BIOMEDICAL INFORMATION TECHNOLOGY

BIOMEDICAL INFORMATION TECHNOLOGY

EDITED BY

DAVID DAGAN FENG

PROFESSOR, SCHOOL OF INFORMATION TECHNOLOGIES
UNIVERSITY OF SYDNEY

AND

CHAIR-PROFESSOR OF INFORMATION TECHNOLOGY
HONG KONG POLYTECHNIC UNIVERSITY

AMSTERDAM • BOSTON • HEIDELBERG • LONDON
NEW YORK • OXFORD • PARIS • SAN DIEGO
SAN FRANCISCO • SINGAPORE • SYDNEY • TOKYO

Academic Press is an imprint of Elsevier

Academic Press is an imprint of Elsevier

30 Corporate Drive, Suite 400, Burlington, MA 01803, USA
525 B Street, Suite 1900, San Diego, California 92101-4495, USA
84 Theobald's Road, London WC1X 8RR, UK

This book is printed on acid-free paper. ♾

Library of Congress Cataloging-in-Publication Data
Application submitted

British Library Cataloguing-in-Publication Data
A catalogue record for this book is available from the British Library.

ISBN 978-0-12-373583-6

For information on all Academic Press publications
visit our Web site at www.books.elsevier.com

Printed in the United States of America
07 08 09 10 11 9 8 7 6 5 4 3 2 1

Contents

Acknowledgments

The editor would like to take the opportunity to express his sincerely appreciation to all of the contributors of this book for making it possible to have such a comprehensive coverage of the most current information in this very dynamic field, to Dr. Fu for helping with formatting this book, to the support from the University of Sydney and Hong Kong Polytechnic University, and to the support from ARC and PolyU/UGC grants.

About the Editor

(David) Dagan Feng received his M.E. in Electrical Engineering & Computing Science (EECS) from Shanghai Jiao Tong University in 1982, M.Sc. in Biocybernetics and Ph.D. in Computer Science from the University of California, Los Angeles (UCLA) in 1985 and 1988, respectively. After briefly working as an Assistant Professor at the University of California, Riverside, he joined the University of Sydney at the end of 1988 as a Lecturer, Senior Lecturer, Reader, Professor, Head of Department of Computer Science, and the Head of School of Information Technologies. He is currently an Associate Dean (International IT) of Faculty of Science at the University of Sydney; Honorary Research Consultant, Royal Prince Alfred Hospital, the largest hospital in Australia; Chair-Professor of Information Technology, Hong Kong Polytechnic University; Advisory Professor and Chief Scientist of Med-X, Shanghai Jiao Tong University; Guest Professor, Northwestern Polytechnic University, Northeastern University and Tsinghua University. His research area is Biomedical & Multimedia Information Technology (BMIT). He is the Founder and Director the BMIT Research Group. He has published over 400 scholarly research papers, pioneered several new research directions, made a number of landmark contributions in his field with significant scientific impact and social benefit, and received the Crump Prize for Excellence in Medical Engineering. More importantly, however, is that many of his research results have been translated into solutions for real-life problems and have made tremendous improvements to the quality of life for those involved. He is a Fellow of the Australian Academy of Technological Sciences and Engineering, ACS, HKIE, IEE, and IEEE. Professor Feng is a Special Area Editor of *IEEE Transactions on Information Technology in Biomedicine*, Editorial Board Advisor or member for *The Visual Computer (International Journal of Computer Graphics)*, *Biomedical Signal Processing and Control, Control Engineering Practice, Computer Methods and Programs in Biomedicine, The International Journal of Image and Graphics (IJIG)*, and is the current Chairman of IFAC-TC-BIOMED.

Contributors

Professor Jorge R. Barrio, Ph.D.
Department of Molecular and Medical Pharmacology,
David Geffen School of Medicine,
University of California, Los Angeles (UCLA)

Dr. Alessandra Bertoldo, Ph.D.
Department of Information Engineering,
University of Padova

Professor Zaver M. Bhujwalla, Ph.D.
Director of the JHU In Vivo Cellular and Molecular Imaging Center,
Director of the Cancer Imaging Resource,
Departments of Radiology and Oncology,
The Johns Hopkins University School of Medicine

Dr. Tom Weidong Cai, Ph.D.
Biomedical & Multimedia Information Technology (BMIT) Research Group,
School of Information Technologies,
University of Sydney

Professor Ewart Carson, D.Sc., Ph.D, CEng, FIET, FIEEE, FAIMBE, FIAMBE
Professor of Systems Science,
Centre for Health Informatics,
City University, London, UK

Josephine Chen
Research Center for Genetic Medicine,
Children's National Medical Center

Dr. Sirong Chen, Ph.D.
Department of Diagnostic Radiology,
Hong Kong Sanatorium & Hospital, and
Honorary Associate,
Biomedical & Multimedia Information Technology (BMIT) Research Group,
School of Information Technologies,
University of Sydney

Professor Wesley W. Chu, Ph.D., FIEEE
Distinguished Professor,
Computer Science Department,
University of California, Los Angeles (UCLA)

Professor Claudio Cobelli, Ph.D., FIEEE
Department of Information Engineering,
University of Padova

Dr. Ruth E. Dayhoff, M.D.
Director, VistA Imaging System Project,
Health Provider Systems, VA Office of Information, Los Angeles (UCLA)
U.S. Department of Veterans Affairs (VA)

Professor Piet C. De Groen, M.D.
Mayo Clinic

Dr. Stefan Eberl, Ph.D.
Principal Scientist, Department of PET and Nuclear Medicine,
Royal Prince Alfred Hospital, and
Adjunct Associate Professor,
Biomedical & Multimedia Information Technology (BMIT) Research Group,
School of Information Technologies,
University of Sydney

Professor David Dagan Feng, Ph.D., FACS, FATSE, FHKIE, FIEE, FIEEE
Director, Biomedical & Multimedia Information Technology (BMIT) Research Group,
Professor, School of Information Technologies,
University of Sydney,
Honorary Research Consultant,
Royal Prince Alfred Hospital, Sydney, and
Chair-Professor of Information Technology,
Centre for Multimedia Signal Processing,
Department of Electronic & Information Engineering,
Hong Kong Polytechnic University

Dr. Catherine A. Foss, Ph.D.
Departments of Radiology and Oncology,
The Johns Hopkins University School of Medicine

Professor Matthew T. Freedman, Ph.D.
Department of Oncology and Lombardi Cancer Center,
Georgetown University

Professor Michael Fulham, M.D.
Director, Department of PET and Nuclear Medicine,
Royal Prince Alfred Hospital, Sydney,
Adjunct Professor,
Biomedical & Multimedia Information Technology (BMIT) Research Group,
School of Information Technologies,
University of Sydney, and
Clinical Professor,
Faculty of Medicine,
University of Sydney

Professor Maryellen L. Giger, Ph.D., SMIEEE, FAAPM, FAIMBE
Professor of Radiology, the Committee on Medical Physics, and College,
Chair, Committee on Medical Physics
Vice-Chair for Basic Science Research, and Section Chief, Radiological Sciences,
Department of Radiology,
University of Chicago

Professor Kristine Glunde, Ph.D.
Departments of Radiology and Oncology,
The Johns Hopkins University School of Medicine

Dr. Yetrib Hathout, Ph.D.
Assistant Professor, Research Center for Genetic Medicine,
Children's National Medical Center

Professor Doan B. Hoang, Ph.D.
Director, ARN Networking Research Laboratory,
Faculty of Information Technology,
University of Technology, and
Honorary Associate,
Biomedical & Multimedia Information Technology (BMIT) Research Group,
School of Information Technologies,
University of Sydney

Professor Eric P. Hoffman, Ph.D.
Clark Professor of Pediatrics, Biochemistry and Molecular Biology, Neuroscience, & Genetics,
School of Medicine and Health Sciences,
George Washington University, and
Director, Research Center for Genetic Medicine,
Children's National Medical Center

Professor H. K. Huang, D.Sc., FRCR(Hon.)
Professor and Director of Imaging Informatics,
Department of Radiology,
Keck School of Medicine,
University of Southern California,
Chair Professor of Medical Informatics
The Hong Kong Polytechnic University, and
Honorary Professor,
Shanghai Institute of Technical Physics,
The Chinese Academy of Sciences,

Professor Sung-Cheng Huang, D.Sc.
Department of Molecular and Medical Pharmacology,
David Geffen School of Medicine,
University of California, Los Angeles (UCLA)

Dr. Emil Jovanov, Ph.D.
Electrical and Computer Engineering Department,
University of Alabama, Huntsville

Professor Peter Kazanzides, Ph.D.
Assistant Research Professor of Computer Science,
The Johns Hopkins University

Dr. Jinman Kim, Ph.D.
Biomedical & Multimedia Information Technology (BMIT) Research Group,
School of Information Technologies,
University of Sydney

Mr. Peter M. Kuzmak, M.S.B.M.E.
Biomedical Engineer and Senior VistA Imaging System Developer,
Health Provider Systems, VA Office of Information,
U.S. Department of Veterans Affairs (VA)

Dr. Eugene Y. S. Lim, Ph.D.
Hospital Scientist, Departments of PET and Nuclear Medicine,
Royal Prince Alfred Hospital, Sydney, and
Honorary Associate,
Biomedical & Multimedia Information Technology (BMIT) Research Group,
School of Information Technologies,
University of Sydney

Professor Brent J. Liu, Ph.D.
Deputy Director of Imaging Informatics,
Departments of Radiology and Biomedical Engineering,
Keck School of Medicine and Viterbi School of Engineering,
University of Southern California

Dr. Zhenyu Liu, Ph.D.
Computer Science Department,
University of California, Los Angeles (UCLA)

Dr. Gladys Goh Lo, M.D.
Department of Diagnostic Radiology,
Hong Kong Sanatorium & Hospital

Professor Murray Loew, Ph.D.
Department of Electrical and Computer Engineering,
George Washington University

Dr. Wenlei Mao, Ph.D.
Computer Science Department,
University of California, Los Angeles (UCLA)

Mr. Kevin Meldrum
Senior Architect and Computerized Patient Record System Developer
Health Provider Systems, VA Office of Information
U.S. Department of Veterans Affairs (VA)

Dr. Aleksandar Milenković, Ph.D.
Electrical and Computer Engineering Department,
University of Alabama, Huntsville

Dr. Javad Nazarian, Ph.D.
Research Center for Genetic Medicine,
Children's National Medical Center

Dr. Dejan Rašković, Ph.D.
Electrical and Computer Engineering Department,
University of Alaska

Erica Reeves
Research Center for Genetic Medicine,
Children's National Medical Center

Dr. Abdul Roudsari, Ph.D.
Director, Centre for Health Informatics,
City University, London, UK

Dr. Andrew J. Simmonds, Ph.D.
Assistant Director, ARN Networking Research Laboratory,
Faculty of Information Technology,
University of Technology, Sydney

Professor Nadine Smith, Ph.D.
Department of Bioengineering,
Penn State University

Dr. Seu Som, Ph.D.
Principal Medical Physicist
Department of Nuclear Medicine & PET,
Liverpool Hospital

Dr. Kenji Suzuki, Ph.D., SMIEEE
Assistant Professor of Radiology,
Department of Radiology,
University of Chicago

Dr. Damian M. Tan, Ph.D.
School of Electrical and Computer Engineering,
Science, Engineering & Technology Portfolio,
RMIT University

Professor Russell H. Taylor, Ph.D. FIEEE
Director, NSF Engineering Research Center for CISST
Professor of Computer Science, with joint appointments in Mechanical Engineering, Radiology, and Surgery,
The Johns Hopkins University

Dr. Xiu Ying Wang, Ph.D.
Biomedical & Multimedia Information Technology (BMIT) Research Group,
School of Information Technologies,
University of Sydney, and
School of Computer Science,
Heilongjiang University

Professor Yue Wang, Ph.D.
Director, Computational Bioinformatics and Bio-imaging Lab
Departments of Electrical, Computer, and Biomedical Engineering,
Virginia Polytechnic Institute and State University

Professor Yu-Ping Wang, Ph.D.
Department of Computer Science and Electrical Engineering,
University of Missouri—Kansas City

Dr. Zhiyong Wang, Ph.D.
Biomedical & Multimedia Information Technology (BMIT) Research Group,
School of Information Technologies,
University of Sydney

Dr. Zuyi Wang, Ph.D.
Research Center for Genetic Medicine,
Children's National Medical Center

Professor Andrew Webb, Ph.D., FIMBE
Director, Huck Institute Magnetic Resonance Centre,
Department of Bioengineering,
Penn State University

Dr. Peter Weller, Ph.D.
Senior Lecturer in Medical Informatics,
Centre for Health Informatics,
City University, London, UK

Dr. Lingfeng Wen, Ph.D
Biomedical & Multimedia Information Technology (BMIT) Research Group,
School of Information Technologies,
University of Sydney

Professor Anna M. Wu, Ph.D.
Department of Molecular and Medical Pharmacology,
David Geffen School of Medicine,
University of California, Los Angeles (UCLA)

Professor Hong Ren Wu, Ph.D.
Professor of Visual Communications Engineering,
Discipline Head, Computer and Network Engineering,
School of Electrical and Computer Engineering,
Science, Engineering & Technology Portfolio,
RMIT University

Professor Chris Wyatt, Ph.D.
Departments of Electrical, Computer, and Biomedical Engineering,
Virginia Polytechnic Institute and State University

Dr. Kai-Ming Au Yeung, FRCR
Department of Diagnostic Radiology,
Hong Kong Sanatorium & Hospital

Dr. Xiaofeng Zhang, Ph.D.
Department of Bioengineering,
Penn State University

Dr. Qinghua Zou, Ph.D.
Computer Science Department,
University of California, Los Angeles (UCLA)

Introduction

We have all witnessed the revolutionary changes in recent years brought about by the development of information technology. These changes have been key to modernizing many disciplines and industries, and biomedicine is no exception. The importance of biomedical information technology has been widely recognized and its application has expanded beyond the boundary of health services, leading to the discovery of new knowledge in life sciences and medicine. In the meantime, life sciences and medicine are becoming an important driving force for the further development of information technology and related disciplines. Many emerging areas have recently been developed, including health informatics, bioinformatics, imaging informatics (or even medical imaging informatics; see Chapter 13 of this book), medical biometrics, systems physiology, systems biology, and biocybernetics. This book aims to provide readers with a comprehensive and up-to-date overall picture of information technology in biomedicine.

This book is divided into two major parts: technological fundamentals and integrated clinical applications. The technological fundamentals cover key medical imaging systems: Electronic Medical Record (EMR) standards and systems; image data compression; content-based medical image retrieval; modeling and simulation; techniques for parametric imaging; data processing and analysis; image registration and fusion; visualization and display; data communication and transmission; security and protection for medical image data; and biological computing. The integrated clinical applications include picture archiving and communication systems (PACS) and medical imaging informatics (MII) for filmless hospitals; a knowledge-based digital library for retrieving scenario-specific medical text documents; integrated multimedia patient record systems; computer-aided diagnosis (CAD); clinical decision support systems (CDSS); medical robotics and computer-integrated interventional medicine; functional techniques for brain magnetic resonance imaging; molecular imaging in biology and pharmacology; the evolution of e-health systems; and smart medical home. Most of the chapters include over 100 references and comprehensively summarize the most recent cutting-edge research in these areas.

This book is a well-designed research handbook instead of a collection of research papers, and is intended for scientific and clinical researchers and practitioners. It is also well-suited for use as a textbook for senior undergraduate and junior postgraduate students with exercises at the end of each chapter to facilitate a better understanding of the comprehensive knowledge covered by this book. Ten chapters are contributed from our Biomedical & Multimedia Information Technology (BMIT) Research Group, School of Information Technologies, University of Sydney and Centre for Multimedia Signal Processing, Department of Electronic and Information Engineering, Hong Kong Polytechnic University, including from our BMIT Group senior members Professor Michael Fulham, who is an Adjunct Professor in the School of Information Technologies and Clinic Professor in the Faculty of Medicine, University of Sydney, Director of PET and Nuclear Medicine Departments, Royal Prince Alfred (RPA) Hospital, Clinical Director for Medical Imaging Service Central Sydney Area Health Services, Chairman of RPA PACS Steering Committee, and the winner of the U.S. NIH Outstanding Performance in Research Award and Australian Eccles Lectureship Award; and Professor Doan B. Hoang, who is an Honorary Associate of the School of Information Technologies, University of Sydney, Professor of Computer Networks and Director of the ARN Networking Research Laboratory, Faculty of Information Technology, University of Technology, Sydney; as well as our BMIT regular research collaborator and Chapter 3 co-author, Professor Henry Wu, who is a Professor of Visual Communications Engineering and Discipline Head of Computer and Network Engineering at the School of Electrical and Computer Engineering, RMIT University, Melbourne, Australia. The following 13 chapters are purposely reserved for contributions from other external international top-leading research groups headed by the world's authorities in their respective areas. These international research leaders who contributed to the remaining 13 chapters are introduced in the following paragraphs.

Chapter 1: "Medical Imaging" is contributed by Professor Andrew Webb, Director of Huck Institute Magnetic Resonance Centre, and his team in the Department of Bioengineering at Penn State University. Professor Webb's main research program is in high field applications of magnetic resonance imaging and spectroscopy, with an emphasis on applications to small animal imaging and microimaging. He has been a full professor since 2003 and has published over 130 journal articles in peer-reviewed publications. He is also the author of a widely used textbook *Introduction to Biomedical Imaging* (Wiley, 2003). Professor Webb is a Fellow of the American Institute for Medical and Biological Engineering, as well as having been awarded a Wolfgang Paul Prize from the Humboldt Foundation from 2001 to 2004.

Chapter 5: "Data Modeling and Simulation" is contributed by Professor Claudio Cobelli and his colleague Dr. Alessandra Bertoldo at the Department of Information Engineering, University of Padova, Italy. Professor Cobelli's main research subject, the field of modeling of endocrine-metabolic systems, has received competitive research grants from

MIUR-MURST, EU and the U.S. National Institutes of Health. He has been a full professor in bioengineering since 1981, and has published over 228 papers in well-established internationally refereed journals. He has also published a number of international leading books in his area and is co-author of *Carbohydrate Metabolism: Quantitative Physiology and Mathematical Modeling* (Wiley, 1981), *The Mathematical Modeling of Metabolic and Endocrine Systems* (Wiley, 1983), *Modeling and Control of Biomedical Systems* (Pergamon Press, 1989), *Modeling Methodology for Physiology and Medicine* (Academic Press, 2000), *Tracer Kinetics in Biomedical Research: from Data to Model* (Kluwer Academic/Plenum Publishers, 2001), etc. Professor Cobelli, Fellow of IEEE, is an active research leader, the founding Chairman of the International Federation of Automatic Control (IFAC), Technical Committee on Modeling and Control for Biomedical Systems (including Biological Systems), and is currently an Associate Editor of *IEEE Transactions on Biomedical Engineering* and of *Mathematical Biosciences* and on the Editorial Board of the *American Journal of Physiology: Endocrinology and Metabolism.*

Chapter 7: "Data Processing and Analysis" is contributed by Professor Yue Wang's group and his collaborators at the Virginia Polytechnic Institute and State University, University of Missouri, Georgetown University, and George Washington University. Professor Wang has also worked closely with the Johns Hopkins Medical Institutions. His research focuses on computational bioinformatics and bio-imaging for diagnosis and molecular analysis of human diseases, with an emphasis on the strategic frontier between statistical machine learning and systems biomedical science. He leads a multidisciplinary and multi-institutional research effort to improve the outcome for patients with cancers, muscular dystrophies, and cardiovascular diseases, an initiative supported by the U.S. National Institutes of Health and Department of Defense. His work has also advanced the broad scientific fields of pattern recognition, signal processing, statistical information visualization, and machine learning. Professor Wang is an elected Fellow of the American Institute for Medical and Biological Engineering (AIMBE), and is currently an Associate Editor for the *International Journal of Biomedical Imaging, EURASIP Journal on Bioinformatics and Systems Biology,* and *IEEE Signal Processing Letters.* Professor Wang is on the ISI (Web of Knowledge) list of highly cited authors in the Category of Engineering.

Chapter 12: "Biologic Computing" is contributed by Professor Eric P. Hoffman and his team at the Research Center for Genetic Medicine, Children's Medical Center, Washington D.C. Dr. Hoffman is a Professor of Pediatrics, Biochemistry and Molecular Biology, Neuroscience, and Genetics at the George Washington University School of Medicine and Health Sciences, and the Director of the Research Center for Genetic Medicine, Children's National Medical Center, Washington D.C. He received his Ph.D. degree in biology (genetics) from Johns Hopkins University in 1986 and subsequently worked as a post-doctoral research fellow at the Harvard Medical School and Children's Hospital for two years. His laboratory is the top contributor of Affymetrix microarray data in the public domain, and he has focused bioinformatics methods developments on quality control and standard operating procedures, signal/noise balance, and public access databases, including the popular PEPR resource (http://pepr.cnmcresearch.org). His laboratory has enjoyed an impressive research grant track record from NIH and Department of Defense, as well as outstanding publication track record in the area of biological computing in well-recognized journals, for example, *Nature, Cell, Nature Medicine, Neuron, Neurology, Brain, Journal of Cell Biology, Journal of Biological Chemistry,* and *Bioinformatics.* Dr. Hoffman is among the most highly cited scientists (more than 12,000 citations to date).

Chapter 13: "PACS and Medical Informatics for Filmless Hospitals" is contributed by Professor H. K. (Bernie) Huang, Director, and Professor Brent J. Liu, Deputy Director of Informatics, Department of Radiology, Keck School of Medicine, University of Southern California. He is also the Chair Professor of Medical Informatics at Hong Kong Polytechnic University and an Honorary Professor at the Shanghai Institute of Technical Physics and at the Chinese Academy of Sciences. Professor Huang has pioneered PACS research, developed the PACS at UCLA in 1991, and developed the hospital-integrated PACS at UCSF in 1995. He has authored and co-authored seven books, published over 200 peer-reviewed articles, and received several patents. His book: *PACS and Imaging Informatics*, published by John Wiley & Sons in 2004, is the only textbook in this field. During the past 25 years, Professor Huang has received over 21 million U.S. dollars in PACS, medical imaging informatics, tele-imaging, and image-processing–related research grants and contracts. He has mentored 22 Ph.D. students and over 30 post-doctoral fellows from around the world. Professor Huang has been a consultant for many national and international hospitals, imaging manufacturers in the design and implementation of PAC systems, and enterprise level EPR with image distribution. He has been a Visiting Professor in many leading universities around the world and Board Member in leading medical imaging manufacturers.

Chapter 14: "KMeX: A Knowledge-Based Digital Library for Retrieving Scenario-Specific Medical Text Documents" is contributed by Professor Welsey W. Chu and his team in the Computer Science Department, University of California (UCLA), Los Angeles. Professor Chu is a UCLA Distinguished Professor and former chairman of the department. He received his Ph.D. from Stanford University in 1966, worked with IBM and Bell Laboratories from 1964 to 1966 and 1966 to 1969, respectively, and has joined UCLA since 1969. During the first two decades, he has made fundamental contributions to the understanding of statistical multiplexing and did pioneering work in file allocation, as well as directory design for distributed databases and task partitioning in real-time distributive systems, for which he was elected as an IEEE Fellow. During the past decade, his research interests have evolved to include intelligent information systems and knowledge acquisition for large information systems. Professor Chu led the development of CoBase, a cooperative database system for structured data, and KMed, a knowledge-based multimedia medical image system. CoBase has been successfully used in logistic applications to provide approximate matching of objects. Together with the medical school staff, the KMed project has been extended to the development of a medical digital library, which consists of structured data, text documents, and images. The system provides approximate content-matching and navigation and serves as a cornerstone for future paperless hospitals. In addition, Professor Chu conducts research on data mining of large information sources, knowledge-based text retrieval, and extending the relaxation methodology to XML (CoXML) for information exchange and approximate XML query answering in the Web environment. In recent years, he also researches in the areas of using inference techniques for data security and privacy protection (ISP). Professor Chu has received best paper awards at the 19th International Conference on Conceptual Modeling in 2000 for his work on XML/Relational schema transformation. He and his students have received best paper awards at the American Medical Information Association Congress in 2002 and 2003 for indexing and retrieval of medical free text, and have also been awarded a Certificate of Merit for the Medical Digital Library Demo System at the 89th Annual Meeting of the Radiological Society of North America in 2003. He is also the recipient of the IEEE Computer Society 2003 Technical Achievement Award for his contributions to intelligent information systems.

Chapter 15: "Integrated Multimedia Patient Record Systems" is contributed by Dr. Ruth E. Dayhoff and her Multimedia Medical Record group, which is part of the Office of Information of the U.S. Department of Veterans Affairs (VA). This organization is responsible for the software and systems used by the clinicians and staff at 156 VA hospitals and almost 900 clinics, the largest health care network in the United States. The VA's software, called Veterans Health Information System & Technology Architecture (VistA) is developed by the VA's Office of Information. Initial work started over 25 years ago, and over 60 different hospital information system modules are in use. VistA Imaging, the multimedia patient record component, has grown and evolved over the past 16 years. Dr. Ruth Dayhoff, M.D., is a physician and early pioneer in medical informatics. She directs the VistA Imaging development team. The team participates in integrating the Healthcare Enterprise initiatives and other major health care standards. The VistA System is undergoing a major data standardization effort necessitated by the new capabilities to view and filter a patient's entire record, including information stored at remote sites. This work involves domains such as orders, progress note titles, problems, and imaging procedures. Another major focus within the VA is monitoring the quality of health care that is provided. Software plays a major role in this effort, and is constantly enhanced to provide additional reminders to clinicians and monitoring tools for the organization. As a result, the Department of Veterans Affairs has recently been recognized by multiple authorities as providing the highest quality health care in the United States.

Chapter 16: "Computer-Aided Diagnosis" is contributed by Professor Maryellen L. Giger and her colleague Kenji Suzuki at the University of Chicago. Dr. Giger is a Professor of Radiology and resides on the Committee on Medical Physics at the University of Chicago, is the Director of the Graduate Programs in Medical Physics, and oversees her research lab of 12 members, including post-doctoral trainees, research associates, and graduate students. She also serves as Chief of the Radiological Sciences Section and Vice Chair for Basic Science Research in the Department of Radiology, University of Chicago. Dr. Giger received her Ph.D. in medical physics from the University of Chicago in 1985. Dr. Giger is recognized as one of the pioneers in the development of computer-aided diagnosis. She has authored or co-authored more than 240 scientific manuscripts (including 120 peer-reviewed journal articles), is inventor/co-inventor on approximately 25 patents, and serves as a reviewer for various granting agencies, including the NIH and the U.S. Army. Dr. Giger is an Associate Editor for *Medical Physics* and *IEEE Transactions on Medical Imaging*. She is an elected fellow of the American Institute for Medical and Biological Engineering (AIMBE) and the American Association of Physicists in Medicine (AAPM), and serves on various scientific program committees. During recent years, she has been invited to give presentations on CAD at SPIE, BIROW, SCAR, IWDM, CARS, AAPM, and RSNA, as well as presentations at various workshops and conferences of the NCI. Her research interests include digital radiography and computer-aided diagnosis in multi-modality breast imaging, chest/CT imaging, cardiac imaging, and bone radiography.

Chapter 17: "Clinical Decision Support Systems" is contributed by Professor Ewart Carson and his colleagues, Dr. Abdul Roudsari, and Dr. Peter Weller at the Centre for Health Informatics, City University, London, UK. Professor Carson is a Professor of Systems Science, and for many years was the Director of the Centre for Measurement and Information in Medicine at City University, which has now been restructured as the Centre for Health Informatics. He served as the Director of the Institute of Health Sciences from 1993 to 1999. His areas of research interest and expertise include modeling in physiology and medicine; modeling methodology for health resource management; clinical decision support systems; development and evaluation of model-based decision support systems; evaluation methodologies with particular application in telemedicine; and integrated policy modeling for ICT enhanced public health care. He has led a range of major research projects funded by UK and European agencies, and has successfully supervised some 40 Ph.D. students. Publications include some 13 authored and edited books and more than 300 journal papers and book chapters. Dr. Carson is a member of the Executive Team of the Healthcare Technologies Professional Network of the IEEE, Associate Editor of Computer Methods and Programs in Biomedicine, a Technical Board member of the International Federation of Automatic Control (IFAC), and Chairman of the IFAC Coordinating Committee for Biological and Ecological Systems. He is an Honorary Member of the Royal College of Physicians (London), Fellow of IEEE, Fellow of the American Institute of Medical and Biological Engineers, and Fellow of the International Academy of Medical and Biological Engineering. Due to his exceptional outstanding contributions in his field, he received the 2005 IEEE Engineering in Medicine and Biology Career Achievement Award.

Chapter 18: "Medical Robotics and Computer-Integrated Interventional Medicine" is contributed by Professor Russell H. Taylor and Dr. Peter Kazanzides from Johns Hopkins University. Professor Taylor received a B.E.S. degree from The Johns Hopkins University in 1970 and a Ph.D. in Computer Science from Stanford in 1976. He joined IBM Research in 1976, where he developed the AML robot language and various other projects, managed robotics and automation technology research activities from 1982 to 1988, led the team that developed the first prototype for the Robodoc® system for robotic hip replacement surgery from 1988 to 1989, and served as the Manager of Computer Assisted Surgery from 1990 to 1995. In September 1995, Dr. Taylor moved to Johns Hopkins University as a Professor of Computer Science, with joint appointments in Radiology, Surgery, and Mechanical Engineering. He is the Director of the NSF Engineering Research Center for Computer-Integrated Surgical Systems and Technology and is also currently on the Scientific Advisory Board of Integrated Surgical Systems for IBM, where he subsequently developed novel systems for computer-assisted craniofacial surgery and robotically-augmented endoscopic surgery. At Johns Hopkins, he has worked on all aspects of CIIM systems, including modeling, registration, and robotics in areas including percutaneous local therapy, microsurgery, and minimally-invasive robotic surgery. He is Editor Emeritus of the *IEEE Transactions on Robotics and Automation,* Fellow of IEEE and AIMBE. In February, 2000 he received the Maurice Müller award for excellence in computer-assisted orthopaedic surgery. Dr. Kazanzides received a Ph.D. in electrical engineering from Brown University in 1988, and began work on surgical robotics in March 1989 at IBM Research with Dr. Russell Taylor. Dr. Kazanzides co-founded Integrated Surgical Systems (ISS) in November, 1990 to commercialize the robotic hip replacement research performed at IBM and the University of California, Davis. As Director of Robotics and Software, he was responsible for the design, implementation, validation, and support of the ROBODOC® hardware and software. In 2002, Dr. Kazanzides joined the NSF Engineering Research Center for Computer-Integrated Surgical Systems and Technology (CISST ERC) at Johns Hopkins University.

Chapter 20: "Molecular Imaging in Cancer" is contributed by Professor Zaver M. Bhujwalla and her colleagues, Dr. Kristine Glunde and Dr. Catherine A. Foss in the Departments of Radiology and Oncology at the Johns Hopkins University School of Medicine. Professor Bhujwalla joined the Department of Radiology at the Johns Hopkins University School of Medicine in 1989 after completing her Ph.D. from the University of London and has built an internationally-recognized cancer functional and molecular imaging program at Johns Hopkins. She is currently the Director of the JHU In Vivo Cellular and Molecular Imaging Center (JHU ICMIC Program), and Director of the Cancer Imaging Resource of the Sidney Kimmel Comprehensive Cancer Center at Johns Hopkins. Over the past decade, Dr. Bhujwalla's work has focused on the application of imaging technology to promote the understanding of cancer. These studies encompass studying cancer from the sub-cellular to the clinical stage with imaging, with a strong impact on both basic scientific research and clinical applications.

Chapter 21: "Molecular Imaging in Biology and Pharmacology" is contributed by Professor Henry Sung-Cheng Huang and his colleagues in the Department of Molecular and Medical Pharmacology, David Geffen School of Medicine, UCLA. Professor Huang has pioneered the quantification of PET images and was involved in the tomography reconstruction of early PET scanners in the early 1970s. He has investigated a series of radioactivity quantification issues in PET imaging, including photon attenuation correction scheme for PET, that have had a lasting impact on all biomedical imaging fields. He is a pioneer in using compartmental models to model the kinetic behavior of positron-labeled tracers (started in the late 1970s). His modeling papers on FDG in 1979 and 1980 have shaped the way glucose utilization rates in local tissue are currently measured *in vivo.* He has expanded its application from brain tissue to myocardium and to tumors, and from research to the clinical setting. His early papers are still frequently quoted in the literature, and the model continues to be used in the field. In addition to the FDG modeling, Dr. Huang has developed models and study methodologies for many other PET tracers as well, including O-15 water/oxygen, N-13 ammonia, C-11 Palmitate, C-11 acetate, FESP, and FDOPA. In conjunction with biologists/physicians, Professor Huang has demonstrated the value of quantitative biomedical imaging and has advanced our understanding of the biological/physiological changes in diseases. He has also made exceptional outstanding contributions in many related areas and has over 800 peer-reviewed publications (including 293 full journal papers in well-established journals, 488 peer-reviewed short papers/abstracts in well-established journals and keynote/invited/special presentation articles, 22 book chapters, and three U.S. patents and software copyrights) with frequent citations. He has served as Deputy Chief Editor, Associate Editor or editorial board consultant for major journals in his areas, for example, *Cerebral Blood Flow and Metabolism, Molecular Imaging and Biology*, and *Journal of Nuclear Medicine*. He has received numerous prestigious awards, such as George Von Hevesy Prize, Award of Excellence for Best Paper, and Outstanding Scientist Award.

Chapter 22: "From Telemedicine to Ubiquitous M-Health: The Evolution of E-Health Systems" is contributed by Dr. Dejan Rašković from the University of Alaska, Fairbanks, who leads the DIA-sponsored Laboratory for Energy and Performance Profiling of Wireless Sensor Networks and performs research in wireless sensor networks, battery-aware processing, and embedded systems architecture. Contributing authors include Dr. Piet C. De Groen from the Mayo Clinic, who is a Professor of Medicine and former Program Director of Mayo Clinic/IBM Computational Biology Collaboration at the Mayo Clinic, Rochester, and Drs. Aleksandar Milenković and Emil Jovanov from the University of Alabama in Huntsville. The wearable health monitoring group at the University of Alabama has been developing wireless intelligent sensors and wearable health sensors for more than seven years (http://www.ece.uah.edu/~jovanov/whrms/). The group has pioneered the concept of the wireless body area network of intelligent sensors (WBAN) for ambulatory health monitoring, and developed a few dozens different sensors and systems for wearable health monitoring. Their wireless distributed system for stress monitoring has been used at the Navy Aviation Medical Research Lab at Pensacola, Florida for more than four years. They have established a collaboration with the Mayo Clinic in Rochester, MN, and currently work on wearable ambulatory monitoring.

Professor David Dagan Feng
Professor, School of Information Technologies,
University of Sydney, and
Chair-Professor of Information Technology,
Hong Kong Polytechnic University

I

Technological Fundamentals

1

Medical Imaging

Dr. Xiaofeng Zhang,
Prof. Nadine Smith, and
Prof. Andrew Webb
Penn State University

1.1 Introduction

Medical imaging forms a key part of clinical diagnosis, and improvements in the quality and type of information available from such images have extended the diagnostic accuracy and range of new applications in health care. Previously seen as the domain of hospital radiology departments, recent technological advances have expanded medical imaging into neurology, cardiology, and cancer centers, to name a few. The past decade, in particular, has seen many significant advances in each of the imaging methods covered in this chapter. Since there are a large number of texts (see Bibliography) that deal in great detail with the basic physics, instrumentation, and clinical applications of each imaging modality, this chapter summarizes these aspects in a succinct fashion and emphasizes recent technological advances. State-of-the-art instrumentation for clinical imaging now comprises, for example, 64-slice spiral computed tomography (CT); multi-element, multidimensional phased arrays in ultrasound; combined positron emission tomography (PET) and CT scanners; and rapid parallel imaging techniques in magnetic resonance imaging (MRI) using large multidimensional coil arrays. Furthermore, on the horizon are developments such as integrated diffuse optical tomography (DOT)/MRI. Considered together with significant developments in new imaging contrast agents—so-called "molecular imaging agents"—the role of medical imaging looks likely to continue to expand in modern-day health care.

1.2 Digital Radiography

Planar X-ray imaging has traditionally been film-based and is used for diagnosing bone breaks, lung disease, a number of gastrointestinal (GI) diseases (fluoroscopy), and conditions of the genitourinary tract, such as kidney stones (pyelography). Increasingly, images are being formed and stored in digital format for integration with picture archiving and communication systems (PACSs), ease of storage and transfer, and image manipulation in, for example, digital subtraction angiography. Many of the components of conventional film-based systems (X-ray source, collimators, anti-scatter grids) are essentially identical to those in digital radiography, the only difference being the detector itself.

1.2.1 Formation and Characteristics of X-rays

A schematic of an X-ray source is shown in Figure 1.1 (a). A potential difference, termed the accelerating voltage (kVp), typically between 90 and 150 kV, is applied between a small helical cathode coil of tungsten wire and a rotating anode consisting of a tungsten target embedded in a rotating copper disc. When an electric current is passed through the cathode, electrons are emitted via thermionic emission and accelerate toward the anode target; X-rays are then created by the interaction of these electrons with the target: This electron flow is termed the *tube current* (mA). X-rays then pass through a "window" in the X-ray tube. In order to create the desired thin X-ray beam, a negatively charged focusing cup is placed around the cathode. A broad spectrum of X-ray energies is emitted from the X-ray tube, as shown in Figure 1.1 (b). Characteristic lines are produced when the accelerated electrons knock out a bound electron in the K-shell of the tungsten anode, with the resulting hole being filled by an electron from the L-shell, and the difference in binding energy of the two electrons being transferred to an X-ray. The broad "hump" component of the X-ray spectrum arises from "general radiation," which corresponds to an accelerated electron losing part of its kinetic energy when it passes close to a tungsten atom in the target and this energy being emitted as an X-ray. Overall, the number of X-rays produced by the source is proportional to the tube current, and the energy of the X-ray beam is proportional to the square of the accelerating voltage.

The collimator, also termed a *beam restrictor,* consists of lead sheets that can be slid over one another to restrict the beam dimensions to match those of the area of the patient to be imaged.

1.2.2 Scatter and Attenuation of X-rays in Tissue

The two dominant mechanisms for the interaction of X-rays with tissue are photoelectric absorption and Compton scattering. Photoelectric interactions in the body involve the energy of an incident X-ray being absorbed by an atom in tissue, with a tightly bound electron emitted from the K- or L-shell: The incident X-ray is completely absorbed and does not reach the detector. The probability (P_{photo}) of photoelectric absorption occurring is given by:

$$P_{\text{photo}} \propto \frac{Z_{\text{eff}}^3}{E^3}, \tag{1.1}$$

where Z_{eff} is the effective atomic number, and E is the X-ray energy. Since there is a large difference in the values of Z_{eff} for bone ($Z_{\text{eff}} = 20$ due to the presence of Ca) and soft tissue ($Z_{\text{eff}} = 7.4$), photoelectric absorption produces high contrast between bone and soft tissue.

Compton scattering involves the transfer of a *fraction* of an incident X-ray's energy to a loosely bound outer shell of an atom in tissue. The X-ray is deflected from its original path but typically maintains a substantial component of its original energy. The probability of Compton scattering is essentially independent of the effective atomic number of the tissue, linearly proportional to the tissue electron density, and weakly

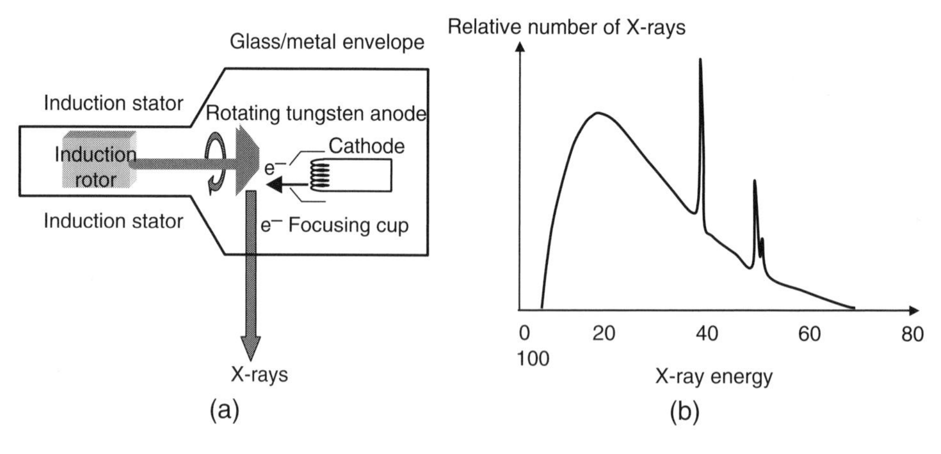

FIGURE 1.1 (a) Schematic of an X-ray tube. (b) Typical energy spectrum from a tungsten anode with an accelerating voltage of ∼100 kVp.

dependent on the X-ray energy. Since the electron density is very similar for bone and soft tissue, Compton-scattered X-rays result in very little image contrast.

Attenuation of the intensity of the X-ray beam as it travels through tissue can be expressed mathematically by:

$$I_x = I_0 e^{-(\mu_{\text{Compton}} + \mu_{\text{photoelectric}})x}, \qquad (1.2)$$

where I_0 is the intensity of the incident X-ray beam, I_x is the X-ray intensity at a distance x from the source, and μ is the linear attenuation coefficient of tissue, measured in cm^{-1}.

The contribution from photoelectric interactions dominates at lower energies, whereas Compton scattering is more important at higher energies. X-ray attenuation is often characterized in terms of a mass attenuation coefficient, equal to the linear attenuation coefficient divided by the density of the tissue. Figure 1.2 plots the mass attenuation coefficient of fat, bone, and muscle as a function of the incident X-ray energy. At low-incident X-ray energies, bone has by far the highest mass attenuation coefficient. As the incident X-ray energy increases, the probability of photoelectric interactions decreases greatly, and the value of the mass attenuation coefficient becomes much lower. At X-ray energies greater than about 80 keV, Compton scattering is the dominant mechanism, and the difference in the mass attenuation coefficients of bone and soft tissue is less than a factor of 2. At incident X-ray energies greater than around 120 keV, the mass attenuation coefficients for bone and soft tissue are very similar.

In cases in which there is little contrast—for example, between blood vessels and surrounding tissue—X-ray contrast agents can be used. There are two basic classes of contrast agents: those based on barium and those based on iodine. Barium sulphate is used to investigate abnormalities such as ulcers, polyps, tumors, or hernias in the GI tract. Since barium has a K-edge at 37.4 keV, X-ray attenuation is much higher in areas where the agent accumulates. Barium sulphate is administered as a relatively thick slurry. Orally, barium sulphate is used to explore the upper GI tract, including the stomach and esophagus (the so-called "barium meal"). As an enema, barium sulphate can be used either as a single or "double" contrast agent. As a single agent, it fills the entire lumen of the GI tract and can detect large abnormalities. As a double contrast agent, barium sulphate is introduced first, followed usually by air: The barium sulphate coats the inner surface of the GI tract, and the air distends the lumen. This double agent approach is used to characterize smaller disorders of the large intestine, colon, and rectum.

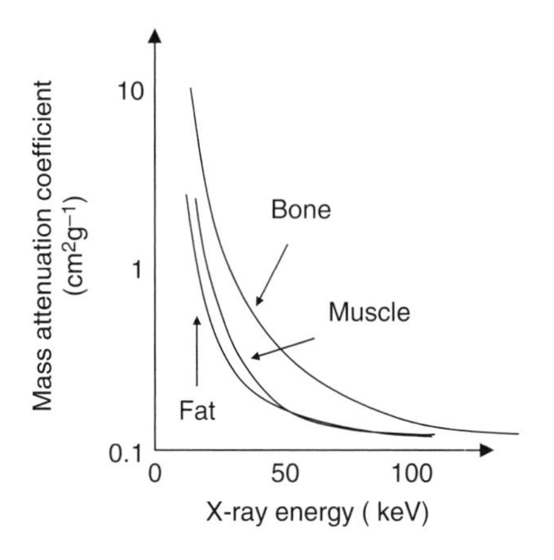

FIGURE 1.2 Mass attenuation coefficient for bone, muscle, and fat as a function of incident X-ray energy.

Iodine-based X-ray contrast agents are used for a number of applications, including intravenous urography, angiography, and intravenous and intra-arterial digital subtraction angiography. An iodine-based agent is injected into the bloodstream, and since iodine has a K-edge at 37.4 keV, X-ray attenuation in blood vessels is enhanced compared with the surrounding soft tissue. This makes it possible to visualize arteries and veins within the body. Digital subtraction angiography (DSA) is a technique in which one image is taken before the contrast agent is administered and a second is taken after injection of the agent, and the difference between the two images is computed. DSA gives very high contrast between the vessels and the tissue and can produce angiograms with extremely high spatial resolution, resolving vessels down to ~100 μm in diameter.

1.2.3 Instrumentation for Digital Radiography

The detector placed on the opposite side of the patient to the X-ray source consists of an anti-scatter grid and recording device. The role of the anti-scatter grid is to minimize the number of Compton-scattered X-rays that reach the detector, since these reduce image contrast. The grid consists of thin strips of lead spaced by aluminium for structural support. The grid ratio, the length of the lead strips divided by the interstrip distance, has values between 4:1 and 16:1, and the strip line density ranges from 25 to 60 per cm.

Digital radiography has largely replaced the use of X-ray film for recording the image. A large-area (41 × 41 cm) flat-panel detector (FPD) consists of an array of thin-film transistors (TFT). The FPD is fabricated on a single monolithic glass substrate. A thin-film amorphous silicon transistor array is then layered onto the glass. Each pixel of the detector consists of a photodiode and associated TFT switch. On top of the array is a structured thallium doped cesium iodide (CsI) scintillator, which consists of many thin, rod-shaped crystals (approximately 6–10 μm in diameter) aligned parallel to one another. When an X-ray is absorbed in a CsI rod, the CsI scintillates and produces light. The light undergoes internal reflection within the fiber and is emitted from one end of the fiber onto the TFT array. The light is then converted into an electrical signal by the photodiodes in the TFT array. This signal is amplified and converted into a digital value for each pixel using an analog-to-digital (A/D) converter. Each pixel typically has dimensions of 200 × 200 μm.

1.3 Computed Tomography

1.3.1 Principles of Computed Tomography

CT acquires X-ray data at different angles with respect to the patient and then reconstructs these data into images. The basic scanner geometry is shown in Figure 1.3. A wide X-ray "fan-beam" and large number of detectors (typically between 512 and 768) rotate synchronously around the patient. The detectors used are ceramic scintillators based on Gd_2O_2S, with different companies adding trace amounts of various elements to improve performance characteristics. Behind each scintillator is a silicon photodiode to convert light into current flow. The current is amplified and then digitized. The combined data represent a series of one-dimensional projections.

Prior to image reconstruction, the data are corrected for the effects of beam hardening, in which the effective energy of the X-ray beam increases as it passes through the patient due to the greater degree of attenuation of lower X-ray energies. Corrections are also made for imbalances in the sensitivities of individual detectors and detector channels. Reconstructing a 2D image from a set of projections—$p(r,\phi)$, acquired as a function of r, the distance along the projection, and the rotation angle ϕ of the X-ray source and detector—is performed using filtered backprojection. Each projection $p(r,\phi)$ is Fourier-transformed along the r-dimension to give $P(k,\phi)$, and then $P(k,\phi)$ is multiplied by $H(k)$, the Fourier transform of the filter function $h(r)$, to give $P_{\text{filt}}(k,\phi)$. The filtered projections, $P_{\text{filt}}(k,\phi)$, are inverse-Fourier-transformed back into the spatial domain and backprojected to give the final image, $\hat{f}(x,y)$:

$$\hat{f}(x,y) = \sum_{j=1}^{n} \mathbb{F}^{-1}\left\{P_{filt}\left(k,\phi_j\right)\mathrm{d}\varphi\right\}, \tag{1.3}$$

where $\mathbb{F}^{-1}$ represents an inverse Fourier transform and n is the number of projections. The filter is typically a lowpass cosine or generalized Hamming function. The reconstruction algorithm assumes that all of the projections are parallel. However, Figure 1.3 shows that in the case of an X-ray fan-beam, this is not the case. The backprojection algorithm is adapted by multiplying each projection by the cosine of the fanbeam angle, with the angle also incorporated into the filter. After reconstruction, the image is displayed as a map of tissue CT numbers, which are defined by:

$$\mathrm{CT_o} = 1000\frac{\mu_o - \mu_{\mathrm{H_2O}}}{\mu_{\mathrm{H_2O}}}, \tag{1.4}$$

where $\mathrm{CT_o}$ is the CT number and μ_o is the linear attenuation coefficient of the tissue.

1.3.2 Spiral and Multislice Computed Tomography

Spiral CT acquires data as the patient table is moved continuously through the scanner, with the trajectory of the X-ray beam through the patient tracing out a spiral pattern, as shown in Figure 1.4. This technique enables very rapid scan times, which can be used, for example, for a complete chest and abdominal study during a single breath-hold. Full three-dimensional vascular imaging data sets can be acquired very shortly after injection of an iodinated contrast agent. The instrumentation for spiral CT is very similar to that of conventional third-generation CT scanners, but with multiple slip-rings being used for power and signal transmission.

The spiral trajectory is defined in terms of parameters such as the spiral pitch, (p), defined as the ratio of the table feed (d) per rotation of the X-ray source to the collimated slice thickness (S). Due to the spiral trajectory of the X-rays throughout the patient, modification of the backprojection reconstruction algorithm is necessary in order to form images that correspond closely to those that would have been acquired using a

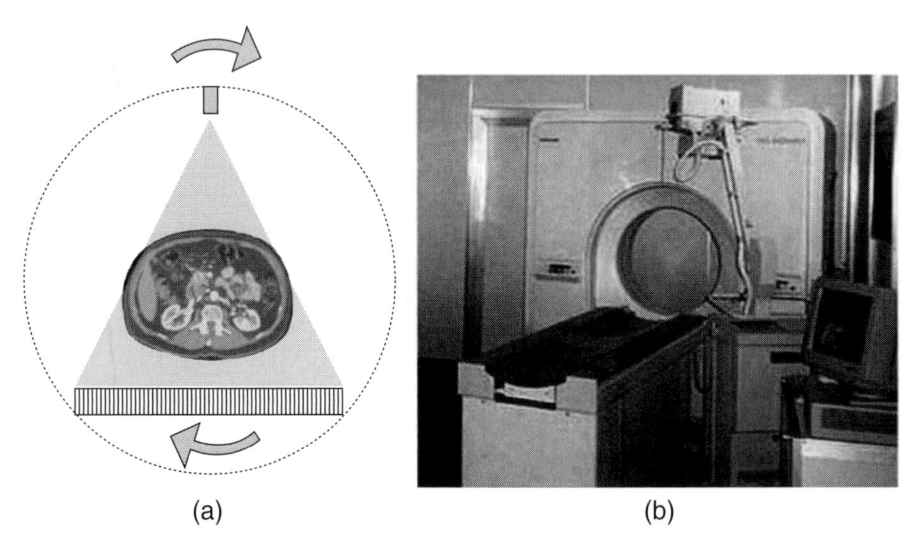

FIGURE 1.3 (a) Schematic of the operation of a third-generation CT scanner. (b) Photograph of a CT scanner with patient bed.

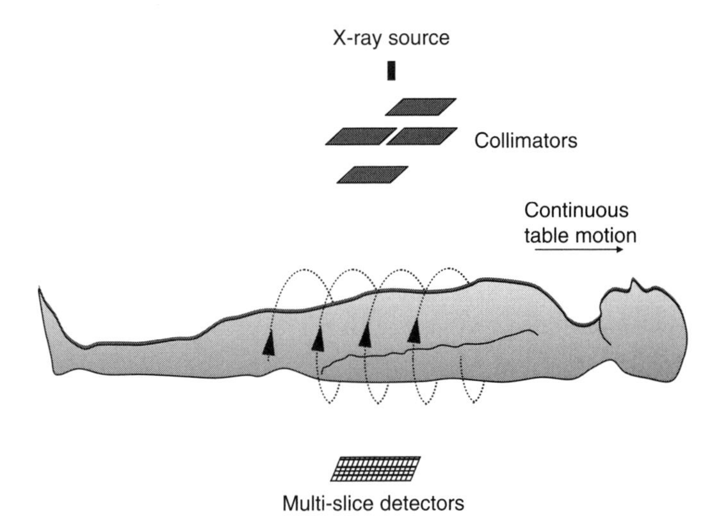

FIGURE 1.4 Continuous motion of the patient while the X-ray source and detectors rotate causes the X-rays to trace out a helical trajectory through the patient. Multi-slice detectors (not shown to scale) enable very thin slice thicknesses to be acquired.

single-slice CT scanner. Images are usually processed in a way that results in considerable overlap between adjacent slices. This has been shown to increase the accuracy of lesion detection, for example, since with overlapping slices there is less chance that a significant portion of the lesion lies between slices.

The vast majority of new CT scanners are multislice scanners; that is, they incorporate an array of detectors in the direction of table motion, as shown in Figure 1.4, in addition to spiral data acquisition. Multislice spiral CT can be used to image larger volumes in a given time, or to image a given volume in a shorter scan time compared with conventional spiral CT. The collimated X-ray beam can also be made thinner, giving higher-quality three-dimensional scans, with slice thicknesses well below 1 mm. Sixty-four–slice machines are now offered by all vendors, which allow very high resolution images to be acquired, as shown in Figure 1.5.

1.4 Nuclear Medicine

1.4.1 Radioactive Nuclides in Nuclear Medicine

In contrast to X-ray, ultrasound, and MRI, nuclear medicine imaging techniques do not produce an anatomical map of the body, but instead image the spatial distribution of radioactive materials (radiotracers) that are introduced into the body. Nuclear medicine detects early biochemical indicators of disease by imaging the kinetic uptake, biodistribution, and clearance of very small amounts (typically nanograms) of radiotracers, which enter the body via inhalation into the lungs, direct injection into the bloodstream, or oral administration. These radiotracers are compounds consisting of a chemical substrate linked to a radioactive element. Abnormal tissue distribution or an increase or decrease in the rate at which the radiopharmaceutical accumulates in a particular tissue is a strong indicator of disease. Radiation in the form of γ-rays is detected using an imaging device called a *gamma camera.* The vast majority of nuclear medicine scans are performed using technetium-containing radiotracers. ^{99m}Tc exists in a metastable state and is formed from ^{99}Mo according to the following scheme:

$$^{99}_{42}\text{Mo} \xrightarrow{\tau_{1/2}\ 66\ \text{hours}} \beta + {}^{99m}_{43}\text{Tc} \xrightarrow{\tau_{1/2}\ 6\ \text{hours}} {}^{99g}_{43}\text{Tc} + \gamma.$$

The energy of the emitted γ-ray is 140 keV, which is high enough for a significant fraction to pass through the body without being absorbed, and low enough not to pentrate the

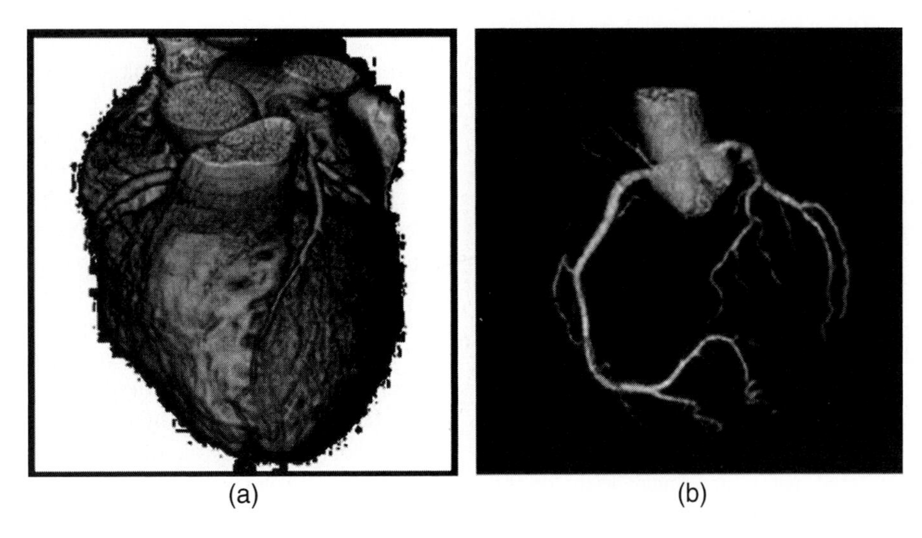

FIGURE 1.5 (a) Three-dimensional volume rendering of the cardiac surface with data from a multislice spiral CT system. (b) Three-dimensional cardiac angiogram.

collimator septa used in gamma cameras to reject scattered γ-rays. Tc-based radiotracers are produced from an on-site technetium generator, which can be replenished on a weekly basis. The generator comprises an alumina ceramic column with radioactive ^{99}Mo absorbed onto its surface in the form of ammonium molybdenate. The column is housed within a lead shield for safety considerations. ^{99m}Tc is obtained by flowing an eluting solution of saline through the generator. The solution washes out the ^{99m}Tc, which binds very weakly to the alumina, leaving the ^{99}Mo behind. The ^{99m}Tc eluted from the generator is in the form of sodium pertechnatate, $NaTcO_4$. The majority of radiotracers, however, are prepared by reducing the pertechnetate to ionic technetium (Tc^{4+}) and then complexing it with a chemical ligand that binds to the metal ion. Examples of ligands include diphosphonate for skeletal imaging, diethylenetriaminepentaacetic acid (DTPA) for renal studies, hexamethylpropyleneamineoxime (HMPAO) for brain perfusion, and macroaggregated albumin for lung perfusion.

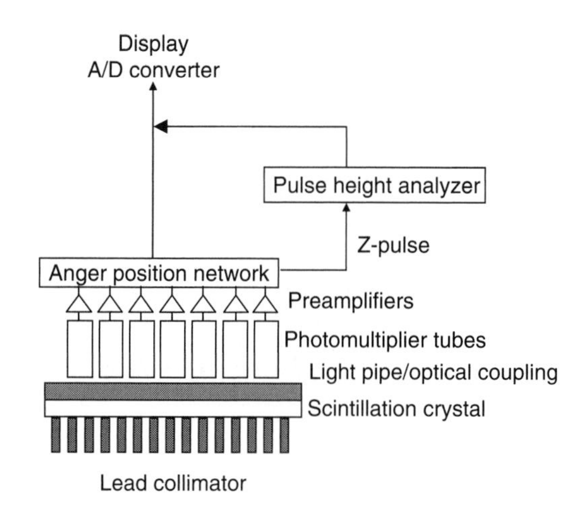

FIGURE 1.6 Schematic of an Anger gamma camera used for planar nuclear medicine.

1.4.2 Nuclear Medicine Detectors

The gamma camera is based on a large scintillation crystal that transduces the energy of a γ-ray into light. In front of the crystal is a lead collimator, usually of a hexagonal "honeycomb" structure, which minimizes the contribution of Compton scattered γ-rays, analogous to the setup described previously for X-ray imaging. The crystal itself is made of thallium-activated sodium iodide, NaI(Tl), which converts the γ-ray energy into light at 415 nm. The intensity of the light is proportional to the energy of the incident γ-ray. The light emission decay constant, which is the time for the excited states within the crystal to return to equilibrium, is 230 ns, which means that count rates of 10^4–10^5 γ-rays per second can be recorded accurately. The linear attenuation coefficient of NaI(Tl) is 2.22 cm^{-1}, and so 90% of the γ-rays that strike the scintillation crystal are absorbed in a 1-cm thickness. Approximately 13% of the energy deposited in the crystal via γ-ray absorption is emitted as visible light. The only disadvantage of the NaI(Tl) crystal is that it is hygroscopic and so must be sealed hermetically.

The light photons emitted by the crystal are detected by hexagonal-shaped (sometimes square) photomultiplier tubes (PMT), which are closely coupled to the scintillation crystal via light pipes. Arrays of 61, 75, or 91 PMTs, each with a diameter of between 25 and 30 mm, are typically used. The output currents of the PMTs pass through a series of low-noise preamplifiers and are digitized. The PMTs situated closest to a particular scintillation event produce the largest output current. By comparing the magnitudes of the currents from all of the PMTs, the location of individual scintillations within the crystal can be estimated using an Anger logic circuit (Figure 1.6). In addition, the summed signal from all the PMTs, termed the *z*-signal, is sent to a pulse-height analyzer (PHA), which compares the *z*-signal with a threshold value that corresponds to that produced by a γ-ray with energy 140 keV. If the *z*-signal is significantly below this threshold, it is rejected as having originated from a Compton-scattered γ-ray. A range of values of the *z*-signal is accepted, with the energy resolution of the system being defined as the full-width half maximum (FWHM) of the photopeak; typically, it is about 14 keV (or 10%) for most gamma cameras. The narrower the FWHM of the system, the better it is at discriminating between unscattered and scattered γ-rays.

1.4.3 Single Photon Emission Computed Tomography

The relationship between single photon emission CT (SPECT) and planar nuclear medicine is exactly the same as that between CT and planar X-ray imaging. In SPECT, two or three gamma cameras are rotated around the patient in order to obtain a set of projections that are then reconstructed to produce a two-dimensional image (Figure 1.7). Adjacent slices are produced from separate rows of PMTs in the two-dimensional array. SPECT uses similar instrumentation and radiotracers as does planar scintigraphy, and most SPECT machines can also be used for planar scans. Projections can be acquired either in a "stop-and-go" mode or during continuous rotation of the gamma camera. Image reconstruction can be performed either by filtered backprojection, as in CT, or by iterative methods. In either case, attenuation and scatter correction of the data are required prior to image reconstruction.

Attenuation correction is performed using either of two methods. In the first, the attenuation coefficient is assumed to be uniform in the tissue being imaged. A patient outline is formed by fitting an ellipse or circle from the acquired data. This approach works well when imaging homogeneous tissues such as the brain. However, for cardiac applications, for example, a spatially variant correction must be applied based

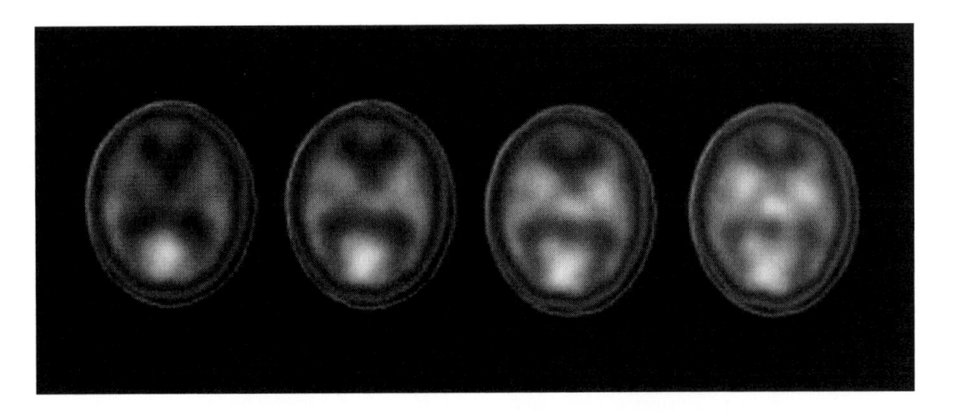

FIGURE 1.7 SPECT images of the brain.

on direct measurements of tissue attenuation using a transmission scan with tubes of known concentration of radioactive gadolinium (^{153}Gd), which emits ~100 keV γ-rays, placed around the patient. The transmission scan can be performed with the patient in place before the actual diagnostic scan, or it can be acquired simultaneously with the diagnostic scan. Since the attenuation coefficient is measured for 100 keV γ-rays, a fixed multiplication factor is used to convert these numbers to 140 keV. The attenuation map is calculated from the transmission projections using filtered backprojection.

The second step in data processing is scatter correction, which must be performed on a pixel-by-pixel basis, since the number of scattered γ-rays is not spatially uniform. The most common method uses dual-energy window detection: One energy window is centered at 140 keV with a fractional width (W_m) of ~20%, and a "subwindow" is centered at 121 keV with a fractional width (W_s) of ~7%. The main window contains contributions from both scattered and unscattered γ-rays, but the subwindow has contributions from only scattered γ-rays. The true number of primary γ-rays, C_{prim}, can be calculated from the total count, C_{total}, in the main window and the count, C_{sub}, in the subwindow:

$$C_{prim} = C_{total} - \frac{C_{sub} W_m}{2W_s}. \qquad (1.5)$$

Along with filtered backprojection, iterative reconstruction methods are also available on commercial machines. These iterative methods can often give better results than filtered backprojection, since accurate attenuation corrections based on transmission source data can be built into the iteration process, as can the overall modulation transfer function (MTF) of the collimator and gamma camera. Typically, the initial estimate of the distribution of radioactivity can be produced using filtered backprojection. Projections are then calculated from this initial estimate and the measured attenuation map, and these are compared with the projections actually acquired. The differences (errors) between these two data sets are computed and the estimated image correspondingly updated. This process is repeated a number of times to reach a predetermined error threshold. The most commonly used iterative methods are based on *maximum-likelihood expectation maximation* (ML-EM), with the particular implementation being the *ordered subset expectation maximum* (OSEM) algorithm. Potential instability in the reconstruction from noisy data normally necessitates applying a filter, such as a two- or three-dimensional Gaussian filter with an FWHM comparable to the intrinsic spatial resolution of the data.

1.4.4 Positron Emission Tomography

Radionuclides used in PET scanning emit positrons, which travel a short distance in tissue before annihilating with an electron resulting in the formation of two γ-rays, each with an energy of 511 keV. The two γ-rays travel in opposite directions to one another and are detected by a ring of detectors placed around the patient (Figure 1.8). The location of the two crystals that detect the two anti-parallel γ-rays defines a line along which the annihilation occurred. This process is referred to as *annihilation coincidence detection* (ACD) and forms the basis of signal localization in PET. The spatial distribution, rate of uptake, and rate of washout of a particular radiotracer are all quantities that can be used to distinguish diseased from healthy tissue. Radiotracers for PET have very short half-lives (e.g., $^{11}C = 20.4$ minutes; $^{15}O = 2.07$ minutes; $^{13}N = 9.96$ minutes; $^{18}F = 109.7$ minutes) and must be synthesized on-site using a cyclotron. After production, they are incorporated via rapid chemical synthesis into structural analogues of biologically active molecules, such as ^{18}F-fluorodeoxyglucose (FDG) and ^{11}C-palmitate. Robotic units are available commercially to synthesize ^{18}FDG, $^{15}O_2$, $C^{15}O_2$, $C^{15}O$, and $H_2^{15}O$.

The individual scintillation crystals used in PET are either bismuth germanate (BGO: $Bi_4Ge_3O_{12}$) or, increasingly commonly, lutetium silicon oxide (LSO: Lu_2SiO_5: Ce). The advantages of LSO are its short decay time (allowing a short coincidence time, reducing accidental coincidences, as will be described), a high emission intensity, and an emission wavelength close to 400 nm, which corresponds to maximum sensitivity for standard PMTs. Multislice capability can be

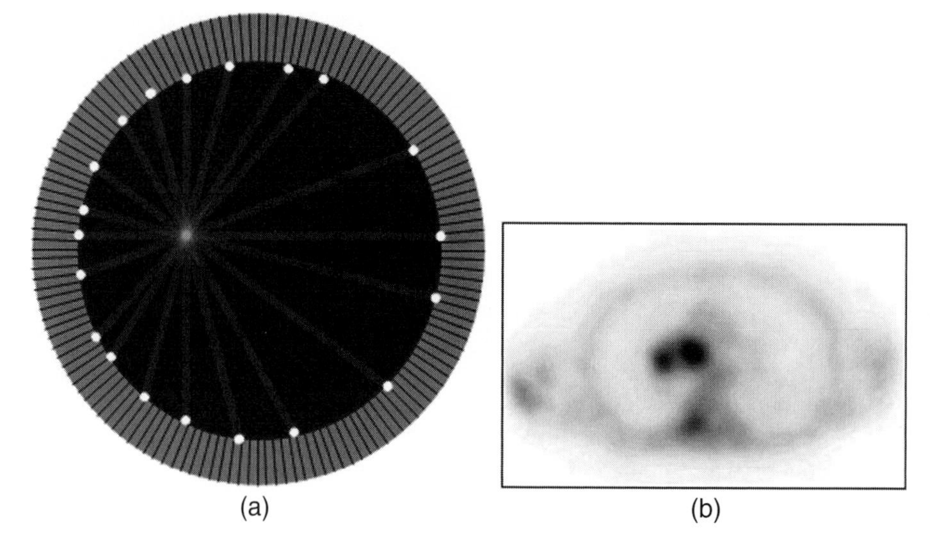

FIGURE 1.8 (a) Image formation using PET. Anti-parallel γ-rays strike pairs of detectors that form a line integral for filtered backprojection. (b) Abdominal PET study using FDG with hot spots indicating the presence of small tumors.

introduced into PET imaging, as it can for CT, by having a number of detector rings stacked adjacent to one another. Each ring typically consists of 16 "buckets" of 8×8 blocks of scintillation crystals, each block coupled to either 16 (BGO) or 4 (LSO) PMTs. The number of rings in a high-end multislice PET scanner can be up to 48. Retractable septa (lead or tungsten) are positioned within each ring: These can be retracted for imaging in three-dimensional mode.

When a γ-ray interacts with a particular detector crystal, it produces a number of photons. These photons are converted into an amplified electrical signal, at the output of the PMT, which is fed into a PHA. If the electrical signal is above a certain threshold, then the PHA generates a "logic pulse," which is sent to a coincidence detector. Typically, this logic pulse is 6–10 ns long. When the next γ-ray is detected, a second logic pulse is sent to the coincidence detector, which adds the logic pulses together and passes the summed signal through a separate PHA. If the logic pulses overlap in time, then the system accepts the two γ-rays as having evolved from one annihilation and records a line integral between the two crystals. The PET system can be characterized by its "coincidence resolving time," which is defined as twice the length of the logic pulse, or 12–20 ns in this case.

Prior to reconstruction, the data must undergo attenuation correction and must have accidental and scattered coincidences removed. Prior to the development of dual CT/PET scanners (see the next section), an external ring source of positron emitters, usually containing germanium-68, was used for a transmission-based calibration. However, with the advent of CT/PET scanners, anatomical information from the CT scan, together with knowledge of tissue attenuation factors, is used for attenuation correction. *Accidental coincidences* refer to events in which the line integral formed by the detection of the two γ-rays is assigned incorrectly. These occur due to the finite coincidence resolving time of the system, γ-rays passing through the crystal and not being detected, and the presence of background radiation. The most common method of estimating accidental coincidences uses additional parallel timing circuitry, which splits the logic pulse from one of the detectors into two components. The first component is used in the standard mode to measure the total number of coincidences. The second component is delayed well beyond the coincidence resolving time so that only accidental coincidences are recorded. The accidental coincidences are then removed from the acquired data. Image reconstruction used either filtered backprojection or iterative methods.

Due to the detection of two γ-rays, the *point spread function* (PSF) in PET is essentially constant through the patient. The PSF is limited by three factors:

1. The finite distance that the positron travels before annihilation with an electron ($\sim$1 mm for ^{18}F)
2. The statistical distribution ($180 \pm -0.3°$), which characterizes the relative trajectories of the two γ-rays, meaning that a 60-cm–diameter ring has a spatial resolution of 1.6 mm, whereas a 100-cm–diameter ring has a resolution of 2.6 mm
3. The size of the detection crystal; one-half of the crystal diameter is often assumed

The most common clinical application of PET is in tumor detection using ^{18}F-FDG. In the body, the radiopharmaceutical FDG is metabolized in exactly the same way as naturally occurring 2-deoxyglucose. Once injected, FDG is actively transported across the blood–brain barrier (BBB) into the cells in brain tissue. Inside the cell, FDG is phosphorylated by glucose hexokinase to give FDG-6-phosphate. This chemical is trapped inside the cell, since it cannot react with G-6-phosphate dehydrogenase, which is the next step in the glycolytic cycle. The amount of intracellular FDG is, therefore, proportional to both the rate of initial glucose transport and subsequent intracellular phosphorylation. Malignant cells, in general, have higher rates of aerobic glucose metabolism than healthy cells; and therefore, in PET scans using FDG, the tumors show up as areas of increased signal intensity, as seen in Figure 1.8.

Future technical advances in PET technology seem likely to be based on time-of-flight (TOF) PET scanners, which can potentially increase the signal-to-noise ratio significantly over today's scanners. If the PET detectors have good time resolution, then the actual location of the annihilation can be estimated by measuring the difference in the arrival times of the two γ-rays. In its original implementation in the early 1980s, the only scintillator that was sufficiently fast was BaF_2, which had a timing resolution of <0.8 ns, corresponding to a blurring of ± 6 mm. However, recently, LSO crystals with much higher detection sensitivity have been used at the detectors. Although not widespread within the clinical community, commercial products using this technology do exist, including the Philips Gemini TF and CPS Hi-Rez systems.

1.4.5 Combined Positron Emission Tomography/Computed Tomography Scanners

The development of dual-modality PET/CT scanners has evolved rapidly from the research laboratory in the late 1990s to clinical practice today. Indeed, essentially all PET scanners are now commercially available only as combined PET/CT systems, and SPECT/CT systems are becoming increasingly common. The two separate scanners are installed adjacently and share a common patient bed. There are two major reasons for using the combined approach:

1. The anatomical information obtained from CT is complementary to the functional information from PET or SPECT and can be used to remove false positives, as will be described.
2. The information from CT can be used for accurate attenuation correction algorithms for the PET or SPECT data to allow better quantitation of the kinetics of biodistribution of the particular agent.

In particular, CT/PET is widely used for imaging of the most commonly used PET agent, ^{18}FDG. Although FDG does accumulate in tumors, it also distributes in regions defined by tissue necrosis and/or inflammation, in addition to biodistribution in many healthy tissues. CT provides the anatomical information that can aid in removing false positives corresponding to these cases.

1.5 Ultrasonic Imaging

Ultrasound is non-ionizing, real-time, portable, and inexpensive compared with other clinical imaging modalities. However, images can be difficult to interpret, requiring expert training. In addition, organs such as the brain located beneath bone cannot be imaged clearly. Nevertheless, ultrasound is particularly functional for obstetrics (fetal imaging) and quantification of blood flow using Doppler measurements.

Clinical ultrasound imaging uses frequencies in the range of 1–15 MHz. Unlike X-rays, mechanical sound propagation requires a medium to support transmission. Ultrasound is a sinusoidal pressure wave that causes the molecules to become displaced from their equilibrium position. A one-dimensional representation of this interaction can be used to simplify this description. Figure 1.9 shows one wavelength of a sinusoidal pressure wave propagating in the x-direction. The pressure oscillates between a maximum (compressional, P_c) and a minimum (rarefractional, P_r) value about an ambient pressure as it moves through the medium. Within the medium, molecules move closer together due to the compressional pressure and spread apart due to the rarefactional pressure. Wave propagation also depends on other parameters, such as density, particle displacement, temperature, attenuation, and other variables that will be covered in this section.

1.5.1 Fundamentals of Ultrasound

Sound waves traveling through a fluid medium cause a periodic change in the density, pressure, and temperature as a function of time. The speed at which the wave travels though a material is given by $c = f\lambda$, where c is the speed of sound (in m/s) through the medium, λ is the wavelength (m) and f is the frequency (s^{-1} or Hz). For water at 20°C, the speed of sound is

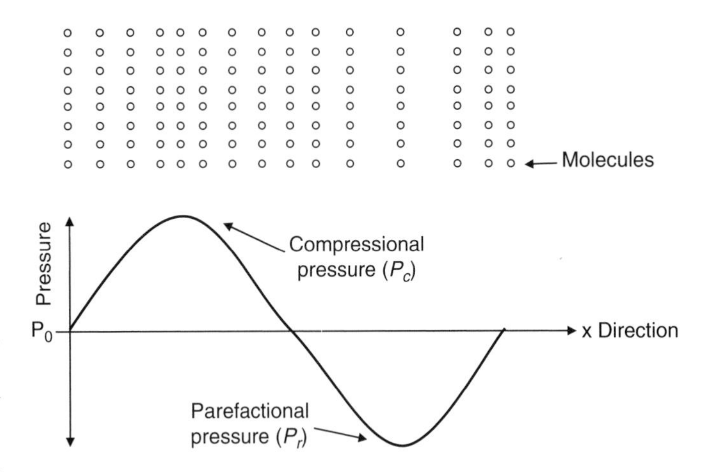

FIGURE 1.9 Schematic of molecular motion within tissue imposed by the passage of an ultrasound wave.

TABLE 1.1 Acoustic properties of biological tissues and matter at temperatures 20–25°C

	Value of $Z \times 10^6$ (kg/[m^{2}s])	Speed of Sound (m/s)
Air	0.0004	330
Blood	1.61	1550
Bone	7.8	3500
Fat	1.38	1450
Brain	1.58	1540
Muscle	1.7	1580
Liver	1.65	1570
Kidney	1.62	1560

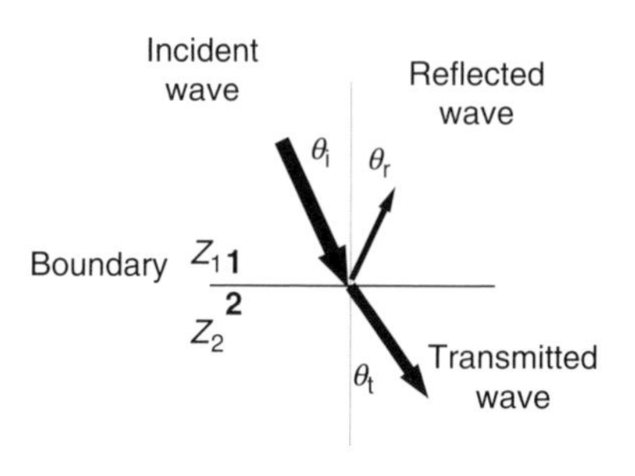

FIGURE 1.10 Behavior of an ultrasound beam incident upon a boundary between two tissues with characteristic impedances Z_1 and Z_2.

1,481 m/s. The speeds of sound in various tissues have values in the range of 1,450 to 1,580, as listed in Table 1.1. The relatively small variations among the different values are due to differences in specific tissue constituents, such as the percentage of protein, collagen, fat, and water. In contrast, bone has a much higher speed of sound. The relationship between density and speed of sound in fluids is given by:

$$c = \sqrt{\frac{B}{\rho}}, \tag{1.6}$$

where ρ is the density (in kg/m^3) and B is the adiabatic bulk modules (Pa, Pascals or N/m^2).

The wave equation describes the propagation of a wave in a lossless medium and is developed from the equations of state and motion and the continuity equation. Changes in density related to changes in pressure are described by the equation of state. The continuity equation is based on the conservation of mass and describes the motion of particles that produces a change in density. Variations in pressure are related to change in particle displacement through the equation of motion or Newton's law of motion. Additionally, the density, pressure, and temperature of a medium vary periodically when a sound wave is passed through the fluid, thereby affecting the speed of sound. Combining the equations of continuity and motion gives the one-dimensional linear, lossless wave equation:

$$\frac{\partial^2 P}{\partial x^2} = \frac{1}{c^2}\frac{\partial^2 P}{\partial t^2}. \tag{1.7}$$

The wave equation explicitly shows the direct relationship between the pressure wave as a function of space (distance traveled) and time. The characteristic impedance, Z, of a material is defined as:

$$Z = \rho c, \tag{1.8}$$

in which Z has units of kg/(m^{2}s). Table 1.1 lists the characteristic impedances for air, water, and selected tissues. Acoustic impedance implies resistance to the propagating ultrasound wave. As a wave travels through different layers of tissue, it encounters different specific acoustic impedances, and therefore a certain fraction of the intensity of the wave is transmitted, with the remainder being reflected at the interface between the different tissues. Figure 1.10 shows an ultrasound wave traveling through a medium with impedance Z_1 to another medium with impedance Z_2. The pressure reflection coefficient (R_p) and transmission reflection coefficient (T_p) are given by:

$$R_p = \frac{p_r}{p_i} = \frac{Z_2\cos\theta_i - Z_1\cos\theta_t}{Z_2\cos\theta_i + Z_1\cos\theta_t},$$

$$T_p = \frac{p_t}{p_i} = \frac{2Z_2\cos\theta_i}{Z_2\cos\theta_i + Z_1\cos\theta_t}, \tag{1.9}$$

where p_r and p_i are the reflected and incident pressures, respectively. Equation 1.9 shows that the reflected wave will undergo a 180° phase shift from the incident if the wave travels from material of low acoustic impedance to one of high impedance; that is, $Z_1 < Z_2$. Snell's law governs the refracted wave at the boundary of fluids:

$$\frac{\sin\theta_i}{\sin\theta_t} = \frac{c_1}{c_2}, \tag{1.10}$$

where c_1 and c_2 are the speeds of sound in fluids 1 and 2, respectively. In the case of a tissue/air interface, almost all of the ultrasound energy is reflected ($R_p = -0.99$). Clinically, this demonstrates why lung imaging is difficult, given that almost all of the energy is reflected at the boundary.

Attenuation of the ultrasound wave as it passes through tissue is comprised of two effects: absorption and scattering. The absorption mechanism consists of viscous losses, heat conduction, and relaxation processes, while scattering occurs when acoustic energy is deflected or redirected from its normal propagation. Recalling that sound waves in a medium cause expansions and contractions (Figure 1.10), we note that fluids exhibit resistance to the distortion, which is known as *viscosity* (η). Thus, the relative motion between adjacent parts of the medium caused by expansions and compressions leads to a viscous loss or frictional loss. Thermal losses result from conduction of thermal energy between higher-temperature compressions and lower-temperature rarefactions. Taking into account both viscous and thermal conductivity losses through the medium gives rise to the classical absorption coefficient. *Relaxation* refers to the dynamics of the disturbance of the structure of a fluid due to a propagating wave, and different mechanisms are characterized by different relaxation times. An example of how a wave is attenuated and the significance of the relaxation time is exhibited when the period of the acoustic cycle is greater than the time required for a portion of the compression energy of the fluid to be converted into internal energy of molecular vibration. During the expansion cycle, some of this energy will be delayed in its restoration, resulting in a tendency toward pressure equalization and an attenuation of the wave.

When a sound wave encounters a small (relative to the ultrasound wavelength) solid obstacle, a fraction of the wave is scattered. *Scattering* can be defined as the change of amplitude, frequency, phase velocity, or direction of propagation as the result of an obstacle or nonuniformity in the medium. Different behavior is seen for a scattering volume consisting of a single scatterer or a statistical distribution of scatterers. The degree and directionality of scattering are affected by the physical properties of the scatterer, such as its density, compressibility, roughness, and thermal conductivity.

1.5.2 Transducers and Beam Characteristics

When polarized crystalline or ceramic materials are subjected to mechanical stress, they produce an electrical voltage. The converse is also true, such that an oscillating electrical voltage causes the material to vibrate, thereby producing a pressure wave in a medium in direct contact with the material. This phenomenon, known as the *piezoelectric effect,* forms the basis of an ultrasound transducer. Transducers are usually made from polarized ferroelectric ceramics such as lead zirconate titanate (PZT). The resonance frequency, f_o, of the transducer is defined as:

$$f_o = \frac{c_{crystal}}{2t}, \tag{1.11}$$

where $c_{crystal}$ is the speed of sound in the piezoceramic ($\approx$ 4000 m/s for PZT) and t is the ceramic thickness (Figure 1.11). The ceramic itself is often represented as a disk that is electrically driven by silver-coated electrodes attached to opposite faces of the disk. Applying a sinusoidal voltage at frequency f_o causes the disk to vibrate and produce a pressure wave at f_o. Since the spatial resolution in the axial direction is proportional to the length of the pulse in tissue, the transducer is mechanically damped to produce a short pulse of energy.

The radiation or spatial intensity field from a circular piston is a complicated three-dimensional pattern (Figure 1.11). Close to the face of the transducer, the pressure field oscillates between a series of maxima and null points. The final oscillation is known as the last axial maximum, located at

$$\text{last axial maximum} \cong \frac{a^2}{\lambda}. \tag{1.12}$$

For a plane piston, this location also forms the boundary between the near field ("Fresnel zone") and the far field ("Fraunhofer zone") of the transducer. Beyond the far field, the beam diverges at an angle $\theta = \sin^{-1}(0.61\lambda/a)$. Similar to a radiating radio antenna, the off-axis field pattern also has a series of much smaller pressure field lobes and nulls (not shown). The null between the main lobe and the first side lobe is at $\theta = \sin^{-1}(0.61\lambda/a)$.

Image formation (see the next section) using a single-element transducer requires mechanical movement over the region of interest. The vast majority of transducers used in clinical practice are transducer arrays, consisting of a large number of much smaller elements, which can be driven independently. These arrays can be one dimensional or two dimensional, as shown in Figure 1.11.

1.5.3 Image Acquisition and Display

Single lines of pulse–echo ultrasound are termed *A-mode lines.* Knowing the speed of sound in tissue, the time delay between transmission and signal reception defines the depth of the reflected or backscattered signal. The beam can be swept through the region of interest by varying the excitation times

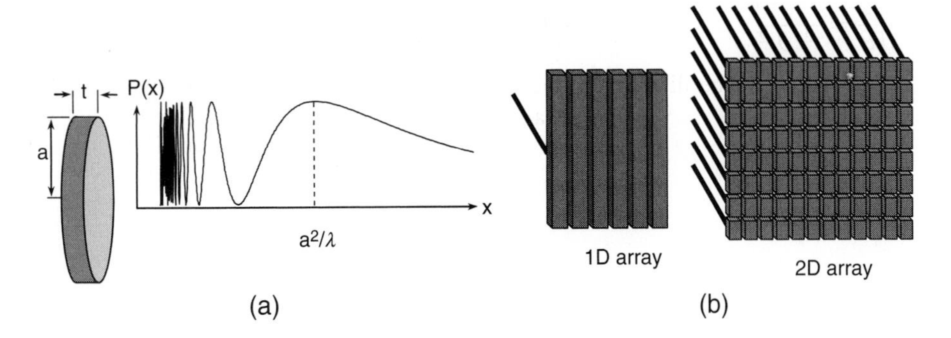

FIGURE 1.11 (a) Plot of the pressure produced by a plane-piston single-element transducer as a function of distance from the transducer. (b) Schematics of one- and two-dimensional transducer arrays.

of individual elements in a transducer array to form a B-mode (brightness) image. Although array systems are electrically complex, their overriding advantages are ease of focusing and multidimensional image acquisition. B-mode imaging can be used to examine stationary organs such as the kidney, breast, and liver, or moving objects such as the beating heart or the flow of blood in the carotid artery. Linear arrays are designed for use in conventional high-resolution imaging of musculoskeletal or superficial vascular features and can also be used in compound scanning (SonoCT) and Doppler blood velocity determination. Two-dimensional arrays can have up to 2,400 elements and produce full-volume images for cardiology applications. Arrays are attractive because they can be used to focus on an object or organ within the body by varying the transmit and to receive signals (phasing) to the elements. Three-dimensional volume imaging can be acquired by mechanically scanning a phased-array transducer perpendicular to the plane of each B-mode scan. Figure 1.12 shows a conventional two-dimensional image of a fetus compared with a three-dimensional volume image of a fetus.

Ultrasound images often contain "artifacts," which can be misinterpreted unless a skilled technician is interpreting the images. In fact, these artifacts contain valuable information if understood. Examples of image artifacts include reverberations, acoustic shadows, and speckle. *Reverberations* are the appearance of equally spaced repeating lines in an image, caused by the transducer being located near a strong reflector. *Acoustic shadows* occur when the sound field is transmitted and reflected through a highly attenuating object or organ. In the image, the shadow appears as a dark area behind the object of interest. The appearance of light and dark spots in a homogeneous material such as liver is called *speckle*. This pattern arises from the constructive and destructive interference of waves as a result of scattering from small structures. One of the most recent advances in ultrasound imaging is the use of compound scanning (also known as *SonoCT*) to overcome many of the image artifacts found in conventional scanning. Compound imaging adjusts the phasing of the array elements to obtain multiple image views and planes at several angles. These tomographic images are combined in real time into a single averaged image. The acquisition of these averaged images at multiple angles suppresses artifacts such as speckle, noise, and shadows and reinforces real structures and organs.

Ultrasound can also be used to measure blood flow using the well-known Doppler effect. A continuous wave (CW) Doppler system consists of a probe with two transducer elements (one for transmit, the other for receive) and the ultrasound beam aligned at an angle θ to the blood vessel. The change in the ultrasound frequency, Δf, or the Doppler shift frequency, compared with the incident transmit frequency, f_i, is given by:

$$\Delta f = f_i - f_r = \frac{2v\cos\theta}{c} f_i, \quad (1.13)$$

where c is the speed of sound in blood, v is the blood flow velocity, and f_r is the frequency measured at the receive element. In contrast, flow velocity measurements from a pulsed Doppler system used a single transducer operating in pulse–echo mode. Here, the transducer sends a short ultrasound pulse that is backscattered from the moving blood, and the signal is detected by the same tranducer. The advantage to pulsed Doppler is that the pulse–echo signals can be gated to acquire flow information within a specific region of interest, defined by a minimum and maximum depth:

$$depth_{\min} = \frac{c(t_d - t_p)}{2} \qquad depth_{\max} = \frac{c(t_d + t_g)}{2}, \quad (1.14)$$

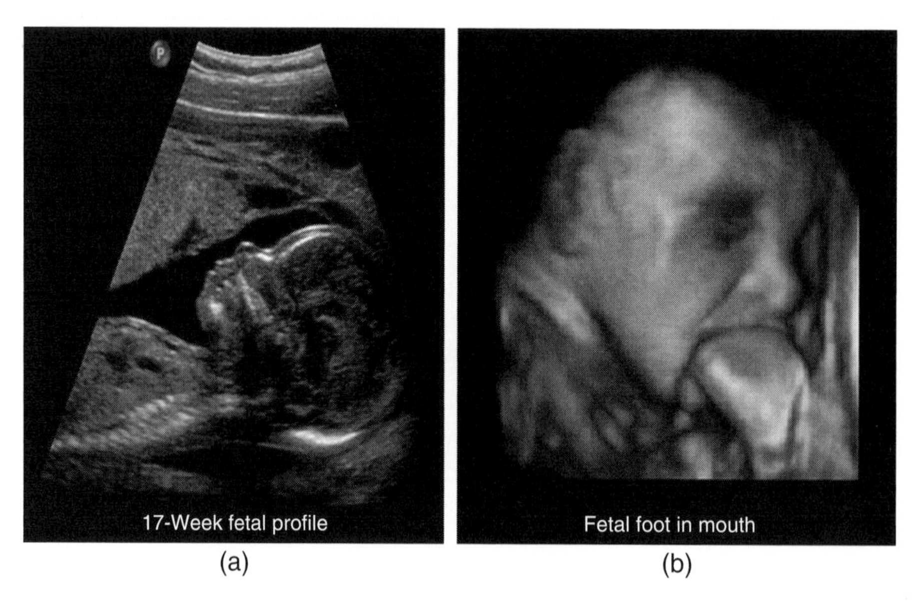

FIGURE 1.12 (a) Two-dimensional B-mode ultrasound image of the fetus *in utero*. (b) Three-dimensional ultrasound image.

where t_p is the duration, in seconds, of the transmitted pulse; t_d is the time delay (s) between the end of the transmitted pulse and the receiver gate being opened; and t_g is the time (s) during which the receiver gate is on to detect the return echo from the moving blood. Compared with CW Doppler, one disadvantage is that there is a limit to the highest blood velocity, v_{max}, that the system can determine, given by

$$v_{max} = \frac{c^2}{8f_i depth_{max}}. \quad (1.15)$$

This limit is based on the Nyquist criterion that the sampling rate must be greater than twice the highest frequency present in the signal.

1.6 Magnetic Resonance Imaging

MRI is a non-ionizing technique with excellent soft-tissue contrast and high spatial resolution (~1 mm). The temporal resolution is typically much slower than for ultrasound or CT, with scans lasting several minutes. The cost of MRI scanners is relatively high, and the large superconducting magnet requires special housing in clinical environments. The major uses of MRI are in the areas of brain disease, spinal disorders, angiography, cardiac assessment, and musculoskeletal damage.

1.6.1 Basis of Magnetic Resonance

The first requirement for MRI is to produce a strong, temporally stable and spatially homogeneous magnetic field within the patient. The majority of magnets use superconductor technology to produce the magnetic field. The superconducting wire must be able to carry a large current, which limits the material to certain alloys, particularly niobium-titanium, which is formed into multistranded filaments within a copper conducting matrix. This superconducting matrix is housed in a stainless steel can containing liquid helium at a temperature of 4.2 K. This can is surrounded by a series of radiation shields and vacuum vessels, with an outer container of liquid nitrogen being used to cool the outside of the vacuum chamber and the radiation shields. The most common fields for clinical scanning are 3-tesla systems, although systems operating at 7 tesla now exist for experimental human investigations.

When protons are placed in a strong external magnetic field, the interaction between their magnetic moments and the magnetic field means that they can align in two different configurations, commonly termed "parallel" and "anti-parallel" states, shown in Figure 1.13. The number of protons in each state is given by the Boltzmann distribution:

$$\frac{N_{\text{anti-parallel}}}{N_{\text{parallel}}} = \exp - \left[\frac{\Delta E}{kT}\right] = \exp - \left[\frac{\gamma h B_o}{2\pi kT}\right], \quad (1.16)$$

where B_0 is the strength of the magnetic field, k is Boltzmann's constant, h is Plank's constant, ΔE is the energy gap between the two states, and T is the temperature in Kelvin. The size of the MRI signal is proportional to the difference in populations between the two energy levels:

$$N_{\text{parallel}} - N_{\text{anti-parallel}} = N_s \frac{\gamma h B_0}{4\pi kT}, \quad (1.17)$$

where N_s is the total number of protons in the body. Despite large static magnetic fields, Equation 1.17 shows that at an operating magnetic field of 3 tesla, for every one million protons, there is a population difference of only approximately ten protons between the parallel and anti-parallel orientations. In order to stimulate transitions between energy levels, electromagnetic energy has to be applied at a frequency (ω) corresponding to the difference between the two levels:

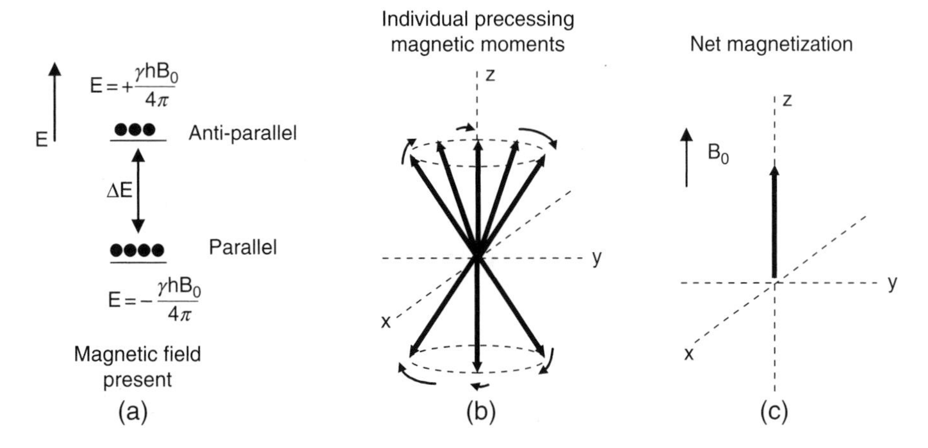

FIGURE 1.13 (a) Zeeman splitting of the proton energy levels induced by application of a static magnetic field. (b) Precession of all of the proton magnetic moments about the applied magnetic field. (c) Net magnetic moment at equilibrium aligned in the direction of the magnetic field.

$$\frac{h\omega}{2\pi} = \Delta E = \frac{\gamma h B_0}{2\pi} \Rightarrow \omega = \gamma B_0. \quad (1.18)$$

If one considers each magnetic moment as a vector (Figure 1.13), then the equilibrium condition is characterized by the z-component of magnetization (M_z) being M_0 (the total magnetization of the patient), with the transverse component (M_{xy}) equal to zero. After a pulse of radiofrequency (RF) energy has been applied, the magnetization is tipped from the z direction (Figure 1.13) into the transverse plane and precesses around the direction of the applied magnetic field at the *Larmor frequency*, given by $\omega = \gamma B_0$. After spatial encoding using magnetic field gradients (see next section), the signal is detected via Faraday induction using an RF coil. Often, the same coil is used to transmit the RF energy and to detect the signal. There are many forms of coil, depending upon whether the RF field produced should be homogeneous over a large volume of the patient or only a small localized volume is to be investigated. Since Faraday's law states that voltage is proportional to the time-dependent rate of magnetic flux, a higher B_0 field gives a higher precessional frequency and hence a higher signal voltage. Overall, therefore, the measured MRI signal is proportional to the square of the B_0 value, providing a major impetus to the ever-increasing static magnetic fields.

Absorption of electromagnetic energy by the spin system results in a non-Boltzmann distribution of the population levels, equivalent to a nonequilibrium value of the M_z and M_{xy} components of magnetization. The return to thermal equilibrium is governed by two different relaxation times: T_1 determines the return of M_z to M_0, and T_2 the return of M_{xy} to zero. Different tissues have quite different values of T_1 and T_2, as shown in Table 1.2, and these differences can be used to introduce contrast into MR images.

1.6.2 Magnetic Field Gradients

In order to introduce spatial information into the MR signal and thereby form images, magnetic field gradients are used to make the proton precessional frequency spatially dependent. Three separate gradient coils are required to encode the three spatial dimensions within the body. Since only the z-component of the magnetic field interacts with the proton magnetic moments, it is the spatial variation in the z-component of the magnetic field (B_z) that is important. Image reconstruction is simplified considerably if the magnetic field gradients are linear over the region to be imaged; that is,

$$\frac{\partial B_z}{\partial z} = G_z \quad \frac{\partial B_z}{\partial x} = G_x \quad \frac{\partial B_z}{\partial y} = G_y. \quad (1.19)$$

By convention, for human studies, the z direction lies along the head-to-foot axis; the y-axis corresponds to the vertical (spine to abdomen) direction, and the x-axis goes from side to side (right to left). The magnetic field, B_z, experienced by all nuclei with a common coordinate z, is:

$$B_z = B_0 + zG_z, \quad (1.20)$$

where G_z has units of tesla per meter. The corresponding precessional frequencies (ω_z) of the protons, as a function of their position in z, are given by:

$$\omega_z = \gamma B_z = \gamma(B_0 + zG_z). \quad (1.21)$$

Analogous expressions can be obtained for the spatial dependence of the resonant frequencies in the presence of the x- and y-gradients. The requirements for gradient coil design are that the gradients be as linear as possible over the region being imaged, that they be efficient in terms of producing high gradients per unit current, and that they be fast in switching times for use in rapid imaging sequences. Copper is used as the conductor, with chilled-water cooling to remove the heat generated by the current. The simplest configuration for a coil producing a gradient in the z direction is a *Maxwell pair*, shown in Figure 1.14 (a), which consists of two separate loops of multiple turns of wire, each loop containing equal currents flowing in opposite directions. The magnetic field produced by this gradient coil is zero at the center of the coil

TABLE 1.2 Tissue relaxation times at 1.5 tesla

Tissue	T_1 (ms)	T_2 (ms)
Fat	260	80
Muscle	870	45
Brain (gray matter)	900	100
Brain (white matter)	780	90
Liver	500	40
Cerebralspinal fluid	2400	160

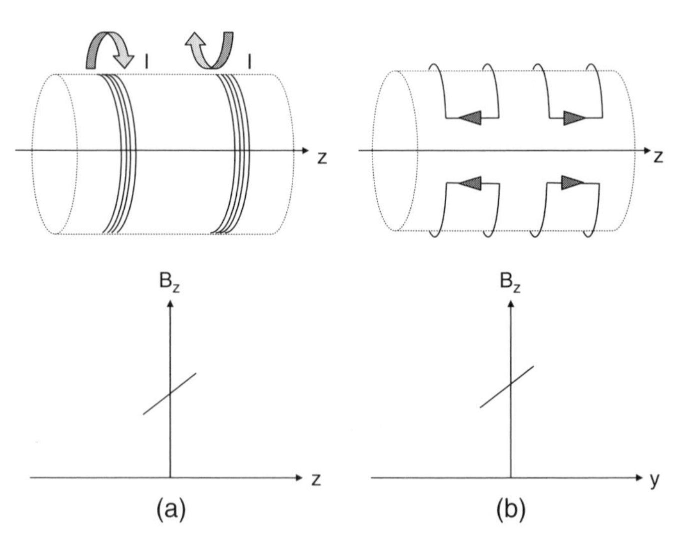

FIGURE 1.14 (a) Maxwell pair used to produce a linear magnetic field in the z direction. (b) Four-element Golay coils used to produce a linear magnetic field in the y direction.

and is linearly dependent upon position in the z direction over about one-third of the separation of the two loops. The x- and y-gradient coils are completely independent of the z-gradient coils: The usual configuration is to use four arcs of wire, as shown in Figure 1.14 (b).

When the current in the gradient coils is switched rapidly, eddy currents can be induced in nearby conducting surfaces, such as the radiation shield in the magnet. These eddy currents, in turn, produce additional unwanted gradients that may decay only very slowly, even after the original gradients have been switched off. All gradient coils in commercial MRI systems are now "actively shielded" to reduce the effects of eddy currents. Active shielding uses a second set of coils placed outside the main gradient coils, the effect of which is to minimize any stray gradient fields.

1.6.3 Fourier Imaging Techniques

Acquisition of the data required for conventional MRI comprises three independent components: slice selection, phase encoding, and frequency encoding. The combination of a frequency-selective RF pulse and the slice-select gradient excites protons only within a thickness given by $\Delta\omega/\gamma G_{\text{slice}}$, where $\Delta\omega$ is the frequency bandwidth of the pulse; protons outside this slice are not excited. Application of the phase-encoding gradient G_{phase} for a time τ_{pe} prior to data acquisition imparts a spatially dependent phase shift into the signal given by:

$$\phi(G_y, \tau_{pe}) = \omega_y \tau_{pe} = \gamma G_y y \tau_{pe}, \tag{1.22}$$

where y is denoted as the phase-encoding direction. During signal acquisition, the frequency-encoding gradient G_{freq} generates a spatially dependent precessional frequency in the acquired signal. Overall, ignoring relaxation effects, the detected signal is given by:

$$s(G_y, \tau_{pe}, G_x, t) \propto \int_{\text{slice}} \int_{\text{slice}} \rho(x, y) e^{-j\gamma G_x x t} e^{-j\gamma G_y y \tau_{pe}} dxdy, \tag{1.23}$$

where $\rho(x,y)$ is the proton density (that is, the number of protons at a given (x,y) coordinate) and x is the frequency-encoding dimension. If two variables are defined:

$$k_x = \frac{\gamma}{2\pi} G_x t, \quad k_y = \frac{\gamma}{2\pi} G_y \tau_{pe}, \tag{1.24}$$

then the acquired MRI signal can be expressed as:

$$S(k_x, k_y) \propto \int_{\text{slice}} \int_{\text{slice}} \rho(x, y) e^{-j2\pi k_x x} e^{-j2\pi k_y y} dxdy. \tag{1.25}$$

Image reconstruction is obtained by an inverse two-dimensional Fourier transform:

$$\rho(x, y) = \int_{-\infty}^{\infty} \int_{-\infty}^{\infty} S(k_x, k_y) e^{+j2\pi(k_x x + k_y y)} dk_x dk_y. \tag{1.26}$$

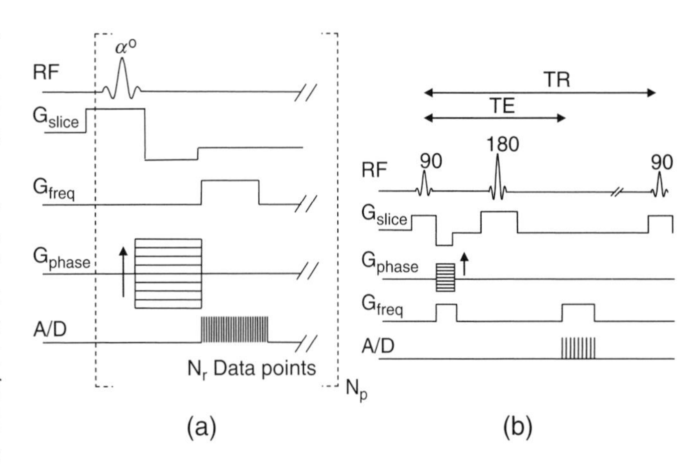

FIGURE 1.15 (a) Gradient–echo imaging sequence. (b) Spin–echo imaging sequence.

The two most commonly used sequences are shown in Figure 1.15. The gradient–echo sequence is used for rapid imaging, whereas the spin–echo sequence has a higher intrinsic sensitivity. Each imaging sequence is repeated N_p times, with the phase-encoding gradient incremented for each repetition. This results in N_p lines being acquired in the k_y direction, and N_r points in the k_x direction. Two delays are defined and can be altered by the operator:

- TE = the echo time, which is defined as the delay between the middle of the initial RF pulse and the center of the data acquisition time.
- TR = the repetition time, defined as the time between successive applications of the sequence.

When the effects of T_1 and T_2 relaxation are taken into account, it can be shown that in a gradient–echo sequence, the image intensity $I(x,y)$ is given by:

$$I(x, y) \propto \frac{\rho(x, y)\left(1 - e^{-TR/T_1}\right) e^{-TE/T_2^*} \sin\alpha}{1 - e^{-TR/T_1} \cos\alpha}, \tag{1.27}$$

where T_2^* is the spin–spin relaxation time, including the effects of magnetic field inhomogeneity. For a spin-echo imaging sequence, the corresponding expression is:

$$I(x, y) \propto \rho(x, y)\left(1 - e^{-TR/T_1}\right) e^{-TE/T_2}. \tag{1.28}$$

The times TR and TE within the imaging sequence can be chosen to give different contrasts in the image. For example, Figure 1.16 shows the effects of increasing the TE on a simple brain scan acquired with a spin–echo sequence.

One of the most important techical developments in the past few years has been the introduction of parallel imaging, in which a degree of spatial encoding is performed by an array of small RF coils. Using this type of technology, the number of phase-encoding steps can be reduced up to a theoretical limit of the number of RF coils, thus speeding up data acquisition

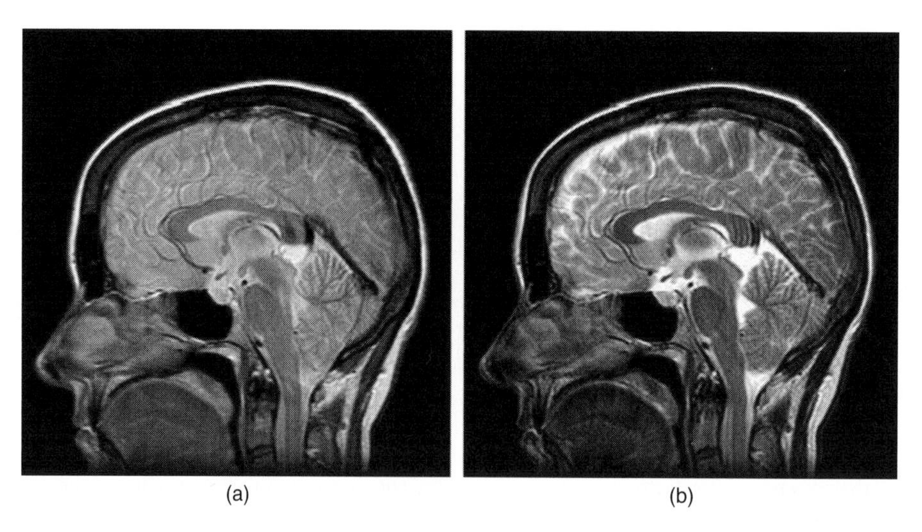

FIGURE 1.16 Sagittal images through the human brain with less (a) and more (b) T_2 weighting.

considerably. Most commercial systems now offer this capability under various acronyms, with acceleration factors up to an order-of-magnitude having been shown in developmental systems.

1.6.4 Magnetic Resonance Imaging Contrast Agents

As with many imaging modalities, contrast agents can be used to improve contrast in MR images. There are two general types of agents used in MR:

1. "Positive" MR contrast agents—those that produce high intensity on images—are extensively used in tumor diagnosis and MR angiography. These paramagnetic agents are not detected per se, unlike the tracers used in nuclear medicine, but work by reducing the T_1 value of the water protons that either transiently bind to or diffuse close to the agent: These two mechanisms are termed "inner sphere" and "outer sphere," respectively. These agents are therefore used in conjunction with so-called T_1-weighted sequences. The most commonly used agents are gadolinium chelates, since the Gd^{3+} ion has seven unpaired electrons, and these cause very efficient T_1 relaxation of neighboring protons in water molecules. Commonly used agents are Gd-DTPA (trade name Magnetvist), Gd-DTPA-bis(methylamide) (Gd-DTPA-BMA, trade name Omniscan), and ($\pm$)-10 (2-hydroxypropyl)-1,4,7,10-tetraazacyclodecane-1,4,7-triacetatogadolinium[III] (Gd-HP-DO3A, trade name Prohance).
2. "Negative" MR contrast agents are based on small ferromagnetic iron particles, with various types of coating and size distributions. Ferridex is a liver imaging agent approved by the U.S. Food and Drug Administration that consists of dextran-coated superparamagnetic iron oxide (SPIO) particles with diameters in the range of 80–100 nm. These agents reduce the T_2 value of the water protons by causing inhomogeneities in the local magnetic field and therefore producing areas of signal void in T_2-weighted sequences. Since the particles accumulate in healthy regions of the reticuloendothelial system (liver, spleen, lymph node, bone marrow), comparisons of images before and after administration of the agent reveal diseased regions with unchanged signal intensity.

One of the most recent developments is the design of *molecular imaging agents*. These have so far been used only in animal studies, but they hold immense promise for the future. True molecular imaging agents can be used, for example, to detect the presence of different types of enzymes. Figure 1.17 shows one such example in which the contrast agent is in an "inactive state" (a) in the absence of the enzyme (since all the coordinate sites around the Gd are filled). In the presence of the particular enzyme (b), one of the coordinate sites becomes vacant, and water can undergo very efficient inner-sphere relaxation.

1.7 Diffuse Optical Imaging

Near infrared (NIR) imaging methods are characterized by their noninvasive nature (milliwatt-levels of energy), chemical specificity (capable of resolving concentrations of oxy- and deoxyhemoglobin), and good temporal resolution (typically on the order of 10 ms per measurement). In addition, NIR image systems are portable and inexpensive and therefore make bedside application feasible. Despite being a relatively "young" imaging technique, NIR methods have already found a number of *in vivo* biomedical applications, including

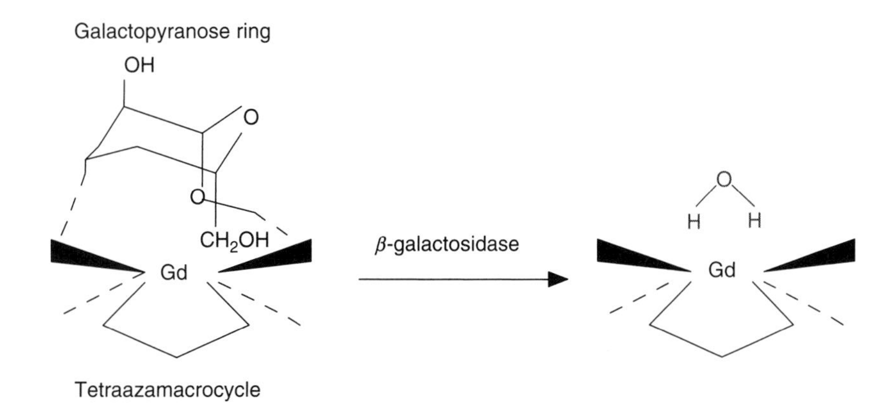

FIGURE 1.17 Schematic of the operation of a molecular imaging agent sensitive to the presence of β-galactosidase. Activation of the contrast agent involves cleaving of the chemical bonds to the galactopyranose ring, which opens up a coordination site for water to interact with the central gadolinium ion.

mammography and real-time monitoring of blood oxygenation levels of patients during medical procedures.

1.7.1 Propagation of Light Through Tissue

For biomedical applications, the NIR spectrum of interest spans approximately 650 to 950 nm. The absolute concentrations of blood constituents (e.g., oxy- and deoxyhemoglobin, lipid) are of great interest. Light propagating in tissue is subject to not only absorption but to scattering processes. Within the window of the NIR spectrum, the absorption and scattering properties of tissue allow a measurable amount of light to pass through a clinically useful quantity of tissue. Below 650 nm, the absorption of hemoglobin increases to a point that no measurable amount of light can travel through tissue. Above 950 nm, the absorption of water is so significant that tissue becomes practically opaque, as shown in Figure 1.18.

Light propagation in a medium of arbitrary geometry is most commonly modeled using the Boltzmann transport equation, alternatively known as the *radiative transfer equation* (RTE). Although the RTE ignores electromagnetic wave properties such as polarization, and particle properties such as inelastic collisions, it is generally sufficient to describe the interaction of photons with tissue for medical imaging. The

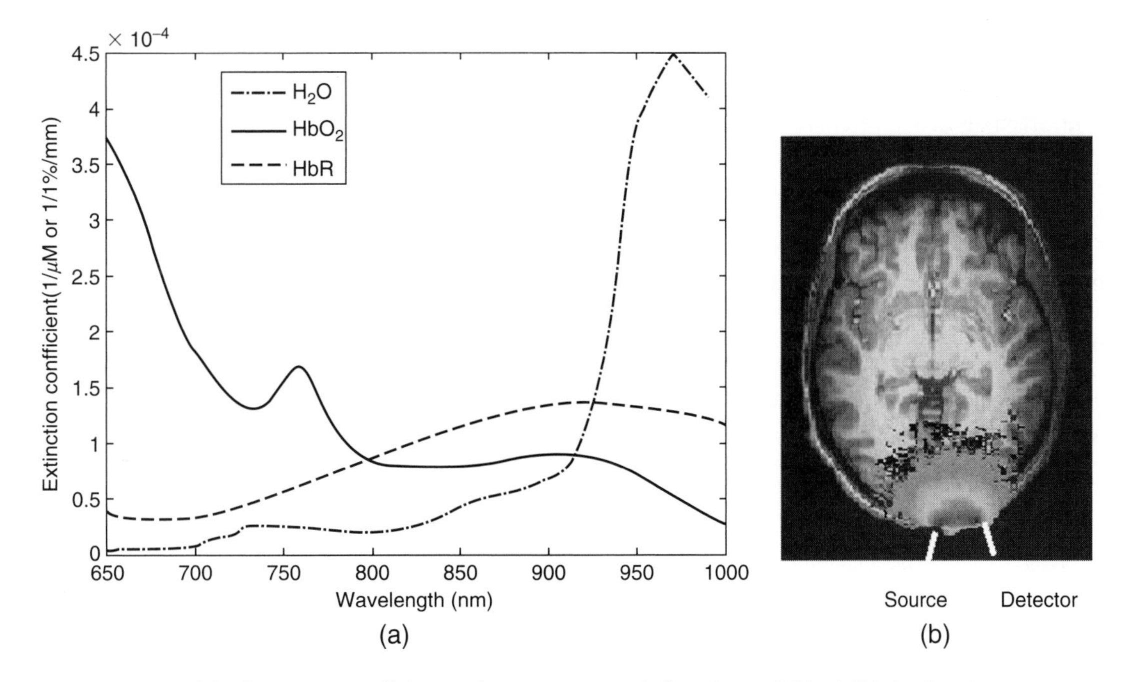

FIGURE 1.18 (a) Absorption coefficients of water, oxy-, and deoxyhemoglobin (Hb) in the NIR spectrum (650–1000 nm). (b) The normalized intensity (log scale) of the "banana-shaped" light bundle (modulation frequency 150 MHz) of a particular optical channel through the human head.

RTE can be reduced to a much simpler form, known as the diffusion equation:

$$\left(\frac{1}{c(r)}\frac{\partial}{\partial t} - \nabla\cdot\frac{1}{3\left(\mu_a(r)+\mu_s'(r)\right)}\nabla + \mu_a(r)\right)\Phi(r,t) = q_0(r,t), \quad (1.29)$$

where Φ is the radiant energy fluence, μ_a is the absorption coefficient, μ_s' is the reduced scattering coefficient, c is the speed of light in the medium, q_0 is a point light source, t is time, and r represents position. The frequency-domain counterpart of equation (1.29) is given by:

$$\left(\frac{i\omega}{c(r)} - \nabla\cdot\frac{1}{3\left(\mu_a(r)+\mu_s'(r)\right)}\nabla + \mu_a(r)\right)\Phi(r,\omega) = q_0(r,\omega). \quad (1.30)$$

One can use these equations to model photon migration paths through different tissue types. For example, Figure 1.18 shows results from a Monte Carlo simulation of light passage through the human brain, superimposed on a structural MRI scan.

Retaining only the first-order expansion of a perturbation series of the diffusion equation is referred to as the *Born approximation.* If only the change in the absorption coefficient is being studied (e.g., in hemodynamic functional studies), this results in:

$$\delta\Phi(r_s,r_d) = -\int_{\Omega}\delta\mu_a(r)\Phi_0(r_s,r)G_0(r,r_d)d^3r, \quad (1.31)$$

where $\delta\Phi$ is the difference of fluence radiated from the light source and measured by the detector; $\delta\mu_a$ is the change in absorption coefficient; and the integral kernel $\Phi_0(r_s,r)G_0(r,r_d)$ is the product of the fluence (evaluated at r in response to a source at r_s) and Green's function (evaluated at r in response to a source placed at the detector position r_d). The integral kernel is evaluated using the optical properties at the resting state (baseline). The physical significance of Equation (1.31) is that the change in the absorption coefficient (assuming constant scattering coefficient and refractive index) is related to the change in the optical signal (the difference between the emitted and detected lights), given the baseline condition of the medium.

1.7.2 Measurement of Blood Oxygenation

In order to relate the measured absorption and scattering of light to the underlying physiology, the starting point is the modified Beer-Lambert's law:

$$OD = -\ln\left(\frac{I}{I_0}\right) = \varepsilon C d\sigma + G, \quad (1.32)$$

where OD is the optical density; I_0 and I are the input and output light intensities, respectively; ε is the extinction coefficient, which is a function of the type of absorbers and the wavelength of light; C is the concentration of the absorber; d is the optical source-detector distance; σ is the differential pathlength factor (DPF), which is a function of the wavelength of light and the type of tissue and represents the increment of effective pathlength of light due to scattering; and G is a factor accounting for measurement geometry and contact loss. Physiological *changes* can be measured via:

$$\Delta OD = -\ln\left(\frac{I_{final}}{I_{initial}}\right) = \varepsilon\Delta C d\sigma = \Delta\mu_a d\sigma, \quad (1.33)$$

where $\Delta\mu_a$ is the change in absorption coefficient. Absorption due to different absorbers can be superimposed. Considering only the absorption due to oxy- (HbO_2) and deoxyhemoglobin (HbR) at a given wavelength λ, Equation 1.33 can be rewritten as

$$\Delta OD^{\lambda} = \left(\varepsilon^{\lambda}_{HbO_2}\Delta[HbO_2] + \varepsilon^{\lambda}_{HbR}\Delta[HbR]\right)\sigma^{\lambda} d. \quad (1.34)$$

In Equation 1.34, d is determined by the geometry of the optical probe and $\varepsilon^{\lambda}_{HbO_2}$, $\varepsilon^{\lambda}_{HbR}$, and σ^{λ} are specified by the optical properties of tissue and the wavelength of light. For a given system, $\varepsilon, \sigma, \lambda$, and d are constants. Therefore, there are only two variables: $\Delta[HbO_2]$ and $\Delta[HbR]$, which can be determined by taking measurements at two different wavelengths

$$\Delta[HbR] = \frac{\varepsilon^{\lambda_2}_{HbO_2}\Delta\mu_a^{\lambda_1} - \varepsilon^{\lambda_1}_{HbO_2}\Delta\mu_a^{\lambda_2}}{\left(\varepsilon^{\lambda_1}_{HbR}\varepsilon^{\lambda_2}_{HbO_2} - \varepsilon^{\lambda_2}_{HbR}\varepsilon^{\lambda_1}_{HbO_2}\right)},$$
$$\Delta[HbO_2] = \frac{\varepsilon^{\lambda_1}_{HbR}\Delta\mu_a^{\lambda_2} - \varepsilon^{\lambda_2}_{HbR}\Delta\mu_a^{\lambda_1}}{\left(\varepsilon^{\lambda_1}_{HbR}\varepsilon^{\lambda_2}_{HbO_2} - \varepsilon^{\lambda_2}_{HbR}\varepsilon^{\lambda_1}_{HbO_2}\right)}, \quad (1.35)$$

where $\Delta\mu_a^{\lambda}$ is the change in absorption coefficient at wavelength λ. In terms of other physiologically interesting parameters, the sum of $\Delta[HbO_2]$ and $\Delta[HbR]$ is the change in total hemoglobin concentration $\Delta[HbT]$, which is proportional to the change in regional cerebral blood volume (ΔrCBV). In addition, the ratio of $\Delta[HbO_2]$ to $\Delta[HbT]$ or $\Delta[HbR]$ is a good indicator of the change in tissue oxygenation level.

1.7.3 Image Reconstruction

If only very few optical sources and detectors are used in measurements, then very limited spatial localization of the signal can be achieved. Increasingly, systems capable of using dozens of sources and detectors are being developed. Even with this number, the process of image reconstruction with a reasonable spatial resolution represents a considerable challenge, since the problem is mathematically underdetermined and ill-posed and requires sophisticated mathematical treatment. Predicting optical signals based on the knowledge of the optical sources and detectors (such as their positions, sizes, orientations, etc.) and the optical properties of the medium (such as the absorption and scattering coefficients) is referred to as the *forward problem.* The process of reconstructing the optical properties of the medium, knowing the relevant parameters of the optical sources and detectors as well as the measured optical signals, is consequently called the *inverse problem.*

As an example, suppose that one is measuring the change in absorption coefficient associated with functional brain activity. The measured optical signal is recorded as a function of time for all source–detector combinations, or optical channels. The time-series data acquired for each optical channel are termed a measurement. It follows that for m measurements in a field of view (FOV) defined by n voxels, one can formulate the image reconstruction problem as a generalized matrix

$$AX = b. \tag{1.36}$$

The m-by-n coefficient matrix A is the so-called Jacobian matrix, where

$$a_{ij} = -\Phi_0(r_s,r)G_0(r,r_d)\Delta V. \tag{1.37}$$

The measurements b are represented by an m-by-one vector, called the measurement vector, where

$$b_i = \Delta\Phi(r_{s,i},r_{d,i}). \tag{1.38}$$

The solution X is a one-by-n vector that contains all voxels in the image, where

$$x_j = \Delta\mu_a(r_j). \tag{1.39}$$

If A is a singular or near-singular matrix, the matrix is noninvertible. If A is not a square matrix ($m \neq n$), the direct inversion of matrix A is not well defined. Under both conditions, singular value decomposition (SVD) can be used to solve the linear equations. This produces a nonnegative diagonal matrix S and unitary matrices U and V, such that

$$A = USV^T, \tag{1.40}$$

and the solution is given by

$$X = VS^{-1}U^Tb. \tag{1.41}$$

Since S is diagonal, its inversion is trivial. In this way, the inversion of an ill-conditioned A is performed via simple matrix multiplications. Variations of the SVD method have been developed to address the problem of instability due to singular values, which effectively amplifies the noise.

An alternative approach is to use backprojection. Backprojection does not rely on any particular photon migration model in solving the forward problem. Although diffuse optical tomography (DOT) differs from CT in that tissue is highly scattering to NIR light, the simplicity of backprojection algorithms has gained some users in DOT, assuming that the depth of the physiological activation can be assumed or ignored. A number of linear iterative reconstruction techniques can also be used for image reconstruction. In particular, the simultaneous iterative reconstruction technique (SIRT) was designed to overcome the problem of "noisy" image reconstruction typical of many algebraic reconstruction techniques. Nonlinear optimization methods, such as congugate gradients, have also been used extensively.

In addition to the above mathematical techniques, applying physiological and spatial *a priori* knowledge to constrain the solution space is another important and effective technique that helps to solve such ill-posed and underdetermined inverse problems. For example, MRI can be used to provide structural information in solving the forward problem, which is an important incentive for developing multimodality imaging techniques.

1.7.4 Measurement Techniques

An NIR system consists of three major modules: optical sources, photodetectors, and the data-acquisition system. Laser diodes (LDs) and light-emitting diodes (LEDs) are the dominant choices for the optical sources in modern systems. LDs have the advantages in intensity, directionality, and spectral width. Nonetheless, LEDs have some attractive features, such as simplicity of use, lower power consumption, and lower risk to the eyes. The choices of the detectors are somewhat more flexible: PMTs, avalanche diodes, photodiodes, and charge-coupled devices (CCDs) can all be used. Each has its unique advantages and characteristics. The choice of a specific device as the photodetector is dependent on the particular application and specific design aims. In most designs, the optical sources (particularly if using LDs) and photodetectors are coupled to the surface of the patient via fiber optics. It should be noted that in some designs the sources and detectors (e.g., using LEDs and photodiodes) can be attached to the surface directly. The data-acquisition hardware is usually part of the computer where data processing and image reconstruction are performed. However, not all systems are connected to a personal computer, especially those that are designed to be compact and highly portable.

There are three major modes of data acquisition: CW, frequency domain, and time domain. In the CW method, the intensity of light emitted from the optical sources is usually amplitude-modulated at slightly different very low frequencies (typically 1–10 kHz) so that different simultaneous light sources can be distinguished. In the frequency-domain method, the light source is amplitude-modulated at a much higher frequency, typically 100–500 MHz, using a sinusoidal function

$$S = I_0 + I(\omega)\sin(\omega t + \theta(\omega)), \tag{1.42}$$

where S is the intensity of the light, I_0 is termed the direct-current (DC) component, I is the alternating-current (AC) component, and θ is the phase. The detected light is converted into an electrical signal via photodetectors and decomposed into the same sinusoidal form: AC, DC, and phase signals. The AC and DC signals are determined primarily by the absorption coefficient, while the phase is more sensitive to the scattering coefficient. Both the AC and phase signals are also functions of the modulation frequency. It is worth noting that the DC

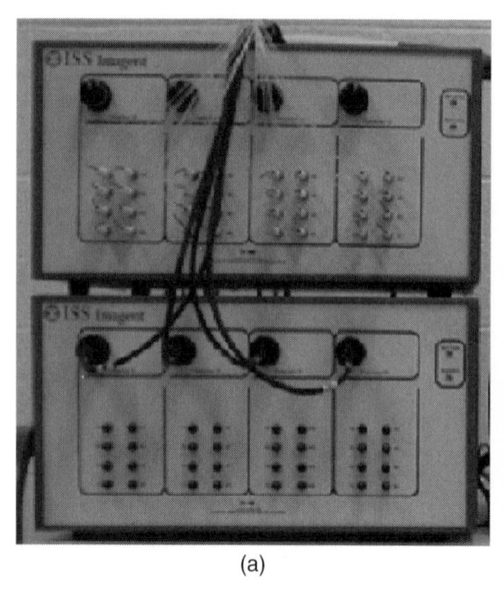
(a)

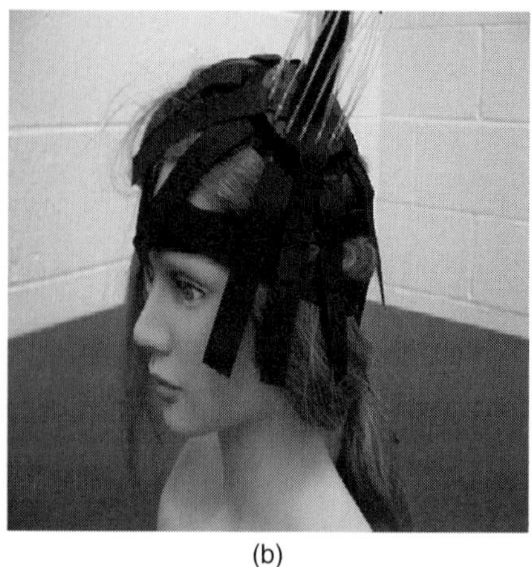
(b)

FIGURE 1.19 (a) A representative frequency-domain DOT system. (b) Custom-built optical probe for studying the primary motor cortex.

signal is in principle equivalent to the signal measured in the CW method. In this sense, the frequency-domain method provides more information than the CW method. However, the measurement signal-to-noise ratio using this method is typically significantly lower than that using the CW method. A representative frequency-domain system is shown in Figure 1.19. This type of instrumentation is often used in studies of brain activation, in combination with a multisensor optical probe, as also shown in Figure 1.19.

In the time-domain method, a very short (~50 ps) pulse is transmitted by the optical source. The detector measures the intensity of the light as a function of time (typical delay 0.1–1.0 ns for neuroimaging depending on the source–detector distance and the geometry). In practice, however, because of the limited sensitivity of the photodetector, this function is discrete in time and is measured in terms of the numbers of accumulated photons over short periods of time. From the point of view of a measurement system, if one can generate a delta-pulse (i.e., infinitely short pulse) as the optical input and the system is linear-time–invariant, the measured optical signal is the impulse response of the system. In this sense, the time-domain method is equivalent to taking frequency-domain measurements using an infinite number of modulation frequencies. In reality, the measurement sensitivity (number of photons captured by the photodetector within a very short period of time) is a major limiting factor with current technology. No time-domain system is available commercially at the time of this writing, although there are a few providers of specialized modules, such as Becker & Hickl (Berlin, Germany).

The most well studied and perhaps most mature application of DOT is mammography. Optical sources and detectors are usually positioned so that the measurement channel is across the breast, and they are often interleaved as well to increase the efficiency of measurement, as shown in Figure 1.20. A hand-held scanning device has been developed as an in-home-use self-screener for the early discovery of breast tumors. This device uses a photodiode as the detector and a pair of modulated LEDs as optical sources operating 180° out of phase. If the effective differential pathlength factors (DPF) of the two channels are equal, the detected optical signal has a zero phase, which suggests a homogeneous medium or a perfectly symmetric medium. By scanning the device over the breast, inhomogeneities larger than a certain threshold size can be detected eventually by causing asymmetry to the DPF of the two measurement channels (Figure 1.20), which triggers an alarm.

Developing DOT for clinical purposes is an active research area. One promising application is to monitor postsurgery recovery of brain tissue oxygenation, which is of major concern for patients after brain surgery. DOT potentially enables early diagnosis of tissue oxygen deprivation to minimize brain damage, instead of having to wait until patients regain consciousness. Another important application is the noninvasive diagnosis of cortical areas affected by stroke, which is difficult to diagnose using MRI or CT before permanent tissue damage has occurred. In the case of breast cancer screening, for example, a highly optically absorbing abnormality in the breast suggests a highly concentrated vasculature and might be good evidence for the physician to make a decision on a biopsy.

1.8 Biosignals

In addition to the information from the imaging modalities described in this chapter, there are a number of other clinically diagnostic patient measurements that are routinely acquired and analyzed. These include electrocardiography (ECG/EKG) and electroencephalography (EEG). The "shape," timing, and

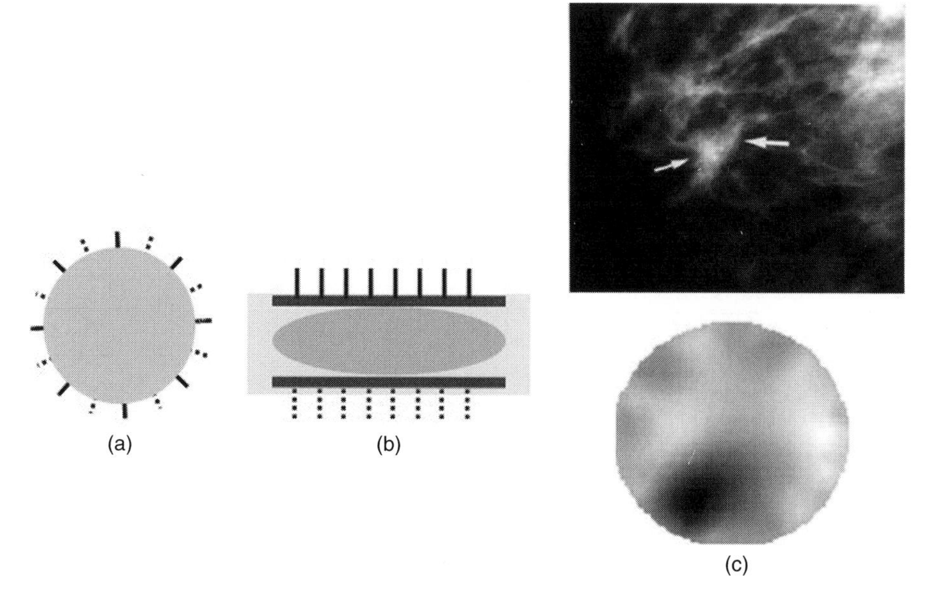

FIGURE 1.20 (a) Topology of optical sources (solid) and detectors (dotted) in a typical setup for a mammography experiment. (b) Breast placed between two plates where the sources and detectors are mounted (optical matching fluid can be used to improve optical contact). (c) A DOT image (bottom) compared with an MRI mammogram (top). (S. Nioka, B. Chance, Technology in Cancer Research and Treatment, vol. 4, pp. 497, 2005.)

frequency data contained in these typically one-dimensional, time-series measurements contain information useful for diagnosis. These types of measurements are not spatially localized and therefore are not referred to as imaging but as *biosignals,* although one should note that high-density EEG arrays do allow the possibility of localizing the "source" of the neural activity. A very brief description of EKG and EEG signals is presented here.

1.8.1 Electroencephalography

EEG involves measurements made on the scalp of the patient corresponding to the electric fields produced by spontaneous neural activity. The amplitude of the signals is approximately 100 μV, with frequency components below 50 Hz. Typically, 21 electrodes are located on the scalp surface, with the exact positions referenced to the anatomy of the particular patient. Additional electrodes can also be placed in intermediate positions. The electrodes can be unipolar or bipolar. The former involves referencing the potential measured by each electrode to either the average of all the electrodes or to a "neutral" electrode; the latter arrangement measures the difference in voltage between a pair of electrodes.

The EEG signal contains components corresponding to alpha, beta, delta, and theta waves. Alpha waves originate from the occipital region in awake patients with eyes closed; beta waves can be measured over the frontal and parietal lobes; and delta and theta waves are present when the patient is asleep. The approximate frequency ranges of the waves are: 8–13 (alpha), 13–30 (beta), 0.5–4 (delta), and 4–8 (theta). Clinical use of EEG relates to brain injury, with abnormal waveforms reflecting different areas of pathological disease or injury. EEG recordings are also extensively used in epileptic patients, where the onset and duration of an epileptic event can be detected via hyperactivity of the electrical signal, with severe spikes, as shown in Figure 1.21.

1.8.2 Electrocardiograms

The normal rhythmic cardiac electrical impulse originates in pacemaking cells in the sinoatrial (SA) node, which is located at the junction of the right atrium and superior vena cava. The electrical pulse passes through conducting tracts to activate first the right atrium and then the left atrium. The impulse is delayed at the atrioventricular (AV) node before continuing to the bundle of His, the left and right bundle branches, and the Purkinje network.

Prior to excitation, the ventricular wall has a resting potential of –90 mV. Rapid depolarization is followed by an equally

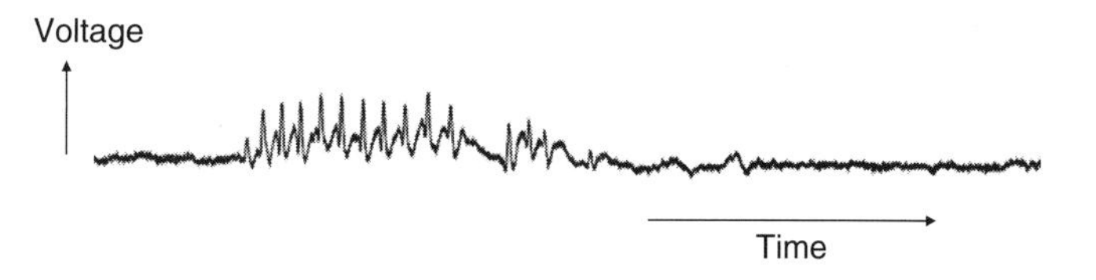

FIGURE 1.21 An indication of an epileptic seizure is apparent in the increased electrical activity of the brain as measured by EEG.

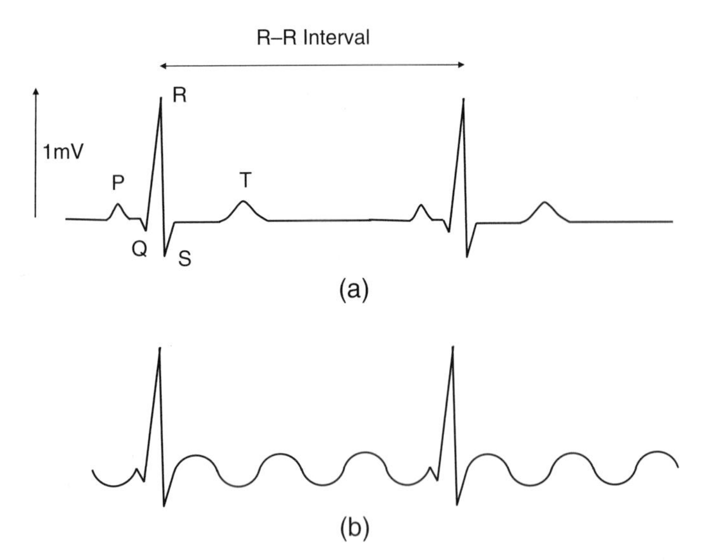

FIGURE 1.22 (a) Standard EKG trace from a healthy patient. (b) Trace corresponding to atrial flutter, with the source of the pacemaker impulse traveling in a circular trajectory within the atria.

rapid repolarization for ~200–300 ms, before final repolarization restores the membrane potential to the resting state.

Measurements of the electrical activity of the heart are made by using electrodes placed on the body surface. The measured signal can be thought of as the sum of different spatially localized electrical activities. Typical features are the P wave, the QRS complex, and the T wave, as shown in Figure 1.22, with the intervals between and durations of these features being clinically diagnostic. The P wave corresponds to atrial depolarization; the QRS complex is associated primarily with ventricular depolarization; and the T wave is most associated with ventricular repolarization. The P–R interval corresponds to conduction delay in the AV node, and the S–T interval to the duration of the repolarization plateau.

A number of electrodes are used for the measurements. A typical arrangement is to have one each on the right arm, left arm, left leg, and right leg (for grounding). A typical EKG is shown in Figure 1.22 (a). Different diseases often display very characteristic changes in the EKG signal. For example, in the case of a disease in which the AV node is diseased, the AV nodal delay is greatly increased, corresponding to a much longer P–R interval. Figure 1.22 (b) shows another example, that of atrial flutter.

1.9 Appendix

A.1 Fourier Transforms

The forward Fourier transform of a time-domain signal $s(t)$ is given by:

$$S(f) = \int_{-\infty}^{\infty} s(t) e^{-j2\pi ft} dt. \tag{A.1}$$

The inverse Fourier transform of a frequency-domain signal $S(f)$ is represented by:

$$s(t) = \frac{1}{2\pi} \int_{-\infty}^{\infty} S(f) e^{+j2\pi ft} df. \tag{A.2}$$

Similarly, the forward Fourier transform of a spatial-domain signal $s(x)$ has the form:

$$S(k) = \int_{-\infty}^{\infty} s(x) e^{-j2\pi kx} dx. \tag{A.3}$$

The corresponding inverse Fourier transform of a spatial frequency-domain signal $S(k)$ can be expressed as:

$$s(x) = \int_{-\infty}^{\infty} S(k) e^{+j2\pi kx} dk. \tag{A.4}$$

Signals are often acquired in more than one dimension, and the corresponding multidimensional Fourier transformations are given by:

$$S(k_x,k_y,k_z) = \int_{-\infty}^{\infty}\int_{-\infty}^{\infty}\int_{-\infty}^{\infty} s(x,y,z) e^{-j2\pi\left(k_x x + k_y y + k_z z\right)} dx dy dz \tag{A.5}$$

$$s(x,y,z) = \int_{-\infty}^{\infty}\int_{-\infty}^{\infty}\int_{-\infty}^{\infty} S\left(k_x,k_y,k_z\right) e^{+j2\pi\left(k_x x + k_y y + k_z z\right)} dk_x dk_y dk_z. \tag{A.6}$$

A.2 Filtered Backprojection

The problem of reconstructing a two-dimensional image from a series of one-dimensional projections, denoted by $p(r,\phi)$ is common to a number of imaging modalities. Backprojection assigns an equal weighting to the pixels contributing to each point in a particular projection. This process is repeated for all of the projections, and the pixel intensities are summed to give the reconstructed image, $\hat{f}(x,y)$. Mathematically, $\hat{f}(x,y)$ can be represented as:

$$\hat{f}(x,y) = \sum_{j=1}^{n} p(r,\phi_j) d\phi, \tag{A.7}$$

where n is the number of projections. Simple backprojection results in a number of image artifacts and blurring, the remediation of which is the rationale for using filtered backprojection. In this process, each projection $p(r, \phi)$ is multiplied by a spatial filter, $h(r)$, prior to backprojection. Commonly used functions are Shepp–Logan, lowpass cosine, and generalized Hamming filters. For computational efficiency, the process is performed in the spatial frequency domain. Each projection is Fourier-transformed along the r-dimension to give $P(k,\phi)$, and then $P(k,\phi)$ is multiplied by $H(k)$, the Fourier transform of $h(r)$, to give $P'(k,\phi)$:

$$P'(k,\phi) = P(k,\phi)H(\mathrm{k}). \tag{A.8}$$

The filtered projections, $P'(k,\phi)$, are inverse–Fourier-transformed back into the spatial domain and backprojected to give the final image, $\hat{f}(x,y)$:

$$\hat{f}(x,y) = \sum_{j=1}^{n} \mathbb{F}^{-1}\left\{P'\left(k,\phi_j\right)\right\} d\phi, \tag{A.9}$$

where $\mathbb{F}^{-1}$ represents an inverse Fourier transform.

A.3 Iterative Image Reconstruction

As described previously, PET and SPECT images can be reconstructed using analytical inversion techniques such as filtered backprojection of the line integral signals. However, this very simple line integral model of the data is not exact. For example, in SPECT there is a depth-dependent point spread function, and in PET there exist variations in different detector-pair sensitivities as well as Compton scatter in the detectors. The other major issue, not accounted for using analytical techniques, is the statistical variability, governed by the Poisson distribution, in the data; this is an especially important factor when the number of counts is low.

Iterative techniques are used extensively for image reconstruction in PET and SPECT. The goal is to estimate the spatial dependence of the radiotracer distribution that "agrees best" with the acquired data. Iterative methods can model the detection process without assuming explicitly a line-integral between source and detector. They also allow explicit consideration of the statistical nature of the measurement noise. There are a large number of different algorithms, which can be characterized broadly in terms of three parameters: (1) the choice of cost function and model for the data, (2) the optimization procedure; that is, the particular algorithm used to either maximize or minimize the cost function, and (3) the computational "cost" of the algorithm; that is, its rate of convergence and stability.

One can describe the reconstruction problem as solving the set of equations $\mathbf{y} = \mathbf{Pf}$, where $\mathbf{y}$ is the measured projection data, $\mathbf{P}$ is a projection matrix, and $\mathbf{f}$ is the unknown source distribution. The earliest iterative method incorporating a Poisson model was the *maximum likelihood* (ML) method. Provided that detection of individual radioactive decay events is independent, the conditional probability for $\mathbf{y}$ is given by:

$$p(y|f) = \prod_i e^{-\bar{y}_i} \frac{\bar{y}_i^{y_i}}{y_i!}. \tag{A.10}$$

Now the ML estimate of $\mathbf{f}$ is computed by maximizing the above equation (in practice the log of the likelihood estimate is used). Use of this algorithm in PET assumes that random and scattered coincidences have been estimated. A related general approach to solving ML problems uses an *expectation maximization* (EM) algorithm.

The ML-EM method converges very slowly; hence, in practice it is implemented using the *ordered subsets EM* (OSEM) method, in which the acquired data are broken up into a number of subsets and the EM algorithm is applied sequentially to each set in turn (this has obvious analogies with the fast Fourier transform implementation of the discrete Fourier transform algorithm). In all of these algorithms, the voxel values are updated simultaneously during one iteration; an alternative approach is to update the value of a single voxel for each iteration, as in iterated coordinate-ascent methods.

It is also possible to introduce prior knowledge into the iteration process. For example, smoothness priors are commonly used, with a smoothing penalty added to the likelihood function. This allows selection between sets of equivalent solutions. This also addresses the issue that ML methods in PET and SPECT are inherently ill-conditioned. Alternative approaches to the problem of ill-conditioning include post-reconstruction filtering and stopping the iterative process before convergence occurs using prespecified conditions.

1.10 Exercises

1. Two X-ray images of the hand are shown in the following figure. One corresponds to an X-ray beam with an effective energy of 140 keV, and the other to an effective energy of 50 keV. Explain which is which, and the reasons for the differences in image contrast and signal intensity.

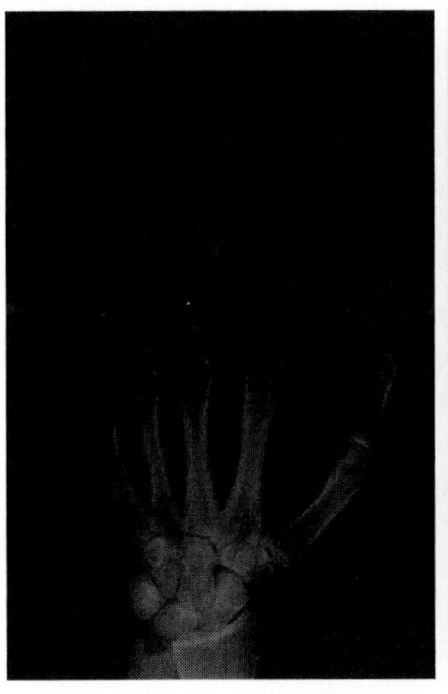
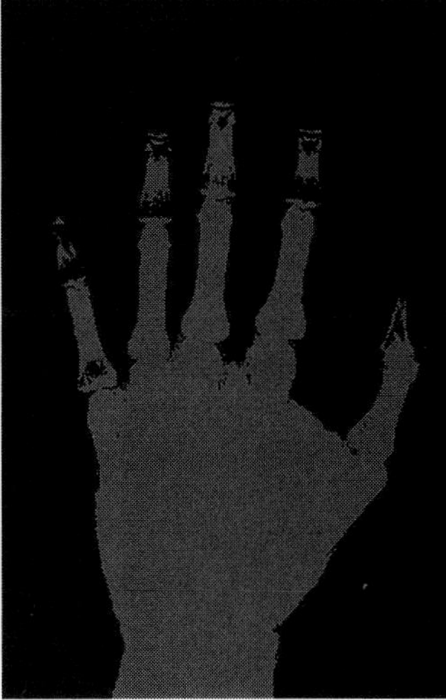

FIGURE 1.23 Two planar X-ray images of the hand—see Exercise 1.

2. In mammography, should the intensifying screen be placed in front of the X-ray film or behind the film in order to achieve the highest spatial resolution? Explain the reasons for your answer.

3. A SPECT scan is taken of a patient, and an area of radioactivity is found at the position marked by the "x." The sinogram from the SPECT scan is shown in the figure. Assuming that the radioactivity is uniform within the targeted area, what is the shape of the area of radioactivity? (For ease, assume that the 0 degree projection corresponds to the SPECT detector being directly below the patient.)

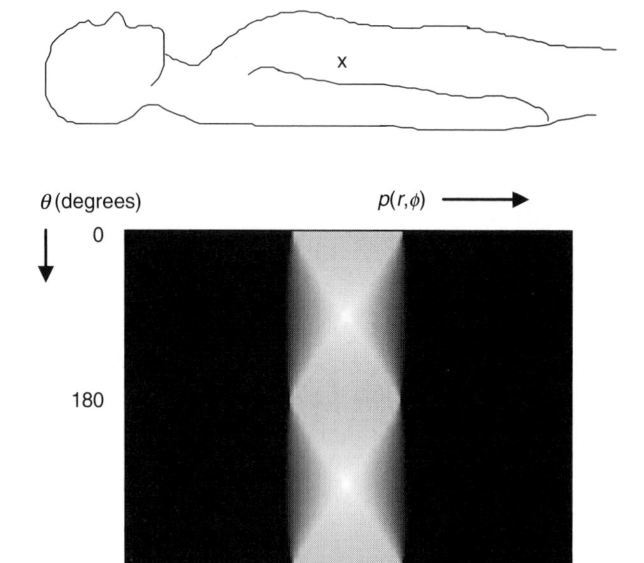

FIGURE 1.24 Sinogram obtained from a SPECT scan of the biodistribution of a radionuclide centered at position x—see Exercise 3.

4. Match up the following image contrast agents with their respective clinical protocols with a one-sentence explanation of why the particular agent is used.
   (a) Tri-iodinated–based compounds
   (b) ^{18}F-fluorodeoxyglucose
   (c) ^{99m}Tc-sulfur colloid ($\sim$2 μm diameter)
   (d) Barium sulphate
   (e) ^{133}Xe gas
      (i) SPECT liver imaging
      (ii) PET tumor imaging
      (iii) SPECT lung imaging
      (iv) X-ray fluoroscopy
      (v) Digital subtraction X-ray angiography
5. The schematic in the figure shows the setup for a SPECT scan of the heart. Three areas are outlined: soft tissue (gray), lungs (white), and the radionuclide distributed in the heart (black). Sketch to scale the one-dimensional projections at each detector position 1–4 in the following cases:
   (a) No attenuation correction is applied to the data.
   (b) A uniform attenuation coefficient equal to that of soft tissue is assumed, and the data are correspondingly corrected.
   (c) Attenuation correction based on a transmission scan with an external source is used.

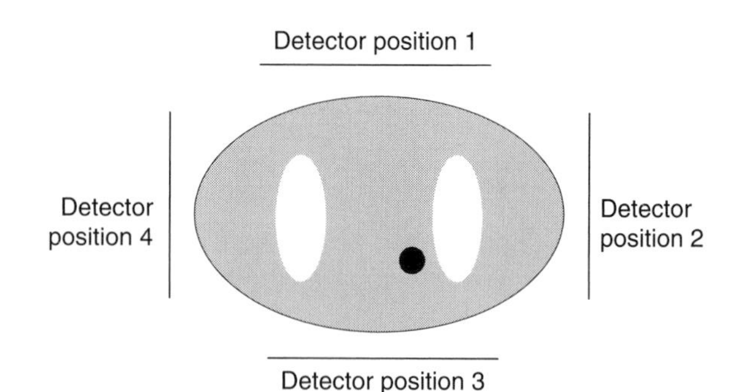

FIGURE 1.25 Schematic of the setup used for a SPECT scan of the heart—see Exercise 5.

6. For the object shown in the figure, qualitatively sketch the B-mode ultrasound image obtained from a 1 μs pulse of ultrasound. Acoustic impedances: muscle 1.61, cyst 1.52, fat 1.38, liver 1.5 (all × 10^5 g/cm^{2}s). Attenuation coefficients (dB/cm/MHz): muscle 1.0, fat 0.8, liver 1.0, and cyst 1.0. Speeds of sound (m/s): muscle 1540, fat 1540, liver 1540, and cyst 3080.

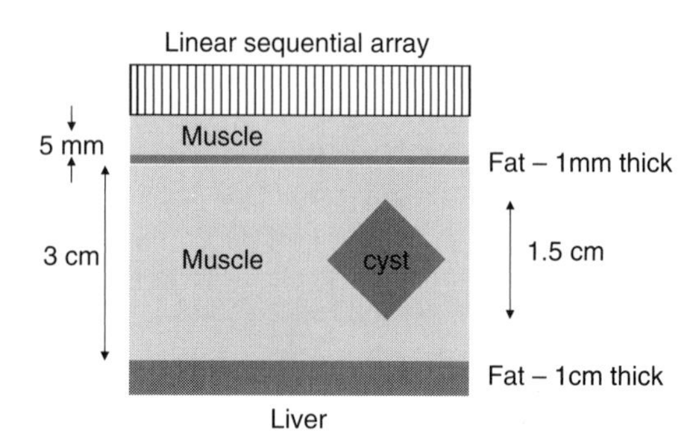

FIGURE 1.26 Geometric representation of different aspects of the body used to generate a B-mode ultrasound scan—see Exercise 6.

7. The B-mode image in the figure shows what is known as a "mirror artifact." Explain what might give rise to this effect.

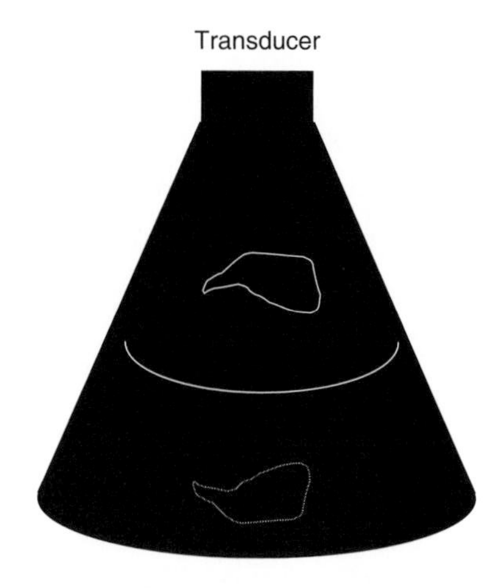

FIGURE 1.27 Illustration of a mirror artifact produced in B-mode ultrasound—see Exercise 7.

8. Suppose you are an ultrasound engineer working for a transducer manufacturing company. A customer needs a plane-piston transducer that resonates at 5 MHz and has a radius of 2.5 cm. Sketch the axial beam shape for the customer as the normalized intensity versus the axial distance, and indicate the location of the last axial maximum in water at 20°C. What is the thickness of the transducer?
9. Imagine that flow information is desired from a vessel that is 4 mm in diameter and lies at a depth of 5 cm below the skin. If a train of ultrasound pulses is sent out (each pulse consisting of five cycles of ultrasound at a frequency of 5 MHz), determine the length of the ultrasound pulse, the time delay between the end of the transmitted pulse and the receiver gate being opened, and the time for which the receiver gate is open (assuming an ultrasound velocity of 1540 m/s in tissue). What is the highest velocity that can be determined?
10. A region of the brain to be imaged contains areas corresponding to tumor, normal brain, and fat. The relevant MRI parameters are:

$$\rho(\text{tumor}) = \rho(\text{fat}) > \rho(\text{brain})$$
$$T_1(\text{fat}) > T_1(\text{tumor}) > T_1(\text{brain})$$
$$T_2(\text{fat}) > T_2(\text{tumor}) > T_2(\text{brain})$$

Which type of weighted spin–echo sequence should be run in order to get contrast between the three different tissue types? Explain your reasoning, including why the other two types of weighting would not work.

1.11 References and Bibliography

General Imaging

J. T. Bushberg et al. *The Essential Physics of Medical Imaging.* Lippincott, Williams and Wilkins, 2001.
W. R. Hendee and E. R. Ritenour. *Medical Imaging Physics,* 4th ed. Wiley, 2002.
A. G. Webb. *Introduction to Biomedical Imaging.* IEEE Press Series in Biomedical Engineering. IEEE Press, 2003.

X-ray and Computed Tomography

J. McNamara (Ed.). *Digital Radiography and Three-Dimensional Imaging.* Needham Press, 2006.
M. F. Reiser et al. (Eds.). *Multislice CT.* Springer, 2004.

Nuclear Medicine

P. E. Valk et al. *Positron emission tomography: Basic Science and Clinical Practice.* Springer-Verlag, 2003.
F. A. Mettler and M. J. Guiberteau. *Essentials of Nuclear Medicine Imaging,* 5th ed. Saunders, 2005.

Ultrasonic Imaging

F. W. Kremkau. *Diagnostic Ultrasound: Principles and Instruments,* 7th ed. Saunders, 2005.
C. R. Hill, J. C. Bamber, and G. R. ter Haar. *Physical Principles of Medical Ultrasonics,* 2nd ed. John Wiley & Sons, 2002.

Magnetic Resonance Imaging

Z.-P. Liang and P. C. Lauterbur. *Principles of Magnetic Resonance Imaging. A Signal Processing Perspective.* IEEE Press Series in Biomedical Engineering. IEEE Press, 2000.
M. A. Brown and R. C. Semelka. *MRI: Basic Principles and Applications,* 2nd ed. Wiley-Liss, 1999.
E. M. Haake et al. *Magnetic Resonance Imaging: Physical Principles and Sequence Design.* Wiley-Liss, 2000.

Diffuse Optical Imaging

R. Frostig (Ed.). *In Vivo Optical Imaging of Brain Function.* CRC Press, 2002.

Diffuse Optical Imaging Review Papers

A. P. Gibson, J. C. Hebden, and S. R. Arridge. *Physics in Medicine and Biology.* 50:R1-R43, 2005.

Biosignals

John G. Webster. *Medical Instrumentation: Application and Design,* 3rd ed. Wiley, 1997.

Electroencephalography: Basic Principles, Clinical Applications, and Related Fields

E. Niedermeyer and F. L. da Silva. *Electroencephalography: Basic Principles, Clinical Applications, and Related Fields,* 5th ed. Lippincott, Williams & Wilkins, 2004.
J. Malmivuo and R. Plonsey. *Bioelectromagnetism.* Oxford University Press, 1995.

2

Electronic Medical Records

Dr. Eugene Y. S. Lim,[1,2]
Prof. Michael Fulham,[1,2] and
Prof. David Dagan Feng[1,3]
[1] *University of Sydney*
[2] *Royal Prince Alfred Hospital*
[3] *Hong Kong Polytechnic University*

2.1 Introduction

2.1.1 Background

Medical record, health record, patient record, patient medical record, and medical chart are often used interchangeably to describe a systematic documentation of medical information on a particular patient [1]. This patient medical record (PMR) has traditionally been a paper record, which contains diverse data, accumulated over a single or multiple episodes where the patient interacted with the health system. The paper-based medical record (PBMR) is generally bundled into a single (occasionally multiple) file(s) and kept in a central repository often called the Medical Record Department. The individual file is identified by a coding system that links patient demographics to a unique number (e.g., a medical record number [MRN]). Staff in the Medical Record Department are responsible for filing, updating, retrieving, and maintaining the PBMR [2–4]. The PBMR will often contain details of hospital inpatient and outpatient admissions; hand-written file notes from medical, nursing, and allied health staff; results of laboratory, pathology, and imaging investigations; operation reports; and copies of correspondence sent to referring and local medical officers. In some instances, paper copies of imaging tests, operative findings, and endoscopy results will also be included. However, in the current complex medical environment encountered by patients in their medical journey, the PBMR will often be incomplete because many departments in a large hospital maintain their own databases with pertinent data kept within the department. Or the PBMR may be impossibly cumbersome when countless pages of meaningless data (e.g., entire printed notes from an intensive care admission) are filed in it. Another consideration is that while departmental data will be necessary for the daily functioning of that department, there will be duplication of some data that are kept in the central repository. Parts of the record may be illegible, but perhaps most importantly, the data may not be immediately available where and when they are needed. Thus, in the current medical environment, which is increasingly reliant upon electronic media for access to patient information, the PBMR is a suboptimal solution.

Improvements in information technology coupled with the increasing volume of data that need to be stored in a medical record and the need for these data across multiple sites where different medical services are accessed by patients have renewed interest in seeking an electronic solution to manage these critical data. To state the obvious, effective management, processing, and communication of patient data in medical records to the relevant staff improves the quality of healthcare [5–8]. If a digital medical record, called an electronic medical record (EMR), were able to overcome the limitations of a PBMR, it should follow that the quality of care should also improve. Benefits of an EMR include ready access, rapid searching, secure storage, and safe transmission of patient data. However, an EMR is not a trivial undertaking. Important considerations are the quantity and complexity of the data, the marked diversity in the information infrastructure and databases within and across hospitals (public, private, and university-based) and the community (local general practice, medical center, rural or community center). Homogeneity in system and infrastructure, which is found in some large institutions such as within the Veterans Hospitals in the United States, is an exception rather than the rule, and what is needed is a statewide or nationwide solution. Heterogeneity across the various systems is also likely to grow as patients live longer with more medical and surgical co-morbidities [9]. It is not uncommon for patients to have multiple specialists look after part of their medical condition, such as a renal physician, a cardiologist, a diabetologist, and a neurologist for the patient with vascular disease. The greater emphasis on and demand for an evidence-based approach to medicine is a logical and seamless fit for an EMR. Rapid electronic access to patient data and to large medical databases will allow decisions to be made on the best available data relevant to a particular patient or problem.

At present, there is not a single EMR approach that provides all things for all patients and all health care personnel; however, there have been some recent useful attempts [6]. Many of these new approaches use advanced network technologies. Telemedicine, the use of telecommunications technology for medical diagnosis and patient care, has enabled the exchange of medical, imaging and surgical information [10–12]. The broader concept of eHealth has emerged as a new paradigm for providing health care using telecommunication technologies [13]. mHealth, where mobile devices and wireless networks are coupled to provide real-time health data, offers tremendous advantages [14]. Research in these areas continues to be encouraged because of the high expectations of modern society and governments [15]. The major challenges for these approaches are to adapt to and cater to the heterogeneity in the targeted environment, whether it is patient- or system-based [9, 16, 17]. Table 2.1 lists some of the diverse systems. These are available in large hospitals, where the most serious illnesses continue to be managed and treated. We suggest that customized system design is an essential element to the greater penetration and acceptance of electronic approaches by health care users [6]. Any new system should be sufficiently easy to use [18–21] and adaptable within the local workflow or setting [20, 22–24].

2.1.2 Overview of Electronic Medical Records

In this chapter, we address the structure and design issues of an EMR. In the earliest phase of an EMR, the main effort is to convert patient records into digital format for archiving by scanning reports, letters, and other parts of the record. A more advanced form of EMR requires the application of processing and analytic methods [25, 26]. However, before this can be carried out, it is essential that a standard terminology, classification, and rules for communication be developed. A longstanding issue in health care is the lack of a uniform terminology for even common disorders. A number of organizations are continuing attempts to develop standard terminologies, but as yet there is not a single agreed-upon approach that can be applied across the whole of medicine. In part, this relates to the proliferation of medical knowledge and the extent and degree of specialization over the past decade.

Nevertheless, the first step in the development of a more sophisticated EMR is the introduction of common terminologies and communication standards. However, this must occur

TABLE 2.1 Examples of data management systems in large hospitals

Systems	Description
HIS (Hospital information system)	Admission, discharge, and billing
RIS (Radiology information system)	Patient tracking, workflow management, report generation
PACS (Picture archive and communication system)	Accesses and distributes medical images and reports
Anesthesia system	Administration/recording of anesthetic agents
Order entry system	Electronic ordering of tests, treatment
Pharmacy system	Administration/dispensation of pharmaceuticals
Surgery scheduling system	Scheduling operating time in the operating room

simultaneously with a thorough understanding of the environment in which the EMR will be used and the intended outcomes from the end users. While it is self-evident that the true measure of success of any EMR is its integration into the targeted environment, such integration is not trivial. We emphasize that the targeted workflow must be considered early rather than late in the planning phase [27]. Although usability studies can be employed in the early stages of development of an EMR, there is no substitute for extensive collaboration and consultation with the users of the final product [28].

2.2 Medical Data and Patient Records

2.2.1 Paper-Based Medical Records

As mentioned earlier, data contained in a PMR are complex and diverse. At a simple level, these data can be separated into demographic and historic information. Demographic data include details such as the patient's name, date of birth, the unique MRN or patient identification number (PID), address and contact details, next-of-kin, mother's maiden name, and names/address/contact details of the referring doctors and/or local medical practitioners. Meanwhile, the historic data include information relevant to the medical domain such as: the current clinical diagnosis, medical history, medications, allergies, examination findings, treatment plan, results of investigations, nursing observations, and treatment plans and notes from other allied health professionals including physiotherapists, occupational therapists, and social workers. Over time, the historic data contain sequential events that chronicle the development of various diseases and the investigations, complications, and treatment of these disease states. Although reports of various investigations may be found appended to the PMR, in the past it was not possible to include images from the various tests that were performed. For example, departments of medical imaging would maintain their own physical archive of previous X-rays, which were linked to the PMR by the MRN. Figure 2.1 shows a typical example of de-identified notes from a patient file.

These pages demonstrate some of the limitations mentioned earlier [29]. Over two pages and in less than 24 hours, there are six different sets of handwriting. All have varying degrees of legibility, abbreviations, and author identification, but overall the handwritten notes are difficult to read. Further, the Institute of Medicine suggests that handwritten reports or notes, manual order entry, nonstandard abbreviations, and poor legibility are sources of medical error [30]. The typed notes, meanwhile, are much easier to follow. Authors of the notes are clearly identified as is the time the entry was made, but the history is documented once again. The most appropriate tool to capture medical data is still to some extent controversial. Interpretation of information from paper was ranked higher than that from a computer display [2]. Some studies suggest that a paper-based approach is a more effective method of communication between clinicians and there is high satisfaction level, but this is surely tempered by the quality and legibility of the notes [31]. Another consideration is that a paper-based approach is still regarded in legal circles as the primary data source, and insurance companies rely on a PBMR to evaluate appropriateness of admission and length of stay [31]. However, the PBMR remains a discrete entity that cannot be shared or distributed and appropriately is jealously guarded by medical records departments. The PBMR follows the patient but generally only within a single hospital. If a patient enters another hospital system, health care personnel are generally reliant on the patient's memory rather than on robust data about the medical condition.

2.2.2 Electronic Medical Records

A variety of terms have been used to describe an EMR. To some extent, these terms have arisen from their use in different environments.

- The Personal Health Record (PHR): An individual's own account of his or her medical history in a digital format
- The Electronic Medical Record (EMR): A provider-based electronic medical record that includes health documentation for a patient covering all services provided within an enterprise
- The Electronic Patient Record (EPR): A patient-centered system containing patient documentation

There are certain common attributes of an EMR. The digital nature of an EMR allows data contained within it to be searched and retrieved. Other attributes include system quality (e.g, accessibility, usability), information quality (e.g. readability, accuracy), and decision support (e.g. data analysis). Accessibility describes the degree to which a system is reliable. Although the EMR has the potential to be accessed by multiple users at multiple locations, these same users are then dependent upon the electronic medium for critical data. An unreliable system could lead to medical errors [32, 33]. Usability describes the ability of the EMR to be integrated into the clinical workflow in a seamless manner [7]. Everyone is familiar with poor software that does not deliver the desired outcomes or delivers them at such a cost in time and frustration that users no longer use the software. Legibility is not an issue with an EMR although scanned documents can be problematic. However, a criticism remains that complex data are often easier to read from a piece of paper rather than from a computer monitor. Data storage via electronic media is far less space- and weight-expensive. Accuracy is difficult to define in this context but can be defined as the degree (measured as a percentage) of correctness, completeness, and inclusiveness in a dataset [34]. Computer-defined fields, when appropriately designed for data entry, provide a mechanism for an EMR to be more comprehensive than a PBMR.

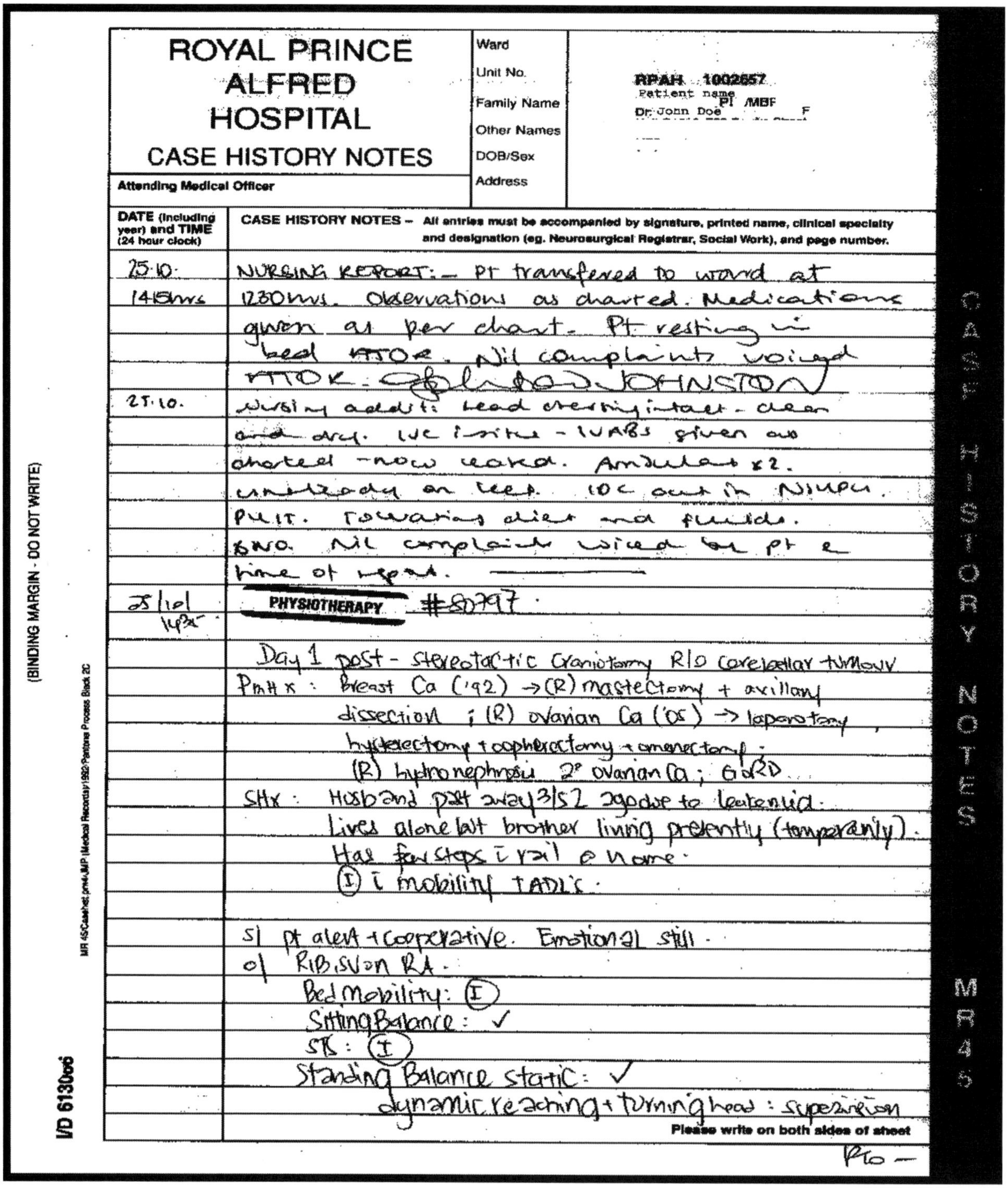

ROYAL PRINCE ALFRED HOSPITAL
CASE HISTORY NOTES
Attending Medical Officer
Ward
Unit No.
Family Name
Other Names
DOB/Sex
Address
RPAH 1002657
Patient name
PI /MBF
Dr John Doe F
DATE (including year) and TIME (24 hour clock)
CASE HISTORY NOTES – All entries must be accompanied by signature, printed name, clinical specialty and designation (eg. Neurosurgical Registrar, Social Work), and page number.

25.10. 1415hrs NURSING REPORT:– Pt transfered to ward at 1230hrs. Observations as charted. Medications given as per chart. Pt resting in bed NTOR. Nil complaints voiced NTOR. [signature] JOHNSTON

25.10. Nursing addit: head dressing intact – clean and dry. IVC in situ – IVABs given as charted – now ceased. Ambulant x2. unsteady on feet. IDC out in NIMPU. PU IT. Tolerating diet and fluids. SNO. Nil complaints voiced by pt at time of report.

25/10/ 1430 PHYSIOTHERAPY #80797.

Day 1 post – stereotactic Craniotomy R/O cerebellar tumour
PMHx: Breast Ca ('92) → (R) mastectomy + axillary dissection; (R) ovarian Ca ('05) → laparotomy, hysterectomy + oopherectomy + omenectomy;
(R) hydronephrosis 2° ovarian Ca; GORD.
SHx: Husband past away 3/52 ago due to leukemia. Lives alone but brother living presently (temporarily). Has few steps c̄ rail @ home.
(I) c̄ mobility + ADL's.

S/ pt alert + cooperative. Emotional still.
O/ R/B, SV on RA.
Bed Mobility: (I)
Sitting Balance: ✓
STS: (I)
Standing Balance static: ✓
dynamic reaching + turning head: supervision

Please write on both sides of sheet
PTO –

(BINDING MARGIN - DO NOT WRITE)
MR 45/Casehist.pm4/JMP (Medical Records) 1992/Pantone Process Black 2C
I/D 6130ec
CASE HISTORY NOTES MR45

(a)

FIGURE 2.1 Data from a typical hospital medical record. In (a) and (b) there are multiple hand-written notes.

Family Name	Other Names	DOB	Unit No

DATE (including year) and TIME (24 hour clock)	CASE HISTORY NOTES – All entries must be accompanied by signature, printed name, clinical specialty and designation (eg. Neurosurgical Registrar, Social Work), and page number.
25/10	physio continue
14^25	O/ Mobility: leaning to (R) + slightly ataxic gait.
	1x mod (A), need to stand on (R) side of pt.
	Rx/ standing balance training
	Mobilised 20m a/a.
	Returned to bed
	A/ 1x mod (A) for mobility, standing on (R) side of pt.
	OK to sit in (N) high back blue chair.
	P/ will R/v 1/7 to work on mobility + balance
26/10	Nursing: Patient observed for sleep for long periods.
05^00 AM	Mobilised out to toilet X 2 with assistance of 1,
	remains unsteady on feet. Head dressing dry and intact
	NIL complaints voiced. BNO o/night.
06^15 AM	Observations stable @ 06^00 AM. Medications given as
Addit:	charted. GCS 15/15 PEARL
26/10	PHYSIOTHERAPY # 80797
1030	Day 2 post stereotactic craniotomy e/o cerebellar tumours
	Nursing staff report pt quite teary + upset today
	re husband's death etc. Pt not wanting to do
	much today.
	P. R/v 1/7
26/10	[illegible] (medical oncology) 88/21
	Progress noted
	Stereotactic Craniotomy and removal
	of cerebellar metastases 24/10/06
	CT Post-op ⇒ no evidence of residual
	tumour
	Frozen Section → adenocarcinoma
	1 lesion vascular / 1 lesion avascular
	? 2 different pathologies

(BINDING MARGIN - DO NOT WRITE)

(b)

FIGURE 2.1 Cont'd

(Continued)

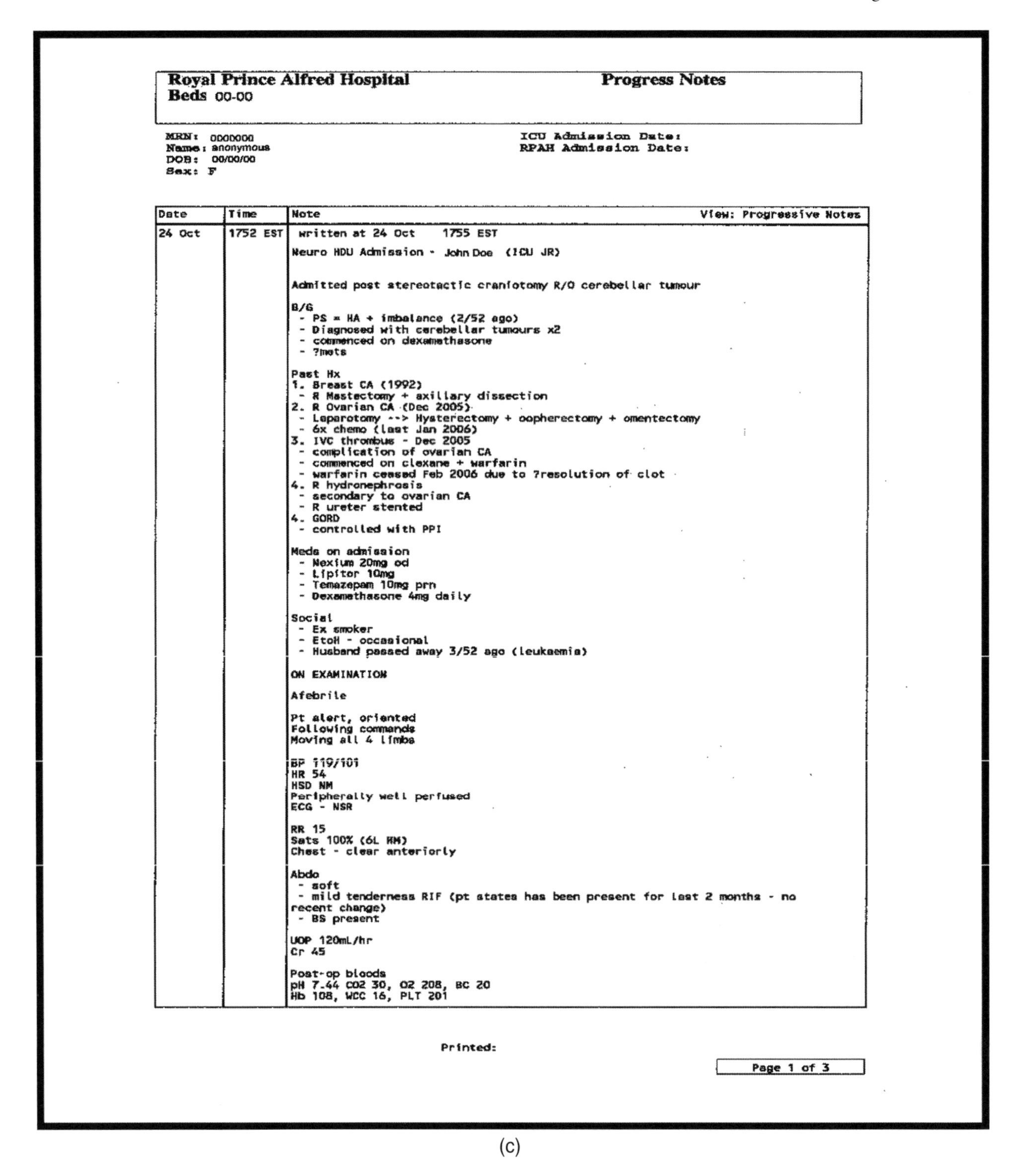

Royal Prince Alfred Hospital **Progress Notes**
Beds 00-00

MRN: 0000000
Name: anonymous
DOB: 00/00/00
Sex: F

ICU Admission Date:
RPAH Admission Date:

Date	Time	Note View: Progressive Notes
24 Oct	1752 EST	Written at 24 Oct 1755 EST Neuro HDU Admission - John Doe (ICU JR) Admitted post stereotactic craniotomy R/O cerebellar tumour B/G - PS = HA + imbalance (2/52 ago) - Diagnosed with cerebellar tumours x2 - commenced on dexamethasone - ?mets Past Hx 1. Breast CA (1992) - R Mastectomy + axillary dissection 2. R Ovarian CA (Dec 2005) - Laparotomy --> Hysterectomy + oopherectomy + omentectomy - 6x chemo (last Jan 2006) 3. IVC thrombus - Dec 2005 - complication of ovarian CA - commenced on clexane + warfarin - warfarin ceased Feb 2006 due to ?resolution of clot 4. R hydronephrosis - secondary to ovarian CA - R ureter stented 4. GORD - controlled with PPI Meds on admission - Nexium 20mg od - Lipitor 10mg - Temazepam 10mg prn - Dexamethasone 4mg daily Social - Ex smoker - EtoH - occasional - Husband passed away 3/52 ago (leukaemia) ON EXAMINATION Afebrile Pt alert, oriented Following commands Moving all 4 limbs BP 119/101 HR 54 HSD NM Peripherally well perfused ECG - NSR RR 15 Sats 100% (6L HM) Chest - clear anteriorly Abdo - soft - mild tenderness RIF (pt states has been present for last 2 months - no recent change) - BS present UOP 120mL/hr Cr 45 Post-op bloods pH 7.44 CO2 30, O2 208, BC 20 Hb 108, WCC 16, PLT 201

Printed:

Page 1 of 3

(c)

FIGURE 2.1 In (c) and (d), there are type-written notes from an electronic database that is used in the intensive care unit.

2.3 Terminology Standards—Vocabulary and a Clinical Coding System

One of the major challenges for an EMR is standardization of the terminology that is to be used by a variety of different and perhaps geographically separated health care personnel [35]. This relates to the clinical vocabulary as well as to the coding system for various diseases. A clinical vocabulary can be defined as a collections of words or terms that represent the conceptual information that makes up a given knowledge

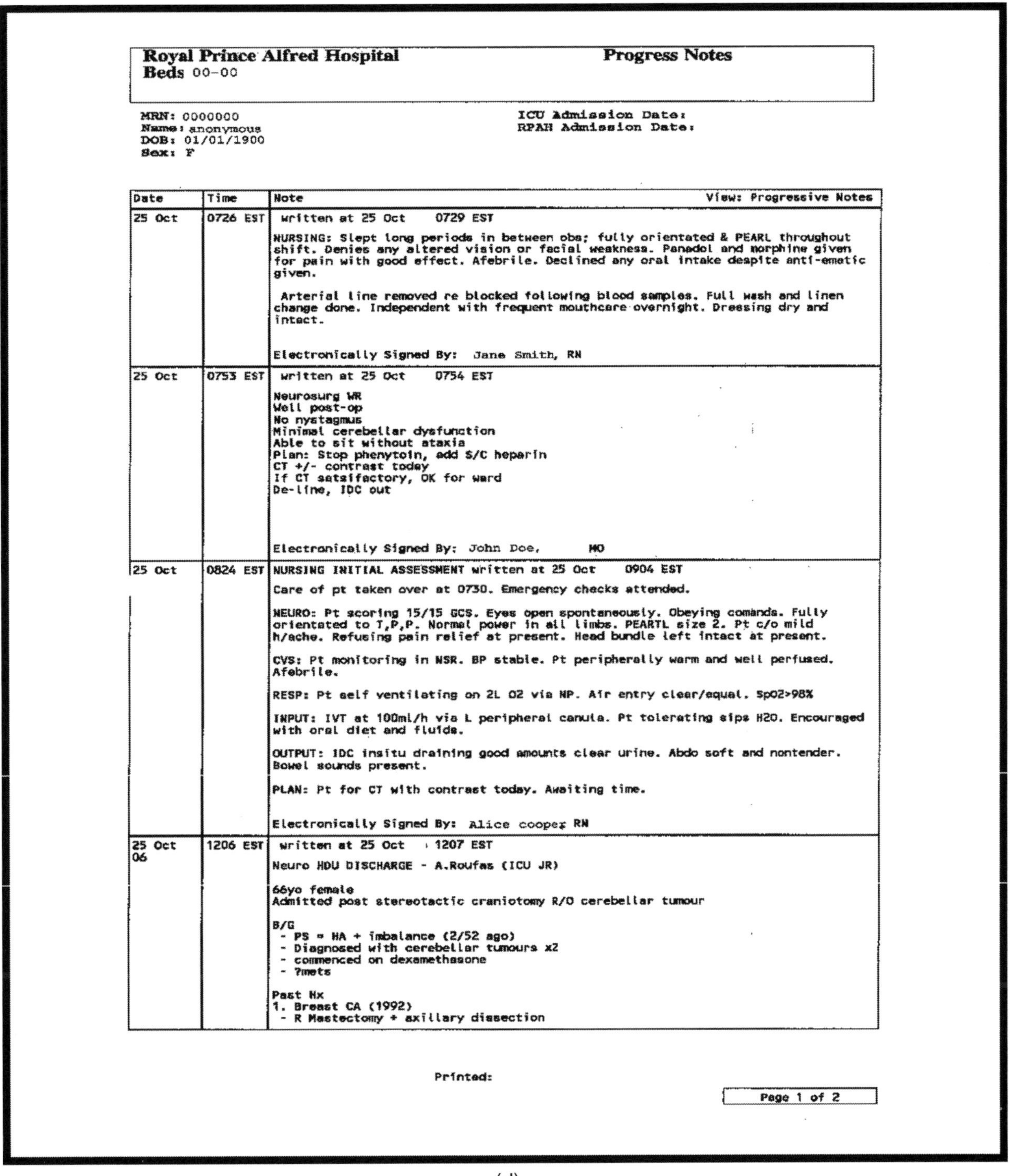

Royal Prince Alfred Hospital **Progress Notes**
Beds 00-00

MRN: 0000000
Name: anonymous
DOB: 01/01/1900
Sex: F

ICU Admission Date:
RPAH Admission Date:

Date	Time	Note (View: Progressive Notes)
25 Oct	0726 EST	written at 25 Oct 0729 EST NURSING: Slept long periods in between obs; fully orientated & PEARL throughout shift. Denies any altered vision or facial weakness. Panadol and morphine given for pain with good effect. Afebrile. Declined any oral intake despite anti-emetic given. Arterial line removed re blocked following blood samples. Full wash and linen change done. Independent with frequent mouthcare overnight. Dressing dry and intact. Electronically Signed By: Jane Smith, RN
25 Oct	0753 EST	written at 25 Oct 0754 EST Neurosurg WR Well post-op No nystagmus Minimal cerebellar dysfunction Able to sit without ataxia Plan: Stop phenytoin, add S/C heparin CT +/- contrast today If CT satsifactory, OK for ward De-line, IDC out Electronically Signed By: John Doe, MO
25 Oct	0824 EST	NURSING INITIAL ASSESSMENT written at 25 Oct 0904 EST Care of pt taken over at 0730. Emergency checks attended. NEURO: Pt scoring 15/15 GCS. Eyes open spontaneously. Obeying comands. Fully orientated to T,P,P. Normal power in all limbs. PEARTL size 2. Pt c/o mild h/ache. Refusing pain relief at present. Head bundle left intact at present. CVS: Pt monitoring in NSR. BP stable. Pt peripherally warm and well perfused. Afebrile. RESP: Pt self ventilating on 2L O2 via NP. Air entry clear/equal. SpO2>98% INPUT: IVT at 100ml/h via L peripheral canula. Pt tolerating sips H2O. Encouraged with oral diet and fluids. OUTPUT: IDC insitu draining good amounts clear urine. Abdo soft and nontender. Bowel sounds present. PLAN: Pt for CT with contrast today. Awaiting time. Electronically Signed By: Alice cooper RN
25 Oct 06	1206 EST	written at 25 Oct 1207 EST Neuro HDU DISCHARGE - A.Roufas (ICU JR) 66yo female Admitted post stereotactic craniotomy R/O cerebellar tumour B/G - PS = HA + imbalance (2/52 ago) - Diagnosed with cerebellar tumours x2 - commenced on dexamethasone - ?mets Past Hx 1. Breast CA (1992) - R Mastectomy + axillary dissection

Printed:

Page 1 of 2

(d)

FIGURE 2.1 Cont'd

domain. The terms are generally defined as actual events or entities and mapped to their cognitive representations called concepts [36]. A number of other descriptors are used in the literature to describe clinical vocabularies. These include health [37], structured clinical [38], and controlled medical vocabularies as well as reference terminologies, but they can be used interchangeably [39]. Unfortunately, the modern medical clinical vocabulary is vast and replete with jargon and abbreviations, some of which are generalized, such as stat (from *statim*, immediately), but most may be specific to a particular area or subspecialty. Some individual vocabularies have been identified by standards organizations as candidates for specific uses [40]. The criteria vary, but the core requirements for a comprehensive clinical vocabulary include an accurate representation of clinical detail, effective storage and retrieval, translation and compatibility between coding methods, and effective management of the data. It is also important to mention the term *synonymy*, which has a central role in clinical vocabularies. Synonymy can be defined as a semantic relation where two words (or concepts) have the same meaning [41]. Synonyms are important because they are often used to describe different terms in records that refer to the same concept [42]. It is logical that a meaningful and unambiguous vocabulary will enable a smooth translation from natural medical language to the more structured representations required by application programs. Yet as noted before, this is not a trivial undertaking, and there is not a single vocabulary that is accepted as a universal standard for the representation of clinical concepts.

There are three commonly used clinical coding systems: International Classification of Diseases (ICD), Systematized Nomenclature of Human and Veterinary Medicine Clinical Terms (SNOMED CT), and Medical Subject Headings (MeSH). The major aim is to establish consistency across the medical spectrum so that communication is facilitated. Figure 2.2 shows how these three systems classify idiopathic Parkinson's disease (IPD), which is a progressive neurodegenerative disorder of unknown cause that is characterised by a tremor, typically of the limbs; slowness in initiating movement; increased tone in the limbs; a stooped posture; a gait with small steps; and sometimes dementia. The major structures that are involved in the brain are the basal ganglia (a series of deep nuclei) and the extrapyramidal motor system.

2.3.1 International Classification of Diseases

The ICD is a classification, terminology, or vocabulary introduced by the World Health Organization. The most recent version is the 10th Edition (1992), hence ICD-10. The ICD chapters are divided by major anatomic systems or cause. The previous version, ICD-9 with clinical modification (ICD-9-CM) is still widely used as a system of assigning codes to diagnoses and procedures in many countries like the United States. In ICD-9-CM, coding employs a three-digit number with an optional fourth digit separated by a decimal point (e.g., 332 for Parkinson's disease and 332.1 for Parkinsonism) while code G20 is used in ICD-10. The implementation of ICD-10 is not yet expected to phase out the use of ICD-9-CM.

2.3.2 Systemized Nomenclature of Medicine Clinical Terms

SNOMED CT is a universal health care terminology developed by a division of the College of American Pathologists (CAP). SNOMED consists of a set of axes, each of which serves as a taxonomy for a specific set of concepts (e.g., organisms, disease, procedures). Coding of patient information is accomplished by combining terms from multiple axes (postcoordination) to represent complex terms. SNOMED has several synonyms for Parkinson's disease as shown in Table 2.2.

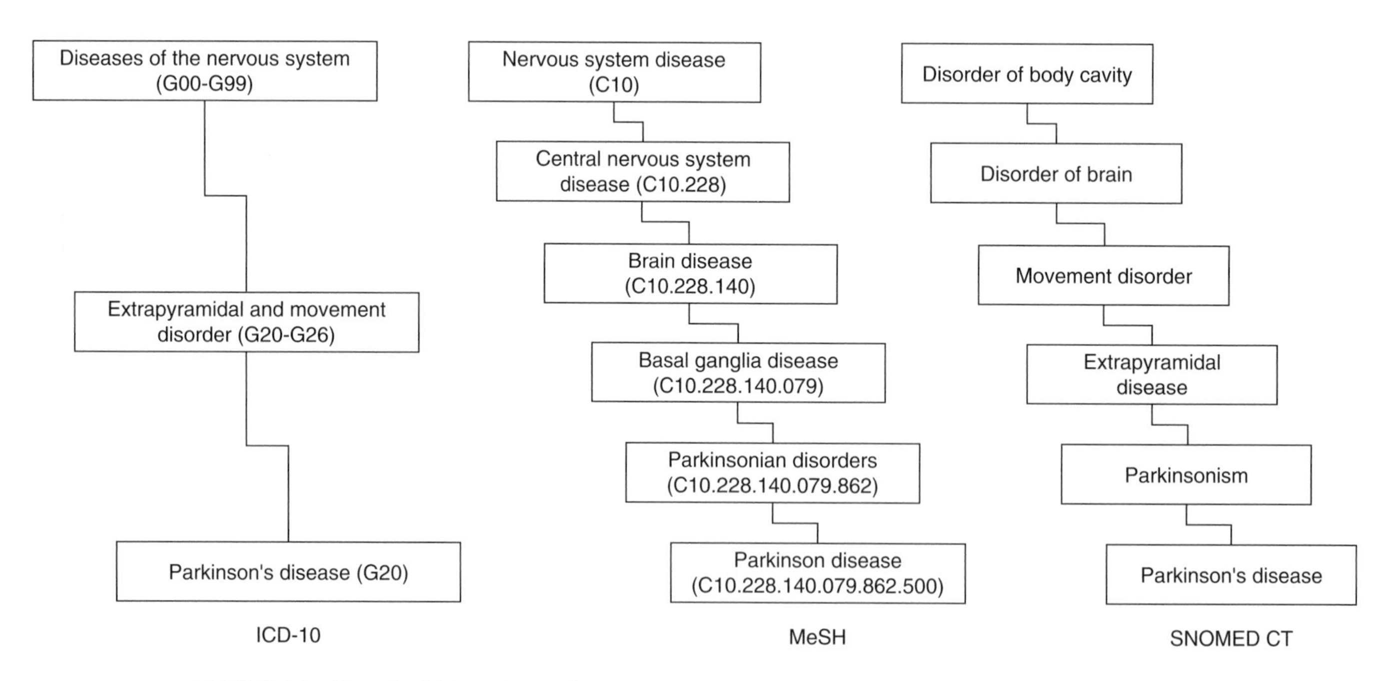

FIGURE 2.2 How the ICD, MeSH, and SNOMED-CT approach the classification of Parkinson's disease.

TABLE 2.2 SNOMED CT synonyms for Parkinson's disease

Preferred description	Synonyms
Parkinson's disease	Paralysis agitans Idiopathic parkinsonism Primary parkinsonism Shaking palsy Idiopathic Parkinson's disease Parkinson disease PD–Parkinson's disease

2.3.3 Medical Subject Heading

MeSH is a large controlled vocabulary for indexing journal articles and books in the life sciences. It was created and is maintained and updated by the National Library of Medicine (NLM) [43]. It is used by the MEDLINE article database and by the NLM catalog of book holdings. The vocabularies and their supporting informatics systems were designed to be used both by indexing professionals and by medical staff who have various degrees of computer experience.

2.3.4 Unified Medical Language System

The Unified Medical Language System (UMLS) is a large number of clinical vocabularies and classifications used to map structures between the vocabularies and the classifications. This integration can provide interconcept relationships from the established vocabularies, including SNOMED, ICD, and MeSH. There is a component within the UMLS called Metathesaurus®. The primary purpose of this component is to map between coding systems to enable information exchange between different clinical databases and systems. A semantic network maps categories to medical concepts defined in the Metathesaurus. There are 135 semantic types and 54 semantic network relationships available [44]. Figure 2.3 shows a portion of the network, where the super type is biologic function, and it has two children: physiologic and pathologic functions. Each of these has server children in the hierarchy with an "is-a" relationship.

Another component in the UMLS called the SPECIALIST Lexicon contains syntactic, morphologic, and orthographic information for biomedical and common words in the English language. The Lexicon and its associated lexical resources are used to generate the indices to the Metathesaurus and have wide applicability for natural language processing applications in biomedicine.

2.3.5 Logical Observation Identifiers Names and Codes

Logical Observation Identifiers Names and Codes (LOINC) is a terminology database for identifying laboratory observations that was developed by and is maintained by the Regenstrief Institute, an internationally recognized nonprofit medical research organization. The aim of LOINC is to standardize laboratory and clinical codes for use in health care, outcomes management, and research [45]. It is authorized by the American Clinical Laboratory Association and the CAP and is one of the standards used in U.S. government systems for the electronic exchange of clinical health information.

It is likely to become a Health Insurance Portability and Accountability Act (HIPAA) [46] standard. HIPAA is a federal regulation that establishes national standards for health care

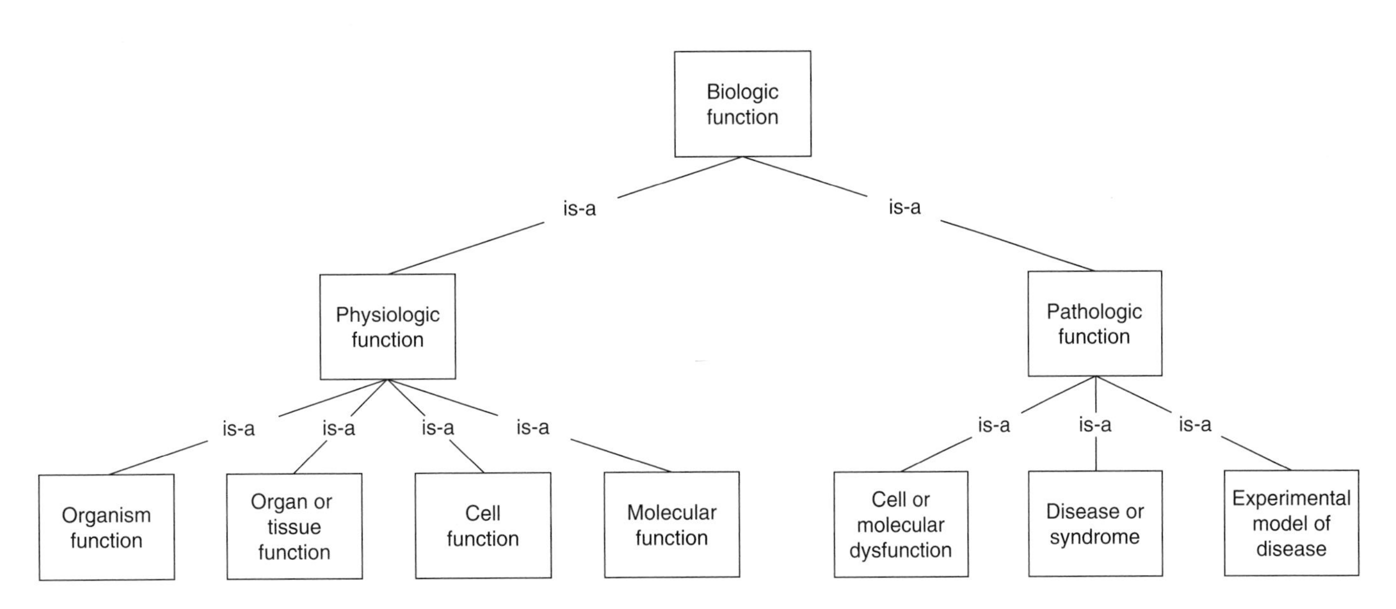

FIGURE 2.3 Part of UMLS network of biologic function.

information in the United States. LOINC applies universal code names and identifiers to medical terminology related to the EMR. A unique code (format: nnnnn-n) is assigned to each entry upon registration. There are six main parts in the codes: an analytic component, a property measured, a time aspect of the property measured, system/sample types, scale of the measurement, and type of method used [45].

2.4 Information Exchange Standards

The main aim of the medical information standards is to ensure that all data needed to accomplish a specific communication can be transmitted from one system to another [47].

2.4.1 Health Level 7

Health Level 7 (HL7) is a standards-developing organization that is accredited by the American National Standards Institute [48]. The domain of HL7 is in clinical and administrative data. The HL7 standard is a specification, not software, for information exchange between medical applications, and it includes a protocol for data exchange. It defines the format and content of the messages that applications must use when exchanging data. HL7 standards are based on the reference information model [49]. The current standard is HL7 version 2.4. Version 2.5 and version 2.6 were further constrained by the previous standard. Recently, version 3 made fundamental changes to the existing HL7 messaging approaches that introduced clinical document architecture (CDA) [50]. A feature of the CDA is its ability to be viewed in a browser using a single style sheet based on XML. Messages with CDA are, by definition, able to include text, images, sounds, and other multimedia content. A major advantage of HL7 is system independence resulting from open system architecture. Open system architecture discloses its specifications, and, by following appropriate protocols, add-on components can be developed without regard to vendor specifications. This avoids the need to develop an entire infrastructure simultaneously. Each component can be developed and integrated individually.

2.4.2 Digital Imaging and Communication in Medicine

The move to digital images in radiology and their anticipated dissemination electronically prompted the American College of Radiology (ACR) and the National Electrical Manufacturers Association (NEMA) to form a joint committee in 1983 to create a standard format for storing and transmitting medical images [51]. The committee published the original ACR-NEMA standard in 1985, which has been revised since then. In 1993, the standard was renamed to digital imaging and communications in medicine (DICOM version 3.0) and permitted the transfer of medical images in a multivendor environment. The DICOM standard contains the network components required for file exchange. It uses transmission control protocol/Internet protocol to communicate between systems. DICOM files can be exchanged between two entities that have the capability to receive the information, both image and patient data. The patient data are encoded in a header, including the patient name, type of scan, image dimensions, and other information related to the scan. The concept of DICOM arose from a desire to integrate scanners, servers, workstations, and network hardware from multiple vendors into a picture archive and communication system. For picture archive and communications system imaging devices, servers and workstations have DICOM conformance statements that identify the DICOM classes that are supported. DICOM has been widely adopted by hospitals because of its interchangeability. HL7 and DICOM require the definition of all data elements that are sent. In many cases, the content requires a specific vocabulary that can be interpreted between the systems. At present, there is heterogeneity between the standards. To improve the electronic interchange of clinical data, universal standard formats should be defined to include images, signals, multimedia, and text. In the absence of such a standard, a number of interpretation-based methods have been proposed; one of these uses a middleware system as an interpreter [52, 53].

2.5 Usability Issues in Electronic Medical Records

Usability encompasses appropriateness for the intended purpose, ease of use, and ease of learning. There are four main components in usability: learnability, time efficiency, user error rates, and user satisfaction [54]. Hartson [55] emphasized the importance of user interaction in usability. Interaction is the user's perception of the impact of the system with the workflow. In this context, it is referred to as human–computer interaction (HCI). HCI itself has a number of components: design, evaluation, and implementation. Usability is an important element in any system, but the ramifications of design-induced error can be catastrophic in a medical environment and may lead to severe morbidity and even death [56]. To increase acceptability of any EMR-based system, an extensive usability study is necessary in the development process. In the following section, a brief theoretical background of HCI is discussed.

2.5.1 Theoretical Background—Human–Computer Interaction

The main theories used in HCI are activity [57, 58], task-analysis [59, 60], and cognitive [61, 62] theory. Given that the main aim of computer system design is to assist users in their professional activities, the user interface should enable the user to complete tasks and achieve goals that are associated with the chosen activity. Activity theory builds on the work of Vygotsky and is a framework that allows the study of different forms of human practice [57]. The basic tenet of activity theory is that human (work) activity is driven by needs. The activity is performed using one or more tools. The tools are viewed in terms of usefulness and as mediators of activity rather than as inanimate objects [63].

Activity theory provides a paradigm to describe and understand the way that humans interact with computers in the user's environment [58]. The human element gives rise to the need for the activity. Activities consist of distinct actions or series of actions, which in turn consist of operations. Operations are ways of executing actions that enable the goal to be achieved. An operation may begin as a conscious act but may become routine and almost subconscious with practice and repetition. This approach to the analysis of human activities can also draw on processes such as interactive development and evaluation between users and developers seen with contextual inquiry/design [64, 65]. Contextual inquiry identifies the issues of concern, the tasks to observe, the questions to ask, and the criteria to use for screening participants.

Task analysis is used to design system functionalities and interfaces. It is essential for the developer to clearly understand the tasks to be done using the system and the way each task is performed to meet the needs of the users [60]. This process of describing tasks and their relationships is called task-analysis. Annett and Duncan introduced hierarchical task-analysis to evaluate the training needs of an organization [66].

Cognitive theory stems from cognitive psychology in which humans can be modeled as cognitive information processors [61, 67]. This approach describes man as a sensory, effectors, and central processing system (perception, cognition, storage). The goals, operator, methods, and selection model are typical examples used to explain the nature and structure using cognitive theories [68].

2.5.2 Workflow Analysis

Efficient work processes are essential to any successful organization. One method to achieve improved efficiency is based on workflow analysis. Workflow can be defined as a set of tasks comprised of coordinated computer-based and human activities [69]. A workflow model or schema is a formal computerized representation of work procedures that controls the sequence of the tasks that are performed [70]. Ancillary benefits of better efficiency can be improved cost-effectiveness, productivity, communications, and user satisfaction. The value of workflow analysis techniques is gradually being recognized in the health care industry [71]. Processes and their associated workflows can be modeled in various ways using different tools, which are mainly based on object-oriented modeling methods. The four main approaches are as follows [72]:

1. Informational modeling: The focus is on the flow, structure, and interrelationships in the information.
2. Functional modeling: Attention is directed to tasks being performed and the related information.
3. Organizational modeling: The focus is on agents (humans or computers) and resources involved in each task; communication between the agents is analyzed.
4. Transactional modeling: Issues of timing (sequencing) and control are examined, both within and between the tasks in the process.

These modeled workflows can be hierarchically decomposed into subworkflows [73]. Activities and tasks can be allocated to one or more agent. The term agent can also refer to organizational units or roles. In health care, there are two main types of workflow: ad hoc and procedural [74]. Ad hoc workflows are processes in which the outcome cannot be predicted. Procedural workflows have a predefined structure that is implemented each time the workflow is performed. For example, in a medical imaging department, the performance of a chest computed tomography (CT) scan is procedural and routine, and the outcome can be predicted. The sequence of procedures follows a predefined protocol. After the scan is completed, the transmission of imaging report to the requesting physician is an ad hoc process as the timing and receipt by the requesting physician cannot be predicted. In a workflow model, the outcome of each process is recorded. A description of who performed the process, how it was performed, and the tools and other resources used is analyzed. These results are then specified as they relate to the current state of the workflow process, workflow-relevant data, organizational structure, and the available technology.

The health care environment poses its own problems for any workflow analysis as the work activities tend to be very heterogeneous, range from simple to exceedingly complex, and can be composed of ad hoc elements as well as repetitive and well-defined processes.

2.5.3 User Involvement

There is no question that user involvement improves user satisfaction with the end result [75, 76]. In this context, user involvement refers to participation in the system development process by the target user group, which tends to be medical professionals. However, it is also important to emphasize that user buy-in generally requires the user to see benefits from the proposal.

2.6 User Interface

The design principles of the user interface have been discussed by a number of investigators [59, 77]. Some key principles in the development of a user interface for medical applications are shown in Table 2.3.

A number of principles must also be followed when data are displayed and perceived effectively (Table 2.4). Well-designed information presented in a logical manner can lead to a higher degree of usability. Conversely, an illogical and poorly designed display reduces performance and effectiveness of the system.

2.6.1 Synoptic Presentation

Part of the EMR can be displayed in a synoptic (brief or condensed) format. A synoptic report simplifies a complex medical problem. Records following one of the clinical vocabularies are readily organized and classified. There has been some by-in for surgery and pathology reports where there is a relatively high degree of repetition and use of standard procedures [78, 79]. Nevertheless, a synoptic report may cause some ambiguity, which once again emphasizes the importance of design and consultation with the user.

2.6.2 Narrative Presentation

Narrative text reports are the major component of a PBMR. Admission notes, progress notes, imaging reports, and discharge summaries are collected and stored as narrative text. An example of a short medical imaging report is shown in Figure 2.4. The example shown is a short report, but the succinctness and value of such reports often depend on the reader. Unnecessary data provided for the referring physician clutter many such reports, and in this setting, a synoptic report offers advantages.

To convert a narrative report to a coded format remains a continuing challenge and is an area of very active research in the field of natural language processing (NLP) [80–82]. NLP is a synthetic area between artificial intelligence and linguistics dealing with problems of automatic generation and understanding of human language. The foundation of NLP is parsing of sentences and of terms used in narrative text. Natural language can be converted into a structured representation for computers, and information from computers can be converted to natural language. The major advantage of NLP is that it can be applied to existing narrative reports. NLP can be combined with the clinical vocabularies discussed previously to enhance medical coding.

TABLE 2.3 User interface design principles

Principle	Description
Consistency	Interface uses consistent terms and concepts.
Ease of use	Interface is easy to understand and use.
Robustness	Interface is robust and reliable.
User control	Override is available in an unexpected situation.
Flexibility	Various input tools are available.
Feedback and help	A context-sensitive user guide and assistance are available.

TABLE 2.4 Data display principles

Principle	Description
Grouping	Logically related data are grouped.
Standard	Consistent display of data according to standards.
Highlight	Data are highlighted according to importance.
Graphics	Appropriate graphics can increase effectiveness.
Colors and fonts	Appropriate fonts and color enhance perception.

2.6.3 Graphic User Interface and Standard Data Entry

A graphic user interface (GUI) contains a graphic component in addition to text and accepts input from devices such as a keyboard and mouse. The GUI displays a graphic output on the computer monitor in response to the input. There are a number of different principles used for GUI design, and these include object-oriented user interfaces and function-oriented design [83].

The graphic component of a GUI can be designed and arranged in a structured way called structured data entry (SDE) [84–86]. This approach creates patient records that can be edited and searched and that can have decision support functions [87, 88]. It has been suggested that this type of approach can improve completeness and reduce ambiguity. The structure of the data must be reflected in the GUI design such that the user can choose from a number of options or templates [87]. For a given medical specialty, a limited dataset may only need to be specified and collected. Advantages of SDE include uniform and consistent patient data. However, an SDE requires careful design. Long lists of values (which require scrolling and a rigid hierarchy) should be avoided [89]. A very important consideration is the time required to enter the data. A patient history and physical examination can be a tedious exercise to enter into a computer when compared to scribbling a quick note that is legible to the person transcribing it.

2.6.4 Web-Based Interface (Static and Dynamic)

Another approach is to use a Web-based interface [90–92]. The immediate and obvious advantages are that it provides ready access and is operating system–independent. Information is accessed with a Web browser over a network. The interface can be updated and maintained centrally, which is also attractive. Although there may be some functional limitations on a Web interface, client-side scripting can add functionalities. Most application-specific functionalities can be implemented

Name:	Sample patient	**Referring Doctor:**	John Doe
DOB:	01/01/2000	**MRN:**	0000000
Age:	6	**Patient #:**	Q0000 I PID 00000
Sex:	Male	**Study Date:**	01/01/2006

PET WB FDG - Melanoma pre Six

Technical data:

The patient was scanned on the PET-CT scanner (LSO Biograph) after the intravenous injection of 351 MBq of FDG.

Report:

The patient was scanned with arms above the head. There are no large mass lesions in the cerebrum. There is mild misregistration between the anatomical and functional data due to head movement.

There is a slightly irregular, markedly glucose avid lesion in the soft tissues of the left inguinal region consistent with the known site of disease. This lesion appears to be solitary. There is mild diffuse FDG uptake medial to the main abnormality in the lower abdominal wall, which may represent the site of previous surgery. There are no other abnormalities in the pelvis. The paraaortic nodes and abdominal viscera are clear. In the thorax, I cannot identify any glucose avid foci, but the small nodules which are referred to on the request form are approximately 5 mm in size, which may be beyond the resolution of the scanner. There are no abnormalities in the mediastinum or in the rest of the study to indicate other sites of disease.

Conclusion:

There is a slightly irregular, markedly glucose avid lesion in the soft tissues of the left inguinal region consistent with active high-grade tumors. But the lesion appears to be solitary and there are no other abnormal foci of increased FDG uptake in the rest of the study.

FIGURE 2.4 A sample medical imaging report.

using Java and JavaScript. Web-based communication systems have been set up for transmitting medical images and for online consultation [93]. However, a dynamic interaction between client servers has been lacking. Most of the data are statically transmitted when the Web page is refreshed. Recently, dynamic and interactive Web interfaces have been introduced by using server-side technologies such as Hypertext Preprocessor and Asynchronous JavaScript and Extensible Markup Language [XML] (AJAX) [94]. AJAX enables faster, more responsive Web applications through a combination of JavaScript, the document object model, and XML. In the AJAX model, JavaScript calls to the server can update a single element in the user interface with data retrieved from a server. The application is much more responsive than a static interface because the full screen is not reloaded.

2.6.5 Alternative–Input Interfaces

A number of investigators have reported on alternative methods for data input [95–99]. Instrumentation improvements now allow handwriting and voice recognition input to the EMR. Accuracy, speed, and portability remain continuing challenges for handwriting and voice recognition (VR) in a busy hospital/ward setting. Voice technologies have penetrated a number of areas in a hospital, but the need for greater computer power has generally limited its use to relatively quiet environments. Although VR is readily accomplished on a laptop, it would be desirable to have the same functionality on a handheld device. Newer areas of research and development are the tactile interfaces. A tactile interface supplements or replaces other forms of output with haptic (hand) feedback [95]. Tactile sensation, a sense of roughness and friction, is being used in the field of HCI. Touch screens can be used in medical imaging to manipulate three-dimensional whole body, vascular, and neurologic data. In another approach, the tangible user interface [98, 99], the environment is in three-dimensional space; this has been used in neurosurgery to aid surgical planning [100].

2.6.6 Advanced Computing Devices

Still other approaches utilize mobile computing devices and a wireless network [101, 102]. Mobile devices are ideally suited to a busy hospital because there is no requirement to find a free computer terminal, and the devices are carried to wherever the healthcare worker is required. Other obvious benefits are real-time data uploads and downloads to the device. These data can include patient lists and the associated laboratory and imaging results as well specific protocols for patient care. Electronic versions of drug references and abridged versions of medical textbooks have been available on mobile devices for some time, and these are widely used by resident medical staff [24, 62, 103–108].

2.7 Evaluation

Any new intervention should have an appropriate evaluation to measure its impact, which may be positive or negative. The three common areas of interest relevant to the EMR are system quality, information quality, and user satisfaction [4]. Some studies include the impact of the system on individuals and organizations [16, 109]. System quality is often measured with the emphasis on time efficiency, including response time and time savings [3, 110–113]. Time is a key criterion for measuring the integration of the system into clinical workflow. A system can be considered effective if it reduces time required for documentation and access time, even if the time efficiency does not translate into better patient care [114]. Attributes of information quality can include accuracy, completeness, and legibility. Depending on the application of the EMR, single or multiple attributes can be included in the design of the evaluation [109, 111, 115]. User satisfaction can refer to the system itself or to its content [4]. Other than the attributes mentioned above, general evaluation methods widely used in software engineering can be applied to an EMR [116].

Formal methods used in cognitive science are generally needed to evaluate usability [117]. Observation is one such technique. Observation simply involves watching individuals as they use a system; this can be done physically or by means of a video. The users can also be interviewed as the tasks are performed—a so-called *think aloud*, which provides an opportunity for user feedback. Questionnaires and surveys are also commonly used. A questionnaire generally asks users to rate a system according to the attributes and dedicated functionalities. To obtain reliable results from the survey, questions need to be precise rather than general. It is also informative to assess the background of participating users, such as their experience and familiarity with the working environment and system. Another approach is to assess and evaluate the new system as it is developed in a process referred to as a *cognitive walk-through*. Assessors evaluate the component tasks of the system sequentially. The basic steps that are involved include the following:

1. A description of the system and its functionalities
2. A description of the task the user is to perform using the system
3. A complete, detailed list of actions required to accomplish the task
4. A profile of the target users and their backgrounds

2.8 Electronic Medical Records System—A Case Study: A Web-Based Electronic Record for Medical Imaging

In our Department of Molecular Imaging, we specialize in positron emission tomography (PET) CT scans that are carried out mainly in the assessment of patients with a variety of malignancies. We currently perform between 22 and 24 patient PET/CT studies a day. The scanning procedure is explained to the patient, and then pertinent clinical details including diagnosis, treatment, results of other investigations, and medications are recorded, and informed consent for the procedure is obtained [118]. The record of the interview was originally paper-based and hand-written; these data were then dictated using a Dictaphone, and secretarial staff would type the dictated history into the department information system (IS). However, we identified issues with legibility of handwritten notes leading to delays in transcribing taped data into the IS, and we found it increasingly difficult to use this approach with our large number of patients. We sought to use recent advances in information technology to improve our efficiency. We used a Web-based approach called Web-based imaging electronic patient history (WEBI-EPH), a personal digital assistant (PDA), and a wireless network to access the IS within the department.

2.8.1 The Clinical Workflow

A profile of the clinical workflow in the department was outlined before designing the new system. A contextual/enquiry method based on activity theory was applied in the system development and interface design [65]. The profile of the traditional history-taking process was as follows:

1. User (consultant or registrar) interviews the patient. He or she explains the procedure, obtains informed consent, and records pertinent historic data on paper.
2. The paper-based history is dictated using a Dictaphone.
3. The dictated material is typed into the IS by secretarial staff.

The elements of the new system are as follows (Figure 2.4):

1. Prior to interviewing the patient, user creates a patient record.
2. User interviews the patient and records the data on the mobile PDA as the interview proceeds.
3. The mobile device displays a series of fields that are appropriate to the underlying condition (e.g., lymphoma, lung cancer, bowel cancer).
4. Input is automatically saved.
5. Input is automatically updated in real time to the IS.
6. The patient undergoes the scan.
7. Updated data are available to the reporting clinician before the scan is completed.

Fields were created to capture the patient's current treatment, medical history, medications taken, and social/environmental factors. Each component creates queries to the middleware tier, as shown in Figure 2.5, which are then updated to the IS.

2.8.2 Web-Based Imaging Electronic Patient History

The WEBI-EPH was designed to accommodate the sequence of events that occurs when a patient is assessed prior to undergoing a scan, which is shown schematically in Figure 2.5. The fields were designed, tailored, and piloted for two common medical conditions where PET-CT scanning provides unique information: lung cancer and the evaluation of lymphomas (a type of blood cancer). A look-up list was generated for expected answers to common questions, and the design reflected the usual sequence in which these questions were asked. The graphic component was used to define basic units. A basic unit was a dataset where the elements had distinctive properties that were described separately. Each unit consisted of several descriptive elements including properties and values. In the classification of current symptoms, a basic unit was pain, and it included information about the location and duration as properties and values. Data input was achieved by selecting the relevant item from a list of values provided by the template or by typing/writing using a stylus. In cases requiring more complicated input, multiple forms were designed in a hierarchic structure. For each template, the first layer was shown immediately. For more detail, the logical second layer of descriptive elements appeared and was inserted into the first-layer values. This was done in such a way that details relevant to the scan interpretation were not lost and inclusiveness of data was maintained.

The WEBI-EPH design included demographic data, indication for the scan, current symptoms, medical history,

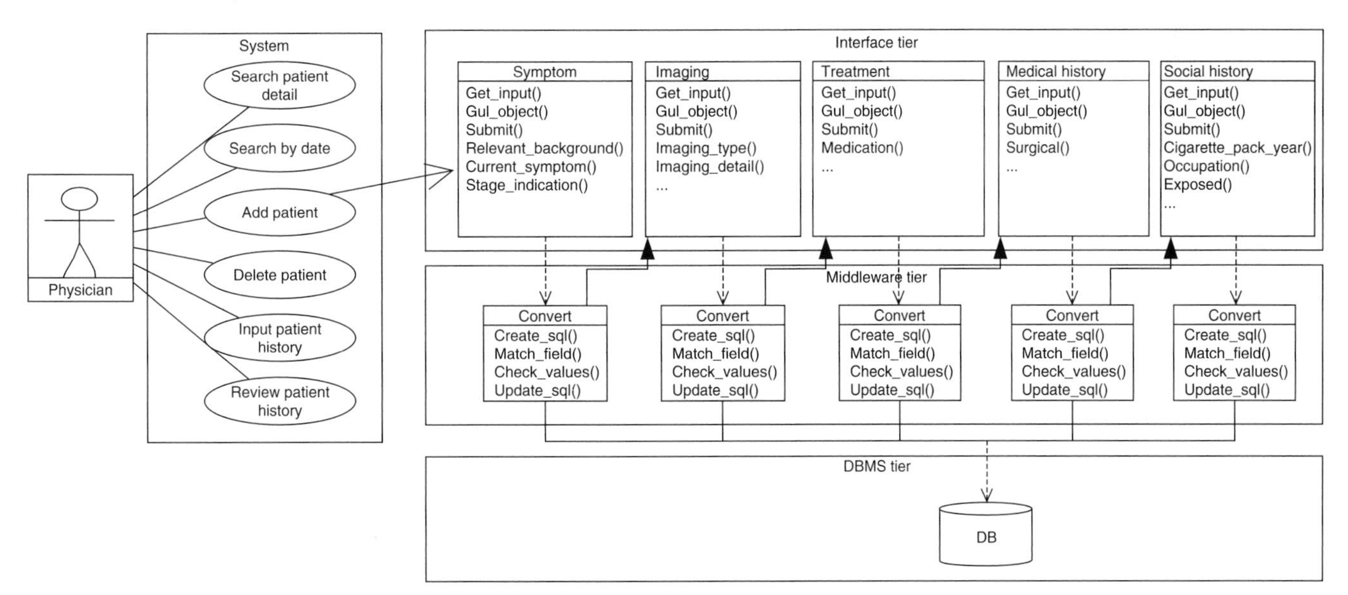

FIGURE 2.5 Sequence using unified modeling language to create a WEBI-EPH. User selects a function from the system interface, which initiates the processes. The input process has five stages.

Department of P.E.T & Nuclear Medicine
at Royal Prince Alfred Hospital

PATIENT DATA SHEET

Patient Name:
PID Number:
MRN: nil

Referring Dr:
Study date: 6/9/2005
Patient DOB: 10/6/1943

Indication for study: Restaging.

Diagnosis: NHL

Relevant History: 62 yo. ♂ R.H.
Retired sheet metal worker.
BHx:
1. 1999 – incidental finding of Mantle cell NHL in abdominal LN's after bowel resection for multiple benign polyps. Apparently disease limited to abdomen initially.
Mx: CHOP × 6 cycles.
In CR for 3½ yrs.
2. 2003 – recurrence, ? apparently causing spinal cord compression.
Mx: mabthera × 4 cycles
– cyclophosphamide
3. 2004 – another recurrence involving mediastinum
Mx: mabthera
RTX
4. Further recurrence in Jan 05
Mx: cyclophosphamide
F/up CT (01/06/05) showed progression causing airway compromise
Mx: Hyper CVAD (last dose 7/12 ago)
last dose 3/52 ago.

Other investigations:

Medications / Doses: Losec, Allopurinol, Bactrim.

Karnofsky Performance Scale (KPS)
Normal, no complaints 100
Able to carry on normal activities, minor signs or symptoms of disease 90
Normal activity with effort 80
Cares for self. Unable to carry on normal activity or to do active work 70
Requires occasional assistance but able to care for most of his / her needs 60
Requires considerable assistance and frequent medical care 50
Disabled. Requires special care and assistance 40
Severely disabled. Hospitalisation indicated though death not imminent 30
Very sick. Hospitalisation necessary, active supportive treatment necessary 20
Moribund 10
Dead 0

KPS = 80.

FO, stem cell TP.

If Woman of Reproductive Age (15 - 50 yrs)
* Has the patient been asked regarding possibility of pregnancy ☐ Yes ☐ No
* Is a pregnancy test required ☐ Yes ☐ No
If Yes, Result:

Study and scanning data (General Nuclear Medicine Studies only)

CXR – nil
ETOH – nil

Name (Printed):
Signature:
Date: 06/09/05

FIGURE 2.6 The PBMR used in our department shows typical acronyms and the difficulty with hand-written data.

management plan, and medications taken. A form-input–based Web-interface was designed using JavaServer Pages in a Web server. Data from the Web interface were transmitted from the PDA to the IS using Java Database Connectivity in real time via a wireless network. Fields were designed to minimize handwriting. We piloted the EPH in 180 patients with lung cancer and 130 patients with lymphoma over a 3-month period in 2005. We measured the time required to complete and update a PBMR and the WEBI-EPH to the IS and the time for the dictated history to appear in the patient's file in the IS. We also rated the WEBI-EPH and the PBMR for legibility, inclusiveness, and relevance to the scan interpretation. A PBMR of patient history is shown in Figure 2.6. The Web interface WEBI-EPH is shown in Figure 2.7.

The time taken to complete the WEBI-EPH ranged from 2 to 59 minutes, and the average was 10.14 minutes ($\pm$ 4.2); the outlying value was the result of an interruption in the interview because of an unforeseen event when the registrar had to attend to another sick patient. All 310 records were available immediately in the IS. For the PBMR, the average time to completion was 24 minutes; all records were dictated, but none of those had appeared in IS over the period surveyed; the average time for the dictated data to appear in the IS was 10 months. The EPH also received higher evaluation scores than the PBMR in all categories: legibility 2.3 (PBMR), 5 (EPH); inclusiveness 3.7 (PBMR), 4.7 (EPH); and relevance to scan 3.3 (PBMR), 4.3 (EPH). EPH data were legible, inclusive, and immediately available without loss of accuracy. Our preliminary data indicate that a custom-designed EPH is a marked improvement over a PBMR.

2.9 Summary

The theoretic advantages of an EMR are widely appreciated, but as yet an effective, robust, routine EMR that is integrated into health care remains elusive. Nevertheless, such an EMR is imminent and will be achievable with recent advances in information technology. However, it should be emphasized that user input in the design is essential, the workflow where the EMR is to be used should be carefully mapped, the medical data that are to be captured should be standardized as much as possible, and the input method should be easy to use and fast. Further reading from the reference list is recommended.

Acknowledgements

The authors are grateful to the support from Australian Research Council and Hong Kong Polytechnic University/University Grant Committee grants.

2.10 Exercises

1. Discuss the main advantages of using EMR with standard terminologies.
2. Discuss emerging computer interfaces and possible applications in the development of an EMR.
3. List the three major factors for evaluation of an EMR-based system and its attributes.

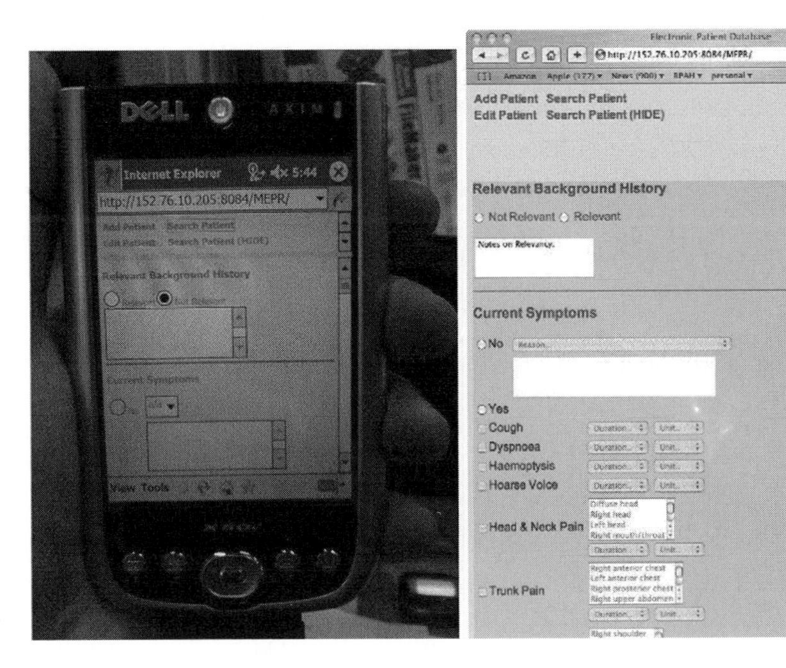

FIGURE 2.7 PDA. Web-based interface on PDA is on the left and screen capture from desktop browser is on the right.

2.11 References and Bibliography

1. U.S. General Accounting Office. *Automated Medical Records: Leadership Needed to Expedite Standards Development.* U.S. Government Accounting Office; IMTEC-93-17, 1993.
2. J. Stausberg et al. Comparing paper-based with electronic patient records: lessons learned during a study on diagnosis and procedure codes. *J. Am. Med. Inf. Assoc.* 10:470–477, 2003.
3. G. Makoul, R. H. Curry, and P. C. Tang. The use of electronic medical records: Communication patterns in outpatient encounters. *J. Am. Med. Inf. Assoc.* 8(6):610–615, 2001.
4. M. Van der Meijden et al. Determines of success of inpatient clinical information systems: A literature review. *J. Am. Med. Inf. Assoc.* 10:235–243, 2003.
5. O. Ratib, M. Swiernik, and J. M. McCoy. From PACS to integrated EMR. *Computerized Medical Imaging and Graphics.* 27(2–3):207–215, 2003.
6. D. P. Lorence and R. Churchill. Clinical knowledge management using computerized patient record system: Is the current infrastructure adequate? *IEEE Trans. Inf. Technol. Biomed.* 9(2):283–288, 2005.
7. L. Poissant et al. The impact of electronic health records on time efficiency of physicians and nurses: A systematic review. *J. Am. Med. Inf. Assoc.* 12:505–516, 2005.
8. A. Van Ginneken. The computerized patient record: Balancing effort and benefit. *Int. J. Med. Inf.* 65:97 –119, 2002.
9. S. T. C. Wong et al. Workflow-enabled distributed component-based information architecture for digital medical imaging enterprises. *IEEE Trans. Inf. Technol. Biomed.* 7(3):171–183, 2003.
10. A. James, Y. Wilcox, and R. N. G. Naguib. A telematics system for oncology based on electronic health and patient records. *IEEE Trans. Inf. Technol. Biomed.* 5:16–17, 2001.
11. M. Krol. Telemedicine. *IEEE Potentials.* 16(4):29–31, 1997.
12. S-C. Hwang and M-H. Lee. A Web-based telePACS using an asymmetric satellite system. *IEEE Trans. Inf. Technol. Biomed.* 4(2):212–215, 2000.
13. G. Eysenbach. What is e-health? *J. Med. Internet Res.* 3(2): e20, 2001.
14. R. Istepanian. M-health emerging mobile health systems. In: R. S. H. Istepanian, S. Laxminarayan, C. S. Pattichis (Eds.). *Biomedical Engineering International Book Series.* Springer, 2005.
15. D. W. Bates et al. A proposal for electronic medical records in U.S. primary care. *J. Am. Med. Inf. Assoc.* 10(1):1–10, 2003.
16. G. Southon, C. Sauer, and K. Dampney. Lessons from a failed information systems initiative: Issues for complex organisations. *Int. J. Med. Inform.* 55(1):33–46, 1999.
17. J. Aarts, H. Doorewaard, and M. Berg. Understanding implementation: The case of a computerized physician order entry system in a large Dutch university medical center. *J. Am. Med. Inform. Assoc.* 11(3):207–216, 2004.
18. K. Dansky et al. Electronic medical records: are physicians ready? *J. Health Manage.* 44(6):440, 1999.
19. M. van der Meijden et al. Development and implementation of electronic medical records in the United States. *J. Am. Med. Inform. Assoc.* 12(1):3, 2005.
20. E. Ammenwerth et al. Factors affecting and affected by user acceptance of computer-based nursing documentation: Results of a two-year study. *J. Am. Med. Inform. Assoc.* 10(1):69, 2003.
21. D. R. Kaufman et al. Usability in the real world: Assessing medical information technologies in patients' homes. *J. Biomed. Inform.* 36(1–2):45–60, 2003.
22. S. T. C. Wong et al. Workflow-enabled distributed component-based information architecture for digital medical imaging enterprises. *IEEE Trans. Inf. Technol. Biomed.* 7(3):171–183, 2003.
23. M. C. Beuscart-Zephir et al. Integrating users' activity modeling in the design and assessment of hospital electronic patient records: The example of anesthesia. *Int. J. Med. Inform.* 64:157–171, 2001.
24. C. Pappas et al. A mobile e-health system based on workflow automation tools. In *The 15th IEEE Symposium on Computer-based Medical Systems* (CBMS 2002). 2002; Maribor, Slovenia.
25. D. Scalise. Evidence-based medicine. An overview for hospital leaders. *Hosp. Health Network.* 79(9):8, 2005.
26. W. D. Sacket. The need for evidence based medicine. *J. R. Soc. Med.* 88(11): 620–624, 1995.
27. H. Laerum, T. H. Karlsen, and A. Faxvaag. Effects of scanning and eliminating paper-based medical records on hospital physicians' clinical work practice. *J. Am. Med. Inform. Assoc.* 10(6):588–595, 2003.
28. N. Rodriguez et al. A usability study of physicians' interaction with a paper-based patient record system and a graphical-based electronic patient record system. In *Proc. AMIA Symp.*, 2002.
29. B. Webb and K. Powell. From a paper to an electronic medical record. *Inform. Healthc Aust.* 5(3):97–100, 1996.
30. L. T. Kohn, J.M. Corrigan, and Molla S. Donaldson (Eds.). Committee on Quality of Health Care in America, Institute of Medicine. *To Err is Human: Building a Safer Health System.* TNA Press, 1999.
31. H. Tange. The paper-based patient record: Is it really so bad? *Comput. Methods Programs Biomed.* 48(1–2): 127–131, 1995.
32. R. Koppel et al. Role of computerized physician order entry systems in facilitating medication errors. *J. Am. Med. Inform. Assoc.* 293(10):1197–1203, 2005.
33. R. Brodell et al. Prescription errors: Legibility and drug name confusion. *Arch. Fam. Med.* 6(3):296–298, 1997.
34. W. Hogan, and M. Wagner. Accuracy of data in computer-based patient records. *J. Am. Med. Inform. Assoc.* 4:342–355, 1997.

35. C. G. Chute. Clinical classification and terminology: Some history and current observations. *J. Am. Med. Inform. Assoc.* 7:298–303, 2000.
36. K. E. Campbell et al. Representing thoughts, words, and things in the UMLS. *J. Am. Med. Inform. Assoc.* 5:421–431, 1998.
37. B. L. Humphreys et al. Planned NLM/AHCPR large-scale vocabulary test: Using UMLS technology to determine the extent to which controlled vocabularies cover terminology needed for health care and public health. *J. Am. Med. Inform. Assoc.* 3:281–287, 1996.
38. J. L. Kannry et al. Portability issues for a structured clinical vocabulary: Mapping from Yale to the Columbia Medical Entities Dictionary. *J. Am. Med. Inform. Assoc.* 3:66–79, 1996.
39. K. A. Spackman, K. E. Campbell, and R. A. Cote. SNOMED-RT: A reference terminology for health care. In *Proc. Ann. AMIA*. Washington, DC, 1997.
40. C. J. McDonald. The barriers to electronic medical record system and how to overcome them. *J. Am. Med. Inform. Assoc.* 4(3):213–221, 1997.
41. J. Lyons. *Linguistic Semantics. An Introduction.* Cambridge University Press, 1997.
42. K. Fung et al. Integrating SNOMED CT into the UMLS: An exploration of different views of synonymy and quality of editing. *J. Am. Med. Inform. Assoc.* 12:486–494, 2005.
43. S. J. Nelson et al. The MeSH translation maintenance system: Structure, interface design, and implementation. In *Proceedings of the 11th World Congress on Medical Informatics.* IOS Press, 2004.
44. C. Tilley and J. Willis. *Unified Medical Language System Basics.* National Library of Medicine, 2004.
45. A. W. Forrey et al. Logical observation identifier names and codes (LOINC) database: A public use set of codes and names for electronic reporting of clinical laboratory test results. *Clin. Chem.* 42(1):81–90, 1996.
46. Health and Human Services Web site. The health insurance portability and accountability (HIPAA) act of 1996. Available at: http://www.hhs.gov/ocr/hipaa/. *U.S. Office for Civil Rights.* 0991-AB29:20223–20258, 2005.
47. E. H. Shortliffe et al. *Medical Informatics: Computer Applications in Health Care and Biomedicine.* Springer-Verlag, 2003.
48. Health Level Seven, I. Health Level-7 Standards Web site. Available at: http://www.hl7.org/.
49. R. H. Dolin et al. The HL7 clinical document architecture. *J. Am. Med. Inform. Assoc.* 8(6):552–569, 2001.
50. R. H. Dolin et al. HL7 clinical document architecture, release 2. *J. Am. Med. Inform. Assoc.* 13(1):30–39, 2006.
51. National Electrical Manufacturers Association. Digital imaging and communication in medicine strategic document version 4.0. Available at: http://medical.nema.org/.
52. I. Foster and R. L. Grossman. Data integration in a bandwidth-rich world. *Commun. ACM.* 46(11):50–57, 2003.
53. K. Schelfthout and T. Holvoet. Coordination middleware for decentralized applications in dynamic networks. In *DSM '05: Proceedings of the 2nd International Doctoral Symposium on Middleware.* ACM Press, 2005.
54. H. R. Hartson and D. Hix. Human-computer interface development: Concepts and systems for its management. *ACM Comput. Surv.* 21(1):5–92, 1989.
55. H. R. Hartson. Human-computer interaction: Interdisciplinary roots and trends. *Journal of Systems and Software.* 43:103–118, 1998.
56. J. J. van Bemmel. Medical data, information, and knowledge. *Methods Inform. Med.* 27:109–110, 1998.
57. L. Vygotsky. The instrumental method in psychology. In J. Wertsch (Ed.). *The Concept of Activity in Soviet Psychology.* Sharpe, 1981.
58. S. Bodker. *Through the Interface: A Human Activity Approach to User Interface Design.* Lawrence Erlbaum Associates, 1991.
59. J. Hackos and J. Redish. *User and Task Analysis for Interface Design.* John Wiley & Sons, Inc., 1998.
60. D. Diaper. *Task Analysis for Human-Computer Interaction.* Prentice Hall, 1990.
61. S. K. Card, T. P. Moran, A. Newell. *The Psychology of Human-Computer Interaction.* Lawrence Erlbaum Associates, 1983.
62. A. E. Carroll, S. Saluja, P. Tarczy-Hornoch. The implementation of a personal digital assistant (PDA) based patient record and charting system: Lesson learned. In *Proceedings of the American Medical Infomatics Association Symposium,* 2002.
63. J. J. Preece et al. *Human–Computer Interaction.* Addison-Wesley, 1994.
64. J. Simonsen and F. Kensing. Using ethnography in contextual design. *Commun. ACM* 40(7):82–88, 1997.
65. H. Beyer and K. Holtzblatt. *Contextual Design: Defining Customer-Centered Systems.* Morgan Kaufmann, 1998.
66. J. Annett and J. Duncan. Task analysis and training design. *Occupational Psychology.* 12:211–221, 1967.
67. P. H. Lindsay and D. A. Norma. *Human Information Processing.* Academic Press, 1972.
68. D. E. Kieras. Towards a practical GOMS model methodology for user interface design. In M. Helander (Ed.). *Handbook of Human–Computer Interaction.* North-Holland Publishing, 1998.
69. K. Lei and M. Singh. A comparison of workflow metamodels. In *Proceedings of the ER-97 Workshop on Behavioral Modeling and Design Transformations: Issues and Opportunities in Conceptual Modeling,* 1997.
70. J. E. Bardram. Plans as situated action: An activity theory approach to workflow systems. In *Proceedings of the 1997 European Conference on Computer Supported Cooperative Work (ECSCW'97).* 1997.

71. Mueller, M. et al. Workflow analysis and evidence-based medicine: towards integration of knowledge-based functions in hospital information systems. In *Proceedings of the AMIA Symposium*, 1999.
72. B. Curtisl, M. I. Kellner, and J. Over. Process modeling. *Commun. ACM*, 35(9):75–90, 1992.
73. P. Lawrence. *Workflow Handbook*. John Wiley & Sons Ltd., 1997.
74. S. Gräber. The impact of workflow management systems on the design of hospital information systems. In *Proceedings of the Annual Fall Symposium of the American Medical Informatics Association (AMIA)*. 1997.
75. J. D. McKeen, T. Guimaraes, and J. C. Wetherbe. The relationship between user participation and user satisfaction: An investigation of four contingency factors. *MIS Quarterly*. 18(4):427–451, 1994.
76. B. Ives and M. H. Olson. User involvment and MIS success: A review of research management science. 30(5):586–603, 1984.
77. D. Stone et al. *User interface design and evaluation*. Elsevier, 2005.
78. I. Edhemovic et al. The computer synoptic operative report—A leap forward in the science of surgery. *Ann. Surg. Oncol.* 11(10):941–947, 2004.
79. K. Leslie and J. Rosai. Standardization of the surgical pathology report: Formats, templates, and synoptic reports. *Semin. Diagn. Pathol.* 11(4):253–257, 1994.
80. P. Spyns. Natural language processing in medicine: An overview. *Methods Inform. Med.* 35:285–301, 1996.
81. C. Friedman et al. A general natural language text processor for clinical radiology. *J. Am. Med. Informatics. Assoc.* 1:161–174, 1994.
82. C. Friedman et al. Representing information in patient reports using natural language processing and the extensible markup language. *J. Am. Med. Inform. Assoc.* 6(1):76–87, 1999.
83. E. Gamma et al. *Design patterns: Elements of Reusable Object-Oriented Software*. Addison-Wesley, 1995.
84. F. Keller et al. Standardized structure and modular design of a pharmacokinetic database. *Comput. Methods Programs Biomed.* 55(2):107–115, 1998.
85. Y. Matsumura et al. Method of structured data entry in electronic patient record. *Jpn. J. Med. Inform.* 17(3):193–201, 1997.
86. P. Moorman et al. A model for structured data entry based on explicit descriptional knowledge. *Methods Inf. Med.* 33(5):454–463, 1994.
87. Y. Matsumura et al. Dynamic viewer of medical events in electronic medical record. *Studies in Health Technology and Informatics*. 84(Part 1): 648–652, 2001.
88. R. K. Los et al. Row modeling applied to generic structured data entry. *J. Am. Med. Inform. Assoc.* 11:162–165, 2004.
89. I. M. Kuhn et al. Automated ambulatory medical record systems in the US. In B.I. Blum (Ed). *Information Systems for Patient Care*. Springer, 1984.
90. W. D. Cai D. Feng, and R. Fulton. Web-based digital medical images. *IEEE Comput. Graph. Appl.* 44–47.
91. E. Bellon et al. Web-access to a central medical record to improve cooperation between hospital and referring physicians. *Stud. Health Technol. Inform.* 93:145–153.
92. C. Lau et al. Asynchronous Web-based patient-centered home telemedicine system. *IEEE Trans. Inf. Technol. Biomed.* 49(12):101–107, 2002.
93. Y. Lim, T. Cai, and D. Feng. A web-based collaborative system for medical image analysis and diagnosis. In *Proceedings of the 2001 Visual Information Processing Conference*, 2002.
94. J. J. Garrett. *AJAX: A New Approach to Web Applications*. 2005.
95. H. Iwata, S. Sugano. *Whole-body covering tactile interface for human robot coordination*. 2002.
96. G. Z. Thomas et al. A hand gesture interface device. *SIGCHI Bull.* 17(SI):189–192, 1987.
97. C. R. Diego, K. Krasimir, and K. Oussama. The haptic display of complex graphical environments. In *Proceedings of the 24th Annual Conference on Computer Graphics and Interactive Techniques*. ACM Press/Addison-Wesley Publishing Co. 1997:345–352.
98. M. G. Gorbet, M. Orth, and H. Ishii. Triangles: Tangible interface for manipulation and exploration of digital information topography. In *Proceedings of the SIGCHI Conference on Human Factors in Computing Systems, Los Angeles, CA*. ACM Press/Addison-Wesley Publishing Co. 1998:49–56.
99. G. W. Fitzmaurice, H. Ishii, and W. A. S. Buxton. Bricks: Laying the foundations for graspable user interfaces. *Proceedings of the SIGCHI Conference on Human Factors in Computing Systems, Denver, CO*. ACM Press/Addison-Wesley Publishing Co. 442–449, 1995.
100. K. Hinckley et al. Passive real-world interface props for neurosurgical visualization. In *Proceedings of the SIGCHI Conference on Human Factors in Computing Systems: Celebrating Interdependence*. ACM Press. 452–458, 1994.
101. Y-C. Lu et al. A review and a framework of handheld computer adoption in healthcare. *Int. J. Med. Inform.* 74(5):409–422, 2005.
102. R. Istepanian, E. Jovanov, and Y. Zhang. Guest editorial introduction to the special section on M-health: Beyond seamless mobility and global wireless health-care connectivity. *IEEE Trans. Inf. Technol. Biomed.* 8:405–414, 2004.
103. M. Bojovic and D. Bojic. MobilePDR: A mobile information system featuring update via Internet. *IEEE Trans. Inf. Technol. Biomed.* 9(1):1–5, 2005.

104. G. Pour et al. Web service-oriented enterprise architecture for mobility support in prescription management. *Proc. IEEE International Conference on e-Technology, e-Commerce and e-Service*, 2005.
105. R. Andrade et al. Wireless and PDA: A novel strategy to access DICOM-compliant medical data on mobile devices. *Int. J. Med. Inform.* 71:157–163, 2003.
106. A. Krause et al. Mobile decision support for transplantation patient data. *Int. J. Med. Inform.* 73:461–464, 2004.
107. D. E. Malter, and T. J. Davis. Handheld computers making the rounds with physicians. Devices put medical know-how-literally in the palm of your hand. *Mich. Med.* 100:30–31, 2001.
108. E. Ammenwerth. Mobile information and communication tools in the hospital. *Int. J. Med. Inform.* 57:21–40, 2000.
109. E. Ammenwerth et al. A randomized evaluation of a computer-based nursing documentation system. *Methods Inf. Med.* 40(2):61–68, 2001.
110. J. M. Overhage et al. Controlled trial of direct physician order entry: Effects on physicians' time utilization in ambulatory primary care internal medicine practices. *J. Am. Med. Inform. Assoc.* 8(4):361–371, 2001.
111. C. Lovis et al. Evaluation of a command-line parser-based order entry pathway for the Department of Veterans Affairs electronic patient record. *J. Am. Med. Inform. Assoc.* 8(5):486–498, 2001.
112. S. S. Warshawsky et al. Physician use of a computerized medical record system during the patient encounter: A descriptive study. *Comput. Methods Programs Biomed.* 43(3-4):269–273, 1994.
113. M. Weiner et al. Contrasting views of physicians and nurses about an inpatient computer-based provider order-entry system. *J. Am. Med. Inform. Assoc.* 6(3):234–244, 1999.
114. M. Apkon and P. Singhaviranon. Impact of an electronic information system on physician workflow and data collection in the intensive care unit. *Intensive Care Med.* 27:122–130, 2001.
115. F. C. G. Southon, C. Sauer, and C. N. G. Dampney. Information technology in complex health services: organizational impediments to successful technology transfer and diffusion. *J. Am. Med. Inform. Assoc.* 4(2):112–124, 1997.
116. I. Sommerville. *Software Engineering.* 7th ed. Addison Wesley, 2004.
117. G. P. Peter et al. Cognitive walkthroughs: A method for theory-based evaluation of user interfaces. *Int. J. Man-Mach. Stud.* 36(5):741–773, 1992.
118. E. Y. Lim et al. A medical imaging Web-based electronic patient history (EPH) via a personal assistant (PDA) in a wireless environment. *J. Nucl. Med.* 47(5):367, 2006.

3

Image Data Compression and Storage

Prof. Hong Ren Wu,[1]
Dr. Damian M. Tan,[1]
Dr. Tom Weidong Cai,[2] and
Prof. David Dagan Feng[2,3]
[1] *RMIT University*
[2] University of Sydney
[3] *Hong Kong Polytechnic University*

3.1 Introduction

Picture compression is an important tool in the modern digital world. Over the years, the move toward the digital media has led to the proliferation of digital picture compression systems. This is particularly noticeable in the entertainment industry, consumer electronics, and security/surveillance systems. However, data compression is becoming increasingly important in biomedical imaging applications as well, due to the increased popularity of digital biomedical imaging systems, the constant improvement of image resolution, and the practical need for online sharing of information through networks. Picture compression came about following the advent of analog television broadcasting. Initial methods of limiting signal bandwidth for transmission were relatively simple. They included subsampling to lower picture resolution and/or interlacing of television pictures into alternate fields in alternate picture frames. With the introduction of color television, subsampling was extended to the color channels as well [1]. The digital picture is seen as a natural migration from the analog picture. Hence, it is felt that digital picture compression is, in many ways, a natural evolution of analog compression. While this is true in some sense, the nature of digital signals and of analog signals is quite different. Consequently, the methods for compressing digital and analog pictures are distinct from one another. This chapter will first provide a basic introduction to digital picture compression, focusing on general concepts and methods, then introduce some advanced data compression techniques. These techniques are used in noisy medical image data sets with high compression ratios (CRs) and improved image quality, which have pioneered in biomedical diagnostically lossless data compression research. An extended discussion of classical data compression can be found in [2–8], while the new research in diagnostically lossless data compression can be found in the references given in the later sections.

3.2 Picture Compression

A basic compression system has both an encoding and a decoding component. The encoder is responsible for compression, while the decoder handles the reverse process; that is, decompression. The goal of any compression system is to reduce the size of signal data while maintaining information integrity, or a certain degree of it. In the context of compression, it is important to note the difference between *data* and *information*. Data are the individual samples of signal, while information conveys the content of all data samples. From this perspective, compression can be seen as a function of data versus information, as expressed by Shannon's theory of entropy [9]. Before moving on to information entropy, a brief overview is presented first of the concepts and terminologies in picture coding.

3.2.1 Picture Coding Concepts and Terminologies

Pictorial data in general are representative of both two-dimensional (image) and three-dimensional (video) spaces; but in the biomedical domain, use is often made of 3D volume images (e.g., computed tomography [CT], magnetic resonance imaging [MRI]) and 4D time-varying 3D volume images (e.g., from positron emission tomography [PET] and single photon emission computed tomography [SPECT]). In image processing terminology, an image is made up of individual *pixels* (picture elements) arranged by rows and columns (or *voxels* for volume images arranged by rows and volumns for each image plane). Video data consist of multiple static images, commonly referred to as *frames,* arranged along the temporal/spatial[1] axis (Figure 3.1). Pixels in a picture have a common depth that determines the number of discrete luminance and/or chrominance levels. This pixel depth is measured in *bits,* following the binary system utilized in existing computer architectures. Most common images have bit depths of 8 bits per pixel per channel, while medical images may have between 10 and 16 bits per pixel. Each bit, within the continuous string of bits, is a binary number with value of 0 or 1. An 8-bit pixel,

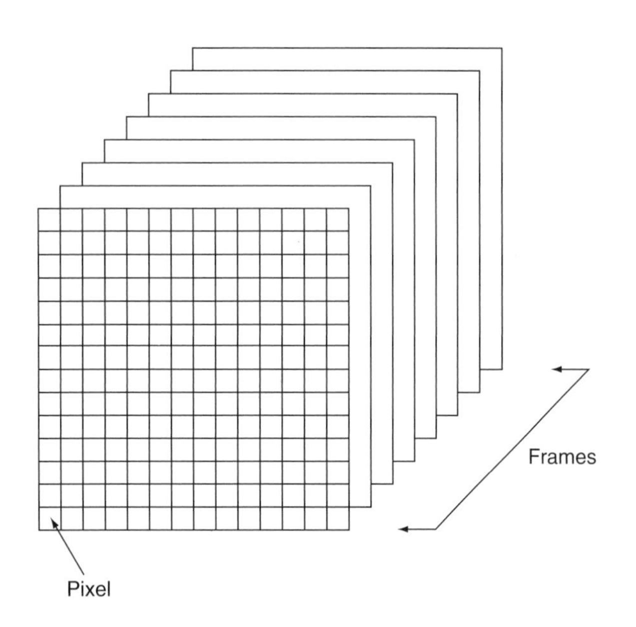

FIGURE 3.1 Illustration of pixels and frames.

[1] Some volumetric images, such as biomedical images, are three-dimensional spatial projection data or reconstructed projection data.

therefore, may range from $0000\ 0000_2$ to $1111\ 1111_2$, which in decimal terms is 0 to 255, having 2^8 (256) discrete levels. For any unsigned[2] binary number, the leftmost bit in a binary sequence is called the *most significant bit* (MSB), having the highest order of magnitude. The rightmost bit is the *least significant bit* (LSB), with the lowest order of magnitude.

Natural pictures may be grayscale or color. Grayscale pictures have a single luminance channel, while color pictures have three separate channels. Typically, color channels are arranged according to the primary colors of red, green, and blue (the RGB color space). However, it is also a common practice to represent color channels in the component color domain with one luminance and two chrominance channels (YCrCb color space) [1, 10].

The uncompressed representation of symbols/pixels[3] appears in the form of *fixed length binary codes.* The minimum length, l_m, of these codes is dependent on the size of a given symbol set,[4] n, governed by

$$l_m = \lceil \log_2 (n) \rceil . \tag{3.1}$$

Here, $\lceil\ \rceil$ denotes rounding up to the nearest integer.[5] For digital pictures, this code length is dependent on pixel depth. For example, an 8-bit pixel uses an 8-bit fixed length code. Although compressed pictures are often quoted in terms of their CRs, the more conventional measurement is in *bit rate,* taken in units of *bits per pixel* (bpp). The CR, C_{ratio}, defined as

$$R_c = \left(\frac{bpp_{uncompressed}}{bpp_{compressed}} \right) : 1, \tag{3.2}$$

is a measurement of compression gain, whereas bit rate is a measurement of the size of compressed data.

3.2.2 Shannon's Theory of Entropy

Shannon's theory of entropy [9], in short, stipulates that a minimum amount of data is necessary to represent a certain amount of information. If the amount of data falls short of the entropy, then information cannot be fully represented, leading to information loss. Conversely, if the entropy level is surpassed, then an excessive amount of data has been used to represent information, thus some redundancies exist. Given a symbol set $\boldsymbol{A}$, the discrete form of information entropy H, in bits, is written as

$$H = \sum_{i=1}^{n} p(i) \log_2 \left(\frac{1}{p(i)} \right) = - \sum_{i=1}^{n} p(i) \log_2 p(i), \tag{3.3}$$

[2] Nonnegative number.

[3] *Symbol* is a reference to generic data (e.g., text, audio), whereas *pixel* is specifically associated with picture data.

[4] The number of different symbols. For pixels, it is the number of discrete magnitude levels.

[5] Rounding is needed to remove fractions, since the fixed length binary codes have integer representation (quantization).

where $p(i)$ is the probability associated with the occurrence of symbol $A(i)$ such that $\sum p(i) = 1$, for $i=\{1,2,\ldots,n\}$ and $A(i) \in \boldsymbol{A}$. It is clear from Equation 3.3 that H is a summation of all the information contained within each symbol in $\boldsymbol{A}$, relative to the probability of occurrence. This self-contained information in each symbol is referred to as *self-information* and is defined as

$$\iota(i) = -\log_2 p(i). \tag{3.4}$$

Redundancy in data, as defined by Shannon [9], is a proportional difference between the entropy, H, and the actual size of the source data, l_m, as defined in Equation 3.1, relative to 1. It is formulated as

$$\rho = 1 - \left(\frac{H}{l_m}\right). \tag{3.5}$$

Example 1: Given a set of symbols, $\boldsymbol{A}$, with the probability distribution shown in Table 3.1, the entropy is computed as:

$$\begin{aligned} H(A) &= -[p(1)\cdot \log_2(p(1)) + p(2) \\ &\quad \cdot \log_2(p(2)) + \ldots + p(6)\cdot \log_2(p(6))] \\ H(A) &= -[0.3\cdot \log_2(0.3) + 0.1\cdot \log_2(0.1) \\ &\quad + \ldots + 0.2\cdot \log_2(0.2)] \\ H(A) &= 2.4087 \text{ bits per symbol.} \end{aligned}$$

For most uncompressed natural audio and picture signals, the amount of data used to represent the information is above the entropy threshold. Consequently, it is possible to compress these signals. For example, according to Equation 3.1, the symbol set $\boldsymbol{A}$ from Table 3.1, with symbol size of six ($n=6$), required 3 bits to denote each symbol with fixed length codes. However, the information entropy associated with $\boldsymbol{A}$ is 2.4087 bits per symbol. From Equation 3.5, the redundancy in $\boldsymbol{A}$ is calculated as

$$\begin{aligned} \rho(A) &= 1 - \left(\frac{2.4087}{3}\right) \\ &= 0.1971. \end{aligned}$$

3.2.3 Entropy Coding

Entropy coding is a reference term for information lossless compression at or near the data entropy. Its general operation consists of two stages: modeling and coding. Modeling is performed to identify and describe data redundancies. It is carried out through statistical analysis of the data set in order to capture the *probability distribution of symbols* (PDS) in the set. The coding phase then encodes the information in data, based on the description of data derived during the modeling phase, by assigning a distinct code to each symbol.[6] The size of each code, measured in number of bits, is dependent on the probability of occurrence of its respective symbol. Generally, the most common symbol will have the smallest code size, while the least common symbol will have the largest code size. The manner in which these codes are generated is dependent on the coding algorithm. There are a variety of entropy coding algorithms available [8]. The most common of these is the Huffman code [11], which generates its alphabet of codewords through recursive sorting of source symbols, of a given probability distribution, into a binary tree. In each iteration, the Huffman coding algorithm performs the following operations:

1. The PDS is sorted in descending order.
2. The two symbols with the lowest probability of occurrence are then grouped together to build a tree branch and generate an updated symbol table. For every tree branch, each of the two merged symbols is assigned a binary digit, 0 for the last and 1 for the second last. With each successive iteration, the merged symbols will then progressively build up their binary codeword (BCW).
3. The process is iterated until the root of the tree is reached. This occurs when no more tree branches can be formed; that is, when there is only one symbol remaining. The number of iterations, i, required to generate the entire tree is one less than the total number of symbols, n, i.e., or $i=n-1$.

Example 2: Given the PDS in Table 3.1, the Huffman code is generated as follows:
Iteration 1:

$$\begin{aligned} \text{probability: } p(a) &= 0.3,\ p(b) = 0.1,\ p(c) = 0.2,\ p(d) \\ &= 0.05,\ p(e) = 0.15,\ p(f) = 0.2 \end{aligned}$$

1. Sorting symbols:

$$A = \{a,b,c,d,e,f\} \Rightarrow \{a,c,f,e,b,d\}$$

2. Group the lowest two symbols and build the tree branch:

$$\begin{aligned} z &= \{b,d\},\ b = 1,\ d = 0, \\ A_1 &= \{a,c,f,e,z\} \\ \text{binary codeword} &: b = 1, d = 0 \end{aligned}$$

3. $\text{size}(A_1) > 1$

TABLE 3.1 Example probability distribution of symbol set $\boldsymbol{A}$

i	1	2	3	4	5	6
$A(i)$	a	b	c	d	e	f
$p(A(i))$	0.3	0.1	0.2	0.05	0.15	0.2

[6] Codes may also be assigned to a sequence of symbols, as in the case of arithmetic coding.

Iteration 2:

probability: $p(a) = 0.3,\ p(c) = 0.2,\ p(f) = 0.2,\ p(e) = 0.15,\ p(z) = 0.15$

1. Sorting symbols:

$$A_1 = \{a,c,f,e,z\} \Rightarrow \{a,c,f,e,z\}$$

2. Group the lowest two symbols and build the tree branch:

$$y = \{e,z\}, e = 1, z = 0,$$
$$A_2 = \{a,c,f,y\}$$

binary boundary : $e = 1,\ b = 01,\ d = 00$

3. $\text{size}(A_2) > 1$

Iteration 3:

probability: $p(a) = 0.3,\ p(c) = 0.2,\ p(f) = 0.2,\ p(y) = 0.3$

1. Sorting symbols:

$$A_2 = \{a,c,f,y\} \Rightarrow \{a,y,c,f\}$$

2. Group the lowest two symbols and build the tree branch:

$$x = \{c,f\}, c = 1, f = 0,$$
$$A_3 = \{a,y,x\}$$

binary boundary : $c = 1,\ f = 0,\ e = 1,\ b = 01,\ d = 00$

3. $\text{size}(A_3) > 1$

Iteration 4:

probability: $p(a) = 0.3,\ p(y) = 0.3,\ p(x) = 0.4$

1. Sorting symbols:

$$A_3 = \{a,y,x\} \Rightarrow \{x,a,y\}$$

2. Group the lowest two symbols and build the tree branch:

$$w = \{a,y\},\ a = 1,\ y = 0,$$
$$A_4 = \{x,w\}$$

binary codeword: a=*1*, c=*1*, f=*0*, e=*01*, b=*001*, d=*000*

3. $\text{size}(A_4) > 1$

Iteration 5:

probability: $p(x) = 0.4,\ p(w) = 0.6$

1. Sorting symbols:

$$A_4 = \{x,w\} \Rightarrow \{w,x\}$$

2. Group the lowest two symbols and build the tree branch

$$v = \{w,x\},\ w = 1,\ x = 0,$$
$$A_5 = \{v\}$$

binary codeword : $a = 11,\ c = 01,\ f = 00,\ e = 101,\ b = 1001,\ d = 1000$

3. $\text{size}(A_5) = 1$: Stop iteration.

This operation is visualized in Figure 3.2. Once the Huffman tree is generated, the encoding process replaces all source symbols in a data set with code symbols to produce a coded binary sequence. For decoding, one simply tracks the binary decision level, in the coded sequence, from the root of the Huffman tree to the leaves. When a binary decision reaches a leaf, a source symbol is decoded and tracking of the next source symbol begins anew at the root of the Huffman tree.

It is important to note that each individual code generated in Figure 3.2 is distinct and therefore uniquely decodable. For example, the encoded binary sequence **000111011000** can be decoded only into symbols *f*(**00**), *c*(**01**), *a*(**11**), *c*(**01**), *d*(**1000**).

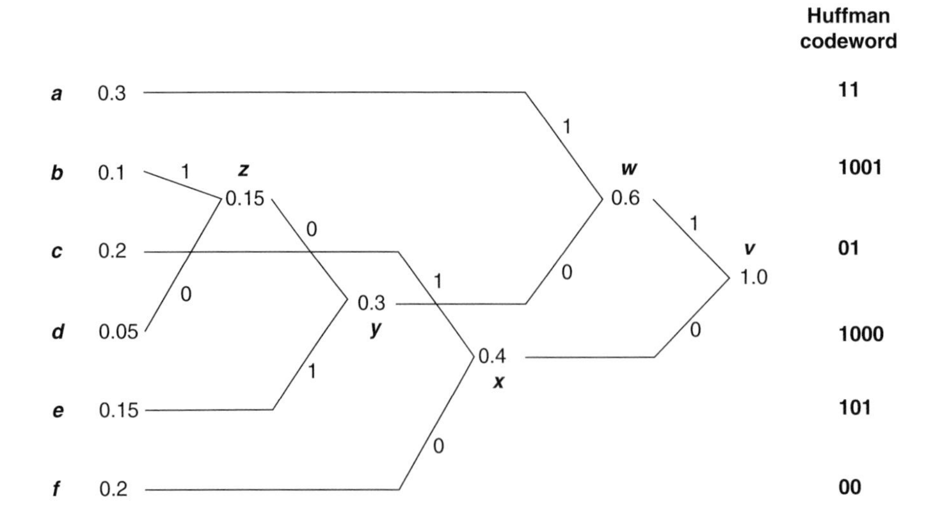

FIGURE 3.2 Huffman tree corresponding to the PDS in Table 3.1. The Huffman codewords are generated by tracing the binary numbers along each branch from the root to each leaf of the tree.

The compression rate, r_H, associated with the Huffman codes in Figure 3.2 can be computed as follows:

$$r_H = \sum_{i}^{N} p(i) \cdot b(i), \tag{3.6}$$

where $p(i)$ and $b(i)$ are, respectively, the probability of occurrence and the codeword length of the ith Huffman codeword. The expected compression rate of the Huffman codes in Figure 3.2 is

$$r_H = [p(1) \cdot b(1) + p(2) \cdot b(2) + \ldots + p(6) \cdot b(6)]$$

$$r_H = [0.3 \times 2 + 0.1 \times 4 + 0.2 \times 2 + 0.05 \times 4 + 0.15 \times 2 + 0.2 \times 3]$$

$$r_H = 2.45 \text{ bits per symbol.}$$

The Huffman code is generally quite efficient, compression-wise. It also has simple and fast implementation. However, its integer codeword may be less efficient in situations in which a fractional codeword[7] occurs. Take the example of symbol set $\boldsymbol{A}$ (Table 3.1). A comparison of its Huffman code size, S_{hc}, with its optimum code size, $S_{oc} = \log_2(\boldsymbol{A})$, given in Table 3.2, demonstrates some inefficiencies of the Huffman code for symbols ***a, c,*** and ***f.*** While the Huffman codes for symbols ***b, d,*** and ***e*** are more efficient than the optimal code, these symbols appear with less frequency than ***a, c,*** and ***f.***

TABLE 3.2 Symbol set *A*. Codeword sizes (bps) for the optimum code (S_{oc}) and the Huffman code (S_{hc}). The Huffman code size is shown in Figure 3.2

i	1	2	3	4	5	6
$A(i)$	a	b	c	d	e	f
$p(A(i))$	0.3	0.1	0.2	0.05	0.15	0.2
S_{oc} (i)	1.737	3.3219	2.3219	4.3219	2.737	2.3219
S_{hc} (i)	2	4	2	4	3	2

Therefore, on average, the Huffman codes will require 2.45 bits to code one symbol, whereas the optimum codes[8] require 2.4087 bits per symbol.

In contrast to Huffman coding, arithmetic coding [12, 13] can indirectly handle fractional codeword and consequently has become the preferred choice for entropy coding in a number of applications [14, 15]. Instead of encoding each symbol separately, arithmetic coding encodes a group of symbols. The arithmetic codeword used to represent a sequence of symbols is formed through successive cascades of the probability interval of source symbols. Figure 3.3 illustrates the formation of the cascaded intervals. A step-by-step approach for encoding and decoding is shown in Tables 3.3 and 3.4, respectively.

The encoding procedure for arithmetic coding has two steps in each iteration:

1. Identity the interval range, R, of the symbol to be coded:

$$R = H - L, \tag{3.7}$$

where H and L are the upper and the lower bound, respectively, of the interval range.

2. Update both the upper and lower bounds relative to the encoded symbol:

$$H = L + R \times S_H \tag{3.8}$$

$$L = L + R \times S_L, \tag{3.9}$$

where S_H and S_L are the upper and lower interval bounds, respectively, of the encoded symbol.

The decoding phase is also carried out in two steps per symbol:

1. Locate the symbol interval for codeword, C_a: $S_L < C_a < S_H$.

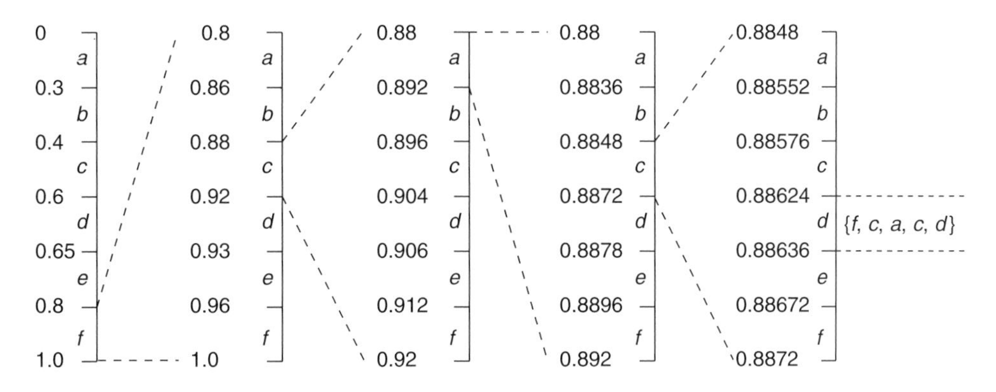

FIGURE 3.3 Arithmetic coding with cascaded intervals for symbols {*f, c, a, c, d*} from PDS in Table 3.1.

[7] Codeword with fractional part.

[8] Optimum code encodes at the entropy level.

TABLE 3.3 Arithmetic encoding of symbol sequence {*f*, *c*, *a*, *c*, *d*} from PDS in Table 3.1

Symbol	*R*	*L*	*H*
	1.0	0.0	1.0
f	0.2	0.8	1.0
c	0.04	0.88	0.92
a	0.012	0.88	0.892
c	0.0024	0.8848	0.8872
d	0.00012	0.88624	0.88636

TABLE 3.4 Arithmetic decoding of symbol sequence {*f*, *c*, *a*, *c*, *d*} from PDS in Table 3.1

Code	Symbol	S_L	S_H	R_s
0.88625	*f*	0.8	1.0	0.2
0.43125	*c*	0.4	0.6	0.2
0.15625	*a*	0.0	0.3	0.3
0.52083	*c*	0.4	0.6	0.2
0.60417	*d*	0.6	0.65	0.05

2. Update the codeword for decoding the next symbol:

$$C_a = (C_a - S_L)/R_s, \tag{3.10}$$

where $R_s = S_H - S_L$ is the symbol range.

The initial coding intervals are taken as the boundary intervals of the PDS. As each symbol is encoded, the coding intervals are refined. These refined intervals are proportional to the PDS but limited to within the interval range of the coded symbol. For example, in Figure 3.3 (also Table 3.3), after the first symbol (*f*) is encoded, the intervals for the proportional distribution of symbols are limited within the range of 0.8 and 1.0. Encoding the symbol sequence {***f*, *c*, *a*, *c*, *d***} in Figure 3.3 (Table 3.3) will result in the final coding interval (0.88624, 0.88636). Any arithmetic code number, C_a, that resides within this range can be used to represent the encoded symbol sequence. Generating the cascaded probability intervals is a trivial task. The real challenge for arithmetic coding is how to efficiently encode the number between two given interval points. With each successive cascade, the range of the probability interval decreases. As a result, the arithmetic precision needed to represent the code number increases. This in turn dictates the number of bits needed to encode the arithmetic code number. In arithmetic coding, there is no fixed codeword for any particular source symbol, since symbols are coded in groups rather than individually. Furthermore, as a consequence of group encoding, individual symbols may be coded indirectly in fractions. Take the example in Figure 3.3: The symbol sequence {***f*, *c*, *a*, *c*, *d***} may be represented by the code number $C_a = 0.88628$. Its 8-bit integer representation is 226 ($(2^8 - 1) \times 0.88628$).[9] Since an 8-bit codeword[10] is used to encode five symbols, it is possible to only extrapolate the average code size per symbol, l_a, by

$$l_a = \frac{\text{codeword length}}{\text{no. of symbols}} = \frac{8}{5} = 1.6 \text{ bits per symbol.}$$

However, this does not imply that only 1.6 bpp are needed to encode symbols with the probability distribution given in Table 3.1. The example given here, taken from Figure 3.3, covers only the first five symbols. In addition, the integer representation of the arithmetic code number, C_a, is an overly simplistic view of arithmetic coding. The full description of arithmetic coding is given in Witten et al. [110]. A general comparison between arithmetic and Huffman coding is covered in Sayood [8].

Entropy coding may operate in an *adaptive* or a *static* manner [8]. The difference between these two modes is that the former is able to adapt to changes in data, while the latter does not. Adaptive entropy coding, also referred to as dynamic entropy coding, utilizes a predetermined PDS in its initial operation. This initial PDS is generally acquired, heuristically, through generic sample data. During the encoding process, the PDS is continually updated with coded data to reflect the statistical profile of the data set more accurately. Adaptive coding is particularly useful in instances where the *a priori* profile of the PDS is unknown[11] or where different segments of the source data exhibit different PDS profiles. In addition to this, adaptive coding is often employed in situations where the statistical analysis of data is impractical due to time constraints. This is envisaged when dealing with volumetric data. In instances where data volume or time constraints are not an issue, then static entropy coding may be employed. Static entropy coding generally models each data set independently to capture the exact PDS. However, coding with the exact PDS incurs some overheads, since the PDS used for encoding is also needed for decoding. For example, in order to decipher the Huffman codes at the decoding end, a copy of the Huffman tree generated at the encoding end must be made available. This is possible only if the PDS is sent to the decoder. Consequently, an overhead is incurred for storing/transmitting the PDS needed to generate the Huffman tree.

While the entropy for any given set of data may be determined analytically, it is understood that under normal circumstances, existing entropy coding techniques can never reach the entropy threshold. For static entropy coding, this

[9] An 8-bit binary integer number has 256 discrete levels, ranging from 0 to 255 ($2^8 - 1 = 255$).

[10] In practice, the codeword size must be selected in a fashion that allows distinct representation of all possible combinations of symbol sequences. Here, an 8-bit codeword is used only as an example to illustrate fractional codes.

[11] Real-time encoding where data is encoded as soon as it is digitized.

is due to the practical limitation of compression overheads, i.e., the PDS. In order to maximize the compression performance of entropy coders, it is important to have an accurate PDS of pictures. Since the PDS is dependent on picture content and thus varies from picture to picture, it is necessary, in the interest of having an accurate PDS, for static entropy coding to generate and transmit a copy of the PDS from the encoder to the decoder. In the case of adaptive entropy coding, transmission of the PDS is unnecessary, since a default PDS already exists. However, for adaptive coding, the problem lies in the adaptation process, which infers inefficiency, hence the need to adjust. Though the adaptation approaches the entropy, it would never reach the entropy threshold, since the entropy is progressively estimated from causal data samples, not all data samples. Coding at the entropy threshold is possible only if there is no variation in data statistics and if the initial PDS is optimal, a most unusual situation if it does occur.

3.2.4 Classification of Picture Compression

Picture coding has traditionally been classified into two general categories: information *lossless* and information *lossy*. Lossless coding maintains information integrity. Lossy coding, in contrast, degrades information integrity to a certain extent. However, any information lost is offset by a higher CR. Therefore, lossy compression is a balancing act between information quality and compression performance as dictated by the *rate versus distortion* (R–D) curves of pictures (Figure 3.4). Ideally, lossless compression would be the approach of choice for encoding picture data. In

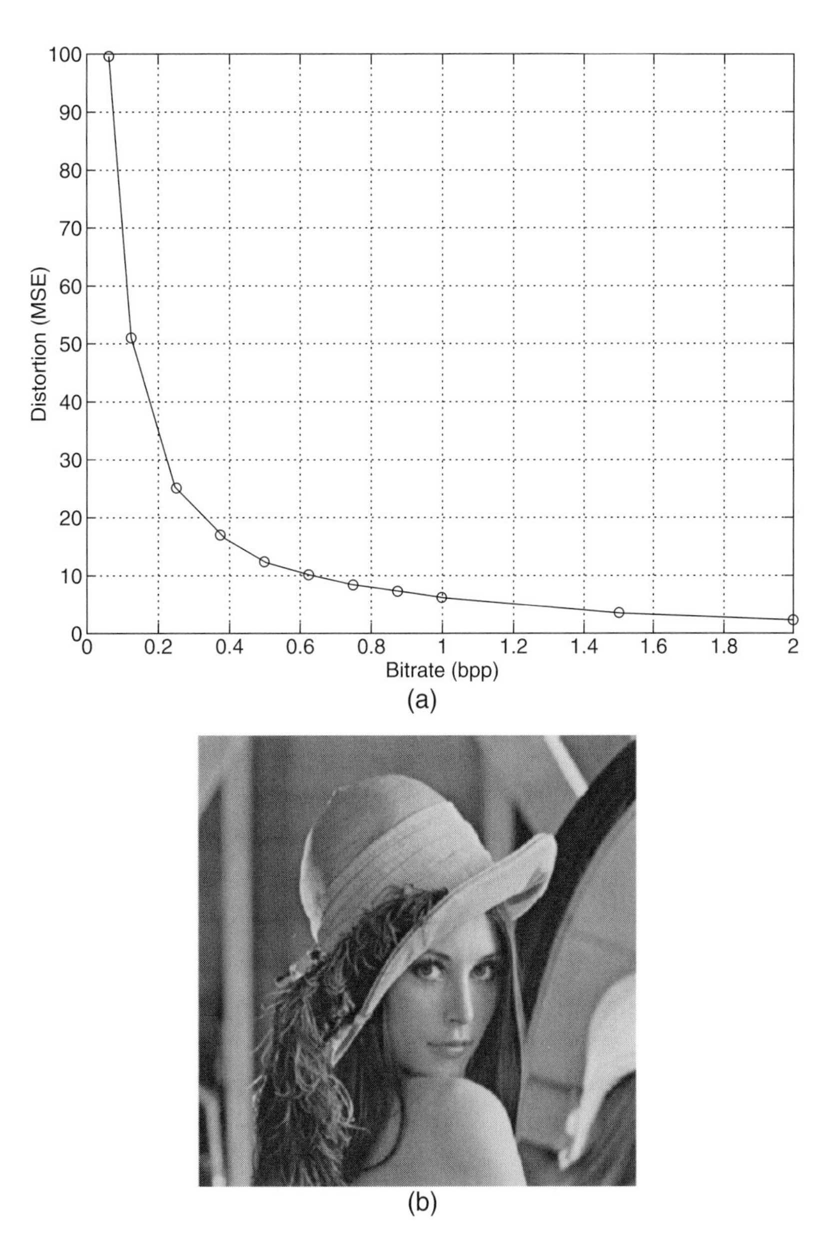

FIGURE 3.4 Example of R–D curve (a) for the *Lena* image (b) coded at different bit rates with the JPEG2000 coder [15]. As the bit rate increases, the distortion decreases. An increase in bit rate corresponds to a lower CR.

practice, however, lossless compression is unable to deliver the necessary CR required for most consumer applications. The limitation of lossless compression becomes obvious when dealing with video or volumetric data. Consequently, there is preference for lossy coding of digital video, as reflected in existing industrial standards [16].

3.2.5 Lossless Picture Coding

Lossless picture coding in its most basic form is equivalent to entropy coding. However, picture data, being two-dimensional in nature, generally have strong correlations between adjacent pixels. Consequently, it is common practice to employ predictive coding prior to entropy coding to further improve compression performance. Utilizing predictive coding such as *differential pulse code modulation* (DPCM)[12] [17, 18] has the effect of reshaping the PDS. For natural images, predictive coding leads to sharper PDS, usually centered about the zero prediction with a Gaussian or a Laplacian profile. This translates to better compaction of data, since the probability distribution is concentrated on fewer numbers of symbols, as evident in Figure 3.5.

A simple DPCM encodes the difference between two pixels such that

$$r[n] = x[n] - x[n-1], \tag{3.11}$$

where $r[n]$ is the predicted residue, and $x[n]$ and $x[n-1]$ are the current and previous pixels, respectively. It is, however, more common in predictive coding to use multiple *causal* samples to generate the predicted residue. Causality in this context refers to samples that have been encoded; that is, $x[i]$ for $i < n$. The convention in image coding is to read pixels left to right and top to bottom. Thus, any pixel that is above or to the left of the current pixel is considered causal (Figure 3.6).

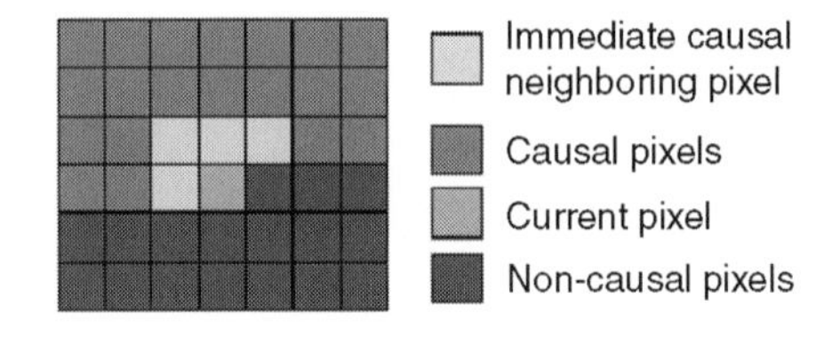

FIGURE 3.6 Causal samples for predictive coding.

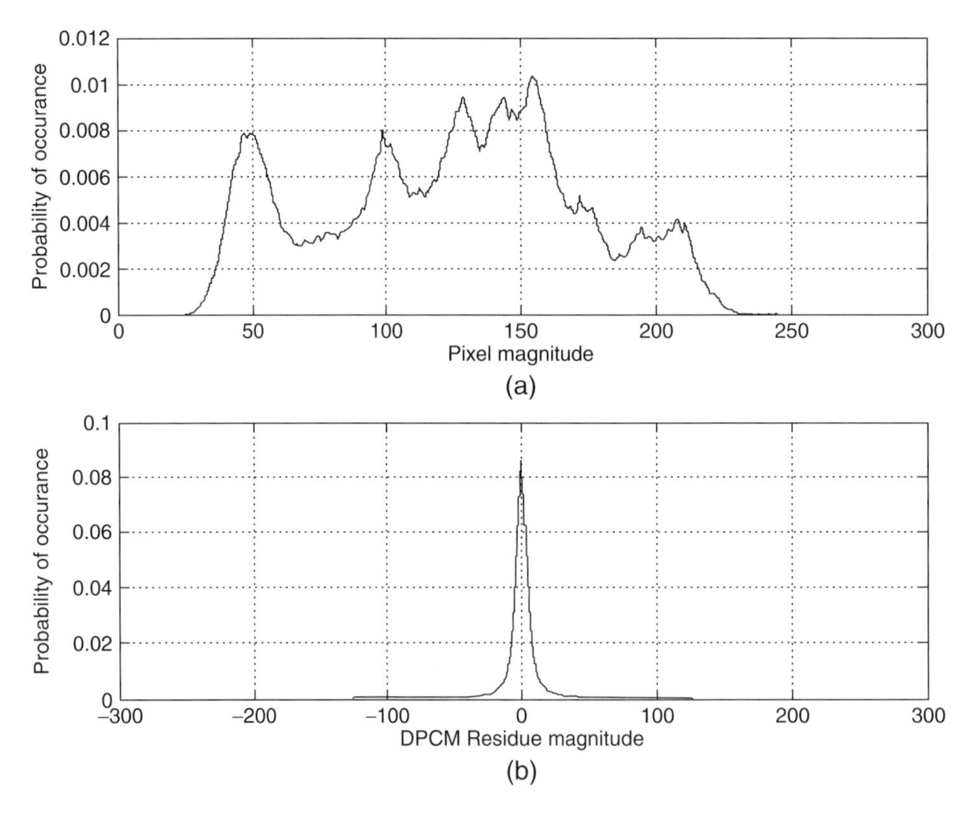

FIGURE 3.5 PDS of the *Lena* image. (a) PCM (entropy: 7.4456 bpp). (b) Row-wise DPCM (entropy: 5.0475 bpp).

[12] Pulse code modulation (PCM) [19] encodes analog signals in digital waveforms.

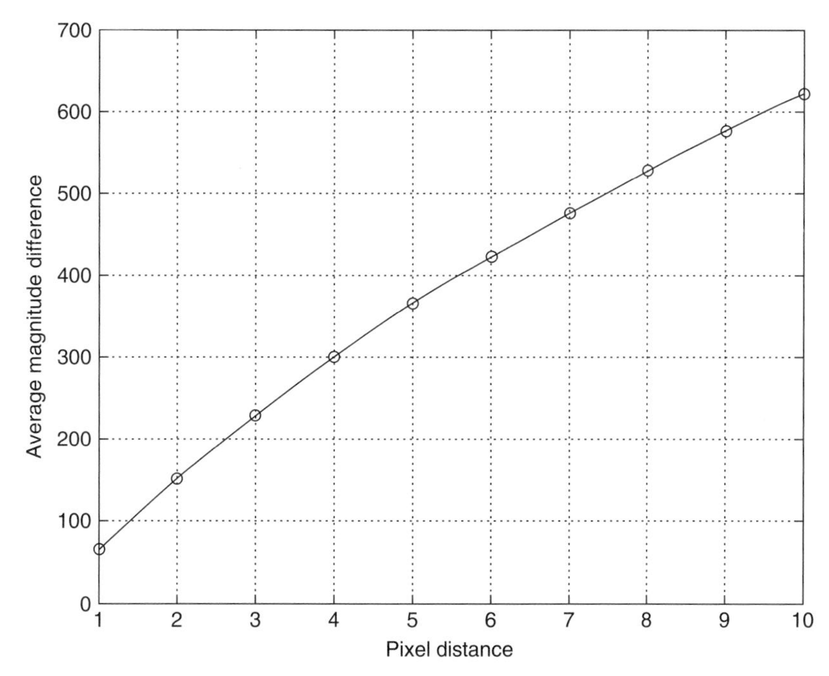

FIGURE 3.7 Correlation relative to pixel distance measured according to the average difference between pixels of the *Lena* image.

One common approach to predictive coding is to utilize all four immediate neighboring causal samples [20]. Other combinations that extend beyond the immediate neighbors are also possible [14].

Generally, having more causal pixels would improve prediction, since correlation between adjacent pixels or pixels in close proximity is typically high. However, not all adjacent pixels are correlated. Further, pixel correlation degrades relative to distance, as evident in Figure 3.7.

3.2.5.1 Context-Based Coding

In text compression, it is simple to envisage that the occurrence of a letter in an English word[13] is dependent on previous letters in that word [21]. For example, if the letter Q is encountered, the probability of the next letter being U is almost certain. Similarly, if the two previous letters in a word are QU, it is likely that the next letter will be a vowel. In light of this, it would be most prudent to adjust the coding operation to take advantage of this conditional correlation. Context-based coding operates by selecting the most suitable context for encoding data based on the behavior of past data samples.[14] Each context within a context coder maintains a separate PDS, independent of other contexts, for entropy coding [8]. For image compression, context selection is dependent on structural characteristics of neighboring pixels. Typically, these characteristics are measured in terms of horizontal, vertical, and/or diagonal edges. Context coding employs multiple conditional predictors for residue coding. Again, the predictor selection for coding any given pixel is also dependent on structural characteristics. However, it is important to distinguish between context selection and predictor selection. The purpose of context selection is to match the actual residue produced by the selected predictor to a suitable probability distribution. The predictor, on the other hand, is selected in order to minimize the prediction residue. Having multiple predictors increases computation complexity and cost. However, it usually leads to better prediction, hence a more favorable probability distribution for entropy coding [8]. The manner in which context coding is employed varies with different lossless algorithms. This is illustrated in the next two sections with two different coders: the low-complexity (LOCO) coder and the context arithmetic lossless image coder (CALIC).

3.2.5.2 Low-Complexity Coder

The LOCO coder [20] is the core of the JPEG-LS [22] coding standard. It relies on the four causal neighboring pixels for prediction and employs a conditional predictor given by

$$\hat{x} = \begin{cases} \min(W,N), & NW \geq \max(W,N) \\ \max(W,N), & NW \leq \min(W,N), \\ W + N - NW, & \textit{otherwise} \end{cases} \tag{3.12}$$

where $\hat{x}$ is the predicted pixel, with W, N, and NW the neighboring pixels, illustrated in Figure 3.8. The residue is the difference between the original pixel, x, and the predicted pixel, $\hat{x}$.

[13] Insofar as the natural English language is concerned.

[14] It is perhaps more accurate to differentiate between multicontext and unicontext coding, as opposed to context and noncontext coding, since all coding schemes have at least one context by default.

The context selection process is dependent on three difference measures, D_i for $i=\{1, 2, 3\}$, with simple horizontal and vertical edge detection. These difference measures are defined as

$$\begin{aligned} D_1 &= NE - N \\ D_2 &= N - NW \\ D_3 &= NW - W. \end{aligned} \tag{3.13}$$

The response of each difference measure is partitioned into $2T + 1$ equal distance interval, z_i, such that

$$z_i = \begin{cases} -T, & D_i \le -T \\ -T+1, & -T < D_i \le -T+1 \\ \dots & \\ 0, & D_i = 0 \\ \dots & \\ T-1, & T-1 < D_i \le T \\ T, & T < D_i \end{cases} \tag{3.14}$$

The equal distance interval is chosen to simplify implementation. Ideally however, the intervals should be optimized according to structural statistics of images. The total number of contexts, C, is then taken as the number of possible permutations for the given numbers of intervals and difference measures, $C = (2T+1)^3$. The number of contexts may be reduced by merging symmetric intervals in Equation 3.14 such that $-T \le z_i \le T \rightarrow 0 \le |z_i| \le T$ and thus resulting in $C_{sym} = ((2T+1)^3 + 1)/2$ contexts. The LOCO coder encodes the residue with Golomb codes [23] under regular operation. However, if flat regions are encountered, adaptive run-length coding[15] is used.

3.2.5.3 CALIC

CALIC [14] is moderately sophisticated. Compared with the LOCO coder, it has more developed edge detection, residue prediction, and context modeling functions. Coding operation begins with two edge-sensitive gradient estimation functions based on neighboring pixels (Figure 3.8):

$$\begin{aligned} d_h &= |W - WW| + |N - NW| + |NE - N| \\ d_v &= |W - NW| + |N - NN| + |NE - NNE|. \end{aligned} \tag{3.15}$$

These gradient estimates determine the predicted pixel, $\bar{x}$, under the following conditions:

$$\bar{x} = \begin{cases} N, & d_h - d_v > 80 \\ W, & d_v - d_h > 80 \\ (\tilde{x} + N)/2, & d_h - d_v > 32 \\ (\tilde{x} + W)/2, & d_v - d_h > 32 \\ (3\tilde{x} + N)/4, & d_h - d_v > 8 \\ (3\tilde{x} + W)/4, & d_v - d_h > 8, \end{cases} \tag{3.16}$$

$$\text{where } \tilde{x} = (N + W)/2 + (NE - NW)/4. \tag{3.17}$$

[15] See Capon [24] for a description of run-length coding.

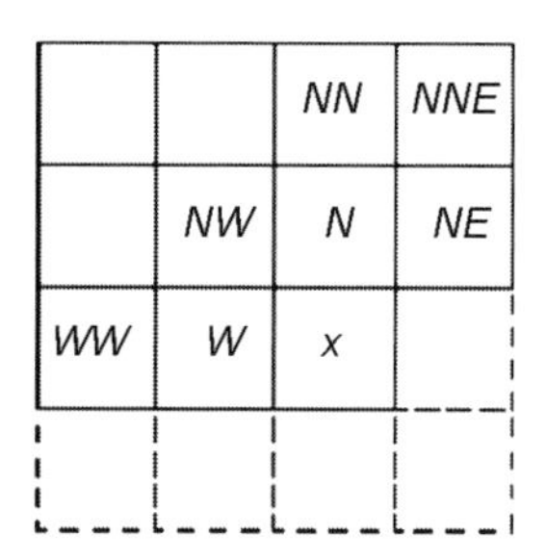

FIGURE 3.8 Neighboring samples used for prediction for LOCO and CALIC coders.

Context modeling is based on prediction error and texture. The prediction error, e, is modeled as

$$e = ad_h + bd_v + c|e_w|, \tag{3.18}$$

where $e_w = W - \overline{W}$ is the previous prediction error, with $\overline{W}$ being the prediction of W. The coefficients a, b, and c are parameters to be optimized. For efficient operation, $a=b=1$ and $c=2$ have been suggested [14]. In the basic CALIC implementation, e is optimally quantized into eight intervals. The boundaries of these intervals were obtained through dynamic programming and found to be

$$\xi_i = \{5,15,25,42,60,85,140\},\ 0 \le i \le 7.$$

The actual intervals are

$$\phi = \{0 < e \le \xi_1, \xi_1 < e \le \xi_2, \dots, \xi_6 < e \le \xi_7, \xi_7 < e\}.$$

Texture context, being dependent on the activity of neighboring pixels, is modeled under eight separate factors, C_t, defined as

$$\begin{aligned} C_t &= \{x_0, x_1, \dots, x_6, x_7\} \\ &= \{N, W, NW, NE, NN, WW, 2N - NN, 2W - WW\}, \end{aligned} \tag{3.19}$$

where $t=\{1, 2, \dots, 7\}$. Individual factor, x_k, is measured against the prediction derived in Equation 3.12 as follows:

$$b_k = \begin{cases} 0, & x_k \ge \bar{x} \\ 1, & x_k < \bar{x} \end{cases}, \quad k = \{0,1,\dots,7\}. \tag{3.20}$$

The combined measurement of all factors, $B=\{b_7, b_6, \dots, b_0\}$, determines the behavior of the pixel neighborhood and forms the number of texture contexts, $2^8 = 256$. However, due to dependencies[16] between factors in C_t, the actual number of texture contexts is reduced to 144. The overall number of contexts is a combination of texture and prediction error. For the basic CALIC implementation, the prediction error context is reduced from 8 to 4 to form a total of $144 \times 4 = 576$ coding contexts. Once the context has been determined through Equations 3.18 and 3.19, the predicted pixel, $\bar{x}$, is entropy-coded with context arithmetic coding.

[16] See Wu and Memon [14] for further details.

3.2.5.4 Near-Lossless Compression

A major limitation of lossless compression has always been its compression performance. To address this deficiency, the near-lossless coding scheme has been proposed [20]. Near-lossless coding allows for controlled degradation of picture quality in order to improve compression gain. This is carried out through the quantization of the prediction residue, r, prior to entropy coding.[17] This quantization operation is performed in a manner that guarantees that the quantized residue will not deviate beyond a certain point, $\pm\Delta$, as defined by users. The quantized residue, $\bar{r}$, is derived as follows:

$$\bar{r} = \text{sign}(r)\left\lfloor\frac{|r| + \Delta}{2\Delta + 1}\right\rfloor, \tag{3.21}$$

where Δ determines the maximum magnitude of deviation, and sign(r) is defined as

$$\text{sign}(r) = \begin{cases} -1, & r < 0 \\ 1, & r \geq 0 \end{cases}. \tag{3.22}$$

Since a near-lossless coding scheme appends only an additional quantization stage to the coding system, it can be readily adapted to practically any lossless coding algorithm.

3.2.6 Transform-Based Lossy Picture Coding

In order to attain a higher CR that surpasses that of lossless coding schemes, it has become a necessity to tolerate some information loss. Transform-based coding is employed generally for this purpose. It consists of a transformation operator followed by quantization and bitstream coding (Figure 3.9).

3.2.6.1 Transformation

The transformation operation is intended to rearrange data, in pictures, in a manner that facilitates compression. The most common form of data transformation applied in picture coding is frequency based. Nonfrequency transforms such as fractal [25, 26] have also been studied. A frequency transform projects data from the time domain, **x**, to the frequency domain, **X**, relative to a given filter set, s:

$$\mathbf{X} = T_s(\mathbf{x}). \tag{3.23}$$

The transformation process, T, is carried out through a convolution operator ($*$) defined by

$$\mathbf{X} = T_s(\mathbf{x}) = \mathbf{S} * \mathbf{x} = \sum_{i}^{I} \mathbf{S}[i] \cdot \mathbf{x}[n-1], \tag{3.24}$$

where S is the transform filter of length I, $n=\{1, 2, \ldots, N\}$, with N the number of data samples. In regard to filter length, most signal processing applications, including picture compression, typically employ *finite impulse response* (FIR) filters. FIR filters have a finite number of filter taps, or finite filter length. This is desirable for practicability. The alternative *infinite impulse response* (IIR) filters, while superior to FIR filters in terms of frequency selectivity [27], require infinite sampling. Infinite sampling may be carried out through recursive filtering [28, 29]. For invertible systems with forward (analysis) and inverse (synthesis) transformation, the synthesis IIR operation can be realized only with signals of finite samples (e.g., still images) [27].

The process of frequency transformation decomposes pictorial data into different subband images. In this respect, frequency transformation is referred to as *subband transform.* Subband transforms are classified according to their decomposition structure and filters. The decomposition structure determines the way in which transform coefficients are arranged. The nature of this arrangement may be subband [30], block based [31], or hierarchical [32, 33] (Figure 3.10). Hierarchical decomposition is also known as multiresolution or *wavelet transform.*

Image transformation is generally two-dimensional, since images are 2D data. For video, there is a natural extension from 2D to 3D transform to account for the temporal dimension. However, due to the prevailing coding philosophy [10, 16], most video coders adhere to the 2D transform for an individual picture frame.

Filters are the core of the transform and dictate transformation characteristics such as frequency response. The principles of filter design originated from Fourier analysis [34]. Consequently, a basic and commonly employed tool in

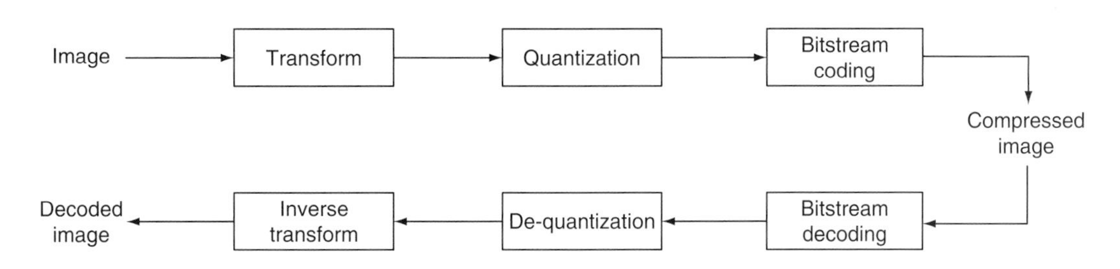

FIGURE 3.9 Principal components of a transform-based coding structure.

[17] Only the quantized residue, $\bar{r}$, is entropy-coded, not the residue, r.

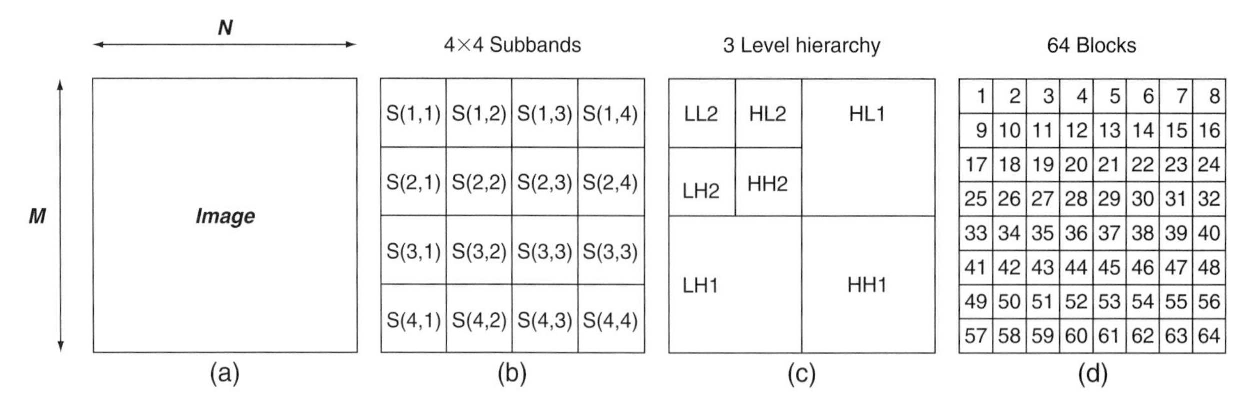

FIGURE 3.10 Transform decomposition structures. (a) An image of $M \times N$ dimension. (b) A 4×4 subband decomposition. Each band $S(k,l)$ with k, $l = \{1, 2, 3, 4\}$ has $M/4 \times N/4$ transform coefficients. (c) A two-level dyadic hierarchical decomposition. Each successive level is a quarter of the resolution of the previous. (d) Block-based transform with 64 blocks. Each block is of $M/8 \times N/8$ dimension.

signal analysis is the *discrete Fourier transform* (DFT) [3, 31], defined as

$$X_{DFT}[k] = \sum_{m=0}^{M-1} x[m] \cdot e^{\frac{-2\pi j}{M}km}$$
$$= \sum_{m=0}^{M-1} x[m] \cdot \left(\cos\left[\frac{2\pi j}{M} \cdot km\right] - j\sin\left[\frac{2\pi j}{M} \cdot km\right]\right), \tag{3.25}$$

where M is the filter length, j the imaginary unit, and k the frequency band of the selected filter. In picture compression, the application and performance of various transformation kernels have been explored [2, 4]. The more popular among these is the *discrete cosine transform* (DCT) [31], formulated as

$$X_{DCT}[k] = \sum_{m=0}^{M-1} x[m] \cdot \cos\left[\frac{\pi}{M}\left(m + \frac{1}{2}\right)k\right]. \tag{3.26}$$

The DCT is a derivation of the real component of the DFT, as is evident in Equation 3.25. It offers near-optimal decorrelation performance, second only to that of the Karhunen–Loève transform (KLT) [35], with respect to the first-order Markov random process [3, 7]. Decorrelation has been seen as an important feature for filters, since it corresponds to energy packing ability. Energy packing leads to compaction of statistical redundancy, whereby a substantial amount of pixel energy is contained with a small number of transform coefficients. The popularity of the DCT is in no small part credited to the availability of fast DCT algorithms, so that its practical application has proliferated in digital communication equipment and services.

Separable filters such as the DFT and the DCT may be extended from one dimension, as in Equations 3.25 and 3.26, to two dimensions through separate horizontal (φ_h) and vertical (φ_v) transformation operations performed in two stages [2, 27]. For the DCT, this is formalized as

$$X_{DCT}[k,l] = \sum_{m=0}^{M-1} \varphi_v(k,m) \cdot \sum_{n=0}^{N-1} x[m,n] \cdot \varphi_h(l,n) \tag{3.27}$$

$$\text{with } \varphi_v(k,m) = \cos\left[\frac{\pi}{M}\left(m + \frac{1}{2}\right)k\right] \tag{3.28}$$

$$\varphi_h(l,n) = \cos\left[\frac{\pi}{N}\left(n + \frac{1}{2}\right)l\right]. \tag{3.29}$$

Filters are designed to address specific characteristics of the data they operate upon. They have various properties, which are discussed in detail in Vetterli and Kovačević [27]. For picture coding purposes, there are a number of highly desirable properties that filters should have, such as phase linearity and orthogonality, among others. Complete recovery of data is allowed by *perfect reconstruction*, defined as

$$\mathbf{x} = T^{-1}(T(\mathbf{x})), \tag{3.30}$$

where $\mathbf{x}$ is both the input and output signal, T and T^{-1} are the forward and reverse transforms, respectively. Invertibility, while not strictly necessary for lossy coding, has become desirable in light of the move toward *scalable* coding [36]. For full scalability in picture quality, picture coders must have the flexibility to encode from lossy to lossless quality, such as in the case of a JPEG2000-compliant image coder [15].

Filters that cater to critical sampling—that is, sampling at the *Nyquist frequency* [2]— are particular useful, since they ensure that the number of input samples prior to transformation equals the number of output samples after transformation. Any transform that has more output than input samples effectively expands data. While such *overcomplete* transforms have been applied in pictorial compression, they are challenging and complex to implement [37]. The Nyquist frequency, F, is defined as the

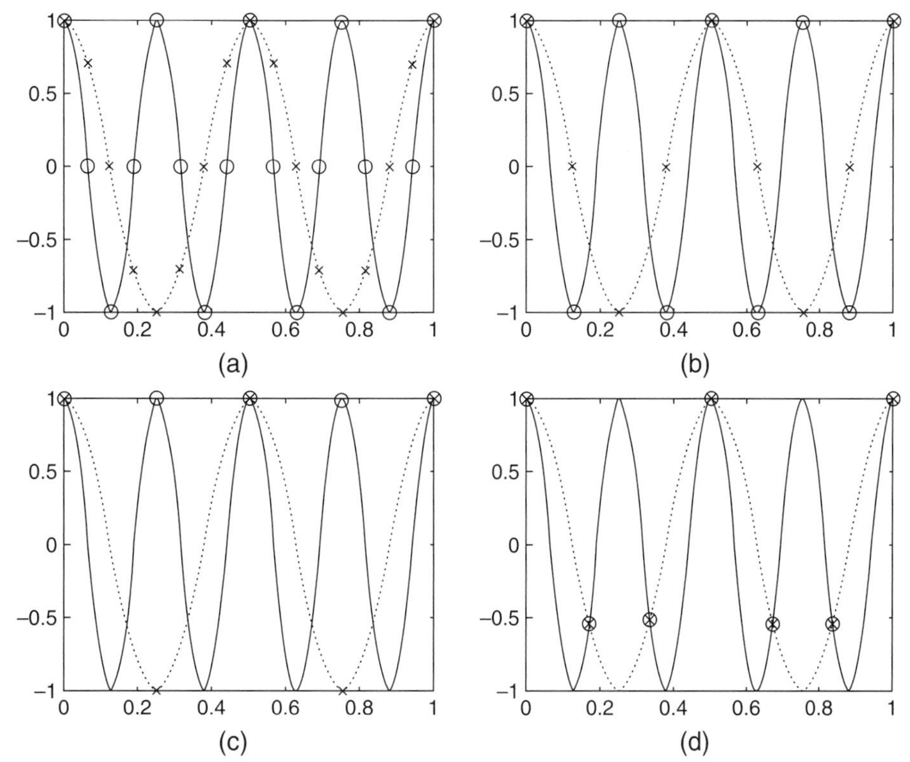

FIGURE 3.11 Two periodic signals $S(F_1)$ and $S(F_2)$ with respective frequencies F_1 (dashed line) and F_2 (solid line) are depicted. Here, $F_2 = 2F_1$. The sampling points associated with the sampling frequency, F_s, are denoted by circles (○) and crosses (×). (a) $F_s = 4F_2$. (b) $F_s = 2F_2$. (c) $F_s = F_2$. (d) $F_s = 2/3F_2$. In both (a) and (b), the subsampled signal at F_2 retains its periodic structure and is distinct from F_1. In (c) and (d), the subsample signal at F_2 has lost its original periodic structure. Further, in (d) it is impossible to distinguish between F_1 and F_2 from the subsampled signal. Note that in (c), $S(F_1)$ is unaffected by aliasing, since $F_s \geq 2F_1$.

bandwidth (i.e., maximal frequency) of a signal [2]. The Nyquist rate, F_N, is defined as twice the Nyquist frequency. Sampling below the Nyquist rate results in signal aliasing [28]. When aliasing occurs, signal components with frequencies F_a above half of the sampling frequency F_s, also known as the *folding frequency* F_f, are indistinguishable from those with frequencies F_b below the folding frequency F_f, such that $F_a = F_b + n \times F_f$, where n is an integer with $0 < F_b \leq F_f$ and $F_a > F_f$. This is illustrated in Figure 3.11 with time-domain sinusoidal signals. For both auditory and pictorial signals, the appearance of aliasing typically results in high-frequency signal distortions [2].

Aliasing in the transform domain occurs when downsampling is carried out on filters with overlapping frequency responses (Figure 3.12) [27, 29]. Typically, filters are designed with multiple overlapped response bands to cover the entire range of possible frequencies. Overlapping is a practical necessity, since idealized filters,[18] requiring infinite filter taps, are impossible to implement [29]. When these overlapped filters are downsampled, that is, sampled at the critical frequency, aliasing is induced in the overlapped regions.

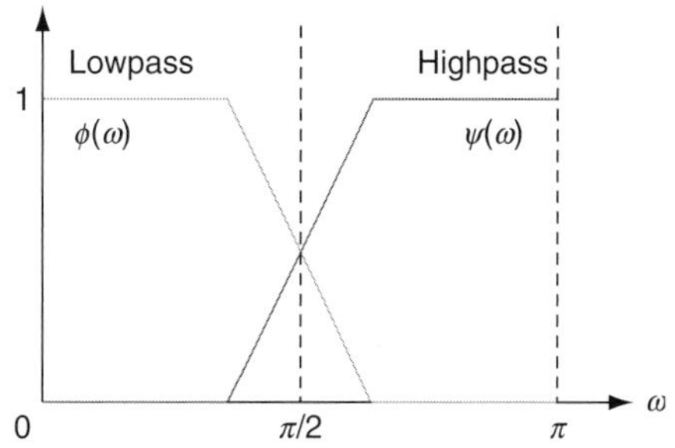

FIGURE 3.12 Overlapping frequency responses of a two-band system with a lowpass ($\varphi(\omega)$) and a highpass ($\varphi(\omega)$) filter. The overlapped region is the triangular area that centers around midfrequency, $\pi/2$.

Two classes of invertible filters that allow critical sampling and address the aliasing factor are orthogonal[19] and bi-orthogonal filters. In these filters, aliasing is dealt with through alias cancellation in the reconstruction/synthesis process [27, 38]. Therefore, while aliasing still exists in transformed data, these aliasing components are effectively eliminated during the

[18] With perfect (nonoverlapped) frequency localization.

[19] Orthogonal filters with unity gain are referred to as orthonormal filters [27].

inverse transform. However, total alias cancellation is possible only if the transform data remain undamaged/unaltered—a condition that is impossible to satisfy if the quantization operation is undertaken after transform [27].

Differences between orthogonal and bi-orthogonal filters can be seen in structural terms. Orthogonal filters allow for m number of frequency response bands, whereas bi-orthogonal filters have dyadic response bands, one highpass and one low-pass. In addition, orthonormal systems such as the DFT and DCT are inherently critically sampled and thus preserve vector length[20] [27]. Bi-orthogonal filters, on the other hand, do not preserve vector length. Consequently, they have more inputs than outputs. For image compression applications, this disparity may be resolved through the extension of input signal samples [15, 36]. In terms of filter properties, bi-orthogonal filters have a number of advantages over orthogonal filters, such as phase linearity and regularity. These properties are discussed further in [27].

3.2.6.2 Quantization

In order to achieve an acceptable level of compression, it is necessary to limit the possible range of symbols to be coded after transformation. The quantization operation in image coders effectively remaps the transform coefficient from a larger to a smaller set of discrete numbers. Quantizers may be scalar or vector based. In addition, scalar quantizers may be uniform or nonuniform (Figure 3.13). Scalar quantization is usually carried out through the division operation

$$X_q[n] = \left\lfloor \frac{X[n]}{q} \right\rfloor, \tag{3.31}$$

where $X_q[n]$ is the scalar quantized symbol, $X[n]$ is the transformed coefficient, q is the quantization step size, and $\lfloor \ \rfloor$ denotes rounding down to the nearest integer. The methodology for designing optimal quantizers is discussed in Lloyd [39] and Max [40].

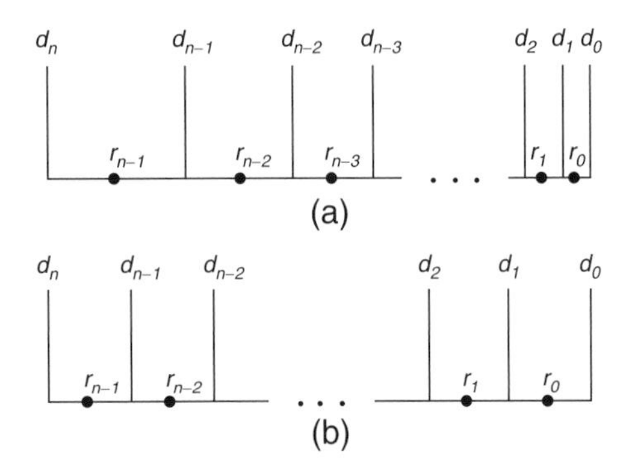

FIGURE 3.13 Nonuniform (a) and uniform (b) scalar quantization. Each set of decision levels, d, corresponds to a quantized response, r. For uniform quantization, the decision levels are equal distances apart. Nonuniform quantization has variable distance intervals between decision levels.

Quantization introduces quantization errors, e, defined as the difference between the dequantized, $X_r[n]$, and unquantized data samples:

$$e = X[n] - X_r[n], \tag{3.32}$$

where $X_r[n] = X_q[n] \times q$. Quantization error leads to the irrecoverable information loss in lossy coding. While scalar quantization is quite effective in shaping a favorable PDS for entropy coding, it has been shown that *vector quantization* (VQ) is generally more effective [41]. In fact, scalar quantization is seen as a subset of VQ, where the vector length equals unity. VQ operates by approximating a group of transformed coefficients,[21] $\chi[u] \in X$, from the transformed space, X, with a vector symbol, $V[k] \in V$. The vector symbol is chosen from a vector codebook, V, of size M based on some distance minimization criteria, f_{DMC}, such as the *mean squared error* (MSE). With the MSE criterion, this process is formulated as

$$\begin{aligned} V[k] &= f_{DMC}(\chi[u], V) \\ &= \min(\mathrm{MSE}(\chi[u], V[m])), \quad \forall m \in M. \end{aligned} \tag{3.33}$$

The MSE is defined as

$$\mathrm{MSE(A,B)} = \frac{1}{N}\sum_{n}^{N}(\mathrm{A}[n] - \mathrm{B}[n])^2. \tag{3.34}$$

The effectiveness of VQ is ultimately dependent on the size of the vector codebook, M. Having a large codebook with more vector codes would lead to better approximation. Unfortunately, as the size of the codebook grows, the overhead associated with storing and transmitting the codebook increases. Additionally, the computation for the vector-matching operation also increases. Thus, there is a practical limitation that curtails the effectiveness of VQ.

Another realization of quantization is recursive quantization[22] associated with progressive bitplane coding. This approach breaks the transformed coefficient into its constituent bit components and encodes each bit component progressively from the MSB to the LSB. The attraction of this method is that it is particularly suited for scalable coding.

3.2.6.3 Bitstream Coding

The final stage of a transform coder is concerned with two things: first, entropy coding of quantized transformed data; second, the efficient arrangement of an entropy-coded data stream. General details of entropy coding have been covered previously (Section 3.2.3). For bitstream coding, it is common

[20] The number of input samples equals the number of output samples.

[21] Current discussion centers on frequency-domain transformed coefficients. However, VQ may also operate on time-domain data.

[22] Also referred to as progressive bitplane quantization.

to adapt various combinations of coding techniques to suit the nature of the data being coded. For example, the JPEG still-image coder employs run-length [2] coding prior to Huffman coding in an effort to reduce the number of quantized coefficients to be coded. This is also true for standardized hybrid video coders [42]. Similarly, for the JPEG2000 coder, run-length coding has also been utilized, albeit conditionally. Further, the JPEG2000 coder employs a context binary arithmetic coder for entropy coding to match the progressive bitplane quantization strategy. The formation of the final data stream arranges the order in which entropy-coded data are stored. This is especially important where scalability is concerned. For example, if a picture is coded with resolution scalability in mind, then the final bitstream should be formed in order of picture resolution; that is, from low resolution to high resolution.

3.2.6.4 Video and Image Coding

One approach to encoding digital video is to encode each video frame independently, as seen in motion JPEG2000 [43]. While this method does have some useful applications, particularly in video editing, its compression performance is unimpressive compared with three-dimensional transform coding [44–46, 81] and hybrid video coding techniques [10, 16]. Conventional hybrid video coding strategy (Figure 3.14) operates in two modes: *intraframe* and *interframe.*

The intraframe mode, for all intents and purposes, behaves as a still-image coder and is intended for encoding *reference* frames only. The interframe mode complements this by encoding the *difference* frames. While the physical encoding operation in the inter mode is equivalent to that of the intra mode, its primary component is a prediction engine used to perform the *displace frame difference* (DFD) operation [10]. The DFD generates the difference frame based on the reference frame (Figure 3.15). The reference frame is then encoded in a similar manner to the intraframe mode.

In order to account for motion in video data for DFD operations, *motion estimation* (ME) and *motion compensation*

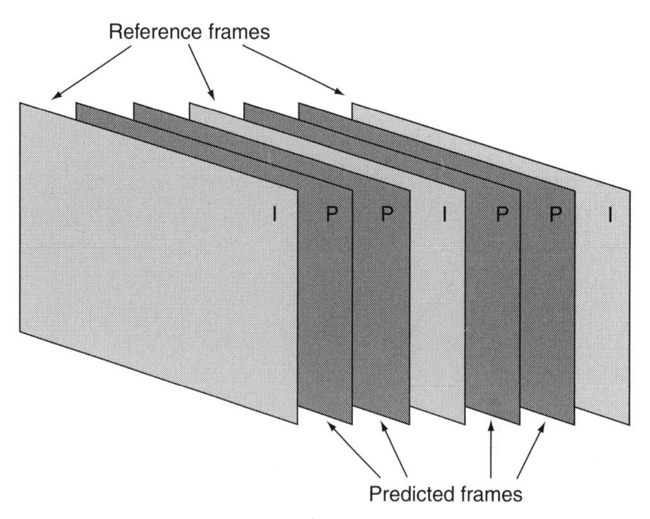

FIGURE 3.15 Arrangement of reference (**I**) and difference (**P**) frames for hybrid video coding.

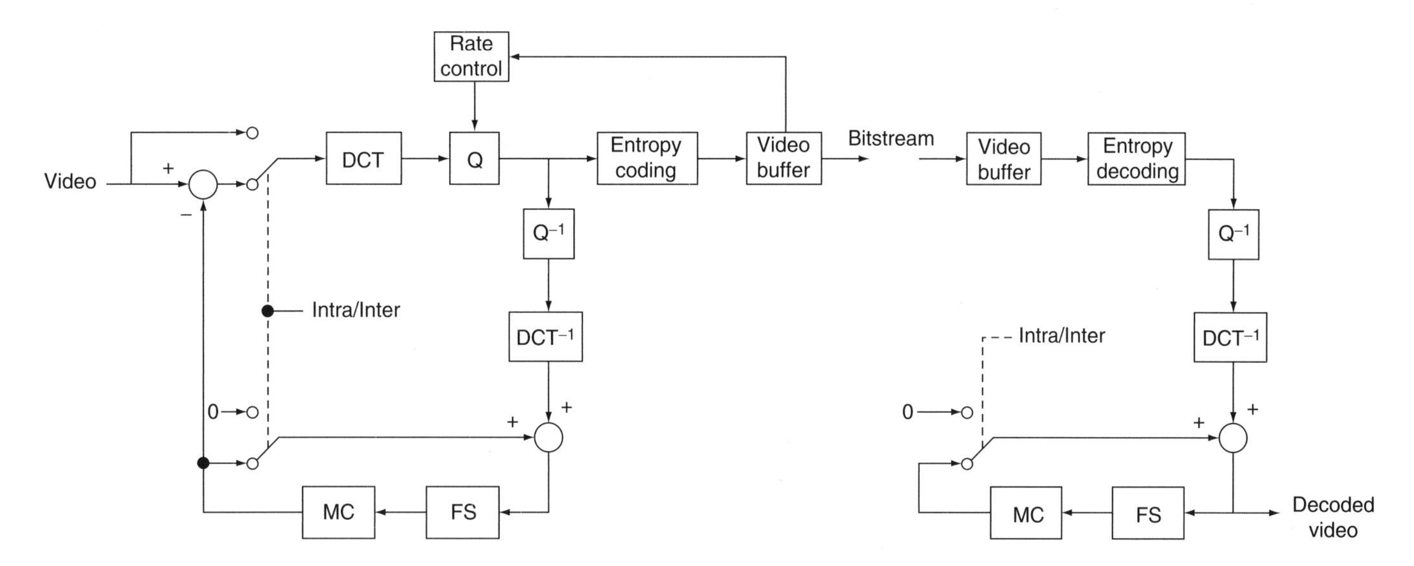

FIGURE 3.14 A simplified hybrid video coding structure (MPEG) that switches between intra- and interframe coding. The difference between these two modes is the additional prediction operation for the interframe mode. In the intraframe mode, DCT is performed over video data followed by quantization (Q). The quantized data are then entropy-coded to form the compressed bitstream. The entropy-coding process also feeds into a rate control function that attempts to maintain coding at a desirable bit rate. For the interframe mode, the reconstruction of previously encoded frames is needed in order to generate the prediction for the current frame with motion compensation (MC). Consequently, a reverse process with dequantization (Q^{-1}) and inverse DCT (DCT^{-1}) is encapsulated in the encoder. Frame storage (FS) is employed to store these previously encoded reference frames for MC operations. The decoding end mirrors the encoder, but in the reverse order.

(MC) [5, 7] functions are used to track moving objects in the temporal field. This has the effect of reducing difference or residue errors from DFD operations. However, even with MC and ME, motion mismatch may still occur, resulting in high residue errors. Three-dimensional transform coders have no standardized structure. Some of these coders are direct 3D extensions of their 2D counterparts [47, 81]. Others, however, adopt the hybrid coding structure with ME/MC [44, 46, 48]. While 3D transform coding is focused primarily on general digital video coding applications, it has also been proposed for coding of volumetric medical images [49].

3.2.6.5 Scalability

Scalability has been an issue of intense interest in picture coding in recent years. It is a flexibility feature that enables some degree of control over the variations between the encoding and decoding ends with respect to picture quality, resolution, and, in the case of video, frame rate [10]. It is envisaged that scalability would be most useful in instances where pictures are coded to the optimum scale[23] but decoded at different scales, subject to the requirements and limitations at the decoding end [10]. An example of this may be a central picture repository, such as an art collection or archive of medical images, having high-quality pictures that are accessed remotely. Scalability is also particularly useful in transcoding, where digital videos are reencoded to different bit rates, qualities, or resolutions [50]. To maintain scalability in picture coding, it is necessary to arrange the bitstream of the coder into appropriate layers that reflect the scale used. For example, if pictures are scaled according to quality, then the layers should be arranged according to layers of decreasing degree of quality improvement [51, 52]. That is, the first layer should have the highest degree of quality improvement, and the last layer should have the least.

3.2.7 Perceptual Picture Coding

The impact on picture quality in lossy coding has always been an area of concern, particularly for high-quality images. Traditional methods for quantifying distortions and picture quality do not consider the human factor [53]. It is recognized that picture quality is dependent on perceived picture content. As a result, it has become increasingly common for picture coders to incorporate, at the very least, some aspects of the *human vision system* (HVS) into the coding system. HVS-based coding may operate at two levels, above or below threshold vision—the former is perceptually lossy, while the latter is perceptually lossless. The conceptual view of perceptual coding and traditional lossless and lossy coding is depicted in Figure 3.16.

There are two main issues in perceptual coding. The first is the modeling of the HVS, for which a detailed treatment is given in Wandell [55]. Human vision has a physical and a psychological aspect. The psychological aspect is concerned with the human mind—specifically, what the mind perceives based on memories and experiences. The physical aspect deals with the physiology of the HVS, which broadly consists of the human eye, the visual pathways, and the visual cortex. The levels of understanding among these three physical components vary. For example,

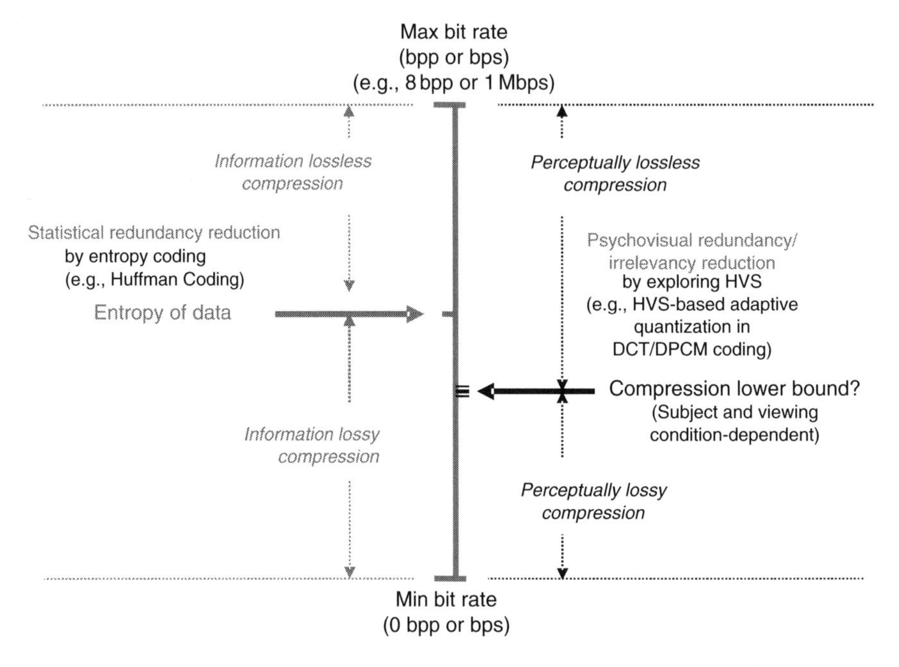

FIGURE 3.16 Conceptual view of picture coding philosophy from Hwang et al. [54].

[23] Scale in reference to quality, bit rate, and resolution.

literary knowledge of the human eye is sufficiently well developed to allow accurate modeling of the visual acuity of the eyes and color sensitivity. However, the more exact mechanical operation of the visual pathways and, specifically, the visual cortex is less developed. Hence, the model of the visual cortex only approximates basic primitive functions of the real visual cortex [56]. Nevertheless, this model does offer a credible behavioral approximation of the HVS. The basic physiological characteristics of the HVS can be summarized as follows:

- It is sensitive to the frequency and orientation of visual stimuli.
- It operates in a relative rather than an absolute manner. More specifically, the HVS sees the contrast between the luminance of two visual stimuli rather than the absolute luminance level of the two stimuli.
- The visibility of visual stimuli may be affected by masking and facilitation. Masked signals would have diminished visibility, while facilitated signals increase their visibility.

The *contrast gain control* (CGC) model [57] (Figure 3.17) provides a reliable generic description of the HVS. Functionally, it evaluates the perceived similarity/difference between two images, a reference, and an altered copy of the reference. The CGC model consists of four main parts: contrast estimation, filtering, masking, and pooling. Contrast estimation translates images from the absolute scale to the relative scale via the *contrast sensitivity function* (CSF).

This is followed by the filtering operation, which projects images from the space domain into an oriented frequency domain. There are a number of realizations for the filtering operation, including the steerable pyramid transform [58], cortex transform [59], and Gabor array [57], to name a few. Masking then attenuates individual coefficient samples, in the oriented frequency domain, according to the activity of their surroundings. The masking response, R_m, has the general form

$$R_m[l,\theta,i,j] = v_{m,1} \cdot \frac{X[l,\theta,i,j]^{p_m}}{h_m^{q_m}[l,\theta,i,j] + v_{m,2}}, \quad (3.35)$$

where the transform coefficient $X[l,\theta,i,j]$ is masked by an inhibition factor $h_m^{q_m}[l,\theta,i,j]$ of masking domain m; variables l, θ, and (i,j) specify the frequency, orientation, and spatial location, respectively, of the transform coefficient; and p_m, q_m, $v_{m,1}$, and $v_{m,2}$ are parameters to be optimized. Masking may occur in spatial, frequency, and/or orientation domain [57]. Note that Equation 3.35 provides a separable model for quantifying masking in individual domain. An alternative model that concurrently quantifies masking in all domains has been proposed in [57]. Typically, only spatial and/or orientation masking is considered [51, 60, 61]. For spatial masking, the inhibition factor is a measurement of the activity surrounding the target coefficient, $X[l,\theta,i,j]$, given by

$$h_s^{q_s}[l,\theta,i,j] = \sum_u^U \sum_v^V X[l,\theta,u,v]^{q_s}. \quad (3.36)$$

Spatial masking is localized within a windowed area, as specified by U and V. This windowed area is generally centered on spatial location (i, j). Orientation masking measures the activity over the same spatial location, but over orientation, φ. It is written as

$$h_o^{q_o}[l,\theta,i,j] = \sum_\phi^\Phi X[l,\phi,i,j]^{q_o}. \quad (3.37)$$

Separable masking models may be unified into a single measurable quantity by weighted summation during pooling, as in [60]. The pooling stage sums up all the differences between masking responses of the reference and processed image. The Minkowski summation is generally used for this purpose. It is defined as

$$D_{CGC} = g_m \cdot \sum_m^M \left(\sum_l^L \sum_\theta^\Theta \sum_i^I \sum_j^J (R_m[l,\theta,i,j] - \bar{R}_m[l,\theta,i,j])^\beta \right)^{\frac{1}{\beta}}, \quad (3.38)$$

where g_m is the weight for masking domain m, and $R_m[l,\theta,i,j]$ and $\bar{R}_m[l,\theta,i,j]$ are the masking responses of the reference and processed pictures, respectively. The overall distortion spans all resolution levels (L), orientations (Θ), and spatial locations (i,j).

The second issue regarding perceptual coding is the adaptation of the HVS model to picture coders. There are a number of ways to adapt vision models to coding structures. The most common method is through the quantization stage, where

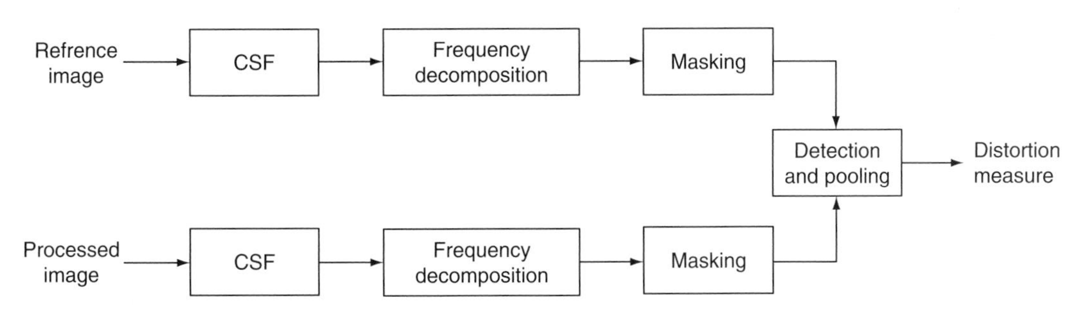

FIGURE 3.17 The contrast gain control model.

vision models regulate the quantization operation to control the level of perceived distortions in coded pictures [62–64]. An unusual method of adapting the HVS to coders is to design transformation filters that are specifically tuned to aspects of the HVS [65]. For scalable coders such as the JPEG2000 still-image coder, vision models may be adapted to the error distortion measure in the R–D function [51, 60, 61] (see Section 3.2.4).

Perceptual coders may be rate driven or quality driven. The purpose of the rate-driven perceptual coder is to encode pictures to the best possible visual quality for a given bit rate. In the quality-driven coder, pictures are encoded to a desired visual quality level at the lowest possible bit rate. A quality-driven coder operating at just below the super-threshold level would provide visually lossless quality coding (Figure 3.18). The *super-threshold level* is defined as the point at which differences between two visual stimuli are just perceptible [55]. It is also commonly referred to as the *just-noticeable-difference* (JND) level.

3.2.8 Standardized Coders

Under the auspices of the International Standards Organization (ISO), the Joint Photographic Experts Group (JPEG) and the Moving Pictures Experts Group (MPEG) have been the primary entities responsible for the development of industry standard picture coders. JPEG is responsible for still-image compression that includes grayscale and color images for both lossy and lossless encoding. JPEG-LS is a lossless coder based on the LOCO coding engine (see Section 3.2.5.2). The

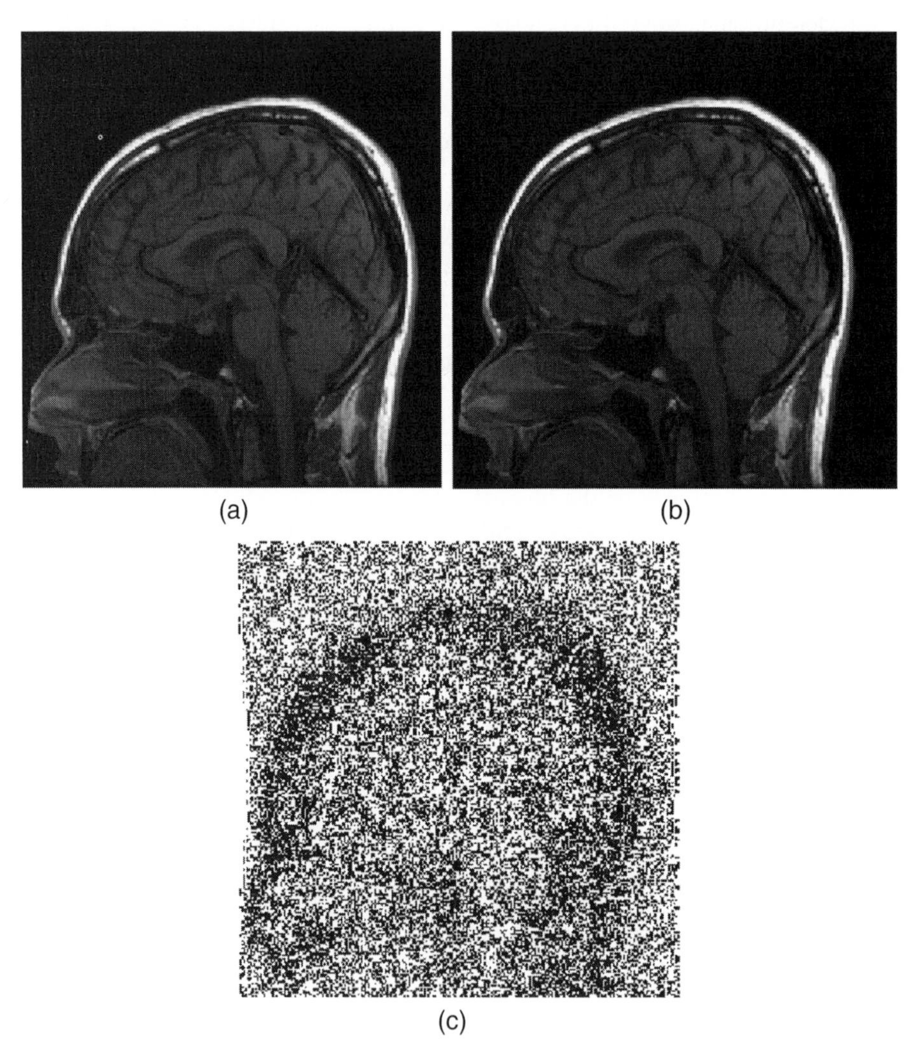

FIGURE 3.18 MRI slice of a brain. (a) Original image. (b) Perceptually lossless coded image [71]. (c) Difference image between original and perceptually lossless coded image. For the difference image, white areas indicate no pixel difference, while dark areas contain some pixel difference. The difference image reveals areas within the image where psychovisual redundancies exist.

LOCO coder was a fairly advanced coding system when it was first introduced. In the years since, more efficient lossless coding algorithms have appeared [66, 67].

The JPEG baseline is lossy image coding using DCT scalar quantization with run-length and Huffman codes for entropy coding [68]. It is considered to be superseded by the recent state-of-the-art JPEG2000 coder. The JPEG2000 still-image coder [36] employs *discrete wavelet transform* (DWT) with progressive bitplane coding. Entropy coding is handled by a moderately sophisticated context-based arithmetic coder. This coding system is scalable in resolution and compression rate. In addition, it can operate in both lossless and lossy modes. While its performance in the lossy mode is superior to the JPEG baseline, its performance in the lossless mode is inferior to JPEG-LS [69].

MPEG is responsible for video compression technology. Its MPEG-x series of coders is closely related to the International Telecommunication Union (ITU) line of H.26x coders. The most recent standardized video coder, the MPEG-4 AVC/H.264 [16], has been developed through joint efforts of the ISO and ITU. Both the MPEG-x and H.26x coders share a similar hybrid DCT/MC coding framework. Over the past decade, the underlying hybrid technology has matured considerably, and substantial performance gains have been made in MC, entropy coding, and postfiltering. The performance of MPEG-4 AVC/H.264 over its predecessors is analogous to the performance of JPEG2000 over JPEG baseline [70].

3.2.9 Applications of Picture Coding to Medical Images

Picture compression has become increasingly important to medical imaging. The shift toward digital media has provided more flexibility in the way in which medical images are taken, transported, and made available for diagnosis (e.g., telemedicine [82]). As in many other applications, digitized medical images require storage space and bandwidth for transportation over communication networks. The problem arises when storage and transmission requirements exceed capacity. While it is possible to increase capacity, it may also be prudent to invest some effort in compression so as to reduce the storage and transmission requirements.

The critical issue for medical images in regard to compression is information integrity. Information loss should be avoided when possible. Thus, in this regard, the medical fraternity sees compression as lossy or lossless. In situations where information loss is unavoidable due to practical reasons, the attention is shifted from prevention to minimization of information degradation [72]. The deterioration of information through compression may be deemed acceptable if the diagnostic value of medical images is preserved.

The question for lossy medical coding, then, is twofold. First, under what conditions could lossy compression be used? Second, what is the error tolerance level for maintaining the diagnostic value of medical images? For example, which pixels in a medical image contain critical diagnostic information? Moreover, what effects do distortions from lossy coding have on diagnostic quality? Currently, there are no formal guidelines for the use of lossy coding in medical images. This may be partly due to legal considerations. The possibility that the loss of some diagnostic information may lead to a drastic misdiagnosis has considerable legal ramifications. Until lossy compression can guarantee the preservation of diagnostic information in medical images, it is likely that medical imaging will focus more on lossless compression [73].

If the diagnostic value of medical images is taken in terms of measurable perceptible quantities, then perceptual coding may be a solution that clinically retains diagnostic information of medical images. To this effect, perceptual lossless coding has been shown to be equally or more effective than the lossless and the near-lossless coding strategies [71]. Ultimately, however, the compression of medical images may be dependent on the nature of the diagnosis and individual situation.

3.3 Compression in the DICOM Standard

This section will provide a brief description of picture coders supported under the image compression component of the Digital Imaging and Communications in Medicine (DICOM) standard (see Chapters 2 and 13 of this book and [74]). The DICOM standard provides a format for collating all information associated with individual medical images. It encapsulates pictures compressed through standardized coders within its structure, thus ensuring modularity. This modular arrangement allows for future introduction and retirement of coders to and from the standard. Once medical images are bound to the DICOM format, the manner in which they are stored and transmitted is covered by the Picture Archiving and Communication System (PACS) (see Chapter 13 and [75, 76]).

3.3.1 DICOM Recommended Coders

DICOM does not necessarily support all features of standardized coders. Additionally, DICOM neither specifies nor recommends under what conditions lossy compression should be used. The decision is left entirely to individual users. Coders that are currently supported in the DICOM standard are:

- JPEG-LS [22] for lossless and near-lossless compression based on the LOCO coder [20].
- JPEG baseline [68, 77] for lossy compression. It implements DCT with scalar quantization and Huffman coding.
- JPEG2000 [15, 36], which supports both lossless and lossy compression through reversible (5/3) and irreversible (9/7) filters, respectively. It also supports scalable coding

and utilizes wavelet transform with bitplane coding and arithmetic coding.
- MPEG-2 MP@ML (main profile at main level) [42] for compressing multiframe images.

3.3.2 Image Modality

Diagnostic imaging falls into three general categories: transmission, reflection, and emission imaging. Transmission imaging such as roentgenography (X-ray) projects particles through a medium to capture specific characteristics within that medium. The acoustic-based reflection techniques send pulse signals into a medium. Information within the medium is then captured from reflected signal pulses. Emission imaging operates by capturing signal emanations from within a medium. These emanations may be induced externally, through the injection of radioactive isotopes, as in the case of nuclear medicine, or may come about naturally in MRI.

The modality of a medical image specifies the method by which the image is captured; that is, MRI, CT, ultrasound, etc. Different modalities are intended to extract different types of information. A list of supported modalities is provided in Chapter 1 and reference [74]. The DICOM standard provides no recommendation as to what type of compression, lossy or lossless, should be used for any particular modality. This decision is left to the individual user. One factor that may affect the choice of compression system is the size of digitized images. Medical images have the tendency to be sizable due to their bit depth, which ranges between 8 and 16 bits, and they have resolution that may exceed 1000 × 1000 pixels [78]. In addition, when dealing with multiframe/sliced images, the amount of storage requirement becomes most noticeable. Therefore, it may be more economical to have high compression with some information loss if the diagnostic value of medical images can be maintained [79, 80].

3.4 Data Compression for Dynamic Functional Images

This section will present the medical data compression for multidimensional dynamic functional images based on various diagnostically lossless coding schemes. A brief background on multidimensional dynamic functional imaging studies with physiological parameter estimation is given, followed by addressing the need for developing efficient multidimensional data compression to reduce the volume of dynamic functional images without affecting physiological parametric estimation and clinical decision making. After that, diagnostically lossless compression techniques for dynamic functional image data are described in three aspects: *compression in temporal domain* based on an optimal image sampling schedule; *compression in spatial domain* with clustering analysis; and *compression in sinogram domain* by a combination of *principal component analysis* (PCA) and a channel-weighted JPEG2000 coding scheme.

3.4.1 Multidimensional Dynamic Functional Imaging Studies

As mentioned in Section 1.4, dynamic functional imaging such as PET in nuclear medicine can provide image-wide quantification of physiological, pharmacological, and biochemical functions within the body and can support the visualization of the distribution of these functions corresponding to anatomical structures. Physiological function can be estimated by observing the behavior of a small quantity of an administered substance "tagged" with radioactive isotopes (tracers). Images are formed by the external detection of gamma rays emitted from the patient when the tracers decay. Since they allow the observation of the effects of physiological processes, where most diseases are functional in nature and structural changes are secondary, functional imaging techniques are invaluable aids to patient diagnosis and treatment [83]. The range of tests that can be performed in functional imaging studies is extremely large and covers all organ systems of the body. In some studies, the time course of tracer redistribution from administration onward must be observed and quantified to enable the calculation of physiological parameters by tracer kinetic modeling. Figure 3.19 illustrates a typical study for processing and analysis of dynamic functional image data and subsequent generation of human brain parametric images using PET with the glucose tracer ^{18}F-fluoro-deoxyglucose (FDG). To estimate physiological parameters and form parametric images, for each cross-section plane, the PET scanner acquires a series of scans at a predetermined rate (not necessarily constant), typically for 20–60 minutes, in which projection views (sinogram data) are acquired at multiple angles and reconstructed into slices that depict regional tracer uptake and function during the study. From these data, a *tissue time–activity curve* (TTAC) can be plotted for each voxel, and the physiological parameter value for that voxel calculated by the application of a tracer kinetic model to the TTAC. If the modeling process is repeated for every plane, then a 3D physiological parametric image can be constructed [84]. More details of generating parametric images can be found in Chapter 6.

These dynamic functional imaging studies, however, are accompanied by a growth in the size of the image data. For example, a routine dynamic PET study using a CTI 951 scanner (CTI Inc., Knoxville, TN) typically involves the acquisition of 31 cross-sectional image planes of 128 × 128 pixels each, at 20 to 30 time points. The resulting four-dimensional data set contains upward of 11 million data points, requiring approximately 22 megabytes of storage for just one study for one bed position. Such a large number of images places a considerable

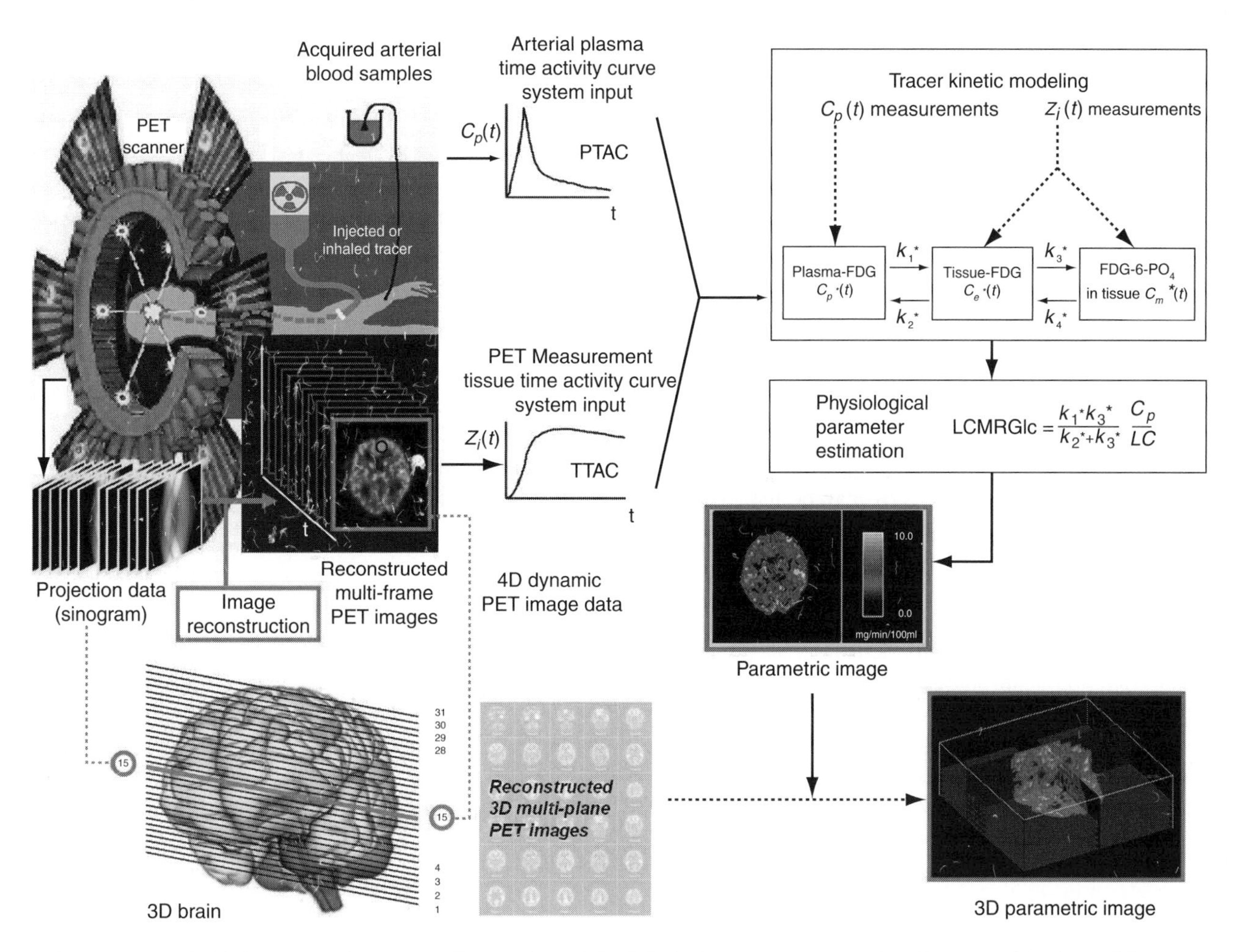

FIGURE 3.19 A brief diagram of the process of generation of physiological parametric images based on tracer kinetic modeling—for example, quantitative estimation of regional glucose metabolic rate with FDG PET. After intravenous injection of the FDG tracer, the time course of the regional radiotracer concentration in the brain is obtained by acquiring a series of images. At the same time, the input function is obtained from a series of blood samples. The physiological parameter of interest, in this case the local cerebral metabolic rate of glucose, is estimated by fitting a compartmental model to the data. Four-dimensional data (three dimensions in space and one in time) are required to construct the three-dimensional parametric image, which depicts regional glucose metabolism quantitatively in quantitative units of mg/100g/min.

load on computer storage space and retrieval, data processing, and transmission time. When the image resolution has to be improved or the scanning procedure has to include the whole body, rather than just a single organ, the demand for space is greatly increased. Therefore, techniques for dynamic functional image data compression are of great interest.

As mentioned in Section 3.2.4, conventional image compression algorithms can be divided into two major categories, lossless and lossy. Lossless compression algorithms allow for perfect reconstruction of the original images from compressed data. These algorithms yield modest CRs, typically between 1.7:1 to 2.1:1 for medical image data. On the other hand, lossy compression can achieve higher CRs. However, the original images can be reconstructed only approximately from compressed data, though the differences may not be distinguishable by the HVS. The challenge in the development of a practical image compression scheme for dynamic medical images is the development of compression algorithms that are lossless for diagnostic purposes; that is, they make no difference to doctors' qualitative and quantitative assessments, yet attain high CRs to reduce storage, transmission, and processing burdens. It should be noted that in the clinical situation, a slight loss of precision in a derived parameter may be undetectable visually and quite insignificant relative to the measurement error.

The conventional compression algorithms are not specifically tailored for the diagnostic use of dynamic functional image data. Therefore, new algorithms have to be developed to fully exploit spatial and temporal redundancies in these data. In addition, the variation of data can be organized in such a novel way as to remove the measurement noise and improve the measurement reliability. In the following subsections, three different diagnostically lossless compression techniques for dynamic functional image data will be reviewed: compression in temporal domain based on an optimal image sampling schedule [85–87], compression in spatial domain with clustering analysis [86–88], and compression in sinogram domain by PCA and a channel-weighted JPEG2000 scheme [89–91]. For simplicity and clarity, the dynamic functional image data in human brain FDG PET studies are used to illustrate the practicality of these compression techniques.

3.4.2 Diagnostically Lossless Compression in Temporal Domain

In dynamic functional imaging studies, the reliability of the temporal frames is directly influenced by the sampling schedule and the duration of each frame. The number of counts and hence the statistical reliability of a frame increases with its duration. However, in order to obtain quantitative information about dynamic processes, a certain number of temporal frames are required. Conventional sampling schedules (CSSs) [92–95] that involve the acquisition of a large number of temporal frame images have been empirically developed but may not be optimal for the extraction of accurate physiological parameter estimates. Most previous studies suggest that a higher sampling frequency should be used over the early stages. This conclusion, however, imposes a considerable burden on the computer image storage space and data processing. To remedy these limitations, an *optimal image sampling schedule* (OISS) has been developed and has been demonstrated to be an effective way to reduce image storage requirements while providing comparable parameter estimates [85, 96]. It was found that if a different cost function for parameter estimation were used—which depends only on the direct PET measurement, rather than the instantaneous measurement—the accuracy of parameter estimation could remain almost unchanged when two neighboring image frames were combined into one.

Finding the optimal image sampling schedule involves minimizing the determinant of the covariance matrix of the estimated parameters **p**, or conversely maximizing the determinant of the Fisher information matrix [97] by rearranging the sample intervals, using the minimum number of required samples. To illustrate the practicality of the OISS algorithm, the five-parameter FDG model [98] for describing the behavior of FDG in brain tissue with the effects of cerebral blood volume is adopted in this section, in which the first four parameters $k_1^* \sim k_4^*$ represent model transport and reaction rate constants, and the fifth parameter *CBV* is used to depict the effects of cerebral blood volume. For the five-parameter FDG model with:

$$\mathbf{p} = \lfloor k_1^*, k_2^*, k_3^*, k_4^*, CBV \rfloor, \tag{3.39}$$

the information matrix **M** with elements m_{ij} can be represented as:

$$\mathbf{M} = [m_{ij}] = \left[\sum_{k=1}^{N} \frac{1}{\delta^2(t_k)} \left(\frac{\partial C_i^*(t_k,\mathbf{p})}{\partial p_i} \right) \left(\frac{\partial C_i^*(t_k,\mathbf{p})}{\partial p_j} \right) \right]. \tag{3.40}$$

Here, $C_i^*(t_k,\mathbf{p})$ is the output function of the five-parameter FDG model:

$$\begin{aligned} C_i^*(t) = {} & \frac{k_1^*(1 - CBV)}{\alpha_2 - \alpha_1} \{ (k_3^* + k_4^* - \alpha_1) e^{-\alpha_1 t} \\ & + (\alpha_2 - k_3^* - k_4^*) e^{-\alpha_2 t} \} \otimes C_p^*(t) + CBV \cdot C_p^*(t), \end{aligned} \tag{3.41}$$

where $\otimes$ is the convolution operator, $C_p^*(t)$ is the FDG concentration in plasma represented by the *plasma time activity curve* (PTAC), and

$$\alpha_{1,2} = \left(k_2^* + k_3^* + k_4^* \mp \sqrt{(k_2^* + k_3^* + k_4^*)^2 - 4k_2^* k_4^*} \right) / 2. \tag{3.42}$$

The required sampling schedule can be adjusted iteratively to maximize the determinant of **M**, det(**M**), using an automatic algorithm [85]. A set of *a priori* parameters has to be provided as the nominal parameters of the model. The optimization procedure starts with an initial sampling schedule $\{I_1, I_2, \ldots, I_N\}$ and then iteratively adjusts the sample intervals. At each iteration, each interval is inspected and adjusted toward the direction in which det(**M**) increases. The optimization procedure always converges, as there are a finite number of intervals to start with, and det(**M**) increases monotonically. It has been shown [85, 99] that the minimum number of temporal frames required is equal to the number of model parameters to be estimated. Therefore, for the five-parameter FDG model, five temporal frames should be sufficient to obtain parameter estimates with similar statistical accuracy and reliability to the conventional technique, which typically requires the acquisition of more than 20 temporal frames. This reduces the number of temporal frames obtained and, consequently, reduces data storage. Furthermore, as fewer temporal frames are reconstructed, the computational burden posed by image reconstruction is reduced. Figure 3.20(a) shows the original 22 temporal frames for the 15th plane from one patient study. Due to the lower tracer concentration and short acquisition time in the first few frames, these images have been scaled to be visible. A set of five temporal frame images derived from the OISS algorithm is illustrated in Figure 3.20(b), where a CR of 4.4:1 is obtained.

3.4.3 Diagnostically Lossless Compression in Spatial Domain

The data compression whose principles were introduced in Section 3.4.2 is used mainly for exploiting temporal

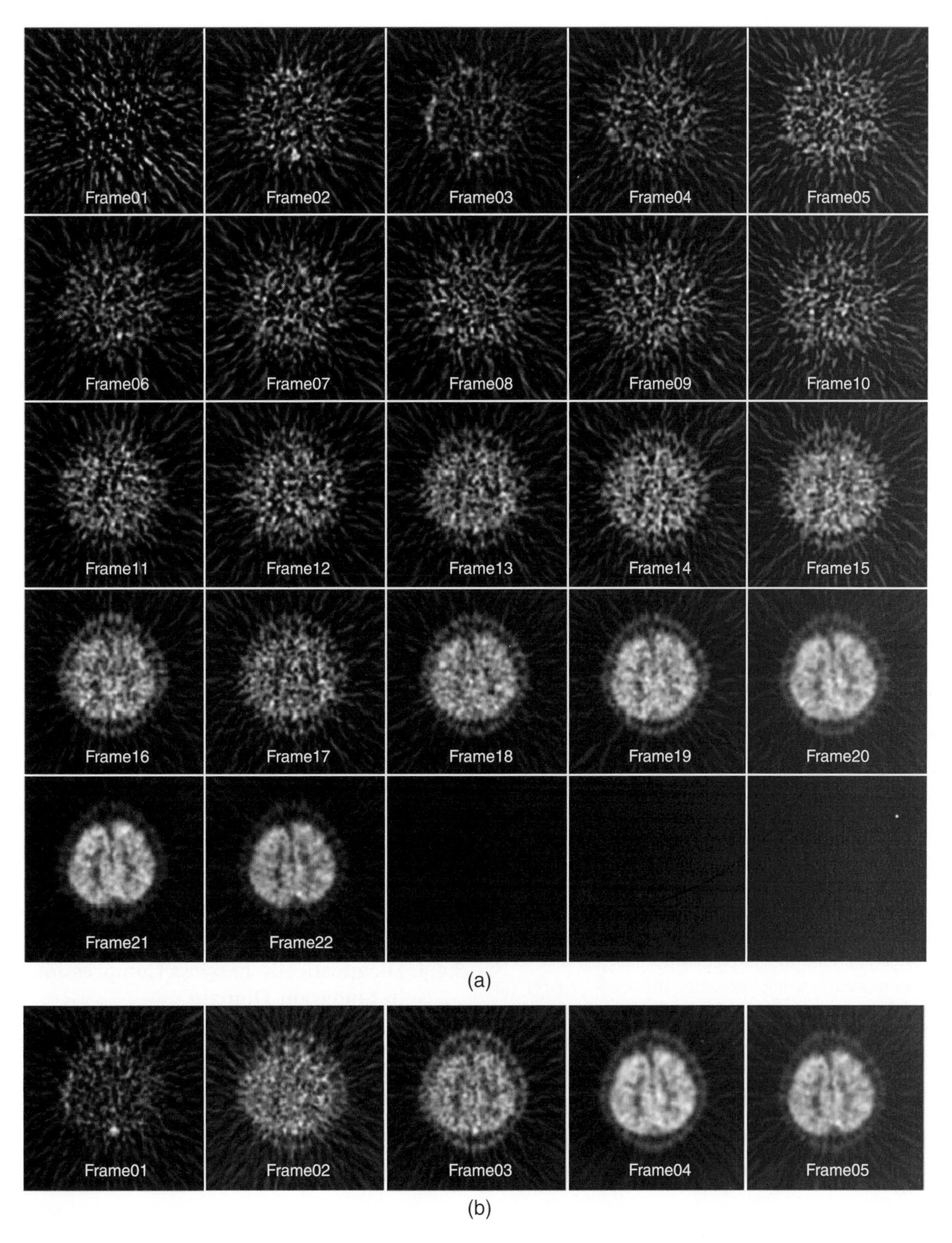

FIGURE 3.20 (a) A set of 22 temporal frame images for the 15th cross-section plane from a patient brain FDG PET study. (b) Results of compressing the images in the temporal domain with the OISS algorithm [87].

redundancies in dynamic functional image data. In terms of removing spatial redundancies in the data, for each cross-section plane, the reduced set of temporal image frames can be further compressed to a single indexed image using cluster analysis. In general, a TTAC can be obtained from each voxel in dynamic PET images. However, many TTAC curves have similar kinetics. Clustering techniques can therefore be adopted to automatically classify pixels into a certain number of typical TTAC types corresponding to different brain regions. The main idea behind clustering algorithms is to group and classify image-wide TTACs, $z_i(t)$ (where $i = 1, 2, \ldots, R$, and R is the total number of image voxels), into S cluster groups C_j (where $j = 1, 2, \ldots, S$, and $S << R$) by measurement of the magnitude of natural association (similarity characteristics). In Li et al. [86], an indirect agglomerative clustering algorithm is used to conduct the data compression in spatial domain, based on the traditional Euclidean distance criterion measure:

$$D^2\left(z_i, \bar{z}_{C_j}\right) = \sum_t \left\lfloor z_i(t) - \bar{z}_{c_j}(t) \right\rfloor^2, \qquad (3.43)$$

where $\bar{z}_{c_j}(t)$ denotes the mean TTAC within each cluster C_j. Here, the clustering analysis technique is applied to further compress the reduced set of temporal frames into:

- A single indexed image that represents a mapping of the cluster groups to their respective pixel time-activity curve (TAC) locations (i.e., the spatial distribution of kinetic behavior)
- An index table that contains the mean TAC for each cluster group

In contrast with other kinds of medical images, dynamic PET images have a consistent general structure, consisting of an approximately oval region containing almost all of the information of interest. Therefore, pixels containing background noise and negative values were suppressed prior to cluster analysis in order to get accurate clustered TTAC results. Using cluster analysis, the reconstructed images have been further compressed in spatial domain, and a CR of 8.6:1 can be gained. Furthermore, PNG (Portable Network Graphics) [100], a well-known standard image lossless compression method, can be used to compress the single indexed image, achieving a further CR of 1.8:1. The PNG coding format was chosen over other lossless image compression methods due to its efficiency, portability, flexibility, and lack of legal encumbrances. In many centers performing clinical dynamic PET studies, the extraction of physiological parameters (i.e., the generation of parametric images) is of major importance. With the proposed compression technique, it can be implemented through:

1. Decompression of the indexed image
2. Tracer kinetic modeling and parameter estimation
3. Pixel-wise mapping

The resultant images obtained from the compressed data correspond to the generated functional images [86]. Figure 3.21 shows the result of applying cluster analysis to the temporal image frames in Figure 3.20(b). Compression does not appear to have degraded image quality and fine structural information of the human brain, while the overall CR obtained for the combined compression approach in temporal and spatial domains is 68.1:1 [87].

In the cluster analysis algorithm pertaining to Figure 3.21, the number of clusters used in dynamic functional image data is a critical issue. A sufficient number of clusters are usually required to ensure that the functional data contained in the dynamic images are adequately represented; however, too many clusters will increasingly reflect the variation in the TTACs due to noise and will increase the heterogeneity in the index image, resulting in increased noise and less scope for compression of the index image. A performance evaluation for compression of dynamic brain FDG PET image data has been conducted in Chen et al. [88], and the optimal number of clusters was shown to be 21~42. For a cluster number of 42, the compression ratio achieved was approximately 87:1, while the minimal practical number of clusters was shown to be 21, which gives a maximum CR of approximately 86:1 [88].

3.4.4 Diagnostically Lossless Compression in Sinogram Domain

Dynamic functional imaging studies using a CSS produce large numbers of temporal image frames that may not provide the

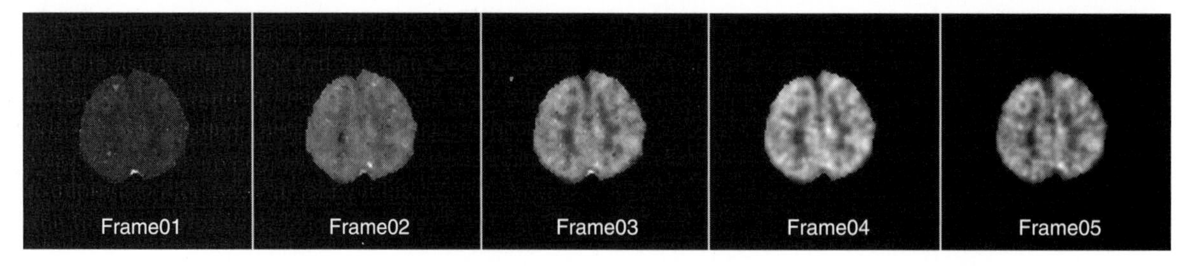

FIGURE 3.21 Results of applying cluster analysis to the temporal image sequence in Figure 3.20(b) [87].

maximum information for the study. In Section 3.4.2, the OISS was shown to greatly compress dynamic functional image data in temporal domain, reducing the number of time samples required to the number of model parameters that are being estimated while providing good parameter estimates. However, the OISS is model dependent and requires an input function, typically from arterial blood sampling, which is complicated to retrieve. Moreover, the OISS is optimized for a specific model rather than for each individual patient data set, and conventional compartment models may not be adequate to describe heterogeneous tumor tissues that require more complex modeling [101, 102]. An alternative, model-independent approach for dimensionality reduction of dynamic PET data involves multivariate data analysis techniques such as PCA [103]. Previous PCA-based temporal compression approaches for dynamic PET data have been applied in the image domain after reconstruction of the sinogram projection data [104, 105]. This requires image reconstruction for every temporal frame in the CSS (typically 22 or more), which imposes a large computational burden and introduces reconstruction errors that can affect the later PCA. An alternative is to apply PCA early to the sinograms, before image reconstruction [89, 106, 107], to reduce the computational cost of image reconstruction and improve quantification.

In Chen et al. [90], a combined temporal and spatial compression technique is proposed for the compression of dynamic functional image data in sinogram domain, including a temporal compression stage based on the application of PCA directly to the sinogram data to reduce the dimensionality of the data, followed by a spatial compression stage using JPEG2000 to each principal-component channel weighted by the signal in each channel. Figure 3.22 illustrates the framework of the combined

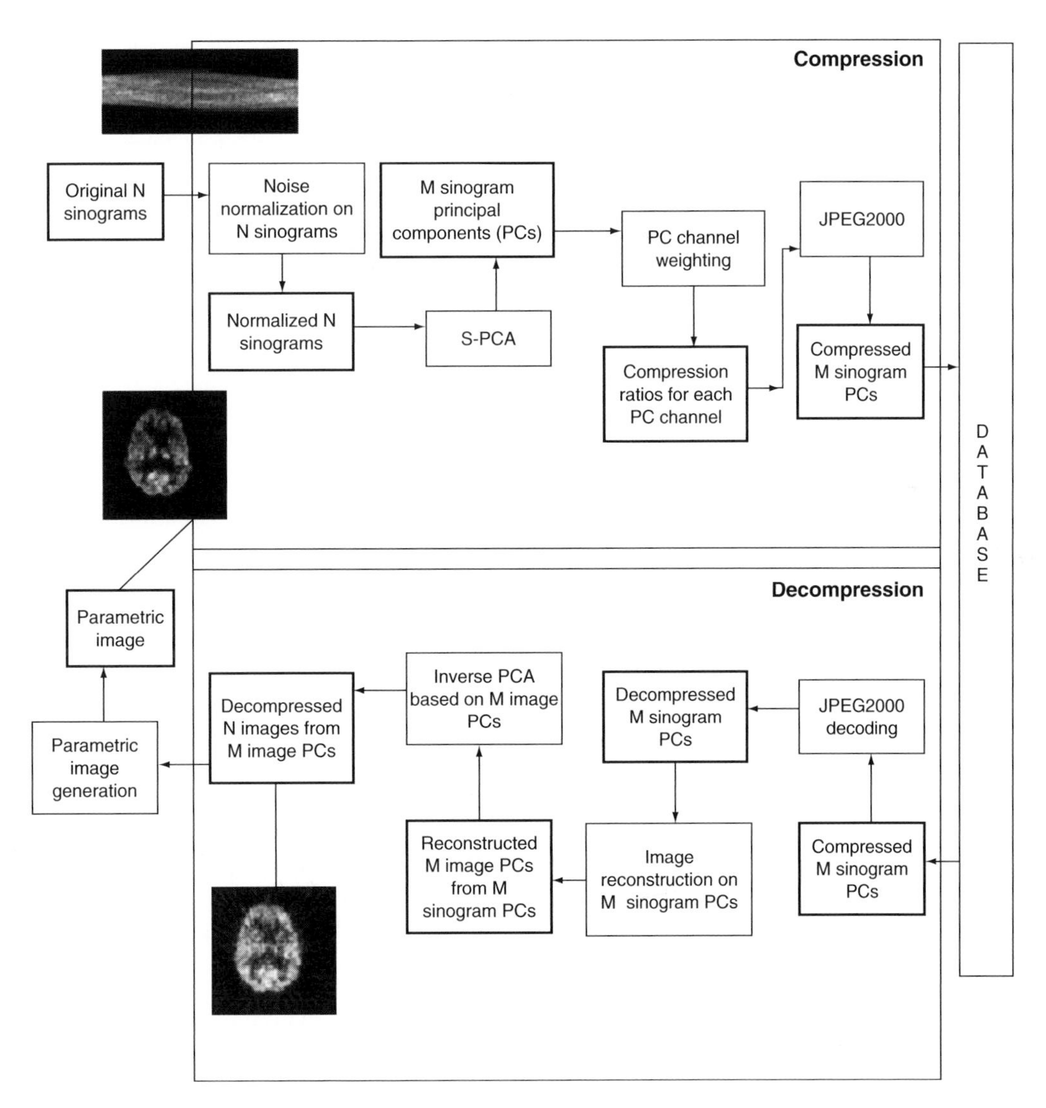

FIGURE 3.22 Dynamic functional image data compression using PCA and channel-weighted JPEG2000 in sinogram domain [90].

data compression technique using PCA and channel-weighted JPEG2000 in sinogram domain, including the following four major coding steps:

1. Sinogram noise normalization
2. Sinogram-domain PCA
3. PCA channel weighting
4. Sinogram-domain JPEG2000 coding, and three major decoding steps:
   a. JPEG2000 decoding
   b. PCA channel reconstruction
   c. Inverse PCA

3.4.4.1 Sinogram Noise Normalization

A PET scanner in 2D mode outputs the initial projection data in the form of N sinograms. These sinogram data are assumed to have been corrected for attenuation, randomness, and differences in detector efficiencies. Since PCA is a data-driven technique that cannot itself distinguish noise from signal, it is necessary to scale or normalize each frame so that each has approximately equal noise variance [108]. Assuming approximate Poisson statistics (attenuation and random-corrected sinograms are no longer exactly Poisson), each temporal sinogram frame is noise-normalized through dividing by $\sqrt{N/\Delta T}$, where N is the total number of detection counts in the sinogram frame, and ΔT is the time duration of the frame. Several other data preprocessing transformations have been developed to normalize the noise in each frame to improve the PCA of the signal [109].

3.4.4.2 Sinogram-Domain PCA

The PCA is applied directly to the time series of N (noise-normalized) sinograms to produce a reduced number of M sinogram principal-component (S-PC) channels and is performed simultaneously on the data from all spatial planes. The objective of PCA is to represent orthogonal maximum variance directions for the analyzed data set. This multivariate image analysis approach is well suited to high-dimensional, highly correlated data such as those of dynamic PET. If there are N frames in the CSS, PCA produces M principal components, where $M \leq N$ and the eigenvalues of the PCA channels are ordered from largest to smallest. Given a random vector population $\mathbf{X}_{sinogram} = (x_1, \ldots, x_n)^T$, $x_1, \ldots, x_n$ in this case represents the individual time samples of the dynamic tomographic study in sinogram domain; the mean vector of the population is defined as $\mu_{sinogram} = E\{\mathbf{X}_{sinogram}\}$; and the covariance matrix is:

$$C = E\{(\mathbf{X}_{sinogram} - \mu_{sinogram})(\mathbf{X}_{sinogram} - \mu_{sinogram})^T\}. \quad (3.44)$$

After eigen-analysis of C, the eigenvalue-eigenvector pairs $(\lambda_1, e_1), (\lambda_2, e_2), \ldots, (\lambda_n, e_n)$ are ordered by eigenvalues in descending order. To reduce the data set, only the first M eigenvectors of the covariance matrix are used to represent the data. Let $\mathbf{A}_{sinogram}$ be a matrix consisting of the first M eigenvectors of the covariance matrix as the row vectors. The transformation of data vector $\mathbf{X}_{sinogram}$ is then given by:

$$\mathbf{P}_{sinogram} = \mathbf{A}_{sinogram}(\mathbf{X}_{sinogram} - \mu_{sinogram}), \quad (3.45)$$

where $\mathbf{P}_{sinogram}$ is in an orthogonal coordinate system defined by the eigenvectors.

3.4.4.3 Principal Components Analysis Channel Weighting

It is required that each principal component in the M sinogram be compressed with different qualities according to its priority (importance of the signal) in the set of principal components. The principal-component channel with the higher priority requires less CR.

3.4.4.4 Sinogram-Domain JPEG2000 Coding

JPEG2000 is applied to each M sinogram principal-component channel with a weighted CR to produce compressed M sinogram principal components. As noted in Section 3.2.8, JPEG2000 is based on the DWT [36] rather than the DCT, which provides significant improvements over JPEG, including progressive decoding by image quality and improved compression efficiency.

3.4.4.5 JPEG2000 Decoding

Decoding is performed on the M JPEG2000 compressed sinogram principal-component channels to regenerate M decompressed sinogram principal-component channels.

3.4.4.6 Principal Components Analysis Channel Reconstruction

A set of M image-domain principal-component channels is reconstructed from the M decompressed sinogram-domain principal components by using an image reconstruction algorithm such as filtered backprojection (FBP) or ordered subset expectation maximization (EM). Define this reconstructed space as R_{image}. An advantage of applying PCA before the image reconstruction stage is that the lowered noise levels in the PCA channels allow for reduced filtering in the FBP algorithm, which reduces blurring and partial volume effects in the final result.

3.4.4.7 Inverse Principal Components Analysis

The inverse of the PCA is performed on the M image principal-component channels to regenerate a time series of N image frames. During sinogram-domain PCA, as described in Section 3.4.4.2, PCA of the sinograms created a transformation matrix

$\mathbf{A}_{sinogram}$ and a mean vector $\mu_{sinogram}$. The inverse PCA transformation in the image domain is then:

$$\mathrm{Inv}(\mathbf{R}_{image}) = (\mathbf{A}_{sinogram})^T \times \mathbf{R}_{image} + \mu_{sinogram}/N_{proj}, \quad (3.46)$$

where N_{proj} is the total number of projection angles in the tomograph. Finally, the inverse of the noise-normalization weighting is performed.

The results of performance evaluation demonstrate that noise-normalized PCA can give equivalent CRs to OISS (to approximately five frames) but with approximately twice the precision. PCA in sinograms avoids introducing image reconstruction errors into the analysis, decreases the computational burden of image reconstruction, and gives similar quantitative accuracy to OISS, as well as better accuracy than image-domain PCA. Figure 3.23 shows the results of a dynamic reconstructed PET series using different data compression methods, in which the signal-to-noise ratio (SNR) of the reconstructed images from both PCA-only and the combined approaches is significantly improved. The improved SNR can be achieved because, after applying noise normalization, the PCA can separate the signal from the noise. It is noted that the reconstructed images from the combined approach are slightly less noisy than those from the PCA-only approach, because JPEG2000 itself has a denoising effect. The results indicate that the combined approach not only can reduce the quantity of data in dynamic PET, but can also improve the image quality of PET. Overall, the combined temporal and spatial compression technique for the compression of dynamic functional image data in sinogram domain can achieve a CR as high as 129:1 while simultaneously reducing noise, improving physiological parameter estimation compared with the uncompressed data, and preserving the sinogram data for later analysis [90].

3.5 Summary

The basics of picture compression were reviewed in this chapter. Information entropy dictates the minimum amount of data needed to carry a certain amount of information. Therefore, any datum that exceeds the entropy level contains

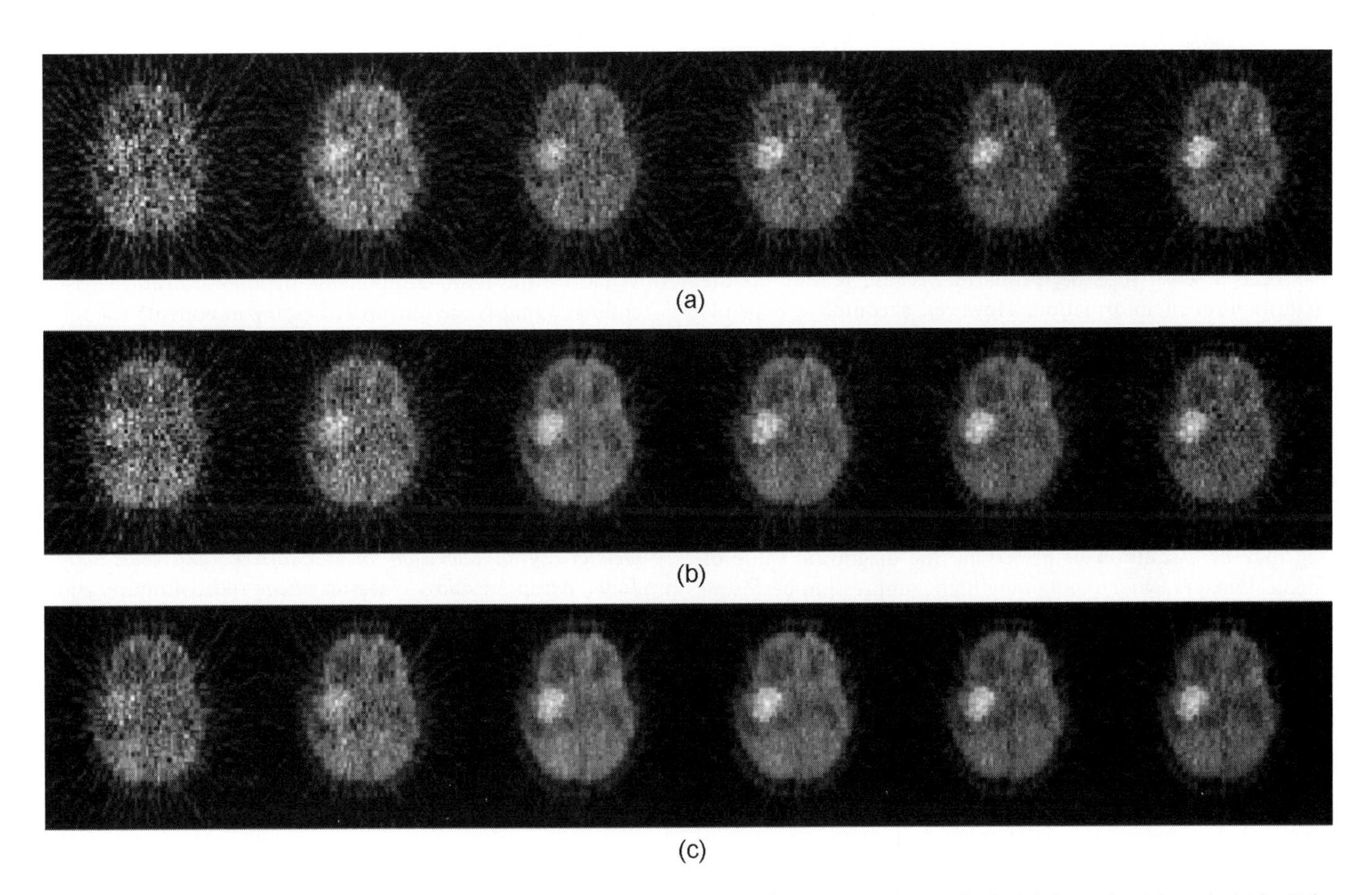

FIGURE 3.23 The results of dynamic reconstructed PET series using different data compression methods: (a) the 11th, 13th, 15th, 17th, 19th, and last frames of the original 22 temporal frames of the CSS; (b) the corresponding reconstructed images from the PCA-only compression approach; (c) the corresponding reconstructed images from the combined compression approach based on PCA and channel-weighted JPEG2000 [90].

redundancies. The purpose of compression is to remove these redundancies. The most rudimentary method for compressing data is through entropy coding. Entropy coders compress data to their source entropy using a variety of algorithms, such as Huffman or arithmetic codes. Most competitive compression strategies employ predictive coding and context-based coding prior to entropy coding to further enhance compression performance.

Compression of pictures is performed either in lossless or lossy manner. Lossless compression is desirable, since it maintains information integrity. However, it has limited CRs. Consequently, it has been more practical in many instances to apply lossy compression. While it has superior compression performance over its lossless counterpart, lossy compression does not preserve all information contained within pictures. In addition, the loss of information is a function of the compression rate as governed by the R–D function. Hence, for lossy coding, it is important to maintain a balance between compression and distortion/quality. The fidelity of pictures is dependent on how much perceptible information is contained within them. The purpose of perceptual coding is to compress pictures based on perceptible information. Consequently, the performance of perceptual coders is generally superior to that of nonperceptual coders in terms of visual quality. Perceptual coders operating at just above the super-threshold vision level produce compressed pictures of the quality level which is deemed imperceptible to their respective original pictures by removing only visually redundant information.

Compression of natural images is a straightforward task that juggles CR and quality. For medical images, lossy compression has been a sensitive issue, primarily because it leads to the deterioration of information. However, a complete dependency on lossless compression is impractical due to its limited compression performance. To mitigate between information loss and compression performance, the idea of preserving diagnostic information as opposed to all information is gaining support. The challenge of this approach, then, is to devise a method for identifying diagnostic information. If diagnostic information is dependent on perception, then perceptual coding may be a solution to preserving the diagnostic value of medical images while maintaining high compression performance. It is important to note that in medical compression, it is impossible to rely entirely on either lossless or lossy coding alone. Therefore, advanced data compression techniques used in noisy medical image data sets with high CRs and improved image qualities were discussed in this chapter. These techniques have pioneered in biomedical diagnostically lossless data compression research. In this chapter, three examples were used: compression in temporal domain, in spatial domain, and in sinogram domain. These new techniques have demonstrated a substantial improvement in image quality while providing a significant reduction in storage space. Additionally, these methods filter out the measurement noise and thus provide more reliable and compact data sets, ready for accurate diagnosis.

Acknowledgements

The authors are grateful to the support from ARC, PolyU/UGC grants. The authors thank Mr. David Wu for providing illustrations of compressed medical images.

3.6 Exercises

1. Given the following probability distribution for alphabet A:

i	1	2	3	4	5	6	7
$A(i)$	a	b	c	d	e	f	g
$p(A(i))$	0.07	0.2	0.1	0.05	0.25	0.13	0.2

   a. Calculate the entropy of A.
   b. Calculate the fixed length code, and compute data redundancy.
   c. Generate the Huffman tree and Huffman codes.
   d. Generate the arithmetic code number for sequence $\{a, e, g, c\}$.
2. What are the advantages and disadvantages of lossless and lossy coding?
3. What is the purpose of predictive coding and context coding?
4. What are the basic components of lossy transform-based coding? What is the purpose of each component?
5. How does aliasing occur?
6. What is perceptual coding?

3.7 References and Bibliography

1. ITU-R Recommendation BT.601-5. *Studio Encoding Parameters of Digital Television for Standard 4:3 and Wide-Screen 16:9 Aspect Ratios.* International Telecommunication Union–Telecommunications Standardization Sector (ITU-T), 1995.
2. N. S. Jayant and P. Noll. *Digital Coding of Waveform: Principles and Applications to Speech and Video.* Prentice Hall, 1984.
3. R. Clarke. *Transform Coding of Images.* Academic Press, 1985.
4. A. K. Jain. *Fundamentals of Digital Image Processing.* Prentice Hall, 1989.
5. R. Clarke. *Digital Compression of Still Images and Video.* Academic Press, 1995.
6. A. N. Netravali and B. G. Haskell. *Digital Pictures—Representation, Compression and Standards.* 2nd ed. Plenum Press, 1995.

7. K. R. Rao and J. J. Hwang. *Techniques and Standard for Image, Video and Audio Coding.* Prentice Hall, 1996.
8. K. Sayood. *Introduction to Data Compression.* 3rd ed. Morgan Kauffman Publishers, 2006.
9. C. E. Shannon. A mathematical theory of communication. *Bell System Technical Journal.* 27:379–423, 623–656, 1948.
10. B. G. Haskell. A. Puri, and A. N. Netravali. *Digital Video: An Introduction to MPEG–2.* Chapman and Hall, 1997.
11. D. A. Huffman. Method for the construction of minimum redundancy codes. *Proceedings of the IRE.* 40(9): 1098–1101, 1952.
12. J. Rissanen. Generalised Kraft inequality and arithmetic coding. *IBM Journal of Research and Development.* 20:198–203, 1976.
13. A. Moffat, R. Neal, and I. H. Witten. Arithmetic coding revisited. *ACM Transaction on Information Systems.* 16(3):256–294, 1998.
14. X. Wu and N. Memon. Context-based adaptive lossless image coding. *IEEE Trans. Comm.* 45(4):437–444, 1997.
15. ISO/IEC JTC 1/SC 29. *Information Technology—JPEG 2000 Image Coding System—Part 1: Core Coding System.* ISO/IEC 15444–1:2000, 2000.
16. ITU–T Recommendation H.264/ISO IEC 11496–10. *Advance Video Coding.* International Telecommunication Union–Telecommunications Standardization Sector (ITU-T)/International Organization for Standardization (ISO), 2002.
17. C. C. Cutler. *Differential Quantization of Communication Signals.* US Patent No. 2,605,361, 1952.
18. C. W. Harrison. Experiment with linear prediction in television. *Bell Systems Technical Journal.* 29:764–783, 1952.
19. W. M. Goodall. Television by pulse coding modulation. *Bell Systems Technical Journal* 28:33–49, 1951.
20. M. Weinberger, G. Seroussi, and G. Sapiro. The LOCO–I lossless image compression algorithm: Principles and standardization into JPEG–LS. *IEEE Trans. Image Proc.* 9(8):1309–1324, 2000.
21. C. E. Shannon. Prediction and entropy of printed English. *Bell System Technical Journal* 30:50–64, 1951.
22. ISO/IEC JTC 1/SC 29. *Information Technology—Lossless and Near-Lossless Compression of Continuous-Tone Still Images: Baseline.* ISO/IEC 14495–1:1999, 1999.
23. S. W. Golomb. Run–length encodings. *IEEE Trans. Inform. Theory.* IT–12:399–401, 1966.
24. J. Capon. A probabilistic model for run-length coding of pictures. *IRE Trans. Inform. Theory.* IT–5:157–163, 1959.
25. R. Dansereau and W. Kinsner. Perceptual image compression through fractal surface interpolation. *Canadian Conf. Elect. Comp. Eng.* 2:899–902, 1996.
26. J. E. Jacquin. A novel fractal block-coding technique for digital images. *Proc. IEEE Int. Conf. Acoustics, Speech and Signal Processing.* 4:2225–2228, 1990.
27. M. Vetterli and J. Kovačević. *Wavelets and Subband Coding.* Prentice Hall, 1995.
28. A. V. Oppenheim, R. W. Schafer, and J. R. Buck. *Discrete-Time Signal Processing.* Prentice Hall, 1999.
29. S. K. Mitra. *Digital Signal Processing: A Computer-Based Approach.* 2nd ed. McGraw-Hill, 2001.
30. J. W. Woods and S. D. O'Neil. Subband coding of images. *IEEE Trans. Acoustics, Speech and Signal Processing.* 34(5):1278–1288, 1986.
31. K. R. Rao and P. Yip. *Discrete Cosine Transform—Algorithms, Advantages, and Applications.* Academic Press, 1990.
32. P. J. Burt and E. H. Adelson. The Laplacian pyramid as a compact image coder. *IEEE Trans. Comm.* 31(4):532–540, 1983.
33. S. G. Mallat. A theory for multiresolution signal decomposition: The wavelet representation. *IEEE Trans. Pattern Anal. Mach. Intel.* 11(7):674–693, 1989.
34. J. Fourier. *Théorie Analytique de la Chaleur.* 1822.
35. V. R. Algazi and D. J. Sakrison. On the optimality of the Karhunen-Loève expansion. *IEEE Trans. Inform. Theory,* IT–15:319–321, 1969.
36. D. Taubman and M. Marcellin (Eds). *JPEG2000: Image Compression Fundamentals, Standards and Practice.* Springer, 2002.
37. S. G. Mallat and Z. Zhang. Matching pursuits with time-frequency dictionaries. *IEEE Trans. Sig. Proc.* 41:3397–3415, 1993.
38. S. Mitra and J. F. Kaiser. *Handbook of Digital Signal Processing.* John Wiley & Sons, 1993.
39. S. P. Lloyd. Least square quantization in PCM. *IEEE Trans. Inform. Theory,* IT–28(2):127–135, 1982.
40. J. Max. Quantizing for minimum distortion. *IRE Trans. Inform. Theory* 6(1):7–12, 1960.
41. A. Gersho and R. M. Gray. *Vector Quantization and Signal Compression.* Kluwer Academic Publishers, 1992.
42. ISO/IEC JTC 1/SC 29. *Information Technology—Generic Coding of Moving Pictures and Associated Audio Information: Video.* ISO/IEC 13818–2:2000, 2000.
43. ISO/IEC JTC 1/SC 29. *Information Technology—JPEG 2000 Image Coding System—Part 3: Motion JPEG 2000.* ISO/IEC 15444–3:2002, 2002.
44. J. R. Ohm. Three-dimensional subband coding with motion compensation. *IEEE Trans. Image Proc.* 3(5): 559–571, 1994.
45. D. Taubman and A. Zakhor. Multirate 3D subband coding of video. *IEEE Trans. Image Proc.* 3(5):572–588, 1994.
46. J. Xu et al. 3D embedded subband coding with optimal truncation (3D ESCOT). *Applied and Computational Harmonic Analysis: Special Issue on Wavelet Applications in Engineering.* 10:290–315, 2001.
47. J. Hua, Z. Xiong, and X. Wu. High-performance 3D embedded wavelet video (EWV) coding. *Proc. IEEE 4th Workshop on Multimedia Signal Processing.* 569–574, 2001.

48. S.-T. Hsiang and J. W. Woods. Embedded video coding using invertible motion compensated 3-D subband/wavelet filter bank. *Sig. Proc.: Image Communication.* 16(8): 705–724, 2001.
49. M. Benetiere et al. Scalable compression of 3D medical datasets using a (2D+T) wavelet video coding scheme. *International Symposium on Signal Processing and its Applications.* 2:537–540, 2001.
50. J. Xin, C. W. Lin, and M. T. Sun. Digital video transcoding. *Proc. IEEE* 93(1):84–97, 2005.
51. D. Taubman. High performance scalable image compression with EBCOT. *IEEE Trans. Image Proc.* 9(7): 1158–1170, 2000.
52. J. R. Ohm. Advances in scalable video coding. *Proc. IEEE* 93(1):42–56, 2005.
53. B. Girod. What's wrong with mean-squared error. In A. B. Watson (Ed). *Digital Images and Human Vision.* MIT Press, 1993.
54. J. J. Hwang, H. R. Wu, and K. R. Rao. Picture coding and human visual system fundamentals. In H. R. Wu and K. R. Rao (Eds). *Digital Video Image Quality and Perceptual Coding.* CRC. 3–43, 2006.
55. B. A. Wandell. *Foundations of Vision.* Sinuar, 1995.
56. G. E. Legge and J. M. Foley. Contrast masking in human vision. *J. Opt. Soc. Am.* 70(12):1458–147l, 1980.
57. A. B. Watson and J. A. Solomon. Model of visual contrast gain control and pattern masking. *J. Opt. Soc. Am.* 14(9):2379–2391, 1997.
58. E. P. Simoncelli et al. Shiftable multi-scale transforms. *IEEE Trans. Inform. Theory.* 38(2):587–607, 1992.
59. A. B. Watson. The cortex transform: Rapid computation of simulated neural images. *Computer Vision, Graphics, and Image Processing.* 39:311–327, 1987.
60. D. M. Tan, H. R. Wu, and Z. Yu. Perceptual coding of digital monochrome images. *IEEE Sig. Proc. Letters.* 11(2):239–242, 2004.
61. Z. Liu, L. J. Karam, and A. B. Watson. JPEG2000 encoding with perceptual distortion control. *IEEE Trans. Image Proc.* 15(7):1763–1778, 2006.
62. J. Limb. On the design of quantisers for DPCM coders—A functional relationship between visibility, probability, and masking. *IEEE Trans. Comm.* 26:573–578, 1978.
63. A. Netravali and B. Prasada. Adaptive quantisation of picture signals using spatial masking. *Proc. IEEE.* 65(4):536–548, 1977.
64. A. B. Watson. DCT quantization matrices visually optimized for individual images. *Proc. Human Vision, Visual Processing, Digital Display IV.* 202–216, 1993.
65. T. P. O'Rourke and R. L. Stevenson. Human visual system based wavelet decomposition for image compression. *J. Vis. Commun. Image Rep.* 6(2):109–121, 1995.
66. D. Shkarin. PPM: One step to practicability. *Proc. Data Comp. Conf.* 202–211, 2002.
67. I. Matsuda, H. Mori, and S. Itoh. Lossless coding of still images using minimum-rate predictors. *Proc. IEEE Int. Conf. Image Proc.* 1:132–135, 2000.
68. W. B. Pennekaker and J. L. Mitchell. *JPEG Still Image Data Compression Standard.* Van Nostrand Reinhold, 1993.
69. D. Santa-Cruz, R. Grosbois, and T. Ebrahimi. JPEG2000 performance evaluation and assessment. *Sig. Proc. Image Commun.* 17(1):113–130, 2002.
70. G. J. Sullivan. Video compression—From concepts to the H.264/AVC standard. *Proc. IEEE* 19(1):18–31, 2005.
71. D. Wu et al. Perceptually lossless medical image coding. *IEEE Transactions on Medical Imaging.* 25(3):335–344, March 2006.
72. J. H. C. Reiber et al. Angiographic and intravascular ultrasound. In *Handbook of Medical Imaging.* J. Beutel, H. L. Kundel, and R. L. van Metter (Eds.)., 2nd ed. SPIE Press, 2004.
73. D. A. Clunie. Lossless compression of grayscale medical images: Effectiveness of traditional and state-of-the-art approaches. *Proc. SPIE Med. Imaging.* 3980:74–84, 2000.
74. NEMA PS 3. *Digital Imaging and Communications in Medicine (DICOM)—Part 5: Data Structures and Encoding.* National Electrical Manufacturers Association, 2006.
75. A. J. Duerinckx and E. J. Pisa. Filmless picture archiving and communication system (PACS) in diagnostic radiology. *Proc. SPIE* 318:9–18, 1982. Reprinted in *IEEE Computer Society Proceedings of PACS'82*, order No 388.
76. S. J. Dwyer III. A personalized view of the history of PACS in the USA. *Proceedings of the SPIE: Medical Imaging 2000: PACS Design and Evaluation: Engineering and Clinical Issues.* 3980:2–9, 2000.
77. ISO/IEC JTC 1/SC 29. *Information Technology—Digital Compression and Coding of Continuous-Tone Still Images: Requirements and Guidelines.* ISO/IEC 10918–1:1994, 1994.
78. S. H. Becker and R. L. Arehson. Costs and benefits of picture archiving and communication system. *J. Amer. Med. Informatics Assoc.* 1(5):361–371, 1994.
79. N. C. Phelan and J. T. Ennis. Medical image compression based on a morphological representation of wavelet coefficients. *Med. Physics.* 26(8):1607–1611, 1999.
80. O. Kocsis et al. Visually lossless threshold determination for microcalcification detection in wavelet compressed mammograms. *European Radiology.* 13(10):2390–2396, 2003.
81. B.-J. Kim and W. A. Pearlman. An embedded wavelet video coder using three-dimensional set partitioning in hierarchical trees (SPIHT). *Proc. Data Comp. Conf.* 251–260, 1997.
82. M. Krol. Telemedicine. *IEEE Potentials* 16(4):29–31, 1997.
83. D. Feng et al. Techniques for functional imaging. In *Medical Imaging Techniques and Applications*, C. T. Leondes (Ed.).

Gordon and Breach International Series in Engineering, Technology and Applied Science. Gordon and Breach Science Publishers, 1997.

84. D. Feng. Information technology applications in biomedical functional imaging. *IEEE Trans. Inform. Tech. Biomed.* 3(3):221–230, 1999.
85. X. Li, D. Feng, and K. Chen. Optimal image sampling schedule: A new effective way to reduce dynamic image storage space and Functional image processing time. *IEEE Trans. Med. Imaging.* 15(5):710–719, 1996.
86. D. Ho, D. Feng, and K. Chen. Dynamic image data compression in spatial and temporal domains: Theory and algorithm. *IEEE Trans. Inform. Tech. Biomed.* 1(4):219–228, 1997.
87. D. Feng, W. Cai, and R. Fulton. Dynamic image data compression in the spatial and temporal domains: Clinical issues and assessment. *IEEE Trans. Inform. Tech. Biomed.* 6(4):262–268, 2002.
88. Z. Chen et al. Performance evaluation of functional medical imaging compression via optimal sampling schedule designs and cluster analysis. *IEEE Trans. Biomed. Eng.* 52(5):943–945, 2005.
89. Z. Chen et al. Temporal processing of dynamic positron emission tomography via principal component analysis in the sinogram domain. *IEEE Trans. Nuclear Science.* 51(5):2612–2619, 2004.
90. Z. Chen, D. Feng, and W. Cai. Temporal and spatial compression of dynamic positron emission tomography in sinogram domain. *Inter. J. Image Graphics.* 5(4):839–858, 2005.
91. Z. Chen. *Biomedical Functional Imaging Data Compression and Analysis.* Ph.D. thesis, University of Sydney, 2003.
92. J. Delforge, A. Syrota, and B. M. Mazoyer. Experimental design optimization: Theory and application to estimation of receptor model parameters using dynamic positron emission tomography. *Phy. Med. Biol.* 34:419–435, 1989.
93. R. A. Hawkins, M. E. Phelps, and S.C. Huang. Effects of temporal sampling, glucose metabolic rates, and disruptions of the blood–brain barrier on the FDG model with and without a vascular compartment: studies in human brain tumours with PET. *J. Cereb. Blood Flow Metab.* 6:170–183, 1986.
94. S. Jovkar et al. Minimization of parameter estimation errors in dynamic PET: Choice of scanning schedules. *Phys. Med. Biol.* 34:895–908, 1989.
95. B. M. Mazoyer et al. Dynamic PET data analysis. *J. Comp. Assist. Tomog.* 10:645–653, 1986.
96. X. Li and D. Feng. Toward the reduction of dynamic image data in PET studies. *Computer Methods and Programs in Biomedicine.* 53:71–80, 1997.
97. D. Z. D'Argenio. Optimal sampling times for pharmacokinetic experiments. *J. Pharmacokinet. Biopharm.* 9:739–756, 1981.
98. D. Feng, W. Cai, and R. Fulton. An optimal image sampling schedule Design for cerebral blood volume and partial volume correction in neurologic FDG–PET studies. *Australia New Zealand J. Med.* 28(3):361, 1998.
99. D. Feng, X. Li, and W. C. Siu. Optimal sampling schedule design for positron emission tomography data acquisition. *Control Eng. Practice.* 5(12):1759–1766, 1997.
100. L. D. Crocker. PNG: The portable network graphic format. *Dr. Dobb's J.* July:36–49, 1995.
101. T. Thireou et al. Performance evaluation of principal component analysis in dynamic FDG-PET studies of recurrent colorectal cancer. *Comp. Med. Imag. Graph.* 27:43–51, 2003.
102. A. H. Andersen, D. M. Gash, and M. J. Avison. Principal component analysis of the dynamic response measured by fMRI: A generalized linear systems framework. *Mag. Res. Imag.* 17(6):795–815, 1999.
103. B. S. Everitt and G. Dunn. *Applied Multivariate Data Analysis.* 2nd ed. Arnold. 2001:48–65, 2001.
104. V. Chameroy and R. D. Paola. High compression of nuclear medicine dynamic studies. *Int. J. Card. Imag.* 5:261–269, 1990.
105. T. Kao, S. H. Shieh, and L. C. Wu. Dynamic radionuclide images compression based on principal components analysis. *Eng. Med. Biol. Soc.* 3:1227–1228, 1992.
106. C. M. Kao, J. T. Yap, and M. N. Wernick. High-resolution reconstruction of dynamic PET image sequences using a low-order approximation. *Proc. SPIE.* 2622:796–801, 1995.
107. M. N. Wernick, E. J. Infusion, and M. Milosevic. Fast spatio-temporal image reconstruction for dynamic PET. *IEEE Trans. Med. Imag.* 18:185–195, 1999.
108. F. Pedersen, M. B. E. Bengtsson, and B. Langstrom. Principal component analysis of dynamic PET and gamma camera images: A methodology to visualize the signals in the presence of large noise. *IEEE Conference Record Nuclear Science Symposium and Medical Imaging Conference.* 3:1734–1738, 1993.
109. M. Samal et al. Experimental comparison of data transformation procedures for analysis of principal components. *Phys. Med. Biol.* 44:2821–2834, 1999.
110. H. Witten, R. M. Neal, and J. G. Clear. Arithmetic coding for data compression. *Comm. ACM* 30(6):520–540, 1987.

4

Content-Based Medical Image Retrieval

Dr. Tom Weidong Cai,[1]
Dr. Jinman Kim,[1] and
Prof. David Dagan Feng[1,2]
[1] *University of Sydney*
[2] *Hong Kong Polytechnic University*

4.1 Introduction

In the past three decades, various medical imaging techniques, as introduced in Chapter 1, have advanced rapidly, providing powerful tools for patient diagnosis, treatment planning, medical reference, and training. Medical image data have been expanded rapidly in quantity, content, and dimension—due to an enormous increase in the number of diverse clinical exams performed digitally and to the large range of image modalities available [1–3]. This development has therefore led to an increased demand for efficient medical image data retrieval and management. In current medical image databases, images are indexed and retrieved mainly by alphanumeric keywords, classified by human experts. However, purely text-based retrieval methods are unable to sufficiently describe the rich visual properties or features of the image content and therefore pose significant limitations on medical image data retrieval [4–6]. The ability to search by medical image content is becoming increasingly important, especially with the current trend toward evidence-based practice of medicine [6, 7].

In this chapter, we present an overview of current techniques of *content-based medical image retrieval* (CBMIR). We first give an introduction to typical generic *content-based image retrieval* (CBIR), including its key components: image feature extraction, similarity comparison, indexing scheme, and interactive query interface; followed by a short review of the major image visual features, such as color, texture, shape, and spatial relationships (Section 4.1.2). Then we briefly address the need for CBMIR and related challenges in Section 4.1.3. The major techniques used in CBMIR are reviewed in detail with four different categories: retrieval based on physical visual features in Section 4.2, retrieval by geometric spatial features in Section 4.3, retrieval by combination of semantic and visual features in Section 4.4, and retrieval based on physiologically functional features in Section 4.5. Conclusions are drawn in Section 4.6.

4.1.1 Fundamentals of Content-Based Image Retrieval

The recent escalating use of digital images in diverse application areas such as medicine, education, remote sensing, and entertainment has led to enormous image archives and repositories that require management and retrieval of effective image data [8–12]. This development is similar to the rapid increase in the amount of alphanumeric data during the early days of computing, which led to the development of database management systems (DBMSs) by organizing data into interrelated collections for convenient information retrieval and storage [11]. Early work on image retrieval was still based on the textual annotation of images; that is, images were first manually annotated by keywords or descriptive text and organized by topical or semantic hierarchies in traditional DBMSs to facilitate easy access based on standard Boolean queries. However, such purely text-based methods posed significant limitations on image retrieval. Manual annotation is subjective, time-consuming, and prohibitively expensive, and the sheer content volume of very large image databases is simply beyond the manual indexing capability of human experts. Furthermore, many visual features in images, such as irregular shapes and jumbled textures, are extremely difficult to describe in text. Such a text-based approach also limits the scope of searching to that predetermined by the author of the system and leaves no means for using the data beyond that scope.

In contrast to traditional text-based approaches that performed retrieval only at a conceptual level, the recently developed CBIR methods support full retrieval by visual content/properties of images, by retrieving image data at a perceptual level with objective and quantitative measurements of the visual content and integration of image processing, pattern recognition, and computer vision [9, 10, 13, 14]. A typical CBIR system is depicted in Figure 4.1.

Firstly, the visual contents for each image in the image database are extracted. This content consists of a set of distinguishing features (a *multidimensional feature vector*) precomputed via an *offline feature extraction process.* The feature vector is then stored in a *feature metadata repository.* To retrieve images, the user submits a *query example image* to the system, and the example image is then converted into an *internal feature vector* via an *online feature extraction process.* The similarity or closeness between the feature vector of the user's query image and the feature metadata items is calculated and ranked during a *similarity comparison* stage. Retrieval is

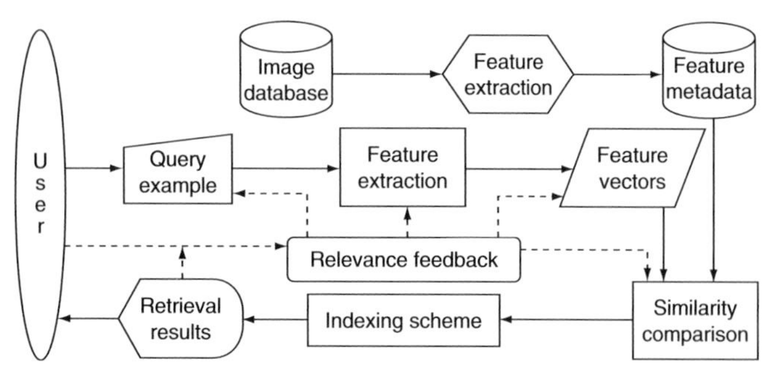

FIGURE 4.1 General architecture of the CBIR system.

performed by applying an *indexing scheme* that can be used to support fast retrieval and to make the system scalable to large image databases. If the *retrieval results* in response to the query are not fully satisfactory, the user can give some positive or negative feedback to the system, and the modified query can be resubmitted via the interactive *relevance feedback.* Such a feedback/retrieval cycle can be repeated until the user is satisfied with the retrieval results. Among these various components, the four key issues in any CBIR system are feature extraction, similarity comparison, the indexing scheme, and the interactive query interface.

4.1.1.1 Feature Extraction

Feature extraction is the basis and most important component of the CBIR system. In a broad sense, features may include both those that are text based (keywords, annotations) and those that are visual (color, texture, shape, spatial relationships). Since there already exists a rich literature on text-based feature extraction in the traditional DBMS and information retrieval research communities, this chapter will be confined to general visual feature extraction. Representation of images needs to consider which features are most useful for representing the contents of images and which approaches can effectively code the attributes of the images. Due to perception subjectivity, there exists no single best representation for any given feature—instead, multiple representations are used to characterize the feature from different perspectives. It is well acknowledged that a good visual feature representation should be invariant to the accidental variance introduced by the imaging process (e.g., the variation of the illuminant of the scene taken in the image). There is a trade-off, however, between the invariance and the discriminative power of the visual feature, since a very wide class of invariance may lose the ability to discriminate among essential differences [13]. A short review of the major image visual features used in CBIR will be given in the next section.

Moreover, a visual feature can be either global or local. If the feature extraction is applied on the whole image, the derived content features then become *global* features. In order to obtain more selective features at a finer resolution, the image is often divided into parts (subareas or homogeneous regions) before features are computed from each part, and this is *local* feature extraction. The easiest way to divide an image is to partition it into equal-size blocks or tiles [13]. Such simple partition, however, does not generate perceptually meaningful or salient regions. A better method is to divide the image into homogeneous regions according to some criterion using segmentation algorithms [14]. This is particularly true in medical imaging studies. It is uncommon that a condition or disease will alter an image over its entire spatial extent; more often than not, diagnostic features of interest manifest themselves in local regions. In a screening situation, the radiologist scans the entire image and searches for features that could be associated with disease;

in a diagnostic situation, however, the medical expert concentrates on the region of suspected abnormality and examines its characteristics to determine whether such a *region of interest* (ROI) exhibits signs related to a particular disease [15].

4.1.1.2 Similarity Comparison

To measure similarity, the general approach is to represent the image features as multidimensional vectors. Selection of metrics has a direct impact on the performance of a CBIR system. A similarity-comparison function maps between pairs of feature vectors and a positive real-valued number, which is chosen to be representative of the visual similarity between two images. Therefore, the retrieval result is not a single image but a list of images ranked by their similarities with the query image. Various similarity-comparison approaches have been developed for image retrieval based on empirical estimates of the distribution of features, including:

- Minkowski form distance (a generic form of the well-known Euclidean distance)
- Mahalanobis distance
- Quadratic form (QF) distance
- Proportional transportation distance
- Earth mover's distance
- Kullback-Leibler divergence (KLD)
- Jeffrey divergence (JD)

Readers are referred to Long et al. [13] for a detailed review.

4.1.1.3 Indexing Scheme

When the number of images in the database is small, a sequential linear search can provide a reasonable performance. However, with large image collections, indexing support for similarity-based queries becomes necessary and can help in avoiding sequential scanning. Index structures ideally filter out all irrelevant images by checking image attributes with the user's query and retain only relevant images without analyzing the entire database. Retrieved images are ranked in order of similarity to a query. Some commonly used multidimensional indexing approaches include:

- Linear quad trees [16]
- k-dimensional binary search (K-D-B) trees [17]
- Holey brick (hB) trees [18]
- Rectangle (R) trees [19, 20] and their variants $R+$-trees [21] and R^*-trees [22]
- X-trees (having X as their leaves) [23]
- Telescopic vector (TV) trees [24]
- Similarity search (SS) trees [25]

Most of these indexing approaches perform reasonably well for a small number of dimensions, but they explore exponentially with increasing dimensionality and eventually reduce to sequential searching. One of the methods commonly used for addressing this problem is the application of dimension reduction techniques, such as *principal components analysis* (PCA), an optimal technique that linearly maps input data to a coordinate space such that the axes are aligned to maximally reflect the variations in the data [26]. Some very good reviews and comparisons of various indexing techniques can be found in White and Jain [27] and Ng and Sedighian [28].

4.1.1.4 Interactive Query Interface

An interactive retrieval interface allows the user to formulate and modify queries. The ability of users to express their search needs accurately and easily is crucial in any CBIR system. The most appealing paradigm in many ways is *query by example*: providing a sample of the kind of output desired and asking the system to retrieve further examples of the same kind. Virtually all current CBIR systems now offer such searching, where the user submits a query image and the system retrieves and displays thumbnails of some closest-matching images in the database. However, the user will not always have an example image on hand. Several alternative query formulation approaches have been proposed, such as Aslandogan and Yu [29]:

- Category browsing
- Simple visual feature query
- Feature combination query
- Localized feature query
- Query by sketch
- User-defined attribute query
- Object relationship query
- Concept query

The ability to refine searches online in response to user indications of relevance—known as *relevance feedback*—is particularly useful for improving the effectiveness of CBIR systems interactively [30, 31]. The main idea of relevance feedback is to use positive and negative examples from the user to improve system performance. For a given query, the retrieval system first returns a list of ranked images based on predefined similarity metrics. The user then labels the retrieved images that are relevant to the query as positive examples and those not relevant to the query as negative examples. The system will subsequently refine the retrieval results based on the user's feedback by a certain learning algorithm and return a new ranked list of images to the user. This process can continue to iterate until the user is satisfied.

Relevance-feedback strategies help to alleviate the semantic gap [14] between low-level visual features and high-level semantic features, since it allows CBIR systems to learn users' image perceptions. The learning algorithm usually deals with small training samples (typically less than 20 per round of interaction), asymmetry in training samples (a few negative examples are normally fed back to the system), and a requirement that the algorithm be fast enough to support user interaction in real time [32, 33]. Commonly used relevance-feedback learning algorithms include:

- Genetic algorithms [34]
- Weight-based learning approaches [35]
- Bayesian probabilistic methods [36]
- Support vector machines (SVMs) [37]

4.1.2 Image Features in Content-Based Image Retrieval

Clearly, the key to the CBIR framework lies in feature extraction, in which the quantitative image features, computed automatically, are used to characterize image content. In essence, image features can be classified into (1) general visual features and (2) domain-specific semantic features. *General visual features* typically include primitive image information that refers to the constituents and composition of an image, such as color, texture, shape, and spatial relationships. *Domain-specific semantic features,* on the other hand, are application dependent and consist mainly of abstract information that refers to the "meaning" of an image, describing high-level image semantic content in specialized domains.

This section will briefly introduce the general visual features that can be used in most CBIR applications. Readers are referred to [8–10, 12, 14, 29] for detailed reviews on visual feature extraction. Typically, general visual features can be further classified into (a) *physical visual features,* including color and texture, and (b) *geometric spatial features,* such as shape and spatial relationships. (The domain-specific semantic features, which can be obtained either by textual annotation or by complex inference procedures based on image visual content, will be covered in Section 4.4, concentrating on the medical domain application, CBMIR.)

4.1.2.1 Color

Color is the most frequently used general visual feature for CBIR due to its invariance with respect to image scaling, translation, and rotation and to its three color-component values (e.g., red/green/blue [RGB], hue/saturation/value [HSV], CIE $L^*a^*b^*$ or CIE $L^*u^*v^*$ [luminance/chrominance values set by the Commission Internationale d'Éclairage]), which make its discrimination potential superior to the single-gray intensities of images [8, 9, 13]. There are many color features that have been developed for CBIR, such as:

- *Color histogram:* A most effective color representation, with the distribution of the number of pixels for each quantized color bin located in three different color components [38, 39]
- *Color moments:* Very compact color representations, with three low-order moments (mean, variance, and skewness) for each color component [26, 40]
- *Color coherence vectors* (CCVs): Incorporating spatial information into the color histogram (histogram refinement) [41]
- *Color correlogram:* A color descriptor characterizing both color distributions of pixels and the spatial correlation of pairs of colors [42]
- *HDS-S (hue/diff/sum–structure):* A color structure descriptor for capturing local color image structure based on the MPEG-7 HMMD (hue-min-max-difference) color space [43, 44] (see Section 4.2.1)

4.1.2.2 Texture

Texture is a powerful discriminating visual feature that has been widely used in pattern recognition and computer vision for identifying visual patterns with properties of homogeneity that cannot result from the presence of only a single color or intensity [45]. Texture presents almost everywhere in nature. The size of the image patch and the number of distinguishable gray-level primitives and the spatial relationships between them are all interrelated elements that characterize a texture pattern [10]. Some commonly used texture features are:

- *Co-occurrence matrices,* with 14 texture descriptors for capturing the spatial dependence of gray levels [46]
- *Tamura features,* with six visual texture components designed in accordance with psychological studies of the human perception of texture [47]
- *Run-length matrices,* for quantifying the coarseness of texture in specified directions [48]
- *Wavelet transform coefficients,* representing frequency properties of texture patterns, including pyramid-structured and tree-structured wavelet transforms [49, 50]
- *Gabor filters,* as orientation and scale tunable edge and bar/line detectors [51, 52]
- *Wold decomposition,* providing perceptual properties with three components: harmonic (repetitiveness), evanescent (directionality), and indeterministic (randomness) [53, 54]
- Markov random field (MRF) [55–57]
- Fourier power spectrum [58]
- Fractal dimension [59]
- Shift-invariant principal components analysis (SPCA) [60]

4.1.2.3 Shape

Shape can be used to identify an object or region as a meaningful geometric form. To humans, perceiving a shape means capturing prominent/salient elements of the object or region [10]. Therefore, shape features in an image are normally represented after that image has been segmented into objects or regions. Due to the difficulties in fully automated image segmentation [13] and the variety of ways a given 3D object can be projected into 2D shapes in 2D images, CBIR based on shape features is considered to be one of the most challenging tasks and has usually been limited to specific applications where objects or regions are readily available [14]. In general, shape representation techniques fall into two broad categories: (1) boundary-based and (2) region-based approaches. *Boundary-based* approaches work on the outer boundary of the shape, and the shape descriptors in this category include:

- A *Fourier descriptor,* which describes the shape of an object with the Fourier transform of its boundary [61, 62]
- A *turning function,* for comparing both convex and concave polygons [63]
- A *finite element method* (FEM), with a stiffness matrix and its eigenvectors [64]
- A curvature scale space (CSS) [65, 66]
- Chord-length statistics [67]
- Chain encoding [68]
- Beam angle statistics (BAS) [69]
- A wavelet descriptor [70]

Shape descriptors commonly used in *region-based* approaches include:

- Invariant moments, a set of statistical region-based moments [71, 72]
- Zernike moments [68]
- Generalized complex moments [73]
- Morphological descriptors [74]

4.1.2.4 Spatial Relationships

Spatial relationships between multiple objects or regions in an image usually capture the most relevant and regular part of the information in the image content [10, 13] and are very useful for image retrieval and searching. Spatial relationships can be divided into (1) directional (or orientation) relationships and (2) topological relationships. *Directional* relationships capture relative positions of objects with respect to each other, such as "left," "above," and "front," and are usually calculated through objects' centroids or barycenters. *Topological* relationships describe neighborhood and incidence between objects, like "disjoint," "adjacency," and "overlapping," and are calculated via objects' shapes. The most commonly used approach to describe spatial relationships is the *attributed relational graph* (ARG) [75, 76], in which objects are represented by graph nodes, and the relationships between objects are represented by arcs between such nodes. Another well-known approach, called the *2D strings* method, is based on symbolic projection theory and allows a bidimensional arrangement of a set of objects to be encoded into a sequential structure [77]. In addition to the ARG and 2D strings methods, *spatial quad trees* [78] and *symbolic images* [79] are used to represent spatial relationships.

4.1.3 Content-Based Image Retrieval in the Medical Domain

In medicine, the majority of acquired medical images are currently stored with a limited text-based description of their content. As image databases expand, it is becoming increasingly apparent that these simple text-based descriptions are inadequate for the proper search and retrieval of medical images. As a consequence, valuable diagnostic and prognostic information in such databases remains unusable, and the demand increases for efficient retrieval techniques that can tap the expertise contained in these databases [80]. In previous sections of this chapter, CBIR has been shown to be a viable alternative to text-based image retrieval, with the ability to search for an image depending on metrics for comparing image with visual/spatial properties that can match human judgments of similarity. It is therefore very natural to apply CBMIR by retrieving medical images according to their domain-specific image features, providing an important alternative and complement to traditional text-based retrieval.

The potential benefits of CBMIR range from clinical decision support to medical education and research [6]. Clinical knowledge has shown that visual characteristics of medical images have strong effects on diagnosis [81]. Therefore, diagnosis by comparing past and current medical images associated with pathological conditions has become one of the primary approaches in case-based reasoning or evidence-based medicine [7], while the clinical decision-making process can be beneficial to find other images of the same modality or the same anatomical region of the same disease [6]. CBMIR can aid in such diagnoses in the following way: After observing an abnormality in a diagnostic image, a physician can query a database of known cases to retrieve images (and associated textual information) that contain regions with features similar to that observed in the image of interest. With the knowledge of disease entities that match features of the selected region, the physician can be more confident of the diagnosis and may be able to expand the differential diagnosis to include pathological entities not previously considered. Here, CBMIR is the source of relevant supporting evidence from prior known cases, providing the physician trained in its use with set examples that are close to his/her decision boundary, along with the correct class labels (proven pathology) for these examples. The less-experienced practitioner can also benefit from this expertise in that the retrieved images, if visually similar, can serve the role of an expert consultant [82–84]. CBMIR could be used to present cases that are not only similar in diagnosis, but also similar in appearance, and in cases with visual similarity but different diagnoses. It would therefore be useful as a training tool for medical students, residents, and researches to browse and search large collections of disease-related illustrations using their visual attributes [6]. The success of CBMIR will open up many new vistas in medical services and research, such as disease tracking, differential diagnosis, noninvasive surgical planning, clinical training, and outcomes research [85].

Albeit the need is clearly identified, developing CBMIR systems imposes several distinct challenges compared with CBIR for general images, and demands a thorough understanding of the nature and requirements of medical images. The following are ways in which this understanding is most useful [2, 5, 6, 85–88]:

1. Medical image data are *heterogeneous* in how they are collected, distributed, and displayed. Images are acquired from different modalities and in different settings in terms of position, resolution, contrast, and signal-to-noise ratio. Inside one modality, the tuning of an imager may lead to significantly varying images—for example, a magnetic resonance imaging (MRI) scan may be used for acquisition of completely different anatomical and functional information, or an imaging scan by positron emission tomography (PET) or single photon emission computed tomography (SPECT) may be conducted for different organ studies with different tracer settings. A hospital can generate thousands of diverse images of different modalities for different clinical studies every day. Retrieval from these heterogeneous images is much more complex compared with a single image modality, and the existing CBIR approaches hardly address all medical image properties.
2. Medical images, except histological, dermatoscopic, and endoscopic, and tongue color images, are intensity-only images, represented in grayscale. The color features frequently used in CBIR thus cannot be applied.
3. Medical images are usually of low resolution and high noise and are therefore difficult to automatically analyze for extracting visual features.
4. Many human organs are made of soft tissue or are nonrigid bodies. The diseased tissue often has no regular shape, and there is no clear boundary with respect to the surrounding healthy part. Automatic and accurate image segmentation of these organs or lesion regions is difficult to achieve.
5. A large fraction of medical images capture human anatomy, which is three-dimensional and thus provides additional information not available in 2D images. Therefore, an additional, complex registration procedure is required before implementation of volumetric image comparison and image feature extraction.
6. Staging of the disease state and the monitoring of patient progress over time are fundamental to diagnostic and therapeutic decisions and to outcome assessment in long-term follow-up. The CBMIR system is required to have the ability to define and track *temporal relation* in the medical image sets of a patient taken at different periods, together with the medical history of that patient.
7. Careful treatment of medical images is required due to issues of patient privacy and other legal constraints. Such security and administrative barriers hinder CBIR research within the medical domain [85].

In recent years, various CBMIR approaches have been developed and integrated mainly in research prototypes. In a broad sense, these retrieval techniques can be classified into four different categories according to the key image features used:

- Retrieval based on physical visual features such as color and texture
- Retrieval based on geometric spatial features such as shape, 3D volumetrics, and spatial relationships
- Retrieval by combination of semantic and visual features, including semantic pathology interpretation approaches and generic model-based methods
- Retrieval based on physiologically functional features such as the dynamic activities of glucose metabolism in human brain images

Details of these techniques will be given in subsequent sections.

4.2 Content-Based Medical Image Retrieval by Physical Visual Features

4.2.1 Retrieval Based on Color

As mentioned in Section 4.1.2, color is the most extensively used low-level feature for CBIR. However, since the majority of medical images are intensity-only images carrying less information than color images, color-based retrieval would be applicable to medical images based only on light photography, where color is an inherent feature and any deviations or changes in the color of a particular sample from a normal sample can have significant medical implications [4, 89]. Medical images that can benefit from color-based CBMIR include:

- Histological images
- Dermatoscopic images
- Endoscopic images
- Tongue images, in which the well-known color moments and color histogram approaches are often adopted

Histological images are taken via light microscopy and can be used for assisting the pathologist to observe and analyze the fine details of biological cells and tissues. These images usually possess unique color signatures, including various subtle changes in color, such as in jaundice, congestion, and pigmentation; and in exudation and effusion [4, 90]. In Mattie et al. [80], the mean optical density and color moments to RGB components were calculated separately for the cytoplasm, nucleus, and nucleolus and adopted as specific local color descriptors in a content-based cell image retrieval system. As a significant color descriptor, the histogram of color distribution has been used in retrieving cytohistological breast carcinoma images [91], breast cancer biopsy slide images [92], and microscopic pathology images of the prostate, liver, and heart [93]. In Tang et al. [94], the local RGB color histograms were applied as the coarse features for prepartitioned high-resolution histological images of the gastrointestinal (GI) tract. Since color characteristics in stained tissue images are prominent within these coarse structures, the extracted color histograms make themselves an ideal coarse detector for iconic content analysis and retrieval, especially when an image database contains a large number of high-resolution images [94–97].

Dermatoscopic imaging is a technique that allows microscopic examination of skin lesions and has already proven to be an effective tool for analysis of skin erythema [98], evaluation of wound status [99], and early detection of skin cancers [100]. Skin color is produced by a combination of complex mechanisms and is used as vital information in dermatology to interpret the characteristics of a lesion and its depth in skin [101]. In Rahman et al. [100], a CBMIR system was developed as a diagnostic aid to dermatologists for skin cancer recognition. A 64-dimensional color histogram consisting of four uniformly quantized bins for each color channel and two color moments (mean and variance) is extracted from the segmented skin lesion and used to provide specific local features for retrieval of color variation.

Endoscopic images, on the other hand, are taken by a lighted tubelike instrument with a camera, which is placed in the GI tract for viewing abnormalities such as bleeding, growth of tumors, polypoid lesions, and ulcerations [102]. A content-based endoscopic image retrieval (CBEIR) system with color clustering was recently presented in Xia et al. [103]. In order to reduce the color sensitivity to noise and the color histogram dimension, and considering that endoscopic images generally contain only a few dominant colors (such as red, yellow, and purple) and HSV color space is most approximate to human perception, the original 24-bit RGB color images are converted into HSV color space, in which the HSV components have been nonuniformly quantified into six, four, and eight levels, respectively. Therefore, color features of endoscopic images can be represented by a clustered 192-bin ($6 \times 4 \times 8$) histogram. Based on some preliminary studies of upper GI tract endoscopic images, Kim et al. [104] proposed an endoscopic image analysis system based on a domain-specified color model and the color variations in the endoscopic images of the stomach for detecting and retrieving abnormal regions such as sites of early gastric cancer and inflammatory changes in the surroundings caused by rubor, erosion, intestinal metaplasia, or atrophic gastritis [104, 105]. Since most endoscopic images have redness on the whole due to the influence of hemoglobin (a predominant pigment in the GI mucosa), the detailed colors of an endoscopic image can be determined by the amount of hemoglobin; that is, the distribution of blood flow in the mucous membrane, using IHb (index of hemoglobin) [106, 107]. Such IHb can be calculated from the values of the original RGB channels (Vr, Vg, and Vb) [107]: $\text{IHb} = 32\{\log_2(\text{Vr}/\text{Vg})\}$ and used as a unique color feature descriptor of the GI endoscopic images.

Similarly, Tjoa and Krishnan [108] proposed a new color feature extraction approach for the classification of colon status from colonoscopic images. In general, colonoscopic images contain rich color information associated with various lesions; for example:

1. Malignant tumors are normally inflated and inflamed, and the inflammation is usually reddish and more severe in color than the surrounding tissues.
2. Benign tumors exhibit less intense hues.
3. Redness may specify bleeding.
4. Blackness may be treated as indication of deposits due to laxatives.
5. Green may mean the presence of fecal material.
6. Yellow relates to pus formation [108].

Based on these properties, some local features are extracted from the chromatic and achromatic histograms of the image. In the histograms of each image, certain lower- and upper-threshold values of the ROIs are selected for the extraction of the quantitative parameters. The color features are defined as follows:

$$\beta_C = \sum_{i=L_1}^{L_2} Hist_C(i) \Big/ \sum_{i=0}^{L-1} Hist_C(i), \tag{4.1}$$

where β_C is a set of the color features for various image components $C = \{\text{I(Intensity), R(Red), G(Green), B(Blue), H(Hue), S(Saturation)}\}$; $Hist_C(i)$ is the histogram amplitude at level i of a particular color component C; L is the number of gray levels; and L_1 and L_2 are the lower- and upper-threshold values of the histogram of the region [108].

Tongue diagnosis—that is, inspection of the tongue—is one of the most important methods in traditional Chinese medicine (TCM), where a physician visually examines the color and other properties of the substance and coating of the tongue. It has been shown to be a simple and noninvasive way to identify the body condition/state and symptoms of the patient and can be further combined with the other three major TCM diagnostic methods—listening/smelling, inquiry, and palpation—for determining the actual disease or deficiency in the patient [89, 109–111]. However, tongue diagnosis in TCM is usually based on the capacity of the physician's human vision for detailed discrimination. Environmental factors such as the difference of light sources and surrounding illuminations may affect the physician in making accurate diagnoses via observation of the tongue. Therefore, it is vital to build an objective and quantitative tool/system—*computerized tongue characterization*—for acquiring true-color tongue images that are invariant to illumination changes and for supporting automatic analysis of tongue properties [111–113]. Various tongue diagnostic systems have been recently developed to support acquisition, processing, analysis, storage, and retrieval of tongue images and to be useful aids in standardization and automation of tongue diagnoses [111, 114–118]. The key components of these systems normally include a color charge-coupled device (CCD) camera, standardized light sources, a semi-closed box/chest with a head/face supporting mechanism, and a computer system with image processing tools and image databases. Some sample tongue images captured from the systems are shown in Figure 4.2 [111, 119, 120].

According to the general principles of TCM, the tongue body can be divided into five subregions, each of which corresponds to the health conditions of different organs of the human body, as shown in Figure 4.2(c): Subregions A1 (root) and A2 (middle/center) represent the health conditions of the stomach and spleen; subregion A3 (tip) corresponds to the heart and lungs; and subregions A4 (right side) and A5 (left side) indicate the health information of the liver and gall bladder [111, 117].

Given the fact that there are large variations in the colors of tongue substance and coating, in Shen et al. [111] a total of 15 color categories are set up to run color feature extraction—the colors of the tongue substance are divided into six different classes (Figure 4.3a), while the coating colors are divided into nine different classes (Figure 4.3b) [111, 119]. Based on subregions A1 to A5 and the 15 color categories, the local color features of tongue images were extracted by estimating the distribution of substance (and coating) of each color category in each subregion of the segmented tongue body; these have been used for tongue image analysis and retrieval in a *tongue image analysis instrument* (TIAI) system.

Moreover, in order to obtain color features that are more consistent and correspond to human vision, color reproduction was conducted for image capture, transfer, and display [121]. Some other systems for tongue image analysis and retrieval include:

- A tongue-computing model (TCoM) for the diagnosis of appendicitis using a total of 22 local metrics in four color spaces (RGB, CIE Y xy, CIE Luv, and CIE Lab) with color moments (mean and variance) for different subregions of the tongue body [114]
- A vision-based tongue diagnosis system using the local (block-size) RGB color mean metric [115]
- A tongue diagnosis supporting system based on quantized color class labels [116]
- A computerized tongue examination system (CTES) using color relaxation with decision boundaries for HSV color space [117]
- A digital tongue inspection system (DigiTIS) based on an RGB color histogram [118]

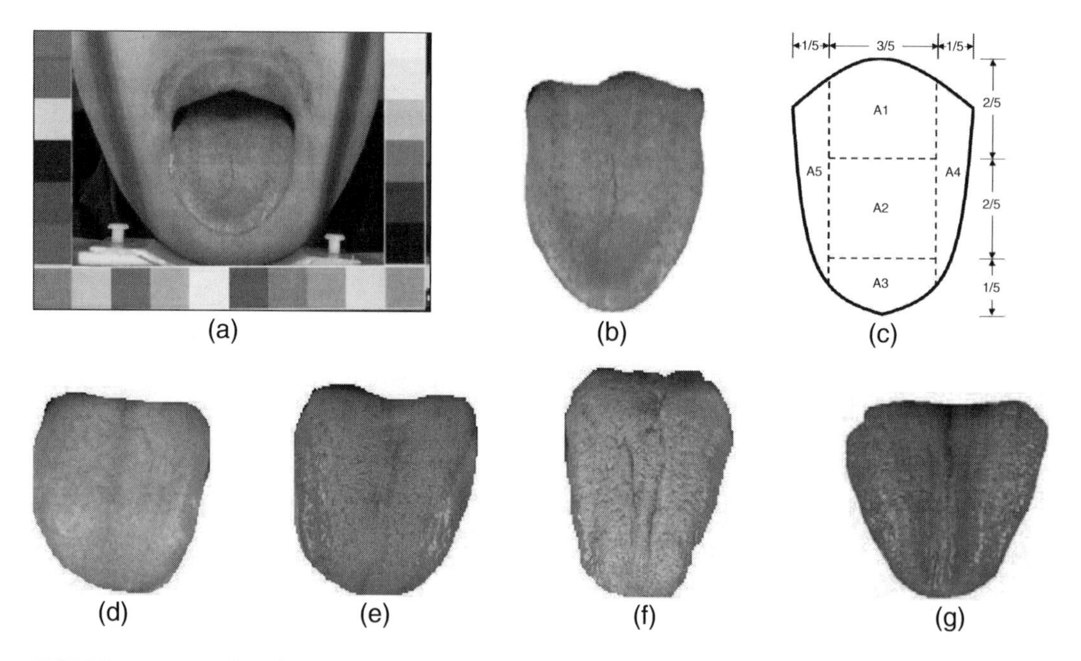

FIGURE 4.2 Examples of tongue images after color calibration. (a) An original tongue image with standard color bars for color calibration. (b) A segmented normal tongue body image. (c) The partition of a tongue: five subregions corresponding to the health condition of different organs. (d) A light red tongue with thin white coating. (e) A dark purple tongue with thin white coating. (f) A light white tongue with yellow thick coating. (g) A deep red tongue with brown coating. Courtesy of Prof. L. S. Shen, Beijing University of Technology, China.

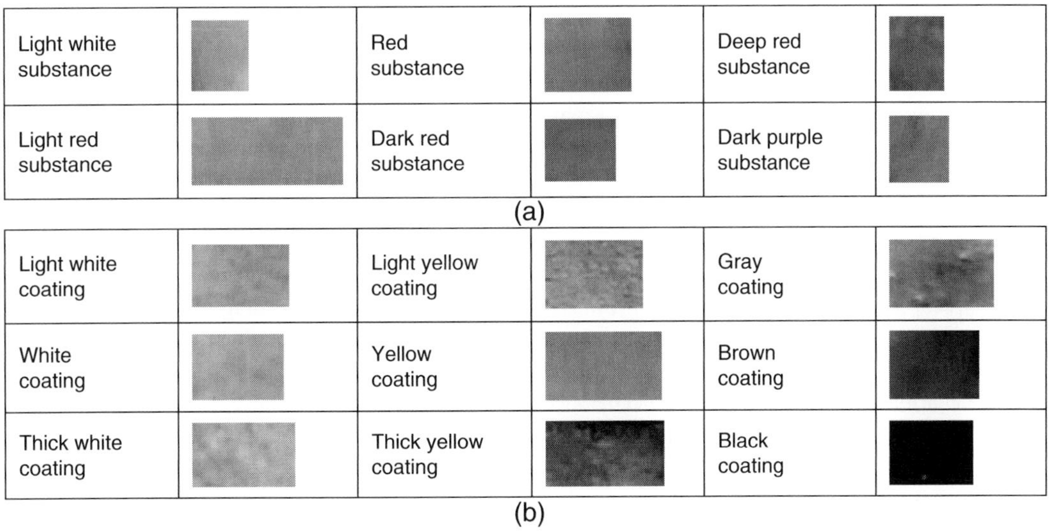

Light white substance		Red substance		Deep red substance	
Light red substance		Dark red substance		Dark purple substance	

(a)

Light white coating		Light yellow coating		Gray coating	
White coating		Yellow coating		Brown coating	
Thick white coating		Thick yellow coating		Black coating	

(b)

FIGURE 4.3 Sample images of 15 different color categories: (a) six color classes of the tongue substance; (b) nine color classes of the tongue coating. Courtesy of Prof. L. S. Shen, Beijing University of Technology, China.

For generic CBMIR—that is, retrieving varied medical images acquired from various image modalities, including the color medical images mentioned, color-based retrieval is vital and needs to be included in the systems. [122] simply used a different quantization of the HSV space with six hues, three saturations, and three values for retrieving color medical images in CasImage [123]. Howarth et al. [43], on the other hand, proposed a new color structure descriptor, HDS-S (see Section 4.1.2.1), for retrieval of varied medical images from the ImageCLEFmed [124] collection. In the HMMD color model, the hue component is from HSV, and the min and max components are taken from the minimum and maximum values, respectively, in the RGB model. The diff component is the difference between min and max. Based on the MPEG-7 standard [44], the HMMD space was unevenly quantized into 184 bins in the three-component HDS (hue, diff, sum) color space. Then an HDS-S color structure descriptor was extracted from a quantized HDS histogram based on an 8×8 sliding window—each color bin contains the number of window positions for which there is at least one pixel falling into the bin under consideration. Since the proposed HDS-S can discriminate between images with the same global color distribution but differing local color structures, it can benefit the retrieval of color medical images that tend to have similar overall color but differing color structure corresponding to intrinsic image content [43].

4.2.2 Retrieval Based on Texture

Most medical images acquired and displayed in grayscale are often highly textured, and consequently, examination of medical images usually requires interpretation of tissue appearance; that is, the local intensity variations, based on different texture properties, such as smoothness, coarseness, regularity, and homogeneity [1, 125]. Since texture acquires such distinguished importance, it is becoming one of the most commonly used characteristics in medical image analysis, classification, and retrieval [1, 6, 126]. Among the various texture descriptors introduced in Section 4.1.2, mainly co-occurrence matrices and Gabor filters are adopted in CBMIR.

The well-known co-occurrence matrices approach for texture feature representation explores the gray-level spatial dependencies of texture by constructing co-occurrence matrices based on different orientations and distances among image pixels and extracting meaningful statistics from the matrices as texture representations [8]. In the co-occurrence matrices approach, given a distance d at an orientation angle θ, $p_{(d,\theta)}(l_1, l_2)$, the (l_1, l_2) coefficient of the corresponding matrix $\mathbf{P}_{(d,\theta)}$ is the co-occurrence count or probability of going from a gray level l_1 to another gray level l_2 with an intersample spacing of d along the axis making an angle θ with the x axis. If the number of distinct gray levels in the quantized image is L, then the co-occurrence matrix $\mathbf{P}$ will be of size $L \times L$. For computational efficiency, the number of gray levels can be reduced by binning, which is a simple procedure in which the total range of values is divided by a smaller amount—the required number of bins—thus "shrinking" the co-occurrence matrix. Different co-occurrence matrices can be constructed by mapping the gray-level probabilities based on the spatial relations of pixels at different angular directions specified by θ while scanning the image according to the distance d [15, 46, 87, 127]. On their own, these co-occurrence matrices do not provide any measure of texture that can easily be used as descriptors. The information in the matrices needs to be further extracted as a set of feature values. In Haralick et al. [46], a total of 14 second-order statistical quantities called *Haralick texture features* are computed out of the coefficients. In order to allow images of different sizes to be compared, before extracting these features, all co-occurrence matrices are normalized by dividing each coefficient in a matrix by the sum of all elements. The formulas for calculating these texture features are listed in Table 4.1.

The co-occurrence matrices approach, which provides enough discrimination power for various texture appearances, has been frequently used in CBMIR [4–6]. Tsang et al. [128] proposed a texture-based image retrieval system for normal anatomical regions presented in CT studies of the chest and abdomen by using local and global co-occurrence texture descriptors. To extract global-level features, 20 co-occurrence matrices for the presegmented images were first constructed for four different orientations (θ as horizontal 0°, vertical 90°, and two diagonals 45° and 135°) and five displacements ($d = 1, 2, 3, 4, 5$). Then 10 Haralick features were computed for each of the 20 matrices. The obtained 20 values for each feature were further averaged and recorded as a mean-based feature vector for the corresponding image. For local (pixel)-level feature extraction, the co-occurrence matrix construction was based on setting a 5×5 neighborhood window for each pixel within the segmented region without the orientation and displacement settings; that is, only a single co-occurrence matrix was produced for each pixel rather than for each choice of θ and d. Therefore, for each co-occurrence matrix (each pixel), the same set of Haralick features was calculated. In Raicu et al. [127], the above 10 Haralick features and 11 texture descriptors extracted by run-length matrices [48, 125] were calculated and stored in a texture dictionary for classification and retrieval of images of various human organ tissues, including the backbone, heart, kidney, liver, and spleen.

In Orphanoudakis et al. [129], various texture descriptors were computed based on a gray-tone co-occurrence matrices method associated with regions of interest and were used as key image content descriptors in a so-called AttributeMatch retrieval system for MRI head scans—for instance, for the retrieval and tracking of images of Alzheimer's disease [130]. Haralick descriptors have also been shown to be effective for the characterization of intrinsic texture features in high-resolution CT (HRCT) images of lung [82]. In Felipe et al. [131], a tool for texture extraction called TextEx was developed

TABLE 4.1 The 14 Haralick texture features

Feature	Equation	Feature	Equation
F_1: (energy) Angular second moment (ASM)	$\sum_{l_1=0}^{L-1}\sum_{l_2=0}^{L-1}[p(l_1,l_2)]^2$	F_8: Sum entropy	$-\sum_{k=0}^{2(L-1)} p_{x+y}(k)\log[p_{x+y}(k)]$
F_2: Contrast	$\sum_{k=0}^{L-1}k^2 \sum_{l_1=0}^{L-1}\sum_{l_2=0,\ \lvert l_1-l_2\rvert=k}^{L-1} p(l_1,l_2)$	F_9: Entropy	$-\sum_{l_1=0}^{L-1}\sum_{l_2=0}^{L-1} p(l_1,l_2)\log[p(l_1,l_2)]$
F_3: Correlation	$\dfrac{\sum_{l_1=0}^{L-1}\sum_{l_2=0}^{L-1} l_1 l_2 p(l_1,l_2)-\mu_x\mu_y}{\sigma_x\sigma_y}$	F_{10}: Difference variance	Variance of p_{x-y}
F_4: Sum of squares (variance)	$\sum_{l_1=0}^{L-1}\sum_{l_2=0}^{L-1}\left(l_1-\mu_f\right)^2 p(l_1, l_2)$	F_{11}: Difference entropy	$-\sum_{k=0}^{L-1} p_{x-y}(k)\log[p_{x-y}(k)]$
F_5: Inverse difference moment	$\sum_{l_1=0}^{L-1}\sum_{l_2=0}^{L-1}\frac{1}{1+(l_1-l_2)^2}p(l_1, l_2)$	F_{12}: Information measure A of correlation	$\frac{HXY-HXY_1}{\max\{HX, HY\}}$
F_6: Sum average	$\sum_{k=0}^{2(L-1)} kp_{x+y}(k)$	F_{13}: Information measure B of correlation	$(1-\exp[-2(HXY_2-HXY)])^{1/2}$
F_7: Sum variance	$\sum_{k=0}^{2(L-1)}(k-F_6)^2 p_{x+y}(k)$	F_{14}: Maximal correlation coefficient	$\sqrt{\text{second largest eigenvalue of } Q}$

Notations: μ_x, μ_y, σ_x, and σ_y are the means and standard deviations of C_x and C_y, respectively;

$$p_x(l_1) = \sum_{l_2=0}^{L-1} p(l_1,l_2);\ p_y(l_2) = \sum_{l_1=0}^{L-1} p(l_1,l_2);\quad p_{x+y}(k) = \sum_{l_1=0}^{L-1}\sum_{\substack{l_2=0\\ l_1+l_2=k}}^{L-1} p(l_1,l_2)\ \text{where } k = 0, 1, 2, \ldots, 2(L-1);$$

$$p_{x-y}(k) = \sum_{l_1=0}^{L-1}\sum_{\substack{l_2=0\\ \lvert l_1-l_2\rvert=k}}^{L-1} p(l_1,l_2)\ \text{where } k = 0, 1, 2, \ldots, L-1;\ Q(l_1,l_2) = \sum_{k=0}^{L-1}\frac{p(l_1,k)p(l_2,k)}{p_x(k)p_y(k)};$$

$$HXY = F_9;$$

$$HXY_1 = \sum_{l_1=0}^{L-1}\sum_{l_2=0}^{L-1} p(l_1,l_2)\log[p_x(l_1)p_y(l_2)];\quad \text{and } HXY_2 = -\sum_{l_1=0}^{L-1}\sum_{l_2=0}^{L-1} p_x(l_1)p_y(l_2)\log[p_x(l_1)p_y(l_2)]$$

to index and retrieve CT and MRI scans based on six texture descriptors extracted by the co-occurrence matrices method, supporting tissue identification of brain, spine, heart, lung, breast, adiposity, muscle, liver, and bone.

Besides the applications in texture-based retrieval of CT and MRI scans, the co-occurrence matrices approach has also been used in retrieving ultrasound images that contain various granular texture layouts [132, 133], mammography images with benign and malignant masses [134], tongue images with various texture layouts at different subregions of the tongue body [114,117,121], and dermatoscopic images with differential texture structures specific to skin lesions [100]. For texture-based retrieval of color endoscopic images, a color co-occurrence model [135] was set up to integrate color and texture characteristics for efficient image retrieval. The original RGB color space was first converted to HSV space and rebinned to 192 levels (H: 6 × S: 4 × V: 8). Then 12 color co-occurrence matrices with size 192 × 192 were constructed for four different orientations (0, $\pi/4$, $\pi/2$, $3\pi/4$) and three different distances (1, 3, 9). Finally, two statistics (mean and standard derivation) of four Haralick descriptors (CONtrast, ENerGy, CORrelation, and ENTropy) were computed from each matrix to form a textual feature vector $< \mu_{CON}, \sigma_{CON}, \mu_{ENG}, \sigma_{ENG}, \mu_{COR}, \sigma_{COR}, \mu_{ENT}, \sigma_{ENT} >$ [135].

Howarth et al. [43] and Rahmann et al. [88] reported the use of the co-occurrence matrices approach for retrieving large collections of medical images from ImageCLEFmed and from World Wide Web medical image atlases, such as X-rays of the chest with enlarged heart; sagittal or frontal views of head MRIs; chest CTs with micronodules; abdominal CTs with liver blood vessels; angiograms of aortas; and microscopic images of leukemia, Alzheimer's disease, bacterial meningitis, skin lesions, etc. Unlike another well-known method, that of Tamura features [47], which provides only visually meaningful texture properties, the co-occurrence matrices approach allows detection of some abnormalities in medical images that are beyond human appreciation of complexity and are otherwise difficult to determine by other texture extraction methods, and it provides valuable information about medical images that may not be visible to the human eye [127]. In these CBMIR systems, a subset of Haralick texture features is normally selected and adopted, based on experiments, and these features have exhibited the best performance among the total of 14 features. The most efficient and frequently used descriptors in the CBMIR are: Energy, Entropy, Contrast, Inverse Difference Moment, Correlation, and Variance.

Gabor filters have been commonly adopted as powerful edge/line/bar detectors with orientation and scale (frequency) tunable properties, and their statistics in the image or image parts (regions) are often used to characterize the underlying texture information while achieving minimum joint 2D uncertainties in both spatial and frequency domains [52]. Since Gabor filters allow one to choose arbitrary orientation and scale, and considering that textural images are usually distinguishable with orientation and scale features, Gabor filters have also been widely used to extract texture features from images for CBIR [51, 52, 136]. A 2D Gabor function is a Gaussian-modulated sinusoid defined as:

$$g(x, y) = \left(\frac{1}{2\pi\sigma_x\sigma_y}\right) \exp\left[-\frac{1}{2}\left(\frac{x^2}{\sigma_x^2}+\frac{y^2}{\sigma_y^2}\right) + 2\pi jW_x\right], \tag{4.2}$$

where σ_x and σ_y are the standard deviations of the Gaussian-shape envelop along the x and y directions, respectively, and W_x is the modulation frequency of the filter. A set of self-similar Gabor filters is then generated via scaling (m) and orientation (n):

$$g_{mn}(x, y) = a^{-m} g(x', y'), \tag{4.3}$$

where $a > 1$; $x' = a^{-m}(x \cos\theta + y \sin\theta)$; $y' = a^{-m}(-x \sin\theta + y \cos\theta)$; $\theta = n\pi/K$; $m = 0, 1, \ldots, S-1$; $n = 0, 1, \ldots, K-1$; and S and K are the total number of scales and of orientations, respectively. Based on the obtained Gabor filters, given an image $I(x,y)$, its Gabor transform is defined to be:

$$W_{mn}(x, y) = \int I(x, y) g^*_{mn} (x - x_1, y - y_1) dx_1 dy_1, \tag{4.4}$$

where * indicates the complex conjugate. Then the mean μ_{mn} and the standard deviation σ_{mn} of the magnitude $|W_{mn}|$ form a feature vector, $< \mu_{00}, \sigma_{00}, \ldots, \mu_{mn}, \sigma_{mn}, \ldots, \mu_{S-1\,K-1}, \sigma_{S-1\,K-1} >$, representing the texture features of the image. Here, the whole image is decomposed at S scales and K orientations by using the Gabor filters and is then ready for texture-based indexing and retrieval.

The application of Gabor filters in CBMIR has been reported and covered in various medical image categories such as cardiac MRIs [87]; CT liver images [137]; histological images of 10 organs [138] and the GI tract [95]; mammography images [134]; and very large collections of various other medical images [43, 122]. In Glatard et al. [87], a bank of 42 Gabor filters with angular bandwidths of 30° and frequency bandwidths of one octave were used to compute the mean and standard deviation of the magnitude response of each heart MRI scan for extracting feature vectors of myocardium texture, supporting CBMIR queries such as: "Given one vertical slice, find slides corresponding to a given instant in time along one cardiac cycle (in particular the end of systole or the end of diastole)." The basis of such myocardium texture-based retrieval lies in the fact that the contraction of the myocardium is correlated to the fineness of its texture presented in the image: (1) The more the myocardium is contracted, the finer its texture will be; and (2) the contraction of the myocardium corresponds to a reduction of its volume, and its fibers then lie more closely together [87].

A CBMIR system for CT liver images based on Gabor texture is presented in Zhao et al. [137], allowing retrieval of different types of CT findings and manifestations such as low attenuation with infiltration, lipiodol retention, and multifocal nodular types. However, all hepatic pathology-bearing patches and other abnormal regions need to be manually delineated before Gabor feature extraction. For high-resolution histological image retrieval, since texture patches in histological images are normally not homogeneous, it is difficult to directly use global Gabor features to image indexing and retrieval. Instead, the original image is usually divided into subimages or blocks, on the hypothesis that texture patterns within a block are homogeneous. Gabor filters are then used to extract texture features for these patterns. Finally, a histological image can be represented with a finite number of feature vectors, each corresponding to one block [95, 138]. In Lam et al. [95], a total of 15 semi-fine Gabor features based on a 64×64 window size were used for retrieving histological images of six GI tract organs. For the CBMIR system presented in Zhao et al. [138], a Gabor filter bank has 18 filters of three scales and six orientations, which are calculated as texture features to index and retrieve histological images from 10 organ categories: adrenal gland, heart, kidney, liver, lung, pancreas, spleen, testis, thyroid, and uterus.

Gabor features have also been adopted for generic CBMIR; that is, retrieving medical images acquired from various image modalities. Muller et al. [122] investigated the potential of using a bank of real circularly symmetric Gabor filters with three scales and four orientations for retrieving varied medical images from CasImage teaching files collections. The resultant 12 Gabor filters have been shown to give good coverage of the frequency domain and little overlap between filters. However, the performance of these Gabor features has not yet been compared with other characteristics, such as those based on the co-occurrence matrices approach or wavelet filters. Another generic CBMIR study [43] concluded that a Gabor filter bank with two scales and four orientations gave the best performance for retrieving varied medical images from the ImageCLEFmed collection.

4.3 Content-Based Medical Image Retrieval by Geometric Spatial Filters

4.3.1 Retrival Based on Shape

Shapes found in medical images express different characteristics for different parts of the anatomy. Some possess readily identifiable shapes, such as those of the brain, heart, lungs, kidneys, and several bones, while others can be arbitrary, such as those of lesions or tissues. Disease processes can affect the structure of organs and cause deviation from their expected shapes. Even abnormal entities tend to demonstrate differences in shape between benign and malignant conditions [15]. Furthermore, shapes can undergo nonrigid deformation over time, over the progression of disease, or from patient to patient. Therefore, shape information becomes one of the most important and effective criteria in characterizing many pathologies identified by medical experts. Medical image retrieval by shape features appears promising for quickly finding the same anatomical region of the same disease and can be beneficial in supporting certain disease diagnoses. Many CBIR techniques based on shape features that exhibit and retain different shape characteristics have been developed, as mentioned in Section 4.1.2. Some of them seem to be well fitted to certain CBMIR applications, while several new approaches have been proposed for particular shapes that populate the databases in some specific medical imaging studies.

The well-known *Fourier descriptors* (FDs) represent the shape in a frequency domain with the Fourier transform of its boundary signatures and can be used to discriminate different shapes. The lower-frequency FDs contain information about the general shape, while the higher-frequency FDs hold information about smaller details of the shape [62]. The most general form of representation of a contour (2D shape) can be just a closed sequence of N successive boundary points or pixels (x_i, y_i), where $i = 0, 1, 2, \ldots, N-1$. Three main signatures of the contour are defined: (1) curvature, (2) centroidal distance, and (3) complex coordinate functions.

The curvature function at a point i along the contour is defined as the rate of change in tangent direction of the contour: $C_i = d\theta_i/d_i$, where θ_i is the turning function of the contour, that is, $\theta_i = \arctan(y_i'/x_i')$, here $y_i' = dy_i/d_i$ and $x_i' = dx_i/d_i$. The centroidal distance function expresses the distance of the boundary points from the centroid (x_c, y_c) of the shape: $R_i = [(x_i - x_c)^2 + (y_i - y_c)^2]^{1/2}$. And the complex coordinate function can be obtained by simply representing the coordinates of the boundary points as complex numbers: $Z_i = (x_i - x_c) + j(y_i - y_c)$. Through Fourier transforms of these three functions, three sets of complex coefficients (i.e., FDs) can then be generated. In order to achieve rotation invariance (since contourencoding is irrelevant to the choice of the reference point), only the amplitudes of the complex coefficients are used, and the phase components are discarded. Scale invariance is achieved by dividing the amplitudes of the coefficients by the amplitudes of the discrete cosine (DC) descriptor or the first nonzero frequency coefficient. And the translation invariance is obtained directly from the contour representation [13, 62]. The FDs of the curvature are

$$FD_C = \left[|F_1|,|F_2|,\ldots,|F_{M/2}|\right], \tag{4.5}$$

and the FDs of the centroidal distance are

$$FD_R = \left[\frac{|F_1|}{|F_0|},\frac{|F_2|}{|F_0|},\ldots,\frac{|F_{M/2}|}{|F_0\ |}\right], \tag{4.6}$$

where F_i denotes the *i*th component of the Fourier transform coefficients. The FDs of the complex coordinate are:

$$FD_Z = \left[\frac{|F_{-(M/2-1)}|}{|F_1|}, \ldots, \frac{|F_{-1}|}{|F_1|}, \frac{|F_2|}{|F_1|}, \ldots, \frac{|F_{M/2}|}{|F_1|}\right], \quad (4.7)$$

where F_1 is the first nonzero frequency component used for normalizing the transform coefficients.

The study in Antani et al. [139] showed that the FD-based retrieval approach could achieve better performance for spine X-ray images with vertebral shapes, and effectively described various pathologies identified by medical experts as being consistently and reliably found in the image collection. The shapes of cervical and lumbar vertebrae were first segmented from the digitized spine X-ray images, and each shape was then outlined by the coarse radiologist-marked nine-point model that is widely used in the vertebra morphometry community. The nine-point shapes were further made dense to 36 points by linear interpolation. The resulting dense segmented vertebra boundary shapes could be used to extract Fourier descriptors as shape attributes. The nature of spine X-ray images in grayscale, offering very little in terms of texture for the anatomy of interest, makes this an ideal application area for CBMIR by FDs.

Brodley et al. [82] presented a CBMIR system for lung HRCT images, in which FDs were extracted from different *pathology-bearing regions* (PBRs) with seven types of diseases: centrilobular emphysema, paraseptal emphysema, sarcoid, invasive aspergillosis, broncheitasis, eosinophilic granuloma, and idiopathic pulmonary fibrosis identified by radiologists. The local shape attributes using FDs have been demonstrated to significantly improve retrieval performance in the domain of HRCT images of the lung over a purely global approach. Shape-based retrieval by FDs has also been proven to be an effective tool for pathology images and can assist physicians in differential diagnosis of lymphoproliferative disorders [140]. Various nucleus shapes segmented from original leukocyte images were characterized through FDs and used for efficient retrieval of different types of leukocytes (band neutrophils, lymphocytes, monocytes, and polymorphonuclear leukocytes) and three particular disorders: chronic lymphocytic leukemia, follicular center cell lymphoma, and mantle cell lymphoma [140].

Recalling the character of FDs, lower-frequency FDs describe the general shape property, while higher-frequency FDs reflect shape details; that is, FDs will become larger if the object shape becomes more complex and rough. It appears that such FDs will be ideal in helping to perform retrieval of mammography images and distinguish accurately between benign masses and malignant tumors, due to the following observation: Benign masses are normally round or macrolobulated in appearance and are well circumscribed with smooth contours; malignant tumors, on the other hand, usually possess rough, jagged, or irregular boundaries, including microlobulations, spicules, and concavities [141, 142]. Rangayyan et al. [141] investigated the use of various shape descriptors such as FDs, compactness, invariant moments, acutance measure, and chord-length statistics to discriminate between benign and malignant breast tumors. Given that some benign lesions may have a speculated appearance and that round/well-defined malignant lesions do exist, an evaluation study of the use of the different shape factors we have discussed to distinguish between circumscribed and speculated tumors was also conducted in Rangayyan et al. [141]. The results showed that the FD method gave higher accuracy for circumscribed/speculated classification, but less accuracy than *acutance measure* for benign/malignant classification. Overall, acutance measure achieved the best performance for breast cancer classification and retrieval. As presented in Rangayyan et al. [141] and Rangayyan and Elkadiki [143], acutance measure is a descriptor of the sharpness or change in density across a mass margin and can be obtained by first computing the sum of differences $d(j)$ along the normal to each boundary point $j = 0, 1, \ldots, N-1$, where N is the number of boundary points of the region:

$$d(j) = \sum_{i=1}^{n_j} \frac{f(i) - b(i)}{2i}. \quad (4.8)$$

Here, $f(i)$ and $b(i)$ are pixels along the normal inside and outside, respectively, of the region, and $i = 1, 2, \ldots, n_j$, where n_j is the number of pixel pairs along the normal used to calculate the differences for the *j*th boundary pixel and is limited to a predefined maximum value $n_{\max}$. The acutance measure A can then be derived by normalizing $d(j)$ over all boundary pixels:

$$A = \frac{1}{d_{\max}}\left[\frac{1}{N}\sum_{j=0}^{N-1}\frac{d^2((j))}{n_j}\right]^{1/2}, \quad (4.9)$$

where $d_{\max}$ is a normalization factor dependent upon the maximum gray-level range and $n_{\max}$, such that A is within the range (0, 1]. In Rangayyan et al. [141], acutance measure was used as a descriptor of edge strength or diffusion of a breast tumor or mass into the surrounding regions, in which low values indicated malignant tumors, while high values implied benign masses.

Unlike FDs and acutance measure, *invariant moments* can be computed from a region's boundary or silhouette. The former is more sensitive to high-frequency edge details, while the latter is less sensitive to noise and is an indicator of gross shape [141]. The 2D $(p+q)$th-order central moments μ_{pq} of a density distribution function $f(x,y)$ are derived as:

$$\mu_{pq} = \sum_x \sum_y (x - x_c)^p (y - y_c)^q f(x, y), \quad (4.10)$$

where (x_c, y_c) is the center of the region, and the summation is over all pixels in the region boundary. The central moments can be further normalized for scale invariance: $\eta_{pq} = \mu_{pq}/\mu_{00}^r$, where $\gamma = (p+q+2)/2$. A set of seven low-order central moments invariant to translation, rotation, and scale can be obtained [144, 145] as follows:

$$
\begin{aligned}
M_1 &= \eta_{20} + \eta_{02} \\
M_2 &= (\eta_{20} - \eta_{02})^2 + 4\eta_{11}^2 \\
M_3 &= (\eta_{30} - 3\eta_{12})^2 + (3\eta_{21} - \eta_{03})^2 \\
M_4 &= (\eta_{30} + \eta_{12})^2 + (\eta_{21} + \eta_{03})^2 \\
M_5 &= (\eta_{30} - 3\eta_{12})(\eta_{30} + \eta_{12}) \\
&\quad \left[(\eta_{30} + \eta_{12})^2 - 3(\eta_{21} + \eta_{03})^2\right] \\
&\quad + (3\eta_{21} - \eta_{03})(\eta_{21} + \eta_{03}) \\
&\quad \left[3(\eta_{30} + \eta_{12})^2 - (\eta_{21} + \eta_{03})^2\right] \\
M_6 &= (\eta_{20} - \eta_{02})\left[(\eta_{30} + \eta_{12})^2 - (\eta_{21} + \eta_{03})^2\right] \\
&\quad + 4\mu_{11}(\eta_{30} + \eta_{12})(\eta_{21} + \eta_{03}) \\
M_7 &= (3\eta_{21} - \eta_{03})(\eta_{30} + \eta_{12}) \\
&\quad \left[(\eta_{30} + \eta_{12})^2 - 3(\eta_{21} + \eta_{03})^2\right] \\
&\quad - (\eta_{30} - 3\eta_{12})(\eta_{21} + \eta_{03}) \\
&\quad \left[3(\eta_{30} + \eta_{12})^2 - (\eta_{21} + \eta_{03})^2\right].
\end{aligned}
\tag{4.11}
$$

The invariant moments have been adopted for retrieving HRCT lung images [82] and mammography images [134]. The evaluation study in Alto et al. [134] showed that shape-based retrieval by invariant moments could be used to classify breast tumors and achieve similar performance as FDs. In Zhu and Schaefer [146], a shape-based retrieval system for thermal medical images was developed by using a set of combinations of invariant moments:

$$
\begin{aligned}
\beta_1 &= \frac{\sqrt{M_2}}{M_1}, \quad \beta_2 = \frac{M_3\mu_{00}}{M_1 M_2}, \quad \beta_3 = \frac{M_4}{M_3}, \\
\beta_4 &= \frac{\sqrt{M_5}}{M_4}, \quad \beta_5 = \frac{M_6}{M_1 M_4}, \quad \beta_6 = \frac{M_7}{M_5}.
\end{aligned}
\tag{4.12}
$$

Such combinations can be found to achieve invariance not only to translation, rotation, and scale, but also to contrast [147]. Each thermal image is then characterized by the six invariant moments in Equation 4.12: $\Phi = [\beta_1, \beta_2, \beta_3, \beta_4, \beta_5, \beta_6]$. Image retrieval is performed by finding those images whose invariant moments are closest to the ones calculated for a given query image, and the results have shown that this approach is very robust for thermal images of the arms, neck, lower back, dorsal view, and legs [146, 148].

Other shape-based retrieval techniques used in CBMIR include:

- Retrieval of MRI heart scans based on the simple turning-function approach [149]
- Retrieval of angiograms and MRI scans with mass/centroid/dispersion features [150]
- Retrieval of dental radiographic images using the finite element method (FEM) and eigendecomposition approach [151]
- Retrieval of mammogram images using Zernike moments [152], as well as morphological features for fast search of tumor shapes [153]
- Retrieval of pathology images based on integrated region matching (IRM) [154]
- Retrieval of functional-MRI (fMRI) brain scans using concentric circle (CC) features extracted from wavelet-decomposed ROI subsets [81]
- Retrieval of varied medical images using MPEG-7's contour shape descriptor [155], salient point detector [156], optimal dynamic time warping (DTW) approach [157], convex hull model [158], and similarity flooding approach [159]

4.3.2 Retrieval by 3D Volumetric Features

The majority of medical images capture human anatomy that is 3D in structure. Three-dimensional medical images can be used for nondestructive inspection of the body and its component regions *in vivo* and *in vitro*, supporting execution and monitoring of interventions and providing quantitative measurements to determine whether an abnormality is present by comparison with normal controls in diagnosis and treatment planning [1, 3, 160]. However, feature extraction and retrieval of 3D medical images in most of the existing CBMIR systems are still performed based on 2D slices—simply following the way of conversional image retrieval in the 2D domain and departing from their originally obtained 3D form. It is believed that more accurate CBMIR, with more discriminating power, can be achieved if we can take full advantage of the information available in the 3D spatial domain by performing retrieval of these medical images based on their 3D volumetric features [161].

Liu et al. [162] proposed a CBMIR approach based on 3D *ideal midsagittal plane* (iMSP) features for retrieving 3D CT neuroimages of hemorrhage (blood), bland infarct (stroke), and normal brain. A basic observation from neuroradiologic imaging is that normal human brains exhibit an approximate bilateral symmetry that is often absent in pathological brains. Given this observation, a symmetry detector was first constructed to automatically extract an iMSP—a virtual geometric plane about which the 3D anatomical structure captured in a brain image presents maximum bilateral symmetry [163]. The basic idea is to find where the MSP is supposed to be for a given 3D brain image, especially for pathological brains, where the anatomical MSP is often distorted (shifted or bent) due to large lesions. Automatically locating and retrieving possible lesions (e.g., bleeds, stroke, tumors) can therefore be conducted by detecting asymmetrical regions with respect to the extracted iMSP.

After the iMSP is aligned in the middle of each 3D volumetric image, a stack of cross-sectional 2D slices are temporarily derived as basic image units, from which three types of asymmetry features are then calculated to quantify and capture the statistical distribution difference of various brain asymmetries: (1) global statistical properties, (2) measures of asymmetry of halved and quartered brains, and (3) local asymmetrical region-based properties. These features are calculated from:

- The original image unit with its iMSP aligned
- The difference image of the original image and its mirror reflection with respect to the iMSP
- The thresholded difference image
- The original image masked by the thresholded binary image

For each image unit, a feature vector with 48 image features is constructed, including the means, the variances, and the X and Y gradients of the intensity images at different regions and under various Gaussian smoothing, scaling, and thresholding [162].

This retrieval approach, although taking advantage of the 3D iMSP characteristics, provides global query and access on entire 2D-slice images, instead of local retrieval based on segmented 3D lesions. Since the 3D medical images are usually acquired from different modalities in different hospitals, each 3D neuroradiology scan may start and end at different portions of the brain with different angles or along different axes. This requires that 3D image data sets be properly registered and segmented before existing CBMIR techniques can be directly used for meaningful results [164–166].

Guimond and Subsol [166] developed a content-based retrieval of 3D brain MRIs with the user-selected 3D anatomical structure defined as the *volume of interest* (VOI) in a given reference image. The retrieval process was turned into a deformable registration process using global-to-local affine followed by free-form transformations for the best matching between the user-defined VOI and 3D images in the database, and correlation was used as a measure for morphological differences. This retrieval method, however, relies on the VOI definition to be a simple local subdivision of the image, rather than the segmented region.

Megalooikonomou et al. [167] proposed a CBMIR approach for 3D fMRI images using specific 3D *concentric sphere* features extracted from pathological VOIs. Since tumors or lesions in medical images are often considered to be homogeneous regions for simplicity, the VOIs formed by the voxels with the same value were first segmented, followed by a three-step feature extraction process of these VOIs. In the first step, the center of mass V of the VOI was computed. In step 2, using V as the center, a series of $1 \ldots k$ concentric spheres in the 3D domain were constructed with regular increments of their radii. In the last step, for each sphere, the k-dimensional feature vectors $\mathbf{F}_s$ and $\mathbf{F}_r$ were constructed, representing the fraction of the sphere occupied by the VOI and the fraction of the VOI occupied by the sphere, respectively. The obtained 3D concentric sphere feature vectors actually mapped the entire VOI to a specific point in k-dimensional space. To project the characterization vector to a space of lower dimensionality, the well-known Karhunen-Loève (KL) transform or the closely related *singular value decomposition* (SVD) can be adopted. Furthermore, for a given training set and classes of VOIs, experimentation can be conducted as an optimal classification process to obtain an appropriate value for each increment of radius of the concentric sphere [81, 167].

In Kim et al. [161], a 3D VOI-based retrieval approach for multidimensional dynamic brain PET images was proposed and integrated into a prototype VOI-based *functional image retrieval* system (VOI-FIRS). For each dynamic image data set, VOIs that contained various physiological characteristics related to local cerebral glucose consumption were first segmented based on a domain knowledge-based classification process; then a set of VOI functional and physiological features were extracted (more details can be found in Section 4.5). For the extraction of VOI visual features, an anatomical standardization procedure of the 3D stereotactic surface projection transformation in the NEUROSTAT package [168–170] was adopted to deform the 3D image into a standard stereotactic atlas by linear scaling of the image to correct individual brain sizes and by nonlinear warping to minimize regional anatomical variations. A set of transformation library files was then created for the dynamic PET images and applied to the corresponding segmentation results for warping the segmented images into the same standardized image frame of reference. After that, the centroid moment for each VOI and its corresponding volume (i.e., the total number of voxels in the VOI) was estimated and indexed into the database. The VOI volumetric features can be retrieved by measuring the absolute differences between the two volumes. In the similarity measure of the VOI location, the digital *Zubal phantom* [171] with fully labeled phantom was transformed into the standard coordinate system, as with dynamic PET images. The similarity of the VOI location was measured based on the 3D spatial distance (in voxels) between the user-defined point from the Zubal atlas and the centroids of the VOIs indexed in the database.

The ability of physicians to express their search needs and to navigate their search results accurately and easily is crucial in CBMIR systems [172], especially for retrieval of multidimensional medical images by volumetric features. In Kim et al. [161], the VOI-FIRS was responsive to user interaction in the retrieval process. As one of the most important interface functionalities of query components, the "query by VOI visual features" graphic user interface (GUI) (see Figure 4.4) can support user selection of the VOI location from the transformed Zubal phantom that is used as the standard brain atlas (center window in the figure).

The user is allowed to navigate in the 3D viewing space (rotation, scaling, and translation) and to change the viewing planes of sagittal, coronal, and transaxial slices, providing both

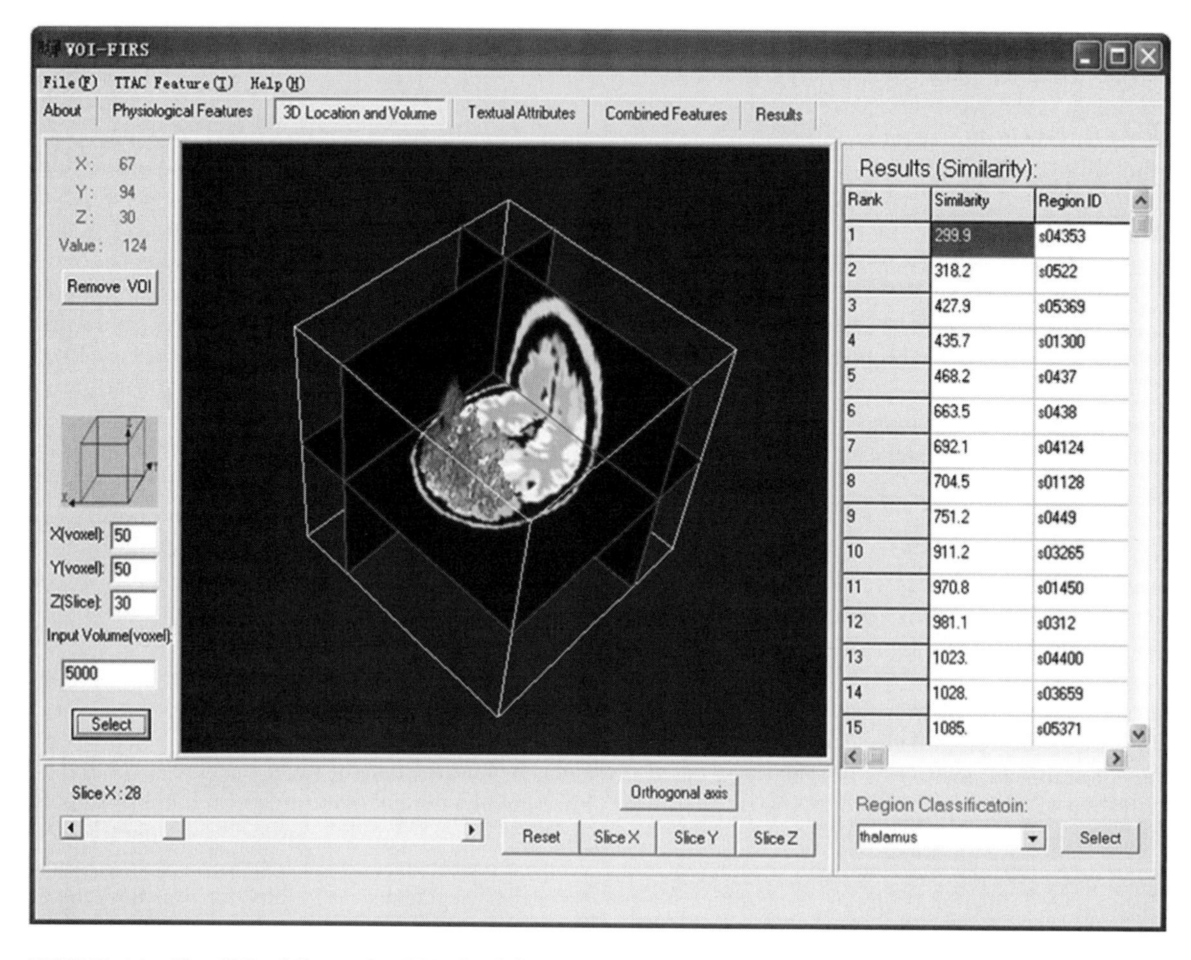

FIGURE 4.4 The GUI of "query by VOI visual features" exemplified by the rendered surface corresponding to the user-selected "prefrontal lobes" in the orthogonal Zubal slices [161].

conventional and 3D orthogonal views. The location feature can be initiated by selecting a point either on the Zubal phantom in the center display zone (setting voxel coordinates numerically via an input panel at the left middle) or on the labeled Zubal structure at the bottom right. The volume of the VOI can also be set numerically, which works independently of the location feature. The GUI of the output folder/window for display of retrieved results is illustrated in Figure 4.5. Each thumbnail result in the left part [Figure 4.5(a–d)] is presented as an *active display zone*, instead of a still index image, and can be individually navigated in orthogonal planes. The similarity indices from the retrieved VOIs are listed below the images. Any of the retrieved images can be enlarged and the VOI surface rendered, as shown in the right display zone in Figure 4.5(e). The segmentation result from which the features were extracted can also be retrieved from the database for inspection of the segmented VOIs. In this example, the combination query was formulated to search for images having similar dynamic functional behavior and size apparent in a malignant brain tumor case. The query was formulated to include a kinetic curve as a functional feature derived from an existing tumor VOI in the database, of 2,000 voxels as the volume parameter, with equal weights applied to the two selected features. The highest-ranked result shown in Figure 4.5(a) and enlarged in Figure 4.5(e) with the VOI surface rendered is of a patient study with a prominent tumor (indicated by the arrows). The result shows that the query successfully identified VOIs with high kinetic behavior and user-defined volume, which is in agreement with the similarity indices for the query features shown below the images [161].

4.3.3 Retrieval by Spatial Relationships

Spatial relationships are an important piece of medical knowledge, since a physician's mental model of the patient includes understanding not only the shape, size, and boundaries of organs or lesions, but also his or her spatial extents in relation to adjacent structures in the human body. Medicine thus critically relies on knowing where body structures are located and their locations relative to other structures [173, 174]. Given that

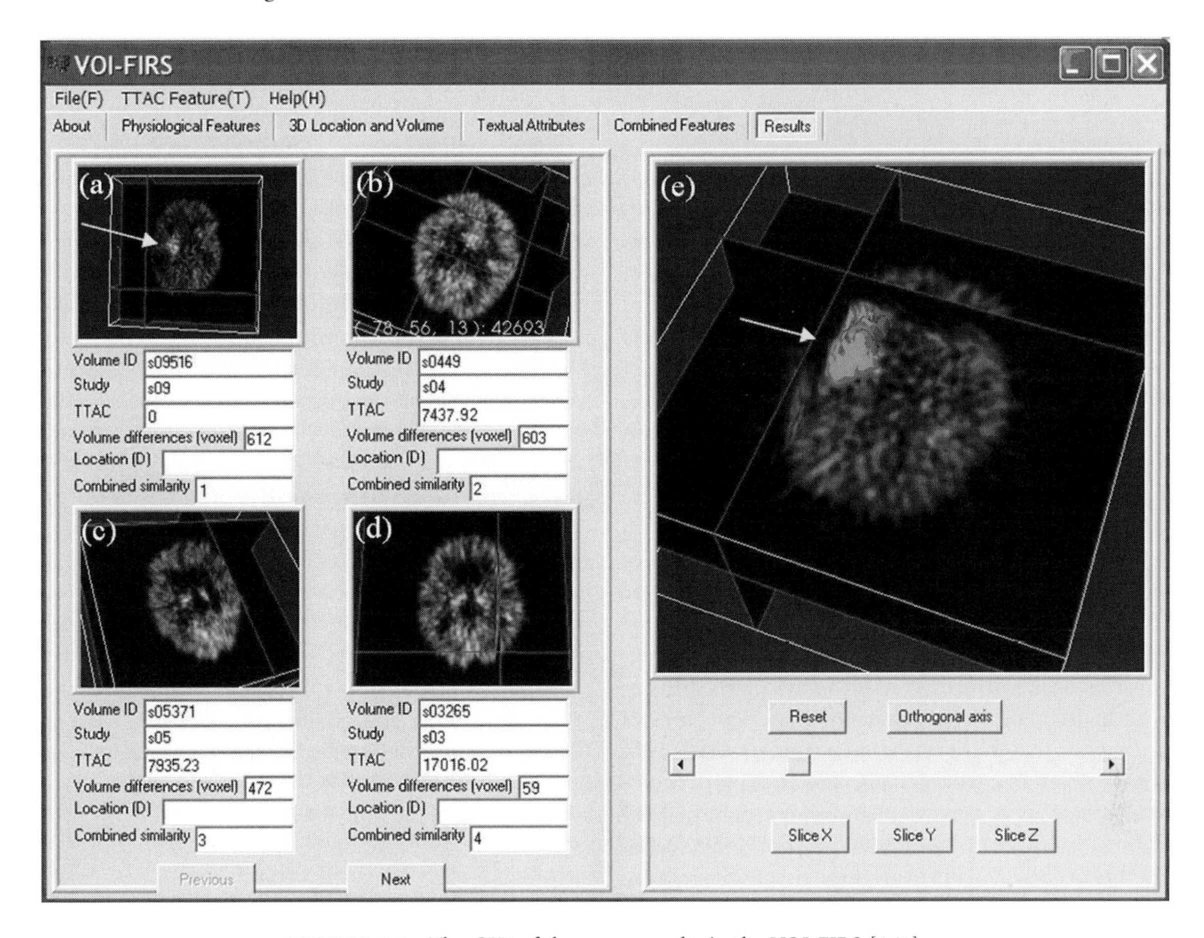

FIGURE 4.5 The GUI of the query results in the VOI-FIRS [161].

abnormalities are defined as gross deviations from anatomical models, the spatial relationships of body structures are often critical to the diagnosis, prognosis, and mechanics of human disease. For example, spatial content in terms of relationships in surgical or radiation therapy of brain tumors is very decisive, because the location and related adjacent structures of a tumor have profound implications on a therapeutic decision [175]. Low-level features cannot always capture or describe these complex scenarios. Various spatial relationships have therefore been modeled and used as content features for CBMIR to support complex image queries like "Find all image cases demonstrating the invasion of an adenoma into the sphenoid sinus" [173] and "Retrieve all images having a dangerous tumor on the left lung above and approximately touching another tumor" [176], as well as to assist the physician to understand and integrate the complex relationships between patient symptoms, diagnostic image features, and underlying disease pathology [174, 176–179].

In Petrakis [177], a CBMIR system for MRI scans and CT images was developed based on ARGs (see Section 4.1.2.4), which represent features of objects (regions) and relationships between them. In an ARG descriptor, the objects are represented by graph nodes, and the relationships between objects are constructed by arcs between such nodes. Both nodes and arcs are labeled by the attribute values of the object's features and relationship properties, respectively. The individual objects are described by four attributes:

- *Size* (s): computed as the size of the area the object occupies
- *Perimeter* (p): calculated as the length of the object's bounding contour
- *Roundness* (r): computed as the ratio of the smallest to the largest second moment
- *Orientation* (o): defined as the angle between the horizontal direction and the aixis of elongation that is the axis of least second moment

The spatial relationships between objects are described by three properties:

1. *Relative distance* (rd): calculated as the minimum distance between surrounding contours
2. *Relative orientation* (ro): defined as the angle with the horizontal direction of line connecting the centers of mass of the objects

3. *Relative position* (*rp*): defined by the values (−1, 0, 1) corresponding to objects that are:
   a. The first inside the second (−1)
   b. Outside each other (0)
   c. The second inside the first (1)

In Petrakis [177], the images were first subjected to presegmentation to obtain polygonal approximations of object contours, followed by shape corrections or deletion of insignificant segments by experts. The segmented objects were further classified into predefined classes corresponding to normal or abnormal anatomical structures, such as ventricle, hematoma, and tumor. The individual objects were then represented by 5D feature vectors of the normalized form <*s*, *p*, *r*, 1+cos(*o*), 1+sin(*o*)>, while relative orientations were defined by <*rd*, 1+*rp*, 1+cos(*o*), 1+sin(*o*)>.

Spatial-relationship models can also be used to retrieve microscopic tissue images [174]. Conventional region- or image-level search algorithms always assume that regions or images consist of uniform pixel feature distributions. However, complex tissue images normally contain many pixels and regions that have different feature characteristics. Two images with similar regions may have very different interpretations if the regions have different spatial arrangements [1, 174]. Therefore, a visual grammar of spatial relationships may help in describing these scenarios. In Aksoy et al. [174], the spatial relationships between all region pairs in a tissue image were represented by an $n \times n$ region relationship matrix:

$$R = \{\{r_{ij}, d_{ij}, \theta_{ij}\} | i, j = 1, \ldots, n, \forall i \neq j\}, \tag{4.13}$$

where $r_{ij} = \pi_{ij}/\pi_i$ is the ratio of the common perimeter to the perimeter of the first region; π_i and π_j are the perimeters of the first and second region, respectively, and π_{ij} is the common perimeter between two regions; $d_{ij} = \|v_i - v_j\|$ is the distance between the centroids; v_i and v_j are the centroids of the first region and second region, respectively; and θ_{ij} is the angle between the horizontal (column) axis and the line joining the centroids. Here, each region pair was assigned a degree of their spatial relationship using fuzzy class membership functions Ω_c with $c \in$ {DIS, BOR, INV, SUR, NEAR, FAR, RIGHT, LEFT, ABOVE, BELOW}. Then the degree of membership of regions i and j to class c could be defined as $\Omega_c(r_{ij}, d_{ij}, \theta_{ij})$. These class functions were divided into three relationship groups:

1. Perimeter class: DISjoined, BORdering, INVaded_by, and SURrounded_by
2. Distance class: NEAR and FAR
3. Orientation class: RIGHT, LEFT, ABOVE, and BELOW, since multiple relationships may be used to represent a region pair; for example, BORdering from ABOVE, or INVaded_by from LEFT

Based on these second-order region relationships, the higher-order region relationships were constructed for more complex combined relationship representation.

Since medical image content is very rich in properties, characteristics, salient objects, and spatial relationships that heavily relate to medical knowledge, it is of great interest to retrieve medical images by combining spatial relationships with high-level medical domain knowledge. In Chu et al. [178], a knowledge-based approach to retrieve medical images by feature and content with spatial and temporal constructs was proposed with a four-layered semantic image model representing the spatial, temporal, and evolutionary nature of medical objects. Chbeir et al. [176] developed the *medical image management system* (MIMS), based on a global description of a medical image to achieve an efficient retrieval process integrating a high level of precision (especially in terms of spatial relations) required by the medical domain. A spatial knowledge-based model (SKM) was integrated into the MIMS to provide coherent and effective objectivity of interpretation at different facets (or views) of medical images. Notwithstanding that spatial-relationship approaches are included in Chu et al. [178] and Chbeir et al. [176], considering the overall focus of these works on the integration of high-level medical domain knowledge in the CBMIR, more details of these two studies will be given in Section 4.4.2.

4.4 Content-Based Medical Image Retrieval by Combination of Semantic and Visual Features

4.4.1 Retrieval by Semantic Pathology Interpretation

The ultimate goal for CBMIR is to find medically meaningful similar cases. However, these cases may or may not present medical images that are similar in the usual visual sense. Sometimes similar visual features in images may not imply similar diagnoses or symptoms, and vice versa. It thus appears that combining visual features in the image with domain knowledge to reach the right subset of relevant cases in the database is a key to the success of CBMIR.

Sometimes medical images derived from a specific organ are similar visually and differ only in small details that can be missed by untrained eyes, but such domain-specific subtle differences may be of pathological significance [94]. One such domain is medical radiology, for which clinically useful information consists of gray-level variations in highly localized regions of the image [180]. For example, for HRCT images of the lung, the number of pathology-bearing pixels as a fraction of all the image pixels is so small that global signatures cannot be used for image characterization. Moreover, the unclear boundaries between the pathology-bearing pixels and those bearing the image of the surrounding healthy part are difficult to discern, and such PBRs are unlikely to be automatically extracted. Since medical images frequently give rise to ambiguity in interpretation and in diagnosis, current retrieval techniques using primitive image characteristics—such as color, texture, and shape—are likely to be insufficient for some domain-specific CBMIR systems. One approach to remedy such limitation is to associate high-level semantic information with low-level visual image data, albeit such an approach is still controversial in terms of subjectiveness [6].

In Shyu et al. [180], a *physician-in-the-loop* approach for content-based retrieval of HRCT lung images was developed and integrated into a CBMIR system called ASSERT (Automatic Search and Selection Engine with Retrieval Tools). To archive an image into the database, a physician spent a few seconds to delineate the PBRs and any relevant anatomical landmarks. The information regarding the pathology of the lung may reside as much in the location of each PBR with respect to the anatomical markers as it does in the characteristics of the PBR. Four major perceptual categories used by the physician for detecting pathology in the image included:

- Linear and reticular opacities
- Nodular opacities
- Diffuse regions of high attenuation (high-density areas)
- Diffuse regions of low attenuation (low-density areas)

After the extraction of PBRs, a lung region extraction algorithm was used to determine the boundary of the lungs. For feature extraction of the image, the system computed attribute vectors that characterized the PBRs individually and the part of the image that consisted of just the entire lung region. Each PBR was characterized by a set $\mathbf{A}_{PBR}$ of commonly used physical and spatial features such as texture, shape, and other gray-level attributes. However, more importantly, the PBR also included another set $\mathbf{B}_{PBR}$ of attributes that measured the perceptual categories used by physicians for identifying and interpreting pathology in HRCT lung images. For $\mathbf{A}_{PBR}$, a large number of general-purpose attributes were calculated; in total, 255. This did provide the user an exhaustive characterization of a PBR, but in light of efficient indexing and searching, only a small subset of these attributes could be used for database indexing and retrieval. Therefore, a greedy search algorithm called SFS (sequential forward selection) was applied to reduce the dimensionality of the feature space while retaining the ability to accurately classify each image as belonging to its associated disease pattern, resulting in only 12 general features in $\mathbf{A}_{PBR}$, such as the area, contrast, entropy, and edginess histogram of the PBR. For $\mathbf{B}_{PBR}$, 14 pathology attributes for measuring four major perceptual categories, such as shape of the bronchial wall, curvature of an adjacent fissure, and adjacent artery size, were computed. Finally, a 26-dimensional mixture pathology feature vector $\mathbf{F}_{PBR} = < \mathbf{A}_{PBR}, \mathbf{B}_{PBR} >$ could be formed as a PBR feature descriptor that was maximally discriminatory with regard to the different diseases [180].

In this PBR pathology interpretation approach, the physician is an integral part of the whole CBMIR system, in the sense that it is the physician who delineates the PBRs with semantic pathology interpretation. Taking into account the physician's perceptual categories, which correlate strongly with the various lung diseases, [181] proposed a better alternative to this method. It extracted only those general features that measured the presence or the absence of the various perceptual categories that the physicians used for disease diagnosis. That is, general features were used to describe the perceptual categories and discriminate among them. After determining what perceptual categories were present in a PBR, the user could determine the disease of the PBR. An advantage of such a hierarchical approach is that it is now easier to decide what features are needed to characterize the PBRs [181]. To support semantic integration and knowledge exchange in the medical radiology domain, a CBMIR framework called an "evolutionary system for semantic exchange of information in collaborative environments" ("Essence") was developed and reported in Barb et al. [182], extracting and managing visual content of lung pathologies. In Essence, an XML-based shared ontology was developed based on the common knowledge from expert radiologists and information from two well-known references, [183, 184]. Physicians were able to build their personalized semantic search criteria by customizing the degrees of satisfaction of features to existing semantic terms and by adding new semantic terms to existing perceptual categories. The system also supported refining the shared ontology by adapting the assignment of semantic terms to image features based on individuals' preferences [182].

Tang et al. [94] presented an intelligent I-Browse system that combined low-level image processing technology with high-level semantic interpretation for retrieving histological images of the GI tract—a range of histological images originating from the esophagus, stomach, small intestine (small bowel), large intestine (large bowel), appendix, and anus. Before semantic features extraction, two sets of relevant histological features (also called semantic labels) in GI tract images were first defined by consulting with histopathologists:

1. The 15 coarse feature labels represented by different uppercase letters—such as lumen (L), mucosa (M), submucosa (S), muscularis externa (E), and serosa/adventitia (A), providing an overall structural description of the image content that is important for the later reasoning procedures
2. The 63 fine feature labels indicated by different numbers (1, 2,..., 63), such as adipose tissue #1, lumen #33, stomach–junction of lumen and foveolae #63, for distinguishing different visual appearances within each coarse region

To enable automatic analysis and interpretation, the original microscopy image was first partitioned as a set of subimages, each of 64 × 64 pixels. For each subimage, via the *visual feature detector* (VFD), a set of coarse features (normalized color and gray-level histograms) were extracted and passed to a three-layer neural network of multilayer perceptrons (MLPs) for classifying the subimage into one of the 15 coarse feature classes, by assigning a coarse feature label to the subimage. Using the VFD, a set of semi-fine features (means, standard deviations of the gray and color levels, and Gabor filters) were also computed and used to classify the subimage into one of the 63 fine feature classes using a Bayes minimum risk classifier via marking a fine feature label to the subimage. Since each fine class actually corresponded to a coarse class, another coarse label result could be obtained for the subimage from these semi-fine features, whose classified fine class result could be matched into one of the coarse labels. Therefore, each subimage was with two letters and one number, representing two coarse feature labels and one fine feature label, respectively, and the whole microscopy image could be presented with three matrices (the label map). With these label matrices, the *semantic analyzer* (SA) went through an iteration process to refine and correct the fine histological feature label of each subimage according to the histological context in the *knowledge base* (KB) and was able to produce a set of hypotheses on the labels associated with subimages by the *hypothesis generator* (HG), if those labels were deemed erroneously detected by the extracted coarse and semi-fine features. Based on the hypotheses, a number of fine features were invoked to extract and confirm the visual features within the suspected regions. Such analysis-and-detection cycles iterated until the semantic analyzer found a coherent result and no further change was needed. The refined final label map was then used to construct the semantic content representation structure called *Papillon,* a codename used in the I-Browse system, and could be used to automatically generate the textual annotation for the image in the database with the *annotation generator* (AG).

When the query was submitted by free text (natural language), the *free text analyzer* (FTA) would extract the information in the query and convert it into the Papillon. The Papillon actually bridged information from different media (image and text), linking together the SA, AG, and FTA components in the system. Therefore, when the query was issued by sample image or free text, their semantic content in the Papillon could be used for the retrieval [94, 95, 185–189]. Some other semantic description-based CBMIR systems include a *property concept frame* (PCF) approach for retrieval of histopathology images [190] and a brain CT image retrieval system based on the hierarchical medical image description model [191].

4.4.2 Retrieval Based on Generic Models

In light of the heterogeneous forms presented in medical images, most CBMIR systems are task specific; that is, they are limited to a particular modality, organ, or diagnostic study and, hence, are usually not directly transferable to other medical applications [4, 6, 192]. Medical knowledge arises from anatomical and physiological information, requiring regional features to support diagnostic queries. However, interpretation of medical images depends on both image and query context. Since the context of queries is unknown when images are entered into the database, the number and kind of image features can be especially subject to continuous evolution, and the CBMIR scheme must be generic and flexible [4, 192, 193].

To conduct efficient medical image retrieval, there is a need for building comprehensive data models capturing the structured abstracts about images, and supporting more sophisticated query predicates based not only on primitive image characteristics, but also on generic semantic features with the inclusion of knowledge. Chu et al. [178] developed a CBMIR system called KMeD, in which a knowledge-based semantic temporal image model was included containing four layers: (1) the raw data layer (RDL), with image collection; (2) the feature and content layer (FCL), with different features extracted from the image content, including shape, spatial-relationship characteristics, and temporal features; (3) the schema layer (SL), representing entities and relationships (spatial, temporal, and evolutionary) among objects based on the features in the previous layer; and (4) the knowledge layer (KL), containing hierarchical structures called *type abstraction hierarchies* (TAHs) for classifying shape and spatial-relationship features.

The FCL contains shape features such as type, area, volume, diameter, length, and circumference, and spatial-relations features between a pair of objects, including orthogonal relations (i.e., east, south, southeast, etc.) and containment relations (i.e., invades, contains, etc.). In SL, the image objects can be represented as visual entities with textual attributes and visual attributes, while multiple versions of an object over a period of time (e.g., the stages undergone by a tumor during the cancer process of a particular patient) can be linked to form a stream entity for that time period. Moreover, the evolutionary object constructs for evolution, fusion, and fission, while the temporal-relation object constructs the temporal relationships between peer objects and between an object and its super or aggregated type. The TAHs in the KL have been designed as hierarchical structures so that at higher levels of abstraction, more generalized concepts are specified (i.e., a wider range of feature values are used), and at lower levels of abstraction, more specific concepts are described (i.e., a narrower range of feature values) [173, 178, 194]. In the KMeD system, the knowledge-based temporal, evolutionary, and spatial features extracted from the images are classified and captured in the image data model and stored in tables. For example, the query "Retrieve all image cases demonstrating a pituitary gland microadenoma that evolved into a macroadenoma with suprasellar extension pressing against the optic chiasm" can be given by searching a pituitary gland–microadenoma containment relationship table, then following the evolutionary path that leads to a macroadenoma and selecting the instances from the outside contact relationship table between a macroadenoma and the optic chiasm [173].

Existing CBIR systems hardly address all medical image properties, since medical images have different acquisition parameters and modalities and specific noise characteristics for each imaging system. As such, the development of global features that can represent an entire medical image database seems to be practically infeasible. Aiming to provide efficient retrieval of generic medical images with coherent and effective objectivity of interpretation at different facets (or views) of medical images, Chbeir and Favetta [195] proposed a global description of medical images in which a hyperspaced image data model was constructed [175, 196]. The data model was structured as a multispaced form in which each space contained a set of features (contextual, physical, spatial, and semantic), and considered the medical image as a composition of contextual and content feature spaces. The contextual space collects the general data attached to the image without taking its visual content into account. It does this through three components:

1. *The independent context* (e.g., the medical specialty, patient name, acquisition date) has no impact on the image description and, due to patient privacy and other legal constraints, needs careful treatment and can be managed separately.
2. *The pseudo-independent context* (e.g., the patient's age and gender, the image quality) is vital for CBMIR, since it contains very important background knowledge and may help in determining methods to be used to construct and compare image content features. For example, the age of the patient is a determinant factor when considering organ shapes.
3. *The dependent context* (e.g., image type, incidence [sagittal/coronal/axial/others], the scene described as a triplet <title, organ, alteration>, diagnostic report, voice report) can significantly help in the image description. For instance, with help from natural language processing or voice segmentation, the diagnostic report can be used to clarify missing factors in some situations, such as "Describing a lung X-ray of a person remains incomplete and insignificant if we ignore that he smokes."

The content space, on the other hand, provides a global image description and can be used for various query types. In general, a medical image is considered to be composed of a set of *salient image objects* in three different forms:

1. The *anatomical organ* (AO) presents the medical organs found in the image, such as the brain, lungs, hands, etc., and gathers a set of medical regions. It is also called the *organ of interest* (OOI).
2. The *medical region* (MR) describes the internal structure of the AO, such as the left ventricle and the right lobe. It allows one to locate any anomaly and is synonymous with the ROI.
3. The *medical sign* (MS) concerns either medical anomalies (such as tumors, fractures, and lesions) identified and detected by physicians, or unidentified (or variable) objects found in the image. Sometimes it is referred to as the PBR.

Each salient object (AO, MR, or MS) is projected on the following subspaces:

- *The physical subspace* contains low-level physical properties of the image content, such as various global or local color and texture features that can be extracted manually, semi-automatically, or automatically, depending on the contextual space (image type, format, quality, etc.) and may be used later to analyze other subspaces. Moreover, the physical analysis can be achieved based on the pseudo-independent and dependent contexts. For example, the patient's age is a determinant factor when considering the medical organ shape. The type of the image determines the appropriate color extraction approach.
- *The spatial subspace* holds middle-level geometric features of salient objects such as the shape and spatial-relationship features.
- *The semantic subspace* concerns high-level semantic properties of salient objects. The objective of the semantic subspace is to integrate high-level features of objects and relations judged primordial by medical users for image description. However, such semantic feature analysis may require human intervention, since explicit semantic objects must be recognized. The semantic subspace is usually described manually by the user due to the fact that the medical domain is very complex, and each term may have several meanings depending on the context. Medical signs can be codified by some existing, albeit controversial, labeling codes for disease classification such as the ICD-10 (International Classification of Diseases, 10th Revision) [197] or the Unified Medical Language System (UMLS) [198].

The hyperspaced image data model previously discussed has been integrated into the MIMS prototype [176, 196, 199–201].

Lehmann et al. [192] presented a general structure for content-based image retrieval in medical applications (IRMA) based on a generic multistep approach including categorization of the entire image, registration with respect to prototypes, extraction and query-dependent selection of local features, hierarchical blob/object representation, and image retrieval. To cope with the complexity of medical knowledge, IRMA split the whole retrieval process into seven consecutive steps, with each step representing a higher level of image abstraction, reflecting an increasing level of image content understanding:

1. *Image categorization* (based on global features). This can determine the imaging modality and its body orientation as well as the examined body region and biological system for each image entry with a detailed hierarchical coding scheme [202] to supplement the existing standard (e.g., DICOM).
2. *Image registration* (in geometry and contrast). Diagnostic inferences derived from medical images are normally deduced from an incomplete but continuously evolving model of normality [4]. Therefore, registration is based on prototype images defined for each category by experts with prior medical knowledge or by statistical analysis, in which the prototypes can be used for determination of parameters for rotation, translation, scaling, and contrast adjustment.
3. *Feature extraction* (using local features). This derives various local image descriptions with either a category-free or a category-specific approach. Like the global features for categorization, the number of local feature images is extensible.
4. *Feature selection* (category and query dependent). The separation of feature selection from feature extraction enables the former task to be retrieval dependent. It can integrate both the image category and the query context into the abstraction process with a precomputed set of adequate features. For instance, the retrieval of radiographs with respect to bone fractures or tumors can be conducted using a shape-based or texture-based feature set, respectively.
5. *Indexing* (multiscale blob representation). This provides an abstraction of the previously generated and selected image features, resulting in a compact image description via clustering of similar image parts into regions described by invariant moments as "blobs." Thereafter, the blob representation of the image is adjusted with respect to the parameters determined in the registration step, yielding a multiscale "blob tree."
6. *Identification* (incorporating prior knowledge). This provides the link between medical *a priori* knowledge and certain blobs generated during the indexing step. Therefore, it is the fundamental basis for introduction of high-level image understanding by analysis of regional or temporal relationships between blobs.
7. *Retrieval* (on abstract blob level). This is performed by searches in the hierarchical blob structures. This retrieval step requires online computations, while all other steps can be performed automatically in batch mode at image entry time (offline computation).

The above multistep approach has been applied to the IRMA database of radiographs (consisting of medical images of six major body regions taken from daily routine), narrowing the gap between the semantic imprint of images and any alphanumeric description, which is always incomplete [192, 193].

Some other CBMIR systems that provide varied medical image retrieval include I^2C (Image Indexing by Content) [203], COBRA (Content-Based Retrieval Architecture) [204], ImageEngine [205], and MedGIFT [122, 206, 207]. MedGIFT in particular, with its integration of the GNU Image Finding tool (GIFT) [208], the Multimedia Retrieval Markup Language (MRML) [209], and CasImage, provides an open-source framework of reusable components for a variety of CBMIR systems to foster resource sharing and avoid costly redevelopment.

4.5 Content-Based Medical Image Retrieval by Physiologically Functional Features

The CBMIR techniques introduced in this chapter so far are designed mainly for anatomical images that capture human anatomy at different levels and provide primarily structural information. Unlike those anatomical images, functional/molecular images such as PET and SPECT allow the *in vivo* study of physiological and biochemical processes, providing functional information previously not available. This is what most distinguishes medical images from general images [86, 210, 211]. Physiological function can be estimated at the molecular level by observing the behavior of a small quantity of an administered substance tagged with radioactive atoms. Images are formed by the external detection of gamma rays emitted from the patient when the radioactive atoms decay. Glucose metabolism, oxygen utilization, and blood flow in the brain and heart can be measured with compounds labeled with carbon (^{11}C), fluorine (^{18}F), nitrogen (^{13}N), and oxygen (^{15}O), which are the major elemental constituents of the body.

Existing CBMIR approaches may not be optimal when applied to functional images due to the latter's unique characteristics with regard to the inherent knowledge of the disease state as it affects the physiological and biochemical processes before the morphological change of the body. Such quantitative physiological information inside the functional image content is unlikely to be retrieved by common image retrieval techniques using color, texture, and shape features. Color is not captured in the imaging process of functional features, whose images are usually acquired and displayed in grayscale or pseudo-color. Therefore, the color feature is unlikely to be applicable to functional images. Texture is likely to be confounded by the statistical noise in functional images. Shape is also unlikely to be relevant to function. Indeed, function is likely to result in changes in apparent shape during acquisition as the tracer redistributes. It appears that the development of CBMIR for functional images should take into account specific physiologically functional features [161, 212].

An early study on content-based retrieval of dynamic PET functional images was reported in Cai et al. [212]. Based on this work, Kim et al. [161] recently developed a new VOI-based retrieval system for multidimensional dynamic functional $[^{18}F]$2-fluoro-deoxy-glucose (FDG) brain PET images, which are widely used to determine the local cerebral metabolic rate of glucose (LCMRGlc) and depicts the glucose consumption and energy requirements of various structural and functional components in the human brain. In dynamic functional imaging studies, prior knowledge has the form of a tracer kinetic model to a time series of PET tracer uptake measurements. Such functional information can be defined in terms of a mathematical model $\mu(t|p)$ (where $t = 1,2,\ldots,T$ are discrete sampling times of the uptake measurements, while the number of conventional scan time intervals T is 22, and p is a set of the model parameters), whose parameters describe the delivery, transport, and biochemical transformation of the tracer. The input function for the model is the *plasma time activity curve* (PTAC) obtained from serial blood samples. Reconstructed PET images provide the *tissue time activity curve* (TTAC), or the output function, denoted by $\mathbf{f}(t)$ for every voxel in the image.

Application of the model on a voxel-by-voxel basis to measured PTAC and TTAC data using certain rapid parameter estimation algorithms [213, 214] yields physiological parametric images. In Kim et al. [161], a four-dimensional fuzzy c-means cluster analysis [215, 216] was used to construct VOI functional groups consisting of voxels that have similar kinetic behaviors. The physiological TTACs were first extracted for each of the N nonzero voxels in the image to form the kinetic feature vector comprising the voxel values at the dynamic time sequence of tracer uptake measurements. After applying the optimal image sampling schedule (OISS) technique [217, 218] for the dynamic FDG brain PET image study based on the five-parameter FDG model, the dimension of TTAC vectors was reduced from 22 to 5, while the signal-to-noise ratio of the individual image frames was increased for betterclustering output. The fuzzy c-means cluster analysis was then applied to assign each of the N feature vectors to a set number C of distinct cluster groups and minimized the objective function J:

$$J = \sum_{i=1}^{N} \sum_{j=1}^{C} u_{ij}^{P} D_{ij}^{2}, \tag{4.14}$$

where P $(1 \leq P \leq \infty)$ is a weighting exponent on each fuzzy membership, which determines the amount of fuzziness of the resulting classification, and u_{ij} is the membership degree of the ith feature vector in the cluster j. The similarity measure between the ith feature vector $\mathbf{f}_i(t)$ and the cluster centroid $\bar{\mathbf{f}}_{c_j}(t)$ of the jth cluster group c_j was computed using the Euclidean distance:

$$D_{ij} = \left[\sum_{t=1}^{T} s(t) \left(\mathbf{f}_i(t) - \bar{\mathbf{f}}_{c_j}(t) \right)^2 \right]^{1/2}, \tag{4.15}$$

where $s(t)$ is a scale factor of time point t equal to the duration of the tth frame divided by the total dynamic acquisition time. The scale factor gives more weight to the later frames with longer scan time durations, which contain more reliable data. The minimization of J was achieved by iteratively updating u_{ij}:

$$u_{ij} = \left[\sum_{k=1}^{C} \left[\frac{D_{ij}}{D_{ik}} \right]^{\frac{2}{P-1}} \right]^{-1} \tag{4.16}$$

and the cluster centroids $\bar{\mathbf{f}}_{c_j}(t)$:

$$\bar{\mathbf{f}}_{c_j}(t) = \frac{\sum_{i=1}^{N} u_{ij}^{P} \mathbf{f}_i(t)}{\sum_{i=1}^{N} u_{ij}^{P}}. \tag{4.17}$$

Therefore, a probabilistic weighting was assigned to every voxel i, representing it to be likely a member of each cluster j. For any voxel, the sum of the assigned membership degrees was 1.0. The procedure was terminated when the convergence inequality

$$\max_{ij} \left\{ \left| u_{ij}^{m+1} - u_{ij}^{m} \right| \right\} > \varepsilon \tag{4.18}$$

was satisfied, where m was the iteration step and $0 < \varepsilon < 1$. Upon convergence, a cluster map was created by assigning to each voxel a value equal to the cluster number for which it had the highest degree of fuzzy membership. From the derived clustered results, the region-growing algorithm [145] was applied to the voxels in each cluster to construct the VOIs for grouping the voxels that were spatially connected and separating the different structures that may have been classified into a cluster due to the similarity of the voxels' kinetic behavior. The TTAC feature vectors extracted from the VOIs were indexed as physiologically functional features and used as a key query method in the proposed VOI-FIRS [161].

Figure 4.6 shows the GUI of the query component "query by functional and physiological features" in VOI-FIRS. The user is allowed to manually draw the TTAC feature curve with the labeled grid or to select from a list of predefined sample TTACs if needed. Once the selection has been made, the TTAC curve can be manually adjusted for individual TTAC sampling points. As the TTAC curve is concentrated in the early temporal frames, the drawn curve can be zoomed for closer inspection.

An example of dynamic image retrieval based on physiologically functional features is illustrated in Figure 4.7. The sample TTAC, which approximates a pattern found in gray matter of dynamic brain FDG-PET images (as shown in Figure 4.7(a)), and the 3D location of the "right thalamus" selected from the labeled structures in the Zubal phantom panel (Figure 4.7(b)) were set as the query features. Weighting was set to 50% for the functional feature and 50% for the 3D volumetric location feature (see Section 4.3.2). The highest-ranked retrieved VOI is shown in Figure 4.7(c), where the query identified a VOI representing the right thalamus. Figure 4.7(d) presents the top-ranked result from changing the location feature to "left thalamus." The result demonstrated that by retrieving based

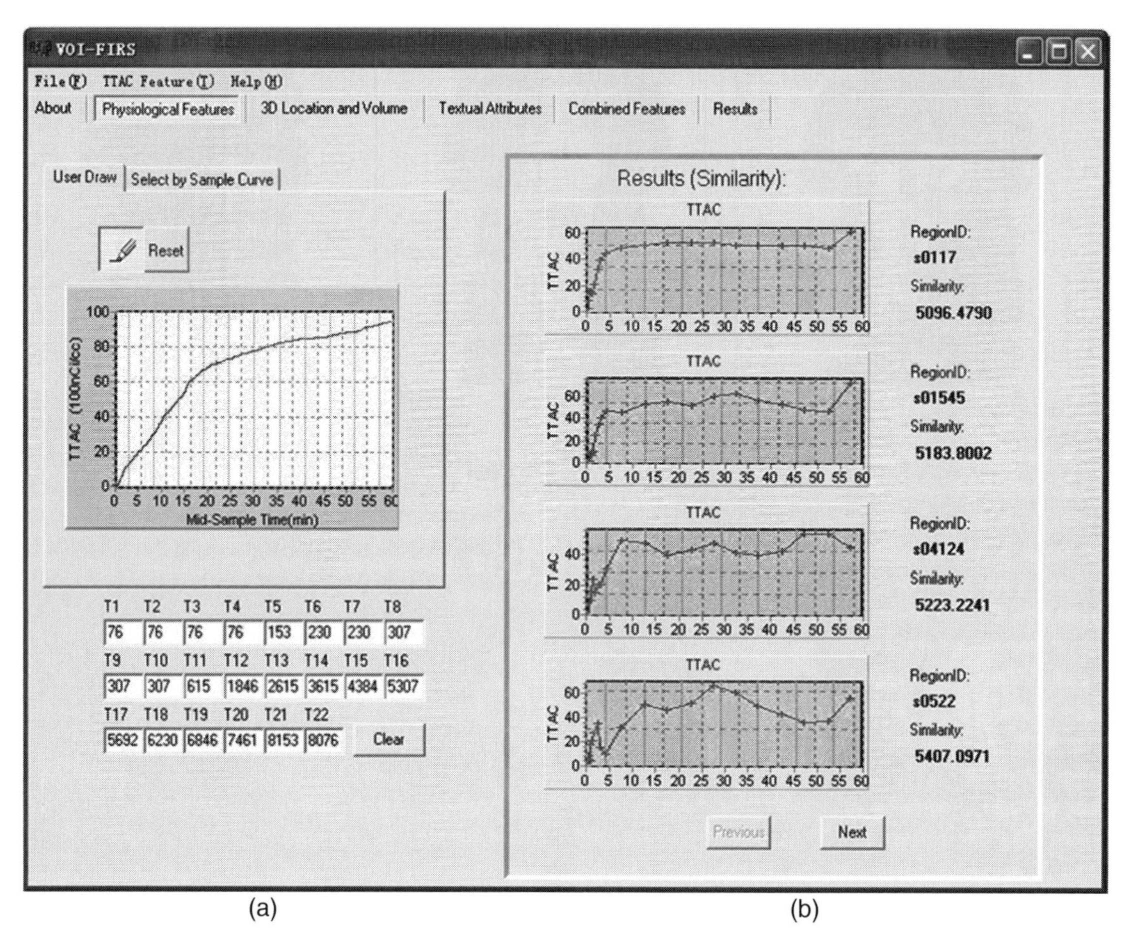

FIGURE 4.6 The GUI of "query by physiological features" shows the user-drawn TTAC curve (a) and the retrieved VOIs with their TTAC curves and similarity indices (b) [161].

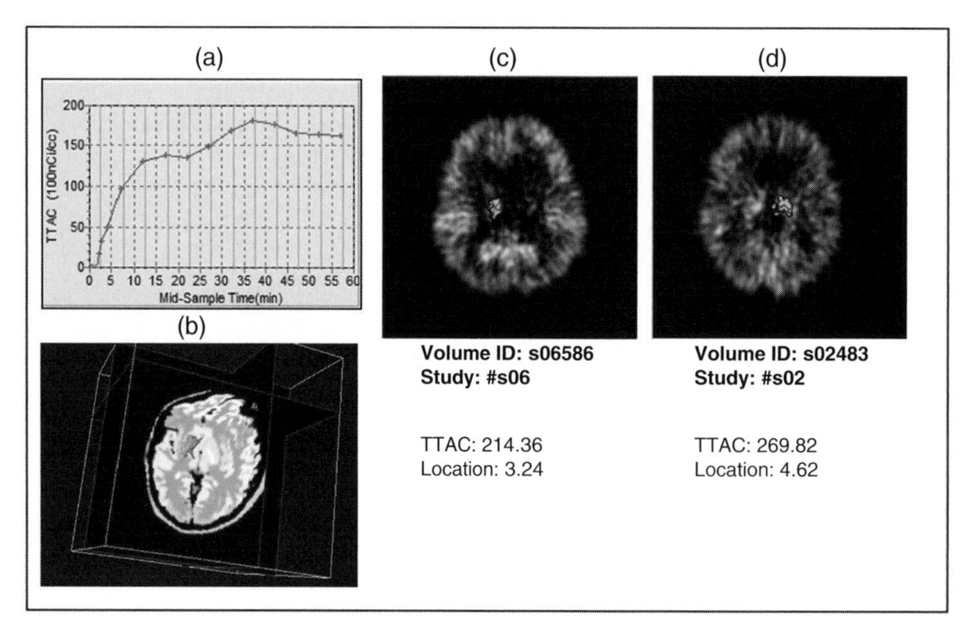

FIGURE 4.7 (a) The sample TTAC curve; (b) selection of a location (right thalamus) in the standard atlas (Zubal) panel; (c) retrieved result with the combination query features of (a) and (b); (d) retrieved result with (a) and a different location feature (left thalamus) as the query feature [161].

on the combination of functional features and the spatial properties of the dynamic PET images in the 3D volumetric location feature, VOIs with the user-defined kinetic TTAC characteristics could be successfully identified, which might not have been possible from the functional feature alone [145, 196].

4.6 Summary

This chapter introduced CBIR and its key components, including image feature extraction, similarity comparison, indexing scheme, and interactive query interface. The need for CBIR in the medical domain (CBMIR) and its related challenges were discussed, followed by a detailed review of the current major CBMIR techniques in four different categories: retrieval based on physical visual features (color and texture); retrieval based on geometric spatial features (shape, 3D volumetric features, and spatial relationships); retrieval by combination of semantic and visual features (semantic pathology interpretation and generic models); and retrieval based on physiologically functional features. The success of CBMIR can open up many new vistas in medical services and research, such as in disease tracking, differential diagnosis, noninvasive surgical planning, clinical training, and outcomes research.

Acknowledgments

The authors are grateful to the support from ARC and PolyU/UGC grants.

4.7 Exercises

1. Describe the mechanism of CBIR.
2. What are the primary differentiating factors between CBIR and CBMIR?
3. Texture as a visual feature has been successfully applied in numerous CBMIR systems (e.g., MRI head scans, HRCT images of the lung). What is an image texture, and what are its attributes that enable content-based retrieval?
4. Why can 3D volumetric features be used in CBMIR? What are the advantages and disadvantages of 3D volumetric features versus 2D shape features?
5. Give an example of a CBMIR application in clinical decision support.
6. What are the advantages and disadvantages of combining semantic and visual features in CBMIR? How does combining these two components exceed expected results from using just one?

4.8 References and Bibliography

1. I. Bankman (Ed.). *Handbook of Medical Imaging Processing and Analysis.* Academic Press, 2000.
2. H. K. Huang. *PACS and Imaging Informatics: Basic Principles and Applications*, 2nd ed. Wiley-Liss, 2004.
3. J. Duncan and N. Ayache. Medical image analysis: Progress over two decades and the challenges ahead. *IEEE Trans. PAMI.* 22(1):85–106, 2000.
4. H. D. Tagare, C. Jaffe, and J. Duncan. Medical image databases: A content-based retrieval approach. *J. Am. Med. Inform. Assoc.* 4(3):184–198, 1997.
5. L. H. Y. Tang, R. Hanka, and H. H. S. Ip. A review of intelligent content-based indexing and browsing of medical images. *Health Informatics J.* 5:40–49, 1999.
6. H. Müller et al. A review of content-based image retrieval systems in medical applications—Clinical benefits and future directions. *Int. J. Med. Info.* 73:1–23, 2004.
7. J. Boissel et al. Getting evidence to pre-scribers and patients or how to make EBM a reality. *Proc. Med. Info. Europe Conf.* France, 2003.
8. Y. Rui, T. S. Huang, and S.-F. Chang. Image retrieval: Past, present, and future. *Proc. Int. Symp. Multimedia Inform. Proc.* Taipei, Taiwan, 1997.
9. A. W. M. Smeulders et al. Content-based image retrieval at the end of the early years. *IEEE Trans. PAMI.* 22(12): 1349–1380, 2000.
10. A. D. Bimbo. *Visual Information Retrieval.* Morgan Kauffman Publishers, 1999.
11. V. Castelli and L. D. Bergman (Eds.). *Image Databases: Search and Retrieval of Digital Imagery.* John Wiley & Sons, 2002.
12. D. Feng, W. C. Siu, and H. J. Zhang (Eds.). *Multimedia Information Retrieval and Management: Technological Fundamentals and Applications.* Springer, 2003.
13. F. Long, H. Zhang, and D. Feng. Fundamental of content-based image retrieval. In D. Feng, W. C. Siu, and H. Zhang (Eds). *Multimedia Information Retrieval and Management: Technological Fundamentals and Applications.* Springer, 1–26, 2003.
14. Y. Rui, T. S. Huang, and S. F. Chang. Image retrieval: Current techniques, promising directions and open issues. *J. Vis. Comm. Image Rep.* 10:39–62, 1999.
15. R. M. Rangayyan. *Biomedical Image Analysis.* CRC Press, 2005.
16. J. Vendrig, M. Worring, and A. Smeulders. Filter image browsing: Exploiting interaction in retrieval. *Proc. Vis. '99 Information and Information Systems*, 1999.
17. J. Robinson. The k–d–B–tree: A search structure for large multidimensional dynamic indexes. *Proc. SIGMOD Conf.* Ann Arbor, 1981.
18. D. Lomet and B. Salzberg. A robust multimedia-attribute search structure. *5th Int. Conf. Data Eng.* 296–304, 1989.

19. T. Brinkhoff, H. Kriegel, and B. Seeger. Efficient processing of spatial joins using R-trees. *Proc. ACM SIGMOD.* 237–246, 1993.
20. A. Guttman. R-tree: A dynamic index structure for spatial searching. *Proc. ACM SIGMOD.* 47–57, 1984.
21. T. Sellis, N. Roussopoulos, and C. Faloutsos. The R+-tree: A dynamic index for multidimensional objects. *Proc. 12th VLDB.* 507–518, 1987.
22. N. Beckmann et al. The R*-tree: An efficient and robust access method for points and rectangles. *Proc. ACM SIGMOD.* 322–331, 1990.
23. S. Berchtold, D. A. Keim, and H.-P. Kriegel. The X-tree: An index structure for high-dimensional data. *Prod. 22nd Int. Conf. Very Large Data Bases.* 28–39, 1996.
24. K.-I. Lin, H. V. Jagadish, and C. Faloutsos. The TV tree: An index structure for high-dimensional data. *VLDB J.* 3(4):517–549, 1994.
25. D. A. White and R. Jain. Algorithms and strategies for similarity retrieval. *Technical Report VCL–96–01.* Visual Computing Laboratory, University of California, San Diego, 1996.
26. M. Flickner et al. Query by image and video content: The QBIC system. *IEEE Computer.* 28(9):23–32, 1995.
27. D. White and R. Jain. Similarity indexing: Algorithms and performance. *Proc. SPIE Storage and Retrieval for Image and Video Databases,* 1996.
28. R. Ng and A. Sedighian. Evaluating multi-dimensional indexing structures for images transformed by principal component analysis. *Proc. SPIE Storage and Retrieval for Image and Video Databases,* 1996.
29. Y. A. Aslandogan and C. T. Yu. Techniques and systems for image and video retrieval. *IEEE Trans. KDE.* 11(1):56–63, 1999.
30. Y. Rui et al. Relevance feedback: A power tool for interactive content-based image retrieval. *IEEE Trans. Circ. Sys. Video Tech.* 8(5):644–655, 1998.
31. J. Huang, S. Kumar, and M. Metra. Combining supervised learning with color correlograms for content–based image retrieval. *Proc. ACM Multimedia '95.* 325–334, 1997.
32. R. Torres and A. Falcao. Content-based image retrieval: Theory and applications. *Revista de Informatica Teorica e Aplicada.* 13(2):161–185, 2006.
33. X. S. Zhou and T. S. Huang. Relevance feedback in image retrieval: A comprehensive review. *Multimedia Systems* 8:536–544, 2003.
34. C. Lopez-Pujalte, V. Bote, and F. Anegon. Order-based fitness functions for genetic algorithms applied to relevance feedback. *J. Am. Soc. Info. Sci. and Tech.* 54(2):152–160, 2003.
35. Y. Rui et al. A power tool in interactive content-based image retrieval. *IEEE Trans. Cir. Sys. Video Tech.* 8(5):644–655, 1998.
36. I. Cox et al. The Bayesian image retrieval system, PicHunter: Theory, implementation, and psychophysical experiments. *IEEE Trans. Image Processing.* 9(1):20–37, 2000.
37. S. Tong and E. Chang. Support vector machine active learning for image retrieval. *Proc. 9th ACM Int. Conf. Multimedia.* 107–118, 2001.
38. J. Huang et al. Spatial color indexing and applications. *Int. J. Computer Vision* 35(3):245–268, 1999.
39. G. Pass and R. Zabith. Comparing images using joint histograms. *Multimedia Systems.* 7:234–240, 1999.
40. W. Niblack et al. Querying images by content using color, texture, and shape. *SPIE Conf. Storage and Retrieval for Image and Video Database.* 1908:173–187, 1993.
41. G. Pass and R. Zabith. Histogram refinement for content-based image retrieval. *IEEE Workshop on Applications of Computer Vision.* 96–102, 1996.
42. J. Huang et al. Image indexing using color correlogram. *IEEE Int. Conf. Computer Vision and Pattern Recognition* (Puerto Rico). 762–768, 1997.
43. P. Howarth et al. Medical image retrieval using texture locality and colour. *CLEF2004–LNCS* 3491. 740–749, 2005.
44. B. S. Manjunath and J. R. Ohm. Color and texture descriptors. *IEEE Trans. Cir. Sys. Video Tech.* 11:703–715, 2001.
45. J. R. Smith and S.-F. Chang. Automated binary texture feature sets for image retrieval. *Proc. ICASSP* (Atlanta), 1996.
46. R. M. Haralick, K. Shanmugam, and I. Dinstein. Texture features for image classification. *IEEE Trans. Sys., Man., and Cyb.* SMC-3(6):610–621, 1973.
47. H. Tamura, S. Mori, and T. Yamawaki. Texture features corresponding to visual perception. *IEEE Trans. Sys., Man., and Cyb.* SMC-8(6):460–473, 1978.
48. X. Tang. Texture information in run-length matrices. *IEEE Trans. Image Processing* 7(11):1602–1609, 1998.
49. A. Laine and J. Fan. Texture classification by wavelet packet signatures. *IEEE Trans. PAMI.* 15(11):1186–1191, 1993.
50. T. Chang and C. C. J. Kuo. Texture analysis and classification with tree-structured wavelet transform. *IEEE Trans. Image Processing.* 2(4):429–441, 1993.
51. J. G. Daugman. Complete discrete 2D Gabor transforms by neural networks for image analysis and compression. *IEEE Trans. ASSP.* 36:1169–1179, 1988.
52. B. S. Manjunath and W. Y. Ma. Texture features for browsing and retrieval of image data. *IEEE Trans. PAMI.* 18(8):837–842, 1996.
53. J. Francos. Orthogonal decompositions of 2D random fields and their applications in 2D spectral estimation. In N. K. Bose and C. R. Rao (Eds.), *Signal Processing and its Applications.* North Holland. 20–27, 1993.
54. F. Liu and R. W. Picard. Periodicity, directionality, and randomness: Wold features for image modeling and retrieval. *IEEE Trans. PAMI.* 18(7):722–733, 1996.

55. P. P. Ohanian and R. C. Dubes. Performance evaluation for four classes of texture features. *Pattern Recognition.* 25(8):819–833, 1992.
56. C. Shyu et al. ASSERT: A physician-in-the-loop content-based retrieval system for HRCT image databases. *Comput. Vis. Image Understand.* 75(1–2):111–132, 1999.
57. J. Mao and A. K. Jain. Texture classification and segmentation using multiresolution simultaneous autoregressive models. *Pattern Recognition.* 25(2):173–188, 1992.
58. J. Weszka, C. Dyer, and A. Rosenfild. A comparative study of texture measures for terrain classification. *IEEE Trans. Sys., Man., and Cyb.* SMC–6(4):269–285, 1976.
59. A. P. Pentland. Fractal-based description of natural scenes. *IEEE Trans. PAMI.* 6(6):661–674, 1984.
60. C. Chatfield and A. Collins. *Introduction to multivariate analysis.* Chapman & Hall, 1983.
61. E. Persoon and K. Fu. Shape discrimination using Fourier descriptors. *IEEE Trans. Sys., Man., and Cyb.* 7:170–179, 1977.
62. H. Kauppinen, T. Seppnaen, and M. Pietikainen. An experimental comparison of autoregressive and Fourier-based descriptors in 2D shape classification. *IEEE Trans. PAMI.* 17(2):201–207, 1995.
63. E. M. Arkin et al. An efficiently computable metric for comparing polygonal shapes. *IEEE Trans. PAMI.* 13(3), 1991.
64. A. Pentland, R. W. Picard, and S. Sclaroff. Photobook: Content-based manipulation of image databases. *Int. J. Computer Vision*, 1996.
65. S. Abbasi, F. Mokhtarian, and J. Kittler. Enhancing CSS-based shape retrieval for objects with shallow concavities. *Image and Vision Computing.* 18(3):199–211, 2000.
66. F. Mokhtarian and S. Abbasi. Shape similarity retrieval under Affine transforms. *Pattern Recognition* 35(1):31–41, 2002.
67. Z. You and A. K. Jain. Performance evaluation of shape matching via chord length distribution. *Computer Vision, Graphics, Image Processing.* 28:185–198, 1984.
68. A. K. Jain. *Fundamentals of Digital Image Processing.* Prentice Hall, 1986.
69. N. Arica and F. Vural. BAS: A perceptual shape descriptor based on the beam angle statistics. *Pattern Recognition Letters.* 24(9–10):1627–1639, 2003.
70. G. C.-H. Chuang and C.-C. J. Kuo. Wavelet descriptor of planar curves: Theory and applications. *IEEE Trans. Image Processing.* 5(1):56–70, 1996.
71. M. K. Hu. Visual pattern recognition by moment invariants, computer methods in image analysis. *IRE Trans. Info. Theory* 8, 1962.
72. L. Yang and F. Algregtsen. Fast computation of invariant geometric moments: A new method giving correct results. *Proc. IEEE Int. Conf. on Image Processing.* Austin, Texas, 1994.
73. Y. S. Kim and W. Y. Kim. Content-based trademark retrieval system by using visually salient feature. *Proc. IEEE Conf. Computer Vision and Pattern Recognition.* 307–312, 1997.
74. L. Prasad. Morphological analysis of shapes. *CNLS Research Highlights.* Los Alamos National Laboratory, 1997.
75. D. H. Ballard and D. M. Brown. *Computer Vision.* Prentice Hall, 1982.
76. S.-K. Chang. *Principles of Pictorial Information Systems Design.* Prentice Hall Int'l Editions, 1989.
77. S. K. Chang, Q. Y. Shi, and C. Y. Yan. Iconic indexing by 2D strings. *IEEE Trans. PAMI.* 9(3):413–428, 1987.
78. H. Samet. The quadtree and related hierarchical data structures. *ACM Computing Surveys.* 16(2):187–260, 1984.
79. V. N. Gudivada and V. V. Raghavan. Design and evaluation of algorithms for image retrieval by spatial similarity. *ACM Trans. Info. Sys.* 13(2):115–144, 1995.
80. M. E. Mattie et al. PathMaster: Content-based cell image retrieval using automated feature extraction. *J. Am. Med. Informatics Assoc.* 7(4):404–415, 2000.
81. Q. Wang, V. Megalooikonomou, and D. Kontos. A medical image retrieval framework. *Proceedings of the IEEE Workshop on Machine Learning for Signal Processing (MLSP'05).* 233–238, 2005.
82. C. Brodley et al. Content-based retrieval from medical image databases: A synergy of human interaction, machine learning and computer vision. *Proc. the 10th National Conf. on Artificial Intelligence.* 760–767, 1999.
83. A. Marchiori et al. CBIR for medical images—An evaluation trial. *Proc. IEEE Workshop on Content-Based Access of Image and Video Libraries.* 89–93, 2001.
84. I. el-Naqa et al. A similarity learning approach to content-based image retrieval: Application to digital mammography. *IEEE Tran. Med. Imaging.* 23(10):1233–1244, 2004.
85. S. T. C. Wong. CBIR in medicine: Still a long way to go. *Proc. IEEE Workshop on Content-Based Access of Image and Video Libraries.* 114, 1998.
86. S. T. C. Wong and H. K. Huang. Design methods and architectural issues of integrated medical image data base systems. *Computerized Medical Imaging and Graphics.* 20(4):285–299, 1996.
87. T. Glatard, J. Montagnat, and I. E. Magnin. Texture based medical image indexing and retrieval: Application to cardiac imaging. *Proc. 6th ACM SIGMM Int. Workshop on Multimedia Info. Retrieval.* 135–142, 2004.
88. M. M. Rahman, P. Bhattacharya, and B. C. Desai. A framework for medical image retrieval using machine learning and statistical similarity matching techniques with relevance feedback. *IEEE Trans. Info. Tech. Biomed.* 11(1):58–69, 2007.
89. C. H. Li and P. C. Yuen. Regularized color clustering in medical image database. *IEEE Trans. Med. Imag.* 19(11):1150–1155, 2000.
90. S. Tamai. The color of digital imaging in pathology and cytology. *Digital Color Imaging in Biomedicine*, no. 2:61–66, 2001.

91. H. K. Choi et al. Design of the breast carcinoma cell bank system. *Proc. 6th Int. Workshop on Enterprise Networking and Computing in Healthcare Industry (HEALTHCOM '04)*. 88–91, 2004.
92. F. Schnorrenberg et al. Content-based retrieval of breast cancer biopsy slides. *Tech. and Health Care.* 8:291–297, 2000.
93. L. Zheng et al. Design and analysis of a content-based pathology image retrieval system. *IEEE Trans. Info. Tech. Biomed.* 7(4):249–255, 2003.
94. H. L. Tang, R. Hanka, and H. S. Ip. Histological image retrieval based on semantic content analysis. *IEEE Trans. Info. Tech. Biomed.* 7(1):26–36, 2003.
95. R. W. K. Lam et al. An iconic and semantic content based retrieval system for histological images. *VISUAL 2000–LNCS.* 1929:384–395, 2000.
96. L. H. Tang et al. Automatic semantic labelling of medical images for content-based retrieval. *Proc. Int. Conf. Artificial Intelligence, Expert Systems and Applications (EXPERSYS 1998).* 77–82, 1998.
97. L. H. Tang et al. Extraction of semantic features of histological images for content–based retrieval of images. *Proc. IEEE Symp. Computer-Based Medical Systems (CBMS 2000).* Houston, Texas, 2000.
98. M. Nischik and C. Forster. Analysis of skin erythema using true-color images. *IEEE Trans. Med. Imaging.* 16(6):711–716, 1997.
99. G. L. Hansen et al. Wound status evaluation using color image processing. *IEEE Trans. Med. Imaging.* 16(1):78–86, 1997.
100. M. M. Rahman, B. C. Desai, and P. Bhattacharya. Image retrieval–based decision support system for dermatoscopic images. *Proc. 19th IEEE Symp. on Computer-Based Medical Systems.* 285–290, 2006.
101. M. Nishibori. Problems and solutions in medical color imaging. *Proc. Sec. Int. Symp. Multispectral Imaging and High Accurate Color Reproduction.* 9–17, 10–11, 2000.
102. B. V. Dhandra et al. Analysis of abnormality in endoscopic images using combined HIS color space and watershed segmentation. *18th Int. Conf. Pattern Recognition*, 2006.
103. S. Xia et al. An endoscopic image retrieval system based on color clustering method. *Third Int. Symp. Multispectral Image Processing and Pattern Recognition, Proc. SPIE.* 5286:410–413, 2003.
104. K. B. Kim, S. Kim, and G. H. Kim. Analysis system of endoscopic image of early gastric cancer. *IEEE Trans. Fundamentals of Electronics, Communications Computer Sciences.* E89-A(10):2662–2669, 2006.
105. U. Honmyo et al. Mechanisms producing color change in flat early gastric cancers. *Endoscopy.* 29:366–371, 1997.
106. T. Ogihara et al. Display of mucosal blood flow function and color enhancement based on blood flow index (IHb color enhancement). *Clinical Gastroenterology.* 12: 109–117, 1997.
107. S. Tsuji et al. Functional imaging for the analysis of the mucosal blood hemoglobin distribution using electronic endoscopy. *Gastrointest. Endosc.* 34:332–336, 1988.
108. M. P. Tjoa and S. M. Krishnan. Feature extraction for the analysis of colon status from the endoscopic images. *BioMedical Engineering Online.* 2(9):1–17, 2003.
109. Z. L. Chen. *Research on Tongue Diagnosis.* Shanghai Science and Technology Publishing House, 1982.
110. G. Maciocia. *Tongue Diagnosis in Chinese Medicine.* Eastland Press, 1995.
111. L. S. Shen et al. Image analysis for tongue characterization. *Chinese J. Electronics.* 12(3):317–323, 2003.
112. N. M. Li et al. The contemporary investigations of computerized tongue diagnosis. In *Handbook of Chinese Tongue Diagnosis.* Shed-Yuan Publishing. 1315–1317, 1994.
113. B. G. Wei et al. Recent progresses in analysis of tongue manifestation for traditional Chinese medicine. *Chinese J. Biomed. Eng.* 14(2): 55–64, 2005.
114. B. Pang, D. Zhang, and K. Wang. Tongue image analysis for appendicitis diagnosis. *Info. Sci.* 175:160–176, 2005.
115. C. H. Li and P. C. Yuen. Tongue image matching using color content. *Pattern Recognition.* 35:407–419, 2002.
116. Y. G. Wang et al. An image analysis system for tongue diagnosis in traditional Chinese medicine. *CIS 2004, LNCS 3314.* 1181–1186, 2004.
117. C. C. Chiu. A novel approach based on computerized image analysis for traditional Chinese medical diagnosis of the tongue. *Computer Methods and Programs in Biomedicine.* 61:77–89, 2000.
118. J. H. Jang et al. Development of the digital tongue inspection system with image analysis. *Proc. Second Joint EMBS/BMES Conf.* 1033–1034, 2002.
119. A. M. Wang. *Research on Image Analysis for Tongue Characterization.* Ph.D dissertation, Beijing Polytechnic University, 2001.
120. X. F. Zhang. *The Primary Study of Classification and Recognition on Tongue Manifestation in Traditional Chinese Medicine.* Ph.D dissertation, Beijing University of Technology, 2006.
121. B. G. Wei. *Research on Color Reproduction and Analysis of Texture, Shape and State of Tongue for Traditional Chinese Medicine.* Ph.D dissertation, Beijing University of Technology, 2005.
122. H. Muller et al. Integrating content-based visual access methods into a medical case database. *Proc. Medical Informatics Europe Conf.* 480–485, 2003.
123. A. Rosset et al. Casimage project—A digital teaching files authoring environment. *J. Thoracic Imaging.* 19(2):1–6, 2004.
124. P. Clough, H. Muller, and M. Sanderson. The CLEF 2004 cross-language image retrieval track. *Proc. 5th Workshop Cross-Language Evaluation Forum, CLEF 2004.* 3491:597–613, 2004.

125. D. Xu et al. Run-length encoding for volumetric texture. *Proc. VIIP,* 2004.
126. T. M. Lehmann et al. Similarity of medical images computed from global feature vectors for content-based retrieval. *Proc. KES2004-LNAI.* 3214:989–995, 2004.
127. D. S. Raicu et al. A texture dictionary for human organs tissues'classification. *Proc. 8th World Multiconf. on Sys. Cyb. and Info.(SCI2004).* 2004.
128. W. Tsang et al. Texture-based image retrieval for computerized tomography databases. *Proc. 18th IEEE Symp. Computer-Based Medical Systems (CBMS'05).* 593–598, 2005.
129. S. Orphanoudakis, C. Chronaki, and D. Vamvaka. I2Cnet: Content-based similarity search in geographically distributed repositories of medical images. *Computerized Medical Imaging and Graphics.* 20(4):193–207, 1996.
130. P. A. Freeborough and N. C. Fox. MR image texture analysis applied to the diagnosis and tracking of Alzheimer's disease. *IEEE Trans. Med. Imag.* 17(3):475–479, 1998.
131. J. C. Felipe, A. J. M. Traina, and C. Traina. Retrieval by content of medical images using texture for tissue identification. *Proc. 16th IEEE Symposiu. on Computer-Based Medical Systems (CBMS'03).* 175–180, 2003.
132. D. M. Kwak et al. Content-based ultrasound image retrieval using a coarse to fine approach. *Ann. N.Y. Acad. Sci.* 980:212–224, 2002.
133. M. A. Sheppard and L. Shih. Efficient image texture analysis and classification for prostate ultrasound diagnosis. *Proc. IEEE Conf. Computational Systems Bioinformatics.* 7–8, 2005.
134. H. Alto, R. M. Rangayyan, and J. E. Leo Desautels. Content-based retrieval and analysis of mammographic masses. *J. Electronic Imaging.* 14(2):023016–1–17, 2005.
135. S. Xia et al. A content-based retrieval system for endoscopic images. *Proc. 27th Annual Conf. IEEE EMBS.* 1720–1723, 2005.
136. A. K. Jain and F. Farroknia. Unsupervised texture segmentation using Gabor filters. *Pattern Recognition.* 24(12):1167–1186, 1991.
137. C. G. Zhao et al. Liver CT-image retrieval based on Gabor texture. *Proc. 26th Annual Conf. IEEE EMBS.* 1491–1494, 2004.
138. D. Zhao, Y. Chen, and H. Correa. Statistical categorization of human histological images. *Proc. IEEE Int. Conf. Image Processing (ICIP'05).* 628–631, 2005.
139. S. Antani et al. Evaluation of shape similarity measurement methods for spine X-ray images. *J. Vis. Commun. Image R.* 15, 285–302, 2004.
140. D. Comaniciu, D. Foran, and P. Meer. Shape-based image indexing and retrieval for diagnostic pathology. *Proc. 14th Int. Conf. Pattern Recognition.* 902–904, 1998.
141. R. M. Rangayyan et al. Measures of acutance and shape for classification of breast tumors. *IEEE Trans. Med. Imag.* 16(6):799–810, 1997.
142. S. Ciatto, L. Cataliotti, and V. Distante. Nonpalpable lesions detected with mammography: Review of 512 consecutive cases. *Radiol.* 165(1):99–102, 1987.
143. R. M. Rangayyan and S. G. Elkadiki. Algorithm for the computation of region-based image edge profile acutance. *J. Electron. Imag.* 4(1):62–70, 1995.
144. M. K. Hu. Visual pattern recognition by moment invariants. In J. K. Aggarwal, R. O. Duda, and A. Rosenfeld. *Computer Methods in Image Analysis.* IEEE Computer Society, 1977.
145. R. C. Gonzalez and R. E. Woods. *Digital Image Processing.* Prentice Hall, 2002.
146. S. Zhu and G. Schaefer. Thermal medical image retrieval by moment invariants. *Int. Symp. Biol. Med. Data Anal. LNCS* 3337:182–187, 2004.
147. S. Maitra. Moment invariants. *Proc. IEEE* 67:697–699, 1979.
148. G. Schaefer, S. Y. Zhu, and S. Ruszala. Visualisation of medical infrared image databases. *Proc. 27th Annual Conf. IEEE EMBS.* 634–637, 2005.
149. G. P. Robinson et al. Medical image collection indexing: Shape-based retrieval using KD-trees. *Comput. Vis. Graphics Image Proces.* 20(4):209–217, 1996.
150. A. J. M. Traina et al. Content-based image retrieval using approximate shape of objects. *Proc. 17th IEEE Symposium on Computer-Based Medical Systems (CBMS'04).* 91–96, 2004.
151. W. Zhang et al. Shape-based indexing in a medical image database. *Proc. IEEE Workshop Biomed. Image Analysis.* 221–230, 1998.
152. J. Felipe et al. Effective shape-based retrieval and classification of mammograms. *Proc. 2006 ACM Symp. Applied Computing.* 250–255, 2006.
153. P. Korn et al. Fast and effective retrieval of medical tumor shapes. *IEEE Trans. KDE.* 10(6), 1998.
154. J. Z. Wang. Pathfinder: Multiresolution region-based searching of pathology images using IRM. *J. Am. Med. Informatics Asso. (AMIA)–Proc. AMIA Annual Symp., 2000 Symposium Suppl.* Los Angeles, CA. 883–887, 2000.
155. C. Ng and G. Martin. Content–description interfaces for medical imaging. *Technical Report CS–RR–383.* Coventry, U.K. 2001.
156. W. Liu and Q. Tong. Medical image retrieval using salient point detector. *Proc. 2005 IEEE EMBS 27th Annual Conference.* Shanghai, China. 6352–6355, 2005.
157. S. Chu, S. Narayanan, and C.-C. Kuo. Efficient rotation invariant retrieval of shapes with applications in medical databases. *Proc. 19th IEEE Symp. Computer-Based Medical Systems (CBMS'06).* 673, 678, 2006:.
158. P. A. Mlsna and N. M. Sirakov. Intelligent shape feature extraction and indexing for efficient content-based

medical image retrieval. *Proc. 6th IEEE Southwest Symp. Image Analysis Interpretation.* 172–176, 2004.
159. B. Fischer et al. Content-based image retrieval by matching hierarchical attributed region adjacency graphs. *Proc. SPIE–Medical Imaging: Image Processing.* 5370:598–606, 2004.
160. J. K. Udupa and G. T. Herman. *3D Imaging in Medicine.* CRC Press, 2000.
161. J. Kim et al. A new way for multidimensional medical data management: Volume of interest (VOI)–based retrieval of medical images with visual and functional features. *IEEE Trans. Info. Tech. Biomed.* 10(3):598–607, 2006.
162. Y. Liu et al. Semantic-based biomedical image indexing and retrieval. In L. Shapiro, H. Kriegel, and R. Veltkamp (Eds.). *Trends and Advances in Content-Based Image and Video Retrieval.* Springer, 2004.
163. Y. Liu, R. T. Collins, and W. E. Rothfus. Robust midsagittal plane extraction from normal and pathological 3D neuroradiology images. *IEEE Trans. Med. Imag.* 20(3):175–192, 2001.
164. Y. Liu, W. E. Rothfus, and T. Kanade. Content-based 3D neuroradiologic image retrieval: Preliminary results. *Proc. IEEE Workshop Content-based Access of Image and Video Libraries, in Conjunction with Int. Conf. Computer Vision.* 91–100, 1998.
165. J. Declerck et al. Automatic retrieval of anatomical structures in 3D medical images. *Technical Report 2485,* INRIA. Sophia-Antipolis, France, 1995.
166. A. Guimond and G. Subsol. Automatic MRI database exploration and applications. *Int. J. Pattern Recog. Artif. Intell.* 11, 1997:1345–1365.
167. V. Megalooikonomou, H. Dutta, and D. Kontos. Fast and effective characterization of 3D region data. *Proc. IEEE Int. Conf. Image Processing (ICIP'02).* 421–424, 2002.
168. S. Minoshima et al. An automated method for rotational correction and centering of three-dimensional functional brain images. *J. Nucl. Med.* 33:1579–1585, 1992.
169. S. Minoshima et al. Automated detection of the intercommissural line for stereotactic localization of functional brain images. *J. Nucl. Med.* 34:322–329, 1993.
170. S. Minoshima et al. Anatomic standardization: Linear scaling and nonlinear warping of functional brain images. *J. Nucl. Med.* 35:1528–1537, 1994.
171. I. G. Zubal et al. Computerized three-dimensional segmented human anatomy. *Med. Phys.* 21:299–302, 1994.
172. T. M. Lehmann et al. Extended query refinement for content-based access to large medical image databases. *Proc. SPIE–Medical Imaging: PACS and Imaging Informatics.* 5371(15):90–98, 2004.
173. W. W. Chu, A. F. Cardenas, and R. K. Taira. KMED: A knowledge-based multimedia medical distributed database system. *Info. Sys.* 19(4):33–54, 1994.
174. S. Aksoy et al. Interactive classification and content-based retrieval of tissue images. *Proc. SPIE–Applications of Digital Image Processing XXV.* 4790:71–81, 2002.
175. R. Chbeir and F. Favetta. A global description of medical image with high precision. *Proc. IEEE Int. Symp. Bio-Informatics and Biomedical Engineering (BIBE'2000).* 289–296, 2000.
176. R. Chbeir, Y. Amghar, and A. Flory. MIMS: A prototype for medical image retrieval. *Proc. 6th Int. Conf. Content Based Multimedia Information Access,* RIAO2000. 846–861, 2000.
177. E. Petrakis. Content-based retrieval of medical images. *Int. J. Comput. Res.* 11(2):171–182, 2002.
178. W. W. Chu et al. Knowledge-based image retrieval with spatial and temporal constructs. *IEEE Trans. KDE.* 10(6):872–888, 1998.
179. P. M. Willy and K. H. Küfer. Content-based medical image retrieval (CBMIR): An intelligent retrieval system for handling multiple organs of interest. *Proc. 17th IEEE Symp. Computer-Based Medical Systems (CBMS'04).* 103–108, 2004.
180. C. R. Shyu et al. ASSERT: A physician-in-the-loop content-based retrieval system for HRCT image databases. *Computer Vision and Image Understanding.* 75(1/2): 111–132, 1999.
181. C. R. Shyu et al. Using human perceptual categories for content-based retrieval from a medical image database. *Computer Vision and Image Understanding.* 88:119–151, 2002.
182. A. S. Barb, C. R. Shyu, and Y. P. Sethi. Knowledge representation and sharing using visual semantic modeling for diagnostic medical image databases. *IEEE Trans. Info. Tech. Biomed.* 9(4):538–553, 2005.
183. E. Stern and S. Swensen. *High Resolution CT of the Chest: Comprehensive Atlas.* 2nd ed. Lippincott, Williams & Wilkins, 2000.
184. W. Webb, N. Muller, and D. Naidich. *High-Resolution CT of Lung.* Lippincott-Raven, 1996.
185. K. K. T. Cheung et al. An object-oriented framework for content-based image retrieval based on 5-tier architecture. *Proc. Asia-Pacific Software Eng. Conf. 99* 174–177, 1999.
186. R. W. K. Lam et al. A multi-window approach to classify histological features. *Proc. Int. Conf. Pattern Recognition* 2:259–262, 2000.
187. K. K. T. Cheung et al. A software framework for combining iconic and semantic content for retrieval of histological images. *VISUAL 2000–LNCS* 1929:488–499, 2000.
188. L. H. Tang et al. An intelligent system for integrating semantic and iconic features for image retrieval. *Proc. Computer Graphics International.* 240–245, 2001.
189. L. Tan et al. Integration of intelligent engines for a large scale medical image database. *Proc. 13th IEEE Symp. Computer-Based Medical Systems.* 235–240, 2000.

190. M. C. Jaulent et al. A property concept frame representation for flexible image content retrieval in histopathology databases. *Proc. Annual Symp. Am. Soc. Med. Informatics (AMIA)*. 379–383, 2000.
191. H. Shao, W. C. Cui, and L. Tang. Medical image description in content-based image retrieval. *Proc. 27th Annual Conf. IEEE EMBS*. 6336–6339, 2005.
192. T. M. Lehmann et al. Content-based image retrieval in medical applications. *Methods Inf. Med.* 4:354–361, 2004.
193. D. Keysers et al. A statistical framework for model-based image retrieval in medical applications. *J. Electronic Imaging*. 12(1):59–68, 2003.
194. C. Hsu, W. Chu, and R. Taira. A knowledge-based approach for retrieving images by content. *IEEE Trans. KDE.*, 1996.
195. R. Chbeir and F. Favetta. A global description of medical imaging with high precision. *IEEE Trans. Sys., Man, Cyb.–Part B: Cybernetics*. 33(5):752–757, 2003.
196. R. Chbeir et al. A hyper-spaced data model for content and semantic-based medical image retrieval. *Proc. ACS/IEEE Int. Conf. Computer Systems and Applications*. 161–167, 2001.
197. International Classification of Diseases, 10th Revision. http://www.who.int/classifications/icd/en
198. B. L. Humphreys (Ed.). *UMLS Knowledge Sources–First Experimental Edition Documentation*. National Library of Medicine, 1990.
199. R. Chbeir, Y. Amghar, and A. Flory. System for medical image retrieval the MIMS model. *Proc. 3rd Int. Conf. Visual (VISUAL'99), LNCS 1614*. 37–42, 1999.
200. S. Atnafu, R. Chbeir, and L. Brunie. Content-based and metadata retrieval in medical image database. *Proc. 15th IEEE Symp. Computer-Based Medical Systems (CBMS 2002)*. 327–332, 2002.
201. R. Chbeir, S. Atnafu, and L. Brunie. Image data model for an efficient multi-criteria query: A case in medical databases. *Proc. 14th Int. Conf. Scientific Statistical Database Management (SSDBM'02)*. 165–174, 2002.
202. T. Lehmann et al. The IRMA code for unique classification of medical images. *Proc. SPIE*. 5033:440–451, 2003.
203. S. C. Orphanoudakis, C. Chronaki, and S. Kostomanolakis. I2C: A system for the indexing, storage, and retrieval of medical images by content. *Med. Informatics*. 19(2):109–122, 1994.
204. E. Ei-Kwae, H. Xu, and M. Kabuka. Content-based retrieval in picture archiving and communication systems. *J. Digital Imaging*. 13(2):70–81, 2000.
205. H. Lowe et al. Automated semantic indexing of imaging reports to support retrieval of medical images in the multimedia electronic medical record. *Meth. Info. Med.* 38(303–7), 1999.
206. H. Muller, C. Lovis, and A. Geissbuhler. The medGIFT project on medical image retrieval. *Proc. 15th IEEE Symp. Computer-Based Medical Systems (CBMS2002)*. 321–326, 2002.
207. H. Muller et al. Comparing feature sets for content-based image retrieval in a medical case database. *Proc. SPIE 2004*, 2004.
208. http://ww.gnu.org/software/gift
209. http://www.mrml.net
210. D. Feng. Information technology applications in biomedical functional imaging. *IEEE Trans. Info. Tech. Biomed.* 3(3):221–230, 1999.
211. D. Feng et al. Techniques for functional imaging. In C. T. Leondes (Ed.). *Medical Imaging Techniques and Applications*. Gordon and Breach International Series in Engineering, Technology and Applied Science. Gordon and Breach Science Publishers. 85–145, 1997.
212. W. Cai, D. Feng, and R. Fulton. Content-Based Retrieval of Dynamic PET Functional Images. *IEEE Trans. Info. Tech. Biomed.* 4(2):152–158, 2000.
213. S. C. Huang et al. Non-invasive determination of local cerebral metabolic rate of glucose in man. *Amer. J. Physiol.* 238: E69–E82, 1980.
214. D. Feng et al. An evaluation of the algorithms for determining local cerebral metabolic rates of glucose using positron emission tomography dynamic data. *IEEE Trans. Med. Imag.* 14:697–710, 1995.
215. J. Bezdek. *Pattern Recognition with Fuzzy Objective Function Algorithm*. Kluwer, 1981.
216. J. Kim et al. An objective evaluation framework for segmentation techniques of functional positron emission tomography studies. *IEEE NSS-MIC Conf.* 5:3217–3221, 2004.
217. X. Li, D. Feng, and K. Chen. Optimal image sampling schedule: A new effective way to reduce dynamic image storage and functional image processing time. *IEEE Trans. Med. Imag.* 15(5):710–719, 1996.
218. D. Feng, W. Cai, and R. Fulton. An optimal image sampling schedule design for cerebral blood volume and partial volume correction in neurologic FDG-PET studies. *Aust. N.Z. J. Med.* 28:361, 1998.

5

Data Modeling and Simulation

Dr. Alessandra Bertoldo and
Prof. Claudio Cobelli
University of Padova

5.1 Introduction

In vivo imaging techniques like positron emission tomography (PET) and magnetic resonance imaging (MRI) are providing crucial functional information at the organ/tissue level of the human body. However, this functional information is not directly available in quantitative terms by simply looking at the images. This usually requires an interpretation of the image with a mathematical model of the underling physiologic process. Various classes of models (e.g., models of data or input-output, models of system, graphic models) have been proposed to interpret PET and MRI data. Focus here is on a specific class of model of system, compartmental modeling, which is the most frequently used.

Compartmental models are widely employed to solve a broad spectrum of physiologic and clinical problems related to the distribution of materials in living systems. The governing law of these models is conservation of mass, and the models describe the events in the system by a finite number of variables; that is, they are described by ordinary differential equations. These characteristics make them very attractive to users because they formalize physical intuition in a simple and reasonable way. Their usefulness in research, especially in conjunction with tracer experiments, has been demonstrated at whole-body, organ, and cellular levels. Examples and references can be found in several books [1–5]. Purposes for which compartmental models have been developed are various, but the most relevant here are the following:

1. Identification of system structure. Such models examine different hypotheses regarding the nature of specific physiologic mechanisms.
2. Estimation of unmeasurable quantities. These quantities might include the estimation of internal parameters and other variables of physiologic interest.
3. Simulation of the intact system behavior where ethical or technical reasons would not allow direct experimentation on the system itself.

In this chapter, we will first briefly review some fundamentals on compartmental models focusing on tracee and tracer kinetics. Subsequently, we discuss some aspects of model identification and parameter estimation. Finally, we will show the power of compartmental modeling methodology in interpreting PET and nuclear magnetic resonance (NMR) functional imaging data.

5.2 Compartment Models

Before discussing the theory of compartmental models, we first need to give some definitions. A compartment is an amount of material that acts as though it is well mixed and kinetically homogeneous. A compartmental model consists of a finite number of compartments with specified interconnections among them. These interconnections represent a flux of material which physiologically represents transport from one location to another or a chemical transformation or both. An example is shown in Figure 5.1. Control signals arising in neuro-endocrine systems can also be described. In this case, there can be two separate compartmental models: one for the hormone and one for the substrate, and they interact via control signals. An example is shown in Figure 5.2.

Given our introductory definitions, it would be useful to discuss possible candidates for compartments before explaining what we mean by a compartment's well-mixed and kinetic homogeneity. Consider the notion of a compartment as a physical space. Plasma is a candidate for a compartment. A substance such as plasma glucose could be a compartment, and zinc in bone could be a compartment also as could insulin in β-cells. In some experiments, several different substances in

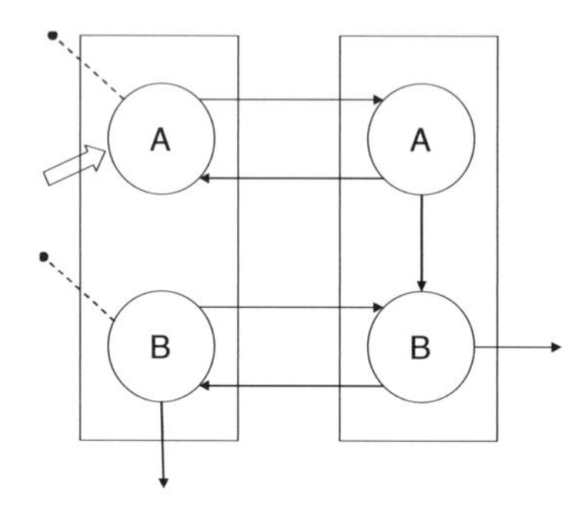

FIGURE 5.1 The compartmental system model showing the interconnections among compartments. The administration of material into and sampling from the accessible pools are indicated by the input arrow and measurement symbols (dotted line), respectively. The solid arrows represent the flux of material from one compartment to another.

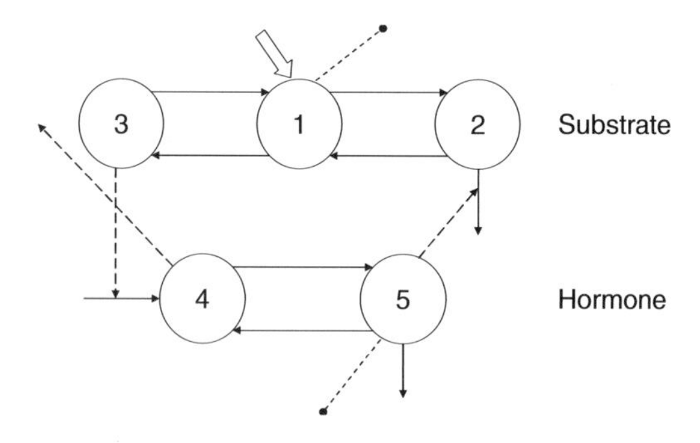

FIGURE 5.2 An example of a multicompartmental model of an endocrine-metabolic control system. The top and bottom multicompartmental models describe the metabolism of the substrate and hormone, respectively. The dotted arrows represent control signals. For example, the dotted arrow from compartment 3 to the input arrow into compartment 4 indicates that the amount of material in compartment 3 controls the input of material into compartment 4.

plasma can be followed, such as glucose, lactate, and alanine. Thus, there can be more than one plasma compartment in the same experiment, one for each of the substances being studied. This notion extends beyond plasma. Glucose and glucose-6-phosphate might need to be shown by two different compartments, depending on whether they are found in liver or muscle tissue. Thus, a single physical space or substance may actually be represented by more than one compartment, depending on the components measured or their location.

In addition, one must distinguish between compartments that are accessible and nonaccessible for measurement. Researchers often try to assign physical spaces to the nonaccessible compartments. This is a very difficult problem which is best addressed by the recognition that a compartment is actually a theoretic construct, one which may in fact combine material from several different physical spaces. To equate a compartment with a physical space depends upon the system under study and the assumptions made about the particular model.

With these notions of what might constitute a compartment in mind, it is easier to define the concepts of well-mixed and kinetic homogeneity. Well-mixed means that any two samples taken for a compartment at the same time would have the same concentration of the substance being studied and therefore would be equally representative. Thus, the concept of well-mixed relates to the uniformity of the information contained in a single compartment.

Kinetic homogeneity means that every particle in a compartment has the same probability of taking any pathway leaving the compartment. When a particle leaves a compartment, it does so because of metabolic events related to transport and utilization, and all particles in the compartment have the same probability of leaving due to one of these events.

This process of combining material with similar characteristics into collections that are homogeneous and that behave identically is what allows one to reduce a complex physiologic system into a finite number of compartments and pathways. The number of compartments required depends both on the system being studied and on the richness of the experimental configuration. A compartmental model is clearly unique for each system studied because it incorporates known and hypothesized physiology and biochemistry specific to that system. It provides the investigator with insights into the system's structure and is only as good as the assumptions that are incorporated into its structure.

5.2.1 Tracee Model

In this section, we will discuss the definition of the tracee model using Figure 5.3. This is a typical compartment, the ith compartment. The tracee model can be formalized by defining precisely the flux of tracee material into and out of this compartment and can be establishing the measurement

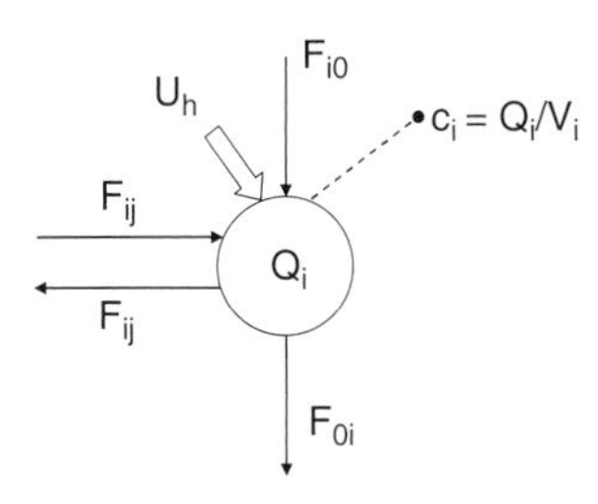

FIGURE 5.3 The ith compartment of an n-compartmental model showing fluxes into and out of the compartment, inputs, and measurements.

equation if this compartment is accessible for sampling. Once this is understood, the process of connecting several such compartments together into a multicompartmental model and writing the corresponding equations is easy.

Let Figure 5.3 represent the ith compartment of an n-compartment model of the tracee system with Q_i denoting the mass of the compartment. The arrows represent fluxes into and out of the compartment. The input flux into the compartment from outside the system, the de novo synthesis of material, is represented by F_{i0}; the flux to the environment and therefore out of the system by F_{0i}; the flux to and from compartment j by F_{ji} and F_{if}, respectively; and finally, $U_h(h = 1,\ldots,r)$ denotes an exogenous input. All fluxes F_{ij} $(i = 0,1,\ldots,n;\ j = 0,1,\ldots,n;\ i \neq j)$ and masses $Q_i(i = 1,2,\ldots,n)$ are ≥ 0. The dashed arrow with a bullet indicates that the compartment is accessible to measurement. This measurement is denoted by $C_1(1 = 1,\ldots,m)$, where we assume it is a concentration, $C_1 = Q_i/V_i$ where V_i is the volume of compartment i. As already noted, usually only a small number of compartments are accessible to test inputs and measurements.

By using the mass balance principle, one can write for each compartment

$$\dot{Q}_i(t) = -\sum_{\substack{j=0\\ j\neq i}}^{n} F_{ji}(Q_1(t)\cdots,Q_n(t)) + \sum_{\substack{j=1\\ j\neq i}}^{n} F_{ij}(Q_1(t),\cdots,Q_n(t))$$
$$+ F_{i0}(Q_1(t),\cdots,Q_n(t)) + U_h(t) \tag{5.1}$$
$$C_1(t) = \frac{Q_1(t)}{V_i} \qquad Q_i(0) = Q_{i0},$$

where $\dot{Q}_i(t) = \dfrac{dQ_i(t)}{dt}$ and $t > 0$ is time, the independent variable. All the fluxes F_{ij}, F_{i0}, and F_{0i} are assumed to be functions of the compartmental masses Q_i.

If one writes the generic flux $F_{ji}(j = 0,1,\ldots,n;\ i = 1,2,\ldots,n;\ j \neq i)$ as

$$F_{ij}(Q_1(t),\cdots,\ Q_n(t)) = k_{ij}(Q_1(t),\cdots,Q_n(t))Q_i(t) \tag{5.2}$$

where $k_{ji}(\geq 0)$ denotes the fractional transfer coefficient between compartment i and j, Equation 5.1 can be rewritten as:

$$\dot{Q}_i(t) = -\sum_{\substack{j=0\\ j\neq i}}^{n} k_{ij}(Q_1(t),\ldots,\ Q_n(t))Q_j(t)$$
$$+\sum_{\substack{j=1\\ j\neq i}}^{n} k_{ji}(Q_1(t),\ldots,Q_n(t))Q_j(t) \tag{5.3}$$
$$+ F_{i0}(Q_1(t),\ldots,Q_n(t)) + U_h(t)$$
$$C_1(t) = \frac{Q_1(t)}{V_i} \qquad Q_i(0) = Q_{i0}.$$

Equation 5.3 describes the nonlinear compartmental model of the tracee system. To make the model operative, one has to specify how the k_{ij} and F_{i0} depend upon the Q_i. This obviously depends upon the system being studied. Usually the k_{ij} and F_{i0} are functions of one or a few of the Q_i. Some possible examples include the following:

- k_{ij} are constant, and thus do not depend upon any Q_i.

$$k_{ij}(Q_1(t),\ldots,Q_n(t)) = k_{ij} = \text{ constant}. \tag{5.4}$$

- k_{ij} are described by a saturative relationship such as Michaelis-Menten.

$$k_{ij}(Q_j(t)) = \frac{V_M}{K_m + Q_j(t)} \tag{5.5}$$

or the Hill equation:

$$k_{ij}(Q_j(t)) = \frac{V_M Q_j^{m-1}(t)}{K_m + Q_j^m(t)}. \tag{5.6}$$

Note that when $m = 1$ in the above, Equation 5.6 becomes Equation 5.5.

- k_{ij} is controlled by the arrival compartment, such as by a Langmuir relationship.

$$k_{ij}(Q_j(t)) = \alpha\left(1 - \frac{Q_i(t)}{\beta}\right). \tag{5.7}$$

- k_{ij} is controlled by a remote compartment different from the source (Q_j) or arrival (Q_i) compartments. For example, using the model shown in Figure 5.2, one could have:

$$k_{02}(Q_5(t)) = \gamma + Q_5(t) \tag{5.8}$$

or a more complex description such as:

$$k_{02}(Q_2(t),\ Q_5(t)) = \frac{V_m(Q_5(t))}{K_m(Q_5(t)) + Q_2(t)}, \tag{5.9}$$

where now one has to further specify how V_m and K_m depend on the controlling compartment, Q_5.

The input F_{i0} can also be controlled by remote compartments. For example, for the model shown in Figure 5.2, one can have:

$$F_{30}(Q_4(t)) = \frac{\delta}{\varepsilon + Q_4(t)} \quad (5.10)$$

$$F_{40}(Q_3(t)) = \eta + \lambda Q_3(t) + \mu \dot{Q}_3(t). \quad (5.11)$$

The nonlinear compartmental model given in Equation 5.3 permits the description of a physiologic system in nonsteady state under very general circumstances. Having specified the number of compartments and the functional dependencies, there is now the problem of assigning a numerical value to the unknown parameters that describe them. Some of them may be assumed to be known, but some need to be tuned to the particular subject studied. Often, however, the data are not enough to arrive at the unknown parameters of the model, and a tracer is employed to enhance the information content of the data.

5.2.2 Tracer Model

In this section, we will formalize the definition of the tracer model using Figure 5.4. This parallels exactly the notions introduced above, except now we follow the tracer, denoted by lowercase letters, instead of the tracee. The link between the two, the tracee and tracer models, is given in the following section.

Suppose an isotopic radioactive tracer is injected (denoted by u_h) into the ith compartment, and denote q_i its tracer mass at time t (Figure 5.4). Assuming an ideal tracer, the tracer–tracee indistinguishability ensures that the tracee rate constants k_{ij} also apply to the tracer. Again as with the tracee, the measurement is usually a concentration, $y_l(t) = q_i(t)/V_i$.

The tracer model, given the tracee model (Equation 5.3), is:

$$\begin{aligned}\dot{q}_i(t) = &-\sum_{\substack{j=0\\ j\neq i}}^{n} k_{ji}(Q_1(t),\ldots,Q_n(t))q_j(t)\\ &+\sum_{\substack{j=0\\ j\neq i}}^{n} k_{ji}(Q_1(t),\ldots,Q_n(t))q_j(t)\\ &+u_h(t) \quad q_i(0)=0\end{aligned} \quad (5.12)$$

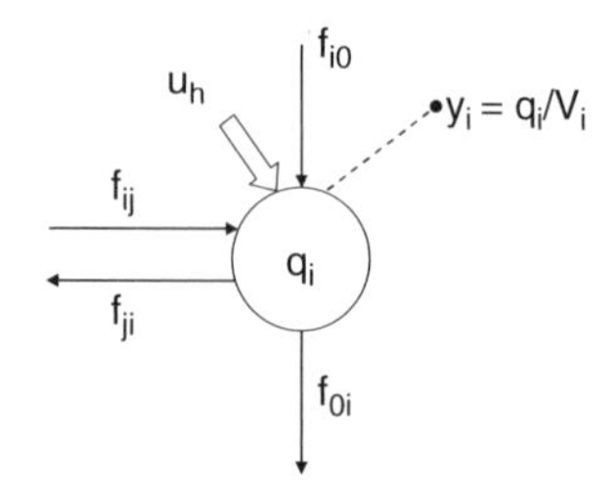

FIGURE 5.4 The ith compartment of an n-compartmental tracer model showing fluxes into and out from the compartment, inputs, and measurements.

$$y_l(t) = \frac{q_i(t)}{V_i}.$$

Note that the endogenous production term F_{i0} in Equation 5.3 does not appear in Equation 5.12; this is because this term applies only to the tracee.

5.2.3 Linking Tracer and Tracee Models

The model necessary to arrive at the nonaccessible system properties is obtained by linking the tracee and tracer models to form the tracer–tracee model; this model is described by Equations 5.3 and 5.12. The problem one wishes to solve is how to use the tracee data $C_l(t)$ and the tracer data $y_l(t)$ to obtain the unknown parameters of the model. In the general setting, with the tracee system in nonsteady state, the problem is complex. The difficulty is reduced considerably when the tracee system is in steady state. Because this situation is also the experimental protocol most frequently encountered in PET and NMR functional imaging studies, in the following we will consider this important special case.

If the tracee is in a constant steady state, the exogenous input U_h is zero, all the fluxes F_{ij} and masses $Q_i(t)$ in the tracee model (Equation 5.1) are constant, and the derivatives $Q_i(t)$ are zero. As a result, all the fractional transfer coefficients k_{ij} are constant.

The tracee and tracer models given in Equations 5.3 and 5.12, respectively, thus become

$$0 = -\sum_{\substack{j=0\\ j\neq i}}^{n} k_{ji}Q_i(t) + \sum_{\substack{j=1\\ j\neq i}}^{n} k_{ij}Q_j + F_{i0} \quad Q_i(0) = Q_{i0} \quad C_l = \frac{Q_i}{V_i} \quad (5.13)$$

$$\dot{q}_i(t) = -\sum_{\substack{j=1\\ j\neq i}}^{n} k_{ji}q_i(t) + \sum_{\substack{j=1\\ j\neq i}}^{n} k_{ij}q_j(t) + u_h(t) \quad q_i(0) = 0 \quad (5.14)$$

$$y_l(t) = \frac{q_i(t)}{V_i}.$$

This is an important result: The tracer compartmental model is linear and time-invariant if the tracee is in a constant steady state, irrespective of whether it is linear or nonlinear. The modeling machinery for Equations 5.13 and 5.14 is greatly simplified with respect to the nonlinear models shown in Equations 5.3 and 5.12. The strategy is to use the tracer data to arrive at the k_{ij} and the accessible pool volume V_i of Equation 5.14, and subsequently use the steady-state tracee model of Equation 5.13 to solve for the unknown parameters F_{i0} and the remaining Q_i.

5.3 Model Identification

With the tracer-tracee model described by Equations 5.13 and 5.14, we can now proceed to model identification, the process by which we can arrive at a numeric value of the unknown model parameters from the tracer (and tracee) actual measurements. Let's assume that measurement error is additive; thus, tracer actual measurements (assume the scalar case) are described at sample time t_i:

$$z(t_i) = y(t_i) + v(t_i) \quad i = 1, \ldots, N, \tag{5.15}$$

where $v(t_i)$ is the tracer measurement error. The error is usually given a probabilistic description in that it is assumed to be independent and often Gaussian. With Equation 5.15 and the model Equations 5.13 and 5.14, the compartmental model identification problem can now be defined. We can begin to estimate the unknown model parameters from the $z(t_i)$ noisy data contained.

Before solving this problem, however, we must deal with a prerequisite issue for the well-posedness of our parameter estimation. This is the issue of *a priori* identifiability. As seen below, this requires reasoning that uses ideal noise-free data, (i.e., Equations 5.13 and 5.14).

5.3.1 *A priori* Identifiability

A priori identifiability is a key step in the formulation of a structural model in which parameters are going to be estimated from a set of data. The question *a priori* identifiability addresses is the following: Do the data contain enough information to estimate all of the unknown parameters of the postulated model structure? This question is usually referred to as the *a priori* identifiability problem. It is set in the ideal context of an error-free model structure and noise-free, continuous time measurements and is an obvious prerequisite for well-posedness of parameter estimation from real data. In particular, if it turns out in such an ideal context that the postulated model structure is too complex for the particular set of data (i.e., some model parameters are not identifiable from the data), there is no way in a real situation, where there is error in the model structure and noise in the data, that the parameters can be identified. The *a priori* identifiability problem is also referred to as the structural identifiability problem because it is set independently of a particular set of values for the parameters. For the sake of simplicity, in what follows, only the term *a priori* will be used to qualify the problem.

Only if the model is *a priori* identifiable is it meaningful to use the techniques to estimate the numeric values of the parameters from the data that will be discussed later. If the model is *a priori* nonidentifiable, a number of strategies can be considered. One would be to enhance the information content of the experiment by adding, when feasible, inputs and/or measurements. Another possibility would be to reduce the complexity of the model by simplifying its model structure (e.g., by lowering the model order) or by aggregating some parameters. These simple statements allow one to foresee the importance of *a priori* identifiability also in relation to qualitative experiment design (e.g., definition of an experiment), which allows one to obtain an *a priori* identifiable model with the minimum number of inputs and measurements.

Before discussing the problem in depth and the methods available for its solution, it is useful to illustrate the fundamentals through some simple examples. Then, some formal definitions will be given, using these simple examples where the identifiability issue can be easily addressed.

5.3.1.1 Examples

5.3.1.1.1 Example 1 Consider the single compartmental model shown in the left side of Figure 5.5, where the input is a bolus injection of a tracer given at time zero and the measured variable is the tracer concentration. The model and measurement equations are:

$$\dot{q}(t) = -k \cdot q(t) + u(t) \qquad q(0) = 0 \tag{5.16}$$

$$y(t) = \frac{q(t)}{V}, \tag{5.17}$$

where $u(t) = d \cdot \delta(t)$; that is, d is the magnitude of the bolus dose. The unknown parameters for the model are the rate constant k and the volume V. Equation 5.17 defines the observation on the system in an ideal context of noise-free and continuous-time measurements. In other words, the

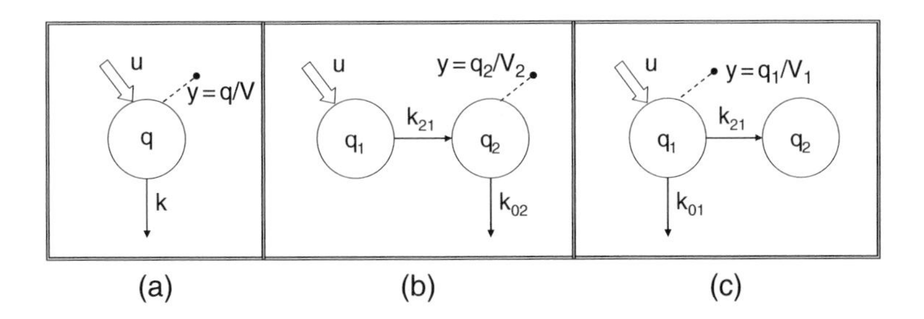

FIGURE 5.5 A single-compartment (a), a two-compartment (b), and a two-compartment model in which the irreversible loss is from compartment 1 (c). For all the models, the tracer input u(t) is a bolus injection of dose d given at time zero. The compartments are characterized by a volume V, and y is the measured tracer concentration.

model output is describing what is measured continuously and without errors; it does not represent the noisy discrete times measurements. To see how the experiment can be used to obtain estimates of these parameters, note that the solution of Equation 5.16 is the mono-exponential

$$q(t) = de^{-kt}. \quad (5.18)$$

The model output y(t) is thus given by

$$y(t) = \frac{d}{V} e^{-kt} \equiv Ae^{-\lambda t}. \quad (5.19)$$

The model output or ideal data are thus described by a function of the form $Ae^{-\lambda t}$, and the parameters that can be determined by the experiment are A and λ. These parameters are called the observational parameters. What is the relationship between the unknown model parameters k and V, and the observational parameters A and λ? From Equation 5.19, one sees immediately:

$$A = y(0) = \frac{d}{V} \quad (5.20)$$

$$\lambda = k. \quad (5.21)$$

In the example above, the unknown parameters k and V of the model are *a priori* uniquely or globally identifiable from the designed experiment because they can be evaluated uniquely from the observational parameters A and λ. Because all model parameters are uniquely identifiable, the model is said to be *a priori* uniquely or globally identifiable from the designed experiment.

So far, we have analyzed the identifiability properties of the model by inspecting the expression of the model output and deriving the relationships between the observational parameters and the unknown model parameters. The method is easy to understand because it requires only some fundamentals of differential calculus. However, the approach is not practicable in general as it works easily only for some simple linear models of order one and two. For linear models of higher order, the method becomes quite cumbersome, and its application is practically impossible.

A simpler method to derive the desired relationships between observational parameters and unknown model parameters consists of writing the Laplace transform for the model output and is known as the transfer function method. Briefly, the advantage of the Laplace transform method is that there is no need to use the analytical solution of the system of linear differential equations. By writing the Laplace transform of the state variables such as masses and then of the model outputs such as concentrations, one obtains an expression that defines the observational parameters as a function of the unknown model parameters. This gives a set of nonlinear algebraic equations in the original parameters.

For the model of Figure 5.5, the Laplace transforms of Equations 5.16 and 5.17 are, respectively:

$$s \cdot Q(s) = -k \cdot Q(s) + d \quad (5.22)$$

$$Y(s) = \frac{Q(s)}{V}, \quad (5.23)$$

where s is the Laplace variable, and the capital letter denotes the Laplace transform of the corresponding lower case letter variable.

The transfer function is

$$H(s) \equiv \frac{Y(s)}{U(s)} = \frac{Q(s)/V}{d} = \frac{[d/s + k]/V}{d} = \frac{1/V}{s+k} \equiv \frac{\beta}{s+\alpha}. \quad (5.24)$$

The coefficients α and β are determinable from the experiment; that is, they are the observational parameters, and thus one has:

$$\beta = \frac{1}{V} \quad (5.25)$$

$$\alpha = k. \quad (5.26)$$

That is, the model is *a priori* uniquely identifiable.

For this simple model, the advantage of the Laplace transform method is not evident, but its power will be appreciated when we consider the next example.

5.3.1.1.2 Example 2 Consider the two-compartment model shown in the Section (b) of Figure 5.5, where a bolus injection of tracer is given into compartment 1. The accessible compartment is compartment 2. Assume the measured variable is the tracer concentration

$$y(t) = q_2(t)/V_2.$$

The equations describing this model, assuming a bolus input, are:

$$\dot{q}_1(t) = -k_{21}q_1(t) + u(t) \qquad q_1(0) = 0 \quad (5.27)$$

$$\dot{q}_2(t) = k_{21}q_1(t) - k_{02}q_2(t) \qquad q_2(0) = 0 \quad (5.28)$$

$$y(t) = \frac{q_2(t)}{V_2}, \quad (5.29)$$

where $u(t) = d \cdot \delta(t)$. The unknown model parameters are k_{21}, k_{02}, and V_2. To see how the experiment can be used to obtain estimates of these parameters, one can use either the time domain solution of Equation 5.28 (a sum of two exponentials) or the transfer function method, which is much more straightforward. The transfer function is

$$H(s) = \frac{Y(s)}{U(s)} = \frac{k_{21}/V_2}{(s+k_{21})(s+k_{02})} \equiv \frac{\beta_1}{s^2 + \alpha_2 s + \alpha_1}, \quad (5.30)$$

where the coefficients α_1, α_2, β_1 are the observational parameters (known from the experiment) linked to unknown model parameters by:

$$\beta_1 = k_{21}/V_2 \quad (5.31)$$

$$\alpha_2 = k_{21} + k_{02} \quad (5.32)$$

$$\alpha_1 = k_{21}k_{02}. \quad (5.33)$$

Equations 5.31, 5.32, and 5.33 are nonlinear, and it is easy to verify that it is not possible to obtain a unique solution for the unknown parameters. In fact, from equations 5.32 and 5.33, parameters k_{21} and k_{02} are interchangeable, and thus each has two solutions, say k_{21}^{I}, k_{21}^{II} and k_{02}^{I}, k_{02}^{II}. As a result, from equation 5.31, V_2 has two solutions also, V_2^{I} and V_2^{II}. The two solutions provide the same expression for the model output y(t). When there is a finite number of solutions (more than one; two in this case), the unknown parameters are said to be *a priori* nonuniquely or locally identifiable from the designed experiment. When all the model parameters are identifiable (uniquely or nonuniquely) and there is at least one of the model parameters that is nonuniquely identifiable (in this case, all three are), the model is said to be *a priori* nonuniquely or locally identifiable.

It is also worth noting that in this case one has parameters that are *a priori* uniquely identifiable, but these are not the original parameters of interest. They are combinations of the original parameters, in particular $k_{21}k_{02}$, $k_{21} + k_{02}$ and k_{21}/V_2. To achieve unique identifiability of this nonuniquely identifiable model, one could design a more complex experiment or, if available, exploit additional independent information available on the system. In this particular case, knowledge of V_2, or a relationship between k_{21} and k_{02}, allows one to achieve unique identifiability of all model parameters.

5.3.1.1.3 Example 3 Consider the two-compartment model shown in Section (c) of Figure 5.5, where a bolus injection of a tracer is given at time zero and where the measured variable is the concentration of drug in plasma. The equations describing this model are:

$$\dot{q}_1(t) = -(k_{01} + k_{21})q_1(t) + u(t) \qquad q_1(0) = 0 \tag{5.34}$$

$$\dot{q}_2(t) = k_{21}q_1(t) \qquad q_2(0) = 0 \tag{5.35}$$

$$y(t) = \frac{q_1(t)}{V_1}. \tag{5.36}$$

The unknown model parameters are k_{21}, k_{01} and V_1.

To see how the experiment can be used to obtain estimates of these parameters, one notes that the transfer function is:

$$H(s) = \frac{Y(s)}{U(s)} = \frac{1/V_1}{s + k_{21} + k_{01}} \equiv \frac{\beta}{s + \alpha}, \tag{5.37}$$

and thus

$$\beta = 1/V_1 \tag{5.38}$$

$$\alpha = k_{21} + k_{01}. \tag{5.39}$$

It is easy to see that, whereas V_1 is uniquely identifiable, k_{21} and k_{01} have an infinite number of solutions lying on the straight line $\alpha = k_{21} + k_{01}$.

When there are an infinite number of solutions for a parameter, one says the parameter is *a priori* nonidentifiable from the designed experiment. When at least one of the model parameters is nonidentifiable (in this case, there are two), the model is said to *a priori* nonidentifiable.

As with the previous example, one can find a uniquely identifiable parameterization; that is, a set of parameters that can be evaluated uniquely). In this case, the parameter is the sum $k_{01} + k_{12}$ (V_1 has been seen to be uniquely identifiable). Again, to achieve unique identifiability of k_{01} and k_{21}, either a more informative experiment is needed (e.g., measuring also in compartment 2), or additional information on the system, such as a relationship between k_{01} and k_{21}, is required.

5.3.1.2 Definitions

The simple examples have allowed understanding of the importance of the *a priori* identifiability problem and have provided a means of introducing some basic definitions. Below, we will give some generic definitions that also hold for more general models such as the nonlinear compartmental model (Equations 5.3 and 5.12). Consider the model (Equation 5.14). Define with $\mathbf{p} = [p_1, p_2, \ldots, p_M]^T$ the set of M unknown model parameters (i.e., the k_{ij} and either V_i). So the model (Equation 5.14) can be written as $y_l = g_l(t, \mathbf{p})$. Define now the observational parameter vector $\boldsymbol{\Phi} = [\phi_1, \ldots, \phi_R]^T$ having the observational parameters ϕ_j, $j = 1, \ldots, R$ as entries. Each particular input–output experiment will provide a particular value $\hat{\boldsymbol{\Phi}}$ of the parameter vector $\hat{\boldsymbol{\Phi}}$; that is, the components of $\hat{\boldsymbol{\Phi}}$ can be estimated uniquely from the data by definition. Moreover, the observational parameters are functions of the basic model parameters p_i, which may or may not be identifiable:

$$\boldsymbol{\Phi} = \boldsymbol{\Phi}(\mathbf{p}). \tag{5.40}$$

Thus to investigate the *a priori* identifiability of model parameters p_i, it is necessary to solve the system of nonlinear algebraic equations in the unknown p_i obtained by setting the polynomials $\boldsymbol{\Phi}(\mathbf{p})$ equal to the observational parameter vector $\hat{\boldsymbol{\Phi}}$:

$$\boldsymbol{\Phi}(\mathbf{p}) = \hat{\boldsymbol{\Phi}}. \tag{5.41}$$

These equations are called the exhaustive summary.

Examples of this have already been provided in working out Examples 1, 2, and 3 in equations 5.20–5.21 and 5.25–5.26, 5.31–5.33, and 5.38–5.39, respectively.

One can now generalize definitions. Let us give them first for a single parameter of the model, and then for the model as a whole.

The single parameter p_i is *a priori* uniquely (globally) identifiable if and only if the system of equations (Equation 5.41) has one and only one of solution:

- Nonuniquely (locally) identifiable if and only if the system of equations (5.41) has for p_i more than one but a finite number of solutions
- Nonidentifiable if and only if the system of Equations (5.41) have for p_i infinite solutions

- The model is *a priori*
- Uniquely (globally) identifiable if all of its parameters are uniquely identifiable
- Nonuniquely (locally) identifiable if all of its parameters are identifiable either uniquely or nonuniquely and at least one is nonuniquely identifiable

5.3.1.3 The Transfer Function Method

The problem now is to assess, solely on the basis of knowledge of the assumed model structure and of chosen experimental configuration, whether the model is *a priori* nonidentifiable, nonuniquely identifiable, or uniquely identifiable. The most common method to test *a priori* identifiability of linear dynamic models is the transfer function method. Assuming we have r inputs and m outputs, the approach is based on the analysis of the r x m transfer function matrix:

$$H(s,p) = [H_{ij}(s,p)] = \frac{Y_i(s,p)}{U_j(s)}, \qquad (5.42)$$

where each element H_{ij}, of H is the Laplace transform of the response in the measurement variable at port i, $y_i(t,p)$ to a unit impulse at port j, $u_j(t) = \delta(t)$. The transfer function approach makes reference to the coefficients of the numerator and denominator polynomials of each of the m x r elements $H_{ij}(s,p)$ of the transfer function matrix, respectively, $\beta_1^{ij}(p),\ldots,\beta_n^{ij}(p)$ and $\alpha_1^{ij}(p),\ldots,\alpha_n^{ij}(p)$. These coefficients are the $2n \times r \times m$ observational parameters ϕ_ℓ^{ij}. Therefore, the exhaustive summary can be written as:

$$\begin{aligned} \beta_1^{11}(p) &= \phi_1^{11} \\ &\vdots \\ \alpha_n^{11}(p) &= \phi_{2n}^{11} \\ &\vdots \\ \beta_1^{rm}(p) &= \phi_1^{rm} \\ &\vdots \\ \alpha_n^{rm}(p) &= \phi_{2n}^{rm}. \end{aligned} \qquad (5.43)$$

This system of nonlinear algebraic equations needs to be solved for the unknown parameter vector p to define the identifiability properties of the model.

We have discussed the Laplace transform method to generate the exhaustive summary of the models. The method is simple to use, even for system models of order greater than two. What becomes more and more difficult is the solution; that is, to determine which of the original parameters of the model are uniquely determined by the system of nonlinear algebraic equations. In fact, one has to solve a system of nonlinear algebraic equations that is increasing both in number of terms and in degree of nonlinearity with the model order. In other words, the method works well for models of low dimension (e.g., order two or three) but fails when applied to relatively large models because the system of nonlinear algebraic equations becomes too difficult to be solved.

To deal with the problem in general, there is the need to resort to computer algebra methods. In particular, a tool to test *a priori* identifiability of linear compartmental models of general structure that combines the transfer function method with a computer algebra method is needed; therefore, the Grobner basis algorithm has been developed [6]. Finally, it is worth noting that for some classes of linear compartmental models (i.e., catenary and mamillary models) and for the general two- and three-compartmental models, explicit identifiability results are available [4].

From the above considerations, it follows that *a priori* unique identifiability is a prerequisite for well-posedness of parameter estimation and for the reconstructability of stated variables in nonaccessible compartments. It is a necessary step that, because of the ideal context where it is posed, does not guarantee a successful estimation of model parameters from real input–output data.

5.3.2 Parameter Estimation

At this stage, a model has been formulated, and the parameter estimation is well-posed. In this section, we describe how to obtain numerical estimates of the unknown parameters from the noisy experimental data and how to judge the quality of parameter estimation; that is, is the model able to describe the data and what is the precision with which the unknown model parameters are estimated? We will only consider Fisher estimation techniques and, particularly, weighted least-squares (WLS). This is probably the most widely used parameter estimation technique. For its connection to Maximum Likelihood estimation as well as for Bayesian estimation techniques, the reader is referred to reference [5].

5.3.2.1 Weighted Least Squares

A model of the system has now been formulated. The model contains a set of unknown parameters to which we would like to assign numeric values from the data of an experiment. We assume that we have checked its *a priori* identifiability. The experimental data are also available. In mathematical terms, the ingredients we have are the model output which can be written as:

$$y(t) = g(t,p), \qquad (5.44)$$

where g(t,p) is related to the model of the system and the discrete-time noisy output measurements, z_i,

$$z(t_i) = z_i = y(t_i) + v(t_i) = g(t,p) + v_i \quad i = 1,\ldots,N, \qquad (5.45)$$

where v_i is the measurement error of the ith measurement.

The problem is to assign a numeric value of p from the data z_i. Regression analysis is the most widely used method to adjust the parameters, characterizing a model to obtain the best fit to a set of data. The weighted residual sum of squares (WRSS) is a good and commonly used measure of how good the fit to the data is. It is given by:

$$\mathrm{WRSS} = \sum_{i=1}^{N} w_i (z_i - y_i)^2, \tag{5.46}$$

where N is the number of observations and $(z_i - y_i)$ the error between the observed and predicted value for each sample time t_i; w_i is the weight assigned to the ith datum. WRSS can be considered as a function of the model parameters: WRSS = WRSS(p). The idea is to minimize WRSS with respect to the parameter values characterizing the model to be fitted to the data. It is natural to link the choice of weights to what is known about the precision of each individual datum. In other words, one seeks to give more credibility, or weight, to those data in which precision is high, and less credibility, or weight, to those data in which precision is small.

The measurement error is v_i in Equation 5.44. It is a random variable, and assumptions about its characteristic must be made. The most common assumption is that the sequence of v_i is a random process with zero mean (i.e., no systematic error) independent samples and variance known. What this means can be formalized in the statistical setting using the notation E, Var and Cov to represent, respectively, mean, variance, and covariance.

Then:

$$E(v_i) = 0 \tag{5.47}$$

$$\mathrm{Cov}(v_i, v_j) = 0 \text{ for } t_i \neq t_j \tag{5.48}$$

$$\mathrm{Var}(v_i) = \sigma_i^2. \tag{5.49}$$

Equation 5.47 means the errors v_i have zero mean; Equation 5.48 means they are independent, and Equation 5.49 means the variance is known. A standardized measure of the error is provided by the fractional standard deviation (FSD) or coefficient of variation (CV):

$$\mathrm{FSD}(v_i) = \mathrm{CV}(v_i) = \frac{\mathrm{SD}(v_i)}{z_i}, \tag{5.50}$$

where SD is the standard deviation of the error

$$\mathrm{SD}(v_i) = \sqrt{\mathrm{Var}(v_i)}. \tag{5.51}$$

The FSD or CV is often expressed as a percentage; that is, the percentage fractional standard deviation or percentage coefficient of variation, by multiplying $\mathrm{SD}(v_i)/z_i$ in Equation 5.50 by 100.

We have considered the case where the variance is known (Equation 5.49). However, one can also easily handle the case where the variance is known up to a proportionality constant, $\mathrm{Var}(v_i) = b_i\sigma^2$ with b_i known and σ^2 unknown. We shall not consider this case explicitly in the following in order not to make the presentation too heavy. For more details, the reader is referred to Cobelli et al. [4].

Knowing the error structure of the data, how are the weights w_i chosen? The natural choice is to weight each datum according to the inverse of the variance:

$$w_i = \frac{1}{\sigma_i^2}. \tag{5.52}$$

It can be shown that this natural choice of weights is optimal in the linear regression case. Therefore, it is very important to have correct knowledge of error of the data and to weight each datum according to this error.

The problem now is how to estimate the error variance. Ideally, one would like to have a direct estimate of the variance of all sources of error. This is a difficult problem. For instance, the measurement error is just one component of the error; it can be used as an estimate of the error only if the investigator believes that the major source of error arises after the sample is taken. To have a more precise estimate of the error, the investigator should have several independent replicates of the measurement z_i at each sampling time t_i from which the sample variance σ_i^2 at t_i can be estimated. If there is a major error component before the measurement process, for instance an error related to drawing a plasma sample or preparing a plasma sample for measurement, then it is not sufficient to repeat the measurement *per se* on the same sample several times. In theory, in this situation it would be necessary to repeat the experiment several times. Such repetition is not often easy to handle in practice. Finally, there is the possibility that the system itself can vary during the different experiments.

In any case, because the above-mentioned approach estimates the variance at each sampling time t_i, it requires several independent replicates of each measurement. An alternative more practical approach consists of postulating a model for the error variance and estimating its unknown parameters from the experimental data.

A flexible model that can be used for the error variance is:

$$\sigma_i^2 = \alpha + \beta(y_i)^\gamma, \tag{5.53}$$

which can be approximated in practice by:

$$\sigma_i^2 = \alpha + \beta(z_i)^\gamma, \tag{5.54}$$

where σ, β, and y are non-negative model parameters relating the variance associated with an observation to the value of the observation itself. Values can usually be assigned to these parameters, or they can also be estimated from the data themselves.

5.3.2.2 Linear Regression Let us consider a linear model with M parameters and put the regression problem in compact matrix-vector notation. The measurement equation can be written as:

$$\mathrm{z} = \mathrm{y} + \mathrm{v} = \mathrm{G} \cdot \mathrm{p} + \mathrm{v} \tag{5.55}$$

$$\begin{bmatrix} z_1 \\ z_2 \\ \vdots \\ z_N \end{bmatrix} = \begin{bmatrix} y_1 \\ y_2 \\ \vdots \\ y_N \end{bmatrix} + \begin{bmatrix} v_1 \\ v_2 \\ \vdots \\ v_N \end{bmatrix} = \begin{bmatrix} g_{11} & g_{12} & \cdots & g_{1M} \\ g_{21} & g_{22} & \cdots & g_{2M} \\ \cdots & \cdots & \cdots & \cdots \\ g_{N1} & g_{N2} & \cdots & g_{NM} \end{bmatrix} \begin{bmatrix} p_1 \\ p_2 \\ \vdots \\ p_M \end{bmatrix} + \begin{bmatrix} v_1 \\ v_2 \\ \vdots \\ v_N \end{bmatrix} \tag{5.56}$$

with

$$p = \left[p_1, p_2, \cdots, p_M\right]^T \tag{5.57}$$

$$G = \begin{bmatrix} g_{11} & g_{12} & \cdots & g_{1M} \\ g_{21} & g_{22} & \cdots & g_{2M} \\ \cdots & \cdots & \cdots & \cdots \\ g_{N1} & g_{N2} & \cdots & g_{NM} \end{bmatrix}. \tag{5.58}$$

The measurement error v, assuming a second-order description (mean and covariance matrix [NxN]), is:

$$E[v] = 0 \tag{5.59}$$

$$E[v\ v^T] = \Sigma_v, \tag{5.60}$$

and, since independence is assumed, Σ_v is diagonal:

$$\Sigma_v = \mathrm{diag}\left(\sigma_1^2, \sigma_2^2, \ldots, \sigma_N^2\right). \tag{5.61}$$

Now, if we define the residual vector r:

$$r = z - G \cdot p, \tag{5.62}$$

the weighted residual sum of squares is

$$WRSS(p) = \sum_{i=1}^{N} \frac{r_i^2}{\sigma_i^2} = r^T \Sigma_v^{-1} r = (z - G \cdot p)^T \Sigma_v^{-1} (z - G \cdot p). \tag{5.63}$$

The WLS estimate of p is that which minimizes WRSS(p)

$$\hat{p} = \arg\min_{p}\ WRSS(p) = \arg\min_{p} (z\text{-}G \cdot p)^T \Sigma_v^{-1} (z\text{-}G \cdot p). \tag{5.64}$$

After some calculations, one has:

$$\hat{p} = \left(G^T \Sigma_v^{-1} G\right)^{-1} G^T \Sigma_v^{-1} z. \tag{5.65}$$

It is also possible to obtain an expression of the precision of $\hat{p}$. Because data z are affected by a measurement error v, one has that $\hat{p}$ is also affected by an error that we call estimation error. It can be defined as:

$$\tilde{p} = p - \hat{p}. \tag{5.66}$$

$\tilde{p}$ is a random variable because $\hat{p}$ is random. The covariance matrix of $\tilde{p}$ is

$$\Sigma_{\tilde{p}} = \mathrm{cov}(\tilde{p}) = E\left[\tilde{p} \cdot \tilde{p}^T\right] = \Sigma_{\hat{p}}, \tag{5.67}$$

and one can show that

$$\Sigma_{\hat{p}} = \left(G^T \Sigma_v^{-1} G\right)^{-1}. \tag{5.68}$$

The precision of the estimate $\hat{p}_i$ of p is often expressed in terms of standard deviation, the square root of the variance $\mathrm{Var}(\hat{p})$:

$$SD(\hat{p}_i) = \sqrt{\mathrm{Var}(\hat{p}_i)}. \tag{5.69}$$

It can be given in terms of FSD or CV, which measures the relative precision of the estimate

$$FSD(\hat{p}_i) = CV(\hat{p}_i) = \frac{SD(\hat{p}_i)}{\hat{p}_i}. \tag{5.70}$$

As noted previously, FSD and CV can be expressed as a percent by multiplying by 100.

From Equations 5.65 and 5.68, one sees that both $\hat{p}$ and $\Sigma_{\hat{p}}$ depend upon the σ_i^2. This is why it is essential that the investigator appreciate the nature of the error in the data.

Up to this point, the assumption has been made that the model is correct. In this case, from the comparison between the equation describing the data $z = G \cdot p + v$ and the definition of the residual $r = z - G \cdot \hat{p}$, one can immediately conclude that residuals r must reflect the measurement errors v. For this in fact to be true, two conditions must be met: The correct model or functional description of the data has been selected, and the parameter estimation procedure has converged to values close to the true values. The sequence of residuals can thus be viewed as an approximation of the measurement error sequence.

One can check whether the above two conditions hold by testing the assumptions made regarding the measurement error on the sequence of residuals. Usually, the measurement error is assumed to be a zero mean, independent random process having a known variance. These assumptions can be checked on the residuals by means of statistical tests.

Independence of the residuals can be tested visually using a plot of residuals versus time. It is expected that the residuals will oscillate around their mean, which should be close to zero, in an unpredictable way. Systematic residuals—that is, a long run sequence of residuals above or below zero—suggest that the model is an inappropriate description of the system because it is not able to describe a nonrandom component of the data.

A formal test of nonrandomness of residuals in the run test. A run is defined as a subsequence of residuals having the same sign (assuming the residuals have zero mean); intuitively, a very small or very large number of runs in the residuals sequence is an indicator of nonrandomness; that is, of systematic errors in the former and of periodicity in the latter case. For details and examples, we refer the readers to Cobelli et al. [4].

In WLS estimation, a specific assumption on the variance of the measurement errors has been made. If the model is correct, the residuals must reflect this assumption.

Because

$$\mathrm{Var}\left(\frac{v_i}{\sigma_i}\right) = \frac{1}{\sigma_i^2} \mathrm{Var}(v_i) = 1. \tag{5.71}$$

If we define, the weighted residuals as

$$\text{wres}_i = \frac{\text{res}_i}{\sigma_i}, \tag{5.72}$$

they should be a realization of a random process having unit variance. By plotting the weighted residual versus time, it is thus possible to test visually the assumption on the variance of the measurement error: Weighted residuals should lie in a −1, +1 wide band. A typical plot of weighted residuals is shown in Figure 5.6.

A pattern of residuals different from that which was expected indicates either the presence of errors in the functional description of the data or that the model is correct but that the measurement error model is not appropriate. In this case, it is necessary to modify the assumptions on the measurement error structure. Some suggestions can be derived by examining the plot. As an example, consider the case where the variance of the measurement error is assumed to be constant. The residuals are expected to be confined in a −1, +1 wide region. If their amplitude tends to increase in absolute value with respect to the observed value, a possible explanation is that the variance of the measurement error is not constant, thus suggesting a modification of the assumption on the measurement error variance.

5.3.2.3 Nonlinear Regression Let us now turn to the nonlinear model of equations:

$$y(t) = g(t,p) \tag{5.73}$$

$$z_i = y_i + v_i = g(t_i,p) + v_i \quad i = 1,2,\ldots,N. \tag{5.74}$$

Let us put the model in the compact matrix-vector notation like we did with the linear model. One has:

$$y = G(p) \tag{5.75}$$

$$z = y + v = G(p) + v \tag{5.76}$$

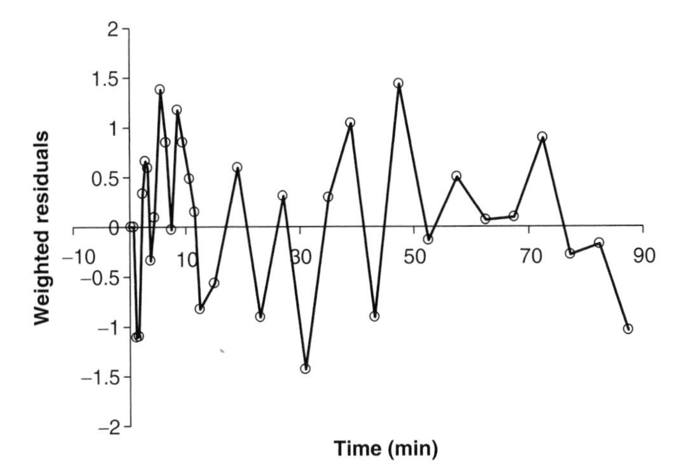

FIGURE 5.6 Plot of weighted residuals versus time.

where

$$z = [z_1,z_2,\ldots,z_N]^T \tag{5.77}$$

$$y = [y_1,y_2,\ldots,y_N]^T \tag{5.78}$$

$$p = [p_1,p_2,\ldots,p_M]^T \tag{5.79}$$

$$G[p] = [g(t_1,p),g(t_2,p),\ldots,g(t_N,p)]^T \tag{5.80}$$

$$v = [v_1,v_2,\ldots,v_N]^T \text{ with } \Sigma_v = \text{cov}(v). \tag{5.81}$$

The WLS estimate of p is the one that minimizes

$$\text{WRSS}(p) = [z - G(p)]^T \Sigma_v^{-1} [z - G(p)]. \tag{5.82}$$

It can be easily shown, using the simple nonlinear model $y(t) = Ae^{-\alpha \cdot t}$, that an explicit analytic solution for p analogous to Equation 5.65 is not possible.

To arrive at an estimate of p, one possible strategy is based on iterative linearization of the model, the Gauss-Newton method. Let us go back to the model Equation 5.73 and consider the expression of y(t) obtainable through its Taylor series expansion around a specific value of p, say

$$p^0 = [p_1^0,p_2^0,\ldots,p_M^0]^T \tag{5.83}$$

by neglecting the terms that contain derivatives of second order and higher

$$y_i = g(t_i,p) \cong g(t_i,p^0) + \left[\frac{\partial g(t_i,p^0)}{\partial p_1} \; \frac{\partial g(t_i,p^0)}{\partial p_2} \cdots \frac{\partial g(t_i,p^0)}{\partial p_M}\right] \begin{bmatrix} p_1 - p_1^0 \\ p_2 - p_2^0 \\ \cdots \\ p_M - p_M^0 \end{bmatrix}, \tag{5.84}$$

where the derivatives are evaluated at $p = p^0$. Notice that this equation is now linear in p. The data relate to y_i as Equation 5.74. Thus, using Equation 5.84 and moving to vector notation one has:

$$\begin{bmatrix} z_1 - g(t_1,p^0) \\ z_2 - g(t_2,p^0) \\ z_N - g(t_N,p^0) \end{bmatrix} = \begin{bmatrix} \frac{\partial g(t_1,p^0)}{\partial p_1} & \frac{\partial g(t_1,p^0)}{\partial p_2} & \cdots & \frac{\partial g(t_1,p^0)}{\partial p_M} \\ \frac{\partial g(t_2,p^0)}{\partial p_1} & \frac{\partial g(t_2,p^0)}{\partial p_2} & \cdots & \frac{\partial g(t_2,p^0)}{\partial p_M} \\ \cdots & \cdots & \cdots & \cdots \\ \frac{\partial g(t_N,p^0)}{\partial p_1} & \frac{\partial g(t_N,p^0)}{\partial p_2} & \cdots & \frac{\partial g(t_N,p^0)}{\partial p_M} \end{bmatrix} \begin{bmatrix} p_1 - p_1^0 \\ p_2 - p_2^0 \\ \cdots \\ p_M - p_M^0 \end{bmatrix} + \begin{bmatrix} v_1 \\ v_2 \\ \vdots \\ v_N \end{bmatrix}, \tag{5.85}$$

and thus

$$\Delta z = S \cdot \Delta p + v \tag{5.86}$$

with obvious definition of $\Delta\mathbf{z}$, $\mathbf{S}$ and $\Delta\mathbf{p}$ from Equation 5.85. Now, because $\Delta\mathbf{z}$ is known ($\mathbf{p}^0$ is given, z is measured) and $\mathbf{S}$ can be computed, one can use WLS to estimate $\Delta\mathbf{p}$ with the linear machinery by using Equation 5.65 with the correspondence $\mathbf{z} \leftrightarrow \Delta\mathbf{z}$, $\mathbf{p} \leftrightarrow \Delta\mathbf{p}$ and $\mathbf{G} \leftrightarrow \mathbf{S}$:

$$\Delta\hat{\mathbf{p}} = \left(\mathbf{S}^T\Sigma_v^{-1}\mathbf{S}\right)^{-1}\mathbf{S}^T\Sigma_v^{-1}\Delta\mathbf{z}. \tag{5.87}$$

Hence, a new estimate of p can be obtained as

$$\mathbf{p}^1 = \mathbf{p}^0 + \Delta\hat{\mathbf{p}}. \tag{5.88}$$

Now, with $\mathbf{p}^1$, which is by definition a better estimate than $\mathbf{p}^0$ because $\mathrm{WRSS}(\mathbf{p}^1) < \mathrm{WRSS}(\mathbf{p}^0)$, the process can restart. The model is linearized around $\mathbf{p}^1$; a new estimate $\mathbf{p}^2$ is obtained, and so on until the cost function stops decreasing significantly, for example, when two consecutive values of WRSS(p) are within a prescribed tolerance.

Once $\hat{\mathbf{p}}$ has been obtained, by paralleling the linear case, one can obtain the covariance of the parameter estimates as:

$$\Sigma_{\hat{\mathbf{p}}} \cong \left(\mathbf{S}^T\Sigma_v^{-1}\mathbf{S}\right)^{-1} \tag{5.89}$$

with

$$\mathrm{S} = \begin{bmatrix} \frac{\partial g(t_1,\hat{\mathbf{p}})}{\partial p_1} & \frac{\partial g(t_1,\hat{\mathbf{p}})}{\partial p_2} & \cdots & \frac{\partial g(t_1,\hat{\mathbf{p}})}{\partial p_M} \\ \frac{\partial g(t_2,\hat{\mathbf{p}})}{\partial p_1} & \frac{\partial g(t_2,\hat{\mathbf{p}})}{\partial p_2} & \cdots & \frac{\partial g(t_2,\hat{\mathbf{p}})}{\partial p_M} \\ \cdots & \cdots & \cdots & \cdots \\ \frac{\partial g(t_N,\hat{\mathbf{p}})}{\partial p_1} & \frac{\partial g(t_N,\hat{\mathbf{p}})}{\partial p_2} & \cdots & \frac{\partial g(t_N,\hat{\mathbf{p}})}{\partial p_M} \end{bmatrix}. \tag{5.90}$$

Residuals and weighted residuals are defined as for the linear case above.

The linear machinery has been used to solve the nonlinear case. However, it is worth remarking that the nonlinear case is more complex to handle than the linear case. This is true, not only from a computational point of view, but also conceptually as a result of the presence of local minima of WRSS(p) and the necessity of specifying an initial estimate of p, p^0. To illustrate graphically this additional complexity, let us consider the scalar case with WRSS(p) as a function of p shown in Figure 5.7. There is more than one minimum for WRSS, and this is distinctly different from the linear case, where there is only one (unique) minimum. The minima shown in Figure 5.7 are called local minima. The difference then between the linear and nonlinear case is that in linear regression there is a unique minimum for WRSS whereas in the nonlinear case there may be several local minima for WRSS. Among the local minima, the smallest is called the global minimum. This has obvious implications for the choice of p^0. Generally, to be sure one is not ending up at a local minimum, several tentative values of p^0 are used as starting points.

The steps of nonlinear least squares estimation have been illustrated using the Gauss-Newton iterative scheme. This outlines the principles of that class of algorithms which require the computation of derivatives contained in matrix S. This is usually done numerically, using, for example, central difference methods, although other strategies are also available such as the sensitivity system [4]. This class is referred to as gradient-type (derivative) algorithms. Numerically refined and efficient algorithms such as the Levenberg-Marquardt technique, based on the Gauss-Newton principle, are available and are implemented in many software tools.

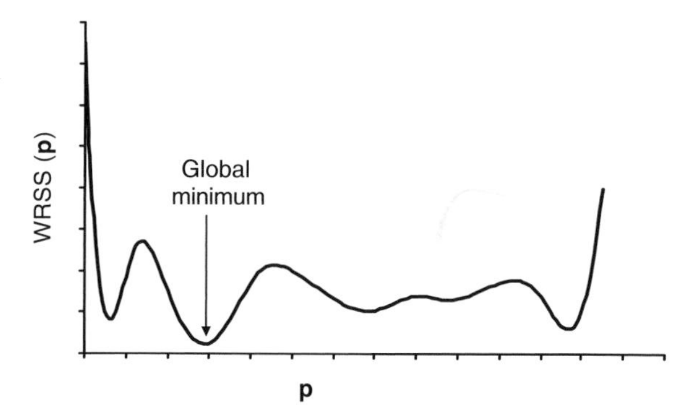

FIGURE 5.7 WRSS as a function of p.

Another category of algorithms for minimizing WRSS that has been applied in physiologic model parameter estimation is one that does not require the computation of the derivatives. These algorithms are known as direct search methods, and both deterministic and random search algorithms are available and implemented in software tools. An efficient deterministic direct search algorithm is the simplex method. It is worth emphasizing that with a direct search method, the computation of the derivatives is not required; a direct comparison of gradient versus direct search methods is difficult and may be problem-dependent. Available experience in physiologic model parameter estimation tends to favor the gradient-type methods.

5.3.2.4 Test of Model Order

Up to this point, only the problem of testing whether or not a specific model is an appropriate description of a set of data has been examined. Consider now the case where different candidate models are available, and the problem is to select the model that provides the best description of the data. For example, when performing multiexponential modeling of a decay curve:

$$y(t) = \sum_{i=1}^{n} A_1 e^{-\lambda_i \cdot t} \tag{5.91}$$

the model order—that is, the number n of exponentials—is not known *a priori*. A mono-, bi-, and tri-exponential model are usually fitted to the data, and the results of parameter estimation are evaluated to select the optimum order or the best value for n.

Relying solely on WRSS and an examination of the weighted residuals to determine the optimum model order is not appropriate because as the model order increases WRSS will decrease. For example, in dealing with a tracer decay curve following a bolus injection, each additional exponential term added to the sum of exponentials will decrease WRSS. Similarly, the pattern of residuals will become more random. However, each time an exponential term is added, two parameters are added, and the degrees of freedom are decreased by two. Thus intuitively, when comparing different model structures, both WRSS and the degrees of freedom should be evaluated. This is to check whether or not the reduction of WRSS truly reflects a more accurate representation of the data or whether it is merely the result of the increase in the number of parameters. Hence, additional tests are required.

The two tests that are frequently used to compare model structures are the F-test and tests based on the principle of parsimony. We briefly describe below only the latter and refer the reader for illustration of the F-test to Cobelli et al. [4].

The most commonly used tests that implement the principle of parsimony—that is, that choose the model that is best able to fit the data with the minimum number of parameters—are the Akaike information criterion (AIC) and the Schwarz criterion (SC). More than two models can be compared, and the model that has the smallest criterion is chosen as the best.

If one assumes that errors in the data are uncorrelated and Gaussian, with a known measurement error variance, then the criteria are:

$$AIC = WRSS + 2 \cdot M \quad (5.92)$$

$$SC = WRSS + M \cdot \ln N, \quad (5.93)$$

where M is the number of parameters in the model, and N is the number of data. Although AIC and SC have different derivations, they are similar because they are made up of a goodness-of-fit measure plus a penalty function proportional to the number of parameters M in the model. Note that in SC, M is weighted ln(N), with large N; this may become important.

5.4 Model Validation

It is not difficult to build models of systems—the difficulty lies in making them accurate and reliable in answering the question asked. For the model to be useful, one has to have confidence in the results and the predictions that are inferred from it. Such confidence can be obtained by model validation. Validation involves the assessment of whether a compartmental model is adequate for its purpose. This is a difficult and highly subjective task when modeling physiologic systems because intuition and an understanding of the system, among other factors, play an important role in this assessment. It is also difficult to formalize related issues such as model credibility or the use of the model outside its established validity range. Some efforts have been made, however, to provide formal aids for assessing the value of models of physiologic systems. Validity criteria and validation strategies for models of physiologic systems are available that take into account both the complexity of model structure and the extent of available experimental data [7]. Of particular importance is the ability to validate a model-based measurement of a system parameter by an independent experimental technique. A model that is valid is not necessarily a true one; all models have a limited domain of validity, and it is hazardous to use a model outside the area for which it has been validated.

5.5 Simulation

Suppose one wishes to see how the system behaves under certain stimuli, but it is inappropriate or impossible to carry out the required experiment. If a valid model of the system is available, one can perform an experiment on the model by using a computer to see how the system would have reacted. This is called simulation. Simulation is thus an inexpensive and safe way to experiment with the system. Clearly, the value of the simulation results depends completely on the quality or the validity of the model of the system.

Having derived a complete model, including estimating all unknown parameters and checking its validity in relation to its intended domain of application, it is now possible to use it as a simulation tool. Computer simulation involves solving the model (i.e., the equations that are the realization of the model) to examine its output behavior. This might typically be the time course of one or more of the system variables. In other words, we are performing computer experiments on the model. In fact, simulation can be used either during the process of model building or with a complete model. During model building, simulation can be performed to clarify some aspects of behavior of the system or part of it to determine whether a proposed model representation seems to be appropriate. This would be done by comparison of the model response with experimental data from the same situation. Simulation, when performed on a complete, validated model, yields output responses that provide information regarding system behavior. Depending on the modeling purpose, this information assists in describing the system, predicting behavior, or yielding additional insights (explanation).

Why carry out computer simulation? The answer is that it might not be possible, appropriate, convenient, or desirable to perform particular experiments on the system (e.g., it cannot be done at all, it is too difficult, it is too expensive, it is too dangerous, it is not ethical, or it would take too long to obtain results). Therefore, we need an alternative way of experimenting with the system. Simulation offers such an alternative that overcomes the above limitations. Such experimenting can provide information that is useful in relation to our modeling purpose.

To perform computer simulation, we first need a mathematic model that is complete in terms of all its parameters being specified and that has initial conditions specified for all the variables. If the model is not complete in the sense of there being unspecified parameter values, then formal parameter estimation techniques must be employed to obtain such estimates. The model is then implemented on the computer. This assumes that the model equations cannot be, or are not being, solved analytically and that a numeric solution of the system is needed.

The model is solved on the computer; this solution process produces the time course of the system variables. In technical terms, the computer implementation is done either using a standard programming language such as Fortran or C or by using a specialist simulation package such as MATLAB®.

5.6 Case Study

To illustrate the methodologic points we have been making, consider the following set of data, which will be described by a sum of exponentials. In our previous discussion, we have focused on compartmental models. However, when one is starting from scratch, it is often wise to fit the data to a sum of exponentials because this gives a clue as to how many compartments will be required in the model.

Consider the data given in Table 5.1; these data are radioactive tracer glucose concentrations measured in plasma following an injection of tracer at time zero. The time measurements are minutes, and the plasma measurements are disintegrations per minute per milliliter. The experiment was performed in a normal subject in the basal state [8]. To select the order of the multi-exponential model that is best able to describe these data, one-, two- and three-exponential models can be considered:

$$y(t) = A_1 e^{-\lambda_1 \cdot t} \tag{5.94}$$

$$y(t) = A_1 e^{-\lambda_1 \cdot t} + A_2 e^{-\lambda_2 \cdot t} \tag{5.95}$$

$$y(t) = A_1 e^{-\lambda_1 \cdot t} + A_2 e^{-\lambda_2 \cdot t} + A_3 e^{-\lambda_3 \cdot t}. \tag{5.96}$$

The measurement error is assumed to be additive

$$z_i = y_i + v_i, \tag{5.97}$$

where the errors v_i are assumed to be independent, Gaussian with a mean of zero, and an experimentally determined standard deviation of

$$SD(v_i) = 0.02 \cdot z_i + 20. \tag{5.98}$$

These values are shown associated with each datum in Table 5.1. The three models are to be fitted to the data by applying weighted nonlinear regression with the weights chosen equal to the inverse of the variance. The plots of the data and the model predictions together with the corresponding weighted residuals are shown in Figure 5.8, and the model parameters are given in Table 5.2.

Examining Table 5.2, we can see that all parameters can be estimated with acceptable precision in the one- and two-exponential models, while some parameters of the three exponential model are very uncertain. This means that the three-exponential model cannot be resolved with precision from the data. In fact, the first exponential is so rapid, $\lambda_1 = 4.6 min^{-1}$, that it practically has vanished by the time of the first available datum at 2 minutes. The other two exponential terms have values similar to those obtained for the two-exponential model. In addition, the final estimates of A_1 and λ_1 are also dependent upon the initial estimates; that is, starting from different initial points in parameter space, the nonlinear regression procedure yields different final estimates while producing similar values of the WRSS. Therefore, the three-exponential model is not numerically identifiable and can be rejected at this stage.

One can now compare the fit of the one- and two-exponential models. Nonrandomness of the residuals for the one-exponential model is evident because the plot reveals long runs of consecutive residuals of the same sign. The run test allows one to check the independence formally, and from the values of Z, one can conclude that the residuals of the two-exponential model is consistent with the hypothesis of independence because the Z value lies within the 5% region of acceptance (−1.96,1.96), or equivalently, the *P* value is high. Conversely, the Z value for the one-exponential model indicates that the hypothesis of independence is to be rejected, with a $P < 0.5\%$.

Most residuals for the two-exponential model lie between 1 and 1, which indicates they are compatible with the assumptions on the variance of the measurement error. On the other hand, only a few of the residuals of the one-exponential model

TABLE 5.1 Plasma data from a tracer experiment

Time	Plasma		Time	Plasma	
2	3993.50	99.87	28	2252.00	65.04
4	3316.50	86.33	31	2169.50	63.39
5	3409.50	88.19	34	2128.50	62.57
6	3177.50	83.55	37	2085.00	61.70
7	3218.50	84.37	40	2004.00	60.08
8	3145.00	82.90	50	1879.00	57.58
9	3105.00	82.10	60	1670.00	53.40
10	3117.00	82.34	70	1416.50	48.33
11	2984.50	79.69	80	1333.50	46.67
13	2890.00	77.80	90	1152.00	43.04
14	2692.00	73.84	100	1080.50	41.61
15	2603.00	72.06	110	1043.00	40.86
17	2533.50	70.67	120	883.50	37.67
19	2536.00	70.72	130	832.50	36.65
21	2545.50	70.91	140	776.00	35.52
23	2374.00	67.48	150	707.00	34.14
25	2379.00	67.58			

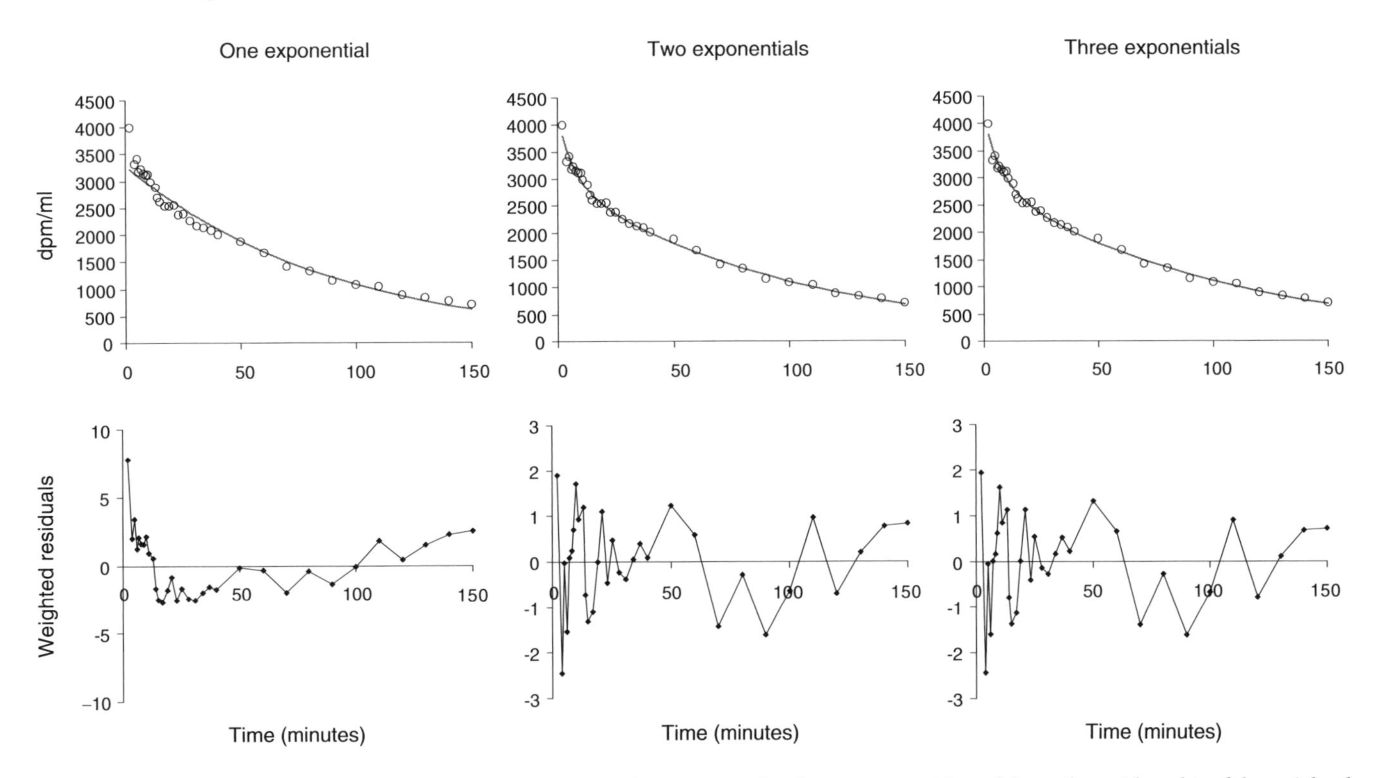

FIGURE 5.8 The best fit of the data given in Table 5.1 to a single-, a two-, and a three-exponential model together with a plot of the weighted residuals for each case. The exponential coefficients and eigenvalues for each model are given in Table 5.2.

TABLE 5.2 One-, two-, three-exponential model parameter estimates (see text for explanation)

		1 Exponential	2 Exponentials	3 Exponentials
A_1		3288 (1%)	1202 (10%)	724866 (535789%)
λ_1		0.0111 (1%)	0.1383 (17%)	4.5540 (7131%)
A_2			2950 (2%)	1195 (14%)
λ_2			0.0098 (3%)	0.1290 (14%)
A_3				2925 (2%)
λ_3				0.0097 (3%)
Run test:	Z value	−5.13	−1.51	
	5% region	[−1.96,1.96]	[−1.96,1.96]	
	P value	<0.5%	>6%	
χ^2 **test**	WRSS	167.1	32.98	
	5% region	[16.8,47.0]	[16.0,45.7]	
	P value	<0.5%	>20%	
F test	F ratio	2 vs 1: 29.59		
		5% region [0,3.33]		
		P value <0.5%		
AIC		171.10	40.98	
SC		174.09	46.97	

fall in this range. To test formally whether the weighted residuals have unit variance, as expected if the model and/or assumptions on the variance of the measurement error are correct, the X^2-test can be applied. For the one-exponential model, the degrees of freedom $df = \mathrm{N} - \mathrm{M} = 31$, and for the level of significance equal to 5%, the region of acceptance is 16.8, 47.0. Because the WRSS is greater than the upper bound 47.0, the assumption of unit variance of the residuals has to be rejected with $P < 0.5\%$. For the two-exponential model, the P values are higher, and the WRSS lies within the 5% region

(16.05, 45.72), indicating that the residuals are consistent with the unit variance assumption. The WRSS decreases, as expected, when the number of parameters in the model increases.

The F test indicates that the two-exponential model reduces the WRSS significantly when compared with the one-exponential model because the F value is greater than $F_{max} = 3.33$ (evaluated for a 5% level of significance from the $F_{29.59}$ distribution). Similar conclusions can be derived from the AIC and SC, which assume their lower values for the two-exponential model.

5.7 Quantification of Medical Images

Having put the compartmental modeling methodology on firm ground, we can now move into discussing its application to quantification of functional brain PET and MRI imaging data.

5.7.1 Positron Emission Tomography

We discuss below some classical PET compartmental models that require an arterial blood or plasma input function (i.e., the measured radioactivity concentration in the blood supply is used as input to the system). For receptor-binding studies, there are less invasive alternatives, the so-called reference tissue models such as those described by Lammertsma, et al. [9, 10], but for the sake of space, we will not discuss them here.

5.7.1.1 Blood Flow

The model used is shown in Figure 5.9.

The model equation is:

$$\frac{dC_t(t)}{dt} = K_1 C_p(t) - k_2 C_t(t) \qquad C_t(0) = 0, \tag{5.99}$$

where C_t is the tracer concentration in tissue, C_p is the tracer concentration in arterial plasma, and K_1 and k_2 are two first-order kinetic rate constants. The tracer concentrations are often measured in [nCi/ml] or in [kBq/ml], and the acquisition time in [min]. Consequently, K_1 is in [ml_{plasma}/ml_{tissue}/min] or, shortly, in [ml/ml/min] or [ml/100gr/min] and k2 in [min^{-1}].

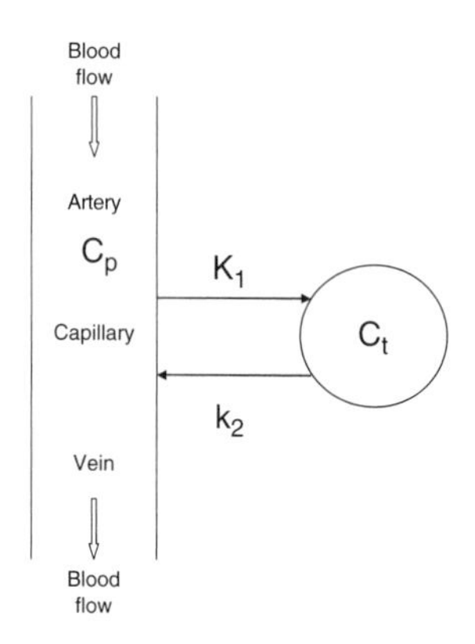

FIGURE 5.9 The one-tissue compartmental model.

Following the introduction of the PET tracer, the total concentration of radioactivity measured by a PET scanner, C(t), is the sum of the tissue activity in a region of interest (ROI), or in a voxel, and a certain fraction of blood tracer concentration:

$$C(t) = (1 - V_b) \cdot C_t(t) + V_b \cdot C_b(t), \tag{5.100}$$

where V_b is the fraction of the measured volume occupied by blood, and C_b is the tracer concentration in the whole blood. The three model parameters are *a priori* uniquely identifiable and can be estimated by weighted nonlinear least squares.

In PET neuroimaging, because blood volumes in the human brain are small, 2%–4% [11] the blood volume term is often omitted.

This model structure is used to quantify blood flow with the freely diffusible $[^{15}O]H_2O$ PET tracer and by assuming:

$$K_1 = F \tag{5.101}$$

$$k_2 = \frac{F}{\lambda}, \tag{5.102}$$

where F is the rate of blood flow per unit mass of tissue, and λ is the tissue:blood partition coefficient. The partition coefficient was originally defined as the ratio of the tissue to venous blood concentration. In PET, the partition coefficient refers to the concentration ratio between a tissue compartment and arterial plasma at equilibrium, $\lambda = C_t/C_p|_{equilibrium} = K_1/k_2$, and it is more correctly called volume of distribution, generally symbolized by V_d. However, traditionally in $[^{15}O]H_2O$ PET quantification, the ratio $\lambda = K_1/k_2$ is named partition coefficient.

In quantitative $[^{15}O]H_2O$ PET imaging, plasma tracer activity is normally obtained by automatic blood sampling. In this case, delay and dispersion affect the measured plasma activity and consequently the parameter estimates. However, the bias introduced by delay and dispersion can be taken into account by assuming [12]:

$$C_{p\_measured}(t) = C_p(t+\Delta t) \otimes \frac{1}{d} e^{-\frac{1}{d} \cdot t} = \frac{1}{d} \int_0^t C_p(\tau + \Delta t) e^{-\frac{1}{d}(t-\tau)} d\tau, \tag{5.103}$$

where Δt is the delay in [min], and d is the dispersion value in [min]. Solving Equation 5.103 by using known value for d and Δt, one can derive the corrected $C_p(t)$ time activity curve not affected for delay and dispersion and obtain unbiased parameter values for blood flow and partition coefficient. Because the discrepancy between C_p_measured and C_p is negligible after a few minutes, delay and dispersion correction is not affecting parameter estimates for PET studies that have a duration of 60 minutes or more. Thus, in this case, discrepancy between C_p_measured and C_p is ignored.

5.7.1.2 Glucose Metabolism

The two-tissue compartment model proposed by Sokoloff in 1977 [13] was originally developed for autoradiographic studies in the brain with 2-[^{14}C]deoxyglucose and subsequently used for PET studies of glucose utilization with [^{18}F]fluorodeoxyglucose, [^{18}F]FDG. [^{18}F]FDG is an analog of glucose that crosses the blood–brain barrier by a saturable carrier-mediated transport process and competes with glucose for the same carrier. Once in the tissue, [^{18}F]FDG, like the glucose, can be either be transported back to plasma or phosphorylated. The model structure is shown in Figure 5.10, where C_p is [^{18}F], FDG plasma arterial concentration, C_e[^{18}F] FDG is cerebral tissue concentration, C_m[^{18}F]FDG-6-P is cerebral concentration in tissue, K_1[ml/100gr/min or ml/ml/min] and k_2[$\min^{-1}$], respectively, are the rate constants of [^{18}F]FDG forward and reverse transcapillary membrane transport, and k_3[$\min^{-1}$] is the rate constant of [^{18}F] FDG phosphorylation.

The model equations are:

$$\frac{dC_e(t)}{dt} = K_1C_p(t) - (k_2 + k_3)C_e(t) \quad C_e(0) = 0 \tag{5.104}$$

$$\frac{dC_m(t)}{dt} = k_3C_m(t) \quad C_m(0) = 0. \tag{5.105}$$

The total concentration of radioactivity measured by a PET scanner, C(t), is the sum of the tissue activity in an ROI or in a voxel, and a certain fraction of blood tracer concentration:

$$C(t) = (1 - V_b) \cdot (C_e(t) + C_m(t)) + V_b \cdot C_b(t), \tag{5.106}$$

where V_b is the fraction of the measured volume occupied by blood, and C_b is the tracer concentration in whole blood. All four parameters, [K_1,k_2,k_3,V_b], are *a priori* uniquely identifiable. Of note is that model parameters reflect [^{18}F] FDG kinetics and not glucose kinetics. However, from [^{18}F] FDG parameter estimates one can also derive the cerebral metabolic rate of glucose utilization, CMR_{glu} from:

$$CMR_{glu} = \frac{K_1k_3}{k_2 + k_3}\frac{C_p^{glu}}{LC}, \tag{5.107}$$

where C_p^{glu} is the arterial plasma glucose concentration [mg/dl], and LC is the lumped constant [unitless] (i.e., the factor that describes the relation between the glucose analog [^{18}F] FDG and glucose itself). LC is given by:

$$LC = \frac{E^{FDG}}{E^{GLU}}, \tag{5.108}$$

where E^{FDG} and E^{GLU} are, respectively, the extraction of [^{18}F] FDG and glucose. LC value in brain in humans under normal physiology has a value in the range of 0.85 [14,15]. From model parameter estimates, it is possible to also derive the distribution volume for C_e compartment by:

$$V_d = \frac{K_1}{k_2 + k_3}[\text{ml}/100\text{gr}] \text{ or } [\text{ml}/\text{ml}]. \tag{5.109}$$

5.7.1.3 Receptor Binding

PET allows the study of receptor density and radio ligand affinity in the brain. Quantification of the ligand–receptor system is of fundamental importance not only to understand how the brain works (e.g., how it performs the various commands and reacts to stimuli), but also in the investigation of the pathogenesis of important diseases like Alzheimer's disease and Parkinson's disease. In recent years, PET has become an increasingly used tool to quantitate important parameters like the receptor density. The most–used compartmental model is the two-tissue model shown in Figure 5.11, where C_{f+ns} is the free and the nonspecifically bound tissue concentration of the PET ligand, and C_s is the tissue concentration of specifically bound ligand. Parameters K_1 [ml $\text{ml}^{-1}\min^{-1}$] and k_2 [$\min^{-1}$] represent rate constant of ligand transfer from plasma to the tissue and vice versa, while k_3 [$\min^{-1}$] represents the transfer

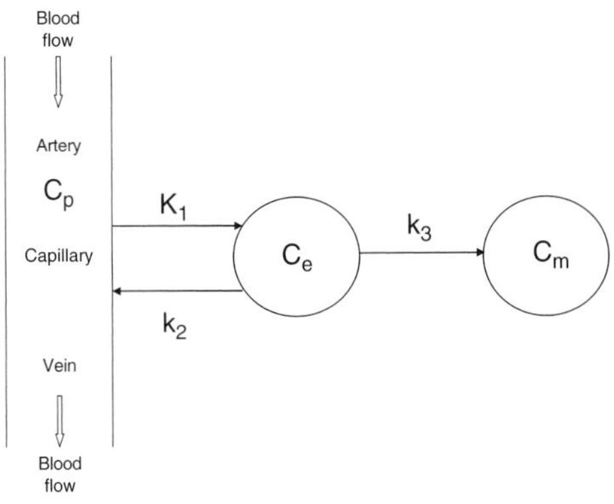

FIGURE 5.10 The two-tissue compartmental model.

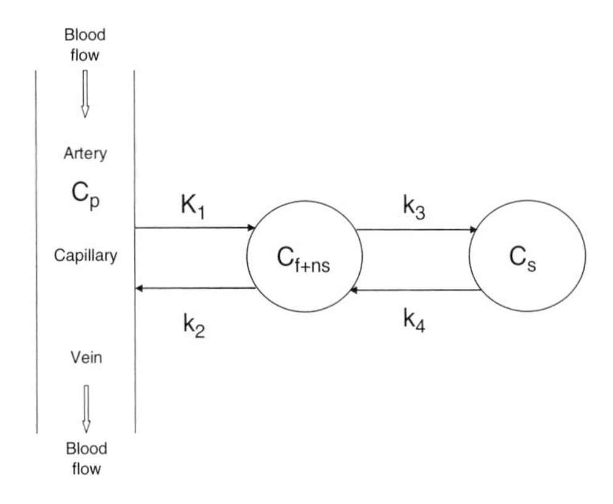

FIGURE 5.11 The two-tissue compartmental model.

of tracer to the specifically bound compartment, and k_4 [min^{-1}] is the return from the specifically bound compartment to the free and nonspecifically bound compartment. In truth, the comprehensive compartmental model for PET ligand studies requires three compartments, as shown in Figure 5.12, where C_p is the arterial plasma concentration corrected for metabolites, C_f is the concentration of free ligand, C_{ns} is the concentration of nonspecifically bound ligand, and C_s is the concentration of specifically bound ligand. The model equations are:

$$\begin{aligned} \frac{C_f(t)}{dt} &= K_1C_p(t) - (k_2 + k_3 + k_5)C_f(t) + k_4C_s(t) + k_6C_{ns}(t) \\ \frac{C_s(t)}{dt} &= k_3C_f(t) - k_4C_s(t) \\ \frac{C_{ns}(t)}{dt} &= k_5C_f(t) - k_6C_{ns}(t), \end{aligned} \tag{5.110}$$

with initial conditions $C_f(0) = C_s(0) = C_{ns}(0) = 0$ and where K_1 [ml/ml/min] is the rate constant of transfer from plasma to free ligand tissue compartment, and k_2, k_3, k_4, k_5, k_6[min^{-1}] are the rate constants of ligand transfer from tissue to plasma and inside the tissue.

To better understand the physiologic meaning of parameters k_3 and k_4, let's assume that the binding of the ligand to the receptor site is describable as a bimolecular reaction:

$$L + R \underset{k_{off}}{\overset{k_{on}}{\rightleftarrows}} LR, \tag{5.111}$$

where L represents the ligand, R is the receptor site, LR is the binding product, k_{on} is the association rate of the ligand with the receptor sites, and k_{off} is the dissociation rate of the specifically bound reaction product. In the notation of Figure 5.12, one has that C_f and C_s represent L and LR, respectively. Thus:

$$\frac{dC_s(t)}{dt} = k_{on}C_f(t)C_r(t) - k_{off}C_s(t), \tag{5.112}$$

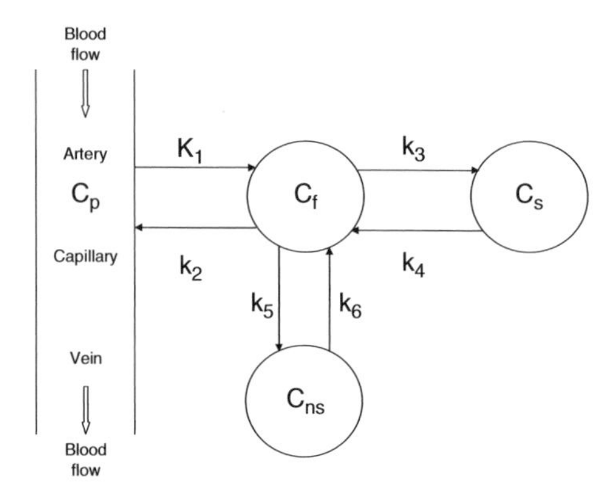

FIGURE 5.12 The three-tissue compartmental model.

where C_r denotes the concentration of receptors. If B_{max} is the total number of available reactions sites, then

$$B_{max} = C_s + C_r, \tag{5.113}$$

and, if the ligand is present in tracer concentration, the concentration C_s is negligible, and thus:

$$B_{max} \approx C_r. \tag{5.114}$$

Equation 5.112 becomes:

$$\frac{dC_s(t)}{dt} = k_{on}B_{max}C_f(t) - k_{off}C_s(t) = k_3C_f(t) - k_4C_s(t), \tag{5.115}$$

with $k_3 = k_{on}B_{max}$ and $k_4 = k_{off}$.

An important parameter is also the equilibrium binding constant K_d, which is defined with the ligand–receptor reaction in steady state as

$$K_d = \frac{C_s}{C_rC_f} = \frac{k_{on}}{k_{off}}. \tag{5.116}$$

The PET measurement is the result of the tracer present in the tissue and of that present in the blood of the ROI. Consequently, the measurement equation for the three-tissue compartment model is:

$$C(t) = (1 - V_b)(C_f(t) + C_{ns}(t) + C_s(t)) + V_bC_b(t), \tag{5.117}$$

where C_b is whole-blood tracer concentration, and V_b is the fraction of the measured volume occupied by blood. However, the three-tissue model is *a priori* only nonuniquely identifiable; in particular, it admits two solutions for each parameter. To ensure unique identifiability, it is usually assumed that the exchange rates between the free tissue and nonspecific binding pools are sufficiently rapid (compared with the other rates of the model) so that the three-tissue compartment model reduces to the two-tissue model where $C_{f+ns}(t) = C_f(t) + C_{ns}(t)$ is the free and nonspecific binding tracer concentration. For this reason, the two-tissue model is the most used to quantify receptor binding studies. The model equations for the two-tissue model are:

$$\begin{aligned} \frac{C_{f+ns}(t)}{dt} &= k_1C_p(t) - (k_2 + k_3)C_{f+ns}(t) + k_4C_s(t) \\ \frac{C_s(t)}{dt} &= k_3C_{f+ns}(t) - k_4C_s(t), \end{aligned} \tag{5.118}$$

with initial conditions $C_{f+ns}(0) = C_s(0) = 0$ and with

$$k_3 = k_{on}B_{max}\ f, \tag{5.119}$$

where f is given by:

$$f \equiv \frac{C_f}{C_{f+ns}} = \frac{C_f}{C_f + C_{ns}} = \frac{C_f}{C_f\left(1 + \dfrac{C_{ns}}{C_f}\right)} = \frac{1}{1 + \dfrac{k_5}{k_6}}. \tag{5.120}$$

The measurement equation becomes:

$$C(t) = (1 - V_b)(C_{f+ns}(t) + C_s(t)) + V_bC_b(t). \tag{5.121}$$

The model now is *a priori* uniquely identifiable, and in addition to k_1,k_2,k_3,k_4,V_b, it is also possible to estimate the binding potential (BP):

$$BP = f \cdot \frac{B_{max}}{K_d} = \frac{k_3}{k_4} \quad (5.122)$$

and the distribution volumes:

$$V_{d-C_{f+ns}} = \left.\frac{C_{f+ns}}{C_p}\right|_{equilibrium} = \frac{K_1}{k_2} \quad (5.123)$$

$$V_{d-CS} = \left.\frac{C_s}{C_p}\right|_{equilibrium} = \frac{K_1 k_3}{k_2 k_4} \quad (5.124)$$

$$V_d = \left.\frac{C_s + C_{f+ns}}{C_p}\right|_{equilibrium} = \frac{K_1}{k_2} + \frac{K_1 k_3}{k_2 k_4} = \frac{K_1}{k_2}\left(1 + \frac{k_3}{k_4}\right). \quad (5.125)$$

When a region is void of specific receptors, the two-tissue compartmental mode collapses into a one-tissue compartmental one because, in this case, $k_3 = k_4 = 0$ and, its total distribution volume Vd is given by $V_d^{void} = K_1/k_2$. In this case—that is, in presence of regions void of specific receptors—BP in the region or voxel having specific receptors can be estimated from:

$$BP = \frac{V_d - V_d^{void}}{V_d^{void}} = \frac{\frac{K_1}{k_2}\left(1 + \frac{k_3}{k_4}\right) - \frac{K_1^{void}}{k_2^{void}}}{\frac{K_1^{void}}{k_2^{void}}} \quad (5.126)$$

$$= [\text{assuming } \frac{K_1}{k_2} = \frac{K_1^{void}}{k_2^{void}}] = \frac{k_3}{k_4}.$$

To obtain unbiased model parameter estimates, it is not only required to choose the more adequate model for the particular PET ligand, but one must be particularly careful with the handling of the possible presence of metabolites in plasma measures. Unlike [^{18}F]FDG and [^{15}O]H2O plasma data, the metabolism of the other tracers, particularly those used for binding studies, often leads to labeled metabolites in plasma. In this case, using the model of Equation 5.118, it must be assumed that the metabolites do not enter into the tissue and that metabolites must be removed from the blood measurements before identifying model parameters.

5.7.2 Arterial Spin Labeling–Magnetic Resonance Imaging

Quantification of arterial spin labeling (ALS) images allows cerebral blood flow (CBF) to be obtained [16, 17]. ASL is based on the idea of magnetically labeling blood flowing into the slices of interest [16, 17]. Because blood exchanges with tissue water, altering the tissue magnetization, a perfusion-weighted image can be generated by the subtraction of an image in which inflowing spins have been labeled from an image in which spin labeling has not been performed. In the imaged slice, the evolution of the tissue magnetization can be described by the following modified Bloch equation:

$$\frac{dM_t(t)}{dt} = \frac{M_t^0 - M_t(t)}{T_{1t}} + CBF[m_a(t) - m_v(t)], \quad (5.127)$$

where M_t is the longitudinal magnetization of the brain tissue per unit mass, M_t^0 is the magnetization of fully relaxed tissue per unit mass, T_{1t} is the longitudinal relaxation time of brain tissue water, CBF is blood flow (perfusion) (mL 100 g-1 min-1), m_a is the magnetization of water in the inflowing arterial blood per unit volume of blood, and m_v is the equivalent term for the outflowing venous blood. At equilibrium:

$$m_v^0 = m_a^0 = m_a(t) = \frac{M_t^0}{\lambda}, \quad (5.128)$$

where λ, is the blood–brain partition coefficient, which represents the difference between water concentrations in blood and in tissue. By assuming full exchange between blood and tissue water:

$$m_v(t) = \frac{M_t(t)}{\lambda}. \quad (5.129)$$

From which:

$$\frac{dM_t(t)}{dt} = \frac{M_t^0 - M_t(t)}{T_{1t}} + rCBF\left[\frac{M_t^0}{\lambda} - \frac{M_t(t)}{\lambda}\right] = \left(\frac{1}{T_{1t}} + \frac{rCBF}{\lambda}\right)M_t^0 - \left(\frac{1}{T_{1t}} + \frac{rCBF}{\lambda}\right)M_t(t). \quad (5.130)$$

By defining:

$$T_{1tapp} = \left[\frac{1}{T_{1t}} + \frac{rCBF}{\lambda}\right]^{-1}, \quad (5.131)$$

one has

$$\frac{dM_t(t)}{dt} = \frac{1}{T_{1app}}M_t^0 - \frac{1}{T_{1app}}M_t(t). \quad (5.132)$$

Assuming that a slice-selective inversion pulse has been applied at $t = 0$ on the tissue magnetization, Equation 5.132 yields:

$$M_t(t) = M_t^0\left[1 - 2e^{-t/T_{1app}}\right]. \quad (5.133)$$

Once estimated T_{1app}, CBF can be obtained from:

$$CBF = \frac{\lambda}{T_{1app}}\left(1 - \frac{M_t^{lab}}{2M_t^0}\right), \quad (5.134)$$

where M_t^{lab}/M_t^0 is the ratio of magnetization with and without arterial spin saturation. In particular, M_t^{lab} is the tissue magnetization after reaching the steady state.

5.7.3 Dynamic Susceptibility Contrast–Magnetic Resonance Imaging

The model used for cerebral hemodynamics quantification from dynamic susceptibility contrast (DSC)–MRI images is based on the principles of tracer kinetics for nondiffusible tracers [18]. It relies on the assumptions that the contrast agent is totally nondiffusible, that there is no recirculation of the contrast agent, that the contrast-agent is confined to the intravascular space, and that the system is in steady state during the experiment. In particular, in the case of intact blood–brain barrier (BBB), the model makes it possible to relate CBF to the concentration time curve of tracer within a given VOI, C_{VOI}, by convolution between the residue function, R (the fraction of tracer still present in the VOI at time t following an ideal bolus injection) and the arterial input function, C_{AIF}:

$$C_{VOI}(t) = \frac{\rho}{k_H} CBF \int_0^t C_{AIF}(\tau) R(t-\tau) d\tau. \quad (5.135)$$

R(t) is an unknown, dimensionless, positive, and decreasing function in time for which:

$$R(0) = 1. \quad (5.136)$$

C_{VOI} in Equation 5.135 is assumed known, but it needs to be obtained from the DSC–MRI signal. In DSC–MRI the contrast agent amount present within a voxel locally perturbs the total magnetic field, thus decreasing relaxation time constants and influencing the detected $T2^*$-weighted signal, S(t), from the voxel, as follows:

$$S(t) = S_0 e^{-\Delta R_2^*(t) \cdot T_E}, \quad (5.137)$$

where:

1. $S_0 = S(0)$ is the signal value from water protons at time $t = 0$, when no contrast agent is yet present.
2. $\Delta R_2^*(t) = R_2^*(t) - R_2^*(0)$ is the change in transverse relaxation rate; that is, the difference between water proton $T2^*$-relaxation rate $R2^*(t) = 1/T2^*(t)$ and its value at $t = 0$.
3. T_E is the echo-time; that is, a time parameter specific to the particular gradient–echo sequence adopted.

Within frequently used low-dosage ranges of contrast agents at common field B_0 strength, a linear relationship between the change in transverse relaxation rate and tracer concentration $C_{VOI}(t)$ within the voxel of volume can be assumed [19]:

$$C_{VOI}(t) = \kappa_{VOI} \Delta R_2^*(t), \quad (5.138)$$

where κ_{VOI} is an unknown proportionality constant depending on the tissue, the contrast agent, the field strength, and the pulse sequence. From Equations 5.137 and 5.138, one can derive:

$$C_{VOI}(t) = -\frac{\kappa_{VOI}}{T_E} \ln\left(\frac{S(t)}{S_0}\right), \quad (5.139)$$

which is the fundamental equation of DSC-MRI, relating the tracer concentration profile within a voxel to the measured signal produced by the perturbed water protons spin-½ system. Equation 5.139 is used to convert both arterial and tissue DSC–MRI measured signals. Because of the complexity of the relaxation mechanism underling the DSC–MRI signal generation and the consequent difficulty in retrieving the correct κ_{VOI}, value for each voxel, the same proportionality constant $\kappa = \kappa_{VOI}$ is usually assumed for both tissue and arterial concentration. However, this assumption can affect the correct quantification of the derived parameter below [20].

5.7.3.1 Cerebral Blood Flow

To obtain CBF, one needs to deconvolve (see below) Equation 5.135 to calculate $R'(t) = CBF \cdot R(t)$, and, subsequently, CBF from $R'(t)$ value at time $t = 0$:

$$CBF = R'(0). \quad (5.140)$$

In general, an analytic solution is not possible, and several techniques can be used to compute an approximate solution. One simple method to solve the inverse problem is using the convolution theorem of Fourier transform, which states that the transform of two convolved function equals the product of their individual transforms:

$$F\{CBF \cdot R(t) \otimes C_{AIF}(t)\} = F\{C_{VOI}(t)\}. \quad (5.141)$$

From Equation 5.141, one obtains:

$$CBF \cdot R(t) = F^{-1}\left\{\frac{F\{C_{VOI}(t)\}}{F\{C_{AIF}(t)\}}\right\}, \quad (5.142)$$

where F^{-1} denotes the inverse of the Fourier transform F. The Fourier transform approach has the attraction of being theoretically very easy to implement and insensitive to delays between the arterial input function and the tissue. However, its use is not without problems, and discordant results have been reported in literature. For instance, Ostergaard et al. showed that the Fourier transform approach biases CBF, in particular, underestimating it in case of high flow [21, 22]. On the other hand, other researchers found satisfactory estimates of CBF using the Fourier transform compared with other more sophisticated deconvolution techniques [23].

Another method to solve Equation 5.135 is to resort to a linear algebraic approach. More precisely, assuming that tissue and arterial concentrations are measured at equidistant times $t_i = t_{i-1} + \Delta t$ and choosing Δt so that $CBF \cdot R(t)$ is reasonably approximated by a staircase function, a discrete version of Equation 5.135 can be written in matrix form:

$$C_{VOI}(t_j) \approx CBF \cdot \Delta t \cdot \sum_{i=0}^{j} C_{AIF}(t_i) \cdot R(t_j - t_i), \quad (5.143)$$

and thus

$$\begin{pmatrix} C_{VOI}(t_1) \\ C_{VOI}(t_2) \\ \cdots \\ C_{VOI}(t_N) \end{pmatrix} = CBF \cdot \Delta t \cdot \begin{pmatrix} C_{AIF}(t_1) & 0 & \cdots & 0 \\ C_{AIF}(t_2) & C_{AIF}(t_1) & \cdots & 0 \\ \cdots & \cdots & \cdots & \cdots \\ C_{AIF}(t_N) & C_{AIF}(t_{N-1}) & \cdots & C_{AIF}(t_1) \end{pmatrix} \cdot \begin{pmatrix} R(t_1) \\ R(t_2) \\ \cdots \\ R(t_N) \end{pmatrix} \quad (5.144)$$

or, in compact form:

$$C_{VOI} = CBF \cdot \Delta t \cdot C_{AIF} \cdot \mathbf{R}, \quad (5.145)$$

where $C_{VOI} \in \Re^{Nx1}$, $C_{AIF} \in \Re^{NxN}$, $\mathbf{R} \in \Re^{Nx1}$, and Δt is the sampling interval. Equation 5.145 is a standard matrix equation that can be inverted to yield CBF·R if $\det(C_{AIF}) \neq 0$:

$$CBF \cdot \Delta t \cdot \mathbf{R} = C_{AIF}^{-1} \cdot C_{VOI}. \quad (5.146)$$

This approach has been termed *raw deconvolution* in the literature [24]. Albeit appealing for its simplicity, it is well known to perform poorly because it is extremely sensitive to noise.

A widely used approach to solve Equation 5.145 that overcomes the limitations of raw deconvolution is singular value decomposition (SVD). SVD was introduced as a method to estimate R(t) by Ostergaard et al. [21, 22]. SVD constructs matrices V, W, and U^T so that the inverse of C_AIF can be written:

$$C_{AIF}^{-1} = \mathbf{V} \cdot \mathbf{W} \cdot \mathbf{U}^T, \quad (5.147)$$

where W is a diagonal matrix, and V and U^T are orthogonal and transpose orthogonal matrices, respectively. Given this inverse matrix, CBF · R is found as:

$$CBF \cdot \mathbf{R} = \frac{\mathbf{V} \cdot \mathbf{W} \cdot \mathbf{U}^T \cdot C_{VOI}}{\Delta t}. \quad (5.148)$$

SVD has been shown to be a reliable technique for deconvolution because it reduces the effect of noise on R(t) estimation. This is achieved by setting to zero the elements in the diagonal matrix W obtained by SVD when they are smaller than a threshold value given beforehand.

SVD represents the most-used approach to quantify bolus tracking MRI data. However, this method does have limitations [25–27].

5.7.3.2 Cerebral Blood Volume

In case of intact BBB, one can also measure cerebral blood volume (CBV) by means of DSC–MRI images, using the ratio of the areas under the concentration time curve of tracer within a given VOI, C_{VOI}, and the arterial concentration C_{AIF} and normalizing it to the density of brain tissue ρ:

$$CBV = \frac{k_H}{\rho} \frac{\int_0^\infty C_{VOI}(\tau)d\tau}{\int_0^\infty C_{AIF}(\tau)d\tau}, \quad (5.149)$$

where k_H accounts for the difference in hematocrit between large vessels (LV) and small vessels (SV) because only the plasma volume is accessible to the tracer, $k_H = (1 - H_{LV})/(1 - H_{SV})$.

5.7.3.3 Mean Transit Time

An additional parameter that can be derived is the mean transit time (MTT). MTT can be calculated by using the central volume theorem of indicator dilution theory [28] for which MTT is the ratio of CBV to CBF in the VOI:

$$MTT = \frac{CBV}{CBF}. \quad (5.150)$$

5.8 Exercises

1. Analyze the *a priori* identifiability of the following model:

$$\dot{C}_e(t) = K_1 C_p(t) - (k_2 + k_3) C_e(t) \qquad C_e(0) = 0$$

$$\dot{C}_m(t) = k_3 C_m(t) \qquad C_m(0) = 0$$

$$y(t) = (1 - V_b)(C_e(t) + C_m(t)) + V_b \cdot C_b(t),$$

   where $C_p(t)$ and $C_b(t)$ are known functions.
2. Explain why compartmental modeling is widely used in quantification of medical imaging.
3. Explain why it is important to weight the measured data to correctly estimate model parameters.
4. Explain under which hypothesis the two-tissue can be used instead of the three-tissue compartment model to estimate the binding potential with PET receptor images.
5. How can cerebral blood flow be estimated when the brain-blood barrier is not intact.
6. Explain why deconvolution is needed to quantify DSC–MRI images.

5.9 References and Bibliography

1. E. R. Carson, C. Cobelli, and L. Finkelstein. *The Mathematical Modeling of Metabolic and Endocrine Systems.* John Wiley, 1983.
2. K. R. Godfrey. *Compartmental Models and Their Application.* Academic Press, 1983.
3. J. A. Jacquez. *Compartmental Analysis in Biology and Medicine.* 3rd ed. Biomedware, 1996.

4. C. Cobelli, D. M. Foster, and G. Toffolo. *Tracer Kinetics in Biomedical Research: From Data to Model.* Kluwer Academic/Plenum Publishers, 2002.
5. C. Carson and C. Cobelli. *Modeling Methodology for Physiology and Medicine.* Academic Press, 2001.
6. S. Audoly et al. Global identifiability of linear compartmental models—A computer algebra algorithm. *IEEE Trans. Biomed. Eng.* 45:36–47, 1998.
7. C. Cobelli et al. Validation of simple and complex models in physiology and medicine. *Am. J. Physiol.* 246:R259–R266, 1984.
8. C. Cobelli and G. Toffolo. Compartmental vs. noncompartmental modeling for two accessible pools. *Am. J. Physiol.* 247:R488–R496, 1984.
9. A. A. Lammertsma et al. Comparison of methods for analysis of clinical [11C]raclopride studies. *J. Cereb. Blood Flow Metab.* 16:42–52, 1996.
10. A. A. Lammertsma and S. P. Hume. Simplified reference tissue model for PET receptor studies. *Neuroimage.* 4:153–158, 1996.
11. D. E. Kuhl et al. Local cerebral blood volume determined by 3-dimensional reconstruction of radionuclide scan data. *Circ. Res.* 36:610–619, 1975.
12. H. Iida et al. Error analysis of a quantitative cerebral blood flow measurement using H215O autoradiography and positron emission tomography, with respect to the dispersion of the input function. *J. Cereb. Blood Flow Metab.* 6:536–545, 1986.
13. L. Sokoloff et al. The [14C]deoxyglucose method for the measurement of local cerebral glucose utilization: theory, procedure, and normal values in the conscious and anesthetized albino rat. *J. Neurochem.* 28:897–916, 1977.
14. M. M. Graham et al. The FDG lumped constant in normal human brain. *J. Nucl. Med.* 43:1157–1166, 2002.
15. S. G. Hasselbalch et al. The (18)F-fluorodeoxyglucose lumped constant determined in human brain from extraction fractions of (18)F-fluorodeoxyglucose and glucose. *J. Cereb. Blood Flow Metab.* 21:995–1002, 2001.
16. L. M. Parkes. Quantification of cerebral perfusion using arterial spin labeling: Two-compartment models. *J. Magn. Reson. Imaging.* 22:732–736, 2005.
17. E. L. Barbier, L. Lamalle, and M. Decorps. Methodology of brain perfusion imaging. *J. Magn. Reson. Imaging.* 13:496–520, 2001.
18. K. L. Zierler. Theoretical basis of indicator-dilution methods for measuring flow and volume. *Circ. Res.* 10:393–407, 1962.
19. B. R. Rosen et al. Contrast agents and cerebral hemodynamics. *Magn. Reson. Med.* 19:285:292, 1991.
20. V. G. Kiselev. On the theoretical basis of perfusion measurements by dynamic susceptibility contrast MRI. *Magn. Reson. Med.* 46:1113–1122, 2001.
21. L. Ostergaard et al. High resolution measurement of cerebral blood flow using intravascular tracer bolus passages. Part I: Mathematical approach and statistical analysis. *Magn. Reson. Med.* 36:715–725, 1996a.
22. L. Ostergaard et al. High resolution measurement of cerebral blood flow using intravascular tracer bolus passages. Part II: Experimental comparison and preliminary results. *Magn. Reson. Med.* 36:726–736, 1996b.
23. A. M. Smith et al. Whole brain quantitative CBF and CBV measurements using MRI bolus tracking: comparison of methodologies. *Magn. Reson. Med.* 43:559–564, 2000.
24. G. De Nicolao, G. Sparacino, and C. Cobelli. Nonparametric input estimation in physiological systems: Problems, methods, and case study, *Automatica.* 33:851–870, 1997.
25. K. Murase, M. Shinohara, and Y. Yamazaki. Accuracy of deconvolution analysis based on singular value decomposition for quantification of cerebral blood flow using dynamic susceptibility contrast-enhanced magnetic resonance imaging. *Phys. Med. Biol.* 46:3147–3159, 2001.
26. R. Wirestam et al. Assessment of regional cerebral blood flow by dynamic susceptibility contrast MRI using different deconvolution techniques. *Magn. Reson. Med.* 43:691–700, 2000.
27. M. R. Smith et al. Removing the effect of SVD algorithmic artifacts present in quantitative MR perfusion studies. *Magn. Reson. Med.* 51:631–634, 2004.
28. P. Meier and K. L. Zierler. On the theory of the indicator-dilution method for measurement of blood flow and volume. *J. Appl. Physiol.* 6:731–744, 1954.

6

Techniques for Parametric Imaging

Prof. David Dagan Feng,[1,2]
Dr. Lingfeng Wen,[1,3] and
Dr. Stefan Eberl[1,3]
[1]*University of Sydney*
[2]*Hong Kong Polytechnic University*
[3]*Royal Prince Alfred Hospital*

6.1 Introduction

6.1.1 Structural and Functional Imaging

Medical imaging has become an important component of modern medicine in many areas, such as detecting abnormalities, staging progression of disease, and guiding surgery, through noninvasive visualization of internal structures or functional changes in the human body. Medical imaging can be classified into two broad categories: structural and functional imaging. X-ray computed tomography (CT) and magnetic resonance imaging (MRI) are regarded as structural imaging because they provide detailed visualization of anatomic boundaries and morphologic variations. In contrast, positron emission tomography (PET) and single photon emission computed tomography (SPECT) visualize *in vivo* physiologic or biochemical processes and are thus referred to as functional imaging. They are capable of detecting subtle changes of biochemical and physiologic function like glucose metabolism and blood flow at an early stage of the disease, before marked changes become apparent on structural imaging.

The radioactive drug is generally referred as a tracer because it "traces" a functional process. After the tracer is administrated, its specific chemical and biologic properties lead to differential uptake and clearance between normal and abnormal tissues, giving rise to characteristic spatial and temporal distributions in the human body. PET relies on the detection of the two opposing 511keV photons from positron annihilation to determine the tracer distribution in the patient. A typical PET scanner consists of hundreds of detectors forming a detector ring, which detects a large number of these opposing photon pairs in coincidence followed by reconstruction to form images of the 3D tracer distribution. These unique features of PET facilitate quantitative measurement of *in vivo* function and metabolism. More recently, PET scanners have been combined with CT scanners into the one instrument [1]. The combined PET-CT scanner can measure co-registered CT images, providing more accurate attenuation correction and an anatomic frame of reference for the interpretation of the low-resolution PET data. The recent introduction of time-of-flight PET-CT further improves image quality by accurately measuring and using the time difference of two annihilation photons reaching the two opposing detectors [2]. The most attractive feature of

PET is its diversity of tracers, which not only visualize functional change, but also provide early diagnosis of some disorders with disorder-specific tracers. Some clinical PET tracers are listed in Table 6.1.

TABLE 6.1 Some PET tracers and their applications

Isotope name	Tracer name	Purpose
Fluorine-18	^{18}F-FDG	Glucose metabolism
Fluorine-18	^{18}F-Fluoride	Bone metabolism
Carbon-11	^{11}C-Flumazenil	Brain epilepsy
Carbon-11	^{11}C-PIB	Early diagnosis of Alzheimer's disease
Oxygen-15	^{15}O-Water	Blood flow/perfusion
Oxygen-15	^{15}O-O_2	Oxygen consumption
Nitrogen-13	^{13}N-Ammonia	Myocardial perfusion

Modern SPECT scanners employ one or more gamma cameras rotating around the patient, detecting emitted photons from the decay of the tracer. These features lead to poorer image quality of SPECT with high level of random noise as compared with PET. Advances in gamma camera technology and correction algorithms have resulted in SPECT potentially providing quantitative measures of functional processes [3]. A diverse range of SPECT tracers is also available in the clinic, with a small sample of SPECT tracers listed in Table 6.2.

TABLE 6.2 Some SPECT tracers and their purposes in the clinic

Isotope name	Tracer name	Purpose
Technetium-99m	^{99m}Tc-MIBI	Myocardial perfusion scan
Technetium-99m	^{99m}Tc-MDP	Bone scan
Technetium-99m	^{99m}Tc-MAA	Perfusion of lung
Iodine-123	^{123}I-MIBG	Neuroendocrine tumors
Thallium-201	^{201}Tl-chloride	Myocardial perfusion

Although functional imaging suffers from relatively low spatial resolution compared with structural imaging, as shown in Figure 6.1, PET and SPECT are still attractive for their unique abilities to detect early functional changes in disease before anatomic changes are apparent on anatomic imaging. This is particularly important where early treatment is essential before damage becomes irreversible, such as in some forms of dementia [4]. Limited functional imaging can also be performed with the structural imaging modalities, such as brain activation measurements with functional MRI using the blood-oxygenation-level–dependent (BOLD) method [5]. However, in this chapter, functional imaging refers to PET and SPECT.

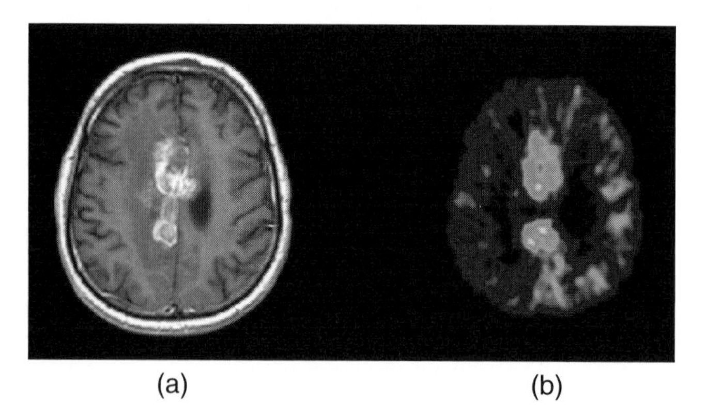

FIGURE 6.1 Difference between structural imaging (a, MRI) and functional imaging (b, PET) in the human brain.

6.1.2 Kinetic Modeling and Parametric Images

In functional imaging, the injected tracer is transported throughout the body by the circulation, following the behavior of a targeted dynamic process without affecting the normal physiologic process. The tracer is accumulated or released over time from normal and abnormal tissues, based on its specific physiologic and biochemical properties. The distribution of the tracer at any time point can be determined from the external detection of the photons emitted in the body by the radioactive decay of the tracer. Kinetic modeling is a highly versatile tool for analyzing experiments in living systems, with application in many branches of biology [6]. The compartment model is most commonly used in functional imaging for its simplicity and its ability to model a large range of physiologic and biochemical processes [7]. Figure 6.2 demonstrates an example of a three-compartment and four-parameter compartment model for 2-[^{18}F]fluorodeoxyglucose (FDG), a radioactive analog of glucose [8]. The compartment in kinetic modeling can be a physical or a chemical compartment, such as two compartments respectively for FDG in plasma and tissue, and one compartment for the phosphorylated form 2-[^{18}F]fluorodeoxyglucose-6-phosphate (FDG-6-PO_4) in tissue.

The fractional transfer rates between compartments are nonlinear in many biologic systems, such as facilitated diffusion and receptor binding systems. The corresponding fractional transfer rates are functions of the amount of substance in the compartments and may be saturated at a maximum. However, if the dynamic system is in steady state and is time-invariant over the study duration, the administration of a small amount of tracer will not disturb the steady state or underlying transfer rates [9]. Therefore, linear fractional transfer rates can be assumed for the tracer in the compartment model analysis, which allows taking advantage of the attractive mathematical properties of linear compartment models. The compartment model in Figure 6.2 is thus assumed to be a linear model, whose arrows indicate the direction of transfer governed by the rate constant. Rate constants describe the fraction of substance in one compartment being transported to another compartment per unit time. For example, if K_1 is equal to 0.1/min, then 10% of FDG mass in plasma is transported to the tissue per minute.

Figure 6.3 shows a flow chart for estimating cerebral metabolic rate of glucose (CMRGlc) with FDG and PET. The time

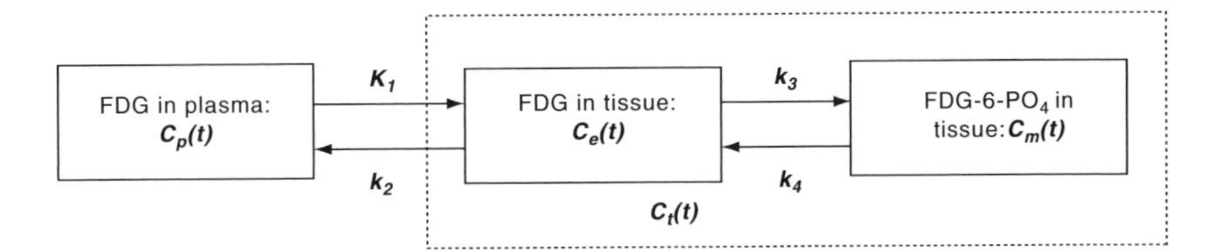

FIGURE 6.2 A three-compartment and four-parameter model for FDG.

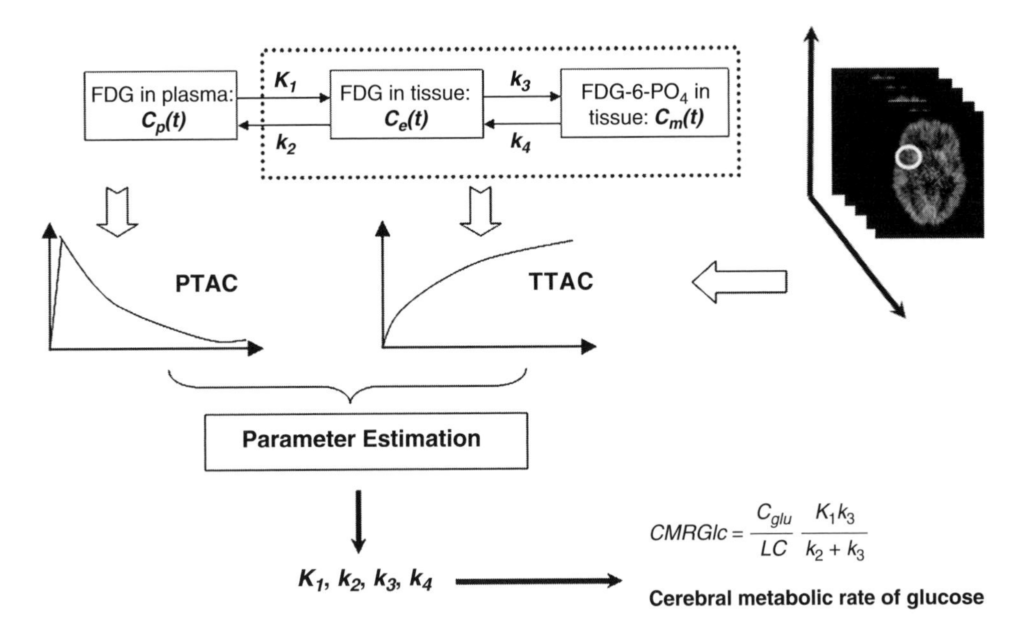

FIGURE 6.3 Flow chart of estimating cerebral metabolic rate of glucose (CMRGlc) using ROI.

course of FDG uptake in the brain is followed by a dynamic PET study that collects a series of images over predetermined time intervals. A region of interest (ROI) is defined to generate a tissue time activity curve (TTAC) of the average activity concentration within the ROI, which represents the sum of FDG and the phosphorylated FDG-6-PO_4 in the tissue. Arterial blood samples are measured to obtain a plasma time activity curve (PTAC) over the whole scan period, which is used as input function (IF) to the kinetic model. The microparameters of K_1, k_2, k_3, and k_4 of the kinetic model are estimated from the model equations based on the measured TTAC and PTAC. CMRGlc can then be estimated as shown in Figure 6.3 based on the rate constants, the endogenous glucose concentration (C_{glu}), and the lump constant (LC), which accounts for differences between endogenous glucose and FDG.

If parameter estimation is performed for TTACs derived from each voxel of the image volume, then images of CMRGlc covering the three-dimensional image volume can be generated as shown in Figure 6.4. Parametric images form a three-dimensional image volume whose voxels represent quantitative functional parameters. This avoids operator-dependent manual delineation of ROI, and also avoids the need for prior knowledge on the spatial distribution for a newly developed tracer. Parametric images encapsulate the detailed information contained in the dynamic scans and can visualize functional changes such as metabolism, blood flow, and receptor binding.

6.1.3 Compartment Model Parameter Estimation

Parameter estimation is a discipline that provides tools for mathematically modeling phenomena and the estimation of the constants appearing in these models [10]. Diverse approaches are available for parameter estimation. These are optimized in terms of the tracer, compartment model, and imaging implementation. A compartment model for regional cerebral blood flow (rCBF) is used to demonstrate two basic methods of parameter estimation. In contrast to the indirect measurement of metabolic rate of glucose with FDG, kinetic modeling of blood flow directly measures rCBF transported into the tissue, as shown in Figure 6.5, based on the Kety-Schmidt single compartment model [11, 12].

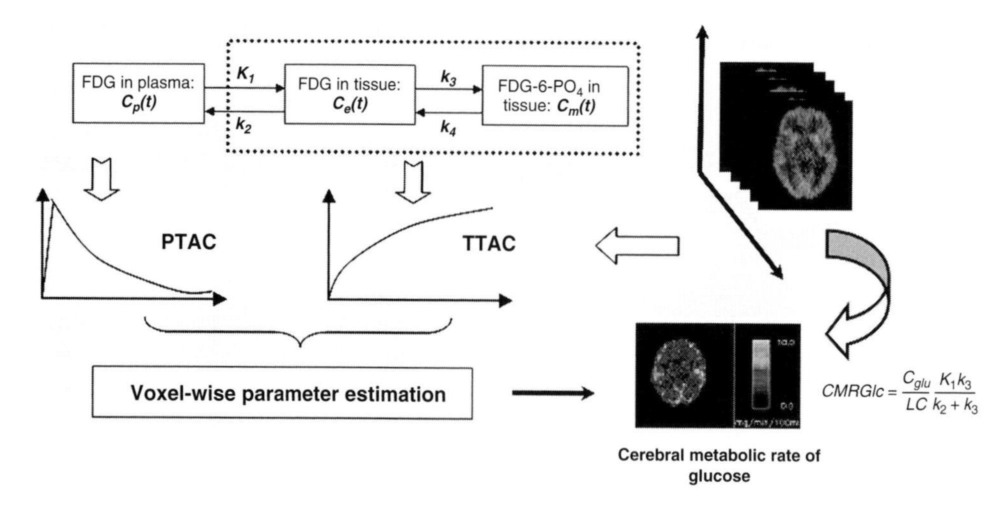

FIGURE 6.4 Flow chart of generating parametric image of cerebral metabolic rate of glucose (CMRGlc).

The model yields the following differential equation for the radioactivity concentration (Equation 6.1):

$$\frac{dC_t(t)}{dt} = rCBF \cdot C_p(t) - \left(\frac{rCBF}{V_d} + \lambda\right) \cdot C_t(t), \tag{6.1}$$

where rCBF is the local perfusion flow, $C_p(t)$ represents the tracer concentration in arterial blood, V_d is the volume of distribution for water, and λ is the radioactive decay constant of the tracer. Equation 6.1 can be simplified to Equation 6.2 by substituting $rCBF/V_d + \lambda$ with K:

$$\frac{dC_t(t)}{dt} = rCBF \cdot C_p(t) - K \cdot C_t(t). \tag{6.2}$$

Equation 6.2 for rCBF study can be transformed into Equation 6.3 by the Laplace transform:

$$C_t(s) = \frac{rCBF}{s+K} C_p(s). \tag{6.3}$$

Taking the inverse Laplace transform, Equation 6.3 can be reorganized into Equation 6.4:

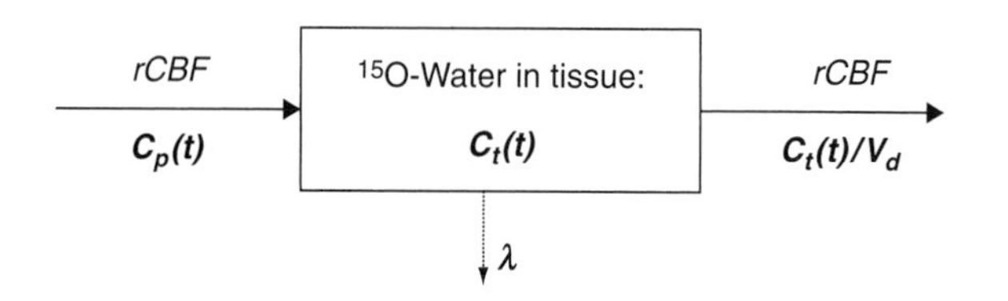

FIGURE 6.5 A single-compartment model for regional cerebral blood flow (rCBF).

$$\begin{aligned} C_t(t) &= rCBF \cdot \int_0^t C_p(\tau) e^{-K(t-\tau)} d\tau \\ &= rCBF \cdot C_p(t) \otimes e^{-Kt}, \end{aligned} \tag{6.4}$$

where $\otimes$ represents the mathematical operation of convolution, $C_t(t)$ is the measured tissue activity, and $C_p(t)$ is the tracer concentration in the plasma from arterial blood samples. Even for this simple compartment model, the solution to the differential equation is nonlinear, requiring a nonlinear fitting routine to estimate f and K from the measured data. Nonlinear least-squares fitting, discussed in the next section, is the method of choice for fitting the nonlinear equations resulting from compartment models.

6.1.3.1 Nonlinear Least-Squares Fitting

Nonlinear least-squares (NLS) analysis comprises a group of numeric procedures that find the "optimal estimates" of the parameters for the experimental data [13]. In this context, the NLS method is frequently called NLS fitting, since it is used to iteratively fit the model equation to the TTAC for a given PTAC. As shown in Figure 6.6, the NLS fitting consists of initial parameters and an iterative procedure, adjusting estimated parameters to derive a better approximation until sufficient convergence has been reached. Within the iterations, the improved parameter estimates derived from the previous step are used as a new set of initial parameters in the present step. The objective function is chosen to evaluate how well the estimated model TTAC approximates the measured TTAC. In general, the objective function minimizes the nonweighted or weighted sum of squared differences between the estimated and measured TTAC.

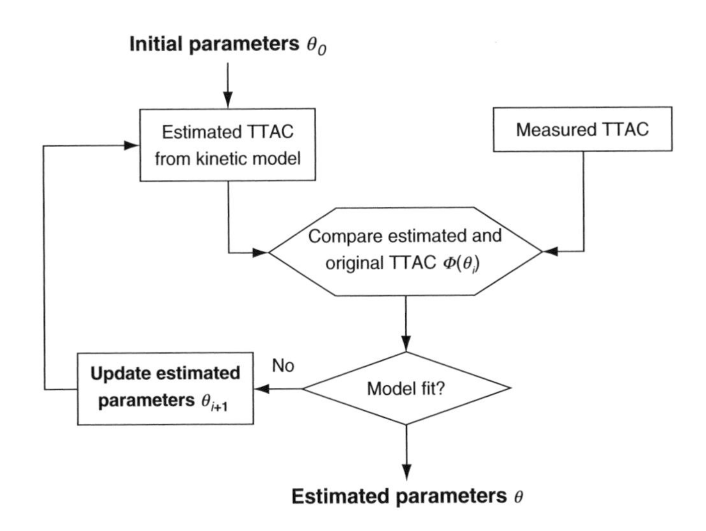

FIGURE 6.6 Diagram illustrating iterative process for fitting model to measured data. TTAC: tissue time activity curve.

Given a set of initial parameters $\theta_0 = [rCBF,\ K]$, the estimated TTAC can be derived according to Equation 6.4 and the measured $C_p(t)$. The sum of squared differences between the estimated TTAC and measured TTAC forms the objective function in Equation 6.5:

$$\Phi(\theta_i) = \sum_{j=1}^{N} [C^E(t_j) - C^M(t_j)]^2, \tag{6.5}$$

where θ_i are the estimated parameters in the ith iteration, N is the number of sampling frames, $C^E(t_j)$ is the estimated activity for the jth sampling point using Equation 6.4, and $C^M(t_j)$ is the measured activity for the jth sampling point. If $\Phi(\theta_i)$ hasn't reached its minimum, the parameters θ_i are adjusted to θ_{i+1} using multidimensional optimization techniques such as the Newton-Gauss or Levenberg-Marquardt algorithms [14, 15]. The updated θ_{i+1} is then used in the next iteration. Once the minimum is reached or estimation has converged to within a set tolerance, the whole process is terminated, and the final estimates are regarded as the results of the NLS method.

Because noise exists in PET and SPECT data, with particularly low signal to noise ratio (SNR) in the early frames, a modified objective function Equation 6.6 can be adopted using a weighted least-square distance.

$$\Phi_W(\theta_i) = \sum_{j=1}^{N} w_j[C^E(t_j) - C^M(t_j)]^2, \tag{6.6}$$

where the weight function, $W = [w_1, w_2, \ldots, w_N]$, can be chosen to be proportional to the frame durations with higher weights for frames with longer collection times, or proportional to the inverse of variance of the measurements [16]. This is then referred to as the weighted nonlinear least-square (WNLS) fitting method.

The NLS and WNLS methods are considered the methods of choice for providing parameter estimates for TTACs with optimum statistical accuracy and reliability [8]. The practical implementations of NLS and WNLS are dependent on the choice of initial values. If poor initial values are given, the methods may converge very slowly, fail to converge, or be trapped in local minima [17]. Because of the high computational burden, NLS and WNLS are not well suited for the generation of parametric images. However, NLS methods are often used as the gold standard to assess the performance of other methods for fitting ROI TTACs to a model.

6.1.3.2 Steady-State Technique

Steady-state technique is also called the equilibrium method. Constant infusion of tracer leads to a steady state eventually being reached with a constant tissue concentration. For example, the time to reach equilibrium state is approximately 10 minutes for $C^{15}O_2$. In steady state, $dC_t(t)/dt = 0$, thus rearranging Equation 6.1 yields:

$$C_t(T) = \frac{rCBF \cdot C_p(T)}{\lambda + rCBF/V_d}, \tag{6.7}$$

where T is the time when the concentration reaches a steady state. The estimation of rCBF can be derived by Equation 6.8:

$$rCBF = \frac{\lambda}{C_p(T)/C_t(T) - 1/V_d} = \frac{\lambda \cdot C_t(T) \cdot V_d}{C_p(T) \cdot V_d - C_t(T)}. \tag{6.8}$$

Since the $C_p(T)$ and $C_t(T)$ are known from the measurements, the derivation of rCBF depends on the values of λ and V_d. rCBF has a nonlinear relationship with the tissue concentration $C_t(T)$, with the degree of nonlinearity depending on λ. A tracer with a short half-life (i.e., larger λ values) can improve the linearity. This fact and the requirement that equilibrium be reached in a reasonable short time restrict the steady-state technique to tracers with short half-lives, such as ^{15}O with a half-life of 2 minutes [9].

The steady-state technique requires only one scan and one blood sample according to Equation 6.8. However, estimation of rCBF is sensitive to measurement error of the steady-state plasma concentration. A 10% error in $C_p(T)$ may give rise to $20 \sim 30\%$ error in the rCBF [18]. Thus, a few blood samples are usually collected over the scanning period to obtain an average measurement of $C_p(T)$, improving the accuracy of the estimation [19, 20].

6.2 Parametric Image Estimation Methods

FDG is the most commonly used tracer in PET imaging because of its availability (half-life of 109.7 minutes) and wide application in the assessment of glucose metabolism for neurology and the diagnosis, staging, and monitoring of treatment in oncology. Several reasonable assumptions are generally made for the FDG kinetic model [8, 21]. For example, it is assumed that the compartment is homogenous with respect to blood flow, transport rate, and substance mass and that the arterial plasma concentration of FDG is approximately equal to the capillary plasma concentration.

These assumptions and knowledge lead to the three-compartment and four-parameter model for FDG shown in Figure 6.2. The corresponding differential equations of activity concentration in tissue for FDG and FDG-6-PO_4, respectively, are shown in Equations 6.9 and 6.10:

$$\frac{d}{dt}C_e(t) = K_1 C_p(t) - (k_2 + k_3)C_e(t) + k_4 C_m(t) \quad (6.9)$$

$$\frac{d}{dt}C_m(t) = k_3 C_e(t) - k_4 C_m(t), \quad (6.10)$$

where $C_p(t)$ represents the FDG concentration as a function of time in plasma, which is measured by arterial blood sampling. $Ce(t)$ represents free FDG concentration in tissue; $C_m(t)$ represents the concentration of FDG-6-PO_4 in tissue. Because PET measures the total activity in the tissue space, the measured $C_t(t)$ is the sum of $C_e(t)$ and $C_m(t)$. Solving Equations 6.9 and 6.10 can yield the differential equation for $C_t(t)$ shown in Equation 6.11:

$$\frac{d^2C_t(t)}{dt^2} = K_1\frac{dC_p(t)}{dt} + K_1(k_3 + k_4)C_p(t) - (k_2 + k_3 + k_4)\frac{dC_t(t)}{dt} - k_2k_4C_t(t). \quad (6.11)$$

The solutions for $C_e(t)$ and $C_m(t)$ *can be derived as shown in Equations 6.12 and 6.13:*

$$C_e(t) = \frac{K_1}{\alpha_2 - \alpha_1}[(k_4 - \alpha_1)\cdot e^{-\alpha_1 t} + (\alpha_2 - k_4)\cdot e^{-\alpha_2 t}] \otimes C_p(t) \quad (6.12)$$

$$C_m(t) = \frac{K_1 k_3}{\alpha_2 - \alpha_1}(e^{-\alpha_1 t} - e^{-\alpha_2 t}) \otimes C_p(t), \quad (6.13)$$

where $\otimes$ denotes the mathematical convolution, $\alpha_{1,2} = \frac{k_2 + k_3 + k_4 \mp \sqrt{(k_2 + k_3 + k_4)^2 - 4k_2k_4}}{2}$. Once the rate constants are obtained from parameter estimates, the CMRGlc is calculated as defined by Equation 6.14 [7, 8]:

$$CMRGlc = \frac{1}{LC}\cdot\frac{K_1 k_3}{k_2 + k_3}\cdot C_{glu}, \quad (6.14)$$

where C_{glu} is the endogenous glucose concentration in the plasma, and LC denotes the lumped constant accounting for the difference in transport and phosphorylation between glucose and deoxyglucose. C_{glu} can be obtained from analyzing blood samples, and a prior value can be used for LC—for example, 0.52 provided in Reivich et al [22].

6.2.1 Autoradiographic Technique

Usually a dynamic study is required to generate the TTACs that allow estimation of the rate constants (K_1, k_2, k_3, and k_4) and hence *in vivo* quantitation of CMRGlc. The autoradiographic (ARG) technique requires only one static study (one frame) and thus substantially reduces the complexity of the study and computational burden compared with the NLS and WNLS methods.

Given that $C_t(t) = C_e(t) + C_m(t)$, Equation 6.14 can be rewritten as Equation 6.15:

$$CMRGlc = \frac{C_{glu}}{LC}\cdot\frac{K_1 k_3}{k_2 + k_3}\cdot\frac{C_t(t) - C_e(t)}{C_m(t)}. \quad (6.15)$$

As the accuracy of CMRGlc estimation is rather insensitive to variation of the rate constants, their values can be assumed to be constant in the brain [8]. The estimated CMRGlc then approximates its true value when the rate constants (K_1, k_2, k_3, *and* k_4) are based on the mean kinetic values ($\overline{K}_1$, $\overline{k}_2$, $\overline{k}_3$, and $\overline{k}_4$) from a population of control subjects [21]. Therefore, the formula of the ARG method is presented in Equation 6.16 with $C_e(t)$ and $C_m(t)$ substituted by estimates derived from the population rate constant and the measured input function:

$$CMRGlc \approx \frac{C_{glu}}{LC}\cdot\frac{C_t(t) - \frac{\overline{K}_1}{\alpha_2 - \alpha_1}[(\overline{k}_4 - \alpha_1)\cdot e^{-\alpha_1 t} + (\alpha_2 - \overline{k}_4)\cdot e^{-\alpha_2 t}]\otimes C_p(t)}{\frac{\overline{k}_2 + \overline{k}_3}{\alpha_2 - \alpha_1}(e^{-\alpha_1 t} - e^{-\alpha_2 t})\otimes C_p(t)}, \quad t > T. \quad (6.16)$$

With the prior knowledge of *LC*, the required measurements for the ARG technique are a single image scan at time T after injection. However, the tracer concentration of FDG in plasma, $C_p(t)$, still has to be determined from the time of injection to the end of the image acquisition. As before, endogenous glucose concentration has to be measured at least once during the study. Equation 6.16 provides good estimates of CMRGlc for the scan times, *T*, between 30 minutes and 2 hours. The advantages of the ARG method are its simplicity and computational efficiency, and it has thus become the method of choice for routine clinical neurologic studies where the assumptions are sufficiently satisfied.

However, in pathologic states such as tumor or stroke, the rate constants can differ significantly from the assumed population means derived from brain studies, leading to high bias in CMRGlc estimates [8, 21]. For example, negative values were obtained in patients with stroke [8, 21]. Alternative formulations have been investigated to improve the accuracy of the ARG method estimates [23]. One approach assumes $C_t(t)$ is proportional to CMRGlc because Equations 6.12 and 6.13 demonstrate that $C_e(t)$ and $C_m(t)$ are linearly dependent to the influx rate K_1. For the variations of K_1 and k_2, better estimation of CMRGlc is provided by Equation 6.17 because of the normalization of K_1 [21, 23–25]:

$$CMRGlc \approx \frac{C_{glu}}{LC}\cdot\frac{\overline{K}_1\overline{k}_3}{\overline{k}_2 + \overline{k}_3}\cdot\frac{C_t(t)}{\overline{C}_e(t) + \overline{C}_m(t)}. \quad (6.17)$$

If the tissue's free FDG is assumed to be small compared to the total radioactivity in the tissue, ($C_e(t) << Ct(t)$), Equation 6.18 provides a further simplified formula for CMRGLC at later time ($T > 45$ min) [26]:

$$CMRGlc \approx \frac{C_{glu}}{LC} \cdot \frac{C_t(T)}{\int_0^T C_p(t) \cdot dt}. \tag{6.18}$$

6.2.2 Standardized Uptake Value Method

Although the ARG method is attractive, a series of blood samples is still required to obtain the plasma time activity curve. A popular alternative approach is the standardized uptake value (SUV) method, which has been widely applied in differentiating malignant from benign tumors and assessing the efficacy of therapy [27], especially for whole-body FDG-PET.

If the time integral of $C_p(t)$ is assumed proportional to the injected tracer dose divided by body weight of the scanned subject with a proportional constant b (that is, $\int_0^T C_p(t) \cdot dt = b \frac{Dose}{Weight}$) Equation 6.18 can be reorganized into Equation 6.19:

$$MRGlc = \frac{C_{glu}/100}{LC \cdot b/100} \cdot \frac{C_t(T)}{dose/Weight} = \frac{C_{glu}/100}{LC \cdot b/100} \cdot SUV, \tag{6.19}$$

where $SUV = C_t(T)/(dose/Weight)$, and MRGlc is the metabolic rate of glucose for the whole body rather than for the brain. If LC and b are both constants, SUV is proportional to MRGlc [27] if C_{glu} is also constant in Equation 6.19. This is the theoretical background of the SUV method. Therefore patients should fast before and during PET scans to keep plasma glucose concentration relatively stable for the duration of the PET study [28].

There are many factors affecting accuracy of SUV values, such as patient size, standardized measurement time, plasma glucose level, and partial volume effects. For example, SUV in a lung tumor was observed to vary by 40% from 5.5 to 7.7 between 30 and 60 minutes [29]. The use of lean body mass [30] or body surface area [31] has been proposed to provide more accurate estimation for the SUV method instead of traditional body weight. Because of its computational efficiency and clinical practicality, the SUV method is the method of choice for providing a semiquantitative measure of MRGlc in routine clinical studies, despite its shortcomings.

6.2.3 Integrated Projection Method

The integrated projection (IP) method was developed to address the issue of computational cost with the NLS or WNLS methods. The IP method was proposed to derive rCBF from dynamic studies in terms of time integrals of the tissue and blood activity concentrations [32, 33]. For the Kety-Schmidt compartment model in Figure 6.5, if blood and tissue measurements have been corrected for decay, the differential Equation 6.1 can be converted to Equation 6.20:

$$\frac{dC_t^*(t)}{dt} = rCBF \cdot C_p^*(t) - \frac{rCBF}{V_d} \cdot C_t^*(t), \tag{6.20}$$

where $C_p^*(t) = C_p(t) \cdot e^{\lambda t}$ and $C_t^*(t) = C_t(t) \cdot e^{\lambda t}$ are radioactive decay corrected terms. Integrating Equations 6.1 and 6.20 over the scan duration from time 0 and T, the estimates of $rCBF$ and V_d can be solved in Equation 6.21 if zero initial conditions are assumed:

$$\widehat{rCBF} = \frac{\int_0^T C_t^*(\tau)d\tau \cdot [\lambda \int_0^T C_t(\tau)d\tau + C_t(T)] - C_t^*(T) \cdot \int_0^T C_t(\tau)d\tau}{\int_0^T C_p(\tau)d\tau \cdot \int_0^T C_t^*(\tau)d\tau - \int_0^T C_t(\tau)d\tau \cdot \int_0^T C_p^*(\tau)d\tau}$$

$$\hat{V}_d = \frac{\int_0^T C_t^*(\tau)d\tau \cdot [\lambda \int_0^T C_t(\tau)d\tau + C_t(T)] - C_t^*(T) \cdot \int_0^T C_t(\tau)d\tau}{\int_0^T C_p^*(\tau)d\tau \cdot [\lambda \int_0^T C_t(\tau)d\tau + C_t(T)] - C_t^*(T) \cdot \int_0^T C_p(\tau)d\tau}, \tag{6.21}$$

where $\widehat{rCBF}$ and $\hat{V}_d$ are the estimates of $rCBF$ and V_d using the IP method. If the integral time, T, is chosen to be long enough so that $C_t(T) << \lambda \int_0^T C_t(\tau)d\tau$, the equations of the original integrated method may be simplified further into Equation 6.22:

$$\widehat{rCBF} = \frac{\lambda \int_0^T C_t^*(\tau)d\tau \cdot \int_0^T C_t(\tau)d\tau - C_t^*(T) \cdot \int_0^T C_t(\tau)d\tau}{\int_0^T C_p(\tau)d\tau \cdot \int_0^T C_t^*(\tau)d\tau - \int_0^T C_t(\tau)d\tau \cdot \int_0^T C_p^*(\tau)d\tau}$$

$$\hat{V}_d = \frac{\lambda \int_0^T C_t^*(\tau)d\tau - C_t^*(T)}{\lambda \int_0^T C_p^*(\tau)d\tau - C_t^*(T) \cdot \int_0^T C_p(\tau)d\tau / \int_0^T C_t(\tau)d\tau}. \tag{6.22}$$

The advantage of the integrated method is the simple computation for parameter estimation based on the direct reconstructing of the time integrals of the projection data. Because reconstruction is a linear operation, integration can be performed either prior to reconstruction or postreconstruction [34]. Performing the integration prior to reconstructions avoids the need to reconstruct the dynamic data.

6.2.4 Weighted Integrated Method

The weighted integrated method (WIM) is an optimized version of the IP method to minimize the noise effect on estimates of rCBF and improve statistical accuracy and reliability of estimates [35, 36]. The method was originally proposed by introducing weighted integrals using two arbitrary weighting functions and their differentials into the IP method [34]. It was then extended using piecewise continuous weighting functions instead of the differentials [37]. The WIM method was later extended for CMRGlc using the two-compartment and three-parameter Sokoloff model (assuming $k_4 = 0$) [38].

For the two-compartment and four-parameter model for FDG (Figure 6.2), integrating Equation 6.11 twice from 0 to a time, t, derives the linearized equation for $C_t(t)$ given by Equation 6.23 [39]:

$$C_t(t) = P_1 \int_0^t C_p(\tau)d\tau + P_2 \int_0^t \int_0^t C_p(\tau)d\tau^2 + P_3 \int_0^t C_t(\tau)d\tau + P_4 \int_0^t \int_0^t C_t(\tau)d\tau^2, \quad (6.23)$$

where $P_1 = K_1$, $P_2 = K_1(k_3 + k_4)$, $P_3 = -(k_2 + k_3 + k_4)$, and $P_4 = -k_2 k_4$. After the four weighting functions, $W_1(t), W_2(t)$, $W_3(t), W_4(t)$ are multiplied with the corresponding integral terms in Equation 6.23, followed by integration over the scan duration of T, the obtained simultaneous equations can be solved in matrix form Equation 6.24:

$$A = B\theta_{WIM} + \varepsilon, \quad (6.24)$$

where $\theta_{WIM} = \begin{bmatrix} P_1 \\ P_2 \\ P_3 \\ P_4 \end{bmatrix}$, $A = \begin{bmatrix} \int_0^T W_1(t)C_t(t)dt \\ \int_0^T W_2(t)C_t(t)dt \\ \int_0^T W_3(t)C_t(t)dt \\ \int_0^T W_4(t)C_t(t)dt \end{bmatrix}$, $\varepsilon = \begin{bmatrix} \varepsilon_1 \\ \varepsilon_2 \\ \varepsilon_3 \\ \varepsilon_4 \end{bmatrix}$

ε corresponds to the error terms, and $B =$

$$\begin{bmatrix} \int_0^T W_1(t) \int_0^t C_p(\tau)d\tau dt, & \int_0^T W_1(t) \int_0^t \int_0^t C_p(\tau)d\tau^2 dt, & \int_0^T W_1(t) \int_0^t C_t(\tau)d\tau dt, & \int_0^T W_1(t) \int_0^t \int_0^t C_t(\tau)d\tau^2 dt \\ \int_0^T W_2(t) \int_0^t C_p(\tau)d\tau dt, & \int_0^T W_2(t) \int_0^t \int_0^t C_p(\tau)d\tau^2 dt, & \int_0^T W_2(t) \int_0^t C_t(\tau)d\tau dt, & \int_0^T W_2(t) \int_0^t \int_0^t C_t(\tau)d\tau^2 dt \\ \int_0^T W_3(t) \int_0^t C_p(\tau)d\tau dt, & \int_0^T W_3(t) \int_0^t \int_0^t C_p(\tau)d\tau^2 dt, & \int_0^T W_3(t) \int_0^t C_t(\tau)d\tau dt, & \int_0^T W_3(t) \int_0^t \int_0^t C_t(\tau)d\tau^2 dt \\ \int_0^T W_4(t) \int_0^t C_p(\tau)d\tau dt, & \int_0^T W_4(t) \int_0^t \int_0^t C_p(\tau)d\tau^2 dt, & \int_0^T W_4(t) \int_0^t C_t(\tau)d\tau dt, & \int_0^T W_4(t) \int_0^t \int_0^t C_t(\tau)d\tau^2 dt \end{bmatrix}.$$

The estimates of WIM can be derived by solving Equation 6.24:

$$\hat{\theta}_{WIM} = B^{-1}A \quad (6.25)$$

Once the estimates for $\hat{\theta}_{WIM}$ are obtained, the microparameters can be calculated according to Equation 6.26.

$$\hat{K}_1 = \hat{P}_1, \hat{k}_2 = -\frac{\hat{P}_2}{\hat{P}_1} - \hat{P}_3, \hat{k}_3 = -\hat{P}_3 - \hat{k}_2 - \hat{k}_4, \quad \hat{k}_4 = -\frac{\hat{P}_4}{\hat{k}_2}. \quad (6.26)$$

The advantage of WIM is that the computational cost is substantially reduced through the linear calculations. Whereas a range of time-weighting functions has been proposed, the choice of weighting function still affects reliability of the parameter estimates. Thus, optimization of the weighting function can keep noise effects in parameter estimates to a minimum, but it may require considerable computational time to achieve the objective for each pixel [9, 34].

6.2.5 Spectral Analysis

In kinetic modeling, the numbers of compartments and corresponding interconnections are assumed to be known *a priori*. However, in some experiments such as the study of a new drug or tracer, prior knowledge may not be available about the behavior of the tracer. In addition, the assumption of homogenous tracer distribution and behavior within the ROI may not be accurate for regions in tumors [40]. The spectral analysis (SA) method identifies the unit impulse tissue response function with no prior assumptions of a specific kinetic model [41, 42].

The SA method solves a linear problem by modeling TTAC as a linear combination of a series of basis function $C_j(t)$ in Equation 6.27:

$$C_t(t) = \sum_{j=1}^{N} \alpha_j C_j(t), \quad (6.27)$$

where α_j denotes non-negative coefficients. Typically the basis functions are exponentials convolved with the PTAC, $C_j(t) = C_p(t) \otimes \exp(-\beta_j t)$. β_j is the parameter defining basis function, which is predetermined and fixed to cover an appropriate spectral range. Usually, N is chosen to be equal to 100 in order to generate a large number of basis functions.

The application of spectral analysis to PET data without decay correction provides useful constraints on the range of values for β_j. For example, for the FDG study with a decay constant $\lambda = 0.000105\ s^{-1}$, a suitable range of β_j is from λ to 1 s^{-1}. 1 s^{-1} represents the fastest spectral component that can be realistically extracted in practice [41].

The fit coefficients α_j are constrained to be non-negative by using non-negative least-square fitting, such as a simplex method [15], in the curve fitting. The impulse response function (IRF) is then given by summation of the basis functions as given by Equation 6.28:

$$h(t) = \sum_{j=1}^{N} \alpha_j \cdot \exp(-\beta_j - \lambda)t. \quad (6.28)$$

An important feature of the process is that most α_j coefficients are returned as zero, with only two to three α_j being non-zero because of the imposed constraints. The parameters of interest then can be derived from the obtained IRF in Equation 6.28. The delivery rate constant K_1 can be estimated from IRF at time t=0 and is given by $h(0) = \sum_{j=1}^{N} \alpha_j$ [9]. The volume of distribution V_d is defined by $V_d = \int_0^\infty h(t)dt = \sum_{j=1}^{N} \frac{\alpha_j}{\beta_j - \lambda}$ [40, 41].

The spectral analysis can be directly applied to projection data so that the reconstruction burden may be reduced. The efficiency of spectral analysis is well suited to incorporating into iterative parametric image reconstruction algorithms [40, 43]. This is further described in Section 6.2.8.

6.2.6 Graphic Analysis Methods

Graphic analysis techniques transform the data to allow estimates of the parameters of interest to be determined from a linear plot (hence graphic analysis) with linear regression analysis. The slope and intercept of the graphic analysis are related to parameters of interests in the kinetic modeling. The graphic analysis methods are computationally efficient, and the estimates are usually highly reliable in contrast to other methods.

6.2.6.1 Patlak Graphic Analysis

One of the best-known graphic analysis techniques is the Patlak graphic analysis (PGA) [44, 45]. It estimates the influx rate constant $K_i = (K_1k_3)/(k_2 + k_3)$, which is a more appropriate measure of FDG uptake than SUV. The Patlak plot was originally proposed for a two-compartment and three-parameter Sokoloff model with one irreversible compartment for FDG-6-PO_4 in tissue. For the compartment model with $k_4 = 0$ in Figure 6.2, the differential equations for the mass concentration in tissue are quite similar to Equations 6.9 and 6.10, except terms related to k_4 are removed in Equation 6.29:

$$\begin{aligned} \frac{d}{dt}C_e(t) &= K_1C_p(t) - (k_2 + k_3)C_e(t) \\ \frac{d}{dt}C_m(t) &= k_3C_e(t). \end{aligned} \tag{6.29}$$

The total tracer concentration in the tissue can be solved from Equation 6.29 as:

$$\frac{dC_t(t)}{dt} = \frac{K_1 \cdot k_3}{k_2 + k_3} \cdot C_p(t) + \frac{k_2}{k_2 + k_3}\frac{dC_e(t)}{dt}. \tag{6.30}$$

After a sufficiently long time post tracer administration $(t > t^*)$, it can be assumed that an equilibrium has been reached between the plasma and free tissue concentrations, $C_e(t)$. In other words, $C_e(t)/C_p(t)$ tends to a constant when $t > t^*$. Thus, integrating Equation 6.30 and dividing the equation by the plasma concentration of $C_p(t)$ yields Equation 6.31:

$$\begin{aligned} \frac{C_t(t)}{C_p(t)} &= \frac{K_1 \cdot k_3}{k_2 + k_3} \cdot \frac{\int_0^t C_p(\tau)d\tau}{C_p(t)} + \frac{k_2}{k_2 + k_3}\frac{C_e(t)}{C_p(t)} \\ &\approx K_i \cdot \frac{\int_0^t C_p(\tau)d\tau}{C_p(t)} + \frac{k_2}{k_2 + k_3}Const,\ t > t^*, \end{aligned} \tag{6.31}$$

where $K_i = (K_1k_3)/(k_2 + k_3)$, and Const is a constant. Therefore, plotting $\frac{\int_0^t C_p(\tau)d\tau}{C_p(t)}$ vs. $\frac{C_t(t)}{C_p(t)}$ will then generate a straight line with slope of K_i and intercept of $k_2Const/(k_2 + k_3)$ in Figure 6.7b. As K_i is directly related to CMRGlc (Equation 6.14), voxel-by-voxel parameter estimates of K_i using PGA can readily derive parametric images of CMRGlc. In the clinical application, the value of t^* ranges from approximately 15 to 60 minutes post tracer administration for FDG studies [46].

The advantages of PGA are its computational efficiency, reliability, and simplicity. PGA does not necessarily require full dynamic scan protocols. For example, four scans of 5 minutes duration are sufficient to achieve adequate accuracy for parametric images of CMRGlc [46]. The drawback of PGA is its inability to estimate individual microparameters. The assumptions of $k_4 = 0$ will underestimate the value of CMRGlc if there is appreciable dephosphorylation.

6.2.6.2 Logan Graphic Analysis

The Logan graphic analysis (LGA) extends graphic analysis methods to reversible compartments [47, 48]. Based on the differential Equations 6.9 and 6.10, the total tissue tracer concentration $C_t(t)$ can be derived by Equation 6.32:

$$\begin{aligned} C_t(t) = \frac{K_1}{k_2} \cdot \frac{k_3 + k_4}{k_4} \cdot C_p(t) - \frac{k_3 + k_4}{k_2k_4} \cdot \frac{dC_t(t)}{dt} - \frac{1}{k_4} \\ \cdot \frac{dC_m(t)}{dt}. \end{aligned} \tag{6.32}$$

Integrating Equation 6.32 and dividing the equation by the tissue tracer concentration of $C_t(t)$ yields:

$$\begin{aligned} \frac{\int_0^t C_t(\tau)d\tau}{C_t(t)} = \frac{K_1}{k_2} \cdot \frac{k_3 + k_4}{k_4} \cdot \frac{\int_0^t C_p(\tau)d\tau}{C_t(t)} - \frac{k_3 + k_4}{k_2k_4} \\ \cdot -\frac{1}{k_4} \cdot \frac{C_m(t)}{C_t(t)}. \end{aligned} \tag{6.33}$$

The assumption of the Logan plot is that when equilibrium is reached, the ratio between $C_m(t)$ and $C_t(t)$ becomes constant. So if $t > t^*$, the Equation 6.33 can be reorganized into Equation 6.34:

$$\frac{\int_0^t C_t(\tau)d\tau}{C_t(t)} = V_d \cdot \frac{\int_0^t C_p(\tau)d\tau}{C_t(t)} + Int, \tag{6.34}$$

where V_d is volume of distribution, which is $K_1(k_3 + k_4)/(k_2k_4)$, Int is a constant. Thus, when plotting $\frac{\int_0^t C_p(\tau)d\tau}{C_t(t)}$ vs. $\frac{\int_0^t C_t(\tau)d\tau}{C_t(t)}$, the slope of the LGA plot is the volume of distribution as shown in Figure 6.7c. If $C_m(t)$ is assumed to be far less than $C_t(t)$ at steady state, $Int \approx -(k_3 + k_4)/(k_2k_4)$ and the influx rate of K_1 would be derived from the slope and intercept of the Logan plot (i.e., $K_1 \approx -V_d/Int$). Usually, t^* is assumed to be about 40 minutes for FDG studies.

The volume of distribution has been used extensively to monitor changes related to receptor-ligand binding in neuroreceptor studies [48]. For other studies, the Logan plot is still attractive because of its computational efficiency and stable estimates of V_d. K_1 is potentially underestimated because of the underlying assumption detailed above. Despite V_d being based on a compartment model in this section, one advantage of the Logan method is that it does not require prior knowledge of the tracer's kinetics. Parametric images derived by the Logan plot give rise to direct quantitative intrasubject and intersubject comparisons.

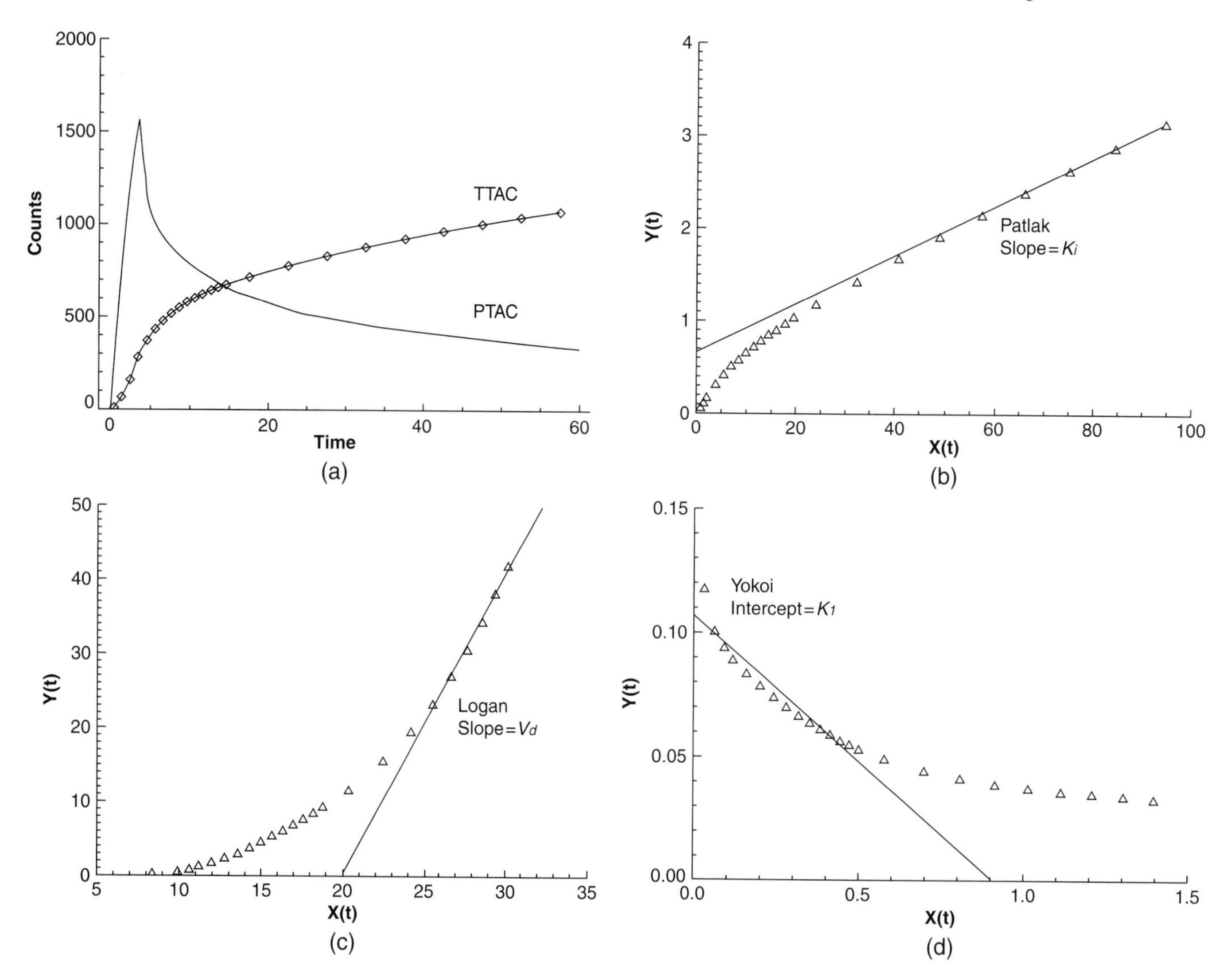

FIGURE 6.7 Graphic analysis methods (a) PTAC and TTAC used in the demonstration. (b) Patlak plot for three-compartment and three-parameter model. (c) Logan plot for three-compartment and four-parameter model. (d) Yokoi plot for two-compartment and two-parameter model.

6.2.6.3 Yokoi Plot

The Yokoi plot was proposed to estimate cerebral blood flow for a two-compartment and two-parameter model [49]. The kinetics of cerebral blood flow can be written by a first-order differential equation (6.35):

$$\frac{dC_t(t)}{dt} = K_1 C_p(t) - k_2 C_t(t), \tag{6.35}$$

where $C_t(t)$ is the tracer concentration in tissue, and $C_p(t)$ is the tracer concentration in plasma. The integration of Equation 6.35, divided by the integral of $C_p(t)$, derives Equation 6.36:

$$\frac{C_t(t)}{\int_0^t C_p(\tau)d\tau} = K_1 - k_2 \frac{\int_0^t C_t(\tau)d\tau}{\int_0^t C_p(\tau)d\tau}. \tag{6.36}$$

When plotting $\frac{\int_0^t C_t(\tau)d\tau}{\int_0^t C_p(\tau)d\tau}$ vs. $\frac{C_t(t)}{\int_0^t C_p(\tau)d\tau}$, the slope of the Yokoi plot is $-k_2$, while the intercept of line is K_1, as shown in Figure 6.7(d). Because K_1 dominates the early part of the kinetics, when $t < t^*$, the Yokoi plot provides a more accurate estimate of K_1. The advantages of the Yokoi plot are its rapid computation and simplicity to derive parametric image of K_1. However, lower SNR at early time frames gives rise to unreliable estimates using the Yokoi plot. Thus, low-pass filter may need to be applied to improve SNR before the Yokoi plot is applied.

6.2.7 Linear Least-Squares Method

The linear least-square method is similar to the weighted integrated method, and it estimates parameters from the matrices of linearized equations. The advantage of linear least-square method and extended methods is that they avoid the need for specifying initial parameters required for the NLS method and determination of optimal weights in the WIM method.

6.2.7.1 Linear Least-Squares

The linear least-squares (LLS) method is a special case of the more general nonlinear least square method [13]. For parameter estimation for a general compartmental model for FDG, all the available dynamic information from measurements of plasma and imaging data is used to derive reliable estimates of CMRGlc [39, 50].

The differential Equation 6.11 can be discretized by the second-order integration for m imaging durations, whose mid scan times range from t_1 to t_m. The algebraic equations are shown by Equation 6.37:

$$\begin{aligned}
C_t(t_1) &= P_1\int_0^{t_1} C_p(\tau)d\tau + P_2\int_0^{t_1}\int_0^{t_1} C_p(\tau)d\tau^2 \\
&\quad + P_3\int_0^{t_1} C_t(\tau)d\tau + P_4\int_0^{t_1}\int_0^{t_1} C_t(\tau)d\tau^2 + \varepsilon_1 \\
C_t(t_2) &= P_1\int_0^{t_2} C_p(\tau)d\tau + P_2\int_0^{t_2}\int_0^{t_2} C_p(\tau)d\tau^2 \\
&\quad + P_3\int_0^{t_2} C_t(\tau)d\tau + P_4\int_0^{t_2}\int_0^{t_2} C_t(\tau)d\tau^2 + \varepsilon_2 \\
&\vdots \\
C_t(t_m) &= P_1\int_0^{t_m} C_p(\tau)d\tau + P_2\int_0^{t_m}\int_0^{t_m} C_p(\tau)d\tau^2 \\
&\quad + P_3\int_0^{t_m} C_t(\tau)d\tau + P_4\int_0^{t_m}\int_0^{t_m} C_t(\tau)d\tau^2 + \varepsilon_m,
\end{aligned} \tag{6.37}$$

where $P_1 = K_1$, $P_2 = K_1(k_3 + k_4)$, $P_3 = -(k_2 + k_3 + k_4)$ and $P_4 = -k_2k_4$, $\varepsilon = [\varepsilon_1,\varepsilon_2,\ldots,\varepsilon_m]^T$ are the equation error terms. Those equation errors are not independent but are correlated because of the involved integrations of measurements of PTAC and TTAC, even though the measurement errors are independent [51]. Rearranging Equation 6.37 into a matrix form yields Equation 6.38:

$$y = X\theta_{LLS} + \varepsilon, \tag{6.38}$$

where $y = [C_t(t_1), C_t(t_2), \ldots, C_t(t_m)]T$, $\theta_{LLS} = [P_1,P_2,P_3,P_4]^T$, $\varepsilon = [\varepsilon_1,\varepsilon_2,\ldots,\varepsilon_m]^T$, and

$$X = \begin{bmatrix}
\int_0^{t_1} C_p(\tau)d\tau, & \int_0^{t_1}\int_0^{t_1} C_p(\tau)d\tau^2, & \int_0^{t_1} C_t(\tau)d\tau, & \int_0^{t_1}\int_0^{t_1} C_t(\tau)d\tau^2 \\
\int_0^{t_2} C_p(\tau)d\tau, & \int_0^{t_2}\int_0^{t_2} C_p(\tau)d\tau^2, & \int_0^{t_2} C_t(\tau)d\tau, & \int_0^{t_2}\int_0^{t_2} C_t(\tau)d\tau^2 \\
\vdots & & & \\
\int_0^{t_m} C_p(\tau)d\tau, & \int_0^{t_m}\int_0^{t_m} C_p(\tau)d\tau^2, & \int_0^{t_m} C_t(\tau)d\tau, & \int_0^{t_m}\int_0^{t_m} C_t(\tau)d\tau^2
\end{bmatrix}.$$

Thus, the LLS estimator of θ can be solved by Equation 6.39:

$$\hat{\theta}_{LLS} = (X^TX)^{-1}X^TY. \tag{6.39}$$

Once $\hat{\theta}_{LLS}$ is derived, the estimated microparameters can be obtained according to Equation 6.26, followed by the calculation of parameters of interests, like MRGlc and V_d. The LLS method doesn't require any optimization or prior initial parameters. The statistically non-independent error terms result in potentially biased estimation for the LLS method.

6.2.7.2 Generalized Linear Least-Squares

The generalized linear least-squares (GLLS) method was proposed to address the biased estimation in the LLS method [51, 52]. The equation error terms in Equation 6.37 are not independent with respect to time because of the overlapping integration, even though the measurement errors are independent. For example, the equation error term ε_1 contains the measurement error $E(t_1)$, whereas ε_2 contains the measurement errors $E(t_1)$ and $E(t_2)$. In practice, the equation error term ε_m may contain certain other errors in addition to the measurement errors, such as the approximation errors from the numerical integration method [34].

Reconsidering Equation 6.11, the Laplace transform converts the equation into the frequency domain in Equation 6.40:

$$s^2C_t(s) = sP_1C_p(s) + P_2C_p(s) + sP_3C_i(s) + P_4C_t(s), \tag{6.40}$$

where $P_1 = K_1$, $P_2 = K_1(k_3 + k_4)$, $P_3 = -(k_2 + k_3 + k_4)$, and $P_4 = -k_2k_4$. Rearranging Equation 6.40, with a white measurement noise term of E(s) added, can yield Equation 6.41:

$$C_t(s) = \frac{sP_1C_p(s) + P_2C_p(s)}{s^2 - sP_3 - P_4} + E(s). \tag{6.41}$$

Equation 6.41 can be reorganized into Equation 6.42, which depicts the noise term colored by the equation term of $s^2 - sP_3 - P_4$, even if the measurement noise $E(s)$ is white. That is the reason why the estimates of the LLS method are biased.

$$\begin{aligned}
s^2C_t(s) &= sP_1C_p(s) + P_2C_p(s) + sP_3C_t(s) + P_4C_t(s) \\
&\quad + (s^2 - sP_3 - P_4)E(s)
\end{aligned} \tag{6.42}$$

The principle of the GLLS method is to prewhiten the equation noise term to obtain unbiased estimates. Both sides of Equation 6.42 are divided by an autoregressive filter $F(s) = s^2 - s\hat{P}_3 - \hat{P}_4$ as show in Equation 6.43:

$$\begin{aligned}
\frac{s^2C_t(s)}{s^2 - s\hat{P}_3 - \hat{P}_4} &= \frac{sP_1C_p(s)}{s^2 - s\hat{P}_3 - \hat{P}_4} + \frac{P_2C_p(s)}{s^2 - s\hat{P}_3 - \hat{P}_4} \\
&\quad + \frac{sP_3C_t(s)}{s^2 - s\hat{P}_3 - \hat{P}_4} + \frac{P_4C_t(s)}{s^2 - s\hat{P}_3 - \hat{P}_4} \\
&\quad + \frac{(s^2 - sP_3 - P_4)}{s^2 - s\hat{P}_3 - \hat{P}_4}E(s).
\end{aligned} \tag{6.43}$$

If $F(s)$ is close to the term $s^2 - sP_3 - P_4$, the equation noise in Equation 6.43 is whitened, and the estimates would become unbiased. Therefore, Equation 6.43 can be transformed into the temporal domain by the inverse Laplace transform in Equation 6.44:

$$C_t(t)+\hat{P}_3\psi_1 \otimes C_t(t)+\hat{P}_4\psi_2 \otimes C_t(t) = \hat{P}_1\psi_1 \otimes C_p(t)+\hat{P}_2\psi_2 \otimes C_p(t)+\hat{P}_3\psi_1 \otimes C_t(t)+\hat{P}_4\psi_2 \otimes C_t(t), \quad (6.44)$$

where $\psi_1 = \frac{\lambda_2 e^{-\lambda_2 t} - \lambda_1 e^{-\lambda_1 t}}{\lambda_2 - \lambda_1}$ and $\psi_2 = \frac{e^{-\lambda_1 t} - e^{-\lambda_2 t}}{\lambda_2 - \lambda_1}$. The values of λ_1 and λ_2 are determined by $\hat{P}_3$ and $\hat{P}_4$ with $\lambda_{1,2} = -\frac{\hat{P}_3 \pm \sqrt{\hat{P}_3 + 4\hat{P}_4}}{2}$. Therefore, the estimates of the GLLS method can be solved in the matrix form from m measurements ($m \geq 4$ for a four-parameter model) in Equation 6.45:

$$\hat{\theta}_{GLLS} = (Z^T Z)^{-1} Z^T r, \quad (6.45)$$

where $\hat{\theta}_{GLLS} = \begin{bmatrix} \hat{P}_1 \\ \hat{P}_2 \\ \hat{P}_3 \\ \hat{P}_4 \end{bmatrix}$, $r = \begin{bmatrix} C_t(t_1)+\hat{P}_3\psi_1 \otimes C_t(t_1)+\hat{P}_4\psi_2 \otimes C_t(t_1) \\ C_t(t_2)+\hat{P}_3\psi_1 \otimes C_t(t_2)+\hat{P}_4\psi_2 \otimes C_t(t_2) \\ \vdots \\ C_t(t_n)+\hat{P}_3\psi_1 \otimes C_t(t_n)+\hat{P}_4\psi_2 \otimes C_t(t_n) \end{bmatrix}$,

$$Z = \begin{bmatrix} \psi_1 \otimes C_p(t_1), \psi_2 \otimes C_p(t_1), \psi_1 \otimes C_t(t_1), \psi_2 \otimes C_t(t_1) \\ \psi_1 \otimes C_p(t_2), \psi_2 \otimes C_p(t_2), \psi_1 \otimes C_t(t_2), \psi_2 \otimes C_t(t_2) \\ \vdots \\ \psi_1 \otimes C_p(t_n), \psi_2 \otimes C_p(t_n), \psi_1 \otimes C_t(t_n), \psi_2 \otimes C_t(t_n) \end{bmatrix}.$$

In a manner similar to the LLS method, the estimated $\hat{\theta}_{GLLS}$ can derive the microparameters in terms of Equation 6.26. Theoretically, Equation 6.45 needs to be performed iteratively until unbiased estimates are achieved or the termination criterion is reached [39, 51]. In the implementation, estimates from the LLS method are used as the parameters in the autoregressive filter for the first iteration. In practice, two or three iterations provide satisfactory results in most cases, and further iterations are typically not necessary for PET studies.

The GLLS method is a computationally efficient algorithm for generating parametric images without having to specify initial parameters. The microparameters in the kinetic model can be estimated by the GLLS method as can macroparameters such as volume of distribution and CMRGlc.

6.2.7.3 Improved Versions for Generalized Linear Least-Squares Methods

The GLLS method has been successfully applied to PET data in the brain [51], heart [53], and liver [54]. However, despite the potential of advanced SPECT providing quantitative information, the high level of noise intrinsic in SPECT still gives rise to unsuccessful fits using GLLS, especially for voxel-by-voxel TTACs used to generate parametric images [55]. The unsuccessful fits may result in negative parameter estimates, which are physiologically meaningless. Efforts have been made to improve the reliability of GLLS for challenging SPECT data.

One approach uses a prior volume of distribution in the curve fitting to reduce the number of estimated parameters. V_d is a relatively stable functional parameter and is predominantly influenced by late time frames with relative higher SNR than earlier time frames in SPECT. The approach is referred to as the V_d-aided GLLS method.

According to Equation 6.28, the definition of V_d can be rewritten in Equation 6.46.

$$V_d = \frac{K_1}{k_2}\left(1 + \frac{k_3}{k_4}\right) = -\frac{P_2}{P_4}. \quad (6.46)$$

Because V_d is assumed to be constant in the V_d-aided GLLS, the differential equation for $C_t(t)$ can be rearranged into Equation 6.47, with three unknown parameters for the three-compartment and four-parameter model:

$$\frac{d^2}{dt^2} C_t(t) = P_1 \frac{d}{dt} C_p(t) + P_3 \frac{d}{dt} C_t(t) + P_4[C_t(t) - V_d C_p(t)]. \quad (6.47)$$

The parameter estimates of the V_d-aided GLLS can then be derived by solving the matrix Equation 6.48:

$$\theta_{Vd-GLLS} = (Z_{Vd}^T Z_{Vd})^{-1} Z_{Vd}^T \cdot r, \quad (6.48)$$

where:

$$\hat{\theta}_{GLLS} = \begin{bmatrix} \hat{P}_1 \\ \hat{P}_3 \\ \hat{P}_4 \end{bmatrix}, r = \begin{bmatrix} C_t(t_1)+\hat{P}_3\psi_1 \otimes C_t(t_1)+\hat{P}_4\psi_2 \otimes C_t(t_1) \\ C_t(t_1)+\hat{P}_3\psi_1 \otimes C_t(t_2)+\hat{P}_4\psi_2 \otimes C_t(t_2) \\ \vdots \\ C_t(t_1)+\hat{P}_3\psi_1 \otimes C_t(t_m)+\hat{P}_4\psi_2 \otimes C_t(t_m) \end{bmatrix},$$

$$Z_{Vd} = \begin{bmatrix} \psi_1 \otimes C_p(t_1), \psi_1 \otimes C_t(t_1), \psi_2 \otimes C_t(t_1) - V_d\psi_2 \otimes C_p(t_1) \\ \psi_1 \otimes C_p(t_2), \psi_1 \otimes C_t(t_2), \psi_2 \otimes C_t(t_2) - V_d\psi_2 \otimes C_p(t_2) \\ \vdots \\ \psi_1 \otimes C_p(t_n), \psi_1 \otimes C_t(t_n), \psi_2 \otimes C_t(t_n) - V_d\psi_2 \otimes C_p(t_n) \end{bmatrix}.$$

The microparameters are then obtained by Equation 6.49:

$$\hat{K}_1 = \hat{P}_1, \hat{k}_2 = \frac{\hat{V}_d \cdot \hat{P}_4}{\hat{P}_1} - \hat{P}_3, \hat{k}_3 = -\hat{P}_3 - \hat{k}_2 - \hat{k}_4, \quad \hat{k}_4 = -\frac{\hat{P}_4}{\hat{k}_2}. \quad (6.49)$$

V_d can be predetermined from the GLLS method for a lower-order compartment model [56]. However, because of the potentially biased estimation from the different underlying model, the optimum V_d has to be searched over a reasonable range of values determined from the lower-order compartment model estimate of V_d, which substantially increases computational cost. Thus, V_d derived from the Logan plot was used directly as the prior for the V_d-aided GLLS in recent investigations [55]. Although reliability of estimates is improved for noisy SPECT data, the V_d-aided GLLS is still not completely immune from the negative parameter estimates for extremely noisy curves.

The statistical resampling technique solves the problem by incorporating the bootstrap Monte Carlo (BMC) method into the fitting routine. The BMC approach is a well-established, robust statistical method with no underlying assumption about the noise level and distribution [15]. The only assumption is that the sequential order of the data points is not important in the data processing.

Given a curve of n data points, n points are randomly selected from the original curve with some points duplicated and others not selected. The points that are duplicated and not selected will vary randomly for each iteration [57]. For one given TTAC, a number of resampled curves are generated, followed by the GLLS fitting for each resampled curve. The mean parameters, derived from successful fits of the resampled curves, are used as the results. This is referred as the BMC-aided GLLS. This method has been found to overcome the problem of unsuccessful fits in GLLS-derived parametric images from dynamic SPECT studies [58].

For the BMC-aided GLLS, the fitting part can be substituted by the V_d-aided GLLS, which is the integration of the two improved methods (BMC–V_d-aided GLLS). This method can further improve the reliability of parameter estimates for noisy SPECT data. Because each derived BMC curve requires fitting using the GLLS method, the BMC-aided GLLS and the BMC–V_d-aided GLLS have substantially increased the computational burden over the standard GLLS method.

6.2.8 Parametric Image Reconstruction Method

In general, the data for each frame of a dynamic study are first reconstructed, followed by parameter estimation from either ROI-based or voxel-by-voxel TTACs derived from the reconstructed data. This approach tends to reduce SNR because particularly the early, short frames contain a very high level of noise [59]. Furthermore, the requirement for mechanically rotating the detectors around the patient with SPECT can lead to inconsistent projections and introduce bias and artifacts, particularly for tracers with fast kinetics [60, 61]. Parametric image reconstruction (PIR) methods can overcome these problems by carrying out parameter estimation in the projection data space prior to reconstruction.

Typically, the PIR method consists of two steps. One is the expression of the kinetic model in the projection domain. The other is incorporation of the parameter estimation into the reconstruction. As described in Section 6.2.5, the kinetic model can be expressed as a linear function of the unknown parameters. Such a time-dependent tracer distribution can be directly incorporated into a reconstruction with a temporal component.

PIR can be classified into non-iterative and iterative approaches. The non-iterative approaches were implemented to solve the time-dependent Radon transform [60, 62], for example, to estimate the exponential factors using linear time-invariant system theory for the kinetic model, followed by estimating parameters using a linear estimation technique [60]. It is more attractive incorporating the kinetic model into an iterative algorithm, such as ordered subset expectation maximization (OS-EM) [61], because the temporal component and noise properties can then be explicitly modeled and included in the reconstruction.

Nonlinear parameter estimation provides more accurate estimates because linear estimation relies on the model linearization, which may introduce bias [59]. However, performing nonlinear parameter estimation with iterative reconstruction is computationally intensive. The standard acceleration technique can be used to reduce reconstruction time [61]. Alternatively, a multiresolution reconstruction scheme can be applied to overcome slow convergence of maximum *a posteriori* reconstruction at low spatial frequency [59].

6.3 Noninvasive Methods

Parameter estimates for parametric imaging usually require frequent blood sampling to obtain tracer concentration in blood or plasma as an input function for kinetic analysis. Arterial blood sampling is considered the gold standard, providing the most accurate measures. However, arterial sampling may cause patients discomfort and carries some risks such as arterial thrombosis, arterial sclerosis, and ischemia. The arterialized-venous method using a heated limb avoids the discomfort of arterial puncture [8, 63], but requires frequent blood sampling and prolonged warming to achieve arterial-venous shunting. In addition, it also exposes personnel to additional radiation and risks associated with handling blood. A number of approaches have been proposed to reduce the need for blood sampling.

6.3.1 Image-Derived Input Function

The image-derived input function (ID-IF) method relies on a sufficiently large vascular structure, such as the left ventricle or major artery, to be in the imaging field of view. An ROI is then drawn over the vascular structure to derive the concentration of the tracer in blood. The use of a manually defined ROI within the left ventricle is well established for cardiac studies [64]. However, spill-over from tracer uptake in adjacent structures as a result of limited spatial resolution can introduce large bias in the blood activity concentration estimates, particularly at late time frames when blood concentration is low. Thus, careful placement of ROI and correction of spill-over are essential for the ID-IF method from the left ventricle images [65, 66].

As shown in Figure 6.8, assuming $C_{bROI}(t)$ is the measured blood time activity curve (BTAC) for a ROI in the left ventricle and $C_{tROI}(t)$ is the measured TTAC from an ROI over adjacent myocardium, the spill-over effect is demonstrated in Equation 6.50 [66]:

$$\begin{aligned} C_{bROI}(t) &= C_b(t) + F_{mb} \cdot C_t(t) \\ C_{tROI}(t) &= C_t(t) + F_{bm} \cdot C_b(t), \end{aligned} \tag{6.50}$$

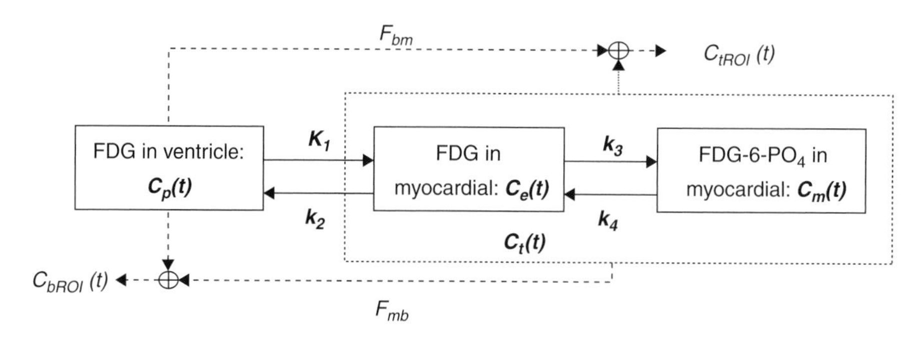

FIGURE 6.8 The modified compartment model of FDG with spill-over effect in the heart.

where $C_b(t)$ is the corrected BTAC, $C_t(t)$ is the corrected TTAC for adjacent myocardium, F_{mb} is the spill-over fraction from myocardial to the blood pool in the ventricle, and F_{bm} is the spill-over fraction from the blood pool to myocardium. The expected input function $C_p(t)$ can be resolved by Equation 6.51:

$$C_p(t) = \frac{C_b(t) - F_{mb} \cdot C_t(t)}{1 - F_{mb} \cdot F_{bm}}. \tag{6.51}$$

Substituting $C_p(t)$ from Equation 6.51 in Equations 6.12 and 6.13 leads to the unknown parameters for estimating $C_t(t)$ consisting of K_1, k_2, k_3, k_4, F_{mb}, and F_{bm}. Thus, the original invasive parameter estimates for four microparameters are transformed into the noninvasive estimates for four microparameters and two spill-over factors. These six unknown parameters can be estimated by the NLS or other methods.

When the heart is outside the scanner field of view, other large vessels can be used to derive IF from imaging data, such as the venous sinus [67], cardiac aorta [68], and abdominal aorta [69, 70]. If the spatial resolution and system sensitivity are sufficient, carotid arteries can potentially yield ID-IF for cerebral studies [71]. The ID-IF method relies on manually drawn ROIs on the early imaging data, when blood activity is highest and tissue activity is low, for optimum delineation of vascular structure. For example, the initial 36-second images are used to delineate the carotid artery [71]. Manual drawing of ROIs is time-inefficient and subjective, which increases variability and may introduce bias. The clustering technique can address the problem and aid automatic identification of ID-IF by classifying TTACs into clusters. Because the time course of BTAC is quite distinct from a TTAC, it is easy to identify ID-IF automatically using the cluster method [72]. The simultaneous estimation (SIME) method avoids the choice of ROI placement by simultaneous estimation of kinetic parameters for the compartment model and parameters for a mathematically defined IF during the NLS fitting of the TTACs [73, 74]. To provide accurate estimates, several distinct TTACs are required from different tissues to provide sufficient information to estimate the additional IF parameters.

6.3.2 Reference Tissue Model

The binding potential (BP), which reflects the densities of transporters or receptors in a brain ROI [17], has been used extensively to evaluate changes in receptor density and binding [48]. The volume of distribution has also been used to estimate receptor binding [75]. The reference tissue model, originally proposed for neuroreceptor studies, assumes the existence of a tissue region with a negligible concentration of specific binding sites. As shown in Figure 6.9, K_1, k_2, K_1', and k_2' describe the exchange of tracer between plasma and free (nonspecifically bound) ligand compartments in the ROI and the reference region, respectively; and k_3, k_4 represent the exchange of the tracer between the free compartment and a specifically bound ligand compartment [76]. $C_{REF}(t)$ is the tissue tracer concentration in the tissue devoid of receptors; $C_t(t)$ is the measured tissue tracer concentration in the receptor-rich region.

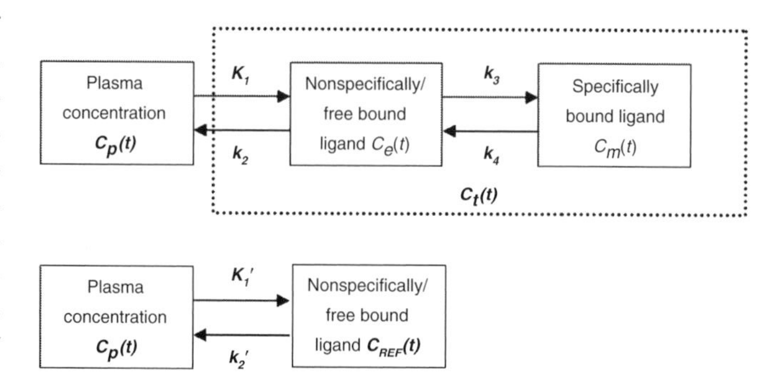

FIGURE 6.9 The three-compartment model and two-compartment model as reference models in the neuroreceptor study.

The simplified reference tissue model (SRTM) method assumes that the volume of distribution for nonspecific binding is the same for the reference model and the receptor-rich model (i.e., $K_1/k_2 = K_1'/k_2'$). The kinetics of the receptor-rich region can be described as a function of the reference region without requiring blood sampling, as shown in Equation 6.52 [77]:

$$C_t(t) = R_I \cdot C_{REF}(t) + \left[k_2 - \frac{R_I \cdot k_2}{1 + BP}\right] \cdot C_{REF}(t) \otimes e^{-[k_2/(1+BP)+\lambda]t}, \tag{6.52}$$

where RI is relative rate of delivery of $RI = K_1/K_1'$, BP is binding potential for the receptor-rich region, and $BP = k_3/k_4$, λ is the physical decay constant of the radiotracer. The three unknown parameters (R_1, BP, and k_2) in Equation 6.52 can be estimated by fitting Equation 6.52 to the measured $C_{REF}(t)$ and $C_t(t)$ data. The LLS method has been applied to generate voxel-by-voxel parametric images by introducing a set of basis parameters in the SRTM method [76]. If the clearance rate of reference region, k_2', is constant across brain regions, a two-step simplified reference tissue model (SRTM2) can improve the quality of neuroreceptor functional images using a global value of k_2', which improves reliability of the estimated parameters [78].

The reference tissue model method can be applied with graphic analysis as well. The noninvasive Logan plot can estimate the distribution volume ratio (DVR), which is the ratio of V_d in a receptor-rich region to V_d in a non-receptor region [75]. For example, Equation 6.53 gives the equation of the Logan plot for the reference non-receptor region in Figure 6.9:

$$\frac{\int_0^t C_{REF}(\tau)d\tau}{C_{REF}(t)} = V_d' \cdot \frac{\int_0^t C_p(\tau)d\tau}{C_{REF}(t)} - \frac{1}{k_2'}, \qquad (6.53)$$

where $V_d' = K_1'/k_2'$. Solving the integral term of $C_p(t)$ and substituting the term in Equation 6.34 yields Equation 6.54 [75, 79]:

$$\begin{aligned}\frac{\int_0^t C_t(\tau)d\tau}{C_t(t)} &= \frac{V_d}{V_d'} \cdot \frac{\int_0^t C_{REF}(\tau)d\tau}{C_t(t)} + \frac{V_d}{k_2'V_d'} \cdot \frac{C_{REF}(t)}{C_t(t)} + Int \\ &= DVR \cdot \frac{\int_0^t C_{REF}(\tau)d\tau}{C_t(t)} + Int', \end{aligned} \qquad (6.54)$$

where $DVR = V_d/V_d'$, $Int = \frac{V_d}{k_2'V_d'} \cdot \frac{C_{REF}(t)}{C_t(t)} + Int$. If $\frac{C_{REF}(t)}{C_t(t)}$ becomes constant after $t > t^*$, the plot of $\frac{\int_0^t C_{REF}(\tau)d\tau}{C_t(t)}$ vs. $\frac{\int_0^t C_t(\tau)d\tau}{C_t(t)}$ becomes linear, and the slope of the Logan plot is equal to DVR. DVR is widely used in neuroreceptor studies to assess the receptor binding with tracers for which reference tissue regions exist.

The noninvasive Logan plot requires $C_{REF}(t)/C_t(t)$ to become constant during the scanning period, which depends on the characteristics of the radioligand [17]. The multilinear reference tissue model (MRTM) method can estimate binding potentials without such assumptions. The MRTM method assumes $K_1/k_2 = K_1'/k_2'$ as for the SRTM method. The same degree of nonspecific binding in the reference tissue and receptor-rich region leads to $V_d/V_d' = 1 + k_3/k_4$. Substituting $DVR=1+BP$ in Equation 6.54 yields,

$$\begin{aligned}\frac{\int_0^t C_t(\tau)d\tau}{C_t(t)} &= (1+BP) \cdot \frac{\int_0^t C_{REF}(\tau)d\tau}{C_t(t)} + \frac{1+BP}{k_2'} \cdot \frac{C_{REF}(t)}{C_t(t)} + Int \\ &= a \cdot \frac{\int_0^t C_{REF}(\tau)d\tau}{C_t(t)} + b \cdot \frac{C_{REF}(t)}{C_t(t)} + c. \end{aligned} \qquad (6.55)$$

Equation 6.55 is a multilinear equation with partial regression coefficients of a, b, and c, which may be estimated by multiple regression analysis [80]. Once the voxel-by-voxel parameters are obtained, parametric images of binding potentials can be calculated from the relationship $BP = a - 1$. As for the SRTM2 method, the number of parameters to be estimated can be reduced from three to two in the updated two-parameter MRTM (MRTM2) method with a fixed k_2' [81]. The incorporation of fixed k_2' substantially reduces variability of the binding potential by a factor of two or three over the MRTM or SRTM method.

6.4.3 Population-Based Input Function and Cascaded Modeling Approaches

The image-derived method and the reference tissue model both rely on dynamic scans to provide sufficient information for noninvasive parameter estimates. However, the required long scan durations for dynamic studies are often impractical in a busy clinical practice. For example, a dynamic FDG neurologic study requires at least 60 minutes of scanning compared with 5 ~ 10 minutes for a static study. If the shape of IF can be assumed to be similar across subjects, an individual IF may be derived from a population-based standard arterial input curve.

The standard IF (SIF) can be derived by averaging the actual arterial blood curves for several subjects [82]. To avoid SIF's noise effect on generation of CMRGlc images, a smooth curve can be derived by normalizing individual IF to a standard 370 MBq/kg injection and curve-fitting with linear interpolation. Only two arterial blood samples taken at around 10 and 45 minutes postinjection are required to calibrate the SIF in the population-based input function method. Arterialized venous blood sampling has also been used for the calibration of the SIF [83]. The SIF is given for a 3-minute infusion protocol using the arterialized-venous method in Equation 6.56:

$$C_p(t) = \begin{cases} 0.009094 + 7.8720 \cdot t - 0.5666 \cdot t^2, \; 0 \le t < 3.5\,\text{min} \\ 5.646 \cdot e^{-1.5680(t-3.5)} + 6.581 \cdot e^{-0.1438(t-3.5)} \\ \quad +7.848 \cdot e^{-0.0109(t-3.5)}, \; t \ge 3.5\,\text{min}. \end{cases} \qquad (6.56)$$

The Equation 6.56 calibrated with two arterialized venous blood samples has been shown to achieve highly correlated values of CMRGlc compared with frequent blood sampling ($r \ge 0.992$)[83]. The number of blood samples for calibration may be further reduced to one at around 40 minutes postinjection with little loss in accuracy [84]. Furthermore, SIF was also investigated without any blood samples [85], relying only on the SIF normalization by injected dose and body mass in Equation 6.57:

$$SIF(t) = \frac{\sum_{i=1}^{N} \frac{BM_i}{ID_i} \cdot C_p(t)_i}{N}, \qquad (6.57)$$

where N is total number of subjects, BM is the body mass, and ID is the injected dose. Thus, the individual IF can be estimated from Equation 6.58:

$$C_p(t) = \frac{ID}{BM} SIF(t). \quad (6.58)$$

Using body surface area instead of BM may provide better estimation of IF [85]. Other approaches have been investigated to normalize SIF for better estimation, such as the use of initial distribution of volume of tracer through statistically minimizing difference among the individual IF and derived standardized IF [86]. However, the population-based input function without any blood sample calibration potentially suffers from more than 10% bias in the generated CMRGlc because of the many factors that can influence an individual IF.

Alternatively, the PTAC can be estimated by a mathematical model that preserves the shape of individual IFs while reducing the effect of random noise in blood sampling on the kinetic parameter estimates. A four-compartment model of blood pools is used to describe the tracer distribution over time in the circulatory system, as shown in Figure 6.10 [87].

The compartment of "Pool 1" includes the tracer in the veins where the tracer is introduced, the right cardiac chambers, and part of the pulmonary system. The second compartment of "Pool 2" encompasses the arteries where blood samples are taken, including "arterialized" veins. The vascular tissue and interstitial space make up most of the compartment of "Pool 3", while the fourth compartment describes the cellular space of the tracer. The model input, u(t), is the bolus FDG injection, which is an impulse function of the model. Random noise, including measurement error, is described by e(t). The PTAC measurement, $C_p(t)$, is the model output response function. Clinical data have validated that the model in Equation 6.59 is the most suitable PTAC model, consisting of a three-exponential function with a pair of repeated eigenvalues [73, 87]:

$$C_p(t) = (A_1 t - A_2 - A_3) \cdot e^{\lambda_1 t} + A_2 \cdot e^{\lambda_2 t} + A_3 \cdot e^{\lambda_3 t}. \quad (6.59)$$

The parameters of A_1, A_2, and A_3 can be estimated by the curve-fitting. The curve derived by the PTAC model efficiently

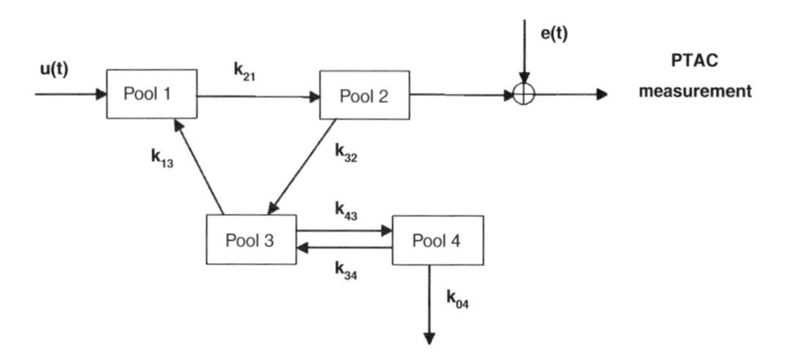

FIGURE 6.10 The four-compartment model of blood pools describing tracer distribution in the circulatory system.

improves SNR and has been shown to facilitate generation of functional images, both in computer simulation and in clinical studies [88]. For example, the PTAC model has been applied to study the effect of input function sampling schedule [89] and to model the SIF, which facilitates calibration with the selected number of blood samples [83].

Furthermore, based on the PTAC model in Equation 6.59, the PTAC can also be noninvasively estimated by the SIME method from dynamic data of an individual patient [73, 74], as described in Section 6.3.1. Because several TTACs are used in the estimation, a cascaded modeling technique in Figure 6.11 is applied for the simultaneous estimation of parameters of one PTAC and a number of kinetics. The cascaded modeling converts noninvasive estimation into solving solutions of multiple systems with a single input and multiple outputs for several known TTACs. The success of the SIME method is relying on the appropriate choices of ROIs for distinct TTACs.

6.4 Clinical Applications of Parametric Images

6.4.1 Blood Flow Parametric Images

The circulatory system is essential to maintain basic physiologic function in the living body. Blood flow in the circulatory system delivers metabolic substances and clears waste to support metabolism in the tissue. Rate of blood flow is dependent on vascular resistance and regional functional variation, and any abnormality may be relevant to disorders. Measurement of blood flow has become a clinical routine to evaluate the variability of tissue function.

Perfusion studies can be performed for the evaluation of rCBF in the brain and myocardial blood flow (MBF) in the heart using perfusion tracers such as ^{15}O-water and ^{13}N-ammonia for PET, ^{201}Tl, and ^{99m}Tc-sestamibi for SPECT. Parametric images of rCBF have been used in many areas, for example, diagnosing neurodegenerative disorders [90], localizing epileptic focus [91], assessing stroke recovery [92], and investigating cognitive or behavioral functional specialization [93].

^{15}O-water is recognized as an ideal tracer for measuring blood flow because it freely diffuses across cell membranes. When the heart is in the field of view, the input function can be derived noninvasively using the ID-IF method. The ROI for the left ventricle (LV) is defined within the LV chamber in the early scan images. In contrast, the other ROI for myocardial tissue is placed within the myocardial wall in the late images, as shown in Figure 6.12 [94].

Once the spill-over fractions are generated through analyzing the derived time activity curves from the ROIs, the input function can be estimated according to Equation 6.51. The autoradiographic method for ^{15}O-water is then used to

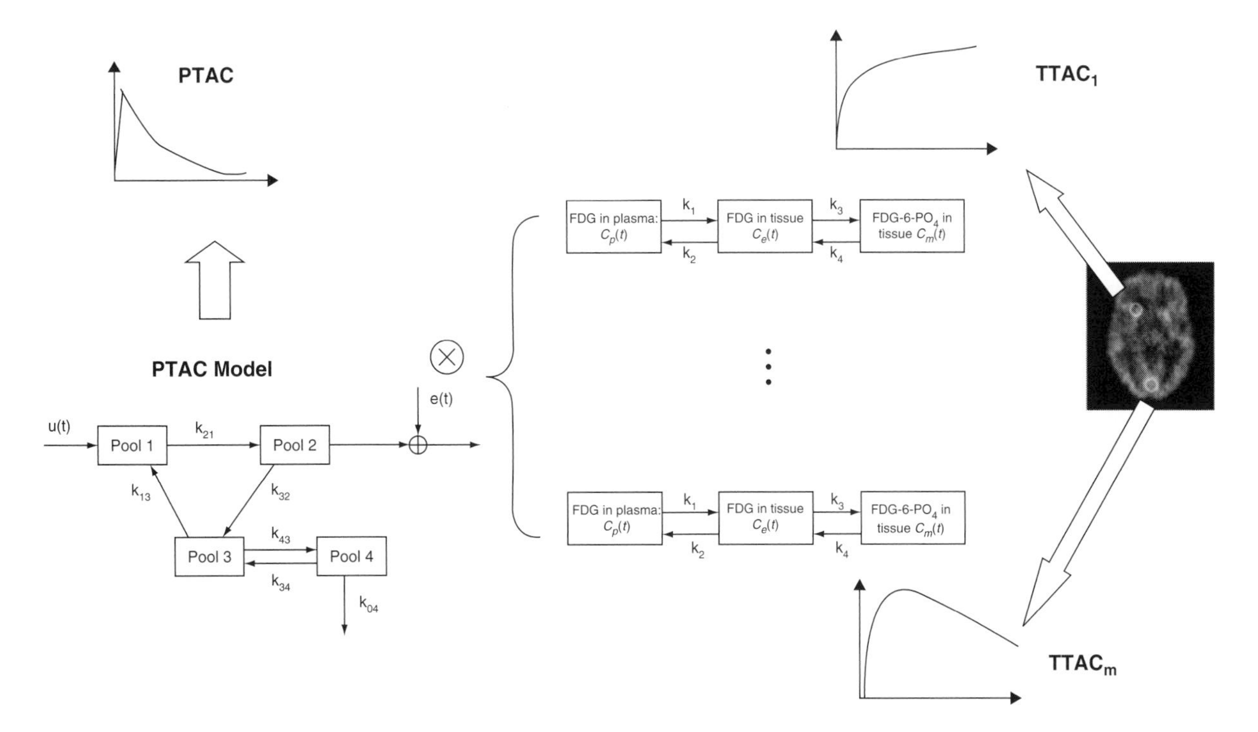

FIGURE 6.11 Flow chart of the cascaded modeling for the simultaneous estimation method, where ⊗ is the convolution operator.

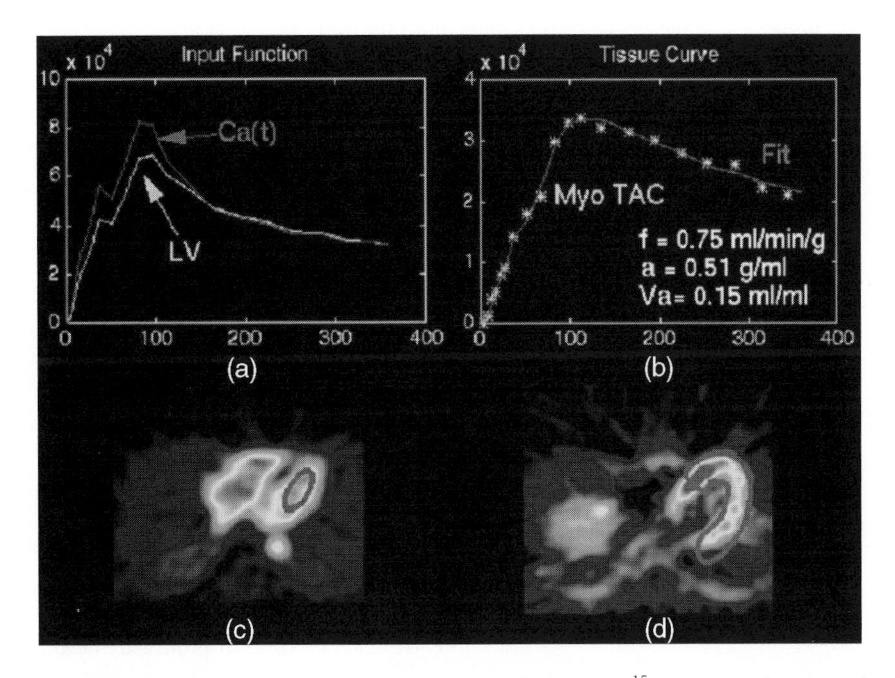

FIGURE 6.12 Image-derived input function for the heart in ^{15}O-water study. The red lines on (c) and (d) show the outlines of the ROIs for left ventricular chamber (c) and for adjacent myocardial region (d). LV is the time activity curve for left ventricular (a); Myo TAC is the curve for myocardial tissue (b), Ca(t) is the estimated input function (a). Courtesy of Dr. H. Iida, Akita, Japan. For a more detailed view of this figure, please visit our companion site at: http://books.elsevier.com/companions/9780123735836.

generate the parametric images of rCBF using the image-derived IF (Figure 6.13).

In the heart, measurement of MBF is used to assess coronary artery disease and LV dysfunction [95, 96]. Because of the anatomy and motion of the heart, spill-over correction between myocardium and LV is essential [97]. The electrocardiograph-gated image acquisition is capable of reducing false-positive perfusion studies by removing the confounding effects of heart motion [98]. Automated quantification has been used to derive ejection fraction (EF) and regional myocardial wall motion and thickening from the gated study as shown in Figure 6.14, all of which contribute valuable information to the diagnosis of the patient.

Although the perfusion images are usually interpreted visually and qualitatively, voxel-by-voxel quantitative images can provide more objective and accurate results. ^{201}Tl is a SPECT tracer that can assess both myocardial perfusion and cellular viability. Figure 6.15 shows parametric images from a ^{201}TlCl study in a patient. The Yokoi plot is applied to process the voxel-by-voxel TTACs from dynamic reconstructed data that have been processed by low-pass filtering to reduce noise. As the two-compartment and two-parameter model for blood flow is applied, the volume of distribution is obtained according to the equation, $V_d = K_1/k_2$.

Quantitative parametric images can not only identify regional differences but also can assess global changes that are not apparent on nonquantitative images. The apparently higher values of K_1 with thicker wall (Figure 6.15c) indicate contamination from an adjacent liver area rather than increased blood flow in this myocardial region because the limited spatial resolution caused spill-over from the high blood flow of the liver. In contrast, V_d of myocardium is less affected by the liver because of the lower V_d of liver compared to that of the myocardium [9].

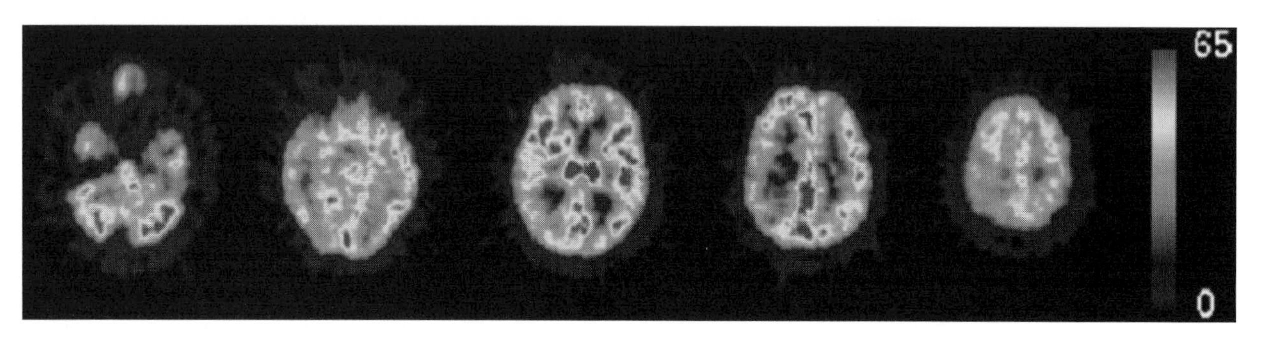

FIGURE 6.13 rCBF images using the image-derived input function from the ^{15}O-water study. Courtesy of Dr. H. Iida, Akita, Japan. For a more detailed view of this figure, please visit our companion site at: http://books.elsevier.com/companions/9780123735836.

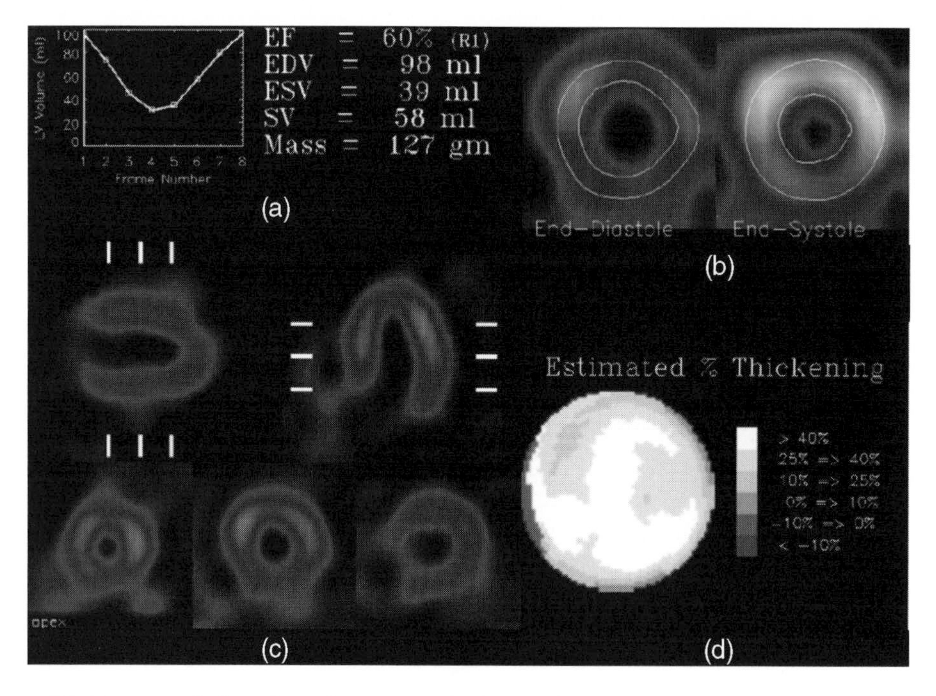

FIGURE 6.14 Display of gated SPECT with the automated quantified parameters (b). LV volume time curve over cardiac cycle (a), estimated variation of thickening (d). For a more detailed view of this figure, please visit our companion site at: http://books.elsevier.com/companions/9780123735836.

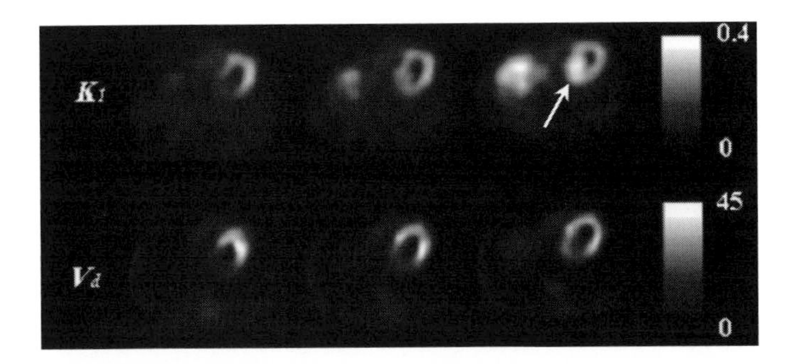

FIGURE 6.15 Parametric images of K_1 (a) and V_d (b) using the Yokoi plot for a ^{201}Tl study. For a more detailed view of this figure, please visit our companion site at: http://books.elsevier.com/companions/9780123735836.

6.4.2 Oxygen Consumption Parametric Images

Oxygen use is another measure of brain function because the cerebral metabolism is dependent on the transported glucose and oxygen. Increased metabolism in the brain requires greater consumption of oxygen, which may result in an increase of rCBF. However, because of the physical reasons why rCBF may not increase, the extra supply of oxygen would rely on the increased delivery from the hemoglobin to the surrounding tissue. That is the reason why the regional cerebral metabolic rate of oxygen consumption ($rCMRO_2$) provides a more accurate measure of the cerebral metabolism than rCBF.

$rCMRO_2$ is related to regional cerebral blood flow (rCBF), regional cerebral oxygen extraction fraction (rCOEF), and the arterial oxygen concentration (CaO_2) [99], as shown in Equation 6.60:

$$rCMRO_2 = rCBF \cdot rCOEF \cdot CaO_2. \tag{6.60}$$

In general, a dual-tracer study is taken to derive $rCMRO_2$ using the steady-state method [100]. Two steps are involved in the process to derive $rCMRO_2$. ^{15}O-water is administrated via constant infusion until the steady state is reached, followed by generation of rCBF images using the steady-state method. The second step involves measurement of rCOEF through the inhalation of $^{15}O_2$, again using the steady-state method. The value of CaO_2 in Equation 6.60 is measured from the arterial blood samples.

The two-step method was later found to overestimate rCOEF because it neglected vascular ^{15}O activity [101]. A correction method using regional cerebral blood volume (rCBV) was then proposed to account for the intravascular ^{15}O activity in the estimation of rCOEF. rCBV is usually measured by inhaling $C^{15}O$ [102]. Therefore, the two-step method was extended to a three-step sequential study with rCBF from ^{15}O-water, rCOEF from $^{15}O_2$, and rCBV from $C^{15}O$ infusions.

The three-step method requires a prolonged data-acquisition period. A relatively high level of radiation exposure is also required to reach steady state for each scan. Thus, an alternative method has been developed to substitute the steady-state method with the autoradiographic method [103], and improvements have been made to simplify and optimize the derivation of $rCMRO_2$ [99, 104, 105]. Recently, a two-step autoradiographic method has been proposed for fast generation of parametric images of $rCMRO_2$ using a look-up table procedure with a dual-scan study of $^{15}O_2$, and $H_2{}^{15}O$ [106]. Figure 6.16 demonstrates an example of parametric images respectively for rCBF, rCOEF, and $rCMRO_2$.

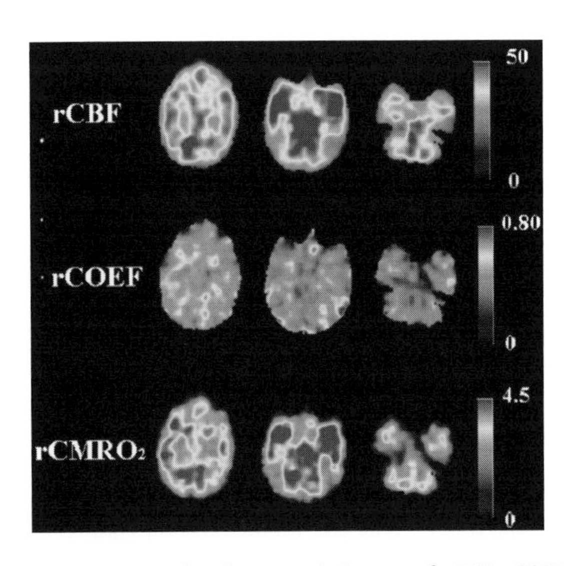

FIGURE 6.16 Example of parametric images of rCBF, rCOEF and $rCMRO_2$. Courtesy of Dr. H. Iida, Akita, Japan. For a more detailed view of this figure, please visit our companion site at: http://books.elsevier.com/companions/9780123735836.

^{11}C-acetate, which can study the tricarboxylic acid cycle activity, has been frequently used as an indirect measurement of regional myocardial metabolic rate of oxygen consumption ($rMMRO_2$) [107, 108]. However, the lack of an appropriate model that accurately describes the complex tissue kinetics of ^{11}C-acetate and the lost correlation with $rMMRO_2$ under some conditions have prevented absolute quantification of $rMMRO_2$ using ^{11}C-acetate [109, 110]. The direct, accurate quantification of $rMMRO_2$ can be obtained with $^{15}O_2$ inhalation [109, 111]. The three-step method can be used to derive regional myocardial blood flow (rMBF), regional myocardial oxygen extraction fraction (rMOEF), and regional myocardial blood volume (rMBV) from a sequential study of continuous inhalation of $C^{15}O$ and $^{15}O_2$ and administration of $H_2{}^{15}O$ with spill-over correction.

6.4.3 Glucose Metabolism Parametric Images

Physiologic processes in the living body require energy, which is released through metabolism. Glucose metabolism is the most common metabolism and in the presence of oxygen occurs via oxidative phosphorylation in most living cells. Hypoxia, such as can be encountered in tumors, may drive the glucose metabolism switch from oxidative phosphorylation to simple glycolysis as a means of energy generation [112]. Thus, the measurement of glucose metabolism is important to accurately assess the tissue metabolism correlated to energy release, especially for oncology studies in most tissues.

2-[^{18}F]fluorodeoxyglucose (FDG), which is a glucose analog undergoing similar uptake as glucose, is the most commonly used tracer in PET. FDG is carried by glucose transporters into the tissue, where it is phosphorylated to yield FDG-6-PO_4. Because FDG-6-PO_4 cannot be further metabolized and dephosphorylation occurs only slowly, FDG-6-PO_4 becomes trapped in the tissue and accumulates at a rate proportional to glucose metabolism. The increased glucose consumption of cancer cells leads to high FDG uptake, which can be visualized with PET studies. This feature has enabled FDG-PET to gain widespread clinical acceptance in oncology for detecting, staging, and assessing treatment response in patients with cancer.

Analysis of FDG-PET data consists of visual interpretation, semiquantitative evaluation, and quantitative assessment. The visual interpretation relies on the physician's knowledge and identification based on the tumor's region, boundary, and contrast against background. This approach is a subjective, qualitative evaluation of images. The SUV method is a semiquantitative method, assuming that the tumor uptake, normalized to the injected dose and a measure of total volume of distribution, is related to the metabolic rate of glucose.

The SUV method does not require prolonged data acquisition and blood sampling. For example, SUV > 5.0 is associated with a worse prognosis for lung cancer [113]. SUV has been shown to provide an independent prognostic value for primary non–small cell lung cancer (NSCLC) for SUV > 7, to supplement other factors in the choice of appropriate treatment [114], while SUV $= 7$ is the discriminative point of prognosis for small-cell lung cancer (SCLC) [115]. Although SUV is dependent on patient size, uptake time, and plasma glucose concentration, the SUV method has been frequently used as a measure of FDG uptake to assess differences between scans [112].

Figure 6.17 demonstrates SUV images for staging and evaluating response to radiotherapy for one patient. The parametric images were derived by dividing the corresponding FDG images by the ratio of the injected dose and body weight. The body weight of the patient was 77 kg. For the first study (a), 378 MBq of FDG was injected in contrast to 491 MBq of FDG in the second study. The focal abnormality near the central mediastinum with the maximum value of SUV above 5 (arrow in Figure 6.17) is prominent and obvious. The lesion was diagnosed as stage IIIA NSCLC. Thus, radiotherapy was chosen to treat the tumor. After 3 months, the second study (b) showed almost complete resolution of the focal abnormality, demonstrating a marked response to treatment.

Because the cortical regions in the brain have high glucose metabolism, the SUV method is less applicable in the brain. A kinetic study is often used to measure the glucose metabolic rate instead. The Patlak graphic plot can be applied on a pixel-by-pixel basis, generating parametric images of K_i, which improves image contrast and gives a direct view of glucose consumption [116]. If population-based kinetic constants are assumed for all brain regions, a static study can be used with the autoradiographic technique to estimate CMRGlc (see Section 6.2.1), avoiding a prolonged image acquisition.

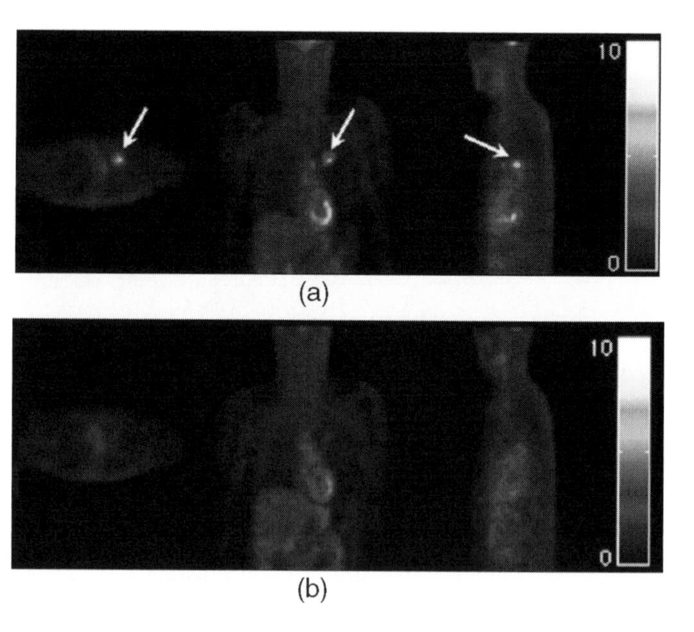

FIGURE 6.17 SUV images for whole body FDG-PET. (a) Initial staging of NSCLC. (b) Response to treatment. For a more detailed view of this figure, please visit our companion site at: http://books.elsevier.com/companions/9780123735836.

The quantitative values of CMRGlc are mainly dependent on the area under the curve of the input function when the autoradiographic method is used. Thus, the population-based input function can further decrease the required number of blood samples to only two without affecting the accuracy of CMRGlc [83]. Figure 6.18 depicts images of CMRGlc derived by the autoradiographic method for a neurologic study of FDG-PET.

A number of cortical regions such as the parietal lobes and the occipital association cortex are observed with hypometabolism (blue area in Figure 6.18); that is, decreased CMRGlc. The changes of quantitative values and morphologic features of the images are consistent with Alzheimer's disease.

6.4.4 Neuroreceptor Binding Parametric Images

FDG is not a tumor-specific tracer, and other pathologic states such as inflammation and infection may have high uptake that can result in false positives in the diagnosis of cancer. Receptor-specific tracers provide efficient imaging of receptor binding, for instance, to investigate neurologic disorders [117]. Continued research into development of diverse receptor-specific tracers holds great promise for future advances and application in functional imaging.

Receptors have a prominent role in the brain. Imaging of the distribution, density, and activity of receptors provides insight into the organization of functional networks in the brain, which cannot be achieved by the structural imaging or imaging of blood flow, oxygen, and glucose metabolism [118]. Neuroreceptor studies have been widely used to evaluate the effects of novel drugs in humans through specifically targeted receptors such as dopamine transporter, serotonin receptor, muscarinic receptors, and nicotinic receptors [119, 120]. Parametric

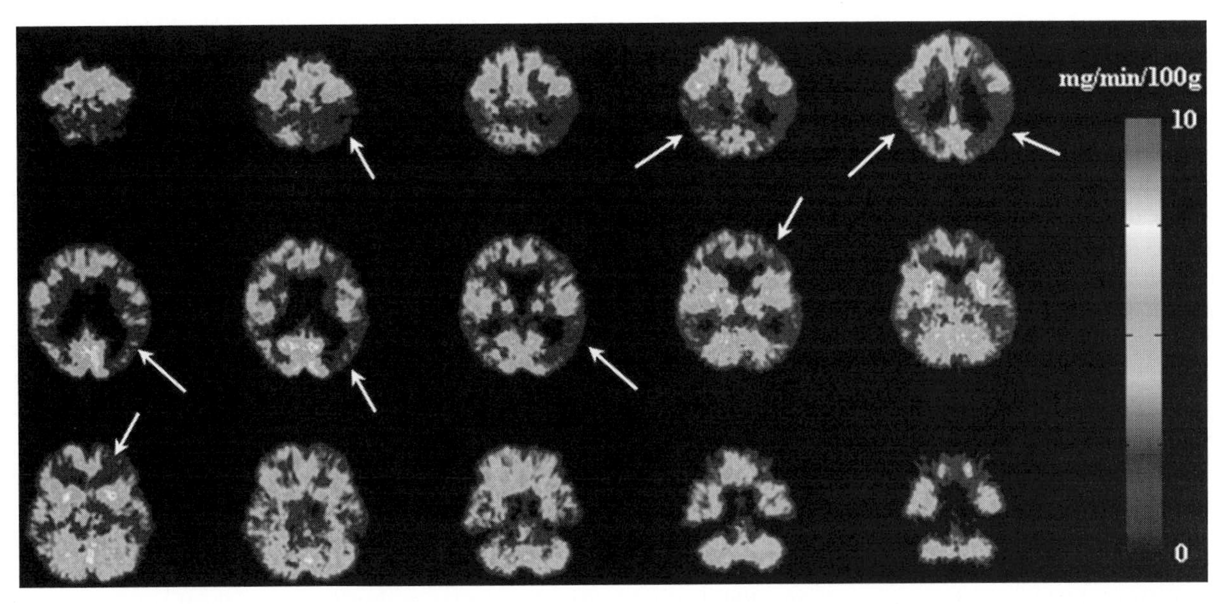

FIGURE 6.18 Parametric images of rCMRGlu for a neurologic study using FDG-PET. The input function is derived by the population-based IF with two calibration points. For a more detailed view of this figure, please visit our companion site at: http://books.elsevier.com/companions/9780123735836.

images of specially targeted tracers enhance early diagnosis and assessment of disease progression. For example, ^{11}C-PIB, which is targeted at amyloid plaque deposits, is showing great promise in the early differential diagnosis of Alzheimer's disease [121] and in the efficacy of treatments targeting plaque deposits.

Graphic methods are often used in the generation of parametric images for receptor-specific studies because of their computational efficiency and relative immunity to noise. For example, in a baboon study for neuronal nicotinic acetylcholine receptors (nAChRs), which are implicated in various neurodegenerative disorders, the Logan method was applied to process reconstructed SPECT data. The tracer of 5-[^{123}I]-iodo-A-85380 is an nAChRs-specific tracer. The corresponding parametric images of V_d are shown in Figure 6.19. The baseline

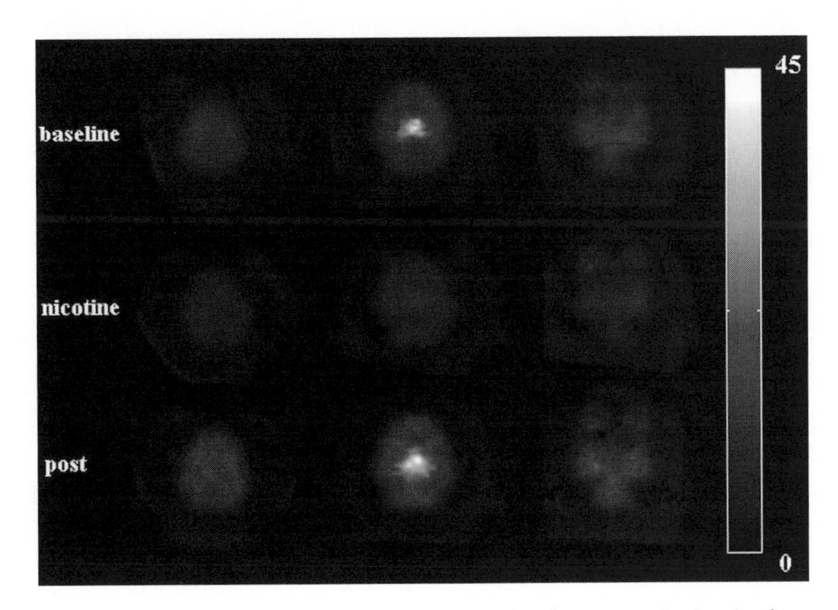

FIGURE 6.19 Parametric images of V_d for the baseline state, nicotine implantation, and post-nicotine infusion, respectively. The middle images at each row demonstrate normal, blocking, and upregulation of nAChRs, respectively. For a more detailed view of this figure, please visit our companion site at: http://books.elsevier.com/companions/9780123735836.

images (top row) show higher V_d in the thalamus, which has been found to be rich in nAChRs receptors. After a chronic dose of nicotine was infused for 2 weeks, no uptake was observed in the thalamus (middle row) because of the larger concentration of infused nicotine blocking the binding of the tracer. A significant increase in V_d was observed in the thalamus (bottom row) after cessation of nicotine infusion as a result of upregulation of nAChRs caused by the exposure to nicotine.

In addition to the graphic methods, a number of other methods have been applied to neuroreceptor studies, including the LLS method. For example, the GLLS method and improved GLLS methods for three-compartment and four-parameter model can simultaneously generate parametric images for multiple parameters from the nAChRs data. Figure 6.20 depicts the parametric images of K_1, V_d, and BP.

In Figure 6.20, the voxels with nonphysiologic rate constant estimates (negative rate constants or > 1), were set to zero. The original GLLS suffered from a large number of unsuccessful fits because of the high level of noise in SPECT (first column). Prior knowledge of V_d indeed improves the reliability of parametric images with slightly smoother images of V_d and BP (second column); however, an appreciable fraction of voxels still could not be successfully fitted. The bootstrap resampling enables the BMC-aided GLLS to eliminate unsuccessful fits even for noisy SPECT data (third column). The limitation of the BMC-aided GLLS is the overestimation of V_d and BP because the BMC resampling favors the bootstrap curves with higher V_d. The BMC-V_d–aided method improves the reliability of GLLS for SPECT further, as shown in the fourth column for K_1, V_d, and BP.

6.5 Summary

Parametric images derived from functional tracer studies enable quantitative estimates of physiologic or biochemical processes in the living body. Functional imaging not only provides unique information related to *in vivo* physiologic processes but is also capable of deriving quantitative functional parameters that can be used in the diagnosis of disorders, the assessment of treatment, and the evaluation of a novel drug. The process of parameter estimates usually involves a kinetic model that abstracts the complex physiologic process and describes the tracer distribution in the tissue.

Some of the major techniques for parametric imaging have been described in this chapter with tracer kinetic models for blood flow, glucose metabolism, and neuroreceptor studies. The appropriate methods of parameter estimates may be chosen according to the tracer, kinetic model, and scan duration. Noninvasive methods, which avoid frequent and invasive blood sampling, were also covered. Examples of

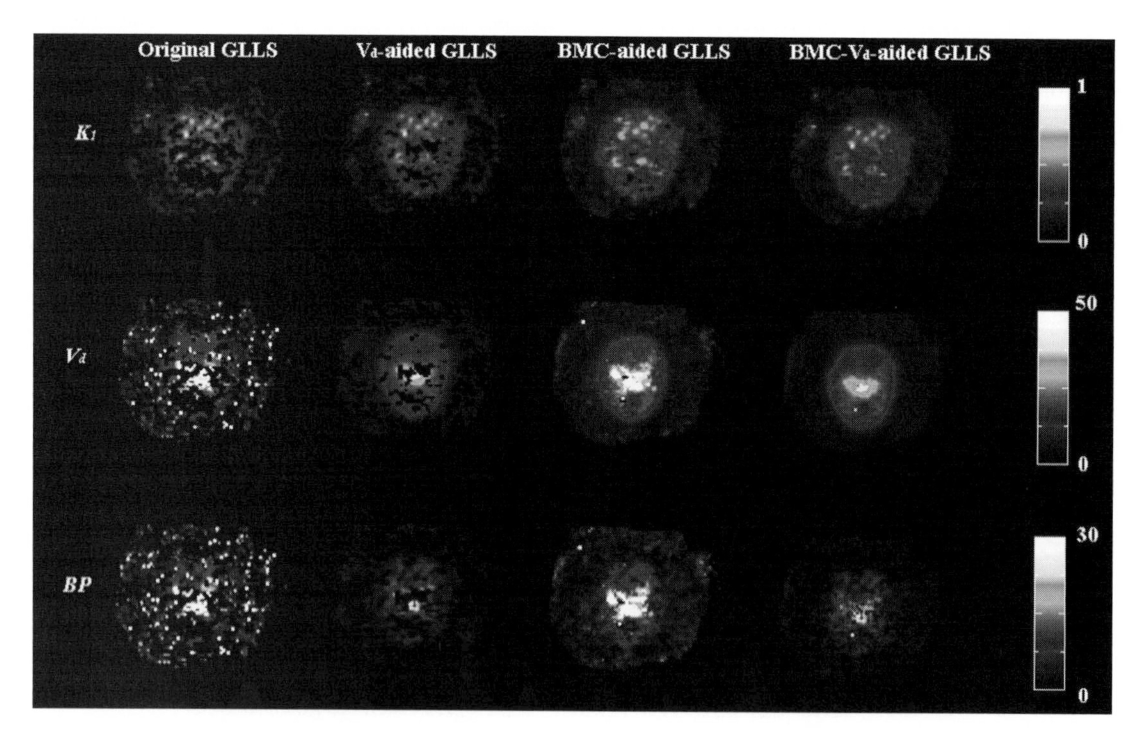

FIGURE 6.20 Parametric images for influx rate: K_1; volume of distribution: $V_d = K_1/k_2\ (1 + k_3/k_4)$; and binding potential: $BP = K_1\ k_3/k_2\ k_4$, respectively. Voxel values with nonphysiologic rate constant estimates (negative rate constants or > 1) were set to zero. For a more detailed view of this figure, please visit our companion site at: http://books.elsevier.com/companions/9780123735836.

parametric images have been presented for blood flow, oxygen consumption, glucose metabolism, and neuroreceptor binding studies for both clinical and research applications.

With the development of new tracers targeting specific receptors, the present techniques for parameter estimates will need to be validated for new physiologic processes. Innovative methods will be developed as well to address the kinetics of the new tracers and provide more reliable parametric images for the developed tracers. Complex physiologic processes still pose a challenging task for parametric imaging techniques.

Acknowledgments

The authors are grateful to the support from ARC, PolyU/UGC grants.

6.6 Exercises

1. State the differences between structural and functional imaging.
2. Why is the SUV considered to be proportional to metabolic rate of glucose? What are the necessary conditions in this case?
3. Give the equation of Logan plot for two-compartment and two-parameter model. Note: $V_d = K_1/k_2$.
4. State why the GLLS method can provide unbiased estimation compared with the LLS method.
5. List the major noninvasive input functions and compare their merits and limitations.

6.7 References and Bibliography

1. T. Beyer et al. A combined PET/CT scanner for clinical oncology. *J. Nucl. Med.* 41:1369–1379, 2000.
2. J. S. Karp et al. Benefit of time-of-flight information for whole-body FDG PET. *J. Nucl. Med.* 47(suppl):529, 2006.
3. P. Almeida et al. Absolute quantitation of iodine-123 epidepride kinetics using single-photon emission tomography: Comparison with carbon-11 epidepride and positron emission tomography. *Eur. J. Nucl. Med.* 26:1580–1588, 1999.
4. D. H. S. Silverman. Brain ^{18}F-FDG PET in the diagnosis of neurodegenerative dementias: Comparison with perfusion SPECT and with clinical evaluations lacking nuclear imaging. *J. Nucl. Med.* 45:597–607, 2004.
5. J. J. Pekar. A brief introduction to functional MRI. *IEEE Eng. Med. Biol. Mag.* 35:24–26, 2006.
6. C. Burger and A. Buck. Requirements and implementation of a flexible kinetic modeling tool. *J. Nucl. Med.* 138:1818–1823, 1997.
7. S.–C. Huang and M. E. Phelps. Principles of tracer kinetic modeling in positron emission tomography and autoradiography. In *Positron Emmission Tomography and Autoradiography: Principles and Applications for the Brain and Heart.* 287–346. Raven Press, 1986.
8. S.–C. Huang et al. Noninvasive determination of local cerebral metabolic rate of glucose in man. *Am. J. Physiol.* 238:E69–E82, 1980.
9. S. Eberl. *Quantitative Physiological Parameter Estimation from Dynamic Single Photon Emission Computed Tomography (SPECT).* Sydney, Australia: University of New South Wales, 2000.
10. Z. Zhang. Parameter estimation techniques: A tutorial with application to conic fitting. *Image and Vision Computing Journal.* 15:59–76, 1997.
11. S. S. Kety. The theory and applications of the exchange of inert gas at the lungs and tissues. *Pharmacol. Rev.* 3:1–41, 1951.
12. S. S. Kety. Measurement of local blood flow by the exchange of an inert, diffusible substance. *Methods Med. Res.* 8:228–236, 1960.
13. M. L. Johnson and L. M. Faunt. Parameter-estimation by least-squares methods. *Methods Enzymol.* 210:1–37, 1992.
14. R. Fletcher. *Practical Methods of Optimization.* John Wiley & Sons, 1981.
15. W. H. Press et al. *Numerical Recipes in C: The Art of Scientific Computing.* 2nd ed. Cambridge University Press, 1992.
16. R. E. Carson. Parameter estimation in positron emission tomography. In *Positron Emission Tomography and Autoradiography: Principles and Applications for the Brain and Heart.* Raven Press. 347–390, 1986.
17. M. Ichise, J. H. Meyer, and Y. Yonekura. An introduction to PET and SPECT neuroreceptor quantification models. *J. Nucl. Med.* 42:755–763, 2001.
18. A. A. Lammertsma et al. Accuracy of the oxygen-15 steady state technique for measuring rCBF and $rCMRO_2$. *J. Cereb. Blood Flow Metab.* 1(suppl):S3–S4, 1981.
19. E. Meyer and Y. L. Yamamoto. The requirement for constant arterial radioactivity in the $C^{15}O_2$ steady-state blood flow model. *J. Nucl. Med.* 25:455–460, 1984.
20. J. A. Correia et al. Analysis of some errors in the measurement of oxygen extraction and oxygen consumption by the equilibrium inhalation method. *J. Cereb. Blood Flow Metab.* 5:591–599, 1985.
21. K. Wienhard. Measurement of glucose consumption using [^{18}F]fluorodeoxyglucose. *Methods.* 27:218–225, 2002.
22. M. Reivich et al. Glucose metabolic rate kinetic model parameter determination in humans: The lumped constants and rate constants for [^{18}F]fluorodeoxyglucose and [^{11}C]deoxyglucose. *J. Cereb. Blood Flow Metab.* 5:179–192, 1985.

23. G. D. Hutchins et al. Alternative approach to single-scan estimation of cerebral glucose metabolic rate using glucose analogs with particular application to ischemia. *J. Cereb. Blood Flow Metab.* 4:35–40, 1984.
24. R. A. Brooks. Alternative formula for glucose utilization using labeled deoxyglucose. *J. Nucl. Med.* 23:583–539, 1982.
25. C. M. Clark, E. W. Grochowski, and W. Ammann. A method for comparing different procedures of estimating regional glucose metabolism using fluorine-18-fluorodeoxyglucose. *J. Nucl. Med.* 33:157–160, 1992.
26. C. G. Rhodes et al. *In vivo* disturbance of the oxidative metabolism of glucose in human cerebral gliomas. *Ann. Neurol.* 14:614–626, 1983.
27. S.-C. Huang. Anatomy of SUV. *Nucl. Med. Biol.* 27:643–646, 2000.
28. P. Lindholm et al. Influence of the blood glucose concentration on FDG uptake in cancer—A PET study. *J. Nucl. Med.* 34:1–6, 1993.
29. J. W. Keyes Jr. SUV: Standard uptake or silly useless value. *J. Nucl. Med.* 36:1836–1839, 1995.
30. K. R. Zasadny and R. L. Wahl. Standardized uptake values of normal tissues at PET with 2-[fluorine-18]-fluoro-2-deoxy-D-glucose: Variations with body weight and a method for correction. *Radiology.* 189:847–850, 1993.
31. C. K. Kim et al. Standardized uptake values of FDG: Body surface area correction is preferable to body weight correction. *J. Nucl. Med.* 35:164–167, 1994.
32. S.-C. Huang, R. E. Carson, and M. E. Phelps. Measurement of local blood flow and distribution volume with short-lived isotopes: a general input technique. *J. Cereb. Blood Flow Metab.* 2:99–108, 1982.
33. S.-C. Huang et al. Quantitative measurement of local cerebral blood flow in humans by positron computed tomography and ^{15}O-water. *J. Cereb. Blood Flow Metab.* 3:141–153, 1983.
34. D. Feng et al. Techniques for functional imaging. In C. T. Leondes and D. Feng (Eds.). *Medical Imaging Systems Techniques and Applications.* Gordon and Breach Science Publishers. 85–145, 1997.
35. N. M. Alpert et al. Strategy for the measurement of regional cerebral blood flow using short-lived tracers and emission tomography. *J. Nucl. Med.* 4:28–34, 1984.
36. R. E. Carson, S.-C. Huang, and M. V. Green. Weighted integration method for local cerebral blood flow measurements with positron emission tomography. *J. Cereb. Blood Flow Metab.* 6:245–258, 1986.
37. T. Yokoi et al. A new approach of weighted integration technique based on accumulated image using dynamic PET and $H_2{}^{15}O$. *J. Cereb. Blood Flow Metab.* 11:492–501, 1991.
38. H. Iida et al. Effect of real-time weighted integration system for rapid calculation of functional images in clinical positron emission tomography. *IEEE Trans. Med. Imaging* 14:116–121, 1995.
39. D. Feng et al. An evaluation of the algorithms for determining local cerebral metabolic rates of glucose using positron emission tomography dynamic data. *IEEE Trans. Med. Imaging* 14:697–710, 1995.
40. S. R. Meikle et al. Pharmacokinetic assessment of novel anti-cancer drugs using spectral analysis and positron emission tomography: a feasibility study, *Cancer Chemother. Pharmacol.* 42:183–193, 1998.
41. V. J. Cunningham and T. Jones. Spectral analysis of dynamic PET studies. *J. Cereb. Blood Flow Metab.* 13:15–23, 1993.
42. F. Turkheimer et al. The use of spectral analysis to determine regional cerebral glucose utilization with positron emission tomography and [^{18}F]fluorodeoxyglucose: Theory, implementation, and optimization procedures. *J. Cereb. Blood Flow Metab.* 14:406–422, 1994.
43. J. Matthews et al. The direct calculation of parametric images from dynamic PET data using maximum-likelihood iterative reconstruction. *Physics Med. Biol.* 42:1155–1173, 1997.
44. C. S. Patlak, R. G. Blasberg, and J. D. Fenstermacher. Graphical evaluation of blood-to-brain transfer constants from multiple-time uptake data. *J. Cereb. Blood Flow Metab.* 3:1–7, 1983.
45. C. S. Patlak and R. G. Blasberg. Graphical evaluation of blood-to-brain transfer constants from multiple-time uptake data: Generalizations. *J. Cereb. Blood Flow Metab.* 5:584–590, 1985.
46. K. Wienhard et al. Estimation of local cerebral glucose utilization by positron emission tomography of [^{18}F] 2-fluoro-2-deoxy-D-glucose: A critical appraisal of optimization procedures. *J. Cereb. Blood Flow Metab.* 5:115–125, 1985.
47. J. Logan et al. Graphical analysis of reversible radioligand binding from time-activity measurements applied to [N-^{11}C-methyl]-(–)-cocaine PET studies in human subjects. *J. Cereb. Blood Flow Metab.* 10:740–747, 1990.
48. J. Logan et al. Effects of blood flow on [^{11}C]raclopride binding in the brain: model simulations and kinetic analysis of PET data. *J. Cereb. Blood Flow Metab.* 14:995–1010, 1994.
49. T. Yokoi et al. A new graphic plot analysis for cerebral blood flow and partition coefficient with iodine-123-iodoamphetamine and dynamic SPECT validation studies using oxygen-15-water and PET. *J. Nucl. Med.* 34:498–505, 1993.
50. W. D. Heiss et al. *Atlas of Positron Emission Tomography of the Brain.* Springer-Verlag, 1985.
51. D. Feng et al. An unbiased parametric imaging algorithm for nonuniformly sampled biomedical system parameter estimation. *IEEE Trans. Med. Imaging* 15:512–518, 1996.
52. D. Feng, Z. Wang, and S.-C. Huang. A study on statistically reliable and computationally efficient algorithms for generating local cerebral blood flow parametric image

with positron emission tomography. *IEEE Trans. Med. Imaging.* 12:182–188, 1993.
53. K. Chen et al. Generalized linear least-squares method for fast generation of myocardial blood flow parametric images with N–13 ammonia PET. *IEEE Trans. Med. Imaging.* 17:236–243, 1998.
54. H.-C. Choi et al. Fast parametric imaging algorithm for dual-input biomedical system parameter estimation. *Comput. Methods Programs Biomed.* 81:49–55, 2006.
55. L. Wen et al. Improved generalized linear least squares algorithms for generation of parametric images from dynamic SPECT. *J. Nucl. Med.* 47(suppl):61, 2006.
56. H.-C. Choi et al. Methods for improving reliability of GLLS for parametric image generation. Presented at 6th IFAC Symposium on Modelling and Control in Biomedical Systems, Reims, France, 2006.
57. L. Wen et al. Effect of reconstruction and filtering on kinetic parameter estimation bias and reliability for dynamic SPECT: A simulation study. *IEEE Trans. Nucl. Sci.* 52:69–78, 2005.
58. L. Wen, S. Eberl, and D. Feng. Enhanced parameter estimation with GLLS and the Bootstrap Monte Carlo method for dynamic SPECT. Presented at 28th IEEE EMBS Annual International Conference, 2006.
59. M. E. Kamasak et al. Direct reconstruction of kinetic parameters images from dynamic PET data. *IEEE Trans. Med. Imaging* 24:636–650, 2005.
60. R. H. Huesman et al. Kinetic parameter estimation from SPECT cone-beam projection measurements. *Physics Med. Biol.* 43:973–982, 1998.
61. D. J. Kadrmas and G. T. Gullberg. A maximum a posteriori algorithm for the reconstruction of dynamic SPECT data. Presented at IEEE Nuclear Science Symposium and Medical Imaging Conference, 1998.
62. G. T. Gullberg et al. Dynamic cardiac single-photon emission computed tomography. In *Nuclear Cardiology: State of the Art and Future Directions.* Mosby-Year Book Inc. 137–187, 1998.
63. M. E. Phelps et al. Tomographic measurement of local cerebral glucose metabolic rate in humans with (F-18)2-fluoro-2-deoxy-D-glucose: Validation of method. *Ann. Neurol.* 6:371–388, 1979.
64. S. S. Gambhir et al. Simple noninvasive quantification method for measuring myocardial glucose utilization in humans employing positron emission tomography and fluorine-18 deoxyglucose. *J. Nucl. Med.* 30:359–366, 1989.
65. H. Iida et al. Use of the left ventricular time-activity curve as a noninvasive input function in dynamic oxygen-15-water positron emission tomography. *J. Nucl. Med.* 33:1669–1677, 1992.
66. K.-P. Lin et al. Correction of spillover radioactivities for estimation of the blood time-activity curve from the image LV chamber in cardiac dynamic FDG PET studies. *Physics Med. Biol.* 40:629–642, 1995.
67. L. D. Wahl, M. C. Asselin, and C. Nahmias. Regions of interest in the venous sinuses as input functions for quantitative PET. *J. Nucl. Med.* 40:1666–1675, 1999.
68. T. Ohtake et al. Noninvasive method to obtain input function for measuring tissue glucose utilization of thoracic and abdominal organs. *J. Nucl. Med.* 32:1432–1438, 1991.
69. G. Germano et al. Use of the abdominal aorta for arterial input function determination in hepatic and renal PET studies. *J. Nucl. Med.* 33:613–620, 1992.
70. V. Dhawan et al. Quantitative brain FDG/PET studies using dynamic aortic imaging. *Physics Med. Biol.* 39:1475–1487, 1994.
71. K. Chen et al. Noninvasive quantification of the cerebral metabolic rate for glucose using positron emission tomography, ^{18}F-fluoro-2-deoxyglucose, the Patlak method, and an image-derived input function. *J. Cereb. Blood Flow Metab.* 18:716–723, 1998.
72. M. Liptrot et al. Cluster analysis in kinetic modelling of the brain: a noninvasive alternative to arterial sampling. *Neuroimage.* 21:483–493, 2004.
73. D. Feng et al. A technique for extracting physiological parameters and the required input function simultaneously from PET image measurements: Theory and simulation study. *IEEE Trans. Inf. Technol. Biomed.* 1:243–254, 1997.
74. K. P. Wong et al. Simultaneous estimation of physiological parameters and the input function—*In vivo* PET data. *IEEE Trans. Inf. Technol. Biomed.* 5:67–76, 2001.
75. J. Logan et al. Distribution volume ratios without blood sampling from graphical analysis of PET data. *J. Cereb. Blood Flow Metab.* 16:834–840, 1996.
76. R. N. Gunn et al. Parametric imaging of ligand-receptor binding in PET using a simplified reference region model. *Neuroimage* 6:279–287, 1997.
77. A. A. Lammertsma and S. P. Hume. Simplified reference tissue model for PET receptor studies. *Neuroimage* 4:153–158, 1996.
78. Y. Wu and R. E. Carson. Noise reduction in the simplified reference tissue model for neuroreceptor functional imaging. *J. Cereb. Blood Flow Metab.* 22:1440–1452, 2002.
79. J. Logan. Graphical analysis of PET data applied to reversible and irreversible tracers. *Nucl. Med. Biol.* 27:661–670, 2000.
80. M. Ichise et al. Noninvasive quantification of dopamine D2 receptors with iodine-123-IBF SPECT. *J. Nucl. Med.* 37:513–520, 1996.
81. M. Ichise et al. Linearized reference tissue parametric imaging methods: Application to [^{11}C]DASB positron emission tomography studies of the serotonin transporter in human brain. *J. Cereb. Blood Flow Metab.* 23:1096–1112, 2003.
82. S. Takikawa et al. Noninvasive quantitative fluorodeoxyglucose PET studies with an estimated input function derived from a population-based arterial blood curve. *Radiology.* 188:131–136, 1993.

83. S. Eberl et al. Evaluation of two population-based input functions for quantitative neurological FDG PET studies. *Eur. J. Nucl. Med.* 24:299–304, 1997.
84. K. Wakita et al. Simplification for measuring input function of FDG PET: investigation of 1-point blood sampling method. *J. Nucl. Med.* 41:1484–1490, 2000.
85. T. Tsuchida et al. Noninvasive measurement of cerebral metabolic rate of glucose using standardized input function. *J. Nucl. Med.* 40:1441–1445, 1999.
86. T. Shiozaki et al. Noninvasive estimation of FDG input function for quantification of cerebral metabolic rate of glucose: optimization and multicenter evaluation. *J. Nucl. Med.* 41:1612–1618, 2000.
87. D. Feng, S.-C. Huang, and X. Wang. Models for computer simulation studies of input functions for tracer kinetic modeling with positron emission tomography. *Int. J. Biomed. Comput.* 32:95–110, 1993.
88. D. Feng. Information technology applications in biomedical functional imaging. *IEEE Trans. Inf. Technol. Biomed.* 3:221–230, 1999.
89. D. Feng, X. Wang, and H. Yan. A computer simulation study on the input function sampling schedules in tracer kinetic modeling with positron emission tomography (PET). *Comput. Methods Programs Biomed.* 45:175–186, 1994.
90. K. Hirao et al. The prediction of rapid conversion to Alzheimer's disease in mild cognitive impairment using regional cerebral blood flow SPECT. *Neuroimage.* 28:1014–1021, 2005.
91. R. C. Knowlton et al. Ictal SPECT analysis in epilepsy--Subtraction and statistical parametric mapping techniques. *Neurology.* 63:10–15, 2004.
92. J. M. Mountz, H. G. Liu, and G. Deutsch. Neuroimaging in cerebrovascular disorders: measurement of cerebral physiology after stroke and assessment of stroke recovery. *Semin. Nucl. Med.* 33:56–76, 2003.
93. N. Shuke et al. Demonstration of positional posterior circulation cerebral ischemia on cerebral blood flow SPECT. *Clin. Nucl. Med.* 26:559–560, 2001.
94. H. Iida et al. Noninvasive quantification of cerebral blood flow using oxygen-15-water and a dual-PET system. *J. Nucl. Med.* 38:1789–1798, 1998.
95. D. A. Sivaratnam, R. O. Bonow, and V. Kalff. Assessment of myocardial viability in dysfunctional myocardium. In P. J. Ell and S. S. Gambhir (Eds.). *Nuclear Medicine in Clinical Diagnosis and Treatment.* Vol. 2. 3rd ed. Churchill Livingstone. 1159–1170, 2004.
96. R. H. J. A. Slart et al. Imaging techniques in nuclear cardiology for the assessment of myocardial viability. *Int. J. Cardiovasc. Imaging* 22: 2006.
97. J. Nuyts et al. Three-dimensional correction for spillover and recovery of myocardial PET images. *J. Nucl. Med.* 37:767–774, 1996.
98. M. R. Mansoor and G. V. Heller. Gated SPECT imaging. *Semin. Nucl. Med.* 24:271–278, 1999.
99. N. Hattori et al. Accuracy of a method using short inhalation of $^{15}O-O_2$ for measuring cerebral oxygen extraction fraction with PET in healthy humans. *J. Nucl. Med.* 45:765–770, 2004.
100. R. S. Frackowiak et al. Quantitative measurement of regional cerebral blood flow and oxygen metabolism in man using ^{15}O and positron emission tomography. *J. Comput. Assist. Tomog.* 4:727–736, 1980.
101. A. A. Lammertsma and T. Jones. Correction for the presence of intravascular oxygen-15 in the steady-state technique for measuring regional oxygen extraction ratio in the brain: 1. Description of the method. *J. Cereb. Blood Flow Metab.* 3:416–424, 1983.
102. W. R. Martin, W. J. Powers, and M. E. Raichle. Cerebral blood volume measured with inhaled $C^{15}O$ and positron emission tomography. *J. Cereb. Blood Flow Metab.* 7:421–426, 1987.
103. M. A. Mintun et al. Brain oxygen utilization measured with O-15 radiotracers and positron emission tomography. *J. Nucl. Med.* 25:177–187, 1984.
104. H. Iida, T. Jones, and S. Miura. Modeling approach to eliminate the need to separate arterial plasma in oxygen-15 inhalation positron emission tomography. *J. Nucl. Med.* 34:1333–1340, 1993.
105. M. Shidahara et al. Evaluation of a commercial PET tomography-based system for the quantitative assessment of rCBF, rOEF and $rCMRO_2$ by using sequential administration of ^{15}O-labeld compounds. *Ann. Nucl. Med.* 16:317–327, 2002.
106. N. Kudomi et al. Rapid quantitative measurement of $CMRO_2$ and CBF by dual administration of ^{15}O-labled oxygen and water during a single PET scan—A validation study and error analysis in anesthetized monkeys. *J. Cereb. Blood Flow Metab.* 25:1209–1224, 2005.
107. M. Brown et al. Delineation of myocardial oxygen utilization with carbon-11-labeled acetate. *Circulation* 76:687–696, 1987.
108. J. J. Armbrecht et al. Regional myocardial oxygen consumption determined noninvasively in humans with [1-^{11}C]acetate and dynamic positron tomography. *Circulation.* 80:864–872, 1989.
109. H. Ukkonen et al. Use of [^{11}C]acetate and [^{15}O]O_2 PET for the assessment of myocardial oxygen utilization in patients with chronic myocardial infraction. *Eur. J. Nucl. Med.* 28:334–339, 2001.
110. P. G. Camici et al. Positron emission tomography. In P. J. Ell and S. S. Gambhir (Eds.). *Nuclear Medicine in Clinical Diagnosis and Treatment.* 3rd ed. Churchill Livingstone. 2:1075–1091, 2004.
111. H. Iida et al. Noninvasive quantification of regional myocardial metabolic rate of oxygen by use of $^{15}O_2$ inhalation

and positron emission tomography: Theory, error analysis, and application in humans. *Circulation.* 94:792–807, 1996.

112. G. J. Kelloff, J. M. Hoffman, and B. Johnson. Progress and promise of FDG-PET imaging for cancer patient management and oncologic drug development. *Clin. Cancer Res.* 11:2785–2808, 2005.
113. J. Guo et al. *In vitro* proton magnetic resonance spectroscopic lactate and choline measurements, 18F-FDG uptake, and prognosis in patients with lung adenocarcinoma. *J. Nucl. Med.* 45:1334–1339, 2004.
114. J. F. Vansteenkiste et al. Prognostic importance of the standardized uptake value on ^{18}F-fluoro-2-deoxy-glucose-positron emission tomography scan in non–small-cell lung cancer: An analysis of 125 cases. *J. Clin. Oncol.* 17:3201–3206, 1999.
115. I. Brink et al. Small cell lung cancer (SCLS): Glucose metabolism and prognostic significance. *J. Nucl. Med.* 47(suppl):163–164, 2006.
116. C. J. Hoekstra et al. Monitoring response to therapy in cancer using [^{18}F]-2-fluoro-2-deoxy-D-glucose and positron emission tomography: An overview of different analytical methods. *Eur. J. Nucl. Med.* 27:731–743, 2000.
117. J. Passchier et al. On the quantification of [^{18}F]MPPF binding to 5-HT$_{1A}$ receptors in the human brain. *J. Nucl. Med.* 42:1025–1031, 2001.
118. W. D. Heiss and K. Herholz. Brain receptor imaging. *J. Nucl. Med.* 47:302–312, 2006.
119. A. van Waard. Measuring receptor occupancy with PET. *Current Pharm. Des.* 6:1593–1610, 2000.
120. C.–M. Lee and L. Farde. Using positron emission tomography to facilitate CNS drug development. *Trends Pharmacol. Sci.* 27:310–315, 2006.
121. C. C. Rowe et al. Visual analysis of ^{11}C-PIB PET and ^{18}F-FDG PET for detection of Alzheimer's disease. *J. Nucl. Med.* 47(suppl):75, 2006.

7

Data Processing and Analysis

Prof. Yue Wang[1]
Prof. Chris Wyatt[1]
Prof. Yu-Ping Wang[2]
Prof. Matthew T. Freedman[3] and
Prof. Murray Loew[4]
[1] *Virginia Polytechnic Institute and State University*
[2] *University of Missouri—Kansas City*
[3] *Georgetown University*
[4] *George Washington University*

7.1 Introduction

Biomedical data processing and analysis has become a major component of biomedical research and clinical applications. To successfully detect and diagnose disease, it is vital for clinicians to properly apply the latest data processing and analysis technologies. Because of either the nature or volume of available biomedical data, early or obscured signs of disease can go undetected or can be misinterpreted. To combat these inaccuracies, biomedical researchers and clinicians have come to rely on advanced data processing and analysis techniques and software.

Biomedical data processing and analysis have been woven into the fabric of the signal processing and pattern analysis community. Initially, the efforts in this area were seen as applying pattern analysis and computer vision techniques to another interesting dataset. However, over the last two to three decades, the unique nature of the problems presented within this area of study has led to the development of a new discipline in its own right.

The focus of Chapter 7 is on data processing and analysis. It is an essential reference that details the primary methods, techniques, and approaches used to improve the quality of biomedical data visualization and interpretation as well as quantitative detection and diagnostic decision aids. This information is presented by the contributing authors, who are at the forefront of biomedical data processing and analysis. This comprehensive volume illustrates analytical techniques such as medical image enhancement, medical image segmentation, medical image feature extraction, computer-aided diagnosis, and data-driven medical decisions by clinicians.

7.2 Medical Image Enhancement

Because of limited capability of a display system, the optical imaging noises, and many other factors, the acquired medical images usually have poor quality. Image enhancement is the procedure used to alter the appearance of an image or the subset of the image for better contrast or visualization of certain features and to facilitate the subsequent image-based medical diagnosis. For example, in the X-ray mammogram imaging for the breast cancer diagnosis, image enhancement is usually used to improve the visibility of microcalcifications, masses, and soft tissues. The design of a good image enhancement system should consider the specific features of the medical images and understand the imaging procedure of a particular imaging modality [1]. In brain imaging using

X-ray computed tomography (CT), the bony structure and soft tissues display different contrast and geometric features, while in positron emission tomography (PET) brain images, there is little structural information. Therefore, the filtering algorithm should be different for improving the soft tissue contrast and enhancing anatomic structures. Another example is that in microscopic imaging, the images are usually acquired at different focal planes and at different time intervals and spectral channels. The design of the enhancement algorithm should fully take advantage of this multidimensional information.

There has been a variety of image enhancement algorithms available. They are usually categorized into two types: Spatial-domain– and transform-domain–based methods. The spatial-domain methods include image operations on a whole image or a local region based on the image statistics. Histogram equalization, image averaging, sharpening of images using edge detection and morphology operators, and nonlinear median filtering all belong to this category. The other class is a transform-domain–based method because the image operations are performed in the transform domain, such as in the Fourier and wavelet domain. The frequency transform methods facilitate the extraction of certain image features that cannot be derived from the spatial domain. One can manipulate the transformation coefficients in the frequency domain and then recover the image in the spatial domain to highlight interested image contents. As one of powerful image transforms, wavelet approaches have been used in recent years for medical image analysis. We will introduce several wavelet approaches for image contrast enhancement. Finally, we will discuss how to evaluate the performance of enhancement algorithms and use chromosome image enhancement as an example to compare among different image enhancement approaches.

7.2.1 Spatial-Domain Methods Using Pixel Operations

When display devices have a limited range of gray level over which the image features are most visible, one can use the global method to adjust all the pixels in the image to ensure that the features of interest fall into the visible range of display. This technique is also called contrast stretching [2]. For example, if I_1 and I_2 define the intensity range of interest, a scaling transformation can be introduced to map the image intensity f to the image g with the range of 0 to $I_{\max}$ as $g = \frac{I - I_1}{I_2 - I_1} \cdot I_{\max}$. This mapping is a linear stretch. A number of nonlinear monotonic pixel operations exist [2,3]. The image intensity scaling is usually used for contrast stretch and clipping, display calibration, etc. Because no *a priori* information may be available to identify useful intensity bands, it is useful to distribute the intensity information uniformly over the available intensity bands. This technique is called histogram equalization [2].

When there is more than one image available, image averaging is a simple way to enhance the signal-to-noise ratio. For example, in microscopic imaging, one usually has multiple images at different focal planes. For motion images, images of the same scene are acquired at different time points. These multiple images can be properly registered and averaged to reduce noise. Image subtraction is usually performed when two images of the same object are obtained at different imaging conditions and have significant similarities [3]. The image subtraction will enhance the differences between the two images. An application is the background subtraction or correction. In the microscopic imaging of human chromosomes, the image is contraindicated by slowly varying the background shading pattern. One can move the microscope stage to obtain an empty field and use this image as the background. Then the background is subtracted from the image that contains chromosomes to remove the shading.

7.2.2 Local Operations

Because image properties vary at different pixels, the operation of spatial filtering is usually performed in a local neighborhood. An example of local filtering is the local area histogram equalization [2], obtained with a modification of the histogram equalization. The local histogram equalization adapts the histogram equalization technique from whole image to small and overlapping areas of the image [2], which takes into account the image local features. However, this nonlinear filtering is computationally intensive. There are a number of variants on the image histogram transforms by considering the local properties of the image. A local histogram transform was performed based on local standard deviation [4]. A wavelet transform-based histogram equalization [5] was also introduced for the enhancement of gastric sonogram images.

Median filter is a well-used nonlinear filter that replaces the original gray level of a pixel by the median of the gray values of pixels in a specific neighborhood. The median filter is also called the order-specific filter [3] because it is based on statistics derived from ordering the elements of a set rather than taking the means. This filter is popular for reducing noise without blurring edges of the image [6]. The noise-reducing effect of the median filter depends on two factors: the spatial extent of the neighborhood and the number of pixels involved in the median calculation. An adaptive neighborhood can be used for preserving edges while smoothing noise [2]. In adaptive neighborhood filtering, a region centered at a pixel is grown until a prespecified criterion of region growing is satisfied. Mathematical morphology approach is another type of nonlinear operation, where the filters are represented as the combination of two simple set operations: dilation and erosion. Based on the properties of morphologic filters, an algorithm was designed for enhanced segmentation and extraction of suspicious mass areas from mammographic images [7]. The morphologic filters are also used in combination with active contours for the automated extraction of foreign objects in radiographic images [8].

The unsharp masking technique is commonly used for sharpening the edges of the image. This filtering method involves the convolution of an image with a specific filter such as the high-frequency filtering mask. In general the unsharp masking operation is represented by:

$$g(x,y) = f(x,y) + \lambda e(x,y), \tag{7.1}$$

where $\lambda > 0$, and e(x,y) is often taken to be the gradient of the image. The operation is equivalent to adding the gradient information to the image. A commonly used gradient function is the discrete Laplacian filter [6]. Other gradient filters such as the Sobel filter can be used to compute the first-order gradient in the x- and y-directions. The directional filters such as the steerable filter [9] can extract directional information of the image to sharpen the image at a specific direction. Examples using local derivative information for medical image enhancement can be found in references [10–12].

7.2.3 Frequency Domain Methods

In many cases, filtering in frequency domain is more straightforward than in spatial domain when reducing noises because noises can be easily identified in frequency domain. When an image is transformed into the Fourier domain, the low-frequency components usually correspond to smooth regions or blurred structures of the image, whereas high-frequency components represent image details, edges, and noises. Thus, one can design filters according to image frequency components to smooth images or remove noise [1, 13]. Low-pass filtering will usually smooth images by attenuating high-frequency components, and high-pass filtering will emphasize the image edges or sharp details by attenuating low-frequency components.

The Wiener filter is an optimal filter derived under a minimum of mean-squared error criterion [1, 3, 6]. When the noisy image is obtained as the sum of the image and stationary noise $g(x,y) = f(x,y) + n(x,y)$, where the noise is assumed to be spectrally white with the zero mean and variance σ^2, the Wiener filter is derived as follows [2]:

$$H(u,v) = \frac{P_f(u,v)}{P_f(u,v) + \sigma^2}, \tag{7.2}$$

where P_f is the power spectrum of the signal. The conventional Wiener filter has limitations [3]. If the signal is a realization of a non-Gaussian process such as in natural images, the Wiener filter is outperformed by nonlinear estimators. A number of variants of Wiener filter exist by considering the spatial variant characteristics of the signal and noise [2]. An approach to make the filter spatially variant is by using a local spatially varying model of the noise parameter σ_n. This filter changes from pixel to pixel. A variant of this filter is the noise-adaptive Wiener filter by Lee et al. [14], which models the signal locally as a stationary process. The filter is given by the following:

$$\tilde{f}(x,y) = m_f(x,y) + \frac{\sigma_f^2(x,y)}{\sigma_f^2(x,y) + \sigma_n^2(x,y)}(g(x,y) - m_f(x,y)), \tag{7.3}$$

where m_f is the local mean of the signal f, and σ_f^2 is the local signal variance. This filter is similar to the unsharp filtering (Equation 7.1).

The Wiener filter only relates to second-order statistics of the input image. By introducing nonlinearity in the image, some limitations can be overcome. Abramatic and Silverman [15] proposed a filter that is a linear combination of the stationary Wiener filter H and the identity map:

$$H_\alpha = H + (1 - \alpha)(1 - H). \tag{7.4}$$

The modified adaptive filter equals the Wiener filter for $\alpha = 1$, and for $\alpha = 0$ the filter becomes the identity map. From the study of the human visual system, Knutsson et al. introduced an anisotropic component in the model of Abramatic and Silverman [16]:

$$H_{\alpha,\gamma} = H + (1 - \alpha)(\gamma + (1 - \gamma)\cos^2(\varphi - \theta))(1 - H), \tag{7.5}$$

where the parameter Y controls the degree of anisotropy, φ is the angular direction of the filter, and θ defines the orientation of local image structure. The more dominant the local orientation, the smaller the Y value and the more anisotropic the filter. The local direction and the level of anisotropy can be estimated with three oriented Hilbert transform pairs. The weighting function $\cos^2(\varphi - \theta)$ was imposed by its ideal interpolation properties, and the directed anisotropy filter can be implemented as a steerable filter by Freeman and Adelson [9].

7.2.4 Wavelet Domain Methods

Human visual perception occurs at multiple scales; hence, edges of an image can be extracted from Laplacian of Gaussian (LoG) operators, as proposed by Maar-Hildreth [6]. Wavelets are developed in applied mathematics for the analysis of multiscale image structures [17]. Wavelet functions are distinguished from other transformations such as Fourier transform because they not only dissect signals into their component frequencies but also vary the scale at which the component frequencies are analyzed. As a result, wavelets are exceptionally suited for applications such as data compression, noise reduction, and singularity detection in signals. The ability to vary the scale of the function as it addresses different frequencies also makes wavelets better suited to signals with spikes or discontinuities than traditional transformations such as the Fourier transforms. The application of wavelets to medical image enhancement has been extensively studied. We introduce two types of wavelet transforms and the enhancement algorithms based on these transforms. The use of other image transforms such as the Steerable [9] and Gabor filtering [18] transforms can be found elsewhere.

Mathematically, using wavelet transforms, a signal can be decomposed into the low-frequency and high-frequency components at dyadic scales 2^j. A typical orthogonal/bi-orthogonal wavelet transform is shown in Figure 7.1. It uses low-pass and high-pass filters $\{h\}$ and $\{g\}$ [17]. After each level of decomposition, the number of wavelet coefficients becomes half of the previous decomposition. The low-pass and high-pass components of the image are contained in the coefficients {c} and {d}, respectively.

A more general family of wavelets suitable for image enhancement is the non-orthogonal wavelet transforms or frames. Zhong and Mallat [19] proposed a family of non-orthogonal wavelet transforms. The edge information can be extracted from the zero-crossings and/or extrema of the wavelet transforms. These wavelet transforms are translation-invariant and outperform the orthogonal wavelet transforms when reducing ringing effects at the signal edges [20]. A more general family of differential wavelets was proposed [21, 22]. If we define the smoothing and wavelet transform of an image f as $S_{2^j}f$ and $W_{2^j}f$, we can compute the wavelet transforms using a fast algorithm

$$\begin{cases} S_{2^j}f = S_{2^{j-1}}f * h_{\uparrow 2^{j-1}} \\ W_{2^j}f = S_{2^{j-1}}f * g_{\uparrow 2^{j-1}} \end{cases}, 1 \leq j \leq J, \tag{7.6}$$

where $\{h\}$ and $\{g\}$ are the low-pass and high-pass filters, and $\uparrow 2^{j-1}$ is the up-sampling operation by putting $2^{j-1} - 1$ zeroes between two samples in the filter. The values of these filters can be found in Wang [21].

One can manipulate the wavelet transform coefficients to magnify coefficients to enhance the authentic signals while suppressing noise. The modification of the wavelet coefficients results in a nonlinear mapping from the wavelet transform coefficients x to a new value. The hard-thresholding and soft-thresholding functions proposed by Donoho and Johnstone [23] are such nonlinear functions. For example, the soft-thresholding function was given by:

$$\theta(x) = \begin{cases} x - T, & if\ x \geq T \\ x + T, & if\ x \leq -T \\ 0, & if\ |x| \leq T \end{cases}. \tag{7.7}$$

The threshold T is usually chosen to be $T = \sigma\sqrt{2\log_2 N}$, where N is the length of the signal. Coefficients below the threshold T or above –T are shrunk to a nearly zero value. The thresholding Equation 7.7 is usually applied to the orthogonal/bi-orthogonal wavelet transform domain. A translation invariant wavelet transform such as in Equation 7.6 is more favorable [20]; the translation-invariant procedure reduces ripple effects when estimating discontinuous signals. A variety of enhancement schemes [24, 25, 26] were proposed based on the non-orthogonal transforms. The nonlinear mapping function used in these schemes acts as a multiscale unsharp mask.

The differential wavelet transforms (Equation 7.6) facilitate the extraction of edges at multiple scales. Because the edge patterns are correlated spatially, we have used this property to identify edges and subsequently amplify them. We have adopted multiscale point-wise products (MPP) to measure the cross-scale correlation [27]. The MPP is defined:

$$P_K(n) = \prod_{j=1}^{K} W_{2^j}f(n), \tag{7.8}$$

where $\{W_{2^j}f\}$ are the wavelet transforms defined in Equation 7.6. This criterion was used for detection and localization [28], denoising [19], and filtering of magnetic resonance images (MRIs)[29]. In fact, even before the advent of wavelet transform, the MPP had been used to enhance multiscale signal peaks while suppressing noise by exploiting the multiscale correlation of desired signals [30]. Because the maxima of $W_{2^j}f(n)$ tend to propagate across scales because of edges in the signal f(n), whereas the maxima caused by noise does not, $P_K(n)$ reinforces the response of the signal rather than the noise. Analysis of edge patterns indicates that the multiscale product has an inherent ability to suppress isolated and narrow impulses while preserving the edge responses across different scales [27]. A more detailed analysis of the probability distribution function of the MPP can be found in Sadler and Swami [28]. Based on this observation, we have proposed the following nonlinear mapping function using the MPP as a criterion to modify the wavelet coefficients [27]

$$\theta(x) = \begin{cases} \lambda x, & if\ |x| \geq \mu \\ 0, & otherwise \end{cases}, \tag{7.9}$$

where λ is an adjustable constant corresponding to the scale, and its choice can have different degrees of enhancement. The threshold parameter μ can be set by users. Larger values of μ result in a high denoising effect, and vice visa. The choice of μ

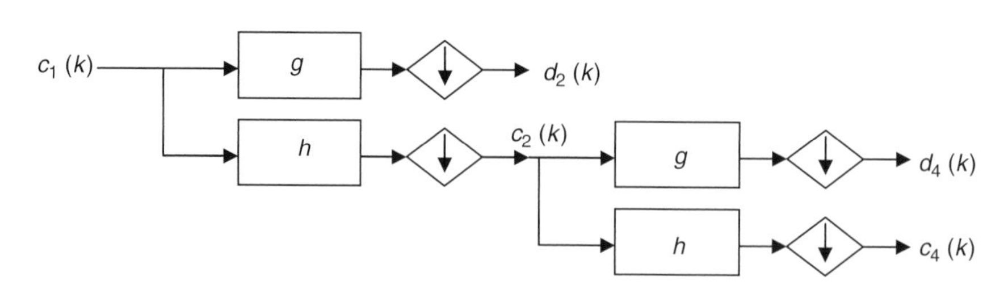

FIGURE 7.1 Multiresolution wavelet pyramid decompositions at two levels.

depends on the noise level in the image, which is similar to the hard thresholding method used in Coifman and Donoho [20].

7.2.5 Evaluation of Image Enhancement

Quantification of contrast enhancement is generally difficult [31]; there is no universal measure for specifying either the objective or the subjective performance of the enhancement algorithm. Contrast is often defined as the difference in mean luminance between an object and its surrounding. There are many measures of contrast. For example, in the definition proposed in Gordon and Rangayyan [32], the local contrast is defined as the difference of the mean values in two rectangular windows centered on a pixel. Specifically, the local contrast $c(x,y)$ is defined as:

$$c(x,y) = \frac{|p - a|}{|p + a|}, \qquad (7.10)$$

where p and a are the average values of gray levels in the center window and surrounding window of the pixel location (x,y), as illustrated in Figure 7.2. It gives the contrast measure c in the range [0,1]. The performance measure, contrast improvement ratio (CIR), is defined as the ratio of the enhanced image and the original image within the region of interest (ROI) R,

$$CIR = \frac{|\sum_{(x,y)\in R} |c(x,y) - \tilde{c}(x,y)|^2}{\sum_{(x,y)\in R} |c(x,y)|^2}, \qquad (7.11)$$

where c and $\tilde{c}$ are the local contrast values of the original and the enhanced images, respectively. Figure 7.2 illustrates the center and surrounding regions. The CIR can be used as an objective criterion to evaluate different enhancement techniques.

We use chromosome enhancement as an example to compare the wavelet approach [27] with several different approaches. These are the adaptive contrast stretch (ACS), the adaptive contrast enhancement (ACE), and the contrast gain transform (CGT) [4]. The CGT and ACE parameters used in the article by Chang and Wu [4] are 25 and 2.0 in the experiments. The parameters λ (Equation 7.9) used in our proposed method for the three scales are 5, 2, and 2, respectively. In addition, we also compare our method with the multiscale contrast enhancement (MCE) approach proposed in Boccignone and Ferraro [33]. Figure 7.3 shows one example of the spread image enhancement using different enhancement methods. One can see that the proposed wavelet method produces the best visualization effect after enhancing the band patterns. The adaptive contrast enhancement (ACE) method and the CGT method, on the other hand, both cause blurring at the edges. Chromosome images are used in routine cytogenetic diagnosis and cancer research. Image enhancement is desired for high-resolution display and visualization of the chromosome band patterns [31]. A set of 21 human chromosome images, including 10 chromosome metaphase spread images and 11 karyotype images, was tested in the experiments. The test results, in terms of the average CIRs measured from the spread and karyotype images, are tabulated in Wang et al. [27]. Among all the methods tested, the wavelet approach consistently yielded the highest CIRs.

A more objective evaluation of an enhancement algorithm is determined by the subsequent application. For example, in the chromosome image enhancement, the ultimate purpose is to

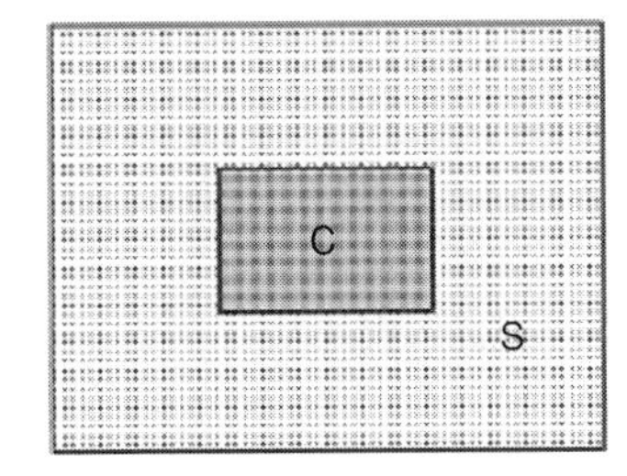

FIGURE 7.2 The local contrast is defined as the measure between the center region and the surrounding region. The size of the center and surrounding window are 3 and 7, respectively.

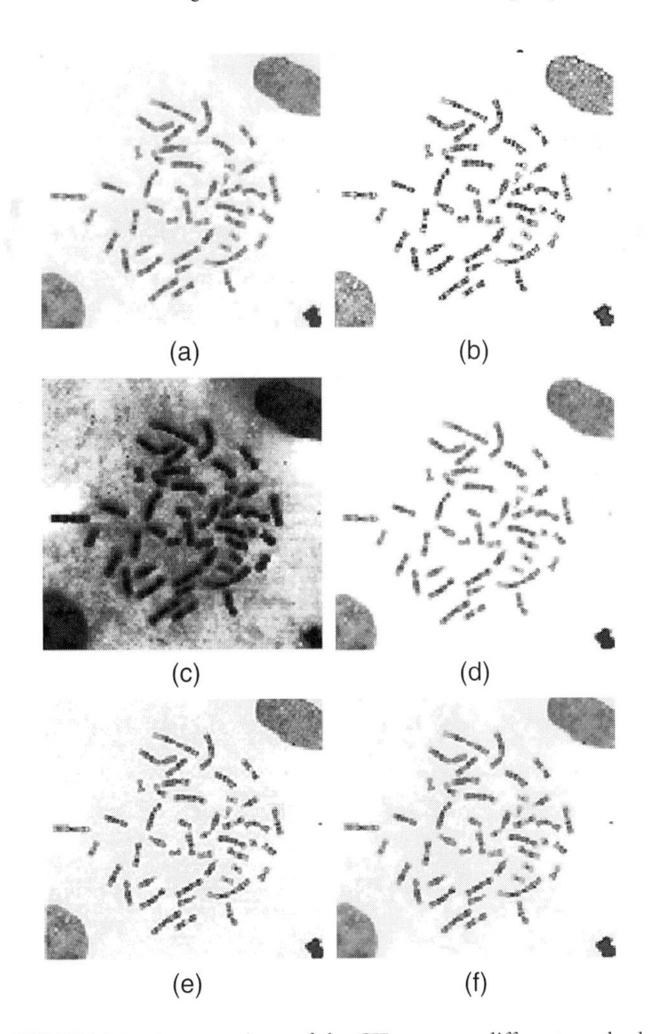

FIGURE 7.3 A comparison of the CIRs among different methods for enhancing chromosome images. (a) Original image, (b) Enhancement using the proposed method, (c) MCE, (d) CS, (e) ACE, (f) CGT.

improve the classification accuracy of the chromosome images. The enhanced image using the wavelet approach results in the highest classification accuracy among the enhancement algorithms that we compared [27].

7.3 Medical Image Segmentation

Note: In the following, references to pixels also include volumetric elements (voxels), with some noted exceptions.

Image segmentation is the process of partitioning an image into sets of pixels corresponding to regions of physiologic interest. For example, Figure 7.4(a) is a single slice from an MRI. Figure 7.4(b) shows a segmentation of the image into brain and nonbrain tissues as a collection of pixels with an anatomic label (red) for brain tissues.

The segmented image can be used to make measurements such as brain volume, to detect abnormalities, or to visualize areas such as the brain surface in Figure 7.4(c). Image segmentation has been an active area of research in image processing and computer vision from the outset of digital imaging. As a result, there is a wide array of segmentation approaches for medical images.

7.3.1 Intensity and Texture

A major class of segmentation methods form regions with similar intensities or textures. The motivation behind this approach is that these regions correspond to different objects or features in the original image. The various methods differ in their exact definition of a region and the way pixels are assigned to those regions. It is difficult to define what a meaningful region is in the general sense, since this is often application dependent. However, a basic definition of a region is as follows:

- Regions contain pixels that are similar with respect to a homogeneity criteria.
- Regions should be topologically simple and void of small holes and non-uniform edges.
- Adjoining regions should be characteristically different as determined by the homogeneity criteria.

Mathematically, this can be considered as partitioning the image space into disjoint sets that satisfy the homogeneity criteria.

$$R_i = \{x \in R^N | T_i < H[f(x)] < T_{i+1}\}, \tag{7.12}$$

where R_i is region number i, x is a vector whose length depends on the image dimensions, and H is the homogeneity criteria applied to the image *f*. Each region corresponds to a different set of values constrained by the threshold functions T_i and T_{i+1}. In general, T can be a complicated function based on prior information. The requirement that the sets be disjoint may be relaxed during the region formation process, but the final result should contain mutually exclusive sets. The homogeneity criteria have been defined in many different ways and are used to measure the degree to which the pixel belongs to the region.

The simplest formulation of the homogeneity criteria is to use the intensity at the pixel location itself or the output of an image processing filter; for example, smoothing or a texture measure. The thresholds, T_i, then correspond to intensity limits, and the regions are groups of pixels with intensities in the given limits. This technique is referred to as *thresholding* and has two basic parameters: the number of regions and the threshold values. If only two regions are considered, a single threshold is used. This rarely describes real scenes, so multiple thresholds are often required, giving rise to multivalued thresholding. The number of regions and thresholds can be determined from *a priori* knowledge of expected image

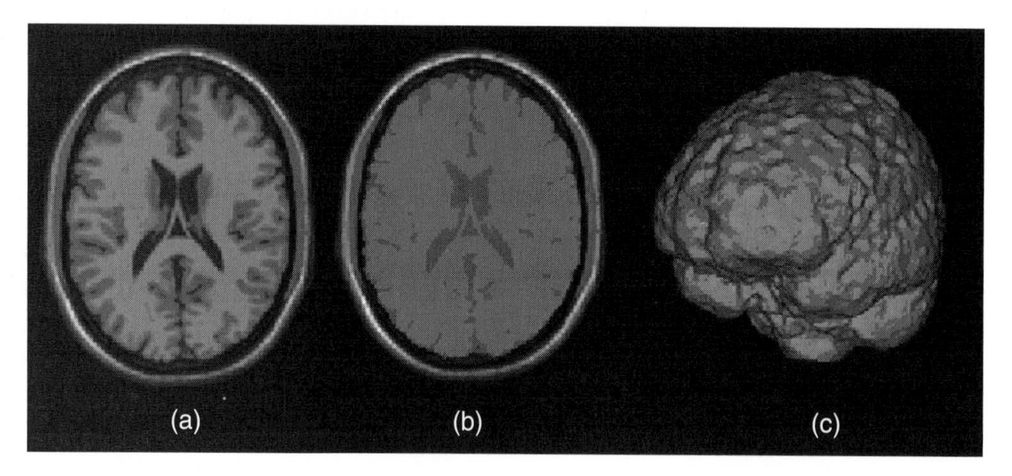

FIGURE 7.4 Example of image segmentation used to identify pixels belonging to brain tissue. (a) Slice from a T2-weighted image. (b) Region identified as brain highlighted in red. (c) Surface rendering of the brain region from the entire image volume. The MR images are from the BrainWeb database (http://www.bic.mni.mcgill.ca/brainweb). For a more detailed view of this figure, please visit our companion site at: http://books.elsevier.com/companions/9780123735836.

intensities or estimated from histogram analysis. The histogram analysis may locate peaks and subdivide the pixel range accordingly or, as is more often the case, model the pixel intensities using probability distributions. The parameters of the distributions can then be estimated from the image histogram and used to form linear discriminate functions that specify the threshold values.

The histogram analysis approach is a subset of a larger class of region-based segmentation algorithms that model the image as a stationary random process and individually classify pixels based on a set of features calculated from the image. These classification or clustering techniques are a subset of tools from pattern recognition and can use a wide variety of information from the image. In the case of classification, the goal is to find the most probable segmentation given the measured image. This is accomplished by maximizing the posterior distribution to obtain the most likely segmentation (the maximum *a posteriori* or MAP estimate). The clustering approaches use measures of intraclass and interclass homogeneity, like the Fisher criteria, to determine the best segmentation. Methods such as hierarchical clustering [34] and nearest neighbor algorithms (c-means or k-means) can also be used to group pixels into regions. When the intensity of a pixel and its neighbors is used, the classification can be considered a region-based technique.

The results obtained by thresholding and classification of individual pixels often contain many small holes or gaps in the regions, especially in noisy situations. Also, the results must be processed to obtain pixels that are spatially isolated using a labeling algorithm. A more focused approach to region segmentation is known as pixel aggregation or region growing. This technique starts with a specified point in the image, called the seed. It then collects all the pixels that are adjacent to the region that meet the homogeneity criteria. Because only a single pixel is added to the region at a time, restrictions as to region size and shape can be made to guide the region formation process. Using this approach, a single connected region with certain topologic characteristics can be extracted without examining the entire image domain. In addition to selecting the threshold values, the seed point must also be located. This is often specified by an operator, but may also be automatically determined in some applications.

Alternatively, the entire image domain can be iteratively subdivided until the combination of pixels in each region meets the homogeneity criteria. The process may also be reversed, combining individual regions, until no more can be merged and each region meets the homogeneity criteria. These two approaches form a hierarchical-directed graph structure. In the case of splitting, the graph has a single root corresponding to the image and terminal nodes that correspond to individual regions. In the case of merging, the graph has a root for each pixel and a smaller set of terminal nodes that correspond to individual regions. The homogeneity criteria used in split and merge methods use either the maximum and minimum of an image characteristic or a statistical test to determine when to stop the split or merge process. As with the case for pixel aggregation, constraints of size and shape can be applied to restrict the region topology.

The region-growing and split-merge segmentation approaches are sequential in nature. They make decisions about inclusion or exclusion from the region based on a previous estimate of the region. A more generalized sequential approach that modifies the decision criteria per iteration is called relaxation [35]. In relaxation, the likelihood of each pixel belonging to a particular class or region is measured using probabilities or fuzzy membership. These individual likelihoods are updated iteratively using the likelihood of neighboring pixels at the previous iteration. Relaxation bears a striking similarity to a statistical approach to segmentation based on Markov random fields (MRF) and is used with several different algorithms including simulated annealing [36], iterated conditional modes [37], and in several medical applications [38].

Another approach to region-based segmentation, the watershed, is derived from the field of mathematical morphology. Using a geographic analogy, consider the gradient of the image as a topological map. The map is gradually flooded, and regions separated by peaks or crests in the gradient are artificially separated by a barrier called the watershed. When the map is completely flooded, the set of points within a watershed forms the individual regions. Whereas this technique does use edge information, the flooding process is much like the region-growing approaches described above. The watersheds are difficult to construct in two and three dimensions, and the technique tends to over-segment the image. These issues have been addressed somewhat by using first-in first-out breadth-first techniques to perform the flooding [39] and morphological postprocessing to correct over segmentation [40].

One of the major difficulties in applying the region-based approaches is the process of parameter estimation. For thresholding, classification, clustering, and MRF approaches, the number of partitions of the features may be difficult to determine. It is often the case that *a priori* knowledge of the number of classes does not produce a useful segmentation, and automatic methods are ad hoc in nature. In general, region-based methods tend to over-segment the images because of blurring and noise. In terms of computational complexity, the region-growing techniques are the fastest region-oriented techniques because they only examine a subset of the total number of pixels in the image. Thresholding, classification, and relaxation approaches have linearly increasing complexity with the size of the image because they must examine each pixel. The MAP and MRF approaches are often computationally expensive because they must extremize a complicated functional.

7.3.2 Edges

To identify an object from a two- or three-dimensional image, you can, as in the previous section, locate the interior of the object itself. Alternately, you can identify only the boundaries of the objects and interpret the interior as taking on values from the original image or some average of those values. This technique, referred to as edge finding or edge detection, is a useful tool for segmentation.

A natural definition of an edge is the location where objects are separated in the image. From this definition of an edge, the process of edge finding can be considered using a differential operator. It is well known that the process of numeric differentiation is ill posed in the sense of Hadamard [41]. Thus, the output of the system does not depend continuously on the data, and noise is amplified. As a result, one of the major aspects of edge detection research has been to regularize the differentiation process.

The first attempts to obtain a discrete differential operator were based on simple finite difference approximations to the continuous derivatives. These simple operators can then be applied to the image to obtain an approximation to each directional derivative. Each directional derivative can then be combined to form the gradient, ∇f. To address the issue of noise, the finite difference operators were combined with simple filters to produce the classic edge detection operators of Sobel and Prewitt [42]. Both of these approaches combine a simple low-pass filter with the finite difference operator to obtain edge filters that are less affected by noise.

A more formal approach to the regularization of the differential operators can be obtained from the theory of inverse systems. Here, the linear image formation model is explicitly used to obtain a least mean square solution to the inverse problem by finding a minimum of a cost. The differential operator smoothes the solution and thus provides the required stability. The smoothing effect of the regularization operator can be obtained either explicitly through filtering or implicitly by approximation. The filtering approach led to the Gaussian regularization filter introduced by Marr [43] and has connections to scale-space approaches [44–47]. The Marr-Hildreth operator combines a filter with the selection of gradient peaks by means of the second derivative. The filter used is the Gaussian, which can be shown to minimize the uncertainty between the spread in the spatial domain and the spread in the frequency domain from infinite impulse response and finite impulse response filters [41]. The Marr-Hildreth operator is then the Laplacian of Gaussian or Mexican hat filter. The approximation approach to regularization led to the Haralick edge operator [48]. This technique first approximates the image locally using a least-squares fit polynomial. The edges are then located using the zero crossings of the second directional derivative of the polynomial, steered in the gradient direction.

The filtering approach to regularization of differential operators can be combined with desirable edge operator characteristics in a variational framework. This idea is normally attributed to John Canny, who proposed a set of criteria that a good edge detector should obey [49]. Canny proposed three principle characteristics an edge detector should have. First, the differential operator should respond maximally to a true edge and minimally to noise. Second, the operator should locate edges with the highest possible precision. Third, the operator should provide a unique estimate of an edge.

The noise used in the Canny model is additive, stationary, white noise with known or estimated statistics. The detection criteria can be considered the maximization of the signal-to-noise ratio (SNR). The optimal operator that maximizes the SNR is a matched filter. The localization criteria can be quantified by the dislocation of the measured edge location from the true edge location. In contrast, Deriche assumed the optimal operator to have an infinite spatial extent, resulting in a simpler form that can be implemented recursively. Spacek used criteria similar to those of Canny but incorporated the expression for the average distance between local maxima directly into the criteria function. A comparison of the performance relative to the three optimal criteria shows that, in general, the Deriche and Spacek operators provide more accurate detection and localization of edges than the Canny operator [41]. In practice, however, the Canny operator can be approximated by a first derivative of Gaussian filter, which is simple conceptually and has an efficient implementation.

The algorithms discussed thus far have been concerned with filters that implement the derivative operation in a stable and well-defined manner. The output of these filters must be examined to explicitly locate the edges, as a chain of points in two dimensions or as a mesh of points in three dimensions. This postprocessing step is called edge linking and is a difficult part of the edge-based approach. The simplest method is to threshold the gradient magnitude image and use morphological operators and labeling to identify edge chains. The main difficulty with this approach is the selection of the threshold, which can produce gaps and is ad hoc in nature. Graph searching and the Hough transform are two examples of methods for bridging the gaps produced by thresholding the gradient response. The graph search techniques use both greedy algorithms and heuristic search to fill gaps, while the Hough transform is used to locate parametrically defined lines and curves.

A more principled approach is to locate zero-crossings of the second derivative, like those produced by the Marr-Hildreth operator. Although this is well-defined numerically, it does not guarantee closed contours and does not handle noise responses well. A set of rules can be established for the selection of edge points, such as those proposed by Canny in conjunction with his optimal operator. These rules include nonmaximal suppression and hysteresis thresholding and have been applied to many other edge operators.

7.3.3 Geometric Models

Deformable models begin with an estimate of the object shape to be segmented and its approximate location in relation to the image. This estimate is then modified iteratively until it matches the local image data using mathematical techniques from variational calculus and optimization. The application of deformable models is appropriate when *a priori* information about the target object's location and shape are known or can be provided manually, often the case in medical image segmentation.

Deformable models are themselves a subset of variational segmentation techniques and implicitly include correlations between pixels (or voxels) by using a geometric representation for the objects to be segmented. The usefulness of the various deformable models stems from the focus on local information, controlled in a global manner by a functional. The geometric variational approaches include the distributed image space (DIS) models and lumped parameter space (LPS) models. Examples of the DIS models are the physically based deformable models [50–53] and the geometric deformable models [54–57]. LPS models include the active shape models [58–60] and the statistical shape models [61]. DIS models attempt to minimize a functional that is only defined on a curve or surface of the image. The application of variational calculus to the functional gives rise to a partial differential equation (PDE) that determines the optimal deformation of the model parameters. In the case of the physically based deformable models, the PDE is a necessary condition for a minimum called the Euler-Lagrange equation and is used to determine the motion of the curve. In the case of the geometric deformable models, the first variation of the functional is used to derive the Hamilton-Jacobi equation, giving the steepest descent direction toward a functional minima. LPS models typically optimize over a much smaller set of parameters than the DIS models and are much more compact representations. Each of the deformable model approaches requires initialization by a contour or surface estimate, either by hand or from some other high level method (i.e., a database of shapes).

The LPS models use compact, parameterized shape models to deform and register the shape to the measured image by minimizing a functional. A model of each shape is required, and care must be taken to ensure that the model can fully capture the expected variation in the object. One example of a shape model is that of Fourier descriptors, a representation of the parameterized shape as a finite combination of basis functions [59]. These basis functions are often sinusoids with the coefficients determined using an optimization strategy. Another LPS approach uses figural shape and statistical models [58] to represent the object and its local deformations. These and related techniques have shown promise in medical image analysis for low-contrast segmentation where variability in the target shape(s) is low.

In general, LPS models cannot change their topology from the initial shape, and they require better description of the shape itself than other deformable models. Thus, they are closely related to object recognition tasks. However, because LPS models rely heavily on the *a priori* knowledge of the shape itself, they are often more robust solutions to contrast-limited problems for which topological flexibility is not necessary.

Distributed image space models are physically based deformable models, sometimes referred to as snakes or active contour models. They generally attempt to minimize a function by combining an internal spline energy of the model, a potential term derived from the intensity and gradient of the image along the contour, and a term including external forces that can be applied to the model given *a priori* knowledge. Computer implementation requires the sampling of the curve into snaxels (snake elements) and iterative solution of the Euler-Lagrange equations [50] or other minimization strategies [62, 63]. The physically based model is also applicable to three-dimensional surfaces, referred to as balloons [64], and to multiscale implementations [52, 65].

The main criticism of physically based deformable models is the difficulty in tracking the parameterized curve or surface, especially in areas where the surface changes topology. These issues have been addressed by adapting the physically based approach (T snakes or surfaces [51]) and an implicit representation of the deformable model. The T snakes and T surfaces of McInerney and Terzopoulos [51] are able to change topology by a re-parameterization of the contour at each iteration using an implicit formulation. The physically based deformable models also require weighting coefficients to adjust the interaction of the forces on the contour. These weights are difficult to determine in a principled manner and are adjusted experimentally.

The geometric deformable models embed the curve or surface in a two-dimensional or three-dimensional space termed the level function and use numeric schemes introduced by Osher and Sethian [66] to compute the Hamilton-Jacobi equation for a length or area preserving functional. The result of minimization of this functional and embedding in a level set is a gradient flow equation that uses an image-dependent speed to control the contour movement. The main advantage of the above geometric approach is that the surface is defined by the level set such that changes in topology are possible without further intervention, a powerful motivation for their use.

The difficulty with the above geometric deformable model is the lack of any physical principle to drive it. The contour moves toward a minimal length curve with an image dependent speed; however, there is no physical interpretation of stopping or selection of the speed term. The addition of image-independent forces [54] adds to the parameters that must be chosen, and while improving noise independence, this reduces the sensitivity to low-contrast and high-curvature boundaries.

The work by Caselles et al. [67] and Kichenassamy et al. [68] unites the physical energy minimization and implicit form for the contour evolution into the geodesic deformable model. Subsequent analysis and application in both two dimensions and three dimensions have shown the approach to be useful for segmentation in medical images.

7.4 Medical Image Feature Extraction

Classification, comparison, or analysis of images is performed almost always in terms of a set of features extracted from the images. Usually this is necessary for one or more of the following reasons [69]:

1. Reduction of dimensionality. An 8-bit-per-pixel image of size 256×256 pixels has $256^{65536} \approx 10^{157826}$ possible realizations. Clearly it is worthwhile to express structure within and similarities between images in ways that depend on fewer, higher-level representations of their pixel values and relationships.
2. Incorporation of cues from human perception. Much is known about the effects of basic stimuli on the visual system. In many situations, moreover, we have considerable insight into how humans analyze images (essential, for example, in the training of radiologists and photo-interpreters). Use of the right kinds of features would allow for the incorporation of that experience into automated analysis.
3. Transcendence of the limits of human perception. Notwithstanding the great facility that humans have in understanding many kinds of images, there are properties (e.g., some textures) of images that we cannot perceive visually but which could be useful in characterizing them. Features can be constructed from various manipulations of the image that make those properties evident.
4. The need for invariance. The meaning and utility of an image often are unchanged when the image itself is perturbed in various ways. Changes in one or more of scale, location, brightness, and orientation, for example, and the presence of noise, artifact, and intrinsic variation are image alterations to which well-designed features (depending on the application) are wholly or partially invariant. Many of the examples of features presented below exhibit invariance in at least one of those ways.

Features can be based on individual pixels (e.g., the number having an intensity greater than x; the distance between two points), on areas or volumes (the detection of regions having specific shapes), on time (the flow in a vessel, the change in an image since the last examination), and on transformations (wavelet, Fourier, and many others) of the original data. The assumption made here is that feature extraction is automated, and we describe only those features that can be computed without user interaction. That process generally is easier if the image has been segmented; that is, divided into regions each of which is internally homogeneous and different from its neighbors. Segmentation is discussed elsewhere in this chapter and is sometimes alternated with feature extraction.

7.4.1 Feature Extraction Across Space, Time, and Frequency

The distribution of gray levels (intensity values) or of color levels (e.g., of red, green, and blue) can be of great value in describing an image. For example, if each picture element (pixel) could take on one of only two widely separated values, the image would have high contrast, and essentially no variation in tone or shade would be evident. At the other extreme, if all intensities were represented equally often among the pixels, the image could look washed out, having little contrast. A way to describe the distribution of levels is the histogram, a plot of the relative frequency of occurrence of each gray level (or color intensity). In the cases mentioned above, the first histogram would have two spikes, and the second would be flat: a horizontal line.

Often it is useful, therefore, to characterize the histogram's shape [70, 71]. Some of the commonly used descriptors are the mean and the mode (location parameters); the central moments (e.g., variance, skewness, and kurtosis), which describe rough shape; energy (sum of squares of the intensity values, emphasizing the larger values); and entropy (a measure of nonuniformity).

Regions correspond to areas that are homogeneous in some characteristic(s). The shape of a subimage may be described in terms of its boundary (contour-based) and/or its interior (region-based).

Descriptors of contour include the following: (1) The chain code [71], which uses a connected sequence of straight-line segments, all of a specified length, oriented at angles that are multiples of 45 degrees. The code number indicates the line's orientation, and a sequence of numbers corresponds to the sequence of edges bounding a region. (2) A set of values of the curvature of the surface at each point [72]. (3) Radial edge-gradient analysis [73] (e.g., for describing spiculations in mammography.(4) A signature, a one-dimensional representation of the boundary. Some property of the boundary (e.g., distance from the centroid of the region) is plotted as a function of angle or arc length. (5) Segments obtained from decomposition of the boundary; this may be based, for example, on concavities in the boundary. These can be detected by finding the convex hull [71] (the smallest convex region that contains the original region where A convex region is one in which any two points can be connected by a line that lies entirely within the region) and noting those subregions where the original boundary departs from the convex hull. (6) If the coordinates of points on a digitized boundary are taken as the components of complex numbers, then the sequence of those numbers will describe a closed path in a complex plane. The Fourier transform of that sequence yields a set of coefficients, of which a subset can be used in an inverse Fourier transform to reconstruct the contour approximately. The subset of coefficients can serve as useful features.

Useful descriptors of shape that are derived from the entire object include the following: (1) Effective diameter (the diameter of circle that has the same area as the region); it equals 2 $(A/\pi)^{1/2}$, where A is the area. (2) Circularity (a circle has value 1); it equals $4\pi A/P^2$, where P is the perimeter. (3) Compactness (minimum for a circle): P^2/A. (4) Projections define the cumulative intensity of a region as measured along a set of parallel rays cast through the region in any direction. Most commonly the projections are in the horizontal and vertical directions. If a region has pixel value $f(x,y)$ at location (x,y), the horizontal and vertical projections p_h are, respectively:

$$p_h(i) = \sum_{j=1}^{n} f(i,j) \text{ and } p_h(j) = \sum_{i=1}^{m} f(i,j), \qquad (7.13)$$

where m and n are the overall sizes (in pixels) of the image or area of interest. These measures can be useful for measuring the height and width of an object (zero values of the projections indicate the end of a region) and also the homogeneity of a region (inferred from the variation in the projection values across their extent). (5) The skeleton of a region (also called the medial axis) can be taken as the set of points maximally distant from the boundary, as the locus of centers of maximal disks inside the region. It thus describes the orientation and approximate curvature of the region as a whole and can indicate the presence of branches in the object. Numerous algorithms exist [70] for the calculation of skeletons from gray-scale and binary images. (6) Topological descriptors include the number C of connected components (a connected component is one in which all member pixels have a path to all other member pixels; the path's pixels must lie in the component; connectivity of the path is defined as being either 4- or 8-adjacent), the number H of holes (connected nonregion pixels lying within the region), and the Euler number C-H.

Texture does not have a formal definition, though usually it is taken to refer to regularities, smoothness, and roughness in images. The perception of texture depends on scale and sometimes orientation, so a description must allow the explicit incorporation of those factors.

One way to describe relationships among pixels is to choose a relationship and examine the image to determine the ways in which the relationship appears. Let P be a relationship operator, and let $\mathbf{A}$ be an $L \times L$ matrix, where L is the number of possible gray levels. The operator P can be viewed as a displacement vector $P = (\delta x, \delta y)$ that specifies the direction and spacing from a given pixel to another. Each element a_{ij} of $\mathbf{A}$ contains the count of the number of times that such a pair of pixels occurs related in space by P and having, respectively, gray levels b_i and b_j.

Let n be the number of point pairs in the image that satisfy P. If a matrix $\mathbf{C}$ is normalized by dividing every entry of $\mathbf{A}$ by n, then C_{ij} is an estimate of the joint probability that a pair of points satisfying P will have values (b_i,b_j). The matrix thus defined will in general not be symmetric because of the directionality of the relationship between the pixels. The matrix $\mathbf{C}$ is called a gray-level co-occurrence matrix [70, 72].

An understanding of the properties of $\mathbf{C}$ may be developed by considering possible values of P. If the texture in the image is coarse—that is, δ is smaller than the texture element's dimension—then pixels separated by δ will have similar gray levels, and there will be many counts along the main diagonal of the matrix. Conversely, fine variations within the image (δ comparable to texture-element size) will appear in $\mathbf{C}$ as substantial numbers of counts located far from the diagonal, making the overall matrix more uniform.

In practice, $\mathbf{C}$ (or $\mathbf{A}$) is computed for several values of P. One way to define the relationship more formally is to specify the angle φ and distance d from the first to the second pixel [70]. Using $\mathbf{C}_{\varphi,d}(i,j)$ or $\mathbf{A}_{\varphi,d}(i,j)$ to denote the entries in the matrix for gray levels i and j, we can extract several features from the co-occurrence matrix that will give insight into the textural nature of the image [70, 74]. Because $\mathbf{C}$ is a histogram, some of the features used for it are applicable here also.

Sequences of two-dimensional images arise in many applications. A set of CT or MRI slices, for example, may be viewed as a sequence. A pixel in a given slice could be compared to the corresponding pixel in the next slice, and so on, yielding a one-dimensional plot of intensity versus slice number for each pixel. The variation in size of the slices means that some pixels will not have a neighbor in the adjacent slice, and thus the lengths of the one-dimensional plots will differ in general. The sequence could occur over time rather than space. Examples of this include a series of images of a beating heart and images of drug uptake in the brain. Those slices in time often can be treated in the same way as the CT and MRI slices.

Pixels that are adjacent in a given slice are likely to have similar intensities (because they are likely to be describing the same tissue type or phenomenon), and this similarity is likely also to hold across time. The extent of that persistence in space and/or time can be measured in several ways. Correlation is a well-known process by which one signal is shifted relative to the other, and at each value of shift we compute the sum of the products of corresponding points' values. When the two signals are the same, we have the autocorrelation, and its value as a function of shift is a measure of the amount of information retained in later values of the signal. It is a measure of persistence of the structure of the signal and can be computed for each of the one-dimensional plots defined above, as a function of shift in time or space (slice). The correlation length is a feature that the user can define as a measure of the maximum shift that can occur before the autocorrelation drops below a given value.

Equally, correlation can be calculated within a slice. In two dimensions, an image or a section of it is shifted relative to another image (or itself), and the sum of products is computed. This can provide a measure of spatial homogeneity.

Autocorrelation is one example of a tool from time series analysis [75]. Probability models for time series provide a richer set of tools that help describe structure. Autoregressive processes use past values of a signal to estimate the next value; the coefficients of such regressions may be useful descriptors and allow comparison of one-dimensional functions among adjacent pixels.

The more general analysis of the multidimensional problem (two-dimensional or three-dimensional versus time) is addressed by hyperspectral imaging [76–78] for which a variety of tools and methods exist.

Images usually are represented in the two-dimensional space domain, which displays intensity as a function of position in (x,y). The Fourier transform provides an additional representation in the two-dimensional spatial frequency domain. The two axes represent spatial frequency (in units of cycles per unit length or angle), as manifest in orthogonal directions in the space domain, the (u,v) space. A regular pattern of bars that are parallel to the x-axis and that alternate between black and white along the y-axis, for example, will have a zero-frequency component in u (because there is no variation) and a single value of frequency in v (because of the regularity).

As an object rotates in (x,y), its Fourier transform also rotates about the origin in (u,v). This means that properly extracted features from (u,v) will be invariant to rotation in (x,y), which usually is a desirable characteristic. Examples include rings and disks that span regions of spatial frequency. Their circular symmetry ensures that an object's representation (its measure of power in a given spatial frequency range, which can for example, measure texture coarseness and scale) is insensitive to rotation [71]. Conversely, power measured within a given radial wedge can be used to detect orientation.

One application of Fourier-related features arises in efforts to model human perception. As described elsewhere in this chapter, the human observer's detection and recognition tasks are difficult for the observer to describe directly. Measurements of observer eye movement [79] during the performance of those tasks, however, provide some insight into the mechanisms of detection of important structures in an image.

A perceptually correlated metric has been demonstrated [80] that quantifies the conspicuity of local, low-level (or bottom-up) visual cues and identifies those spatial frequencies that are most distinct and perhaps most relied upon by radiologists for decision-making. The goal is to provide a holistic, or top-down, measure of conspicuity that would quantify the variability in perceptual pop-out that occurs as a result of case differences (anatomical background), lesion size, location, or signal-to-noise, for example. A measure of feature conspicuity derived from visually inspired filters has been used to show [81] that ROIs obtained from eye-position data exhibited statistically significant differences in the conspicuity associated with different class types (true positive and true negative), reader experience levels, and cases. The Gabor filter examines ROIs and extracts information from each at a variety of spatial frequencies and orientations; a new salience measure [80–82] combines those Gabor responses into a single value. The features contained in each pass band and the average salience measure led to a measure that is useful for lesion detection and image categorization [81].

7.4.2 Characterization and Selection of Features

Features are extracted and selected on the basis of one or more criteria, which may overlap and sometimes conflict. Those criteria include the ability of a set of features to group samples of a similar type (clustering), to separate samples of different types (classification), to convey intrinsic information about the samples (description), to capture properties that humans find important for a task (e.g., salience), and to remain useful in the presence of noise, changes in scale and orientation, and measurement error (invariance).

Minimizing the number of features used in a given classification or clustering problem generally is desirable because, when constructed based on a limited amount of design (training) data, classifiers that use a large number of features will not perform well on new data [83]. This curse of dimensionality leads to the need for better features and for ways to select the best subsets of candidate features. Both of those subjects have been addressed extensively in the literature [83] and will not be discussed here; they should, however, be considered carefully in the design of any practical system.

7.4.3 Applications of Feature Extraction

Much of current automation in medical imaging aims to produce a system that can automatically assign the correct label (e.g., normal or abnormal) to a given image; this is the classification problem. The goal is to achieve both a high true-positive rate and a low false-positive rate. There is inevitably a trade-off between the two; this is made explicit by the receiver operating characteristic (ROC) curve, described elsewhere [84], and often summarized by the single statistic A_z, the area under the curve, which we seek to maximize. One way to measure classifier performance is with A_z.

Classifier design usually requires a set of labeled training data, from which we extract features that we evaluate and select as indicated above. A separate set of data (the testing set) then is used to assess the overall effectiveness of the combination of classifier and feature set, often using the ROC area as the criterion.

In some circumstances, the available data samples do not have class labels. Analysis in this unsupervised case is called clustering, and there are several reasons that this situation can be of interest [83]. They include the following: (a) One could design a classifier with large amounts of (presumably less-expensive) unlabeled data, and then use labeled data to fine-tune the classifier; (b) slowly changing characteristics of the data (e.g., as in gradual appearance of disease, or in time- or geography-based epidemiology studies) may be identified and tracked by a classifier operating in the unsupervised mode; (c) candidate features can be evaluated for their ability to induce natural clusters in the data.

The interaction between feature selection and clustering can provide insights to the structure of the data and offers the opportunity to incorporate whatever side information we may have about the problem, such as the number of classes, the prior probabilities of occurrence of various classes, or the probabilistic nature of the features for each of the hypothesized or actual classes.

Instead of attempting to deal explicitly with each of the very large number of possible images that can be created by varying their gray-scale structures, we can use features to capture the essential information for a given task that is contained in a given image or a set of images. In that case, when a set of n features is extracted from a given image, it then is represented as a point in n-dimensional feature space, and well-separated points in that space would then make evident the differences among the contents of the set of images. The goal is for images that are similar (as defined by the task) to cluster in feature space. Here we also see the interaction with clustering methods.

Queries to an image storage and retrieval system can be formulated explicitly in terms of features, or as query-by-example. If the user knows the characteristics of the desired images in terms of the extracted features (size, shape descriptors, etc.), then it is straightforward to search the database for images that match either exactly or approximately. In the second case, the user submits an image for which a match is sought. Features are extracted from that query image and used as search terms in the database. Clearly, one can iterate on this process, by selectively deleting features from the set of search terms.

Feature extraction in medical images is part science and part art. It is very much task-specific. Descriptions by the expert of the diagnostic process provide the scientist with starting points for the definition of candidate features to be used then for rigorous mathematical formulation and analysis. The set of candidates, and the subsequent elimination of some of them, will be driven in part by the next step (usually classification or clustering), and the task-specificity thus also enters at this point. The user must define carefully the nature of the source images and what conclusions are expected at the end of the process. That will yield a principled basis for feature definition and selection.

7.5 Medical Image Interpretation

This section explains how radiologists work when they look at and interpret a medical image. Radiologists work, to some degree, in image space rather than verbal space. They, therefore, may have great difficulty explaining what they do, how they process the data in the image to create a diagnosis, and why they are so certain (if they are) that they are correct. The data in an image are ambiguous, and the radiologist has innate skills or has learned the skill of assembling ambiguous information into patterns and those patterns into diagnoses. To some degree, identifying these findings depends on gestalt processes, but how these processes work within the mind is poorly understood. Gestalt provides names for what occurs but not explanations.

Radiologists are inconsistent in what they identify on an image, disagreeing with themselves and with other radiologists with some medium frequency when the image has difficult-to-interpret findings. When you discuss the findings on an image with a radiologist, he or she may have difficulty explaining the thought processes that led to the identification of disease and the selection of a diagnosis. Diagnoses based on images vary in their certainty, and the degree of uncertainty may be included in the radiologist's report on an image. For research purposes, knowing ground truth is important—one would want to train and test one's algorithm on the most definite cases, but only a limited number of definite cases are available. Knowledge of how certain the diagnosis is should be conveyed along with the algorithms that have been developed or tested on cases.

7.5.1 How Radiologists Work

Radiologists are diverse in their training and skills and use somewhat different methods for image interpretation. Some are more skillful in subconsciously detecting unusual findings on images and do so very rapidly and consistently; others reach equal subconscious skill levels by carefully defined search methods that they have trained themselves to follow. In general, radiologists will have difficulty describing what they do when they are interpreting an image. Often, they will point to the abnormality but cannot describe how they found it—it is just so obvious to them that they cannot tell you what they did, or if they missed a finding, why they missed it. The following is a systematic description of what radiologists appear to do. This is based on findings learned through systematic but still incomplete investigations. Remember, however, that most radiologists will not understand the processes they follow or how the sections that follow relate to the things they actually do as they interpret medical images.

7.5.2 Search, Detection, Description, Diagnosis

There appear to be four processes in image interpretation: search, detection/rejection, description, and diagnosis. Search is the nonsystematic or systematic visual review of the image during which attention is focused on specific areas for more intensive review. Detection/rejection is the process of evaluating each of these areas of attention to decide whether they are indeed abnormalities requiring further analysis or are spurious findings that can be rejected. Description is the processes of defining an area of detection that may or may not involve the use of words. Diagnosis is assigning a category to the process detected: normal or abnormal. If abnormal, what is it likely to represent.

There are three methods commonly used for search. Search can represent a quick review of the image looking for areas that are most conspicuous. Search can represent a systematic search attempting to review all areas of the image. Search can be based on the expectation of finding features that point toward a single suspected disease or several suspected diseases. For example, Figure 7.5 shows a mammogram in which there is an abnormality, a mass. The mass is of high contrast and is quite conspicuous. It stands out from the background. Figure 7.6 shows a mammogram in which there are microcalcifications. Although each of these calcifications is of high contrast, they are small features. They can be found by a systematic search that surveys the entire breast tissue image, for example, by scanning up and down over the entire image, often with the use of magnification. If one is looking at a mammogram, one knows that there are only a few features of cancer. These include a mass, microcalcifications, and architectural distortion. Thus, one searches the mammogram in expectation that if one or more of these findings is present, cancer may be present. Thus, the expectation of what might be found guides the search. There is a risk that a computer program for disease detection may do quite well detecting conspicuous features but would fail with more subtle findings. Sometimes, the most obvious finding on a mammogram will represent scar, whereas the cancer is quite hard to see. In addition, if the computer program is to help radiologists with detection, it is the subtle findings that they need help with.

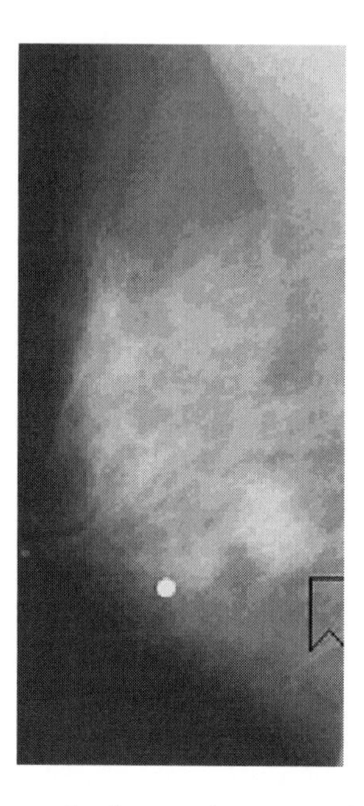

FIGURE 7.5 An example of a conspicuous mass in the lower portion of a woman's breast (arrow). The small, round, white object is a metal marker indicating that the mass could be felt during breast palpation.

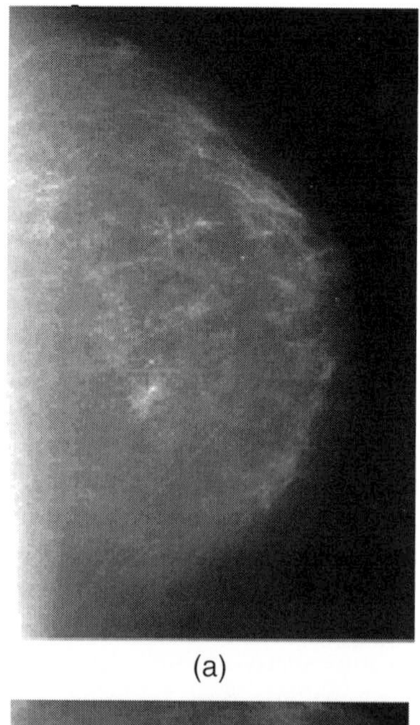

(a)

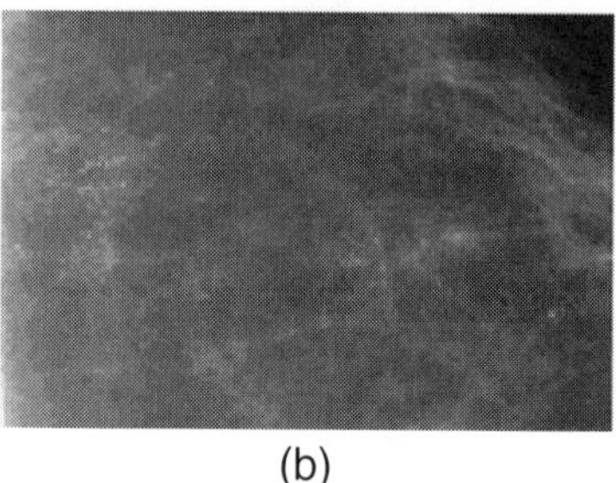

(b)

FIGURE 7.6 (a) This mammogram demonstrates microcalcifications of a type that indicate malignancy. On the view of a large portion of the breast, the microcalcifications are not well seen. (b) This magnified view of a portion of the same breast shows the microcalcifications and white dots and short white lines.

Sometimes search involves only a single image; sometimes it involves a comparison of images obtained with the same modality (several mammograms, several CTs). Other times, the comparison is made of images of several imaging modalities (comparing a CT to an MRI, for example).

Normal breast patterns tend to be symmetrical, so when something is observed on one view of a mammogram that could be normal or abnormal, it is common for a radiologist to compare the finding to the opposite side. In evaluating the breasts, symmetry is good, asymmetry can point to disease. Sometimes, the only sign of cancer on a mammogram is the finding of more tissue in one breast than the other. Each breast appears normal, but it is the asymmetry that indicates that one is abnormal. Many parts of the body are right–left symmetrical.

When two images are obtained at different times, the search pattern may be directed by knowledge of what is seen on the other image. On a chest radiograph that appears initially normal, one may sometimes see an area of pneumonia on the prior film. This then directs the radiologist's attention to the same area of the current film to see whether any pneumonia can still be seen.

When a small area of abnormal attenuation is shown on a liver CT, an MRI may be performed to aid in diagnosis. The radiologist, guided by the liver CT, will look especially at the area on the MRI that represents the same area shown to be abnormal on the CT.

When the radiologist has data available that are not image data, that may guide him or her to look for certain patterns of disease. Thus, if the radiologist knows that the patient has smoked cigarettes for many years, the radiologist may search more fully for signs of lung cancer and emphysema. To some extent, the radiologist will see what he or she expects to see. This has good and bad features. To some degree, it enhances the likelihood that disease will be detected and the correct diagnosis will be made. It has the detrimental feature, however, of what is called assignment or attribution error in diagnosis. If a finding on the image is consistent with the expected diagnosis, the radiologist is more likely to assign a finding as caused by that expected diagnosis, when it may actually have another cause.

Once the radiologist identifies an area of interest, he or she applies a critical analysis of this, either accepting it as an area that requires further attention or as an area of normality that can be dismissed. The initial process of discrimination that occurs is subconscious, and thus its exact mechanism is not known. It is presumed that it is related to an image memory bank that each radiologist develops that represents the patterns of normality and the patterns of disease. Many such analyses occur when an image is viewed, and the process is extremely fast. The usual viewing method often starts with a global overview followed by a very quick review of many small areas on the image. Only a few of these areas of attention reach the level of conscious evaluation. If they reach the level of conscious evaluation, then the radiologist makes a conscious effort to separate the image pattern into features that may represent disease or normality. To some extent, this is also related to that individual radiologist's stored memories of normal and abnormal patterns, and to some extent, it is an extrapolation from these known patterns looking for similarities. Radiologists quickly detect what doesn't fit the pattern of normal.

Knowledge of normal patterns is quite important to the radiologist because the variation of normal is quite wide. When a radiologist wants to learn how to interpret a new type of imaging, he or she can learn extensively from looking at normal cases. Once the range of normal appearances is learned and is active at the subconscious level, then variations from the various normal appearances can be rapidly detected.

Another major component of this discrimination is a series of image interpretation patterns called *gestalt*. To varying degrees, all image patterns are somewhat ambiguous rather than definitive. Gestalt (German, for the way things have been put together, their shape or form) is a group of poorly understood processes by which the brain processes image data to form the shapes that result in interpretations. To a large extent these are subconscious, but decisions based on them can rise to the conscious level. The four main gestalt patterns are emergence, closure, multistability, and invariance. These are basic patterns of seeing that affect medical image interpretation [85–88].

Emergence occurs when the structure of an object in an image is not well defined. The brain can insert the margins of the object so that the brain can interpret it. An alternative interpretation is grouping. The brain can group ambiguous images features to create an object that can be interpreted and described.

Closure occurs when the actual visible object has incomplete margins. The brain can complete the margins so that the object can be recognized.

Multistability allows one to interpret an incompletely defined object in different ways—thus, an ambiguous object could be one of two structures and can be alternately interpreted as either of these objects. This is commonly used with ambiguous structures when the radiologist reassembles an object in different ways to determine whether it is real and/or what it might represent. Radiologists can rapidly change groupings of ambiguous data to form different potential combinations to form an image, selecting those that are possible findings of disease and excluding those that are not likely.

Invariance is the ability to recognize an object when it is seen from a different perspective. One can learn to recognize the shape of an object and then continue to recognize it when it is rotated, changed in size, or warped in shape. Certain types of images are obtained as stacks of consecutive adjacent slices. Radiologists viewing these can quickly understand the relationship of one slice to the next and can visualize structural relationships across images.

Recognition that radiologists use gestalt in the analysis of images is important. Radiologists do this completion very quickly. Currently, computer image pattern recognition programs have difficulty with gestalt tasks.

Invariance recognition can be programmed to some extent, but invariance incorporating warping of the shape of a structure can be quite difficult for a computer. Emergence, closure, and multistability represent difficult problems for the computer to solve and are probably only partially solved with current technology.

Radiologists seldom describe the findings on images in detail. The long-standing trend has been for radiologists to switch from extensive description to diagnosis; the same change has occurred in pathology. Years ago, radiologists were trained in how to describe images by having one trainee describe in words the findings on a radiograph to another trainee who was not allowed to look at the image. The second trainee would then have to diagnosis the patient from the verbal description. This is no longer done.

Engineers developing computer analysis systems for medical images often request descriptions of findings, reflecting a desire to translate into computer code the observations of radiologists; these descriptions by radiologists are often brief and incomplete. To obtain a description that helps an engineer, it may be necessary for the engineer to ask specific questions of the radiologist, reflecting what it is that the engineer can do, rather than relying on what the radiologist describes.

Radiologists may request from the engineer things that cannot be achieved by engineering, or even if they can, the radiologists may not recognize the amount of work that the engineer must do to create what seems to the radiologist a simple thing. A thoughtful radiologist may know what is needed to make his or her work easier and be able to prioritize needs. The engineer can understand what is easy and what is hard to achieve. In a dialogue, a decision can often be arrived at that combines what is most important with what is most easily achievable. Radiologists often want everything all at once and will usually not accept that that which is so easy for them to visualize may be difficult to achieve with current computer methods.

Radiologists vary in the way in which they formulate diagnoses. There are three main patterns:

1. Providing a diagnosis that is really an abstract description. For example, saying that there is an infiltrate in the lungs. This means the lungs are abnormal, but it does not classify the abnormality into a specific disease. Infiltrates can be caused by pneumonia, heart failure, pulmonary emboli, and a group of other diseases, so a diagnosis of infiltrate is really a description. Similarly, a mass can be caused by cancer, infection, inflammation without infection, and several other diseases.
2. Providing a diagnosis with indication of uncertainty. For example, there is an infiltrate likely caused by pneumonia. To some extent, patterns can indicate that a process is most likely caused by a certain disease.
3. Providing a diagnosis, for example, pneumonia, and not indicating the uncertainty of that diagnosis. This is done on the assumption that the physician who reads the report knows that there is uncertainty in many radiology diagnoses and wants to know what is most likely, not several or all the possibilities. If the radiologist's diagnosis does not fit clinically, then the other physician may ask for additional possibilities. That is, when what is probable, based on the image, appears to be wrong, the physician will start to think of other possible diagnoses.

In general, radiologists are trained to view the image without knowledge of the patient's symptoms or history. This is to help the radiologist avoid assignment or attribution error (see above). If one is more likely to see what one expects to see, then foreknowledge of the symptoms or suspected diagnosis may result in the radiologist underestimating the importance of findings that do not agree with the expected diagnosis or that fail to explain the symptoms. This is part of the problem of satisfaction of search, a process that is described below.

Once the radiologist considers that everything on the image has been seen, and often after having decided on a temporary diagnosis, he or she will then look at the clinical information and prior reports to see whether this information fits the initial radiology diagnosis. Whether it does or not, the radiologist will then usually look again at the image to see whether there are any findings that could further explain the clinical concern or contradict the clinical impression. [89] The radiologist's role in this case is to decide whether findings on the image explain the clinical findings, support the diagnosis, or indicate that another diagnosis is more likely.

Every so often, the radiologist will appear to rapidly claim that a patient has an unexpected diagnosis, often with a speed that startles the observer who is not a radiologist. Radiologists can have incredible image memory skills, and, if the pattern is unique for one rare disease, they may recognize it in an instant. In the radiologist's mind, the correct diagnosis leaps out of without thought or censorship.

In other cases, the radiologist will rapidly take what appears to be unfocused information about a patient and, combined with the image, rapidly assemble these items into an unexpected diagnosis.

In the first case, the radiologist, if asked, will often say either "it is this" or "I just remembered it." The radiologist will usually not understand how the surprising diagnosis was made.

Radiologists are thinking in image space rather than verbal space. For many, it is quite difficult to explain in words how they reached their conclusion, though they have learned to describe in words what they have found.

7.5.3 Transforming Data to Information

Radiologists face certain problems when they interpret images. These problems result in inconsistency of diagnosis [90–92]. A radiologist can disagree with himself or herself when viewing an image a second time. For difficult cases, the rate of change is approximately 20%. Two radiologists can disagree with each other. For difficult cases, the rate is approximately 30%. The radiologist looking at an image may find something that is obvious and explains the patient's symptoms and stop looking. This is called satisfaction of search. Many images that radiologists view are ambiguous. The radiologist must assemble the finding out of ambiguous details. There is often diagnostic ambiguity: The same findings can have more than one reasonable diagnosis and can have a few rare diagnoses as well.

Intraobserver variability is a well-recognized problem that is not fully understood. If you provide a radiologist with the same image twice and ask for a diagnosis, the radiologist may or may not provide the same description and diagnosis. When a case is easy, such as a major fracture of a large bone, the radiologist most likely will see the same finding twice, but the more ambiguous the finding on the radiograph, the greater the chance that the diagnosis will change. Part of this is the result of the variability of identifying things through gestalt processes. Poorly defined or incomplete structures may not always be interpreted the same—the gestalt may be different at different times. The structure of many poorly defined objects has to emerge from the ambiguity of the actual densities in the image, and at different times, different things may emerge. When radiologists are interpreting difficult images, intraobserver variability can be in the range of 20%, meaning that 10% of abnormal findings previously seen will not be seen the second time, and 10% of those things not seen previously can emerge.

In the same way as with intraobserver variability, if you show two or more radiologists the same image, they may have different interpretations. The difference between the interpretations of two or more radiologists is called interobserver variability. If the findings are obvious to the radiologists, they are likely to agree, but if the findings are ambiguous or difficult to see, they may disagree with each other in as much as 30% of cases. As with intraobserver variability, sometimes one radiologist will recognize one group of findings, and the other radiologist will recognize a different, but overlapping, group of findings. Fortunately for radiologists and patients, most things on images are relatively obvious to a radiologist, and so the intraobserver and interobserver variability, overall, is much less [92].

For the engineer working to create a computer program for image analysis, this intraobserver and interobserver variability creates problems in knowing what ground truth to use. If all the radiologists who look at an image come to the same diagnosis, the case is likely too simple to benefit from a computer program to help with detection. If the case is difficult enough so that a radiologist will disagree with herself or himself or that two radiologists will disagree with each other, it becomes difficult to know what ground truth to use. On the other hand, it is precisely in these cases where radiologists have difficulty and where a computer detection and diagnosis program could be of the greatest help to radiologists.

The task of a radiologist is to review images and come to a final impression or diagnosis as quickly as possible. In doing this, there is a risk that the radiologist may see a finding that, for that person, mentally completes the task so that he or she can move on to the next case. Deciding that enough has been seen means that the search has been satisfied. Radiologists satisfy their search on every image they view; at some point they stop looking and move on to the next case. The problem of satisfaction of search is that, sometimes, there is something else on the image that the radiologist should see. The missed finding is then ascribed as having been missed because of satisfaction of search [93–95].

Each radiologist decides on each image when the search has been satisfied. There are no rules or criteria to know when enough time and effort have been spent; rather, if the radiologist has missed something, it may be because the search was not satisfied or because the ambiguity in the image was assembled differently and the finding was not seen and would not be seen even with more time spent. Indeed, some studies have shown that radiologists who spend too much time on an image may talk themselves out of a finding that is real and important [96, 97].

Many medical images are somewhat ambiguous in their findings. Perhaps the most common ambiguity is edge ambiguity. The easiest way to demonstrate this is to have several radiologists outline the same finding. There will almost always be some differences in the edges drawn. The second common form of ambiguity is of inclusion and exclusion. When a lesion is seen, there is often some heterogeneity of the finding. When a radiologist views this, some parts may appear to be appropriate for inclusion (they are part of the finding) and others appropriate for exclusion (they are not part of the finding). This image ambiguity can result in different radiologists drawing different margins or providing different sizes of an object. A third ambiguity is contrast ambiguity. When a finding is of low contrast, one radiologist may identify the finding as present and another may decide that there is nothing there. This problem is greater on images with more image noise and decreased on images with lesser amounts of noise.

There are several thousand medical diagnoses, and there are books for coding most of them. Most patients, however, have common diseases. The findings on an image that point toward a diagnosis are often not specific to that diagnosis but can occur in several different diseases. In general, radiologists will either provide a descriptive diagnosis (as with infiltrate, as discussed above) or, if they want to list a specific disease, will list the two or three that best correspond to the image findings and other things they know about the patient. A disease diagnosis coming solely from an image should be considered a probability statement. With a fracture of the femur, the diagnosis is almost always correct. With a mass in the kidney, there can be a high statistical probability of diagnosis of mass, but not 100% certainty from the image alone as to whether that mass is cancer. With lung infiltrates, the degree of diagnostic uncertainty increases. The engineer designing a diagnostic program has to consider the degree of uncertainty of diagnosis of the images used to train the system and in any claims made for its accuracy.

7.5.4 Postdiagnostic Recommendations

Most radiologists at the end of their official report of an image or series of images will provide either a descriptive or disease diagnosis. In general, if the radiologist considers the diagnosis likely, he or she will not indicate any uncertainty in the wording of the report. It will be assumed that the patient's primary clinician understands that there is uncertainty of diagnosis based on image findings alone. In some cases, the report will contain some estimate of the likelihood of the diagnosis or likelihood of an alternative diagnosis. Words commonly used to indicate uncertainty are rare, possible, likely, or probable. Although these terms do not have precise usages by radiologists, one can interpret them as follows: rare means $<$ 1%

likelihood; possible means about 5% likelihood; likely means about 50% likelihood; probable means about 95% likelihood; and definite means about 99% likelihood.

For mammogram interpretation, a standard set of names has been developed. The Breast Imaging Reporting and Data System (BI-RADS) was developed by radiologists and is used commonly to describe the likelihood that a mammogram is normal or shows cancer [98]. BI-RADS 1 is assigned to a normal mammogram. BI-RADS 2 is assigned to a mammogram that shows some abnormality but that abnormality does not represent cancer. BI-RADS 3 indicates that there is a finding that most likely is not cancer, but cancer is possible, and therefore the patient should have some additional follow-up or evaluation. BI-RADS 4 is assigned to a mammogram that shows a finding that is of moderate suspicion for cancer. BI-RADS 5 is assigned to a mammogram that shows a finding of probable cancer.

After the diagnosis, some radiologists may include a statement recommending the next step or steps that could be used to confirm the radiologist's diagnosis. If there is any such statement, there will usually be a statement that the recommended method is only one of several appropriate methods for further establishing the diagnosis.

Radiologists and other physicians in general are used to dealing with uncertainties in diagnosis. They know they do not see everything on every image and that their interpretation of what they see may not be consistent were they to look at the image again themselves or with other radiologists. Engineers working with radiologists should expect some degree of uncertainty in what radiologists detect and diagnose.

Establishing ground truth or the gold standard—the absolute truth so that one can develop computer systems based on truth—is a major and difficult task and is probably not fully achievable. For this reason, it is perhaps better to think of how to deal with the uncertainty of diagnosis and accept that one will have to work with images with different degrees of ground truth.

Different methods are used for establishing truth for different types of problems where truth needs to be known either for training or validating a computer algorithm; the higher the standard of proof, the fewer the number of cases likely to be available for engineering applications. Because the engineer may not be able to obtain a sufficient number of cases with the highest levels of proof, he or she should always incorporate into the description information on the standard of proof used for developing and testing the program.

Proof comes at several levels. The highest level of proof normally requires a pathologist's diagnosis, providing the pathologist with accurate information about the location from which the tissue was sampled. Thus, an image-guided biopsy can be used to confirm that the tissue came from the area of interest on the image. Even with this, some uncertainty remains because pathologists also encounter the same types of problems as radiologists when they interpret the pathology images: image ambiguity, intraobserver and interobserver variability, and diagnostic ambiguity. In addition, there are problems of sampling error, where the area of interest on the image is not the area where the tissue came from There are also changes in tissue that occur when the tissues are processed for the pathology examination.

The next highest level of proof would be one where the findings on the image match the findings of the pathologist, even though the actual location used for sampling the tissue is not exactly known. In this level of proof, and in others, it is common to use a consensus panel (see below) to confirm the image findings and, in some cases, the pathology findings.

When there is no pathologic proof available, the next lower level of proof is obtained by the use of a consensus panel. A consensus panel is a group (usually) of recognized experts or specialists in evaluating that type of image and disease. Consensus means that all of the experts agree on the image interpretation. The members of a consensus panel can evaluate a case as a group, providing one final consensus or as individuals each of whom provides a diagnosis. These separate individual diagnoses can be combined in several ways because there will almost always be some degree of disagreement. Results may indicate the number of experts who agreed compared to those who disagreed; so, for example, five of five means there were five radiologists and all five agreed; three of five means that only three agreed. If all radiologists are required to see a finding for the finding to count, then harder-to-see lesions may be excluded because not all of the radiologists saw the subtle lesion. Thus, when a full consensus is reached (e.g., five of five), it may be harder to show the benefit of a system such as a computer-aided detection system because harder-to-see lesions may be excluded. If one allows cases where a fewer number of radiologists need to agree, then one may include cases where the diagnosis of one or two radiologists is incorrect. However, if correct, the cases are likely to be harder, and it would, therefore, be better to show the benefit of a computer system if the computer algorithm detected them.

7.6 Summary

This chapter provided an introduction of various biomedical data processing and analysis methods, most of them aimed at assisting data visualization and diagnostic decision-making. Most of the reviewed work revolves around computerized medical image processing and analysis, ranging from the description of the basic steps a clinician takes in medical decisions to a more elaborate exposition of specific methods such as image enhancement, segmentation, feature extraction, image interpretation.

Most of the mathematical and technical concepts presented in this chapter exhibit a general level of sophistication. To fully understand some of the discussed techniques, consultation to the cited references and other supplementary materials may be required.

7.7 Exercises

1. Compare the difference between image spatial smoothing, frequency filtering, and wavelet filtering methods in image enhancement. Compare both the advantages and disadvantages of these methods.
2. What is the difference between the orthogonal, bi-orthogonal, and translation invariant wavelet transform?
3. Add the Gaussian white noise to a medical image. Perform the wavelet-based soft and hard thresholding algorithms, and calculate the signal-to-noise ratio (SNR) improvement using these two de-noising approaches.
4. What is the difference between image smoothing and image sharpening? How do you evaluate these two different approaches?
5. Briefly describe the four types of gestalt processes that radiologists use in the interpretation of images. How might these affect computer image analysis? What types of computer vision approaches could one apply to mimic gestalt processes?
6. What is assignment or attribution error? What is its potential effect on the diagnoses that radiologists make? How might attribution error adversely affect a computer program for image diagnosis?
7. What is satisfaction of search? What is (are) the problem(s) associated with satisfaction of search? Can a computer detection system have a problem with satisfaction of search? Can a computer program for image analysis be deleterious because of the problem of satisfaction of search?
8. What is intraobserver variability? What is interobserver variability? If you are designing a computer system to aid radiologists in diagnosis, how will these two types of variability affect your design and testing? Can a computer system have intramachine variability? Can two computer systems show intermachine variability? What are the implications of this for computer programs for image analysis?
9. What is the BIRADS? Why should computer systems provide a BIRADS-equivalent measure of certainty or uncertainty? Or, should they not provide an equivalent measure of certainty or uncertainty?
10. What is ground truth (also called gold standard)? How can it be established? How might a computer detection or diagnosis system positively or negatively affect the determination of ground truth?
11. How might you deal with uncertainties in ground truth in evaluating computer systems for detection or diagnosis of abnormalities?

7.8 References and Bibliography

1. A. P. Dhawan. Medical image analysis. In *IEEE Press Series on Biomedical Engineering*. John Wiley & Sons, Inc, 2003.
2. R. B. Paranjape. Fundamental enhancement techniques. In Issac N. Bankman (Ed.). *Handbook of Medical Imaging*. Academic Press. 2–18, 2000.
3. K. Castleman. *Digital Image Processing*. Prentice Hall, 1996.
4. D.-C. Chang and W.-R. Wu. Image contrast enhancement based on a histogram transformation of local standard deviation. *IEEE Trans. Med. Imaging*. 17(4):518–531, 1998.
5. H. C. Lien, J. C. Fu, and S. T. C. Wong. Wavelet-based histogram equalization enhancement of gastric sonogram images. *Comput. Med. Imaging Graphics*. 24:59–68, 2000.
6. A.K. Jain. *Digital Image Processing*. Prentice-Hall, 1989.
7. H. Li et al. Computerized radiographic mass detection–Part I: Lesion site selection by morphological enhancement and contextual segmentation. *IEEE Trans. Med. Imaging*. 20(4):289–301, 2001.
8. J. Xuan et al. Automatic detection of foreign objects in computed radiography. *J. Biomed. Opt.* 5(4):425–431, 2000.
9. W. T. Freeman and E. H. Adelson. The design and use of steerable filters. *IEEE Trans. Pattern Anal. Mach. Intell.* 13(9):891–806, 1991.
10. K. S. Song et al. Adaptive mammographic image enhancement using first derivative and local statistics. *IEEE Trans. Med. Imaging* 16(5):495–502, 1997.
11. A. Polesel, G. Ramponi, and V. J. Mathews. Image enhancement via adaptive unsharp masking. *IEEE Trans. Image Process.* 9(3):505–510, 2000.
12. A. Beghdadi and A. Le Negrate. Contrast enhancement technique based on local detection of edges. *Comp. Vis. Graph. Image Proc.* 46:162–174, 1989.
13. C. B. Ahn, Y. C. Song, and D. J. Park. Adaptive template filtering for signal-to-noise ratio enhancement in magnetic resonance imaging. *IEEE Trans. Med. Imaging* 18(6):549–556, 1999.
14. J. S. Lee. Digital image enhancement and noise filtering by local statistics. *IEEE Trans. Pattern Anal. Mach. Intell.* 2:165–168, 1980.
15. J. F. Abramatic and L. M. Silverman. Nonlinear restoration of noisy images. *IEEE Trans. Pattern Anal. Mach. Intell.* 4(2):141–149, 1982.
16. H. Knutsson, R. Wilson, and G. H. Granlund. Anisotropic non-stationary image estimation and its applications—Part I, Restoration of noisy images. *IEEE Trans. Communications*. COM–31(3):388–397, 1983.
17. S. Mallat. A theory for multiresolution signal decomposition: Wavelet representation. *IEEE Trans. Pattern Anal. Mach. Intell.* 11(7):674–693, 1989.
18. D. A. Clausi and H. Deng. Design-based texture feature fusion using Gabor filters and co-occurrence probabilities. *IEEE Trans. Image Process.* 14(7):925–936, 2005.

19. S. Mallat and S. Zhong. Characterization of signals from multiscale edges. *IEEE Trans. Pattern Anal. Machine Intell.* 14(7):710–732, 1992.
20. R. R. Coifman and D. L. Donoho. Translation-invariant de-noising. In A. Antoniadis and G. Oppenheim (Eds.). *Wavelets and Statistics.* Springer-Verlag, 1995.
21. Yu-Ping Wang. Image representations using multiscale differential operators. *IEEE Trans. Image Process.* 8(12):1757–1771, 1999.
22. Y-P. Wang and S. L. Lee. Scale-space derived from B-splines. *IEEE Trans. Pattern Anal. Mach. Intell.* 20(10):1050–1065, 1998.
23. D. Donoho and I. Johnstone. Ideal spatial adaptation via wavelet shrinkage. *Biometrika.* 81:425–455, 1994.
24. A. Laine, X. Zong, and E. Geiser. Speckle reduction and contrast enhancement of echocardiograms via multiscale nonlinear processing. *IEEE Trans. Med. Imaging.* 17(4):532–540, 1998.
25. A. Laine, I. Koren, and F. Taylor. Enhancement via fusion of mammographic features. *International Conference on Image Processing.* 2:722–726, 1998.
26. A. Laine and W. Huda. Enhancement by multiscale nonlinear operators. In Issac N. Bankman (Ed.). *Handbook of Medical Imaging.* Academic Press. 33–56, 2000.
27. Y.-P. Wang et al. Chromosome image enhancement using multiscale differential operators. *IEEE Trans. Med. Imaging.* 22(5):685–693, 2003.
28. B. M. Sadler and A. Swami. Analysis of multiscale products for step detection and estimation. *IEEE Trans. Inform. Theory.* 45(3):1043–1051, 1999.
29. Y. Xu et al. Wavelet domain filters: A spatial selective noise filtration technique. *IEEE Trans. Image Process.* 3(11):747–757, 1994.
30. A. Rosenfeld. A nonlinear edge detection technique. *Proc. IEEE.* 814–816, 1970.
31. Q. Wu et al. The effect of image enhancement on biomedical pattern recognition. *Second Joint IEEE EMBS–BMES Conference,* 2002.
32. R. Gordon and R. M. Rangayyan. Feature enhancement of film mammograms using fixed and adaptive neighborhoods. *Appl. Opt.* 23(4):560–564, 1984.
33. G. Boccignone and M. Ferraro. Multiscale contrast enhancement. *Electron. Lett.* 37(2):751–752, 2001.
34. R. O. Duda, P. E. Hart, and D. G. Stork. *Pattern Classification.* 2nd ed. John Wiley & Sons, Inc., 2001.
35. A. Rosenfeld and A. C. Kak. *Digital Picture Processing.* Vol. 2. Academic Press, 1982.
36. D. Geman and S. Geman. Stochastic relaxation, Gibbs distributions, and the Bayesian restoration of images. *IEEE Trans. Pattern Anal. Mach. Int.* 6:721–741, 1984.
37. J. Besag. On the statistical analysis of dirty pictures. *J. Royal Stat. Soc.* B 48:259–302, 1986.
38. K. Held et al. Markov random field segmentation of brain MR images. *IEEE Trans. Med. Imaging.* 16:878–886, 1997.
39. L. Vincent and P. Soille. Watersheds in digital spaces: An efficient algorithm based on immersion simulations. *IEEE Trans. Pattern Anal. Mach. Intell.* 13:583–598, 1991.
40. L. Najman and M. Schmitt. Geodesic saliency of watershed contours and hierarchical segmentation *IEEE Trans. Pattern Anal. Mach. Intell.* 18:1163–1173, 1996.
41. O. Faugeras. *Three-Dimensional Computer Vision: A Geometric Viewpoint.* MIT Press, 1996.
42. R. C. Gonzalez and R. E. Woods. *Digital Image Processing.* 3rd ed. Addison Wesley Publishing Company, 1993.
43. D. Marr. *Vision.* W. H. Freeman and Co., 1982.
44. T. Lindeberg. Scale-space for discrete signals. *IEEE Trans. Pattern Anal. Mach. Intell.* 12:234–254, 1990.
45. Y. Lu and R. C. Jain. Behavior of edges in scale space. *IEEE Trans. Pattern Anal. Mach. Intell.* 11:337–356, 1989.
46. J. Sporring et al. *Gaussian Scale Space Theory.* Kluwer Academic Publishers, 1997.
47. A. P. Witkin. Scale space filtering. *Proc. Intl. Joint Conf. on AI, 1983.* 1019–1022, 1983.
48. R. Haralick. Digital step edges from zero crossing of second directional derivative. *IEEE Trans. Pattern Anal. Mach. Intell.* 6:58–68, 1984.
49. J. Canny. A computational approach to edge detection. *IEEE Trans. Pattern Anal. Mach. Intell.* 8:679–698, 1986.
50. M. Kass, A. Witkin, and D. Terzopoulos. Snakes: Active contour models. *Int. J. Computer Vision.* 1:321–331, 1988.
51. T. McInerney and D. Terzopoulos. Topology adaptive deformable surfaces for medical image volume segmentation. *IEEE Trans. Med. Imaging.* 18:840–850, 1999.
52. J. A. Schnabel and S. R. Arridge. Active shape focusing. *Image and Vision Computing* 17:419–428, 1999.
53. C. Xu and J. L. Prince. Snakes, shapes, and gradient vector flow. *IEEE Trans. Image Process.* 7:359–369, 1998.
54. R. Malladi, J. A. Sethian, and B. C. Vemuri. Shape modelling with front propagation: A level set approach. *IEEE Trans. Pattern Anal. Mach. Intell.* 17:158–175, 1995.
55. W. J. Niessen, B. M. t. H. Romeny, and M. A. Viergever. Geodesic deformable models for medical image analysis. *IEEE Trans. Med. Imaging.* 17:634–641, 1998.
56. K. Siddiqi et al. Area and length minimizing flows for shape segmentation. *IEEE Trans. Image Process.* 7:433–443, 1998.
57. A. Yezzi et al. A geometric snake model for segmentation of medical imagery. *IEEE Trans. Med. Imaging.* 16:199–209, 1997.
58. S. M. Pizer et al. Segmentation, registration, and measurement of shape variation via image object shape. *IEEE Trans. Med. Imaging* 18:851–865, 1999.
59. L. H. Staib and J. S. Duncan. Model-based deformable surface finding for medical Images. *IEEE Trans. Med. Imaging.* 15:720–731, 1996.
60. B. C. Vemuri and Y. Guo. Snake pedals: Compact and versatile geometric models with physics-based control. *IEEE Trans. Pattern Anal. Mach. Intell.* 22:445–459, 2000.

61. T. F. Cootes et al. Use of active shape models for locating structure in medical images. *Image and Vision Computing.* 12:355–365, 1994.
62. L. D. Cohen and R. Kimmel. Global minimum for active contour models: A minimal path approach. *Int. J. Comput. Vision.* 24:57–78, 1997.
63. J. Park and J. M. Keller. Snakes on the watershed. *IEEE Trans. Pattern Anal. Mach. Intell.* 23:1201–1205, 2001.
64. L. D. Cohen and I. Cohen. Finite-element methods for active contour models and balloons for 2D and 3D images. *IEEE Trans. Pattern Anal. Mach. Intell.* 15:1131–1147, 1993.
65. J. A. Schnabel. Multi-scale active shape description in medical imaging. University College London, 1997.
66. S. J. Osher and J. A. Sethian. Fronts propagating with curvature dependent speed: Algorithms based on Hamilton-Jacobi formulation. *J. Computational Physics.* 79:12–49, 1988.
67. V. Caselles, R. Kimmel, and G. Sapiro. Geodesic active contours. *Proc. 5th Int. Conf. Computer Vision.* MIT Press. 694–699, 1995.
68. S. Kichenassamy et al. Conformal curvature flows: From phase transitions to active vision. *Arch. Rational Mech. Anal.* 134:275–301, 1996.
69. M. H. Loew. Feature extraction. In M. Sonka and J. M. Fitzpatrick (Eds.). *Medical Image Processing and Analysis.* SPIE Press. 273–341, 2000.
70. M. Sonka, V. Hlavac, and R. Boyle. *Image Processing, Analysis, and Machine Vision.* Pacific Grove, CA: Brooks/Cole, 1999.
71. R. C. Gonzalez and R. E. Woods. *Digital Image Processing.* 2nd ed. Prentice Hall, 2002.
72. R. M. Haralick and L. G. Shapiro. *Computer and Robot Vision.* Addison-Wesley, 1992.
73. H. Zhimin et al. Analysis of spiculation in the computerized classification of mammographic masses. *Med. Phys.* 22:1569–1579, 1995.
74. R. M. Haralick. Statistical and structural approaches to texture. *Proceedings of the IEEE.* 67:786–804, 1979.
75. C. Chatfield. *The Analysis of Time Series: An Introduction.* 6th ed. Chapman and Hall, 2003.
76. D. Landgrebe. Hyperspectral image data analysis. *IEEE Signal Processing Magazine.* 19:17–28, 2002.
77. J. Freeman et al. Multispectral and hyperspectral imaging: Applications for medical and surgical diagnostics. Presented at the *19th Annual International Conference of the IEEE Engineering in Medicine and Biology Society.* 1997.
78. B.-C. Kuo et al. Regularized feature extractions for hyperspectral data classification. Presented at *2003 IEEE International Geoscience and Remote Sensing Symposium, IGARSS '03.*, 2003.
79. H. L. Kundel. Visual search in medical images. In *Physics and Psychophysics.* H. L. Kundel and R. Van Metter (Eds.). *Handbook of Medical Imaging,* vol. 1. SPIE Press. 837–858, 2000.
80. P. Perconti and M. H. Loew. Salient features in mammograms using Gabor filters and clustering. Presented at *Medical Imaging 2004: Image Perception, Observer Performance, and Technology Assessment.* 2004.
81. P. Perconti and M. H. Loew. Analysis of parenchymal patterns using conspicuous spatial frequency features in mammograms and applied to the BI–RADS density rating scheme. Presented at *SPIE Medical Imaging 2006: Image Perception, Observer Performance, and Technology Assessment.* 2006.
82. P. Perconti and M. H. Loew. An objective measure for assembling databases used to train and test mammogram CAD algorithms. Presented at *Proc. 3rd IEEE International Symposium on Biomedical Imaging: From Nano to Micro.* 2006.
83. R. O. Duda, P. E. Hart, and D. G. Stork. *Pattern Classification.* 2nd ed. Wiley-Interscience, 2000.
84. H. H. Barrett and K. J. Myers. *Foundations of Image Science.* Wiley-Interscience, 2004.
85. S. Lehar. Gestalt isomorphism and the primacy of subjective conscious experience: A gestalt bubble model. *Behavioral & Brain Sciences.* 26:375–444, 2004.
86. http://en.wikipedia.org/wiki/Gestalt_psychology
87. C. Torrans and N. Dabbagh. Gestalt and Instructional Design. http://chd.gmu.edu/immersion/knowledgebase/strategies/cognitivism/gestalt/gestalt.htm.
88. J. W. Oestmann et al. Chest "gestalt" and detectability of lung lesions. *Eur. J. Radiol.* 16:154–157, 1993.
89. K. S. Berbaum and W. L. Smith. Use of reports of previous radiologic studies. *Acad. Radiol.* 5:111–114, 1998.
90. J. Yerushalmy. The statistical assessment of the variability in observer perception and description of roentgenographic pulmonary shadows. *Radiol. Clin. North Am.* 7:381–392, 1969.
92. M. T. Freedman and T. Osicka. Reader variability: What we can we learn from computer-aided detection experiments. *J. Am. Coll. Radiol.* 3:446–455, 2006.
93. R. L. Siegle et al. Rates of disagreement in imaging interpretation in a group of community hospitals. *Acad. Radiol.* 5:148–154, 1998.
93. K. S. Berbaum et al. Can order of report prevent satisfaction of search in abdominal contrast studies? *Acad. Radiol.* 12:74–84, 2005.
94. K. S. Berbaum et al. Can a checklist reduce SOS errors in chest radiography? *Acad. Radiol.* 13:296–304, 2006.
95. K. S. Berbaum et al. Role of faulty visual search in the satisfaction of search effects in chest radiography. *Acad. Radiol.* 5:9–19, 1998.
96. T. Osicka et al. Computer-aided detection of lung cancer on chest radiographs: Differences in the interpretation time of radiologist's showing vs. not showing improvement

with CAD. Proc. of SPIE: Image Perception, Observer Performance, and Technology Assessment. 5034:483–494, 2003.

97. K.S. Berbaum Role of faculty visual search in the satisfaction of search effect in chest radiography. *Acad. Radiol.* 5:9–19, 1998.

98. American College of Radiology. *Illustrated Breast Imaging Reporting and Data System (BI–RADS)*. 3rd ed. American College of Radiology, 1998.

8

Data Registration and Fusion

Dr. Xiu Ying Wang,[1,2]
Dr. Stefan Eberl,[1,3]
Prof. Michael Fulham,[1,3]
Dr. Seu Som,[4] and
Prof. David Dagan Feng[1,5]
[1] *University of Sydney*
[2] *Heilongjiang University*
[3] *Royal Prince Alfred Hospital*
[4] *Liverpool Hospital*
[5] *Hong Kong Polytechnic University*

8.1 Introduction

In current clinical practice, large amounts of imaging data, acquired from different imaging devices, over multiple time points, are used for the accurate diagnosis and management of patients with a variety of diseases. Anatomical imaging modalities such as magnetic resonance imaging (MRI), computed tomography (CT), and X-ray depict mainly detailed morphological structure. Functional imaging modalities such as positron emission tomography (PET) and single photon emission computed tomography (SPECT) reveal information primarily about underlying biochemical and physiological changes. Recently, combinations of functional and anatomical imaging technologies into a single device, PET/CT and SPECT/CT scanners, have widened the array of medical imaging approaches and offered new challenges in the assimilation of imaging data. Further, each of these imaging technologies can have its own inherent value for patient management, and ideally all such imaging data would be available for the one individual when they are required. However, the seamless integration of such diverse data acquired on different scanners at different times poses substantial challenges.

Medical image registration is an important step in maximizing the information embedded in imaging datasets. Registration aims to spatially match datasets that may differ in time of acquisition, imaging device, and acquisition angle. After registration, spatial correspondence between functional information and anatomical structure can be achieved. Data fusion often follows the registration procedure to combine complementary information from multiple image datasets and represent these heterogeneous data in a common coordinate system. Information that was not apparent in an individual dataset can be extracted by accurate registration and fusion. For example, PET data have poorer spatial resolution compared with CT data. On the other hand, it can be difficult to accurately localize abnormalities in CT data, whereas these are identified in PET. Registration followed by fusion of these multimodal data allows functional abnormalities that are evident in PET, but not in CT, to be accurately localized and allows a more complete insight into the underlying problem.

The increase in diagnostic accuracy provides for better patient management; in the context of a patient with non–small cell lung cancer (NSCLC), registered data may mean the difference between surgery aimed at cure and a palliative approach by the ability to better stage the patient (see Figure 8.1). Further, registration of multiple studies from the same imaging technique (monomodality) performed over multiple time intervals is critical in the longitudinal assessment of certain processes, such as serial studies in patients with lymphoma, where scans are done over several months to gauge the effectiveness of chemotherapy.

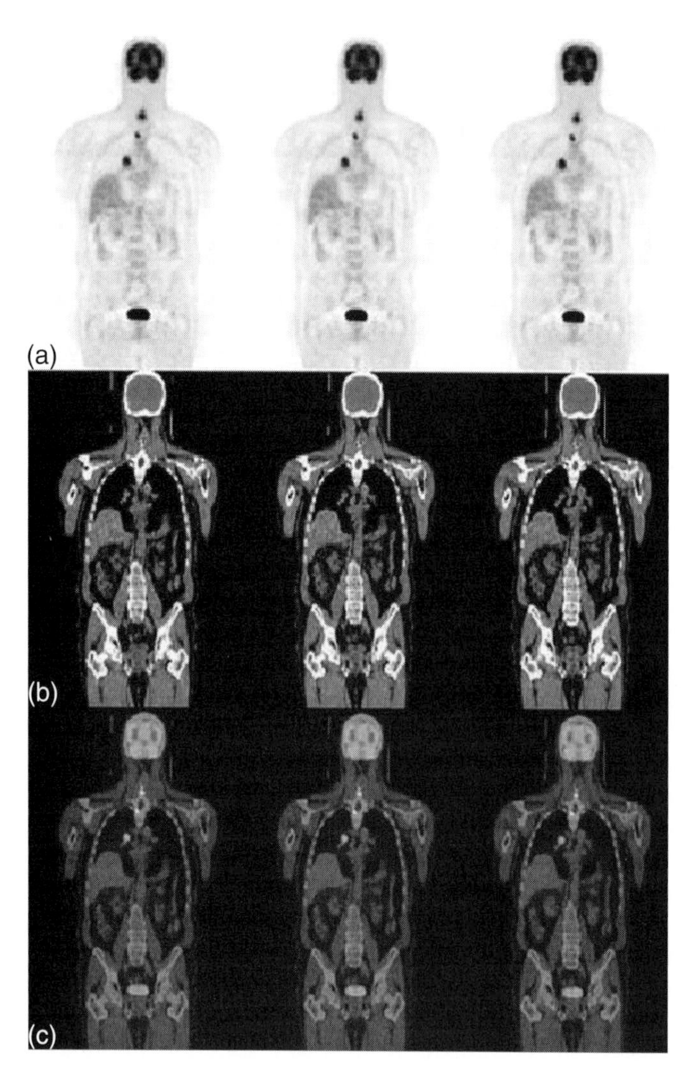

FIGURE 8.1 Coronoal PET (a), CT (b), and fused images (c) of ^{18}F-fluoro-deoxyglucose (FDG) PET/CT images in a 69-year-old man with NSCLC involving the right main bronchus. PET detected additional sites of disease in the bony skeleton, clearly shown in the fused images. These findings mean that the patient has not been offered surgery because there is evidence of extensive disease outside the chest.

Medical image registration and fusion are playing an increasingly important role in improving the quality and effectiveness of health care through the provision of imaging data in a usable format. There are a wide range of clinical applications, including radiation therapy planning and assessment, disease progress monitoring, detection of dynamic structural and functional changes, and *image-guided surgery* (IGS). In IGS, registration and fusion of preoperative data to intraoperative findings provides an important navigational tool to precisely localize the area of interest and limit interference with normal structures. Image registration is also essential in building statistical atlases to capture and encode morphological or functional variability over a large population. The population-based atlas can be used in the automatic labeling, segmentation, and interpolation of structures and tissues, while the disease-specific atlas can assist the detection of pathology.

However, research into image registration is not new and dates back to the 1980s [1]. After decades of intensive research, numerous algorithms have been proposed, and a number of reviews, surveys, and books have been published in this area [2–12]. In these surveys, registration methodologies are classified into up to nine different categories [5], based on criteria such as image dimensionality, registration feature space, transformation, similarity measure, image modalities, optimization, interaction, and subjects involved. These criteria may overlap with each other in the registration procedure, so that any registration scheme can be the combination of different choices of these criteria [2].

In addition to a large number of software-based registration algorithms, more advanced imaging devices, such as combined PET/CT and SPECT/CT scanners, provide hardware-based solutions for registration and fusion by performing functional and anatomical imaging in the one imaging session with the one device. Although combined PET/CT scanners can reduce misregistration between PET and CT data by obtaining these data at one imaging session, software-based registration may still be required to correct misregistration caused by patient motion between the PET scan and the CT scan.

This chapter aims to impart an understanding of registration and fusion fundamentals, major methodologies and techniques, and applications of registration and fusion in the clinical environment.

8.2 Fundamentals of Biomedical Image Registration and Fusion

8.2.1 Registration Definition

Image registration is a primary tool in comparing or combining images acquired from multiple sensors, at different times, or at different viewpoints for analysis or visualization. The main task of the registration algorithm is to determine a mapping to spatially relate the image sets so that these images can be represented in a common coordinate system. Mathematically [2], image registration can be expressed as:

$$I_R(X_R) = g(I_S(T(X_S))), \tag{8.1}$$

where

- I_R and I_S are the reference-image and study-image sets, respectively, indexed by their spatial coordinates.
- $T\colon (X_S) \to (X_R)$ is the spatial transformation of coordinates of study image X_S to the coordinate system of reference image X_R.
- $g\colon (I_s) \to (I_R)$ is the one-dimensional intensity transformation.

8.2.2 Main Components of Registration

As shown in Figure 8.2, the study-image set is compared with the reference-image set using a similarity measure. Based on the similarity measure, updated transformation parameters are estimated to provide an improved spatial match between the two image sets. The study images are interpolated and transformed with the updated parameters and again compared with the reference images to allow further improvement in transformation parameters by the optimization step. This procedure is repeated until the optimum transformation parameters are found, which are then used to register the study-image set to the reference-image set.

8.2.3 Classification of Registration Based on Input Data

According to the dimensionality of the input images, registration can be classified as 2D, 3D, 2D-3D (when one image is two-dimensional and the other is three-dimensional), or multidimensional (when time domain is added as the fourth dimension) registration categories. Comparatively, 2D registration is easier and faster because of the smaller data volume and the fewer transformation parameters to be computed. Applications of 2D registration include image mosaics that provide a whole view of a sequence of partially overlapped images and atlas construction. 3D medical image registration is required for most clinical applications. However, the lack of computational efficiency and automation of 3D registration techniques limits their application in the routine clinical setting, particularly for large datasets such as 3D whole-body data. In applications such as IGS, 2D-3D registration is required to align 3D preoperative images (e.g., CT, MRI) to 2D intraoperative images (e.g., ultrasound, X-ray), with the aim of achieving a safer and less-invasive surgical result. Computational efficiency or speed is one of the main concerns in this registration application scenario. Multidimensional registration is used to register a series of medical images acquired at different times for applications such as tumor growth monitoring, cancer staging, and treatment response assessment.

The input images to be registered may be from the same imaging modality (e.g., CT-CT, PET-PET, MRI-MRI) or different imaging modalities (e.g., CT-PET, CT-MRI, MRI-PET), and accordingly, registration can be cataloged as *monomodal* and *multimodal.* Monomodal registration is used mainly to detect changes over time due to disease progression and treatment. Multimodal registration is used to combine the complementary information from modalities (e.g., CT, MRI) to optimally visualize both soft tissue information and bone structure.

8.2.4 Registration Transformations

8.2.4.1 Basic Concept of Transformation

In any biomedical imaging procedure, many factors can lead to distortions and deformations in the images. For instance, different underlying physics principles of imaging sensors are primary causes of differences among the multimodal images to be registered; in addition, even for monomodal images, such factors as intersubject differences, voluntary and involuntary motion of the subject during imaging, and differences in positions and poses of the subject in different studies can lead to significant differences in spatial orientation of structures and organs of interest [7]. To align medical image data with these differences and deformations, a registration transformation, T, which may be linear or nonlinear, must be determined. A transformation is linear if for any two images X_1, X_2 and two scalars α, β, it satisfies:

$$T(\alpha X_1 + \beta X_2) = \alpha T(X_1) + \beta T(X_2). \tag{8.2}$$

The number of parameters used to describe the transformation is known as the number of *degrees of freedom*.

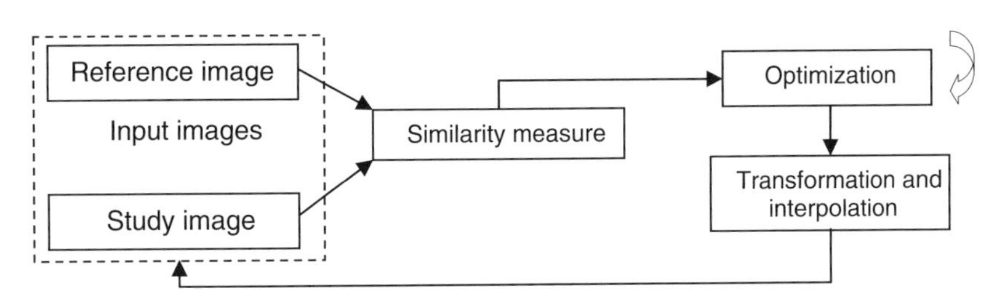

FIGURE 8.2 Image registration framework.

8.2.4.2 Homogeneous Coordinates

By introducing an additional dimensional vector, homogeneous coordinates not only include Cartesian coordinates and transformation in one matrix, but importantly, make the combination of linear transformations simple.

In homogeneous coordinates, a 3D point (x, y, z) is represented by $(x, y, z, 1)$, and 3D transformations of rotation, translation, and scaling are represented by 4 * 4 matrices. A series of these linear transformations can be combined into a single matrix by the multiplication of their corresponding matrices. In homogeneous coordinates, the transformation composed by two consecutive transformations T_1 and T_2 can be computed by:

$$\begin{bmatrix} x_1 \\ y_1 \\ z_1 \\ 1 \end{bmatrix} = T_1 \begin{bmatrix} x \\ y \\ z \\ 1 \end{bmatrix}; \begin{bmatrix} x_2 \\ y_2 \\ z_2 \\ 1 \end{bmatrix} = T_2 \begin{bmatrix} x_1 \\ y_1 \\ z_1 \\ 1 \end{bmatrix} = T_2{}^* T_1 \begin{bmatrix} x \\ y \\ z \\ 1 \end{bmatrix}. \quad (8.3)$$

For instance, T_r is the translation matrix, R_x is the rotation matrix of rotating α degrees about the x-axis, and S is the scaling matrix:

$$T_r = \begin{bmatrix} 1 & 0 & 0 & t_x \\ 0 & 1 & 0 & t_y \\ 0 & 0 & 1 & t_z \\ 0 & 0 & 0 & 1 \end{bmatrix}; R_x = \begin{bmatrix} 1 & 0 & 0 & 0 \\ 0 & \cos\alpha & -\sin\alpha & 0 \\ 0 & \sin\alpha & con\,\alpha & 0 \\ 0 & 0 & 0 & 1 \end{bmatrix};$$

$$S = \begin{bmatrix} s_x & 0 & 0 & 0 \\ 0 & s_y & 0 & 0 \\ 0 & 0 & s_z & 0 \\ 0 & 0 & 0 & 1 \end{bmatrix}. \quad (8.4)$$

The transformation T, including rotation about the x-axis and then scaling and translation, can be expressed as:

$$\begin{bmatrix} x_1 \\ y_1 \\ z_1 \\ 1 \end{bmatrix} = T \begin{bmatrix} x \\ y \\ z \\ 1 \end{bmatrix} = T_r{}^* S^* R_x \begin{bmatrix} x \\ y \\ x \\ 1 \end{bmatrix}$$

$$= \begin{bmatrix} s_x & 0 & 0 & t_x \\ 0 & s_y\cos\alpha & -s_y\sin\alpha & t_y \\ 0 & s_z\sin\alpha & s_z\cos\alpha & t_z \\ 0 & 0 & 0 & 1 \end{bmatrix} \begin{bmatrix} x \\ y \\ z \\ 1 \end{bmatrix}. \quad (8.5)$$

8.2.4.3 Rigid-Body Transform

Rigid transformation is a linear transformation that has six degrees of freedom in three dimensions, including three rotations around three axes and three translations in three directions. After rigid transformation, the distance between points and the angle between lines will not be changed. Rigid transformation can be used to cope with simple differences caused by position changes of the subject and is generally used in brain image registration due to the rigid structure of the skull.

All rigid transformations can be considered as special cases of the affine transformation, which is composed of rotations, translations, scalings, and shears. Parallel lines remain parallel after affine transformation. Affine distortions may be present in medical images—for instance, through scaling displacements in MRI scans due to the miscalibration of MR gradients, or skewing errors that appear in CT images if the gantry is tilted. Affine transformations, with their 12 degrees of freedom in 3D, can be used to correct these displacements in the images and are usually used as global transformations to provide good initial estimates for complex nonlinear registrations.

8.2.4.5 Elastic Transformation

Linear registration approaches are limited outside the brain because of organ motion and deformation (in heart, lungs, liver, bowel, etc.) that occur as a result of normal physiology and the effects of a disease process. Registration outside the brain thus requires complex, nonlinear transformations. To cope with these more complicated changes, more degrees of freedom are needed in nonlinear registration. The complex nonlinear transformation cannot be simply represented as a 4 * 4 matrix. Instead, it is usually represented as a displacement field D, which is composed of displacement vectors v_i. One displacement vector is defined for each individual point in the images as the difference between point positions p_{Ri} and p_{Si}:

$$D = \{v_i | v_i = p_{Ri} - p_{Si}, p_{Ri} \in X_R \;\&\; p_{Si} \in X_S\}. \quad (8.6)$$

After nonlinear transformation, straight lines will not be preserved. Compared with rigid-body transformations, the additional degrees of freedom of nonrigid transformations will inevitably increase the registration complexity and slow down the registration speed. Nonrigid image registration thus remains an active and challenging research area.

8.2.5 Interpolation

Interpolation is an essential component of image registration and is required whenever the image needs to be transformed. When the points in an image are mapped to nongrid positions after transformation, interpolation is performed to approximate the values for these transformed points. Interpolation also compensates for resolution differences among the images to be registered. For instance, interpolation is needed to compensate for the difference between the intraslice resolution and the interslice resolution. Since images from different imaging modalities have different resolutions, lower-resolution images in a multimodal image registration are often interpolated to the sample space of the higher-resolution images. A survey of interpolation methods in medical image processing is presented in Lehmann et al. [13].

The most frequently used interpolation methods include:

- Nearest neighbor
- Linear
- Bilinear
- Trilinear
- Bicubic
- Tricubic
- Quadrilinear
- Cubic convolution

The more complex the interpolation method, the more surrounding points are used, and the slower the registration speed [12]. For instance, the nearest-neighbor interpolation is a fast technique because only the nearest grid point is taken into account, and there is no need for a floating-point computation. In contrast, when using tricubic interpolation, 64 points are needed to estimate a new value. To speed up the registration procedure, computationally low cost interpolation techniques are often preferred.

Linear interpolation is one of the most popular techniques. The value of a certain point will be determined by the weighted combination of its neighbors, and the weights will depend on the distances from the neighbors to the point. Because of its good trade-off between accuracy and computational complexity, the bilinear interpolation, which needs four points to get an interpolated value, is frequently employed [14]. According to the research of [15], in cardiac and thorax image registration, trilinear interpolation, in which eight points are involved to calculate an interpolated value, can be used to achieve good registration performance [12].

Other interpolation techniques such as sinc and window sinc are also used in image registration. Some special interpolation techniques such as partial-volume interpolation [16] and stochastic interpolation [17] were proposed for mutual information (MI)-based registration. The interpolation effects on MI-based image registration are analyzed in Ji et al. [18].

8.2.6 Optimization

Almost every registration procedure requires an optimization algorithm, which searches for the optimal transformation to minimize a cost function (or maximize the similarity measure). For a given optimization algorithm, the registration procedure can be expressed mathematically as:

$$T_{optimal} = \underset{(T)}{\arg\min}\, f(T(X_S), X_R), \tag{8.7}$$

where T is the registration transformation, and f is the cost function to be minimized.

8.2.6.1 Gradient-Based Methods

Gradient-based optimization methods [26] are often used to determine the search direction in which the value of the cost function will be decreased locally. The gradient vector of cost function $f(\vec{x})$, where $\vec{x}$ is an n-dimensional vector $\vec{x} = [x_1, x_2, \ldots, x_n]^T$, can be expressed as:

$$\nabla f(x) \equiv g(x) \equiv \left[\frac{\partial f}{\partial x_1}, \frac{\partial f}{\partial x_2}, \ldots, \frac{\partial f}{\partial x_n}\right]^T,$$

and the second-order partial derivatives can be represented by a Hessian matrix:

$$\nabla^2 f(x) \equiv H(x) \equiv \begin{bmatrix} \dfrac{\partial^2 f}{\partial^2 x_1} & \cdots & \dfrac{\partial^2 f}{\partial x_1 \partial x_n} \\ \vdots & & \vdots \\ \dfrac{\partial^2 f}{\partial x_1 \partial x_n} & \cdots & \dfrac{\partial^2 f}{\partial^2 x_n} \end{bmatrix}^T. \tag{8.8}$$

Function $f(x)$ can be approximated by its Taylor series expansion about x_k:

$$\begin{aligned} f(\vec{x}_k + \vec{x}) &\approx f(\vec{x}_k) + (\nabla f(\vec{x}_k))^T \vec{x} + \frac{1}{2}\vec{x}^T \nabla^2 f(\vec{x}_k)\vec{x} \\ &= f(\vec{x}_k) + g(\vec{x}_k)^T \vec{x} + \frac{1}{2}\vec{x}^T H(\vec{x}_k)\vec{x}. \end{aligned} \tag{8.9}$$

8.2.6.1.1 General Gradient-Based Optimization Algorithm.

The gradient-based optimization is performed iteratively:

Step 1: Initialize. Set iteration $k = 0$ and, initialize vector $\vec{x}_k$. Compute $f(\vec{x}_k)$.
Step 2: Check the convergence criterion. If the convergence criterion is satisfied, the optimization procedure will be stopped, and $\vec{x}_k$ is the solution.
Step 3: Compute a search direction. The vector $\vec{p}_k$, which defines the search direction, will be computed.
Step 4: Compute a step length l_k. A positive scalar l_k will be determined so that $f(\vec{x}_k + l_k\vec{p}_k) < f(\vec{x}_k)$.
Step 5: Update variables. Set $\vec{x}_{k+1} = \vec{x}_k + l_k\vec{p}_k$, compute $f(\vec{x}_{k+1})$, set $k = k + 1$, and return to Step 2.

Computing the search direction $\vec{p}_k$ and finding the step length l_k are two major issues of gradient-based optimization algorithms, and different methods of calculating search direction will generate various gradient-based methods.

8.2.6.1.2 The Steepest Descent Method.

In the steepest descent method, the first-order Taylor series is used to approximate the function [26]:

$$f(\vec{x}_k + \vec{p}_k) \approx f(\vec{x}_k) + (\nabla f(\vec{x}_k))^T \vec{p}_k. \tag{8.10}$$

As a result, the gradient vector at each point is used as the search direction, which is also the steepest descent direction at that point:

$$\vec{p}_k = -\nabla f(\vec{x}_k). \tag{8.11}$$

8.2.6.1.3 Newton's Method and Quasi-Newton Methods [26]. In Newton's method, a second-order expansion is used to approximate the function, and the search direction can be obtained as a solution to the equation

$$H(\vec{x}_k)\vec{p}_k = -g(\vec{x}_k).$$

Quasi-Newton methods approximate and update the Hessian matrix at each iteration. There are two efficient implementations of quasi-Newton methods: One is the Dividon-Fletcher-Powell (DFP) algorithm, in which the inverse Hessian is calculated, and the other is the Broyden-Fletcher-Goldfarb-Shanno (BFGS) method, in which the approximation of Hessian is used.

Christensen et al. [19] reported on gradient descent optimization, and a number of investigators have used the quasi-Newton method [20, 21]. Gradient-based optimization methods are computationally efficient, but their performance is highly dependent on the initial estimation, and they are prone to being trapped in local optima.

8.2.6.2 Powell Algorithm

For the Powell algorithm [22], derivatives are not required in choosing the successive searching directions, which can reduce the computational cost. The Powell method performs a succession of one-dimensional optimizations, finding the best solution for each variable, and the single-variable optimizations are used to determine the new search direction. The algorithm is repeated until it is unable to find a new solution that is a major improvement over the current solution. This algorithm has been used frequently as an optimization strategy in image registration [16, 23–25].

8.2.6.3 Downhill Simplex Method

The Nelder-Mead downhill simplex algorithm [26] also requires no derivatives. However, unlike the Powell algorithm, downhill simplex is a multidimensional optimization method starting with an initial simplex. In n-dimension, the simplex is an $(n + 1)$-point geometrical figure. The simplex method searches the optimum solution downhill through a complex n-dimensional topology by operations of reflection, expansion, contraction, and multiple contractions. At each step of the search:

1. The function value at these $n + 1$ points is evaluated.
2. The points with highest value, second-highest value, and lowest value are determined.
3. A new point is generated, and one of the existing points may be replaced by the new point to generate a new simplex.
4. Step 3 is repeated until the difference between the highest value and the lowest is less than the predefined tolerance.

Downhill simplex is not as efficient as the Powell algorithm due to the larger number of evaluations involved, but it is more robust. Downhill simplex has been used, for example, in Hill et al. [27]. Rohlfing et al. [28] adopted a variant of the downhill simplex algorithm restricted to the direction of the steepest ascent.

8.2.6.4 Global Optimization

The algorithms previously discussed are local optimization techniques and might be trapped in local optima as a result of a good local similarity measure or improper implementation factors such as interpolation and changes of overlap between the images [29]. To achieve good registration, global optimization methods are required.

8.2.6.4.1 Quasi-Exhaustive Optimization. The quasi-exhaustive searching method was adopted as optimization strategy for medical data registration [e.g., 4, 30]. Because of its high computational complexity, the exhaustive method was used for only simple transformations (e.g., translation). It is not an efficient choice and becomes impractical for searching for global optimization when transformations become more complex.

In addition to cost-expensive quasi-exhaustive optimization, the search for a global optimal registration transformation can be performed by genetic algorithms [31], the simulated annealing method [32], and the particle swarm technique [33].

8.2.6.4.2 Genetic Algorithm. The genetic algorithm (GA) [31] is an interesting optimization technique based on the Darwinian concept of survival of the fittest.

1. The GA starts with the initialization of the population of n random solutions to the similarity measure to be optimized. The solution consists of transformation parameter values (genes) that are encoded as chromosomes made of bits and connected as a single string, called the *individual.*
2. Then, the fitness of each individual is estimated by the similarity function that can be calculated from the genes stored in the individual.
3. According to survival of the fittest, the pairs of fit individuals are selected to recombine their genes to produce offspring.
4. The current generation will be replaced by the offspring.
5. The population of solutions evolves through operations of selection, crossover, and mutation, in order for generations to produce the fittest solution.

Multiresolution optimization schemes have been used widely to escape from a local optimum and speed-up registration [e.g., 15, 25]. More recently, a new derivative-free global optimization for medical image registration was proposed in Wachowiak and Peters [34].

8.3 Feature-Based Medical Image Registration

Registration methods seek to optimize values of a cost function or similarity measure that define how well two image sets are registered. The similarity measures can be based on the distances between certain homogeneous features and differences of gray values in the two image sets to be registered. Accordingly, biomedical image registration can be classified in terms of feature-based or intensity-based methods.

In feature-based registration, which involves corresponding landmarks and features identified in the datasets to be registered, the transformation that is required to spatially match the features can be computationally efficiently determined and applied to the image datasets. Feature-based medical image registration methods can be classified into point-based approaches [35], curve-based algorithms [36, 37], and surface-based methods [1, 38]. A preprocessing step is usually needed to extract the features manually or semi-automatically, which makes this an operator-intensive and -dependent approach. An automated approach to feature extraction and definition is desirable because it would avoid intensive operator interaction and still take advantage of the computational efficiency of feature-based registration.

8.3.1 Landmark-Based Registration

Landmark-based registration involves identifying corresponding landmark points, matching the landmarks, and estimating the image transformation from the locations of the landmarks. The corresponding points are also called homologous landmarks, to emphasize that they should present the same feature in the different images. These points can be anatomical features or markers attached to the patient that can be identifiable in both image modalities.

8.3.1.1 Extrinsic Landmarks (Fiducial Markers)

Extrinsic landmarks refer to the artificial markers attached to the subject. These landmarks can be noninvasive, such as a mold, frame, dental adapter, or skin markers [5]. However, because of the elasticity of human skin, skin markers do not provide an accurate registration result. Although extrinsic landmark screw markers [39] and stereotactic frames [40] provide a robust basis for registration and a "gold standard" for brain image registration, they are uncomfortable and invasive.

Since fiducial markers can usually be easily detected in the images, registration based on these extrinsic landmarks is often automated and can be used in monomodal and multimodal image registration. Once the corresponding fiducial landmarks have been extracted, complex optimization and computation of registration parameters are not needed. The end result is fast registration. This scheme can be used in IGS, where registration efficiency is one of the primary concerns [39].

8.3.1.2 Intrinsic Landmarks

Intrinsic landmarks can be anatomically or geometrically salient points extracted from the patient images. Such landmarks should be uniquely localized and scattered evenly over the image volume and should carry substantial and characteristic information of the image. In registration based on anatomical landmarks, intensive user interaction is usually involved to manually identify the corresponding morphological feature points, and the accuracy of the registration result is highly dependent on the experience of the user. Geometric landmarks such as corner points, intersection points, and local extrema [41] can be segmented automatically; however, the accuracy of registration based on such landmarks may depend on the precision of the segmentation algorithms.

8.3.1.2.1 Iterative Closest Point Algorithm. Once the landmarks have been determined, the iterative closest point (ICP) algorithm can be used to register the images [e.g., 5, 43]. The ICP method proposed by [42] can be used with seven presentations of geometric data, including landmarks, free-form curves, and surfaces [42]. Furthermore, no prior knowledge about correspondence between the features is required, which eases the registration procedure greatly.

For two landmark sets,

$$P_R = \{p_{Ri} | p_{Ri} \in X_{Reference}, i = 0,1, \cdots m-1\} \text{ and}$$
$$Q_S = \{q_{Si} | q_{Si} \in X_{Study}, i = 0,1, \cdots n-1\},$$

from the reference and the study image, respectively. The ICP algorithm repeats the following steps until the mean-square difference falls below the predefined threshold:

Step 1: Determine the closest reference points for each study landmark point.
Step 2: Find the mean-square distance matrix and the transformation.
Step 3: Apply the transformation on the study landmark points, and redetermine the new closest point set.

A more efficient technique to speed up registration can be found in Kapoutsis et al. [44].

8.3.1.2.2 Thin-Plate Splines. The ICP technique is usually suitable for rigid-body registration, while thin-plate splines (TPSs) can be used in elastic registration. The TPS interpolation was first introduced by Bookstein [45, 46]. Because it can produce a smooth spline interpolation, has high computation speed, and can correct local elastic deformations, it is a commonly used elastic registration method [47]. Rohr [8] provided a good example of elastic registration based on TPSs.

The transformation function $f(p_i) = q_i$, $i = 1,2,\ldots,n$ is to be determined to minimize the energy function, which reflects the amount of variation. The least bending energy of the TPS function is:

$$E = \iint_{R^2} \left(\left(\frac{\partial^2 f}{\partial x^2} \right)^2 + 2\left(\frac{\partial^2 f}{\partial x \partial y} \right) + \left(\frac{\partial^2 f}{\partial y^2} \right)^2 \right) dxdy. \quad (8.12)$$

The TPS function is expressed as:

$$f(x,y) = a_1 + a_x x + a_y y + \sum_{i=1}^{n} w_i U(\| (x,y) - p_i \|), \quad (8.13)$$

where

- The coefficients a_1, a_x, and a_y define the affine part of the transformation, whereas the coefficient w defines the elastic deformation.
- p_i is the ith landmark.
- $U(\| (x,y) - p_i \|) = \| (x,y) - p_i \|^2 \log(\| (x,y) - p_i \|^2)$ is the radial basis function.

In order to ensure that $f(x,y)$ has square integratable second derivatives, the following conditions must be satisfied:

$$\sum_{i=1}^{n} w_i = 0 \text{ and} \sum_{i=1}^{n} w_i x_i = \sum_{i=1}^{n} w_i y_i = 0.$$

The coefficient vector $a = (a_1, a_x, a_y)^T$ and $w = (w_1, w_2, \ldots, w_n)^T$ can be computed through the following linear equations:

$$\begin{cases} kw + Pa = v \\ P^T w = 0 \end{cases}, \quad (8.14)$$

where

- v represents column vectors of landmarks.
- $k_{ij} = U_i(p_j) = U(\| (x_i, y_i) - (x_j, y_j) \|)$.
- $(1, x_i, y_i)$ is the ith row in P.

These two vector equations can be solved by:

$$\begin{cases} w = K^{-1}(v - Pw) \\ a = (P^T K^{-1} P)^{-1} P^T K^{-1} v \end{cases}. \quad (8.15)$$

8.3.2 Line-Based Registration

Multimodal (CT/MRI) registration can make use of line features such as edges, which have local maximum gradient magnitude, ridges, or crest lines [37, 48]; representations of the skull [36, 49]; and boundaries [50]—all extracted from images.

"Snakes," or active contours, first proposed by Kass [51], provide effective contour extraction techniques and have been widely applied in image segmentation and shape modeling [52, 53], boundary detection and extraction [54], motion tracking and analysis [55], and deformable registration [56–58]. Active contours are energy-minimizing splines, which can detect the closest contour of an object. The shape deformation of an active contour is driven by both internal energy and external energy:

$$E^*_{snake} = \int_0^1 E_{snake}(v(s))ds$$
$$= \int_0^1 E_{int}(v(s)) + E_{image}(v(s)) + E_{con}(v(s))ds. \quad (8.16)$$

An active contour can be represented by a curve $v(s) = [x(s), y(s)]$. The contour coordinates (x, y) can be expressed as the function of arc length s. The snakes are influenced by internal forces, image forces, and external constraint forces. The classic active contour model previously shown is flexible, since it maintains the shape as a curve, and the final form of the contour can be influenced by feedback from a higher-level process. However, this classic snake is sensitive to the initial contour guess and cannot deal with concavity. Some approaches such as the balloon model [59] and gradient vector flow (GVF) [60] have been proposed to solve these problems. More recently, there have been proposals for "united snakes," which unify finite element formulations, Hermitian shape functions, and B-spline functions in a consistent finite element formulation, and these have been tested in segmentation and dynamic chest image analysis [61].

In the balloon model, the initial contour does not need to be close to the target contour, as in the original version of the snake. The original snake is modified by using an inflation force so that the curve reacts like a balloon. The contour will be inflated and pass the weak edges but will be stopped if the edge is strong with respect to the inflation force.

The GVF, on the other hand, is defined to move the boundary into concavities. The GVF model is less sensitive to the initial contour than is the traditional snake. GVF has thus gained considerable popularity and has been used in elastic biomedical registration and fusion of multimodal cardiac [62], thoracic, and abdominal data [63].

Rather than solely depending on the interface, as in the parametric active contour presented by the snake, the geometric active contour presented by a *level set,* which was first introduced in Osher and Sethian [64], fits the initial contour into a surface. Compared with the snake, the level set's advantages include the ability to (1) easily calculate the intrinsic geometric properties and (2) curve evolution. In addition, topological adaptations such as splitting and merging can be handled naturally [65]. Due to these advantages, level sets have attracted more research attention [66] and have been applied in image segmentation. Their properties make them suitable for 3D medical image registration. But there are very few registration methods based on it, since it has only relatively recently been proposed for registration by Vemuri et al. [67].

8.3.3 Surface-Based Registration

In surface-based registration, the corresponding structure surfaces, which can be extracted automatically by various segmentation algorithms, are used as distinct features in the registration procedure.

The *head and hat* algorithm [1] is one example of a successful surface fitting technique for multimodal image registration. In this method, two equivalent surfaces are identified in the images. The first surface, extracted from the higher-resolution images, is represented as a stack of discs and is referred to as the "head." The second surface, referred to as "hat," is represented as a list of unconnected 3D points extracted from the lower-resolution image volume. The registration is determined by iteratively transforming the hat surface with respect to the head surface, until the closest fit of the hat onto the head is found. Because the segmentation task is comparatively easy and the computational cost is relatively low, this method remains popular. However, it is prone to error for convoluted surfaces.

Surface-based elastic registration can be used in intersubject applications and atlas registration. Elastic surface-based registration approaches include, among others, the elastic matching approach proposed by Bajcsy and Kovacic [69] and the finite-element model (FEM) technique proposed by Terzopoulos and Metaxas [70]. More details of surface-based registration algorithms can be found in the review in Audette et al. [71].

8.4 Intensity-Based Registration

In addition to the registration schemes based on the features extracted from the images, registration can directly utilize all the image intensity information without requiring segmentation or extensive user interactions. Intensity-based registrations, which can be fully automated, have attracted significant research attention, and numerous registration methods have been proposed. These include correlation-based methods, Fourier-based approaches, moment and principal axes methods [72], minimization of variance of intensity ratios [27, 74], and mutual information methods [23, 75].

In intensity-based registration, a cost function or similarity measure, which is based on the raw image content and is sensitive to misregistration, is defined, and the image datasets are iteratively transformed until the cost function is optimized. There are several well-established intensity-based similarity measures used in biomedical image registration.

8.4.1 Intensity Difference and Ratio Similarity Measures

8.4.1.1 Registration Measures Based on Intensity Differences

Methods of minimizing the intensity difference include *sum of squared differences* (SSD) and *sum of absolute differences* (SAD), which exhibit a minimum in the case of perfect matching [2]:

$$SSD = \sum_{i}^{N} (I_R(i) - T(I_S(i)))^2 \qquad (8.17)$$

$$SAD = \frac{1}{N} \sum_{i}^{N} |I_R(i) - T(I_S(i))|, \qquad (8.18)$$

where

- $I_R(i)$ is the intensity value at position i of reference image R.
- $I_S(i)$ is the corresponding intensity value in study image S.
- T is geometric transformation.

SSD and SAD are suitable for monomodal image registration only when intensity differences among the data are sufficiently small. In spite of their high computational efficiency, these cost functions can lead to false monomodal data registration when intensity changes are significant—for instance, due to an operation. This undesirable property obviously restricts the application of this category of registration criteria.

8.4.1.2 Variance of Intensity Ratios Algorithm

Introduced by Woods [73] for PET data registration, the algorithm derived from the variance of intensity ratios (VIR) is based on the assumption that when images are accurately registered, the value of a pixel in one image can be related to the value of a pixel in another image by a single factor. This strict assumption is seldom true for the data of different modalities, and this algorithm is suitable for only monomodal image registration.

1. In the VIR algorithm, the image intensity ratio $R(i)$ is calculated by dividing each reference-pixel value $I_R(i)$ by each study-pixel value $I_S(i)$:

   $$R(i) = I_R(i)/T(I_S(i)).$$

2. The standard deviation of the ratio is calculated by: $\delta_R = 1/N \sum_i (R(i) - \bar{R})$, and N is the number of the pixels.
3. The registration is achieved by minimizing the normalized standard deviation $\delta_{std} = \delta_R/\bar{R}$ of the ratio image.

8.4.1.3 Partitioned Intensity Uniformity

To overcome the limitation of registration based on minimizing the intensity difference, Woods and his colleagues [74] introduced an intrasubject, cross-modal registration for MRI and PET brain scans, known as the *partitioned intensity uniformity* (PIU) algorithm.

Based on an ideal assumption that "all pixels with a particular MR pixel value represent the same tissue type so that values of corresponding PET pixels should also be similar to each other" [74], the PIU algorithm divides the MR image into different partitions according to their intensity values, and then the uniformity of PET values for each partition is maximized by minimizing a certain standard deviation.

In the PIU algorithm, for any MRI voxel located at position i with value j, the value of the corresponding PET voxel is a_{ij}; for all voxels i with an MRI voxel value of j and with n_j as the total number of voxels in the brain region with MRI value j:

$a'_j = \frac{1}{n_j}\sum_{i=1}^{n_j} a_{ij}$ is the mean of a_{ij}.

$\delta_j = \frac{1}{n_j-1}\sum_{i=1}^{n_j}(a_{ij}-a'_j)^2$ is the standard deviation of a_{ij}.

$\delta'_j = \delta_j / a'_j$ is the normalized standard deviation.

$\delta''_j = \sum_{j=1}^{N}\delta'_j\frac{n_j}{N}$ is the weighted average of the measures of normalized standard deviations of the various MRI voxel values j, and registration can be achieved by minimizing this standard deviation.

Although manual removal of scalp, skull, and meninges is required before this rigid registration, the introduction of this algorithm has stimulated enthusiastic research into multimodal medical image registration based on image intensity information.

8.4.2 Correlation Techniques

Correlation techniques were proposed to register medical data from multimodal imaging sensors [36, 49]. In correlation methods, the registration is obtained by maximizing the similarity between images of the same object that may be different due to, for instance, different acquisition conditions. The cross-correlation technique has also been used for rigid motion correction of SPECT cardiac images [11, 76]. However, because these correlation methods are based on the assumption of a linear dependence between the image intensities, which usually is not true for complex multimodal images, these correlation techniques cannot always achieve reliable registration results.

The normalized cross correlation is defined as:

$$CR = \frac{\sum_i (I_R(i)-\bar{I}_R)(I_S^T(i)-\bar{I}_S)}{\sqrt{\sum_i (I_R(i)-\bar{I}_R)^2}\sqrt{\sum_i (I_S^T(i)-\bar{I}_S)^2}}, \tag{8.19}$$

where

- $I_R(i)$ is the intensity value at position i of reference image R.
- $I_S^T(i)$ is the corresponding intensity value in the transformed study image S.
- $\bar{I}_R$ and $\bar{I}_S$ are the mean intensity values of the reference and study images, respectively.

8.4.3 Registration Based on Information Theory

8.4.3.1 Information-Theoretic Concepts and Definitions

Techniques based on information theory play an important role in multimodality medical image registration. *Shannon entropy* is widely used as a measure of information in many branches of engineering. It was originally developed as a part of information theory in the 1940s and describes the average information supplied by a set of symbols $X = \{x\}$ whose probabilities are given by $\{p(x)\}$. The entropy of a random variable X is defined as:

$$H = -\sum_x p(x)\log p(x)\cdot \tag{8.20}$$

Entropy will be maximal when all symbols have equal probability. Entropy will be minimal when one symbol has a probability of 1 and all others have a probability of zero.

Image registration aims to increase the correspondence between the information in the datasets to be registered and to decrease the information in the combined image. Joint entropy provides a method to measure the amount of information in the combined image. If the two datasets are completely independent, their joint entropy will be a maximum value and will be the sum of the entropies of the individual dataset; the more related the images are, the lower their joint entropy is.

The joint entropy $H(X,Y)$ of pairs of random variables (X,Y) is defined as:

$$H(X,Y) = -\sum_{x\in X}\sum_{y\in Y} p_{XY}(x,y)\log p_{XY}(x,y). \tag{8.21}$$

Joint entropy can be calculated using the joint histogram of images involved in the registration. For a registration transformation T, the intensity i in the first image X paired with intensity j in the second image Y, the joint histogram is the probability:

$$p_{i,j}(T) = |\{k\colon X(x_k) = i \wedge T(Y(x_k)) = j\}|. \tag{8.22}$$

The axes in the joint histogram are the intensities of the images to be registered, and the value in the histogram represents the number of occurrences of intensity value pairs. The joint histogram becomes a measure of the degree of statistical dependence between the intensities in the two images. In image registration, when the images are correctly aligned, the joint histograms have tight clusters, and the joint entropy is minimized. These clusters disperse as the images become less well registered, and correspondingly, the joint entropy is increased [23]. Because minimizing the entropy does not require that the histograms be unimodal, the joint entropy is generally applicable to multimodality registration, and segmentation of images is not needed.

8.4.3.2 Registration Based on Maximization of Mutual Information

Mutual information (MI) was simultaneously and independently introduced by two research groups, Collignon et al. [23] and Viola and Wells [75], as a registration criterion for multimodal images. Maes and Suetens [78] and Pluim et al. [29] give good summaries of the development history of the MI criterion for medical image registration.

8.4.3.2.1 Definition. MI measures the degree of statistical dependence of two random variables and can be used to measure how well one image explains the other. For two datasets $X=\{x\}$ and $Y=\{y\}$, MI is defined as:

$$I(X,Y) = \sum_{x,y} p_{XY}(x,y) \log \frac{p_{XY}(x,y)}{p_X(x) \cdot p_Y(y)}, \tag{8.23}$$

where $p_{XY}(x,y)$ is the joint distribution of the intensity pair (x,y), and p_X (x) and p_Y (y) are the marginal distributions of x and y.

MI can be calculated by entropy:

$$\begin{aligned} I(X,Y) &= H(X) + H(Y) - H(X, Y) \\ &= H(X) - H(X|Y) \\ &= H(Y) - H(Y|X), \end{aligned} \tag{8.24}$$

where $H(Y \mid X)$ is the conditional entropy, which is the amount of uncertainty left in Y when X is known [16]. $H(Y \mid X)$ is defined as:

$$H(Y|X) = \sum_{x \in X} \sum_{y \in Y} p_{XY}(x,y) \log p_{Y|X}(y|x). \tag{8.25}$$

Therefore, MI measures the amount of information that one image contains about the other. If X and Y are completely independent, $p_{XY}(x,y) = p_X(x) \cdot p_Y(y)$ and $I(X,Y) = 0$ reaches its minimum; if X and Y are identical, $I(X,Y) = H(X) = H(Y)$ arrives at its maximum. Registration can be achieved by searching the transformation parameters that maximize the mutual information.

If both $p_X(x)$ and $p_Y(y)$ will not change with registration transformations, then entropies $H(X)$ and $H(Y)$ will be constants, and hence, the registration can be computed by minimizing the joint entropy $H(X,Y)$. If only one of the marginal distributions changes with transformations, then registration based on MI can be achieved by minimizing the conditional entropy $H(Y \mid X)$ or $H(X \mid Y)$. However, in most cases, because registration transformation will change the overlaps between the images, both $p_X(x)$ and $p_Y(y)$ will change, and both $H(X)$ and $H(Y)$ will vary accordingly. In implementation, the joint entropy and marginal entropies can be estimated by normalizing the joint and marginal histograms of the overlap sections of the images.

Algorithm Description

1. Allocate an $N_X \times N_Y$ array for bins in the joint histogram, with N_X and N_Y as the number of intensities in the images.
2. Compute the joint intensity histogram $HIS(X,Y)$ of the overlap part $X \cap Y$ in the images by binning the intensity pairs. For every pixel $i \in X \cap Y$, if the intensity value in X is $X(i) = x$, and the corresponding value in Y is $Y(i) = y$, then $HIS(x,y) = HIS(x,y) + 1$.
3. Calculate the normalized joint intensity histogram $PDF(x,y) = \dfrac{HIS(x,y)}{\sum_{x,y} HIS(x,y)}$.
4. Calculate the marginal intensity distributions by the sums of the rows and the columns of the normalized joint histogram: $p_X(x) = \sum_y PDF(x,y)$ and $p_Y(y) = \sum_x PDF(x,y)$.
5. Calculate the mutual information.

Because no assumption is made regarding the nature of this dependence and no limit constraints are imposed on the image content, maximization of MI is a general and powerful criterion and is suitable for multimodality medical image registration [29]. Various implementations and extensions of MI-based registration techniques have been proposed in [16, 77, 79, 80–82].

8.4.3.2.2 Registration Based on Normalized Mutual Information. Maximization of MI sometimes may generate false registrations—for instance, MI may increase with increasing misregistration due to changing overlap between the image data [80]. Studholme [83] proposed a normalized MI-based registration approach, which is less sensitive to changes in image data overlap:

$$NMI(X,Y) = \frac{H(X) + H(Y)}{H(X,Y)}. \tag{8.26}$$

Maes et al. [16] proposed the entropy correlation coefficient (ECC) technique to get a better registration result:

$$ECC(X,Y) = 2 \cdot \frac{I(X,Y)}{H(X) + H(Y)}. \tag{8.27}$$

Although the MI criterion has been widely used in registration of multimodal medical images of MRI/CT, MRI/PET, CT/PET, and CT/SPECT, as pointed out by Maes and Suetens [78], registration directly based on maximizing the MI criterion may not be able to get robust alignment of thorax CT and PET images. Use of the MI criterion for thorax and abdomen images may not be as feasible as for brain images, mainly because of reduced anatomical information in PET images of such areas [78].

8.4.3.3 Medical Image Registration Based on Other Information Measures

In addition to MI, researchers have explored other information measures in image registration. Zhu [84] proposed a cross-entropy optimization based on a volume image registration method. A divergence measure (Jensen divergence) was used as a similarity measure for image registration by He et al. [85]. Measures of F (Fisher)-information (e.g., V-, I_α-, and x^α-information) were applied in image registration, and the performance, robustness, and accuracy of rigid registration were studied by Pluim et al. [86], who showed that some measures have the potential to achieve more accurate registration than MI, even though they may be more difficult to optimize.

As mentioned, intensity-based methods can be automatic and accurate. However, they are also computationally expensive and are inefficient because the image contents have to be included in spatial transformation at each optimization iteration to allow calculation of an updated cost function. Furthermore, this category of schemes does not make use of *a priori* knowledge of the organ structure.

8.5 Hybrid Registration and Hierarchical Registration

In addition to feature-based registration and the intensity-based scheme, hierarchical and hybrid image registration have been explored to increase computational efficiency, achieve automation, and find better solutions.

By combining intensity-based techniques with feature-based methods, hybrid registration approaches try to exploit the merits of both and at the same time avoid their disadvantages. Hybrid registration can be automatic, more accurate, and faster than either of its registration components used separately.

By decomposing the datasets into multiple resolutions and performing the registration from low resolutions to high resolutions, hierarchical registration methods have the potential to increase registration speed, avoid local minima, and therefore improve registration performance.

8.5.1 Hybrid Registration

As mentioned previously, because feature-based registration methods are efficient, they can be used to tackle more complex nonrigid deformations. However, it is hard to generate one-to-one correspondence for regions at some distance from the extracted registration features. Hybrid registration approaches were proposed to overcome this limitation.

Johnson and Christensen [87] proposed consistent hybrid registration combining a landmark technique with an intensity-based method. Hybrid registration was composed of two main steps: (1) the landmark-based step, where TPSs were used to obtain the exact correspondence at the landmarks and (2) the intensity-based step, where the regions away from landmarks were matched by minimizing the intensity differences in these regions without affecting the matched landmark regions. The landmark-based step provided a global registration, while the intensity-based step was used to refine the registration locally. Experiments on 2D MRI scans demonstrated that this hybrid registration could produce better registration than using the intensity or the landmark method alone.

Hellier and Barillot [88] proposed a hybrid registration that reversed the order of Johnson and Christensen [87]. In this method, global registration was obtained by optical flow–based photometric similarity, while local elastic registration was achieved by a sparse landmarks-based technique. Based on their experiments, the authors claimed that the algorithm was efficient and was capable of coping with functional variability.

A hybrid deformable registration for intersubject cortical structures in the brain was proposed by Borgetors [38]. The hybrid method combined the advantages of a volumetric approach and a surface warping method. As a first step, the volumetric method was used to reduce the variations between the model cortical structure and the individual cortical structure and to provide initial estimation for the surface-based registration in the second step. However, accurate registration may depend on topologically correct cortical surface reconstruction, which is a prerequisite step in this hybrid method.

A more recent hybrid retinal image registration method, which combined an area-based (intensity-based) method with a feature-based technique, was proposed by Chanwimaluang et al. [89]. The method consisted of three steps:

1. Binary vascular trees were extracted based on local entropy-based thresholds.
2. Zero-order translation was estimated by using MI techniques.
3. Feature points were used to estimate higher-order transformation.

The hybrid method was effective and robust for retinal image registration.

8.5.2 Hierarchical Registration and Fusion

To achieve fast registration and fusion for clinical applications, it is necessary to decompose the very large datasets into manageable sizes for registration, which can significantly relieve the heavy computational burden and complexity. Hierarchical biomedical image registration schemes have been proposed to produce registration with both increased computational efficiency and the ability to find better solutions [15, 38, 69, 90–92].

In hierarchical strategies, the datasets to be registered are divided into multiple resolution levels to compose registration pyramids, and then the registration procedure is carried out from low-resolution scales to high-resolution scales. This multiscale scheme has several merits. The initial registration of the global information provided by the low-resolution scales helps to avoid being trapped in local minima and hence contributes to improved registration performance. This initial registration estimation also improves computational speed. Based on this estimation, the registration precision of high-resolution scales can be further improved by adding more registration information or features. Because most registration iterations occur on low-resolution scales, with fewer required in high-resolution scales, efficient registration can be achieved through hierarchical registration methods.

One of the key issues in hierarchical registration schemes is the building of registration pyramids. Several categories of hierarchical registration have been proposed [6], such as Gaussian pyramids [90, 92], spline pyramids [15], and wavelet-based registration. In these multiresolution registration approaches in general, registration pyramids are usually created by successively filtering and then downsampling the datasets.

8.5.2.1 Spline-Based Hierarchical Registration

Systematic research into B-spline–based hierarchical registration has been carried out by [15, 20, 91, 94, 95]. In these spline-based registration methods, pyramids were constructed by prefiltering and then downsampling the images. The prefilters can be B-splines [15] or cubic splines [20]. B-splines are piecewise polynomials and can be defined recursively. Based on B-splines, Thévenaz and Unser [15] proposed a faster and more accurate multiresolution registration for multimodal images by using MI as a similarity measure. However, as pointed out by Unser et al. [20], the underlying spline image is not shift invariant, due to the downsampling operation, which leads to small residual error even for an "exact alignment." This effect might be overcome by using higher-order splines. In addition, these spline-based hierarchical schemes concentrated mainly on rigid-body registration.

Due to its global-to-local influence, coarse-to-fine matching, and computational efficiency, hierarchical B-splines in the form of free-form deformation were applied in multiscale registration by Xie and Farin [92]. The hierarchical B-splines were used in point-based registration, surface-based registration, and intensity-based registration.

8.5.2.2 Wavelet-Based Hierarchical Image Registration

Wavelets lend themselves naturally to the separation of image data into different frequencies and resolutions while preserving information at different resolutions. The wavelet decomposition is a key component for automated extraction of features, which then allows efficient rigid, as well as nonrigid, transformation. Pajares and de la Cruz [96] provides a tutorial for wavelet-based image fusion.

Multiresolution analysis (MRA) Mallat [97] is important for the construction of fast two-dimensional wavelets from one-dimensional ones. For a given 2D image of size $2^{m*}2^{n}$, the wavelet-based image decomposition can be achieved by convolving the wavelet lowpass filter $h_\varphi(n)$ and the wavelet highpass filter $h_\psi(n)$ and downsampling by a factor of 2 along rows and columns independently (Figure 8.3).

Mathematically, the series of filtering and downsampling operations used to compute $W_\Psi(j,m,n)$ and $\{W_\Psi^i(j,m,n)|i = D,V,H\}$ can be expressed as:

$$\begin{aligned}
W_\Psi(j,m,n) &= h_\varphi(m)^*[h_\varphi(n)^* W_\varphi(j-1,m,n)|_{n=2k,k\geq 0}]|_{m=2k,k\geq 0}\\
W_\Psi^H(j,m,n) &= h_\psi(m)^*[h_\varphi(n)^* W_\varphi(j-1,m,n)|_{n=2k,k\geq 0}]|_{m=2k,k\geq 0}\\
W_\Psi^V(j,m,n) &= h_\varphi(m)^*[h_\psi(n)^* W_\varphi(j-1,m,n)|_{n=2k,k\geq 0}]|_{m=2k,k\geq 0}\\
W_\Psi^D(j,m,n) &= h_\psi(m)^*[h_\psi(n)^* W_\varphi(j-1,m,n)|_{n=2k,k\geq 0}]|_{m=2k,k\geq 0},
\end{aligned} \tag{8.28}$$

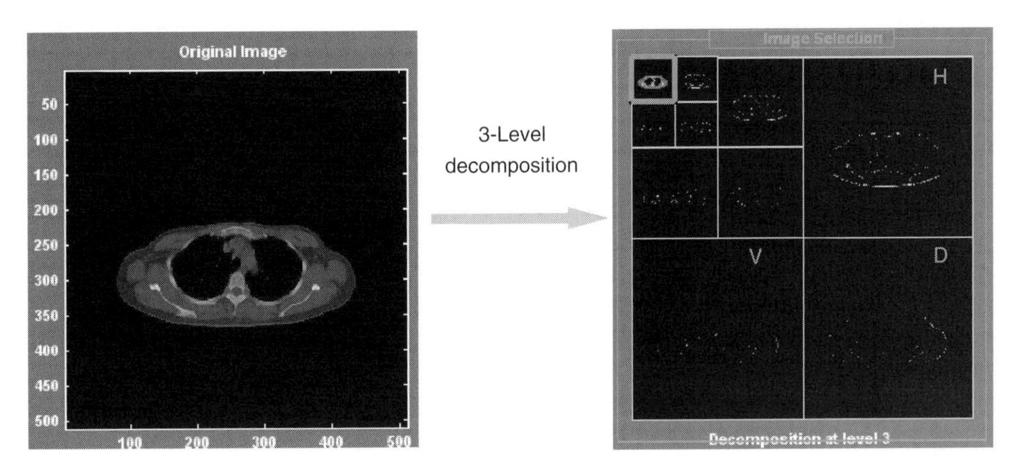

FIGURE 8.3 Wavelet decomposition of original image into quarter-sized subimages.

where

- $W_\Psi(j,m,n)$ are approximation coefficients used to represent global (low-frequency) information.
- $W_\Psi^H(j,m,n), W_\Psi^D(j,m,n)$, and $W_\Psi^V(j,m,n)$ are horizontal coefficients, diagonal coefficients, and vertical coefficients, respectively, which represent local (high-frequency) information.

Fast wavelet transform can be obtained by MRA and used in hierarchical image registration. However, coefficients of the shifted and rotated versions of the same dataset may be distributed differently. The lack of translation invariance and rotation invariance, which are the basic requirements for an image registration procedure, makes discrete wavelet-based registration difficult, especially for intensity- or coefficient-based registration.

Research efforts have been devoted to breaking the barrier of translation invariance and rotation invariance, and innovative techniques such as steerable pyramids [98] and translation invariant wavelets [99] have been proposed. A multiresolution registration method based on a steerable filter was also proposed [100]. To cope with rotation displacements in wavelet-based matching, a rotation-invariant pattern-matching method was introduced by Tsai and Chiang [101] and adopted by Xue et al. [102]. However, because the lack of translation invariance is neglected, this rotation-invariant pattern-matching method is not necessarily robust in dealing with more significant and complex displacements.

Feature-based methods offer a potential solution to the problem of rotation invariance and translation invariance in wavelet-based multiresolution registration. A wavelet-based coarse-to-fine matching method using "points of interest" as feature space was proposed by You and Bhattacharya [103] and Wang and Feng [104], and a surface alignment approach using multiresolution wavelet representation was introduced by Gefen et al. [105].

8.6 Hardware Registration

Outside the brain, software-based image registration algorithms face many challenges due to differences in patient positioning and movement and changes of internal organs between imaging sessions performed on different devices [106, 107]. Despite continued progress in software image registration algorithms, these techniques can still be labor intensive and have limited accuracy and are thus impractical to apply routinely on a patient-by-patient basis [106]. Functional and anatomical images provide complementary data, and a routine clinical combination of information from these modalities can offer:

- Improved lesion detection on both the functional and anatomical images
- Improved localization of the lesion or foci seen with functional imaging, resulting in better differentiation between physiological and pathological uptake
- Precise localization of the abnormal foci, such as in bone versus soft tissue, which can aid in guiding biopsy, treatment planning, etc. [107]

Hardware registration largely overcomes the current limitations of software-based techniques. In hardware registration, the functional imaging device, such as PET, is combined with an anatomical imaging device, such as CT, in the one instrument. Functional and anatomical imaging are then performed in the one imaging session on the same imaging table, which minimizes differences in patient positioning and in the locations of internal organs between the scans. It also ensures that both sets of registered data are ready for reporting almost as soon as the study is completed.

A prototype of a combined PET/CT system to be used clinically was first described by Beyer et al. [108]. This early system consisted of a single-slice spiral CT scanner mounted on the same rotating assembly as a low-end, partial-ring PET scanner. The addition of the CT scanner to a PET scanner brought two main advantages:

- CT provided fast and low-noise transmission data for attenuation correction of the PET data. Previous methods based on radioactive transmission sources added substantially to the total study time and increased the noise in the reconstructed PET images.
- The CT data provided exquisite anatomical detail, which enabled the PET data to be put into anatomical context and thus aided the interpretation of PET, as detailed in the previous section.

For these reasons, PET/CT has gained rapid acceptance in the clinical setting, and currently over 95% of PET scanners sold are PET/CT systems. Current-generation PET/CT scanners combine high-end multislice CT systems with high-end full-ring PET systems (Figure 8.4). Although combined into a single unit, the gantries of the CT and PET are separate units; this allows advantage to be taken of progress in the technology of both PET and CT scanners, which may occur at difference paces.

As the CT and PET units are mounted on separate gantries, transformation matrices have to be determined to map between the two coordinate systems. This is achieved by a calibration scan with sources visible both on CT and on PET arranged in a suitable geometry. The transformation factors mapping between the coordinate systems can then be calculated using similar techniques as those applicable to fiducial markers or stereotactic frame registration techniques (see also Section 8.3). Manufacturers have paid considerable attention to the design of the patient scanning bed. It is a requirement that the bed deflection for a particular region in the patient be the same during the CT scan as it was when the patient passed through the

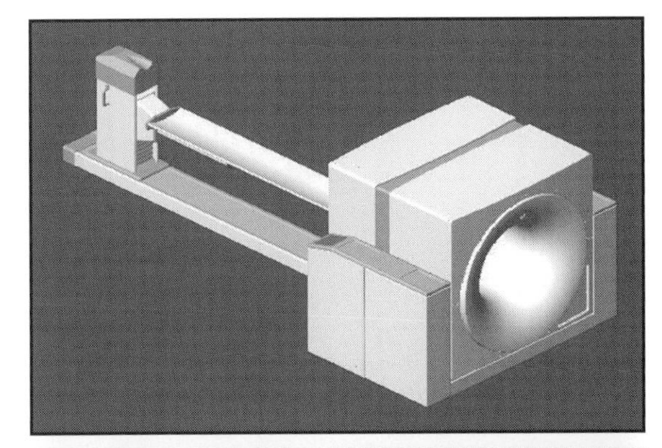

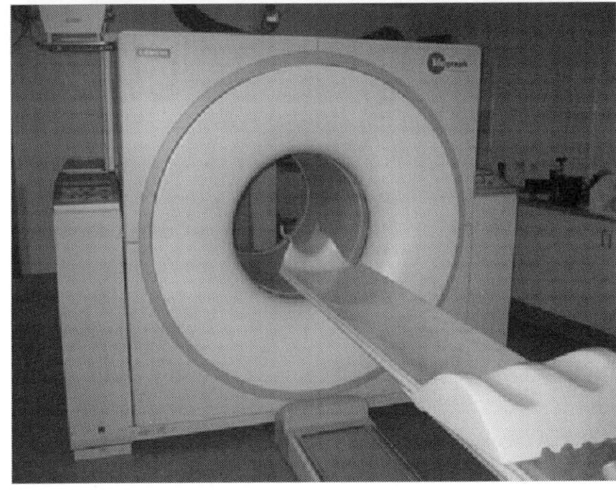

FIGURE 8.4 Illustration and photo of commercial PET/CT system. A multislice CT is combined with a high-end, ring-type PET system to form a hybrid scanner. The CT is mounted at the front of the PET unit. The gantries can be separated for servicing.

PET gantry, irrespective of patient weight. This has been achieved through a number of different designs, including bed supports between the CT and PET gantries or a cantilevered bed, where the fulcrum and support move as shown in Figure 8.4.

The mechanical design and calibration procedures ensure that the CT and PET data are inherently accurately registered if the patient does not move. However, patient motion can be encountered, as illustrated in Figure 8.5, where there is considerable movement in the patient's head between the CT and the PET. This results in not only incorrect anatomical localization, but also artifacts from the attenuation correction based on the misaligned CT data. Misregistration between the PET and CT data can also be due to involuntary motion, such as from respiratory or cardiac processes. An example of artifacts due to respiratory motion is shown in Figure 8.6. The errors introduced due to involuntary motion have been assessed [109, 110]. Voluntary patient motion can be minimized by careful patient position, restraints, and scanning protocol. Effects of involuntary (e.g., respiratory) motion can be reduced through suitable breathing protocols. More recently, considerable interest has focused on acquiring respiratory-gated CT and PET data, to determine and correct for respiratory motion.

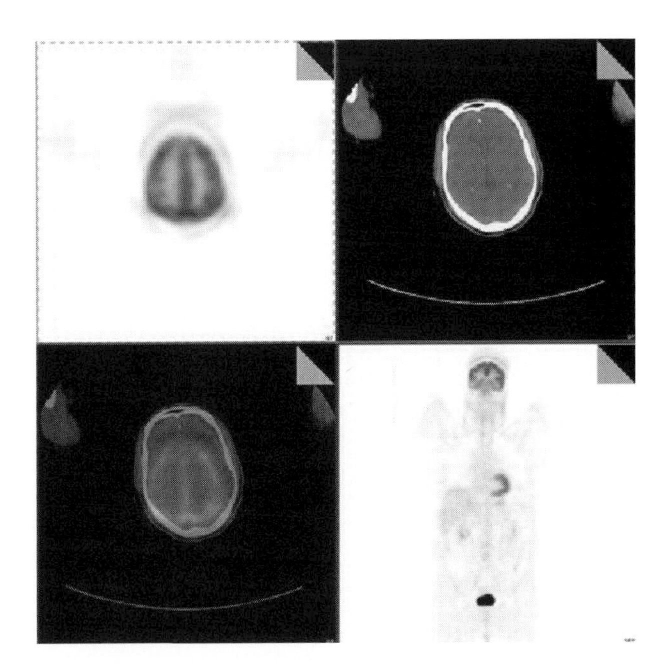

FIGURE 8.5 Patient moved the head between acquisition of CT and PET data in this whole-body PET/CT study. This causes not only loss of anatomical localization but also artifacts in the PET data due to misaligned CT-based transmission data used for attenuation correction.

Despite the limitations in registration accuracy between CT and PET in hybrid systems due to voluntary and involuntary patient motion, PET/CT has proven very valuable in the routine clinical setting. There is, however, a definite role for software registration in the PET/CT setting to remove the misregistrations due to patient motion. However, to be acceptable in the routine clinical setting, it has to be fast and convenient to perform and should not unduly slow down the workflow of PET/CT studies. For this, PET/CT studies offer the advantage that CT and PET imaging are performed in the same session, using the same setup, such as the patient bed, which makes the registration task less demanding and potentially more efficient.

8.7 Assessment of Registration Accuracy

To be clinically useful, a registration method must be sufficiently accurate and robust. Accuracy and robustness are two important criteria for assessing the performance of a registration approach. Accuracy can be measured by the difference between the optimal solution obtained from an algorithm and the real correct solution. Robustness is used to measure how frequently the algorithm can achieve an optimum solution regardless of deformations and image contents and modalities. However, the assessment of registration accuracy and robustness can be difficult due to the strong impact of other factors, such as similarity measures, interpolation techniques, and

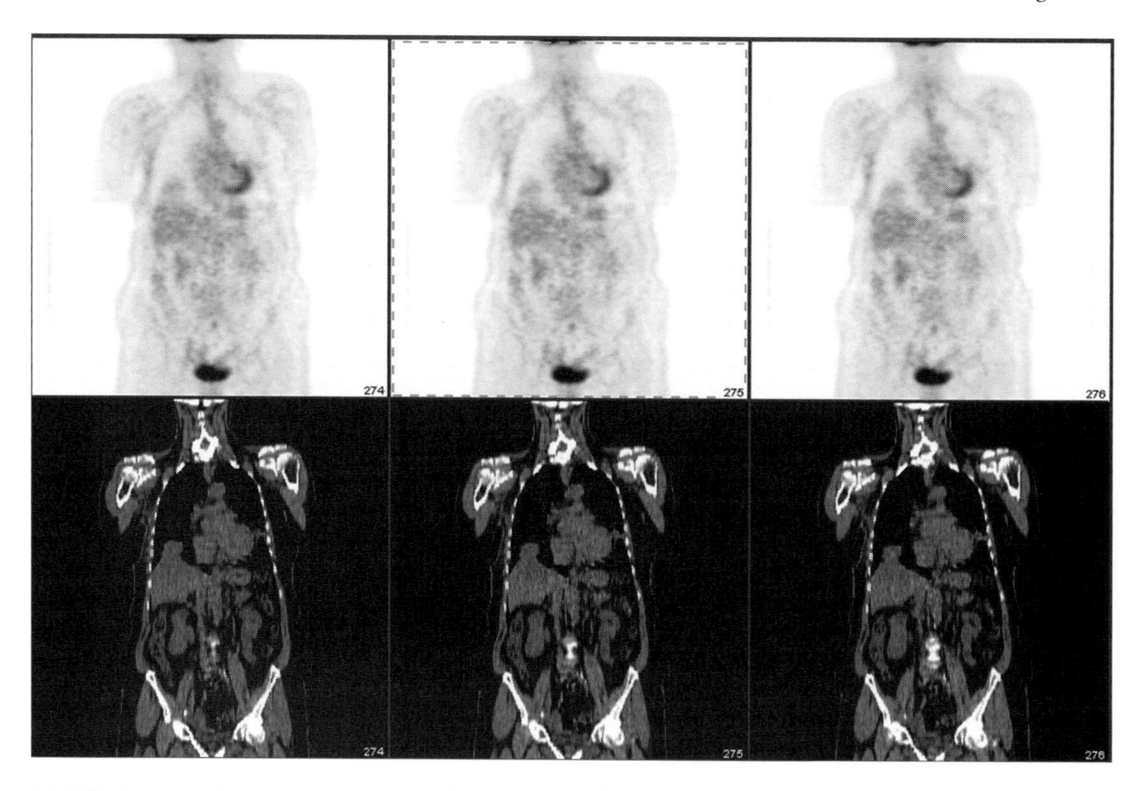

FIGURE 8.6 Respiratory motion during CT resulting in a "mushroom" artifact on the CT at the top of the liver, which results in artificially increased activity in the PET images in the affected region.

optimization algorithms [111], as well as the lack of a "gold standard." Frequently used validation methods include simulation and phantom studies, fiducial markers, and visual inspection.

8.7.1 Simulations and Phantoms

Simulation and phantom studies are important for the estimation of registration accuracy because the phantom is motionless in the imaging scenarios, and the displacement information or artificial transformations are known. Simulation- and phantom-based validations provide measures for computational complexity of different registration methods. In particular, they are useful for estimating the accuracy of intramodality registration methods [27, 111, 112]. However, assessment based on simulations and phantoms may not take all issues and imaging characteristics into account, and the simulated deformations might not be entirely realistic. To get a realistic model for elastic registration, Schnabel et al. [113] presented a validation method for nonrigid registration for breast MRI scans that took into account tissue properties and deformations by using the FEM.

8.7.2 Fiducial Markers

Fiducial markers have been devised to assess registration accuracy. To a certain extent, assessment using fiducial markers provides a gold standard for medical image registration [112]. A basic assumption of fiducial markers is that the motion experienced by the markers is the same as that of the organ of interest. This is not necessarily the case for fiducial markers attached to the skin, since skin can move independently of the organ of interest, such as the brain. To obtain gold standard transformation parameters for registration, skull-implanted fiducial markers have been used [35], with which the accuracy of different registration methods can be compared. But this invasive method is applicable to only rigid-body registration.

8.7.3 Visual Inspection

Visual inspection is a qualitative assessment and is the most intuitive method for the evaluation of registration accuracy. This assessment may involve the inspection of subtraction images, contour overlays, or anatomical landmarks [35]. It has been used widely in both rigid [9] and nonrigid registration [21, 114] assessment. Visual inspection is an important technique by which to assess a registration method for clinical use, but it is subjective and cannot provide a robust, reproducible measure of registration accuracy.

8.7.4 Consistency Measures

Internal consistency measures of transformations provide an elegant method for assessing registration accuracy for suitable image datasets [27, 87, 115–117]. For N images of the same subject, the number of transformations is $p(N,2) = N^* (N-1)$. For three images X_A, X_B, and X_C, if we can apply three transformations $T_{A \to B}$, $T_{B \to C}$, and $T_{C \to A}$ in turn to build a close transformation circuit, then the transformation composition $T_{A \to B} \circ T_{B \to C} \circ T_{C \to A}$ would transform image X_A to itself in the perfect registration process.

Holden et al. [27] proposed the assessment method, which adapted the hierarchical optimization scheme to accelerate the computational speed and avoid local minima. Johnson and Christensen [87] presented a hybrid consistency measurement that fully utilized the advantages of both landmark-based and intensity-based registration methods. However, consistency measures are largely limited to assessing monomodal, rigid registrations.

8.7.5 Investigations of Accuracy Assessment

Researchers have been developing novel and practical validation techniques. Wang et al. [118] proposed a novel automatic method to estimate confidence intervals of the resulting registration parameters and allow the precision of registration results to be objectively assessed for 2D and 3D medical images. Influence of implementation parameters on MRI and SPECT registration by MI was investigated by Zhu and Cochoff [25], and the effect of interpolation artifacts on registration was studied by Tsao [119].

A number of investigations into the accuracy of registration similarity measures have been presented. Penny et al. [112] carried out a comparison of six intensity-based similarity measures for 2D-3D CT registration. The similarity measures were used to register phantom spine images, and their accuracy was compared with the gold standard obtained from fiducial markers. To further test the accuracy and robustness of these similarity measures in the presence of soft-tissue structures and other clinically encountered features, the researchers simulated more realistic gold standard data by overlaying clinical image features on the phantom images.

An assessment of eight similarity measures for rigid-body registration of 3D MRI scans was performed by Hill et al. [27]. These similarity measures were assessed by using simulated brain images and internal consistency measures. Based on their investigation, the authors ranked the accuracy of these registration similarity measures, and their results also showed that image noise had a significant effect on registration accuracy.

A validation protocol for selecting the most suitable similarity measure and corresponding optimization method for a certain application has been proposed by Škerl et al. [111]. Nine similarity measures have been tested on multimodal registration, including MRI/CT and MRI/PET of spine phantom images and brain images. Extensive discussions and summaries of performance validation methods for medical image registration are provided in [9, 11, 14]. However, validation of registration accuracy for clinical applications remains a difficult task, and objective performance validation is still a challenge in the field of biomedical image registration.

8.8 Applications of Biomedical Image Registration and Fusion

8.8.1 Applications in the Brain

As with most developments in medical imaging, the first site of research was the brain. It has always been easier to study the brain because it is relatively stationary with respect to the skull, and tomographs with small apertures were cheaper to build. This research also applies to registration, which has been applied to the localization of tumors, eloquent cortex, and regions of dysfunction. These advances have been applied to neurosurgery, localization of seizure foci, detection of disease at an early stage, and monitoring of patients' responses to treatments.

8.8.1.1 Disease Progress Monitoring and Atlas Construction

Image registration is an essential step in the monitoring of disease progress and in the automatic construction of brain atlases. Accurate monomodal registration of images obtained over time intervals allows a comparison of serial images. Brain atlases are enabling analytical tools for automatic segmentation, labeling, and interpretation of brain tissue and structures. Brains differ across individuals, so by mapping a large amount of brain image data to a common coordinate system, statistical brain atlases can provide anatomical and functional information and interpretation for a specific population group [120]. Even for the same individual, the brain will change over the lifetime. Dynamic brain atlases reveal brain changes due to such factors as age, gender, or disease. For instance, population-based, disease-specific brain atlases provide a template for early-stage brain disease detection and identification [121].

8.8.1.2 Early Detection of Neurodegenerative Disease

The dementias are the major causes of disability in the elderly population. Alzheimer's disease (AD) is the most common cause of dementia [122]. Early detection of AD holds the promise of early intervention, which may delay or halt its progression when disease-modifying agents become available. The comparison of data from subjects at risk for developing a dementing illness with data from normal subjects will be the main focus of disease-modifying agents in the dementias. Preliminary FDG-PET studies have already shown that subtle reductions in glucose metabolism in the posterior cingulate gyrus herald the onset of AD, and these changes, while difficult to detect on routine FDG-PET scans of the brain, can be robustly demonstrated when the subject's PET data are compared with a normal age- and sex-matched atlas. Registration of longitudinal anatomical MRI studies [123–127] allows the identification of probable AD [122]. Voxel-based morphometry (VBM), which is an automated image analysis tool, has been used in the investigation of the presence and severity of anatomical tissue reduction in early-stage AD [128–131]. Alternatively, Thompson and his team performed research in detecting brain abnormalities by using group-specific brain atlases [121, 132–136].

8.8.1.3 Image-Guided Surgery

Most neurosurgical procedures require highly precise localization of the intended cerebral target (e.g., tumor, malformation) to minimize damage to normal structures [137]. Registration is critical in preoperative lesion identification, planning, intraoperative IGS, and postoperative assessment. Conventionally, IGS is done by surgeons' visual perception and examination of 2D anatomical images of MRI or CT scans and then their mentally relating this preoperative image information to the patient in the operation theater. This procedure is very dependent on the surgeon's expertise and experience. Automatic registration techniques that can assist surgeons to directly visualize the accurately fused structures to guide the surgical procedure are clearly advantageous. Registration is a major component in an IGS system. Operation planning is facilitated by the registration of preoperative images or videos [138], and operation navigation can be achieved by the registration of intraoperative images with the patient, as well as the fusion of segmented anatomical structures with the intraoperative images [139, 140].

8.8.1.4 Radiotherapy Planning

Intensity-modulated radiotherapy (IMRT), in which radiation dose to a tumor is varied across its 3D volume, is becoming increasingly utilized to deliver high doses to the most active regions of a tumor. Such an approach maximizes tumor cell death and minimizes damage to normal tissue [141]. Multimodal image registration such as CT/MRI and CT/PET allows more accurate definition of the tumor volume during the treatment planning phase [142]. These datasets can also be used later to assess response to therapy and evaluate suspected tumor recurrence.

8.8.2 Applications Outside the Brain

As mentioned earlier, registration of brain images is easier to achieve than the registration of abdominal or thoracic organs such as the lung, heart, and liver. Outside the brain, organs move relative to each other as a part of normal physiology—the beating heart, peristalsis in the bowel and ureters, respiratory movement of the lungs—and can deform and change shape during and between imaging sessions (e.g., the bladder becomes larger in FDG-PET images because FDG is excreted by the kidneys during the scan). Thus, simple rigid transformations are no longer sufficient, and more complex and challenging elastic transformations are required.

8.8.2.1 Cardiac Image Registration

In developed countries, heart disease is the main cause of death [143]. Cardiac image registration and fusion provide a noninvasive tool that aids diagnosis and risk stratification in patients with heart disease. Registration is essential for the construction of cardiac atlases, as well as for the modeling of heart motion, which is important in the detection of heart disease [144]. The recently introduced hybrid X-ray and MRI system XMR and 64-slice PET/CT systems provide new possibilities for better cardiac diagnosis and management. Image registration of MRI scans and X-ray images is a crucial step in the XMR-guided cardiovascular intervention, as well as in therapy and treatment planning [145, 146]. However, because of motion, low-image resolution, and the lack of anatomical landmarks, cardiac image registration is more complex than brain image registration. The nonrigid and mixed motion of the heart and the thorax structures makes the task even more difficult [62]. Researchers have been devoting considerable effort to finding good registration approaches for cardiac images [37, 153]. Mäkelä et al. [11] presents a good review of cardiac image registration methods.

8.8.2.2 Breast Image Registration

Breast cancer is one of the major causes of cancer-related death, and early detection is a proven approach to reduce its morbidity and mortality. Registration plays an important role in early breast cancer detection. The medical imaging techniques frequently used in breast cancer detection include X-ray mammography, pre- and postcontrast MRI, and ultrasound. Registration of pre- and postcontrast MRI sequences can effectively distinguish different types of malignant and normal tissues [148]. However, breast image registration is not trivial. Due to the elastic property of the breast tissue, one can observe significant temporal changes of breast tissue and shape and motion differences caused by respiration. Breast image registration can be feature based or intensity based. The former involves features such as boundaries, surfaces, and landmark points. Due to the nonrigid deformations at different imaging times and with various imaging equipment, automated breast registration is an ongoing research direction. Maximization of the MI technique has been claimed to be superior to other methods in the intensity-based registration category [148, 149]. A survey of breast image registration is presented in Guo et al. [150].

8.8.2.3 Whole-Body Registration in Oncology Studies for Assessment of Disease Progression and Treatment Response

Whole-body scanning with PET or PET/CT reveals metabolic information at the molecular level and is critical in cancer detection, staging, disease progress, detection of tumor recurrence, and the assessment of treatment response. Whole-body

CT or MRI scans provide information on anatomical changes. These imaging techniques have different properties, such as resolution, radiation exposure, and examination time [151]. Proper registration and fusion to fully utilize the complementary information of these modalities is thus highly desirable.

Newly introduced advanced hybrid imaging techniques such as PET/CT enhance localization, detection, staging, and diagnosis. But as mentioned in Section 8.6, accurate registration between the two modalities is not guaranteed. Furthermore, software-based registration is still required to address the issues of (1) patient motion between the PET and the CT and (2) PET/CT scans performed on the same patient but on different occasions. PET/CT data raise particular challenges for software-based registration. The large data volumes due to the extended scan range require an efficient registration mechanism that can provide registered datasets in a clinically acceptable time frame. In addition, nonrigid, elastic transformation is required in these studies, adding to the complexity and potential computational expense of the registration.

Automatic whole-body registration and fusion for the images from separate PET and CT operations, as well as from the hybrid PET/CT scanners, have been explored and investigated more recently [152–154]. Once again, maximization of mutual information has shown promise for this registration application. However, elastic registration for complex internal organ deformations remains a challenging and ongoing research topic.

8.9 Summary

Due to advances in medical imaging technologies, diverse imaging modalities play an increasingly important role in improving the quality and effectiveness of health care. Medical imaging is now indispensable to good quality care. However, full utilization of these imaging techniques is in its infancy, and the challenge is to seamlessly integrate these data into a user-friendly environment in a rapid time frame. Image registration is a critical element in integrating complementary and heterogeneous medical data into a common coordinate system. Factors such as different imaging principles, patient motion during the imaging procedure, differences due to disease progression or treatment, and complex deformations of internal organs mean that there are still many challenges ahead for medical image registration. The ever-increasing improvements in instrumentation such as 64-slice PET/CT and 128-slice CT and beyond will provide a wider range of imaging datasets with higher resolution and higher dimensionality. Registration will play a key role in managing these datasets effectively.

Apart from applications in the medical imaging environment, data registration and fusion also have wide applications in remote sensing and multimedia areas. For instance, mosaic construction, image/video compression, motion tracking, and content-based retrieval rely on efficient and effective image registration. Data registration and fusion may also have potential and important applications in the research and industry of bio-informatics and biotechnology, such as in the development of new therapeutic agents, by allowing quantitative and qualitative analysis of patterns and expressions of proteins to reveal their biological functions.

Overall, more efficient and automatic image registration will be needed with these applications in the medical and multimedia environments.

Acknowledgments

The authors are grateful to the support from the ARC, PolyU/UGC grants.

8.10 Exercises

1. Describe four main components of the image registration framework; briefly explain the functions of these components; and give one often-used technique for each of them.
2. Explain why maximization of mutual information is more suitable for multimodal image registration than are other intensity-based registration criteria; describe mutual information algorithm based on histogram calculation.
3. Describe why hybrid and hierarchical registration schemes potentially provide more efficient and robust registrations.
4. Explain why registration may still be necessary for data from hybrid scanners, such as PET/CT systems.
5. Give one application of data registration and fusion both in the brain and for an organ outside the brain.

8.11 References and Bibliography

1. C. Chen et al. Image analysis of PET data with the aid of CT and MR images. In *Information Processing in Medical Imaging.* Plenum. 601–611, 1987.
2. L. G. Brown. A survey of image registration techniques. *ACM Computing Surveys.* 24(4):325–376, 1992.
3. C. R. Maurer and J. M. Fitzpatrick. A review of medical image registration. In R. J. Maciunas (Ed.). *Interactive Image Guided Neurosurgery.* American Association of Neurological Surgeons. 17–44, 1993.
4. P. A. van den Elsen, E. J. D. Pol, and M. A. Viergever. Medical Image matching—A review with classification. *IEEE Eng. Med. Biol.* 12:26–39, 1993.
5. J. B. A. Maintz and M.A. Viergever. A survey of medical image registration. *Med. Image Anal.* 2(1):1–36, 1998.

6. H. Lester and S. R. Arridge. A survey of hierarchical nonlinear medical image registration. *Pattern Recognition.* 32:129–149, 1999.
7. I. Bankman. *Handbook of Medical Imaging: Processing and Analysis.* Academic Press, 2000.
8. K. Rohr. Elastic registration of multimodal medical images: A survey. *Auszug aus: Kunstliche Intelligenz.* 3:11–17, 2000.
9. J. M. Fitzpatrick, D. L. G. Hill, and C. R. Maurer. *Handbook of Medical Imaging.* SPIE Press. 375–435, 2000.
10. J. V. Hajnal, D. L. G. Hill, and D. J. E. Hawkes. *Medical Image Registration.* CRC Press, 2001.
11. T. Mäkelä et al. A review of cardiac image registration methods. *IEEE Trans. Med. Imaging.* 21(9):1011–1021, 2002.
12. X. Wang and D. Feng. Biomedical image registration for diagnostic decision making and treatment monitoring. In R. K. Bali (Ed.). *Clinical Knowledge Management: Opportunities and Challenges.* Idea Group Publishing. 159–181, 2005.
13. T. M. Lehmann, C. Gönner, and K. Spitzer. Survey: Interpolation methods in medical image processing. *IEEE Trans. Med. Imaging.* 18(11):1049–1075, 1999.
14. B. Zitová and J. Flusser. Image registration methods: A survey. *Image and Vision Computing.* 21:977–1000, 2003.
15. P. Thévenaz and M. Unser. Optimization of mutual information for multiresolution registration. *IEEE Trans. Image Proc.* 9(12):2083–2099, 2000.
16. F. Maes et al. Multimodality image registration by maximization of mutual information. *IEEE Trans. Med. Imaging.* 16:187–198, 1997.
17. J. Tsao. Efficient interpolation for clustering-based multimodality image registration. *Proc. 7th International Society for Magnetic Resonance in Medicine.* 3:2195, 1999.
18. J. X. Ji, H. Pan, and Z. Liang. Further analysis of interpolation effects in mutual information-based image registration. *IEEE Trans. Med. Imaging.* 22(9):1131–1140, 2003.
19. G. E. Christensen et al. Synthesis of an individual cranial atlas with dysmorphic shape. In *Mathematical Methods in Biomedical Image Analysis.* IEEE Computer Society Press. 309–318, 1996.
20. M. Unser, A. Aldroubi, and C. R. Gerfen. A multiresolution image registration procedure using spline pyramids. *Proc. SPIE Conference on Mathematical Imaging: Wavelet Applications in Signal and Image Processing.* 2034:160–170, 1993.
21. D. Mattes et al. PET-CT image registration in the chest using free-form deformations. *IEEE Trans. Med. Imaging.* 23(1):120–128, 2003.
22. M. J. D. Powell. An efficient method for finding the minimum of a function of several variables without calculating derivatives. *Comput. J.* 7:155–163, 1964.
23. A. Collignon et al. Automated multimodality image registration based on information theory. In Y. Bizais et al. (Eds.). *Proc. 14th International Conference of Information Processing in Medical Imaging: Computational Imaging and Vision.* 3:263–274, 1995.
24. F. Maes, D. Vandermeulen, and P. Suetens. Comparative evaluation of multiresolution optimization strategies for multimodality image registration by maximization of mutual information. *Med. Image Anal.* 3(4):373–386, 1999.
25. Y. Zhu and S. M. Cochoff. Influence of implementation parameters on registration of MR and SPECT brain images by maximization of mutual information. *J. Nucl. Med.* 43(2):160–166, 2002.
26. W. H. Press et al. *Numerical Recipes in C.* Cambridge University Press, 1992.
27. D. L. G. Hill et al. A strategy for automated multimodality image registration incorporating anatomical knowledge and imager characteristics. In H. H. Barrett and A. F. Gmitro (Eds.). *Proc. 13th int. Conf. Information Processing in Medical Imaging: Lecture Notes in Computer Science.* Springer-Verlag. 182–196, 1993.
28. T. Rohlfing and C. C. Maurer. Nonrigid image registration in shared-memory multiprocessor environments with application to brains, breasts, and bees. *IEEE Trans. Info. Tech. Biomed.* 7(1):16–25, 2003.
29. J. P. Pluim, J. B. A. Maintz, and M. A. Viergever. Mutual-information-based registration of medical images: A survey. *IEEE Trans. Med. Imaging.* 22(8):986–1004, 2003.
30. U. Kjems et al. Enhancing the multivariate signal of ^{15}O water PET studies with a new nonlinear neuroanatomical registration algorithm. *IEEE Trans. Med. Imaging.* 18(4): 306–319, 1999.
31. K. F. Man, K. S. Tang, and S. Kwong. Genetic algorithms: Concepts and applications. *IEEE Trans. Industrial Electronics.* 43(5):519–534, 1996.
32. G. K. Matsopoulos et al. Automatic retinal image registration scheme using global optimization techniques. *IEEE Trans. Info. Tech.* 3(1):47–60, 1999.
33. M. P. Wachowiak et al. An approach to multimodal biomedical image registration utilizing particle swarm optimization. *IEEE Trans. Evolutionary Computation.* 8(3):289–301, 2004.
34. M. P. Wachowiak and T. M. Peters. High-performance medical image registration using new optimization techniques. *IEEE Trans. Info. Tech. Biomed.* 10(2):344–353, 2006.
35. J. M. Fitzpatrick et al. Visual assessment of the accuracy of retrospective registration of MR and CT images of the brain. *IEEE Trans. Med. Imaging.* 17(8):571–585, 1998.
36. J. B. A. Maintz, P. A. van den Elsen, and M. A. Viergever. Evaluation of ridge seeking operators for multimodality medical image registration. *IEEE Trans. PAMI.* 18(4): 353–365, 1996.
37. G. Subsol, J. P. Thirion, and N. Ayache. A scheme for automatically building three-dimensional morphometric

anatomical atlases: Application to a skull atlas. *Med. Image Anal.* 2(1):37–60, 1998.
38. G. Borgefors. Hierarchical chamfer matching: A parametric edge matching algorithm. *IEEE Trans. Pattern. Anal. Mach. Intell.* 10:849–865, 1988.
39. C. R. Maurer et al. Registration of head volume images using implantable fiducial markers. *IEEE Trans. Med. Imaging.* 16(4):447–461, 1997.
40. T. Peters et al. Three-dimensional multimodal image guidance for neurosurgery. *IEEE Trans. Med. Imaging.* 15(2):121–128, 1996.
41. J. P. Thirion. New feature points based on geometric invariants for 3-D image registration. *Int. J. Comput. Vis.* 18(2):121–137, 1996.
42. P. J. Besl and N. D. McKay. A method for registration of 3-D shapes. *IEEE Trans. PAMI.* 14(2):239–256, 1992.
43. C. R. Maurer, R. J. Maciunas Jr., and J. M. Fitzpatrick. Registration of head CT images to physical space using multiple geometrical features. *Proc. SPIE: Medical Imaging 1998: Image Processing.* 3338:72–80, 1998.
44. C. A. Kapoutsis, C. P. Vavoulidis, and I. Pitas. Morphological iterative closest algorithm. *IEEE Trans. Image Proc.* 8(11), 1999.
45. F. L. Bookstein. Principal warps: Thin-plate splines and the decomposition of deformations. *IEEE Trans. Patt. Anal. Mach. Intell.* 11(6):567–585, 1989.
46. F. L. Bookstein. Thin-plate splines and the atlas problem for biomedical images. In *Information Processing in Medical Imaging: Proc. 12th International Conference (IPMI'91).* 326–342, 1991.
47. K. Rohr et al. Landmark-based elastic registration using approximating thin-plate splines. *IEEE Trans. Med. Imaging.* 20(6):526–534, 2001.
48. O. Monga, N. Ayache, and P. T. Sander. From voxel to intrinsic surface features. *Image Vision Computing.* 10:403–417, 1992.
49. P. A. van den Elsen et al. Automatic registration of CT and MR brain images using correlation of geometrical features. *IEEE Trans. Med. Imaging.* 14(2):384–396, 1995.
50. C. Davatzikos, J. L. Prince, and R. N. Bryan. Image registration based on boundary mapping. *IEEE Trans. Med. Imaging.* 15(1):112–115, 1996.
51. M. Kass, A. Witkin, and D. Terzopoulos. Snakes: Active contour models. *Int. J. Computer Vision.* 1988:321–331.
52. D. Shen, E. H. Herskovits, and C. Davatzikos. An adaptive-focus statistical shape model for segmentation and shape modeling of 3-D brain structures. *IEEE Trans. Med. Imaging.* 20(4):257–270, 2001.
53. H. Wang and B. Ghosh. Geometric active deformable models in shape modeling. *IEEE Trans. Image Proc.* 9(2):302–308, 2000.
54. S. Ranganath. Contour extraction from cardiac MRI studies using snakes. *IEEE Trans. Med. Imaging.* 14(2): 328–338, 1995.
55. L. Tsap, D. B. Goldgof, and S. Sarkar. Fusion of physically-based registration and deformation modeling for non-rigid motion analysis. *IEEE Trans. Image Proc.* 10(11): 1659–1669, 2001.
56. M. Moshfeghi, S. Ranganath, and K. Nawyn. Three-dimensional elastic matching of volumes. *IEEE Trans. Image Proc.* 3(2):128–138, 1994.
57. P. King, S. Mitra, and B. Nutter. An automated, segmentation-based, rigid registration system for Cervigram™ images utilizing simple clustering and active contour techniques. *Proc. 17th IEEE Symposium on Computer-Based Medical Systems.* 292–297, 2004.
58. X. Wang and D. Feng. Active contour based efficient registration for biomedical brain images. *J. Cereb. Blood Flow Metab.* 25(Suppl):S623, 2005.
59. I. Cohen and I. Cohen. Finite-element methods for active contour models and balloons for 2-D and 3-D images. *IEEE Trans. Pattern. Anal. Mach. Intell.* 15:1131–1147, 1993.
60. C. Xu and J. L. Prince. Snakes, shapes, and gradient vector flow. *IEEE Trans. Image Proc.* 7:359–369, 1998.
61. J. Liang, T. McInerney, and D. Terzopoulos. United snakes. *Med. Image Anal.* 10:215–233, 2006.
62. T. Mäkelä et al. A 3-D model-based registration approach for the PET, MR and MCG cardiac data fusion. *Med. Image Anal.* 7:377–389, 2003.
63. O. Camara, G. Delso, and I. Bloch. Free form deformations guided by gradient vector flow: A surface registration method in thoracic and abdominal PET-CT applications. In J. C. Gee et al. (Eds.). *Workshop on Biomedical Image Registration (WBIR) 2003,* LNCS 2717:224–233, 2003.
64. S. Osher and J. A. Sethian. Fronts propagating with curvature dependent speed: Algorithms based on Hamilton-Jacobi formulations. *J. Computational Physics.* 79:12–49, 1988.
65. C. Xu, A. J. Yezzi, and J. L. Prince. On the relationship between parametric and geometric active contours. *Proc. 24th Asilomar Conference on Signals, Systems, and Computers.* 483–489, 2000.
66. X. Han, C. Xu, and J. L. Prince. A topology preserving level set method for geometric deformable models. *IEEE Trans. Pattern Anal. Mach. Intell.* 25(6):755–768, 2003.
67. B. C. Vemuri et al. Image registration via level-set motion: Application to atlas-based segmentation. *Med. Image Anal.* 7:1–20, 2003.
68. W. C. Lee, M. Tublin, and B. Chapman. Registration of MR and CT images of the liver: Comparison of voxel similarity and surface based registration algorithms. *Comput. Methods Programs Biomed.* 78(2):101–114, 2005.
69. R. Bajcsy and S. Kovacic. Multiresolution elastic matching. *Comp. Vision Graphics Image Processing.* 46:1–21, 1989.
70. D. Terzopoulos and D. Metaxas. Dynamic 3D models with local and global deformations: Deformable superquadrics. *IEEE Trans. PAMI.* 13(7):703–714, 1991.

71. M. Audette, F. Ferrie, and T. Peters. An algorithm overview of surface registration techniques for medical imaging. *Med. Image Anal.* 4(4):201–217, 2000.
72. N. M. Alpert et al. The principal axes transformation: A method for image registration. *J. Nucl. Med.* 31:1717–1722, 1990.
73. R. P. Woods, S. R. Cherry, and J. C. Mazziotta. Rapid and automated algorithm for aligning and reslicing PET images. *J. Comput. Assist. Tomogr.* 16:620–633, 1992.
74. R. P. Woods, J. C. Mazziotta, and S. R. Cherry. MRI-PET registration with automated algorithm. *J. Comput. Assist. Tomogr.* 19(4):536–546, 1993.
75. P. A. Viola and W. M. Wells. Alignment by maximization of mutual information. *Proc. 5th International Conference of Computer Vision.* 16–23, 1995.
76. M. K. O'Connor et al. Comparison of four motion correction techniques in SPECT imaging of the heart: A cardiac phantom study. *J. Nucl. Med.* 39:2027–2034, 1998.
77. W. M. Wells et al. Multi-modal volume registration by maximization of mutual information. *Med. Image Anal.* 1:35–51, 1996.
78. F. Maes, D. Vandermeulen, and P. Suetens. Medical image registration using mutual information. *Proc. IEEE.* 91(10):1699–1722, 2003.
79. B. Kim et al. Mutual information for automated unwarping of rat brain autoradiographs. *Neuroimage.* 5:31–40, 1997.
80. C. Studholme, D. L. G. Hill, and D. J. Hawkes. An overlap invariant entropy measure of 3D medical image alignment. *Pattern Recognit.* 32:71–86, 1999.
81. J. P. Pluim, J. B. A. Maintz, and M. A. Viergever. Image registration by maximization of combined mutual information and gradient information. *IEEE Trans. Med. Imaging.* 19:809–814, 2000.
82. A. Roche, G. Malandain, and N. Ayache. Unifying maximum likelihood approaches in medical image registration. *Int. J. Imaging Systems and Technology.* 11:71–80, 2000.
83. C. Studholme. *Measures of 3D Medical Image Alignment.* Ph.D thesis. University of London, 1997.
84. Y. M. Zhu. Volume image registration by cross-entropy optimization. *IEEE Trans. Med. Imaging.* 21(2):174–180, 2002.
85. Y. He, A. B. Hamza, and H. Krim. A generalized divergence measure for robust image registration. *IEEE Trans. Sig. Proc.* 51:1211–1220, 2003.
86. J. P. Pluim, J. B. A. Maintz, and M. A. Viergever. F-Information measures in medical image registration. *IEEE Trans. Med. Image Registration.* 23(12):1508–1516, 2004.
87. H. Johnson and G. E. Christensen. Consistent landmark and intensity-based image registration. *IEEE Trans. Med. Imaging.* 21(5):450–461, 2002.
88. P. Hellierand and C. Barillot. Coupling dense and landmark-based approaches for nonrigid registration. *IEEE Trans. Med. Imaging.* 22(2):217–227, 2003.
89. T. Chanwimaluang, G. Fan, and S. R. Fransen. Hybrid retinal image registration. *IEEE Trans. Info. Tech. Biomed.* 10(1):129–142, 2006.
90. S. Kovacic and R. Bajcsy. Multiscale/multiresolution representation. In W. Toga (Ed.). *Brain Warping.* Academic Press. 45–65, 1998.
91. P. Thévenaz, U. E. Ruttimann, and M. Unser. A pyramid approach to subpixel registration based on intensity. *IEEE Trans. Image Proc.* 7(1):1–15, 1998.
92. Y. Xie and G. E. Farin. Image registration using hierarchical B-spline. *IEEE Trans. Visualization and Computer Graphics.* 10(1):8594, 2004.
93. S. Periaswamy and H. Farid. Elastic registration in the presence of intensity variations. *IEEE Trans. Med. Imaging.* 22(7):865–874, 2003.
94. P. Thévenaz, U. E. Ruttimann, and M. Unser. Iterative multi-scale registration without landmarks. *IEEE Int. Conf. Image Proc.* 1995:228–231.
95. P. Brigger et al. Centered pyramids. *IEEE Trans. Image Proc.* 8(9):1254–1264, 1999.
96. G. Pajares and J. M. de la Cruz. A wavelet-based image fusion tutorial. *Pattern Recognit.* 37:1855–1872, 2004.
97. S. Mallat. A theory of multiresolution signal decomposition: The wavelet representation. *IEEE Trans. Pattern Anal. Mach. Intell.* PAMI-11:674–693, 1989.
98. E. P. Simoncelli et al. Shiftable multiscale transforms. *IEEE Trans. Inform. Theory.* 38:587–607, 1992.
99. J. Liang and T. W. Parks. A translation-invariant wavelet representation algorithm with applications. *IEEE Trans. Sig. Proc.* 44:225–232, 1996.
100. A. A. Cole-Rhodes et al. Multiresolution registration of remote sensing imagery by optimization of mutual information using a stochastic gradient. *IEEE Trans. Image Proc.* 12(12):1495–1511, 2003.
101. D. Tsaiand and C. Chiang. Rotation-invariant pattern matching using wavelet decomposition. *Pattern Recogn. Lett.* 23:191–201, 2002.
102. Z. Xue, D. Shenand, and C. Davatzikos. Determining correspondence in 3-D MR brain images using attribute vectors as morphological signatures of voxels. *IEEE Trans. Med. Imaging.* 23(10):1276–1291, 2004.
103. J. You and P. Bhattacharya. A wavelet-based coarse-to-fine image matching scheme in a parallel virtual machine environment. *IEEE Trans. Image Proc.* 9(9):1547–1559, 2000.
104. X. Wangand and D. Feng. An efficient wavelet-based biomedical registration for abdominal images. *J. Nucl. Med.* 46:P161p, 2005.
105. S. Gefen et al. Surface alignment of an elastic body using a multiresolution wavelet representation. *IEEE Trans. Biomed. Eng.* 51(7):1230–1241, 2004.
106. D. W. Townsend, T. Beyer, and T. M. Blodgett. PET/CT scanners: A hardware approach to image fusion. *Semin. Nucl. Med.* 34(3):193–204, 2003.

107. D. Delbeke. Incremental value of imaging structure and function. In P. E. Valk et al. (Eds.). *Positron Emission Tomography: Clinical Practice.* Springer-Verlag, 17–26, 2006.
108. T. Beyer et al. A combined PET/CT scanner for clinical oncology. *Eur. J. Nucl. Med.* 41(8):1369–1379, 2000.
109. Y. Nakamoto et al. Accuracy of image fusion of normal upper abdominal organs visualized with PET/CT. *Eur. J. Nucl. Med. Mol. Imaging.* 30(4):597–602, 2003.
110. T. Beyer et al. Dual-modality PET/CT imaging: The effect of respiratory motion on combined image quality in clinical oncology. *Eur. J. Nucl. Med. Mol. Imaging.* 30(4):588–596, 2003.
111. D. Škerl, B. Likar, and F. Pernuš. A protocol for evaluation of similarity measures for rigid registration. *IEEE Trans. Med. Imaging.* 25(6):779–791, 2006.
112. G. P. Penny et al. A comparison of similarity measures for use in 2D-3D medical image registration. *IEEE Trans. Med. Imaging.* 17(8):586–595, 1998.
113. J. A. Schnabel et al. Validation of nonrigid image registration using finite-element methods: Application to breast MR images. *IEEE Trans. Med. Imaging.* 22(2):238–247, 2003.
114. E. R. E. Denton et al. Comparison and evaluation of rigid and nonrigid registration of breast MR images. *J. Comput. Assist. Tomogr.* 23(5):800–805, 1999.
115. R. P. Woods et al. Automated image registration: I. General methods and intrasubject, intramodality validation. *J. Comput. Assist. Tomogr.* 22:141–154, 1998.
116. E. R. E. Denton et al. The identification of cerebral volume changes in treated growth hormone deficient patients using serial 3-D MR image processing. *J. Comput. Assist. Tomogr.* 24(1):139–145, 2000.
117. G. E. Christensen and H. Johnson. Consistent image registration. *IEEE Trans. Med. Imaging.* 20:568–582, 2001.
118. H. S. Wang et al. Objective assessment of image registration results using statistical confidence intervals. *IEEE Trans. Nucl. Science.* 48:106–110, 2001.
119. J. Tsao. Interpolation artifacts in multimodality image registration based on maximization of mutual information. *IEEE Trans. Med. Imaging.* 22(7):854–864, 2003.
120. D. Rueckert, A. F. Frangi, and J. A. Schnabel. Automatic construction of 3-D statistical deformation models of the brain using nonrigid registration. *IEEE Trans. Med. Imaging.* 22(8):1014–1025, 2003.
121. A. W. Toga and P. M. Thompson. The role of image registration in brain mapping. *Image and Vision Computing Journal.* 19:3–24, 2001.
122. P. J. Nestor, P. Scheltens, and J. R. Hodges. Advances in the early detection of Alzheimer's disease. *Nat. Rev. Neurosci.* 5(Suppl):S34–S41, 2004.
123. N. C. Fox, P. A. Freeborough, and M. N. Rossor. Visualisation and quantification of rates of atrophy in AD. *Lancet.* 348:94–97, 1996.
124. N. C. Fox et al. Using serial registered brain magnetic resonance imaging to measure disease progression in Alzheimer's disease: Power calculation and estimates of sample size to detect treatment effects. *Arch. Neurol.* 57:339–344, 2000.
125. W. R. Crum, R. I. Scahill, and N. C. Fox. Automated hippocampal segmentation by regional fluid registration of serial MRI: Validation and application in Alzheimer's disease. *Neuroimage.* 13:847–55, 2001.
126. W. R. Crum, T. Hartkens, and D. L. G. Hill. Non-rigid image registration: Theory and practice. *Br. J. Radiol.* 77:S140–S153, 2004.
127. R. I. Scahill et al. A longitudinal study of brain volume changes in normal aging using serial registered magnetic resonance imaging. *Arch. Neurol.* 60(7):989–994, 2003.
128. G. B. Frisoni et al. Detection of grey matter loss in mild Alzheimer's disease with voxel based morphometry. *J. Neurol. Neurosurg. Psychiatr.* 73:657–64, 2002.
129. G. B. Frisoni and P. Massimo. Multiple sclerosis and Alzheimer's disease through the looking glass of MR imaging. *Am. J. Neuroradiol.* 26:2488–2491, 2005.
130. G. E. Busatto et al. A voxel-based morphometry study of temporal lobe gray matter reductions in Alzheimer's disease. *Neurobiol. Aging.* 24:221–231, 2003.
131. K. Ishii et al. Voxel-based morphometric comparison between early- and late-onset mild Alzheimer's disease and assessment of diagnostic performance of *z*-score images. *Am. J. Neuroradiol.* 26:333–340, 2005.
132. P. M. Thompson et al. Mathematical/computational challenges in creating deformable and probabilistic atlases of the human brain. *Hum. Brain Mapp.* 9:81–92, 2000.
133. P. M. Thompson et al. Detecting dynamic (4D) profiles of degenerative rates in AD patients, using tensor mapping and a population-based brain atlas. *Proc. Soc. Neurosci.*
134. P. M. Thompson et al. Cortical change in Alzheimer's disease detected with a disease-based atlas. *Cerebral Cortex.* 11:1–16, 2001.
135. P. M. Thompson et al. Structural abnormalities in the brains of human subjects who use methamphetamine. *J. Neurosci.* 24(26):6028–6036, 2004.
136. A. W. Toga and P. M. Thompson. Temporal dynamics of brain anatomy. *Annu. Rev. Biomed. Eng.* 5:119–145, 2003.
137. W. E. L. Grimson. Medical applications of image understanding. *IEEE Expert.* 18–28, 1995.
138. M. J. Clarkson et al. Registration of multiple video images to preoperative CT for image-guided surgery. In K. M. Hanson (Ed.). *Proceedings of SPIE: Medical Imaging 1999: Image Processing.* 3661:14–23, 1999.
139. A. Raabe et al. Laser surface scanning for patient registration in intracranial image-guided surgery. *Neurosurgery.* 50(4):797–803, 2002.
140. M. Audette et al. An integrated range-sensing, segmentation and registration framework for the characterization

of intra-surgical brain deformations in image-guided surgery. *Comput. Vis. Image Underst.* 89:226–251, 2003.
141. J. G. Rosenman et al. Image registration: An essential part of radiation therapy treatment planning. *Int. J. Radiat. Oncol. Biol. Phys.* 40(1):197–205, 1998.
142. C. Scarfone et al. Prospective feasibility trial of radiotherapy target definition for head and neck cancer using 3-dimensional PET and CT imaging. *J. Nucl. Med.* 45(4):543–552, 2004.
143. American Heart Association. *Heart and Stroke Statistical Update*, 2006. http://www.american heart.org
144. A. F. Frangi et al. Automatic construction of multiple-object three-dimensional statistical shape models: Application to cardiac modeling. *IEEE Trans. Med. Imaging.* 21(9):1151–1166, 2002.
145. K. S. Rhode et al. A system for real-time XMR guided cardiovascular intervention. *IEEE Trans. Med. Imaging.* 24(11):1428–1440, 2005.
146. M. Sermesant et al. Simulation of cardiac pathologies using an electromechanical biventricular model and XMR interventional imaging. *Med. Image Anal.* 9:467–480, 2005.
147. K. McLeish et al. A study of motion and deformation of the heart due to respiration. *IEEE Trans. Med. Imaging.* 21(9):1142–1150, 2002.
148. D. Rueckert et al. Non-rigid registration of breast MR images using mutual information. *MICCAI '98 lecture notes in computer science.* Cambridge. 1144–1152, 1998.
149. T. Bruckner. Comparison of rigid and elastic matching of dynamic magnetic resonance mammographic images by mutual information. *Med. Phys.* 27(10):2456–2461, 2000.
150. Y. Guo et al. Breast image registration techniques: A survey. *Med. Bio. Eng. Comput.* 44:15–26, 2006.
151. M. D. Seemann. Whole-body PET/MRI: The future in oncological imaging. *Technol. Cancer Res. Treat.* 4(5):577–582, 2005.
152. R. Shekhar et al. Automated 3-dimensional elastic registration of whole-body PET and CT from separate or combined scanners. *J. Nucl. Med.* 46(9):1488–1496, 2005.
153. P. Slomka et al. Automated 3-dimensional registration of standalone F-FDG whole-body PET with CT. *J. Nucl. Med.* 44(7):1156–1166, 2003.
154. Y. Nakamoto et al. Accuracy of image fusion using a fixation device for whole-body cancer staging. *Am. J. Roentgenol.* 184:1960–1966, 2005.

9

Data Visualization and Display

Dr. Jinman Kim,[1]
Dr. Tom Weidong Cai,[1]
Prof. Michael Fulham,[1,2]
Dr. Stefan Eberl,[1,2] and
Prof. David Dagan Feng[1,3]
[1] *University of Sydney*
[2] *Royal Prince Alfred Hospital*
[3] *Hong Kong Polytechnic University*

9.1 Introduction

This chapter describes data visualization and display techniques for multidimensional biomedical images.

Advances in digital biomedical imaging are enabling unprecedented visualization of the structure, function, and pathology of the human body. These images can be acquired in multiple dimensions and with multiple modalities, including magnetic resonance imaging (MRI) and positron emission tomography coupled with computed tomography (PET/CT). These new techniques have also introduced significant challenges for efficient visualization [1–3]. In line with advances in image acquisition, methods to visualize and display these data have also seen rapid developments and challenges. PET/CT was introduced in 2000 with a PET scanner coupled to an early-generation helical CT scanner, but in the space of five years the leading-edge PET/CT has a 64-slice CT coupled to a PET scanner that has much better resolution and sensitivity than the device of 2000. A major challenge now is to put these vast amounts of imaging data in a readily usable and viewable format for interpretation. Fortunately, there has also been tremendous progress in *three-dimensional volume visualization* of biomedical data. In general, this refers to the ability to interact and navigate the image data in a realistic 3D volumetric display. These volumetric displays are typically constructed from 2D slice images that are acquired in a regular pattern (e.g., one slice every millimeter) and make up a volumetric grid. Rapidly improving capabilities for 3D visualization have made this an attractive method for imaging applications, including those geared toward image-guided surgery (IGS), radiotherapy, and computer-aided diagnosis (CAD) [1, 4–11].

With current visualization technologies, it is possible to perform real-time interactive visualization of multidimensional volumes using low-cost hardware instead of restricting it to expensive high-end workstations [12, 13]. These visualization advances have been accompanied by developments and improvements in display and input control devices. In current diagnostic workstations, it is common to find multiple liquid crystal display (LCD) screens to visualize the imaging data in 2D and 3D views. Specialized display devices, such as stereoscopic screens, enable new approaches to visualize and interact with biomedical data; virtual reality (VR) systems, for instance, are able to provide a realistic image alongside highly interactive control of the visualization process [2, 14]. To provide convenient and efficient navigational control of 3D visualizations, new input devices designed for 3D controls have also been introduced [15–17]. Appropriate biomedical visualization and display, given the vast amount of data, is now mandatory to ensure accurate diagnosis. Moreover, intelligent and innovative new visualization algorithms are necessary in order to overcome the increase in image dimensions that are making conventional viewing approaches inefficient.

This chapter describes data visualization and display techniques for multi-dimensional biomedical images. Section 9.2 discusses 2D visualization techniques that include multiplanar reformatting and oblique/curved sectioning. This is followed by a section on volume rendering techniques (surface-based, direct, and texture-based rendering) for 3D volumetric data and for multivariate data (greater than 3D, i.e., time-varying, dynamic sequence). Section 9.4 introduces methods and devices that are used to navigate through volumetric data. Enhancements in volume visualization are then discussed in Section 9.5, followed by visualization optimization methods (hardware and software approaches) in Section 9.6. A case study on dual-modality PET/CT visualization is presented in Section 9.7, which illustrates how advances in visualization techniques are adapted for use with biomedical data to improve diagnosis and image understanding. Display devices and technologies are presented in Section 9.8, which is followed by a summary of cutting-edge biomedical data visualization projects in Section 9.9. A summary of the chapter is then given in Section 9.10.

9.2 Two-Dimensional Visualization Techniques

Conventional methods of visualizing volumetric biomedical images utilize 2D coronal or transaxial views, with multiple images viewed in *montage* or *slice-by-slice* formats. With improvements in the resolution of imaging devices and the vast amount of data they generate, it is now virtually impossible to rely on the previous approaches for image analysis. This has led to the development of new 2D approaches for biomedical data visualization, such as volume slicing, multiplanar reformatting, and curved sectioning, which provide views complementary to conventional 2D views.

9.2.1 Multiplanar Reformatting

Volumetric data allow the voxels in the volume to be reformatted into different orthogonal orientations, namely, transaxial, sagittal, and coronal views. These orthogonal views are often displayed simultaneously, as shown in Figure 9.1, exemplified by a brain MR dataset [18]. Multiplanar images provide an effective tool to visualize volumetric data with interactive control and are the default visualization setup in many diagnostic applications [5, 12, 19]. By simultaneously displaying the three orthogonal views, they enable rapid observation of the volumetric data. These views are often navigated through the "click and drag" method of manipulating a viewpoint within one of the three views of the volume, causing other views to be reformatted according to the new point position. Multiplanar sectioning, as shown in Figure 9.1(d), is another popular approach toward viewing multiplanar images, visualizing the three orthogonal views stacked perpendicular to each other, thereby creating a 3D visualization.

9.2.2 Oblique and Curved Sectioning

In various clinical applications, the required 2D views may not necessarily lie parallel to the orthogonal orientation of the 3D volume image, as obtained from multiplanar reformatting. It is at times necessary to view cross sections made at arbitrary angles through the volume. Oblique sectioning, also known as volume slicing, is a technique that cuts the volume in any conceivable orientation by a user-defined cutting plane [20].

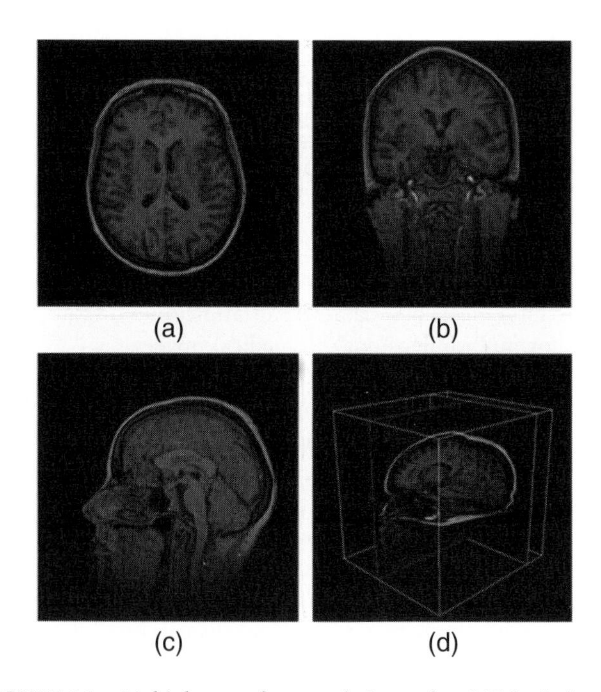

FIGURE 9.1 Multiplanar reformatted views of an MR brain image. (a) Transaxial. (b) Coronal. (c) Sagittal. Multiplanar sectioning shown in (d) is the result of stacking images together.

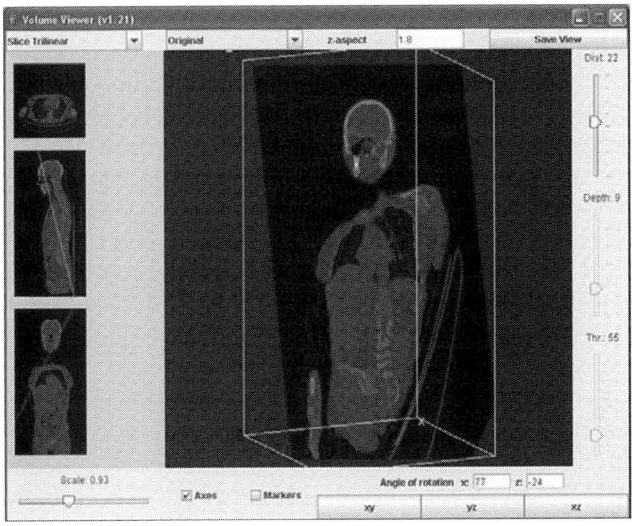

FIGURE 9.2 Oblique sectioning applied to a whole-body CT dataset using ImageJ [22] software with the Volume Viewer plug-in [23]. Thumbnails at left are the cutting planes applied to different orthogonal views, and the resulting 3D volume is in the center. Trilinear interpolation has been applied to volume-rendered images.

This endows users with the ability to depict the inner structure of the volume by removing arbitrary parts of the volume that obscure the primary *volumes of interest* (VOIs). An example of oblique sectioning is shown in Figure 9.2, which demonstrates the removal of occluding sections from a whole-body CT volume. Instead of utilizing a straight line (cutting plane) selection, as in oblique sectioning, curved sectioning traces along an arbitrary path on any orthogonal image to construct an image. Calculation of the voxels on the cutting plane is then essentially performed via an interpolation or resampling operation such as trilinear or nearest-neighbor interpolation [21]. Another approach that has been shown to produce a good selection of viewable data is positioning and resizing a *clipping cube box,* which encapsulates the volume such that only the volumetric data residing inside the box are visible.

9.3 Three-Dimensional Visualization Techniques

In 3D visualization of biomedical data, there are four common techniques that may be used to maximize the visual information, which can be obtained from different types of biomedical data. These visualization techniques—surface rendering, direct volume rendering, texture-based volume rendering, and multivariate rendering—can be used independently or in combination. The first three visualization techniques deal with having different ways to render the data, with each technique having its advantages and limitations. Figure 9.3 compares the three techniques in rendering the bone structure from a CT scan. Rendering was performed using marching-cubes surface rendering [24], direct volume rendering, and texture-based volume rendering, all available through the visualization tool kit (VTK) [25]. In these rendering techniques, appropriate parameters were applied to select the bone structure. In the figure, all three rendering methods appear to be equally matched in visual qualities. However, the marching-cubes method (see the next section) renders only the approximate shapes of the bones (calculating polygonal surfaces from the voxels), whereas the other methods render every voxel belonging to the bones. For multivariate rendering, these techniques are used alongside others to visualize data that are in multiple states, such as time-varying and dynamic data. Differences in rendering techniques are explained in greater detail in the following sections.

9.3.1 Surface Rendering

Surface rendering techniques build a geometrical contour representation of the surface defined by the segmentation of the image volume. The contours of the segmented volume(s) are then extracted with surface tiling techniques [21, 24, 26, 27], which creates polygonal surfaces representing the structure. One of the most well-known algorithms for surface tiling is the *marching-cubes algorithm* [24], which functions by creating

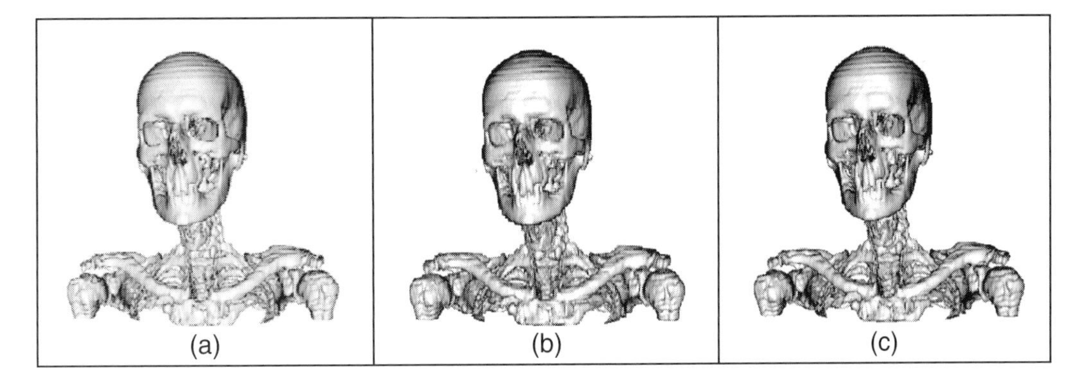

FIGURE 9.3 Rendering of bone structures from a CT dataset using three different rendering algorithms: (a) surface rendering, (b) direct volume rendering, and (c) texture-based volume rendering.

triangulated representations of the surface and has produced many optimizations and variations for different applications. The main attraction that surface rendering offers is its ability to leverage the advances in modern graphics hardware that is specifically designed to process large numbers of polygons, providing the ability for real-time volume rendering. In addition, lighting and shading models can be applied to the surfaces (e.g., Phong shading, Gouraud shading [28, 29]), which aid in depth perception and overall realism of the volume.

Generally, surface rendering techniques require segmentation of the image volume to determine the structures to render. Segmentation as a preprocess is the main disadvantage of surface rendering, due to the complexities in accurate structure delineation and computation. There are numerous segmentation techniques available that are optimized for individual imaging modalities. However, accurate recognition and delineation of all individual structures in an image is likely to be restricted only to controlled environments [2, 30, 31]. Due to the reliance on segmentation results, once preprocessing has been performed, it is computationally inefficient to modify the parameters that generated the surfaces. A further limitation of surface rendering is that only the surface is rendered, and hence potentially important information about structures and pathologies inside the surface-rendered organ or structure is lost.

9.3.2 Direct Volume Rendering

The direct volume method renders every voxel in the image and thus differs from surface rendering in that it does not require surface extraction of the image data to be visualized. Rendering of every voxel allows for natural geometrical structures and the representation of the complete volume data. However, due to the calculations required for every voxel, direct volume rendering is computationally expensive. Real-time rendering performance for direct volume rendering is achievable only by using commodity graphics hardware through the use of specialized rendering optimization techniques. Section 9.6 covers performance optimization for volume rendering visualization. The most versatile approach to direct volume rendering of biomedical data is through *ray casting* (also known as ray tracing) [3, 20, 32], whose basic principle is the casting of rays starting from the viewer's eye through each renderable voxel in the image. It involves sampling, filtering, lighting, and accumulating voxel colors and opacities as the ray passes through the volume. A computationally fast approach to ray casting is *maximum intensity projection* (MIP), which has found several clinical applications [33–35]. MIP works by projecting the voxel with the maximum intensity that falls in the path of the ray. Although computationally efficient, this approach is limited in illustrating depth and orientation, where the MIP of the volume cannot be distinguished between left/right and front/back. In order to improve the sense of 3D with MIP, animations are often created that consist of MIP rendering with rotated viewing angles, which aids in the viewer's perception of the 3D volume. An example of an MIP of whole-body PET data is shown in Figure 9.4, with selection of frame sequences that are used to animate the volume. Other volume rendering techniques that are favorable to biomedical data visualization are *splatting* [36] and *shear warp* [37, 38], which are designed to improve rendering efficiency at the cost of visual quality.

9.3.3 Texture-Based Volume Rendering

Advances in consumer technologies are resulting in the development of graphic cards that are extremely efficient in their texture mapping capabilities. Texture mapping is the process of

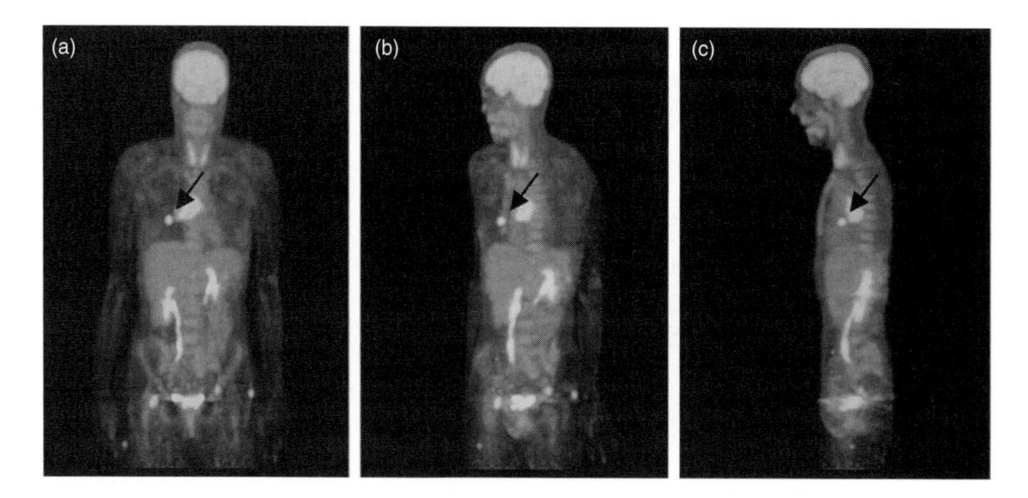

FIGURE 9.4 Sequence of MIP-rendered PET volume data. (a) Coronal view; (b) 45 degrees to the left of (a); and (c) 90 degrees to the left of (a). Animating frames rendered at different angles provide improved perception of the 3D data. These particular data enabled clear identification of the tumor using MIP volume rendering (the bright spot indicated by the arrow).

applying images as textures to geometric objects. This attribute has been utilized in the visualization of biomedical data with real-time interactive capabilities using low-cost consumer hardware [20, 39–41]. The principal concept of texture-based volume rendering is to create parallel planes through the columns of the volume data, in the direction most perpendicular to the user's line of sight. The number of planes to be created is based on the sampling rate, where higher rates equate to better visual quality. In a typical setting, the sampling rate is set proportional to the dimensions of the voxel to be rendered. The parallel planes are converted to polygons that are texture mapped with appropriate 3D texture coordinates [40] that are derived from the volume data and are then drawn in back-to-front order with blending of the voxel colors to result in a smooth image. Texture-based volume rendering produces images that are of comparable quality to direct volume rendering, apart from the noticeable differences during volume manipulations, such as rotation [3]. Due to its image quality and computational efficiency, texture-based volume rendering has become a popular choice in many applications of biomedical data visualization [32, 39, 40, 42, 43].

9.3.4 Multivariate Volume Rendering

Many applications in the field of biomedical visualization require visual outputs of data that contain multiple scalar values at each sample (voxel) point [1, 44–46]. Such data are often referred to as *multivariate data* and can come from numerous applications—for example, multivolume data (dual-modality PET/CT) and time-varying data (ultrasound, flow visualization). In multivolume visualization, two or more complementary datasets are combined and volume-rendered. The combination of these data may be in the form of (1) volume overlay, or overlapping of one volume onto another; or (2) data intermixing, or the calculation of a new voxel value according to two or more values from the same sample. An example of data intermixing of dual-modality biomedical data is discussed in Section 9.7. Time-varying data are frequently used in biomedical applications to visualize the changing scalar values of a selected object over a defined time. Tory et al. [47] introduced *glyphs*, animated arrowhead figures that represent scalar properties (uptake and temperature) with regard to time. Glyphs have been cleverly used to visualize time-varying 3D MRI data for lesion analysis. Thune and Olstad [48], reported animation of 4D ultrasound images (3D spatial and 1D temporal) by direct volume rendering.

9.4 Volume Navigation Interface

In Section 9.2, we introduced 2D approaches to navigate 3D data through the use of a pointer in two dimensions (the *x*- and *y*-axes). In 3D, navigation is based on movement of the volume with an additional *z*-axis. Data with even greater dimensions, such as multivariate data, demand additional interactive features and input devices for efficient control. This section highlights common approaches for the navigation of volume-rendered data.

9.4.1 3D Volume Navigation

Conventional interactive navigational methods for volume rendering consist of scaling, rotation, and translation within the 3D coordinate system. To improve rendering performance during interactive navigation, the sampling rate of the volume can be lowered to accelerate the rendering capabilities [7, 39, 41]. Once the interaction is completed, the sampling rate can be reset to its defaults. Automating the toggle in the sampling rate by listening to changes in mouse interaction allows for progressive refinement during interactive volume rendering. This attribute is particularly important during the rendering of large volume data. Comparison of texture-based rendering performance with regard to varying sampling rates is demonstrated in Figure 9.5.

For 4D or greater-dimensional data, such as time-varying and multimodal volume rendering, additional complexity is added to volume navigation. For time-varying rendering, the volumes are often animated with controls similar to those used for video playback [5]. Such video controls can be used together with 3D navigation, a concept that has found usage in biomedical visualization applications such as those enabling virtual endoscopy [49, 50], which requires movement both in 3D and in the temporal dimension.

9.4.2 Input Devices

In the design of Biomedical data visualization software, due to the complexity in volume navigation, it is often beneficial to customize the input devices. Rosset et al. [12] introduced the use of *jog wheels*, which have been widely adopted in the video and movie industry, thereby providing the ability to control the different dimensions more efficiently than is possible by conventional means, such as a computer mouse. Alternative input devices are readily available that can be used to replace or complement the mouse, such as Hewlett-Packard's 3D SpacePilot [51], consisting of a knob that can be tilted, pushed, pulled, twisted, and turned in a full arc to indicate the direction and velocity with which you wish to move your model. Touch screens have been used as alternative input devices for visualization applications.

With recent developments in and subsequent adaptations of hardware for personal digital assistants (PDAs), tablet personal computers (PCs), and ultramobile PCs (UMPCs), ever greater numbers of visualization applications are taking advantage of stylus and touch-sensitive screens. Figure 9.6 demonstrates the potential use of a touch screen–enabled device for input control to complement a conventional mouse. In this example, the touch-screen input interface was separated from other screens,

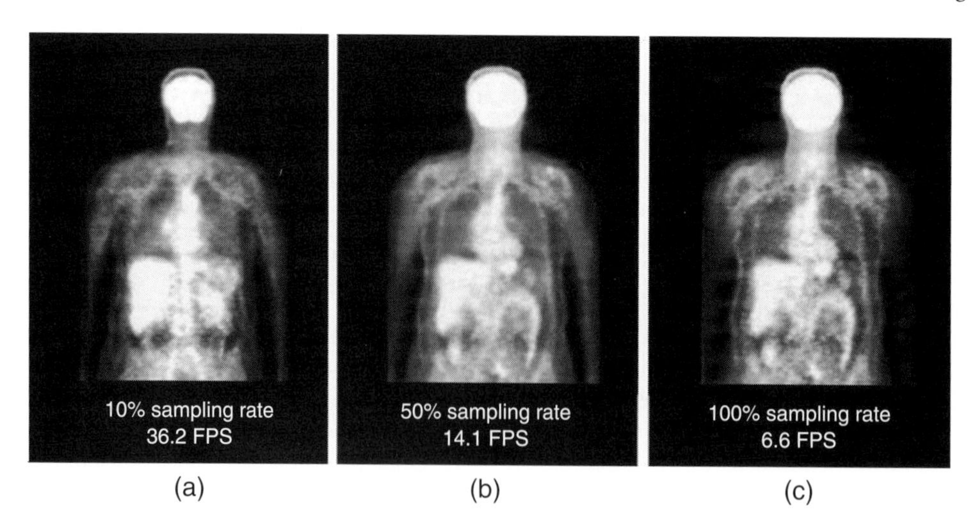

FIGURE 9.5 Different sampling rates applied to whole-body PET data using texture-based volume rendering. The lowest sampling rate has the poorest visual quality, with obvious loss of detail in rendering the internal organ structures. Increasing the sampling rate improves rendering quality. However, there is a decrease in the frames per second (fps) when the volume is interacted with (rotated around all axes). For a more detailed view of this figure, please visit our companion site at: http://books.elsevier.com/companions/9780123735836.

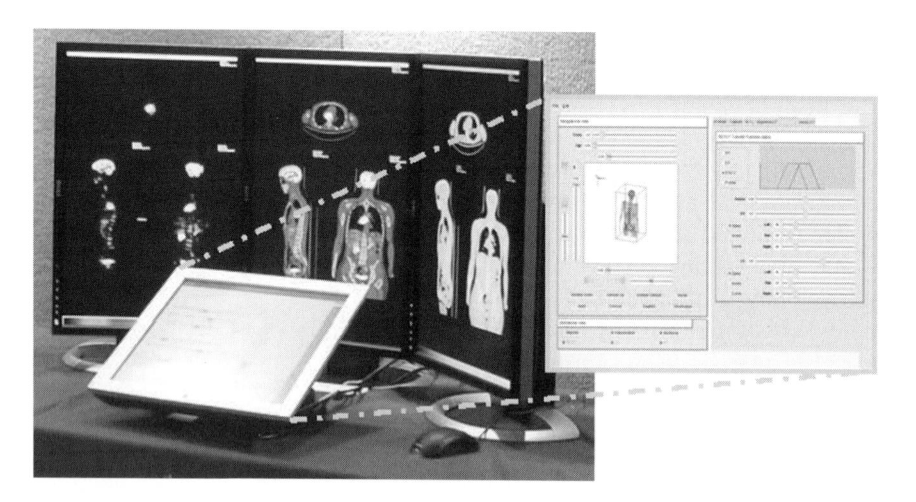

FIGURE 9.6 The touch screen is used to display the input interface that controls the display of data on the other three screens. In this example, the displays are used to visualize PET, fused PET/CT, and CT data, respectively (left to right). By separating the input controls, additional screen space is available to display the images.

thereby freeing the screens to display only images. Such applications can find usefulness for teaching and presentation purposes [15, 52]. Bornik et al. [15] demonstrated a hybrid input device, controlling a stereoscopic projection display (see Section 9.8.1 for stereoscopy) and a desktop display. The input device, named Eye of Ra, is a combination of a wireless mouse, a stylus pen, and motion-tracking sensors. A camera setup is used to track the input device movements, which in turn control the stereo-projected display for rapid and rough control, whereas a stylus pen with mouse buttons is used to control the desktop display for precise control. Kim and Varshney [53] used an eye-tracking device for saliency-guided visualization. Here, based on the region of interest (ROI) on which the eye is focusing, visual emphasis is applied to highlight the ROI. It was shown to be effective in helping users navigate through complex volumetric datasets.

9.5 Volume Enhancement and Manipulation

Enhancement and manipulation techniques applied to 2D slice images, such as filtering, segmentation, and geometrical transformation, can also be extended and applied to 3D image volumes [2, 54]. When extended into the 3D domain, additional information is available, and this therefore generally allows for more accurate (if complicated) methods. An example is the image interpolation that is applied when images are scaled. In 2D data, bilinear interpolation is applied, which calculates a voxel value based on the voxel's eight immediate neighbors. In 3D, trilinear interpolation may be applied instead, which uses an additional 18 neighbors (for a total of 26) in the calculation of a new voxel value, resulting in more accurate interpolation while multiplying the complexity. In volume-rendering visualization, additional enhancements and manipulations are applicable by manipulating 3D attributes such as illumination, shading, and transparency [3, 28, 29].

This section starts with segmentation methods that are fundamental to volume visualization, followed by transfer function techniques that enable the control of visualization attributes for volume enhancements.

9.5.1 Segmentation in Visualization

Segmentation is one of the core components in all image processing, and its outcome dictates the quality of the overall processing [31]. Often in volume rendering, the segmentation of image data is employed in order to separate the VOIs from the entire image. By selecting only the VOIs to be rendered, unnecessary rendering of voxels containing little information of interest can potentially be avoided. In surface rendering, segmentation of VOIs is a mandatory requirement for surface generation. The segmentation in volume rendering is also useful for two-level rendering [13]; that is, a VOI can be rendered using direct volume rendering for optimal quality, while other structures of less importance can be rendered using surface-based techniques.

Recent studies have demonstrated the potential of interactive segmentation in real-time 3D visualization [15, 55–57]. Such methods have the potential to provide a tremendous advantage, since segmentation need not be restricted to being a preprocess, but may also allow us to see resultant rendering as segmentation changes are applied in real time. Furthermore, these methods render only the segmented VOIs, without placing them in the context of surrounding structures. [55] reported a method of correcting segmentation errors from volume-rendered VOIs by adjusting the radius of the viewable volume to reveal the surrounding image, which allows the physician to correct for segmentation errors in volume rendering. The correction was made by rendering only the surrounding voxels within the radius of the VOIs. An interactive segmentation in volume-rendered visualization (texture based) was introduced in [58] for dynamic PET images. Here, the PET images were presegmented using fuzzy c-means cluster analysis, where a fuzzy-logic algorithm assigns probabilistic weightings to every voxel, representing the likelihood that the voxel is a member of a particular segmented cluster (a feature). These weights were then used to interactively adjust the segmentation via computationally efficient thresholding, as shown in Figure 9.7.

9.5.2 Transfer Function Specification

A considerable amount of research has been carried out in the application of transfer functions as a tool for feature selection in volume rendering of biomedical images. A typical feature classification for a particular anatomical structure in volume-

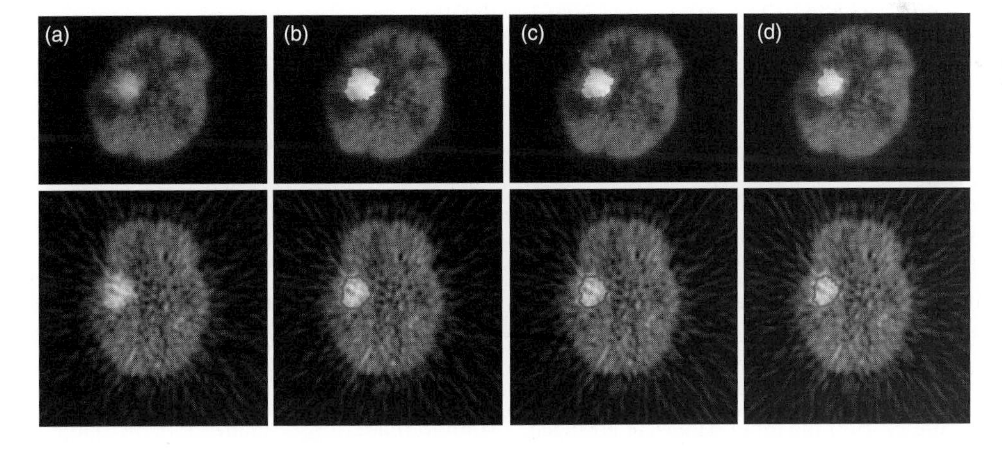

FIGURE 9.7 Interactive segmentation of dynamic PET images, with changes in the segmentation of the brain tumor's definition. Top row is the volume-rendered images of the segmented structures. Bottom row is the 2D representation of these volumes, with segmentation results highlighted in red outline. Image (a) is the original image, and (b) to (d) are the results from varying the segmentation parameter in real-time volume rendering. For a more detailed view of this figure, please visit our companion site at: http://books.elsevier.com/companions/9780123735836.

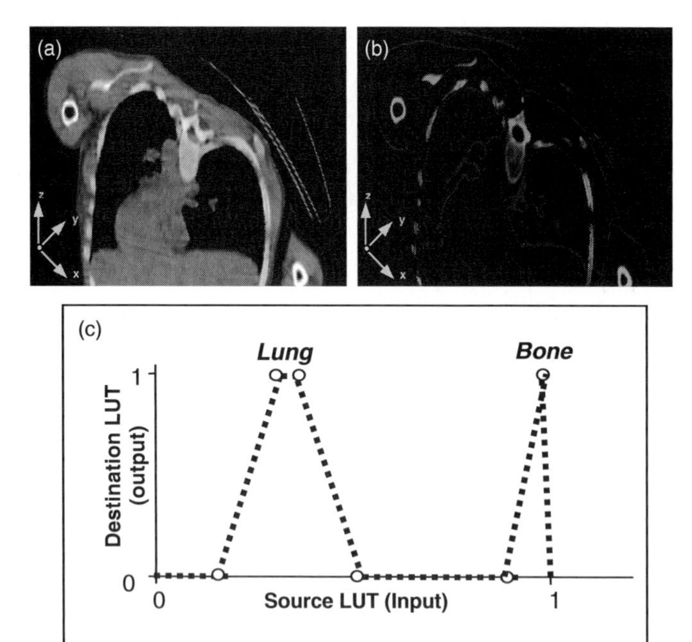

FIGURE 9.8 (a) CT volume of the lung using texture-based rendering; (b) application of transfer function specified in (c) to the volume in (a). A spike-based transfer function appears to produce the most appealing results in revealing structures of interest with gradual fade-out of nearby structures to the selected structures (lung and bone) in CT visualization. For a more detailed view of this figure, please visit our companion site at: http://books.elsevier.com/companions/9780123735836.

rendering visualization can be achieved by a transfer function specification [44, 59, 60]. The transfer function assigns properties such as color, via a lookup table (LUT), and opacity to the voxel data in real time to alter the visualization of the data. Transfer functions are often applied in 1D to the voxels' values from the volume data, as shown in Figure 9.8, which demonstrates the selection of lung nodules and bone in a CT volume.

The capabilities of transfer functions can be extended to multiple dimensions, to selectively visualize data through the control of gradient magnitude and second derivatives of voxels' intensity [44]. Increase in transfer function dimensions also increases complexity. Setting an ideal transfer function specification can be tedious and time consuming, let alone confusing. Automation of transfer function specifications and improvements in user-interface design have the potential to ease their usage and improve efficiency. Konig and Groller [61] introduced a transfer function interface in which variations in data values, color, and opacity attributes from the input data were arranged in thumbnail views for user selection. Meanwhile, Marks et al. [62] reported a design gallery, which presents the user with all possible transfer function variations that are automatically generated and organized based on the input render. A novel and intuitive interface for controlling 3D transfer functions (based on data value, gradient magnitude, and second directional derivatives) was presented by Kniss et al. [44] through a clever use of probing and classification widgets.

9.5.3 Spatial Transfer Function

Section 9.5.1 introduced the importance of incorporating segmentation into the visualization of biomedical data. Instead of rendering the segmentation result, it can be used to specify the transfer function. This allows for the selection of a transfer function based on the spatial properties of the image volume rather than using an LUT, as described in the previous section. An automated approach that removes or suppresses the less important objects of a scene to reveal more important underlying information was reported by investigators [30, 63], who applied different compositing strategies based on the *importance* of the object calculated via image segmentation, which also provides control of the opacity of the voxels. Tzeng and Ma [64] discussed the use of fuzzy-based cluster analysis to segment and classify data into individual objects. This approach enables user manipulation of rendering properties such as LUTs and opacity, as well as fuzzy classification to render spatially related voxels on an object-by-object basis. Zhou et al. [65] reported distance-based volume rendering. Here, the distance of voxels to a focal point was used to control the optical properties of nearby voxels to emphasize the objects of interest and fade out other parts.

9.6 Large Data Visualization and Optimization

Advances in medical scanning technologies are constantly pushing the limits of visualization techniques. Medical data storage commonly exceeds terabytes; such data require advanced computation, are often restricted to dedicated hardware configurations (i.e., parallel rendering), and require software-based optimization. Texture-based volume rendering (Section 9.3.3) is a software-based optimization that permits visualization of large medical data using the ability of modern graphical processing units (GPUs) to render large amounts of textures. Although texture-mapped rendering produces interactive visualization of the data, the capacity of volume rendering is limited to the size of the texture memory in GPUs (currently at 512 MB [megabytes] for a single consumer GPU), and thus it is difficult to maintain an interactive frame rate when large data have to be rendered [39, 66].

This section highlights some of the solutions for optimizing volume-rendering visualization of large biomedical data. A combination of the methods discussed in the subsections that follow can be applied and has been shown to produce particularly good results in texture-based volume rendering [40, 67, 68]. Finally, we show that parallel rendering achieves the computation of a single rendering using several networked computers.

9.6.1 Multiresolution

Multiresolution volume rendering is a common optimization technique that allows applications to interactively render large volume data by assigning multiple *levels of detail* (LODs) in volume rendering [42, 66, 69]. An LOD is used as a controller for the trade-off between quality of rendering and interactive performance and is often applied to data such that when the entire dataset is viewed, a low LOD is applied, which progressively increases in detail as smaller sections of the data are selected for visualization. Multiresolution rendering works by computing data structures that are constructed by decomposition of the data into *bricks* [66], as shown in Figure 9.9, which illustrates variation of rendering resolution according to different LODs. A typical data structure utilized in multiresolution rendering is an *octree*, which is a simple yet efficient algorithm that breaks up the data into uniformly shaped rectangles. The bricks of different LOD levels are combined using interpolation to minimize the visual artifacts that may arise with different LODs.

9.6.2 Empty Space Skipping

At any particular scene during volume rendering, only a portion of the entire data is visualized. *Empty space skipping* is a technique of avoiding the rendering of portions of the volume not being visualized, through the use of precomputed data structures [67, 68]. The principle of empty space skipping is to divide the data into subsections, as in multiresolution rendering. In the rendering phase, visibility testing is performed to determine whether the subsection is visible and should be rendered. There are many data structures that can be used to partition the data, which have optimal performance under varying visualization requirements [67, 70].

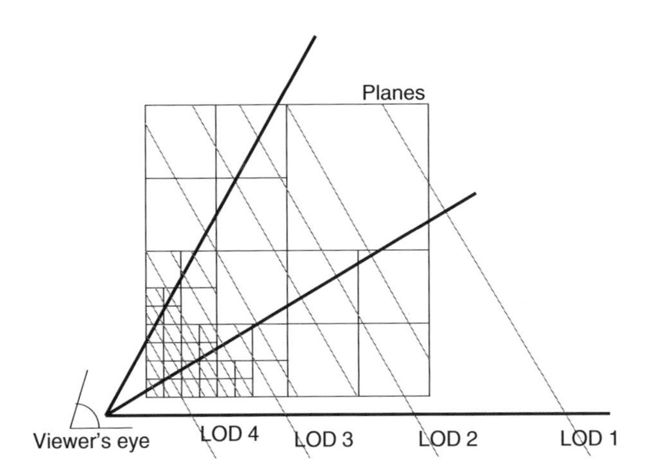

FIGURE 9.9 Multiresolution for single volume data with view-aligned planes. Each brick (square box) in the figure has the same data dimension but a different geometry size; that is, smaller bricks have the same amount of information used in rendering as bigger bricks, thereby increasing the resolution and quality of data nearest to the viewing plane.

9.6.3 Early Ray Termination

Early ray termination is probably the best-known optimization technique in ray-casting volume rendering [68, 70]. As its name implies, this technique is based on the termination of a ray when the contribution from that ray is minimal in the computation of the voxel to display. Early ray termination works only in ray casting that is traversed from front to back.

9.6.4 Parallel Rendering

Whereas the methods described in Sections 9.6.1 to 9.6.3 work on optimizing the computation of data via software-based approaches by trading off between quality and performance, alternate approaches to the rendering of large data harness the computation from multiple graphic cards (hardware) for parallel rendering [40, 71–73]. One way to obtain parallel rendering is to split the rendering processes into several distinct functions that can be applied in series to individual data items. Such a technique is often favored for surface rendering applications [71], which require several separate functional stages: segmentation, surface generation, and then rendering. Instead of computing functions in series, a more common method is to split the data into multiple streams, which can then be operated independently and simultaneously, with each stream responsible for rendering a subsection of the data. All of the subsections are then combined together. Parallel data streaming has been successfully applied on a variety of platforms, ranging from networks of PCs to massively parallel supercomputers for the visualization of large biomedical data via texture-based or direct volume rendering [40, 71].

9.7 Dual-Modality Positron Emission Tomography–Computed Tomography Visualization

In the visualization of dual-modality PET/CT, the PET images provide high sensitivity in tumor detection and tissue characterization, while the coregistered CT data provide the localizations of the anatomical landmarks and boundary definitions of tumors and organs [1, 74]. PET/CT images have led to a new paradigm in biomedical diagnosis and interpretation by enabling visualization of fused anatomical and functional structures. These dual-modality images, which consist of large image dimensions (full-body PET/CT data are typically in the range of $512 \times 512 \times 262$ for CT and $128 \times 128 \times 262$ for PET), have introduced interesting challenges for efficient 3D visualization [74] and stand to benefit significantly from advances in visualization technologies. The following section discusses the application of visualization techniques with PET/CT data.

9.7.1 Dual-Transfer Function

In PET/CT visualization, PET and CT data with independent LUTs are often required to be fused in real time. These volumes may be fused together into a single volume prior to volume rendering [5] or individually rendered using different rendering techniques, such as surface and direct volume rendering, which are then fused together as a preprocess to form a resultant volume [13]. These approaches all rely on the fusion of volumes to be a preprocess prior to volume rendering and are subsequently limited to the application of volume-rendering manipulations, such as window/level and transfer function, to the fused volume and cannot be applied individually to its component scans. Alternatively, fusion of multiple volumes by data intermixing in the rendering phase in direct volume rendering was proposed by Cai and Sakas [75], thus allowing interactive and real-time fusion of the volumes.

The scenario of detecting tumor structures in a lung cancer patient illustrates the requirement to define transfer functions independently in PET/CT visualization. Selecting the VOI from CT (e.g., by transfer function) and adjusting window/level via transfer functions of PET necessitate manipulation of individual volumes, with the ability to visualize the changes in the fused volume in real time. Physicians may benefit from visualizing the functional tumor apparent in the PET scan alongside corresponding anatomical structures for localization from the CT scan. Kim et al. [76] suggested using dual-transfer functions for PET/CT images, thereby providing separate 1D LUT transfer functions independently to the PET and CT volumes to be controlled, with the resultant volumes being fused in real time. Results from use of dual-transfer functions are shown in Figure 9.10. An axial view of a fused PET/CT volume is shown in Figure 9.10(a), with its dual-LUT transfer function in Figure 9.10(b). Equal fusion ratios were applied that were able to visualize the functional tumors (indicated by arrows) with surrounding anatomical structures. Figure 9.10(c) shows the result, using a dual-LUT transfer function in Figure 9.10(d). The transfer function was applied to the PET component for the selection of the LUT range belonging to tumors (high LUT values).

The slope of the transfer function enables gradual increase in the visualization of the LUT values of the voxels as these values reach that of the tumor. Therefore, no voxels belonging to the tumors were erroneously excluded, and they avoided abrupt discontinuation of the selection of structures. For CT, a trapezoidal transfer function was applied to visualize the boundaries of anatomical structures while minimizing spatial overlap with the functional tumor structures. These results demonstrate that the dual-LUT transfer function can be utilized to control the fusion between the PET and CT volumes to avoid overlap and highlight the structure of interest, in this case the tumors, while still retaining the anatomical context provided by the CT data. A different combination of dual-transfer functions is shown in Figure 9.10(e, f). The real-time dual-LUT transfer function applied to the PET/CT volume rendering provides immediate feedback on the accuracy of the feature selection and provides selective rendering of anatomical and functional structures. This permits improved visualization of the anatomical frame of reference and localization of the tumors when compared with the fusion of PET/CT in Figure 9.10(a).

9.7.2 Spatial Transfer Function

Application of the dual transfer function in PET/CT data enables visualization of, for example, tumor structure from PET with surrounding anatomical structures from CT. However, due to the tumor having similar color values as functional organs (e.g., the liver), identification of the tumor using only a transfer function based on the LUT is limited. For such cases, if the location of the tumor is known, a spatial transfer function can be applied that allows control of the viewable range of the segmented structure according to the spatial distance of voxels to the segmented structure [30, 64, 65]. Kim et al. [77] presented an application of spatial transfer to PET/CT data, in which tumors identified on PET were segmented as a preprocess (Figure 9.11(a)). The segmentation result was then used to construct a *distance map* according to spatial distance of voxels to the segmented structure based on the algorithm introduced by Fichtinger et al. [4]. These distances were used to calculate the weights assigned to each individual voxel in the PET, as illustrated in Figure 9.11(b). The weights were then multiplied with the PET, during voxel-by-voxel data intermixing with CT [75]. Several parameters in this algorithm can be fine-tuned. Firstly, the weight applied to the PET data can be controlled in such a way that greater emphasis can be placed on voxels near the tumors. Secondly, cutoffs can be added, such as the red circles in Figure 9.11(b), which can be used to control the application of the weights to only the voxels that reside within these cutoffs. The result of the spatial transfer function is shown in Figure 9.11(c), where tumors are clearly depicted with their surrounding anatomical CT structures. Here, a cutoff of 30 voxels was applied. The spatial transfer function has the advantage of visualizing spatially related voxels to the segmented tumor and enables gradual increase in the transition of the fusion ratio of the voxels as they approach the tumor.

9.7.3 Interactive Segmentation and Volume Interchange in Positron Emission Tomography–Computed Tomography

Instead of using segmentation results for the spatial transfer function, they can be used directly in volume rendering. Rendering the segmented VOI as an independent volume

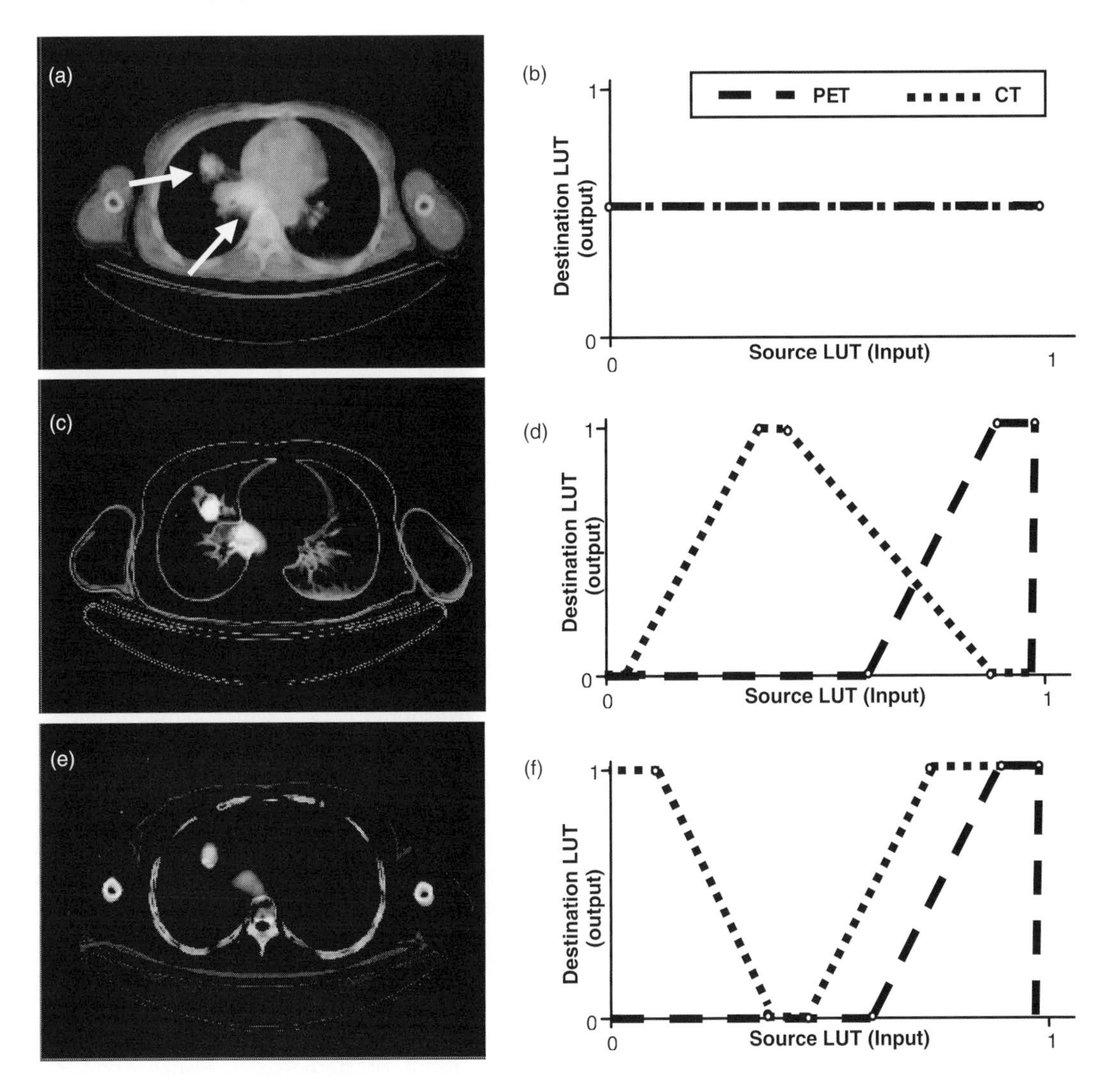

FIGURE 9.10 Application of dual-LUT transfer functions. (a) Axial view of PET/CT with its dual-LUT transfer function in (b) set to equal fusion ratio. Tumors inside the lungs are highlighted by arrows. (c, d) Selection of tumors from PET and the surrounding anatomical lung boundary from CT. (e, f) Identical PET transfer function with inverted and modified CT transfer function. For a more detailed view of this figure, please visit our companion site at: http://books.elsevier.com/companions/9780123735836.

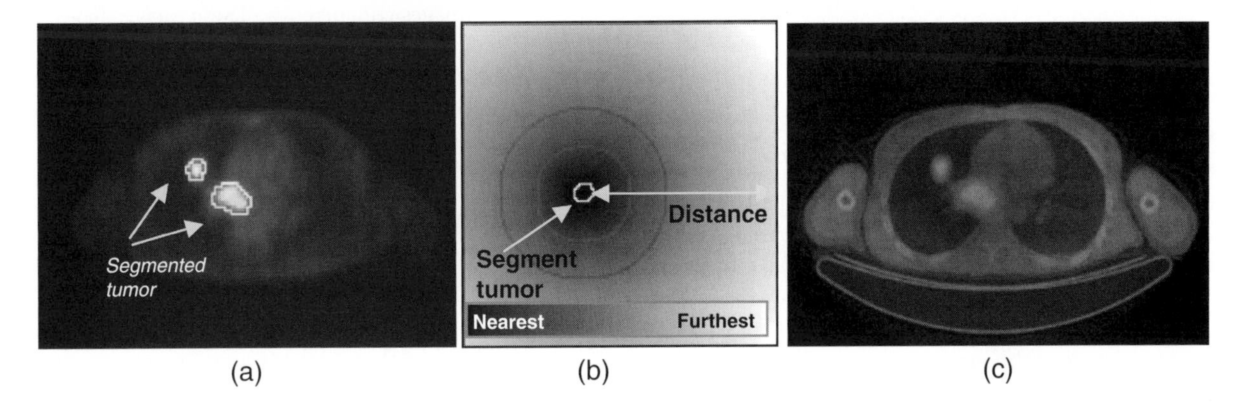

FIGURE 9.11 Overview of the segmentation-based spatial transfer function applied to PET/CT visualization. (a) Original PET image (single axial-view slice) with segmented structures highlighted (arrow); (b) distance map calculated from the segmented result; (c) fused PET/CT result from the spatial transfer function using (b). For a more detailed view of this figure, please visit our companion site at: http://books.elsevier.com/companions/9780123735836.

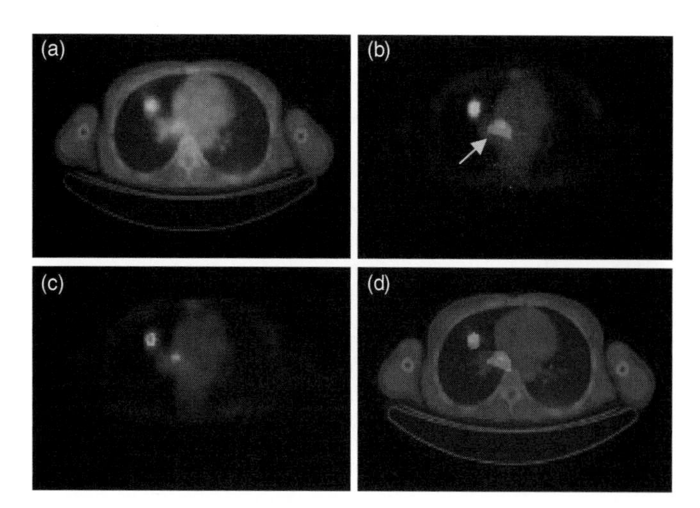

FIGURE 9.12 (a) Volume rendering of PET/CT image using texture-based volume rendering; (b) automated FCM (fuzzy c-means clustering) segmentation result with segmented tumor structures fused with PET; (c) result from varying the PET segmentation parameter to select voxels that more closely resemble the tumor; (d) segmentation result of (b) fused with CT. All volumes have been fused with equal fusion ratios. The transparency level and the LUT of the segmented volumes can be adjusted to reduce obscuration of underlying structures relevant for the interpretation of the images and segmentation results. For a more detailed view of this figure, please visit our companion site at: http://books.elsevier.com/companions/9780123735836.

that is transparent, together with the original PET and CT volumes, is potentially useful in highlighting the VOIs. In PET/CT, such ability can be further extended to allow interchange of the volumes, as shown in Figure 9.12 [41]. Here, the segmented VOIs from the PET data are shown rendered together with either data from PET (Figure 9.12[b]) or CT (Figure 9.12[c]). This was possible through utilization of the high memory bandwidth capacity of modern graphic hardware to rapidly transfer 3D textures from the system memory into the graphic memory, which was used to replace an old volume in the graphic memory with a new volume in real-time volume rendering.

9.8 Data Display Devices

Previous sections of this chapter have discussed algorithms and methods that are used to generate volume-rendered visualizations. The hardware devices that display these visual outputs are equally important. The utilization of digital displays in biomedical data visualization in recent times has been accelerated by the acceptance of picture archiving and communications systems (PACSs) [78] for digital image management, as well as by the additional abilities that digital images provide, such as image processing and volume visualization. The effectiveness of all diagnostic imaging visualization is affected by the quality of the display devices [79]. In the current diagnostic workstation, it is common to find multiple LCD screens for use in visualizing biomedical data in 2D and 3D views. These screens typically have resolutions of 1600×1200 and greater, with pixel depth of 24-bit color, thus providing sufficient display capacity for modern medical imaging modalities. Apart from LCDs, more advanced display devices are being designed and developed for diagnostic rooms and operating theaters. These include stereoscopic displays for virtual reality and depth display devices.

9.8.1 Stereoscopic Visualization

On a 2D screen, true 3D visualization can be obtained only by displaying the data stereoscopically, thereby creating the illusion of depth. This technique has been widely adopted in biomedical visualization for creating virtual reality environments and has found many uses in surgical simulations [14, 50, 80]. Fundamentally, stereoscopy presents slightly different images to the left and right human eyes, such that each eye can see only a single image. The most routinely utilized stereoscopy equipment in biomedical visualization are infrared synchronized shutter glasses, which allow the left/right eye images to be quickly alternated (generally at 60 Hz). Stereoscopic visualization has also been applied to immersive environments, with examples such as Immersadesk [81] and the CAVE virtual-reality environment [82, 83]. As an alternative to shutter glasses, which restrict the viewing capability to a single user per computer screen, the study in [80] proposed polarized filters to separate the stereo information, allowing multiple users to visualize from the same screen with relatively cheap polarized glasses.

As the requirement of wearing glasses was an impediment for adoption of stereoscopic visualization, there has been significant research by display manufacturers (in particular, Sharp [84] and Philips [85]) aiming to allow realistic 3D viewing without the need for 3D glasses. Glasses-free, or *autostereoscopic*, 3D displays work by the device being able to display two separate images concurrently. These images have the same disparity, and the display device uses switchable parallax barrier technology [84] to ensure that each eye sees only one of these images; in doing so, it creates the illusion of depth without the need for glasses to separate the different signals.

9.8.2 Depth Projection

Stereoscopy is a visualization technique that creates the illusion of depth, and therefore the display has no physical depth definition [86]. This has major limitations in creating true depth perception. A device produced by LightSpace Technologies, the DepthCube [87], is a solid-state multiplanar volumetric display that uses a high-speed video projector to display a sequence of slices of the 3D image into multiplanar optical elements. The individual slices in the multiplanar views are than anti-aliased to result in smooth images.

9.9 Applications of Biomedical Visualization

Modern visualization and display technologies are enabling the development of clinical applications that provide new approaches for interacting with and interpreting biomedical data, such as in virtual biopsy [88], motion activity visualization [89], and applications in radiotherapy planning [8]. These applications are capable of providing complete 3D views of the human body that are rendered in real time to volume navigations and manipulations. Furthermore, advances in visualization and display technologies have accelerated the developments of computer-integrated surgery (CIS), IGS, and CAD systems (which are detailed in Chapters 16 and 18).

This section summarizes some of the research that has been made possible by the utilization of cutting-edge biomedical data visualization technologies.

Research efforts in biomedical data visualization have been focused primarily on reconstructing data into 3D volumetric views. However, much other information and many other uses can be obtained from the same data—for example, by tracer kinetic modeling and the modeling of different elements (e.g., bones, cartilage, ligaments, muscles, tendons) and their interrelationships [89–94]. A study by Feng et al. [90, 91] discussed the medical parametric imaging that requires modeling and parameter estimation for certain metabolic, pharmacokinetic, endocrine, or other biochemical systems at voxel-by-voxel level. It is an important technique that provides image-wide quantifications of physiological and biochemical functions and allows the distributions of these functions corresponding to anatomical structures to be visualized. (For more information about parametric imaging, refer to Chapter 6.)

In another study by Magnenat-Thalmann et al. [93], the visualization and understanding of functionalities of human articulation of the shoulders is presented. Here, a generic 3D shoulder was constructed as a physically based model along with other elements and was used to produce a simulation of movements and deformations. This project used the Visible Human Dataset (VHD) [95] and pioneered the innovative use of 3D visual simulation in the medical computing field. The project is currently expanding to the study of hip articulation using dynamic MRI. A study by Benjamin et al. [89] presented the ability to visualize an accurate estimate of joint kinetics for the understanding of joint motion activity. The modeling and simulations of hip articulation were validated using medical data and demonstrated the potential to provide physicians with the benefit of having the technology that enables visualization and observation of motion activity (in 3D) for use in diagnosis of problems of human-body articulations. Further work of this project will investigate the biomechanical model of the hip to simulate and visualize its motion and to understand the possible malfunctioning of the articulation in individual patients. The biomechanical model will rely on the interrelations of the various anatomical elements involved in joint articulation and their influence on the range of motion of the hip [89, 94].

The Visible Human Project (VHP) [95] is an effort to create a detailed dataset of cross-sectional photographs of the human body, in order to facilitate anatomy visualization applications for education, research, and clinical diagnosis. The importance of this project has led to the formation of many research endeavors, such as the motion modeling study previously noted and the development of Korean [96] and Chinese VHPs [97]. The Biomedical Imaging Resource (BIR) at the Mayo Clinic was one of the first laboratories to be involved in VHD-related research, and during the last decade, it has produced many novel 3D visualization and image processing methods (e.g., rendering, registration, segmentation, modeling, classification) and evaluated the effectiveness of these methods for eventual applications in clinical diagnosis and therapy [20, 98, 99]. In the BIR's recent study, Robb et al. [99] developed an anesthesiology training simulation system, in close collaboration with clinicians, using the VHD as the patient. This system enabled the simulation of examinations using 3D visualization of the relationships among the anatomical structures, in addition to its use in needle insertion practice. The training system was built through an immersive environment (a virtual operating theater) created through the use of a head tracking system, a head mounted display, a needle tracking system, and haptic input/feedback. The user may interact with the virtual patient using a wand or haptic feedback devices that provide a sense of touching the patient [100, 101]. This system has been extended to medical simulation used for popliteal nerve (knee) block and prostate brachytherapy training [98] and can also be used with individual patient scans as well as the VHD.

An innovative project that deals with full-body virtual autopsies was recently reported by Ljung et al. [88]. This study reported a procedure based on interactive 3D visualization of large-scale, high-resolution CT data of human cadavers for virtual autopsies, demonstrating its potential to provide key information in guiding criminal investigations. A unique challenge in this project arises from the large volume of data acquired from multidetector CT (MDCT), which captures 8,000 transaxial images for a full-body scan (data size of several gigabytes). In order to interactively visualize these MDCT images in real time, a state-of-the-art volume-rendering pipeline was developed that applied and refined several visualization optimization algorithms, including transfer function–based adaptive LOD [102], interblock interpolation [103], and a single-pass ray casting of multiresolution volumes [104], as well as memory management techniques. Furthermore, this study introduced an extension to the GPU-based ray-casting algorithm that enabled efficient dual transfer function rendering for fast localization of, for example, metal fragments. The described autopsy procedure was evaluated using examples from routine forensic examinations and demonstrated great potential as an imaging application to complement standard autopsies by enabling broad and systematic examination of the full body in forensic investigations.

Diffusion tensor imaging (DTI) is an MRI technique (also known as DT-MRI) that enables the measurement of the diffusion of water in tissues such as bone, muscle, and white matter of the brain. Here, diffusion refers to the ability of water molecules to migrate from one part of a biological system to another, in random molecular motion (moving faster in some directions than in others) [105, 106]. This geometrical feature of DT-MRI enables quantitative characterization and visualization of the local structures in tissues. Due to the large information available in DT-MRI, visualization approaches for it constitute a challenging field, with many alternatives for representing the DT-MRI image. One tool for interpreting the image is the use of a glyph—a parameterized icon that represents the data with its shape, color, texture, location, etc., and can be used to represent the diffusion properties [107] (see Section 9.3.4). In a recent study, Kindlemann and Westin [108] proposed a *glyph packing* method, which would use a particle system with anisotropic potential energy profiles to arrange glyphs, rendered using a texture-based technique, into a dense pattern that would display some of the visual continuity of texture-based visualizations while maintaining the ability to discern the full tensor information at each glyph. This study demonstrated that the use of visualization techniques can complement and also reveal additional information that can aid in the diagnosis and interpretation of DT-MRI images.

9.10 Summary

With continuous development in image acquisition technologies that are resulting in ever-increasing data sizes, the need for visualization of these data in an efficient and intelligent manner is becoming crucial for clinical diagnosis and image understanding. In line with advances in imaging modalities, development of new theories and refinement of techniques in biomedical data visualization are improving the way these data are utilized and interpreted. Furthermore, new discoveries in visualization technologies are enabling alternate approaches to conventional methods. This chapter introduced the visualization and display technologies that are currently being employed in clinical applications, as well as techniques that have the potential to enhance and provide improved diagnostic capabilities.

Starting from 2D and 3D visual methods, core visualization technologies were discussed, including navigation interfaces, volume enhancement and manipulations, and large data visualization. A PET/CT visualization was presented as an example that took advantage of many of the techniques discussed in this chapter. Finally, the next generation of display devices and technologies was briefly covered with regard to their potential clinical applications. Future developments in visualization and displays for biomedical data will carry on the development of new technologies and produce new ways to improve the display and interpretation of biomedical data.

Acknowledgments

The authors are grateful to the support from ARC and PolyU/UGC grants.

9.11 Exercises

1. In Figure 9.3, renders of CT volume data using three different techniques are shown. List the main differences among the rendering techniques and their advantages/disadvantages in biomedical data visualization.
2. Volume-rendering visualization introduces several rendering attributes, such as shadowing and transparency, that create more realistic display of biomedical data. Together with conventional enhancement tools such as window/level, these attributes provide powerful tools for volume enhancement. List and give examples of how these attributes are used for visualization enhancement.
3. Transfer function specification is applicable in 1D, 2D, or greater dimensions. This chapter discussed multidimensional transfer functions used for LUT manipulations. Discuss how multidimensional transfer functions can be applied in the spatial domain.
4. Segmentation is a key component of visualization. However, interactive segmentation is often too computationally intensive for real-time performance in volume rendering. State how this problem can be minimized, in terms of software and hardware solutions.
5. State the basic principles of parallel direct volume rendering and why this technique is favored for surface rendering techniques.
6. Section 9.4.2 presents different input devices that may be used for navigating volumetric biomedical data. Design and analyze a setup of input devices for use in controlling the multidimensional transfer function of multivariate biomedical data.

9.12 References and Bibliography

1. O. Ratib. PET/CT image navigation and communication. *J. Nucl. Med.* 45(1):46S–55S, 2004.
2. I. N. Bankman. *Handbook of Medical Imaging.* Academic Press, 2000.

3. M. Meissner et al. Volume visualization and volume rendering techniques. In *Tutorials 6*. Eurographics, 2000.
4. G. Fichtinger et al. System for robotically assisted prostate biopsy and therapy with intraoperative CT guidance. *Acad. Radiol.* 9(1):60–74, 2002.
5. A. Rosset, L. Spadola, and O. Ratib. OsiriX: An open-source software for navigating in multidimensional DICOM images. *J. Digit. Imaging.* 17(3)205–216, 2004.
6. R. Shahidi, R. Tombropoulos, and R. P. Grzeszczuk. Clinical applications of three-dimensional rendering of medical data sets. *Proc. IEEE.* 86:555–565, 1998.
7. F. Beltrame et al. Three-dimensional visualization and navigation tool for diagnostic and surgical planning applications. *Medical Imaging 2001: Visualization, Display, and Image-Guided Procedures.* SPIE, 2001.
8. I. F. Ciernik et al. Radiation treatment planning with an integrated positron emission and computer tomography (PET/CT): A feasibility study. *Int. J. Radiat. Oncol. Biol. Phys.* 57:853–863, 2003.
9. A. B. Jani, J.-S. Irick, and C. Pelizzari. Opacity transfer function optimization for volume-rendered computed tomography images of the prostate. *Acad Radiol.* 12(6):761–770, 2005.
10. D. T. Gering et al. An integrated visualization system for surgical planning and guidance using image fusion and interventional imaging. *Proceedings of Medical Image Computing and Computer Assisted Intervention.* Springer-Verlag. 809–819, 1999.
11. Y. C. Loh et al. Surgical planning system with real-time volume rendering. *Proc. IEEE Int. Workshop Med. Imag. Augmented Reality,* 2001:259–261.
12. A. Rosset et al. Informatics in radiology (infoRAD): Navigating the fifth dimension: Innovative interface for multidimensional multimodality image navigation. *Radiographics.* 26(1):299–308, 2006.
13. H. Hauser et al. Two-level volume rendering. *IEEE Trans. Vis. Comput. Graph.* 7(3):242–252, 2001.
14. P.-D. Dai et al. A virtual laboratory for temporal bone microanatomy. *IEEE Comput. Sci. Eng.* 7(2):75–79, 2005.
15. A. Bornik et al. A hybrid user interface for manipulation of volumetric medical data. *Proceedings of the 3D User Interface (3DUI '06).* 29–36, 2006.
16. B. Frohlich et al. On 3D input devices. *IEEE Comput. Graph. Appl.* 26(2):15–19, 2006.
17. A. J. Sherbondy et al. Alternative input devices for efficient navigation of large CT angiography data sets. *Radiology.* 234(2):391–398, 2005.
18. *Chapel Hill Volume Rendering Test Data Set.* SoftLab Software Systems Laboratory, University of North Carolina Department of Computer Science.
19. MedView^{TM} [software]. http://www.medimage.com/pet-ct-software.html
20. R. A. Robb. Visualization in biomedical computing. *Parallel Computing.* 25(13–14):2067–2110, 1999.
21. G. Lohmann. *Volumetric Image Analysis.* John Wiley & Sons Ltd., 1998.
22. *ImageJ: Image Processing and Analysis in Java* [homepage]. http://rsb.info.nih.gov/ij
23. U. B. Kai. *Volume Viewer.* Internationale Medieninformatik, 2005.
24. W. E. Lorensen and H. E. Cline. Marching cubes: A high resolution 3D surface construction algorithm. *Comput. Graph.* 21(4):163–169.
25. W. Schroeder, K. Martin, and B. Lorensen. *The Visualization Toolkit: An Object-Oriented Approach to 3D Graphics,* 3rd ed. Kitware, Inc., 2003.
26. J. K. Udupa. *Multidimensional Digital Boundaries.* Academic Press. 311–323, 1994.
27. U. Tiede et al. Investigation of medical 3D-rendering algorithms. *IEEE Comput. Graph. Appl.* 10(2):41–53, 1990.
28. A. H. Watt. *3D Computer Graphics,* 3rd ed. Addison-Wesley, 1999.
29. D. Shreiner et al. *OpenGL(R) Programming Guide: The Official Guide to Learning OpenGL,* 5th ed. Addison-Wesley Professional, 2005.
30. I. Viola, A. Kanitsar, and M. E. Groller. Importance-driven feature enhancement in volume visualization. *IEEE Trans. Vis. Comput. Graph.* 11(4):408–418, 2005.
31. M. Harders and G. Szekely. Enhancing human–computer interaction in medical segmentation. *Proc. IEEE.* 91(9): 1430–1442, 2003.
32. R. Westermann and B. Sevenich. Accelerated volume ray-casting using texture mapping. *Visualization: Proceedings of the Conference on Visualization '01.* 271–278, 2001.
33. G. Kiefer, H. Lehmann, and J. Weese. Fast maximum intensity projections of large medical data sets by exploiting hierarchical memory architectures. *IEEE Trans. Inf. Technol. Biomed.* 10(2):385–394, 2006.
34. H. W. Venema, F. J. H. Hulsmans, and G. J. den Heeten. CT angiography of the Circle of Willis and intracranial internal carotid arteries: Maximum intensity projection with matched mask bone elimination—Feasibility study. *Radiology.* 218:893–898, 2001.
35. M. P. Marks et al. Diagnosis of carotid artery disease: Preliminary experience with maximum-intensity-projection spiral CT angiography. *Am. J. Roentgenol.* 160(6): 1267–1271, 1993.
36. D. Laur and P. Hanrahan. Hierarchical splatting: A progressive refinement algorithm for volume rendering. *Proceedings of the 18th Annual Conference on Computer Graphics and Interactive Techniques.* ACM Press. 285–288, 1991.
37. P. Lacroute and M. Levoy. Fast volume rendering using a shear-warp factorization of the viewing transformation. *Proceedings of the 21st Annual Conference on Computer Graphics and Interactive Techniques.* ACM Press. 451–458, 1994.

38. J. Sweeney and K. Mueller. Shear-warp deluxe: The shear-warp algorithm revisited. *Proceedings of the Symposium on Data Visualisation: ACM International Conference Proceeding Series*. Eurographics Association. 22:95–ff, 2002.
39. P. Bhaniramka and Y. Demange. OpenGL volumizer: A toolkit for high quality volume rendering of large data sets. *Proceedings of the 2002 IEEE Symposium on Volume Visualization and Graphics*. IEEE Press. 45–54, 2002.
40. J. Kniss et al. Interactive texture-based volume rendering for large data sets. *IEEE Comput. Graph. Appl.* 21(4):52–61, 2001.
41. J. Kim et al. Real-time volume rendering visualization of dual-modality PET/CT images with interactive fuzzy thresholding segmentation. *IEEE Trans. Inf. Technol. Biomed.* 11(2):161–169, 2007.
42. W. Manfred et al. Level-of-detail volume rendering via 3D textures. *Proceedings of the 2000 IEEE Symposium on Volume Visualization*. ACM Press. 7–13, 2000.
43. B. F. Tomandl et al. Local and remote visualization techniques for interactive direct volume rendering in neuroradiology. *Radiographics*. 21(6):1561–1572, 2001.
44. J. Kniss, G. Kindlmann, and C. Hansen. Multidimensional transfer functions for interactive volume rendering. *IEEE Trans. Vis. Comput. Graph.* 8(3):270–285, 2002.
45. F. H. Post et al. The state of the art in flow visualisation: Feature extraction and tracking. *Computer Graphics Forum*. 22(4):775–792, 2003.
46. R. S. Laramee et al. The state of the art in flow visualization: Dense and texture-based techniques. *Computer Graphics Forum*. 23(2):203–221, 2004.
47. M. K. Tory, T. Moeller, and M. S. Atkins. Visualization of time-varying MRI data for MS lesion analysis. *Proc. SPIE: Medical Imaging 2001: Visualization, Display, and Image-Guided Procedures*. 4319:590–598, 2001.
48. N. Thune and B. Olstad. Visualizing 4-D medical ultrasound data. *Proceedings of the 2nd Conference on Visualization*. IEEE Computer Society Press. 210–215, 1991.
49. A. Neubauer et al. Advanced virtual endoscopic pituitary surgery. *IEEE Trans. Vis. Comput. Graph.* 11(5):497–507, 2005.
50. A. Gronningsaeter et al. Initial experience with stereoscopic visualization of three-dimensional ultrasound data in surgery. *Surg. Endosc.* 14(11):1074–1078, 2000.
51. HP SpacePilot 3D USB Intelligent Controller [webpage]. http://www.hp.com/sbso/product/workstation/spacepilot.html
52. B. A. Myers. Using multiple devices simultaneously for display and control. *IEEE Personal Communications*. 7(5):62–65, 2000.
53. Y. Kim and A. Varshney. Saliency-guided enhancement for volume visualization. *IEEE Trans. Vis. Comput. Graph.* 12(5):925–932, 2006.
54. N. Nikolaidis. *3-D Image Processing Algorithms*. John Wiley & Sons, 2001.
55. E. Bullitt and S. R. Aylward. Volume rendering of segmented image objects. *IEEE Trans. Med. Imaging*. 21(8): 998–1002, 2002.
56. S.-S. Yoo et al. Interactive 3-dimensional segmentation of MRI data in personal computer environment. *J. Neurosci. Methods*. 112(1):75–82, 2001.
57. L. Vosilla et al. An interactive tool for the segmentation of multimodal medical images. *Proceedings of IEEE EMBS International Conference on Information Technology Applications in Biomedicine*. IEEE Press. 203–209, 2000.
58. J. Kim et al. Interactive fuzzy temporal thresholding for the segmentation of dynamic brain PET images. *J. Cereb. Blood Flow Metab.* 25:S620, 2005.
59. P. Vereda et al. Visualization of boundaries in volumetric data sets using LH histograms. *IEEE Trans. Vis. Comput. Graph.* 12(2):208–218, 2006.
60. M. Levoy. Display of surfaces from volume data. *IEEE Comput. Graph. Appl.* 8(3):29–37, 1988.
61. A. König and E. Gröller. Mastering transfer function specification by using VolumePro technology. *Proceedings of the 17th Spring Conference on Computer Graphics (SCCG '01)*. 279–286, 2001.
62. J. Marks et al. Design galleries: A general approach to setting parameters for computer graphics and animation. *Proceedings of the 24th Annual Conference on Computer Graphics and Interactive Techniques*. ACM Press/Addison-Wesley Publishing, 1997.
63. I. Viola et al. Importance-driven focus of attention. *IEEE Trans. Vis. Comput. Graph.* 12(5):933–940, 2006.
64. F.-Y. Tzeng and K.-L. Ma. A cluster-space visual interface for arbitrary dimensional classification of volume data. *IEEE TVCG Symposium on Visualization*, 2004.
65. J. Zhou, A. Doring, and K. D. Tonnies. Distance-based enhancement for focal region based volume rendering. *Proceedings of Bildverarbeitung fur die Medizin '04*, 2004.
66. S. Guthe et al. Interactive rendering of large volume data sets. *Proc. IEEE Visualization*, 2002.
67. W. Li, K. Mueller, and A. Kaufman. Empty space skipping and occlusion clipping for texture-based volume rendering. *Proc. IEEE Visualization*, 2003.
68. J. Kruger and R. Westermann. Acceleration techniques for GPU-based volume rendering. *Proc. IEEE Visualization*, 2003.
69. W. Ruediger. A multiresolution framework for volume rendering. *Proceedings of the 1994 Symposium on Volume Visualization*. ACM Press, 1994.
70. M. Levoy. *Efficient Ray Tracing of Volume Data*. ACM Press, 1990.
71. T. W. Crockett. An introduction to parallel rendering. *Parallel Computing*. 23(7):819–843, 1997.

72. S. Molnar et al. A sorting classification of parallel rendering. *IEEE Comput. Graph. Appl.* 14(4):23–32, 1994.
73. P. Bhaniramka, P. C. D. Robert, and S. Eilemann. OpenGL multipipe SDK: A toolkit for scalable parallel rendering. *Proc. IEEE Visualization,* 2005.
74. M. N. Wernick and J. N. Aarsvold. *Emission Tomography: The Fundamentals of PET and SPECT.* Elsevier Academic Press, 2004.
75. W. Cai and G. Sakas. Data intermixing and multi-volume rendering. *Computer Graphics Forum.* 18(3):359–368, 1999.
76. J. Kim, S. Eberl, and D. Feng. Visualization of dual-modality anatomical and functional rendered volumes with image fusion using a dual-lookup table transfer function. *IEEE Comput. Sci. Eng.* 9(1):20–25, 2007.
77. J. Kim, S. Eberl, and D. Feng. Multi-modal medical visualization based on spatial transfer function. *Proc. IEEE Visualization* [poster presentation], 2006.
78. H. K. Huang. PACS, image management and imaging informatics. In *Multimedia Information Retrieval and Management.* Springer-Verlag. 476, 2003.
79. A. Badano. AAPM/RSNA tutorial on equipment selection: PACS equipment overview: Display systems. *Radio-Graphics.* 24:879–889, 2004.
80. S. T. Jones, S. E. Parker, and C. C. Kim. Low-cost high-performance scientific visualization. *IEEE Comput. Sci. Eng.* 3(4):12–17, 2001.
81. C. Marek et al. *The ImmersaDesk and Infinity Wall Projection-Based Virtual Reality Displays.* ACM Press, 1997.
82. C. Demiralp et al. CAVE and fishtank virtual-reality displays: A qualitative and quantitative comparison. *IEEE Trans. Vis. Comput. Graph.* 12(3):323–330, 2006.
83. C. Cruz-Neira, D. J. Sandin, and T. A. DeFanti. Surround-screen projection-based virtual reality: The design and implementation of the CAVE. *Proceedings of the 20th Annual Conference on Computer Graphics and Interactive Techniques.* ACM Press. 135–142, 1993.
84. Sharp3D [website]. http://www.codeplex.com/Wiki/View.aspx?ProjectName=Sharp3D
85. Philips N.V. [website]. http://www.research.philips.com
86. K. Balasubramanian. On the realization of constraint-free stereo television. *IEEE Trans. Consum. Electron.* 50(3): 895–902, 2004.
87. A. Sullivan. 3-Deep: New displays render images you can almost reach out and touch. *IEEE Spectrum.* 42(4):30–35, 2005.
88. P. Ljung et al. Full body virtual autopsies using a state-of-the-art volume rendering pipeline. *IEEE Trans. Vis. Comput. Graph.* 12(5):869–76, 2006.
89. B. Gilles et al. Bone motion analysis from dynamic MRI: Acquisition and tracking. *Acad. Radiol.* 12:1285–1292, 2005.
90. D. Feng, Z. Wang, and S.-C. Huang. A study on statistically reliable and computationally efficient algorithms for generating local cerebral blood flow parametric images with positron emission tomography. *IEEE Trans. Med. Imaging.* 12(2):182–188, 1993.
91. D. Feng. Information technology applications in biomedical functional imaging. *IEEE Trans. Inf. Technol. Biomed.* 3(3):221–230, 1999.
92. L. Wen. Fast and reliable estimation of multiple parametric images using an integrated method for dynamic SPECT. *IEEE Trans. Med. Imaging.* 26(2): 179–189, 2007.
93. N. Magnenat-Thalmann and F. Cordier. Construction of a human topological model from medical data. *IEEE Trans Inf Technol Biomed.* 4(2):137–143, 2000.
94. MIRALab [website, Geneva, Switzerland]. http://www.miralab.unige.ch
95. M. J. Ackerman. The visible human project. *Proc. IEEE.* 86(3):504–511, 1998.
96. J. S. Park et al. Visible Korean human: Improved serially sectioned images of the entire body. *IEEE Trans. Med. Imaging.* 24(3):352–360, 2005.
97. S.-X. Zhang, P.-A. Heng, and Z.-J. Liu. Chinese visible human project: Dataset acquisition and its primary applications. *Proc. IEEE EMBC,* 2005.
98. R. A. Robb. Biomedical imaging: Past, present and prediction. *Proc. Symp. Intelligent Assistance in Diagnosis of Multi-Dimensional Medical Images,* 2005.
99. R. A. Robb and D. P. Hanson. Biomedical image visualization research using the Visible Human Datasets. *Clin. Anat.* 19(3):240–253, 2006.
100. D. P. Martin, D. J. Blezek, and R. A. Robb. Simulating lower extremity nerve blocks with virtual reality. *Tech. Reg. Anesth. Pain Manage.* 3:58–61, 1999.
101. A. Pommert et al. Creating a high-resolution spatial/symbolic model of the inner organs based on the visible human. *Med. Image Anal.* 5(3):221–228, 2001.
102. P. Ljung et al. Transfer function based adaptive decompression for volume rendering of large medical data sets. *IEEE Symp. Volume Visualization and Graphics,* 2004.
103. P. Ljung, C. Lundström, and A. Ynnerman. Multiresolution interblock interpolation in direct volume rendering. *Proc. Eurographics/IEEE Symp. Visualization,* 2006.
104. P. Ljung. Adaptive sampling in single pass, GPU-based raycasting of multiresolution volumes. *Proceedings Eurographics/IEEE Workshop on Volume Graphics,* 2006.
105. G. Kindlmann, D. Weinstein, and D. Hart. Strategies for direct volume rendering of diffusion tensor fields. *IEEE Trans. Vis. Comput. Graph.* 6(2):124–138, 2000.
106. C.-F. Westin et al. Processing and visualization for diffusion tensor MRI. *Med. Image Anal.* 6:93–108, 2002.
107. C. Pierpaoli and P. J. Basser. Toward a quantitative assessment of diffusion anisotropy. *Magn. Reson. Med.* 37(6):893–906, 1996.
108. G. Kindlmann and C.-F. Westin. Diffusion tensor visualization with glyph packing. *IEEE Trans. Vis. Comput. Graph.* 12(5):1329–1336, 2006.

10

Data Communication and Network Infrastructure

Prof. Doan B. Hoang[1,2] and
Dr. Andrew J. Simmonds[1]
[1] *University of Technology, Sydney*
[2] *University of Sydney*

10.1 Introduction

The current model for health care is built around hospitals, doctors, nurses, and other medical personnel. This model has, up to now, well served the needs of developed countries and relatively young, healthy, and less mobile communities, by sharing costly infrastructure and medical personnel. However, if we continue with this model, the health care system will be stressed to the point of collapse for a number of reasons, and alternative solutions need to be developed to reduce health care costs while preserving or enhancing the quality of life of a country's citizens. Some of the reasons are:

- Health care is becoming too expensive to deliver. For example, the U.S. Congress, already overburdened with an annual health care bill of more than $1.5 trillion, fears that the health care system will be unable to deal with the increase of potential patients [1].
- Population demographics are changing. The worldwide population of people over the age of 65 is expected to double from 375 million in 1990 to 761 million by 2025 [1]. This implies that preventive and/or assistive care must remove the expensive components of health care. Hospital visits and face-to-face consultations with medical personnel are required only when absolutely necessary.
- Society is increasingly mobile. With the advance of the Internet and mobile technologies, people are becoming extremely mobile, both at work and in their social activities. Health services must follow people on the move, not expect people to come to the service.
- Home care is increasingly preferred. Many people prefer living in their own homes during treatment rather than being in a hospital or nursing home, provided there are satisfactory means for assisting them with home health care services.

Mobile health (m-health) and telemedicine provide alternative and supplementary solutions for coping with new problems in health care. Telemedicine is the use of telecommunications and information technologies for the provision of health care to individuals at a distance and transmission of information to provide that care [2]. M-health is about monitoring the health status of, or providing treatment to, people who are on the move. Telemedicine is often used interchangeably with m-health, but its meaning is different; telemedicine focuses on the transfer of medical data, particularly medical images. However, in both telemedicine and m-health, medical diagnosis takes place remotely, and most m-health applications involve the teletransmission of medical data.

Clearly, communication technologies and network infrastructures play crucial roles in these solutions. Wireless communication is required to support mobile people and medical practitioners. Network infrastructures are needed to support the transfer of medical information and expert advice necessary for treatment. Sensor networks are necessary to keep watch on the vital signs of users who require constant support.

Communication technologies are being used in health care today in a variety of ways—including notification, messaging, web access, videoconferencing, teleradiology, telesurgical consultation, and access to legacy applications. The integration of health care with the Internet, wireless, telecommunication, and mobile technologies has led to increased accessibility to health care providers, more efficient processes, and higher quality of health care services [3–5].

The deployment of communications technologies and network infrastructures aims at eliminating the expensive components of the conventional health care system, providing better health care services to an increasing number of people, and reducing the overall health care cost.

The aim of this chapter is to provide a basic understanding of concepts in data transmission and communications with wired and wireless technologies and a discussion of the role of network architectures and infrastructures for supporting current and emerging health care services. In particular, the chapter focuses on:

- Elucidating the concepts of information and communication channel capacity
- Providing explanations of communication and data transmission techniques
- Discussing the theoretical limitations of a communication channel
- Presenting the Internet architecture and the World Wide Web
- Discussing wireless communication technologies and wireless mobile networks
- Introducing sensors and associated wireless technologies for health monitoring.
- Presenting applications in telemedicine/m-health that require the support of sensors and communication technologies

The chapter is organized as follows: Section 10.2 discusses fundamental concepts of data, data transmission, and the communication channel. This section reviews foundation concepts of information, the nature of data, bandwidth, limitations of a communication channels by Shannon-Hartley law, digital/analog signals, modulation, Asynchronous Digital Subscriber Line (ADSL) technology, and packet switching.

Section 10.3 discusses network layered architecture, the Internet, and the World Wide Web. This section discusses the role of different layers of the architecture of local area and wide area networks. Section 10.4 discusses the essentials of wireless technologies and networks; in particular, it discusses wireless technology from radio frequency to radio transmissions schemes, through narrowband, wideband, and ultra wideband. The section also reviews a number of wireless networks deployed in telemedicine such as mobile cellular networks, wireless local area networks, personal area networks, and satellite communications.

Section 10.5 discusses sensors and their associated wireless infrastructure for health monitoring. This section focuses on sensors, wireless sensor networks and their requirements in supporting health monitoring systems. Section 10.6 presents several relevant applications that illustrate the use of various technologies discussed in earlier sections, as well as some modern applications of wired and wireless networks in telemedicine/m-health which involve emergency, location-based services, tele-training, and tele-operated robot systems.

10.2 Transmission and Communication Technologies

The most important concept in data communications is that the maximum rate at which information is transmitted over a channel (i.e., the channel capacity C with units of bits per second [bps]) is proportional to the available bandwidth. This concept is generally well understood, in that end users today seek high-bandwidth connections to the Internet, preferring asymmetric (or asynchronous) digital subscriber lines (ADSLs) or cable over dial-up connectivity.

Of course, the amount of information transmitted depends on the quality of the link, as well as the technology used. If the link or channel quality is poor and there are high levels of noise, then this will reduce the rate at which information can be transmitted over the link. Before we continue, we need to clarify and define more precisely the terms introduced. We will use real examples of communication technologies to illustrate their principles; unfortunately, this means we will come across many TLAs (three-letter acronyms!).

- **Information (I):** This is a precise term in communications, referring to units of bits. It is actually a measure of uncertainty or surprise. If the content of a message is known before it arrives—that is, if it contains no new information—then the information of the message is 0 bits, regardless of its length. In mathematical terms, if the probability P of a message arriving is 1, then $I=0$. On the other hand, if $P=0$ (i.e., receiving the message is a total surprise, something that should not have happened), then $I=\infty$. In fact, $I = \log_2(1/P)$ bits, where $\log_2(x) = \log_{10}(x)/\log_{10}(2)$.

- **Bandwidth (*B*):** Because B is proportional to channel capacity C, sometimes bps is used as the unit of bandwidth—for example, one might say that the bandwidth of an 802.11-bit wireless link (one of a range of WiFi technologies) is 11 Mbps (megabits per second). However, strictly 11 Mbps is the data rate, and the bandwidth is the amount of frequency spectrum available in the channel, which is measured in hertz (Hz) or, to give it its older name, cycles per second (mains electrical frequency is either 50 or 60 Hz, depending on the country, which means that the polarity of the voltage changes from + to − 50 or 60 times a second).

For example, the WiFi 802.11b wireless technology uses the 2.4 gigahertz (GHz) Industrial, Scientific, and Medicine (ISM) frequency band, which has a frequency range from 2.4 to 2.4835 GHz; that is, a bandwidth of (upper frequency − lower frequency) 0.0835 Ghz, or 83.5 MHz, although it is divided into multiple, overlapping channels of 22 megahertz (MHz) bandwidth each (nonoverlapping channels are channels 1, 6, and 11).

- **Noise:** This is any unwanted signal, ranging from random thermal noise generated in electronic components, to interference from other signals, to noise generated by lightning, etc. Without noise, we could send any amount of information over a channel. A precise voltage level could be assigned as a character representing, say, an encyclopedia; unambiguous reception of this voltage would be equivalent to receiving all the information in the encyclopedia. It is the presence of noise that ultimately limits the capacity of any channel.

We are now in a position to understand equations linking the channel capacity C with the bandwidth B. For a digital technology that uses M characters (e.g., binary communication is a digital technology with $M=2$), the maximum channel capacity or data rate is given by Hartley's law:

$$C = 2B\log_2(M)\text{bps}.$$

Hence, for a binary system, $C=2B$ and (by definition) there is 1 bit per symbol, whereas for a technology with 16 distinct characters (i.e., $M=16$), then $C=8B$ and there are $\log_2(M) = 4$ bits per symbol. So, one approach to increasing data rates is to use more complicated transmission technologies than binary; but first, consider binary signaling.

The binary characters can be different voltages representing 0 and 1, or different frequencies, or even different phases of the same frequency. For a binary system, we use the term *shift keying*—so, if we use voltage levels, then we are using amplitude shift keying (ASK); if we use two different frequencies, we are using frequency shift keying (FSK); and if we use two phases of the same frequency, we are using phase shift keying (PSK). In all cases, we are using broadband communication, where we are modulating a baseband signal (our data) onto a sine wave carrier signal. Mathematically a sine wave is expressed as

$$v(t) = A\sin(2\pi ft + \Phi) \text{ volts},$$

where either the amplitude A (V), the frequency f (Hz), or the phase Φ (radians) can be modulated to carry a signal (or two or more variables can be simultaneously modulated). For example, the early V.21 modem used FSK (two frequencies for the transmit path: one corresponding to binary 0 and the other to binary 1; and a different pair of frequencies on the return path to avoid interference). The chosen frequencies are all in the frequency band of an analog telephone channel, that is from 300 Hz to 3.4 kilohertz (kHz) (with bandwidth of 3.1 kHz). It can transmit 300 bps (yes, only 300 bps!) in full duplex (FDX) mode: transmitting 300 bps and receiving 300 bps simultaneously over the same two-wire (2W) circuit. Note that other transmission modes are half duplex (HDX), where a channel is used alternately for the transmission and reception directions, and a 4W circuit uses two 2W circuits, one for the transmission and one for the reception direction.

For a system with more characters, it is common to combine both amplitude and phase shift keying. For example, eight phases might be distinguished, each with two voltage levels, giving 16 distinct characters. Combining amplitude and phase is called *quadrature amplitude modulation* (QAM): in this case, 16-QAM with 4 bits per symbol. Using an advanced technique called trellis coded modulation (TCM) enables higher data rates to be reached. TCM uses a forward error correction (FEC) code, where the signal contains some redundant information, enabling some characters to be corrected immediately on reception based on information in the received signal. Although sending redundant information increases the amount of data to be sent, the overall effect of the error correction is to enable the data rate to be increased. The FEC code is then modulated using QAM.

The result of TCM is that a V.34 modem can reach speeds of 33.6 kbps FDX over a dial-up connection (with TCM and 1,664-QAM, i.e., 10.7 bits per symbol), this rate being close to the theoretical limit, which is determined by the noise on the channel. The equation for the maximum channel capacity in the presence of noise is called the *Shannon-Hartley law:*

$$C = B\log_2(1 + S/N) \text{ bps}.$$

A channel capacity $C=33.6$ kbps and $B=3.1$ kHz requires a high-quality channel with a signal-to-noise power ratio (S/N) of at least 1,830 (a ratio usually expressed in dB for convenience, where $10\log_{10}(1830) = 32.6$ dB).

Other modems, such as the V.92, which allows a maximum rate of 56 kbps using pulse amplitude modulation (PAM), and ADSL modems (see next section), provide higher speeds than the V.34. But they carry data only over the "last mile," from the local exchange (or central office, in the United States) to the user, whereas the V.34 provides an end-to-end data connection across the telephone network.

10.2.1 Asymmetric Digital Subscriber Lines

In ADSLs, a typical ratio of 10:1 is set between the download rate (from the Internet) and the upload rate (to the Internet), hence the term *asymmetric*. The maximum rates depend on the distance between the user and the telephone exchange, ranging from 8.448 Mbps near the exchange (<3,000 meters) down to 1.554 Mbps farther away (<5,500 m) [6].

The maximum data rate over an analog telephone line using a V.92 modem is only 56 kbps. But ADSLs use the same copper twisted-pair telephone wiring, with one pair carrying signals in both directions (i.e., over the 2W FDX circuit in the local loop between the telephone subscriber and the local telephone exchange)—so how does it achieve such enormous increases in the data rates? The answer is that the line is set up for a digital link, rather than for an analog telephone. What was important for the analog telephone line was that the frequency response should be flat over the 3.1 kHz bandwidth so that voice signals were not distorted; whereas for a digital line, we need only recognize characters, and, as we have seen, it is the bandwidth that is important. The flat frequency response was achieved using "loading coils," which are passive inductors inserted into the line. Removing these greatly increases the available bandwidth.

With the ITU G992.1 ADSL standard, this increased bandwidth is used for multiple 4 kHz bandwidth channels, with each individual frequency slot or channel being modulated very much like the V.34 modem. This is a form of *frequency division multiple access* (FDMA), but the technology for ADSL is called discrete multitone (DMT). There are a maximum of 224 channels for download from the Internet and 25 for upload. So, an ADSL modem can be thought of as 224 QAM modems in one box! Not all of these 224 modems can achieve the full rate simultaneously, but the maximum data rate is still very impressive.

The frequency range for an analog channel (300 Hz to 3.4 kHz) is kept for an existing analog POTS (plain old telephone system!) phone. All the other channels are 4 kHz bandwidth and modulated onto carrier waves spaced 4.3125 kHz apart in the frequency spectrum. The 25 upstream channels are in the frequency range from 30 kHz to 138 kHz, while the 224 downstream channels are in the frequency range from 138 kHz to 1.104 MHz (see Figure 10.1). From 3.4 kHz to 30 kHz is a guard band, to prevent interference between the POTS channel and the digital data.

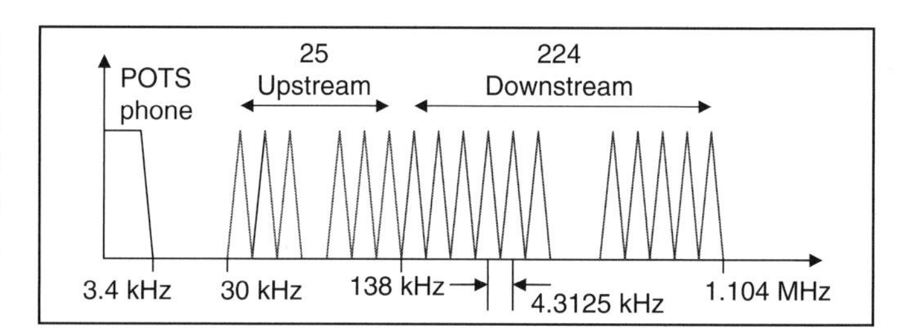

FIGURE 10.1 ADSL frequency spectrum.

A typical ADSL plan provides 1.5 Mbps for download and 256 kbps for upload, while the actual download rate achieved would be typically 1.1 Mbps [7].

10.2.2 The Nature of Data

In the real world, we expect data to be mostly analog; that is, a continuous variable within some range, such as air pressure or vehicle speed. The term *analog* derives from the correspondence between the original signal (e.g., air pressure) and the analog signal (e.g., the voltage output by a pressure transducer, or sensor, which converts air pressure to an electrical signal). But we also deal with digital data, such as a traffic light with its three colors. The process of converting from a continuous analog voltage signal to a binary number can be divided into three stages:

1. **Sampling:** This is a process in which a sample of the signal (e.g., the value of the voltage) is taken at repetitive intervals (so that the signal is no longer continuous). The time between adjacent samples is the sampling period T_s, and the sampling rate or frequency f_s is simply the reciprocal of this; thus, $f_s = 1/T_s$. There is a very important result in communications theory, called the sampling or *Nyquist theorem*, that for a signal with bandwidth B, all the information in the original signal can be recovered if the sampling rate $f_s > 2B$ ($f_s = 2B$ is called the Nyquist rate). Considering our analog telephone channel with $B = 3.1$ kHz, the minimum sampling frequency is just over 6.2 kHz. However, using simple filters to keep costs low means that the sampling frequency needs to be increased, and a value of 8 kHz is chosen as the standard rate.
2. **Quantization**: Consider a simple 4-bit analog-to-digital converter (ADC) that is used to convert voltages in the range from −0.5 V to 15.5 V into a 4-bit number. With 4 bits, we can have 16 possible voltages, ranging from 0000 (corresponding to 0 V) to 1111 (corresponding to +15 V). Any voltage in the range from −0.5 to 0.5 V is converted to 0000, and anything in the range from 14.5 to 15.5 V is converted to 1111, etc. This stage introduces rounding or quantization errors, which, being unwanted,

are actually another form of noise. The maximum quantization error in this example is 0.5 V or (0.5)/16 = 1/32 = 3.125%. In general, for a linear ADC of n bits, the maximum quantization error is $1/(2^{n+1})$, and the resolution is simply stated as n bits.

3. **Serialization:** This is where a rounded (quantized) voltage is converted into a binary number. In this particular example, the analog voltage is converted directly into a binary number. The serialization process is hidden, which is how most ADCs operate.

For an analog telephone channel, an 8-bit number is generated for each speech signal; thus, the digital telephone bit rate is 8 bits × 8 kHz = 64 kbps. Although advantage is taken of the characteristics of speech signals to use nonlinear coding, the resolution is 12 bits for low-amplitude speech but lower resolution for high-amplitude speech.

A significant advantage of digital signals is their generally better performance in the presence of noise. Since a digital signal is selected from a restricted and known set of symbols, it is possible to recover and regenerate the original symbol exactly if the noise is not too great. For example, if the green light in a traffic light is a shade of blue-green, people will still be able to recognize that it should be "green" and will proceed with basic confidence of their safety. If we can correctly recognize the transmitted digital symbol when we regenerate the signal at the receiver, we will remove any noise added in transmission over a link. Hence, the noise in a well-set-up digital communication system is essentially just the noise introduced by the rounding errors of the original ADC process (which by increasing the number of levels can be made arbitrarily small), regardless of the number of links over which the message is transmitted. This is different from what occurs in an analog communication channel, where generally the noise is additive, so the longer the channel, the worse the noise. However, note the qualifier that the channel must be well set up. If the noise is such that we choose the wrong symbol when regenerating the signal at the receiver, then errors are inserted into the data. When a wrong symbol is selected, the noise is actually increased. Hence, above a certain noise threshold, the performance of a digital link tends to collapse, and the signal breaks up, while the analog signal, while degraded, is still meaningful.

10.2.3 Packet Switching

A POTS connection is an example of circuit switching, which means that a physical channel is dedicated to the call for its duration. Nowadays, however, a more usual technique of communicating is packet switching. Sometimes one packet will be enough for the entire message, but in other cases a packet will be part of a call or flow. In packet switching, the physical link is shared among all the packets, so that one packet from one call or flow uses the link, then another from another flow, and so on. Other packets from the flow will follow on the link as and when they are available and as and when the link is free; that is, not necessarily at regular time intervals.

Packet switching is a digital technology, where all information sources are converted to binary data before being sent. This is very useful, as it means that all sources can be transported over the same network, saving management and deployment costs; for example, speech is converted to packets of data and sent using the Voice over Internet Protocol (VoIP). As we have seen, a 64 kbps constant bit rate (CBR) stream is the standard for telephony, but we cannot achieve the CBR with raw traffic over the Internet, as the data are sent in packets. Hence, our 64 kbps data stream must be chopped up and sent in packets over the Internet and reassembled at the destination. As an example of a typical VoIP scenario, for a G.711 codec with two 10 millisecond (ms) samples per packet, the expected packet arrival rate is 50 packets per second, and the packet length is 200 bytes [8] (hence, the total IP data rate is 80 kbps).

There are two types of packet switching: *connectionless* (as in the current Internet), and *connection oriented.* In connectionless packet switching, even though there may be a flow of packets, each packet in principle finds its own way over the network, so packets may end up going different routes, and hence may end up out of sequence (but in general, the Internet is stable enough over a time scale of a week that the packets will all follow the same route unless a fault occurs). In connection-oriented packet switching, all packets follow the same route; and if there is a fault, then the flow is terminated.

10.2.4 Medium Access Schemes

Packet switching is one way of sharing a common medium between multiple channels. Other methods (see Figure 10.2) are:

- FDMA (see Section 10.2.1), which divides the bandwidth of the channel into several subbands of frequencies, and each user is assigned a subband for the total channel time.
- TDMA (time division multiple access), which divides the channel into time slots. Each user is assigned the total channel bandwidth for a time slot on a rotating or on-demand basis.
- CDMA (code division multiple access), which uses *spread spectrum* techniques and assigns a unique digital code rather than radiofrequencies to differentiate among different transmissions. (A discussion of spread spectrum is in Section 10.4.1.2.)

10.3 The Internet and World Wide Web

From home we can connect to the Internet using a V.92 dial-up modem or an always-on ADSL or cable modem. Cable provides access rates of typically 3 Mbps [7] using the feed originally

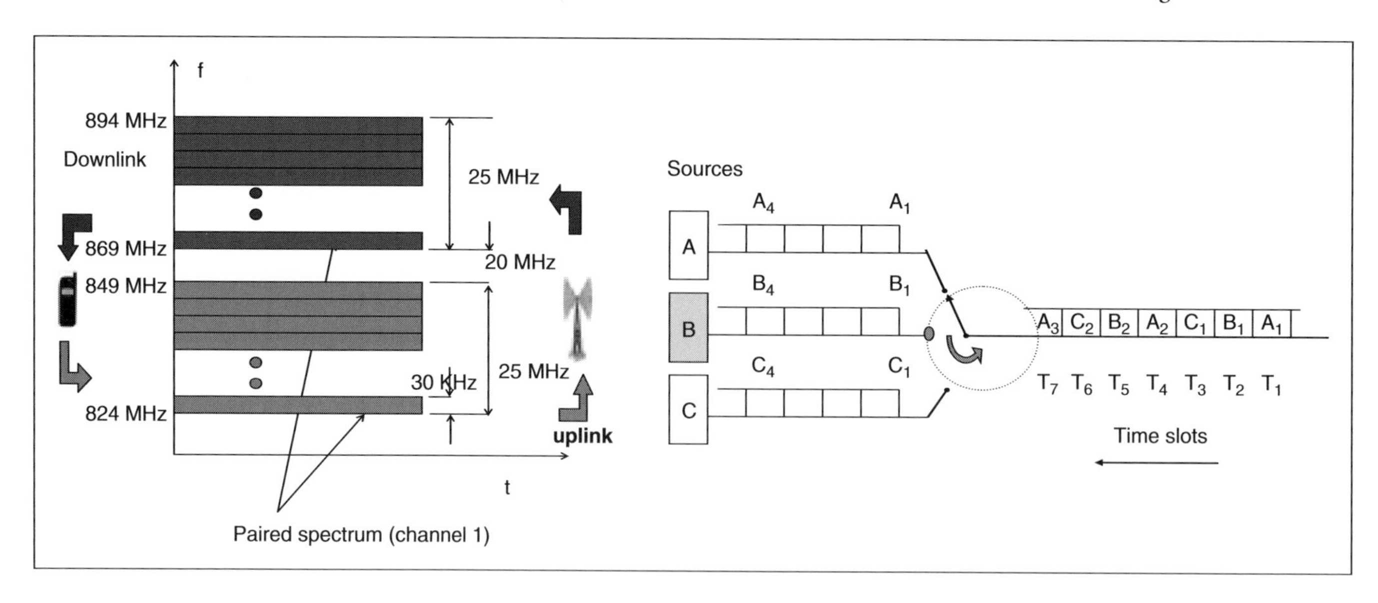

FIGURE 10.2 Multiple access schemes.

designed to support cable TV. Or we can use one of the wireless technologies, as described in Section 10.4. At work we may use an in-house network or connect over a LAN (a local area network), as outlined in Section 10.3.2. In this section we will look at some of the common protocols used to support our use of the Internet, and end up looking at some applications of the World Wide Web (WWW).

10.3.1 Layered Communication Architecture

It has been found to be advantageous to have a layered architecture for networks. The interface between layers is well defined, which enables different technologies and protocols to be "slotted in" as appropriate. And changes to one layer do not cause changes to ripple through the rest of the protocol stack. However, although layered architectures have served us well up to now, thought is being given to linking layers and breaking the model with multilayer services, in order to provide different quality-of-service (QoS) levels to different applications.

There are two common layered architectures: the seven-layer Open Systems Interconnection reference model (OSI RM) of the International Standards Organization (see Figure 10.3), and the TCP/IP (transport control protocol/Internet protocol) stack (see Figure 10.4) [9]. Both architectures have as their bottom two layers a physical layer (Layer 1) and a data-link layer (Layer 2).

Layer 1 deals with physical connectivity such as pin layout, voltage levels, cabling specifications and limits, etc. Often to troubleshoot this layer, all that is needed is to find out whether or not a light-emitting diode (LED) is lit.

Native system		HOST 1 End-User↓		HOST 2 End-User↑	
Host user oriented layers	7	Application	←——→	Application	7
	6	Presentation	←——→	Presentation	6
	5	Session	←——→	Session	5
Host network layers	4	Transport	←——→	Transport	4
	3	Network	Network service layers	Network	3
	2	Data link		Data link	2
	1	Physical		Physical	1
				↑Physical connection	

FIGURE 10.3 The seven-layer OSI RM hierarchy.

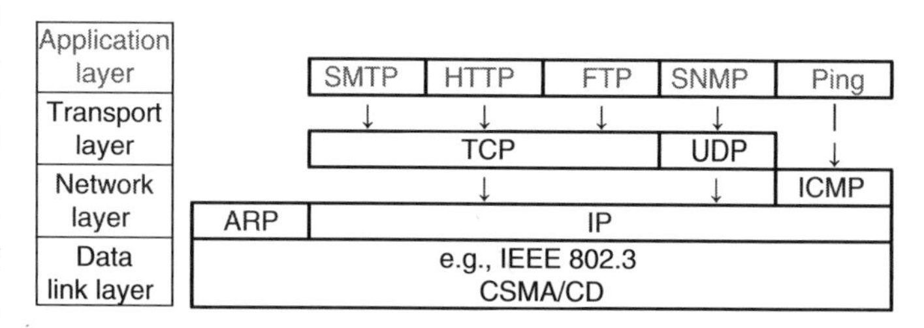

FIGURE 10.4 TCP/IP protocol suite layers. SMTP = simple mail transfer protocol; HTTP = hypertext transfer protocol; FTP = file transfer protocol; SNMP = simple network management protocol; UDP = user datagram protocol; Ping = packet Internet groper; ICMP = Internet control message protocol; ARP = address resolution protocol; IEEE 802.3 CSMA/CD = Institute of Electrical and Electronics Engineers standard for carrier sense multiple access and collision detection.

Layer 2 deals with data transfer between two machines over a physical link. At this layer, both devices have to be running the same protocol at the same speed. The most important standardization at this level comes from Ethernet or, more correctly, 802.3 CSMA/CD (carrier sense multiple access/collision detection). These are essentially the same (minor differences between the two protocols mean that they can run on the same physical network but will not interoperate); however, the old Ethernet protocol proper is so rare now that the term "Ethernet" is often used (as we will from now on) to refer to 802.3 CSMA/CD. (Ethernet is considered at length in Section 10.3.2.1.) Error control is a function of primarily Layer 2, and the data unit is normally called a "frame" in this layer.

Layer 3 is referred to as the network layer in both the OSI RM and the TCP/IP protocol stack. Its function is to ensure that data get from the source to the destination over multiple hops (i.e., over multiple Layer 2 links). However, unlike Layers 1 and 2, the Layer 3 standards are different in the two models: In TCP/IP, the IP provides the mechanism to route packets over a network of networks (i.e., an internetwork, or Internet), while an example of an OSI RM network layer protocol would be X.25. Routing and flow control are primary functions of this layer. The data unit is called a "packet" in this layer.

Layer 4 is the transport layer in both the OSI RM and TCP/IP, although, again, they are not compatible standards. Unlike the network layer, Layer 4 appears only in the end hosts. This layer may be necessary in a gateway between two different network technologies, but really the gateway is acting as a proxy or stand-in for an end host. The function of this layer is to ensure that all packets are received and presented to the next layer in the correct order. The relevant TCP/IP protocol is the TCP, which aims to present to the next layer a service that appears to be connection oriented, even though it actually runs over a connectionless IP service. Sometimes we do not require a connection-oriented service, so the *user datagram protocol* (UDP) is an alternative, lightweight transport protocol in the TCP/IP stack.

In the OSI RM, there is also Layer 5, involving the work session, and Layer 6, dealing with presentation. They are not present in the TCP/IP protocol stack. Layer 5 allows a session to be rolled back to a previously saved common position. In TCP/IP, this function is left to the top layer (the application layer). Layer 6 allows data formats to be adjusted to fit the presentation requirements of the end hosts (e.g., file structures, character sets). TCP/IP again leaves such things to the application layer.

Finally, the top layer deals with the application. This is Layer 7 in the OSI RM or Layer 5 in the TCP/IP protocol stack. In TCP/IP, this could be e-mail via *simple mail transport protocol* (SMTP), the WWW via the well-known http (hypertext transfer protocol) signature, or one of the many other Internet applications. There are similar applications in the OSI RM. Data are segmented into the appropriate size for packets, or are reassembled, at this layer.

Actually, there is another layer even above the application layer: This is the user program, which calls the application layer protocol. It would be better to think of the applications in the top layer as services rather than applications; thus, the user application (e.g., a web browser) uses the http service.

10.3.2 Local Area Networks

LAN protocols are *media access control* (MAC) protocols. Their function is to get data from source to destination host in the same LAN. To do this, they use the MAC address, a unique six-byte address hard-wired (mostly) into every network interface card (NIC) (the subsystem that provides the interface between a computer and a network). The MAC address is expressed in hexadecimal—for instance, 00-13-02-90-71-E2, where the first three bytes identify the manufacturer (Intel, in this case). Try typing "ipconfig/all" into a Windows command shell (from the "start" button in Windows, go to "Run"; type "cmd"; and click "OK") to see the MAC or physical address of your own machine.

10.3.2.1 Ethernet

This protocol has several different versions. The older versions supported a 10 Mbps data rate, but 100 Mbps is common today, and 1 Gbps and 10 Gbps are available. The original Ethernet specification used a copper cable as a "bus"; that is, a common transmission medium connecting all the nodes on the network. Since it was a common medium, if two nodes started to transmit at the same time, there would be a "collision"; that is, both signals would appear together on the bus and each would appear as noise to the other, corrupting each other's signal and preventing any node from being able to recover it. The definition of the 802.3 CSMA/CD standard as "*c*arrier *s*ense *m*ultiple *a*ccess/*c*ollision *d*etection" refers to the way the protocol deals with this problem:

1. CSMA: First check or *sense* if any other node is using the medium. If yes, wait; if no, start to transmit.
2. CD: But some nodes might decide that the medium is free at the same time, causing a collision; so, *detect* it (by recognizing out-of-specification voltage levels), and then recover (by waiting a random time before trying to transmit the frame again). Give up after 15 failed attempts (highly unlikely).

Modifications of this scheme are used for wireless access (since wireless devices also use a common medium), but even if the medium is dedicated, rather than being common, Ethernet is still used for simplicity and flexibility in managing networks.

The general format for describing Ethernet technologies is ***cModex***, where ***c*** refers to the data rate in Mbps; ***Mode*** is either "base," which means that the signal is not modulated (i.e., it is a baseband signal), or "broad," which means that the data are modulated onto a carrier wave to make a broadband signal; and finally, ***x*** either refers to the maximum length of the cable bus to the nearest 100 m or gives (if ***x*** is a character string) a clue as to the medium—for instance, "T" refers to the use of a copper twisted pair, and "F," "LX," or "SX" mean that fiber-optic cable (of various types) is being used. Hence, some Ethernet technologies are 10Broad36, 100BaseT, 10Base5, 1000Base-SX, and 10GBase-T.

10.3.3 The Internet

The Internet is based on the use of the TCP/IP protocol stack. As noted in our introduction of Layer 3 (Section 10.3.1), it is a network of networks, providing a way to link LANs together. Packets are routed from one network to another using IP, while in a LAN the MAC protocol is used to deliver packets (frames at Layer 2) from source to destination host. The *address resolution protocol* (ARP) is used to associate MAC addresses with IP addresses. TCP works in the end hosts to resequence packets and provide flow and congestion control.

The TCP/IP protocol stack and the Internet itself came out of work on ARPANET, the Advanced Research Projects Agency Network developed in the late 1960s for the U.S. Department of Defense, with the first TCP/IP release in 1983. A primary reason for the success of TCP/IP was that IP was engineered to be a very simple network. Because early computing and networking equipment was unreliable, making the network resilient came down to making it connectionless (as opposed to connection oriented), which also helped by making IP simpler. In a connectionless network, even if a next hop router fails, packets can be delivered as long as an alternative route exists. While it satisfies the resiliency constraint, this choice has made QoS provision very difficult, because QoS cannot be guaranteed unless all packets follow the same path—this is a problem the Internet Engineering Task Force (IETF) is currently trying to resolve.

The most widely used IP today is version 4 (IPv4), but there is a problem with the number of available IPv4 addresses. The IPv4 address field is 32 bits, and while theoretically this gives $2^{32} = 4.3 \times 10^9$ addresses, many multiple blocks have been given away, and the number of free addresses is scarce. There are work-arounds for this, such as *network address translation* (NAT), which enables many workstations in an enterprise to use private IP addresses in the enterprise network and share relatively few public IPv4 addresses for connecting to the Internet. However, this suits only organizations that already have IPv4 addresses.

In developing countries, IP version 6 is a preferred solution, since it has what seems now to be a virtually inexhaustible reservoir of addresses. There are 128 bits in the IPv6 address field: 64 bits for the network ID and 64 bits for the host ID. This means that there are a possible 1.8×10^{19} network addresses, used to route packets over the Internet to the destination network—that is 2.8 billion networks for every one of the 6.5 billion people alive today, with each network supporting 1.8×10^{19} host addresses. There are enough bits in the IPv6 host ID to use the 48 bits of the MAC address, which would simplify network administration (although it does raise a privacy issue in that end-user equipment could be tracked). The IPv6 address space contains the IPv4 address space as a subset, so IPv4 networks can work with IPv6 equipment.

10.3.3.1 Internet Protocol Addresses

We have three address formats to consider: hostname, IPv4, and IPv6. The hostname is a more human-friendly label, which really refers to a service rather than a machine. For example, **www.iana.org** is the hostname of a web server for the Internet Assigned Numbers Authority (IANA) (responsible for allocating IP addresses). The Domain Name System (DNS) is used to map between hostnames and the IP address used to route packets across the Internet (e.g., typing "nslookup www.iana.org" into a Windows command shell would show that www.iana.org might have the IPv4 address 192.0.34.162. So, putting "http://www.iana.org" into the uniform resource identifier (URI) field of a web browser has the same effect as typing http://192.0.34.162). The advantage of hostnames, besides being easier to remember, is that they are not machine specific. Hence, the IANA web server might be swapped to another machine, the DNS would be updated, and no user of the hostname would notice any difference, while anyone using the actual IP address would no longer find the web server.

IPv4 addresses are written in a "dotted decimal" notation. They are 4 bytes long, and in decimal a byte is from 0 to 255 bits, so each byte value is written in decimal and separated by a dot from its neighbors. Some addresses are reserved for special use, such as in private or dedicated networks [10].

An IPv6 address consists of 16 bytes, written in groups of two (as four hexadecimal characters) with a colon between them, thus:

2001:0000:0000:0000:1319:8a2e:0370:7334.

Any consecutive groups of four zeroes may be replaced by a double colon, as long as only one double colon is used (two double colons would be ambiguous). Hence, the address in the foregoing example can be shortened to

2001::1319:8a2e:0370:7334.

As mentioned before, IPv6 contains IPv4 addresses, and the format of an IPv4-mapped IPv6 address is

::FFFF:w.x.y.z

(where w.x.y.z is any IPv4 address).

```
C:\Documents and Settings\simmonds>netstat -n

Active Connections

  Proto  Local Address          Foreign Address        State
  TCP    10.50.1.203:1544       192.0.34.162:80        TIME_WAIT
  TCP    10.50.1.203:1545       192.0.34.162:80        TIME_WAIT
  TCP    127.0.0.1:1399         127.0.0.1:1400         ESTABLISHED
  TCP    127.0.0.1:1400         127.0.0.1:1399         ESTABLISHED
```

FIGURE 10.5 Sockets for web browsing.

TCP (or UDP) and IP together provide a "socket" in the end host to identify an application for connecting at the level of the application programming interface (API). With TCP there is a return path, so in this case there is a "socket pair." IP provides the IP address, while TCP provides the port number. An example of a socket used for a web browser is 192.0.34.162:80 (e.g., type "netstat -n" in a Windows shell immediately after loading a new webpage, and look under "Foreign Address"; see Figure 10.5). There are well-known port numbers [11] such as 25 for SMTP and 80 for HTTP.

In Figure 10.5, the establishment of the IPv4 192.0.34.162:80 socket on my PC is caused by opening a web browser and loading the URI **http://www.iana.org/assignments/port-numbers**. This gives the hostname of the server with the page I am trying to find and also specifies that I want to use http (i.e., port 80). A call to a DNS server converts the hostname (www.iana.org) into an IPv4 address. My machine then establishes a socket 192.0.34.162:80 and sends a synchronization (SYN) packet to start the TCP session. The web server at **www.iana.org** replies with a synchronization/acknowledgment (SYN/ACK) packet (using, e.g., port 1544; the port number is not 80, since this is a return path to a browser; i.e., only packets to a web server have the port number 80). My machine then establishes the second socket of the pair (e.g., 10.50.1.203:1544, as I am using a private address with NAT) and sends an ACK back to confirm the TCP session (by this point, both browser and server have acknowledged each other). This SYN to the server, the SYN/ACK back, and the ACK to the server is the three-way TCP "handshake" to establish a TCP session, with the socket pair identifying both ends. The webpage having been delivered, the session is over, so the socket is now in the TIME_WAIT state. The 127.0.0.1 IP address is a special-use address that on any computer refers to itself (i.e., localhost). Here it shows that a looped TCP connection has been established to ensure that the TCP/IP protocol stack is kept in use, in order to avoid any startup delay.

Type "arp -a" in a Windows shell to see the cache (a short-term memory that is rebuilt as needed) of mappings between IP addresses and MAC addresses that your machine has discovered. You should at least see the mapping between the IP and MAC address of your gateway to the Internet.

10.3.4 The World Wide Web

The WWW is actually just one service (using http) that runs on the Internet. It was developed at the Centre Européen de Recherche Nucléaire (CERN) by Tim Berners-Lee, originally as a way of sharing work on physics. The first public release was in 1991. Apart from marrying the Internet with hypertext and a markup language (HTML), it also introduced URIs (also known as URLs [uniform resource locators] and web addresses, although URI is the preferred technical name).

Hypertext is the linking of one or part of a document to another or even to itself, using one-way hyperlinks (e.g., the underlined and generally blue text in a webpage, unless you have clicked on them already, in which case they show a different color). The fact that links are one-way is important, since they do not require changes in two documents in perhaps different web servers to establish a link; hence, the web rapidly grew. On the other hand, it does mean that broken links are to be expected as documents are removed without regard for all the pages linking to them. A hyperlink need not be text—an image can be a hyperlink.

URIs describe the location and method of access of a resource (e.g., a document, an image)—for example, **http://info.cern.ch** is a webpage found on a web server at **info.cern.ch** and is to be accessed using a web browser to display HTML, which is the markup language for the web—the latest version being HTML 4.01 [12]. The markup language is concerned purely with the presentation of information, not with any meaning, although there is a massive effort now to develop a *semantic web* using XML (eXtensible markup language). This would be a web where information could be understood by machines.

A very simple webpage can easily be written:

```
<html>
<head>
<title>an example webpage</title>
</head>
<body>
<p>This is a paragraph with an <a href="http://info.
cern.ch">example hyperlink</a></p>.
</body>
</html>
```

The display instructions in this HTML document are specified by the tags (< >), and a web browser is programmed to display the page appropriately. Note that tags should be paired (a closing tag starts with </) and nested (e.g., the paragraph is completely inside the body that is inside the HTML tags). You may see the code of a webpage in your browser by selecting **View => Source** or **Page Source.** The webpage in our example can be typed in Notepad in Windows and saved as **example.htm.** A web browser can then be used to display the file simply by clicking on it (N.B.: Take care with the " marks).

A key difference between HTML and XML is in the presentation of dates in a document. HTML has no knowledge of dates and will simply display "19/9/2006" as text, leaving it to the reader to recognize that it is probably a date. But associated with an XML document there will be a schema describing what every tag does, allowing new tags to be introduced (hence, the *extensible* in XML)—for instance, **<date>** may be defined so that when a program scans the document, it can recognize **<date>19/9/06</date>** as a date.

For doctors, perhaps the most obvious use of the WWW is by patients looking up their symptoms or their diagnoses so as to be proactive in their own health care. There are many support groups who have webpages giving information and links, doing fund-raising, etc., to help people diagnosed with a particular condition. Or, for people who have symptoms of particular concern, there are many websites that will walk them through self-diagnosis.

As its programs improve, artificial intelligence (AI) will become an essential component in health care. AI programs may become acceptable to patients through the use of *avatars* (i.e., virtual others) such as ALICE (Artificial Linguistic Internet Computer Entity) and Alex [13, 14]. Interestingly, it has been shown that avatars should not be too realistic: The assumption of people that they are talking with a real human being becomes confused due to slight miscues. Of course, the WWW is open to everyone, and there are some snake-oil merchants peddling their wares online. One opportunity to detect those who prey on vulnerable people in need of medical care is when a website says that patients must buy the product or they will end up blaming themselves. A task for health care professionals is to give advice to patients as to which websites are appropriate and can be trusted.

10.4 Wireless and Mobile Technologies in M-Health

10.4.1 Wireless Technology Essentials

10.4.1.1 Radio Waves and the Frequency Spectrum

Conventional wired communications use conductors (twisted-pair copper wires or coaxial cables) or optical fibers to send and receive data, whereas wireless communications utilize electromagnetic waves without relying on wires.

The radiofrequency (RF) spectrum is the lower part of the electromagnetic frequency spectrum. Figure 10.6 shows the RF spectrum and common uses from very low frequency (VLF) to the microwave frequency, EHF (extremely high frequency), and on up through infrared, visible, and ultraviolet (UV) light. RF ranges extend from 10 Hz to over 30 GHz and are divided into 450 different bands. Normally, a license is required from a regulation authority to send and receive on a specific frequency, but unregulated bands are available to use without a license. Two unregulated bands are the ISM band (see Section 10.2) (902–928 MHz, 2.4–2.4835 GHz, and 5.725–5.85 GHz) and the Unlicensed National Information Infrastructure (U-NII) band (5.15–5.35 GHz and 5.725–5.825 GHz).

There are a number of interesting facts associated with electromagnetic waves. With equal power levels, waves with lower frequencies tend to travel farther than higher-frequency waves. And lower-frequency waves tend to penetrate objects better than higher-frequency waves. For example, FM radio

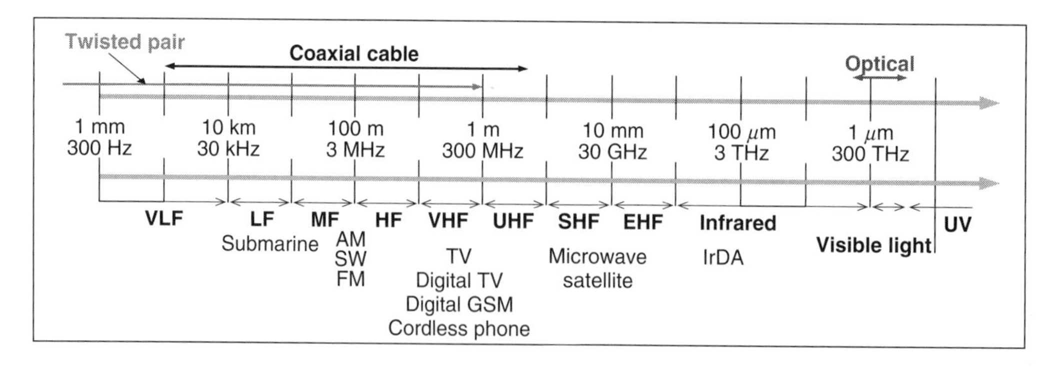

FIGURE 10.6 Radiofrequency spectrum. LF = low frequency; MF = midfrequency; HF = high frequency; VHF = very high frequency; UHF = ultra high frequency; SHF = super high frequency; AM = amplitude modulated; SW = short wave; FM = frequency modulated; GSM = Global System for Mobile Communications; IrDA = Infrared Data Association.

can travel through buildings and other obstacles easily, while higher frequencies such as from GSM phones operating at 1.8 GHz have a harder time penetrating buildings.

In one sense, though, Figure 10.6 is misleading, since it uses a log scale on the *x*-axis (each tick is 10 times the previous tick). Therefore, the bandwidth between ticks is 10 times greater than in the adjacent lower band. Because capacity is proportional to bandwidth, higher frequencies can carry more data.

10.4.1.2 Transmission Schemes

Both infrared and radio waves can be carriers of wireless signals. Infrared transmission utilizes a narrowly focused beam of infrared light that is normally emitted from an LED or semiconductor laser. An infrared signal can be sent to the receiver directly or by reflection to the receiver.

Radio transmission is performed by emitting a sine wave RF carrier into space, the carrier being characterized by amplitude, frequency, and phase, as discussed in Section 10.2. Information (e.g., user data) can be carried by modulation onto one of the carrier's characteristics: its amplitude (as in AM radio), its frequency (as in FM radio), or its phase. A receiver with an appropriate antenna tuned to the right carrier will be able to collect the embedded signal.

Radio signals are *narrowband,* meaning that the information occupies only a narrow band of frequencies centered on the carrier frequency. Figure 10.7 depicts the narrowband radio transmission process using amplitude modulation.

An alternative to narrowband is *spread spectrum* transmission. Spread spectrum is a technique that takes a narrow signal and spreads it over a broader portion of the RF band. This has benefits where, as is often the case, noise occurs at particular frequencies. Spread spectrum is marginally affected by such noise, whereas narrowband RF can be impossible if the noise is in the same frequency band as the narrowband signal.

Two methods are used for spreading the frequency spectrum: frequency hopping spread spectrum (FHSS) and direct sequence spread spectrum (DSSS). With FHSS, spreading is achieved by sending a short burst of user data on one frequency, hopping to a new frequency and sending data for a short period of time on this frequency, then hopping to a new frequency and doing the same, etc., until the transmission is complete. The exact sequence of frequencies used is known as the *hop sequence.* The receiver must know and synchronize with the transmitter hop sequence in order to correctly receive the transmission. DSSS uses an expanded redundant code to transmit each data bit. Figure 10.8 illustrates the transmitting and receiving processes of DSSS. The transmission process involves two stages. In the first stage, the user signal is spread by multiplying itself (XOR ["exclusive or"] operation) with a chipping sequence; and in the second stage, the spread signal modulates an RF carrier before transmission. At the receiving end, the reverse process is performed to recover the user data. Spread spectrum is employed in CDMA, allowing better-quality signal reception and higher (up to 3x) data-carrying capacity compared with TDMA.

Ultra-wideband (UWB) technology refers to a modulation technique based on transmitting very short pulses [12, 15]. The purpose is to spread the information signal over a very large bandwidth so that the signal power can be extremely small for the same channel capacity (Shannon-Hartley law; see Section 10.2). There are two main differences between UWB and other "wideband" systems. First, the bandwidth of UWB systems is more than 25% of the center frequency, or more than 1.5 GHz as defined by the U.S. Federal Communications Commission. This bandwidth is clearly much greater than the bandwidth used by any current scheme (at least an order-of-magnitude wider than current spread spectrum techniques such as FHSS and DSSS in CDMA technology). Second, UWB is typically implemented in a carrierless fashion. UWB implementations can directly modulate an extremely narrow pulse that has very sharp rise and fall times, thus resulting in a waveform that occupies several GHz of bandwidth.

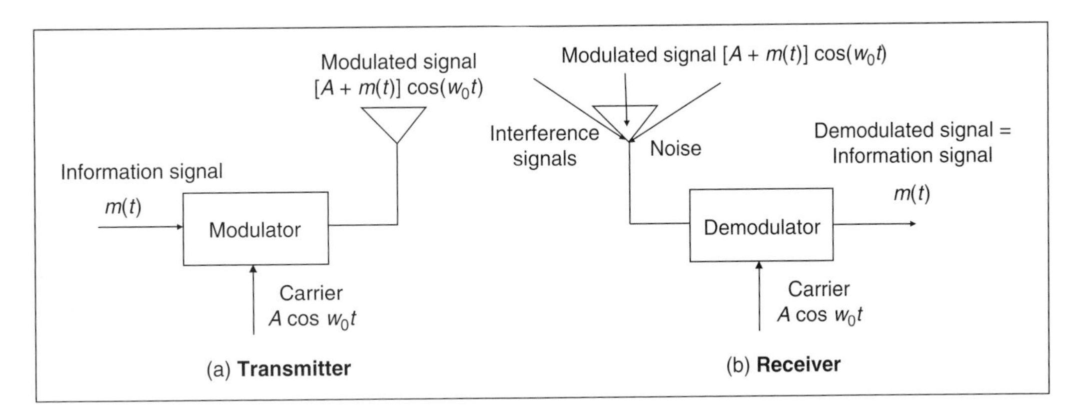

FIGURE 10.7 Narrowband radio transmission.

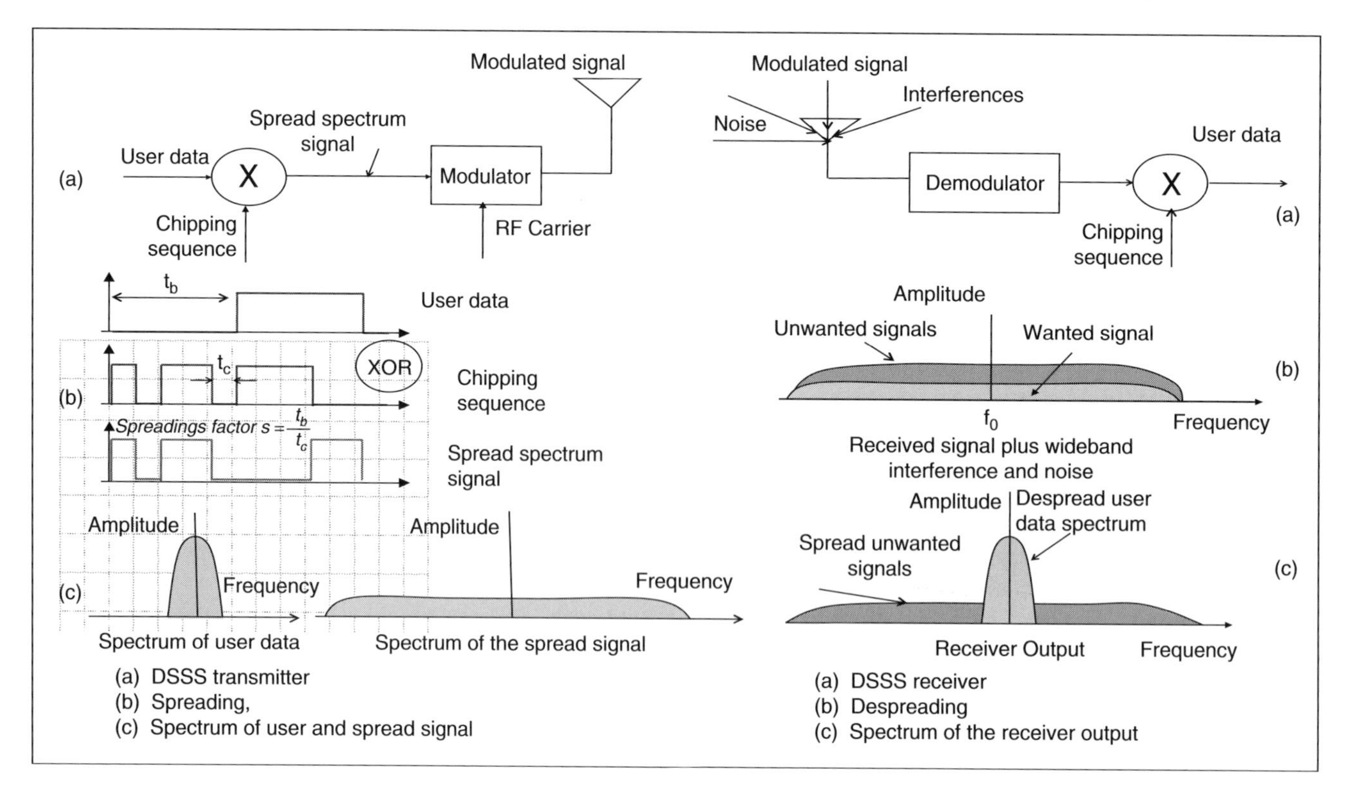

FIGURE 10.8 Spread spectrum transmission.

10.4.2 Wireless Wide Area Networks

The most prominent wireless application of wide area networks (WANs) is the cellular telephone, so called because the coverage area of its network is divided into cells. Cellular telephone networks allow users to move freely while communicating with other users. When a user moves around within the same cell, its associated base station will be responsible for maintaining the connection between the phone and the fixed network. As the user moves beyond the coverage of the current cell to another cell, the cellular network will arrange for a new base station to take over the management of the connection in a hand-off procedure. The user can also roam beyond the coverage of the entire cellular network to another cellular network.

The **first generation** (1G) of wireless cellular technology used analog transmission (Section 10.2) was mainly for voice communication and did not allow roaming.

The **second generation** (2G) uses digital transmission, allows roaming between network operators, and supports data transmission. Examples of 2G networks include the Global System for Mobile Communications (GSM) [16] in Europe and IS-41 in North America. GSM users can send and receive data at rates up to 9,600 bps in normal operation. It is a circuit-switched network. The General Packet Radio Service (GPRS) [16] (the so-called 2.5G) evolved from GSM to transport both voice and data. GPRS uses packet switching for sending data. It enables connection to public and private data networks such as TCP/IP and X.25 networks. The maximum data rate for GPRS networks is 171 kbps.

Third generation (3G) wireless devices use packet-switched networks for both voice and data. The International Telecommunication Union (ITU) has defined two 3G types of networks: WCDMA for GSM networks and CDMA2000 for IS-41 networks. Theoretically, 3G networks can offer a maximum data rate of 144 kbps [16, 17] in rural areas when a user is traveling at 500 km/h; 384 kbps in suburban areas at 120 km/h; and 2 Mbps at 10 km/h.

10.4.3 Wireless Local Area Networks

The IEEE 802.11 standard [18] defines several wireless LANs (WLANs) with different data transmission speeds. The three most important versions of the standard are 802.11a, 802.11b, and 802.11g (also called Wi-Fi5, Wi-Fi, and Wi-FiG, respectively). The most popular one is 802.11b, allowing 11 Mbps at ranges up to 100 m, though the actual throughput of user data ("goodput") is typically 6 Mbps. The 802.11b standard uses DSSS and the ISM 2.4 GHz band for its transmission.

The 802.11 standard supports two modes of operation: ad hoc (or peer to peer) mode and infrastructure mode.

In ad hoc mode, wireless devices can communicate directly between themselves without using an access point (AP). Infrastructure mode requires an AP to relay communications between devices and to connect to other networks. An AP has two functions: to act as a base station for a wireless network and to act as a bridge between wireless and wired networks.

The 802.11b standard uses the *distributed coordinated function* (DCF) access method, which specifies the use of the CSMA procedure for collision avoidance (CA). The CSMA/CA scheme implements a minimum time gap between frames, called DCF-interframe spacing (DIFS). The basic CSMA/CA MAC protocol works as follows. A station wishing to transmit must first "listen" to, or "sense," the radio channel to determine whether another station is transmitting. If the medium is idle for at least one DIFS duration, the station can access the medium at once. If the medium is busy, the station has to wait for one DIFS duration after the end of the current transmission and then enters a contention period. During the contention period, the station selects a random amount of time (called a backoff interval, kept by a backoff timer) to wait before "listening" again to verify a clear channel on which to transmit. If the station does not get access to the medium, it freezes its backoff timer, waits for the medium to be idle again for one DIFS period, and starts the counter again. As soon as the counter expires, the station is allowed to access the medium. This scheme provides fairness, since longer-waiting stations have the advantage over newly entering stations, in that they have to wait only for the remainder of their backoff timer from the previous cycle.

The CSMA/CA protocol provides options for reducing collisions by using Request To Send (RTS) and Clear To Send (CTS) data and ACK transmission frames in a sequential fashion. Communication is established when one of the stations sends a short-message RTS frame to the destination. The RTS frame includes the length of the message. The message duration is known as the *network allocation vector* (NAV). The NAV warns all other stations in the medium to back off for the duration of the transmission. The receiving station issues a CTS frame back to the sender, also including the NAV. If the CTS frame is not received, it is assumed that a collision occurred, and the RTS process starts all over again using CSMA/CA. After the data frame is received, an ACK frame is sent back immediately, verifying a successful data transmission.

10.4.4 Wireless Personal Area Networks

10.4.4.1 Bluetooth

Bluetooth is a wireless technology that uses short-range RF transmissions. It can transmit at a speed of 1 Mbps (2.1 Mbps for the latest version of the standard, v2.0) and has three different power classes for transmitting. Bluetooth operates in the 2.4 GHz ISM band (the same as 802.11b) and uses FHSS for transmission [19].

When Bluetooth devices (maximum of eight) come within range of each other, they automatically connect and form a *piconet.* One device will assume the role of the master, controlling all the communications within the piconet. Others assume the role of slaves and take commands from the master for communications. All devices in a piconet use the same channel for communications. A group of piconets with connections among them is called a scatternet.

10.4.4.2 ZigBee and 802.15.4

The IEEE 802.15.4 standard [20] defines the characteristics of the physical and the MAC layer for low-cost, low-rate, and low-power wireless personal area networks (LR-WPANs).

The ZigBee Alliance [21] is an association of companies working together to develop standards and products for reliable, cost-effective LR-WPANs. ZigBee defines the network layer specifications and provides a framework for application programming in the application layer over the IEEE 802.15.4 standard.

The physical layer uses a DSSS access mode. It supports three frequency bands: a 2,450 MHz band with a maximum data rate of 250 kbps, a 915 MHz band with 40 kbps, and an 868 MHz band with 20 kbps, at ranges up to 10 m. The MAC layer defines two types of nodes: reduced function devices (RFDs) and full function devices (FFDs). FFDs can act as network co-coordinators or network end devices. As network co-coordinators, FFDs can offer synchronization, communication, and network services. RFDs can act as only end devices, may interact with only a single FFD, and are equipped with sensors/actuators.

The 802.15.4 standard supports two types of topology: the star topology and the peer-to-peer topology. In the star topology, a master/slave model is adopted. The master is called the PAN co-coordinator, and this role can be taken by only an FFD; slaves can be RFDs or FFDs and communicate with only the PAN coordinator. In the peer-to-peer topology, an FFD can talk to other FFDs within its radio range and can relay messages to other FFDs outside its range through an intermediate FFD, forming a multihop network. A PAN coordinator is selected to administer the multihop network operation. A particular feature of ZigBee is that it supports a comprehensive set of security measures for WPANs.

10.4.5 Ultra-wideband Communication

UWB technology shows promise for extremely low-cost and low-power sensor networks [22]. UWB can carry a huge amount of data (over 50 Mbps [15]) with very low power over a short distance up to 10 m. UWB radio also has the ability to carry signals through obstacles that tend to reflect narrowband signals. UWB networks can be used in areas too obstacle laden for other wireless protocols to work in. The IEEE standard for UWB, 802.15.3a, is under development.

10.4.6 Satellite Communication

A communication satellite functions as a repeater in the sky. It contains a number of transponders, each of which listens to an uplink channel (Earth station to satellite), amplifies the received signal, and rebroadcasts it on a downlink channel (satellite to Earth station). The most common type of communication satellite today is the geostationary (GEO) satellite, which is positioned at approximately 36,000 km above the equator and hence rotates at the same rate as the Earth. The satellite appears stationary to an Earth station, and communications can always take place without having to track the movement of the satellite. A typical satellite has several transponders, each with a 36–50 MHz bandwidth [23]. Communication between the satellite and Earth stations occurs in the microwave frequency bands: C band at 4–6 GHz and Ku band at 11–14 GHz. A significant fraction of international telephone calls are relayed via the INTELSAT system. Three GEO satellites separated by 120 degrees can cover most of the populated surface of the Earth.

Inexpensive communication using GEO satellites can be achieved with low-cost micro Earth stations called VSATs (very small aperture terminals). These stations can transmit about 1 watt of power and can support an uplink data rate of 19.2 kbps and a downlink rate of 512 kbps. Typical end-to-end delay between two Earth stations is about 270 ms.

The disadvantages of GEO satellite communication include long and noticeable delays and waste of spectrum in point-to-point communications. Satellites characterized as low Earth orbit (LEO) and medium Earth orbit (MEO) have been developed to solve these problems. They circulate at an orbit of around 2,000–12,000 km, resulting in a round-trip propagation delay of less than 50 ms. A user needs only a small GSM-type satellite handset to communicate via these satellites.

10.5 Sensor Networks for Health Monitoring

This section provides a discussion of various issues concerning the deployments of sensors and their associated wireless communication technologies.

10.5.1 Sensors

A sensor is a device that detects the presence and/or the variation of some physical phenomenon, such as voltage or current, and converts the sensed quantity into a useful signal that can be directly measured and processed. An actuator, on the other hand, converts information into actions such as moving itself or initiating actions in other items in its environment. Sensors and actuators often go together as the means of physical interaction between an entity and its surroundings. A *smart sensor* is a sensor that provides extra functions beyond those necessary for generating a correct representation of the sensed quantity. Often, smart sensors possess processing, storage, and decision-making capabilities.

Sensors can measure:

- Physical properties such as pressure, temperature, humidity, and flow
- Motion properties such as position, velocity, angular velocity, and acceleration
- Contact properties such as strain, force, torque, slip, and vibration
- Presence properties by tactility/contact, proximity, distance/range, and motion
- Biochemical properties by biochemical agents
- Identification properties by personal features

Sensors can be conveniently categorized according to one of the following signal domains [24]:

- *Mechanical domain*: Mechanical sensors can convert an applied strain to a change in resistance that can be sensed using electronic circuits (piezoresistive effect), or an applied stress (proportional to force) to a voltage (piezoelectric effect).
- *Thermal domain:* Thermal sensors rely on materials exhibiting thermal expansion on changes in temperature.
- *Magnetic domain*: Magnetic sensors do not require direct physical contact and are useful for detecting proximity effects.
- *Chemical domain*: Chemical and biological transducers are devices that interact with solids, liquids, and gases of all types and convert induced property changes (e.g., mass, resistance) into detectable electrical or optical signals.
- *Radiant domain*: Sensors detect a wide spectrum of electromagnetic radiation, including visible-spectrum and nuclear radiation.

Optical transducers convert light to various quantities that can be detected. These are often based on photoelectric, photoconductive, or photovoltaic effects. In the photoelectric effect, photons (elementary particles of light with zero rest mass) of sufficient energy incident on a charged plate generate a flow of current. In the photoconductive effect, photons generate carriers that lower the resistance of the material. In the photovoltaic effect, photons generate electron-hole pairs in a semiconductor junction that cause current flow. Photovoltaic devices include photodiodes, phototransistors, and solar cells.

For health monitoring, *wearable medical sensors* are of particular interest. These devices are used to monitor a set of key ambulatory parameters in oncology, pediatrics, and geriatrics. Some of these parameters are:

- Heart rate for cardiac function
- Acceleration during walking and running for activity

- Body temperature for illness
- Virtual capacity for severity of airway obstruction in chronic obstructive pulmonary disease
- Blood glucose for vascular or neurological complications
- Electroencephalography (EEG) for seizure disorders, confusion, head injuries, brain tumors, infections, degenerative diseases, and metabolic disturbances that affect the brain
- Electrocardiography (ECG) waveform for cardiac arrhythmias
- Blood pressure
- Arterial oxygen saturation for sleep disorders
- Body weight

The factors that are most important or desirable to be taken into consideration when selecting sensors are:

- **Cost.** In many applications, sensors are most effective when they are used in large numbers, since they provide a comprehensive map or a rich and dense set of information about an environment or health condition that had been unobservable before. However, these applications will be widely deployed only if sensors are affordable.
- **Size.** The size of a sensor is important in many applications. In general, the smaller the size, the more acceptable they appear to be to the user. Some advantages of small and lightweight sensors include ease of attachability to the body, nonintrusiveness, and convenience to users.
- **Power consumption.** Most sensors have to use a battery as their only source of power. It is essential for sensors to consume as little energy as possible so that the battery lasts as long as possible—ideally, for the life of the sensors. Some sensors may be able to harvest power themselves (e.g., solar, wind); however, power-harvesting devices carry with them other unwanted maintenance problems.
- **Mobility.** For health monitoring tasks, it is desirable that sensors be able to communicate with other sensors or control devices while in motion, as they are often attached to a mobile user.
- **Processing capability.** It is often desirable for a sensor to perform some simple data processing such as filtering, removing additive noise before sending useful data to a data collector, etc. This is done to reduce data redundancy and minimize the cost (power) of data transmission.
- **Storage capability.** It is often desirable for a sensor to possess a small amount of storage for critical (or preprocessed) data prior to being passed on to a data collector. For example, the communication channel may be temporarily unavailable or interrupted.

Other factors may include ease of use, clinical approval, durability, and accuracy of measurement. Fortunately, with advances in microchip technology, tiny sensors are becoming available that use very little power, have some processing and storage capability, and can be manufactured in large quantities at relatively low costs.

10.5.2 Monitoring Requirements

There are many factors that need to be considered in a health monitoring system. However, the most important factors regarding wireless technology application to health monitoring are the system security, the bandwidth of the transmission channel, and the power consumption for data transmission.

General *system security* entails authentication, authorization, confidentiality, integrity, availability, and nonrepudiation. However, because of the many constraints, such as limited bandwidth, low power, etc., that are imposed on wireless sensor networks, comprehensive security measures are impractical. But often in a medical setting, there are extra security concerns that must be addressed. With remote monitoring, measurements of a patient's vital signs are often moved from a secure environment in hospitals into the patient's home. This means that there are some additional security issues to consider when putting these systems in use, as they process and transport sensitive information about a person whose privacy must be protected.

Larger *transmitted power* allows wider transmission coverage and provides for better quality of the received signal. However, this is not always possible, because with wireless sensors the power source is always limited. Furthermore, there is a set limit to the amount of radiation that can be emitted by a device, since the radiation may interfere with other critical systems and affect the environment and health of nearby living things. Ideally, one would like to send a lot of data very far, very fast, for many users, all at once. Unfortunately, it is impossible to achieve all these attributes simultaneously.

It is clear from this discussion that desirable properties often conflict with one another in many aspects. For example:

- Small sensors may be desirable for mobility but may not have room for a long-lasting battery or storage capacity.
- Sensors that possess many desirable properties are often expensive and consume a lot of power.
- Sensors require wireless communications for mobility; hence, no permanent power supply can be connected. The communications capability of sensors is thus very limited.
- Sensors may be able to communicate only over a short distance, and hence they have to rely on their neighbors to relay their data to remote data centers. Some forms of self-organizing networks may be necessary.
- Sensors do not often have large storage capacity; hence, they have to upload data frequently to a data center. It is important that the wireless communications technology employed does not drain excessive power from the sensors.

- Sensors do not often have large processing capability. They may not be able to summarize their data before uploading them to the data center; hence, they may have to send a large amount of raw data to their neighbors. It is important to have an efficient wireless communication channel.

It is clear that in order for sensors and sensor networks to be deployed in health monitoring systems, an integrated networking infrastructure is necessary.

10.5.3 Communication in Wireless Sensor Networks

It is thus clear that the choices of sensors, wireless technologies, and applications are interrelated. The choice of wireless technologies should be carefully considered in conjunction with the choice of sensors and the environments of an intended application. Most existing conceptual architectures have three layers, as shown in Figure 10.9(a): The data acquisition layer is responsible for sensing and collecting information concerning health conditions. The data distribution layer is responsible for distributing relevant data to components for analysis. The processing and control layer is responsible for processing, interpreting summarized data, and making appropriate controls or responses. In reality, the functions of a layer are usually carried out by groups of components simultaneously at different topological levels. Figure 10.9(b) shows a typical and practical architecture for health monitoring.

At the lowest level topologically, a group of sensors operate within a confined area (e.g., over the body of a person) and form a wireless *body area network* (BAN) (which is really a PAN wearable on a body) so that they can rely on one another to relay sensed information to a more powerful sensor, which then relays the information (possibly filtered) to a local server. Wireless technology of the 802.15.4 standard is often used to form such a BAN.

The local server acts as a bridge between its sensor network and a central server, which may also serve as a gateway to a WAN such as the Internet or a mobile wireless WAN (WWAN) [25].

The central server is more powerful in terms of its capabilities. It has more power to process and analyze data and extract relevant information for diagnosis and/or initiate appropriate responses. It often communicates with the local server through Bluetooth and/or 802.11 technologies.

At the top layer, the central server may distribute its data to other servers and obtain from them additional data relevant and complementary to the monitored health conditions for forming better responses [26]. Communication between central servers can take place over the wired or the wireless Internet.

From our discussion of sensors, wireless technologies, and health monitoring architectures, several guidelines for the choice of technologies can be recommended:

1. ZigBee technology should be considered if sensors are to be very small and need to communicate over a very short distance, batteries have to last for a long time (a few months), and low data transmission rates are adequate.
2. UWB technology is applicable at high data rates over extremely short distances via small sensors.
3. Bluetooth technology is used as a wire replacement and a bridge between sensor devices and more powerful control devices in WANs. Bluetooth devices can often serve as local servers to coordinate and control wireless sensors.
4. IEEE 802.11 technology is normally deployed as a bridge between sensors and the wired and wireless Internet. It requires more power and is not often used in mobile wireless sensors. However, the 802.20 standard is working on a fully mobile broadband access solution. It will support various vehicular mobility classes up to 250 km/h in a metropolitan area network.
5. Mobile cellular technologies (GSM, GPRS, CDMA2000, WCDMA) are needed to connect devices over WANs. They are often used to connect local servers to global servers that oversee the overall aspects of an application over the mobile Internet.

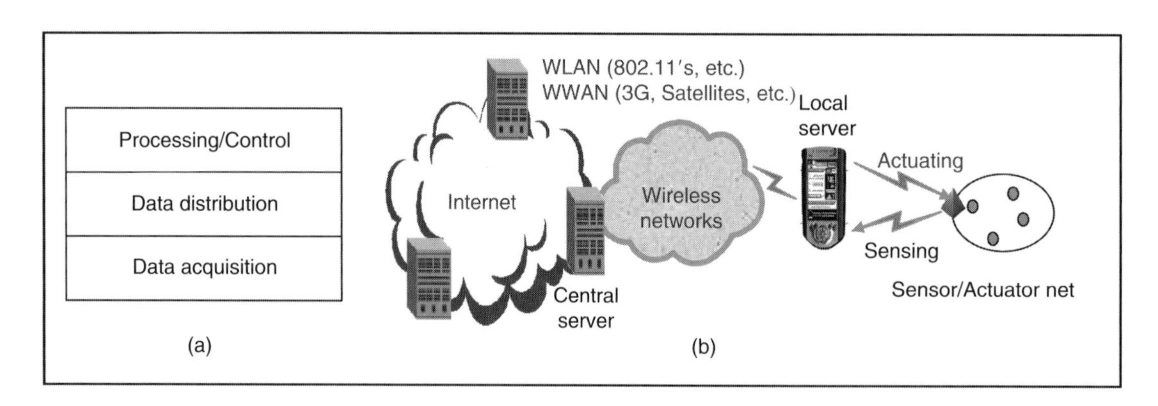

FIGURE 10.9 Monitoring system architecture with wireless sensor networks.

Both the wired and the wireless Internet are integrated for applications that require resource sharing and collaboration across independently managed enterprises.

10.6 Applications of Wireless Technologies in Telemedicine

Health care is about the diagnosis and treatment of illness. Diagnosing an illness involves collecting information about the patient and analyzing that information to decide on the causes of illness and provide appropriate treatment. Telemedicine is the use of telecommunications and information technologies for the provision of health care to individuals at a distance and transmission of information to provide that care [2].

The goal of using communication technologies in medical environments is to provide care when face-to-face physician-to-patient encounters are not possible and to improve the overall quality of health care at an affordable cost. Telemedicine is generally used for remote sensing, decision making, and collaborative arrangements for the real-time management of patients at a distance.

Telemedicine includes diagnosis, treatment, monitoring, and education of patients by using systems that allow ready access to expert advice and patient information, no matter where the patient or relevant information is located. It means that basic patient information must be transferred over computer networks by means of videoconferencing, multimedia, and web-based applications.

In this section, we describe several telemedicine applications that rely heavily on the wired and wireless technologies described in previous sections. The aim is to illustrate the deployment of the transmission and communication technologies within telemedicine applications.

10.6.1 Location-Based Services for Emergency Medical Incidents

In this section, we describe an application, EmerLoc [27], for handling emergency incidents at sites where immediate health support is not available. The system is intended for patients suffering from chronic diseases or requiring continuous monitoring of their physical conditions.

The architecture of the system is depicted in Figure 10.10. Its main components include the patient and his/her portable equipment, the attending doctor and his/her portable equipment, and the central monitoring unit (CMU) and its associated location-based service (LBS). The communication infrastructure includes a wireless sensor network that uses ZigBee/Bluetooth, a WLAN that uses 802.11, and a WWAN that uses GSM/GPRS.

The patient wears a personal device (PD) and a set of sensors. The sensors are attached to the patient's body for measuring biosignals such as ECG, blood pressure, heart rate, breathing rate, and oxygen saturation. The PD collects and processes the patient's vital signals from the sensors. The sensors together with the PD form a PAN that monitors the patient's health status and communicates with the corresponding CMU. The PD transmits alarm signals to the CMU whenever predefined thresholds are exceeded and an emergency situation is imminent. Sensors and PDs are equipped with RF transceivers compatible with Bluetooth or ZigBee. The patient's attending doctor carries a portable doctor device (DD), capable of receiving the alarm signals and the full range of the patient's biosignals.

The CMU controls and coordinates the communication flow among components of the system. It communicates with the PD and the DD over either a WLAN or a WWAN such as GSM or GPRS. The LBS relies on an LSB platform[28] platform that includes a location server of either a mobile network operator or a WLAN position provider.

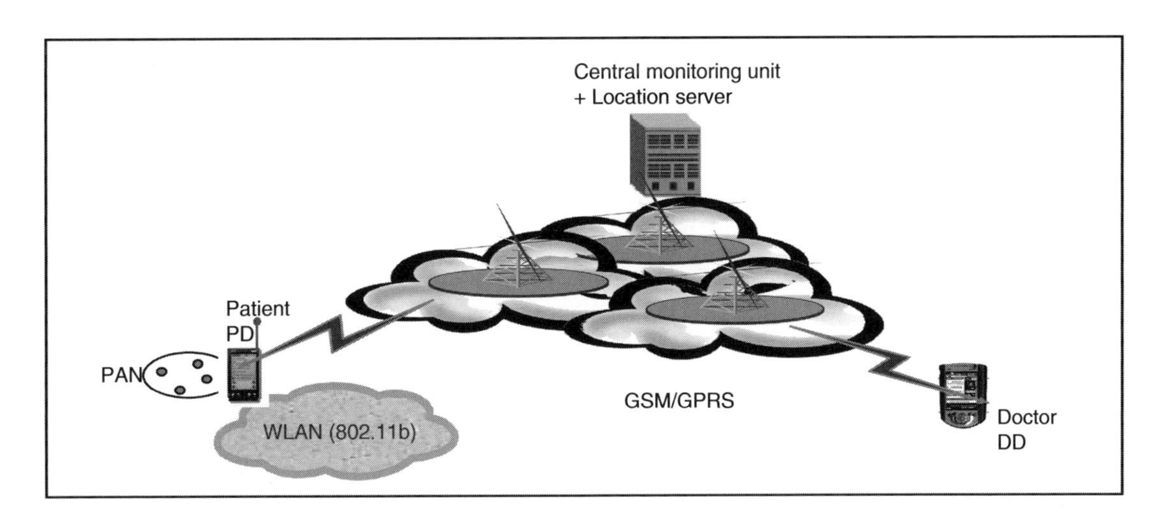

FIGURE 10.10 The overall system architecture of EmerLoc.

10.6.1.1 Operation

When the processed biosignals from the sensors exceed a predefined threshold, the PD generates an alarm signal and relays the emergency information along with its current position to the CMU within its reach through a WLAN or a WWAN.

The CMU acknowledges the receipt of the emergency notification (EN) and uses the patient identification included in the EN to query the medical records database for the appropriate patient's information. The CMU also retrieves the identification of the handling doctor. The doctor's location can be retrieved by the system's location server. Armed with the collected coordinates of the PD and DD, the CMU asks the Geographical Information System (GIS) component for the best routing path from the doctor to the patient. The CMU then sends an alarm message to the DD containing instructions for accessing the emergency information on the CMU. The DD contacts the CMU through a GPRS network to obtain all the relevant information, including a recent CT image along with the routing information.

As the doctor approaches the site of the incident, the DD attaches to the PD's local WLAN. Finally, the DD requests the PD for the full set of recent biosignals and the stored medical records of the patient.

The system was evaluated by 15 physicians in a hospital environment for a period of 10 days. Overall, the results proved "the feasibility of the architecture and its alignment with the widely established practices and standards, while the reaction of potential users who evaluated the system is quite positive" [27].

10.6.2 Mobile Robotic Tele-Ultrasonography Systems

In this section, we describe the advanced medical robotic system called M*o*bile *Tele*-Echography Using an Ultra-*L*ight R*o*bot (OTELLO). Originating as a project funded by the European Information Society Technologies (IST) association, OTELLO developed a fully integrated end-to-end mobile tele-echography system for population groups that are not served locally by medical ultrasound experts.

The system features a fully portable tele-operated robot, allowing a sonography specialist to perform a real-time robotized tele-echography in remote patients. It comprises three subsystems:

1. The **expert station,** where the medical expert interacts with a dedicated pseudo-haptic fictive probe instrumented to control the positioning of the remote robot and emulate an ultrasound probe. Videoconferencing facility is also available.
2. The **communication links:** satellite, 3G wireless, and terrestrial
3. The **patient station,** which comprises a lightweight robotic system with six degrees of freedom (DoF) and its control unit. The robot manipulates an ultrasound probe according to orders sent by the ultrasound expert. The probe retrieves the patient's ultrasound images and sends them to the expert. Videoconferencing facility is also available. Figure 10.11 shows the overall functionality of the OTELLO system.

Three types of critical data are to be transmitted over the OTELLO system: robotic control data, ultrasound stream of images, and medical ultrasound data. The ultrasound medical stream represents the most stringent requirements in terms of data rate and near real-time responses.

The robotic system also has a force feedback mechanism, allowing the expert to move the fictive probe and control the distant probe holder at the remote patient station. The 3G communication link offers a data rate of 144 kbps for a rural outdoor mobile user traveling at a speed of more than

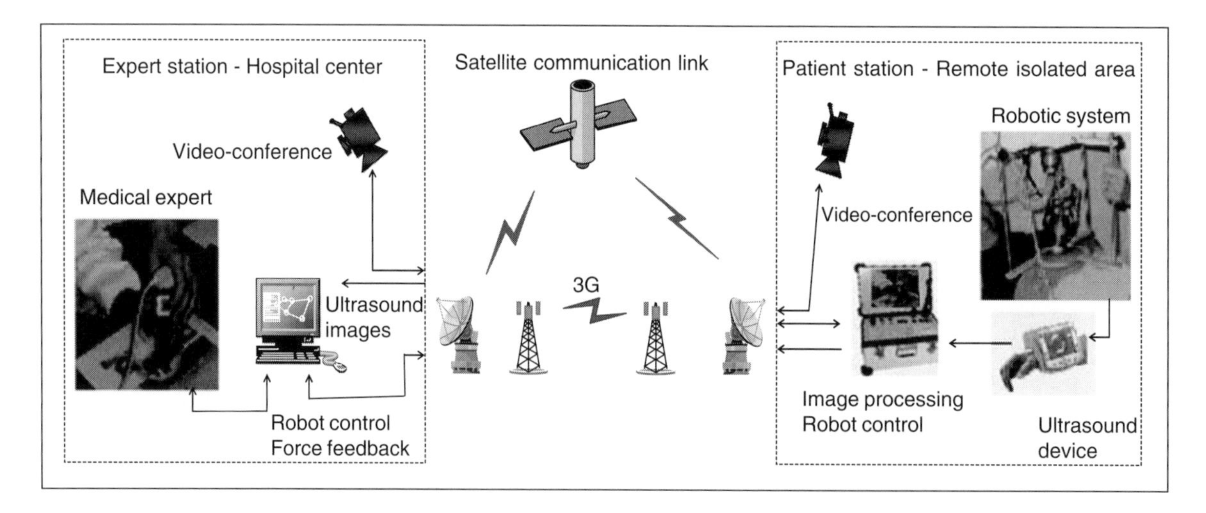

FIGURE 10.11 The overall functionality of the OTELLO system. Adapted from Garawi et al. [29].

120 km/h, a 384 kbps downlink for a pedestrian user traveling at a speed of less than 5 km/hour, and 2 Mbps indoors.

For OTELLO, the patient station sends ultrasound images, ultrasound streams, ambient video, sound, and robot control data, while it receives only robot control, ambient video, and sound from the expert station. The best-suited QoS class for video streaming is the "Streaming" 3G service class that preserves the time relation between information entities of the stream. However, for medical image sequences with real-time requirements, the "Conversational" class would be necessary.

The experimental setup over the 3G/satellite network is shown in Figure 10.11. The ultrasound scanner data are acquired at the rate of 13 frames per second (fps), each frame with a resolution of 320×240 pixels for videoconferencing format. The robotic data flow is generated from the expert station at a rate of 16 bytes every 70 ms. The received robot data stream from the patient station updates the robotic head position continuously. The experiment in Garawi et al. [29] was carried out under different network loading conditions, as reflected by the results. The real-time protocol (RTP) is used for end-to-end real-time streaming, and the UDP/IP protocol is used for the robot data in both directions.

The OTELLO system was tested on a live 3G network (Vodafone, U.K.). The experimental transmissions were carried within London between Kingston University (patient side) and St. George's Medical School (expert side). The test results for transmitting ultrasound streams encoded in the Quarter Common Intermediate Format (QCIF) using the H.263 codec demonstrated successful transmission in 3G real-time environments. The minimum bounds for quality of the received ultrasound information that were clinically acceptable by the medical experts using the OTELLO system for prediagnosis were 5 fps at 35 dB peak signal-to-noise ratio. These results were achieved using 64 kbps at the patient station uplink. It was found that the network delay jitter variations were within the acceptable boundaries of maintaining high-quality real-time interaction for the system: 297 ms compared with a maximum delay of 325 ms. It was concluded that such advanced mobile robotic telemedicine systems could successfully provide clinically acceptable quality ultrasound data using commercial 3G networks.

The demonstration in Garawi et al. [29] also shows that it is possible to overcome the technical difficulties involved in presenting a haptic teaching environment, linking two remote locations across the world.

10.7 Summary

Wired and wireless technologies and communication networks are becoming an integrated part of the health care infrastructure to support current and emerging services and applications in an increasingly mobile and information-driven society, providing better services at reduced costs financially and with fewer human resources.

It is impossible to cover the field of telecommunications and computer networks in one chapter. This chapter attempts to focus on providing a basic understanding of data communications technologies and how they are deployed in various forms in wired and wireless infrastructures to support applications in m-health and telemedicine.

The chapter discussed the TCP/IP stack (see Figure 10.4), in which a TCP/IP application layer service (http) transmits data that are carried in a TCP protocol data unit with a specific port number, which in turn is carried in an IP packet with an IP address used to route the packet across the Internet. The TCP port number and the IP address together form a "socket." The IP packet is finally carried in an Ethernet (802.3 CSMA/CD) frame to the destination host machine using the MAC address. The ARP protocol is used to map between the IP and MAC addresses.

In particular, the chapter discussed the limitations of a communication channel, to make sure that we are aware of them when building an infrastructure for health care. In particular, the chapter discussed aspects of emerging health monitoring that require sensors and wireless sensor networks. The chapter discussed the working of Internet network architecture and other mobile and wireless networks that are being deployed in health care applications.

The chapter identified a large number of open issues and challenges that must be addressed before wireless technologies can be applied on a wide scale within the health care environment.

10.8 Exercises

1. Differentiate between narrowband, wideband, and ultra-wideband transmission schemes.
2. Name some of the limitations of a wireless communication channel.
3. Explain why many networks (wireless sensor networks, WLANs, mobile cellular networks, and the Internet) have to be involved in an application. Describe their role.
4. What do you think are the issues that must be addressed before wireless technologies can be applied on a wide scale within the health care environment?
5. What are the MAC protocol, data rate, and transmission medium of 10GBase-T?
6. What are the main advantages of the IPv6 protocol compared with IPv4?
7. Discuss the opportunities and changes in lifestyle that are likely to occur as all electronic devices in a home (telephone, refrigerator, microwave, etc.) are wirelessly connected to the Internet using IPv6.
8. As their costs fall, RF identification devices will become used to track smaller and smaller unit quantities

(e.g., presently a container, perhaps a crate, later a box, then individual items). What do you think the implications of this trend are for privacy, profiling, and monitoring of patients?
9. Write a simple webpage with links to online medical resources for patients.
10. Discuss the advantages of a digital data network compared with an analog network.
11. Discuss the advantages of a network based on TCP/IP compared with the digital public switched telephone network.

10.9 References and Bibliography

1. E. Dishman. Inventing wellness systems for aging in place. *IEEE Computer.* 37:34–41, 2004.
2. N. F. Guler and E. D. Ubeyli. Theory and applications of telemedicine. *J. Med. Syst.* 26:199–220, 2002.
3. S. E. Kern and D. Jaron. Healthcare technology, economics and policy: An evolving balance. *IEEE Eng. Med. Biol. Mag.* 22:16–19, 2003.
4. R. Holle and G. Zahlmann. Evaluation of telemedical services. *IEEE Trans. Info. Tech. Biomed.* 3:84–91, 1999.
5. P. N. T. Wells. Can technology truly reduce health care costs? *IEEE Eng. Med. Biol. Mag.* 22(1):20–25, 2003.
6. Z. Papir and A. Simmonds. Competing for throughput in the local loop. *IEEE Comm. Mag.* 37, 1999.
7. Australian Communications Authority. *Understanding Your Broadband Quality of Service,* 2004. http://www.acma.gov.au/acmainterwr/aca_home/publications/reports/reports/broadband_quality_of_svce.pdf.
8. Cisco Systems Technology. *Traffic Analysis for Voice over IP* [white paper]. http://www.cisco.com/en/US/tech/tk652/tk701/technologies_white_paper09186a00800d6b74.shtml
9. A. Simmonds. *Data Communications and Transmission Principles: An Introduction.* Macmillan, 1997.
10. Internet Assigned Numbers Authority. *Abuse Issues and IP Addresses* ("Special-Use Addresses"). http://www.iana.org/faqs/abuse-faq.htm#SpecialUseAddresses.
11. Internet Assigned Numbers Authority. *Port Numbers.* http://www.iana.org/assignments/port-numbers.
12. W3C [World Wide Web Consortium]. *XHTML2 Working Group Home Page.* http://www.w3.org/MarkUp.
13. A.L.I.C.E. Artificial Intelligence Foundation [homepage]. http://www.alicebot.org.
14. Lexicle Ltd. *Smart Fridge Centre Stage in Responsive Home Opening* [Alex avatar]. July 13, 2004. http://www.lexicle.com/news/20040713.
15. J. Foerster, E. Green, and S. Somayazulu. Ultra-wideband technology for short- or medium-range wireless communications. *Intel Technology Journal.* Q2, 2001.
16. K. Pahlavan and P. Krishnamurthy. *Principles of Wireless Networks.* Prentice Hall PTR, 2002.
17. Nuntius Systems. *Universal Mobile Telecommunication System (UMTS).* http://www.nuntius.com/solutions23.html.
18. Wikipedia. *IEEE 802.11.* http://en.wikipedia.org/wiki/802.11.
19. Wikipedia. *Bluetooth.* http://en.wikipedia.org/wiki/Bluetooth.
20. J. Gutierrez, E. Callaway, and R. Barrett. *Low-Rate Wireless Personal Area Networks.* IEEE Press, 2004.
21. ZigBee Alliance [homepage]. http://www.zigbee.org
22. Oppermann et al. UWB wireless sensor networks: Uwen—A practical example. *IEEE Comm. Mag.* 42:27–32, 2004.
23. A. Tanenbaum. *Computer Networks*, 3rd ed. Prentice Hall, 1996.
24. F. L. Lewis. Wireless sensor networks. In D. J. Cook and S. K. Das (Eds.). *Smart Environment: Technologies, Protocols, and Applications.* John Wiley, 2004.
25. G. Roussos, A. Marsh, and S. Maglavera. Enabling pervasive computing with smart phones. *IEEE Pervasive Comput.* 4(2):20–27, 2005.
26. C. Otto et al. System architecture of a wireless body area sensor network for ubiquitous health monitoring. *Journal of Mobile Multimedia.* 1:307–326, 2006.
27. I. Maglogiannis and S. Hadjiefthymiades. EmerLoc: Location-based services for emergency medical incidents. *Int. J. Med. Inform.*, 2006.
28. M. Spanoudakis et al. Extensible platform for location based services deployment and provisioning. In D. Katsaros, A. Nanopoulos, and Y. Manolopoulos (Eds.). *Wireless Information Highways.* IRM Press. 339–371, 2004.
29. S. Garawi, R. Istepanian, and M. Abu-Rgheff. 3G wireless communications for mobile robotic tele-ultrasonography systems. *IEEE Comm. Mag.* 44:91–96, 2006.

11

Data Security and Protection for Medical Images

Dr. Eugene Y. S. Lim
Royal Prince Alfred Hospital and University of Sydney

11.1 Introduction

11.1.1 Background

There are ethical and legal obligations for health care providers to preserve the privacy and confidentiality of patient information, which can contain some of the most intimate information conceivable about an individual [1]. Despite the advantages of electronic medical records (EMR), there is a higher chance of disclosure of information to the public compared to other formats such as paper-based records [2]. Fast network and processing tools allow large-scale data searches and retrievals and enable the disclosed data to be transferred. Easy access to networked systems increases the risk of disclosure of the information both intentionally and unintentionally. The actual level of security breaches in current health care systems is not yet significant but is strong enough to raise concerns [3, 4].

In such an environment, the primary risk or danger associated with medical information is the disclosure of content to an unauthorized party, which can result in misuse. The unauthorized alteration of content, whether it is intentional or not, may lead to misdiagnosis and other severe results. The security of medical information is protected by legislation and strict ethics. The Health Insurance Portability and Accountability Act (HIPAA) [5] recently introduced by the U.S. government mandates the development of a national privacy law, security standards, and electronic transactions standards to reduce violations and wrongful disclosures of health information. Three aspects of security arise in relation to dealing with medical images: confidentiality, reliability (integrity, authentication), and availability [6] which can be defined as follows. Confidentiality aims to ensure that only the entitled users have access to the information.

Reliability has two aspects: integrity, which ensures that the information has not been modified by errors in transmission and unauthorized users; and authentication, ensuring the data or information is correctly identified (i.e., is from the correct source and belongs to the correct patient). Availability means that the information is used by the entitled users in the conditions of access and exercise.

Effective distribution and communication of medical images among health care providers are becoming increasingly important. Health care networks are key components in the delivery of cost-effective clinical care. To provide effective communication and exchange of data, the use of public network channels like the Internet is inevitable. Recently, many image-distribution systems that use the Internet have been proposed in the literature such as the Web-based medical image distribution and telemedicine system [7–11]. These systems have great potential because of their cost-effectiveness and easy accessibility.

An important consideration in the implementation of any networked imaging system that uses the Internet is that image data may be susceptible to unauthorized access, disclosure, and alteration. Most clinical picture archive and communication systems (PACS) [12] use the digital imaging and communications in medicine (DICOM) [13] standard image format for medical images. A number of security measures are currently available for PACS and DICOM, such as encrypted transmission, firewall, passwords, and private and public keys, which provide adequate protection for data storage and transmission. However, computer and network speed advance rapidly and may result in security threats to the medical environment. Faster processing speeds increase the vulnerability to brute-force attack and differential cryptanalysis [14]. Extended wide-area networks create possibilities for unauthorized connections to health care networks, resulting in attacks like eavesdropping. Most security tools also have limits. Combining different methods can increase the security of data [15]. In this chapter, some security methods currently being used in practice and new approaches to protect medical images are briefly introduced.

11.1.2 Cryptographic System and Digital Watermarking

A cryptographic system, or cipher, is a mathematical function used for encryption and decryption. At present, a cryptographic system is the security protection for medical information systems such as DICOM, using digital signatures and encryption to improve security [13]. Cryptographic systems are capable of providing confidentiality using encryption and content protection guaranteeing integrity and authenticity using a digital signature. However, there are still some weak aspects in cryptographic systems; for example, no protection is guaranteed after decryption using a legitimate key (which is the most common method of illegally redistributing multimedia content) [6]. The digital signature scheme is a good solution for detecting the intentional or accidental modification of digital contents.

Digital watermarking is a technique to embed encoded information into digital data so that the information is imperceptible but easily decoded by authorized parties [16]. It should be difficult for unauthorized parties to remove the embedded information without the original data. Applications of digital watermarking include copyright protection, authentication, captioning, and tracing of illegal distribution and secure communication. Each application has different requirements. For example, copyright protection requires robust watermarking [17] whereas authentication requires fragile or semifragile watermarking [18]. Robust watermarking is a method in which the embedded watermark should remain the same after any processing technique has been applied, including compression and geometric transformation. This robustness is useful for copyright protection because the ownership of the content can be preserved. Fragile or semifragile watermarking is a method for embedding readily breakable watermarks on the image content such that any processing damages the watermark. The damaged watermark can still be used to authenticate and localize the modification.

The difference between cryptographic systems and digital watermarking is shown in Figure 11.1. A key difference is that a

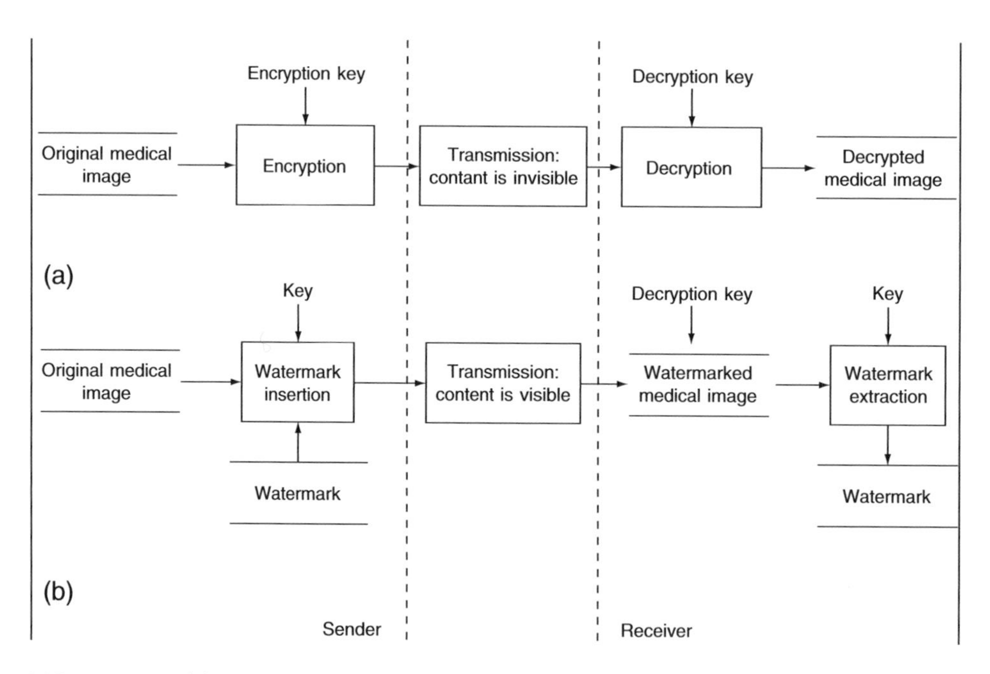

FIGURE 11.1 (a) Cryptographic system for medical images—Image content is hidden during transmission. (b) Embedded digital watermarking—Image is accessible during transmission, but secret information is invisibly or visibly embedded inside the image.

cryptographic system hides the image from unauthorized access by encrypting it while a digital watermark hides secret information within the image content that can be extracted for verification.

11.2 Overview of Cryptographic System

Cryptography originated from the Greek word *kryptós*, hidden, and *gráphein*, to write, meaning secret writing. The encryption process converts a plain text message to cipher text using the cryptographic algorithm. The decryption process is the reverse process of retrieving the original plain text. In general, there are two types of cryptographic algorithms, symmetric (private) key and asymmetric (public) algorithms.

11.2.1 Symmetric Key Cryptography

Symmetric (also called private) key cryptography is based on a shared key that is known only to the communicating parties. Famous symmetric key algorithms are Data Encryption Standard, International Data Encryption Algorithm [19], RC5 [20], and Advanced Encryption Standard.

The level of security provided by the algorithms depends on the secrecy of the key. Both communicating parties have to agree upon the secret key before the message is transmitted. This type of algorithm mainly ensures confidentiality of the message. A major disadvantage of secret key algorithms is that any message can be compromised when the key is disclosed or intercepted. Another issue is the use of a secret key algorithm in a networked environment. When more than two communicating parties or nodes are involved, each pair of parties or nodes is assumed to use separate keys. This is highly computationally complex for large user groups, which is common in the medical environment. On the other hand, an advantage over a public key algorithm is that a symmetric algorithm requires much smaller key sizes for the same level of security. The computations can be faster, and the memory requirements are smaller, which is useful for communication involving small numbers of users.

11.2.2 Asymmetric Key Cryptography

Asymmetric (also called public) key cryptography uses two different keys: a public key and a private key. A public key is available to anyone, and a private key is only known to the key owner. The sender can encrypt the message using the recipient's public key. The encrypted message can only be decrypted with the recipient's private key. Public-key cryptography does not require a shared secret key between communicating parties, thereby solving the problem of key distribution in communication networks. But the overall process of encryption and decryption is much slower than symmetric key cryptography. The key size for public key cryptography has grown very large, requiring large computational processing power and memory size.

When performance is an issue, symmetric key cryptography is preferable to asymmetric key cryptography. To balance performance and effective distribution, a combination of symmetric and asymmetric key cryptography can be used. A temporary random key called a session key can be used. The session key is used to encrypt the message. The session key is then encrypted using the recipient's public key. The public-key encrypted session key is sent along with the encrypted message to the recipient. The session key is retrieved using the recipient's private key. The message is then decrypted using the session key. In medical imaging, because of its large size, this combined approach can improve the performance and security of key management.

11.2.3 Cryptographic Hash Function

Hash functions take any length of input data (e.g., image content) and generate fixed-length strings. The digested string can be appended to the input data during the transmission for error detection. The string is also called parity bits when it is a binary string. Comparing the digest of received data and the appended message digest (also called checksum) checks the error. In general, it detects random error caused by noise during transmission. However, there is a possible threat to the methods such as a malicious attacker intercepting the transmitted data, modifying it, and resending it with the digest recalculated for the modified message.

Cryptographic hash functions using symmetric (private) or asymmetric (public) key cryptography are commonly used for authentication. Authentication is similar to error checking, but it ensures that the origin of data (i.e., authenticity) is correctly identified. Authentication is normally done at the same time as integrity checking. The outcome of the cryptographic hash value (i.e., a binary string) is defined as the digital signature.

To produce a digital signature, the digest of input message M is calculated using hash function $H(x)$. The digested message M' is then encrypted with the key of the sender. The encryption can be either private or public. This encryption of the hash prevents malicious attackers from modifying the input data and recalculating its checksum. The two most widely used cryptographic hash functions are message-digest algorithm 5 and secure hash algorithm. However, the existence of possible security flaws has been reported in both algorithms [21, 22].

Message authentication code (MAC), also known as a cryptographic checksum, is a public function of the input data and a secret key that produces a fixed length value that is used for authentication. MAC algorithms are symmetric-key techniques that provide data origin authentication and data integrity. They have received widespread use in many practical applications such as eCommerce. A keyed hash function h has a secret k-bit key K as a secondary input when used for message authentication.

11.3 Digital Watermarking

Digital watermarking techniques [16, 23–25] can be classified according to their perceptivity, robustness, processing domain, and key types (Table 11.1).

Fragile watermarks do not survive glossy image processing. This fragility enables the detection of modification [18]. Placing the fragile watermark into the perceptually insignificant portions of the data guarantees imperceptibility. Robust watermarks will survive any image processing applied to the content [26]. Robust watermarks are intended to survive common image processing operations such as filtering, scaling, or cropping [27] and are mainly used for security and copyright protection. It is difficult to achieve a single watermark that will survive all image processing techniques. More than one watermarks are embedded in a host image so that at least one watermark survives under different image processing called cocktail embedding [28] have been reported in the literatures.

Cryptography is often used in generating digital watermarks. Either a private or public key can be used. Wong [29] has proposed a public-key watermark that divides an image into blocks and calculates the signature of each block using the public-key hash function. The generated signatures are embedded within the corresponding blocks. This signature can identify the location of modifications up to the base block size. Despite its localization ability and cryptographic key strength, possible methods to attack this method have been reported [26]. The individual signatures depend only on the block so that no interrelationship exists between the signatures. This is called block-wise independency. A possible method to forge the block-wise independent watermarking is called vector quantization (VQ) counterfeiting attack [26]. If an attacker has a sufficient number of watermarked image samples, it is possible for the attacker to generate a new authentic image by adding pieces of different authentic image samples, like a patchwork. Since the introduction of VQ attack, a number of improvements of the existing methods have been proposed [30]: (1) increasing block dimension, (2) including block indices in the signature, (3) including image indices in the signature, and (4) breaking block-wise independency.

TABLE 11.1 Taxonomy of digital watermarking

Classification	Type of watermark
Perceptivity	Visible, invisible
Robustness	Robust, semifragile, fragile
Signal processing domain	Spatial, transform domain
Key types	Private, public key

With regard to the signal processing domain shown in Table 11.1, existing watermarking methods can be classified into transformation domain methods and spatial domain methods [31]. The basic idea of spatial domain methods is to embed watermarks into and extract watermarks from the image in the spatial domain, without transformation. The watermarks are constructed in the spatial domain and embedded directly into an image's pixel data. The most common approach to embedding invisible information is the least significant bit (LSB) substitution method [23, 32, 33]. The fundamental idea is to insert the bits of the hidden message into the LSB of the pixels. Early fragile watermarking systems embedded checksums [34] or pseudo-random sequences in the LSB plane of an image. More recent systems apply more sophisticated embedding mechanisms, including the use of cryptographic hash functions as mentioned in Wong's scheme [35], to aid detection of changes to a watermarked image. In the extraction process, the LSBs of the watermarked image are extracted and concatenated to reconstruct a secret message, then the inverse of the control function is applied. The LSB method can be expanded to many different types of application by using a different kind of control function, depending on its applications.

Compared with spatial domain methods, the basic idea of transform domain methods using transforms like Furrier transform, discrete cosine transform (DCT), and wavelet transform is to embed and extract watermarks in the transform domain. Transform domain methods hide the message in significant areas of the image, making it more robust to attack.

11.4 Medical Image Watermarking

Watermarking techniques enhance existing security measures for distributed data of many kinds including medical images. However, a potential limitation in applying the techniques directly to medical images is that the watermark may alter the original image to an extent that it may not be acceptable for diagnosis or quantitative analysis. The effect of embedded watermarks on the diagnosis and analysis of medical images is a key issue.

In medical imaging, the applications of watermarking [36] are mainly authentication [6], integrity checking [37–42] and metadata embedding [43, 44]. A number of the proposed watermarking methods are for distribution and management of medical images for purposes other than diagnosis [40, 42, 43]. The watermarking methods acceptable for diagnostic analysis are mainly based on either the reversible watermarking schemes [39, 41] or region-based schemes [5, 37, 38], which are discussed next. There are a number of watermarking schemes suitable for compressed medical images in the transform domain. In this chapter, we focus on watermarking methods for noncompressed images. Noncompressed images are accepted for use in diagnosis.

11.4.1 Reversible Watermarking

Watermarking is said to be reversible if the image is deemed authentic and the distortion due to authentication can be completely removed to obtain the original image data. Two principal reversible watermarking techniques are reported in the literature. One is based on robust spatial additive watermarks combined with modulo addition [45], and the second is based on lossless compression and encryption of bit planes [46–48]. Both techniques provide cryptographic strength in verifying the image integrity because the system security is related to a secure cryptographic primitive such as a hash function.

One of the earliest reversible watermarking methods was proposed by Barton [46] and compresses the bits to be affected by the embedding operation. The compression preserves the original data while creating a space for the embedding of secret information. The compressed data and the payload are then embedded into the host image. This method is widely adopted for embedding reversible watermarks [47, 48].

Another reversible data-embedding method [45] uses a spatial additive robust watermarking scheme [49] to embed a watermark pattern W into an 8-bit original image I, where W is calculated from a hash function H (of the original image). A hash function with a secret key K can be used:

$$W = H(p,K), \tag{11.1}$$

where P is payload bits. The generated watermark pattern W is added to the image as follows:

$$I_w = I + W \bmod 256, \tag{11.2}$$

where I_w is the watermarked image. For the recovery of the original image, the payload p is first extracted from the watermarked image, then the watermark pattern W is calculated as above. The embedded data can be removed using subtraction modulo of 256 to obtain the original image I as follows:

$$I = I_w - W \bmod 256. \tag{11.3}$$

The possibility of the above introducing salt-and-pepper noise into the watermarked image when pixels close to zero are flipped to values close to 255 and vice versa in grey scales has been discussed in [50]. The possibility of extracting the payload correctly is reduced when the number of the flipped pixels is too large. This salt-and-pepper distortion in the watermarked image; may be undesirable in some medical imaging applications.

Fridrich et al. [51] proposed a lossless data embedding method for uncompressed image formats. The original image is assumed to be a grey scale image with $M \times N$ pixels including pixel values from the set P, e.g., $P = \{0, \ldots, 255\}$. The pixels in an image are grouped into non-overlapped blocks, each consisting of a number of adjacent pixels. The image is divided into disjoint groups of n adjacent pixels $(x_1, \ldots, x_n)$. A discrimination function F is established to classify the blocks into three different categories: regular (R), singular (S), and unusable (U) groups. An invertible operation F on P called flipping is defined, which is a permutation of grey levels that consists entirely of two cycles. The main idea of the method is that they inspect the image in groups and losslessly compress the bit stream of R and S. The R and S groups are flipped into each other under the flipping operation F, while the U groups do not change status. By assigning a 1 to R and a 0 to S, they embed one message bit in each R or S group. If the message bit and the group type do not match, the flipping operation F is applied to the group to obtain a match. The data to be embedded consist of the overhead and the watermark signal. Although the technique is novel and successful in reversible data hiding, the amount of data that can be hidden is limited. A possible problem with applying the method to medical images is that as the capacity increases, the visual quality drops severely.

Celik et al. [52] presented a high-capacity, low-distortion reversible data-hiding technique. The host signal is quantized, and the residual is obtained in the embedding phase. The context-based, adaptive, lossless image coding (CALIC) compression algorithm [53] is adopted, with the quantized values as side information. The algorithm has three components: prediction, context modeling, and conditional entropy coding. Prediction coding reduces spatial redundancy in the image. The context model examines spatial correlation with different image levels. Conditional entropy coding creates smaller code by translating the correlation. This enables the creation of high capacity for the payload data by compressing the quantization residuals. The compressed residual and the payload data are concatenated and embedded into the host signal using the LSB substitution method. It was reported that the high embedding capacity could be achieved with relatively little distortion.

The major concerns in the development of all lossless embedding algorithms are increasing capacity and minimizing the visual artifact created by embedding the payload.

11.4.2 Region-Based Watermarking

Region-based schemes utilize the spatial redundancy of the image. In most medical images, there exists spatial redundancy such as a background, which has less importance in diagnosis. Coatrieux [37] proposed region-based watermarking for medical image integrity verification. The medical image to be protected is divided into two zones: region of interest (ROI) and non-ROI (NROI), as shown in Figure 11.2.

The ROI is the region needed for the diagnosis, and its integrity is important. The NROI is the peripheral region (outside the ROI) that is not used for diagnosis. The ROI can be defined automatically or delineated semi-automatically by operator. The ROI boundary information is required in the verification process. Signatures generated from cryptographic hash function are calculated from ROI and embedded in the insertion zone. Any image on which the signature extracted from the ROI does not match that stored in the insertion zone indicates possible modification. Thus, any change, whether

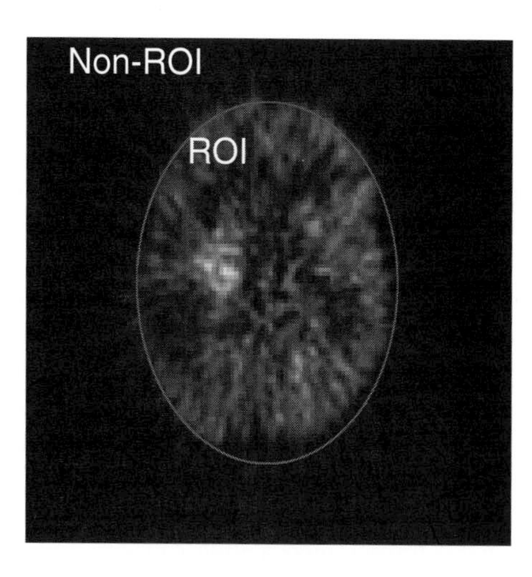

FIGURE 11.2 Region-based embedding scheme. The ROI signature is embedded in the NROI.

due to malicious forging, tampering, casual processing or random errors, can be detected. Using the same ROI and NROI in the embedding and verification process is an important precondition of such methods.

The image ROI can be considered as a binary message M; $|M|$ is the length of the message; its integrity is verified with a signature S_R calculated from cryptographic hash function f. M can be subdivided into N segments m_i of μ_i bits each: $M = m_1, \ldots, m_N$. The signature S_R results from the concatenation of N control words S_i of length $\eta_1, \ldots, \eta_N$ to localize alteration. Each S_i protects the corresponding segment m_i independently. The signature extraction function, f (also called the control function), can be defined as follows.

$$S_R = f(M) = f(m_1, \ldots, m_N) = (h_1, \ldots, h_N). \quad (11.4)$$

In the verification process, the integrity is checked by comparing the embedded signature S_R and the calculated signature from the received image $S'_R = f(M')$.

Any segments in which $S_i \neq S'_i$ are deemed to have been modified.

Cao et al. [5] used a steganographic approach to embed an encrypted digital signature of content and confidential patient information, called a digital envelope (DE). The image is first segmented with the background removed by fitting a minimum rectangle that contains the diagnostically important region. This rectangle separates ROI and NROI. The digital signature (DS) of ROI is produced using the signature extraction function using a private-key such as a cryptographic hash function. The patient information is appended to the digital signature if it is necessary to form DE. The generated DE is embedded within NROI outside of the rectangle. The DE-embedded image is encrypted for transmission.

11.5 Region-Based Reversible Watermarking for Secure Positron Emission Tomography Image Management

A region-based digital watermarking method has been applied to positron emission tomography (PET) images for security and management [54]. Region-based embedding has been used to divide PET images into two regions, ROI and NROI, depending on the diagnostic importance. Patient information is encrypted and embedded within ROI using LSB embedding. The original bits changed because of embedding the encrypted patient information are encrypted and embedded within NROI. The original values of ROI can be restored using data contained in the watermark embedded in the NROI.

A threshold method is used to define ROI and NROI. Pixels in the image with intensity less than the threshold are assigned to NROI. The remaining pixels are assigned to the ROI. It is possible to get misclassified pixels (gaps) in both regions, which are filled using a morphological process [55]. The LSBs of ROI pixels are extracted and concatenated into binary string, which is then encrypted and embedded within the NROI using LSB substitution. Patient information (e.g., name, date of birth) is embedded within ROI (Figure 11.3) The amount of information depends on the capacity of ROI.

The reconstructed images used in the result shown in Figure 11.4 were 128 × 128, unsigned raw data (data were scaled to 0 to 27537 intensity unit). There was no significant visual difference between the original and embedded images (MSE of 0.24 ± 0.03). Decrypting and extracting the embedded message on the background could recover the original ROI. The embedded patient information has many potential uses. Integrating the patient information within the image content offers advantages such as confidentiality of information, compact storage, and fast transmission.

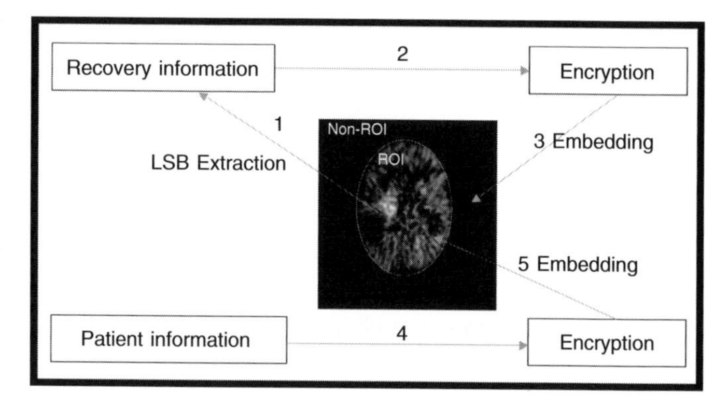

FIGURE 11.3 A conceptual diagram of region-based watermark embedding process with patient information. The data to be changed because of embedding patient information (recovery information) is extracted and embedded within the NROI. Patient information is embedded within the ROI.

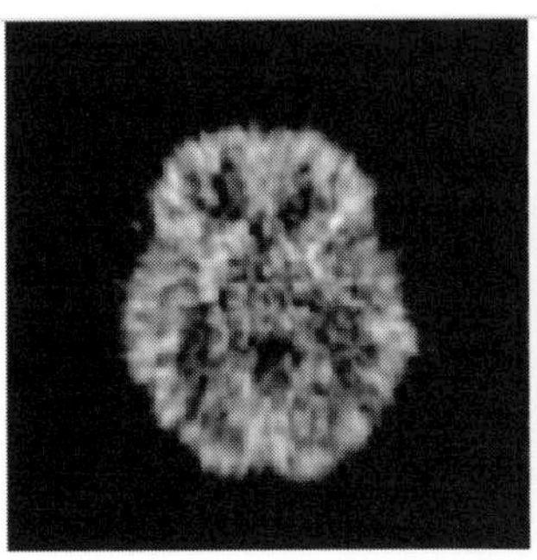
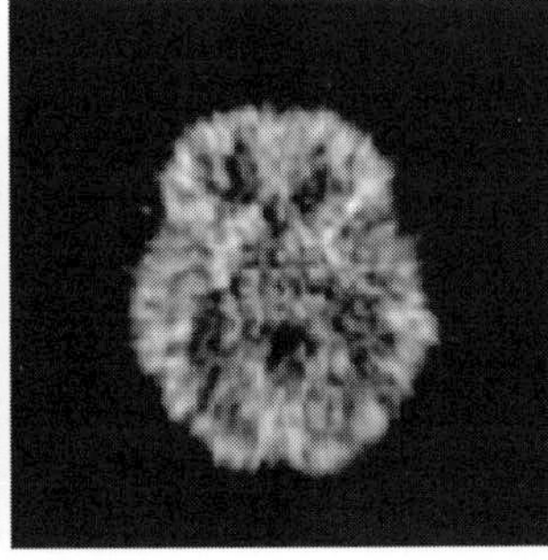
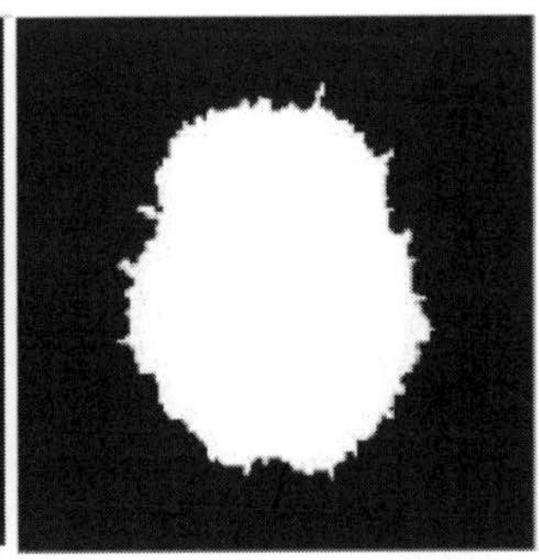

FIGURE 11.4 Original image, watermarked image, and threshold results. This figure illustrates the method used in a clinical fluorodeoxyglucose PET brain scan performed on an ECAT951R, PET scanner (Siemens, Hoffman Estates, IL; CTI, Knoxville, TN) at the Royal Prince Alfred Hospital. There were 31 cross-sectional image planes. A conventional sampling schedule consisting of 22 temporal frames was used to acquire the PET projection data. The reconstruction methods used were filtered-back projection.

11.6 Summary

In this chapter, a number of security methods applicable to medical images are introduced. Currently, cryptographic algorithms are the most widely used security measure, being fast and effective. However, most security tools have their limits and combining different security methods is the best way to protect data. The newly emerging technique of digital watermarking has been introduced as an alternative tool to improve the security of medical images.

Acknowledgments

The author is grateful to the support from ARC and PolyU/UGC grants.

11.7 Exercises

1. Describe the difference between cryptography and digital watermarking.
2. Compare the advantages and disadvantages of the following security measures introduced in this chapter:
   - Private key encryption
   - Public key encryption
   - Reversible watermarking
   - Region-based watermarking
3. Write a pseudo-code using bitwise operation to replace LSB of a given image with a payload (university logo).

11.8 References and Bibliography

1. R. Cushman. Information and medical ethics: Protecting patient privacy. *IEEE Technology and Society Magazine* 15(3):32–39, 1996.
2. J. G. Hodge. Jr., L. O. Gostin, and P. D. Jacobson. Legal issues concerning electronic health information: privacy, quality, and liability. *JAMA*. 282:1466–1471, 1999.
3. J. D. Halamka et al. A WWW implementation of national recommendations for protecting electronic health information. *JAMIA*. 4(6):259–265, 1997.
4. J. Collmann et al. Safe teleradiology: Information assurance as project planning methodology. *JAMIA*. 12(1):84–89, 2005.
5. F. Cao, H. K. Huang, and X. Q. Zhou. Medical image security in a HIPAA mandated PACS environment. *Comput. Med. Imaging Graph.* 27(2–3):185–196, 2003
6. C. Coatrieux et al. Relevance of watermarking in medical imaging. *2000 IEEE EMBS Conf. on Information Technology Applications in Biomedicine*, 2000.
7. J. Zhang et al. Real-time teleconsultation with high-resolution and large-volume medical images for collaborative healthcare. *IEEE Trans. Inf. Technol. Biomed.* 4(4):265–273, 2000.
8. D. Feng et al. Medical image data retrieval and manipulation through the WWW. *Proceedings of the International Symposium on Intelligent Multimedia, Video and Speech Processing*, Hong Kong, 2001.
9. H. Münch et al. Web-based distribution of radiological images from PACS to EPR. *Computer Assisted Radiology and Surgery. Proceedings of the 17th International Congress and Exhibition*, 2003.

10. E. B. Suh et al. Web-based medical image archive system. *Proceedings of SPIE Medical Imaging 2002: PACS and Integrated Medical Information Systems: Design and Evaluation*, 2002.
11. E. Bellon et al. Web-access to a central medical record to improve cooperation between hospital and referring physicians. *Stud. Health Technol. Inform.* 93:145–153, 2003.
12. H. K. Huang. PACS Basic Principles and Applications. Wiley-Liss, 1999.
13. NEMA. Digital imaging and communication in medicine strategic document version 4.0. [Online]. Available: http://medical.nema.org/. 2005: NEMA.
14. E. Biham and A. Shamir. *Differential Cryptanalysis of the Data Encryption Standard.* Springer Verlag, 1993.
15. A. H. Tewfik. Digital watermarking. *IEEE Signal Processing Magazine* 17:17–88, 2000.
16. N. Nikolaidis and I. Pitas. Digital image watermarking: An overview. *ICMCS 99*, 1999.
17. I. J. Cox and M. L. Miller. A review of watermarking and the importance of perceptual modeling. *Proceedings of Electronic Imaging '97*, 1997.
18. E. T. Lin and E. J. Delp. A review of fragile image watermarks. *Proceedings of the Multimedia and Security Workshop at ACM Multimedia '99*, 1999.
19. X. Lai and J. Massey. A proposal for a new block encryption standard. In *EUROCTYPT'91*, 1991.
20. R. Rivest. The RC5 encryption algorithm. *Second International Workshop on Fast Software Encryption*, 1994.
21. W. Xiaoyun and Y. Hongbo. How to break MD5 and other hash functions. *Lecture Notes in Computer Science: Advances in Cryptology, Äì EUROCRYPT 2005.* 19–35, 2005.
22. O. S. Markku-Juhani. Cryptanalysis of block ciphers based on SHA-1 and MD5. *Lecture Notes in Computer Science: Fast Software Encryption.* 36–44, 2003.
23. R. V. Schyndel, A. Z. Tirkel, and C. F. Osborne. A digital watermark. *1st IEEE International Conference on Image Processing*, 1994.
24. E. T. Lin and E. J. Delp. A review of fragile image watermarks. *The Multimedia and Security Workshop (ACM Multimedia '99)*, 1999.
25. X. Kong and R. Feng. Watermarking medical signals for telemedicine. *IEEE Trans. Inf. Technol. Biomed.* 5(3): 195–201, 2001.
26. M. Holliman and N. Memon. Counterfeiting attacks on oblivious block-wise independent invisible watermarking schemes. *IEEE Trans. Image Processing.* 9:432–441, 2000.
27. J. Zhao and E. Koch. Embedding robust labels into images for copyright protection. *Technical report, Fraunhofer Institute for Computer Graphics*, 1994.
28. L. Chun-Shien et al. Cocktail watermarking for digital image protection. *IEEE Transactions on Multimedia.* 2(4):224, 2000.
29. P. W. Wong. A public key watermark for image verification and authentication. *IEEE Inter. Conf. on Image Processing*, 1998.
30. M. Celik, G. Sharma, and E. Saber. A hierarchical image authentication watermark with improved localization and security. *Proc. ICIP 2001*, 2001.
31. M. D. Swanson, B. Zhu, and A. H. Tewfik. Robust data hiding for images. *IEEE Digital Signal Processing Workshop (DSP 96)*, 1996.
32. W. D. Bender, and N. Morimoto. Techniques for data hiding. *IBM Systems Journal.* 35(3/4):131–136, 1996.
33. L. Boney, A. H. Tewfik, and K. N. Hamdy. Digital watermarking for audio signals. *International Conference on Multimedia Computing and Systems*, 1996.
34. O. Bruyndonckx, J.-J. Quisquater, and B. Macq. Spatial method for copyright labeling of digital images. In *Nonlinear Signal Processing Workshop*, 1995.
35. J. Fridrich. Methods for tamper detection in digital images. *Proc. Multimedia and Security Workshop at ACM Multimedia '99*, 1999.
36. F. Mintzer, G. W. Braudaway, and M. M. Yeung. Effective and ineffective digital watermarks. *IEEE ICIP*, 1997.
37. G. Coatrieux, B. Sankur, and H. Maitre. Strict integrity control of biomedical images. *SPIE Conf. 4314: Security and Watermarking of Multimedia Contents III*, 2001.
38. A. Wakatani. Digital watermarking for ROI medical images by using compressed signature image. *Proc. of the 35th Annual Hawaii International Conference on System Sciences (HICSS'02)*, 2002.
39. F. Bao et al. Tailored reversible watermarking schemes for authentication of electronic clinical atlas. *IEEE Trans. Inf. Technol. Biomed.* 9(4):554–563, 2005.
40. Y. S. Lim and D. D. Feng. Multiple block based authentication watermarking for distribution of medical images. *Proceedings of 2004 International Symposium on Intelligent Multimedia, Video and Speech Processing*, 2004.
41. Y. Yang and F. Bao. An invertible watermarking scheme for authentication of electronic clinical brain atlas. *International Conference on Acoustics, Speech, and Signal Processing*, 2003.
42. E. Bertino et al. Privacy and ownership preserving of outsourced medical data. *21st International Conference on Data Engineering*, 2005.
43. D. Anand and U. C. Niranjan. Watermarking medical images with patient information. *Proceedings of the 20th Annual International Conference of the IEEE, Engineering in Medicine and Biology Society*, 1998.
44. Y. S. Lim et al. Interactive invisible captioning for medical image using digital watermarking. *Proc. of International Conference on Image and Vision Computing, Dunedin*, 2001.

45. C. W. Honsinger et al. Lossless recovery of an original image containing embedded data. U.S. patent application, docket no 77102/E/D, 1999.
46. J. M. Barton. Method and apparatus for embedding authentication information within digital data. U.S. patent application, docket no 5 646 997, 1997.
47. J. Fridrich, M. Goljan, and R. Du. Invertible authentication. *Proc. SPIE, Security and Watermarking of Multimedia Contents*, 2001.
48. A. Alattar. Reversible watermark using the difference expansion of a generalized integer transform. *IEEE Trans. Image Processing.* 13(8):1147–1156, 2004.
49. C. W. Honsinger. A robust data hiding technique based on convolution with a randomised phase carrier. In *Proc. PICS'00.*, 2000.
50. J. Fridrich, M. Goljan, and R. Du. Lossless data embedding- new paradigm in digital watermarking. *EURASIP J. Appl. Signal Processing.* 2:185–196, 2002.
51. J. Fridrich, M. Goljan, and R. Du. Lossless data embedding for all image formats. *Proc. SPIE, Security and Watermarking of Multimedia Contents*, 2002.
52. M. Celik et al. Reversible data hiding. *Proc. of International Conference on Image Processing*, 2002.
53. X. Wu. Lossless compression of continuous-tone images via context selection, quantization, and modelling. *IEEE Trans. Image Processing.* 6(5): 656–664, 1997.
54. Y. Lim et al. Web-based functional image display with embedded patient information for security and management. *J. Nucl. Med.* 45(5):492, 2004.
55. M. V. Droogenbroeck and M. Buckley. Morphological erosions and openings: Fast algorithms based on anchors. *Journal of Mathematical Imaging and Vision.* 22(2-3): 121–142, 2005.

12

Biologic Computing

Prof. Eric P. Hoffman,
Erica Reeves,
Dr. Javad Nazarian,
Dr. Yetrib Hathout,
Dr. Zuyi Wang,
and Josephine Chen
Children's National Medical Center

12.1 Introduction

Biologic computing, as defined in this current chapter, focuses on computational and bioinformatics approaches for three fundamental molecules of life: DNA, RNA, and protein. These three building blocks form cells, tissues, and the living human body. DNA is the code that contains the 30,000 or so genes that can be turned on or off in various contexts and sequences in response to development and the environment. DNA databases are focused on the three-billion–letter linear code in humans (smaller in other organisms). The relatively static nature and low dimensionality of DNA allow it to serve as a reference and anchor for most other biologic computing databases. mRNA and protein are more highly dimensional, with dynamic range, modifications, patterns of expression, and interactions with other mRNAs and proteins dictating normal and pathologic conditions. Much of current and future efforts in biologic computing work toward recording changes of mRNA and protein patterns as a function of a specific variable, with the goal of assembling changes into biologically relevant networks and pathways. The long-term goal is to understand and predict the responses of cells, tissues, and the patient to environmental and pathologic conditions; however, the very high dimensionality makes this a formidable task.

12.2 Overview of Genomic Methods

The three types of molecules, DNA, mRNA, and protein, each have their own bioinformatics challenges. DNA has been the most intensively studied and is in many ways the most straightforward. DNA has only four possible components (A, G, T, C), and these are arranged in highly specific linear code to make up the genes and intergenic regions. We will not go over the basic chemistry of DNA, but it is important to understand a few basic concepts to appreciate the existing resources and remaining challenges in computational bioinformatics.

DNA is a double helix with perfectly complementary strands. As such, it is relatively unimportant which strand one databases or visualizes, as the other strand can be easily derived. The three-billion–letter string of DNA is arranged into some superstructure (chromosomes, with telomores and centromeres within each chromosome). However, for the purposes of computational bioinformatics, the chromosomes become a relatively minor feature. More important are the functional units of DNA, the genes or transcript units, that are oriented in both directions within the DNA.

To explain genes and the databases associated with these, we will use one of the most popular Web sites, namely the Genome Browser (http://www.genome.ucsc.edu). The Genome Browser was developed by University California Santa Cruz bioinformatics professor Jim Kent and colleagues in 2000 using MySQL database, with a set of Linux Pentium-class machines acting as Web servers [1]. The Genome Browser uses the genetic code of humans and other organisms as an anchor to which hundreds of other databases are then referenced. Approximately half of the data links are done by UCSC staff, and half are done by remote investigators wishing to have their databases transparently accessible and queried via the Genome Browser site.

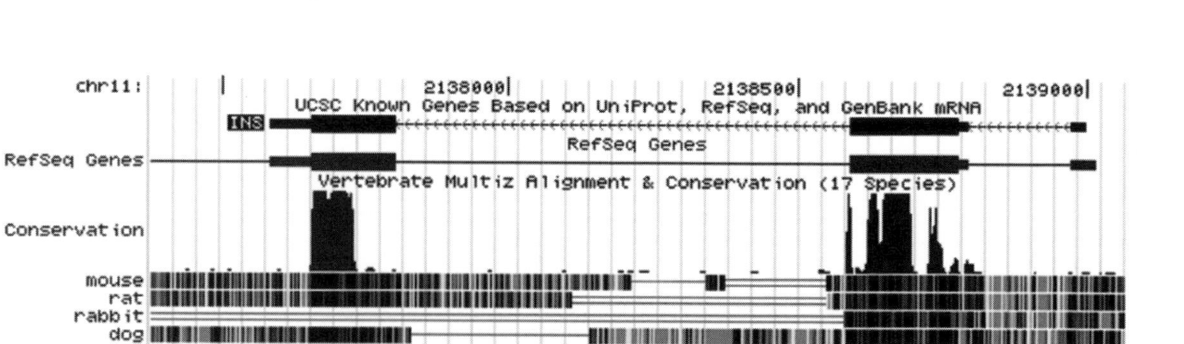

FIGURE 12.1 Genome browser view of the insulin (INS) gene. The position of the gene on the short arm of chromosome 11 (11p15.5) is indicated by the top view, with the base pair coordinates on chromosome 11 indicated (2,138,000 bp region). The INS gene is transcribed in the right-to-left direction (shown by arrows on the INS gene schematic) and contains three exons (heavy lines). The complete gene is about 1,500 bp (1.5 kbp). The initial first exon to the right is small and noncoding (smaller line; 5′ untranslated region of transcript), whereas the second exon contains a small amount of noncoding and a larger amount of coding (amino acid coding; larger bar) sequence. The last exon to the left contains about 2/3 amino acid coding sequence, and the remainder is 3′ noncoding (untranslated). The evolutionary conservation tracks are visualized here, showing that exons 2 and 3 are very highly conserved through evolution, while exon 1 is not. Highly conserved regions imply functional importance of the sequence in that region since there is strong evolutionary pressure to keep these regions the same. The bottom track gives a summary of single nucleotide polymorphisms (SNPs) through the region of the gene. Taken from http://www.genome.ucsc.edu [1].

The size of genes ranges from quite small (perhaps 100–300 bp) (Figure 12.1), to the largest gene known, the dystrophin gene, measuring in at 2.3 million base pairs (Figure 12.2). Overall, there are about 20,000–25,000 transcript units or genes dispersed over the three billion base pairs, but the count depends on how one defines what is considered a gene and what is not.

One quickly then asks, "How is a gene or transcript unit defined?" If a region of DNA functions as a transcript unit, then it should produce an RNA molecule (e.g., be transcribed into RNA). One can then simply take some tissue such as muscle or brain, isolate all RNA, then sequence the RNA fragments. Each RNA must have come from some fragment of DNA somewhere in the genome, so one simply maps the RNA sequence back to the DNA genomic sequence. This process has been done for hundreds of tissues and cells over the last two decades, leading to large databases of expressed sequence tags (ESTs), snippets of RNA sequence that are used as databases to map back to genomic DNA and define transcript units within the genomic DNA.

The process of mapping ESTs back to the linear genomic DNA sequence is a cornerstone of biologic computing. This process relies entirely on the central tenet of biologic computing, namely that the order of bases in DNA and RNA is entirely predictable based on the sequence (e.g., mapping by sequence homology). If an RNA fragment matches to two distinct regions of DNA that are in the same neighborhood, then one can assume that these are two exons separated by an intron. Iteratively doing this process for hundreds of tissues and cells and millions of RNAs leads to the transcript map currently visualized in the genome browser.

There are other aspects of mRNA that are variables in the development and use of biologic computing. As shown in Figure 12.2, the promoters (signals driving transcription at the 5′ end of the gene) and associated first exon can be a variable. An additional variable is alternative splicing, where different exons are used by different cells at different times. Finally, alternative 3′ stop sites can be used. An example where a single transcript unit shows multiple promoters, alternative splicing, and alternative termination sites is shown in Figure 12.3 (tropomyosin 3, TPM3).

It is important to point out that the use of transcript units by cells and tissues depends on many factors, including developmental state, environmental cues, and pathologic states. Indeed, neighboring cells in a tissue may turn different genes on and off, and even when using the same gene may use different variants of that gene. Most importantly, the pattern of gene and protein expression in one cell may alter the pattern of expression in the neighboring cell or in a cell in a distant part of the body. One quickly sees how the neat and invariant order of the three-billion–unit DNA genome is

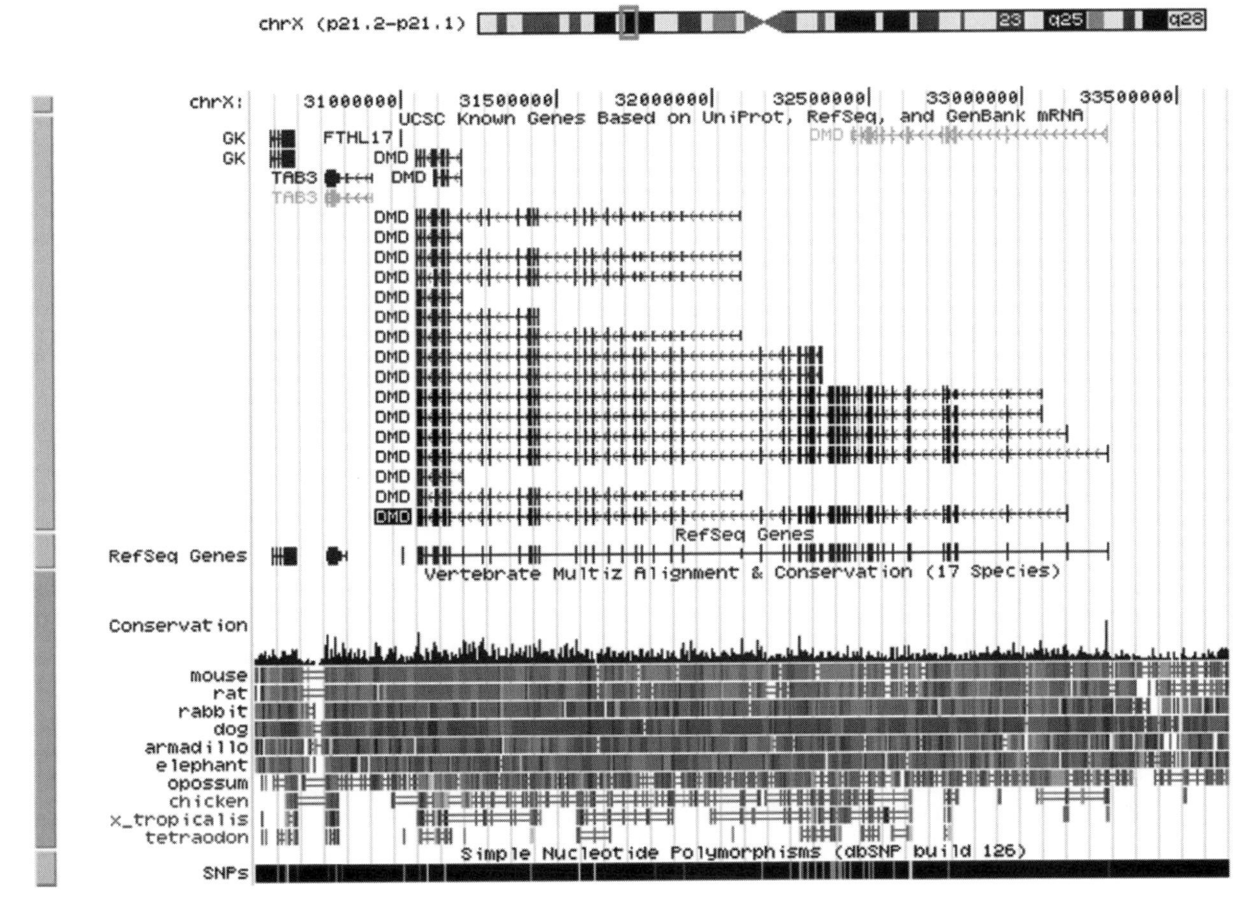

FIGURE 12.2 The Genome browser view of the dystrophin gene, the largest gene known to date. Shown is the dystrophin or DMD gene (Duchenne muscular dystrophy). The gene covers 2.3 million base pairs and includes more than 80 exons. There are multiple alternative start sites of the gene (see alternative exon 1 to the right of the different isoforms listed). The resolution of exons, conservation, and SNPs is all quite compressed in this view, making the information difficult to interpret. There are additional transcript units to the left of the view (GK, TAB3, FTHL17). Comparisons of the relatively simple insulin gene in Figure 12.1 and this highly large and complex gene demonstrates the wide range of transcript units in the human genome. Taken from http://www.genome.ucsc.edu [1].

quickly brought into tremendous complexity at the RNA level. While complex, it remains relatively easy to query, catalogue, and create an RNA database because of the exquisite sensitivity and specificity of computer homology searches and laboratory-based solution hybridization of specific sequences. These are discussed later in this chapter, which also includes a discussion of current bioinformatic tools and challenges.

12.3 Overview of Proteomic Methods

The complexity of RNA patterns and networks pales in comparison to the complexity of protein networks. I typically teach that about 1%–4% of complexity resides in RNA expression; 96%–99% of complexity resides in protein expression and function. Proteins become exponentially more complex at a number of levels. First, there are 20 standard amino acids (building blocks) rather than the four in DNA and RNA. Second, once a protein is translated, a long list of variables influences activity and function of the protein. These include extensive posttranslational modifications (phosphorylation, glycosylation, proteolytic cleavage). Third, folding of the protein dictates activity, as do interactions with additional copies of itself or other proteins. Fourth, subcellular localization and local concentration of substrates and cofactors can dramatically influence protein function.

In addition to the inherent complexity, the understanding of protein networks has been slowed by two technical problems. It is considerably more difficult to purify and sequence proteins, compared to the cloning and automated sequencing technologies available for nucleic acids (DNA, RNA). Also, the exquisitely sensitive and specific sequence-specific hybridization used to query nucleic acids is not available for proteins.

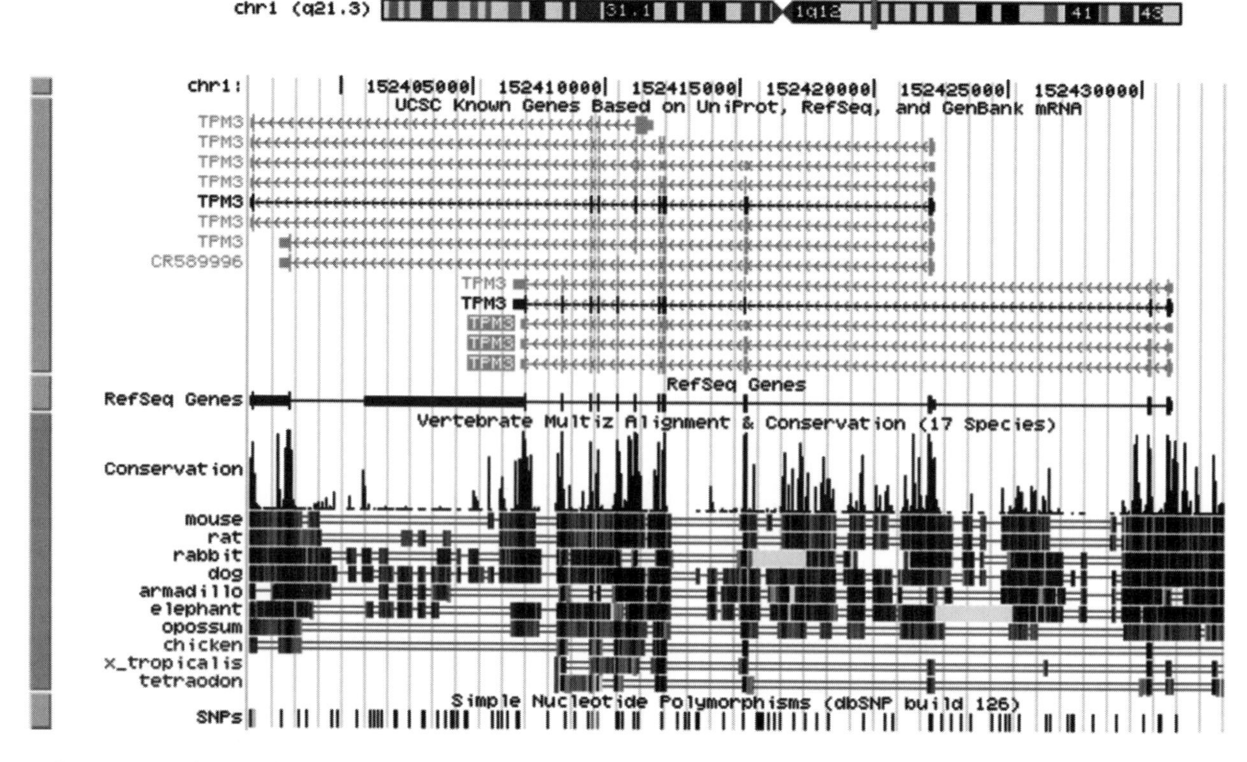

FIGURE 12.3 The complex transcript unit of the tropomyosin 3 gene. The TPM3 gene covers about 35,000 bp (35 kbp) and is transcribed from right to left. At least three different promoter and exon 1 regions are used by different cell types at different times (top transcript uses a promoter and first exon in the center of the gene, while other transcripts are using exons 10–15 kbp to the right). There are also at least three different terminal exons used. The lower transcripts terminate early but also include additional exons not shared with the uppermost transcripts (alternative splicing). This results in a diversity of TPM3 proteins being produced by these transcripts, all from the same parental transcript unit. Taken from http://www.genome.ucsc.edu [1].

To date, it has been impossible to develop probes that can pick a single protein from a complex solution as is possible with hybridization for complex solutions of DNA or RNA.

The lack of sensitive and specific methods of querying proteins in complex solutions has recently changed with the advent of high throughput mass spectrometers (MS), more sensitive protein separation methods, and proteomic profiling using stable isotopes labeling strategies. These are described in a bit more detail later in this chapter. However, a brief description of the principles of high throughput analysis of proteins (proteomics) is given here.

The understanding of the underpinnings of proteomics is critical when considering the bioinformatic aspects of proteomics. The fundamentals of proteomics are as follows:

- In MS, ionized protein fragments moving through a vacuum have a specific mass over charge (m/z) ratio.
- The secondary structure (amino acid sequence) of proteins can be predicted by *in silico* translation of genes (transcript units both known and predicted).
- A protein sequence can be fragmented *in silico* into predicted patterns of m/z fragments.

A pattern of observed m/z fragments detected by MS can then be matched against all theoretical m/z patterns for all known or predicted proteins, and the protein of interest can be assigned and identified.

To begin with the physical and chemical principles behind the use of MS, MS typically have three components: ion source, mass analyzer, and detector (Figure 12.4). The ion source imparts charge to protein fragments (peptides), enabling them to be driven through space by electromagnetic forces within a vacuum. There are a number of methods used for ionization, but the two most frequently used are matrix-assisted laser desorption ionization (MALDI) and electrospray ionization (ESI) (Figure 12.4). The major distinction is that MALDI analyzes samples as dry solids co-precipitated with an ultraviolet laser-absorbing matrix on a probe, while ESI is a solution-based method sprayed through a narrow stainless steel capillary.

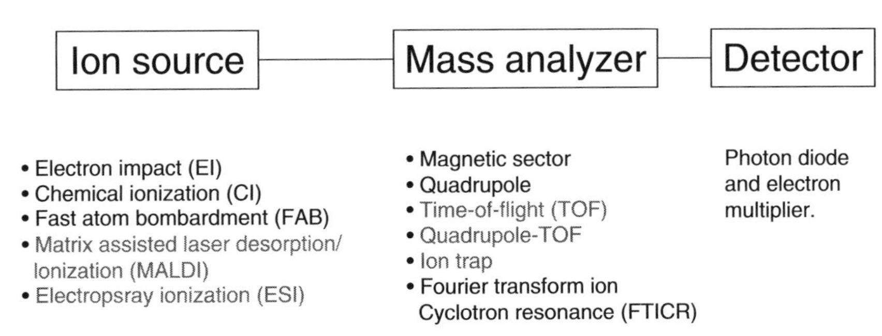

FIGURE 12.4 Overview of mass spectrometer components. The red-font items are the most frequently used ion source and mass analyzer methods. For a more detailed view of this figure, please visit our companion site at: http://books.elsevier.com/companions/9780123735836.

The second component of the MS is the mass analyzer (Figure 12.4). These are vacuum units that manipulate the charged polypeptides in such a way that the m/z ratio is directly related to time at detection by the detector. The three most common mass analyzers are time-of-flight (TOF), where the m/z ratio is related to a linear flight path from the ion source to the detector. Ion traps and quadrupoles have a series of electromagnetic forces that modulate ion frequency in a nonlinear manner. Finally, the detector is where the charged particles are transduced into electrical signals via a photon diode and electron multiplier.

A peptide must be charged (ionized) by the ion source in order to be resolved by the mass analyzer and detected by the detector, but there are a number of possible charged states of a peptide especially (+1 [singly charged], +2 [doubly charged], etc) especially when using ESI as a ionization method. One must know the charge state of the ion in order to determine the m/z value, and thus the precise molecular mass. This is calculated by the presence of naturally occurring stable isotopes, specifically a stable isotope of the carbon atom (Figure 12.5). Carbon typically has an atomic molecular weight of 12 (e.g., ^{12}C). However, the stable (nonradioactive) isotope ^{13}C, with an extra neutron, exists in all natural peptides at about a 1% level (Figure 12.5). This leads to all peptides detected on MS showing a series of peaks rather than a single monoisotopic peak as a result of 1% replacement by ^{13}C (and low levels of other stable isotopes). The monoisotopic mass (no ^{13}C in the peptide) is seen as MW 1296.685 in Figure 12.5, and the addition of one ^{13}C leads to an additional peak to the immediate right, shifted by one neutron (1 mass unit). If the mass difference between the monoisotopic peak (the far left peak at m/z 1297.685/1 – 1296.685/1 = 1), and the peak with one ^{13}C (immediate adjacent peak m/z 1297.685) is equal to one, then the charge state (z = 1) (1296). However, if the peptide is doubly charged, the apparent molecular weight detected on MS is altered due to z = 2 (e.g., 1296.685/2). In this case, the mass difference between the isotopic peaks becomes 0.5 mass units (1297.685/2 – 1296.685/2 = 0.5). The bioinformatics software can quickly determine the charge state of the peptide simply by looking at the distance between adjacent isotopic peaks; if the distance between peaks is 1, then the peptide is singly charged, while a distance between peaks of 0.5 (1/2) indicates a doubly charged state. The improved mass resolution of current MS enabled the accurate detection of stable isotopic variants of peptides, which was a crucial step in the determination of charge state and enabled m/z calculations.

The remaining principles of proteomics depend on the prediction of polypeptide patterns on MS using genomic sequence data (genes, transcript units) and the matching to observed patterns in data from MS. There are three types of polypeptide databases that are often used. The most widely used databases are the SwissProt, a curated protein sequence database that provides a high level of annotation; followed by the National Center for Biotechnology and Information database (NCBI), which contains nonredundant protein and nucleic acid sequences; and the International Protein Index database, which maintains and organizes a large eukaryotic database. Nearly all full-length proteins are too large (high molecular weight) to resolve within most MS. Thus, this database is not used directly for matching to MS data but is used instead as the starting point for *in silico* prediction of polypeptide fragments from the full-length parent protein.

Predicted mass spectra use the molecular mass of each amino acid (Figure 12.6) then calculate the predicted molecular weight of the sum of amino acids of protein fragments (peptides, or polypeptides).

Improved mass resolution can resolve the isotopic peaks for a given peptide.

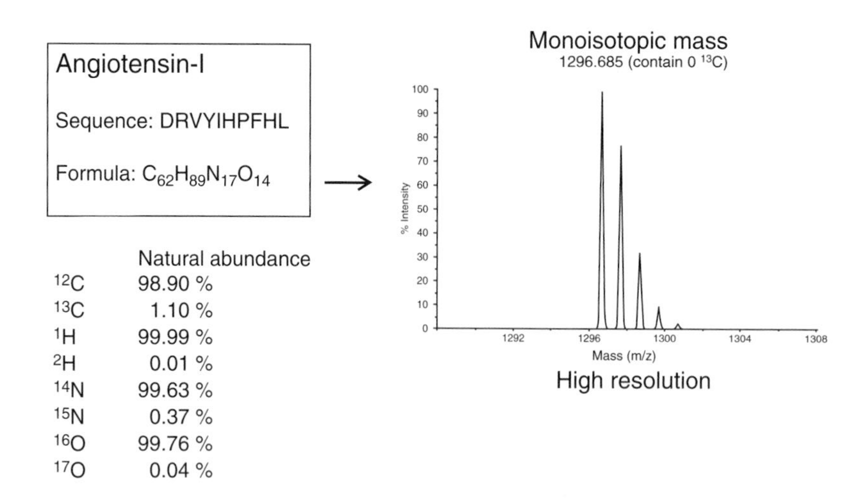

FIGURE 12.5 Isotopic distribution of a peptide analyzed by a high resolution mass spectrometer. The isotopic distribution and intensities are due to the natural abundance of different isotopes of C, H, N, and O.

Amino acid	3-Letter code	1-Letter code	Monoisotopic mass
Glycine	Gly	G	57.021
Alanine	Ala	A	71.037
Serine	Ser	S	87.032
Proline	Pro	P	97.053
Valine	Val	V	99.068
Threonine	Thr	T	101.048
Cysteine	Cys	C	103.001
Leucine	Leu	L	113.084
Isoleucine	Ile	I	113.084
Asparagine	Asn	N	114.043
Aspartic acid	Asp	D	115.027
Glutamine	Gln	Q	128.059
Lysine	Lys	K	128.095
Glutaminic acid	Glu	E	129.043
Methionine	Met	M	131.040
Histidine	His	H	137.059
Phenylalanine	Phe	F	147.068
Arginine	Arg	R	156.101
Tyrosine	Tyr	Y	163.063
Tryptophan	Trp	W	186.079

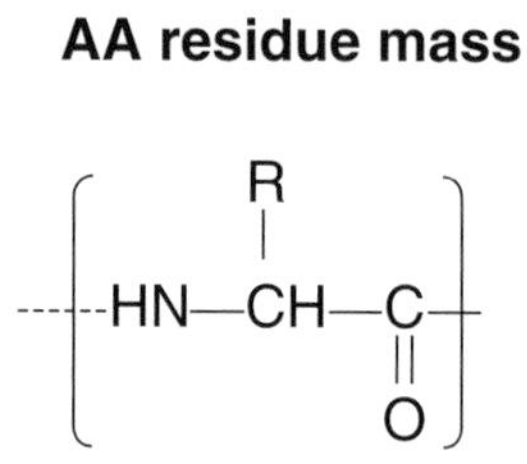

AA residue mass

R
-HN—CH—C-
O

FIGURE 12.6 Amino acids and their corresponding molecular weights (residue mass). The structure of the polypeptide chain with the amide linkages is also shown.

There are two fragmentation patterns that are then assembled into databases: protease fingerprints (e.g., tryptic peptides) and random fragmentation spectra (MS/MS spectra). For protease fingerprints, the most commonly used protease is trypsin, an enzyme that cleaves polypeptide chains immediately after lysine (K) and arginine (R) residue. Peptide fingerprint databases use the primary protein sequence, digest the protein with trypsin *in silico*, create a database of the expected peptide fingerprint, and then match this with the observed peptide fingerprint on MS. If a protein is purified and then digested with trypsin and detected on MS, the peptide fingerprint alone is sufficient to provide an unambiguous protein identification (identification of the parent). Peptide fingerprints require only a determination of the intact peptides emanating from a parent protein (MS).

Two-dimensional electrophoresis of proteins allows resolution of 100–1,000 individual proteins (including post-translationally modified states) (Figure 12.7), where the X axis is charge (isoelectric focusing point) and the Y axis is molecular weight. These spots are purified to the extent that the visible spot is likely to be a single protein. Excision of the spots from the two-dimensional gel, digestion with trypsin, then detection of the peptide map often allow the identification of the protein in each spot.

Additional confidence to a protein identification can be obtained by MS/MS spectra. For MS/MS, specific peptides are gated (isolated in the electromagnetic field in MS), then further fragmented by laser or a neutral collision gas. In Figure 12.8, the 644.4 peptide peak was gated then fragmented, and MS/MS spectrum was obtained as a series of fragment ions that permit determination of the probable amino acid sequence.

Two-dimensional gels have the advantage of identifying proteins with a high degree of confidence. Both peptide maps and MS/MS sequencing of individual peptides should map back to the same parent protein in databases. The disadvantages of two-dimensional gels are their relatively poor sensitivity, limited molecular weight/pI range, and the requirement for relatively large amounts of proteins ($\geq 200\,\mu g$).

A more sensitive method is often called shotgun, where more complex solutions of proteins are digested with trypsin as a mixture then run on solution-based electrospray using multidimensional chromatography coupled to MS. Shotgun

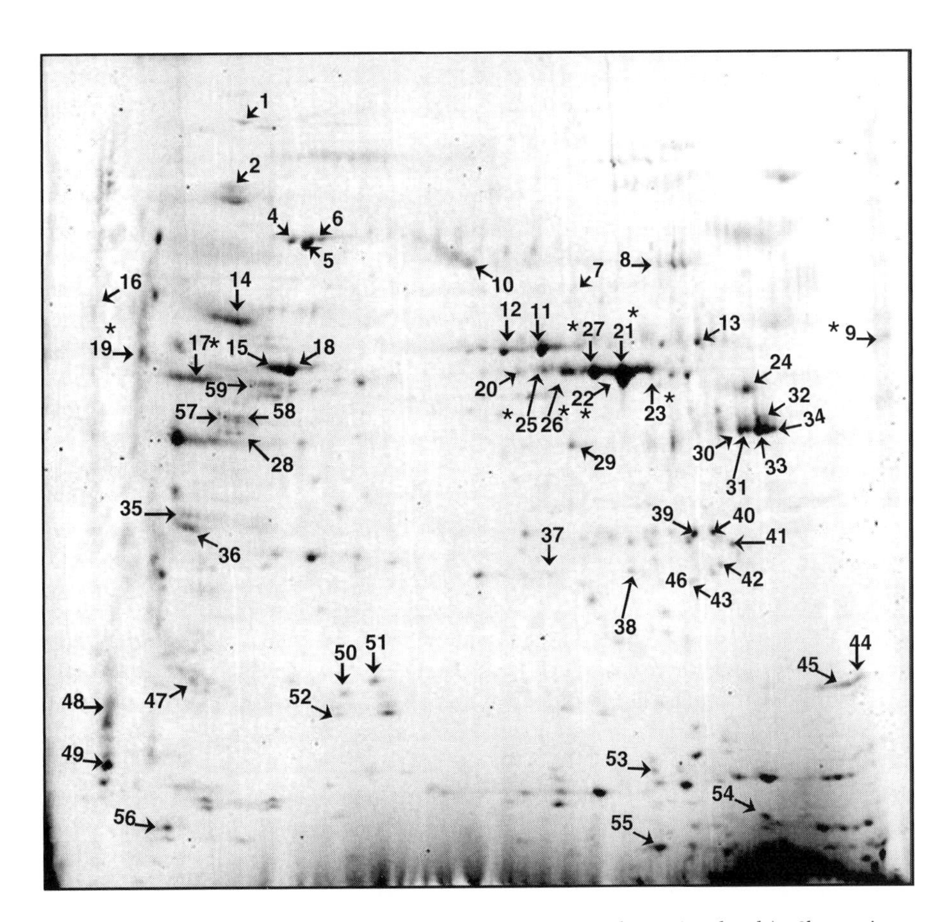

FIGURE 12.7 Two-dimensional gel electrophoresis (two-dimensional gels). Shown is an example of soluble cytoplasmic proteins isolated from the Torpedo fish electric organ, a specialized tissue that is able to generate 200 volts of direct current through water to stun prey. Taken from Nazarian et al. [2].

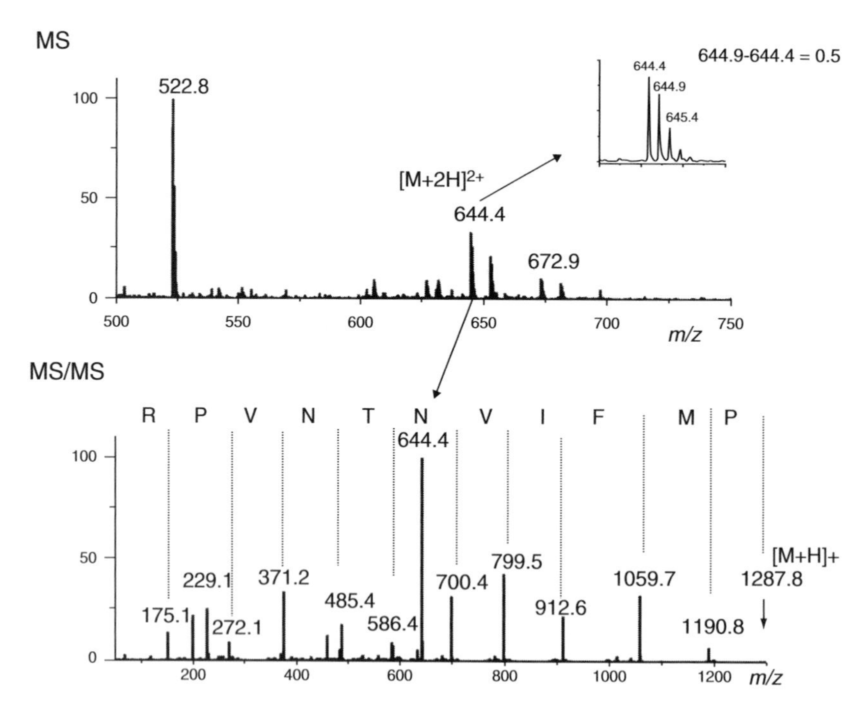

FIGURE 12.8 Gating of a 644.4 peak (MS), fragmentation and detection of MS/MS spectra. The 644.4 peptide is found to be doubly charged as a result of the close spacing (0.5 mass units) of the isotopic variants (upper right blow-up window), hence the designation as $[M + 2H]^{2+}$. The true mono-isotopic weight of this peptide is 1287.8. Resolution of b and y ions (see Figure 12.9) allows the derivation of the amino acid sequence of the peptide.

methods lose the ability to generate peptide fingerprints and rely largely on MS/MS sequence spectra (Figure 12.9). Because the complex mixture of peptides derives from many different parent proteins, it is difficult if not impossible to reconstruct a peptide fingerprint from the highly complex mixture. Also, with shotgun methods, there is typically less coverage of each parent protein, with fewer peptides detected per parent protein. The major advantages of the shotgun method are high throughput and increased proteome coverage.

Statistical significance of parent protein identification becomes more challenging with shotgun MS/MS methods. Typically, a single peptide detected and fragmented is considered insufficient for a robust parent protein identification. Multiple MS/MS identifications of peptides mapping back to the same parent are required before one can confidently conclude that the parent protein was in the original solution under analysis.

Another relatively recent development in proteomics is the ability to compare proteins in two samples in a quantitative and high throughput manner (proteomic profiling). This involves differential labeling of the peptides from one sample relative to a second, unlabeled sample. Many methods are available, but one that is commonly used is metabolic labeling with stable isotopes amino acids. Growth of cells in ^{13}C-labeled arginine and lysine results in each peptide being larger than the same peptide in unlabeled cells (Figure 12.10). Mixture of the labeled and unlabeled solutions results in a doublet for each component peptide in the mixture. Each of the two peaks can be gated and subjected to MS/MS, thus allowing the identification of the peptide, with a relative quantitation of the parent protein in the two original solutions (Figure 12.10).

12.4 Bioinformatics and Information Infrastructure

Bioinformatics and information infrastructure can be done locally (sites at specific labs or universities) for data analysis within a lab or group, or it can be performed by a more centralized data repository such as the NCBI. Intra-lab or intra-university systems are often referred to as a laboratory information management system (LIMS). LIMS can house fairly elaborate methods of tracking and storing information on experimental design, methods, data acquisition, and data interpretation. The more advanced LIMS have a portal for either public access to subsets of the data and/or a portal for export of selected data sets into public repositories. An example of an advanced LIMS that we established for microarray data is described here as one example.

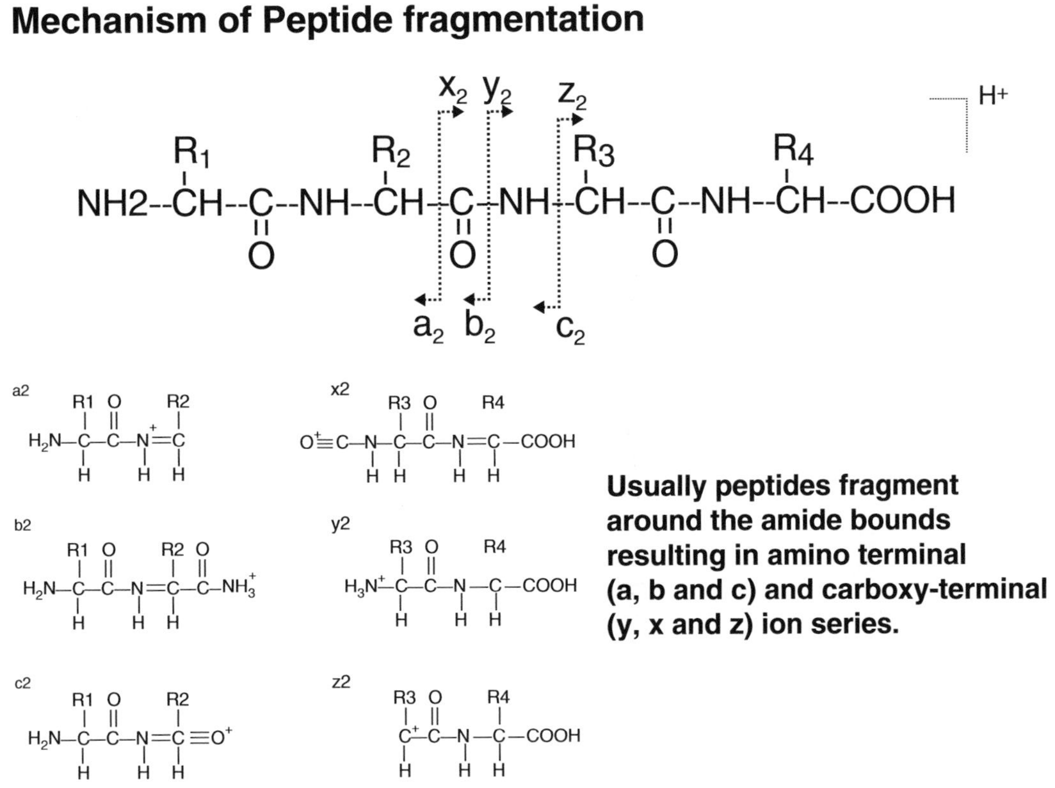

Mechanism of Peptide fragmentation

Usually peptides fragment around the amide bounds resulting in amino terminal (a, b and c) and carboxy-terminal (y, x and z) ion series.

FIGURE 12.9 Mechanisms of peptide fragmentation in MS/MS spectra.

DNA and single nucleotide polymorphism (SNP) databases are relatively straightforward, given the linearity (two-dimensionality) of the genomic data. We will not discuss these in this section of the text, but the reader is referred to outstanding public resources such as the Genome Browser (www.genome.ucsc.edu) and the HapMap project (www.hapmap.org) as examples.

In this section, we focus on the challenging area of mRNA profiling bioinformatics and information infrastructure. mRNA profiling using either Affymetrix microarrays (www.affymetrix.com) or more customized spotted cDNA or oligonucleotide microarrays has been in widespread practice for nearly 10 years, with many tens of thousands of microarray profiles in the public domain. The heavy use of microarrays for mRNA profiling has led to a relatively rich literature on quality control and standard operating procedure, signal derivation, bioinformatics, and statistical analyses (Tumor Analysis Best Practices Working Group 2004).

There are excellent public data repositories for mRNA profiling microarray data (e.g., (ArrayExpress www.ebi.ac.uk/arrayexpress; Gene Expression Omnibus (GEO) http://www.ncbi.nlm.nih.gov/geo/) [3, 4]. Although they can store large numbers of projects and much experimental data, they also have limitations. With so many contributors, it is often difficult to ensure accurate metadata collection process, accessibility, and appropriate data formats. Also, these databases accept many different experimental platforms (e.g., different methods of conducting expression profiling: cDNA arrays, Affymetrix arrays, SAGE), and it becomes difficult or impossible to compare experiments across platforms. One approach to provide some means of comparing data fields across experiments and experimental platforms has been to develop the minimal information about a microarray experiment standards [5, 6]. This does not set a standard for experimental platforms or data but attempts to provide certain data fields that can be mapped across different data sets and databases.

Local LIMS are often able to develop more of an experimental standard and thus provide more power in the data queries and data analysis methods. Two that are commonly used are the Stanford Microarray Database ([7]; http://genome-www5.stanford.edu/) and the Public Expression Profiling Resource (PEPR) at Children's National Medical Center in Washington, D.C. ([8, 9]; http://pepr.cnmcresearch.org).

For PEPR, we focused on improving three aspects of data acquisition and public data analysis:

1. Improving the complete prospective data collection process
2. Improving the application programming interface (API) so that it automatically converts raw mapped image data

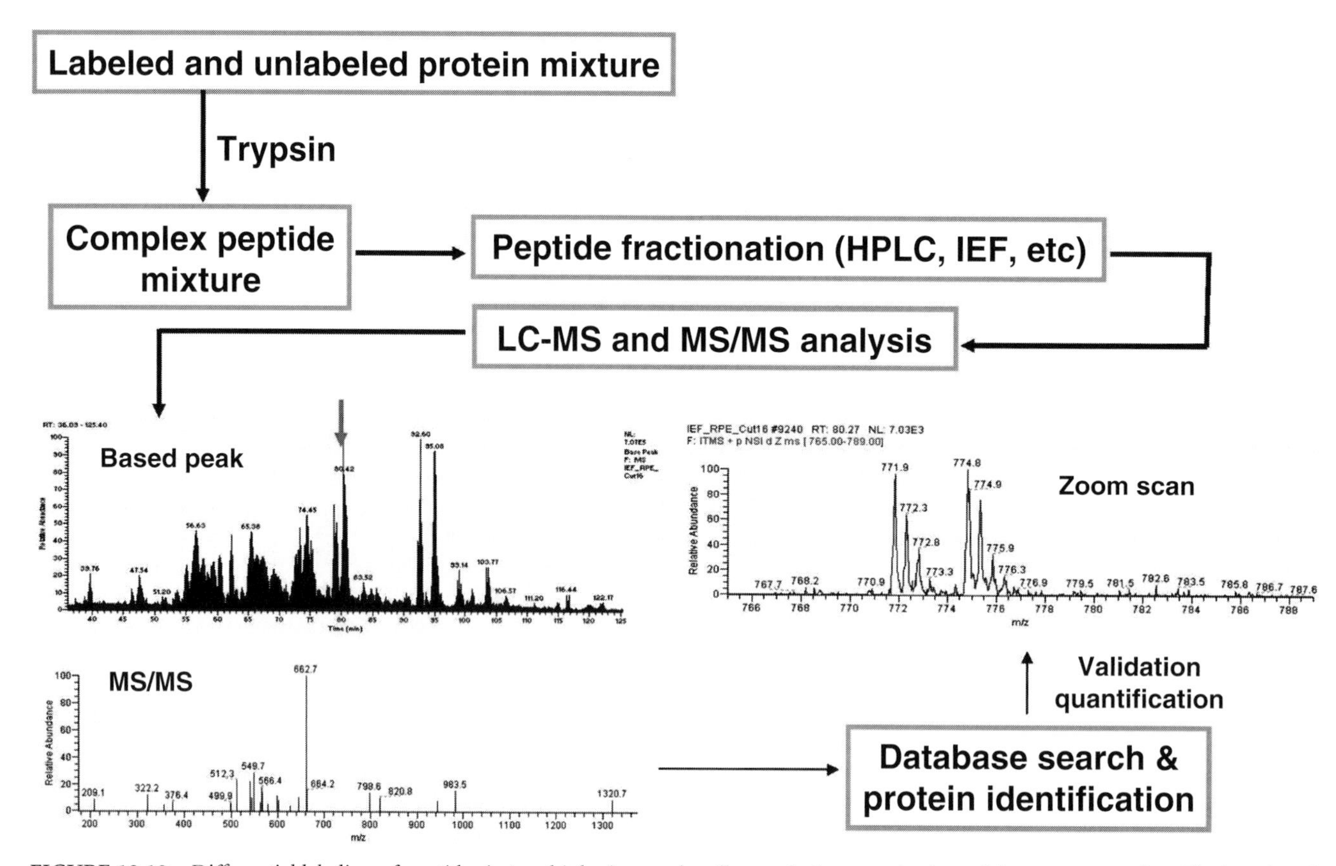

FIGURE 12.10 Differential labeling of peptides in two biologic samples allows relative quantitation of the parent proteins. The based peak shows the peptides detected in a scan of the mass spectrometer (MS). A single peak in the based peak scan (red arrow) is then expanded to greater resolution in the zoom scan, where the pair of peaks representing the same peptide from the two biologic samples (labeled and unlabeled peptides). Isotopic variants of each peptide are obvious in the zoom scan. Individual peptides are then gated and fragmented to provide fragmentation scan representative of the amino acid sequence (MS/MS) (see Figure 12.9). For a more detailed view of this figure, please visit our companion site at: http://books.elsevier.com/companions/9780123735836.

for microarrays (.cel files) to a series of summary signals for each probe set using five probe set algorithms
3. Improving simple public interfaces for dynamic Web-based queries

Regarding the prospective data collection process, we designed and implemented a process where the originator of a microarray project begins interacting with PEPR soon after initial concept of the experiment. This allows implementation of automated checks and balances on data fields and a process of approval at different steps in data generation and public release. A schematic of the design of PEPR is shown in Figure 12.11.

For any bioinformatics and biologic computing development, it is critical to know about issues or concerns regarding the accuracy of data. For mRNA microarray experiments, there is considerable debate regarding how to appropriately interpret hybridization signals on a microarray and derive a normalized signal for each mRNA transcript on the chip and within the project. For Affymetrix microarrays (www.affymetrix.com), each transcript is queried by a probe set with 11 perfect-match 25mers tiled against the last 500 bp of the transcript and 11 paired mismatch 25mers that serve as a possible control for nonspecific binding to the perfect match probe. Thus, there are at least 22 hybridization signals that can be considered for each transcript, and the methods of interpreting signal from noise and normalization methods within and between microarrays is a very active area of research (and hotly debated). The bioinformatic and statistical methods used to move a mapped image file to transcriptional signals for each gene are called *probe set algorithms*, and there are dozens of methods that are available (see Seo and Hoffman [10] for review).

The design of PEPR enables rich metadata search functions (i.e., search by experiment design type or by animal model age or sex), including a Web-interface data input system to capture experiment information prospectively (and remotely such that an investigator in Sweden begins entering metadata and design data prior to initiation of the profiling project in Washington). Unlike other currently utilized profiling packages, our Web-interface data input submission process offers great flexibility

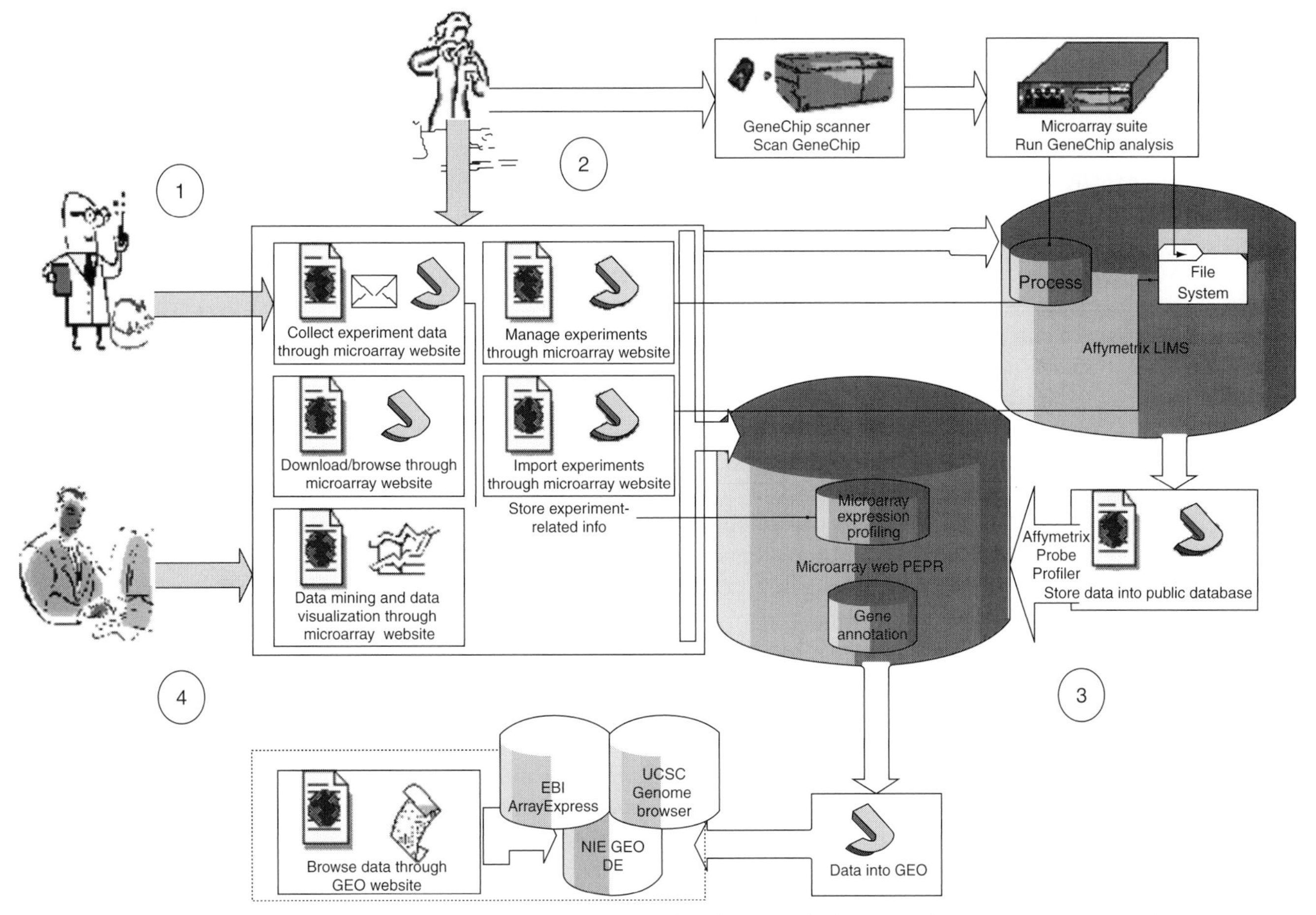

FIGURE 12.11 Public expression profiling resource (PEPR) architecture design. From htpp://pepr.cnmcresearch.org.

to obtain desired experiment metadata (e.g., addition of experiment design type) for analysis and visualization. It provides a mechanism to enforce data input consistency and validation and eliminates the current accessory tables and batch process to filter data. The data consistency expands the search and visualization capabilities. PEPR also utilizes our newly implemented GEO submitted or update APIs to submit new experiments or revised previously published experiment data. PEPR incorporates a custom-designed API for automated conversion of all projects to five probe-set algorithms (MAS5.0, dCHIP PMonly, dCHIP diff, ProbeProfiler PCA, RMA). This API coupled to user interface feature allows the public Web user to rapidly visualize the effect of probe-set algorithm on data interpretation. Finally, PEPR provides off-line batch data exportation that allows the researcher to download/export a series of large data sets while continuing to navigate the site. The generation of .chp, .dat, and .cel data files is done during off-peak hours.

Regarding PEPR process architecture design and implementation, PEPR is a three-tier Java enterprise application composed of a Web tier, a middle tier, and a back-end tier. The Web tier includes a Web server, a Tomcat application server, and various Web components that provide front-end functionalities such as navigation, data browsing, data searching, project submission, project publishing, gene query tool, and user notification. Most Web components interface transparently with PEPR back-end databases. This tier's interface allows users to trigger the middle tier application.

The middle tier is integrated with several third-party services. For some of these, we have purchased enterprise versions of pre-existing software, and for others we wrote or contracted specifically for PEPR (Popchart, Lucene, Affymetrix SDK and Corimbia Probe Profiler SDK). The application is designed to handle time-consuming processes such as Affymetrix data extraction and offline data downloading while allowing a user to navigate the site without waiting for the completion of the process. The middle tier applications require intense computing resources and are responsible for chart visualization generation, offline data download, metadata indexing for keyword search, NCBI GEO data submission, Affymetrix data file extraction and transformation, and Probe Profiler mixture of algorithm data generation.

Most of processes in this tier do not require synchronous response from the PEPR front end. In addition to the conventional Web click-and-wait applications features, PEPR allows users to submit the request without waiting for the completion of the process while the process is guaranteed to be completed. To achieve this asynchronous operation in a reliable manner, an Open JMS queue server is introduced in PEPR implementation, and this serves to enhance the PEPR application functionalities. JMS is designed to handle the message delivery between Web components. When a user submits a request to download a large set of data in PEPR, a Web component in Tomcat application server packages the user's request to a message and drops the message into the JMS queue. The JMS queue is responsible for receiving and delivering the message as a specialized router that looks at the message address and delivers it to the appropriate parties (i.e., offline data download process in the chart). The offline data download process then parses and handles the download request. It continues to search and compress the requested data and then sends the download URL notification to the user. During this process, the user does not have to wait for the lengthy file compression process completion. The JMS queue makes the batch download possible.

The back-end tier is composed of two databases: the PEPR DB and the Affymetrix LIMS DB. PEPR DB stores all sorts of metadata of projects and experiments alone with associated analysis value for real-time data mining purposes. The Affymetrix LIMS DB stores all Affymetrix expression profiling physical data and chipping process information.

The total number of microarrays currently populating our internal LIMS is 7,000, with most from human specimens. The public PEPR resource is populated with 2,827 Affymetrix arrays and shows about 6,000 profile downloads per month from PEPR. There is an automated submission pipeline developed with NCBI GEO, and PEPR is the #1 contributor to GEO, accounting for 12% of all Affymetrix files for vertebrates.

12.5 Data Mining and Large-Scale Biologic Databases

There is an increasing diversity of data sources in the public domain and increasingly flexible tools by which data mining of large-scale biologic databases can be done. Public access data sets have provided a link between the biologists generating the large-scale data and the computer scientist or statistician who wishes access to this data for method development and analysis.

In our experience, there are two common misperceptions that can inhibit the development of biologic computing. First, computer scientists and statisticians tend to assume that the data provided by biologists are as good as they can be (robust, accurate, with good signal/noise balance). This is generally not true, but biologists tend to neglect to inform the quantitative scientists of this. Second, biologists tend to assume that the data analyses provided by computer scientists and statisticians have transformed their data into something that is robust, accurate, with good signal/noise balance. This is also not true, as it counters the classic theorem of "garbage in, garbage out."

The garbage in, garbage out problem is relatively rare with DNA data mining, largely because of the relative simplicity of the problems (low dimensionality) and robust and mature nature of genomics databases. The problem is much more pronounced as the complexity increases and methods become less standardized and robust. A good example is the issue of probe set algorithms for Affymetrix microarrays introduced above. The computer scientist is typically provided a signal for each transcript on each microarray in a project. This signal is a distillation of many composite signals (22 oligonucleotide probe set), with extensive normalization imposed to make transcripts comparable between arrays. Each probe-set algorithm makes many assumptions regarding normality, penalties for hybridization for mismatch probe signal, and background assessments [10–12]. Yet the computer scientist is rarely informed of these assumptions, and the impact on subsequent data interpretation is often not evaluated.

We have studied the probe-set algorithm problem in the setting of signal/noise assessments for different microarray projects [10, 13] and the related question of power calculations for specific species and tissues [14]. Most biologists select one specific probe-set algorithm (e.g., MAS5.0, PLIER, dCHIP, RMA) to convert their microarray project into a set of signals. The choice of the probe-set algorithm is based on the belief that the one chosen out-performs the others not chosen, based on the literature or on their experience. To the contrary, we found that different microarray projects require the use of different probe-set algorithms because each of the probe set algorithms is differentially influenced by uncontrolled sources of biologic and technical noise (confounding variables). Some probe sets algorithms are relatively impervious to extensive noise but then become relatively insensitive. Other probe-set algorithms are exquisitely sensitive, but this results in a high proportion of false positives if confounding variables (noise) become high [10, 14]. We suggest that the method of signal generation in microarrays (selection of probe set algorithm) should be tailored for individual microarray projects.

The selection of a specific probe-set algorithm is clearly a fundamental early step in appropriate data interpretation and analysis. Probe-set algorithms have a profound impact on the data generated, dependent on the intrinsic assumptions of each algorithm. Indeed, if one takes a typical microarray project and queries concordance between two algorithms, one finds only 10%–30% concordance of statistically significant expression differences. Despite the low concordance, the biologist typically selects a single probe-set algorithm, believing one to be best, with little or no knowledge of the underlying assumptions or any assessment of the appropriateness of the selection. The computer scientist and statistician then take the resulting signals from the biologist. They are often unaware of the probe-set algorithm used or the underlying assumptions made in generating the composite signal data. This step is typically done by the biologist, and the resulting signals are provided to the computer scientist, who believes that these are robust and accurate numbers.

Probe-set algorithms are a single step in the path from biologist to computer scientist, but it is exemplary of the relatively deep canyon that exists between these disciplines. As biologic data sets become exponentially larger and more complex, it becomes more and more important to bridge the canyon, with biologists more aware of computational methods and computer scientists more aware of the technical and biologic variables influencing the data provided to them.

One relatively easy way to bridge the divide is for computational researchers to insist on being given raw data rather than processed data. For Affymetrix microarrays, raw data files are the image data from the microarray (.DAT files) and the unprocessed hybridization intensity calculations for each of the 1 million cells (oligonucleotides) on the array (.CEL files). These files are quite large relative to summary tables generated by probe set algorithms. These file types have been available on PEPR for a number of years and are also now available for many projects in both NCBI GEO and ArrayExpress. If the biologic computing specialist accesses the raw data, then a more rational selection of probe-set algorithms (or multiple algorithms) can be done and the resulting data interpreted more sensitively.

Large-scale databases for proteomics data are only beginning to be developed. These make both the biology and the data collection considerably more complex. Whereas RNA experiments are typically conducted by grinding up an entire sample into a pool of transcripts, proteomic experiments often study different subcellular compartments (cytoplasm, membrane, nucleus), adding an additional dimension to data acquisition, storage, and interpretation. RNA microarray experiments typically have one sample = one microarray (image), whereas proteomic experiments often involve deconvolution of complex protein solutions into isoelectric and molecular weight ranges. Also, proteomic data acquisition on MS is an intrinsically linear process over time (as opposed to the highly parallel data generation of hybridization on microarrays). One recent experiment run by our center involved testing the response of patient cells to a drug using proteomic profiling (stable isotope labeling; see above). This experiment involved a time series, using patients and controls who were and were not receiving drug, with analysis limited to a specific subcellular compartment (endoplasmic reticulum). This experiment required a high throughput MS (electrospray ion trap) to run 24 hours a day, 7 days a week for 6 weeks, with both MS and MS/MS (fragmentation) data. The amount of raw data gathered is in the hundreds of gigabytes for this single experiment. How will this be databased and provided to the public? The controversies of probe-set algorithms with RNA microarrays set the stage for the even more complex debate regarding proteomics databasing and data analysis methods. Again, it would be best if computational scientists could have access to the raw MS data, but the logistics of databasing and public access and the complexity of the data become quite challenging.

12.6 Biologic Event-Driven, Time-Driven, and Hybrid Simulation Techniques

The sequence of DNA is relatively static and changes in only certain circumstances. For example, gene mutation and polymorphism are changed from the ground state (normal linear sequence) but can be considered isolated events (unrelated to each other) and relatively easy to characterize, database, and interpret. DNA can be modified by acetylation (transient inactivation of chromatin), and methylation (more permanent inactivation), and the acetylation and methylation state of specific genes likely has both dynamic and static components. For example, one X chromosome in each female cell is inactivated (Barr body; X inactivation) through widespread methylation early in development; this is a static change. On the other hand, the response of a cell to an environmental challenge results in rapid changes in acetylation of specific genes to activate or inactivate transcription (event-driven, time-driven). Methods to access methylation and acetylation status of genes on a genome-wide scale are just beginning to emerge. Once these methods are mature, DNA will take on a dynamic component that will dramatically increase the complexity of databasing and data interpretation.

As described above, existing mRNA profiling and proteomic experiments have intrinsic complexity that makes them challenging to interpret. There are two key approaches that can aid in the interpretation of this highly dimensional data: time series and knowledge networks.

Knowledge networks are computer databases and associated interfaces that assemble preexisting biologic knowledge into a tool that can accept genomics data sets (e.g., mRNA expression profiles) and help interpret the data in the context of preexisting knowledge. There are many types of knowledge networks [15]. One group is gene ontology databases, where genes and encoded proteins are categorized into biochemical- or sequence-defined groups. Another approach is to code existing literature into biologically relevant pathways and networks and then compare genomics data sets to these networks. Again, there are many resources, but two commonly used ones are Ingenuity Pathways Analysis (www.ingenuity.com) and GenMapp (www.genmapp.org). For example, imagine that a biologist has tested the response of some cancer cells to an anticancer drug using mRNA expression profiling (microarrays). The statistically significant transcriptional differences (drug-responsive genes) are uploaded into Ingenuity, and then the program package compares the list of differentially expressed genes with their databases of interacting genes and proteins. The protein networks that show the greatest proportion of changes in the cancer–drug data set are calculated and then ranked by statistical significance. In this case, it is found that there is a previously described protein network governing apoptotic cell death containing 123 proteins, with transcriptional induction of 85 of these proteins by the anticancer drug. The statistical likelihood that 85/123 would be detected by chance is vanishingly small, so the "apoptotic cell death" network is returned to the user as "highly significant altered network." One concludes that the anticancer drug induces apoptotic cell death in the cancer cells studied.

An example of comparison of muscle biopsies from lean and obese subjects is shown in Figure 12.12. In this analysis, expression differences using a threshold of significance were input into Ingenuity software, and a network showing a high proportion of transcripts significantly altered by the obesity state was identified (Figure 12.12). Those proteins indicated by red symbols were significantly up-regulated by obesity, and those with green symbols were significantly down-regulated.

Network analysis software is a highly useful and practical method to condense genomic data sets into biologically relevant pathways deserving further study. The sensitivity and specificity of these packages must be considered tentative at best. Pathways and networks are highly dependent on the cells, tissues, and organisms under study, and these packages generally assume that an interaction described in the literature between two proteins described in mouse eye are relevant to the cancer cells. Additionally, they must be assumed to be a relatively blunt instrument (insensitive), as only a very small proportion of biologic truth in terms of networks and pathways is currently known and published (perhaps 1%).

The second tool to assist in interpretation is the time series. The behavior of a gene or protein as a function of time allows an assessment of biologic plausibility that can help reduce false positives. Most microarray and proteomics experiments are snap shots at only a single point in time (e.g., affected vs. unaffected subjects; see Figure 12.12 as typical example). It is impossible to determine any cause/effect of responses to obesity in the data in Figure 12.12. However, if serial muscle biopsies were taken as a function of time (perhaps after a gastric bypass operation or strict diet), then some of the transcriptional responses seen in Figure 12.12 could be assigned to an earlier time point than others, leading to models of cause and effect.

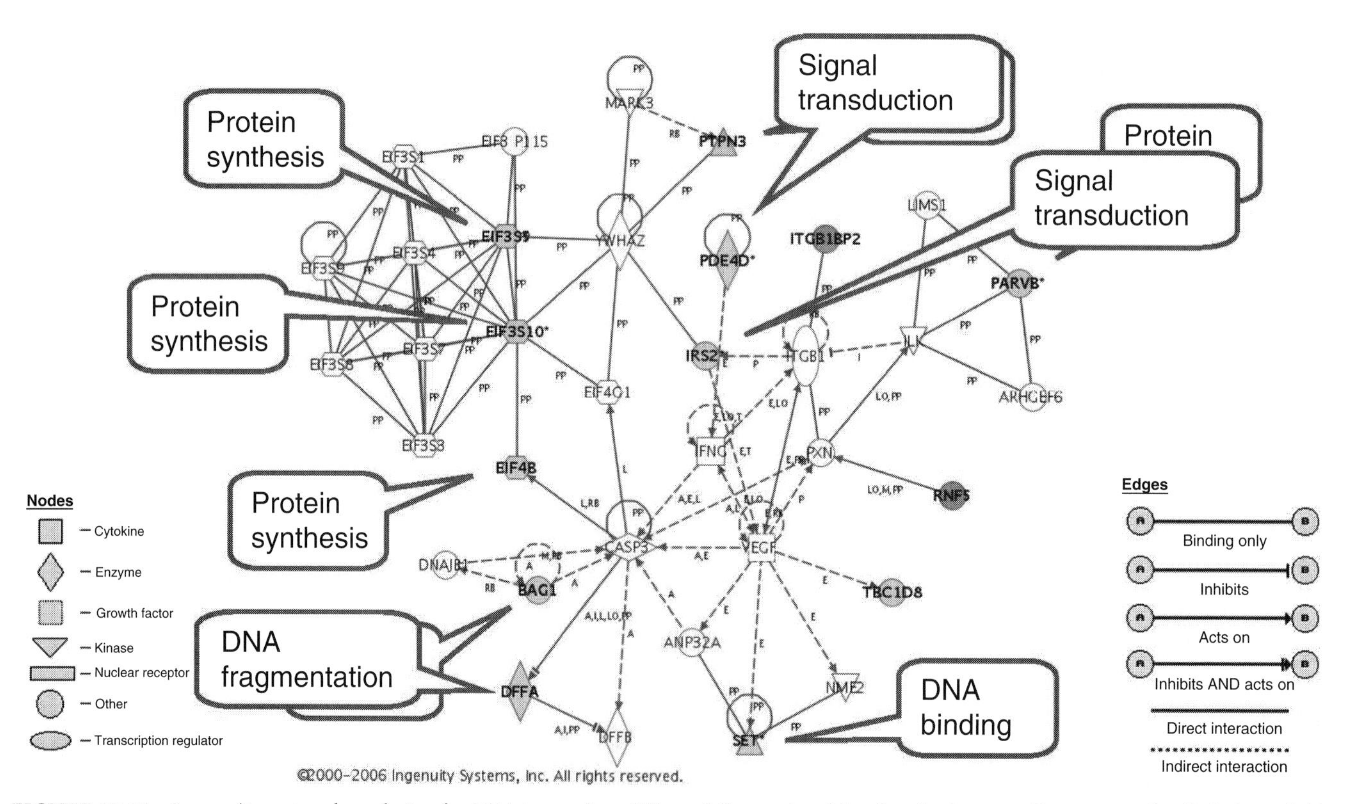

FIGURE 12.12 Ingenuity network analysis of mRNA transcripts differentially regulated by the obesity state. For a more detailed view of this figure, please visit our companion site at: http://books.elsevier.com/companions/9780123735836.

To provide an example of the value of time series data, we have described a 27 time-point muscle regeneration series, where muscle damage was invoked in mouse models and muscle samples were taken as a function of time during recovery from the muscle damage (Figure 12.13). Two transcription factors, myogenin and MyoD, are both seen to be strongly transcriptionally induced around day 3 during recovery. Close inspection of the time series shows the peak of myogenin to be a half-day later than the peak of MyoD: This can establish a hypothesis that MyoD induces myogenin (e.g., MyoD is upstream of myogenin). A cause–effect relationship between these two proteins can thus be established.

Future directions in biologic computing will be focused on combining data from multiple data sets and may include both snap shot (cross-sectional) projects and time series data, possibly done in different species. We recently published an example of a multiproject approach [17, 18]. The first step was to do a snap shot cross-sectional study, where about 125 patient muscle biopsies from 12 disease groups underwent mRNA profiling. Bioinformatic analysis of the profiles led to a relational tree of the different disorders (Figure 12.14). The disease of interest was Emery Dreifuss muscular dystrophy (EDMD), where patients show mutations of components of the nuclear envelope; however, the molecular pathophysiology of the disorder is poorly understood (red box in Figure 12.14).

We then took diagnostic genes from the 6-EDMD node in the tree and queried these in the degeneration/regeneration time series completed in the mouse model (Figure 12.13) and defined a cause–effect transcriptional regulatory pathway that included many of the differentially expressed EDMD-specific transcripts (Figure 12.15) [16]. This defined a model for disease pathogenesis (failure of events during muscle regeneration) that was then tested and validated in a mouse model of EDMD [17].

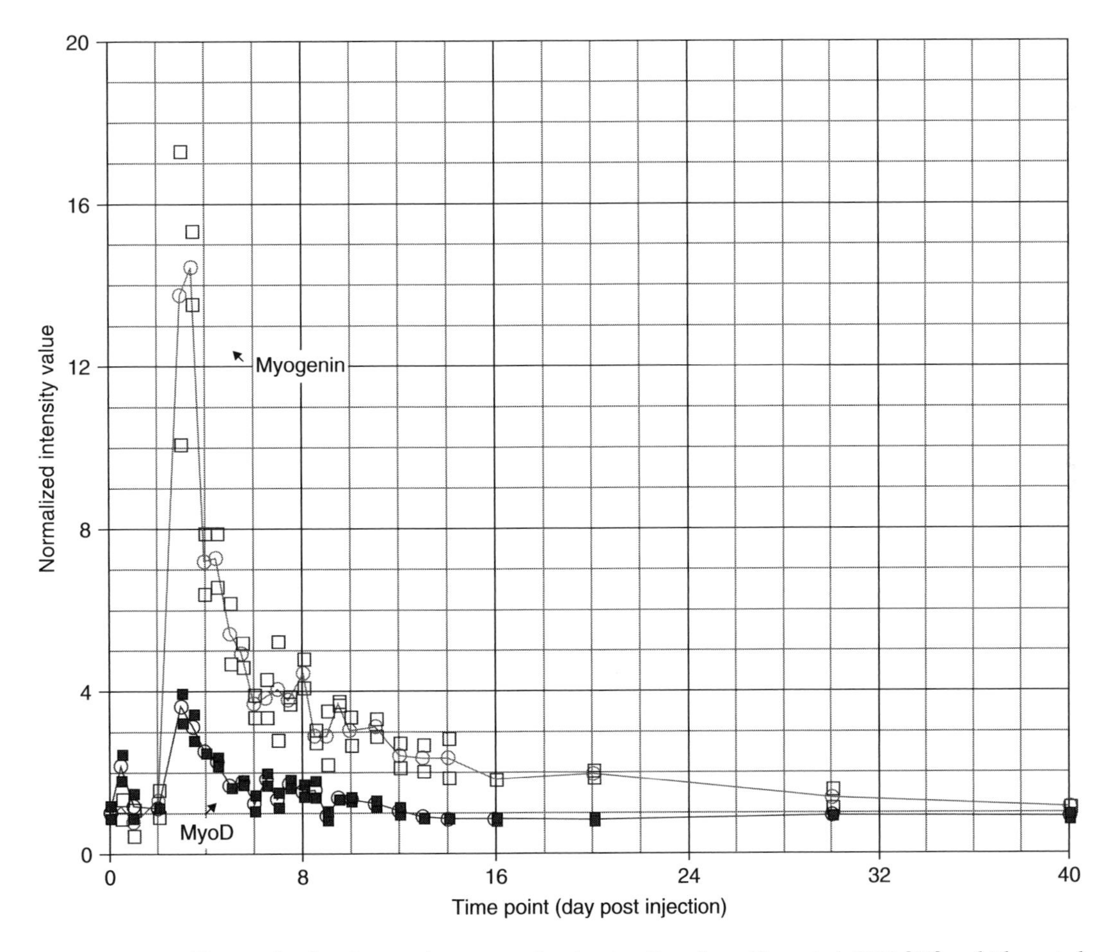

FIGURE 12.13 Time series data in muscle regeneration *in vivo*. Data from Zhao et al. 2002 [18] and Zhao et al. 2003 [19] are publicly available in PEPR (http://pepr.cnmcresearch.org). MyoD shows a peak of expression at 3.0 days, while myogenin shows peak expression 0.5 day later, suggesting that MyoD is upstream of myogenin.

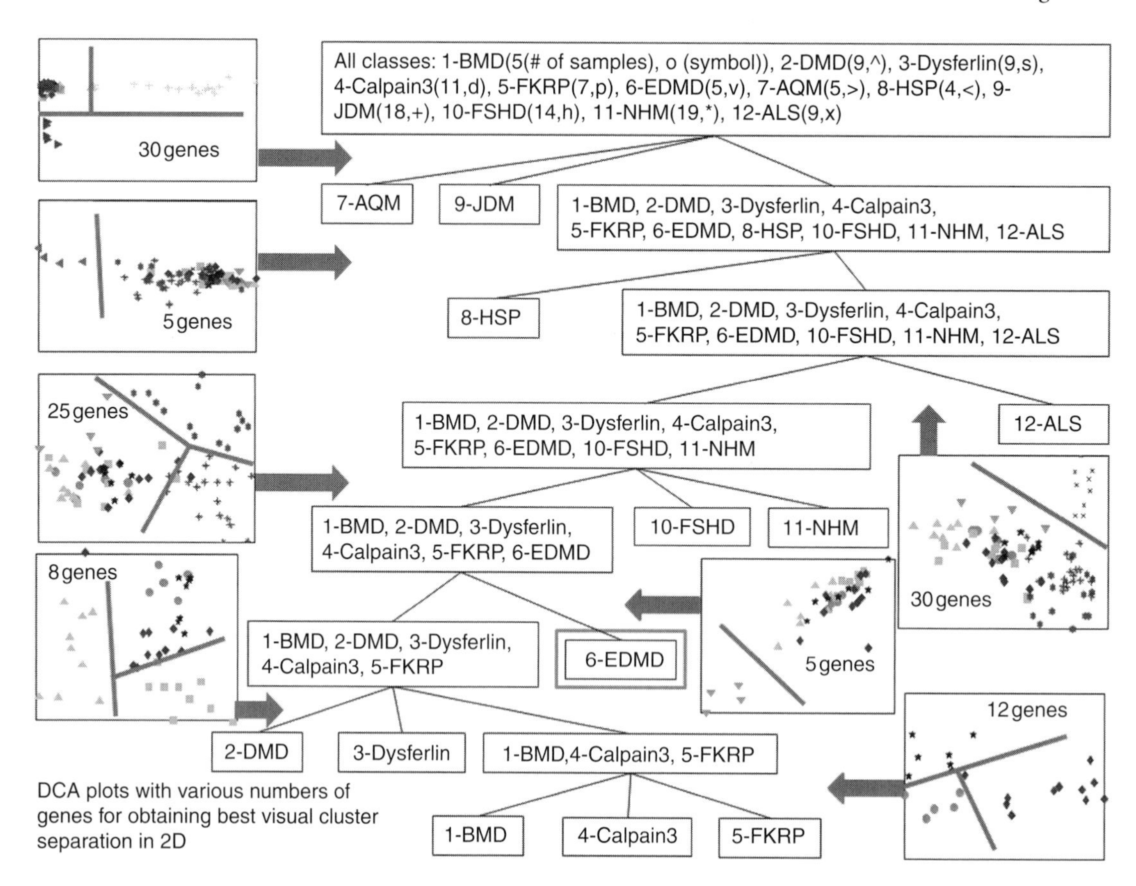

FIGURE 12.14 Diagnostic gene selection in a 12-group data set of muscle biopsies from patients with muscular dystrophy (see Bakay et al. [16]). For a more detailed view of this figure, please visit our companion site at: http://books.elsevier.com/companions/9780123735836.

12.7 Summary

The bioinformatics and biologic computing associated with DNA are quite mature. Hybridization methods allow sensitive and specific identification and quantitation of entire genomes in single microarrays containing millions of features (oligonucleotides at specific addresses on the array). The use of the linear genomic DNA sequence of humans and many other organisms creates an anchor for hundreds of associated databases, such as DNA polymorphisms, mRNA and EST mapping (transcript units, or genes), evolutionary conservation, and others.

The bioinformatics and biologic computing for mRNA is considerably more challenging, because many new variables are introduced such as environmental cues, time, place (in tissue, in body), alternative splicing, and others. There are also different experimental platforms with different amounts of repeated measurement and robust data acquisition and storage intrinsic to each. This makes it much more difficult to define standards or any anchor by which all experiments can be compared. Relatively dense time series data are emerging that begin to define cause/effect pathways, at least with regard to transcriptional regulatory networks.

Proteins are orders of magnitude more complex than mRNA patterns, with posttranslational modifications, subcellular localization, and binding partners all dictating protein activity and function. High throughput proteomics is coming of age with the advent of high resolution MS and associated spectra-matching databases. Proteomic profiling using differentially labeled solutions of peptides is reaching widespread use, but bioinformatics and biologic computing approaches are just beginning to be developed.

Future challenges in biologic computing include defining cell- and tissue-specific pathways and networks and response of networks to environmental and physiologic challenges. A focus will be on integration of DNA, mRNA, and proteomics data sets and databases, with attempts to garner support for established networks while defining new networks through a combination of computational modeling and experimental validation.

Acknowledgments

The authors are grateful for research grant support from the National Institutes of Health (NHLBI Programs in Genomic

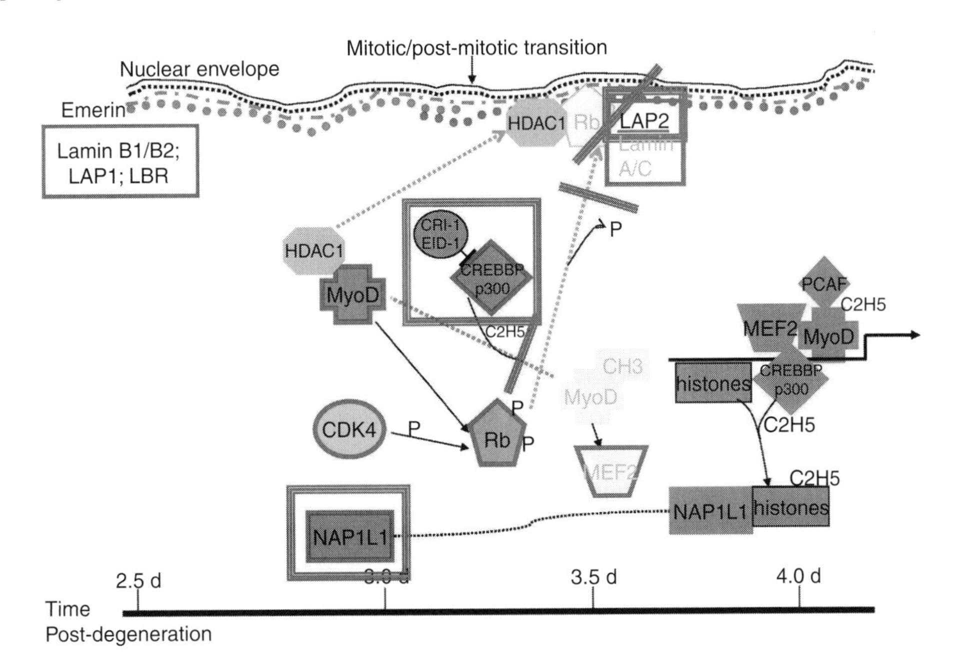

FIGURE 12.15 A model for molecular pathophysiology of a type of muscular dystrophy involving mutations of the nuclear envelope. The X axis (bottom) shows time series data from mouse regeneration (see Figure 12.13). The remainder of the figure shows timed induction of transcriptional pathways during muscle regeneration around the time of the transition from mitotically active cells to postmitotic differentiated myotubes (mitotic/postmitotic transition). The proteins boxed in red are those that are differentially regulated in Emery Dreifuss muscular dystrophy (EDMD) patients, while the red cross-hatches indicate potential blocks in this molecular pathway due to nuclear envelope mutations. Modified from Bakay et al. [16]. For a more detailed view of this figure, please visit our companion site at: http://books.elsevier.com/companions/9780123735836.

Application, NICHD, NINDS, and NIAMS), the Department of Defense CDMRP, and the Foundation to Eradicate Duchenne (FED).

12.8 Exercises

1. How easy is it to study DNA compared to RNA compared to proteins?
2. What factors influence the use of transcript units by cells?
3. Why is it difficult to study proteomics?
4. What are the fundamentals of proteomics?
5. Advantages and disadvantages of 2D gel and shotgun?
6. What are two common misconceptions that can inhibit the development of biological computing?
7. What is a major issue with Affy probe set algorithm with regards to signal/noise? How can one overcome this problem?
8. Describe knowledge networks? Name two commonly used ones.
9. What are some future challenges in biological computing?
10. What is a LIMS system, and what is it used for?

12.9 References and Bibliography

1. J. Kent et al. Genome Browser (http://www.genome.ucsc.edu), 2000.
2. J. Nazarian, Y. Hathout, and E. P. Hoffman. The proteome survey of an electricity-generating organ (Torpedo californica electric organ). *Proteomics.* 7(4):617–627, 2007.
3. H. Parkinson et al. ArrayExpress—A public database of microarray experiments and gene expression profiles. *Nucleic Acids Res.* 35(Database issue):D747–750, 2007.
4. T. Barrett et al. NCBI GEO: Mining tens of millions of expression profiles—Database and tools update. *Nucleic Acids Res.* 35 (Database issue):D760–765, 2007.
5. D. W. Galbraith. The daunting process of MIAME. *Nature.* 444:31, 2006.
6. T. F. Rayner et al. A simple spreadsheet-based, MIAME-supportive format for microarray data: MAGE-TAB. *BMC Bioinformatics.* 7:489, 2006.
7. J. Demeter et al. The Stanford microarray database: Implementation of new analysis tools and open source release of software. *Nucleic Acids Res.* 35(Database issue):D766–770, 2007.

8. J. Chen et al. The PEPR GeneChip data warehouse, and implementation of a dynamic time series query tool (SGQT) with graphical interface. *Nucleic Acids Res.* 32(Database issue):D578–581, 2004.
9. R. R. Almon et al. *In vivo* multi-tissue corticosteroid microarray time series available online at Public Expression Profile Resource (PEPR). *Pharmacogenomics.* 4(6):791–799, 2003.
10. J. Seo and E. P. Hoffman. Probe-set algorithms: Is there a rational best bet? *BMC Bioinformatics.* 7:395–410, 2006b.
11. B. Carvalho et al. Exploration, normalization, and genotype calls of high density oligonucleotide SNP array data. *Biostatistics.* Dec 22, 2006 [Epub ahead of print].
12. R. A. Irizarry, Z. Wu, and H. A. Jaffee. Comparison of affymetrix gene chip expression measures. *Bioinformatics.* 22:789–794, 2006.
13. J. Seo et al. Interactively optimizing signal-to-noise ratios in expression profiling: project-specific algorithm selection and detection p-value weighting in Affymetrix microarrays. *Bioinformatics.* 20(16):2534–2544, 2004.
14. J. Seo, H. Gordish-Dressman, E. P. Hoffman. An interactive power analysis tool for microarray hypothesis testing and generation. *Bioinformatics.* 22:808–814, 2006a.
15. Z. Fang et al. Knowledge guided analysis of microarray data. *J. Biomed. Informatics* 39: 401–411, 2006.
16. M. Bakay et al. Nuclear envelope dystrophies show a transcriptional fingerprint suggesting disruption of Rb–MyoD pathways in muscle regeneration. *Brain* 129(Pt 4):996–1013, 2006.
17. G. Melcon et al. Loss of emerin at the nuclear envelope disrupts the Rb1/E2F and MyoD pathways during muscle regeneration. *Hum. Mol. Genet.* 15:637–651, 2006.
18. P. Zhao et al. Slug is a novel downstream target of MyoD: Temporal profiling in muscle regeneration. *J. Biol. Chem.* 277:20091–20101, 2002.
19. P. Zhao et al. *In vivo* filtering of in vitro expression data reveals MyoD targets. *C. R. Biol.* 326:1049–1065, 2003.

Bibliography

H. Parkinson. Tumor Analysis Best Practices Working Group. Expression profiling—Best practices for data generation and interpretation in clinical trials. *Nat. Rev. Genet.* 5:229–237, 2004.

II

Integrated Applications

13

PACS and Medical Imaging Informatics for Filmless Hospitals

Prof. Brent J. Liu[1] and
Prof. H. K. Huang[1,2,3]
[1] *University of Southern California,*
[2] *Hong Kong Polytechnic University*
[3] *The Chinese Academy of Sciences*

13.1 Introduction

Picture archiving and communication systems (PACS) based on digital, communication, display, and information technologies (IT) have revolutionized the practice of radiology, and in a sense, of the entire clinical continuum in medicine during the past 10 years. This chapter introduces the basic concept, terminology, technology development, implementation, integration, and experiences within the clinical practice. There are many advantages to introducing digital, communications, display, and IT to conventional paper- and film-based operations in radiology and medicine.

13.1.1 The Role of PACS in the Clinical Environment

PACS and IT technologies can be used to improve health care delivery workflow efficiency, resulting in speeding up of health care delivery and reducing operating costs. With all these benefits, the digital, communication, and IT technologies are gradually changing the method of acquiring, storing, viewing, and communicating medical images and related information in the health care industry. One natural development along this line is the emergence of digital radiology departments and the digital health care delivery environment. A digital radiology department has two components: a radiology information management system (RIS) and a digital imaging system. The RIS is a subset of the hospital information system (HIS) or clinical management system (CMS). When these systems are combined with the electronic patient (or medical) record (ePR or eMR) system, which manages selected patient data, the arrival of the total filmless and paperless health care delivery system can become a reality. The digital imaging system, sometimes referred to as a PACS or image management and communication system (IMAC), involves image acquisition, archiving, communication, retrieval, processing, distribution, and display. A digital health care environment consists of the integration of HIS/CMS, ePR, PACS, and other digital clinical systems. The combination of HIS and PACS is sometime referred to as hospital integrated PACS (HI-PACS). The health care delivery system related to PACS and IT is reaching one billion dollars per year (excluding imaging modalities) and continues to grow.

13.1.2 The Role of PACS in Medical Imaging Informatics

PACS originated as an image management system for improving the efficiency of radiology practice. However, it has evolved into a health care enterprise-wide system that integrates information media in multiple forms, including voice, text, medical records, waveform images, and video recordings. To integrate these various data types requires the technology of multimedia: hardware platforms, information systems and databases, communication protocols, display technology, and system interfacing and integration. As the PACS grows in its role within the clinical continuum, it also becomes integrated with these various enterprise-wide media formats and can grow in its richness of content within its database. This wealth of information data becomes the fundamental basis for new approaches in medical research and practice through the discipline of medical imaging informatics, thus ultimately improving the overall health care delivery, research, and education.

13.1.3 General PACS Design

A PACS consists of image and data acquisition, storage, and display subsystems integrated by digital networks and application software. PACS design should emphasize system connectivity. A general multimedia data management system that is easily expandable, flexible, and versatile in its operation calls for both top-down management to integrate various HIS and a bottom-up engineering approach to build a foundation (i.e., PACS infrastructure). From the management point of view, a hospital-wide or enterprise PACS is attractive to administrators because it provides economic justification through a return on investment cost analysis for implementing the system. In addition, proponents of PACS are convinced that its ultimately favorable cost–benefit ratio should not be evaluated as the balance of the resources of the radiology department alone but should extend to the entire hospital or enterprise operation. This concept has gained momentum. Many hospitals and some enterprise-level health care entities around the world have implemented large-scale PACS and have provided solid evidence that PACS improves the efficiency of health care delivery and at the same time saves hospital operational costs. From the engineering point of view, the PACS infrastructure is the basic design concept to ensure that PACS includes features such as standardization, open architecture, expandability for future growth, connectivity, reliability, fault tolerance, and cost-effectiveness. This design philosophy can be constructed in a modular fashion with the infrastructure design described in Section 13.2.

13.1.4 Chapter Overview

This chapter will first describe the PACS infrastructure and its various components in detail. The latter half of the chapter will conclude with implementation and integration strategies for installing a PACS within a health care environment as well as clinical experiences derived from various health care institutions' PACS process.

13.2 PACS Infrastructure

13.2.1 Introduction to PACS Infrastructure Design

The PACS infrastructure design provides the necessary framework for the integration of distributed and heterogeneous imaging devices while supporting intelligent database management of all patient-related information. With this infrastructure, it offers an efficient means of viewing, analyzing, and documenting study results and provides a distribution method for effectively communicating study results to referring physicians. The PACS infrastructure consists of a basic skeleton of hardware components (imaging device interfaces, storage devices, host computers, communication networks, and display systems) integrated with a standardized and robust software system with flexibility for communication, database management, storage management, job scheduling, interprocessor communication, error handling, and network monitoring. The infrastructure as a whole is versatile and can incorporate rules to reliably perform not only basic PACS management operations but also more complex research, clinical service, and educational requests. The software modules of the infrastructure use the ability to handshake and communicate at a system level to permit the components to work together as a system rather than as individual networked computers.

The corresponding hardware components of the general PACS infrastructure include patient data servers, imaging

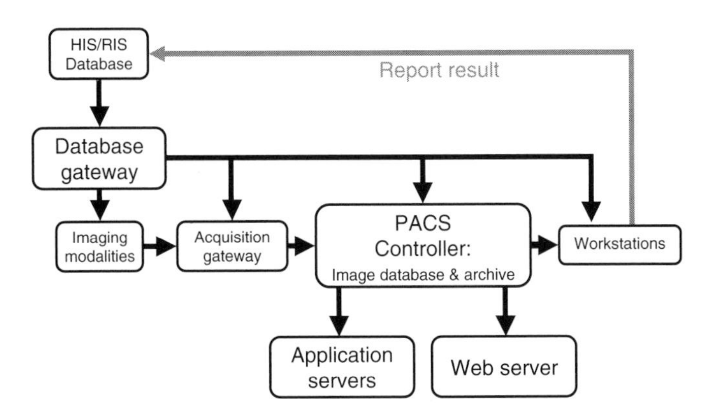

FIGURE 13.1 Generic picture archiving and communication systems components and data flow.

modalities, data/modality interfaces, PACS controllers with database and archive, and display workstations connected by communication networks for handling the data/image flow in the PACS and tuned for a more efficient clinical workflow. Image and data stored in the PACS can be extracted from the archive and transmitted to application servers for various uses. Figure 13.1 shows the basic components and data flow of the PACS. This diagram will be expanded to present additional details in later chapters. The PACS application server concept shown in the bottom of Figure 13.1 broadens the role of PACS in the health care delivery system as a contributor to the advancement of medical imaging informatics field during the past several years. The Web server is optional and is used to distribute PACS studies through wide area networks (WAN) to clinics and physician's offices. Sometimes the Web server is used within the health care enterprise local area network (LAN) to distribute PACS studies throughout the hospital or health care institution.

13.2.2 Industry Standards

Transmission of images and textual information between health care information systems has always been challenging for two major reasons. First, information systems use different computer platforms, and second, images and data are generated from various imaging modalities by made different manufacturers. With the emergent health care industry standards, Health Level 7 (HL7) and Digital Imaging and Communications in Medicine (DICOM), it has become feasible to integrate these heterogeneous, disparate medical images and textual data into an organized system. Interfacing two health care components requires two ingredients, a common data format and a communication protocol. HL7 is a standard textual data format, whereas DICOM includes image and textual data format and communication protocols. In conforming to the HL7 standard, it is possible to share health care information between the HIS, the RIS, and the PACS. By adapting the DICOM standard, medical images generated from a variety of modalities and manufacturers can be interfaced as an integrated health care system. These two standards will be discussed in more detail in the following paragraphs. Integrating health care enterprise (IHE), which is a model for driving the adoption of standards, will also be addressed. IHE combines these available standards with clinical workflow profiles to persuade users and manufacturers to adopt and use this system in daily clinical practice.

13.2.2.1 Health Level 7

HL7, established in March 1987, was organized by a user-vendor committee to develop a standard for electronic data exchange in health care environments, particularly for hospital applications. With the HL7 standard, the level 7 refers to the highest level, which is the application level, in the open systems interconnection (OSI) seven communication levels model. The common goal is to simplify the interface implementation among computer applications from multiple vendors. This standard emphasizes data format and protocol for exchanging certain key textual data among health care information systems, such as HIS, RIS, and PACS. HL7 addresses the highest level (level 7) of the OSI model of the International Standards Organization (ISO), but it does not conform specifically to the defined elements of the OSI's seventh level. It conforms to the conceptual definitions of an application-to-application interface placed in the seventh layer of the OSI model. These definitions were developed to facilitate data communication in a health care setting by providing rules to convert abstract messages associated with real-world events into strings of characters comprising an actual message.

The most commonly used HL7 today is version 2.X, which has many options and is thus flexible. During the past years, version 2.X has been developed continuously, and it is widely and successfully implemented in the health care environment. Version 2.X and other older versions use a bottom-up approach, beginning with very general concepts and adding new features as needed. These new features become options to the implementers so that the standard is very flexible and is easy to adapt to different sites. However, these options and flexibility also make it impossible to have reliable conformance tests of any vendor's implementation. This forces vendors to spend more time in analyzing and planning their interfaces to ensure that the same optional features are used in both interfacing parties. There is also no consistent view of the data when HL7 moves to a new version or when assessing that data's relationship to other data. Therefore, a consistently defined and object-oriented version of HL7 is needed, which is version 3. The initial release of HL7 version 3 was in December 2001. The primary goal of HL7 version 3 is to offer a standard that is definite and testable. Version 3 uses an object-oriented methodology and a reference information model (RIM) to create HL7 messages. The object-oriented method is a top-down method. RIM is the backbone of HL7 version 3, since it provides an explicit representation of the semantic and lexical connections between the information in the fields of HL7 messages. Because each aspect of the RIM is well-defined, very few options exist in version 3. Through object-oriented method and RIM, HL7 version 3 improves many of the shortcomings of previous 2.X versions. Version 3 uses extensible markup language for message encoding to increase interoperability between systems and will include new data interchange formats beyond the American Standard Code for Information Interchange (ASCII) and support of component-based technology such as ActiveX and CORBA. HL7 version 3 will offer tremendous benefits to providers and vendors as well as analysts and programmers, but complete adoption of the new standard will take time and effort.

13.2.2.2 Digital Imaging and Communications in Medicine Standard

ACR-NEMA, formally known as the American College of Radiology and the National Electrical Manufacturers Association, created a committee to develop a set of standards to serve as the common ground for various medical imaging equipment vendors. The goal was for newly developed instruments to be able to communicate and participate in sharing medical image information, in particular within the PACS environment. The committee, which focused chiefly on issues concerning information exchange, interconnectivity, and communications between medical systems, began work in 1982.The first version, which emerged in 1985, specified standards in point-to-point message transmission, data formatting, and presentation and included a preliminary set of communication commands and a data format dictionary. The second version, ACR-NEMA 2.0, published in 1988, was an enhancement to the first release. It included hardware definitions and software protocols as well as a standard data dictionary. However, networking issues were not addressed adequately in either version. For this reason, a new version aiming to include network protocols was released in 1992. Because of the number of changes and additions, it was given a new name: DICOM 3.0. In 1996, a new version was released consisting of 13 published parts that form the basis of future DICOM new versions and parts. Manufacturers readily adopted this version to their imaging products. Currently, the latest version of DICOM has been expanded to 18 parts. Two fundamental components of DICOM are the information object class and the service class. Information objects define the contents of a set of images and their relationships, and the service classes describe what to do with these objects. The service classes and information object classes are combined to form the fundamental units of DICOM, called service-object pairs (SOPs). The next few paragraphs will describe the DICOM data model, which represents the information object, and the DICOM service classes.

13.2.2.3 DICOM Data Model

There are two components relating to the DICOM data model: the DICOM model of the real world and the DICOM file format. The former is used to define the hierarchical data structure from patient to studies, to series, and to images and waveforms. The latter describes how to encapsulate a DICOM file ready for a DICOM SOP service.

The DICOM model of the real world defines several real-world objects in the clinical imaging arena (e.g., patient, study, series, image) and their interrelationships within the scope of the DICOM standard. It provides a framework for various DICOM information object definitions (IOD). The DICOM Model defines four level objects: (1) patient; (2) study; (3) series and equipment; and (4) image, waveform, and structured report document. Each of the above levels can contain several (1–n or 0–n) sublevels. Figure 13.2 shows the DICOM real-world data model. Note the levels with which the above-mentioned four objects reside.

The DICOM file format defines how to encapsulate the DICOM data set of a SOP instance in a DICOM file. Each file usually contains one SOP instance. The DICOM file starts with the DICOM file meta information (optional), followed by the bit stream of the data set and ends with the image pixel data if it is a DICOM image file. The DICOM file meta information includes file identification information. The meta information uses explicit value representations (VR) transfer syntax for encoding. Therefore, the meta information does not exist in the implicit VR-encoded DICOM file. Explicit VR and implicit VR are two coding methods in DICOM. Vendors or implementers have the option of choosing either one for encoding. DICOM files encoded by both coding methods can be processed by most of the DICOM-compliant software. One data set represents a single SOP instance. A data set is constructed of data elements. Data elements contain the encoded values of the attributes of the DICOM object. If the SOP instance is an image, the last part of the DICOM file is the image pixel data.

13.2.2.4 DICOM Service Classes

DICOM services are used for communication of imaging information objects within a device and for the device to perform a service for the object, for example, to store the object or to display the object. A service is built on top of a set of DICOM message service elements (DIMSEs). These DIMSEs are computer software programs written to perform specific functions. There are two types of DIMSEs: one for the normalized objects and the other for the composite objects. DIMSEs are paired in the sense that a device issues a command request and the receiver responds to the command accordingly. The composite commands are generalized, whereas the normalized commands are more specific. DICOM services are referred to as service classes because of the object-oriented nature of its information structure model. If a device provides a service, it is called a

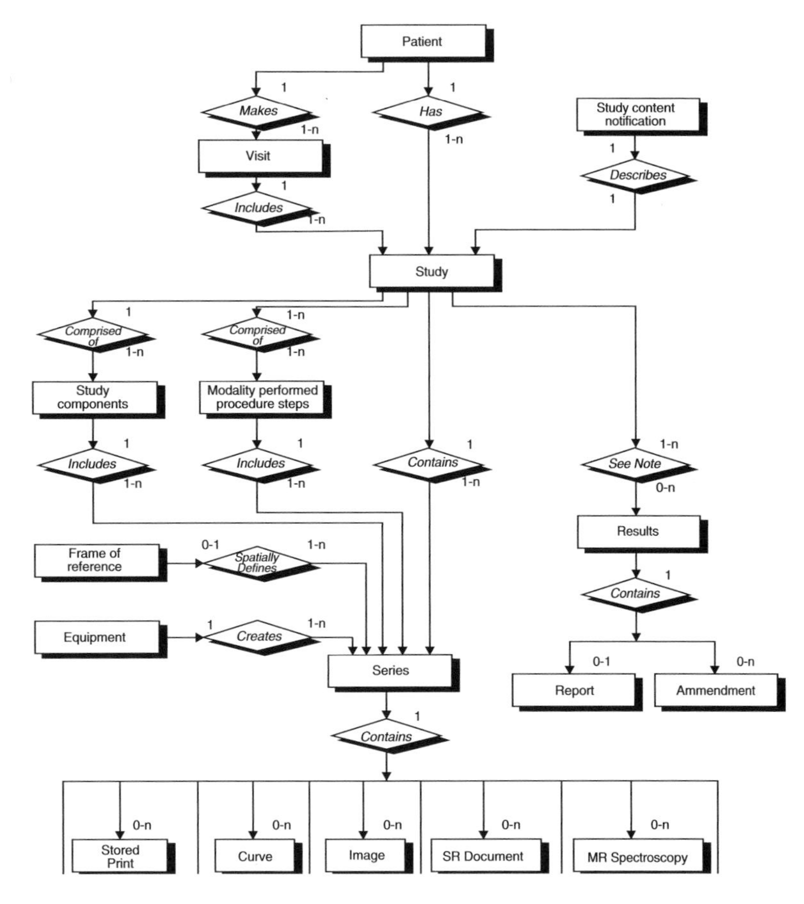

FIGURE 13.2 DICOM model of the real world showing the four main level objects: (1) patient, (2) study, (3) series, and (4) image. Note that there can be multiple instances of each object belonging to the patient.

service class provider; if it uses a service, it is a service class user. Note that a device can be either a service class provider or a service class user or both, depending on how it is used.

DICOM uses existing network communication standards based on the ISO-OSI for imaging information transmission. The ISO-OSI consists of seven layers from the lowest physical (cables) layer to the highest application layer. When imaging information objects are sent between layers in the same device, the process is called a service. When objects are sent between two devices, it is called a protocol. When a protocol is involved, several steps are invoked in two devices; the two devices are referred to as in association using DICOM. If an imaging device transmits an image object with a DICOM command, the receiver must use a DICOM command to receive the information. On the other hand, if a device transmits a DICOM object with a Transmission Control Protocol/Internet Protocol (TCP/IP) communication protocol through a network without invoking the DICOM communication, any device connected to the network can receive the data with the TCP/IP protocol. However, a decoder is still needed to convert the DICOM object for proper use. The most commonly used communication protocol in DICOM is TCP/IP for transmitting DICOM image objects within PACS. To an end user, the two most important DICOM services are (1) send and receive images and (2) query and retrieve images. The query and retrieve services are built on top of the send and receive services.

13.2.2.5 Integrating the Health Care Enterprise

Even with the DICOM and HL7 standards available, there is a need for common consensus on how to use these standards for integrating heterogeneous health care information systems smoothly. IHE is neither a standard nor a certifying authority; instead, it is a high-level information model for driving the adoption of HL7 and DICOM standards. IHE is a joint initiative of RSNA and HIMSS (Health Care Information and Management Systems Society) started in 1998. The mission was to define and guide manufacturers to use DICOM- and HL7-compliant equipment and information systems to facilitate daily clinical workflow operations. The IHE technical framework defines a common information model and vocabulary for using DICOM and HL7 to complete a set of well-defined radiological and clinical transactions for a certain task. These common vocabularies and models would then facilitate health care providers and technical personnel in understanding each other better, which then would lead to smooth systems integration. The first large-scale demonstration was held at the RSNA annual meeting in 1999. Additional presentations were made at RSNA in 2000 and 2001 and at HIMSS 2001 and 2002. In these demonstrations, manufacturers came together to show how actual products could be integrated based on certain IHE protocols. It is the belief of RSNA and HIMSS that with successful adoption of IHE, life would become more pleasant in health care systems integration for both the users and the providers. The IHE integration profiles provide a common language, vocabulary, and platform for health care providers and manufacturers to discuss integration needs and the integration capabilities of products. As of the 2003 implementation, there are 10 integration profiles; this number will grow over time. The 10 implemented IHE profiles are as follows:

1. Scheduled workflow
2. Patient information reconciliation
3. Consistent presentation of images
4. Presentation-grouped procedures
5. Access to radiology information
6. Key image note
7. Simple image and numeric report
8. Basic security
9. Charge posting
10. Postprocessing workflow

13.2.3 Connectivity and Open Architecture

If PACS modules in the same hospital cannot communicate with each other, they become isolated systems, each with its own images and patient information. It would be difficult to combine these modules to form a total hospital-integrated PACS. Open network design is essential, allowing a standardized method for data and message exchange between heterogeneous systems. Because computer and communications technology changes rapidly, a closed architecture would hinder system upgradeability. For example, suppose an independent imaging workstation from a given manufacturer would, at first glance, make a good additional component to a magnetic resonance imaging scanner for viewing images. If the workstation has a closed proprietary architecture design, however, no components except those specified by the same manufacturer can be augmented to the system. Potential overall system upgrading and improvement would be limited. Considerations of connectivity are important even when a small-scale PACS is planned.

13.2.4 Reliability

Reliability is a major concern in a PACS for two reasons. First, a PACS has many components; the probability of a component failing is high. Second, because PACS manages and displays critical patient information, extended periods of downtime cannot be tolerated. The PACS can be considered a mission-critical system within the health care enterprise that should strive for continuous operation 24 hours a day, 7 days a week. In designing a PACS, it is therefore important to use fault-tolerant measures, including error detection and logging software, external auditing programs (i.e., network management processes that check network circuits, magnetic disk space, database status, processer status, and queue status), hardware redundancy, and intelligent software recovery blocks. Some fail-recovery mechanisms that can be used include automatic retry of failed jobs with alternative resources and algorithms and intelligent bootstrap routines (a software block executed by a computer when it is restarted) that allow a PACS computer to automatically continue operations after a power outage or system failure. Improving reliability is costly; however, it is essential to maintain high reliability of a complex system.

13.2.5 Security

Security, particularly the need for patient confidentiality, is an important consideration because of medical-legal issues and Health Insurance Portability and Accountability Act (HIPAA) mandated in April 2003. The violation of data security can be of three different types: physical intrusion, misuse, and behavioral violations. Physical intrusion relates to facility security, which can be handled by building management. Misuse and behavioral violations can be minimized by account control and privilege control. Most sophisticated database management systems have identification and authorization mechanisms that use accounts and passwords. Application programs may supply additional layers of protection. Privilege control refers to granting and revoking user access to specific tables, columns, or views from the database. These security measures provide the PACS infrastructure with a mechanism for controlling access to clinical and research data. With these mechanisms, the system designer can enforce policy as to which persons have access to clinical studies. In some hospitals, for example, referring clinicians are granted image study access only after a preliminary radiology reading has been performed and attached to the image data. An additional security measure is the use of the image digital signature during data communication. If implemented, this feature would increase the system software overhead, but data transmission through open communication channels would be more secure.

13.2.6 Current PACS Architectures

There are three basic PACS architectures: (1) stand-alone, (2) client/server, and (3) Web-based. From these three basic PACS architectures, there are variations and hybrid design types.

13.2.6.1 Stand-Alone PACS Architecture

The three major features of the stand-alone model are as follows:

1. Images are automatically sent to designated reading and review workstations from the archive server.
2. Workstations can also query/retrieve images from the archive server. Workstations have short-term cache storage.
3. Data workflow of the stand-alone PACS model is shown in Figure 13.3.

Following the numerals in Figure 13.3:

1. Images from an examination acquired by the imaging modality are sent to the PACS archive server.
2. The PACS archive server stores the examination images.
3. A copy of the images is distributed to selected end-user workstations for diagnostic reading and review. The server performs this automatically.
4. Historic examinations are prefetched from the server, and a copy of the images is sent to selected end-user workstations.
5. Ad hoc requests to review PACS examinations are made via query/retrieve from the end-user workstations. In addition, if automatic prefetching fails, end-user workstations can query and retrieve the examination images from the archive server.
6. End-user workstations contain a local storage cache of a finite number of PACS examinations.

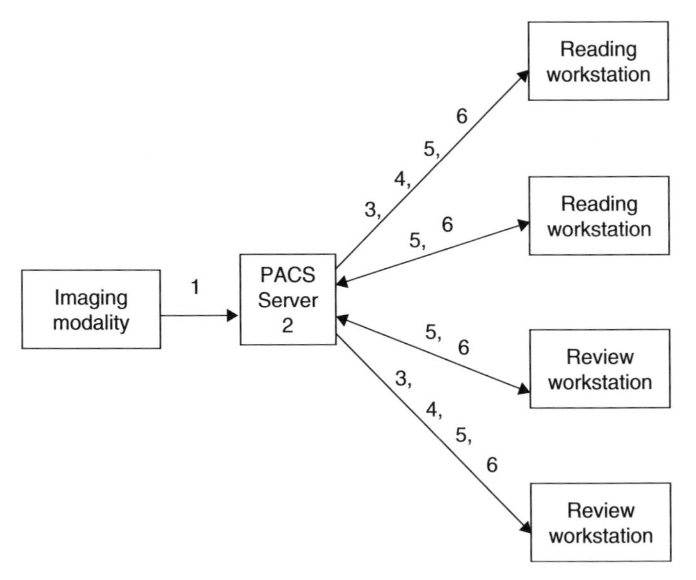

FIGURE 13.3 Stand-alone picture archiving and communication systems architecture and the six workflow steps as described in the text.

The advantages of the stand-alone model are as follows:

- If the PACS server goes down, imaging modalities or acquisition gateways have the flexibility to send directly to the end-user workstation so that the radiologist can continue reading new cases.
- Because multiple copies of the PACS examination are distributed throughout the system, there is less risk of losing PACS data. Some historic PACS examinations will be available in workstations because they have a local storage cache.
- The system is less susceptible to daily changes in network performance because PACS examinations are preloaded onto the local storage cache of end-user workstations and are available for viewing immediately.
- Examination modification to the DICOM header for quality control can be made before archiving.

The disadvantages of the stand-alone system are as follows:

- End-users must rely on correct distribution and prefetching of PACS examinations, which is not possible all the time.
- Because images are sent to designated workstations, each workstation may have a different wordlist, which makes it inconvenient to read/review all examinations at any workstation in one setting.
- End-users depend on the query/retrieve function to retrieve ad hoc PACS examinations from the archive, which can be a complex function compared with the client/server model.
- Radiologists can be reading the same PACS examination at the same time from different workstations because the examination images may be sent to several workstations.

13.2.6.2 Client/Server PACS Architecture

The three major features of the client/server model are:

1. Images are centrally archived at the PACS server.
2. From a single wordlist at the client workstation, an end-user selects images via the archive server.
3. Because workstations have no cache storage, images are flushed after reading.

Data workflow of the client/server PACS model is shown in Figure 13.4.

Following the numerals in Figure 13.4:

1. Images from an examination acquired by the imaging modality are sent to the PACS archive server.
2. The PACS archive server stores the examination.
3. End-user workstations or client workstations have access to the entire patient/study database of the archive server.

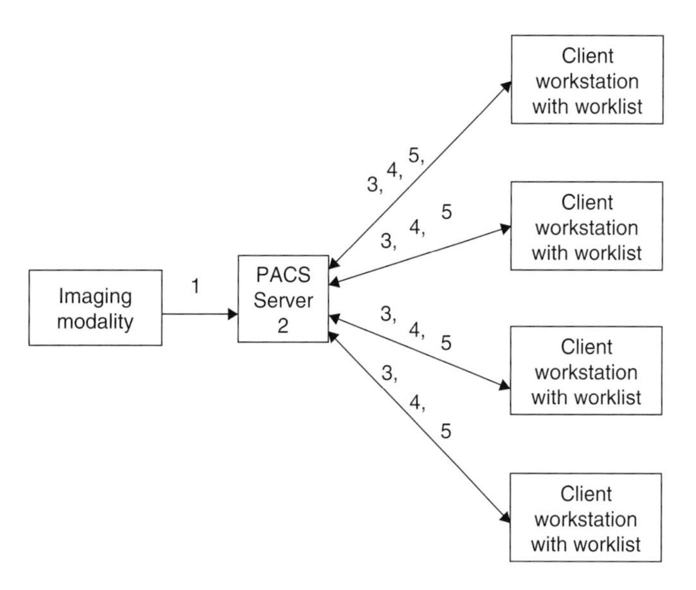

FIGURE 13.4 Client/server picture archiving and communication systems architecture and the five workflow steps as described in the text.

The end-user may select preset filters on the main wordlist to shorten the number of wordlist entries for easier navigation.

4. Once the examination is located on the wordlist and selected, images from the PACS examination are loaded from the server directly into the memory of the client workstation for viewing. Historic PACS examinations are loaded in the same manner.

Once the end-user has completed reading/reviewing the examination, the image data are flushed from memory, leaving no image data in local storage on the client workstation

The advantages of the client/server model are as follows:

- Any PACS examination is available on any end-user workstation at any time, making it convenient to read/review.
- No prefetching or study distribution is needed.
- No query/retrieve function is needed. The end-user just selects the examination from the wordlist on the client workstation, and images are loaded automatically.
- Because the main copy of a PACS examination is located on the PACS server and is shared by the client workstations, radiologists will be aware of when they are reading the same examination at the same time and thus avoid duplicate readings.

The disadvantages of the client/server model are as follows:

- The PACS server is a single point of failure; if it goes down, the entire PACS is down. In this case, end-users will not be able to view any examinations on the client workstations. Newly acquired examinations must be held back from archival at the modalities until the server is back up.
- Because there are more database transactions in the client/server architecture, the system is exposed to more transaction errors, making it less robust compared with the stand-alone architecture.
- The architecture is very dependent on network performance.
- Examination modification to the DICOM header for quality control is not available before archiving.

13.2.6.3 Web-Based Model

The Web-based model PACS is similar to the client/server architecture with regard to data flow. However, the main difference is that the client software is a Web-based application.

Additional advantages as compared with client/server are the following:

- The client workstation hardware can be platform-independent as long as the Web browser is supported.
- The system is a completely portable application that can be used both on-site and at home with an Internet connection.

Additional disadvantages as compared with client/server are as follows:

- The system may be limited in the amount of functionality and performance by the Web browser.

With consistent technology, hardware, and software improvements to database management and performance, clustered and parallel servers, and network communications performance, the client/server and Web-based models have become the architecture of choice for most PACS vendors.

13.3 PACS Components and Workflow

13.3.1 Introduction of Components

This section provides an overview of PACS for two topics. The first topic is the basic concept of PACS and its components, which gives a general architecture and requirements of the system. The second topic is an example of a generic PACS workflow in radiology that highlights the functionalities of these components. As discussed in the previous section, a PACS should be DICOM-compliant. It consists of an image and data acquisition gateway, a PACS controller and archive, and display workstations integrated together by digital networks as shown in Figure 13.1. The following sections introduce these components in more detail.

13.3.2 Image Acquisition Gateway

PACS requires that images from imaging modalities (devices) and related patient data from the HIS and RIS be sent to the PACS controller and archive server. A major task in PACS is to acquire images reliably and in a timely manner from each radiological imaging modality and relevant patient data including study support text information about the patient, a description of the study, and parameters pertinent to image acquisition and processing.

Image acquisition is a major task for three reasons. First, the imaging modality is not under the auspices of the PACS. Many manufacturers supply various imaging modalities, each of which has its own DICOM-compliant statement. Worse, some older imaging modalities may not even be DICOM-compliant. To connect many imaging modalities to the PACS requires tedious and labor-intensive work and the cooperation of modality manufacturers. Second, image acquisition is a slower operation than other PACS functions because patients are involved, and it takes the imaging modality some time to acquire the necessary data for image reconstruction. Third, images and patient data generated by the modality sometimes may contain format information unacceptable to the PACS operation. To circumvent these difficulties, an image acquisition gateway computer is usually placed between the imaging modality(s) and the rest of the PACS network to isolate the host computer in the radiological imaging modality from the PACS. Isolation is necessary because traditional imaging device computers lack the necessary communication and coordination software that is standardized within the PACS infrastructure. Furthermore, these host computers do not contain enough intelligence to work with the PACS controller to recover various errors. The image acquisition gateway computer has three primary tasks: It acquires image data from the radiological imaging device; it converts the data from manufacturer specifications to a PACS standard format (header format, byte ordering, matrix sizes) that is compliant with the DICOM data formats; and it forwards the image study to the PACS controller or display workstations.

Two types of interfaces are used to connect a general-purpose PACS acquisition gateway computer with a radiological imaging modality. With peer-to-peer network interfaces, which use the TCP/IP ethernet protocol, image transfers can be initiated either by the radiological imaging modality (a push operation) or by the destination PACS acquisition gateway computer (a pull operation). The pull mode is advantageous because if an acquisition gateway computer goes down, images can be queued in the radiological imaging modality computer until the gateway computer becomes operational again, at which time the queued images can be pulled and normal image flow resumed. Assuming that sufficient data buffering is available in the imaging modality computer, the pull mode is the preferred mode of operation because an acquisition computer can be programmed to reschedule study transfers if failure occurs (because of failure of the acquisition computer or failure in the radiological imaging modality). If the designated acquisition gateway computer is down and a delay in acquisition is not acceptable, images from the examination can be rerouted to another networked designated backup acquisition gateway computer or workstation.

Although traditionally the image acquisition gateway is a separate computer device within PACS, improvements in server hardware processing speed and memory have provided some manufacturers with the ability to integrate the image acquisition gateway component within the PACS controller or PACS server. Although the image acquisition gateway shares the same hardware as the PACS controller, the main functionalities remain the same as a standalone image acquisition gateway.

13.3.3 PACS Controller and Image Archive

Imaging examinations along with pertinent patient information from the acquisition gateway computer, the HIS, and the RIS are sent to the PACS controller. The PACS controller is the engine of the PACS and consists of high-end computers or servers; its two major components are a database server and an archive system. The archive system consists of short-term, long-term, and permanent storage. These components are explained in more detail in the next section.

The following lists some major functions of a PACS controller:

1. Receives images from examinations via acquisition gateway computers
2. Extracts text information describing the received examination
3. Updates a network-accessible database management system
4. Determines the destination workstations to which newly generated examinations are to be forwarded
5. Automatically retrieves necessary comparison images from a distributed cache storage or long-term library archive system
6. Automatically corrects the orientation of computed radiography images
7. Determines optimal contrast and brightness parameters for image display
8. Performs image data compression if necessary
9. Performs data integrity check if necessary
10. Archives new examinations onto long-term archive library
11. Deletes images that have been archived from acquisition gateway computers

12. Services query/retrieve requests from workstations and other PACS controllers in the enterprise PACS
13. Interfaces with PACS application servers

13.3.4 Display Workstations

A workstation includes communication network connection, local database, display, resource management, and processing software. The fundamental workstation operations are listed in Table 13.1. There are four types of display workstations categorized by their resolutions: (1) high-resolution (2.5K × 2K) liquid crystal display (LCD) for primary diagnosis at the radiology department, (2) medium-resolution (2000 × 1600 or 1600 × 1K) LCD for primary diagnosis of sectional images and at the hospital wards, (3) physician desktop workstation (1K × 768) LCD, and (4) hard-copy workstations for printing images on film or paper. In a stand-alone primary diagnostic workstation, current and historic images are stored in local high-speed magnetic disks for fast retrieval. It also has access to the PACS controller database for retrieving longer-term historic images if needed. Figures 13.5–13.7 show examples of a typical PACS diagnostic workstation displaying various PACS studies. Note the tool set at the bottom of each figure used for manipulating the digital PACS study for case presentation, interpretation, and documentation.

13.3.5 Communications and Networking

A basic function of any computer network is to provide an access path by which end users (e.g., radiologists and clinicians) at one geographic location can access information (e.g., images and reports) at another location. The important networking data needed for system design include location and function of each network node, frequency of information passed between any two nodes, cost for transmission between nodes with various-speed lines, desired reliability of the communication, and required throughput. The variables in the design include the network topology, communication line capacities, and data flow assignments.

TABLE 13.1 Major functions of a picture archiving and communication systems display workstation

Function	Description
Case preparation	Accumulation of all relevant images and information belonging to a patient examination
Case selection	Selection of cases for a given subpopulation
Image arrangement or hanging protocols	Tools for arranging and grouping images for easy review
Interpretation	Measurement tools for facilitating the diagnosis
Documentation	Tools for image annotation, text, and voice reports
Case presentation	Tools for a comprehensive case presentation
Image reconstruction	Tools for various types of image reconstruction for proper display

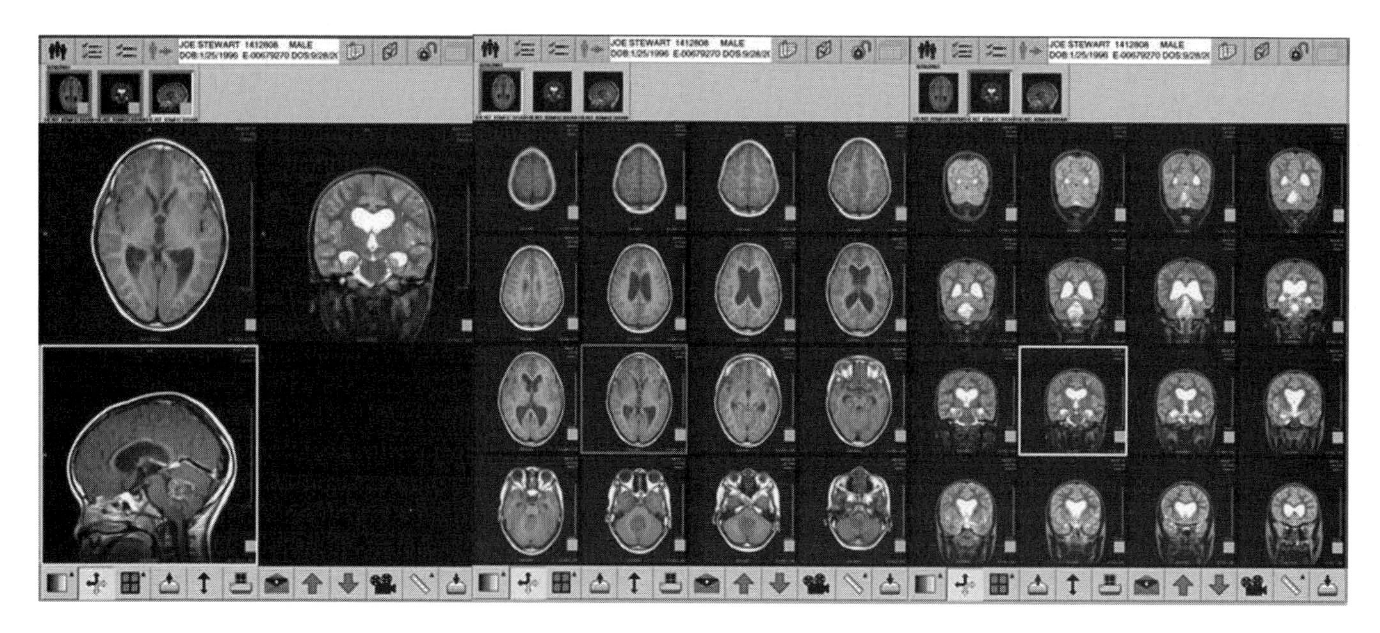

FIGURE 13.5 Example of a picture archiving and communication systems diagnostic workstation displaying a magnetic resonance brain examination.

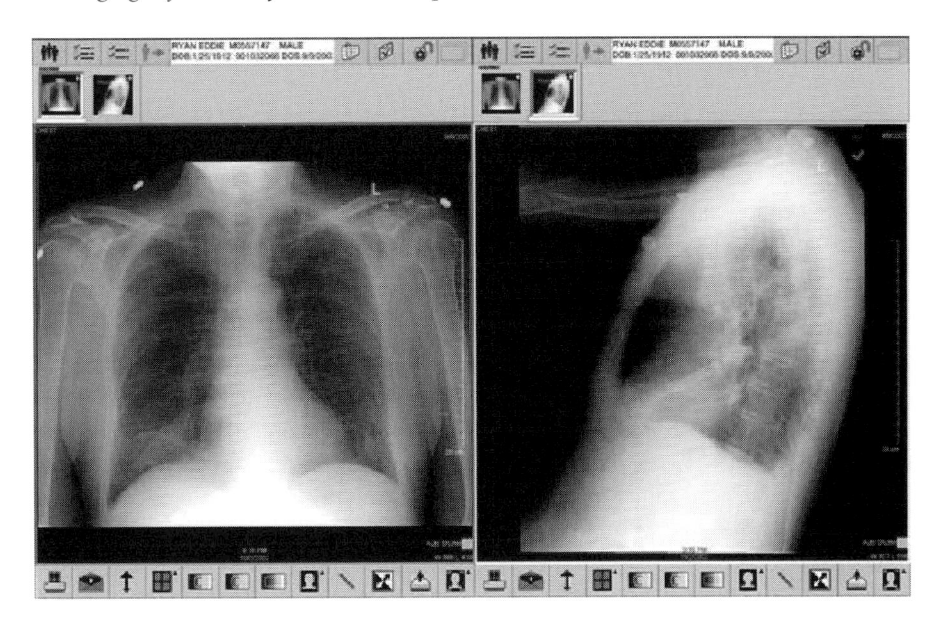

FIGURE 13.6 Example of a picture archiving and communication systems diagnostic workstation displaying a computed radiography chest examination.

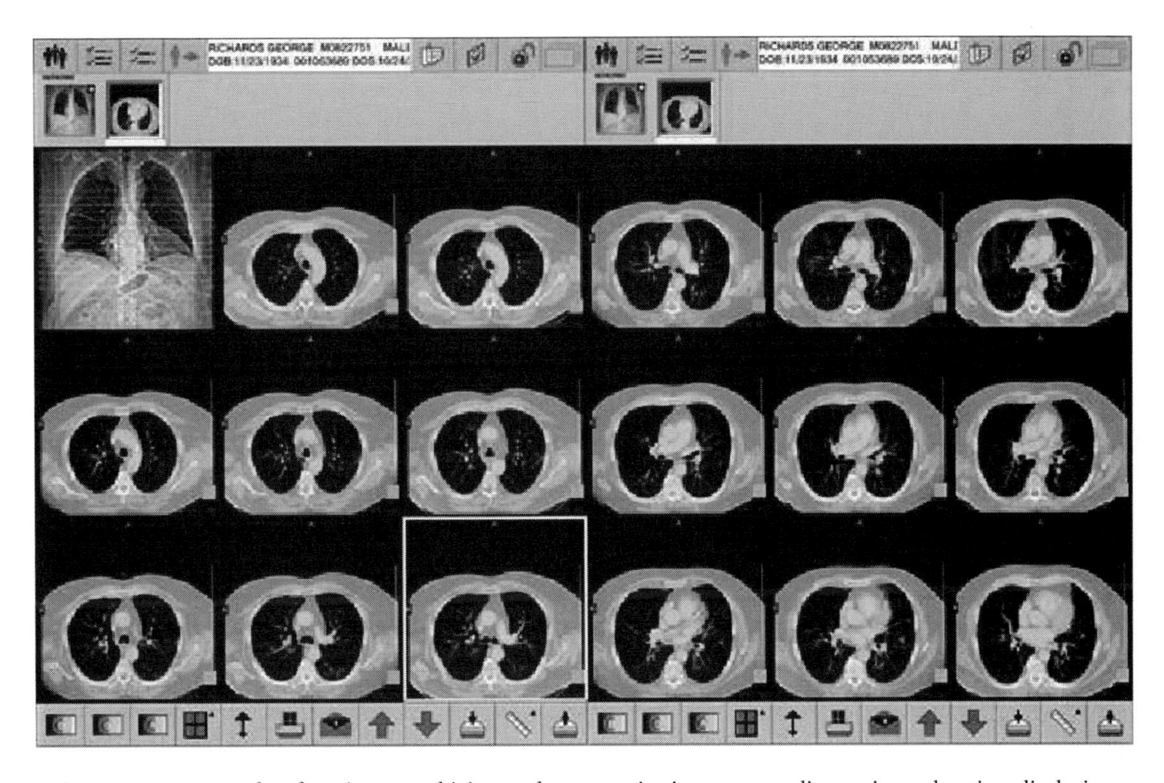

FIGURE 13.7 Example of a picture archiving and communication systems diagnostic workstation displaying a computed radiography chest examination.

At the LAN level, digital communication in the PACS infrastructure design can consist of low-speed Ethernet (10 megabits/s signaling rate), medium-speed (100 megabits/s), or fast-speed (1 gigabit/s) Ethernet, and high-speed asynchronous transfer mode (ATM) technology (155–622 megabits/s and up). In a WAN, various digital service (DS) speeds can be used, which range from DS-0 (56 kilobits/s) and DS-1 (T1, 1.544 megabits/s) to DS-3 (45 megabits/s) and ATM (155–622 megabits/s). There is a trade-off between transmission speed and cost.

The network protocol used should be standard, for example, the TCP/IP and DICOM communication protocol (a higher level of TCP/IP). A low- or medium-speed network is used to connect the imaging modalities (devices) to the acquisition gateway computers because the time-consuming processes during imaging acquisition do not require high-speed connection. Sometimes, several segmented local area Ethernet branches may be used in transferring data from imaging devices to acquisition gateway computers. Medium- and high-speed networks are used on the basis of the balance of data throughput requirements and costs. A faster image network is used between acquisition gateway computers and the PACS controller because several acquisition computers may send large image files to the controller at the same time. High-speed networks are always used between the PACS controller and workstations. It is even more crucial to have high-speed networks to support the client/server PACS architecture because the PACS workstation is highly dependent on data transfer of images from the PACS controller to the PACS workstation's local memory, with performance expectations similar to PACS workstations from a stand-alone architecture where the images are located on the local workstation's hard disk storage.

Process coordination between tasks running on different computers connected to the network is an extremely important issue in system networking. This coordination of processes running either on the same computer or on different computers is accomplished by using interprocessor communication methods with socket-level interfaces to TCP/IP. Commands are exchanged as ASCII messages to ensure standard encoding of messages. Various PACS-related job requests are lined up into disk resident priority queues, which are serviced by various computer system DAEMON (agent) processes. The queue software can have a built-in job scheduler that is programmed to retry a job several times by using either a default set of resources or alternative resources if a hardware error is detected. This mechanism ensures that no jobs will be lost during the complex negotiation for job priority among processes.

13.3.6 PACS Workflow

This section discusses a generic PACS workflow starting from the patient registering in the HIS to the RIS ordering examination, the technologist performing the examination, image viewing, reporting, and archiving. Comparing this PACS workflow with the PACS components and workflow in Figure 13.1 and the general radiology workflow, PACS has replaced many manual steps in the film-based workflow. Figure 13.8 shows the PACS workflow.

13.3.6.1 PACS, Hospital Information Systems, Radiology Information System Workflow

Following the numerals in Figure 13.8:

1. Patient registers in HIS, and a radiology examination is ordered in RIS. An examination accession number is automatically assigned.
2. The RIS outputs HL7 messages of HIS and RIS demographic data to PACS broker/interface engine.
3. The PACS broker notifies the archive server of a scheduled examination for the patient.

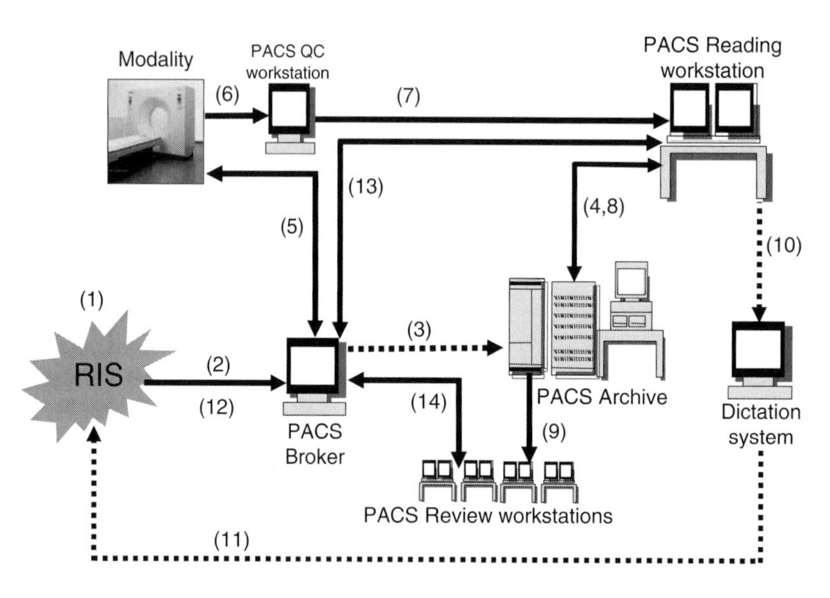

FIGURE 13.8 Generic picture archiving and communication systems workflow showing the numerical steps involved in the clinical workflow as described in the text.

4. Following prefetching rules, historic PACS examinations of the scheduled patient are prefetched from the archive server and sent to the radiologist reading workstation.
5. The patient arrives at examination facility. Modality queries PACS broker/interface engine for DICOM work list.
6. Technologist acquires images and sends PACS examination of images acquired by modality and patient demographic data to QC workstation in DICOM format.
7. Technologist prepares PACS examination and sends to the radiologist reading workstation as prepared status.
8. When the PACS examination arrives at the radiologist reading workstation, it is immediately sent automatically to the archive server. The archive server database is updated, and the PACS examination is marked as prepared status.
9. The archive server automatically distributes the PACS examination to the review workstations in the wards based on patient location received from the HIS/RIS HL7 message.
10. The reading radiologist dictates a report using the examination accession number. The radiologist signs off on PACS examination with any changes. The archive database is updated with changes and marks PACS examination as signed-off status.
11. The transcriptionist retrieves the dictation that corresponds with the examination accession number within RIS and types the report.
12. The RIS outputs HL7 message of results report data along with any previously updated RIS data.
13. The radiologist queries PACS broker/IE for previous reports of PACS examinations on reading workstations.
14. Referring physicians query broker/IE for reports of PACS examinations on review workstations.

13.4 PACS Controller and Image Archive

The PACS central node, considered the engine of the PACS, has two major components: the PACS controller and the archive server. Consisting of both hardware and software architecture, the PACS controller directs the data flow in the entire PACS by using interprocess communication among major processes. The image archive provides a hierarchical image storage management system for short-, medium-, and long-term image archiving.

13.4.1 Image Management and Design Concept

Two major aspects should be considered in the design of the PACS image storage management system: data integrity, which protects against loss of images once they are received by the PACS from the imaging modalities, and system efficiency, which minimizes access time of images at the display workstations. In this section, we only discuss the DICOM-compliant PACS controller and image archive server. To ensure data integrity, the PACS always retains at least two copies of an individual image on separate storage devices until the image has been archived successfully to the long-term storage device (e.g., an optical disk or tape library). This backup scheme is achieved through PACS intercomponent communication among the following PACS components, as shown in Figure 13.1. Figure 13.9 shows the hierarchical image storage management in PACS:

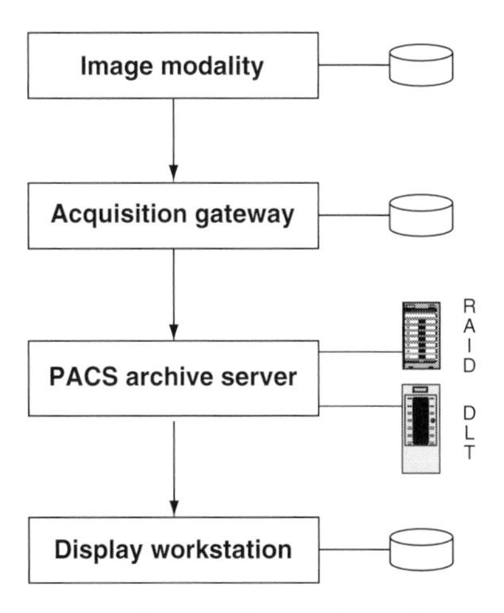

FIGURE 13.9 A diagram showing hierarchic image storage management in picture archiving and communication systems. Note that at least two copies of each picture archiving and communication system image resides on separate storage devices.

1. A copy of the PACS study is stored on the imaging modality until the technologist has verified that the studies have been successfully archived to PACS.
2. A copy of the PACS study is stored on the acquisition gateway computer until the archive subsystem has acknowledged that the study has been received successfully.
3. A copy of the study is retained until the PACS study has been successfully stored to permanent storage (e.g., optical disk or tape library).
4. For stand-alone architecture, a copy of the PACS study is retained on the display workstation until the patient has been discharged or transferred. For client/server architecture, the study is deleted when the review has been completed.

13.4.2 PACS Controller and Archive Server Functions

The PACS controller and the archive server consist of four components: an archive server, a database, a digital linear tape (DLT) library, and a communication network. Attached to the archive system through the communication network are the acquisition computers and the display workstations. Images acquired by the acquisition computers from various radiological imaging devices are transmitted to the archive server, from which they are archived to the DLT library and routed to the appropriate display workstations. The following is a brief description of each of the four subcomponents as well as some of the major functions.

13.4.2.1 The Archive Server

The archive server consists of multiple powerful central processing units (CPUs), small computer systems interface (SCSI) data buses, and network interfaces (Ethernet and ATM). With its redundant hardware configuration, the archive server can support multiple processes running simultaneously, and image data can be transmitted over different data buses and networks. In addition to its primary function of archiving images, the archive server acts as a PACS controller, directing the flow of images within the entire PACS from the acquisition gateway computers to various destinations such as archive, workstations, or print stations.

The archive server uses its large-capacity redundant array of inexpensive disks (RAID) as a data cache, capable of storing several weeks or months or years worth of images acquired from different radiological imaging devices. As an example, a 20-GB disk storage, without using compression, can hold simultaneously up to 500 computed tomography (CT), 1000 magnetic resonance (MR), and 500 computed radiography (CR) studies. Nowadays, very large RAID and SAN technologies are available in the archive server, especially in the client/server model. The magnetic cache disks configured in the archive server should sustain high data throughput for read operation, which provides fast retrieval of images from the RAID.

13.4.2.2 The Database System

The database system consists of redundant database servers running identical reliable commercial database systems, (e.g., Sybase, Oracle) with structured query language (SQL) utilities. A mirror database with two identical databases can be used to duplicate the data during every PACS transaction (not image) involving the server. The data can be queried from any PACS computer via the communication networks. The mirroring feature of the system provides the entire PACS database with uninterruptible data transactions that guarantee no loss of data in the event of system failure or a disk crash. Besides its primary role of image indexing to support the retrieval of images, the database system is necessary to interface with the RIS and the HIS, allowing the PACS database to collect additional patient information from these two health care databases.

13.4.2.3 The Archive Library

The archive library consists of multiple input/output drives (usually DLT, although some older PAC systems may still use optical erasable, WORM disk, optical tape, or CD-ROM) and disk controllers, which allow concurrent archival and retrieval operations on all of its drives. Newer technologies available as archive library solutions include DVD-ROM, large-scale RAID, and storage area network (SAN). The library must have a large storage capacity of terabytes and support mixed storage media if migrating to newer solutions. In this case, most hospitals opt for migrating PACS studies entirely from one data media solution to another to reduce the complexities of managing mixed storage media. Redundant power supply is essential for uninterrupted operation.

13.4.2.4 Communication Networks

The PACS archive system is connected to both the PACS LAN and the WAN. The PACS LAN can have a two-tiered communication network composed of Ethernet and ATM or high-speed Ethernet networks. The WAN provides connection to remote sites and can consist of T1 lines, ATM, and fast Ethernet. The PACS LAN uses the high-speed ATM or Ethernet switch to transmit high-volume image data from the archive server to 1K and 2K display workstations. An Ethernet using 100 MB/s can be used for interconnecting slower-speed components to the PACS server, including acquisition gateway computers, RIS, and HIS, and as a backup of the ATM or the GB/s Ethernet. Failure of the high-speed network automatically triggers the archive server to reconfigure the communication network so that images can be transmitted to the PACS display workstations over slower Ethernet.

13.4.2.5 PACS Controller and Archive Server Functions

In the controller and archive server, processes of diverse functions run independently and communicate simultaneously with other processes using client/server programming, queuing control mechanisms, and job prioritizing mechanisms. Because the functions of the controller and the archive server are closely related, we sometimes use the term archive server to represent both. Major tasks performed by the archive server include image receiving, image stacking, image routing, image archiving, studies grouping, platter management, RIS interfacing, PACS database updating, image retrieving, and image prefetching. The following subsections describe the functionality carried out by each of these tasks. Whenever appropriate, the DICOM standard is highlighted in these processes.

13.4.2.5.1 Image Receiving. Images acquired from various imaging devices in the gateway computers are converted into DICOM data format if they are not already in DICOM. DICOM images are then transmitted to the archive server via the Ethernet or ATM by client/server applications over standard TCP/IP protocols. The archive server can accept concurrent connections for receiving images from multiple acquisition computers. DICOM commands can take care of the send and receive processes.

13.4.2.5.2 Image Stacking. Images that come to the archive server from various gateway computers are stored on its local magnetic disks or RAID (temporary archive) based on the DICOM data model and managed by the database. The archive server holds as many images in its several hundred gigabyte disks as possible and manages them on the basis of aging criteria. During a hospital stay, for example, images belonging to a given patient remain in the archive server's temporary archive until the patient is discharged or transferred. Thus all recent images that are not already in a display workstation's local storage can be retrieved from the archive server's high-speed short archive instead of the lower-speed DLT library. This feature is particularly convenient for radiologists or referring physicians who must retrieve images from different display workstations. In the client/server PACS model, the temporary archive is very large, some have terabytes of capacity with the long-term archive library solution a SAN storage device.

13.4.2.5.3 Image Routing. In the stand-alone (or peer-to-peer) PACS model, images that come to the archive server from various acquisition computers are immediately routed to their destination workstations. The routing process is driven by a predefined routing table composed of parameters including examination type, display workstation site, radiologist name, and referring physician name. All images are classified by examination type (1-view chest, CT head, CT body, etc.) as defined in the DICOM standard. The destination display workstations are classified by location (Chest, Pediatrics, CCU, etc.) as well as by resolution (1K or 2K). The routing algorithm performs table lookup based on the aforementioned parameters and determines an image's destination(s). Images are transmitted to the 1K and 2K workstations over Ethernet, LAN, or ATM and to remote sites over dedicated T1 lines, ATM, or high-speed WAN.

13.4.2.5.4 Image Archiving. Images arriving in the archive server from gateway computers are copied from temporary storage to the archive library for long-term storage. When the copy process is complete, the archive server acknowledges the corresponding acquisition gateway, allowing it to delete the images from its local storage and reclaim its disk space. In this way, the PACS always has two copies of an image on separate magnetic disk systems until the image is archived to the permanent storage. Images from multiple examinations that occur during a patient's hospital stay are scattered temporarily across the archive library.

13.4.2.5.5 RIS and HIS Interfacing and PACS Database Updates. The archive server accesses data from HIS/RIS through a PACS gateway computer. The HIS/RIS relays messages regarding patient admission, discharge, and transfer (ADT) to the PACS only when a patient is scheduled for an examination in the radiology department or when a patient in the radiology department is discharged or transferred. Forwarding ADT messages to PACS not only supplies patient demographic data to the PACS but also provides information the archive server needs to initiate the prefetch, image archive, and studies grouping tasks. Exchange of messages among these heterogeneous computer systems can use the HL7 standard data format running TCP/IP communication protocols on a client/server basis. In addition to receiving ADT messages, PACS receives examination data and diagnostic reports from the RIS. This information is used to update the PACS database, which can be queried and reviewed from any display workstation. Data transactions performed in the archive server, such as insertion, deletion, selection, and update, are carried out by using SQL utilities in the database. Data in the PACS database are stored in predefined tables, with each table describing only one kind of entity. The design of these tables should follow the DICOM data model for operation efficiency. Individual PACS processes running in the archive server with information extracted from the DICOM image header update these tables and the RIS interface to reflect any changes of the corresponding tables.

13.4.2.5.6 Image Retrieving. Image retrieval takes place at the display workstations. The display workstations are connected to the archive system through communication networks. The archive library configured with multiple drives can support concurrent image retrievals from multiple tapes. The retrieved data are then transmitted from the archive library to the archive server via the SCSI data buses. The archive server handles retrieve requests from display workstations according to the priority level of these individual requests. Priority is assigned to individual display workstations and users based on different levels of needs. For example, the highest priority is always granted to a display workstation that is used primary for diagnosis or is in a conference session or at an intensive care unit. Thus, a workstation used exclusively for research and teaching purposes is compromised to allow fast service to radiologists and referring physicians in the clinic for immediate patient care.

13.4.2.5.7 Image Prefetching. The prefetching mechanism is initiated as soon as the archive server detects the arrival of a patient by means of the ADT message from HIS/RIS. Selected historic images, patient demographics, and relevant diagnostic reports are retrieved from the archive library and the PACS database. Such data are distributed to the destination workstation(s) before the completion of the patient's current examination or staged at the short-term RAID storage device.

The prefetch algorithm is based on predefined parameters such as examination type, disease category, radiologist name, referring physician name, location of the workstation, and the number and age of the patient's archived images. These parameters determine which historic images should be retrieved, when they should be retrieved, and where they should go.

13.4.3 Digital Imaging and Communications in Medicine-Compliant PACS Archive Server

The purpose of the DICOM standard is to promote a standard communication method for heterogeneous imaging systems, allowing the transfer of images and associated information among them. By using the DICOM standard, a PACS would be able to interconnect its individual components and allow the acquisition gateways to link to imaging devices. However, imaging equipment vendors often select different DICOM-compliant implementations for their own convenience, which may lead to difficulties for these systems in interoperation. Therefore, it is an important step to perform throughput testing of the entire system from PACS study acquisition to archival to ensure that the system is integrated properly. A well-designed DICOM-compliant PACS server can use two mechanisms to ensure system integration. One mechanism is to connect to the acquisition gateway computer with DICOM providing reliable and efficient processes of acquiring images from imaging devices. The other mechanism is to develop specialized server software allowing interoperability of multivendor imaging systems. Both mechanisms can be incorporated in the DICOM-compliant PACS server.

13.4.4 Hardware and Software Components

The PACS archive server generic hardware components consist of the PACS archive server computer, peripheral archive devices, and fast Ethernet interface and SCSI. For large-scale PACS, the server computer used is a mostly UNIX-based machine. The fast Ethernet interfaces the PACS archive server to the fast Ethernet network, where acquisition gateways and display workstations are connected. The SCSI integrates peripheral archive devices with the PACS archive server. The main archive devices for the PACS server include magnetic disk, RAID, DLT, and CD/DVD (digital video disks) jukeboxes, and newer SAN technologies. RAID, because of its fast access speed and reliability, is extensively used as the short-term archive device in PACS. Because of its large data storage capacity, DLT is mostly used for long-term archiving. Many kinds of storage devices are available for PACS application. In the following we describe the two most popular ones, RAID and DLT, along with a third, newer SAN technology.

13.4.4.1 Redundant Array of Inexpensive Disks

RAID is a disk array architecture developed for fast and reliable data access. A RAID groups several magnetic disks (e.g., eight disks) as a disk array and connects the array to one or more RAID controllers. The size of RAID is usually several hundred gigabytes (e.g., 320 GB for eight disks) to terabytes. With the individual disk size increasing, the size of RAID can also be increased. The RAID controller has a SCSI interface to connect to the SCSI interface in the PACS server. Multiple RAID controllers with multiple SCSI interfaces can avoid the single-point failure in the RAID device.

13.4.4.2 Digital Linear Tape

DLT uses a multiple magnetic tape and drive system housed inside a library or jukebox for large-volume and long-term archive. With current tape drive technology, the data storage size can reach 40 to 200 GB/tape. One DLT can hold from 20 to hundreds of tapes. Therefore, the storage size of DLT can be from one to tens of terabytes, which can hold PACS images from one to several years. DLT usually has multiple drives to read and write tapes. The tape drive is connected to the server through SCSI or fiber-optic connection. The data transmission speed is several megabytes per second for each drive. The tape loading time and data locating time are several minutes. Hence, in general, it takes several minutes to retrieve one CR image from DLT. PACS image data in DLT are usually prefetched to RAID for fast access time.

13.4.4.3 Storage Area Network

A current data storage trend in large-scale archiving is SAN technology. With this new configuration, the PACS server will still have a short-term storage solution in local disks containing unread patient studies. However, for long-term storage, the PACS data are stored in a SAN. This SAN is a stand-alone data storage repository with a single IP address. File management and data backup can be achieved with a combination of digital media (e.g., RAID or DLT) smoothly and with total transparency to the user. In addition, the SAN can be partitioned into several different repositories each storing different data file types. The storage manager within the SAN is configured to recognize and distribute the different clients' data files and store them to distinct and separate parts of the SAN.

13.4.4.4 Archive Server Software

PACS archive server software is DICOM-compliant and supports DICOM storage service class and query/retrieve service class. Through DICOM communication, the archive server receives DICOM studies/images from the acquisition gateway, appends study information to the database, and stores the images in the archive device, including the RAID, DLT, or SAN. It receives the DICOM query/retrieve request from display workstations and sends out the query/retrieve result (patient/study information or images) back to workstations. The DICOM services supported in PACS archive server are C-Store, C-Find, and C-Move. All software implemented in the archive server should be coded in standard programming languages—for example, C and C++ on the UNIX open systems architecture. PACS archive server software is composed of at least six independent components (processes), including receive, insert, routing, send, Q/R-server, and retrieve send. It also includes a PACS database. All of these processes run independently and simultaneously and communicate with other processes through queue control mechanisms.

13.4.5 Disaster Recovery and Backup Archive Solutions

The PACS archive server is the most important component in a PACS, and even though it may have the fault-tolerant feature, chances are it could fail occasionally. A backup archive server is necessary to guarantee its uninterrupted service. Two copies of identical images can be saved through two different paths in the PACS network to two archive libraries. Ideally, the two libraries should be in two different buildings in case of natural disaster. To reduce the cost of redundant archiving, the primary unit can be another DLT library. The backup archive server can be short term (3 months) or long term. The functions of a backup archive server are twofold: maintaining the PACS continuous operation and preventing loss of image data. Data loss is especially troublesome because if a major disaster occurs, it is possible to lose an entire hospital's PACS data. In addition, scheduled downtimes to the main PACS archive have a great impact on a filmless institution. Few current PACS archives feature disaster recovery or a backup archive, and designs are limited at best. Furthermore, current general disaster recovery solutions vary in the approach toward creating redundant copies of PACS data. One novel approach is to provide a short-term fault-tolerant backup archive server using the application service provider (ASP) model at an offsite location. The ASP backup archive provides instantaneous, automatic backup of acquired PACS image data and instantaneous recovery of stored PACS image data, all at a low operational cost because it uses the ASP business model. Figure 13.10 shows the general architecture of an ASP backup archive server. In addition, should the downtime event render the network communication inoperable, a portable solution is available with a data migrator. The data migrator is a portable laptop with a large-capacity hard disk that contains DICOM software for exporting and importing PACS examinations. The data migrator can populate PACS examinations that were stored on the backup archive server directly onto the clinical PACS within hours to allow the radiologists to continue to read previous PACS examinations until new replacement hardware arrives and is installed or until a scheduled downtime event has been completed.

13.5 Large-Scale PACS Implementation

Around the world, because of the need to improve operation efficiency and provide more cost-effective health care, many large-scale health care enterprises have been formed. Each of these enterprises can group hospitals, medical centers, and clinics together as one enterprise health care network. The management of these enterprises recognizes the importance of using PACS and image distribution as a key technology in better-quality and more cost-effective health care delivery at the enterprise level. As a result, many large-scale enterprise-level PACS/image distribution pilot studies, full design, and implementation, are under way. The following are characteristics of these systems:

1. Scale: large enterprise level, from 39 to 399 hospitals and medical centers
2. Complexity: total health care IT integration

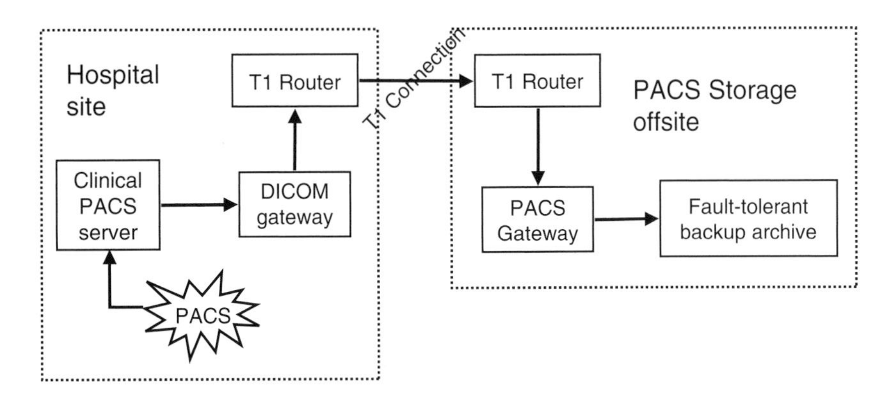

FIGURE 13.10 General architecture of the ASP backup archive server. One DICOM gateway and one picture archiving and communication systems gateway are used as the buffers between the two sites. T1 is used for wide area network (WAN).

3. Goals and Objectives: complete system deployment to the enterprise
4. Costs: extremely expensive
5. Difficulty of Implementation: culture, resources, timeline, and overcoming legacy technologies

13.5.1 Introduction to Hospital Clinical Systems

PACS is a workflow-integrated imaging system designed to streamline operations throughout the entire patient care delivery process. One of its major components, image distribution, delivers relevant electronic images and related patient information to health care providers for timely patient care either within a hospital or in a health care enterprise. Enterprise-level health care delivery emphasizes sharing of enterprise-integrated resources and streamlining operations. In this respect, if an enterprise consists of several hospitals and clinics, it is not necessary for every hospital and clinic to have similar specialist services. A particular clinical service like radiology can be shared among all entities in the enterprise. Under this setup, all patients registered in the same enterprise can be referred to a radiology expert center for examinations. In this scenario, the patient being cared for becomes the focus of the operation. A single index like the patient's name/identification would be sufficient for any health care provider in the enterprise to retrieve the patient's comprehensive record. For this reason, the data management system would not be the conventional HIS, RIS, or other organizational information system. Rather, the ePR or eMR concept will prevail. However, to develop the ePR, the successful integration of the HIS, RIS, and, additionally, voice recognition system, is crucial. The following sections will describe each of the afore-mentioned hospital clinical systems and their interfaces with each other, as well as the ePR concept.

13.5.2 Hospital Information System

The HIS is a computerized management system for handling three categories of tasks in a health care environment:

1. Support clinical and medical patient care activities in the hospital
2. Administer the hospital's daily business transactions (financial, personnel, payroll, bed census, etc.)
3. Evaluate hospital performances and costs and make a long-term forecast

Many clinical departments in a health care center, such as radiology, pathology, pharmacy, clinical laboratories, and other units, have their own specific operational requirements that differ from those of the general hospital operation. For this reason, special information systems may be needed in these departments. Often, these information systems are under the umbrella of the HIS, which maintains their operations. Other departments may have their own separate information systems, and some interface mechanisms are built to integrate data between these systems and the HIS. For example, RIS was originally a component of HIS; later, independent RIS was developed because of the limited support offered by HIS to handle special information required by the radiology departmental operations. However, the integration of these two systems is still extremely important for the health care center to operate as a total functional entity.

Large-scale HIS mostly use mainframe computers. These can be purchased through a manufacturer with certain customization software or homegrown through the integration of many commercial products progressively over years. A homegrown system may contain many reliable legacy components but have out-of-date technology. Therefore, to interface HIS to PACS, caution must be taken to circumvent the legacy problem.

Most HIS are an integration of many information data systems, starting the day the health care data center was established, with older components being replaced by newer ones over many years of operation. In addition to taking care of the clinical operation, the HIS also support hospital and health care center business and administrative functions. They provide automation for such events as patient registration and ADT, as well as patient accounting. They also provide on-line access to patient clinical results (e.g., laboratory, pathology, microbiology, pharmacy, radiology). The system broadcasts in real time the patient demographics and encounters information with HL7 standards to the RIS. Through this path, ADT and other pertinent data can be transmitted to the RIS and the PACS.

13.5.3 Radiology Information System

The RIS is designed to support both the administrative and clinical operation of a radiology department, to reduce administrative overhead, and to improve the quality of radiological examination delivery. Therefore, the RIS manages general radiology patient demographics and billing information, procedure descriptions and scheduling, diagnostic reports, patient arrival scheduling, film location, film movement, and examination room scheduling. The RIS configuration is very similar to the HIS, except that it is on a smaller scale. RIS equipment consists of a computer system with peripheral devices such as RIS workstations (normally no image display), printers, and bar code readers. Most independent RIS are autonomous systems with limited access to HIS. However, some HIS offer embedded RIS as a subsystem with a higher degree of integration.

The RIS maintains many types of patient- and examination-related information, including medical, administrative, patient demographics, examination scheduling, diagnostic reporting, and billing information. The major tasks of the system include:

1. Process patient and film folder records
2. Monitor the status of patients, examinations, and examination resources
3. Schedule examinations
4. Create, format, and store diagnostic reports with digital signatures
5. Track film folders
6. Maintain timely billing information
7. Perform profile and statistics analysis

The RIS interfaces to PACS based on the HL7 standard through TCP/IP over Ethernet on a client/server model using a trigger mechanism. Events such as examination scheduling, patient arrivals, and actual examination begin and end times trigger the RIS to send previously selected information (patient demographics, examination description, diagnostic report, etc.) associated with the event to the PACS in real time.

13.5.4 Voice Recognition System

Typically, radiological reports are archived and transmitted independently from the image files. They are first dictated by the radiologist and recorded on an audiocassette recorder from which a textual form is transcribed and inserted into the RIS several hours later. The interface between the RIS and the PACS allows for sending and inserting these reports into the PACS database, from which a report corresponding to the images can be displayed on the PACS workstation on request by the user. This process is not efficient because the delay imposed by the transcription prevents the textural report from reaching the referring physician in a timely manner. One method is to append the digital voice recordings of the radiologist to the PACS study. The concept of interfacing this method is to have the digital voice database associated with the PACS image database; thus, before the written report becomes available, the referring physician can look at the images and listen to the report simultaneously. The radiologist views images from the PACS workstation and uses the digital Dictaphone system to dictate the report, which converts it from analog signals to digital format and stores the result in the voice message server. The voice message server in turn sends a message to the PACS data server, which links the voice with the images. The referring physicians at the workstation can, for example, in an intensive care unit, request to review certain images and at the same time listen to the voice report through the voice message server linked to the images. Later, the transcriber transcribes the voice by using the RIS. The transcribed report is inserted into the RIS database server automatically. The RIS server sends a message to the PACS database server. The latter appends the transcribed report to the PACS image file and signals the voice message server to delete the voice message.

The ideal method is to use a voice recognition system that automatically translates voice into text. In this case, the voice recognition system is either called within the PACS application or the RIS application. All the necessary fields are populated (e.g., patient name, medical record number, type of study), and the radiologist can begin to dictate. Once the radiologist has completed the dictation, the report can be edited, reviewed, and electronically signed off. It is then ready for distribution. In addition, report templates can be created for common diagnosis results that allow the radiologist to quickly create a report result via voice recognition commands. The report is then sent to RIS via an interface, and RIS can then forward the report to PACS as needed. When the DICOM-structured report standard becomes available, radiologists can directly enter the report through the structured report format while reviewing the images. Thus, the digital voice dictation system may see less use while the voice recognition system will be enhanced with a full set of automatic templates that can be created on demand once the DICOM-structured report becomes more acceptable to radiologists.

13.5.5 Interfacing PACS, Hospital Information Systems, Radiology Information Systems, and Voice Recognition Systems

There are three methods of transmitting data between information systems: through workstation emulation, through database-to-database transfer, and by means of an interface engine.

13.5.5.1 Workstation Emulation

This method allows a workstation of an information system to emulate a workstation of a second system. As a result, data from the second information system can be accessed by the first system. For example, a PACS workstation can be connected to the RIS with a simple computer program that emulates a RIS workstation. From the PACS workstation, the user can perform any RIS function such as scheduling a new examination, updating patient demographics, recording a film movement, and viewing the diagnostic reports. This method has two disadvantages. First, there is no data exchange between RIS and PACS. Second, the user is required to know how to use both systems. Also, a RIS or HIS workstation cannot be used to emulate a PACS workstation because the latter is too specific for HIS and RIS to emulate.

13.5.5.2 Database-to-Database Transfer

The database-to-database transfer method allows two or more networked information systems to share a subset of data by storing them in a common local area. For example, the ADT data from the HIS can be reformatted to HL7 standard and broadcasted periodically to a certain local database in the HIS.

A TCP/IP communication protocol can be set up between the HIS and the RIS, allowing the HIS to initiate the local database and broadcast the ADT data to the RIS through either a pull or push operation. This method is most often used to share information between the HIS and the RIS. A recent trend is the integration between RIS and PACS databases. In this configuration, common elements are shared between both databases, and any changes or modifications made to the patient, study, or image information are updated once without the need to update both databases manually. In addition, at the diagnostic workstation, the RIS application would call the PACS application to display the particular study. An additional monitor is usually used to display the RIS application. The user will navigate through the RIS application to identify and select the particular radiology study to be diagnosed, and the RIS application makes a function call to the PACS application to display the selected PACS study. This method of workflow is called RIS-driven workflow because the RIS is the driver of the diagnostic workflow and the PACS acts as a client in this instance.

13.5.5.3 Interface Engine

The interface engine provides a single interface and language to access distributed data in networked heterogeneous information systems. In operation, it appears that the user is operating on a single integrated database from his or her workstation. In the interface engine, a query protocol is responsible for analyzing the requested information, identifying the required databases, fetching the data, assembling the results in a standard format, and presenting them at the workstation. Ideally, all these processes are done transparently to the user and without affecting the autonomy of each database system. To build a universal interface engine is not a simple task. Most currently available commercial interface engines are tailored to limited specific information systems.

13.5.5.1 Integrating HIS, RIS, PACS, and VR Systems

Another recent trend to streamline the diagnostic workflow and provide as much clinical information as possible to the radiologist has resulted in the integration of HIS/RIS/PACS/ voice recognition (VR) application on one diagnostic workstation. This new integrated workstation is often referred to as the radiology command center and allows the radiologist full access to all available pertinent and historic clinical patient data while making a primary diagnosis. Because this is a fairly recent technology trend, the complexities and challenges of integrating multiple applications on a single workstation have impacted the user with factors such as ease-of-use, reliability, and efficiency. More work in the future is needed to fully realize the potential of such an integrated workstation.

In a hospital environment, interfacing the PACS, RIS, and HIS has become necessary to enhance diagnostic process, PACS image management, RIS administration, and research and training. These are all important aspects to consider when integrating systems.

13.5.6 Electronic Patient Record

The eMR or ePR is the ultimate information system in a health care enterprise. In an even broader sense, if the information system includes the health record of an individual, then it is called the electronic health record (eHR). In this context, we concentrate on ePR. Currently, only small subsets of ePR are actually in clinical operation. One can consider ePR as the big picture of the future health care information system. Although the development of a universal ePR as a commercial product is still years away, its eventual impact on the health care delivery system should not be underestimated. An ePR consists of five major functions:

1. Accepts direct digital input of patient data
2. Analyzes across patients and providers
3. Provides clinical decision support and suggests courses of treatment
4. Performs outcome analysis and patient and physician profiling
5. Distributes information across different platforms and health information systems

HIS and RIS, which deal with patient nonimaging data management and hospital operation, can be considered components of ePR. An integrated HIS-RIS-PACS system, which extends the patient data to include imaging, forms a cornerstone of ePR. Existing ePRs have certain commonalties. They have large data dictionaries with time stamped in their contents and can query and display data flexibly. Examples of successfully implemented EMRs are the Computer-Stored Ambulatory Record developed at Massachusetts General Hospital (in the public domain), the Regenstrief Medical Record System at Indiana University, the Health Evaluation Through Logical Processing system developed at the University of Utah and Latter-Day Saints Hospital, and the Department of Veterans Affairs Health Care Enterprise (VAHE) information system. Among these systems, the VAHE is one of the most advanced systems in the sense that it is being used daily in many of the VA medical centers and it includes images in the ePR. Figure 13.11 shows examples of the ePR from the VAHE system called Veterans Health Information System Technology Architecture (VistA).

As with any other medical information system, the development of the ePR faces several obstacles:

- Finding a common method to input patient examination and related data to the system
- Developing an across-the-board data and communication standard
- Gaining buy-in from manufacturers to adopt the standards
- Gaining acceptance by health care providers

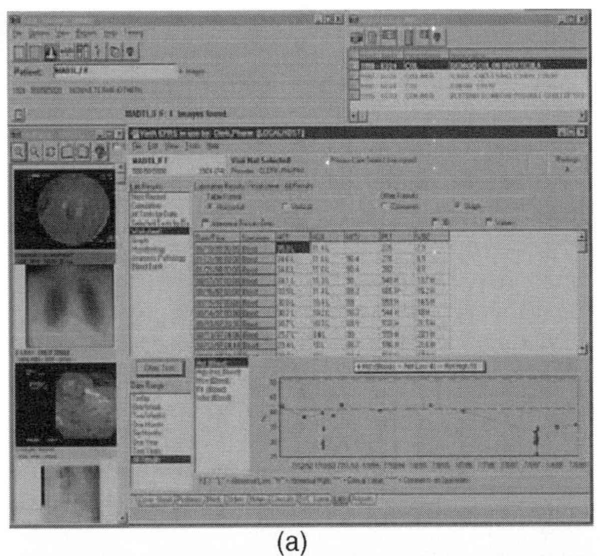

(a)

(b)

FIGURE 13.11 (a) VistA imaging displays the patient record with images. (b) VistA imaging displays thumbnail images, microscopic images, magnetic resonance images, and electrocardiogram. Courtesy of Dr. H. Rutherford; Dayhoff, 2000.

An integrated HIS-RIS-PACS system provides solutions for some of these obstacles.

- DICOM and HL7 standards have been adopted for imaging and text, respectively.
- Images and patient-related data are entered into the system almost automatically.
- The majority of imaging manufacturers have adopted DICOM and HL7 as *de facto* industrial standards.

Therefore, in the course of developing an integrated PACS, one should keep in mind the big picture, the ePR. Anticipation of future connections and the integrated PACS as a subsystem of ePR with images should be considered thoroughly.

13.6 PACS Clinical Experiences

13.6.1 Introduction

In this section, methodology and a road map for PACS implementation and system evaluation within a clinical hospital environment will be discussed. In addition, some examples of clinical experiences and pitfalls will be presented. The philosophy of PACS design and implementation is that, regardless of the scale of the PACS being planned, the strategy should always be to leave room for future expansion, including integration with an enterprise PACS. Thus, if the planning is to have a large-scale PACS now, the PACS architecture should allow its future growth to an enterprise PACS. On the other hand, if only a PACS module is being planned, the connectivity and compatibility of this module with future modules or with a larger-scale PACS are important. The terms we discussed in previous chapters, including open architecture, connectivity, standardization, portability, modularity, and IHE workflow profiles, should all be considered.

13.6.2 PACS Implementation Strategy

When implementing a PACS within a clinical environment, it is very important to recognize some key fundamental concepts that will serve as cornerstones for a successful implementation. First, PACS is an enterprise-wide system or product. It is no longer just for the radiology or imaging department; therefore, careful consideration of all decisions/strategies going forward should include the entire health care continuum, from referring physicians to the radiology department clinical and technical staff to the health care institution's IT department. It is crucial for a successful implementation that some of the key areas within the health care institution have buy-in of the PACS process, including administration, the radiology department, the IT department, and all high-profile customers of radiology (e.g., orthopedics, surgery). Furthermore, a champion or champions should be identified for the PACS process. Usually this is the medical director of radiology, but it can include other physicians as well as IT administrators. Second, PACS is a system with multiple complex components that interact with one another. Each of these components can be an accumulation of multiple hardware components. A general clinical PACS usually includes the archive, the archive server/controller, the DICOM gateway, the Web server, the workstations, and a RIS/PACS interface. Whether considering implementation or acceptance, all components of the system must be assessed. The following sections describe some of the steps involved in implementing a PACS within a health care institution.

13.6.2.1 Risk Assessment Analysis

It is important to perform a risk assessment analysis before implementation so that problem areas and challenges can be mapped out accordingly and timeline schedules can be made to accommodate potential roadblocks. Some areas to focus on are Z the network infrastructure that will be supporting the PACS, the integration of acquisition modality scanners with PACS (e.g., legacy systems, modality work list, quality control workstations), physical space for the PACS equipment, and resource availability. Resource availability is especially crucial because a successful PACS implementation hinges on the support provided by the in-house radiology department. In making risk assessments, it is also helpful to determine areas in which there is a low risk and a high return. These areas are usually departments where there is a high volume of image (film) and a low rate of return of film back to the radiology department (e.g., critical care areas, orthopedics, surgery). These low-risk/high-return areas can help to drive the implementation phase timeline and can also be a good first push in the implementation process.

13.6.2.2 Implementation Phase Development

Implementation of PACS should be performed in distinct phases, which would be tailored based on the risk assessment analysis performed at the health care institution. Usually, the first phase occurs when the main components are implemented such as the archive, archive server/controller, network infrastructure, HIS-RIS-PACS interfaces, workstations, and one or two modality types. The next phases are targeted toward implementing all modality types and a Web server for enterprise-wide and off-site distribution of PACS examinations. The phased approach allows for a gradual introduction of PACS into the clinical environment, with the ultimate goal being the transformation into a filmless department/hospital.

13.6.2.3 Development of Workgroups

Because PACS covers such a broad area within the health care institution, it is important to develop workgroups to handle some of the larger tasks and responsibilities. In addition, a PACS implementation team should be in place to oversee the timely progress of the implementation process. The following are some key workgroups and their responsibilities:

1. RIS-PACS interface and testing: Responsible for integration/testing of RIS/PACS interfaces including the modality work list on the acquisition scanners
2. PACS modalities and system integration: Responsible for the technical integration of modalities with PACS and installation of all PACS devices
3. PACS acquisition workflow and training: Responsible for developing workflow and training for clerical and technical staff and for any construction needed in the clinical areas
4. PACS diagnostic workflow and training: Responsible for developing workflow and training for radiologists and clinicians and for any construction needed in the clinical diagnostic areas (e.g., reading room designs). Figure 13.12 shows an example of the different stages of conversion of a clinical space into a reading room for radiologists.
5. PACS network infrastructure: Responsible for all design and implementation of the network infrastructure to support PACS

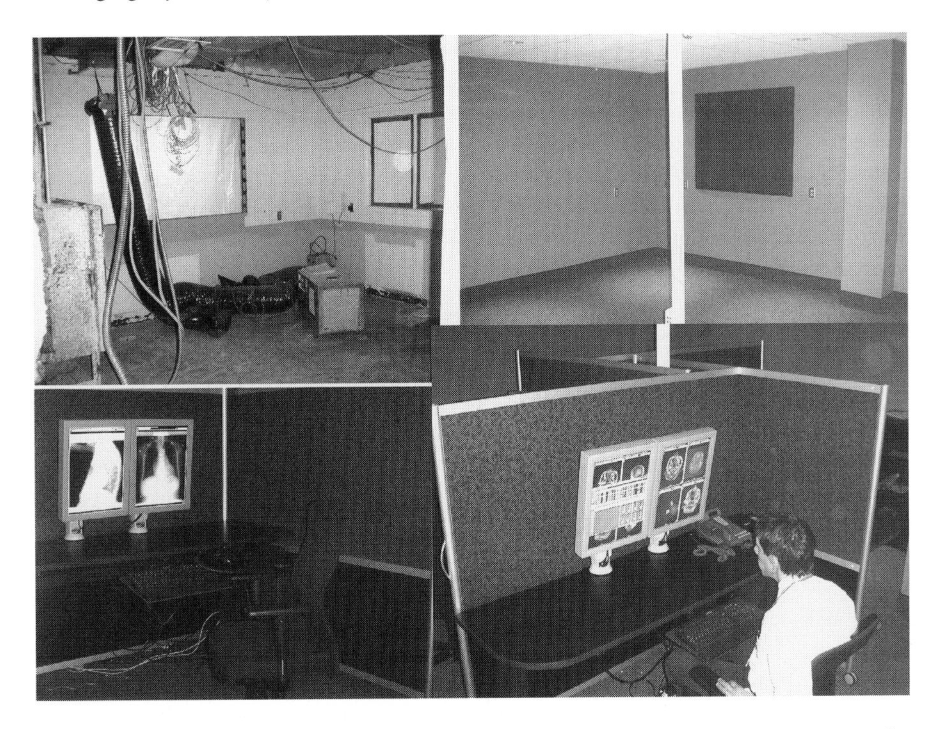

FIGURE 13.12 Different stages of the conversion of a clinical space into a reading room for radiologists. Note the pinwheel-shape design using the center floor of the room. Power and networking are supplied through the column in the center of the floor.

In addition to the above-listed workgroups, a PACS implementation team should be formed to oversee the implementation process. Members should include at least one point person from each workgroup, and additional members should include the PACS implementation manager, the medical director of imaging, the administrative director of imaging, an IT representative, and an engineering/facilities representative. This team should meet at least every 2 weeks and more frequently as the date of live implementation nears. The goals of this team are to update any status items and to highlight any potential stumbling blocks to the implementation process. In addition, this team meeting allows a forum for higher-level administrators to observe the progress of the implementation. It is crucial to identify particular in-house resources for the implementation process. These include a technical supervisor of each modality, a clerical supervisor, a film librarian or film clerk, an RIS support person, and an IT network support person. These resources are an excellent source of information for issues related to PACS such as technologist workflow, clerical workflow, film distribution workflow, design and performance of RIS interface testing with PACS, and overall hospital IT infrastructure.

13.6.2.4 Implementation Management

Developing a schedule and implementation checklist can assist management of the implementation process. This template includes topics such as the task description, the date scheduled for the task, the owner of the task, and a checkmark box to indicate completion of the task. This template allows for finer granularity of the implementation process to protect against overlooked implementation tasks. Input for the checklist can come from the PACS implementation team meetings. Furthermore, the checklist can be broken down into smaller subtask checklists for tracking of issues within each of the workgroups.

13.6.3 System Acceptance

One of the key milestones to system turnover is the completion of the acceptance testing (AT) of PACS. There are a few reasons why AT is important to PACS. First, AT provides vendor accountability for delivering the final product that was initially scoped and promised. It also provides accountability for the in-house administration that there is documentation that the system was tested and accepted. AT also provides a glimpse into determining the characteristics of PACS uptime and whether it will function as promised. Finally, AT provides proof of both PACS performance and functionality as originally promised by the vendor. Most vendors provide their own AT plan; however, usually it is not thorough enough, and the template is not customized to the specific health care institution's needs. The following sections describe some of the steps in designing and developing a robust AT that can be used for final turnover of PACS in the clinical environment.

Acceptance test criteria are divided into two categories. The first category is quality assurance. This includes PACS image quality, functionality, and performance. The second category is technical testing, which focuses on the concept of no single point of failure through the PACS and includes simulation of downtime scenarios. Acceptance criteria should include identifying which PACS components are to be tested. The following are some of the components that should be included are:

1. RIS/PACS interface and/or PACS broker
2. Acquisition gateways
3. Modality scanner(s)
4. Archive server/storage
5. Diagnostic workstation
6. Review workstation
7. Network devices

If the PACS also includes a Web server, then it also should be included within the acceptance testing criteria.

Each of the implementation phases of the PACS process should have an acceptance test performed. Acceptance at each of the phases is also crucial for the vendor because it is only after acceptance that the vendor can collect the remainder of the fee balance, which is negotiated beforehand. The implementation of the AT is a two-phased approach. The first phase should be performed approximately 1 week before the live date. The content of phase one includes the technical component testing focusing on single points of failure, end-to-end testing, contingency solutions for downtime scenarios, and any baseline performance measurements. The second phase should be performed approximately 2 weeks after the live date so that the PACS has stabilized a bit in the clinical environment. The contents of phase two include PACS functional and performance testing as well as any additional network testing, all on a loaded clinical network.

13.6.4 Image/Data Migration

Two scenarios are possible to trigger image/data migration: converting to a new storage technology and increasing data volumes. It is possible for a health care institution to have a dramatic increase in PACS data volumes once it transforms into a filmless institution. This is possible due in part to the continuous image accumulation as well as the integration of new modalities generating mass volumes of PACS data and archiving the large data quantities to PACS. For example, the multislice detector CT scanner is capable of generating up to 1000 images amounting to almost 500 MB of data per examination. It is very likely that a hospital may need to expand the archive storage capacity. Furthermore, most PACS installed in previous years do not have a secondary copy backup of all the archived PACS image data for disaster recovery purposes. It has only been a recent trend for PACS to offer disaster recovery solutions. Therefore, should a hospital decide to upgrade the archive server performance and expand with a higher-capacity data media storage system, there are a few major challenges facing a successful upgrade. One challenge is how to upgrade to a new PACS archive server in a live clinical setting. Another challenge is how to migrate the previous PACS data to a new data media storage system in a live clinical setting.

Some of the issues that surround a migration plan are that the data migration must not hamper the live clinical workflow in any way or reduce system performance. With any migration, it is important that verification be performed to prevent any data loss. Once the data have been successfully migrated to the new data media, the original data media storage system should be removed, which may incur additional downtime of the archive server. Development of a migration plan is key to addressing the surrounding issues and ensuring a data migration that will have the least impact on the live clinical PACS. Because data migration occurs in a live clinical setting, it is important to determine the times at which the data migration will not impact normal clinical workflow. This may include scheduling a heavier data migration rate during off-hours (e.g., nights and weekends) and a lighter rate during operating hours and hours of heavy clinical PACS use. Expert knowledge of the clinical workflow is valuable input toward developing a good schedule data migration. Downtime may be involved both initially and at the end of the data migration process and should be scheduled accordingly with contingency procedures.

It may be necessary to fine-tune the data migration rate because estimates for the migration rate may not be accurate initially. Fine-tuning is very crucial because an aggressive migration rate can adversely affect the performance of the entire clinical PACS. Careful attention to the archive and system performance is especially important during the onset of the data migration. The data migration rate may need to be scaled back. This may be an iterative cycle until an optimal migration rate is achieved that does not adversely affect the clinical PACS.

13.6.5 PACS Clinical Experiences and Pitfalls

The following sections describe an overview of two different PACS clinical experiences from two different sized health care institutions. One is a large-scale health care institution, and the second is a high-profile community-sized hospital. In addition, some PACS pitfalls will be discussed.

13.6.5.1 Clinical Experiences at Baltimore VA Medical Center

The Baltimore VA Medical Center (VAMC) started its PACS implementation in the late 1980s and early 1990s. The VAMC purchased a PACS in late 1991 for approximately $7.8 million, which included $7.0 million for PACS and $800,000 for CR.

The manufacturers involved were Siemens Medical (Erlangen, Germany) and Loral Western Developed Labs (San Jose, CA); the product later changed hands to Loral/Lockheed Martin and then to General Electric Medical Systems. The goals of the project were to integrate with the VA home-grown clinical patient record system (CPRS) and the then to-be-developed VistA imaging system. The project has been under the leadership of Dr. Eliot Siegel, Chairman of the Radiology Department. The system was in operation in the middle of 1993 in the new Baltimore VAMC. This system has since evolved and has been integrated with other VA hospitals in Maryland into a single imaging network, the VA Maryland Health Care System. Four major benefits at the Baltimore VAMC are: changing the operation to filmless, reducing unread cases, reducing retake rates, and drastically improving the clinical workflow.

The two major contributors to the cost of the system are the depreciation and the service contract. The VA depreciates its medical equipment over a period of 8.8 years, whereas computer equipment is typically depreciated over a 5-year time period. The other significant contributor to the cost of the PACS is the service contract, which includes all of the personnel required to operate and maintain the system. It also includes software upgrades and replacement of all hardware components that fail or demonstrate suboptimal performance. This includes replacement of any monitors that do not pass the quality control tests. No additional personnel are required other than those provided by the vendor through the service contract. In the Baltimore VAMC, the radiology administrator, chief technologist, and chief of radiology share the responsibilities of a PACS departmental system administrator. Cost savings attributed to PACS include three areas: (1) film operation costs, (2) space costs, and (3) personnel costs. Films are still used in two circumstances. Mammography examinations are still using films, but they are digitized and integrated to the PACS. Films are also printed for patients who need to have them for hospital or outpatient visits outside the VA health care network. Despite these two uses, film costs have been cut by 95% compared with the figure that would have been required in a conventional film-based department. Additional savings include reductions in film-related supplies such as film folders and film chemistry and processors. The second area in cost savings is space. The ability to recover space in the radiology department because of PACS contributes to a substantial savings in terms of space indirect costs. Finally, the personnel cost savings include radiologists, technicians, and film library clerks. An estimate was made that at least two more radiologists would have been needed to handle the current workload at the VAMC had the PACS not been installed. The efficiency of technologists has improved by about 60% in sectional imaging examinations, which translates to three to four additional technologists had the PACS not been used. Only one clerk is required to maintain the film library and to transport film throughout the medical center.

13.6.5.2 Clinical Experience at Saint John's Health Center

Saint John's Health Center, Santa Monica, CA, has a filmless PACS that acquires approximately 130,000 radiological examinations annually. As the first phase, St. John's implemented the PACS with CR for all critical care areas in April 1999. Phase II, completed in April 2000, included the integration of MR, CT, ultrasound, digital fluorography, and digital angiography within the PACS. Since then, St. John's PACS volumes have increased steadily. The original storage capacity of the PACS archive was a 3.0 TB MOD Jukebox, which would mean that older PACS examinations would have to remain off-line before a year is over. Also, the archive had only a single copy of the PACS data. Therefore, should St. John's encounter a disaster, it might lose all the PACS data because there was no backup. With these considerations, St. John's determined to overhaul its PACS archive system with the following goals:

- Upgrade the archive server to a much larger capacity
- Develop an off-site image/data backup system
- Conduct an image/data migration during the archive system upgrade

These goals were accomplished in late 2001 based on the concepts discussed in this section. With the archive upgrade, all new PACS examinations were archived through a Sun Enterprise 450 platform server with a 270 GB RAID. The examinations were then archived to a network-attached digital tape storage system comprising an additional Sun Enterprise 450 with a 43 GB RAID and a 7.9 TB storage capacity digital tape library. The storage capacity of the tape library technology was forecast to double in the next few years as the tape density doubles, eventually making it a 16 TB library. Figure 13.13 shows the final configuration after the completion of the data migration.

13.6.5.3 PACS Pitfalls

PACS pitfalls are mostly from human error, whereas bottlenecks are due to imperfect design in either the PACS or image acquisition devices. These drawbacks can only be realized through accumulated clinical experience.

Pitfalls resulting from human error are often initiated at imaging acquisition devices and at workstations. Three major errors at the acquisition devices are entering wrong input parameters, stopping an image transmission process improperly, and positioning the patient incorrectly. The errors occur most often at the workstations, where users have to enter many key strokes or click the mouse frequently before the workstation can respond. Other pitfalls at the workstation unrelated to human error are missing location markers in a CT or MR scout view, images displayed with unsuitable lookup tables, and white borders in CR images due to X-ray collimation. Pitfalls created by human intervention can be minimized by implementing a better quality assurance program, providing

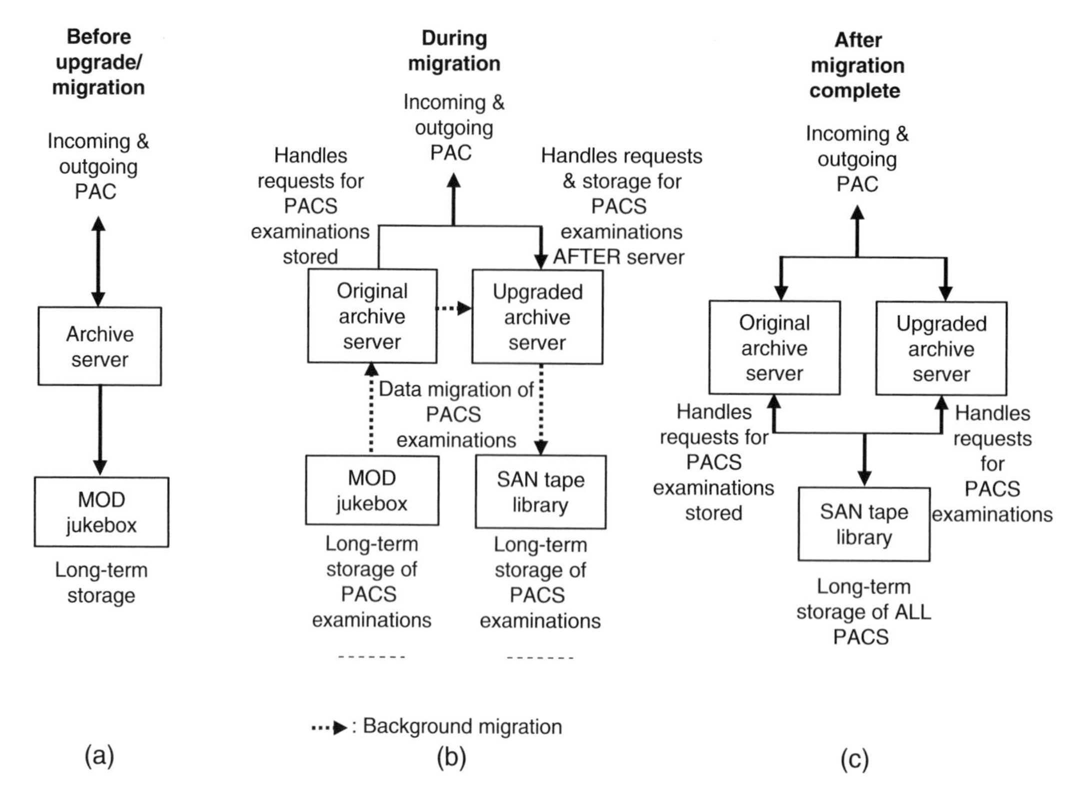

FIGURE 13.13 Before, during, and final archive configuration and process for St. John's Health Center picture archiving and communication system. (a) before, (b) during, (c) after migration with new archive system.

periodic in-service training, and interfacing image acquisition devices directly to the HIS/RIS through a DICOM broker. Bottlenecks affecting the PACS operation include network contention; CR, CT, and MR images stacked up at acquisition devices; slow responses from workstations; and long delays for image retrieval from the long-term archive. Improving the system architecture, reconfiguring the networks, and streamlining operational procedures through a gradual understanding of the PACS clinical environment can alleviate bottlenecks. Utilization of the IHE workflow profiles discussed would also help to circumvent some of the bottleneck problems. During the integration of multivendor PACS components, it should be remembered that, even though each vendor's component may come with a DICOM conformance statement, the components still may not be compatible. These pitfalls can be minimized through the implementation of two DICOM-based mechanisms, one in the image acquisition gateway and the second in the PACS controller, to provide better connectivity solutions for multivendor imaging equipment in a large-scale PACS environment.

13.7 Summary

In this section, various components, terminology, and standards used in PACS were presented and discussed. Integrating the health care enterprise are protocols of image data workflow allowing connectivity of components in PACS from various vendors based on existing standards. The information system used in hospitals is called HIS or CMS, which consists of many clinical databases, like the RIS. These databases are operation-oriented and are designed for special clinical services. The new trend in health care information systems is the ePR, which is patient-oriented (i.e., data goes where the patient goes).

Up-to-date information on these topics can be found in multidisciplinary literature, reports from research laboratories of university hospitals, and medical imaging manufacturers, but not in a coordinated way. Therefore, it is difficult for a radiologist, hospital administrator, medical imaging researcher, radiological technologist, trainee in diagnostic radiology, or student in engineering and computer science to collect and assimilate this information. One major purpose of this section is to provide a brief overview and consolidate PACS-related topics and PACS integration with HIS and ePR. PACS and medical imaging informatics is an ever-growing field that mirrors the ever-changing IT landscape. However, the fundamental concepts remain as important as ever and continue to form the bedrock for this expanding field.

PACS has impacted the health care industry financially and operationally, streamlining clinical workflow and increasing the efficiency of the health care enterprise. Medical imaging informatics infrastructure is an emerging field focused to take advantage of existing PACS resources and image and related data for large-scale horizontal and longitudinal clinical, research, and education applications that could not be performed previously because of insufficient data.

13.8 Exercises

1. Based on the generic PACS basic components diagram and data flow (Figure 13.1) and the components descriptions, identify the single points of failure for both stand-alone and client/server PACS architectures.
2. Describe how the clinical workflow would be impacted for each of the single points of failure.
3. Provide solutions to address the single points of failure identified.
4. Develop a testing script to perform acceptance testing for each of the single points of failure.

13.9 References and Bibliography

R. A. Bauman, G. Gell, and S. J. Dwyer III. Large picture arching and communication systems of the world—Part 1. *J. Digital Imaging.* 9(3):99–103, 1996.

R. A. Bauman, G. Gell, and S. J. Dwyer III. Large picture arching and communication systems of the world—Part 2. *J. Digital Imaging.* 9(4):172–177, 1996.

R. E. Dayhoff, K. Meldrum, and P. M. Kuzmak. Experience providing a complete online multimedia patient record. Session 38. *Health Care Information and Management Systems Society, 2001 Annual Conference and Exhibition. Feb.4–8,* 2001.

R. Dayhoff and E. L. Siegel. Digital imaging within and among medical facilities. In R. Kolodner (Ed.). *Computerized Large Integrated Health Networks—The VA Success.* Springer Publishing, 473–490, 1997.

D. S. Channin. Integrating the health care enterprise: A primer. II. Seven brides for seven brothers: The IHE integration profiles. *RadioGraphics.* 21:1343–1350, 2001.

D. Channin et al. Integrating the health care enterprise: A primer. III. What does IHE do for me? *Radio Graphics.* 21:1351–1358, 2001a.

D. S. Channin et al. Integrating the health care enterprise: A primer.V. The future of IHE. *RadioGraphics.* 21:1605–1608, 2001b.

DICOM Standard 2003. Available at http://medical.nema.org/.

DICOM: *Digital Imaging and Communication in Medicine.* National Electrical Manufacturers' Association, 1996.

A. J. Duerincks. Picture archiving and communication system (PACS). *Proc. SPIE for Medical Applications.* 318, 1982.

HL7. *Health Level Seven. An application protocol for electronic data exchange in health care environments. Version 2.1.* Health Level Seven, Inc., 1991.

Health Level Seven. Available at http://www.hl7.org/.

HL7 Version 3.0: Preview for CIOs, Managers, and Programmers. Available at http://www.neotool.com/company/press/199912_v3.htm#V3.0_preview.

H. K. Huang. Enterprise PACS and image distribution. *Comp. Med. Imaging Graphics.* 27(2–3):241–253, 2003.

H. K. Huang. *PACS and Imaging Informatics: Basic Principles and Applications.* Wiley & Sons, 2004.

H. K. Huang. *PACS: Principles and Applications.* Wiley & Sons, 1999.

H. K. Huang et al. Design and implementation of a picture archiving and communication system: The second time. *J. Digital Imaging.* 9:47–59, 1996.

H. K. Huang, S. T. C. Wong, and E. Pietka. Medical image informatics infrastructure design and applications. *Med. Informatics.* 22(4):279–289, 1997.

B. J. Liu et al. Trends in PACS image storage and archive. *Comp. Med. Imaging Graphics.* 27(2–3):165–174.

B. J. Liu et al. PACS archive upgrade and data migration: clinical experiences. *Proc. SPIE Medical Imaging.* 4685: 83–88, 2002.

B. J. Liu et al. A fault-tolerant back-up archive using an ASP model for disaster recovery. *Proceeding SPIE Medical Imaging.* 4685:89–95, 2002.

C. J. McDonald. The barrier to electronic medical record systems and how to overcome them. *J. Am. Med. Informatics Assoc.* 4(May/June):213–221, 1997.

R. Osman, M. Swiernik, and J. M. McCoy. From PACS to integrated EMR. *Comp. Med. Imaging Graphics.* 27(2–3): 207–215, 2003.

E. L. Siegel, J. N. Diaconis, S. Pomerantz et al. Making filmless radiology work. *J. Digital Imaging.* 8:151–155, 1995.

E. L. Siegel and B. I. Reiner. Filmless radiology at the Baltimore VA medical center: A nine-year retrospective. *Comp. Med. Imaging Graphics.* 27(2–3):101–109, 2003.

E. L. Siegel and D. S. Channin. Integrating the health care enterprise: A primer—Part 1. Introduction. *RadioGraphics.* 21:1339–1341, 2001. www.rsna.org/IHE

F. Yu et al. Some connectivity and security issues of NGI in medical imaging applications. *J. High Speed Networks.* 9:3–13, 2000.

X. Zhou, H. K. Huang. Authenticity and integrity of digital mammography image. *IEEE Trans. Medical Imaging.* 20(8):784–791, 2001.

14

KMeX: A Knowledge-Based Digital Library for Retrieving Scenario-Specific Medical Text Documents

Prof. Wesley W. Chu,
Dr. Zhenyu Liu,
Dr. Wenlei Mao, and
Dr. Qinghua Zou
University of California, Los Angeles (UCLA)

14.1 Introduction

Medical records, such as patient records, lab reports, literature articles, and newsletters, are in free-text form, and oftentimes medical practitioners wish to perform scenario-specific retrieval on these documents. A *scenario* typically refers to a specific health care task, such as searching for treatment methods for a specific disease. Although traditional systems are useful for general *information retrieval* (IR), these systems cannot support scenario-specific IR because:

1. The terms in the query posed by the user may not use a standardized medical vocabulary.
2. There is no effective technique to represent synonyms, phrases, and similar concepts in free text.
3. The terms used in a query and those used in a document for representing the same topic may be mismatched.

In this chapter, we present a new knowledge-based approach (e.g., using the Unified Medical Language System [UMLS]) to mitigate these problems. More specifically, we propose to use the metathesaurus and semantic structure in the UMLS to extract key concepts from the free text for (1) indexing, (2) phrase-based indexing for representing similar concepts, and (3) query expansion to improve the probability of matching query terms with the terms in the document. To do so, the system formulates the query based on the user's input and selects scenario templates such as "disease, treatment" or "disease, diagnosis." Thus, the system is able to retrieve relevant documents for a specific scenario. Furthermore, we propose a topic-oriented directory that is generated based on query-templates and frequently occurring relevant topics in documents. Such a directory system not only selects a set of relevant documents in respect to the query template but also provides cross-references among related topics. These techniques have been implemented in a testbed at the University of California, Los Angeles (UCLA). Using the standard California Office of Statewide Health medical corpus. Our empirical results validate the effectiveness of this new approach over the traditional text retrieval techniques.

Medical information knowledge and clinical data are growing at explosive rates. Ten years ago, medical publications were being added to the world's biomedical journal collections at the rate of approximately 3,000 entries per month. Today, the volume of bibliographic citations is growing at 1,000 per day in Medline alone [1]. Hospitals also generate large amounts of health care data that are stored on computers. Hence, the delivery of quality health care to consumers requires availability and accuracy of IR from large information sources. The demand for the use of evidence-based practices to help improve the quality of care also puts great pressure on health care professionals to regularly access the highest-quality information during health care planning, decision, and delivery. Today, computer-assisted IR and processing are necessary to support quality decision making and to help overcome human cognitive constraints [2].

A medical digital library consists of three types of data: (1) structural data, such as from patient lab results and demographic studies, (2) multimedia data, such as images from magnetic resonance imaging (MRI), and (3) free-text documents, such as patient reports, medical literature, teaching files, and news articles. Previous research focused on the effective retrieval of structural data and image data [3–4]. However, as a rule, medical records are in free-text form and usually require scenario-specific retrieval. For example, a physician may pose the following two queries, one for diagnosis and the other for treatment of a disease:

- Diagnosis scenario: "***diagnosis*** of large-cell lung cancer," from all patient reports
- Treatment scenario: "***treatment*** of large-cell lung cancer," from the collection of medical literature articles (e.g., Medline references).

From this scenario, specific queries cannot be effectively supported by traditional IR systems because of the lack of indexing for free text, ranking the similarity of the content within the document with the query term and a method to resolve the mismatch of the term in the query with that in the document. We developed the following knowledge-based techniques to ameliorate these problems.

14.1.1 Extracting Key Concepts from Free Text

We have developed a new technique of knowledge-based medical extraction to automatically extract key concepts from free text and to permute the set of words in the input free text, thereby generating all valid concepts defined by the controlled vocabulary in a knowledge base (e.g., UMLS). Since the generated valid concept may not be relevant to the query, syntactic and semantic filters are then used to filter out the irrelevant concept. Thus, retrieval efficiency is improved because key concept terms can be used as indices in a free-text directory system, as well as transforming the ad hoc terms in the query into a controlled vocabulary.

14.1.2 Phrase-Based Vector Space Model

Vector space models (VSMs) are commonly used to measure the similarity between a query and a document. Traditional stem-based VSMs cannot match terms in the query with those used in the documents that have similar meanings but different expressions. We developed a knowledge-based/phrase-based VSM [5], which identifies terms with similar meanings and represents them based on both concepts and stems. As a result, this phrase-based VSM yields significantly better retrieval performance than the stem-based VSM.

14.1.3 Knowledge-Based Query Expansion

Queries can be appended with related terms to increase the probability of matching the terms in the query with those of relevant documents. Traditional expansion techniques append all statistically co-occurring terms into the original query, but many of the expanded terms may not be scenario specific. We use a knowledge-based approach that appends the query with only terms related to the scenario of the query.

14.2 Extracting Key Concepts from Documents

14.2.1 The Knowledge Source of the Unified Medical Language System

Since our approach is leveraged on knowledge bases, we shall first briefly describe the UMLS [6] knowledge source, then present an index tool called IndexFinder, which is used for extracting key concepts from free texts. UMLS is a standard medical knowledge source developed by the National Library of Medicine and composed of the UMLS metathesaurus, the SPECIALIST lexicon, and the UMLS semantic network.

The metathesaurus is a central vocabulary component that contains 1.6 million phrases representing over 800,000 concepts from more than 60 vocabularies and classifications in its 2003 edition. We use the metathesaurus as the controlled vocabulary to detect concepts and to derive the conceptual relations using the hyponym relations encoded in it.

A *concept-unique identifier* (CUI) identifies each concept. The metathesaurus encodes "broader-narrower-than" types of relations among the concepts. For example, "lung cancer" is a broader concept than "lung neoplasm." A class of concepts in the metathesaurus is abstracted into one *semantic type* in the semantic network. For example, the concept "lung cancer" belongs to the semantic type "Disease and Syndrome." Each semantic type has several semantic relationships with other types—for example, "Disease and Syndrome" is "treated by" "Therapeutic or Preventive Procedures," "Pharmacological Substance," and "Medical Devices." These semantics are used for knowledge-based query expansion (see Section 14.6).

14.2.2 Indexing for Free-Text Documents

Indexing free text is a difficult task, since its writing does not use a controlled vocabulary. Similar concept terms and synonyms in free text add an additional level of difficulty to such a task. This also applies to ad hoc queries that can be viewed as documents. Unlike medical literature, which provides key words, many free-text documents do not provide such information. To effectively retrieve these free texts, we are motivated to extract the key concepts from these documents. To rapidly retrieve the relevant information/knowledge for a query from a large number of documents, we propose to use a topic-oriented directory system for free text where the document can be obtained based on a set of index terms. Having located a group of documents that satisfy the key concept terms, traditional IR techniques can then be used to rank these documents.

Thus, extracting key concepts from free text is a critical task. Words or word stems are commonly used for indexing, and these indexing techniques do not require any knowledge source. However, synonyms and some morphological differences between the texts in the target documents and the search words used often hamper the search results and are beyond the technological spectrum of word/stem indexing and matching techniques. This issue is particularly problematic in health care, wherein the biomedical language is packed with many interchangeable terms, such as "common cold" and "coryza," "mass" and "lump," "fever" and "pyrexia," "weakness" and "paresis," etc. Therefore, we developed indexing systems based on standard descriptors or dictionaries, such as the UMLS.

TABLE 14.1 Problems with mapping noun phrases individually

Example	Text
1	**Prostate**, right (biopsy) fibromuscular and glandular **hyperplasia**
2	A small **mass** was found in the **left** hilum of the **lung**.

Using search terms generated from standard dictionaries also helps to resolve the differences in synonyms and morphologies and thus reduce user frustration by minimizing the rates of missed hits and failed searches. A significant amount of research has been dedicated to developing effective methods for mapping free text into UMLS concepts. Examples of such efforts include SENSE (SEarch with New SEmantics) [7], MicroMeSH (Micro–Medical Subject Headings) [8], Metaphrase [9], KnowledgeMap [10], PhraseX [11], and MetaMap [12]. Many of these efforts use techniques of *natural language processing* (NLP) to parse passages of free text to generate noun phrases, which are in turn mapped into UMLS phrases. This approach achieves some success; however, it has two major weaknesses:

First, some important concepts cannot be discovered through the identification of noun phrases, because they span multiple noun phrases. Table 14.1 provides examples of texts that reveal the shortcomings of the use of noun phrases.

- Example 1: A word from the first line with a word from the second line forms the key concept, "prostate hyperplasia," which corresponds to concept ID 33577 in the UMLS metathesaurus.
- Example 2: A word from the subject and two words from the location phrase combine to form the key concept, "left lung mass," which corresponds to concept ID 746117 in the UMLS metathesaurus.

Second, NLP requires significant computing resources. As a result, most of the NLP systems work in an offline mode and are not suitable for mapping large volumes of free text into UMLS concepts in real time. To remedy these shortcomings, we developed a new tool called IndexFinder [13] to extract key concepts from free text.

14.2.3 IndexFinder

IndexFinder operates by permuting words in a sentence to generate concept candidates that match the UMLS-controlled vocabulary. Since the generated valid controlled vocabulary and concept terms may contain negative sense and may not be relevant to the query, negation detection is used to identity negative concepts. Further, syntactic and semantic filters based on a specific scenario are used to filter out irrelevant concepts.

14.2.3.1 Text Preprocessing

Since IndexFinder uses the UMLS normalized string table for indexing and also supports certain types of abbreviations, we need to preprocess the input text to normalize words, detect undefined and ambiguous abbreviations, and remove stop words to increase the accuracy of the extraction.

IndexFinder first converts the UMLS controlled vocabulary into an efficient concept indexing structure that resides in the main memory and thus avoids disk access. To detect the concepts embedded in a free-text sentence, IndexFinder scans through the sentence word by word, looks up the indexing structure, and marks every concept where all the words representing that concept have appeared in the sentence. We use the UMLS SPECIALIST lexicon for word normalization and handle synonyms by mapping different wording of the same concept into one entry in the indexing structure. This indexing and matching technique is efficient and able to generate responses in real time for free-text indexing.

14.2.3.2 Negation Detection

Negation detection is an important task in medical document processing, since whether or not a medical symptom has been presented can make for totally different diagnoses of a disease. If a doctor searches for the concept "no cough," returning the concept "cough" is considered to be irrelevant. To handle the negation problem in IndexFinder, we first define a list of terms and negation hues, which carry negative sense for a concept. Then, we identify the UMLS semantic types that can be negated. Finally, for the concepts of these defined semantic types, we combine them with possible negation hues according to certain defined rules. More specifically, IndexFinder relies on the three parts for negation detection:

- *Negation hues list:* Specifies the list of words that tend to negate a concept in a sentence. For example, in medical reports, the words "no," "not," "isn't," etc., are frequently used for negation.
- *UMLS semantic types qualified for negation:* Specifies the list of UMLS semantic types that can be negated. For example, the semantic type T191 (disease/cancer) is qualified for negation, since the concepts related to T191 can appear in patient records in negation form.
- *Rules for negating concepts in a sentence:* Specifies the rules for negating UMLS concepts when negation hues are presented in the same sentence from which the concepts are extracted. For example, "no" tends to negate the immediately succeeding concept; when multiple concepts are qualified for negation, the concept closest to the negation hue is selected for negation.

Figure 14.1 shows the web interface for IndexFinder. The interface has two text panes: The upper text pane takes free text as input, and the lower text pane outputs the identified UMLS concepts. Each line in the output pane shows one identified concept, which contains the concept ID, the concept's phrase string, and the concept's semantic type. Part of the UMLS concepts detected from the input pane is shown in the output pane. Three buttons for adding synonyms, removing inflection, and configuring options, respectively, are at the top of the input window. Results appear when a user clicks the "IFinder Search" button below the input window. Eighteen phrases

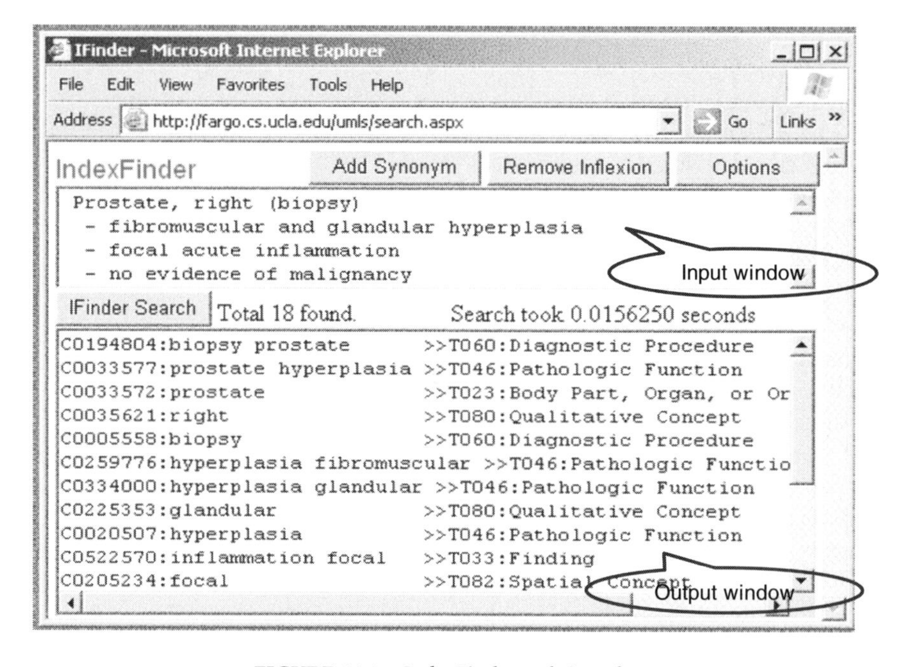

FIGURE 14.1 IndexFinder web interface.

were found when no filters were applied. Each line has a UMLS concept identifier, phrase text, and corresponding semantic type.

14.2.3.3 Syntactic and Semantic Filtering

Although word permutation detects more concept candidates, some concepts may be irrelevant to the original sentence. IndexFinder applies filters that use knowledge source and syntactic or semantic information from the original sentence to filter out irrelevant concepts. For example, if a physician wishes to know what kind of diseases a patient suffers from, it is more desirable to return disease-related UMLS phrases rather than all concepts to the physician. We consider six types of filters, as shown in Figure 14.2.

The first three filters are applied during the mapping process:

- The *symbol type filter* specifies the symbol types of interests. For example, a user who wants to ignore digits (as MetaMap does) can simply not check the "Digits" box, shown in Figure 14.2.
- The *term length filter* specifies the length limitation of candidate phrases.
- The *coverage filter* specifies the coverage condition for a candidate phrase. It has three options: *at least one*, *majority*, and *all*. By default, the *all* option is where every word in a candidate phrase should be present in the input text.

The later three filters are used for further pruning of the candidate phrases:

- The *subset filter* removes phrases if they are subsets of other phrases. For example, if the results are {*lung cancer*} and {*cancer*}, then {*cancer*} will be removed, since it is a subset of the former.
- The *range filter* removes a phrase if the phrase is found from words in the input text to exceed a specific distance.
- The *semantic filter* removes the phrases of semantic types that the user is not interested in. In UMLS, 134 semantic types are defined, and each concept maps to one or several semantic types. For example, as shown in Figure 14.2, the user can select Disease or Syndrome and its two subtypes, so that the resulting phrases will be of these two types. As a result, the filter also eliminates those irrelevant phrases from the set of phrase candidates. Note that the UMLS "is-a" relationship may also be used to filter out more general phrases.

Table 14.2 shows the filtering result for the sample input in Figure 14.1 (also given at the top of Table 14.2). When a subset filter is used, eight phrases are returned. If the Pathologic Function is selected, four answers will be returned. The two terms "prostate" and "focal" will be given if the user wishes to know about body parts or spatial characteristics. Prostate biopsy is the only diagnostic procedure used.

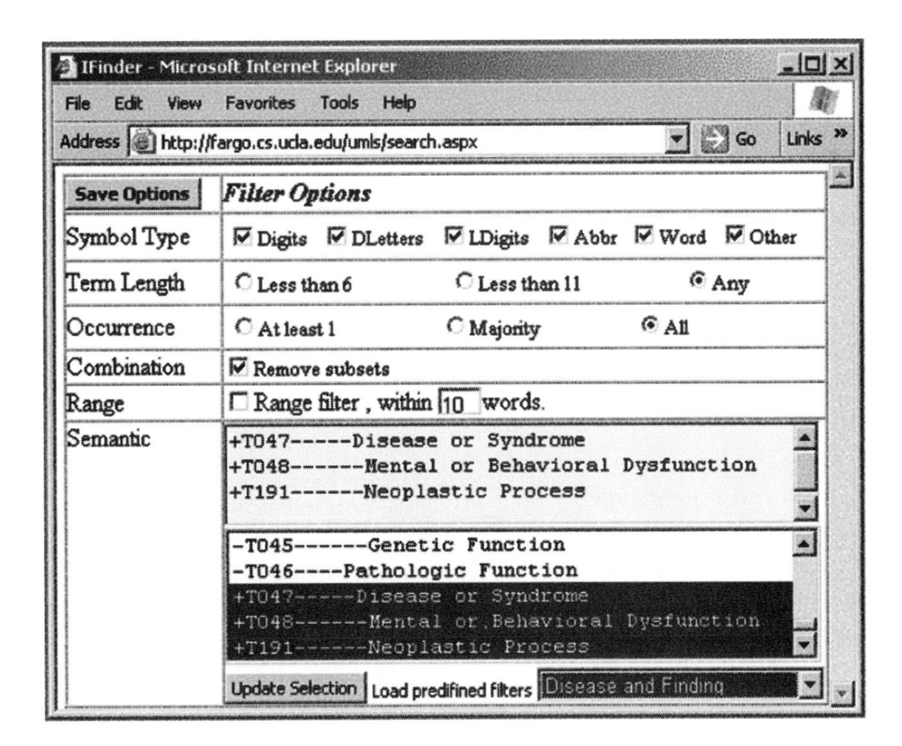

FIGURE 14.2 Filter selection.

TABLE 14.2 Key concepts after filtering

Input: Prostate, right (biopsy)
- fibromuscular and glandular hyperplasia
- focal acute inflammation
- no evidence of malignancy

Filtering	Results
Subset	C0194804: Biopsy prostate C0033577: Prostate hyperplasia C0035621: Right C0259776: Hyperplasia fibromuscular C0334000: Hyperplasia glandular C0522570: Inflammation focal C0333361: Inflammation acute C0391857: No malignancy evidence
Pathologic function (T046)	C0033577: Prostate hyperplasia C0259776: Hyperplasia fibromuscular C0334000: Hyperplasia glandular C0333361: Inflammation acute
Body parts and spatial (T023, T082)	C0033572: Prostate C0205234: Focal
Diagnostic procedure (T60)	C0194804: Biopsy prostate

TABLE 14.3 Comparing results generated by IndexFinder and MetaMap

Input: A small mass was found in the left hilum of the lung.

IndexFinder Results	
C0024873: A mass	>>T190: Anatomical Abnormality
C0700321: Small	>>T080: Qualitative Concept
C0746117: Mass lung left	>>T033: Finding
C0332285: Found	>>T082: Spatial Concept
C0225733: Lung left hilum	>>T029: Body Location or Region
MetaMap Results	
Phrase: "A small mass"	861 Mass, NOS [Anatomical Abnormality] 694 Small [Qualitative Concept]
Phrase: "was"	Meta mappings: <none>
Phrase: "found"	Meta mappings: <none>
Phrase: "in the left hilum"	1000 Left Hilum [Body Part, Organ, or Organ Component]
Phrase: "of the lung"	1000 Lung [Body Part, Organ, or Organ Component] 1000 Lung<3> (Lung diseases) [Disease or Syndrome]

14.2.3.4 Evaluation

IndexFinder is written in C# and runs on a 1.2 GHz personal computer (PC) with 512 MB main memory. We implemented the IndexFinder algorithm as a web-based service that provides web interfaces for users and programs. We tested the web service using 5,783 reports of 128 patients from the UCLA Medical Center. The total size of the documents was 10.8 million bytes. There were 910,000 concepts found in 254 seconds. Therefore, the throughput was about 42,700 bytes per second, which validated that the system could extract key concepts from clinical free texts in real time. Next, we manually examined the mapping results for 100 topic sentences from this set of patient reports. There were a total of 456 UMLS phrases found of the 100 topic sentences. We noticed 18 concepts that were not from a single noun phrase and thus could not be detected by NLP-based methods. Further, we noted that all the concepts detected by IndexFinder were relevant. Filtering was effective in eliminating the irrelevant terms from the validated candidates.

14.2.3.5 Comparison with Natural Language Processing

We performed a comparison study between IndexFinder and MetaMap, which uses the NLP method. We noticed that the NLP tends to break each sentence into small fragments. Conversely, IndexFinder considers all the possible word combinations in the input unit that are valid in UMLS. As a result, NLP does not yield concepts as specific as IndexFinder, as shown in Table 14.3.

We are currently in the process of further evaluating the accuracy of our method. We plan to generate a test dataset by randomly selecting a set of topic sentences from the 5,783 patient reports and then comparing the accuracy of the indexing terms generated by the IndexFinder in terms of the numbers of false negatives and false positives [14].

The key terms extracted by IndexFinder can be used for (1) indexing the free-text documents, which can be used in the directory system for linking the documents with key concepts; (2) formulating scenario-specific queries for content correlation; and (3) transforming the ad hoc query terms to controlled vocabulary, thus increasing retrieval effectiveness.

14.2.3.6 An Example

As a specific clinical application for this research, we have focused on using the IndexFinder to intelligently filter all clinical free text in an electronic medical record for documents that specifically mention brain tumor–related content. It is not uncommon for brain tumor patients to have as many as 50 clinical documents in their medical records. Many of these documents will have nothing to do with the treatment of the brain tumor but are concerned with other health problems. These documents consist of primary care clinical notes, specialist clinical notes, pathology reports, laboratory results, radiology reports, and surgical notes. For instance, free text from a radiology report would read in part:

> The right frontal convexity meningioma is slightly larger now than on the prior examination. The left frontal meningioma is unchanged. There are three other small enhancing nodules seen along the frontal convexities bilaterally, as

TABLE 14.4 Using UMLS semantic type to define interests

Brain tumor characteristics	Relevant UMLS semantic types
Specific cancer	Neoplastic process
Medical intervention	Therapeutic procedure
Anatomical location	Body part, organ, or organ component

TABLE 14.5 Output from IndexFinder for the radiology report excerpt

Semantic descriptor	UMLS code
T191: Neoplastic process	C0025286: Meningioma
T047: Disease or syndrome	C0014068: Encephalomalacia

described above. There are no new lesions seen. There is no mass effect caused by these lesions. There is bifrontal encephalomalacia.

Since our interests focus on brain tumor–related concepts, we can specify a semantic filter work list of pertinent documents based on brain tumor characteristics including cancer type, anatomical location, and medical interventions. These characteristics are then mapped to relevant UMLS semantic types to define semantic filters, as shown in Table 14.4.

A clinician looking for specific documents that address a certain type of brain tumor (e.g., "meningioma") would have to carefully search the individual documents. With IndexFinder, only two key terms, "meningioma" and "encephalomalacia," are returned for the radiology report in our example, as shown in Table 14.5. The two concepts, in fact, are important in the excerpt and thus are good terms for indexing.

14.3 Transforming Similar Queries into Query Templates

Recent studies reveal that users' information requests in a specific domain typically follow a limited number of patterns. In the medical domain [15–18], for example, more than 60% of all the physicians' clinical questions can be classified into 10 frequent categories. We can summarize the frequently asked similar queries and tailor our retrieval system according to the summarized queries. This motivates us to introduce the notion of a *query template.* A query template defines the structure of a group of similar queries that consist of a key concept and scenario concept(s). Filling in the key concept values in a query template results in a specific free-text query.

To find out how to define a query template, we shall investigate a few medical queries presented in Hersh et al. [16] from the Oregon Health and Sciences University medical corpus (OHSUMED).

Q_1: **Lactase deficiency,** *therapy options*
Q_2: **Iron deficiency anemia,** *which test is best*
Q_3: **Thrombocytosis,** *treatment and diagnosis*

where the disease concepts are in boldface, and the scenario concepts are in italics. By inspecting these queries, we note that each focuses on a particular disease concept: "lactase deficiency," "iron deficiency anemia," or "thrombocytosis." Such disease concepts provide the focus of each query. Further, each query asks about a specific scenario related to the disease concept. For example, Q_1 asks about the "treatment" scenario of a disease; Q_2 asks about the "diagnosis" scenario; and Q_3 asks about both.

To generalize these sample queries, we can extract the key concept and scenario concepts (the structural information) and transform them into the following templates. Note that in the templates, we unify the representation of scenario concepts—for instance, mapping "therapy options" to "treatment."

T_1: <Disease and syndrome>, treatment
T_2: <Disease and syndrome>, diagnosis
T_3: <Disease and syndrome>, treatment and diagnosis

Thus, in general, each query template has two essential components:

1. *The key concept.* In the template, we specify only the semantic type of this concept (e.g., "Disease and Syndrome"). The user needs to fill in the concept value to generate a concrete query. For example, filling in "lung cancer" in template T_1 results in a real query of "lung cancer, treatment." Further, the concept must belong to the semantic type defined in the template—for instance, "lung cancer" must be a "Disease and Syndrome" concept.
2. *One or more scenario concepts.* For example, "treatment," "diagnosis," and/or "complication" of some disease concept

In the following sections, we shall illustrate how we use the structural information in query templates to organize the key document features into a topic-oriented directory. Further, the structural information in query templates enables us to expand more scenario-specific terms to the original query and significantly improve the retrieval performance.

14.4 Topic-Oriented Directory

To improve the efficiency of free-text document retrieval in terms of precision and recall of the request documents and to provide cross-reference among related topics, we shall propose to use a directory system that is based on user queries and topic/subtopic hierarchies derived from key features of the documents.

Using our IndexFinder, we are able to automatically extract a set of key features to represent a document. Next we will use data-mining techniques to identify frequently co-occurring key features. Each group of frequent features can be viewed as a directory topic. Since these topics are directly derived from the document content without any generalization, we consider them to be the most specific ones in the directory. Therefore, they are placed at the leaf level of the topic hierarchy. Starting from the most specific topics, we merge these subtopics into more general topics. By continuing this process, a topic hierarchy can be eventually constructed. In order for the merging process to be semantically meaningful, the process is guided by the semantic network in the knowledge base (e.g., UMLS). For the topic hierarchy to be sensitive to directory users, we should reorganize the hierarchy also based on the user querying patterns. One way to achieve this is to adjust the hierarchy so that it corresponds to the frequent browsing patterns from general to specific topics. This can be accomplished by modifying the knowledge hierarchies in the semantic network in accordance with the query granularity to form directory paths and by concatenating these directory paths in the drill-down browsing patterns. As a result, the topics and subtopics in the directory hierarchy are derived based on key features in the documents, as well as on user query patterns.

Such a directory design differs from existing document clustering techniques in the following ways. First, our directory topics are derived from the documents and represented by control vocabulary from a knowledge source. In conventional document clustering, each tree node represents only a subgroup of documents without any semantic meaning. Second, our directory topics are generated from mining the document key features as well as the user queries (query templates). As a result, our directory system can adapt to different types of queries and is user sensitive. Existing document clustering techniques do not consider information related to frequent query patterns and user type. Third, the directory topic hierarchy is organized by the guidance of the semantic network in the knowledge source (e.g., UMLS) and is therefore well defined. The resulting directory structure has more semantic meaning than the statistical approaches and thus is able to provide scenario-specific indexing and improve document retrieval performance.

Let us illustrate the process of organizing a topic-oriented directory system by the following example. Given a large corpus of documents related to disease, we will design a directory system for lung cancer physicians. Based on their interests, most of the query will be related to lung cancer; that is, its diagnosis, treatment, risk factors, etc. As a result, the document collection for these particular physicians can be divided into three topics: lung cancer–related, general cancer–related, and other disease-related documents, as shown in Figure 14.3. Through data mining of the key features of these documents, we are able to derive the following list of topics from the broader topics: lung cancer, diagnosis, treatment, risks of cancer, chemotherapy, surgery, radiation, etc. Such topics and subtopics can be organized with the guidance of the semantic network of UMLS. For example, the topic "lung cancer" can be further divided into various subtopics such as "diagnosis," "treatment," and "risk factors of cancer." Then, based on the knowledge source, the subtopic "treatments" can be organized into the subsubtopics "chemotherapy," "surgery," and "radiation." Since the topics are derived from the key features of the documents, such topics and subtopics can be indexed to represent scenario-specific topics.

Note that the directory is organized based on a given user query (query template), as well as topics and subtopics that derive from the key features of documents. Thus, the directory system not only can provide scenario-specific document retrieval, but can also improve document retrieval performance. Likewise, we can organize the directory system for different user query templates. These different directory systems can be linked and formed into a general directory system for the set of query templates. Nodes in the directory of a query may overlap with nodes in the directories of some other queries. Such overlap provides cross-references of topics and increases the search scope of the nodes (topics). For a given query, the system will navigate according to its directory to retrieve the documents. The overlap nodes may provide cross-references to different scenarios in other directory systems. In order to restrict the cross-reference topics, the user can provide a certain range of topics of interest. As a result, the directory navigator will branch only to these topics. Such focused cross-referencing can increase

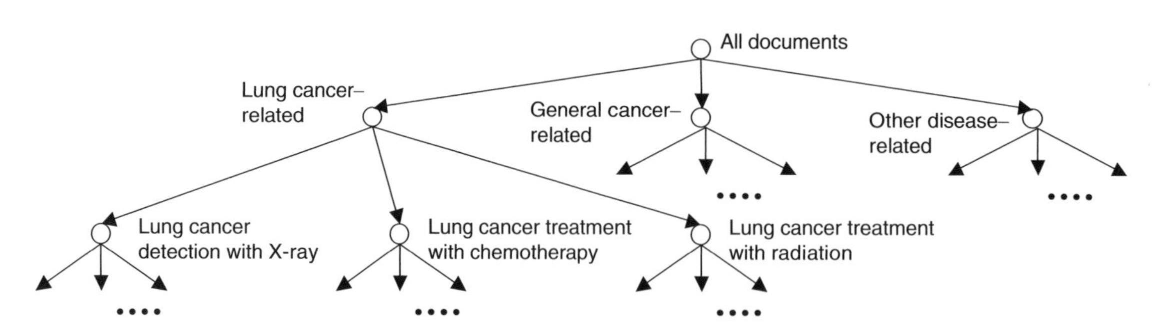

FIGURE 14.3 A sample directory system for a lung cancer physician.

the search scope while providing focused expansion of topics and improving retrieval performance.

14.4.1 Deriving Frequent Directory Topics via Data Mining

Using IndexFinder, we can extract a set of key features from each document. Each feature is a concept defined in the controlled vocabulary of the knowledge source (e.g., the metathesaurus in UMLS). In this section, we will present a data-mining technique to discover topics from document features for directory construction.

A topic can be viewed as a condensed synopsis of a subcollection of documents. For example, "lung cancer and chemotherapy" is a topic that covers all the documents on the treatment of "lung cancer" with "chemotherapy." To capture the meaning of a subset of documents, we typically need multiple concepts, such as "lung cancer" and "chemotherapy." Therefore, a specific topic should consist of multiple concepts. Further, the concepts that belong to one topic should frequently co-occur in the documents. For example, it is meaningless to combine "back pain" and "heart surgery" within a topic, because very few medical documents mention both concepts.

Since a topic is a group of concepts that frequently co-occur in documents, we propose to use frequent item-set mining techniques [19–21] for topic discovery. To map topic discovery into a frequent item-set mining problem, we shall view each document as a market basket, and the concept features extracted from that document as the items in the basket. To use the data-mining techniques for topic discovery, we need to specify a minimum support number. In the topic-discovery context, this minimum support is the minimum number of documents that we want to group under each topic. For example, if any topics in our directory cover at least five documents, then we should set the support level at "5." Table 14.6 further illustrates the mapping between topic discovery and data mining.

If we discover that each topic is a group of frequently co-occurring concepts, any subportion of that group must also be frequent. That is, any subportion of a topic is also a valid topic. For example, if we discovered the topic {"lung cancer," "detection," "biopsy"} as a group of frequent concepts, then a subgroup such as {"detection," "biopsy"} must also be a valid topic. Supergroups of concepts have more specific meanings than subgroups—for instance, {"lung cancer," "biopsy," "detection"} is more specific than {"biopsy," "detection"}. Therefore, it would be desirable to keep only the topics that are supergroups and not those of the subgroups. To efficiently discover these supergroups, we need a specialized data-mining technique called *maximum frequent item-sets* (MFI) mining. We have developed a general-purpose MFI mining algorithm, SmartMiner, which can handle extremely large datasets [21]. We plan to apply this technique to discover topics that have the longest and the most specialized form and use this to construct a more accurate directory system.

TABLE 14.6 Mapping between topic discovery and frequent item-set mining

Topic discovery	Frequent item-set mining
Document	Market basket
Concepts in a document	Items in a basket
Topic as a group of frequently co-occurring concepts	Frequent item set
Minimum number of documents under each leaf topic	Minimum support for frequent item sets

14.4.2 Organizing Topics into a Hierarchical Directory Structure

By mining frequent co-occurring features in the document collection, we obtain a list of topics each with a corresponding set of documents covered by that topic. We shall build a hierarchical structure from these topics for efficient retrieval of relevant documents. Since topics derived from mining frequent document features are the most specific ones in the hierarchy, they are placed at the leaf level. Starting from these most specific topics, we can construct a topic hierarchy by iteratively merging subtopics into more general ones. To construct a scenario-specific and query-sensitive hierarchical directory, we will leverage using a knowledge source, UMLS, and the query templates. The knowledge source organizes its concepts in a general-to-specific fashion. For example, "lung neoplasm" is a more general term than "lung cancer," and "lung cancer" is more general than "non-small-cell lung cancer." This provides useful guidance to determine the general-to-specific relationships among directory topics. For example, "lung cancer with chemotherapy" will be considered more general than "non-small-cell lung cancer with chemotherapy."

UMLS defines multiple hierarchies of concepts. Each UMLS concept hierarchy focuses on one semantic type of concept. For example, the disease-concept hierarchy represents the general-to-specific relationships among all the "Disease and Syndrome" concepts. Similarly, the procedure-concept hierarchy focuses on all "Therapeutic and Preventive Procedure" concepts. The information in query templates can be used to select the appropriate candidate hierarchy.

Let us consider the following example. Suppose that we have discovered four specific topics by mining the key features of the documents:

1. "Lung cancer, surgery"
2. "Lung cancer, radiotherapy"
3. "Heart disease, surgery"
4. "Heart disease, drug therapy"

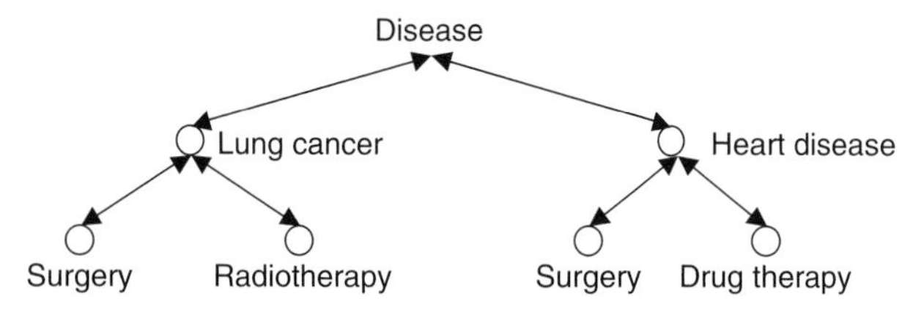

(a) Directory structure derived from the disease-concept hierarchy

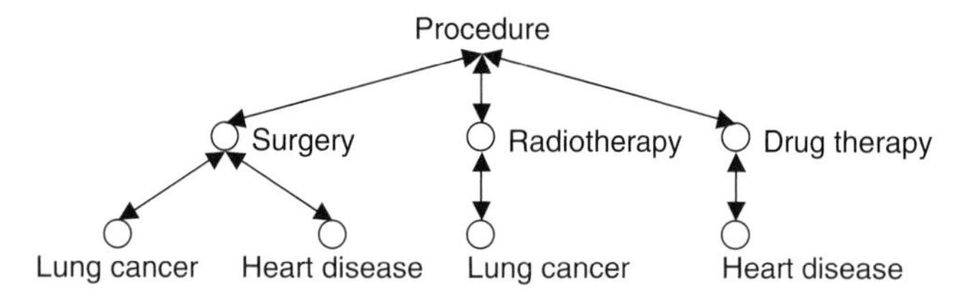

(b) Directory structure derived from the procedure-concept hierarchy

FIGURE 14.4 Different directory structures derived from different sets of query templates.

Following UMLS's disease-concepts hierarchy, entries 1 and 2 and entries 3 and 4 are two pairs of similar topics. At a higher level, both of these two topics fall under a general topic called "disease." The resulting directory structure is shown in Figure 14.4(a). If we use UMLS's procedure-concept hierarchy, the resulting directory structure is shown in Figure 14.4(b). We shall leverage the query template information to select the appropriate candidate structure.

Recall that a query template consists of two parts: a key concept and a set of scenario concepts. For a particular user type, we can identify the set of frequently used query templates. Suppose the key concepts in these frequent templates are "<Disease and Syndrome>"; that is, many templates are seen as "<Disease and Syndrome>, treatment" or "<Disease and Syndrome>, diagnosis," etc. Now consider a sample query constructed by the template "lung cancer, treatment." In the initial step, we want the directory to guide us to a single branch that is all about "lung cancer." Underneath that single branch, we want to further focus on the treatment subtopic. Clearly, the structure in Figure 14.4(a) serves this need better than that in Figure 14.4(b). On the other hand, if the key concept in the query templates is of type "<Therapeutic and Preventive Procedure>," then the structure in Figure 14.4(b) will be preferable.

14.4.3 Navigating the Topic-Oriented Directory

The topic-oriented directory is constructed by the set of topics that are generated by data mining, query templates of a particular user type, and the semantic structure of the knowledge source. With the directory system, identifying a set of relevant documents for a given query is equivalent to selecting a path in the hierarchy to navigate to a leaf node. The path selection should be based on user type and query templates. Further, the directory enables us to easily navigate to broader topics related to the query. For example, if we use the directory in Figure 14.4(a) to answer the query "lung cancer treatment with surgery," we first select the path "disease" → "lung cancer" → "surgery" to reach a subset of documents. Thereafter, we can suggest further reading in the closest path "disease" → "lung cancer" → "radiotherapy."

Multiple directory structures are constructed from the query templates for multiple user types. The commonality in query templates results in overlapping nodes of various directory structures. Such overlapping nodes provide cross-referencing points among multiple directories and enlarge the search scope. For example, the leaf node for "disease" → "lung cancer" → "surgery" in Figure 14.4(a) overlaps with the leaf node "procedure" → "surgery" → "lung cancer" in Figure 14.4(b). Depending on the user's preference, the system may decide the direction and the scope of cross-referencing. For example, a lung cancer oncologist may be interested in the topics of the treatment procedure and in the etiology and development of the patient's disease but not in other diseases such as mental illness. Note that for the most general cases, the navigation path may include generalization (going upward). We propose to use query, user type, and topic hierarchy in the directory to generate and control the navigation path that provides scenario-specific document retrieval.

14.4.4 An Example

We use the 5,000 UCLA medical reports to construct the knowledge hierarchies and a sample topic directory. Directly following the *Parent of* relationships in UMLS, we extract all the possible knowledge hierarchies (or knowledge paths). Such paths cannot be directly used in our directory system for two reasons. First, using all the knowledge paths for our directory system design is not feasible, since the number of knowledge paths in UMLS for a concept can be large. Second, the granularity can be too detailed for a set of documents, and thus we need to simplify the knowledge hierarchies as follows:

- Select a proper source for a knowledge type. The UMLS semantic network defines about 200 knowledge types (or semantic types), such as *disease, treatment, body part*, etc. For each knowledge type, a domain expert can identify the best knowledge source. For example, the ninth edition of the *International Classification of Diseases* can be a good source for *disease* knowledge hierarchies. By applying a source selector for a knowledge type, we significantly reduce the number of knowledge paths for a concept.
- Combine nodes in knowledge paths that contain synonyms. Patient reports may possess synonym concepts.

To reduce the number of knowledge paths, we combine the parents of synonym concepts with a synonym group and assign a concept for the synonym group in the knowledge paths.

- Reduce the number of knowledge paths. Remove the nodes in the knowledge paths that contain only a single child node to the topic concepts. For a specific document set, topics can be derived by data-mining the MFIs. The set of concepts that contains the topics is called topic concepts. For each topic concept, we can extract knowledge paths from UMLS, and we compute the number of descendant nodes in the path. All path nodes with a single child will be removed for simplicity of the topic directory.

Using these three techniques, we can extract knowledge hierarchies from UMLS and simplify them for our directory system design. For example, Table 14.7 shows a portion of the body part knowledge hierarchies we have extracted from UMLS.

There are 875,255 concepts in the UMLS's 2003AA edition. In a real dataset, the number of concepts appearing in the topics of a document set can be much less. Table 14.8 shows the number of concepts for some knowledge types in the set of about 5,000 UCLA medical reports.

TABLE 14.7 Sample disease knowledge hierarchy extracted from UMLS for the UCLA document set

Depth	Disease	CUI
1	Disease	C0012634
	...	
2	. cancer	C0006826
	...	
3	. . respiratory system cancer	C0814136
4	... bronchus cancer	C0345950
4	... lung cancer	C0242379
5	.... small-cell cancer	C0149925
5	.... non-small-cell cancer	C0220601
4	... mediastinum cancer	C0153504
4	... pleural tumor	C0345966
	...	

CUI = concept-unique identifier.

TABLE 14.8 Number of concepts for some knowledge types for the UCLA document set

TUI	Knowledge type	Number of concepts
T191	Disease	181
T184	Finding	171
T061	Treatment	242
T060	Diagnosis	155
T023	Body organ	482

TUI = topic-unique identifier.

TABLE 14.9 Example of the simplified directory paths for some disease concepts

CUI	TUI	Concept name	Knowledge path
C0000735	T191	Abdomen tumor	Disease/cancer/abdomen
C0001418	T191	Adenocarcinoma	Disease/cancer/epithelial/adenocarcinoma
C0001624	T191	Adrenal tumor	Disease/cancer/urological/kidney/adrenal
C0005967	T191	Bone cancer	Disease/cancer/bone
C0006118	T191	Brain tumor	Disease/cancer/neurologic/brain
C0006142	T191	Breast cancer	Disease/cancer/breast
C0006264	T191	Bronchus tumor	Disease/cancer/respiratory/bronchus
...	...	...	...

We obtain knowledge hierarchies for the five types of knowledge as shown in Table 14.8. A portion of the knowledge paths used in our experiment are shown in Table 14.9.

We constructed the topic directory systems by using a set of 50 patient reports from the UCLA Medical Center. Figure 14.5 shows a user giving a usage pattern [disease], after which the system creates a directory for the usage pattern.

Using such a directory, a user is able to obtain patient reports that are organized by disease type + body organ. For example, if a user wants to find reports on cancer/respiratory system/lung, the system returns 33 reports, as illustrated in the upper-left corner of Figure 14.5. When a user clicks on a report, the system will bring the document to the user, as shown in the bottom of the figure.

Such a system provides the user with the capability to generate a topic-oriented directory and to navigate the information that best satisfies the query goals. Such scenario-speific directories generate a set of relevant clinical free-text documents that can then be input for ranking.

14.5 Phrase-Based Vector Space Model for Automatic Document Retrieval

IndexFinder is able to extract key concepts from free text for the directory system. Based on a given query, the directory system is able to identify a group of documents that match the key concepts in the query from a corpus. We need to rank and order this set of documents by their similarity with the target document (query). VSMs are commonly used in IR to perform such ranking. In this section, we shall first present an overview of the phrase-based VSM, a new paradigm to represent documents and to measure document similarities. We then present the performance improvement of this new model and its computation complexity.

Retrieval systems consist of two main processes, *indexing* and *matching*. Indexing is the process of selecting *content identifiers*, also known as *terms* in this setting, to represent a

FIGURE 14.5 Topic directory system experiment.

text. Matching is the process of computing a measure of similarity between two text representations. It is possible for human experts to manually index documents. However, it is more efficient and thus more common to use computer programs to automatically index a large collection of documents.

A basic automatic indexing procedure for English usually consists of (1) splitting the text into words (tokenization), (2) removing frequently occurring words such as prepositions and pronouns (removal of stop words), and (3) conflating morphologically related words to a common word stem (stemming). The resulting word stems could be used as the terms for the given text.

In early retrieval systems, queries were represented as Boolean combinations of terms, and the set of documents that satisfied the Boolean expression was returned in response to the query. Since its inception, the VSM [22] has been the most popular model in information retrieval. In this model, documents and queries are represented by vectors in an n-dimensional space, where n is the number of distinct terms. Each axis in this n-dimensional space corresponds to one term. Given a query, a VSM system produces a ranked list of documents ordered by their similarities to the query. The similarity between a query and a document is computed using a metric on their respective vectors.

14.5.1 The Problem

Although word stems have been shown to be quite effective indexing terms, a recurring question in document retrieval is: What should be used as the basic unit to identify the content in the documents? Or, what is a term?

The problem of using word stems as terms is manifested in several ways:

1. The component words of a phrase sometimes have only a remote, if any, relation with the phrase. For example, separating "photo synthesis" into "photo" and "synthesis" could be misleading.
2. Words can be too general. For example, the individual words "family" and "doctor" are not specific enough to distinguish between "family doctor" and "doctor family."
3. Different words can be used to represent the same thing. For example, both "hyperthermia" and "fever" indicate an abnormal body temperature elevation.
4. The same word can mean different things. For example, "hyperthermia" can indicate an abnormal body temperature elevation as well as a treatment in which body tissue is exposed to high temperature to damage and kill cancer cells.

As a result, many researchers proposed both phrases and concepts in place of words or word stems as content identifiers. However, neither the phrases nor the concepts had been shown to produce significantly better results than word stems in automatic document indexing. On the other hand, through manual indexing, [23] showed the potential of concept-based indexing to produce significant improvements over the stem-based scheme. The high potential shown there and the low performances of current automatic indexing schemes using phrases and concepts led us to the search for such a scheme.

To facilitate discussion, we use the following example query throughout the discussion: "Hyperthermia, leukocytosis, increased intracranial pressure, and central herniation. Cerebral edema secondary to infection, diagnosis and treatment." The first part of the query is a brief description of the patient; the second part is the information desired.

14.5.2 Vector Space Models

14.5.2.1 Stem-Based Vector Space Model

In a stem-based VSM, morphological variants of a word like "edema" or "edemas" are conflated into a single word stem such as "edem" using the Lovins stemmer [24], and the resulting word stems are used as terms to represent the documents. Using the Lovins stemmer, the example query becomes "hypertherm," "leukocytos," "increas," "intracran," "pressur," etc.

Not all word stems are equally important. Authors usually repeat words as they elaborate the major aspects of a subject. Therefore, a frequent word stem in a document is often more important than an infrequent one. On the other hand, a word stem that appears in many documents is less specific than one that appears in only a few. Combining these two aspects, we often evaluate the importance of a word stem following a *term-frequency-inverse-document-frequency* (tf-idf) scheme. We define the weight of stem s in document x as, $w_{s,x} = \tau_{s,x}\iota_s$, where $\tau_{s,x}$ is the number of times s occurs in x, often called the term frequency of s, and ι_s is the inverse document frequency of stem s. One way to compute the inverse document frequency is $\iota_s = \log_2 (N/n_s) + 1$, where N is the number of documents in the collection, and n_s is the number of documents containing stem s, often called the document frequency of s.

To compute the document similarity in the stem-based VSM, we define the *stem-based inner product* between documents x and y as $\langle x,y\rangle^s = \sum_{s\in S} w_{s,x}w_{s,y} = \sum \iota_s^2 \tau_{s,x}\tau_{s,y}$ and define their similarity as the cosine of the angle between their respective document vectors,

$$sim^s(x,y) = \frac{\langle x,y\rangle^s}{\sqrt{\langle x,x\rangle^s \langle y,y\rangle^s}}.$$

14.5.2.2 Concept-Based Vector Space Model

Using word stems to represent documents results in the inappropriate fragmentation of multiword concepts such as "increased intracranial pressure" into their component stems like "increas," "intracran," and "pressur." Clearly, using concepts instead of word stems as content identifiers should produce a VSM that better mimics human thought processes and therefore results in more effective document retrieval.

However, using concepts is more complex than using word stems, because:

1. Concepts are usually represented by multiword phrases.
2. There exist polysemous and synonymous phrases. A phrase is *polysemous* if it can be used to express different meanings, and two phrases are *synonymous* if they can be used to express the same meaning. For example, "fever" and "hyperthermia" are synonyms, since both can be used to denote "an abnormal elevation of the body temperature." On the other hand, "hyperthermia" is polysemous, because it can be used to mean either "fever" or a type of "treatment."
3. Some concepts are related to one another.

Assuming that we can partition the documents into phrases, and ignoring the polysemy, our example query becomes (C0015967), (C0023518), and (C0151740), representing "hyperthermia," "leukocytosis," and "increased intracranial pressure," respectively, where the three strings in the parentheses are CUIs in UMLS.

Not all concepts are equally important, just as not all stems are equally so. We define the weight of a concept c in document x following the tf-idf scheme just like before, $w_{c,x} = \tau_{c,x}\iota_c = \tau_{c,x}(\log_2 (N/n_c) + 1)$, where $\tau_{c,x}$ is the number of times c appears in x, N is the number of documents in the collection, and n_c is the number of documents containing c.

Unlike in the stem-based VSM, where different word stems are considered unrelated, we define the *concept-based inner product* between documents x and y as

$$\langle x,y\rangle^c = \sum_{c\in C}\sum_{d\in C} \iota_c\tau_{c,x}\iota_d\tau_{d,y}s^c(c,d), \qquad (14.1)$$

where we take $s^c(c, d)$, the conceptual similarity between concepts c and d, into consideration. The similarity between documents x and y is defined to be the cosine of the angle between their respective document vectors,

$$sim^c(x,y) = \frac{\langle x,y\rangle^c}{\sqrt{\langle x,x\rangle^c \langle y,y\rangle^c}}.$$

14.5.2.3 Phrase-Based Vector Space Model

Concepts in controlled vocabularies such as UMLS are used in the concept-based VSM. Conceptual similarities needed are often derived from knowledge sources. The qualities of such VSMs therefore depend heavily on the qualities of the controlled vocabularies and the knowledge sources. Some concepts could be missing from the controlled vocabularies. For example, if we detect only concept C0021852 for "small bowel" in the phrase "infiltrative small bowel process" and find no concepts matching either the entire phrase or the fragments "infiltrative" and "process," then we are losing important information when we represent documents using concepts only. Furthermore, the absence of certain conceptual relations in the knowledge sources potentially degrades retrieval effectiveness. For example, treating "cerebral edema" and "cerebral lesion" as unrelated is potentially harmful. Noticing the words "infiltrative" and "process" that match no concepts and the common component word "cerebral" in phrases "cerebral edema" and "cerebral lesion," we propose a phrase-based VSM to remedy the incompleteness of the controlled vocabularies and the knowledge sources.

In the phrase-based VSM, a document is represented as a set of phrases. Each phrase may correspond to multiple concepts (due to polysemy) and consist of several word stems. For example, "infiltrative small bowel process" is represented by phrases (; "infiltr"), (C0021852; "smal," "bowel"), (; "proces"). Our example query now becomes (C0015967, C0203597; "hypertherm"), (C0023518; "leukocytos"), (C0151740; "increas," "intracran," "pressur"), etc.

We use an ordered pair of two sets to represent a phrase $p = (\{(s, \pi_{s,p})\}, \{(c, \pi_{c,p})\})$. The first set, $\{(s, \pi_{s,p})\}$, consists of ordered pairs that indicate the stems and their occurrence counts, $\pi_{s,p}$, in the phrase. The second set, $\{(c, \pi_{c,p})\}$, indicates the concepts and their occurrence counts, $\pi_{c,p}$, in the phrase. We denote the set of all phrases by P. Furthermore, we require that there be at least one stem in each phrase; that is, for each phrase $p \in P$, there exists some stem s such that $\pi_{s,p} \geq 1$. We use a *phrase vector* x^P to represent a document x, $x^P = \{(p, \tau_{p,x})\}$, where $\tau_{p,x}$ is the number of times phrase p occurs in document x. And we define the *phrase-based inner product* as

$$\langle x,y \rangle^P = \sum_{p\in P} \sum_{q\in P} \tau_{p,x}\tau_{q,y} s^P(p,q),$$

where we use $s^P(p,q)$ to measure the similarity between phrases p and q. We call $s^P(p,q)$ the *phrase similarity* between phrases p and q and define it as

$$s^P(p, q) = \max\left(\left(f^s \sum_{s\in S} \iota_s^2 \pi_{s,p}\pi_{s,q}\right), \left(f^c \sum_{c\in C} \sum_{d\in C} \iota_c \pi_{c,p} \iota_d \pi_{d,q} s^c(c,d)\right)\right),$$

where $\iota_s, \iota_c, \iota_d > 0$ are the inverse document frequencies of stem s, concept c, and concept d, respectively, and $s^c(c,d)$ is the conceptual similarity between concepts c and d. As in the concept-based VSM, we ignore polysemy and assume that each phrase expresses only one concept,

$$\pi_{c,p} = \delta_{c,c_p} = \begin{cases} 1 & \text{if } c = c_p \\ 0 & \text{if } c \neq c_p \end{cases},$$

where c_p is the concept that phrase p expresses. Then the phrase similarity is reduced to

$$s^P(p,q) = \max\left(\left(f^s \sum_{s\in S} \iota_s^2 \pi_{s,p}\pi_{s,q}\right), (f^c \iota_{c_p} \iota_{d_q} s^c(c_p,d_q))\right), \tag{14.2}$$

where c_p is the concept phrase p expresses, and d_q is the concept q expresses. Here we use two contribution factors, f^s and f^c, to specify the relative importance of the stem contribution and the concept contribution in the overall phrase similarity. The stem contribution

$$f^s \sum_{s\in S} \iota_s^2 \pi_{s,p}\pi_{s,q}$$

measures the stem overlaps between phrases p and q, and the concept contribution

$$f^c \iota_{c_p} \iota_{d_q} s^c(c_p,d_q)$$

takes the concept interrelation into consideration. Conceptually, when combining the stem contribution and the concept contribution this way, we use stem overlaps to compensate for the incompleteness of the controlled vocabularies in encoding all necessary concepts and the incompleteness of the knowledge sources in describing all necessary concept interrelations. Once again, we define the phrase-based document similarity between documents x and y to be the cosine of the angle between their respective phrase vectors,

$$\text{sim}^P(x,y) = \frac{\langle x,y \rangle^P}{\sqrt{\langle x,x \rangle^P \langle y,y \rangle^P}}.$$

14.5.2.4 Phrase Detection

The building blocks of the concept-based VSM and the phrase-based VSM are phrases. A phrase usually consists of multiple words. Given a controlled vocabulary containing a set of phrases P and a set of documents X, we need to efficiently detect the occurrences of the phrases in P in each of the documents in X. We can achieve this goal by applying indexing methods such as IndexFinder or the Aho-Corasick algorithm.

In our phrase detection, we remove the stop words in the stop list *after* multiword phrase detection. In this way, we correctly detect "secondary to" and "infection" from "cerebral edema secondary to infection." We would incorrectly detect "secondary infection" if the stop words ("to" in this case) were removed before the phrase detection.

14.5.2.5 Conceptual Similarity Evaluation

Among the many possible conceptual relations, we concentrate on the "is–a" relation, also called the *hypernym* relation.

A simple example is that "body temperature elevation" is a hypernym of "fever." Hypernym relations are transitive [25]. For example, "sign and symptom" is a hypernym of "body temperature change," and "body temperature change" is a hypernym of "hyperthermia," so "sign and symptom" is also a hypernym of "hyperthermia." We derive the similarity between a pair of concepts using their relative position in a hypernym hierarchy. For a pair of ancestor-descendant concepts, *c* and *d*, in the hypernym hierarchy, we define their conceptual similarity as

$$s^c(c,d) = \frac{1}{l(c,d)\log_2(D(c)+D(d)+1)}, \qquad (14.3)$$

where $l(c,d)$ is the number of hops between *c* and *d* in the hierarchy, and $D(c)$ and $D(d)$ are the descendant counts of *c* and *d*, respectively.

14.5.2.6 Primitive Word Sense Disambiguation

Polysemy is one of the difficulties people encounter when using concepts. A polysemous phrase can express multiple meanings. As a result, it is necessary to disambiguate polysemous phrases in document retrieval. For example, seeing "hyperthermia," it is necessary to figure out whether it means "fever" or a type of "treatment" using word sense disambiguation [26]. The current accuracy and efficiency of word sense disambiguation algorithms are low. We perform a very primitive word sense disambiguation based on the following observation. UMLS tends to assign a smaller CUI to the more popular sense of a phrase. For example, the CUI for the "fever" sense of "hyperthermia" is C0015967, while the CUI for its "treatment" sense is C0203597. Therefore, we use the concept corresponding to the smallest CUI in the concept-based VSM and the phrase-based VSM.

14.5.3 Experimental Evaluation of the Phrase-Based Vector Space Model

14.5.3.1 Phrase Detection and Conceptual Similarity Derivation via the Unified Medical Language System

In our experiments, we used UMLS as the controlled vocabulary for phrase detection. We also applied the conceptual relations in the metathesaurus to derive conceptual similarities. We are particularly interested in hypernym/hyponym relations. Two pairs of relations in UMLS roughly correspond to the hypernym/hyponym relations: the RB/RN (broader than/narrower than) and the PAR/CHD (parent/child) relations. For example, C0015967 (fever) has a parent concept C0005904 (body temperature change). RB and RN are redundant—for two concepts *c* and *d*, if (*c*, *d*) is in the RB relations, then (*d*, *c*) is in the RN relations, and vice versa. Similarly, PAR and CHD are redundant. As a result, we combine RB and PAR into a single hypernym hierarchy. Hypernymy is transitive. However, the UMLS metathesaurus encodes only the direct hypernym relations, not the transitive closure. We derive the transitive closure of the hypernym relation and use Equation 14.3 to compute the conceptual similarities.

14.5.3.2 The Test Collections

To compare the effectiveness of different VSMs in document retrieval, we need a test collection that provides (1) a set of queries, (2) a set of documents, and (3) the judgments indicating whether a document is relevant to a query.

OHSUMED [16] is a test collection widely used in recent information retrieval tests. OHSUMED contains 106 queries. Each query contains a patient description and an information need. Our example is query 57 in the collection. The document collection is a subset of 348,000 Medline references from 1987 to 1991. Seventy-five percent of the references contain titles and abstracts, while the remainder have only titles. Each reference also contains human-assigned subject headings from the MeSH. Of the references in the document collection, 14,430 are judged by "physicians who were clinically active and were current fellows in general medicine or medical informatics or senior medical residents" to be definitely relevant, possibly relevant, or nonrelevant to each of the 105 queries. The standard recall and precision evaluation that we shall discuss later requires a binary judgment of relevance or nonrelevance. This can be easily achieved by merging the definitely relevant and the possibly relevant documents into a single relevant category.

Another test collection, known as Medlars [27], is based on Medline reference collections from 1964 to 1966. It has been used extensively in document retrieval system comparisons. There are 30 queries and 1,033 references in the collection. The judgments are provided by "a medical school student."

We use both test collections to compare the retrieval effectiveness of different methods. However, based on the qualification of the human experts, the extent, and the up-to-dateness of these collections, we believe that OHSUMED reflects expert judgment better. Therefore, we direct the attention of the reader to the results obtained from the OHSUMED collection in later sections. Table 14.10 compares some statistics of the

TABLE 14.10 Comparison of OHSUMED and Medlars statistics

	OHSUMED		Medlars	
	Query	Document	Query	Document
Number of documents	*105*	*14,430*	*30*	*1,033*
Phrases per document	7.5	112	*11*	90
Stems per phrase	1.34	1.25	1.25	1.14
Concepts per phrase	1.21	1.18	1.27	1.21
Multistem phrases per document	1.96	*21.3*	2.6	*10.8*
Multisense phrases per document	*1.2*	11.3	*2*	9.8

Note: noticeable differences in italics.

two collections. Besides the collection size difference discussed above, other noticeable differences include:

- OHSUMED queries are slightly shorter than those in Medlars.
- OHSUMED documents on average contain more long phrases (those with more than one stem).
- Medlars contains slightly more polysemous phrases (those with multiple senses).

14.5.3.3 Retrieval Effectiveness Measures

The goal of document retrieval is to return documents relevant to a user query before nonrelevant ones. The effectiveness of a document retrieval system is measured by the recall and precision [28–29] based on the user's judgment of whether each document is relevant to a query q. When a certain number of documents are returned, we define *precision* to be the proportion of the retrieved documents that are relevant, and we define *recall* to be the proportion of the relevant documents retrieved so far. More specifically, if we use R_q to represent the set of documents relevant to q, and A to represent the set of retrieved documents, then we define:

$$\text{precision} = \frac{|R_q \cap A|}{|A|} \text{ and recall} = \frac{|R_q \cap A|}{|R_q|}.$$

There are several ways to evaluate retrieval effectiveness using recall and precision. To visually display the change in the precision values as documents are retrieved, we interpolate the precision values to a set of 11 recall points 0, 0.1, 0.2, ..., 1. Averaging the precision values over a set of queries at these recall points illustrates the behavior of a system. Further averaging the 11 average precision values, we arrive at the *average 11-point average precision*, denoted by G_{P11}. Instead of interpolating the precision values to a set of standard recall points, we can also compute the average precision values after each relevant document is retrieved. The average of such a value over a set of queries is the average precision, denoted by G_P.

14.5.3.4 Comparison of the Recall–Precision Curves

Figures 14.6 and 14.7 depict the average precision values of 105 OHSUMED queries and 30 Medlars queries, respectively, at the 11 standard recall points 0, 0.1, 0.2, ..., 1 for five different VSMs.

For the OHSUMED results:

1. "Stems" is the baseline generated by the stem-based VSM. Its average 11-point average precision is $G^s_{P11} = 0.376$.
2. "Concepts Unrelated" is generated by using the concepts as the terms and treating different concepts as unrelated. More specifically, we use $s^c(c,d) = \delta c,d$ in the inner product calculation (Equation 14.1). The average 11-point average precision is $G^{cu}_{P11} = 0.336$, an 11% decrease from the baseline.
3. "Concepts" is similar to case 2, but taking the concept interrelations into consideration, we achieve a significant

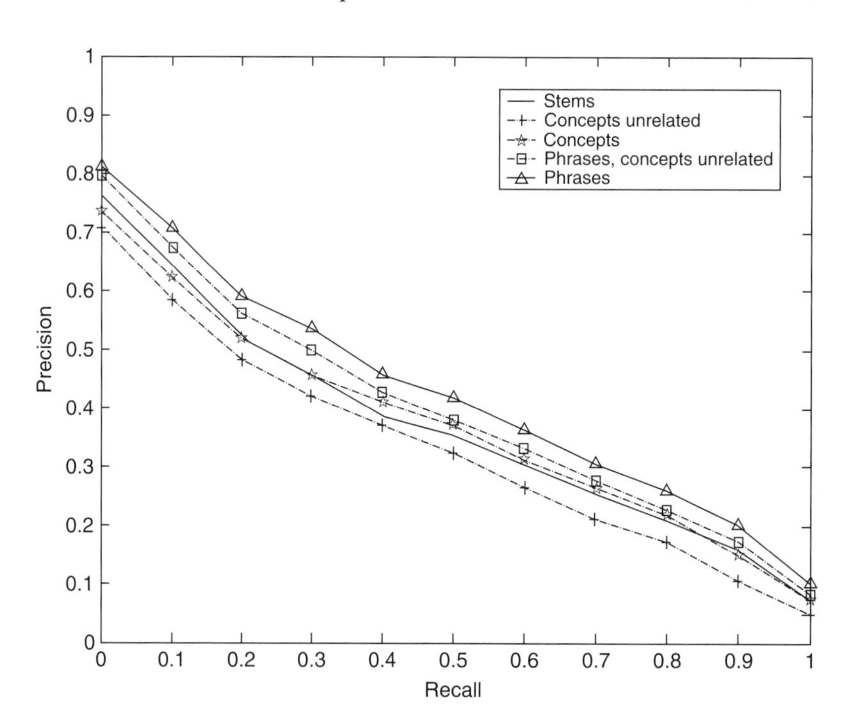

FIGURE 14.6 Comparison of the average recall–precision curves over 105 OHSUMED queries.

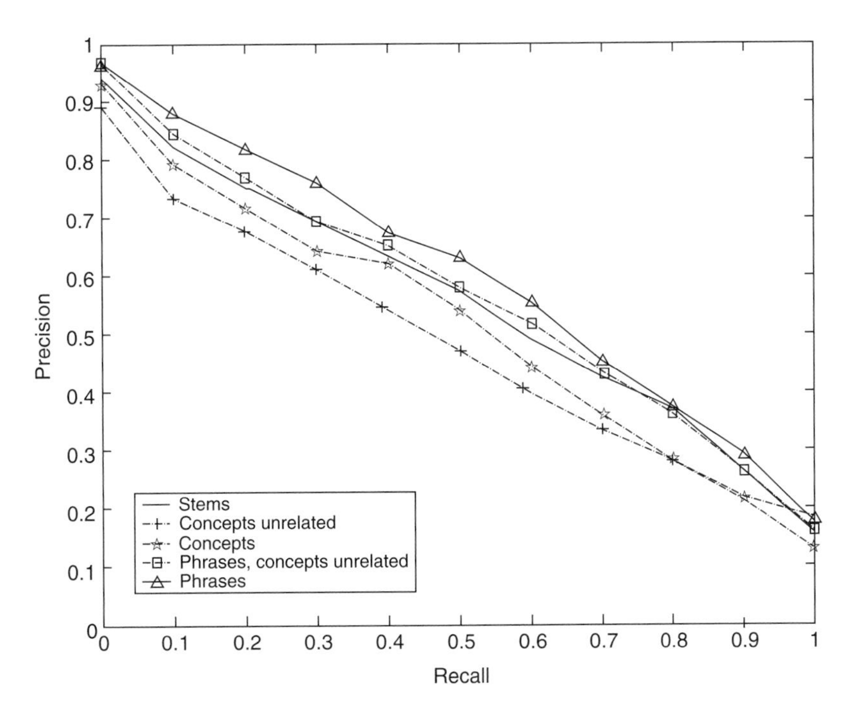

FIGURE 14.7 Comparison of the average recall–precision curves over 30 Medlars queries.

improvement over case 2. The average effectiveness is approximately equal to that of the baseline.

4. "Phrases, Concepts Unrelated" refers to considering contributions from both the concepts and the word stems in a phrase but, once again, treating different concepts as unrelated. By setting $s^c(c_p,d_q)$ in Equation (14.2) to $\delta c_p,d_q$, we achieve significant improvement over the "Concepts Unrelated" case. In fact, its average 11-point average G^{cu}_{P11} is 7.1% better than the baseline.
5. "Phrases" is similar to case 4, but considering the concept interrelations, we achieve an average 11-point average precision of $G^{p}_{P11} = 0.433$, which is a significant 15% improvement over the baseline. In both cases 4 and 5, we used equal weight for the stem and the concept contributions, $f^s = f^c = 1$.

Our experimental results reveal that using only concepts to represent documents and treating different concepts as unrelated can cause the retrieval effectiveness to deteriorate (case 2). Considering concept interrelations (case 3) or relations of different phrases by their shared word stems (case 4) can improve retrieval effectiveness. Measuring the similarity between two phrases using their stem overlaps and the relation between the concepts they represent, the phrase-based VSM (case 5) is significantly more effective than the stem-based VSM.

14.5.3.5 Sensitivity of Retrieval Effectiveness to f^s and f^c

To generate the two sets of recall–precision curves "Phrase, Concepts Unrelated" and "Phrase" in Figures 14.6 and 14.7, we used equal weight, $f^s = f^c = 1$. To study the relative importance of the stem contribution and the concept contribution in the inner product calculation, we vary the weights f^s and f^c and study the change of the average11-point average precision value G_{P11}. The document similarity value depends on the ratio between f^s and f^c, not their absolute values; therefore, we vary the (f^s, f^c) from the stem-only case (1, 0) to the equal-weight phrase case (1, 1) to the concept-only case (0, 1) and study the change of the average 11-point average precision values.

Figure 14.8 depicts the changes of the average 11-point average precision values as the result of the change of f^s and f^c. We observe that the retrieval effectiveness measured by G_{P11} is maximized when f^c is about the same as f^s, and, in this region, the retrieval effectiveness is not sensitive to the change of the relative importance of the stem contribution and the concept contribution.

14.5.3.6 Retrieval Effectiveness Comparison in Cluster-Based Document Retrieval

In the previous section, we showed that the phrase-based VSM is more effective than the stem-based VSM in document retrieval using an exhaustive search. Let us consider a set of N documents. In an exhaustive search system, the similarity values between an incoming query and all the N documents need to be computed *online* before the documents can be returned to the user. Because of the relatively large computation complexity of the VSMs, such an exhaustive search scheme is not feasible for large document collections. Using hierarchical clustering algorithms, we can first construct a

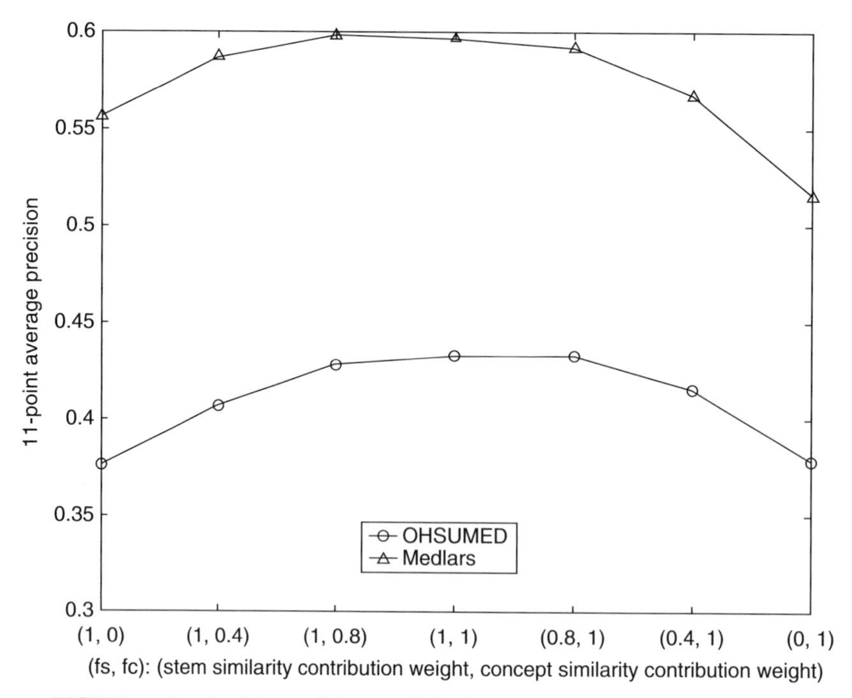

FIGURE 14.8 Sensitivity of G_{P11} to f^s,f^c changes in OHSUMED and Medlars.

document hierarchy using $O(N \log N)$ *offline* document similarity computations and return a ranked list of documents using only $O(\log N)$ *online* comparisons.

We compare the stem-based VSM and the phrase-based VSM using an $O(N\log N)$ spherical *k*-means algorithm that has been shown to produce good clusters in document clustering [30–31]. The resulting document clusters are searched using top-down and bottom-up searching strategies.

Figure 14.9 contains the recall–precision curves of six different searching strategies on the OHSUMED data. They are the result of an exhaustive search of the 14,000 documents in OHSUMED. Their average 11-point average precision values are $G_{11}^s = 0.376$

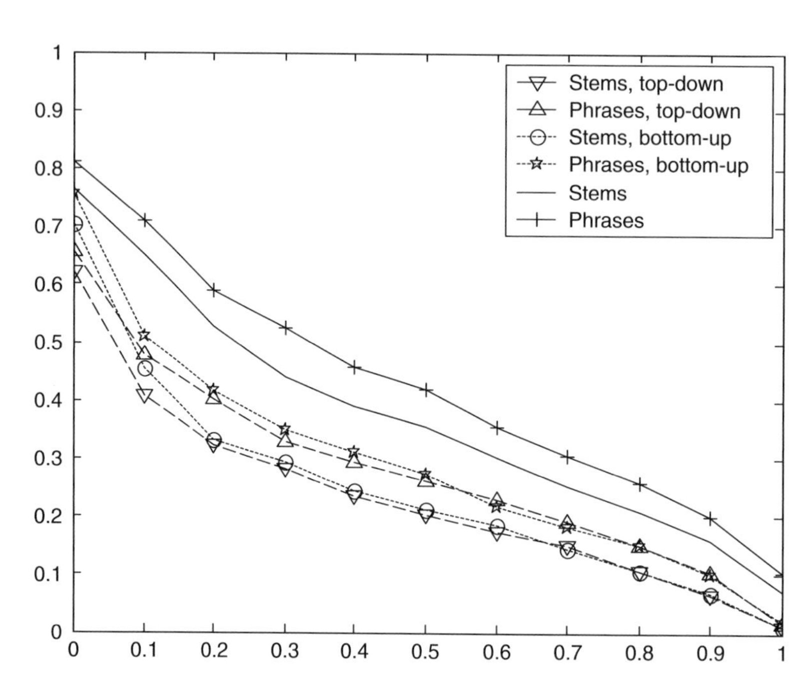

FIGURE 14.9 Retrieval effectiveness comparison of the cluster-based retrieval in OHSUMED.

and $G_{11}^{p} = 0.433$. The other four curves depict the retrieval effectiveness of systems when the document hierarchies are searched. Clearly, the retrieval effectiveness of the cluster-based approaches is lower than that of the exhaustive search–based approaches. That is, by using cluster-based document retrieval, we sacrifice the retrieval effectiveness for more efficient retrieval.

More importantly, using the same searching strategy, we see that the retrieval effectiveness of the phrase-based VSM is always much better than that of the stem-based VSM. For the top-down search, $G_{11}^{s,td} = 0.235$ and $G_{11}^{p,td} = 0.283$; and for the bottom-up search, $G_{11}^{s,bu} = 0.251$ and $G_{11}^{p,bu} = 0.299$. In each case, the phrase-based VSM is about 20% more effective than the stem-based VSM. In IR, if the performance improvement for a new retrieval model exceeds 5% evaluated from 50 queries over an existing model, then it is considered significant enough to warrant using the new retrieval model [23]. In our case, there is a 20% improvement average over 100 queries, representing a significant improvement.

14.5.4 Computation Complexity

The document similarity calculation in the phrase-based VSM is more complex than that in the stem-based VSM. Let us use L to represent the average length of a document. In the stem-based VSM, different word stems are considered unrelated. As a result, by building indexes on the word stems in the documents, an efficient algorithm computes the stem-based similarity between two documents using $O(L \log L)$ time. The time complexity of a straightforward implementation of the phrase-based document similarity calculation is $O(L^2)$. Different phrases in the phrase-based VSM can be related to one another not only because they may share common word stems, but also because the concepts they represent can be related. Therefore, indexing the phrases in the documents does not reduce the time complexity of the phrase-based document similarity calculation to $O(L \log L)$. To reduce the computation complexity, we need to build separate indexes on the concepts and the stems in the documents, keep track of where each stem or concept occurs, and modify the conceptual similarity storage structure. The phrase-based document similarity calculation utilizing such data structure modifications has an $O(L \log L)$ time complexity. For the OHSUMED documents, the improved phrase-based document similarity calculation is about 10 times slower than the stem-based calculation, while the straightforward implementation is over 250 times slower than the stem-based calculation.

Preliminary experimental results show that the number of related concept pairs decreases drastically as the pairwise conceptual similarity value increases. Therefore, we can further reduce the phrase-based computation complexity by treating related concepts with low conceptual similarity values as unrelated. We are currently investigating the trade-off between retrieval effectiveness and computation time complexity when related concepts are treated as unrelated in the phrase-based document similarity calculations.

14.6 Knowledge-Based Scenario-Specific Query Expansion

14.6.1 A Framework for Knowledge-Based Query Expansion

A knowledge-based query expansion and retrieval framework is shown in Figure 14.10. For a given query, *statistical* query expansion (whose scope is marked by the inner dotted rectangle) derives candidate expansion concepts[1] that statistically co-occur with the given query concepts (Section 14.6.2) and assigns weights to each candidate concept according to the statistical co-occurrence. Such weights will be carried through the framework. Based on the candidate concepts derived by statistical expansion, *knowledge-based* query expansion (whose scope is marked by the outer rectangle) further derives the scenario-specific expansion concepts, with the aid of a domain knowledge source such as UMLS (Section 14.6.2). Such knowledge may be incomplete and fail to include all possible query scenarios. Therefore, in an offline process, we apply a *knowledge acquisition and supplementation* module to supplement the incomplete knowledge (Section 14.6.5). After the query is expanded with scenario-specific concepts, we employ a VSM to compare the similarity of the expanded query with each document. Top-ranked documents with the highest similarity measures are output to the user.

14.6.2 Method

Formally, the problem for knowledge-based query expansion can be stated as follows: Given a scenario-specific query with a key concept denoted as c_{key} (e.g., lung cancer, the eye disease keratoconus) and a set of scenario concepts denoted as c_s (e.g., treatment or diagnosis), we need to derive specialized concepts that are related to c_{key}, and the relations should be specific to the scenarios defined by c_s. In this section, we describe how to derive such scenario-specific concepts by presenting existing statistical query expansion methods that generate candidate concepts. We then propose a method that selects scenario-specific concepts from this candidate set with the aid of a domain knowledge source.

14.6.2.1 Deriving Statistically Related Expansion Concepts

Statistical expansion is also referred to as *automatic query expansion* [32–34]. The basic idea is to derive concepts that are statistically related to the given query concepts, where the statistical correlation is derived from a document collection

[1] In the rest of this paper, a *concept* is referred to as a word or phrase that has a concrete meaning in a particular application domain. In the medical domain, concepts in free text can be extracted using existing tools, such as MetaMap [12], IndexFinder [13], etc.

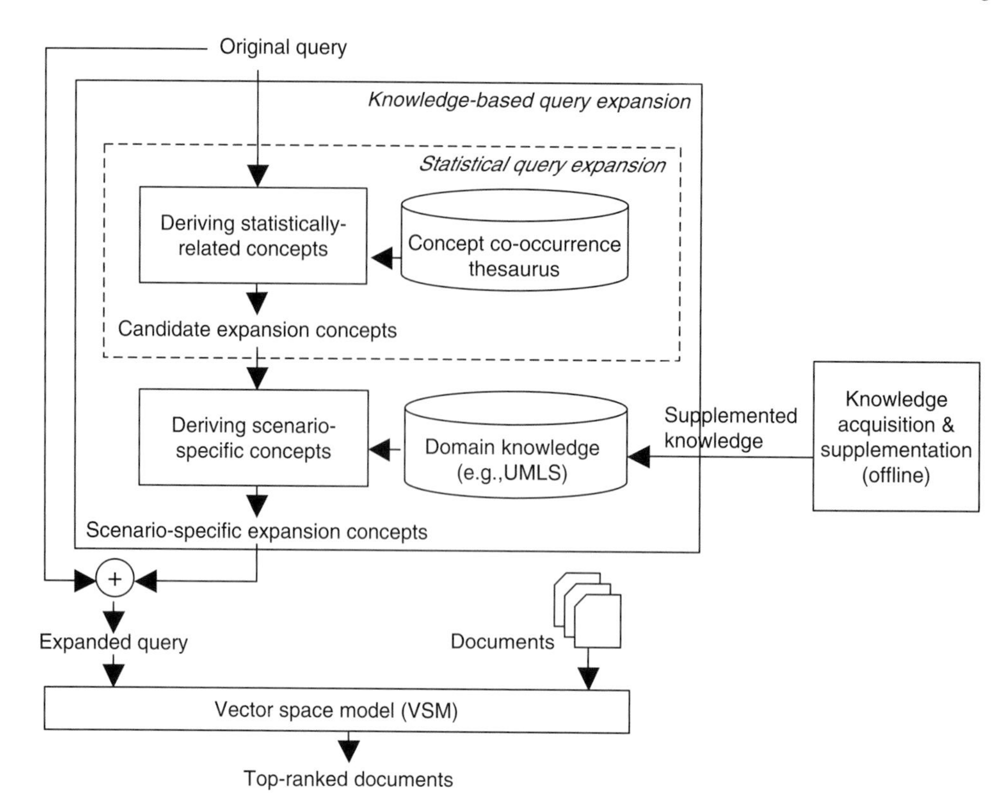

FIGURE 14.10 A knowledge-based query expansion and retrieval framework.

(e.g., OHSUMED [16]). Appending such concepts to the original query makes the query expression more specialized and thus matches relevant documents better. Depending on how such statistically related concepts are derived, statistical expansion methods fall into two major categories:

1. *Co-occurrence thesaurus-based expansion* [35–37]. In this method, a concept co-occurrence thesaurus is first constructed automatically offline. Given a vocabulary of M concepts, the thesaurus is an $M \times M$ matrix, where the $<i,j>$ element quantifies the co-occurrence between concept i and concept j. When a query is posed, we look in the thesaurus to find all concepts that statistically co-occur with concepts in the given query and assign weights to those co-occurring concepts according to the values in the co-occurrence matrix. A detailed procedure for computing the co-occurrence matrix and for assigning weights to expansion concepts can be found in Qiu and Frei [35].
2. *Pseudo-relevance feedback-based expansion* [34, 38–41]. In pseudo-relevance feedback, the original query is used to perform an initial retrieval. Concepts extracted from top-ranked documents in the initial retrieval are considered statistically related and are appended to the original query. This approach resembles the well-known relevance feedback approach, except that instead of asking users to identify relevant documents as feedback, top-ranked (e.g., top 10) documents are automatically treated as "pseudo"-relevant documents and are inserted into the feedback loop. Weight assignment in pseudo-relevance feedback [39] typically follows the same weighting scheme for conventional relevance feedback techniques [38].

We note that the choice of statistical expansion method is orthogonal to the design of the knowledge-based expansion framework (Figure 14.10). In our current experimental evaluation, we used the co-occurrence thesaurus-based method to derive statistically related concepts. For convenience of discussion, we used $co(c_i,c_j)$ to denote the co-occurrence between concept c_i and c_j, a value that appeared as the $<i,j>$ element in the $M \times M$ co-occurrence matrix. Table 14.11 lists the top 15 concepts that are statistically related to keratoconus using the co-occurrence measure. Here, the co-occurrence measure is computed from the OHSUMED corpus.

14.6.2.2 Deriving Scenario-Specific Expansion Concepts

Using a statistical expansion method, we can derive a set of concepts that are statistically related to the key concept, c_{key}, of the given query. Only a subset of these concepts are relevant to the given query's scenario, such as treatment. For example,

TABLE 14.11 Concepts that statistically correlate to keratoconus

Number	Concept
1	Fuchs dystrophy
2	Penetrating keratoplasty
3	Epikeratoplasty
4	Corneal ectasia
5	Acute hydrops
6	Keratometry
7	Corneal topography
8	Corneal
9	Aphakic corneal edema
10	Epikeratophakia
11	Granular dystrophy corneal
12	Keratoplasty
13	Central cornea
14	Contact lens
15	Ghost vessels

the 5th and 8th concepts in Table 14.11 (acute hydrops and corneal) are not related to the treatment of keratoconus. Therefore, in terms of deriving expansion concepts for a query of keratoconus treatment, these two concepts should be filtered out. In this section, we will first describe the type of knowledge structure that enables us to perform this filtering and then present the filtering procedure.

In previous sections, we introduced UMLS and how to apply its subsystems, such as the metathesaurus and the SPECIALIST lexicon, for implementing the IndexFinder and the phrase-based VSM. For the task of knowledge-based query expansion, we apply the subsystem of the semantic network.

The semantic network defines about one hundred semantic types, such as Disease or Syndrome, Body Part, etc. Each semantic type corresponds to a class/category of concepts. The semantic type Disease or Syndrome, for instance, corresponds to 44,000 concepts in the metathesaurus, such as keratoconus, lung cancer, diabetes, etc. Besides the list of semantic types, the semantic network also defines the relations among various semantic types, such as Treatments and Diagnoses. Such relations link isolated semantic types into a graph/network structure. The top half of Figure 14.11 presents a fragment of this network, which includes all semantic types that have a *treats* relation with the semantic type Disease or Syndrome. Relations such as *treats* in Figure 14.11 should be interpreted as follows: Any concepts that belong to semantic type Therapeutic or Preventive Procedure (e.g., penetrating keratoplasty, chemotherapy) have the potential to treat concepts that belong to the semantic type Disease or Syndrome (e.g., keratoconus, lung cancer). However, it is not indicated whether such relations concretely exist between two concepts (e.g., a *treats* relation between penetrating keratoplasty and lung cancer).

Given the knowledge structure in the semantic network, the basic idea in identifying scenario-specific expansion concepts is to use this knowledge structure to filter out statistically correlated concepts that do not belong to the specific semantic types. Let us illustrate this idea through Figure 14.11, using the treatment scenario as an example. We start with the set of concepts that are statistically related to keratoconus. Our goal in applying the knowledge structure is to identify that (1) concepts such as penetrating keratoplasty, contact lens, and griffonia have the scenario-specific relation (i.e., *treats*) with keratoconus and should be kept during expansion, and (2) concepts such as acute hydrops and corneal that do not have the scenario-specific relation with keratoconus are filtered out.

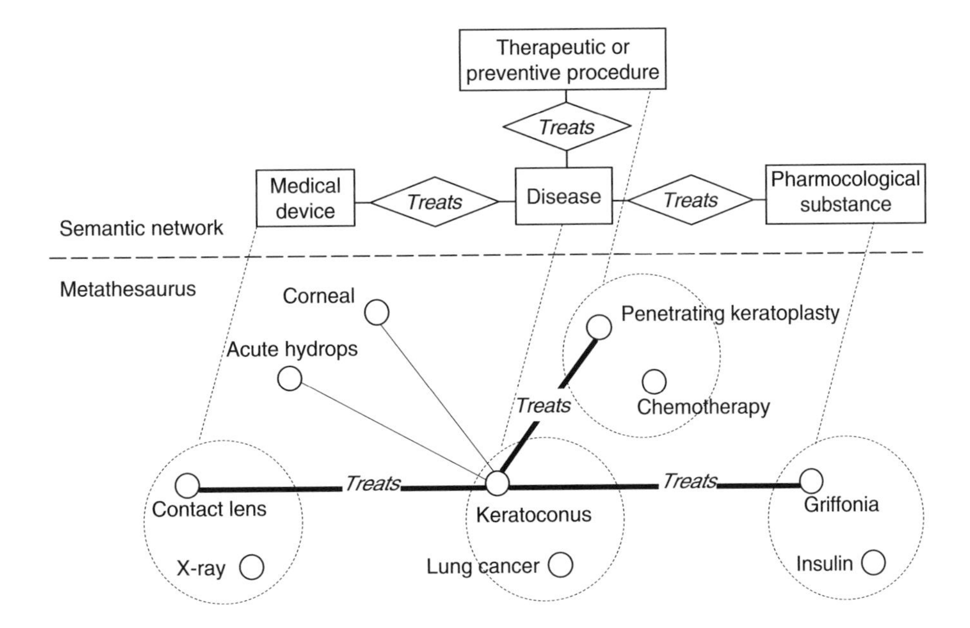

FIGURE 14.11 Using knowledge to identify scenario-specific concept relationships.

Each solid circle in Figure 14.11 represents a single concept, and the solid lines connecting these solid circles indicate strong statistical correlations computed for a pair of concepts—for instance, the solid line between keratoconus and contact lens. A dotted circle represents a class of concepts, and a dotted line links that class of concepts to a corresponding semantic type. For example, the concepts keratoconus and lung cancer are in the class that links to Disease or Syndrome. We identified scenario-specific expansion concepts using the following process: Given a key concept c_{key} of the given query, we first identified the semantic type that c_{key} belongs to. For example, we identified Disease or Syndrome given the key concept keratoconus. Starting from that semantic type, we further followed the relations marked by the query's scenario and reached a set of relevant semantic types. For the previous example, given the query's scenario, treatment, we followed the *treats* relation to reach the three other semantic types, as shown in Figure 14.11. Finally, we identified those statistically related concepts that belonged to the relevant semantic types as scenario specific. We further filtered out other statistically related concepts that did not satisfy this criteria. From the previous example, this final step identified penetrating keratoplasty, contact lens, and griffonia as scenario-specific expansion concepts and filtered out non-scenario-specific ones such as acute hydrops and corneal.

The lists of the concepts for treating and diagnosing keratoconus are shown in Table 14.12. These concepts were derived based on the process we have described and show the top 15 concepts in terms of their correlation with keratoconus. To highlight the effectiveness of the knowledge-based filtering process, we can compare the concepts in Table 14.12 with those in Table 14.11 that are statistically correlated with keratoconus. Five out of these 15 statistically correlated concepts are kept in Table 14.12(a), whereas two are kept in Table 14.12(b).

TABLE 14.12 Concepts that treat or diagnose keratoconus

Number	(a) Concepts that *treat* keratoconus	(b) Concepts that *diagnose* keratoconus
1	Penetrating keratoplasty	Keratometry
2	Epikeratoplasty	Corneal topography
3	Epikeratophakia	Slit lamp examination
4	Keratoplasty	Topical corticosteroid
5	Contact lens	2D Echocardiography
6	Thermokeratoplasty	Transmission electron microscopy (TEM)
7	Button	Interferon
8	Secondary lens implant	Alferon
9	Fittings adapters	Analysis
10	Esthesiometer	Microscopy
11	Griffonia	Bleb
12	Trephine	Tetanus toxoid
13	Slit lamps	Antineoplastic
14	Fistulization	Heart auscultation
15	Soft contact lens	Chlorbutin

This comparison reveals that the knowledge structure is effective in filtering out concepts that are not closely related to the scenario of treatment or diagnosis.

The goal of knowledge-based query expansion is to append specialized terms that appear in relevant documents but not in the original query. Scenario-specific concepts derived from the previous subsection represent a subset of such specialized terms. Another set of highly relevant terms contains hypernyms and hyponyms of the key concept c_{key}.[2] For example, corneal ectasia, a hypernym of keratoconus, is frequently mentioned by documents regarding keratoconus treatment. Therefore, we need to also expand those concepts that are close to c_{key} in the hypernym/hyponym hierarchy.

To expand hypernyms/hyponyms of the key concept to the original query, we again refer to the UMLS knowledge source. The metathesaurus subsystem defines not only the concepts but also the hypernym/hyponym relationships among these concepts. For example, Figure 14.12 shows the hypernyms (parents), hyponyms (children), and siblings of the concept keratoconus. Here we define a concept's siblings as those concepts that share the same parents with the given concept. Through empirical study (which will be discussed later), we have found that expanding the direct parents, direct children, and siblings to the original query generates the best retrieval performance. This is in comparison with expanding parents/children that are two or more levels away from the key concept. Therefore, in the rest of our discussion, we will focus on expanding only the direct parents/children and siblings.

14.6.2.2 Weight Adjustment for Expansion Terms

To match a query and a document using the VSM, we represent both the query and the document as vectors. Each term in the query becomes a dimension in the query vector and receives a weight that quantifies the importance of this term in the entire query. Under this model, any additional term appended to the original query needs to be assigned a weight. An appropriate weight scheme for these additional terms is important because "underweighting" will make the additional terms insignificant compared with the original query and lead to unnoticeable changes in the ranking of the retrieval results. On the other hand, "overweighting" will make the additional terms overly significant and cause a "topic drift" for the original query.

In the past, researchers proposed weighting schemes for these additional terms based on the following intuition: The weight for an additional term c_a should be proportional to its correlation with the original query terms. Thus, the weight for c_a, w_a, is proportional to its correlation with the key concept c_{key}:

$$w_a = co(c_a, c_{key}) \cdot w_{key}, \tag{14.4}$$

[2] A hypernym of concept $c\psi$is a concept with a broader meaning than c, whereas a hyponym is one with a narrower meaning.

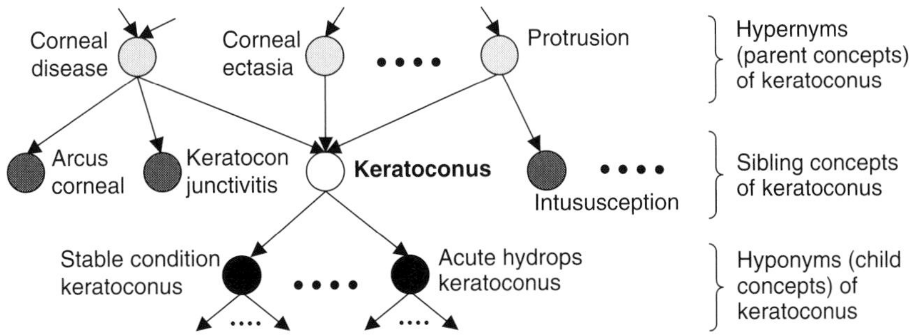

FIGURE 14.12 The direct parents, direct children, and siblings for keratoconus.

where w_{key} denotes the weight assigned to the key concept c_{key}. In Equation (14.4), the correlation between c_a and c_{key}, $co(c_a, c_{key})$ is derived using methods described in Section 14.6.2. In Section 14.6.3, we will further explain how w_{key} is decided according to a common weighting scheme. Given that $co(c_a, c_{key})$ lies in [0, 1], the weight that c_a receives will not exceed that of c_{key}. Using this equation, we compute the weights for the terms that statistically correlate with keratoconus (Table 14.11) and the weights for those that treat keratoconus (Table 14.12a). We list the weights for these terms in Table 14.13(a) and Table 14.13(b), respectively. These weights are computed by assuming that the weight of the key concept (w_{key}) keratoconus is 1.

We will compare the retrieval effectiveness of knowledge-based query expansion with that of statistical expansion. Since the knowledge-based method applies a filtering step to derive a subset of all statistically related terms, the impact created by this subset on retrieval effectiveness will be less than the entire set of statistically related terms. Therefore, weight adjustments are needed to compensate for the filtering. For instance, in our example of keratoconus treatment, the "cumulative weight" for all terms in Table 14.13b is obviously smaller than the cumulative weight of those in Table 14.13a. To increase the impact of the terms derived by the knowledge-based method, we can "boost" their weights by multiplying a linear factor β, so that the cumulative weight of those terms is comparable to those of the statistically related terms. We refer to β as the *boosting factor.* With this factor, we alter Equation (14.4), which assigns the weight for any additional term c_a, as follows:

$$w_a = \beta \cdot co(c_a, c_{key}) \cdot w_{key}. \quad (14.5)$$

We quantify the cumulative weight for both the statistical expansion terms (e.g., those in Table 14.11a) and the knowledge-based expansion terms (e.g., those in Table 14.11b). The former cumulative weight will be larger than the latter. We define β to be the former divided by the latter. In this way, the cumulative weight for the knowledge-based expansion terms equals that of the statistical expansion terms after boosting.

More specifically, we quantify the cumulative weight of a set of expansion terms using the length of the "expansion vector" composed by these terms. Here we define the vector length according to the standard vector space notation: Let $V^{KB} = <w_1^{KB}, \ldots, w_k^{KB}>$ be the augmenting vector consisting solely of terms derived by the knowledge-based method, where w_i^{KB} $(1 \leq i \leq k)$ denotes the weight for the ith term in knowledge-based expansion (Equation 14.4). Likewise, let $V^{stat} = <w_1^{stat}, \ldots, w_l^{stat}>$ be the augmenting vector consisting of all statistically related terms. The process of deriving $\{w_1^{KB}, \ldots, w_k^{KB}\}$ yields $k < l$. Consequently, $\{w_1^{KB}, \ldots, w_k^{KB}\} \subset \{w_1^{stat}, \ldots, w_l^{stat}\}$. Let $|V^{KB}|$ be the length of the vector V^{KB}:

$$|V^{KB}| = \sqrt{(w_1^{KB})^2 + (w_1^{KB})^2 + \ldots + (w_k^{KB})^2}. \quad (14.6)$$

Likewise, let $|V^{stat}|$ represent the length of vector V^{stat}, which can be computed similarly as Equation (14.6). Thus, the boosting factor for V^{KB} is:

$$\beta = \frac{|V^{stat}|}{|V^{KB}|}. \quad (14.7)$$

To study the effects of different levels of boosting, a boosting-level–controlling factor α is introduced to refine Equation (14.7):

TABLE 14.13 Weights for sample expansion concepts

(a) Concepts that statistically correlate to keratoconus	Weight	(b) Concepts that treat keratoconus	Weight
Fuchs dystrophy	0.289	Penetrating keratoplasty	0.247
Penetrating keratoplasty	0.247	Epikeratoplasty	0.230
Epikeratoplasty	0.230	Epikeratophakia	0.119
Corneal ectasia	0.168	Keratoplasty	0.103
Acute hydrops	0.165	Contact lens	0.101
Keratometry	0.133	Thermokeratoplasty	0.092
Corneal topography	0.132	Button	0.067
Corneal	0.130	Secondary lens implant	0.057
Aphakic corneal edema	0.122	Fittings adapters	0.048
Epikeratophakia	0.119	Esthesiometer	0.043
Granular dystrophy corneal	0.109	Griffonia	0.035
Keratoplasty	0.103	Trephine	0.033
Central cornea	0.103	Slit lamps	0.032
Contact lens	0.101	Fistulization	0.030
Ghost vessels	0.095	Soft contact lens	0.026

$$\beta_r = 1 + \alpha \cdot \left(\frac{|V^{stat}|}{|V^{KB}|} - 1 \right), \qquad (14.8)$$

where β_r is the refined boosting factor. The parameter α, ranging within [0, 1], can be used to control the boosting scale. From Equation (14.8), we note that $\beta_r = 1$ when we set $\alpha = 0$, which represents no boosting. β_r increases as α increases. As α increases to 1, β_r reduces to $\frac{|V^{stat}|}{|V^{KB}|}$. Thus, α can be used to experimentally study the boosting sensitivity. (We experimentally evaluated cases of setting $\alpha > 1$. We noted that the retrieval effectiveness in those cases was usually suboptimal compared with cases where α was within [0, 1].)

14.6.3 Retrieval Performance

In this section, we compare the retrieval performance of the knowledge-based query expansion with that of statistical expansion using two standard medical corpuses. We start with the experiment setup and then present the results under selective settings.

14.6.3.1 Testbeds

A testbed for a retrieval experiment consists of three components: (1) a corpus (or a document collection), (2) a set of benchmark queries, and (3) relevance judgments indicating which documents are relevant for each query. Our experiment is based on the following two testbeds:

1. OHSUMED [16]. This testbed was introduced in Section 14.5.3. In the task of evaluating knowledge-based query expansion, we are interested in a subset of the OHSUMED queries that are scenario specific. Among the 106 queries, we have identified a total number of 57 such queries. In Table 14.14, we categorize these 57 queries based on the scenario(s) that each query mentions. The corresponding ID of each query is listed in this table. (The full text of each query is shown in Liu and Chu [42].) Note that a query mentioning multiple distinct scenarios will appear multiple times in this table corresponding to its scenarios.
2. The McMaster University Clinical Hedges Database [43–46]. This testbed was originally constructed for the task of medical document classification instead of free-text query answering. As a result, adaptation is needed for retrieval performance study. We first describe the original dataset and then explain how we adapted it to make it a usable testbed for retrieval performance evaluation.

TABLE 14.14 OHSUMED queries categorized based on their scenarios

Scenario	Query ID
Treatment of a disease	2, 13, 15, 16, 27, 29, 30, 31, 32, 35, 37, 38, 39, 40, 42, 43, 45, 53, 56, 57, 58, 62, 67, 69, 72, 74, 75, 76, 77, 79, 81, 85, 93, 98, 102
Diagnosis of a disease	15, 21, 37, 53, 57, 58, 72, 80, 81, 82, 97
Prevention of a disease	64, 85
Differential diagnosis of a symptom/disease	14, 23, 41, 43, 47, 51, 65, 69, 70, 74, 76, 103
Pathophysiology of a disease	2, 3, 26, 64, 77
Complications of a disease/medication	3, 30, 52, 61, 62, 66, 79
Etiology of a disease	14, 26, 29
Risk factors of a disease	35, 64, 85
Prognosis of a disease	45
Epidemiology of a disease	3
Research of a disease	75
Organisms of a disease	81
Criteria of medication	49, 52, 94
When to administer a medication	33
Preventive health care for a type of patient	96

14.6.3.1.1 Original Dataset. The McMaster Clinical Hedges Database contains 48,000 PubMed articles published in 2000. Each article was classified into one of the following scenario categories: treatment, diagnosis, etiology, prognosis, clinical prediction guide of a disease, economics of a health care issue, or review of a health care topic. Consensus about the classification was drawn from among six human experts [43]. When the experts classified each article, they had access to the hard copies of the full text. However, to construct a testbed for our retrieval system, we were able to download only the title and abstract of each article from the PubMed system. (The full text of each article is typically unavailable through PubMed.)

14.6.3.1.2 Construction of Scenario-Specific Queries. Since the McMaster Clinical Hedges Database is constructed to test document classification, it does not contain a query set. Using the following procedure, we constructed a set of 55 scenario-specific queries and determined the relevance judgments for these queries based on the document classification that can be adapted for them:

- **Step 1.** We identified all the disease/symptom concepts in the OHSUMED query set. We identified such concepts based on their semantic-type information (defined by UMLS). We used these as the key concepts in constructing the scenario-specific queries for the McMaster testbed. In selecting these concepts, we manually filtered out eight concepts (out of an original 90) that we considered too general to make a scenario-specific query (e.g., infection, lesion, carcinoma). After this step, we obtained 82 such key concepts.
- **Step 2.** For each key concept identified in Step 1, four scenario-specific queries were constructed: treatment,

diagnosis, etiology, and prognosis of a disease/symptom. For example, for the concept of breast cancer, we constructed the queries "breast cancer treatment," "breast cancer diagnosis," "breast cancer etiology," and "breast cancer prognosis." Our study was restricted to these four scenarios because UMLS covers only these four.

- **Step 3.** For each query generated in Step 2, we generated its relevance judgments by applying the following simple criterion: A document is considered to be relevant to a given query if (1) experts have classified the document to the category of the query's scenario and (2) the document mentions the query's key concept. This criterion has been our best choice to automate the process of generating relevance judgments on a relatively large scale; however, it may misidentify irrelevant documents as relevant. After we identified the relevant documents for each query, certain queries were filtered out based on the intuition that a query with too few relevance judgments would lead to less reliable retrieval results (especially in terms of precision/recall). For example, for a query with only one relevant document, two similar retrieval systems may obtain completely different precision/recall results if one ranks the relevant document on top and another accidentally ranks it out of the top 10. Following this intuition, queries that had less than five relevant documents were filtered out. After this filtering step, we were left with 55 queries. These queries, together with the scenarios identified for each, are presented in Liu and Chu [42].

14.6.3.2 The Vector Space Model and Indexing

In IR studies, *indexing* refers to the step of converting free-text documents and queries to their respective vector representations [29]. The query and document vectors are then matched based on a VSM. In experimental evaluation of the knowledge-based query expansion method, we focus on results generated by the following two VSMs:

1. *Stem-based VSM* [29]. Using a stem-based VSM, both a query and a document are represented as vectors of word stems. Given a piece of free text, we first removed common stop words such as "a," "the," etc., and then derived word stems from the text using the Lovins stemmer [47]. We further applied the tf-idf weighting scheme (more specifically the $atc \cdot atc$ scheme [48]) to assign weights to stems in documents and the query before expansion. (This weighting process yields the weight for the key concept in Equation (14.1).) Under the stem-based VSM, all terms expanded to a given query need to be in the word-stem format. Thus, for expansion concepts derived from procedures in Section 14.6.2, we applied the following procedure to identify the corresponding word stems: For each expansion concept, we first looked up its string forms in UMLS. We further removed stop words and used the Lovins stemmer to convert the string forms into word stems. Lastly, we assigned weights to these expansion word stems using the method described in Section 14.6.2.
2. *Phrase-based VSM* [5]. Using a phrase-based VSM, both a query and a document are represented as vectors of phrases. We first used the concept extraction method presented in Section 14.2 to identify the concepts appearing in a given query and a set of documents. We further formulated phrase representations of the query and the documents based on the definition of phrases in Section 14.5.2. We applied the weighting method in Section 14.5.2 to assign weights to phrases in the query and the documents. For expansion concepts appended to the original query, we converted them into their corresponding phrase representations and assigned the weights for both concepts and word stems appearing in a phrase using the method described in Section 14.6.2.

14.6.3.3 Evaluation Metrics

We measure the retrieval performance using the following three different metrics:

1. *avgp* = the 11-point precision average (precision averaged over the 11 standard recall points [29])
2. *p@10* = the precision in the top 10 retrieved documents
3. *p@20* = the precision in the top 20 retrieved documents

14.6.3.4 Retrieval Performance Using the Stem-Based Vector Space Model

In the following, we study the performance improvement of knowledge-based expansion as compared with that of statistical expansion.

We used s to denote an expansion size. For a given s, we used both knowledge-based expansion and statistical expansion to expand the top s stems that had the heaviest weights. For knowledge-based expansion, no weight boosting was applied. We computed the three metrics for both methods on the OHSUMED and McMaster testbeds. We further averaged the results over the queries in these two testbeds. Table 14.15 shows the performance comparison of the two methods on both testbeds, which is under the three metrics previously given. The first row in each subtable shows the performance of statistical expansion, whereas the second row shows the performance of knowledge-based expansion and its percentage of improvement over statistical expansion.

In these tables, "s = All" means appending all possible expansion terms that have a nonzero weight (Equation (14.5)) into the original query. Using the knowledge-based method, setting "s = All" led to expanding an average of 1,717 terms to each query on average, with the standard deviation of 1,755;

TABLE 14.15 Performance comparison of the two methods under selected expansion sizes using the stem-based VSM

(a) Performance comparison using the *avgp* metric for the OHSUMED testbed

Expansion size, s	10	20	30	40	50	100	200	300	All
Statistical expansion	0.417	0.424	0.428	0.43	0.429	0.432	0.429	0.43	0.425
Knowledge-based expansion without weight boosting	0.422	0.431	0.430	0.432	0.434	0.438	0.442	0.443	0.445
Knowledge-based expansion with weight boosting	0.428	0.436	0.437	0.437	0.439	0.443	0.446	0.450	0.452

(b) Performance comparison using the *p@10* metric for the OHSUMED testbed

Expansion size, s	10	20	30	40	50	100	200	300	All
Statistical expansion	0.535	0.546	0.549	0.553	0.551	0.567	0.581	0.574	0.567
Knowledge-based Expansion without weight boosting	0.544	0.547	0.554	0.551	0.553	0.572	0.572	0.577	0.588
Knowledge-based expansion with weight boosting	0.552	0.567	0.568	0.577	0.577	0.595	0.586	0.595	0.600

(c) Performance comparison using the *p@20* metric for the OHSUMED testbed

Expansion size, s	10	20	30	40	50	100	200	300	All
Statistical expansion	0.482	0.491	0.493	0.491	0.492	0.496	0.497	0.493	0.496
Knowledge-based expansion without weight boosting	0.483	0.491	0.494	0.496	0.493	0.498	0.496	0.497	0.498
Knowledge-based expansion with weight boosting	0.482	0.496	0.498	0.510	0.509	0.514	0.514	0.513	0.511

(d) Performance comparison using the *avgp* metric for the McMaster testbed

Expansion size, s	10	20	30	40	50	100	200	300	All
Statistical expansion	0.326	0.328	0.325	0.324	0.323	0.319	0.311	0.309	0.295
Knowledge-based expansion without weight boosting	0.325	0.328	0.324	0.326	0.325	0.324	0.321	0.32	0.321
Knowledge-based expansion with weight boosting	0.325	0.326	0.324	0.325	0.323	0.322	0.320	0.315	0.318

(e) Performance comparison using the *p@10* metric for the McMaster testbed

Expansion size, s	10	20	30	40	50	100	200	300	All
Statistical expansion	0.316	0.324	0.324	0.318	0.324	0.311	0.295	0.3	0.293
Knowledge-based expansion without weight boosting	0.322	0.324	0.322	0.325	0.322	0.318	0.315	0.32	0.335
Knowledge-based expansion with weight boosting	0.320	0.322	0.318	0.322	0.320	0.315	0.316	0.313	0.324

(f) Performance comparison using the *p@20* metric for the McMaster testbed

Expansion size, s	10	20	30	40	50	100	200	300	All
Statistical expansion	0.285	0.285	0.285	0.283	0.283	0.281	0.279	0.278	0.279
Knowledge-based expansion without weight boosting	0.285	0.287	0.287	0.291	0.29	0.293	0.286	0.291	0.292
Knowledge-based expansion with weight boosting	0.285	0.289	0.287	0.287	0.289	0.289	0.285	0.287	0.289

using the statistical method, it led to an average of 50,317 terms with the standard deviation of 15,243.

From these experimental results, we observe the following: The performance for knowledge-based expansion generally increases as s increases and usually reaches the peak when $s =$ All. (The only exception is in the case of using the *avgp* metric on the McMaster testbed, in which the performance of the knowledge-based method roughly remains constant as s increases.) On the other hand, the performance of the statistical method degrades as s increases. On the OHSUMED testbed, its performance degrades after s reaches a certain level, such as $s = 100$ (Table 14.15(a)) and $s = 200$ (Tables 14.15(b) and 14.15(c)); on the McMaster testbed, the performance starts degrading almost immediately after s exceeds 20. This is due to the fact that statistical expansion does not distinguish whether an expansion term is scenario specific. As a result, as more terms are appended to the original query, the negative effect of including those non-scenario-specific terms begins to accumulate, and the performance drops after a certain point. In contrast, the knowledge-based method appends scenario-specific terms, and consequently the performance keeps increasing as more "useful" terms are appended.

Our experimental results also revealed that both statistical expansion and knowledge-based expansion consistently outperformed the no-expansion method by more than 5%. On the OHSUMED testbed, for example, the *avgp* of no expansion is 0.382, which is outperformed by the peak performance of statistical expansion at 0.432 and by the peak performance of knowledge-based expansion at 0.452 (Table 14.15(a)). Similarly, the *p@10* and *p@20* of no expansion are 0.532 and 0.470, which are outperformed by the peak performance of statistical expansion at 0.581 and 0.497 and by the peak performance of knowledge-based expansion at 0.600 and 0.514 (Tables 14.15(b) and 14.15(c)).

We evaluated the effectiveness of weight boosting and its impact on retrieval performance. The boosting factor β was computed using Equation (14.8), under the different settings of $\alpha = 0.25, 0.5, 0.75, 1, 1.25, 1.5$. We present the peak performance of weight boosting in the third row of each subtable of Table 14.15. For the OHSUMED testbed, boosting helped improve the performance, and the best performance occurred in the range from $\alpha = 0.5$ to $\alpha = 1.25$. We note that setting $\alpha = 0.5$or $= 0.75$ generally yields the best boosting effect for the *avgp* metric; setting $\alpha = 1$ to 1.25 yields better performances for the *p@10* and *p@20* metrics. For the McMaster testbed, weight boosting failed to yield improvements. Further discussion of weight boosting is presented in Liu and Chu [42, 49].

We further studied how knowledge-based expansion performed for different query scenarios, and experimental results showed that the performance varied depending on the query scenario [42, 49]. More specifically, the method yields more improvements in scenarios such as treatment, differential diagnosis, and diagnosis, whereas it yields fewer improvements in such scenarios as complication, pathophysiology, etiology, and prognosis. An explanation of this lies in the different qualities of the knowledge structures for these scenarios. The knowledge structures (i.e., the fragments of the UMLS semantic network such as Figure 14.11) for the latter four scenarios were originally missing in UMLS and were acquired by ourselves from experts. (See the knowledge acquisition process in Section 14.6.5.) These acquired structures have more semantic types marked as relevant than those for the former three scenarios. As a result, when handling queries with the latter four scenarios, the knowledge-based method keeps more concepts during the filtering step. Thus, the expansion result for the knowledge-based method resembles that of the statistical expansion method, leading to almost equivalent performance between the two methods and less improvements. Further refinement on the clustering and ranking of the knowledge structures for the four scenarios (i.e., complication, pathophysiology, etiology, and prognosis) will increase the improvements in retrieval performance.

14.6.3.4.1 Choice of α for weight boosting. Experimental results revealed that weight boosting was helpful in improving retrieval performance. Further, the performance of weight boosting was sensitive to the query scenario. Certain query scenarios such as treatment and diagnosis are associated with more mature knowledge structures, which require fewer expansion concepts. In these scenarios, setting α between 0.75 and 1.25, which represents more aggressive weight boosting, achieves noticeable improvements. In other scenarios, associated with less mature knowledge structures, such as complication, the difference is insignificant between the set of expansion concepts by our method and those by statistical expansion. As a result, the cumulative weights of the two sets of expansion concepts are close to each other. For such scenarios, our experimental data suggest a more conservative weight boosting with α in the range of 0 to 0.5.

14.6.3.4.2 Comparison with Previous Knowledge-Based Query Expansion Studies. In past studies [50–52], research compared knowledge-based expansion methods against a baseline generated without expansion. Such studies reported an insignificant improvement [51–52] or even degrading performance [50] compared with the no-expansion method. In contrast, our study compares against a baseline generated by statistical expansion. In our experimental setup, this baseline had an observed improvement over the no-expansion method by 5% to 10%.

In Aronson and Rindflesch's study [53], the researchers applied the UMLS metathesaurus to automatically expand synonyms to the original query. In one particular case, their approach achieved a 5% improvement over a previous study [54] that applied statistical expansion on the same testbed. This result indicates the value of knowledge-based query expansion. However, their approach was limited to expanding only synonyms instead of scenario-specific terms. Thus, the improvement was limited.

14.6.3.5 Retrieval Performance Using the Phrase-Based Vector Space Model

In this section, we compare the performance of knowledge-based query expansion with that of statistical expansion by using the phrase-based VSM for query-document matching. The experiments were performed on the 57 scenario-specific queries in OHSUMED. (Similar results were observed on the McMaster testbed and are excluded from this discussion due to space limits.) The results are shown in Table 14.16, under the three metrics, *avgp*, *p@10*, and *p@20*. We present the performance of both knowledge-based query expansion and statistical expansion under selected expansion sizes *s*. We have also provided the retrieval results for the original queries without expansion, as shown in each row and listed under $s = 0$.

From these results, we made the following two major observations:

- With phrase-based VSM, query expansion (both methods) still brings significant improvements for about 10%. For example, both expansion methods yield

TABLE 14.16 Performance comparison of the two methods under various expansion sizes using the phrase-based VSM

(a) Performance comparison using the *avgp* metric

s	0	10	20	30	40	50	100
Statistical expansion	0.440	0.486	0.489	0.483	0.479	0.479	0.460
Knowledge-based expansion	0.440	0.486	0.490	0.487	0.482	0.485	0.475

(b) Performance comparison using the *p@10* metric

s	0	10	20	30	40	50	100
Statistical expansion	0.584	0.612	0.604	0.581	0.579	0.567	0.544
Knowledge-based expansion	0.584	0.612	0.616	0.604	0.600	0.595	0.586

(c) Performance comparison using the *p@20* metric

s	0	10	20	30	40	50	100
Statistical expansion	0.504	0.546	0.540	0.532	0.528	0.525	0.496
Knowledge-based expansion	0.504	0.538	0.546	0.554	0.543	0.542	0.535

a peak *avgp* of 0.49 compared with the *avgp* of the no-expansion method, which is 0.44.

- Both expansion methods achieve the peak performance when expanding 10 to 20 concepts. This makes it desirable to combine query expansion with the phrase-based VSM, since appending 10 to 20 concepts to the original query incurs a small amount of computation overhead. We note that this is in contrast to the case of using the stem-based VSM, in which we need to expand hundreds or thousands of word stems to reach peak performance.

We also noted that the peak performance of the two expansion methods is comparable. That is, expanding 10 to 20 statistically related concepts is almost as good as expanding 10 to 20 scenario-specific concepts identified by the knowledge-based method. This is in contrast to the comparison obtained by using the stem-based VSM, where there is significant difference between the two methods. This is due mainly to the ability of the phrase-based VSM in approximately matching distinct concepts. Recall the fact that expanding all statistically related terms introduces certain heavily weighted terms that are non–scenario specific. Using the stem-based VSM that performs strict matching among terms, the existence of such non-scenario-specific terms promotes the ranking of certain non-scenario-specific documents while demoting the ranking of other, scenario-specific documents. The phrase-based VSM, however, is able to partially match a non-scenario-specific phrase with a scenario-specific one appearing in a relevant document. Subsequently, the existence of certain non-scenario-specific phrases generated by the statistical expansion no longer negatively impacts the retrieval result.

We also note that the precision of using the phrase-based VSM without expansion (the first cell in each row of Table 14.16) is significantly higher than that of using the stem-based VSM (the first cell in each row of Table 14.15). Since the phrase-based VSM is based on UMLS, these improvements can be viewed as the results of a first step in applying human knowledge. On top of this, statistical expansion takes another step and applies statistical knowledge derived from a sample corpus to append statistically correlated concepts. The 5–10% improvement in precision (e.g., an *avgp* of 0.489 for statistical expansion under $s = 20$ compared with an *avgp* of 0.440 for no expansion; Table 14.16(a)) suggests that the statistical knowledge is "additive" to human knowledge to achieve better retrieval results. Knowledge-based query expansion uses statistical expansion as a starting point and attempts to further apply UMLS to refine the query expansion results. Nonetheless, since the same knowledge source has already been applied in the form of the phrase-based VSM, this refinement step yields only a small amount (1–2%) of performance improvement.

14.6.4 Computation Complexity Comparison

The computation complexity of knowledge-based expansion is comparable to that of statistical expansion. In the step of deriving expansion terms, the knowledge-based method requires an additional step of going through all statistically related terms and selecting those that are scenario specific. This step incurs a complexity that is linear to the number of statistically related terms. Since the complexity of identifying all statistically related terms by the statistical method is at least linear to the number of these terms, the additional step in the knowledge-based method does not significantly increase complexity.

In the step of matching an expanded query with documents, the complexity of the knowledge-based method is less than that of the statistical method. The complexity in this step is directly proportional to the number of terms in the expanded query. As revealed by our experiments, knowledge-based expansion requires significantly fewer expansion terms, which reduces the computation complexity.

14.6.5 Knowledge Acquisition

The quality of our knowledge-based method depends largely upon the quality and completeness of the domain-specific knowledge source. The knowledge structure in the UMLS knowledge base is not specifically designed for scenario-specific retrieval. As a result, some frequently asked scenarios (e.g., etiology, complications of a disease) are either undefined in UMLS or defined but with incomplete knowledge. Therefore, we present a methodology that consists of the following two steps:

1. Acquisition of knowledge for undefined scenarios to supplement the UMLS knowledge source
2. Refinement of the knowledge of the scenarios defined in the UMLS knowledge source (including the knowledge supplemented by Step 1)

14.6.5.1 Knowledge Acquisition Methodology

14.6.5.1.1 Knowledge Acquisition for Undefined Scenarios. For an undefined scenario, an incomplete relationship graph, as shown in Figure 14.13, is presented to medical experts. Edges in this relationship graph are labeled with one of the undefined scenarios, such as "etiology." The experts will fill in the question marks with existing UMLS semantic types that fit the relationship. For example, because viruses are related to the etiology of a wide variety of diseases, the semantic type Virus will replace one of the question marks in Figure 14.13. This new relationship graph (etiology of diseases) will be appended to the UMLS semantic network and can be used for queries with the "etiology" scenario.

14.6.5.1.2 Knowledge Refinement Through Relevance Judgments. A relationship graph for a given scenario (either previously defined by UMLS or newly acquired from Step 1) may be incomplete in including all relevant semantic types. A hypothetical example of this incompleteness would be the missing relationship *treats* between Therapeutic or Preventive Procedure and Disease or Syndrome. The basic idea in amending this incompleteness is to explore the "implicit" knowledge embedded in the relevance judgments of an IR testbed. Such a testbed typically provides a set of benchmark queries, and for each query, a prespecified set of relevant documents. To amend the knowledge structure for a certain scenario, such as treatment, we focus on sample queries that are specific to this scenario (e.g., keratoconus treatment). We then study the content of documents that are marked as relevant to these queries. From the content, we can identify concepts that are directly relevant to the query's scenario (e.g., treatment). If the semantic type for those concepts is missing in the knowledge structure, we can then refine the knowledge structure by adding the corresponding semantic types. For example, let us consider a hypothetical case where the type Therapeutic or Preventive Procedure is missing in the knowledge structure of Figure 14.13. If by studying the sample query "keratoconus treatment," we identify quite a few Therapeutic or Preventive Procedure concepts appearing in relevant documents, such as penetrating keratoplasty and epikeratoplasty, we are then able to identify Therapeutic or Preventive Procedure as a relevant semantic type and append it to Figure 14.13.

Given that a typical benchmark query has a long list of relevant documents, it is labor intensive to study the content of every relevant document. One way to accelerate this process is to first apply an incomplete knowledge structure to perform knowledge-based query expansion and conduct retrieval tests based on such expansion. An incomplete knowledge structure leads to an "imperfect" query expansion, which in turn, fails to retrieve certain relevant documents to the top of the ranked list. Comparing this ranked list with the "gold standard" and identifying the missing relevant documents will give us pointers to determine the incomplete knowledge. For example, failure to include Therapeutic or Preventive Procedure in the knowledge structure in Figure 14.11 prevents us from expanding concepts such as penetrating keratoplasty to the sample query of keratoconus, treatment. As a result, documents with a focus on penetrating keratoplasty will be ranked unfavorably

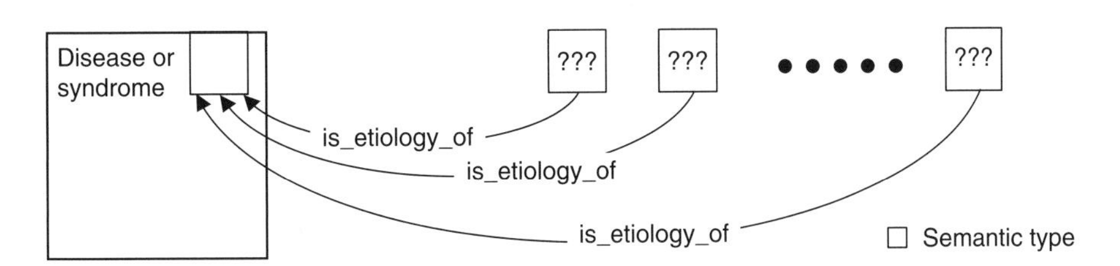

FIGURE 14.13 A sample template to acquire knowledge for previously undefined scenarios.

low. After we identify such documents, we can discover the missing expansion concepts that are contributing to the low rankings and refine the knowledge structure as we have just described.

14.6.5.2 Knowledge Acquisition Process

The 57 scenario-specific queries (Table 14.14) in the OHSUMED testbed were chosen to apply our proposed knowledge acquisition method because of the following considerations:

- The OHSUMED queries are collected from physicians' patients in a clinical setting. Therefore, the OHSUMED query scenarios should be representative of health care, and the knowledge acquired from these scenarios should be broadly applicable.
- The knowledge acquisition methodology also requires the exploration of relevance judgments for a set of benchmark queries. OHSUMED is the largest testbed for medical free-text retrieval that has relevance judgments for knowledge refinement.

We identified 12 OHSUMED scenarios whose knowledge structures were missing in UMLS. We applied the two-step knowledge acquisition method to acquire the knowledge structures for these 12 undefined scenarios and to refine the knowledge structures for all scenarios. During the first step of the acquisition process, we interviewed two intern physicians at the UCLA School of Medicine. During the interview, we first described the meaning of the relationship graphs, as shown in Figure 14.13. Next, we presented the entire list of UMLS semantic types to the experts so that appropriate semantic types were filled into the question marks. We communicated the results from one expert to another until they reached a consensus for each scenario. For the second step of knowledge acquisition, we performed retrieval tests on the OHSUMED testbed using both queries expanded by the knowledge-based method and the method of expanding all statistically related concepts. We focused on 12 queries in which the statistical method outperformed the knowledge-based method in terms of the precision in top 10 results. We further applied the method presented in the previous section to study the content of these top-ranked documents and augmented the knowledge structure for the corresponding scenario with appropriate semantic types.

14.6.5.3 Knowledge Acquisition Results

The acquisition results are shown in Table 14.17. Due to space constraints, we provide only a statistical summary of the results. The scenarios in the first three rows (i.e., treatment, diagnosis, and prevention) are defined in UMLS. The first column in these rows shows the number of semantic types marked as relevant for each scenario (i.e., the number of semantic types that experts have filled into the blank rectangles of Figure 14.13). The second column for these rows is "N/A" because there was no need to acquire knowledge structure from domain experts for these scenarios. The third column shows the number of semantic types added during knowledge refinement (the second step of knowledge acquisition). For example, for the diagnosis scenario, two additional semantic types, Laboratory or Test Result and Biologically Active Substance, were added because of the study on query 97: Iron deficiency anemia, which test is best. These two semantic types were added because their absence prevented the

TABLE 14.17 Knowledge acquisition results

Scenarios	No. of semantic types defined in UMLS	No. of semantic types acquired from experts	No. of additional semantic types through knowledge refinement	Total no. of semantic types after knowledge acquisition
Treatment of a disease	3	N/A	1	4
Diagnosis of a disease	5	N/A	2	7
Prevention of a disease	3	N/A	0	3
Differential diagnosis of a symptom/disease	N/A	10	4	14
Etiology of a disease	N/A	40	1	41
Risk factors of a disease	N/A	40	2	42
Complications of a disease or medication	N/A	15	0	15
Pathophysiology of a disease	N/A	56	0	56
Prognosis of a disease	N/A	15	2	17
Epidemiology of a disease	N/A	13	0	13
Research of a disease	N/A	28	0	28
Organisms of a disease	N/A	7	0	7
Criteria of medication	N/A	26	0	26
When to administer a medication	N/A	5	6	11
Preventive health care for a type of patient	N/A	10	2	12

knowledge-based method from expanding two critical concepts into the original query: serum ferritin and fe iron, each belonging to one of the two semantic types. From the relevance judgment set, we noted that missing these two concepts leads to the low ranking of three relevant documents that heavily use them.

Starting from the fourth row, we list the scenarios for which we need to acquire knowledge structure from domain experts. The first column for these scenarios is "N/A" because these scenarios are originally undefined in UMLS. The second column shows the number of semantic types that experts have filled into the structure template of Figure 14.13. The third column shows the number of additional semantic types from knowledge refinement (the second step of knowledge acquisition), and the last column shows the total number of semantic types after knowledge acquisition.

The proposed knowledge acquisition method on the OHSUMED testbed was shown to be efficient and effective. We finished communicating with domain experts and acquiring the knowledge structures for the 12 scenarios in less than 20 hours and spent an additional 20 hours to refine the knowledge structures by exploring the relevance judgments. The augmented knowledge was applied in our experiments presented in Section 14.6.3 and was effective in improving the retrieval performance of the knowledge-based method over the statistical expansion method.

14.6.6 Study of the Relevancy of Expansion Concepts by Domain Experts

Through experiments on the two standard medical text retrieval testbeds, we observed that under most retrieval settings, knowledge-based query expansion outperformed statistical expansion. Our conjecture was that knowledge-based query expansion selects more specific expansion concepts to the original query's scenario than does statistical expansion. To verify this conjecture, we requested domain experts to manually evaluate the relevancy of expansion concepts.

The basic idea for this study was the following: For each query in a given retrieval testbed, we applied two query expansion methods to generate two sets of expansion concepts. We then prepared an evaluation form that inquired about the relevancy of each expansion concept to the original query. In this form, we presented the query and asked domain experts to judge the relevancy based on the query's scenario(s). For each concept, we provided four scales of relevancy: relevant, somewhat relevant, irrelevant, or do not know. We blinded the method used to generate each concept, and in doing so, we reduced bias that an expert might have had toward a particular method.

To implement this idea, we chose the 57 scenario-specific queries in the OHSUMED testbed. We applied the two expansion methods and derived 40 expansion concepts from each method with the highest weights. We presented the evaluation form consisting of these concepts to three medical experts who were intern doctors at the UCLA School of Medicine. We asked them to make judgments only on those queries that belonged to their area of expertise (oncology, urology, etc.). On average, each expert judged the expansion concepts for 15 queries. Thus, for each expansion method, we obtained 1,600 expansion concepts classified into one of the four categories.

Figures 14.14 and 14.15 present a summary of the results from this human subject study. For the expansion concepts derived from each method, we summarized the results into a histogram. The bins of this histogram were the four scales of relevancy. We noted that 56.9% of the expansion concepts derived by the knowledge-based method were judged as either relevant or somewhat relevant, whereas only 38.8% of expansion concepts by statistical expansion were judged similarly. This represented a 46.6% improvement. The results validate that knowledge-based query expansion derives more relevant expansion concepts to the original query scenario(s) than those by statistical expansion and thus yields improved retrieval performance for scenario-specific queries.

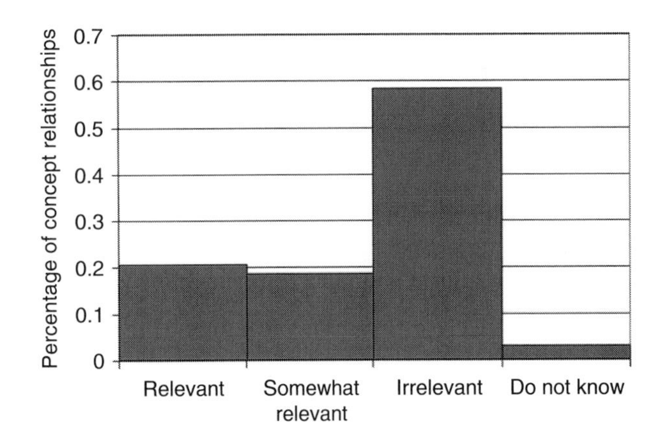

FIGURE 14.14 Relevancy of statistical expansion concepts.

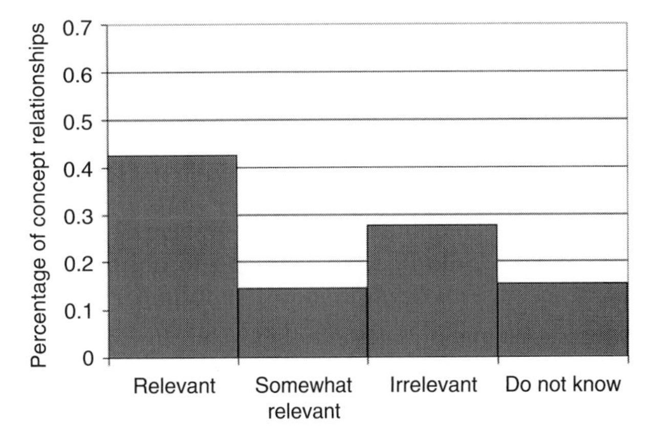

FIGURE 14.15 Relevancy of knowledge-based expansion concepts.

14.7 The KMeX System Architecture for Retrieving Scenario-Specific Free-Text Documents

We have implemented and integrated the three proposed techniques in the KMeX system to provide scenario-specific free-text retrieval (Figure 14.16). This system provides the capability to retrieve many types of medical free-text documents, such as patient clinical reports, medical literature articles, etc. IndexFinder first extracts key concepts and normalize them into standard terms as defined in the knowledge source (e.g., UMLS). Topics and subtopics are then derived by mining the frequently co-occurring features extracted from the documents. With the aid of the knowledge source and the user's query patterns, a topic-oriented directory system can be constructed.

During the retrieval phase, the query expansion module appends the user query with scenario-specific terms. The directory system selects the most relevant topics that match the expanded query. Documents that belong to those topics are submitted to the module that ranks the documents based on their similarity to the query via the phrase-based VSM and returns the documents most similar to the query to users first.

14.8 Summary

We have developed a new knowledge-based approach for retrieving scenario-specific free-text documents, which consists of three integrated components: IndexFinder, phrase-based VSM, and knowledge-based query expansion. IndexFinder extracts key terms from free text, generating conceptual terms by permuting words in a sentence rather than using the traditional techniques based on NLP. Although the generated concepts are matched with the controlled vocabulary in the UMLS and are valid terms, they might not be relevant to the document. Thus, syntactic and semantic filters are used to eliminate the irrelevant candidates. Preliminary evaluation shows that filtering is effective in eliminating irrelevant concepts. Our experimental results show that IndexFinder can process free texts at a speed of about 43,000 bytes of text per second on a PC with Pentium 4. As a result, it is able to extract key UMLS concepts from clinical texts in real time. The extracted concepts can be used for content correlation, document indexing for directory systems, and transforming ad hoc terms in the queries into controlled vocabulary to improve retrieval effectiveness.

The phrase-based VSM has been developed for document retrieval. In this model, we divided each document into a set of

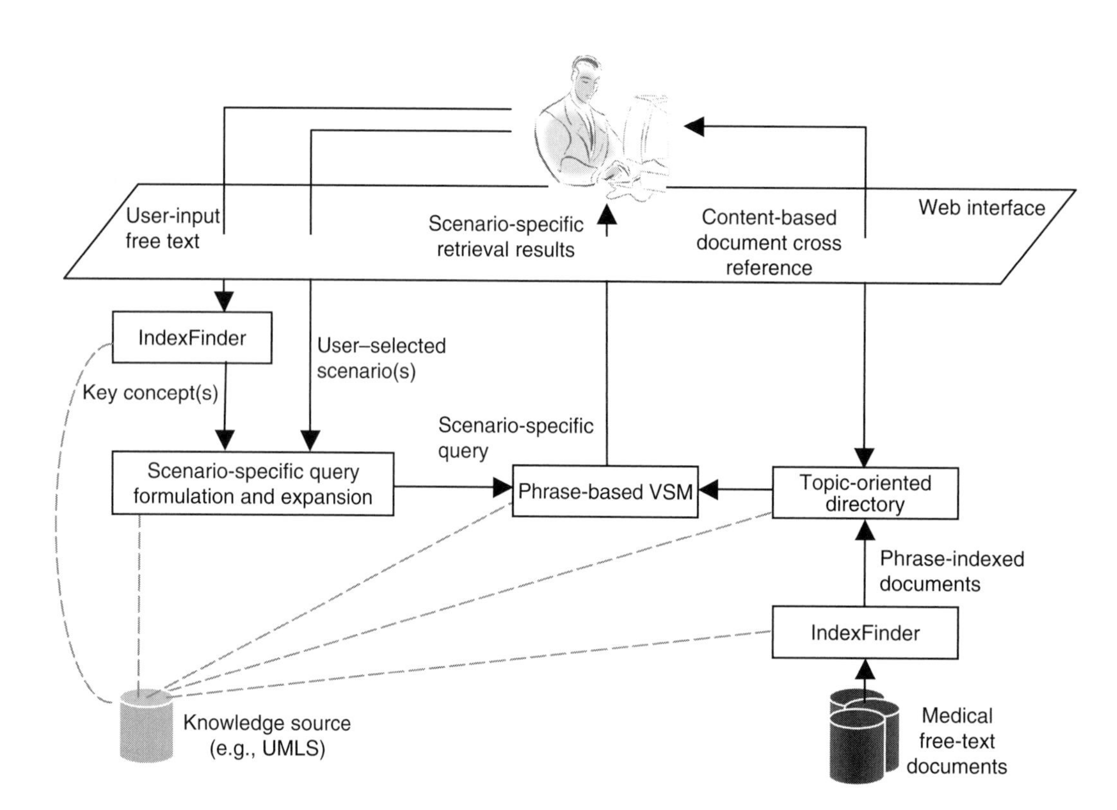

FIGURE 14.16 The KMeX system architecture.

phrases. Each phrase is represented by both a concept defined in a controlled vocabulary and the component word stems. The similarity between concepts is based on the interrelationships of concepts in a knowledge base. The similarity between two phrases is measured by their stem overlaps as well as the similarity between the concepts they represent. The similarity between two documents is defined as the cosine of the angle between their respective phrase vectors.

Using UMLS as both the controlled vocabulary and the knowledge base to derive the conceptual similarities, we demonstrated from different perspectives that the retrieval effectiveness of the phrase-based VSM was significantly higher than that of the current gold standard—the stem-based VSM, because in phrase-based VSM, the stem similarity compensates for the incompleteness of knowledge sources, while the concept similarity compensates for the lack of semantic meaning in the stem similarity. Such a significant increase in retrieval effectiveness was achieved without sacrificing excessive computation efficiency. Knowledge-based query expansion of terms related to the scenario yields 5–10% improvement in precision and recall as compared with the statistical query expansion case. Knowledge-based query expansion can be applied together with the phrase-based VSM. In that case, the peak performance is achieved with very few expansion terms (10 to 20), which is a desirable property.

Topics can be generated from mining document features. Based on query templates and knowledge type hierarchies, free-text documents can be organized into a set of scenario-specific topic-oriented directory systems. In each directory system, the documents are indexed and linked based on the topics. Such topic organization not only improves the retrieval performance for ranking relevant documents but also provides cross-referencing among related topics.

We have implemented a testbed with these three technologies. Using the UCLA Medical Center patient reports as a test set, we have shown that IndexFinder is able to extract features from free-text documents and that data-mining algorithms can be used to organize features into topics and are feasible to construct topic-oriented directory systems. Our knowledge-based query expansion techniques and the phrase-based VSM can be used in conjunction to significantly improve precision and recall. The scenario-specific topic-oriented directory systems further improve retrieval effectiveness and perform content correlation of medical documents.

Acknowledgments

This research is supported in part by NIC/NIH Grant #4442511-33780. We would like to thank Dr. Hooshang Kangarloo, Dr. Denise Aberle, Dr. Suzie El-Saden, Dr. Craig Morioka, Dr. Andrew Chen and Dr. Blaine Kristo from the UCLA School of Medicine for stimulating discussions and insightful comments. We also thank Nancy Wilczynski and Dr. Brian Haynes from the Health Information Research Unit (HIRU) at McMaster University for sharing their valuable dataset for our experimental evaluation.

14.9 Exercises

1. Explain why IndexFinder currently limits word combination within a sentence. Discuss the trade-offs of using other methods of word combination such as phrase, paragraph, or word properties (e.g., part of speech).
2. Discuss why semantic filtering is important for improving the retrieval quality for IndexFinder.
3. Discuss how to handle the negation concept in the IndexFinder.
4. List the reasons why the knowledge-based query expansion technique performs better than the statistical expansion.
5. In what type of queries might the knowledge-based query expansion method not yield significant retrieval performance improvements compared with statistical expansion cases? Suggest ways to improve such queries. (Hint: non-scenario-specific queries).
6. Discuss why the retrieval performance of statistical query expansion improves as the number of expansion terms increases and then degrades with expansion after it reaches a certain size, while the knowledge-based expansion does not exhibit such behavior.
7. Discuss why the phrase-based VSM alone (without applying query expansion) yields similar performance to that of the combination of knowledge-based query expansion and the stem-based VSM.
8. Explain why the expansion size required to reach optimal performance using the phrase-based VSM is much smaller than that using the stem-based VSM.
9. What is the computation complexity of the phrase-based VSM? Suggest methods to reduce the computation complexity.
10. Describe the concept of a topics directory. How does a topics directory complement search techniques to improve document retrieval performance?
11. Discuss the additional tasks and research issues needed to extend the knowledge-based document retrieval methods used in this chapter (i.e., IndexFinder, the phrase-based VSM, knowledge-based query expansion, and topic-oriented directory) to application domains other than medicine and health care.

14.10 References and Bibliography

1. http://www.nlm.nih/gov/pubs/factsheets
2. S. Chu. *Yearbook of Medical Informatics,* International Medical Informatics Association, 2002.
3. W. W. Chu. Cooperative information systems. In J. G. Webster (Ed.). *Encyclopedia of Electrical and Electronic Engineering.* John Wiley & Son, 1998.
4. W. W. Chu et al. Knowledge-based image retrieval with spatial and temporal constructs. *IEEE Trans. Knowl. Data Eng.* 10(6):872–888, 1998.
5. W. Mao and W. W. Chu. Free-text medical document retrieval via phrase-based vector space model. *Proc. American Medical Informatics Association* [AMIA] *Annu. Symp.*, 2002.
6. *UMLS Knowledge Sources,* 14th ed. National Library of Medicine, 2003.
7. Y. L. Zieman and H. L. Bleich. Conceptual mapping of user's queries to medical subject headings. *Proc. AMIA Annu. Symp.*, 1997.
8. P. L. Elkin et al. Mapping to MeSH: The art of trapping MeSH equivalence from within narrative text. *Proc. 12th Symposium on Computer Applications in Medical Care* [SCAMC], 1988:185–190.
9. M. S. Tuttle et al. Metaphrase: An aid to the clinical conceptualization and formalization of patient problems in healthcare enterprises. *Methods Inf. Med.* 37(4–5): 373–383, 1998.
10. J. C. Denny et al. A new tool to identify key biomedical concepts in text documents. *Proc. of AMIA Annu. Symp.*, 2002.
11. S. Srinivasan et al. Finding UMLS metathesaurus concepts in Medline. *Proc. AMIA Annu. Symp.*, 2002.
12. A. R. Aronson. Effective mapping of biomedical text to the UMLS metathesaurus: The MetaMap program. *Proc. AMIA Annu. Symp.*, 2001.
13. Q. Zou et al. IndexFinder: A knowledge-based method for indexing clinical texts. *Proc. AMIA Annu. Symp.*, 2003
14. C. Friedman and G. Hripcsak. Evaluating natural language processors in the clinical domain. *Methods Inf. Med.* 37(4/5):334–344, 1998.
15. R. Haynes et al. Online access to Medline in clinical settings. *Annu. Intern. Med.* 112:78–84, 1990.
16. W. Hersh et al. OHSUMED: An interactive retrieval evaluation and new large test collection for research. *Proc. 17th ACM-SIGIR [Association for Computing Machinery/Special Interest Group on Information Retrieval] Annu. Conf.* 191–197, 1994.
17. J. W. Ely et al. Analysis of questions asked by family doctors regarding patient care. *BMJ.* 319:358–361, 1999.
18. J. W. Ely et al. A taxonomy of generic clinical questions: Classification study. *BMJ.* 321:429–432, 2000.
19. R. Agrawal and R. Srikant. Fast algorithms for mining association rules. *Proc. 20th VLDB [Very Large Data Bases] Conf.* 1994.
20. J. Han, J. Pei, and Y. Yin. Mining frequent patterns without candidate generation. *Proc. 2000 ACM SIGMOD [Special Interest Group on Management Of Data] Int. Conf. Management of Data* (SIGMOD '00). 2000.
21. Q. Zou, W. Chu, and B. Lu. SmartMiner: A depth first algorithm guided by tail information for mining maximal frequent itemsets. *Proc. IEEE Int. Conf. Data Mining.* 2002.
22. G. Salton, A. Wang, and C. S. Yang. A vector space model for automatic indexing. *Comm. ACM.* 18(11):613–620, 1975.
23. Julio Gonzalo et al. Indexing with WordNet synsets can improve text retrieval. *Proceedings of the CoLing/ACL '98 Workshop on Usage of WordNet for NLP.* 1998.
24. J. B. Lovins. Development of a stemming algorithm. *Mechanical Translation and Computational Linguistics.* 11(1):11–31, 1968.
25. J. Lyons. *Semantics.* Cambridge University Press, 1977.
26. N. Ide and J. Veronis. Word sense disambiguation: The state of the art. *Computational Linguistics.* 24(1):1–40, 1998.
27. G. Salton. A new comparison between conventional indexing (Medlars) and automatic text processing (SMART). *J. Am. Soc. Info. Sci.* 23(2):74–84, 1975.
28. C. J. van Rijsbergen. *Information Retrieval.* Butterworths, 1979.
29. G. Salton and M. J. McGill. *Introduction to Modern Information Retrieval.* McGraw-Hill Computer Science Series. McGraw-Hill, 1983.
30. M. Steinbach, G. Karypis, and V. Kumar. A comparision of document clustering techniques. *Proc. KDD [Knowledge Discovery and Data Mining] Workshop on Text Mining,* 2000.
31. Y. Zhao and G. Karypis. *Evaluation of Hierarchical Clustering Algorithms for Document Datasets.* TR 02-022, Dept. of Computer Science, Univ. of Minnesota, 2002.
32. C. Buckley. Automatic Query Expansion Using SMART: TREC 3. in *Proc. 3rd Text REtrieval Conference (TREC-3),* 1994.
33. E. N. Efthimiadis. Query expansion. *Annu. Rev. Info. Sci. Tech.* 31:121–187, 1996.
34. M. Mitra, A. Singhal, and C. Buckley. Improving automatic query expansion. *Proc. ACM SIGIR '98,* 1998.
35. Y. Qiu and H. P. Frei. Concept-based query expansion. *Proc. 16th ACM-SIGIR,* 1993:160–169.
36. Y. Jing and W. B. Croft. An association thesaurus for information retrieval. *Proc. RIAO [Recherche d'Information Assistée par Ordinateur] '94.* 146–160, 1994.
37. J. Xu and W. B. Croft. Query expansion using local and global document analysis. *Proc. 19th ACM-SIGIR.* 4–11, 1996.
38. E. N. Efthimiadis and P. Biron. UCLA-okapi at TREC-2: Query expansion experiments. *Proc. 2nd Text REtrieval Conference* (TREC-2), 1993.

39. C. Buckley et al. Automatic query expansion using SMART: TREC-3. *Proc. 3rd Text REtrieval Conference* (TREC-3), 1994.
40. S. E. Robertson et al. Okapi at TREC-3. *Proc. 3rd Text REtrieval Conference* (TREC-3), 1994.
41. C. Buckley et al. New retrieval approaches using SMART: TREC-4. *Proc. 4th Text REtrieval Conference* (TREC-4), 1995.
42. Z. L. Liu and W. W. Chu. *Knowledge-Based Query Expansion to Support Scenario-Specific Retrieval of Medical Free Text.* Technical Report #060019, Computer Science Department, UCLA, 2006. ftp://ftp.cs.ucla.edu/tech-report/2006-reports/060019.pdf
43. N. L. Wilczynski, K. A. McKibbon, and R. B. Haynes. Enhancing retrieval of best evidence for health care from bibliographic databases: Calibration of the hand search of the literature. *Int. J. Med. Inform.* 10(1):390–393, 2001.
44. S.-L. Wong et al. Developing optimal search strategies for detecting sound clinical prediction studies in Medline. *Proc. AMIA Annu. Symp.* 2003.
45. N. L. Wilczynski and R. B. Haynes. Developing optimal search strategies for detecting sound clinically sound causation studies in Medline. *Proc. AMIA Annu. Symp.*, 2003.
46. V. M. Montori et al. Systematic reviews: A cross-sectional study of location and citation counts. *BMC Med.* 1(2), 2003.
47. J. B. Lovins. Development of a stemming algorithm. *Mechanical Translation and Computational Linguistics.* 11(1–2):22–31, 1968.
48. G. Salton and C. Buckley. Term weighting approaches in automatic text retrieval. *Info. Proc. Manag.* 24(5):513–523, 1988.
49. Z. L. Liu and W. W. Chu. Knowledge-based query expansion to support scenario-specific retrieval of medical free text. *Information Retrieval.* 10(2):173–202, 2007.
50. W. H. Hersh, S. Price, and L. Donohoe. Assessing thesaurus-based query expansion using the UMLS metathesaurus. *Proc. AMIA Annu. Symp.*, 2000.
51. R. M. Plovnick and Q. T. Zeng. Reformulation of consumer health queries with professional terminology: A pilot study. *J. Med. Internet Res.* 6(3), 2004.
52. Y. Guo, H. Harkema, and R. Gaizauskas. Sheffield University and the TREC 2004 genomics track: Query expansion using synonymous terms. *Proc. 13th Text Retrieval Conference* (TREC-13), 2004.
53. A. R. Aronson and T. C. Rindflesch. Query expansion using the UMLS. *Proc. AMIA Annu. Symp.*, 1997.
54. P. Srinivasan. Query expansion and Medline. *Info. Proc. Manag.* 32(4):431–443, 1996.

15

Integrated Multimedia Patient Record Systems

Dr. Ruth E. Dayhoff,
Mr. Peter M. Kuzmak, and
Mr. Kevin Meldrum
*Health Provider Systems,
U.S. Department of Veterans Affairs(VA)*

15.1 Introduction

This chapter will discuss the multimedia patient record as implemented by the U.S. Department of Veterans Affairs (VA), the largest health care network in the United States. The VA's software, called Veterans Health Information System and Technology Architecture (VistA), was developed in-house and is freely available under the U.S. Freedom of Information Act. This automated health care system has allowed the VA to set the national benchmark in quality of health care in the United States. The involvement of frontline providers, use of performance measures, and universal use of electronic health records have enabled the VA to outperform all other sectors of American health care across the spectrum of 294 measures of quality in disease prevention and treatment [1].

The VA's online multimedia patient record includes traditional medical chart information and scanned chart documents, as well as a wide variety of medical images from specialties (see Section 15.2.2). Clinicians perform all medical record activities online, including placing orders, writing progress notes, requesting consults, and viewing and capturing images. Images can also be acquired automatically through standard interfaces with Digital Imaging and Communications in Medicine (DICOM) interfaces. Images are associated with the corresponding patient studies and are thus incorporated directly into the online patient record. Clinicians can easily navigate between chart information and the associated images using the graphical user interface.

The VA's health care enterprise consists of a nationwide network of 156 VA Medical Centers (VAMCs) and approximately 876 outpatient clinics serving a patient population of 7.8 million veterans. Patients are often treated at more than one facility. Clinicians can access their patients' medical information automatically from any facility where care has been provided. Telemedicine is used to share specialty services between facilities and to reduce patient travel.

15.2 Multimedia Patient Record

Complete online patient data, including traditional medical chart information and clinical images, are essential to providing good health care. Information must be available at any location and any time that the patient needs care. A complete multimedia patient record allows health care networks to provide care seamlessly, without repeating studies and delaying treatment.

The contents of the patient record can be divided into four parts:

1. The paper patient chart portion that can be rendered in textual format.
2. The computable data portion, such as laboratory results, that can be used in calculations, graphing, searching, or decision logic to provide additional value to the user.
3. The multimedia portion that traditionally resides in various locations and various departments throughout the medical center. This portion includes radiographs, pathology slides, endoscopy and ultrasound videos, surgery and dermatology photos, and cardiology films. In the past, the disparate media of paper, film, and tape, as well as the sheer volume of these data, have prevented the filing of these elements in the traditional medical record.
4. The remainder of the contents of the paper patient chart includes signed forms, hand-drawn figures, and papers generated by outside institutions. This portion may include graphic elements like diagrams related to progress notes or consults, charts of fluid input and output, and anesthesia records.

It is essential to have all these data available at all workstations in order for clinicians to embrace the electronic medical record. The user interface to an electronic multimedia record can be enhanced to present data in ways not possible with a paper chart or other physical image media (see Figure 15.1). Once the clinical workstation is the most reliable and efficient source of patient information, clinicians will use it enthusiastically. The electronic record will increase clinician productivity, facilitate medical decision making, and improve quality of care [1].

A user logs on to the workstation with a personal hospital information system (HIS) security code. Privileges are verified by the HIS. The user selects a patient and then views the patient's multimedia longitudinal medical record. A number of windows are used to display the patient's image and text data. The online patient chart window allows the user to access traditional chart tabs for the cover sheet, problem list, progress notes, reports, medications, orders, consults, and lab results. Various image windows can be placed on the screen by the user, generally around the chart window. This allows simultaneous viewing of chart and images. The list window (top center) allows the user to filter and sort image studies. The user can view study images by

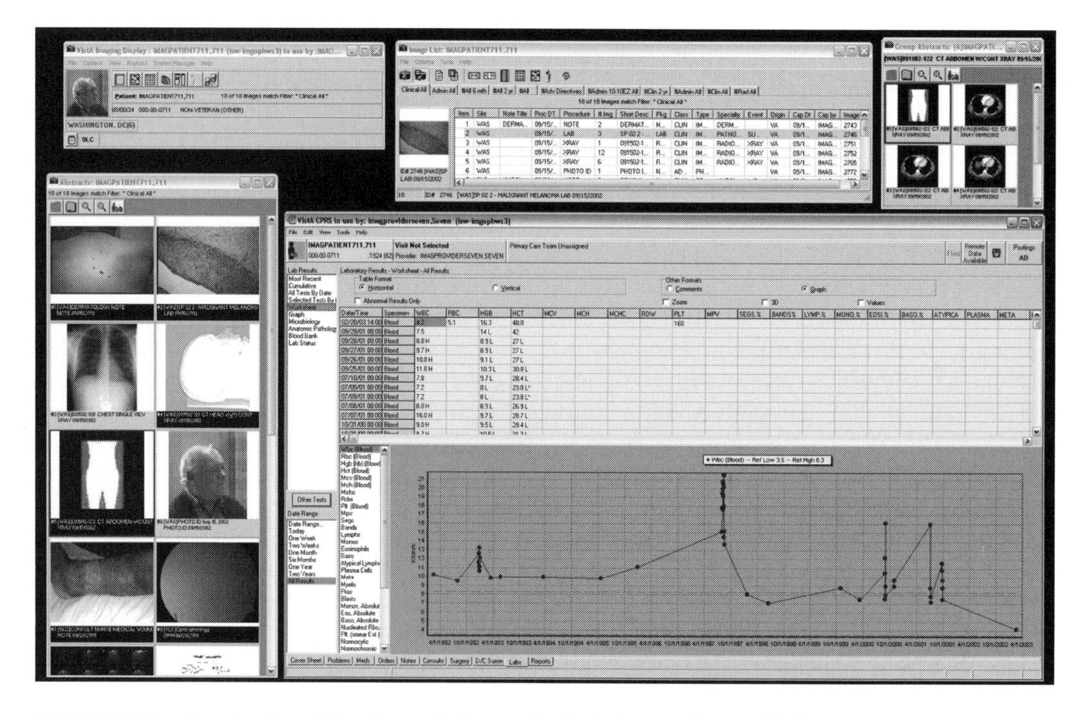

FIGURE 15.1 A patient's multimedia medical record includes images (top and left) integrated with an online computerized patient chart (lower right).

clicking on the list entry or the study thumbnail image (lower left).

15.2.1 User Requirements

The VA has been using a multimedia online patient record for 15 years and has significant experience with medical center use [2, 3]. A number of critical features of a multimedia electronic record system have been identified:

- Single security logon
- Access to all images and data for the patient (from any medical facility in the health care network)
- Easy access to related data (chart and multimedia)
- Ability to search, sort, and filter data and images
- Clear identification of patients and studies, including date and time
- Image/multimedia spatial and color resolution adequate to make medical decisions
- Ability to electronically manipulate and compare images
- Minimal keystrokes or clicks to access key images or data
- Automated capture process that properly documents image attributes
- Delivery of relevant information at the point of care—for example, alerts and notifications of actions required by patient care guidelines
- Ability to annotate diagrams and images
- Conformance to standards for image exchange and retention
- Usable interfaces with consistent and familiar metaphors
- Ease of learning and training for students and residents who rotate through the facilities
- Comprehensive and fully integrated chart with a common presentation
- Ability to customize behavior, rules, and templates to match local policies and practices
- Rapid system response for common user activities
- Overall system reliability

Development of this system has been an evolutionary process, with a cycle of continually adapting the software to accommodate workflow and policy changes as users more fully incorporate information technology into clinical care. Definition of requirements and system design has required close collaboration among the clinical, medical informatics, and information technology professions. Frequent meetings with representative user groups, shadowing of clinicians in their everyday environments, and usability testing have contributed to the refinement of the software. An iterative approach to software development has allowed the design of the system to respond to changes in practices and technology. Advocacy of clinical and administrative leaders has also facilitated the widespread acceptance of an online patient record.

15.2.2 Overview of Multimedia Patient Record Functionality

The VA has combined several major system development efforts to deliver a complete online multimedia patient record containing all of the foregoing elements to users at its hospitals and clinics. The Computerized Patient Record System (CPRS) uses the patient chart paradigm with tabs for cover sheet, problem list, medications, orders, progress notes, consults, lab results, and reports. The clinician may enter all information, including orders, consults, results, notes, and reports using the CPRS software. A number of decision support tools are provided to enhance the patient chart capability. In addition, workflow tracking tools assist in management of the medical record and the institutions' prompt completion of documentation.

The VistA Imaging System brings radiology images, medical photographs, endoscopic pictures, scanned documents, annotated diagrams, and graphical data such as electrocardiograms to the electronic clinician's desktop in an integrated manner. With current technology, the VA is storing over 14 million new medical images per month. Images are distributed along with the electronic patient record using workstations located throughout the hospitals and even across its wide area network [2–5].

The major goal of the multimedia patient record is to provide complete patient data in an integrated manner that facilitates the clinician's decision making. Images and associated text data are available at any time anywhere throughout the hospital and across the VA wide area network on Windows-based workstations that are interfaced to the main HIS. The VA system handles high-quality image data from many specialties (see Figure 15.2), including:

- Anatomical pathology
- Bronchoscopy
- Cardiology
- Dentistry
- Dermatology
- Electrocardiography
- Gastrointestinal endoscopy
- Hematology
- Ophthalmology
- Neurology
- Nuclear medicine
- Nursing
- Podiatry
- Radiology
- Scanned documents
- Surgery
- Textual reports from the HIS and associated medical devices
- Urology
- Vascular care

The VistA Imaging System is being used at all 156 VAMCs, as well as by some other government medical facilities.

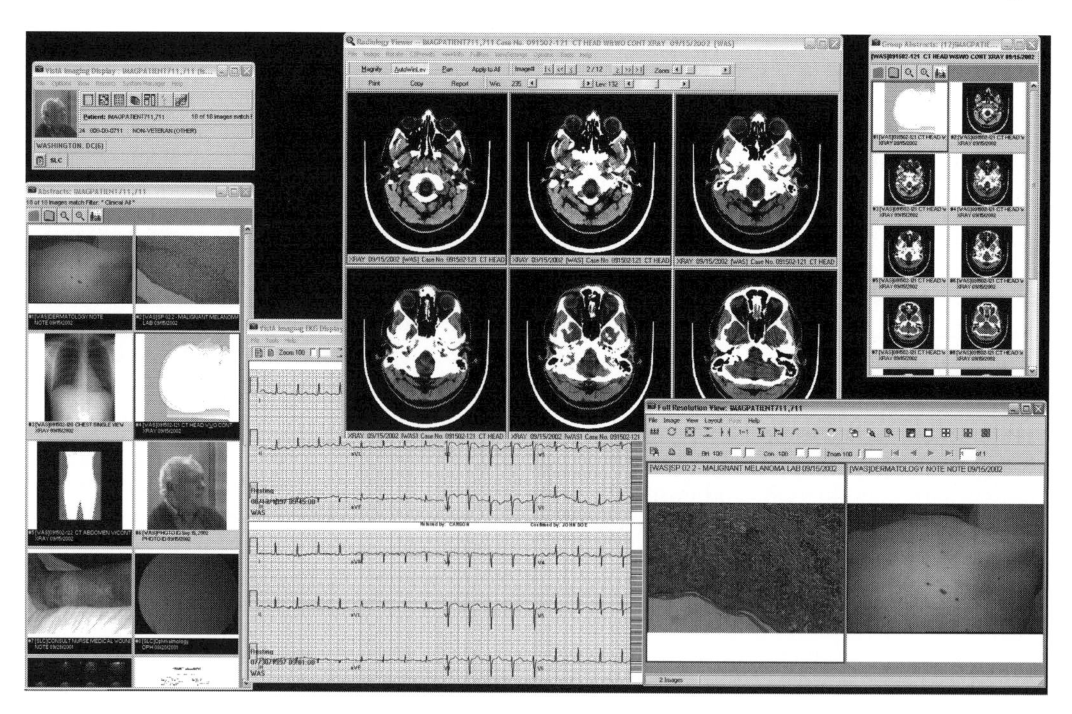

FIGURE 15.2 Multimedia patient record containing dermatology, pathology, nursing, ophthalmology, and radiology images.

It improves the quality of patient care, enhances clinicians' communications, and is used routinely for daily work during patient visits, conferences, morning reports, and ward rounds. The electronic multimedia patient record has played a critical role in allowing the VA to provide the "best care anywhere" in the United States [1].

In Figure 15.2, a menu of small thumbnail images allows the user to select image studies for more detailed viewing (left). Windows show color, grayscale, and document images and electrocardiograms. This "visual chart" capability is typically used by treating clinicians reviewing patients' courses and determining treatment plans. Specialists review procedure data when interpreting studies and writing their reports.

The electronic multimedia patient record developed by the VA demonstrates the breadth of functionality that can be provided with current technology. In addition to serving U.S. veterans, the VA's system provides a model for future multimedia patient record systems.

15.3 Components of the Multimedia Patient Record System Architecture

The multimedia medical record requires a number of interfaced architectural components. The HIS provides database functions for the image registry and medical record text and data. Images are stored in servers by storage management software. Workstations display and capture multimedia medical record images. DICOM gateways capture and save images and other DICOM objects.

15.3.1 Hospital Information System

All VA medical facilities use the same comprehensive VistA HIS, which supports all the clinical services. A client server architecture allows clinical workstation clients to communicate with HIS servers. Modules include Admission-Discharge-Transfer, Laboratory, Pharmacy, Dietetics, Mental Health, and Bar Code Medication Administration, among many others.

The VA has installed a local area network (LAN) at all medical centers to support its client–server architecture, and all facilities are connected by a national wide area network. The LAN at each medical center connects clinical workstations to multiple magnetic and optical disk image file servers, to the VA's HIS, and to commercial medical devices via DICOM gateways. The LAN uses Ethernet (currently up to gigabit/sec). The wide area network carries traffic via TCP/IP (transport control protocol/Internet protocol) to all sites, thus transporting HIS communications, images, and DICOM messages.

A remote procedure call (RPC) architecture allows communication between Windows-based clients and the HIS. All requests for clinical database access are processed using RPC requests that perform the desired operations on the HIS and return the results to the workstation using a TCP/IP message. Security logon and server connections are handled by the HIS in the same manner. Synchronization of applications on the desktop is achieved by conformance to the context management standard from the Health Level Seven (HL7) Clinical Context Object Workgroup [6]. With the proper security privileges, a workstation user may connect to any HIS server on the wide area network and access computerized patient record data through the graphical user interface software. Clinicians may access their patients' remote records to provide health care.

The HIS provides extensive support for VistA Imaging and contains a full image-management infrastructure. Image acquisition, exportation, display, network access, long-term archiving capabilities, and integration with the clinical database are all supported. Hierarchical magnetic and optical disk storage is automatically handled by the HIS image-management software. This provides compliance with required retention periods for images.

15.3.2 Image Database Schema

A common object-oriented approach is used for interfacing all multimedia data to the online patient record. Image management information is stored in the VistA HIS database, in the same manner as all of the rest of the patient text data.

Information about each image is stored in a multimedia object table. Each image entry in the table points to an image file stored on a server outside of the HIS database. The image entry also contains information about the object type and the patient. Figure 15.3 shows three image objects (files A, B, and C) in the multimedia object table. Each image object points to its corresponding image file.

A set of related multimedia objects are collectively joined together into a multimedia group. The multimedia group is then associated with the specific patient study record—for example, a radiology report, an ophthalmology consult, or a progress note. This association allows the user to navigate from the patient to the images and then to the corresponding study report, or from the patient to the study report and then to the corresponding images. In Figure 15.3, the three image objects are joined together to form a multimedia group in the multimedia object table. The multimedia group is associated with a consult request in the consult table. Other studies in the radiology report table and the progress notes table are associated with other multimedia groups in the multimedia object table.

15.3.3 Image Storage

The storage of all document and multimedia data is controlled by an institution's retention schedule, which indicates how long each type of document or multimedia object must be stored by the health care organization. In some cases, retention periods can be in excess of 75 years. Some types of data may have shorter retention periods, depending on local regulations.

Clearly, lengthy retention requirements demand storage formats, devices, media, and software infrastructures that are reliable for a number of years into the future. The storage architecture must provide for redundant copies of data kept in different geographic locations and must also provide rapid recovery in case of disaster. Finally, the system must detect when media or mechanisms are beginning to fail. Migration of data to newer platforms will be required periodically to avoid technological obsolescence.

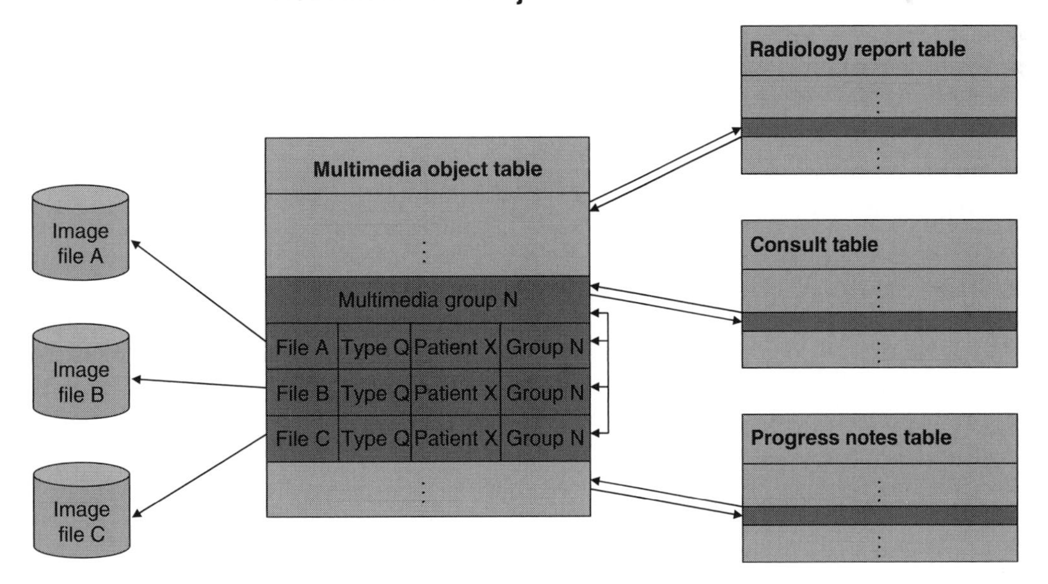

FIGURE 15.3 Multimedia object database schema used for handling images.

Image storage uses a three-tiered approach. The most recently captured images go into fast magnetic storage, typically holding 3–5 years of image data. A second copy of all images is stored on optical media for long-term archiving. In addition, a copy of all images is made on optical or tape media and stored at a different location for disaster recovery. The HIS stores the information about the locations of the patient's images and their association with other patient record data.

The DICOM standard identifies data elements that are to be included in the header section of the image file. Some elements are mandatory, while others are optional. DICOM header data stay with the image file whenever it is stored or communicated to another system. This ensures patient identity. Any changes, such as correction of entry errors or changes in names, require correction in both the database and the image file.

15.3.4 Display Workstations

Clinical workstations are located in most specialty departments and patient care areas. Images are accessed and displayed on the workstation client. The multimedia workstation platform is based on standard Windows computers with a minimum of 1024 × 768 resolution and at least 24-bit color. Some workstations use multiple monitors to allow a larger number of images to be displayed simultaneously. At this time, thin client stations do not fully support manipulations such as window/level modification and are therefore not used for diagnostic purposes. Software running on the workstations was developed by the VA and is written in CodeGear Delphi from Borland, except for integrated commercial off-the-shelf products.

High-resolution radiology workstations allow radiologists to make diagnostic interpretations using VistA Imaging. This diagnostic interpretation software is totally integrated with the radiology HIS module, resulting in a seamless operation and streamlined workflow for the users. These workstations generally include a color monitor and two to four high-resolution grayscale monitors capable of display of three to five megapixels of image data.

15.3.5 Capture Workstations

Capture workstations use the same basic computer hardware as the display workstations, with the addition of a capture device. Frame grab boards are used to capture video input from endoscopes or other video devices. Foot pedals can be added to allow hands-free capturing during procedures. Video capture boards or devices allow capture of video clips. Scanners, either color or grayscale, allow input of photographic prints or film, as well as printed or handwritten medical record pages. Finally, digital cameras, universal serial bus (USB) devices, and magnetic storage attached to the workstation or server can provide images for import.

15.4 Electronic Medical Chart Components

The VA's online patient chart software was designed to resemble a paper chart and includes functional components that are displayed as chart tabs. These tabs are "Cover Sheet," "Problem List," "Medications," "Orders," "Progress Notes," "Consults," "Discharge Summaries," "Surgery," "Labs," and "Reports" (as noted in Section 15.2). Most tabs include browse and search capabilities, in addition to supporting the collection of clinical information. Clinicians may enter progress notes, update the problem list, write orders, enter vital measurements, and record other data that must be collected with each patient encounter.

The online patient chart is always available for access by health care providers and may be used by multiple providers simultaneously. Many clinicians review the status of their patients from their homes via access to the Virtual Private Network (VPN). From remote locations, health care providers can act on abnormal clinical results by entering medication, lab, dietetic, consult, procedure, radiology, and patient-care orders that will be electronically transmitted to the responsible service for immediate action.

The application uses a hierarchically structured set of parameters to allow the behavior of the software to be adapted to specific settings. A baseline set of parameters is provided. Sites may override these by exception. General behavior can be modified for specific hospital locations or even specific clinicians. For example, a baseline set of clinical reminders may be used throughout the hospital while a different subset is used in specific clinics. This capability for fine-tuning is particularly important where the software is used by so many different and varied health care facilities.

Descriptions of the major functional areas follow. These areas are generally represented and accessed by tabs on the patient chart (see Figure 15.1 and figures that follow).

15.4.1 Cover Sheet and Problem List

The cover sheet of the online patient chart facilitates quick orientation to a patient by providing a condensed view of relevant clinical information on one screen. Included are active problems, allergies, immunizations, active medications, recent lab results and vital signs, a list of appointments and admissions, crisis notes, warnings, and reminders.

Clicking on a cover sheet item provides immediate access to a greater level of detail for that item. For example, clicking on an admission displays the discharge summary for that admission, while clicking on a recent lab test displays the results and reference ranges. Allergies and vital measurements may also be updated from the cover sheet. The components that are displayed on the cover sheet are fully customizable and may be tailored to a specific facility or individual.

The problem list of the patient may be maintained through the application. Problems may be added, inactivated, and annotated. Pick lists may be customized to specific user settings to allow easier addition of new problems. There is also a mechanism to automatically update the problem list with a diagnosis when documenting a patient encounter.

15.4.2 Clinical Reminders and Alerts

The clinical reminder system exists to improve preventive health care and to encourage timely clinical interventions to be initiated. Reminders may be viewed on the cover sheet and also during the writing of progress notes. They alert the clinician to certain actions that should be performed. Examples of these actions include examinations, immunizations, patient education, and laboratory tests. Reminders assist in identifying patients who are at risk for hepatitis C, breast cancer, colorectal cancer, hypertension, etc. Some of the actions a clinician may take upon receipt of a reminder include performing an examination, ordering a test, or collecting specific patient information. The information collected in response to a clinical reminder may be automatically appended to the current progress note. Figure 15.4 shows the clinical reminder dialogue that is displayed when tobacco screening is due.

The VA, in cooperation with the Department of Defense and professional organizations, has been developing clinical practice guidelines since the early 1990s. Guidelines have been developed for diabetes mellitus, hypertension, tobacco use cessation, chronic obstructive pulmonary disease (COPD), ischemic heart disease, and depression. The clinical reminder system assists the clinician with following these guidelines. Specific reminders are defined nationally, but other reminders may be defined at each facility to meet additional needs. By allowing the clinician to record the action taken in response to a reminder, the system can monitor conformance to clinical practice guidelines. A variety of reports are available that allow one to view reminders across patients and look for anything that may have been missed. Some reminders are tracked nationally. This allows the VA, as a whole, to evaluate how well it is conforming to clinical practice guidelines and to demonstrate its performance against various quality measures.

Whenever a patient is selected, a list of current alerts is displayed. These notify the clinician of significant events. Some of the areas for which alerts are available include:

- Abnormal results for lab tests and imaging procedures
- Medication orders that are about to expire
- Consults that have been completed or canceled
- Orders that have been flagged for clarification
- Signatures that are required for orders or notes
- Patient movements (admissions, transfers)

Some of the alerts have built-in follow-up actions, allowing the clinician to immediately respond. For example, when an unsigned order notification is received, selection of the notification will display the unsigned orders and allow a signature to be entered. Whenever alerts are selected, the user is taken automatically to the part of the online chart that is relevant to that alert.

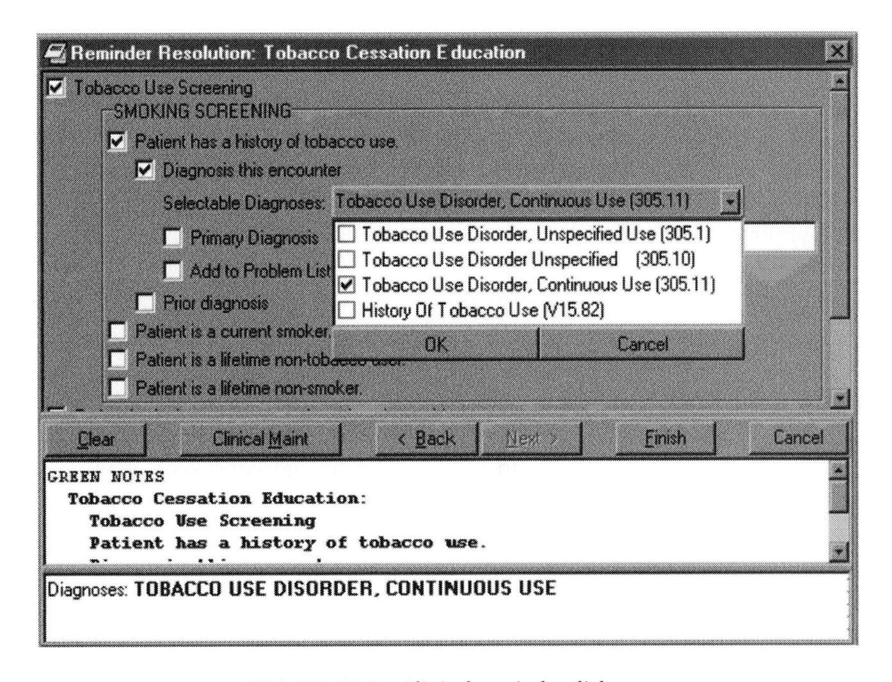

FIGURE 15.4 Clinical reminder dialogue.

15.4.3 Orders and Medications

Orders may be entered and maintained in the online patient chart. Patient orders may be sorted and viewed in a variety of ways. The most common view displays all active orders. Examples of custom views include orders that are about to expire, orders for a specific service, orders that have been written for discharge, orders that have been recently completed, and orders that still require a signature (such as verbal/telephone orders).

Order entry can benefit the clinician by assisting in obtaining all the information necessary to complete an order. If specific information about a patient is required before a procedure may be performed, the clinician is prompted for that information. If additional signatures must be obtained for a particular procedure or treatment, this information may be displayed so that the clinician knows immediately whom to contact. Once an order has been signed, it is transmitted directly to the receiving service. The electronic transmission, completeness, and legibility of the order allow for prompt action. The clinician is able to electronically track the order to completion.

The order entry process may be highly customized to meet the differing needs of clinicians throughout the hospital. Indeed, the degree of customization effort is closely related to the success of an electronic order entry system. The parameters previously described allow customization at the individual clinician level, if necessary. Menus may be used to organize the ordering process in addition to providing some information about policies and protocols for certain areas of ordering. Quick orders may be set up with some or all prompts answered. In situations where there is a large amount of consistency among the orders that are placed, or where certain protocols are in place, quick orders can speed the ordering process, decrease typographical errors, and reduce ambiguities in the order. Order sets are another timesaver, allowing a group of orders to be entered at once and the associated activities to be managed consistently.

Orders can be entered in advance and saved for release to a service at a time in the future. Admission orders may be written weeks ahead. Discharge orders may be written throughout a hospital stay and modified as needed. These orders are easily available for review. Similarly, when writing admission orders, a patient's current outpatient prescriptions are easily viewed and may be transferred to become inpatient orders, if appropriate.

Order checking alerts the clinician during ordering to potential problems that can exist if the order is processed. A message is displayed when a potential problem is detected. If the clinician decides to continue processing the order and if the potential severity of the problem is high, an override reason must be entered. This justification is retained with the order and passed to the service that is processing the order. Some of the order checks that are available include:

- Allergies to contrast media or medication ingredients
- Potential drug–drug interactions
- Duplicate orders or orders for the same drug class
- Laboratory values that may contraindicate an order
- Lab tests that have been ordered too frequently

The order entry system retains a complete, time-stamped history of all orders placed, along with a record of all activity for each order.

A dedicated view of medications is also provided. Separate parts of the screen display outpatient prescriptions, inpatient medications, and medications provided by non-VA sources. Medications may be transferred from the inpatient setting to the outpatient setting and vice versa. This facilitates writing admission and discharge orders.

15.4.4 Progress Notes and Encounter Form

Progress notes may be both viewed and entered through the online patient chart. Extensive facilities exist to define rules for different note types. These rules are based on user roles along with actions taken on a given type of document. It is possible, for example, to restrict who may view, write, or sign a particular type of note. Each type of note has an assigned title. It is possible for sites to create new titles and unique rules for any given title. For example, it is possible to say that a note of a particular title may be viewed by only mental health providers or may be signed by only an attending physician.

A template mechanism is available to reduce typing and to accelerate the entry of notes. A note title may be assigned default (boilerplate) text. Templates may be used throughout the process of note writing. Templates may include elements that are interpreted at run time. These elements expand into text based on previously existing data for the selected patient. For example, the patient's active medication list and recent vital measurements are elements that may be instantiated at run time. Figure 15.5 shows a note in the process of being written using template text. Clinicians have the ability to define their own templates. There are also facilities that allow notes to be uploaded from transcription services.

About 600,000 progress notes are added each workday by clinicians in the VA. Progress notes have been entered online for over ten years. This creates challenges when one needs to locate specific information among a large volume of notes. The viewing of progress notes is facilitated by a variety of filtering and sorting capabilities. Progress notes may be organized by clinic/hospital location, author, title, or time of note. For example, it is possible to list only notes that were created for the Pulmonary Clinic with "COPD" in the subject field. Recently, the capability has been added to allow one to search for specific text within a set of progress notes. Efforts are under way to standardize note titles, so that a list of note titles for a patient who has been seen at multiple VA facilities will be consistent and meaningful.

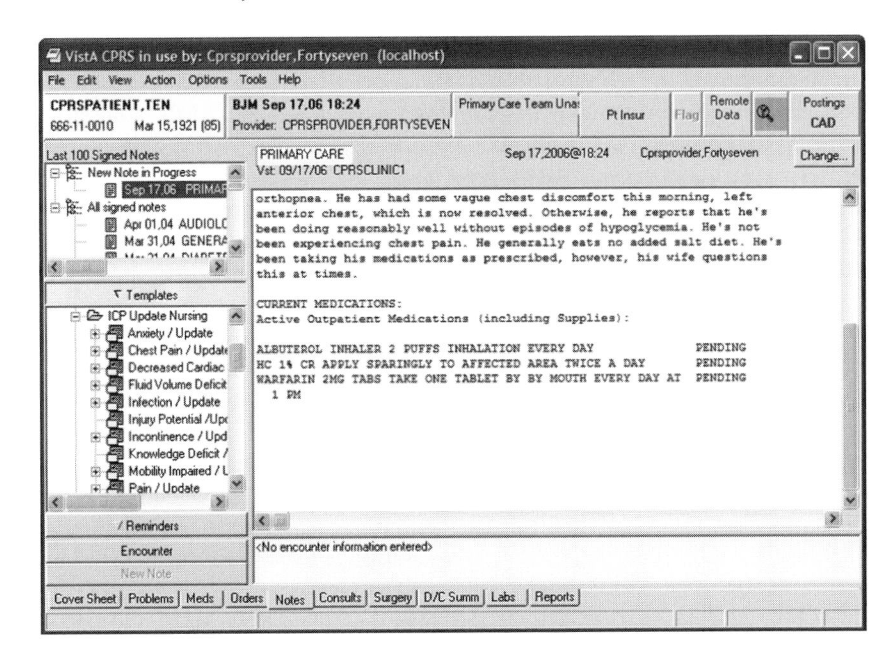

FIGURE 15.5 Progress note in the process of being edited. The "Current Medications" list was prepopulated when the note was started. Any template from the list on the left may be dropped into the note.

An online encounter form is available when writing progress notes. Each clinic has the ability to set up pick lists of encounter-form items that are relevant to that clinic. The encounter form can be used to collect data related to patient education, immunizations, skin tests, health factors, diagnoses, procedures, examinations, and vital signs. The collected encounter-form information is displayed along with any note that matches the same patient visit. The collected data are used for workload measurement and billing. They are also used extensively to support clinical reminders. The encounter-form information may determine whether a particular patient meets the criteria that would cause a specific reminder to be issued. For example, a diagnosis of diabetes would issue a reminder when a foot exam is due.

15.4.5 Discharge Summaries, Consults, Surgery

Discharge summaries, like progress notes, may be viewed and edited online. Discharge summaries share the same mechanisms for sorting and filtering with progress notes. Often, discharge summaries are dictated and become available online after they are transcribed. Alerts that indicate the presence of a newly transcribed discharge summary assist in getting the document electronically signed in a timely manner.

Consults are shown in another specialized view that is similar in capability to progress notes. The consult reports are grouped by consulting service. Additional actions for the consultant or consulting service are available. These actions include:

- Refer a consult to another service.
- Schedule a requested consult.
- Deny a consult request and notify the requestor with the reason.
- Add comments to a consult request.
- Attach related result sets to a consult.
- Record significant findings for a consult.

Using these actions, a consult can be tracked from the initial request until completion.

For surgery, operative reports may be entered directly using the application, but they are often dictated, transcribed, and transmitted into the database separately. Once in the database, the chart application allows the reports to be viewed, verified, updated, and finally signed electronically. The reports are usually grouped by surgical case, but custom views are also available.

15.4.6 Laboratory, Graphing, and Reports

Laboratory results may be viewed in a variety of ways. It is possible to 1) see the most recent results, 2) follow a single set of tests over time, or 3) show abnormal tests only. Clinicians are able to design their own worksheets. This allows them to follow custom panels of tests together over time. Visualization of trends is facilitated by a graphing tool. Graphing extends beyond laboratory results. Nearly any item in the chart can be graphed, and sets of these graphed items may be saved into views to be easily called up again in the future. For example, one may graph specific lab tests in relation to medications prescribed and procedures administered.

A wide variety of reports are available in the online patient chart. These include reports for procedures, medication profiles, order summaries, nutritional assessments, pathology reports, and health summaries. A health summary is a special report that makes available a wide variety of clinical reporting components. These components may be configured to create custom reports that provide relevant information for a particular situation.

Imaging reports contain the narrative for radiology and other imaging procedures. When these reports have associated multimedia objects, an icon is visible next to the report title. Whenever such a report is selected, image displays are synchronized to show the images that are associated with the report. Progress notes may also have associated multimedia and behave in a similar manner.

15.5 Objects Comprised by the Multimedia Patient Record

A number of different image types are part of the multimedia patient record. These include medical color and grayscale images, scanned patient record documents, waveform data such as electrocardiograms and electroencephalograms, sound and video files, stereo image pairs, and three-dimensional reconstructions. The DICOM standard supports these image types and defines header metadata that must be stored with each kind of multimedia object.

Any of these image types can be annotated with text, arrows, measurements, circles, and other markings. Annotations can be stored in a separate file and applied to the original image file, or they can be "burned into" the original image file. The first approach allows multiple annotations to be made available. The second changes the original image and thus is a less appropriate method. The DICOM standard uses an object called "Presentation State" for annotation information.

15.6 Capturing Multimedia Data with a Clinical Workstation

Both text and image data are collected at the point of care. Multimedia data, including clinical images, scanned documents, and motion video, are captured using a clinical capture workstation. Electrocardiogram data are captured on a commercial system and accessed by clinical workstations directly from the commercial storage system, or saved as PDF files. Proper patient and study identification is essential to incorporating data into the electronic patient record. The VA has 15 years of experience acquiring images in clinical disciplines.

Images may be captured by clinical workstations from a number of sources. A frame grab board can be mounted in the workstation, a TWAIN[1] interface can pass images to VistA Imaging, or files may be imported from a disk drive. In addition, images may be transmitted from independent systems using a VistA Imaging–provided application program interface (API). The independent system must implement the interface and pass the images with identifying information to VistA Imaging. Over 50 million images have been acquired using these methods.

Using the VistA Imaging capture workstation graphical user interface, images are collected during medical procedures for diagnostic or follow-up purposes. After logging on to the capture workstation software, the clinical user identifies the patient and the study being performed. Image capture requires only the click of a mouse. The user then enters any pertinent descriptive information related to the image and clicks a button to save it to the patient's record. This simple procedure takes less than 30 seconds, and the result is more informative than a textual description of the image.

Typically, the clinician performing the procedure selects images for capture that are significant to the patient's diagnosis or treatment course. Typically, images are linked to the procedure or consult report in the HIS. The number of images captured per procedure varies by the specialty and is determined by the users themselves. Some specialties have defined the specific set of views to be acquired, while others leave this to the judgment of the clinician.

In some cases, images are captured at the patient's home or nursing home. Images are captured using a digital camera and are later input into the patient's online record.

15.7 DICOM Image Acquisition

The DICOM standard was developed to permit transmission of medical images and their associated information in a multivendor environment. The standard specifies a network protocol, the operation of a variety of service classes, and a mechanism for uniquely identifying information objects across the network. Each information object has a set of attributes. Information objects include images, patient history, studies, reports, and other data. Goals are to achieve compatibility and to improve workflow efficiency between imaging and other health care information systems. A key success factor is that vendors cooperate in testing, and every major diagnostic imaging vendor in the world uses DICOM [8].

Integrating the Healthcare Enterprise (IHE) is an initiative chartered by health care professionals and industry to improve the way computer systems in health care share information. It

[1] Not really an acronym. Nicknamed "*T*echnology *W*ithout *A*n *I*nteresting *N*ame" (http://www.twain.org >> FAQ).

is a multiyear, international effort sponsored by the Radiological Society of North America, the Healthcare Information and Management Systems Society, the American College of Cardiology, the American Academy of Ophthalmology, and several other organizations. IHE promotes the coordinated use of established standards such as DICOM and HL7 to address specific clinical needs in support of optimal patient care. Profiles are defined to specify in greater detail how the standards will be used to meet particular needs. Systems developed in accordance with IHE communicate with one another better, are easier to implement, and enable care providers to use information more effectively [8].

IHE-based standard DICOM interfaces allow the VA to capture images directly from radiology devices such as computed tomography and magnetic resonance imaging scanners, ultrasound systems, computed radiography, and angiography systems, among others. Images can be transferred between systems on portable media such as CDs. Images can also be obtained via DICOM from commercial radiology picture archiving and communication systems (PACS).

Image data may pass through a number of different electronic systems before becoming part of the multimedia patient record. Textual data related to the patient's order and study must be passed from the HIS to the image-producing modalities and the commercial PACS (if present), so that image data are consistent and correctly identified while they traverse multiple systems.

15.7.1 Modalities

In 1999, the IHE initiative defined the methodology for obtaining DICOM objects from radiology image acquisition devices. The IHE Technical Framework Scheduled Workflow Integration Profile defines specific sets of transactions that are essential for robust transfer of images from the acquisition devices (called *modalities*) to a PACS, which might be a commercial radiology imaging system or an image-enabled electronic medical record system such as VistA Imaging [9].

15.7.1.1 Modality Worklist

A DICOM object contains patient and study identification, including a universally unique identifier for each study performed. When the DICOM object is sent to the PACS, this information is used to associate the object to the corresponding study entry in the electronic medical record. It is necessary that this information be exactly correct so that this association can be performed completely automatically.

The DICOM modality worklist service enables the acquisition device to electronically obtain this information. The order placer (e.g., the HIS) first sends an HL7 order message to the PACS, which uses it to populate the modality worklist database. When the patient arrives for the examination, the image acquisition modality performs a query and obtains matching patient and study identification information. Then the examination procedure is ready to begin.

Proper operation of the modality worklist transaction is absolutely essential to the success of the DICOM image acquisition. If patient and study identification must be entered manually, the error rate is so high that the process is unworkable.

15.7.1.2 Modality Performed Procedure Step

At the start of the examination procedure, after the patient and study have been selected from the worklist, the modality sends a message to the PACS indicating that the procedure is beginning. This message identifies the patient, study, and procedure that is being performed. After the procedure is finished, the modality sends a message to the PACS indicating that the procedure has been completed. This termination message lists the DICOM objects that were created during the course of the procedure.

15.7.1.3 Storage

During the course of the examination procedure or soon afterward, the imaging modality will use the DICOM storage service to transmit copies of acquired objects to the PACS image archive.

15.7.1.4 Storage Commitment

Old images on the modality have to be deleted to make room for new ones. The storage commitment transaction allows the modality to verify that the objects previously transmitted to the PACS image archive have been successfully stored. After the examination procedure when the DICOM objects have been transferred to the image archive, the modality sends a storage commitment request transaction to the PACS. The request message contains a list of the DICOM objects that the modality would like to delete. The PACS checks the status of each DICOM object and then sends status information back to the modality in the storage commitment result transaction. The modality can then delete the DICOM objects that it knows have been successfully stored on the PACS.

15.7.1.5 VistA Imaging Modality Interface

In 1996, VistA Imaging implemented the modality worklist and storage functions for radiology. In 2003, this support was extended to the other clinical specialties, primarily dentistry, ophthalmology, and cardiology, but also endoscopy, pathology, and dermatology. Introducing this workflow for clinical specialties has helped foster interoperability standardization efforts in dentistry and eye care [10, 11].

The VA uses a VistA Imaging DICOM gateway to provide DICOM services for the modality. The DICOM gateway provides both the modality worklist and storage services for the DICOM acquisition devices. A VA facility with many modalities may use several DICOM gateways to share the workload.

This VistA Imaging IHE-based interface has been highly successful. There are now about 340 different models of image acquisition devices interfaced to VistA Imaging—about 50 of these are in the clinical specialties outside of radiology. In excess of a half million images per day are acquired across the VA from these modalities.

15.7.2 Portable Media

DICOM objects can be transferred between systems using portable media like CDs. In 2004, IHE Radiology published the Portable Data for Imaging (PDI) integration profile, which specifies how data are to be formatted on CDs using the DICOM standard for interchange between systems [9, 12]. A demonstration at the annual 2004 meeting of the Radiological Society of North America included two dozen vendors who had the capacity to generate and read each other's IHE PDI CDs.

The PDI integration profile, however, did not address how to import these data into a PACS. It was necessary to define a method to resolve the differences between the patient and study identification information on the CD (assigned by an outside system) and that on the local PACS. This was a crucial problem for the VA because imaging studies are often contracted to be performed at outside facilities, and the DICOM images from these studies are always returned on CDs. A solution was developed to treat the DICOM CD import device as an acquisition modality and use the Scheduled Workflow integration profile to handle importation. In the VA, every study that has images must be ordered, whether it is performed within the institution or outside. Since there is always an order, it is easy to use the modality worklist service to provide the CD import device with the local VA system's patient and study identification information. The study on the CD is then matched to the one that was ordered on the local VA system, key data elements in the DICOM headers on the CD are replaced with those from the local VA system, and the objects are then sent to the local VA system.

In 2006, the VA spearheaded the effort to incorporate these requirements into an IHE Radiology integration profile. This resulted in the IHE Radiology Import Reconciliation integration profile, which encompasses the VA requirements and addresses the general issue of importing DICOM objects from portable media on a wider scale [12].

15.7.3 Commercial Picture Archiving and Communication Systems

Commercial PACS are used at some VA facilities for image acquisition and diagnostic radiology reading. VistA Imaging is the VA's official permanent archive for all medical images, and it is necessary that commercial PACSs forward all of their images to VistA Imaging in a timely manner. This assures the VA that all images will be accessible regardless of changes in commercial PACS technology.

The interface between commercial PACSs and VistA Imaging evolved through three generations of standards. The earlier interface versions used the 1993 ACR-NEMA (American College of Radiology–National Electrical Manufacturers Association) and 1997 DICOM standards for all communications between the VistA HIS and the commercial PACS [5, 13]. The VA is moving to another version (2007) that follows IHE Radiology and uses a combination of HL7 for communication of admission, discharge and transfer, orders, and report information and DICOM for handling all aspects of image transfer [14].

The workflow for image acquisition for modalities connected to the commercial PACS will follow the Scheduled Workflow integration profile (see Figure 15.6). The process starts when an HL7 order message is sent to the commercial PACS. This causes the study to be placed on the DICOM modality worklist. The image acquisition device performs the modality worklist query and obtains the patient and study information from the commercial PACS. It then sends a modality performed procedure step (MPPS) message to the commercial PACS indicating that the examination is starting. The images for the examination are acquired and sent to the commercial PACS. At the end of the study, the acquisition device sends an MPPS message to the commercial PACS to inform it that the examination has been completed. This message contains a list of the DICOM objects that were acquired during the course of the examination.

When the commercial PACS receives the completion message from the MPPS, it will check that each of the DICOM objects in the list is stored in its image archive. Once they are all present, the commercial PACS will send an instance availability notification message to VistA Imaging identifying the set of DICOM objects for the examination. VistA Imaging then uses the DICOM retrieve service to obtain copies of them so that they can be placed into the permanent VA image archive (see Figure 15.6). Alternatively, the commercial PACS may send newly received DICOM objects to VistA Imaging so that they can be permanently archived.

15.8 Remote Data and Image Viewing Across the Health Care Network

Because VA patients may receive care at any VAMC, their medical record information may be scattered in different systems. VA has developed two capabilities, remote data views and remote image views, to provide all patient information and images to treating clinicians from the facility where they are stored.

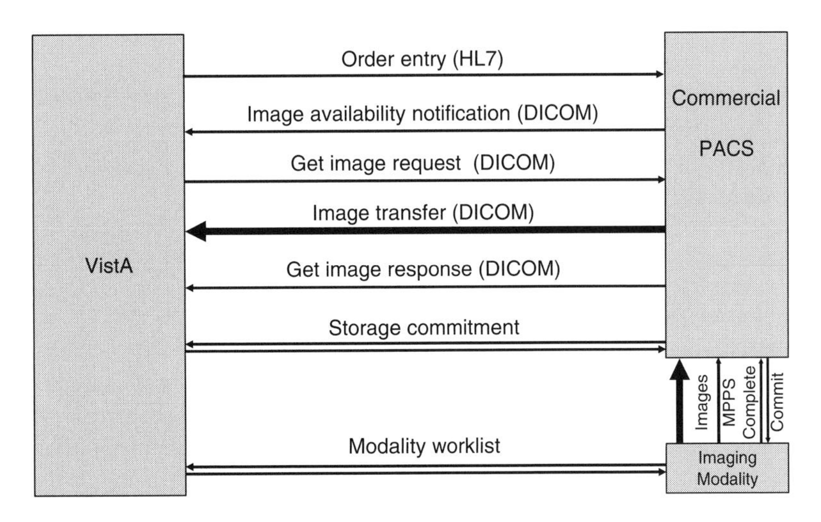

FIGURE 15.6 Messages used for PACS interface.

15.8.1 Remote Data Views

When a patient has been seen at multiple VAMCs, a button at the top right of the screen indicates that remote data are available. Pressing the button reveals a list of sites where the patient has been seen and the date of last activity. Any or all of the sites may be selected (see Figure 15.7), causing data from the selected sites to become available. Each queried site is represented by a tab that appears above the report, allowing reports from each other site to be selected and viewed. Thus, clinicians can directly access data from other sites rather than relying on the patient's memory. This reduces the number of redundant tests and procedures and can provide more comprehensive knowledge of additional treatments, such as medications, that may be provided elsewhere.

In addition to the views of remote data available via the reports tab, the VistA web application (VistAWeb) provides views of remote patient data. VistAWeb integrates data from across VA sites into a single view. It is also used by those who require read-only access to patient data across sites. As a web application, VistAWeb has also provided temporary access to patient charts during emergency situations where hospitals have been evacuated. VistAWeb consists of a simple list of reports available from all VA sites. When a report is selected, all sites that have data for the particular patient are queried, and a table of the returned data is built. This table is displayed

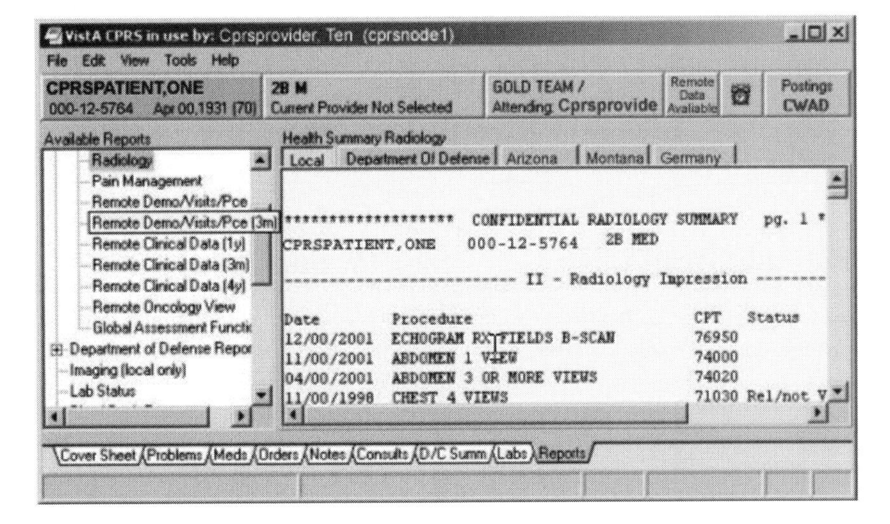

FIGURE 15.7 Report tab providing access to information from other VA sites where the patient has been seen.

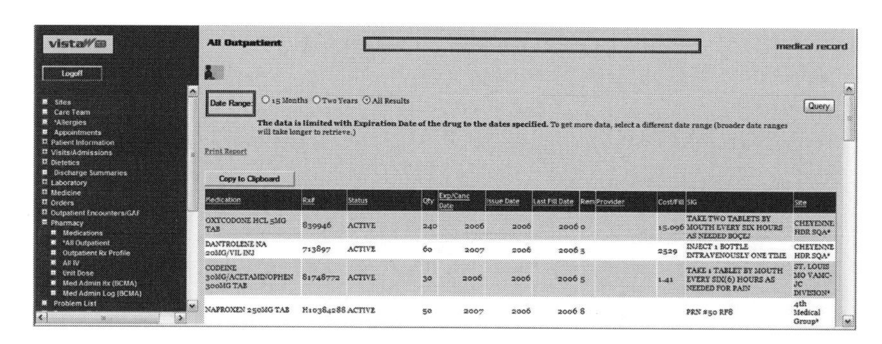

FIGURE 15.8 VistAWeb application providing an integrated view of allergies for a patient, drawing data from multiple VA sites.

on the right side of the VistAWeb screen. An example of an allergy report is shown in Figure 15.8. The user can use hyperlinks to drill down to more specific information.

15.8.2 Remote Image Views

Images are available to clinicians through the similar mechanism of remote image views. A master patient index maintains unique VA identifiers for all patients across the country and the facilities where a patient has been treated. Information related to a particular patient's images can be gathered and used to present a list of available study images to the clinician. Selected images can be displayed on the local workstation for patients under the care of the clinician user. This approach has been called "virtual integration" or "federation" because there is no single enterprise-wide database storage of the information [15].

Benefits of remote image views include:

- Immediate access to images saves time. Access can be from any other point on the VA health care network, without contacting the other facility.
- Redundant testing, which is often done in urgent situations when images and reports are missing, is avoided.
- Patient wait times are reduced because all information is immediately available.

15.9 Impact on Patient Care

The VA currently has over 560 million patient images online. Up to this point in time, there have been 840 million documents (progress notes and discharge summaries) and 1.6 billion orders recorded electronically. Clinicians find that the online multimedia patient record provides a number of benefits to care providers and patients [16], including:

- Ability to provide all patient data, automated reminders, and alerts electronically
- Reductions in errors based on lack of data or nonstandard terminology used to describe images
- Increases in communication among clinicians, improving continuity of care when multiple specialists are involved
- Reductions in physician time spent searching for data
- Improvements to physician education
- Reductions in costs by avoiding film printing
- Decreased hospital admissions, due to rapid availability of electrocardiograms and prior studies
- Reductions in repeat procedures
- Decreased ordering errors through electronic placement of orders and consult requests
- Reduction in medication administration errors through bar-code technology
- Reductions in patient transfers between facilities
- Assistance to patients in understanding their problems
- Reductions in patient wait times
- Increased clinician and technologist productivity

Clinicians use the multimedia patient record during conferences and rounds, on wards, and in the emergency room, operating rooms, intensive care units, clinics, and their own offices. For consultation and medical record access for patients at distant locations, clinicians can view information easily through remote viewing capabilities.

15.10 Summary

The VA's experience indicates that seamless integration of all types of patient data is a critical feature for clinical workstation software. It must be easy and reliable for users to capture patient data in their procedure rooms and view their patients' online multimedia records on workstations anywhere in a medical center. Accurate, synchronized patient identification is essential on all systems that will capture data for inclusion in the online multimedia patient record.

The full range of patient data must be available from a single source—the clinical workstation. This is a key factor in system efficiency, usability, and user acceptance. Because the VistA system is integrated, data that are entered into any part of the system serve all users. No duplicate data entry is needed. This results in a system that contains a "critical mass" of information. Users are more likely to find what they need from the system, and therefore they look first to the system, and find it most effective to place information in it. An institution must reach this critical mass to achieve the maximum benefits from an integrated patient record system.

An online multimedia patient record can present data in ways not possible with a paper chart or other physical media. Data or images can be manipulated on the workstation to present different views. Clinical activities are more efficient, and errors are reduced. Obtaining a critical mass of information online is essential to user satisfaction and efficiency, as well as to achievement of maximum benefits from an integrated patient record system. The involvement of frontline providers in system requirements and design, the use of performance measures in monitoring health care services, and the universal use of the multimedia patient record have enabled the VA to provide the best care anywhere [1].

15.11 References and Bibliography

1. Department of Veterans Affairs. *VistA: Winner of the 2006 Innovations in American Government Award.* Author, 2006. http://www.innovations.va.gov/innovations/docs/InnovationsVistAInfoPackage.pdf#search=%22Harvard%20award%20government%20Dept%20Veterans%20Affairs%20imaging%22
2. R. E. Dayhoff and P. M. Kuzmak. Extending the multimedia patient record across the wide area network. *American Medical Informatics Association (AMIA) Fall Conference.* 1996.
3. R. E. Dayhoff and D. M. Maloney. Exchange of Veterans Affairs medical data using national and local networks. *Extended Clinical Consulting by Hospital Computer Networks.* New Academy of Sciences. 670:50–66, 1992.
4. R. E. Dayhoff. VA's integrated imaging system: A multispecialty, hospital-wide image storage, retrieval and communication system. In E. L.Siegel and R. M. Kolodner (Eds.). *Filmless Radiology.* Springer, 1999.
5. P. M. Kuzmak and R. E. Dayhoff. Success of HIS DICOM interfaces in the integration of the healthcare enterprise at the department of veterans affairs. Medical Imaging 1999: PACS Design and Evaluation/Intl Society for Optical Engineering. *Proc. SPIE.* 3662:44, 1999.
6. HL7 Clinical Context Object Workgroup (CCOW), http://www.hl7.org/special/Committees/ccow_sigvi.htm
7. DICOM standard. http://medical.nema.org/dicom/geninfo/Strategy.pdf
8. *Integrating the Healthcare Enterprise.* http://www.ihe.net
9. *IHE Radiology Technical_Frameworks,* vols. 1, 2, 3, and 4. http://www.ihe.net/Technical_Framework/index.cfm
10. American Dental Association Standards Committee on Dental Informatics. *Technical Report No. 1023.* http://www.ada.org
11. American Academy of Ophthalmology. *IHE Eye Care.* http://www.ihe.net/Eyecare/committees/index.cfm.
12. *Portable Data for Imaging.* http://www.ihe.net/Technical_Framework/upload/IHE_RAD-TF_Suppl_IRWF_TI_2006–04–13.pdf
13. P. M. Kuzmak, G. S. Norton, and R. E. Dayhoff. Using experience with bidirectional HL7–ACR-NEMA interfaces between the federal government HIS/RIS and commercial PACS to plan for DICOM. Medical Imaging 1995 PACS Design and Evaluation: Engineering and Clinical Issues/Intl Society for Optical Engineering. *Proc. SPIE.* 2435, 1995. http://spiedl.aip.org
14. *VistA Imaging Requirements and Specifications.* http://www1.va.gov/imaging/page.cfm?pg=6
15. J. M. Teich. Clinical information systems for integrated healthcare networks. *Proc. AMIA Symp.* 19–28, 1998.
16. B. Kaplan and H. P. Lundsgaarde. Toward an evaluation of an integrated clinical imaging system: Identifying clinical benefits. *Methods Inf. Med.* 35:221–229, 1996.

16

Computer-Aided Diagnosis

Prof. Maryellen L. Giger and
Dr. Kenji Suzuki
University of Chicago

16.1 Introduction

The benefit of a medical imaging exam depends on both the quality of the medical images and the ability of the radiologists who interpret them. Over the years, advances in various image acquisition methods have been made, including the standardization of mammography and development of multislice CT. However, with these acquisition advances come more and more data requiring assimilation and interpretation by the human radiologist. During the past 20 years, developments in computerized image analysis have attempted to improve the interpretation stage of the medical imaging exam by providing a "second opinion" to the radiologist, leading to the use of *computer-aided diagnosis* (CAD) in many breast cancer screening programs. This chapter describes CAD for the detection and diagnosis of diseases, with illustrations on breast cancer, lung cancer, and colon cancer, using examples from the University of Chicago.

16.2 Computer-Aided Diagnosis

16.2.1 Rationale for Computer-Aided Diagnosis

In the clinical interpretation of medical images, limitations are posed by the nature of the human eye/brain visual system, reader fatigue, distraction, the presence of overlapping structures in images, and the vast number of normal cases in screening programs. These limitations provide motivation for the use of CAD with the potential to improve detection, diagnostic performance, and ultimately patient care.

Development and implementation of computer-aided detection and diagnosis involves the application of computer technology in medical image interpretation [1–8]. Radiologists can use the output from a computerized analysis of medical images as a "second opinion" in detecting and characterizing lesions as well as in making diagnostic decisions. It should be noted that the final diagnosis is made by the clinician (e.g., the radiologist). Thus, the computer output needs to be at a sufficient performance level in terms of sensitivity and specificity, and the computer interface should have a user-friendly format for effective and efficient use by the radiologist.

16.2.2 Development of Computer-Aided Diagnostic Methods

Research in CAD has advanced rapidly over the past 20 years—from time-consuming film digitization and computations on a limited number of cases to present-day developments in a variety of medical imaging applications and workstations readied for implementation in the clinical arena. Basic CAD research involves collection of relevant normal and pathological cases, development of a computer algorithm appropriate to the medical interpretation task, validation of the algorithm alone using appropriate cases for performance evaluation and robustness assessment, evaluation of radiologists in the relevant diagnostic task with and without the use of the computer aid, and then ultimate performance evaluation with a clinical trial.

Two general types of systems for CAD are being developed by multiple researchers: CADe for computer-aided detection and CADx for computer-aided diagnosis. CADe involves the use of computer analyses to indicate locations of suspect regions in a medical image. The characterization, diagnosis, and patient management are left to the radiologist. CADx involves the use of computer analyses to characterize a region or lesion, initially located by either a human or a computer, leaving the final diagnosis and patient management to the radiologist. Both CADe and CADx are schematically shown in Figure 16.1.

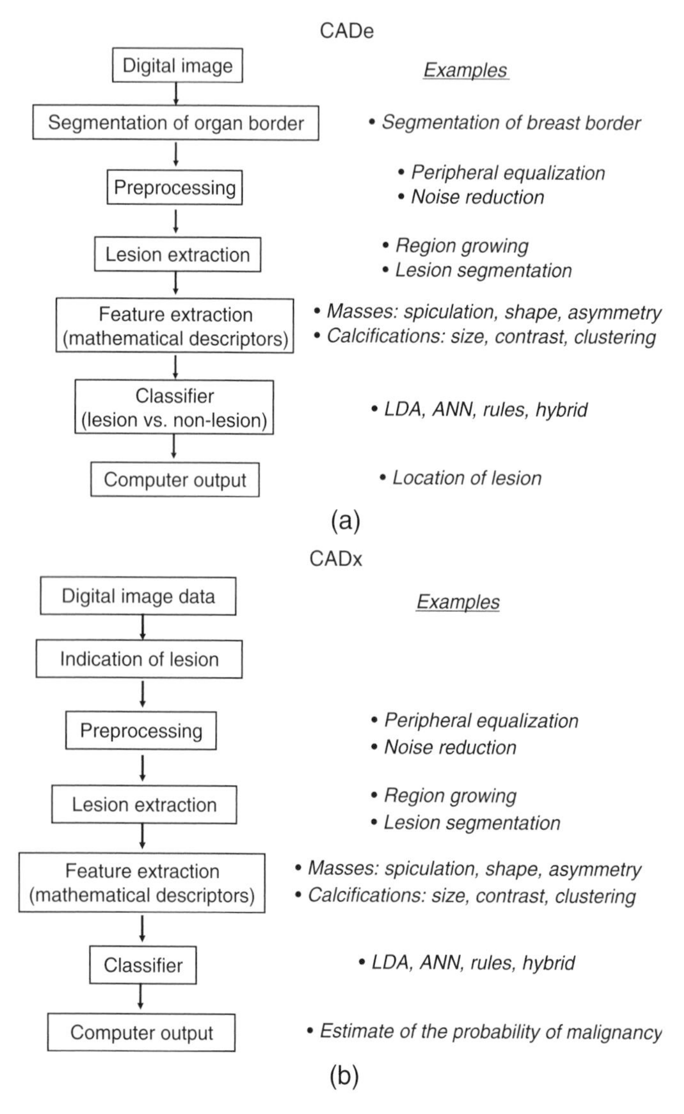

FIGURE 16.1 Schematic diagram illustrating the incorporation of CAD into mammographic interpretation.

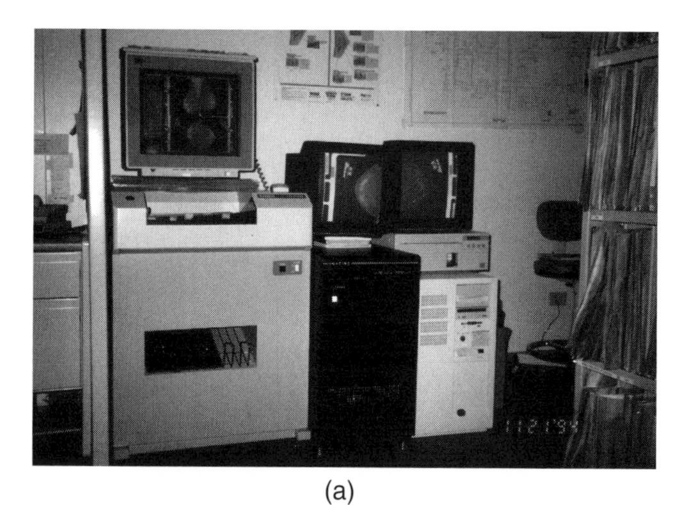

(a)

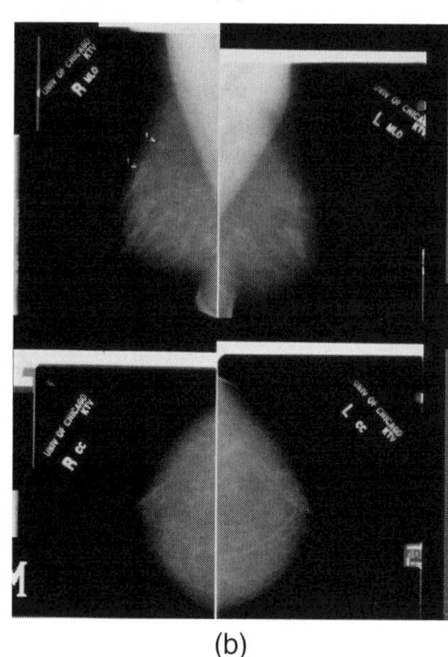

(b)

FIGURE 16.2 First prototype for mammography CAD, developed at the University of Chicago: (a) system and (b) example of computer printout illustrating CADe outputs. Courtesy of M. Giger, University of Chicago.

Figure 16.2 shows the first CAD prototype, which was developed at the University of Chicago. The 1994 system took mammograms as input, and output the computer-determined locations of suspect lesions (clustered microcalcifications and mass lesions) on low-resolution thermal paper. Thus, it is an example of CADe. Image analysis in the system took minutes, while today's systems yield real-time analyses.

16.2.3 Evaluation of Computer-Aided Diagnostic Systems

Discussion of evaluation methods is important at an early part of this chapter in order to help the reader appreciate the difficulties and subtleties of the results given in the various

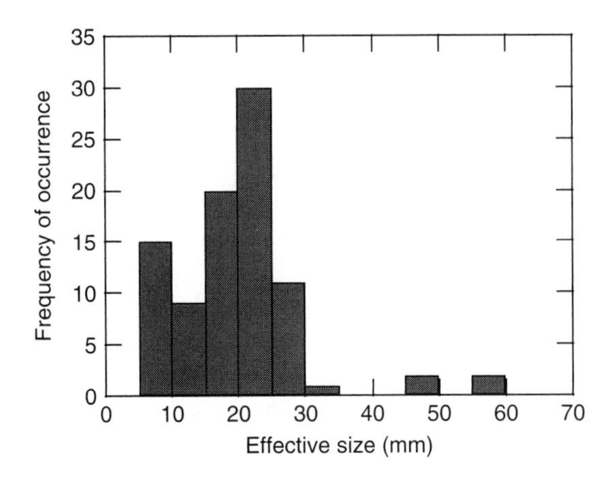

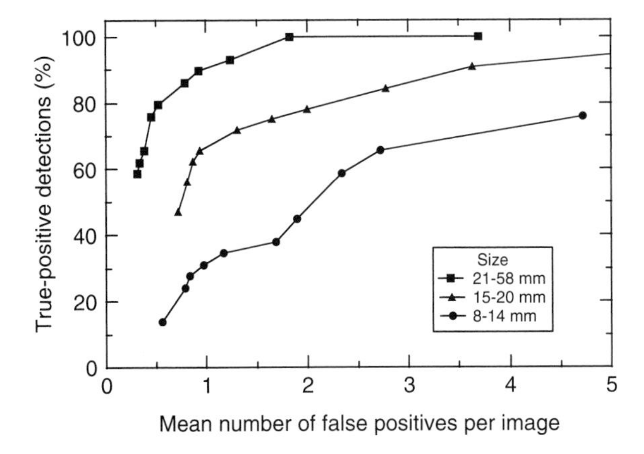

FIGURE 16.3 Example of FROC curves. These demonstrate the effect of database on mass detection performance. The three different databases correspond to images with lesions within three different size ranges. Reprinted with permission from Nishikawa et al. [14].

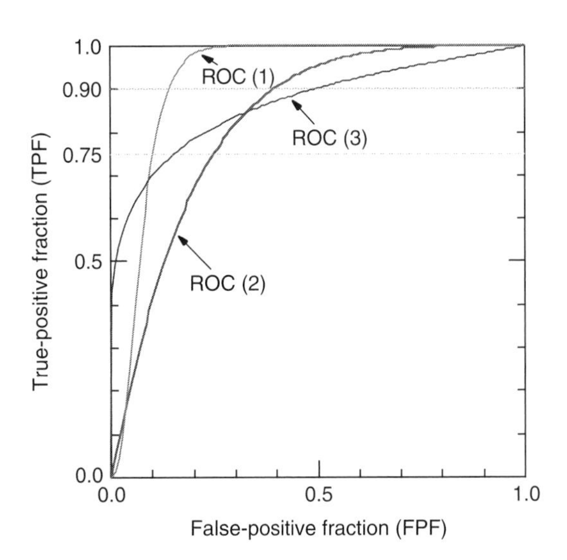

FIGURE 16.4 Three empirical ROC curves used in the computer simulation. The values of the A_z (area under the ROC curve) index were 0.917, 0.825, and 0.862. The values of the partial area indexes were 0.817, 0.484, and 0.261, respectively, for TPF_o (sensitivity threshold of the true-positive fraction) = 0.90; the corresponding values for $TPF_o = 0.75$ were 0.852, 0.606, and 0.523, respectively. Reprinted with permission from Jiang et al. [12].

subsequent sections and in the literature. Evaluations can be given in terms of the computer performance or the human performance during use of the computer output.

Performance levels can be given for (1) lesion detection schemes in terms of the sensitivity (true-positive rate) for detection and the number of false-positive detections (or "false marks") per image, and (2) classification schemes in terms of the sensitivity for classification and the specificity (i.e., 1 minus the false-positive fraction). Figure 16.3 illustrates *free-response receiver-operating-characteristic* (FROC) curves for a mammographic mass detection algorithm. Various performance indices exist for use in the evaluation of computerized methods, such as receiver-operating-characteristic (ROC) [9] and FROC analyses [10, 11]. Appropriate measures are discussed in a paper by Metz [9]. These include the true-positive fraction and false-positive fraction pair and ROC analysis. The area under the ROC curve yields the performance of a computerized classification method [9]. Note that a partial area index from an ROC curve is useful when evaluating a system that requires a high level of sensitivity, such as in the task of determining the likelihood of malignancy of lesions seen on mammograms [12]. See Figure 16.4 for an example of an ROC curve along with indication of the partial area index.

Comparison of different computerized methods in mammography is often not possible because of the use of different databases [13]. That is, it cannot be assumed that a computerized scheme that achieves a high level of performance with one database of mammograms will achieve a similar performance level with another database or with an actual patient population. Nishikawa et al. [14] have shown the effect of the database on mass detection performance using FROC analysis (see Figure 16.3). Use of a database containing very subtle cases will yield a lower performance level. It is possible, for example, that a computerized detection scheme could achieve a sensitivity of 90% at two false positives per image with one database and a sensitivity of 70% at two false positives per image with another database. The characteristics of a database will influence the training (i.e., the development) of a computer method as well as its reported performance level. Databases can be described by objective measures such as lesion size and contrast and by subjective measures such as subtlety for detection. Lesion subtlety is a subjective measure that depends on the particular observer who reports the subtlety rating, the specific task, and the presence or absence of other images and/or information.

Demonstration of robustness is important in order to assess the usefulness and generalizability of a particular computer analysis system across acquisition devices, institutions, and case mixes. A mammographic mass classification scheme was shown to be robust with respect to case mix and digitization technique in Huo [15]. In that study, the investigators failed to show a statistically significant difference between the performance of the computer classification method on the training database in a round-robin evaluation and the performance of the same computer method on an independent database digitized on a different digitizer.

How a database is used will also influence the development and reported performance of a computerized method. For example, in the training and testing of artificial neural networks, it is important that multiple images of the same lesion be kept either in a training set or a testing set but not be spread across both.

Different scoring methods will also vary the performance of a computerized method. For example, in the detection of masses, some investigators use the percent of overlap between the actual lesion and the computer-detected region as a means of determining a true detection. It should be noted that there are different definitions of percent overlap. Some investigators define overlap as the intersection of the two regions, whereas other investigators define overlap as the intersection of the two regions divided by the union of the two regions, the second definition being a stricter criterion [1, 16].

How the actual "truth" is defined for a particular interpretation task will also affect the performance of a computerized method. For example, in the development of computerized nodule detection methods, investigators have trained with cancerous lung nodules, all lung nodules, and/or any "actionable regions." In addition, database collection may be hindered by the rapid growth of the field of computed tomography (CT) imaging, in which helical CT systems have evolved from single-slice to 16-slice to now clinically used 64-slice scanners. In addition, the choice of reconstruction algorithm for a CT scanner will affect the physical image properties and visual appearance of the transaxial images obtained from the CT scanner sinogram data [17]. Thus, it is expected that variations in reconstruction algorithms may also affect the performance of computerized analyses. Investigators are therefore examining the robustness of computerized lung nodule detection schemes across reconstruction algorithms and scanner type.

Computerized detection and diagnosis methods are evaluated both in terms of the computer's own performance as well as in terms of the performance of humans using the computer output as an aid in observer studies. In a sequential observer study, radiologist observers interpret the medical image(s) with just the image data, record their interpretations, and then are shown the computer output. After seeing the computer "interpretation" in terms of a symbol, numerical value, similar images, or graphical presentation, the observer can modify or keep his/her initial interpretation. Analysis of the observer performance before and after use of the computer output can yield information on the benefit (or hindrance) of the computer aid.

16.3 Computer-Aided Diagnosis for Cancer Screening

Focused research on CAD started in the mid-1980s at the University of Chicago with the investigation of CADe methods for the detection of lesions on chest radiographs and mammograms [18, 19]. Computerized detection of lesions (for CADe) involves having the computer locate suspicious regions, leaving the subsequent classification of the lesion (e.g., probability of malignancy) and patient management decisions to the radiologist. In such situations, the computer acts as a second reader or a spell checker in the cancer screening process. Once a possible abnormality is detected, its characteristics must be evaluated by the radiologist in order to estimate a likelihood of malignancy and to yield a decision on patient management.

16.3.1 Breast Cancer: Mammography

While mammography is the best screening method for the early detection of breast cancer, missed lesions do occur. Misses of lesions on radiographic images may be due to the presence of quantum mottle, overlapping normal structures, or radiologists' insufficient search patterns and lapses in perception. In addition, variability in mammographic image interpretation among different radiologists has been reported [20, 21]. In screening programs, the detection of an abnormality is a tedious task, since although most cases are normal, each requires a thorough review by the radiologists. Use of output, however, from a computerized analysis of an image may help the radiologist in detection or diagnostic tasks and potentially improve the overall interpretation of medical images and perhaps overall interpretation time.

One of the earliest published investigations into the computerized analysis of medical images was reported in 1967 by Winsberg et al. [22], which included a computerized method comparing mammographic density patterns in various areas within an individual breast and between right and left breasts. With advances in computer vision, artificial intelligence, and computer technology, along with recognized medical screening needs and the availability of large databases of cases, the field of CAD has grown substantially since the mid-1980s.

As noted earlier, CAD for the detection (i.e., localization) of regions in an image suspected of possessing the disease has been referred to as CADe in order to emphasize the detection, rather than the characterization task. Algorithms for mammographic CADe include detection of mass lesions and clustered microcalcifications, which are two of the primary signs of potential breast cancer. Radiographic mass lesions can be detected and/or characterized by using the mathematical descriptors of various features, including radiating patterns of density (spiculation), margin sharpness, circumscribed configurations, shape, bilateral asymmetries, local textural changes, and temporal stability, as summarized elsewhere [1].

Mathematical descriptors of calcifications are based on the radiographic presentation of individual calcification (e.g., shape, area, brightness) [1, 18, 23–25], the variation of individual features within a cluster [26], the spatial distribution of calcifications within a cluster [26], and the knowledge that clinically significant microcalcifications are clustered [27].

There are currently three computer-aided detection systems approved by the U.S. Food and Drug Administration (FDA): R2/Hologic, ISSI (Intelligent Systems Software Inc.)., and Kodak, which are distributed on multiple vendor systems for screen-film mammography and full-field mammography units. Various studies have shown that mammographically detected cancers are visible in retrospect. Such "missed-lesion" databases have been used by manufacturers to demonstrate their system's performance for the FDA. Many investigators have shown that computer detection can catch 50% to 90% of missed cancers [13, 28, 29]. Websites of the various systems describe performance as well as physical space requirements, workflow aspects, and options.

Various clinical prospective studies of the use of computer-aided detection systems have been performed. It is important to note that proper use of a CADe output requires that any initial human detected region/lesion remain as a detection even if the computer does indicate the location. This is necessary in order to ensure that the radiologists' detection sensitivity remains either constant or improves with the use of the computer aid. There have been, in general, two types of clinical studies of mammography CADe systems:

- Sequential assessment of a case that involves an initial interpretation without computer aid followed by a viewing of the computer output, and a subsequent reinterpretation of the case with the computer aid
- Separate assessment of cases in which computer output is not used within a practice for some set time period and then the computer output is used clinically during a subsequent time period

In one clinical study [30], radiologists interpreted 12,860 mammographic cases using a commercial CAD system alone and then again with the computer output assistance. They showed a 19.5% increase in cancer detection, along with an 18.5% increase in recall rate. Gur et al. [31] investigated the effect of introducing CAD into a radiology practice, and initial results indicated that changes in cancer detection rate and recall rate were not statistically significant. In further analysis of the data, low-volume radiologists showed a 19.7% increase in sensitivity with only a 14.1% increase in call-back rate [30, 31]. To date, additional clinical studies have been performed; these are listed in Table 16.1 [30–38], which also gives the percent change in sensitivity and call-back rate for each study. Note that if the ratio of percent change in sensitivity to the percent change in call-back rate is equal or greater than 1, then one can say that the use of the computer output was beneficial to the interpretation process.

TABLE 16.1 Summary of clinical studies on CADe in mammography*

Study	No. of Studies		% Change	% Change
Sequential	Unaided	Aided	Cancer Detected	Recall Rates
Freer, 2001	12,860	12,860	19.5	18.5
Helvie, 2004	2,389	2,389	10	9.9
Birdwell, 2005	8,692	8,692	7.4	8.4
Khoo, 2005	6,111	6,111	1.3	5.8
Dean, 2006	9,520	9,520	13.3	26
Morton, 2006	21,349	21,349	7.6	10.8
Separate				
Gur & Feig, 2004 (high volume radiologists)	44,629	37,500	−3.32	−4.9
Gur & Feig, 2004 (low volume radiologists)	11,803	21,639	19.7	14.1
Cupples, 2005	7,872	19,402	16.1	8.1

*From different investigators using different CAD systems: either different software versions or different manufacturers, as well as either sequential or separate study designs.

16.3.2 Lung Cancer: Single Projection Radiography and Computed Tomography

Lung cancer continues to rank as the leading cause of cancer deaths in the United States [39, 40], and early detection may allow more timely therapeutic intervention and thus a more favorable prognosis for the patient [41–44]. Chest radiography has been used for detection of lung nodules (i.e., potential lung cancer) because of its low cost, simplicity, and low radiation dose. Radiologists, however, may fail to detect lung nodules in chest radiographs in up to 30% of cases that have nodules visible in retrospect [45, 46]. CAD schemes for nodule detection on chest radiographs are being investigated [19, 47, 48], since studies have shown that use of computer output can improve radiologists' detection accuracy [49–51] by providing potential nodule sites.

A number of researchers have developed image analysis methods for lung nodule detection on chest radiographs. In 1984, a CAD scheme [19, 52, 53] was developed at the University of Chicago, based on a novel difference image approach to initially enhance nodules and suppress the surrounding background (i.e., non-nodules). The performance of the CAD scheme was improved by incorporation of an artificial neural network (ANN) and linear discriminant analysis (LDA) [54], an adaptive thresholding technique [55], and massive-training artificial neural networks (MTANNs) [56]. By use of the MTANNs, the false-positive rate of the CAD scheme

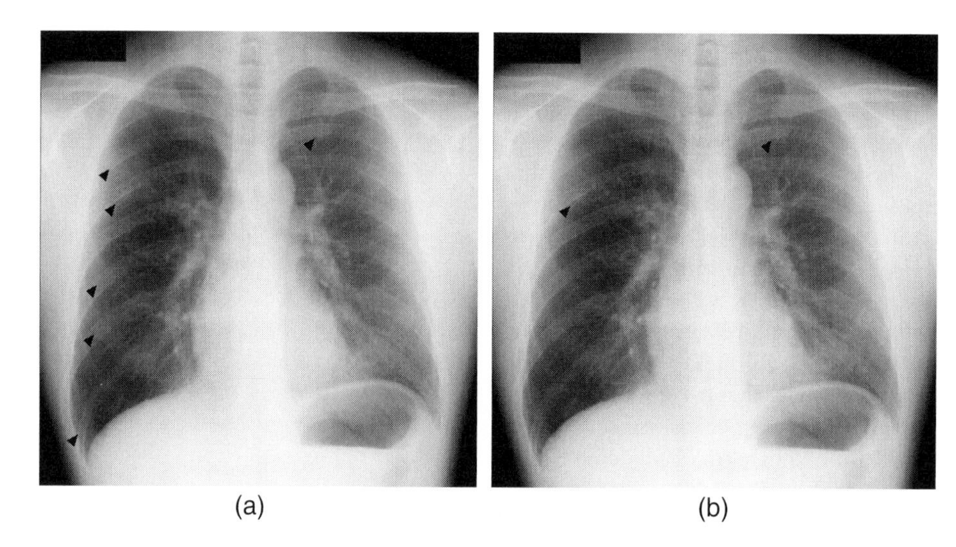

FIGURE 16.5 Illustration of the effect of MTANNs on the false-positive reduction in a CAD scheme for nodule detection in chest radiographs. (a) CAD without MTANNs exhibits one true-positive detection and six false-positive detections (indicated by arrow heads); (b) CAD with MTANNs exhibits one true-positive detection (a small nodule in the left lung) and one false-positive detection. Courtesy of K. Suzuki, University of Chicago.

was improved substantially, while the original sensitivity was maintained, as illustrated in Figure 16.5.

The first commercial CAD system (Riverain Medical, Miamisburg, OH) received FDA approval for clinical use in 2001. The commercial CAD system was tested in a retrospective clinical trial and achieved a sensitivity of 65.0% with 5.3 false positives per image [57]. A study has shown that radiologists' performance was improved with the aid of another commercial CAD system (Mitsubishi Space Software, Japan) having a sensitivity of 73% with 4.0 false positives per image [58].

One of the major challenges in current CAD schemes for chest radiography is to achieve high performance, because there are a large variety of normal structures similar to nodules in chest radiographs [59]. It is difficult to detect nodules overlapping with normal anatomical structures such as ribs and vessels, which account for the majority of false positives [55, 59] and can result in lowering of the sensitivity as well as the specificity of a CAD scheme. To address this issue, an MTANN for suppression of ribs in chest radiographs has been developed [60]. The MTANN suppresses rib opacity in chest radiographs, whereas soft-tissue opacity is maintained (i.e., it produces "soft-tissue" images and "bone" images from standard chest radiographs).

Because CT is more sensitive than chest radiography in the detection of small noncalcified nodules due to lung carcinoma at an early stage [61, 62], lung cancer screening programs are being conducted [61–67] with low-dose helical CT as the screening modality. Thus, an active area of research in computerized image analysis for lung cancer is in the use of computers to aid in the detection of lung nodules on thoracic CT. Even though the use of CT, as opposed to single projection chest radiography, has helped remove much of the camouflaging presence of overlapping structures, the detection of lung nodules is still confounded by the presence of blood vessels. In addition, radiologists' readings of CT images are hindered by the vast number of slices (images) generated by a CT scan, with each requiring careful interpretations. This makes lesion detection a burdensome task that is further complicated by human fatigue and distractions. It is expected that a computerized scheme for the detection of lung nodules on CT images should help radiologists focus their attention on regions suspected of being cancerous. This seems particularly advantageous in potential low-dose CT lung cancer screening procedures, in which most cases will be normal.

Various investigators have been developing methods for the computerized detection of lung nodules on thoracic CT images [17, 68–76]. Currently, methods include both two-dimensional and three-dimensional analyses for the delineation of the lung volume, the segmentation of pulmonary structures indicative of an abnormality, and the extraction of features (i.e., mathematical descriptors of lung nodules). In order to reduce false-positive detections (i.e., computer-indicated locations that in fact do not correspond to lesions), investigators have used ANNs to merge computer-extracted features and distinguish nodules from normal structures such as blood vessels. Information such as continuity between slices can also be used to distinguish the tubelike structure of blood vessels from the approximate spherical nature of nodules. Recently, MTANNs [76] based on linear-output ANN models [77, 78] have been developed for reduction of false positives. Unlike standard ANNs with image features, the MTANN can learn image data

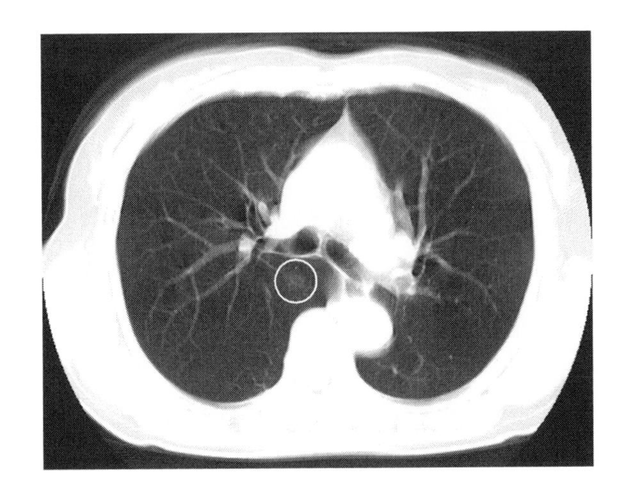

FIGURE 16.6 Illustration of a thoracic CT image with a lung cancer with pure ground-glass opacity that was "missed" in lung cancer screening. CAD incorporating MTANNs correctly detected the lung cancer (indicated by the circle). In an observer study, ten radiologists failed to detect the cancer without CAD, whereas seven radiologists were able to detect the cancer with the aid of CAD. Courtesy of K. Suzuki, University of Chicago.

directly [79]; therefore, inaccuracies in extraction of image features can be avoided. By use of the MTANNs, the false-positive rate of the original CAD scheme was improved from 27.4 to 4.8 false positives per scan, while a relatively high sensitivity of 80% was maintained [76, 80]. One of the advantages of MTANNs is robustness against low-contrast nodules such as ground-glass opacity, which is one of the major sources of false negatives by radiologists [81], as illustrated in Figure 16.6.

CAD schemes are being developed specifically for thin-slice CT [82–84]. In thin-slice CT images, which may range from 1 to 2.5 mm, nodules as well as other normal anatomical structures are more likely to be imaged on multiple contiguous sections on the *z*-axis than those in thick-slice CT images (e.g., slice thickness of 5 or 10 mm). Therefore, three-dimensional volume-processing techniques are applicable to computerized CT analysis methods. A low partial-volume effect is another advantage of thin-slice CT images over thick-slice CT images. Consequently, smaller nodules can be detected more reliably in thin-slice than in thick-slice CT. The performance of these CAD schemes is generally higher than that of CAD schemes for thick-slice CT. As in thick-slice CT, a majority of false positives are caused by lung vessels [84].

16.3.3 Colon Cancer: CT Colonography

While colon cancer is one of the leading causes of cancer deaths, many can be avoided if precursor colonic polyps are detected and removed. CT colonography (virtual colonoscopy) is being examined as a potential screening device (and an alternative to conventional colonoscopy) for the early detection of colonic polyps. When colorectal cancers are detected at an early localized stage, the five-year relative survival rate is 90% [39]. The American Cancer Society recommends that a person at average risk for developing colorectal cancer (beginning at age 50) have colorectal cancer screening. A radiologist's interpretation of a CT colonography exam can be quite time-consuming, due to the potentially large number of axial CT images (400–700 slices). In addition, the overload of image data for interpretation may result in oversight errors. Moreover, the diagnostic performance of CT colonography varies across different clinical trials [85, 86] and depends on the radiologist's experience. Thus, computerized image analysis techniques are being developed to aid in the interpretation of CT colonography images.

Various investigators are developing such computer algorithms for the visual presentation of CT colonography images for human interpretation [85–89] and for computerized image analysis or CAD [90–98] to aid in the detection of colonic polyps. CAD has the potential to (a) increase radiologists' diagnostic accuracy in the detection of polyps, (b) decrease reader variability, and (c) reduce radiologists' interpretation time when CAD is used during the primary read. An improvement in radiologists' detection performance can be achieved because CAD can reduce perceptual errors during the detection of polyps. Decreased inter- and intrareader variability can be achieved because CAD provides objective and consistent results, while the performance of a human reader may be influenced by dependence on skill and experience. Also, reduction of interpretation time can be achieved if radiologists focus mainly on the small number of regions indicated by the CAD scheme and quickly review the large portion of the colon that is likely to be normal. It is important to note that in the combined interpretation process, both the presentation of a three-dimensional image dataset and the format of the computer output are critical for accurate interpretation, user-friendly implementation, and reduction in interpretation times.

The computerized image analysis of CT colonography images has various stages, including segmentation of the colonic wall, detection of polyp candidates, reduction of false-positive detections, and display of final detection output. In the computerized detection of the polyps, geometric features of each voxel of the segmented colon are computed [94]. The geometric model for structures in the lumen of the colon involves the calculation of a shape index in order to help distinguish polyps (which are caplike structures) from folds (which are elongated, ridgelike structures) [90, 94]. Subsequent mathematical descriptors are employed to eliminate false-positive detections that are caused, for example, by the presence of stool in the colon during imaging [96, 97]. Common sources of false positives generated by CAD schemes are haustral folds, residual stool, extracolonic structures such as small bowel and stomach, rectal tubes, and the ileocecal valve. Among these, rectal tubes are relatively "obvious" false

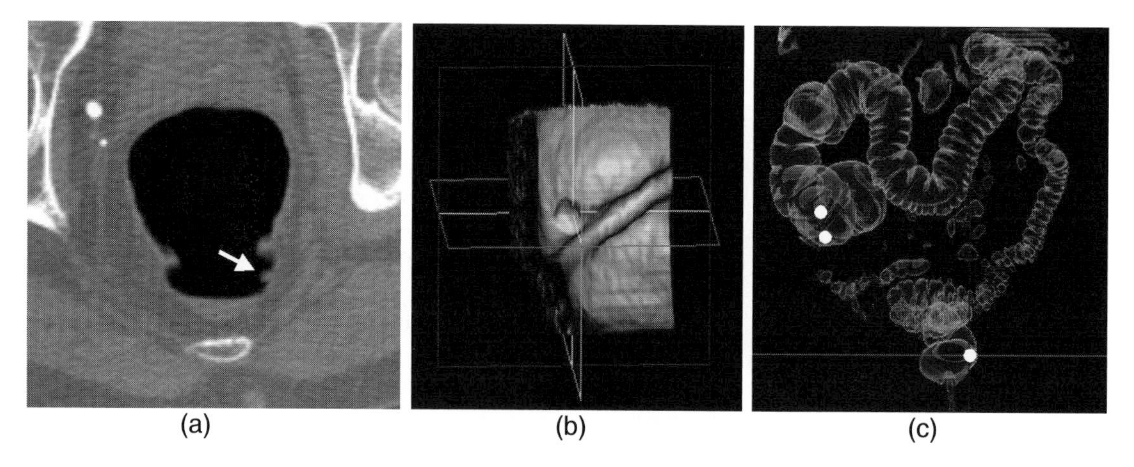

FIGURE 16.7 (a) CAD incorporating MTANNs has correctly detected a small (7 mm) sessile polyp that was missed in a clinical trial and points to it with an arrow. (b) The polyp in the 3D endoluminal view. (c) 3D volume rendering of the colon with three computer outputs indicated by white circles (the one in the rectum is a true-positive detection, and the other two are false-positive detections). Courtesy of K. Suzuki, University of Chicago.

positives; thus, radiologists may lose their confidence in CAD as an effective tool if the CAD scheme consistently generates them. To address this issue, further false-positive reduction techniques employing an MTANN have been developed for elimination of false positives due to rectal tubes [99]. An example of the use of the MTANN is shown in Figure 16.7, in which a polyp that was missed by radiologists in axial CT colonography in a multicenter clinical trial was detected by the computer. This illustrates that by improving the methods for false-positive reduction, one may be able to relax criteria earlier in the algorithm and thus improve sensitivity.

The performance of CAD schemes [90–99] ranges between by-patient sensitivities of 70% and 100% for polyps 6 mm or larger, with two to eight false positives per patient, based on 7–39 polyps in 8–20 patients. A meta-analysis of the reported performance of CT colonography [100] showed that for human readers, the pooled by-patient sensitivities for polyps 10 mm or larger and those between 6 and 9 mm were 88% and 84%, respectively.

16.4 Computer-Aided Diagnosis for Differential Diagnosis

Once a possible abnormality is detected, its characteristics must be evaluated by the radiologist in order to estimate a likelihood of malignancy and to yield a decision on patient management. Characteristics of the lesion may be evaluated further by multiple imaging techniques, including special view mammography, ultrasound, and magnetic resonance imaging (MRI) in order to improve the positive predictive value for biopsy recommendations.

16.4.1 Breast Cancer

The initial investigations into the use of computers in diagnostic mammography involved artificial intelligence techniques to merge observations of image features made by radiologists into useful diagnostic predictions [101, 102]. Ratings from the Breast Imaging Reporting and Data System (BIRADS) [101] provided by humans have been analyzed by computer for lesion characterization. However, in order to eliminate the subjectiveness of human ratings and to more fully automate lesion classification, features extracted using computer vision have been investigated as computerized diagnostic aids [1, 103–105]. Such mathematical descriptors may characterize the lesion using features that radiologists can visually extract, such as mass spiculation or distribution of microcalcifications; or they may characterize the lesion using features that are not so visually apparent to a human observer, such as those extracted using co-occurrence matrices. Computer-extracted features can be obtained from standard mammographic views (cranial–caudal and mediolateral oblique) as well as from special-view mammograms and from prior mammographic exams. Similar to findings vis-à-vis radiologist performance, studies have demonstrated improved computer performance in diagnosing lesions on special-view mammograms as compared with standard views [103] and improved performance when prior mammograms were also analyzed [105].

In observer studies, computerized diagnostic methods have been shown to aid radiologists in the task of distinguishing between malignant and benign lesions [105–107]. Therefore, use of a computer diagnostic aid has the potential to increase sensitivity, specificity, or both in the workup of breast lesions. Investigators have demonstrated that radiologists showed an increase in both sensitivity and specificity in the characterization of clustered microcalcifications and in the associated recommendation for biopsy [106]. In addition, it was shown that an improvement in performance can be obtained by both expert mammographers and community-based radiologists who used CAD information, with the increase greater for the nonexperts [107]. In addition, use of computer output is expected to reduce the variability among radiologists' interpretations [106].

The diagnostic workup of suspect breast lesions may also include imaging with multiple modalities, such as sonography and MRI. Breast sonography is used by many radiologists in distinguishing between solid lesions and cysts, but difficulty exists in distinguishing between benign solid lesions and cancerous lesions. CAD methods in breast ultrasound are being explored by various researchers [108–117]. Mass lesions visible at ultrasound can be classified by computer using a variety of mathematical descriptors of texture, margin, and shape criteria [108]. Observer studies investigating the performance of radiologists using a computerized sonographic analysis method and a multimodality CAD workstation demonstrated that the use of computer-estimated probabilities of malignancy yielded a statistically significant improvement in radiologists' performances in the task of interpreting sonographic breast images and multimodality breast images, respectively [118–120].

Computerized analysis of MRI scans has multiple potential benefits due to the significant variability in the assessment of MRI lesions by radiologists and by the lack of standard imaging protocols for breast MRI. Computer analysis of lesions on MRI scans include morphological features, temporal features, or combinations [121–125]. Temporal features such as uptake or washout rates and time of peak enhancement can be extracted from enhancement curves from either the entire lesion in question or the most enhancing voxels in the lesion (as demonstrated in Figure 16.8). It should be noted that the application of computerized image analysis and CAD to 3D breast MRI has the potential to improve both interpretation performance and interpretation time.

While some features encountered in breast imaging, such as margin sharpness, can be utilized across modalities, others are peculiar to the imaging modality, such as the computer characterization of the posterior acoustic behavior on ultrasound and the inhomogeneity of contrast uptake in breast MRI. Thus, workstations capable of displaying both image data and computer analysis output are necessary (these are discussed in section 16.5).

16.4.2 Lung Cancer

In a screening program with low-dose helical CT in New York, 88% of suspicious lesions were found to be benign nodules on follow-up examinations [126]. In a screening program in Japan, only 10% of the scans with suspicious lesions were diagnosed to be cancer cases [127]. According to findings at the Mayo Clinic, 98.6% of nodules detected by a multidetector CT were benign, and 1.4% of nodules were malignant [128]. Thus, a large number of benign nodules are found with CT, and follow-up examinations such as high-resolution CT (HRCT) and/or biopsy are necessary. Therefore, CAD schemes for classification of nodules as benign or malignant would be useful for reducing the number of "unnecessary" follow-up examinations.

Some investigators have developed CAD schemes for classification of nodules in CT [129, 130]. This type of CAD scheme generally provides radiologists the computer-estimated likelihood of malignancy for assisting them in their task of distinguishing between benign and malignant pulmonary nodules on low-dose helical CT. One such CAD scheme for nodule classification involved the use of nodule segmentation, image feature analysis, and LDA [130]. In a database consisting of 76 primary lung cancers and 413 benign nodules, a comparison of this classification scheme and an MTANN-based classification method was performed, yielding areas under the curve of 0.83 and 0.88, respectively [129, 130]. Similar results were obtained in a study for the distinction between 183 benign and 61 malignant nodules in HRCT, as illustrated in Figure 16.9.

16.5 Intelligent Computer-Aided Diagnostic Workstations: Indices of Similarity and Human/Computer Interfaces

CADe systems output the location of suspect abnormalities to the radiologist by marks on display monitors, hard-copy film, or paper. Note that in the computer output for the detection of potential lesions, both actual lesions and false positives are indicated. Over the years, investigators have worked to increase detection sensitivity while decreasing the number of false marks per image. Investigators have also studied the necessary sensitivity and false-mark rate for computerized detection methods in order to yield improvement in radiologists' performance levels [131]. Sensitivity that is too low or the presence of too many false marks will yield computer output that may be detrimental to radiologist performance.

Radiologists interpret cases rather than individual images; thus, the computerized analysis of images can be case based as opposed to image based. Computer analysis of multiple views or multiple modalities, however, requires effective and efficient displays in order to communicate the multiple images and output to the radiologist. Research has been performed to develop display interfaces, which would better present the computer output to the radiologist [119, 132–134]. For example, the output of a CADx system can be presented in terms of numerical values related to the likelihood of

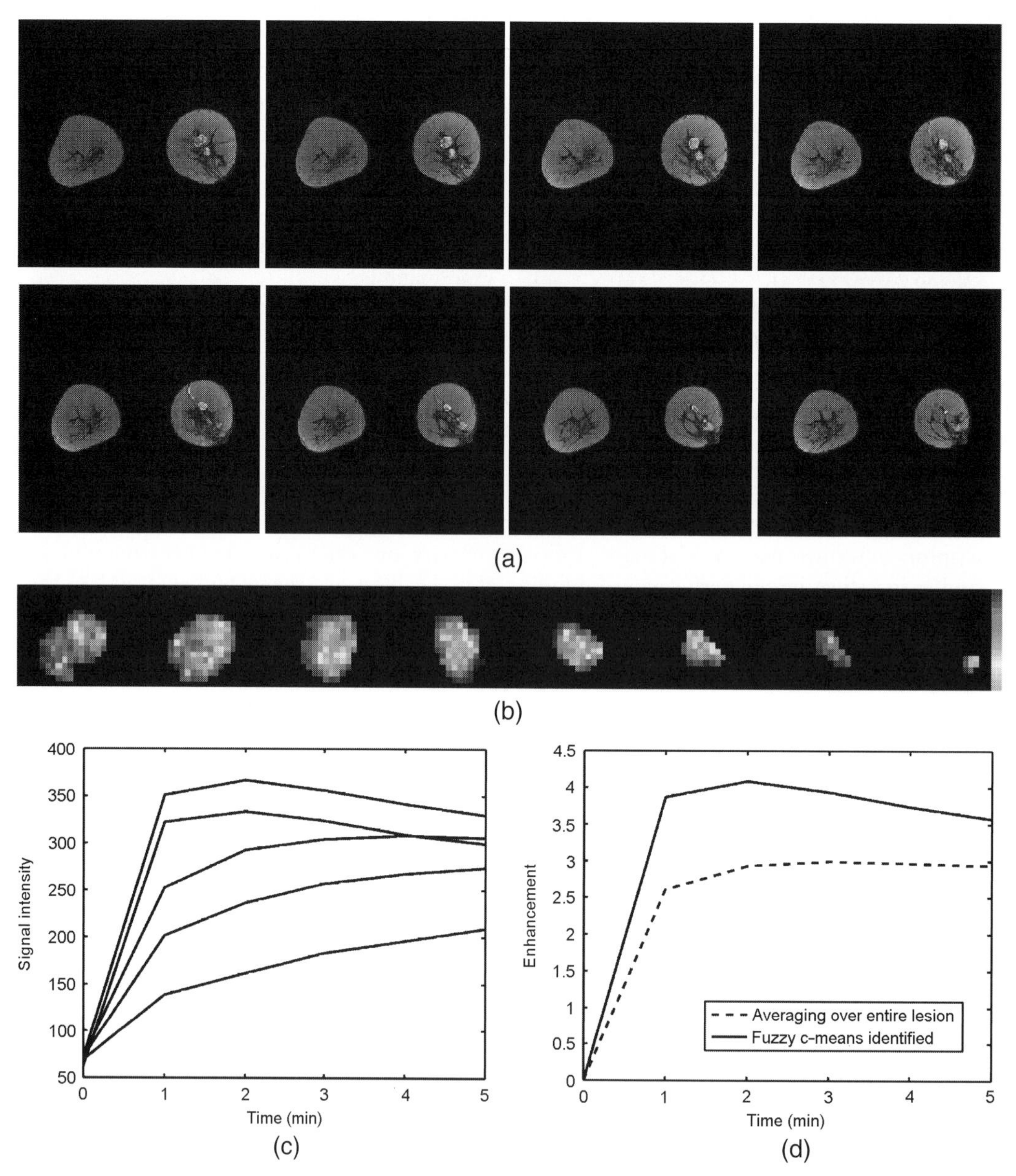

FIGURE 16.8 Illustration of a kinetic analysis for breast MRI CAD using a kinetic curve identification method. (a) 3D breast MRI first postcontrast series displayed as multiple slices, with a malignant mass lesion segmented by a radiologist. (b) Color-encoded membership map from fuzzy cluster-means (FCM) analysis overlapped on the original lesion marking the most enhancing regions in the lesion. (c) Detected prototype curves within the lesion. (d) The characteristic kinetic curve identified by the FCM method (solid line) and the curve obtained by averaging over the radiologist-outlined lesion region (dashed line). Reprinted with permission from Swensen et al. [126].

malignancy, by displaying similar images of known diagnoses or by a graphical representation of the unknown lesion relative to all lesions in a known database (an online atlas) [119, 134]. Searching within an online image atlas can be performed based on individual features, on a likelihood of malignancy, or on psychophysical measures of similarity [119, 134]. Figure 16.10 shows a computer interface that displays similar images and uses color coding to indicate whether the similar images are malignant (red outlines) or benign (green outlines). In addition, the probability of malignancy of the unknown case can be shown relative to the probability distributions of all the malignant and benign cases in the known database. The potential of this interface as an aid in the diagnostic interpretation of lesions by radiologists has been shown for both mammography alone and for a combined mammography and sonography display [119].

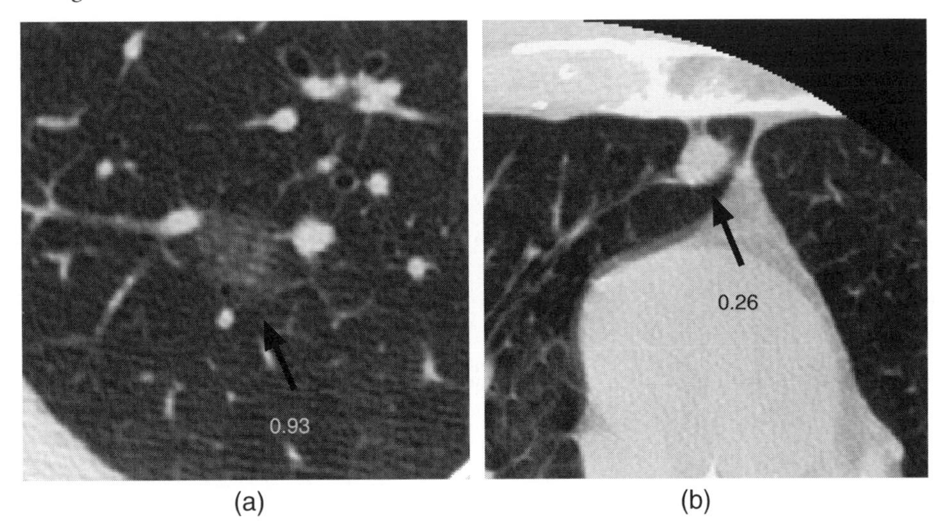

FIGURE 16.9 Illustration of benign and malignant nodules in HRCT with a computer-estimated likelihood of malignancy. (a) Malignant nodule with a high computer-estimated likelihood of malignancy. (b) Benign nodule with a low computer-estimated likelihood of malignancy. It should be noted that average confidence ratings by 16 radiologists in an observer study were 0.49 and 0.46 for (a) and (b), respectively. Courtesy of K. Suzuki, University of Chicago.

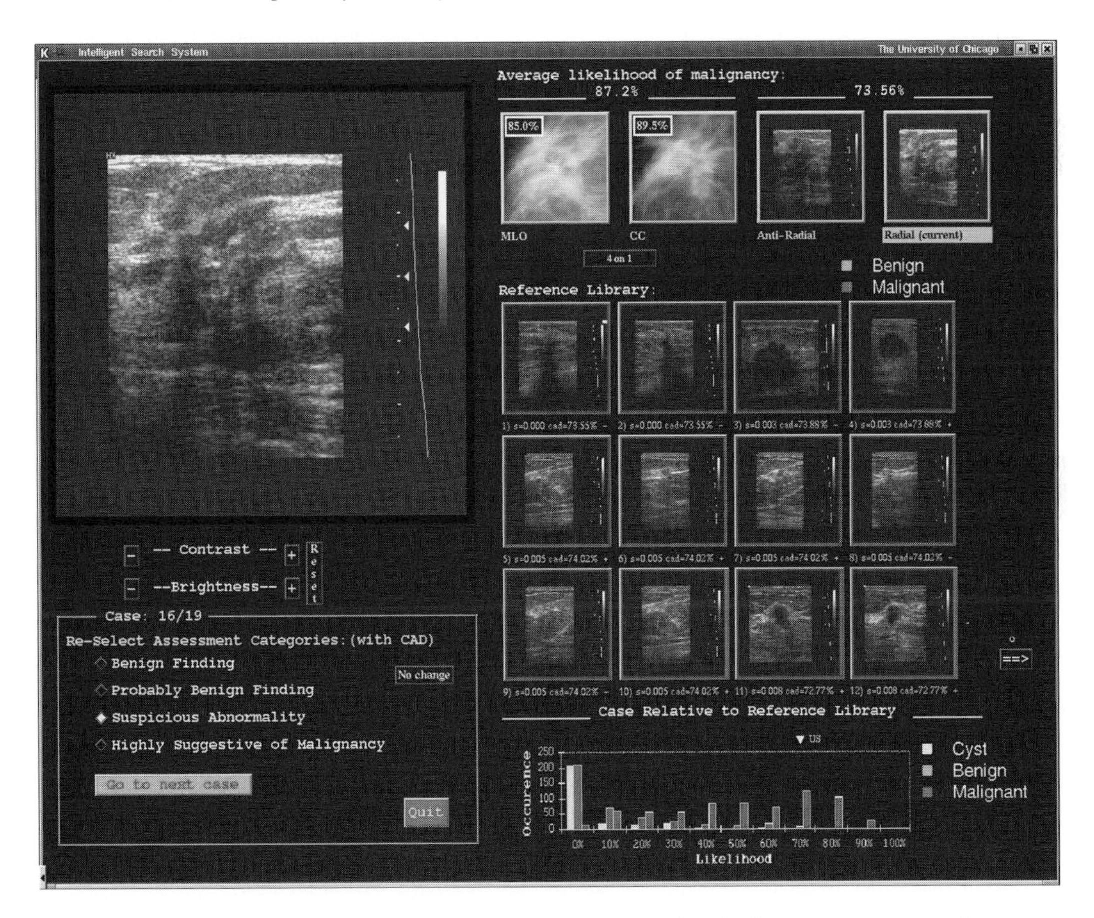

FIGURE 16.10 Display interface for a multimodality workstation that displays computer outputs in numerical, pictorial, and graphical modes for both mammography CADx output and sonography CADx output. Sonography CADx output is shown in the figure. Courtesy of M. Giger, University of Chicago. For a more detailed view of this figure, please visit our companion site at: http://books.elsevier.com/companions/9780123735836.

Acknowledgments

The authors are grateful for the many fruitful discussions with the faculty and research staff in the Department of Radiology, University of Chicago.

M. L. Giger is a stockholder in R2 Technology/Hologic, Sunnyvale, CA. It is the University of Chicago conflict-of-interest policy that investigators disclose publicly actual or potential significant financial interests that may appear to be affected by the research activities.

16.6 Summary

Limitations in the human eye/brain visual system, the presence of overlapping structures in images, and the vast number of normal cases in screening programs provide motivation for the use of computer techniques that have the potential to improve detection and diagnostic performance, and ultimately patient care. The ultimate success of computerized image analyses in the interpretation of medical images depends on both the ability of the computer systems that extract and characterize suspect lesions and the ability of radiologists to incorporate the computer output into their decision making. The clinical usefulness of computer-aided detection for screening mammography is being tested by actual prospective clinical usage of commercial systems. In addition, various observer studies for computer-aided diagnosis indicate the promising role of computer interpretation aids in diagnostic workup. For thoracic CT and CT colonography for the early detection of lung cancer and colon cancer, respectively, effective and efficient displays are also needed to help the radiologist incorporate the computer output into the three-dimensional image data. The development of computerized techniques for aiding in the interpretation of medical images is progressing rapidly, and ultimate incorporation into routine clinical care is expected in the near future.

16.7 Exercises

1. List and discuss three motivations for the use of CADe in cancer screening programs.
2. List four major steps in a computer algorithm for CADe.
3. Discuss the trade-off between true-positive detections and false-positive detections in CADe.
4. Explain the difference between CADe and CADx.
5. Discuss the two types of clinical studies for evaluating the usefulness of CAD.

16.8 References and Bibliography

1. M. L. Giger et al. Computer-aided diagnosis in mammography. In M. Sonka and M. J.Fitzpatrick (Eds.). *Handbook of Medical Imaging,* vol. 2. *Proc. SPIE,* 2000:915–1004.
2. M. L. Giger. Computer-aided diagnosis. In A. G. Haus and M. Yaffe (eds.), *Technical Aspects of Mammography.* RSNA Publications, 1993:283.
3. C. J. Vyborny and M. L. Giger. Computer vision and artificial intelligence in mammography. *Am. J. Roentgenol.* 162:699, 1994.
4. C. J. Vyborny, M. L. Giger, and R. M. Nishikawa. Computer-aided detection and diagnosis. *Radiol. Clin. North Am.* 38:725–740, 2000.
5. K. Doi et al. (Eds.). Computer-aided diagnosis in medical imaging. *Proceedings of the 1st International Workshop on Computer-Aided Diagnosis.* Elsevier, 1999.
6. Y. Jiang. Classification of breast lesions in mammograms. In I. Bankman (Ed.). *Handbook of Medical Imaging, Processing and Analysis.* Academic Press, 2000:341–358.
7. M. L. Giger. Computerized image analysis in breast cancer detection and diagnosis. *Seminars in Breast Disease* 5:99–210, 2002.
8. M. L. Giger and H. MacMahon. Image processing and computer-aided diagnosis. *Radiol. Clin. North Am.* 34:565–596, 1996.
9. C. E. Metz. ROC methodology in radiologic imaging. *Invest. Radiol.* 21:720–733, 1986.
10. J. P. Egan, G. Z. Greenberg, and A. I. Schulman. Operating characteristics, signal detectability, and the method of free response. *J. Acoust. Soc. Am.* 33:993–1007, 1961.
11. P. C. Bunch et al. A free-response approach to the measurement and characterization of radiographic-observer performance. *Journal of Applied Photographic Engineering.* 4:166–171, 1978.
12. Y. Jiang, C. E. Metz, and R. M. Nishikawa. A receiver operating characteristic partial area index for highly sensitive diagnostic tests. *Radiology.* 201:745– 750, 1996.
13. G. M. te Brake, N. Karssemeijer, and H. C. L. Hendriks. Automated detection of breast carcimomas not detected in a screening program. *Radiology.* 207:465–471, 1998.
14. R. M. Nishikawa et al. Effect of case selection on the performance of computer-aided detection schemes. *Med. Phys.* 21:265–269, 1994.
15. Z. Huo. *Computerized Methods for Classification of Masses and Analysis of Parenchymal Patterns on Digitized Mammograms.* Ph.D. thesis, University of Chicago, 1998.
16. M. L. Giger. Current issues in CAD for mammography. In *Digital Mammography '96.* Elsevier, 1996.
17. S. G. Armato et al. Lung cancer: Performance of automated lung nodules detection applied to cancers missed in a CT screening program. *Radiology.* 225:685–692, 2002.

18. H. P. Chan et al. Image feature analysis and computer-aided diagnosis in digital radiography. 1. Automated detection of microcalcifications in mammography. *Med. Phys.* 14:538, 1987.
19. M. L. Giger, K. Doi, and H. MacMahon. Image feature analysis and computer-aided diagnosis in digital radiography. 3. Automated detection of nodules in peripheral lung fields. *Med. Phys.* 15(2):158–166, 1988.
20. C. Beam, P. M Layde, and D. Sullivan. Variability in the interpretation of screening mammograms by U.S. radiologists. *Arch. Intern. Med.* 156:209–213, 1996.
21. J. G. Elmore et al. Variability in radiologists' interpretations of mammograms. *N. Engl. J. Med.* 331:1492, 1994.
22. F. Winsberg et al. Detection of radiographic abnormalities in mammograms by means of optical scanning and computer analysis. *Radiology.* 89:211–215, 1967.
23. Y. H. Chang et al. Identification of clustered microcalcifications on digitized mammograms using morphology and topography-based computer-aided detection schemes: A preliminary experiment. *Invest. Radiol.* 33:746, 1998.
24. N. Ibrahim et al. Automated detection of clustered microcalcifications on mammograms: CAD system application to MIAS database. *Phys. Med. Biol.* 42:2577, 1997.
25. N. Karssemeijer. A stochastic method for automated detection of microcalcifications in digital mammograms. In *Information Processing in Medical Imaging.* Springer-Verlag, 227, 1991.
26. Y. Jiang et al. Automated feature analysis and classification of malignant and benign clustered microcalcifications. *Radiology.* 198:671, 1996.
27. R. M. Nishikawa et al. Computer-aided detection of clustered microcalcifications: An improved method for grouping detected signals. *Med. Phys.* 20:1661–1666, 1993.
28. S. A. Feig. Clinical evaluation of computer-aided detection in breast cancer screening. *Seminars in Breast Disease.* 5:223–230, 2002.
29. L. S. Warren-Burhenne et al. Potential contribution of computer-aided detection to the sensitivity of screening mammography. *Radiology.* 215:554–562, 2000.
30. T. W. Freer and M. J. Ulissey MJ. Screening mammography with computer-aided detection: Prospective study of 12,860 patients in a community breast center. *Radiology.* 222:781–786, 2001.
31. D. Gur et al. Changes in breast cancer detection and mammography recall rates after the introduction of a computer-aided detection system. *J. Nat. Canc. Inst.* 96:185–190, 2004.
32. S. A. Feig et al. Re: Changes in breast cancer detection and mammography recall rates after the introduction of a computer-aided detection system. *J. Nat. Canc. Inst.* 96:1260–1261, 2004.
33. T. Cupples, J. E. Cunningham, and J. C. Reynolds. Impact of computer-aided detection in a regional screening mammography program. *Am. J. Roentgenol.* 185: 944–950, 2005.
34. R. L. Birdwell, P. Bandodkar, and D. M. Ikeda. Computer-aided detection with screening mammography in a university hospital setting. *Radiology.* 236:451–457, 2005.
35. M. Helvie et al. Sensitivity of noncommercial computer-aided detection system for mammographic breast cancer detection: Pilot clinical trial. *Radiology.* 231:208–214, 2004.
36. L. A. Khoo, P. Taylor, and R. M. Given-Wilson. Computer detection in the United Kingdom National Breast Screening Programme: Prospective study. *Radiology.* 237:444–449, 2005.
37. M. J. Morton et al. Screening mammograms: Interpretation with computer-aided detection—Prospective evaluation. *Radiology.* 239:375–383, 2006.
38. J. C. Dean and C. C. Ilvento. Improved cancer detection using computer-aided detection with diagnostic and screening mammography: Prospective study of 104 cancers. *Am. J. Roentgenol.* 187:20–28, 2006.
39. American Cancer Society. *Cancer Facts and Figures 2005.* Author, 2005.
40. B. J. Flehinger, M. Kimmel, and M. R. Melamed. The effect of surgical treatment on survival from early lung cancer. Implications for screening. *Chest.* 101(4):1013–1018, 1992.
41. T. Sobue et al. Survival for clinical stage I lung cancer not surgically treated. Comparison between screen-detected and symptom-detected cases. The Japanese Lung Cancer Screening Research Group. *Cancer.* 69(3):685–692, 1992.
42. R. T. Heelan et al. Non-small-cell lung cancer: Results of the New York screening program. *Radiology.* 151(2): 289–293, 1984.
43. J. H. Austin, B. M. Romney, and L. S. Goldsmith. Missed bronchogenic carcinoma: Radiographic findings in 27 patients with a potentially resectable lesion evident in retrospect. *Radiology.* 182(1):115–122, 1992.
44. P. K. Shah et al. Missed non-small cell lung cancer: Radiographic findings of potentially resectable lesions evident only in retrospect. *Radiology.* 226(1):235–241, 2003.
45. H. MacMahon et al. Computer-aided diagnosis in chest radiology. *J. Thorac. Imaging.* 5(1):67–76, 1990.
46. P. Campadelli, E. Casiraghi, and D. Artioli. A fully automated method for lung nodule detection from postero-anterior chest radiographs. *IEEE Trans. Med. Imaging.* 25:1588–1603, 2006.
47. H. MacMahon et al. Computer-aided diagnosis of pulmonary nodules: Results of a large-scale observer test. *Radiology.* 213(3):723–726, 1999.
48. H. Abe et al. Computer-aided diagnosis in chest radiography: Results of large-scale observer tests at the 1996–2001 RSNA scientific assemblies. *Radiographics.* 23(1):255–265, 2003.

49. T. Kobayashi et al. Effect of a computer-aided diagnosis scheme on radiologists' performance in detection of lung nodules on radiographs. *Radiology.* 199(3):843–848, 1996.
50. M. L. Giger et al. Pulmonary nodules: Computer-aided detection in digital chest images. *Radiographics.* 10(1): 41–51, 1990.
51. M. L. Giger et al. Computerized detection of pulmonary nodules in digital chest images: Use of morphological filters in reducing false-positive detections. *Med. Phys.* 17:861–865, 1990.
52. Y. Wu et al. Reduction of false-positives in computerized detection of lung nodules in chest radiographs using artificial neural networks, discriminant analysis, and a rule-based scheme. *J. Digit. Imaging.* 7:196–207, 1994.
53. X. W. Xu et al. Development of an improved CAD scheme for automated detection of lung nodules in digital chest images. *Med. Phys.* 24(9):1395–1403, 1997.
54. K. Suzuki et al. False-positive reduction in computer-aided diagnostic scheme for detecting nodules in chest radiographs by means of massive training artificial neural network. *Acad. Radiol.* 12(2):191–201, 2005.
55. M. Freedman et al. Computer-aided detection of lung cancer on chest radiographs: Algorithm performance vs. radiologists' performance by size of cancer. *Proc. SPIE Medical Imaging: Image Processing,* 2001:150–159.
56. S. Kakeda et al. Improved detection of lung nodules on chest radiographs using a commercial computer-aided diagnosis system. *Am. J. Roentgenol.* 182(2):505–510, 2004.
57. T. Matsumoto et al. Image feature analysis of false–positive diagnoses produced by automated detection of lung nodules. *Invest. Radiol.* 27(8):587–597, 1992.
58. K. Suzuki et al. Image-processing technique for suppressing ribs in chest radiographs by means of massive training artificial neural network (MTANN). *IEEE Trans. Med. Imaging.* 25:406–416, 2006.
59. S. Miettinen and C. I. Henschke. CT screening for lung cancer: Coping with nihilistic recommendations. *Radiology.* 221(3):592–596, 2001.
60. S. Sone et al. Mass screening for lung cancer with mobile spiral computed tomography scanner. *Lancet.* 351(9111): 1242–1245, 1998.
61. C. I. Henschke et al. Early lung cancer action project: Overall design and findings from baseline screening. *Lancet.* 354(9173):99–105, 1999.
62. C. I. Henschke et al. Early lung cancer action project: Initial findings on repeat screenings. *Cancer.* 92(1): 153–159, 2001.
63. S. J. Swensen et al. Lung cancer screening with CT: Mayo Clinic experience. *Radiology.* 226(3):756–761, 2003.
64. M. Kaneko et al. Peripheral lung cancer: Screening and detection with low-dose spiral CT versus radiography. *Radiology.* 201(3):798–802, 1996.
65. S. Sone et al. Results of three-year mass screening programme for lung cancer using mobile low-dose spiral computed tomography scanner. *Br. J. Cancer.* 84(1): 25–32, 2001.
66. M. L. Giger, K. T. Bae, and H. MacMahon. Computerized detection of pulmonary nodules in computed tomography images. *Invest. Radiol.* 29:459–465, 1994.
67. S. G. Armato III, M. L. Giger, and H. MacMahon. Automated detection of lung nodules in CT scans: Preliminary results. *Med. Phys.* 28:1552– 1561, 2001.
68. M. S. Brown et al. Patient-specific models for lung nodule detection and surveillance in CT images. *IEEE Trans. Med. Imaging.* 20:1242–1250, 200.
69. M. N. Gurcan et al. Lung nodule detection on thoracic computed tomography images: Preliminary evaluation of a computer-aided diagnosis system. *Med. Phys.* 29: 2552–2558, 2002.
70. J. P. Ko and M. Betke. Chest CT: Automated nodule detection and assessment of change over time—Preliminary experience. *Radiology.* 218:267–273, 2001.
71. Y. Lee et al. Automated detection of pulmonary nodules in helical CT images based on an improved template-matching technique. *IEEE Trans. Med. Imaging.* 20: 595–604, 2001.
72. S. G. Armato et al. Automated lung nodule classification following automated nodule detection on CT: A serial approach. *Med. Phys.* 30:1188–1197, 2003.
73. S. G. Armato, M. B. Altman, and P. J. LaRiviere. Automated detection of lung nodules in CT scans: Effect of image reconstruction algorithm. *Med. Phys.* 30:461–472, 2003.
74. K. Suzuki et al. Massive training artificial neural network (MTANN) for reduction of false positives in computerized detection of lung nodules in low-dose computed tomography. *Med. Phys.* 30(7):1602–1617, 2003.
75. K. Suzuki, I. Horiba, and N. Sugie. Neural edge enhancer for supervised edge enhancement from noisy images. *IEEE Trans. Pattern. Anal. Mach. Intell.* 25:1582–1596, 2003.
76. K. Suzuki et al. Extraction of left ventricular contours from left ventriculograms by means of a neural edge detector. *IEEE Trans. Med. Imaging.* 23:330–339, 2004.
77. K. Suzuki and K. Doi. How can a massive training artificial neural network (MTANN) be trained with a small number of cases in the distinction between nodules and vessels in thoracic CT? *Acad. Radiol.* 12:1333–1341, 2005.
78. H. Arimura et al. Computerized scheme for automated detection of lung nodules in low-dose CT images for lung cancer screening. *Acad. Radiol.* 11:617–629, 2004.
79. F. Li et al. Computer-aided detection of peripheral lung cancers missed at CT: ROC analyses without and with localization. *Radiology.* 237:684–690, 2005.
80. R. Wiemker et al. Computer-aided lung nodule detection on high-resolution CT data. *Proc. SPIE Med. Imaging.* 677–688, 2002.

81. T. Oda et al. Detection algorithm of lung cancer candidate nodules on multislice CT images. *Proc. SPIE Med. Imaging,* 2002:1354–1361.
82. M. S. Brown et al. Lung micronodules: Automated method for detection at thin-section CT—Initial experience. *Radiology.* 226(1):256–262, 2003.
83. Hara et al. Detection of colorectal polyps with CT colography: Initial assessment of sensitivity and specificity. *Radiology.* 205:59–65, 1997.
84. A. H. Dachman et al. C colonography with three-dimensional problem solving for detection of colonic polyps. *Am. J. Roentgenol.* 171:989–995, 1998.
85. A. P. Royster et al. CT colonoscopy of colorectal neoplasms: Two-dimensional and three-dimensional virtual-reality techniques with colonoscopic correlation. *Am. J. Roentgenol.* 169:1237–1242, 1997.
86. E. G. McFarland et al. Spiral CT colonography: Reader agreement and diagnostic performance with two- and three-dimensional image-display techniques. *Radiology.* 218:375–383, 2001.
87. M. Macari and A. J. Megibow. Pitfalls of using three-dimensional CT colonography with two-dimensional imaging correlation. *Am. J. Roentgenol.* 176:137–143, 2001.
88. R. M. Summers et al. Automated polyp detection at CT colonography: Feasibility assessment in a human population. *Radiology.* 219:51–59, 2001.
89. H. Yoshida et al. Computerized detection of colonic polyps at CT colonography on the basis of volumetric features: Pilot study. *Radiology.* 222:327–336, 2002.
90. H. Yoshida and J. Nappi. Three-dimensional computer-aided diagnosis scheme for detection of colonic polyps. *IEEE Trans. Med. Imaging.* 20:1261–1274, 2001.
91. G. Kiss et al. Computer-aided diagnosis in virtual colonography via combination of surface normal and sphere fitting methods. *Eur. Radiol.* 12:77–81, 2002.
92. H. Yoshida et al. Computer-aided diagnosis scheme for detection of polyps at CT colonography. *Radiographics.* 22:963–979, 2002.
93. J. Nappi et al. Automated knowledge-guided segmentation of colonic walls for computerized detection of polyps in CT colonography. *J. Comput. Assist. Tomog.* 26:493–504, 2002.
94. J. Nappi and H. Yoshida. Automated detection of polyps with CT colonography: Evaluation of volumetric features for reduction of false-positive findings. *Acad. Radiol.* 9:386–397, 2002.
95. J. Nappi and H. Yoshida. Feature-guided analysis for reduction of false postitives in CAD of polyps for computed tomographic colonography. *Med. Phys.* 30:1592–1601, 2003.
96. J. J. Nappi et al. Computerized detection of colorectal masses in CT colonography based on fuzzy merging and wall-thickening analysis. *Med. Phys.* 31:860–872, 2004.
97. K. Suzuki et al. Massive-training artificial neural network (MTANN) for reduction of false positives in computer-aided detection of polyps: Suppression of rectal tubes. *Med. Phys.* 33(10):3814–3824, 2006.
98. J. Sosna et al. CT colonography of colorectal polyps: A metaanalysis. *Am. J. Roentgenol.* 181(6):1593–1598, 2003.
99. C. J. D'Orsi et al. *Breast Imaging Reporting and Data System (BI–RADS): Mammography,* 4th ed. American College of Radiology, 2003.
100. Y. Wu et al. Artificial neural networks in mammography: Application to decision making in the diagnosis of breast cancer. *Radiology.* 187:81–87, 1993.
101. Z. Huo, M. L. Giger, and C. J. Vyborny. Computerized analysis of multiple-mammographic views: Potential usefulness of special view mammograms in computer-aided diagnosis. *IEEE Trans. Med. Imaging.* 20:1285–1292, 2001.
102. Y. Jiang et al. Improving breast cancer diagnosis with computer-aided diagnosis. *Acad. Radiol.* 6:22, 1999.
103. H. P. Chan et al. Improvement of radiologists' characterization of mammographic masses by using computer-aided diagnosis: An ROC study. *Radiology.* 212:817–827, 1999.
104. Y. Jiang et al. Potential of computer-aided diagnosis to reduce variability in radiologists' interpretations of mammograms depicting microcalcifications. *Radiology.* 220:787–794, 2001.
105. Z. Huo et al. Effectiveness of CAD in the diagnosis of breast cancer: An observer study on an independent database of mammograms. *Radiology.* 224:560–568, 2002.
106. A. T. Stavros et al. Solid breast nodules: Use of sonography to distinguish between benign and malignant lesions. *Radiology.* 196:123, 1995.
107. M. L. Giger et al. Computerized analysis of lesions in US [ultrasound] images of the breast. *Acad. Radiol.* 6:665–674, 1999.
108. B. S. Garra et al. Improving the distinction between benign and malignant breast lesions: The value of sonographic texture analysis. *Ultrason. Imaging.* 15:267–285, 1993.
109. D. R. Chen, R. F. Chang, and Y. L. Huang. Computer-aided diagnosis applied to us of solid breast nodules by using neural networks. *Radiology.* 213:407–412, 1999.
110. R. M. Golub et al. Differentiation of breast tumors by ultrasonic tissue characterization. *J. Ultrasound Med.* 12:601–608, 1993.
111. K. Horsch et al. Automatic segmentation of breast lesions on ultrasound. *Med. Phys.* 28:1652–1659, 2001.
112. K. Horsch et al. Computerized diagnosis of breast lesions on ultrasound. *Med. Phys.* 29:157–164, 2002.
113. B. Sahiner et al. Computerized characterization of breast masses using three-dimensional ultrasound images. *Proc. SPIE.* 3338:301–312, 1998.

114. K. Drukker et al. Computerized detection and classification of cancer on breast ultrasound. *Acad. Radiol.* 11:526–535, 2004.
115. K. Drukker, M. L. Giger, and C. E. Metz. Robustness of computerized lesion detection and classification scheme across different breast ultrasound platforms. *Radiology.* 237:834–840, 2005.
116. K. Horsch et al. Performance of CAD in the interpretation of lesions on breast sonography. *Acad. Radiol.* 11: 272–280, 2004.
117. K. Horsch et al. Multi-modality computer-aided diagnosis for the classification of breast lesions: Observer study results on an independent clinical dataset. *Radiology.* 240:357–368, 2006.
118. B. Sahiner et al. UMich (RSNA 2004).
119. K. G. A. Gilhuijs, M. L. Giger, and U. Bick. Automated analysis of breast lesions in three dimensions using dynamic magnetic resonance imaging. *Med. Phys.* 25:1647, 1998.
120. A. I. Penn et al. Discrimination of MR images of breast masses using fractal interpolation function models. *Acad. Radiol.* 6:156–163, 1999.
121. W. Chen et al. Computerized interpretation of breast MRI: Investigation of enhancement-variance dynamics. *Med. Phys.* 31:1076–1082, 2004.
122. W. Chen, M. L. Giger, and U. Bick. A fuzzy c-means (FCM) based approach for computerized segmentation of breast lesions in dynamic contrast-enhanced MR images. *Acad. Radiol.* 13:63–72, 2006.
123. W. Chen et al. Automatic identification and classification of characteristic kinetic curves of breast lesions on DCE-MRI. *Med. Phys.* 33:2878–2887, 2006.
124. C. I. Henschke et al. Early lung cancer action project: Overall design and findings from baseline screening. *Lancet.* 354(9173):99–105, 1999.
125. F. Li et al. Lung cancer missed at low-dose helical CT screening in a general population: Comparison of clinical, histopathologic, and imaging findings. *Radiology.* 225(3):673–683, 2002.
126. S. J. Swensen et al. Lung cancer screening with CT: Mayo Clinic experience. *Radiology.* 226(3):756–761, 2003.
127. K. Suzuki et al. Computer-aided diagnostic scheme for distinction between benign and malignant nodules in thoracic low-dose CT by use of massive training artificial neural network. *IEEE Trans. Med. Imaging.* 24(9):1138–1150, 2005.
128. M. Aoyama et al. Computerized scheme for determination of the likelihood measure of malignancy for pulmonary nodules on low-dose CT images. *Med. Phys.* 30(3):387–394, 2003.
129. Zheng et al. Soft-copy mammographic readings with different computer-assisted detection cueing environments: Preliminary findings. *Radiology.* 221:633–640, 2001.
130. H. A. Swett et al. Expert system controlled image display. *Radiology.* 172:487–493, 1989.
131. J. Sklansky et al. A visualized mammographic database in computer-aided diagnosis. In *Computer-Aided Diagnosis in Medical Imaging.* Elsevier, 215, 1999.
132. M. L. Giger et al. Intelligent search workstation for computer-aided diagnosis. *Proc. Comput. Assist. Radiol. Surg.* 2000:822–827.

Bibliography

M. L. Giger et al. Results of an observer study with an intelligent mammographic workstation for CAD. In H.-O. Peitgen (Ed.). *Digital Mammography IWDM 2002.* Springer, 297–303, 2003.

C. Muramatsu et al. Investigation of psychophysical measures for evaluation of similar images for mammographic masses: Preliminary results. *Med. Phys.* 32: 2295–2304, 2005.

S. Miettinen. Screening for lung cancer. *Radiol. Clin. North Am.* 38(3):479–486, 2000.

A. Jemal et al. Cancer statistics, 2005. *Cancer J. Clin.* 55(1): 10–30, 2005.

17

Clinical Decision Support Systems

Dr. Peter Weller,
Dr. Abdul Roudsari, and
Prof. Ewart Carson
City University, London, UK

17.1 Introduction

Decision-making is central to the activities of clinical professionals in their dealings with patients. Decisions need to be made when diagnosing the state of the patient, in planning and adjusting therapy, and in monitoring the evolving patient state. Over recent decades, a range of computational methods and tools has been developed to support the clinician in these tasks. In general their purpose is not to replace or substitute for human abilities or skills, but rather to offer support and assistance. Given the increasing importance of information and communications technologies in the infrastructure of health care organization and delivery, it is important that clinicians, allied health care professionals, and others associated with health care be aware of such computational methods and techniques that can support the decision-making processes. This is the motivation for this present chapter.

The aim of the chapter is to provide an overview of the decision-making process, the need for support in such activities, and the broad range of methods and techniques that are now available. Specific objectives include reviewing the nature of the decision-making process, providing a structure in terms of which decision support systems can be classified, and providing examples of the application of a number of such decision support techniques in the clinical setting.

The chapter begins with an overview of the decision-making process and the context of decision support. This is followed by an outline of the nature of human diagnostic reasoning, models of the reasoning process, and issues to be considered in the clinical setting. A structure is then presented for characterizing clinical decision support systems (CDSS). In this section, the broad range of conceptual, methodologic, and technical approaches is reviewed, demonstrating their bases and some of their particular strengths and limitations. A range of applications of decision support systems is then considered: first in hospital and other health care settings and then in relation to medical education. Finally, the major topic of evaluating decision support systems is addressed, including relevant legal and ethical issues.

17.2 Overview of Clinical Decision Support Systems

17.2.1 What Is the Decision-Making Process?

Decision making in the clinical context can be seen as an integral part of the overall care delivery process by considering the simple feedback model shown in Figure 17.1. Here the clinical decision maker receives data and information regarding the state of the patient, and by comparing these current data with past data and a desired state of the patient, using relevant models, determines an appropriate decision action. This decision is then carried out, for example, by other members of the health care team. The result is a change in some attributes of the patient state, either in terms of state of knowledge diagnostically in response to test or investigation, or state of well-being following therapeutic intervention. Feedback, via appropriate information systems, enables the decision maker to review or update the decision as necessary.

Classically, the decision-making process starts with the recognition of the need for a decision, be this diagnostic or therapeutic. The need for a decision then implies that there are a number of alternative courses of action. These alternatives are generated and then assessed. A choice is then made among these alternatives, adopting that which is best in terms of the criteria used to assess worth or value of outcome. The chosen alternative is then implemented. Finally, on the basis of feedback from the result of the decision (as depicted in the feedback loop shown in Figure 17.1), the decision maker is able to learn from the process and hence may be able to enhance future performance. Further elaboration of the decision-making process is presented in section 17.3.

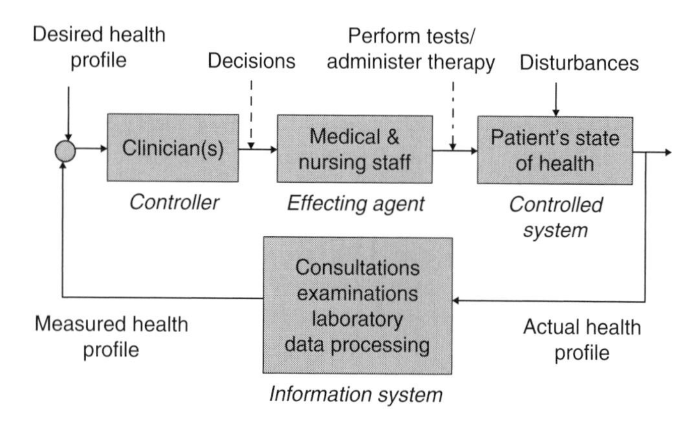

FIGURE 17.1 Information processing and decision-making in health care.

17.2.2 Who Needs Decision Support?

Decision making is central to the organization and delivery of health care. Hence, decision makers are to be found at all levels of health care operation. At the strategic level, policy makers are making investment decisions relating to new health care facilities and provision, including those relating to public health. Operationally, health care professionals, both clinicians and allied health professionals, are making diagnostic and therapeutic decisions so as best to manage the individual patient. In the arena of chronic disease, in particular, increasingly patients both wish to, and are encouraged to, be involved in the management of their condition, as are their family members and care givers.

Decision making is thus an activity that involves health managers and policy makers, health care professionals of all grades, patients, and their care givers among others. Hence all may benefit from support in relation to the decisions needing to be made. Clearly this need for support will vary according to the particular situation and circumstances. However, the domain of decision support is, in principle, something that may be of interest to all involved in the organization and delivery of health care.

17.2.3 What Decisions Need Supporting?

In terms of the needs of the individual patient, clinical decision making includes diagnosis, prognosis, and therapy. Such decisions will relate to different stages in the patient's journey through the health care system. Decisions will also have different degrees of urgency with regard to time. At the one end of the spectrum are life-threatening and traumatic events where almost instantaneous decisions need to be made, while at the other, there is the more relaxed time scale of decisions involved in changing the medication necessary to manage a chronic disease such as diabetes or hypertension. This time factor will influence the extent to which decision support techniques can be employed. When time is of the essence, it may just not be practicable to enjoy the luxury of formal decision support.

Diagnostic, prognostic, and therapeutic decisions relate to managing the individual patient. There are also decisions to be made at policy, management, and resource levels. Can a national health service make a costly new drug routinely available for managing a life-threatening condition? What will be the consequences for clinical outcomes of closing a community hospital? What level of nursing support should be provided in a ward of elderly confused patients? These are decisions where formal decision support techniques can play an equally important role.

In the sections that follow, emphasis will be placed on the role of decision support systems in the management of the individual patient. However, it needs to be borne in mind that many of the methods and techniques discussed are equally relevant to decision making in management, policy making, educational, and research contexts.

17.3 Human Diagnostic Reasoning

17.3.1 Clinical Reasoning and Decision Making

As discussed earlier, one aspect that distinguishes clinical reasoning and decision making from many other arenas is that frequently a decision has to be made, often with incomplete information, in a very limited time frame. The luxury of time to make a well-informed and complete decision is not always an option when a patient's welfare is at stake. In an acute situation, the patient's condition may be rapidly changing. Therefore, any delay in decision making could result in further deterioration of their condition and, in effect, a new decision-making situation. Additionally, once a decision has been made, it is not always easy, or possible, to reverse that decision. This could be the result of the effects of the treatment on the patient, such as drugs or surgery, again creating a new decision scenario. The ingredients of diagnostic reasoning will now be considered.

17.3.2 Deductive, Inductive, and Abductive Reasoning

Deductive reasoning is perhaps the most mechanical reasoning process. It seeks to identify a set of rules which, if followed, will result in the best possible decisions on average or in the long run. For example, condition B is a consequence of symptom A.

Inductive reasoning recognizes that people do not always follow deductive rules and instead use experience and gut feeling to reach a decision or diagnosis. This is particularly true in clinical decision making, where knowledge of previous cases is frequently used to resolve the current situation. An example of this would be that in the clinician's experience most patients with symptom A have condition B.

Abductive reasoning begins with a set of facts and derives the most feasible explanations. The process works in the reverse direction to the previous two forms of reasoning. For example, symptom A is an explanation of condition B.

A more realistic situation for real-world cases would be a combination of the above scenarios, for example a set of rules combined with experience and gut feelings.

17.3.3 Models of Diagnostic Reasoning

The above concepts can be combined to provide a model for diagnostic reasoning as shown below in Figure 17.2 (1). A patient comes to the physician with a set of symptoms, and abductive reasoning is used to choose possible causes that could account for the data. This process may result in a number of possible diagnoses for the patient's illness. Deductive reasoning can then be used to decide which symptoms would be present if the patient had each condition. This may result in further tests (requests for new data) to repeat the loop and test the hypothesis. Inductive reasoning is used to evaluate test results and either confirm or reject the hypothesis. Another possibility is to request further data. An additional aspect of the model is the ability to simulate a possible diagnosis to explore accuracy with the expected data.

Other models of diagnostic reasoning include association-based diagnosis, pathophysiologically based diagnosis, and pattern classification diagnosis. Descriptions of these can be found in Deutsch et al. [1].

17.3.4 Causal Reasoning

An alternative way of expressing reasoning is with causal relationships between events, where one event is caused by the occurrence of another event. An example of this could be the presence of high blood pressure being caused by heart disease. Causal beliefs are expressed using conditional statements with a probability being defined for the connection.

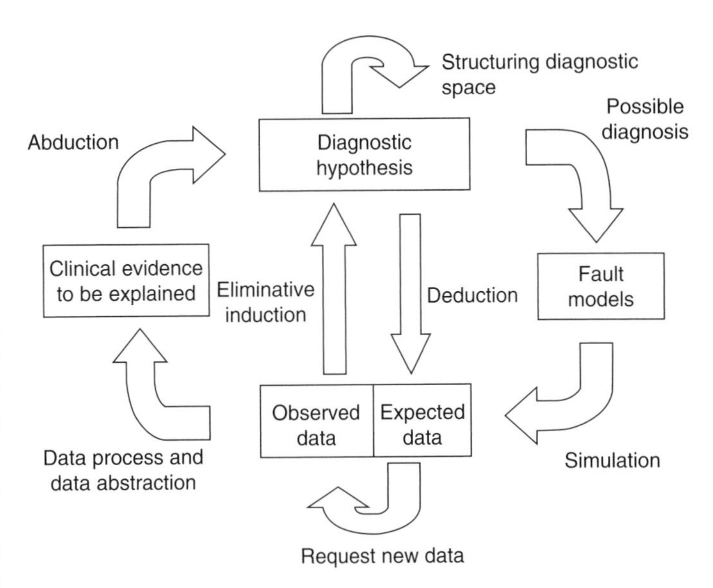

FIGURE 17.2 Model of diagnostic reasoning. Adapted from Deutsch et al. [1].

An interlinked number of these causal beliefs are known as Bayesian networks or causal probabilistic networks. They are popular because they represent uncertainty in medical knowledge in an efficient manner. Variables are represented by nodes, and the causal relationships between these are represented by directed arcs linking the nodes. Uncertainty is represented by probability distributions on the states of each arc. The relationships between nodes are represented by conditional probability tables, and model inference is performed using Bayes' theorem.

In the context of medical decision making, Bayesian networks can be easily integrated with decision theory to yield models for the selection of optimal treatments, or to develop models for health care planning under uncertainty. For a more detailed description of causal reasoning, the reader should consult Pearl [2].

17.3.5 The Decision-Making Process

The decision-making process itself can be broken down into four stages. These will now be introduced and explained. The first stage identifies the need for a decision, clearly defines the problem to be addressed, and then concerns the collection of knowledge relevant to solving the problem. This knowledge collection may include a critical literature review, using evidence-based medicine techniques, or discussions with colleagues.

The second stage involves creating a range of possible decision alternatives to resolve the problem identified in the first stage. The set of possible decisions could be large, so the most realistic candidates are initially considered. It should be noted that making no decision could be classed as a possible strategy.

The third stage involves the evaluation of each of these possible candidates by considering the consequences and potential repercussions involved in following each strategy. The areas that would be considered in this aspect could include ethical, technical, cost, and political issues as well as long-term benefits for both the patient and his or her family. While the detailed consideration of each possible candidate decision may be thought time consuming, it does have the additional merit of allowing the development of suitable contingency plans should circumstances change or if the preferred decision proves to be incorrect.

The last stage of the decision-making process is the making of the actual choice between the possible candidates. The benefits, likelihood of success, risks, and costs are all considered, and a selection is made based on the most appropriate choice of action.

This process has been described as a single loop process, but in some situations it may be possible to revisit the earlier stages and revise the details based on the results from the later stages. For example, on considering the possible alternative strategies, it may be realized that the initial problem was ill-defined and needs to be reconsidered. Figure 17.3 illustrates the process.

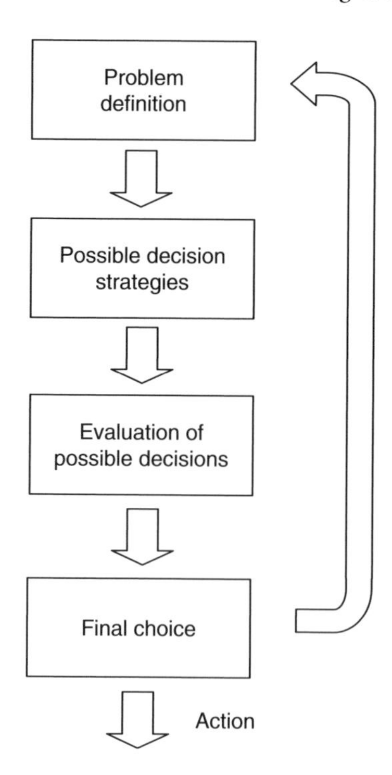

FIGURE 17.3 Decision-making process.

17.3.6 Clinical Judgment, Uncertainty, and Bias

The classical approach to decision making is based on the assumption that the decision maker can identify and evaluate all possible alternatives and their consequences and rationally make a choice as to the most appropriate course of action.

However, the true state of the problem cannot be observed directly. For example the diagnosis of an illness may require testing for further clarification, and this can introduce uncertainty, in terms of the accuracy of the test, to the decision-making process. As a result, the decision maker has to use knowledge that potentially may be imperfect, and so an element of uncertainty is introduced into the process. A fundamental problem with uncertainty is the subjective nature of the terms used. What might be highly probable to one person may be classified as quite unlikely by another. The accepted way to express this uncertainty is in probabilistic terms. For the above example, this would be the specificity and sensitivity of the test. In more general cases, it may be the probability that a patient with a set of symptoms has a certain disease. The subjective nature of general terms for expressing uncertainty remains a concern in most clinical decision making.

A further impediment to the classical decision-making approach is bias, an unfair preference or dislike for a particular choice. Bias can be categorized into a number of different classes:

- *Representative bias*: Bias resulting from the tendency to generalize from a small sample or from a single salient or vivid event or episode

- *Confirmation bias*: Seeking evidence supporting a belief and not data that test competitors
- *Omission bias*: The decision maker feels more responsible for harms caused by action rather than by inaction and as a result tends to omit potentially harmful actions and so avoids potentially beneficial actions in a trade-off.
- *Illusion of control*: Bias resulting from the tendency to overestimate one's ability to control activities and/or events
- *Prior hypothesis bias*: Bias resulting from the tendency to base decisions on strong prior beliefs, even if evidence shows that those beliefs are wrong
- *Groupthink*: Biased decision making that occurs in groups whose members strive for agreement at the expense of accurately assessing the information relevant to the decision being made

17.4 A Structure for Characterizing Clinical Decision Support Systems

Decision support systems for application in the clinical domain have exploited a wide range of conceptual, methodologic, and technical approaches. This section will briefly look at this range of approaches, showing how they can be classified and describing the basis of the major categories of technology that have been adopted in their formulation.

17.4.1 Active, Semi-Passive, and Passive Systems

One classification schema that can be adopted is that of passive, semi-active, and active decision support systems. Active systems include those in which one or more feedback control mechanisms are employed, for instance to administer drug therapy in a closed-loop configuration. Examples have included the administration of sodium nitroprusside to the patient in intensive care following cardiac surgery as a means of controlling blood pressure [3]. The automatic controller in the feedback loop that comprises patient, sensor, controller, and actuator (drug infusion pump) adjusts the rate of drug delivery to achieve a given pattern of control in accordance with rules or algorithms proposed by the expert clinician. In this feedback configuration, the clinician is able to override the automatic mechanisms, so as to retain responsibility for decision making as appropriate.

Semi-active systems include alarms and alerts, indicating to the clinical decision maker that a problem may exist with the patient that requires a decision-making input. Examples span the full clinical spectrum, from those in the context of delivering anesthesia to the patient in the operating room to those now being adopted in situations of home telecare. These could include the patient with diabetes transmitting home-monitored blood glucose data to the clinical center, where automated analysis of such data could flag undesirable patterns of hyperglycemia or hypoglycemia that require a clinical decision regarding changes to the insulin regimen of the patient in question.

Passive systems include those that operate in a consulting mode and those that adopt a critiquing style. Consulting systems are, in essence, advisors. They accept patient-specific data, ask questions, and suggest problem-specific recommendations as requested [1]. In addition to needing patient-specific data, critiquing systems require the clinician to provide the diagnostic conclusion that is thought to be correct or the patient management action that is intended to be taken. The decision support system evaluates the conclusions and proposed decisions and expresses agreement or critique as appropriate. Whatever the outcome, the critiquing system can suggest alternatives.

If the decision support system is linked to the electronic medical record that contains patient data and the management actions being proposed by the clinician, then that system can automatically evaluate patient management. This integration of decision support system and medical record can reduce the need for repeated data entry, because patient data are available from the electronic record or have been entered directly from clinical instruments, for example in the critical care environment [4].

It should be noted that critiquing systems, when linked to electronic records, can also operate in semi-active mode. In this way, they act like watchdogs by providing warnings when they observe specific events or data combinations in the patient's clinical history that are likely to result in unwanted effects. They can also produce reminders to focus the attention of the clinician on problems that might otherwise be overlooked, or to suggest alternative actions if appropriate. A classic example would be the case of a clinical laboratory system that flagged abnormal laboratory values or reminded the physician of the need to retest some laboratory values [5].

17.4.2 Analytic (Mathematical) Models

Analytic models comprise qualitative and quantitative mathematical models that can aid the decision maker by predicting the future state of the patient, and its evolution, based on the present state and a representation of what has passed (system dynamics). Such models include representations of system behavior that allow test signals to be used so that the response of the system to various disturbances can be studied to make predictions of the evolving patient state [6].

Qualitative models can be used to explore time-dependent behavior by representing the patient state trajectory as a set of connected nodes, where the links between the nodes reflect transitional constraints placed upon the system. Classic examples of such qualitative dynamics include the early work of Kuipers [7]. Such models can be used to support decisions involving assessment of patient state and therapy planning.

In the case of diagnostic assessment, the causal mechanisms of the disease process are defined by the precursor nodes and the pathway to the node (decision) of interest. With regard to therapy planning, the desired therapy can be prescribed by investigation of the utility values associated with each link in the disease–therapy relationship. These utility values refer to a cost function, where cost can be defined as the monetary cost of providing the treatment and cost benefit to the patient in terms of efficiency, efficacy, and effectiveness.

Quantitative models typically provide their representation in the form of differential and auxiliary equations. When the parameters of such a model are tuned to match the characteristics of the individual patient, simulation studies enable the model to be run under different scenarios, for example different regimens of drug therapy. In this way, the clinical user can gain insight about the impact of such change in therapy on key physiologic variables, another form of decision support.

An example of such quantitative modeling is that of Coleman and Gay [8] in relation to the human cardiovascular system. The model can predict changes in arterial pressure, cardiac output, and total peripheral resistance in response to several challenges. It is worth noting, however, that when examining the behavior of physiologic systems in the context of decision support, clinicians generally prefer to use symbolic descriptions of continuously varying quantities, such as direction of flow or increasing or decreasing quantities rather than dealing with numeric values as such. In addition, they also tend to replace mathematical equations by qualitative or logical functional constraints that hold among variables and govern their temporal evolution [1].

Quantitative mathematical models can also form an integral part of an active decision support system. One example is that of a nonlinear model-based predictive controller that has been developed to maintain normal levels of blood glucose in subjects with type 1 diabetes during fasting conditions such as an overnight fast [9]. The controller employs a compartmental model which represents the glucoregulatory system. It includes submodels for the absorption of subcutaneously administered insulin and for gut absorption. The controller uses Bayesian parameter estimation to determine time-varying model parameters. The model makes predictions of blood glucose over the following 15-minute interval. These predictions are used to adjust the dosage of insulin administered by pump to the patient to achieve and maintain the desired blood glucose level.

17.4.3 Decision Theoretic Models

Decision theoretic models in the clinical context include algorithms, decision trees, and influence diagrams. Let us consider each of these.

The clinical algorithm is a means of structuring the diagnostic or therapeutic decision problem in the form of a classification tree. The root of the tree represents some initial state, and the branches yield the different options available. In its operation, the choice points of the clinical algorithm are assumed to follow branching logic. Hence, the algorithm consists of a set of questions that must be collectively exhaustive for the particular clinical domain. Equally, the responses available to the clinician at each branch point in the tree must be mutually exclusive. This rigid formalism means that there is only a limited range of well-defined and constrained clinical decision problems that can be addressed, given the lack of flexibility in the formalism. Nevertheless, the approach has been widely applied, with examples ranging from acid–base disorders [10], the diagnosis of mental disorders [11], and at the site of occurrence of major accident and trauma [12].

The decision tree appears in some ways to be of similar structure to the clinical algorithm but is more rigorous as a means of classification. Key features are the calculation of likelihood and cost-benefit for each choice to provide a quantitative measure for each available option. This enables optimization procedures to be used to gauge the probability of success for the correct diagnosis or for a beneficial outcome on the basis of the chosen therapeutic action.

In terms of structure, the decision tree also differs from the clinical algorithm in that the tree contains two types of branching points. At a decision node, the clinician must decide on which choice (branch) is appropriate for the given clinical scenario. At chance nodes, the responses available have no clinician control. For example, the response may be due to patient-specific data, and outcome nodes define the chance nodes at the leaves of the decision tree. That is, they summarize a set of all possible clinical outcomes for the chosen domain [6]. An example of a simple decision tree is shown in Figure 17.4.

The possible outcomes from each chance node must obey the rules of probability and add up to unity; the probability assigned to each branch reflects the frequency of that event occurring in a general patient population. It follows that these probabilities are dynamic (i.e., can be updated), with accuracy increasing as more evidence becomes available. A utility value can be added to each of the outcome scenarios. These utility

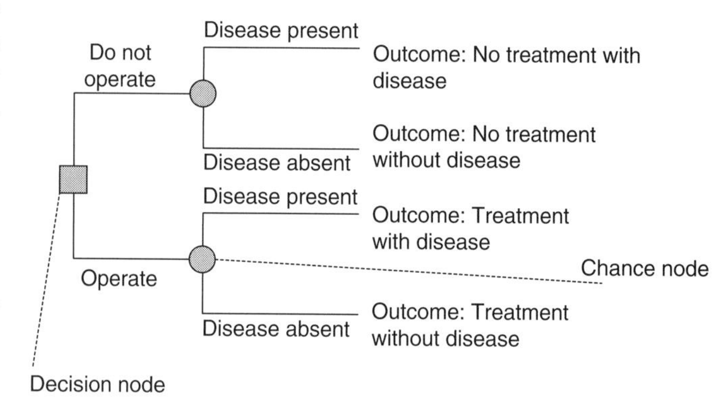

FIGURE 17.4 Simple decision tree.

measures reflect a trade-off between competing concerns, for example, survivability and quality of life, and may be assigned heuristically.

The detailed methodology of this analysis using decision trees is described in many classic works, including those of Sox et al. [13] and Llewellyn and Hopkins [14], which contain many worked examples. However, a very wide range of practical clinical applications is now to be found in mainstream clinical journals, particularly *Medical Decision-Making.* One of the main reasons for the increased interest in this modeling approach has been the desire to contain the costs of medical care while maintaining clinical effectiveness and quality of care. In the United Kingdom, this is reflected in the establishing of the National Institute for Health and Clinical Excellence (NICE) (www.nice.org.uk). Cost-effectiveness analysis is an extension of decision analysis and compares the outcome of decision options in terms of the monetary cost per unit of effectiveness. Thus, it can be used to set priorities for the allocation of resources and to decide between one or more treatment or intervention options. It is most useful when comparing treatments for the same clinical condition. Detailed descriptions of cost-effectiveness analysis and its implications can be found in Gold et al. [15] and Sloan [16].

Influence diagrams provide a further form of representational model as a means of solving decision problems. As a graphic model, they in essence complement the decision tree, providing a different graphic representation of the same mathematical model and operations on it. The classic examples of the role of this modeling approach being applied in the medical domain are to be found in Owens et al. [17] and Nease and Owens [18].

17.4.4 Statistical Models

Statistical models are widely adopted in decision support systems, with the more important approaches including the following.

Database searching: large clinical databases when appropriately examined can yield statistical evidence that is of value diagnostically. In some cases such evidence can form the basis of rule induction for the creation of expert systems. However, the most direct approach for clinical decision making is to determine the relative frequency of occurrence of an entity, or more likely group of entities, in the database of past cases (the frequentist approach). This enables a prior probability measure to be estimated [19]. A drawback of this simple, direct approach to problem solving is that the greater the evidence that is available, the fewer the number of matches in the database that will be found. This is counter to the common wisdom that more evidence leads to an increase in probability of a diagnosis being found. Moreover, such a simple approach lacks any weighting of the individual pieces of evidence. Hence it is not possible to judge which are more significant in relation to patient outcome.

Database searching was one of the earliest statistical approaches to be adopted. However, interest in this approach has been rekindled with the completion of the human genome sequence. For example, methods are being developed for finding data, such as single nucleotide polymorphisms, in the many genetic database resources that are distributed throughout the world [20].

Regression analysis can be used to model the relationship between a response variable of interest and a set of explanatory variables. This involves adjustment of the regression coefficients, which are the parameters of the model, until a best fit to the data set is achieved. This type of model improves on the use of relative frequencies, as logistic regression explicitly represents the extent to which elements of evidence are important in the value of the regression coefficients. Examples can be found in a wide range of clinical applications, with one of the classics being found in the domain of gastroenterology [21].

Statistical pattern analysis is an important tool in support of decision making. In this way, the recognition of patterns in data can be formulated as a statistical problem of classifying the results of clinical findings into mutually exclusive but collectively exhaustive decision regions. This enables physiologic data to be classified as well as the disease states that they give rise to and the therapy options available to treat the disease. The approach has found widespread clinical application, ranging from the analysis of cardiac rhythms to the treatment of head injuries. The methods used to distinguish patterns in data rely on discriminant analysis, which involves obtaining measures of separability between class populations.

The pattern recognition problem can be considered as a two-stage process as shown in Figure 17.5. The pattern vector, P, is an n-dimensional vector derived from the data set used. Let us define the pattern space as Yp, which is the set of all possible values P may assume. The pattern recognition problem can then be formulated as finding a way of dividing Yp into mutually exclusive and collectively exhaustive regions. In the case of analysis of the electrocardiogram, for instance, the complete waveform could be used to perform classifications of diagnostic value. A complex decision function would probably be required in such cases. In some cases, however, it might be appropriate to simplify the pattern vector to investigation of

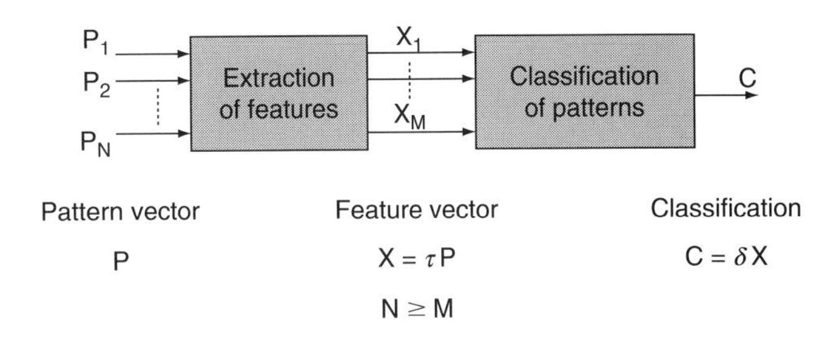

FIGURE 17.5 The basis of pattern recognition.

subfeatures within a pattern. For example, in cardiac arrhythmia analysis, only the R–R interval of the electrocardiogram is required, which allows a much simpler decision function to be used. This may be a linear or nonlinear transformation process:

$$X = \tau P,$$

where X is termed the feature vector and τ is the transformation process [22].

Just as the pattern vector P belongs to a pattern space YP, so the feature vector X belongs to a feature space YX. Because the function of feature extraction is to reduce the dimensionality of the input vector to the classifier, some information is lost. Classification of YX can be achieved using numerous statistical methods including: discriminant functions (linear and polynomial), kernel estimation, k-nearest neighbor, cluster analysis, and Bayesian analysis [22].

Markov models are generally used as a means of representing stochastic processes (random processes that evolve over time). As such, they are particularly suitable to describe the progression of diseases that can be divided into distinct states, with transition probabilities being assigned for transitions between these states. A simple example of a three-state Markov model is depicted in Figure 17.6, showing the possible transitions between the states characterized by the relevant probabilities.

A number of assumptions underpin the Markov model. It is assumed that there are a finite number of distinct health states and that the state transition probabilities remain constant. Moreover, history is ignored, meaning that the probability of moving from one state to another is independent of any history prior to being in that initial state. It is a discrete time model, meaning that transitions between states occur at equal, specified time intervals.

As well as being used to model disease progression in a patient, Markov models can also be used to model hospital stay as a set of daily transitions among the different states. Disease states represent clinically and economically important events in the disease process. Time is divided into discrete periods termed a Markov cycle. Probabilities of transitions refer to one Markov cycle. Suppose that estimates of parameters such as resource use (costs) and health outcome consequences (e.g., quality-adjusted life estimates) are attributed to the states and transitions (e.g., drug effectiveness, discount

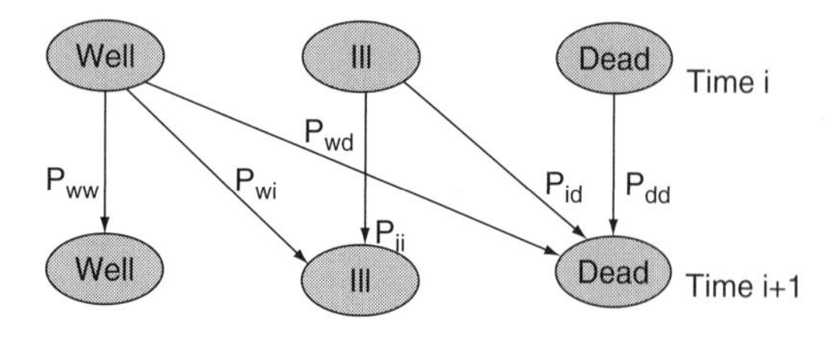

FIGURE 17.6 Three-state Markov model.

rates) in the model. If the model is then run over a large number of cycles, estimates can be made of the long-term costs and outcomes associated with the disease and drug interventions [23]. This is one example of the use of such a model in a decision support role, enabling the consequences of parameter changes to be investigated.

Bayesian analysis is one of the most widely adopted approaches used in the development of decision support systems. In essence, it is an example of a parametric method of estimating class conditional probability density functions. Clinical knowledge is represented as a set of prior probabilities of diseases to be matched with conditional probabilities of clinical findings in a patient population with each disease.

The classification problem becomes one of a choice of decision levels, which minimizes the average rate of misclassification or minimizes the maximum of the conditional average loss function (the so-called minimax criterion) when information about prior probabilities is not available. Bayes' rule is the optimal decision rule that minimizes the average rate of misclassification. It serves as the inference mechanism that allows the probabilities of competing diagnoses to be calculated when patient-specific clinical findings become available.

Bayesian classification is not dependent on the availability of a large clinical database of past cases. Hence, less time is taken in reaching a decision as compared with other database search techniques. Moreover, classification errors resulting from the use of inappropriate clinical inferences are quantifiable. A limitation of the approach, however, is the assumption that the disease states are considered to be complete and mutually exclusive, assumptions which in reality may not be valid.

The early classical clinical application was that of De Dombal and colleagues [24] in relation to the diagnosis of acute abdominal pain. It is worth noting that from the date of the above publication it took a further 20 years for it to be accepted via a multicenter multinational trial.

Bayesian methods continue to find widespread application in the clinical domain. In some instances, this is as a single method. In other cases, it forms part of a multimethod approach, for example, for insulin sensitivity [25]. It is also widely adopted for the analysis of clinical trial data [26, 27].

Bayesian decision theory also finds application in health care technology assessment [22] and is used to analyze economic models that aid decisions about whether new health care technologies should be adopted and the resources that should be allocated [28]. Examples relating to clinical trials and health care technologies more generally can be found in O'Hagan and Stevens [29] and Claxton [30]. These include both the issue of whether a new technology should be adopted immediately and whether there is a need for more research prior to reaching a decision.

The Bayesian approach has a number of advantages over the frequentist approach. It allows knowledge to be accumulated and updated by making use of the prior distribution. It yields more flexible inferences and emphasizes predictions rather than hypothesis testing. It is relatively simple to estimate probabilities involving multiple end points. Finally, it provides a solid theoretical framework for decision analyses and decision support.

17.4.5 Rule-Based and Network-Based Systems

Rule-based systems represent knowledge to inform the decision-making process in the form of production rules. One of the classic early systems was MYCIN developed by Shortliffe [31], and this will be used to illustrate the underlying approach. The application domain in this case was the identification of organisms causing microbial infection and the recommendation of appropriate treatment.

The causal relationships between individual items of factual knowledge held in the knowledge base are represented by production rules of the following form: If premise assertions are true, then consequent assertions are true with confidence weight X.

The measure of certainty attached to each rule aids the reasoning strategy of the system. This measure ranges from -1 corresponding to complete disbelief to $+1$ corresponding to complete belief in the consequent assertions. The assertions can be Boolean combinations of clauses, each of which consists of a predicate statement triple: attribute, object, value. For instance, (Gram's stain, *Escherichia coli*, gramneg) indicates that the Gram's stain of the *E coli* organism is gram-negative.

The formalism employed enables a simple control strategy of goal-directed backward chaining of the production rules to be adopted. In this way, the first rule to be evaluated is the one that contains the highest level goal. For MYCIN, this is to determine whether there are any organisms, or classes of organisms, that require therapy. To deduce the need for therapy involves knowledge of the infections, which is usually unknown initially. Hence, the system tries to satisfy subgoals that originate in the premise of the top goal that will allow the infections to be inferred. Rule chaining is the name given to the process in which the hierarchy of production rules is linked together. The premise portion of each subgoal rule fires a new set of subgoals. This procedure is continued until the most fundamental level of the hierarchy is reached, where the rules become assertions that can only be confirmed or denied by directly questioning the user for the appropriate information.

MYCIN determines the significant microbial infections on the basis of assessing the overall certainty factor that combines the individual degrees of confidence associated with each of the production rules. Then the organisms that account for the infections are found deterministically. Having completed this identification phase, the system can then proceed to recommend an antimicrobial therapy. This involves the MYCIN therapy selector using a description of the infections present and the causal organisms together with a ranking of drugs by their sensitivity and a set of drug-preference categories.

The algorithm used within the therapy selector also calculates the drug dosage needed and contains knowledge to modify the value in light of other complications, for instance renal failure. The therapy selector is able to accept and critique a treatment protocol proposed by the user. Equally, it is able to generate explanation and justification for its treatment selection. Why queries are dealt with by displaying the rule that it is trying to imply. A further why query is answered by ascending the goal-tree hierarchy. How queries are interpreted as the chain of rules that are fired to reach the particular conclusion. Repetition of a how query is answered by descending the goal-tree hierarchy.

The advantage of a production rule system is that each rule is a small quantum of knowledge, with each being independent of all the others. This means that changing or adding knowledge to the MYCIN knowledge base is relatively easy. Also, having a modular data structure facilitates the addition of new rules. A disadvantage lies in the fact that disease states cannot always be adequately described by a rule. Also, it may not always be possible to map a series of desired actions into a set of production rules. Moreover, although new knowledge can be added by inserting a new rule, this may not interact with the existing rule set in the anticipated manner.

In contrast, the semantic network provides a graphic representation of the knowledge base. One of the early classics of this type was CASNET, an expert system for consultation in the diagnosis and treatment of glaucoma [32]. In it, the relevant knowledge for patient-specific reasoning in the decision-thinking process is encompassed in a causal-associational network model of the specific disease process. The semantic network consists of nodes connected by links, where the nodes correspond either to the condition or action part of the rules and the links are inferences between the two. The model of disease is separate from the decision-making strategy, which allows for easier updating of both data structures.

The CASNET model has a descriptive component that includes observations, pathophysiologic states, causal relationships, disease categories, and treatment plans. The observations consist of symptoms and laboratory tests, together forming the direct evidence that a disease is present. The pathophysiologic states describe internal abnormal conditions or mechanisms that can directly cause the observed findings. The causal relations between states are of the following form:

$$n_i \xrightarrow{a_{ij}} n_j,$$

where n_i and n_j are pathophysiologic states and a_{ij} is the causal frequency with which state n_i, when present in a patient, leads to state n_j. Each disease category consists of a pattern of states and observations.

Also included in the CASNET model are the decision rules, which state the degree of confidence with which an inference of a pathophysiologic state can be made from an observed pattern of findings. This translates to the following:

$$t_i \xrightarrow{Q_{ij}} n_j,$$

where t_i is a finding or observation or Boolean combination of findings, n_j is the pathophysiologic state and Q_{ij} is a number in the range -1 to $+1$ representing the confidence with which t_i is believed to be associated with n_j. Q is the certainty factor which indicates the strength of belief that the patient is in state n_j. Dependent upon the threshold value that is set for it, this factor will then confirm or deny the patient as being in that state, or else the situation will remain undetermined. Rules can then be established for connecting disease categories and pathophysiologic states to treatment protocols. These rules take the form of a classification table that consists of ordered triples:

$$(n_1, D_1, T_1), (n_2, D_2, T_2), \ldots, (n_i, D_i, T_i),$$

where n_i is the pathophysiologic state, D_i is the disease process resulting from it, and T_i is the preferred treatment regimen(s) corresponding to that disease state.

The pathogenesis and mechanisms of the disease process are described in terms of cause and effect relationships between pathophysiologic states. Hence, complete or partial disease processes can be characterized by pathways through the network. When a set of cause and effect relationships is specified, the resulting network can be described as an acyclic graph of states. The state network is defined by a four-tuple (S, F, N, X). S is the set of starting states (that is states having no antecedent causes), F is the set of final states, N is the number of states visited between S and F, and X represents the causal relationships between the states visited, in the form of a list.

One of the merits of the CASNET system is its ability to present alternative expertise derived from different clinical consultants. This, taken together with the well-defined nature of the clinical problem being tackled, namely glaucoma, results in the accuracy of diagnosis that led to the relative success of this pioneering system. Evaluation studies revealed a diagnostic accuracy greater than 75% for very difficult cases and an overall figure greater than 90% across the whole spectrum of cases investigated. However, that CASNET's clinical utility was not particularly high limited its adoption in the routine clinical setting. Nevertheless, it has proved to be a powerful research tool in advancing AI in medicine, particularly in catalyzing further advances in causal modeling.

17.5 Decision Support Tools

Software programs written specifically to aid the decision-making process have been in existence since the 1970s, with MYCIN being a prime example as discussed above. While these tools have been applied to a variety of medical domains, they have common essential features that will now be introduced.

17.5.1 Types of Knowledge

The actual knowledge used in a decision support tool will be specific to the application domain, but the source of the knowledge can be from the following:

- *Observations*: Observing the domain experts as they carry out their duties. This approach can often result in the discovery of new knowledge.
- *Academic knowledge*: Printed information from published material in academic publications such as journals and conference proceedings is a useful source of knowledge. Care should always be given to ensure the accuracy of any published information, particularly from unreviewed sources.
- *Experimental*: Results from conducting tests and trials. This is a useful method of collecting knowledge as control is maintained; however, costs and time considerations, in terms of both ethical approvals and collecting a suitable number of subjects, should not be underestimated.

17.5.2 Knowledge Acquisition

The acquiring of knowledge that is both representative of the domain being supported and accurate can be a difficult task but one that is a main contributory aspect to the success of a CDSS. Aspects of knowledge acquisition include the following:

- *Elicitation*: The discovery of aspects of the CDSS domain that are not immediately obvious or a detail so commonly part of the domain process that the experts do not consider mentioning it, yet it has an important role in a CDSS.
- *Domain experts*: The knowledge from experts is a vital ingredient of a CDSS. This can be obtained from interviews, questionnaires, or observations. A nonexpert can play a useful role in extracting knowledge from experts by asking questions to establish the real need for each procedure.
- *Information to users*: Techniques from knowledge engineering and knowledge management can be used to ensure that the final output is of value to the user of a CDSS.

17.5.3 Knowledge Representation

Knowledge representation is one of the key elements of a CDSS system. It is the method by which collected information is stored and presented to the CDSS. A number of methods have been applied for this task, including the following:

- *Heuristic systems*: These are less well defined and can be based on experimental or critical knowledge [33, 34]. A heuristic system can be considered as the knowledge of good practice, good judgment, or plausible reasoning.
- *Artificial neural networks*: Simple models of the human brain can be used to model knowledge. The strength of these tools lies in their ability to be trained for specific domains [35].
- *Expert systems*: A series of if, then rules can be used to define a knowledge base. An inference engine is then used to navigate the knowledge base using defined rules, such as forward to backward chaining [31].
- *Computer languages*: Formal, specialized computer languages, such as PROLOG, that use logic statements to define and navigate a knowledge base.

17.5.4 The Interface

The interface is the medium by which the user interacts with the CDSS. It is a frequently overlooked aspect of CDSS design but is one that has a direct bearing on the acceptance of a system. A CDSS with excellent embedded knowledge and representation will fail to be accepted if the users find the method of accessing the system not to be intuitive and user friendly. The interface should allow the user to both input data and knowledge into the CDSS and to receive the results of the decision-making process fast enough to be of use in the supported domain. The user interface needs to be designed with consideration for the following:

- *Usability*: The ease of entering commands, defining scenarios, and presenting the results in an easily understandable form
- *Acceptability*: The ease of use, with consideration of the environment of application. For example, a CDSS used in a high-dependency unit would require different features than one designed for a General Practicioner (primary care physician) surgery.

17.6 Decision Support Systems in the Hospital and Other Health Care Settings

17.6.1 Examples of Clinical Decision Support Systems in Hospital (Including the Imaging Context), General Practice, and the Community

This section provides some examples of CDSS in the hospital setting. The examples given are not an exhaustive list, but at the time of writing were considered to be a good representation of the current level of implementation of the technology. A detailed list of both current and superseded systems can be found at the open clinical Website (http://openclinical.org/dss.html).

Automedon/SmartCare–Ventilator Management

This CDSS automates and effects clinical guidelines for ventilator management in high-dependency environments [36]. Knowledge engineering techniques have been applied to elicit and model the clinical guidelines with expert system methodologies. These are rule based with forward chaining coupled with temporal reasoning. In addition, software engineering techniques are employed for automatic source code generation.

The system has been approved to meet International Standards Organization standards and successfully implemented in a clinical environment. Further evaluation in a multicentered study is also being undertaken. Dräger is embedding this CDSS in its range of ventilators.

GIDEON–Diagnosis and Treatment of Infectious Diseases

GIDEON is a Web-based decision support tool, using Bayesian modeling, for the diagnosis of global infectious diseases [37]. GIDEON contains four seamlessly integrated, operational modules. Diagnosis uses clinical signs and symptoms to suggest possible infections; epidemiology focuses on the epidemiologic aspects of the disease database; therapy suggests treatment regimens for the disease database; and microbiology can be used for entering laboratory results to assist in the diagnosis process.

LISA–Treatment of Childhood Leukemia

LISA is a Web-based CDSS for providing dosage decision support for the drug regimen in the treatment of acute childhood lymphoblastic leukemia [38]. The system implements the dose adjustment rules as specified in Medical Research Council guidelines. LISA is a rule-based system and uses the PROforma decision engine for determining the course of treatment [39]. Doses have to be constantly monitored, and the treatment regimen is unique for each patient, so this application is ideal for a CDSS.

PERFEX–Expert System for Automated Interpretation of Cardiac Single Photon Emission Computed Tomography Data

PERFEX is a rule-based expert system for automatic interpretation of cardiac SPECT data [40]. This system assists in the diagnosis of coronary artery disease by suggesting the extent and severity of the condition. It provides a patient-specific summary of the state of the main cardiac arteries together with related information. Knowledge-based methods are used to process and map the 3D visual information into symbolic representations, which are used to assess the extent and severity of cardiovascular disease both quantitatively and qualitatively. The knowledge-based system presents the resulting diagnostic recommendations in both visual and textual forms.

17.6.2 Decision Support and Information Systems

An example of the way in which the concepts of information and decision support are converging can be found in the current activity relating to the provision of electronic records. Many countries are developing electronic patient record (EPR) systems to provide a paperless, easy-to-access, location-independent realization of a patient's medical history. A CDSS can interface with the EPR to provide enhanced patient care. The advantages of bringing these elements together include the following:

- Reduction of clinical errors as a result of, for example, incorrect drug doses, patient allergies, or drug interactions
- Reduction of superfluous diagnostic tests and hence savings of patient anxiety, time, and cost
- Enhanced patient care by providing a reminder for clinicians and care givers for repeat prescriptions and screening tests
- Support for clinicians in extended roles

Unfortunately, a number of potential problems are also evident from this approach:

- The CDSS could rely on obsolete knowledge or flawed reasoning. This could lead to potentially serious or even fatal recommendations to the clinician.
- The quality of the advice from a CDSS is difficult to access. Evaluation of a CDSS is not always rigorous, and so the advice from scenarios that do not exactly match the knowledge base may be unpredictable.
- There is currently no framework for approving a CDSS. With the potential availability over the World Wide Web, the number of future systems and potential users could be large.
- High technology applications rapidly become redundant with the progress of computer hardware and software.

17.7 Health Care Education Applications

The ideas and concepts introduced and discussed so far have all been concerned with using patient-related knowledge to provide useful diagnostic information of some form. An alternative use of a CDSS is for training and educating health care professionals and students. In this mode, a CDSS can present scenarios for the student to analyze and provide feedback on performance and accuracy. The following systems are presented to provide examples of applications in this domain.

17.7.1 Simulators for Health Care Students, Health Care Professionals, and Patients

17.7.1.1 High-Dependency Medicine: MacPuf

MacPuf (http://www.chime.ucl.ac.uk/resources/Models/macpuf.htm) is a freely available model of the human respiratory system that allows the student to explore the management of a wide range of medical scenarios. A large number of model variables allow the sophisticated simulation and demonstration of problems in oxygen delivery without harming patients [41].

17.7.1.2 Chronic Diseases: AIDA2

AIDA (www.2aida.net) is a free simulator (both online and downloadable) for the education and teaching of glucose–insulin interaction and insulin dosage and dietary adjustment in diabetes mellitus [42]. The program is based on a model of the human glucose–insulin interaction and can be used for simulating the effects of dietary changes and blood glucose levels for patients with type I diabetes. A knowledge base can be used to identify problems in the included case scenarios.

17.7.2 Provision of Information for the Patient

17.7.2.1 Web-Based Provision–NHS Direct

NHS Direct Online (www.nhsdirect.nhs.uk) is a Web-based CDSS that patients can use to diagnose common health problems and learn about treatments. The interface provides simple pictures or questions to help with the diagnosis. For additional advice, the CDSS is supported by a telephone link to a nurse. In addition, an online health illustrated encyclopedia provides information on a large number of illnesses and conditions, tests, treatments, and operations.

17.7.2.2 The Home Telecare Setting–Decision Support for Lung Transplant Recipients

This work resulted in the development of a computerized rule-based decision support algorithm for the identification of potential acute bronchopulmonary events in lung transplant patients. Weekly home monitoring of forced expiratory volume and respiratory symptoms was used to classify a patient as being in either a stable (improving) state or one that requires watching. The system was well evaluated on 155 patients with sensitivity and specificity greater than 90% and was comparable to the standard human clinical review of the same weekly home-monitoring data [43].

17.8 Verification, Validation, and Evaluation

In essence, the process of evaluating a CDSS is all about assessing its worth or value in relation to the purpose for which it is designed. Because of this relationship between evaluation and purpose, we must start thinking about the process of evaluation from the outset. The very purpose for which the decision support system is designed helps to determine the criteria for assessing the value of the system: Does it meet its intended purpose?

The literature is full of references to the terms verification, validation, and evaluation, and there may be a degree of confusion about their specific meanings.

Verification is the act of checking whether the contents of our decision support system are technically correct. In the case of a rule-based system, this would involve checking that each of the rules is individually correct. Once this is the case, the validity of the system can be assessed. This might typically in the first instance involve examining a series of cases where the outcome is known. The recommendations, say, of a consulting system would be compared with the actual recommendations made by expert clinicians in those clinical cases. A high level of agreement in this retrospective validation test would increase the level of confidence in the validity of the decision support system.

The concept of evaluation is broader than that of verification or validation. Many definitions have been proposed. Examples include: the systematic application of social research procedures to judge and improve the way information resources are designed and implemented; the process of describing the implementation of an information resource and judging its merit and worth; and an overall assessment of the value of a decision support system in relation to its intended purpose. In this context, a clinical decision support system would be an example of an information resource. A good overall review of methods for evaluation can be found in Friedman and Wyatt [44].

17.8.1 A Framework for Evaluation

What is clear is that a proper systemic framework is required for the evaluation process. To begin with, a full evaluation should reflect the totality of the decision support system's intended purpose in terms, for instance, of settings, stakeholders, and perspectives.

Possible settings could range from use in community or primary care environments to use in a hospital clinic, laboratory, or pharmacy. Stakeholders might include some or all of patients, health care workers, health managers, regulatory bodies, public health clinicians, taxpayers, insurance companies, and the health care industry.

Differing perspectives also give rise to different questions being asked as part of the evaluation process. From the patient's perspective, issues would include whether the system was safe and whether it would be helpful in the context of his or her clinical condition. The health care user might also be concerned about whether the system was fast (in relation to the time available for a patient consultation) and accurate. The purchaser of a system would be concerned about its cost benefit and its safety and reliability. For the developer, key issues include whether it works as intended and whether those purchasing it will use it. A further point is that the evaluation issues change over the course of the development and testing processes. This is illustrated in Figure 17.7.

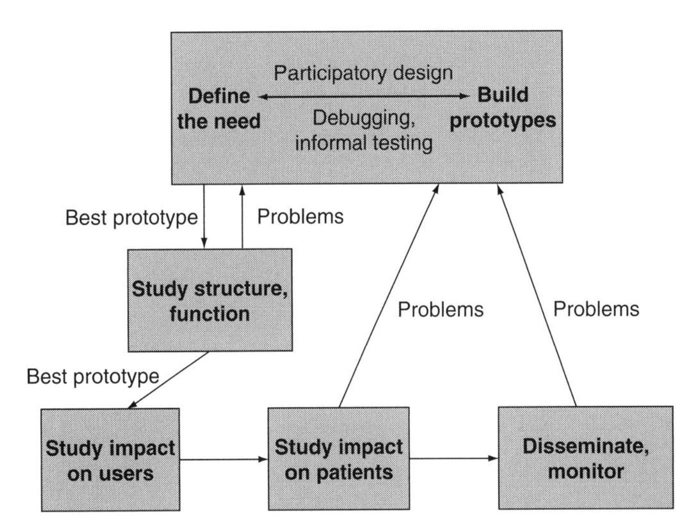

FIGURE 17.7 Changing evaluation issues during the development process.

17.8.2 Evaluability, Formative Evaluation, and Summative Evaluation

One of a number of useful approaches to the evaluation problem that has relevance in the context of decision support is that proposed by Bashshur [45]. The framework that he adopts makes use of the concepts of evaluability, formative evaluation, and summative evaluation.

The first stage of evaluability involves clearly defining the problems and issues to be evaluated; specifying the evaluation criteria, and setting objectives in terms of the stakeholders' expected benefits and costs. Formative evaluation includes a description of system design and implementation, stakeholder analysis of intermediate and short-term effects, where this also provides a focus for additional data collection or modification of the system being evaluated. Summative evaluation involves determining the ultimate effects of the system on health outcomes and making recommendations for modifying the process of health care delivery in the light of the impact of the decision support system.

An example of a model demonstrating the interaction of formative and summative evaluation is depicted in Figure 17.8.

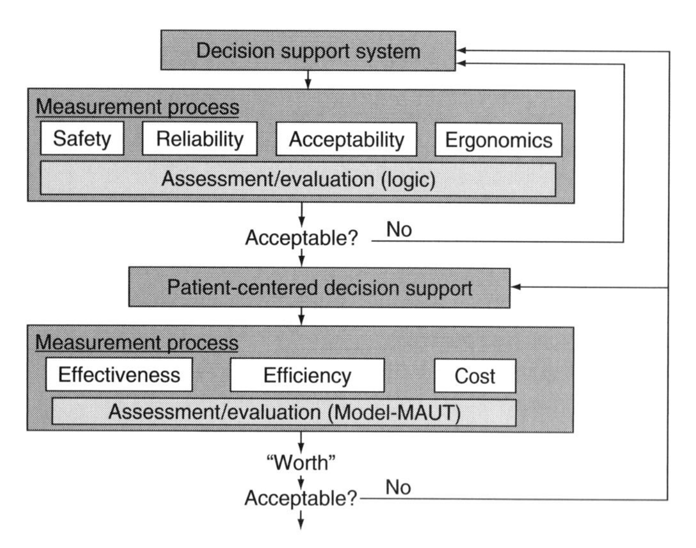

FIGURE 17.8 Model of evaluation.

The first layer of evaluation results from the first measurement process, essentially corresponding to formative evaluation in the terminology of Bashshur [45]. Here a number of criteria need to be satisfied as a necessary precursor to investigating the overall value or worth of a decision support system. For example, necessary requirements to be met might include those of safety, reliability, and usability of hand-held computer implementation that has been designed to offer decision support in the home setting. As such, these would be the criteria upon which assessment and evaluation would be based at this formative level. Since all such criteria must be satisfied, the evaluation in essence corresponds to a logical AND function.

The actual implementation of these criteria will depend on the particular context. For example, one measure of reliability of a system would be the mean time between failures, where in this context failures might represent instances where the decision maker disagreed with the recommendation being proposed by the decision support system.

Once these necessary conditions have been met, evaluation continues via an analysis of the results of subsequent rounds of measurement processes (data gathering exercises). Here the assessment and evaluation processes become more complex. For instance, at any given level the criteria involved might relate to measures of clinical effectiveness, affordability, and organizational impact, or as effectiveness, efficiency, and cost, as shown in Figure 17.8. Hence there is the need to address the problem of defining overall worth or value.

The issue here is not as simple as was the case with formative evaluation. In that case, all the necessary criteria needed to be satisfied. Here we are dealing with a set of criteria that are more difficult to bring together, involving as they do diverse variables relating to both clinical outcomes and financial costs. One possible way of achieving an overall measure of worth at this stage is to adopt a model based on multi-attribute utility theory or something similar. However, this is still an ongoing topic of research.

A number of features of the model presented are worth noting. First, the overall process of evaluation is very much iterative in nature. If at the stage of formative evaluation one of the necessary criteria fails to be met, then it is necessary to return to the design of the system. Similarly, failure at the summative stage will probably also mean a re-examination of the system design. In terms of the formative phase of evaluation, no issues of economic value are addressed (as indicated by the corresponding shaded area in Figure 17.9). These are included in the concerns at the summative stage. On the other hand, all the technical issues are addressed in the formative stage. These features are depicted in Figure 17.9. This figure shows how the

Criteria / Variables	Safety	Reliability	Acceptability of design	Ergonomics	Effectiveness	Efficiency	Cost
Medical	✓	✓			✓	✓	
Economic							✓
Technical	✓	✓	✓	✓			
Organizational	✓	✓	✓			✓	
Social			✓		✓		

FIGURE 17.9 Relationship between evaluation criteria and variables.

various criteria relate to the different types of variables that need to be considered in the evaluation process [46].

What is absolutely clear is that good evaluation practice must be a multifaceted, multidimensional, multi-attribute process. The essence of this is depicted in Figure 17.10. This demonstrates that to gain a measure of the overall worth or value of a decision support system, it is necessary to include a wide range of perspectives. As indicated in this figure, these might typically include the technical, clinical, organizational, and economic, as well as those of the key individuals involved, including both the patient and health care professionals.

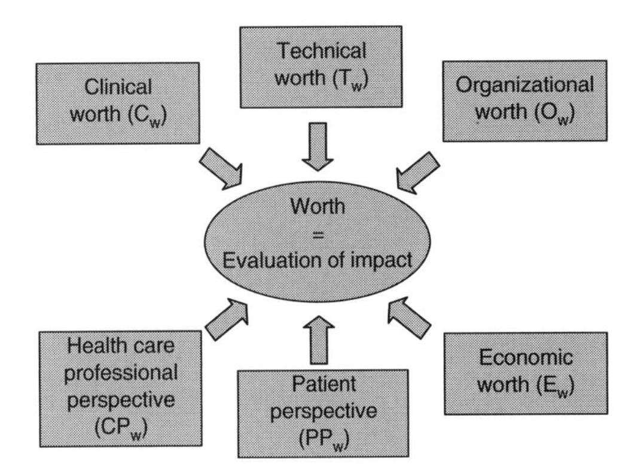

FIGURE 17.10 Evaluation of impact.

17.8.3 The Ethical and Legal Dimensions

Among the impacts that need to be addressed when considering the overall worth of a decision support system are those relating to legal and ethical issues. Here one of the key issues is the differentiation between decision making and decision support.

For the most part, decision support systems are designed to do just that; to support, but not to replace, the decision maker. In other words, the final decision remains with the health care professional. The decision support system can thus be viewed as providing advice that can be either accepted or rejected (or queried if this facility is built into the software). As such, it in some ways parallels the way in which a clinician might seek the advice of a colleague before arriving at a decision.

The systems that come nearest to being decision making rather than decision support are active systems as discussed earlier in section 17.4.1. Examples include closed-loop control of one or more physiologic variable in the context of critical care medicine, for instance blood pressure or blood glucose concentration. However, in such situations there is always provision for override by the clinical decision maker.

Legal liability is a crucial issue in relation to technologic solutions in the decision-making context. How such issues are treated depends very much on the particular national legal framework. For instance, in the United Kingdom, liability would generally be defined in terms of case law; for example, as a result of a case being brought to court following an allegation of negligence by a clinician where decision support technology had been involved. It is worth noting, however, that if technology can be shown to be beneficial, the decision maker would be expected to make appropriate use of it. For instance, in the shipping industry, ship captains have been found to be negligent by failing to make use of navigational decision support systems.

Another issue, ethically, is that of involving the patient. This is important as indicated in Figure 17.10 because it is the patient who is the focus of clinical decision making. Increasingly, the patient is encouraged to participate actively in the decision-making process. Clearly there will be a range of responses to such proposals with some patients still wishing to retain a purely passive role, feeling either unable or unwilling to be involved. However, there are many who welcome the opportunity for active involvement in the decisions that will affect their health. Indeed, a number should be regarded as expert patients, particularly those willing and able to take responsibility for all the day-to-day decisions relating to the management of chronic diseases such as diabetes.

This shift toward greater patient involvement and empowerment has clear implications for decision support and the manner in which it should be used. Increasingly, decision support systems will become an integral component of the patient–clinician encounter. Indeed, one of the authors of this chapter has already experienced participation in decision analysis in a clinical consultation in the area of gastroenterology. Decision support systems are welcomed by many patients who are managing their chronic disease. For example, in the case of diabetes, such systems will increasingly provide advice regarding lifestyle and diet as well as advice relating to day-by-day adjustments in their insulin regimens. All these trends clearly have implications for future developments in clinical decision support in relation to legal and ethical issues.

17.9 Summary

This chapter has provided an overview of clinical decision support systems and the role that they can play in supporting the decision maker in a range of clinically related tasks. First, an overview of the decision-making process has been presented, thereby enabling the context for decision support to be fully appreciated. This was followed by an outline of the nature of human diagnostic reasoning, models of the reasoning process, and issues to be considered in the clinical setting. A structure for characterizing clinical decision support systems was then described. Within this framework, the broad range of conceptual, methodologic, and technical approaches to decision support has been outlined and reviewed, demonstrating their bases and some of their particular strengths and limitations. A range of applications of decision support systems was then considered: first in hospital and other health care settings, and then in relation to medical education. Finally, the major topic of evaluating decision support systems was addressed, including relevant legal and ethical issues.

17.10 Exercises

1. Discuss the advantages and disadvantages of the analytic models presented in the chapter.
2. Describe the types of CDSS systems that have been described in this chapter, and comment on their relative merits in the context of clinical application.
3. Should formal decision-making methods be the normative approach to resolving complex decisions arising in the delivery of health care?

17.11 References and Bibliography

1. T. Deutsch, E. Carson, and E. Ludwig. *Dealing with Medical Knowledge: Computers in Clinical Decision Making.* Plenum, 1994.
2. J. Pearl J. *Probabilistic Reasoning in Intelligent Systems.* Morgan Kaufmann, 1997.
3. J. D. Slate and L. C. Sheppard. Automatic control of blood pressure by drug infusion. *Proceedings of the IEE (Part A).* 129:639, 1981.
4. G. J. Kuperman, R. M. Gardner, and T. A. Pryor. HELP: A Dynamic Hospital Information System. Springer-Verlag, 1991.
5. E. H. Shortlifffe and G. O. Barnett. Medical data: Their acquisition, storage and use. In E. H. Shortliffe and L. H. Perrault (Eds.). *Medical Informatics.* 2nd ed. Addison-Wesley. 41–76, 2001.
6. R. Summers, D. G. Cramp, and E. R. Carson. Non-AI decision making. In J. D. Bronzino (Ed.). *The Biomedical Engineering Handbook.* CRC Press, 2001.
7. B. J. Kuipers. Qualitative simulation. *Artif. Intell.* 29: 289–338, 1983.
8. T. G. Coleman and W. J. Gay. In D. Moller (Ed.). *Advanced Simulation in Biomedicine.* Springer-Verlag. 41–69, 1990.
9. R. Hovorka et al. Nonlinear model predictive control in subjects with type 1 diabetes. *Physiol. Measurement* 25:905–920, 1998.
10. H. L. Bleich. Computer-based consultations: Electrolyte and acid-base disorders. *Am. J. Med.* 53: 285–291, 1972.
11. D. P. McKenzie et al. Constructing a minimal diagnostic decision tree. *Methods Inf. Med.* 32:161–166, 1993.
12. C. D. Newgard, R. J. Lewis, and B. T. Jolly. Use of out-of-hospital variables to predict severity of injury in paediatric patients involved in motor vehicle crashes. *Ann. Emerg. Med.* 39(5):481–491, 2002.
13. H. C. Sox et al. *Medical Decision Making.* Butterworths, 1998.
14. H. Llewellyn and A. Hopkins. *How We Reach Decisions.* Royal College of Physicians, 1993.
15. M. R. Gold et al. (Eds.). Cost-Effectiveness in Health and Medicine. Oxford University Press, 1996.
16. F. A. Sloan (Ed.). *Valuing Health Care.* Cambridge University Press, 1996.
17. D. K. Owens, R. D. Shachter, and R. F. Nease. Representation and analysis of medical decision problems with influence diagrams. *Med. Decis. Making.* 17:241, 1997.
18. R. F. Nease and D. K. Owens. Use of influence diagrams to structure medical decisions. *Med. Decis. Making.* 17:263, 1997.
19. A. Gammerman and A. R. Thatcher. Bayesian inference in an expert system without assuming independence. In M. C. Golumbic (Ed.). *Advances in Artificial Intelligence.* Springer-Verlag. 182–218, 1990.
20. C. A. Goble, R. Stevens, and S. Ng. Transparent access to multiple bioinformatics information sources. *IBM Sys. J.* 40(2):532–551, 2001.
21. D. J. Spiegelhalter and R. P. Knill-Jones. Statistical and knowledge-based approaches to clinical decision-support systems with an application in gastro-enterology. *J. R. Stat. Soc. [Ser A].* 147:35, 1984.
22. R. Summers, D. G. Cramp, and E. R. Carson. Non-AI decision making. In J. D. Bronzino (Ed.). *Medical Devices and Systems (The Biomedical Engineering Handbook, 3rd Edition).* Taylor and Francis, 2006.
23. T. Deutsch, D. Cramp, and E. Carson *Decisions, Computers and Medicines: The Informatics of Pharmacotherapy.* Elsevier, 2001.
24. F. T. De Dombal et al. Computer-aided diagnosis of acute abdominal pain. *B.M.J.* 2:9–13, 1972.
25. O. F. Agbaje et al. Bayesian hierarchical approach to estimate insulin sensitivity by minimal model. *Clin. Sci.* 105:551–560, 2003.
26. R. J. Lewis and R. L. Wears. An introduction to the Bayesian analysis of clinical trials. *Ann. Emerg. Med.* 22(8):1328–1336, 1993.
27. D. J. Spiegelhalter, L. S. Freedman, and M. K. B. Parmar. Bayesian approaches to randomised trials. *J. R. Stat. Soc. [Ser A].* 157:357–416, 1994.
28. G. Parmigiani. *Modeling in Medical Decision Making: A Bayesian Approach.* Wiley, 2002.
29. A. O'Hagan and J. W. Stevens. Bayesian methods for design and analysis of cost-effectiveness trials in the evaluation of health care technologies. *Stat. Methods Med. Res.* 11:469–490, 2003.
30. K. Claxton. The irrelevance of inference: A decision making approach to the stochastic evaluation of health care technologies. *J. Health Econ.* 18:341–364, 1999.

31. E. H. Shortliffe. *Computer-Based Medical Consultations: MYCIN*. Elsevier, 1976.
32. S. Weiss et al. A model-based method for computer-aided medical decision making. *Artif. Intell.* 11:145–172, 1978.
33. C. F. Aliferis and R. A. Miller. On the heuristic nature of medical decision-support systems. *Methods Inf. Med.* 34(1–2):5–14, 1995.
34. S. K. Steginga and S. Occhipinti. The application of the heuristic-systematic processing model to treatment decision making about prostate cancer. *Med. Decis. Making.* 24(6):573–583, 2004.
35. P. J. Lisboa and A. F. G. Taktak. The use of artificial neural networks in decision support in cancer: A systematic review. *Neural Networks* 19(4):408–415, 2006.
36. S. Mersmann and M. Dojat. SmartCare: Automated clinical guidelines in critical care. In R. Lopez de Mantara and L. Saitta (Eds.). *16th European Conference on Artificial Intelligence (ECAI'04).* IOS Press. 745–749, 2004.
37. S. A. Berger and U. Blackman. Computer program for diagnosing and teaching geographic medicine. *J. Travel Med.* 2:2199–2203, 1995.
38. J. Bury et al. A quantitative and qualitative evaluation of LISA, a decision support system for chemotherapy dosing in childhood acute lymphoblastic leukaemia. In M. Fieschi, E. Coiera, and Y.-C. J. Yi (Eds.). *MEDINFO 2004, Proceedings of the 11th World Congress on Medical Informatics.* IOS Press, 197–201, 2004.
39. J. Fox, V. Patkar, and R. Thomson. Decision support for health care: the PROforma evidence base. *Inform. Prim. Care.* 14:49–54, 2006.
40. N. F. Ezquerra et al. PERFEX: An expert system for interpreting 3D myocardial perfusion. In *Expert Systems with Applications.* Pergamon Press, 1992.
41. C. J. Dickinson. A digital computer model to teach and study gas transport and exchange between lungs, blood and tissues (MacPuf). *J. Physiol.*224(1):7P–9P, 1972.
42. P. Tatti and E. D. Lehmann. Utility of the AIDA diabetes simulator as an interactive educational teaching tool for general practitioners (primary care physicians). *Diabetes Technol. Ther.* 3(1):133–140, 2001.
43. S. M. Finkelstein et al. Decision support for the triage of lung transplant recipients on the basis of home-monitoring spirometry and symptom reporting. *Heart Lung.* 34(3):201–208, 2005.
44. C. P. Friedman and J. C. Wyatt. *Evaluation Methods in Biomedical Informatics.* 2nd ed. Springer Science, 2006.
45. R. L. Bashshur. On the definition and evaluation of telemedicine. *Telemed. J.* 1:19–30, 1995.
46. M. Skiadas et al. Design, implementation and preliminary evaluation of a telemedicine system for home haemodialysis. *J. Telemed. Telecare.* 8:157–164, 2002.

18

Medical Robotics and Computer-Integrated Interventional Medicine

Prof. Russell H. Taylor and
Prof. Peter Kazanzides
The Johns Hopkins University

18.1 Introduction

This chapter is concerned with computer-integrated interventional medicine (CIIM). Over the past 50 years, the technology used in interventional medicine increasingly has been computer-based. Medical imaging devices have progressed from simple X-ray units to sophisticated systems combining advanced sensors and computation to provide unprecedented information about a patient's anatomy and physiology. Medical workstations are able to combine information from many sources to help surgeons and other physicians plan interventions and provide real-time information supports in carrying out these plans. Robotic devices and endoscopic cameras enable physicians to perform minimally invasive procedures that would otherwise be impossible. Computer-controlled systems use directed energy to destroy tumors and other malformations inside a patient's body without surgery. Computer-based physiologic monitoring devices are ubiquitous in operating rooms and intensive care units.

This evolution is a natural consequence of the computer's ability to integrate information with action to fundamentally improve treatment processes, in much the same way that computer-integrated systems and processes have affected other sectors of our society, such as manufacturing, transportation, retailing, and agriculture. The basic information loop of interventional medicine is illustrated in Figure 18.1. The process starts with information about the patient, such as images, test results, genetic information, and symptoms. This information is combined with general information about human anatomy and physiology to create a patient-specific model or representation that is used to diagnose the patient's condition and formulate an interventional plan. During the intervention, the virtual reality of the model and plan is registered to the actual reality of the patient and may be coupled to appropriate technology to assist the clinician in carrying out the plan. Further information is typically generated both during and after the intervention to update the model and assess the effect of the intervention.

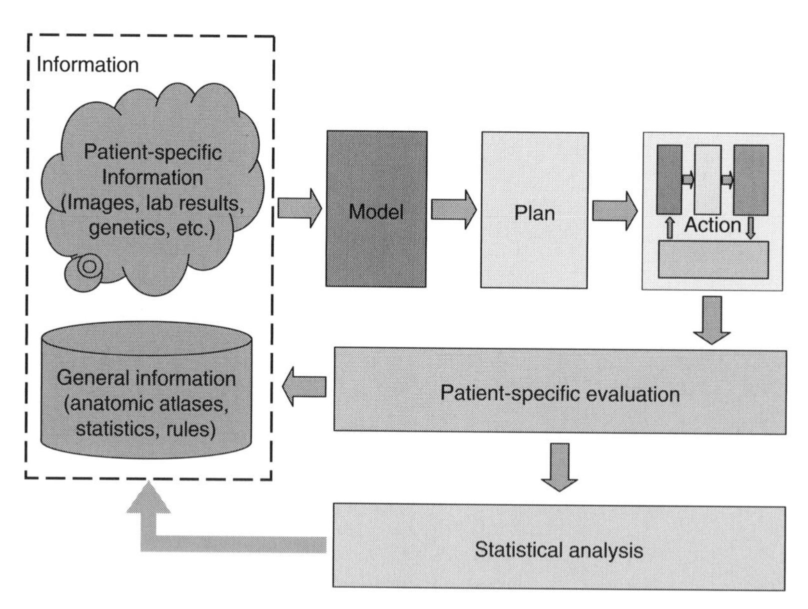

FIGURE 18.1 Computer-integrated interventional medicine (CIIM) as a closed-loop process.

This information may be used subsequently in further treatment of the patient. It may also be analyzed statistically to assess and improve the overall effectiveness of treatment plans and protocols in a manner somewhat analogous to the use of statistical quality control and process learning in manufacturing.

We often refer to this as a closed-loop process of first constructing a patient-specific model and interventional plan, then registering the model and plan to the patient and using technology to assist in carrying out the plan, and finally assessing the result as surgical or interventional computer-aided design/computer-aided manufacturing (CAD/CAM), again emphasizing the analogy between computer-integrated interventional medicine and computer-integrated manufacturing. Of course, it is important to recognize that there are also profound differences between medicine and manufacturing. In particular, our goal is *not* automation of medical interventions. Rather, our goal is to exploit computer-based technology and systems to assist human clinicians in treating patients. Thus, we often refer to these systems as *surgical (or interventional) assistants*, especially when the interventional decisions are highly interactive, as is frequently the case with surgery. However, it is important to remember that these concepts are not incompatible. Although it is often more convenient to think of a CIIM system as being primarily a CAD/CAM or an assistant system, the same underlying concepts and technology are present in both cases. As these systems become more and more sophisticated, the distinction will be harder and harder to make.

18.2 Technology and Techniques

In this section, we will provide a brief overview of key technology components found in CIIM systems, with special attention to surgical navigation and medical robotics. Further discussion may be found in references [1–4].

18.2.1 System Architecture

The overall architecture of CIIM systems is shown in Figure 18.2. Broadly, these systems consist of the following components: (1) computational components that perform a wide variety of image processing, surgical planning, monitoring, and similar tasks; (2) databases of patient-specific information, as well as more generic knowledge bases about human anatomy and physiology, common treatment plans, outcome data, etc.; and (3) devices such as imagers, robots, and human–machine interfaces that relate the virtual reality of computer representations to the actual reality of the patient, interventional room, and clinician.

18.2.2 Registration and Transformations Between Coordinate Systems

Geometric relationships between portions of the patient's anatomy, images, robots, sensors, and equipment in interventional suites are fundamental in all areas of CIIM, and there is an extensive literature on techniques for determining the transformations between the associated coordinate systems

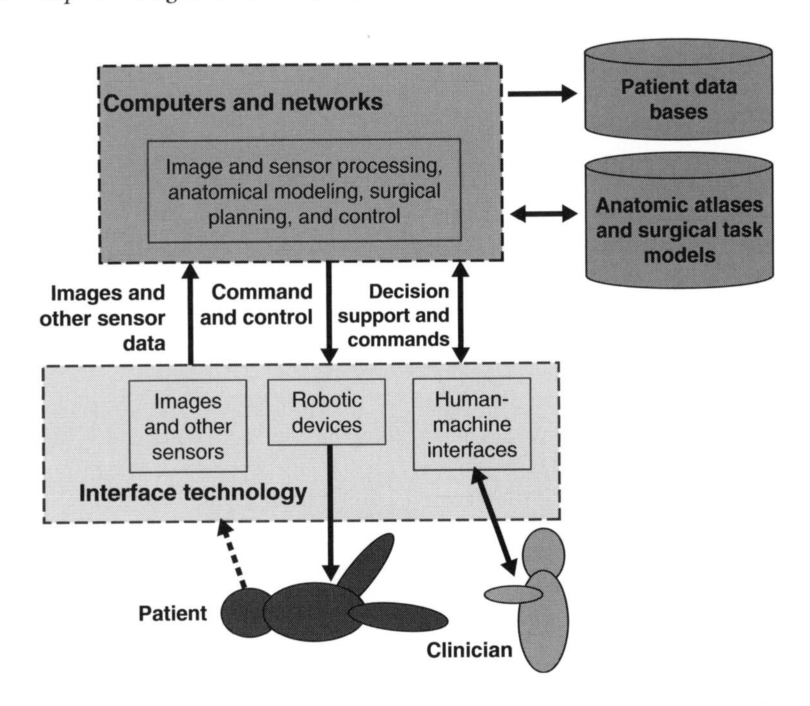

FIGURE 18.2 The architecture of computer-integrated interventional medicine systems.

[5, 6]. The brief discussion below follows the basic framework developed in Lavallee [6]. Given two coordinates $\vec{v}_A = [x_A, y_A, z_A]$ and $\vec{v}_B = [x_B, y_B, z_B]$ corresponding to comparable features in two coordinate systems Ref_A and Ref_B, the process of registration is simply that of finding a function $\mathbf{T}_{AB}(\cdots)$ such that:

$$\vec{v}_B = \mathbf{T}_{AB}(\vec{v}_A).$$

Although nonrigid registrations are becoming more common, $T_{AB}(\cdots)$ is still usually a rigid transformation of the form

$$\vec{v}_B = \mathbf{T}_{AB}(\vec{v}_A) = \mathbf{R}_{AB} \cdot \vec{v}_A + \vec{p}_{AB},$$

where $\mathbf{R}_{AB}$ represents a rotation and $\vec{p}_{AB}$ represents a translation. $\mathbf{R}_{AB}$ is often represented by an axis $\vec{n}$ and angle θ so that

$$\mathbf{R}_{AB}(\vec{n},\theta) = e^{\theta\hat{n}} \text{ where } \hat{n} = \begin{bmatrix} 0 & -n_z & n_y \\ n_z & 0 & -n_x \\ -n_y & n_x & 0 \end{bmatrix}.$$

Thus, if we have two transformations $\mathbf{T}_{AB}$ and $\mathbf{T}_{BC}$, the rotation and displacement components associated with the composite transformation $\mathbf{T}_{AC} = \mathbf{T}_{AB} \cdot \mathbf{T}_{BC}$ will be given by

$$\mathbf{R}_{AC} = \mathbf{R}_{AB} \cdot \mathbf{R}_{BC}$$
$$\vec{p}_{AC} = \mathbf{R}_{AB} \cdot \vec{p}_{BC} + \vec{p}_{AB}.$$

In many cases, $\mathbf{T}_{AB}$ cannot be computed exactly, so the actual transformation $\mathbf{T}^*_{AB}$ is related to the nominal value $\mathbf{T}_{AB}$ by a small perturbation:

$$\mathbf{T}^*_{AB} = \mathbf{T}_{AB} \cdot \Delta\mathbf{T}_{AB}.$$

In this case, we frequently approximate the rotational component of a small rotation $\Delta\mathbf{R}$ by

$$\Delta\mathbf{R} \approx \mathbf{I} + \theta\hat{n},$$

so that

$$\Delta\mathbf{R} \cdot \vec{v} \approx \vec{v} + \theta\vec{n} \times \vec{v}.$$

Further, we often ignore the effects of a small rotation $\Delta\mathbf{R}$ on a sufficiently small translation vector $\Delta\vec{p}$, so that

$$\Delta\mathbf{R} \cdot \Delta\vec{p} \approx \Delta\vec{p}.$$

Thus, if the actual value of a coordinate

$$\vec{v}^*_A \approx \vec{v}_A + \Delta\vec{v}_A,$$

then the actual value of

$$\vec{v}^*_B = \mathbf{T}^*_{AB} \cdot \vec{v}^*_A$$

will be given by

$$\begin{aligned}
\vec{v}^*_B &= \mathbf{T}_{AB} \cdot \Delta\mathbf{T}_{AB} \cdot (\vec{v}_A + \Delta\vec{v}_A) \\
&= \mathbf{T}_{AB} \cdot (\Delta\mathbf{R}_{AB} \cdot \vec{v}_A + \Delta\mathbf{R}_{AB} \cdot \Delta\vec{v}_A + \Delta\vec{p}_{AB}) \\
&\approx \mathbf{T}_{AB} \cdot (\vec{v}_A + \theta\vec{n} \times \vec{v}_A + \Delta\vec{v}_A + \theta\vec{n} \times \Delta\vec{v}_A + \Delta\vec{p}_{AB}) \\
&\approx \mathbf{T}_{AB} \cdot (\vec{v}_A + \theta\vec{n} \times \vec{v}_A + \Delta\vec{v}_A + \Delta\vec{p}_{AB}) \\
&= \mathbf{R}_{AB} \cdot (\vec{v}_A + \theta\vec{n} \times \vec{v}_A + \Delta\vec{v}_A + \Delta\vec{p}_{AB}) + \vec{p}_{AB} \\
&= \vec{v}_B + \mathbf{R}_{AB} \cdot (\theta\vec{n} \times \vec{v}_A + \Delta\vec{v}_A + \Delta\vec{p}_{AB}).
\end{aligned}$$

Thus, the uncertainty in $\vec{v}_B$ will be given by

$$\Delta\vec{v}_B = \mathbf{R}_{AB} \cdot (\theta\vec{n} \times \vec{v}_A + \Delta\vec{v}_A + \Delta\vec{p}_{AB}).$$

There is extensive literature concerning registration methods. Typically, the process involves finding corresponding sets of features $\mathcal{F}_A$ and $\mathcal{F}_B$, and then finding a transformation $T_{AB}(\cdots)$ that minimizes some distance function

$$d_{AB} = distance(\mathcal{F}_B, T_{AB}(\mathcal{F}_A)).$$

Typical features can include artificial fiducial objects (pins, implanted spheres, rods, etc.) or anatomic features such as point landmarks, ridge curves, or surfaces. One very common case involves registration of a set of sample points from an anatomic surface with a computer representation of that surface. In this case, variations of the iterated closest point algorithm of Besl and McKay [7] are commonly used. For example, 3D robot coordinates $\vec{a}_j$ may be found for a collection of points known to be on the surface of an anatomic structure which can also be found in a segmented 3D image. Given an estimate T_k of the transformation between image and robot coordinates, the method iteratively finds corresponding points $\vec{b}_j^{(k)}$ on the surface that are closest to $T_k \cdot \vec{a}_j$ and then finds a new transformation

$$T_{k+1} = \arg\min_{T} \sum_j \|\vec{b}_j^{(k)} - T \cdot \vec{a}_j\|^2 .$$

The process is repeated until some suitable termination condition is reached.

18.2.3 Navigational Trackers

Real time measurement of intraoperative positions and orientations is ubiquitous in CIIM, and a number of technologies are available for this purpose. These include encoded mechanical linkages, ultrasound localizers, electromagnetic localizers, active optical triangulation systems that locate light-emitting diodes, passive optical triangulation systems that locate reflective markers, and more general computer vision systems. Excellent technology surveys may be found in the literature [8–10] and in comparisons of different systems [11–13], although one should be aware that the relative technical capabilities of different technology approaches can change as technology develops.

In recent years, optical systems such as the Optotrak® and Polaris® systems (Northern Digital, Inc., Waterloo, Canada) have been the most widely used option for surgical navigation systems (see Section 18.4.2 and Figure 18.9 later) because of their relatively high accuracy, predictable performance, and insensitivity to environmental variations. However, they do have several limitations. The most serious of these is the requirement that a clear line of sight be maintained between the tracking cameras and the markers being tracked, which can complicate the arrangement of equipment and workflow around the patient. A related drawback is that the markers being tracked must generally be on portions of surgical instruments outside the patient. This approach can lead to inaccuracies in instrument tip position determination and cannot be used with flexible instruments such as catheters.

Electromagnetic trackers were considered for many early surgical navigation applications, but the measurement distortions associated with metal in operating rooms caused them to fall out of favor. More recently, improvements in electromagnetic tracking technology (including reduced distortion and the development of very small sensors) and increased interest in tracking devices inside the patient have led to increased interest in this technology. Current examples include the Aurora® (Northern Digital, Waterloo, Canada), Flock-of-Birds® (Ascension Technology, Burlington, VT), Polhemus Patriot (Polhemus, Inc., Burlington, VT), and proprietary systems used in the Medtronic Axiem® (Medtronic Navigation, Inc. Louisville, CO) and the GE InstaTrak® (General Electric OEC Medical Systems, Salt Lake City, UT).

18.2.4 Robotic Devices

Historically, the term robot as been used for multi-axis machines that are capable of autonomous motion. With this strict definition, the well-known daVinci system would not be classified as a robot, but rather as a teleoperator, because it does not operate autonomously. In fact, this would be true of many of the medical robot systems that have been developed in recent years. Therefore, at least in the medical field, the definition of a robot has been expanded to include virtually any mechanism that provides assistance to the surgeon, whether or not it can operate autonomously. In fact, safety is such a critical concern in medical robotics that it has prompted several researchers to develop robots that are incapable of autonomous motion [14–17]. These systems rely on the surgeon, rather than on motors, to provide sufficient force to create motion. The systems may still contain powered elements (e.g., motors, brakes), but they are only used to constrain motion. Although such systems do not fit the classic definition of a robot, they are considered passive robots in the medical field.

In an industrial setting, the benefit of robotics over fixed automation is that a robot can be programmed to serve in many different capacities. An industrial robot can assemble typewriters, weld car bodies, or debur molded parts. There is, of course, some degree of specialization. A robot that places surface mount components on a printed circuit board is likely to be small and extremely accurate, whereas a robot that installs automobile windshields must be large and powerful. This specialization also applies to medical robots. For example, a robot developed for microsurgery will differ from a robot developed for orthopedic joint reconstruction. Although the field of medical robotics is not yet mature, current experience suggests that medical robots may be more specialized than their industrial counterparts; a robot developed for one medical procedure may not be as easily adapted for other procedures, for reasons outlined below. Some examples of multifunctional medical robots do exist, such as the orthopedic robot systems that assist with hip and knee replacement surgery as well as with ligament repair.

Clearly, there are many similarities between industrial and medical robots. Both (typically) consist of motors, sensors, and articulated links that can be programmed to perform a variety of functions. There are, however, many differences, including their integration in the working environment (factory workcell vs. operating room/interventional suite), the relationship between robot and workpiece (design for manufacturability vs. adapting to human anatomy), and safety systems (keeping humans out of the workspace vs. working alongside and on humans). These issues are discussed in the following paragraphs.

Robots are now commonplace on factory floors, and much experience has been gained in workcell configuration. Workcell design is simplified by the fact that other pieces of equipment, such as conveyer belts and parts feeders, are designed to integrate with robots and other industrial machines. Once a workcell design is completed, the robot and associated equipment are installed and, in most cases, left in place for a long time. In contrast, robots are not (yet) standard equipment in the interventional suite or operating room, where space is limited. Thus, a medical robot must be easily transported in and out of the room or, if permanently installed, should be able to be moved out of the way. In effect, a medical robot must be installed in the medical workcell for each use. This installation includes transporting the robot to the site (e.g., operating room), connecting it to appropriate power sources, and sterilizing it. Because other medical equipment is not designed for compatibility with robotics, the robot must fit in as unobtrusively as possible. It is important to minimize the space requirement around the operating table, since much of this space is needed for the medical team and equipment.

Similarly, the manufacturing industry has widely adopted the principle of design for manufacturability, which means that parts are designed for ease of manufacturing by automated machines, including robots. Furthermore, in an industrial setting, the number of distinct parts is limited, and like parts typically differ only by small manufacturing tolerances. The environment can be further structured using specialized parts feeders to orient or align the parts. In contrast, medical robots must operate on human anatomy, which cannot be redesigned to facilitate robotic procedures and is often not easily accessible from outside the body. Also, although humans have the same types of parts, there are large variations between individuals. A medical robot must be able to sense and adapt to these variations. If sensors alone cannot perform this task (and they often cannot), the clinician should be included in the loop to augment the system's sensing capabilities. This requires a human–machine interface that is easy to use by individuals (clinicians) who do not have robotics backgrounds. In addition, novel kinematic designs are often necessary to operate on the target anatomy without unduly restricting the clinician.

Although the mechanical design of medical robots has many similarities to that of industrial robots, the special requirements associated with interventional procedures (access, workspace, biocompatibility, imaging-device compatibility, etc.) have tended to produce distinct designs. For example, many medical robots are designed to manipulate surgical instruments or needles passed through constrained entry points into the patient's body. This consideration has led many groups [17–20] to develop kinematic structures that decouple tool orientation motions about a remote center of motion (RCM) distal to the robot's structure. In clinical use, the robot is typically positioned so that the RCM point is positioned at the point where the instrument or needle passes into the patient's body (see Figure 18.5 later for an example). Similarly, a number of groups [21–27] have developed robots specifically for use in a magnetic resonance imaging (MRI) environment.

Safety is an important consideration for both industrial and medical robots [28]. In both cases, the goals, in order of priority, are to prevent injury to human beings working near the robot and to prevent the robot from damaging itself, other equipment, or the workpiece. In an industrial setting, safety systems typically involve gates, pressure-sensitive mats, and flashing lights—devices designed to keep people out of the robot's workspace or to shut down the system if a person comes too close. This is especially important when the robot is capable of high speeds or torques. In an industrial robot, high speeds and torques are desirable because they reduce the cycle time, thereby increasing the robot's productivity. In addition, many industrial robots require super-human strength to perform their tasks (e.g., lifting heavy parts). Unfortunately, these desirable attributes increase the potential danger to human beings. In the medical domain, there is little distinction between the two safety goals listed above since the workpiece is a human patient, and other equipment includes life-sustaining medical equipment. Because the medical staff and the patient must be inside the workspace, medical robot safety systems must ensure that they are not harmed, even in the event of a malfunction. The situation is even more challenging when the robot is holding a potentially dangerous device such as a cutting instrument and is supposed to actually contact the patient with this device (in the correct place, of course). As a result, compared to industrial robots, medical robots usually contain more redundancy in hardware and software.

Many classifications systems for medical robotics have been proposed [29]; some of them define systems as being active, semi-active, or passive. There is no universally accepted definition of these terms; some would argue that any robot that is capable of motion (i.e., contains powered actuators) can never be considered passive, whereas others focus on the manner in which the robot is used. This chapter adopts the latter convention, which is an operational definition rather than a mechanical definition. An active robot automatically performs an intervention, such as machining bone. A semi-active robot performs the intervention under the direct control of the surgeon (e.g., a hands-on or cooperative control mode). A passive robot does not actively perform any part of the intervention (e.g., positions a tool guide).

There is some debate whether one class of robots may be better than another class when considering factors such as safety, user acceptance, or regulatory approval. In the latter case, it is likely that the less active a robot is, the more comfortable regulatory agencies will be in granting approval. Regarding safety, although a passive robot may avoid some of the risks inherent with a more active robot, there are still many safety issues that much be considered in all cases. For example, when preparing the bone for a knee prosthesis, regardless of whether the bone is automatically machined by an active robot, cooperatively machined by the surgeon and semi-active robot, or machined by the surgeon using a tool guide positioned by a passive robot, it is critical that the cutting be performed at the correct position and orientation. Therefore, each of these robots must provide a safety system to ensure that sensor failures do not cause them to incorrectly position the cutting tool or tool guide. The question of user acceptance has not yet been answered because currently the difficulty of using medical robots has been a bigger obstacle than whether they are active, semi-active, or passive.

18.2.5 Intraoperative Human–Machine Interfaces

Fundamentally, CIIM systems are intended to work with clinicians, not replace them in the operating room or interventional suite. Consequently, technology and methods for human–machine communication are crucial components in these systems. This communication is two-way, and successful systems must address techniques both for providing information to and for accepting information and direction from the clinician.

Visual display is the most common method for providing information to the clinician. Computer displays relating the positions of surgical instruments to cross-sectional medical images or to X-ray projections are ubiquitous in surgical navigation systems (see Section 18.4.2). The ergonomics of such systems have some serious limitations. Once a procedure has begun, the clinician's attention is necessarily focused on the patient's anatomy, and it is awkward for the clinician to look away from the patient. Consequently, a number of groups have developed systems and devices for superimposing visual information directly on the surgeon's view of the patient.

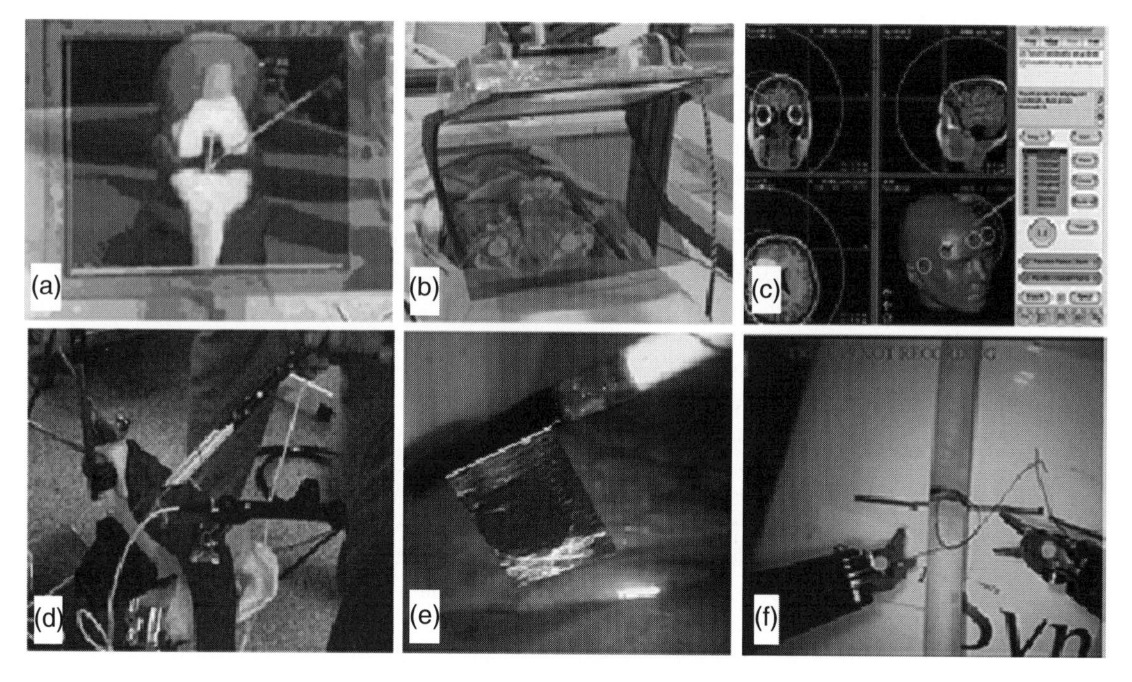

FIGURE 18.3 Visual information display in CIIM systems. (a) CMU image overlay system [42] based on active tracking of surgeon's head, 3D graphics, and semitransparent mirror. (b) Johns Hopkins University image overlay system for simple in-scanner display of scan planes [43, 44]. (c) Typical display from a surgical navigation system. Courtesy Medtronic Surgical Navigation. (d) Osaka/Tokyo laser guidance system [33]. (e) Johns Hopkins University/intuitive surgical overlay of laparoscopic ultrasound onto daVinci surgical robot video monitor [45]. (f) Sensory substitution display of surgical force information onto daVinci surgical robot video monitor [41].

The first such systems [30–32] were designed to inject registered graphic information into a surgical microscope. Subsequently, several groups have developed variations on this theme for use in other environments (Figure 18.3). Some of these systems may use active elements such as laser pointers [33, 34] to help the surgeon achieve a desired alignment. Other forms of feedback used by CIIM systems include auditory feedback, either in the form of computer-generated speech [35] or simple auditory cues [36], haptic (force) feedback [37–40], or visual/auditory representation of tool-tissue interaction forces [41].

There are many ways for a surgeon to provide information or command direction to a CIIM system. The most common are those used with any computer workstation: typed text and mouse-like pointing devices. Intraoperatively, these devices have many limitations, especially because they are difficult to sterilize and they tie up the clinician's hands. One common though clearly limited work-around has been to rely on verbal instructions to technicians operating the equipment. Another has been to rely on computer voice recognition systems [46–49]. Still another method has been to rely on sterile touch screen displays or on the motions of instruments tracked by surgical navigation systems. A few groups have explored video tracking of the clinician's head or eye motions [50].

The motion of surgical robots is frequently commanded through the use of conventional telerobotic master devices, which are essentially powered or unpowered robot manipulators moved by the clinician, or by cooperative control methods in which the robot's motion complies to forces exerted on it by the surgeon (see Section 18.4.3). Other methods, often used in research systems designed for more intelligent assistance to a surgeon, include visual tracking of surgical instruments and target anatomy [51, 52].

18.2.6 Sensorized Instruments

A number of research groups [56–63] have developed sensorized surgical instruments capable of measuring tool-to-tissue interaction forces and providing these results to surgical workstations. Often, these efforts have relied on graphic interfaces to display force data, whether the instrument was manipulated freehand by the clinician or by a robot. For example, Poulose et al. [58, 59] demonstrated that a force-sensing instrument used together with an IBM/JHU LARS robot [51] could significantly reduce both average retraction force and variability of retraction force during Nissen fundoplication. There have also been efforts to incorporate sensed-force information into the control of robotic devices [64–67]. Several researchers [68, 69] have focused on specialized fingers and display devices for palpation tasks requiring delicate tactile feedback (e.g., for detecting hidden blood vessels or cancerous tissue beneath normal tissue). Yet another use of sensorized instruments is in biomechanical studies to measure organ and tissue mechanical properties to improve surgical simulators [70, 71].

There has also been work to integrate nonhaptic sensors into surgical instruments. For example, our group at Johns Hopkins is developing instruments that measure tissue oxygenation as well as force [53–55]. Our plan is to use this information to help surgeons assess tissue viability, avoid ischemic tissue damage during retraction, and distinguish tissue types (see Figure 18.4).

18.2.7 Software and Robot Control Architectures

Figure 18.5 shows the basic control architecture for a robot system. There are two periodic loops: a high-frequency servo loop (typically 1 KHz or higher) that controls the individual motors and a lower-frequency supervisory loop (typically about 100 Hz) that coordinates the individual motors and may also close a loop around an external sensor, such as a force sensor or imaging system. Although the dynamic equations of a robot include coupling between the axes, the standard practice is to perform the servo control of each motor independently. In fact, some robot systems perform the servo control on a distributed network of embedded microprocessors, where each microprocessor is attached to just one or two motor/sensor pairs. Fortunately, most medical robots move rather slowly (often for safety reasons), so the dynamic coupling between joints can be ignored without affecting control performance.

For a typical path-controlled robot, the supervisory control loop consists of a trajectory planner that breaks down a high-level motion command (such as moving at a specified velocity along a straight line) into a set of intermediate setpoints that are sent to the servo control loop(s). Because the high-level motion command is often in a Cartesian coordinate system, this process generally includes the invocation of the robot's inverse kinematic equations to transform the Cartesian coordinates into the robot joint coordinates expected by the servo loop. Although path-controlled robots are common in industrial applications, in the medical field many other supervisory control strategies are often required. One example is a compliant control mode, where robot motion is dictated by the forces and torques applied by the surgeon and measured by a force sensor. The most common approach is to transform the Cartesian velocity to joint velocities using the robot's inverse Jacobian. Alternatively, the Cartesian velocity can be added to the current Cartesian position to obtain a desired position, which can be transformed to joint positions via the robot's inverse kinematics.

Historically, robot manufacturers have provided an interpreted language for programming the robot because this allows the end-user (or systems integrator) to quickly develop new applications or modify existing applications in the field, if necessary. An interpreter environment allows fast implement-test-debug cycles because debugging changes can be made during execution without losing any of the program state

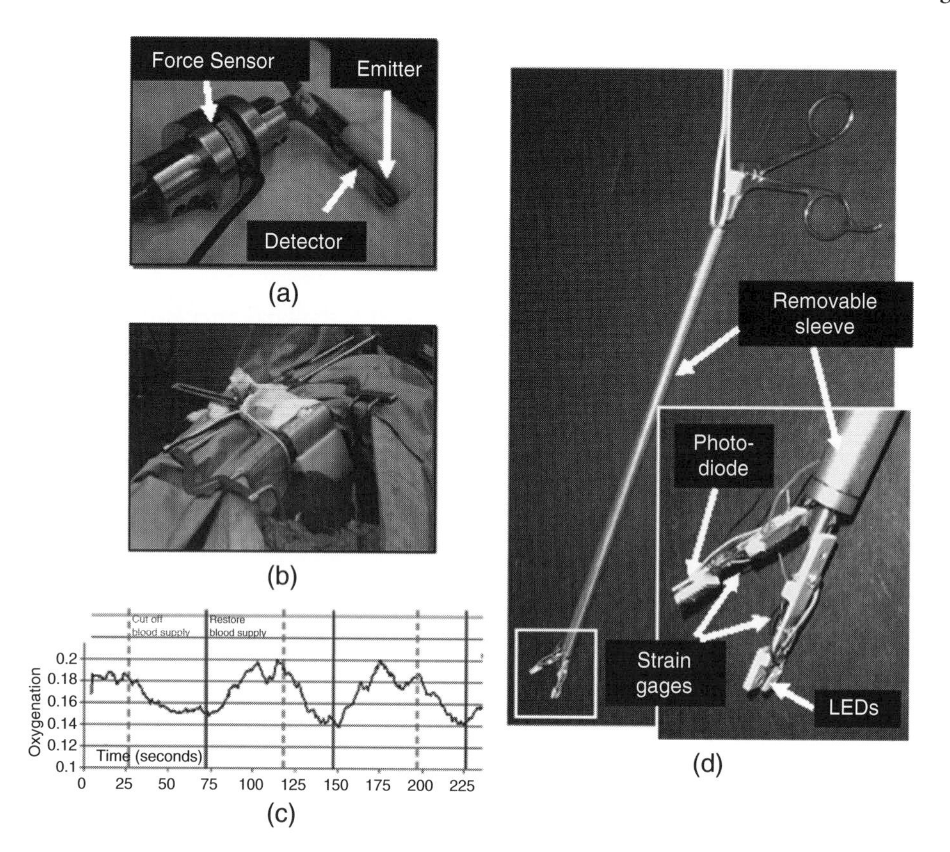

FIGURE 18.4 Sensorized instruments from our laboratory at Johns Hopkins University [53–55]. (a) Liver retractor with integrated force sensor and optical sensors for measuring blood oxygenation. (b) Retraction of pig liver. (c) Sensor readings as blood supply is cut off and restored. (d) Laparoscopic instrument with force and oxygenation sensing fingers.

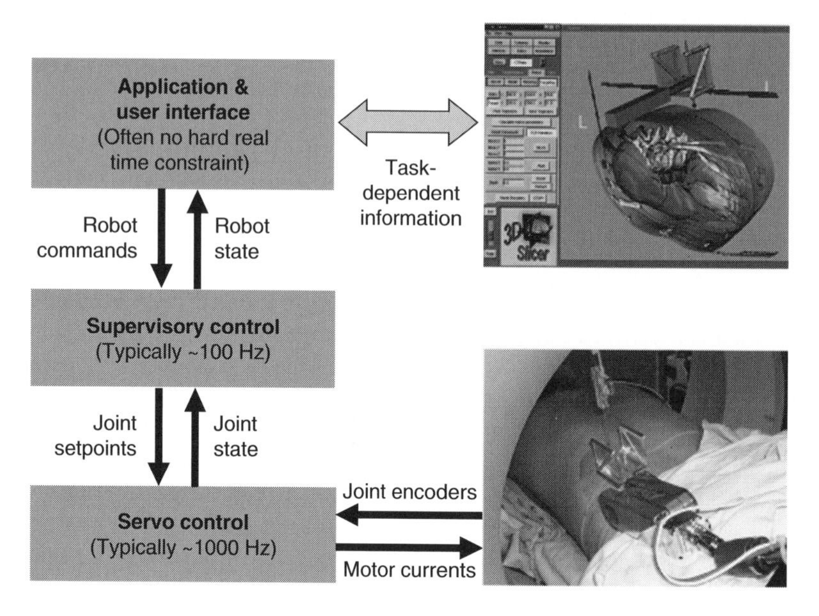

FIGURE 18.5 Architecture of a typical robot system. The robot shown was developed at Johns Hopkins for in-scanner percutaneous needle placement procedures [19, 72, 73]. The screen interface at the top is typical of the sort of research interface commonly developed for similar procedures, although a different interface was used for the kidney biopsy shown at bottom.

such as the robot's position [74]. On the other hand, compiled code runs significantly faster than interpreted code, which is especially important for tasks that require real-time performance, such as closing a loop around an external sensor. For medical robots, regulatory requirements must also be considered. These dictate that a medical device manufacturer must carefully control all software changes (configuration control). This requirement, along with liability considerations, necessitates a system design that prevents inadvertent or unauthorized software modifications, especially in the field. Although it is possible to protect interpreted code from modification (e.g., by encryption of the source code), a compiled language has the advantage of enabling manufacturers to provide end-users with only the executable files. For these reasons (efficiency and security), most medical robots are programmed in a compiled language such as C or C++. For development, however, it is still desirable to have an interactive (interpreted) environment. This can be achieved by wrapping the C/C++ code for use with a standard interpreted language, such as TCL or Python.

The development of standard software and control libraries and application frameworks for medical robotics research represents a significant challenge and opportunity. This goal has been a major research focus at Johns Hopkins University. Figure 18.6 illustrates the software/hardware environment being developed in our laboratory at Johns Hopkins for research on intelligent surgical assistants, which is based on the set of open source software libraries that we are developing [75–77]. The development of this sort of infrastructure can be an important enabler in medical robotics research. See also Section 18.4.6.

18.2.8 Accuracy Evaluation and Validation

Validation of computer-integrated interventional systems is challenging because the key measure is how well the system performs in an operating room or interventional suite with a real patient. Clearly, for both ethical and regulatory reasons, it is not possible to defer all validation until a system is used with patients. Furthermore, it is often difficult to quantify intraoperative performance because there are limited opportunities for accurate postoperative assessment. For example, even though computed tomography (CT) scans are accurate, they may not provide sufficient contrast for measuring the postoperative result, and they expose the patient to additional radiation. For these reasons, most computer-integrated interventional systems are validated using phantoms, which are objects that are designed to mimic (often very crudely) the relevant features of the patient.

One of the key drivers of surgical CAD/CAM is the higher level of accuracy that can be achieved using some combination of computers, sensors, and robots. Therefore, it is critical to be able to evaluate the overall accuracy of such a system. One common technique is to create a phantom with a number of objects whose locations are accurately known, either by precise manufacturing or measurement. If the system uses fiducial-

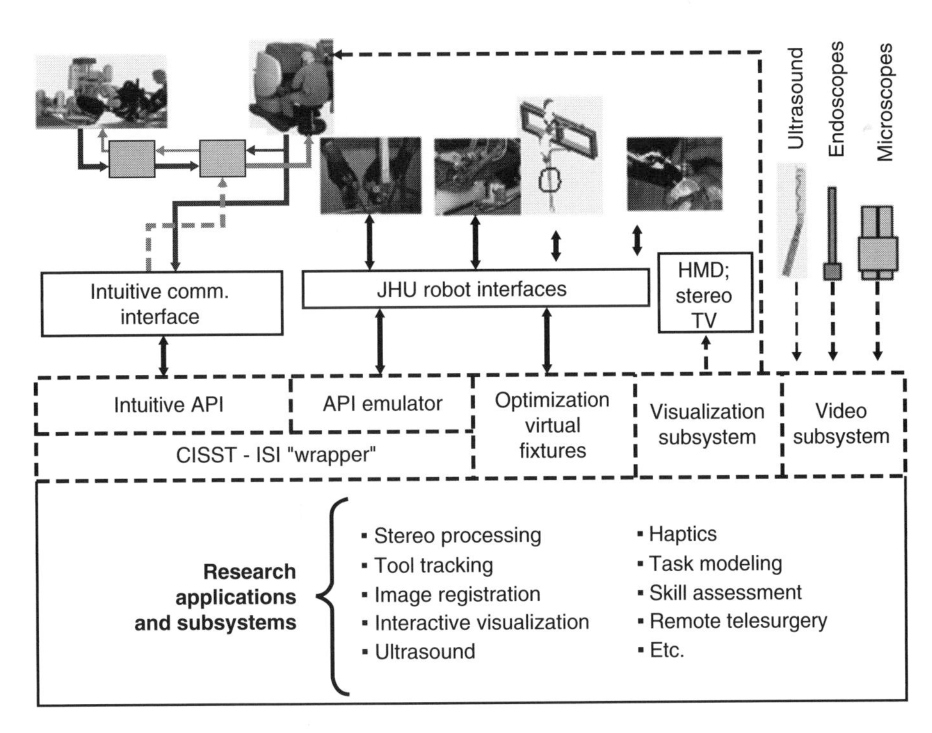

FIGURE 18.6 Modular system environment for robotic surgical assistance research at Johns Hopkins University. Available at http://www.cisst.org/.

based registration, the objects in the phantom should correspond to fiducials. Furthermore, the phantom should contain extra fiducials (not used for registration) or other known features that can be used as targets. If the system uses an anatomic registration, it may still be useful to place a number of fiducials in the phantom so that they can provide a reasonably accurate estimate of the ground truth registration.

The basic technique is to image the phantom, perform the registration, and then locate the target features. Maurer [78] defined the following types of error:

- *Fiducial localization error (FLE)*: The error in locating a fiducial in a particular coordinate system (i.e., imaging system or surgical CAD/CAM system)
- *Fiducial registration error (FRE)*: The root-mean-square (RMS) residual error at the registration fiducials; that is,

$$FRE = \sqrt{\frac{1}{N}\sum_{k=1}^{N}\left\|\vec{b}_k - \mathbf{T}\cdot\vec{a}_k\right\|^2},$$

where $\mathbf{T}$ is the registration transform and $(\vec{a}_k,\vec{b}_k)$ are matched pairs of homologous fiducials ($k = 1, \ldots, N$).

- *Target registration error (TRE)*: The error in locating a feature or fiducial that was not used for the registration; if multiple targets are available, the mean error is often reported as the TRE.

For a robot system, one method for measuring TRE is to locate the targets in the image, transform them to the robot coordinate system (using the registration), and then command the robot to position its instrument at the computed target location. The TRE is given by the difference between the robot's position and the actual position of the target. It may not be practical or convenient to measure this position difference, however, so a common strategy is to manually position the robot at the physical target and then compute the TRE as the difference between the computed position (based on the registration) and the robot's actual position. Essentially, this method uses the robot itself to measure the TRE.

18.2.9 Risk Analysis and Regulatory Compliance

The medical device industry is a heavily regulated industry. In the United States, medical devices must be cleared for market by the Food and Drug Administration (FDA). There are two paths to market. One is via the 510(K) premarket notification process, and the other is via the premarket approval (PMA) process. A manufacturer can obtain a 510(K) clearance if the new device is "substantially equivalent" to an existing device that is already on the market. Otherwise, the PMA application is required. Surprisingly, several medical robots obtained clearance via the 510(K) path, including Aesop (Computer Motion, Inc.), Neuromate (Innovative Medical Machines International, subsequently Integrated Surgical Systems), and daVinci (Intuitive Surgical). In contrast, the earlier ROBODOC System (Integrated Surgical Systems) started down the PMA path and, as of 2006, has not received clearance. ROBODOC was tested in multi-center clinical trials in the United States under an investigational device exemption (IDE), which is the mechanism by which FDA authorizes limited clinical trials to gather supporting data.

In addition to the need for 510(K) or PMA approval, medical device companies must comply with the quality system regulations (QSR) and are periodically audited by the FDA to verify compliance. Initially, the FDA required companies to adhere to good manufacturing practices (GMP), which regulated just the manufacturing phase. For simple devices, this worked well because device failures were primarily due to manufacturing flaws. As devices became more complex, especially with the integration of computers and software, the FDA discovered that a large number of device failures were due to design flaws rather than to manufacturing flaws. The infamous Therac-25 accident, where six patients received massive overdoses of radiation from a computer-controlled medical linear accelerator, is a well-known example [79]. As a result, FDA QSR began to regulate the design phase as well.

In the European market, all products (medical or otherwise) require *Conformité Européenne* (CE) marking. Furthermore, the design and manufacturing processes must comply with International Standards Organization (ISO) 9001 and 9002, respectively (often these are grouped together by the term ISO 9000). The CE marking and ISO 9000 certification are handled by a number of notified bodies, which are independent, nongovernmental entities.

To comply with ISO 9000 and/or FDA QSR, medical device companies must define their development and manufacturing processes and then produce documents (quality system records) that demonstrate adherence to these processes. Although ISO 9000 and FDA QSR are similar, they are not identical, which requires most medical device companies to comply with both of them.

It should be noted that obtaining FDA approval and CE marking and complying with FDA QSR and ISO 9000 are still not enough to guarantee commercial success. Obviously, it is necessary for the device to be marketable (i.e., to provide a favorable cost–benefit ratio). It is perhaps less obvious, however, that the device must also be accepted by the third-party payers in the health care system. In the United States, this consists of Medicare and the health insurance companies. These entities must agree to reimburse for procedures performed with the new technology for that technology to proliferate in the marketplace.

Risk (or hazard) analysis is one of the key elements of a medical device development process and is often a focal point for audits by FDA or notified bodies. A failure modes effects analysis (FMEA) or failure modes effects and criticality analysis (FMECA) are the most common methods [80]. These are bottom-up analyses, where potential component failures are identified and traced to determine their effect on the system. Methods of control are devised to mitigate the hazards associated with these failures. The information is generally presented in a tabular format. The FMECA adds the criticality assessment, which consists of three numerical parameters: the severity (S), occurrence (O), and detectability (D) of the failure. A risk priority number is computed from the product of these parameters; this determines whether additional methods of control are required. The FMEA/FMECA is a proactive analysis that should begin early in the design phase and evolve as hazards are identified and methods of control are developed.

18.3 Surgical CAD/CAM

18.3.1 Example: Robotically Assisted Joint Reconstruction

The relative rigidity of bone and the excellent contrast available in X-ray and CT images make orthopedic procedures, especially joint replacement surgery, natural applications for medical robots, and about 20% of all medical robots surveyed in 2005 were intended for such applications [81]. The authors of this chapter were co-developers of one of the first robotic systems for orthopedic surgery (ROBODOC® [82, 83]), so it is natural for us to use it as an example in discussing surgical CAD/CAM applications. Earlier research using a robot for total knee replacement surgery was performed at the University of Washington [84], and subsequently, a number of other groups also developed systems for similar applications [85–89].

ROBODOC® (Integrated Surgical Systems, Inc., Sacramento, CA), was initially developed for total hip replacement (THR) surgery [90, 91] and was later applied to total knee replacement (TKR) [92]. THR surgery involves preparing an elongated cavity in the femur (thigh bone) and a rounded cavity in the acetabulum (hip socket) to accommodate the two components of a hip prosthesis: the femoral stem (Figure 18.7b) and acetabular cup. Accurate placement of components relative to the patient's bones is very important for achieving a good result. Furthermore, with cementless implants, the bone must be shaped to achieve a close fit between the implant and the bone to encourage the bone to grow into a porous coating on the implant.

For conventional THR surgery, preoperative planning is performed by overlaying templates (outlines) and making measurements on two-dimensional X-rays. Templates are available at different magnification factors so that errors due to X-ray magnification can be minimized. Usually, planning is limited to identifying an approximate range of implant sizes and the approximate desired implant position relative to the bone. During surgery, the bone is prepared using hand-held reamers (drills) and broaches to prepare the desired cavities. Proper execution relies on a significant amount of experience and surgical feel, especially when preparing the femoral cavity. In this case, the surgeon typically begins with the reamer and broach corresponding to the smallest planned implant size. If the cavity feels loose (i.e., insufficient contact with hard cortical bone), the surgeon switches to the next larger size until he or she feels that there is sufficient, but not excessive, cortical contact. If the surgeon chooses a prosthesis that is too large, the femur can fracture either during cavity preparation or during prosthesis insertion. This is one of the most common intraoperative complications associated with THR. Similarly, although the surgeon can plan any desired prosthesis position,

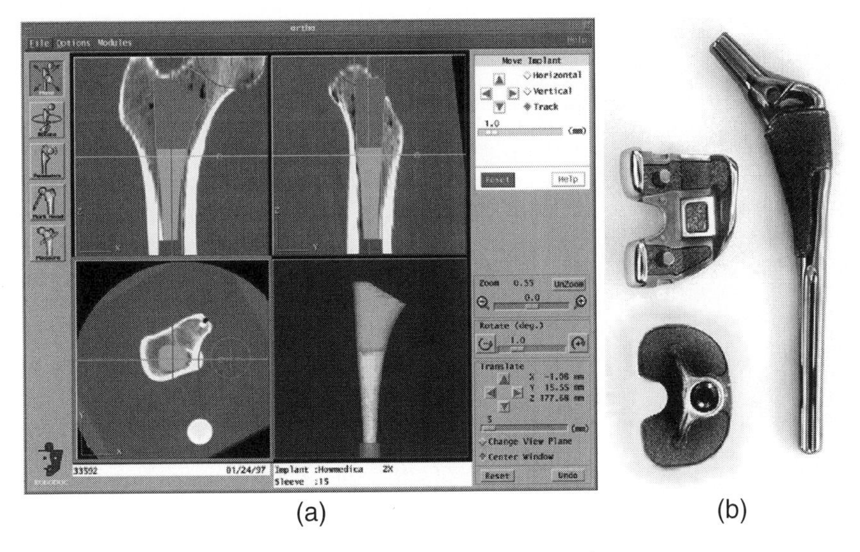

FIGURE 18.7 (a) Typical screen view from ORTHODOC® CT-based planning system for ROBODOC® orthopedic robot. Integrated Surgical Systems, Sacramento, CA. (b) Typical implant components for cementless hip and knee reconstruction surgery.

the actual position is determined mostly by anatomic constraints because the hand-held instruments tend to follow the path of least resistance.

Laboratory tests [93] showed that the conventional method for cavity preparation was inherently inaccurate. The cavities produced were extremely irregular, with large gaps between implant and bone. Further, accurate alignment of the cavity relative to bone was extremely uncertain because the interior surface of the bone could deflect the path of the broach. These considerations led our surgeon colleagues (Drs. Paul and Bargar) to propose the use of a robot to prepare the implant cavity. Expected benefits included adequate and uniform bone ingrowth, uniform stress transfer, reduced stress shielding, less thigh pain, and the elimination of femoral fractures as an intraoperative complication.

The ROBODOC procedure for THR (and TKR) consists of two phases: a preoperative planning phase (ORTHODOC®) and an intraoperative (ROBODOC) phase. The input to ORTHODOC consists of a CT scan of the patient's anatomy, the prosthesis geometry that is supplied by the manufacturers, and clinical decisions made by the surgeon. The surgeon plans the procedure by selecting a prosthesis from the database and positioning it in the CT image. ORTHODOC displays three orthographic views (i.e., orthogonal slices) of the data as well as a 3D model (Figure 18.7). Each joint of the five-axis surgical robot (Figure 18.8a, b) contains two optical encoders for redundant position feedback. The system includes a wrist-mounted six-axis force sensor that monitors the forces applied at the tool. This force information makes it possible to implement functionality such as manual guidance, tactile search, safety checking, and an adaptive cutter feed rate. ROBODOC executes the preoperative plan by machining the specified prosthesis cavity in the femur. This requires the bone to be rigidly attached to the robot. A bone motion monitor is used as a safety sensor to detect motion of the bone relative to the robot. In addition, accurate cavity placement requires a registration between the patient's anatomy in the preoperative plan (i.e., the bones in the CT scan) and the anatomy of the actual patient. The preoperative plan is specified in image (CT) coordinates whereas intraoperative localization of the patient can be obtained in robot coordinates, so registration implies finding the transformation between image and robot coordinates.

Initially, ROBODOC used a "pin-based" registration method, which required the implantation of titanium bone screws (pins) in the femur prior to the CT scan. Registration was accomplished by defining at least three reference points on the pins and then identifying them in both the CT and robot coordinate systems. Because the pins are titanium, the ORTHODOC software could easily locate them in the CT data using image processing techniques. The robot system identified the physical pins via a tactile search, using feedback from its wrist-mounted force sensor [83]. ROBODOC initially used three registration pins, with the centers of the pin heads

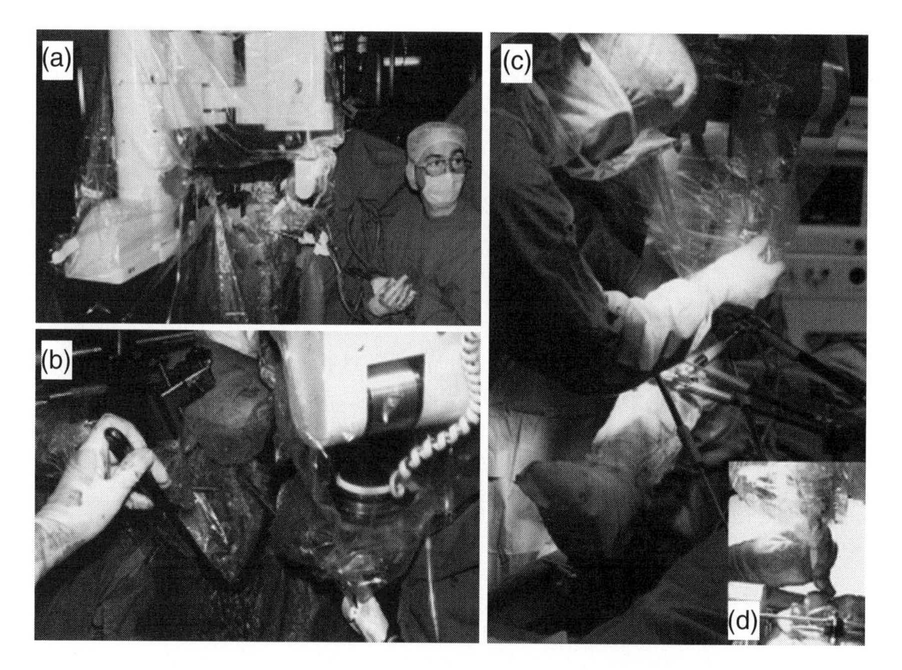

FIGURE 18.8 Clinically applied robots for orthopedic surgery. (a, b) The Robodoc® system for cementless total hip and knee replacement surgery machines bone to match a surgeon-selected implant shape according to a presurgical plan based on patient CT images [83, 94]. (c) The Acrobot system [85] employs cooperative hand guiding with active constraints derived from the implant shape for total knee replacement surgery.

serving as the three reference points. Shortly afterward, it transitioned to a two-pin method, where the third reference point was obtained by creating a virtual pin based on the center and axis of the distal pin. In this case, a longer distal pin was required to enable accurate determination of the pin axis in the CT data.

Although pin-based registration is reliable, it involves an extra (minor) surgery to implant the pins before making the CT scan and was also the source of postoperative knee pain for many patients. This motivated the development of a pinless system [95], which uses anatomic features instead of metal pins as fiducials. Registration is performed using a method similar to the iterated closest point method outlined in Section 18.2.2, using bone surface point positions measured by a small digitizing arm.

Once surgery has begun, ROBODOC provides a visual display of its progress on the computer monitor. As the robot mills the cavity, the monitor displays the CT data overlaid with a model of the prosthesis cavity. The completed portion of the cavity is displayed in one color while the remaining portion is displayed in another. This is similar to the visualization provided by most navigation systems. During surgery, the control software continuously monitors the force sensor and adjusts the cutter feed rate based on the sensed force and on parameters specific to the prosthesis design and cutting tool [96]. This enables the robot to adapt to the patient's anatomy by slowing down in regions of hard cortical bone and speeding up in other regions.

As of December 2006, ROBODOC has been installed in about 50 hospitals around the world and has performed over 10,000 THR surgeries. Use of this system became controversial, especially in Germany, with surgeons and patients reporting both positive and negative results. Two points that both sides seem to agree on are: (1) the robot procedure requires a longer surgery time and has higher surgical costs, compared to the conventional technique; and (2) the robot can execute the preoperative plan more accurately than the conventional technique. There is no consensus, however, on whether the improved accuracy provided by the robot system provides a clinical benefit to the patient.

18.3.2 Example: Needle Placement

Placement of needles or similar devices* is one of the most basic interventional CAD/CAM applications, although there are numerous challenges, depending on the target organ and the operating environment. The problem can be simply stated as placing the tip of the needle at a location specified on an image, typically through an entry point also specified on the image. Both robots and navigation systems have been used to assist with this task. In some cases, the interventional device (robot or navigation system) is used to position a cannula or instrument guide through which the needle is manually advanced. Accurate placement of needles in the brain was one of the first uses of robots in interventional medicine [97–100], and these techniques have since been extended to many parts of the body, including prostate, liver, spine, etc. Further, percutaneous needle placement is a natural application for surgical navigation and image overlay techniques such as those illustrated in Figure 18.3. There is an extensive literature on robotic and nonrobotic systems for needle placement. This section will touch briefly on a few common themes.

When performed using CAD/CAM techniques, the entry and target positions can be identified on preoperative images, intraoperative images, or some combination of the two. In all cases, it is necessary to register the image space to the interventional device (e.g., robot or tracked instrument). When using intraoperative images, this registration can be obtained by placing a calibration object on the robot or patient. The transformation between the calibration object and device coordinate system is known by design, and the transformation between the calibration object and the image coordinate system is computed by locating features of the calibration object in the image, often via image processing techniques. Recent examples from the work of our own group at Johns Hopkins is included in references [101] and [102], although these techniques are widely practiced by many groups.

Another potential issue for CAD/CAM needle placement is target motion. Although this is a relatively minor (though not nonexistent) problem for bone and organs such as the brain (which is encased in the skull), it can be very challenging for soft tissue organs such as liver, kidneys, or prostate, as well as for anatomic targets such as the lungs or spine, which can be affected by respiratory motion or by heartbeats. For this reason, many groups have emphasized placement of needles under direct feedback from imaging modalities such as X-ray fluoroscopy, CT, MRI, or ultrasound. Whether or not direct image feedback is available, it is often important to compensate for motion and/or to register preoperative images with (possibly deformed) intraoperative anatomy or images.

In many cases, needle placement under direct (intraoperative) image guidance is difficult due to patient access issues. This is especially true when the image modality is a closed bore MR scanner, where the patient is placed inside a long cylindrical tube that has a diameter that is not much larger than the patient. Here, the only option for performing needle placement (besides catheter-based methods) is to use a robot that is small enough to fit inside the MRI scanner [27, 103]. The design of MR-compatible robotic devices poses significant challenges as a result of materials and component limitations associated with the high magnetic fields and radiofrequency sensing associated with MR imaging [104]. Even for a robot intended for use with CT or X-ray fluoroscopy, it is generally

* For convenience, we use the term needle placement, but the problem is generic to the placement of any needle-like instrument, including probes, drills, radiation beams, etc.

desirable for the robot's end effector to be as radiolucent as possible to reduce interference with the images used for guidance and targeting.

18.4 Surgical Assistance

18.4.1 Basic Concepts

Interventional procedures, especially those that we think of as surgery, can be highly interactive processes, and many interventional decisions are made in the operating room and executed immediately. The goal of computer-based interventional systems, including medical robots, is not to replace the surgeon or interventionalist* with a machine so much as to provide the surgeon with versatile tools that augment his or her ability to treat patients. Currently, there are three main sub-classes of Assistant Systems, although the distinctions between them are by no means hard and fast.

The first class, intraoperative information support systems, simply provide information to the surgeon, who uses his or her manual dexterity to manipulate the surgical instruments in performing the intervention. An extremely important subclass of these systems (discussed in Section 18.4.2) is surgical navigation systems, which relate surgical instrument positions to medical images and patient anatomy. Interestingly, surgical navigation systems can also be thought of as surgical CAD/CAM because they provide the capability to couple presurgical image-based planning with intraoperative execution.

The second class, surgeon extender robots, are operated directly by the surgeon and augment or supplement the surgeon's ability to manipulate surgical instruments during surgery. Potentially, these systems can give even average surgeons super-human capabilities such as elimination of hand tremor or ability to perform dexterous operations inside the patient's body. The clinical advantages associated with these systems potential include the ability to treat otherwise untreatable conditions, reduced invasiveness and patient morbidity, improved safety and reduced complication rates; and reduced surgeon fatigue. A special subclass of surgeon extender robots is a remote telesurgery system, which permits the surgeon to operate on patients at distances ranging from a few hundred meters to several thousand kilometers.

A third class, auxiliary surgical supports, generally work side by side with the surgeon and perform such functions as endoscope holding, tissue retraction, or limb positioning. These systems typically provide one or more direct control interfaces such as joysticks, head trackers, voice control, or the like. However, there have been some efforts to make these systems smarter to require less of the surgeon's attention during use, for example, by using computer vision to keep the endoscope aimed at an anatomic target or to track a surgical instrument. Although these systems may offer some of the same advantages as surgeon extenders (e.g., reduced tissue damage due to more delicate retraction), their main justification is improved operative efficiency and reduced need for operating room staff.

* For simplicity of discussion, we will use the word "surgeon" throughout the balance of this section, rather than the more inclusive (but awkward) "interventionalist."

18.4.2 Surgical Navigation Systems as Information Assistants

Surgical navigation systems track the positions of surgical instruments and other objects in the operating room and display this information graphically, usually relative to registered images of the patient. Although first developed for neurosurgery [8, 105], they have also been widely adapted to otolaryngological surgery [11, 106], orthopedic surgery [9, 10], craniofacial surgery [107, 108], and other applications placing a high value on precise localization and integration of information from medical imaging systems. There are currently many commercially available systems, and surgical navigation has rather larger acceptance in the interventional systems market than does any form of robotic assistance.

As shown in Figure 18.9, a typical surgical navigation system consists of a navigational tracking device capable of determining the position and orientation of rigid bodies attached to surgical instruments and to the patient's anatomy, together with a computer workstation and display. After a registration step is performed, the workstation is able to compute and display the position of instruments relative to patient images.

18.4.3 Surgeon Extenders

Telesurgical robots are the most widely deployed form of surgeon extender system and have been used extensively for cardiac, prostate, and other minimally invasive laparoscopic procedures. Examples include numerous (dozens) research systems [20, 25, 109–114], as well as commercially deployed systems such the daVinci [115] (Intuitive Surgical, Sunnyvale, CA) and Zeus [116] (formerly marketed by Computer Motion, Goleta, CA).

The architecture of a typical system (here, the daVinci) is shown in Figure 18.10. The system consists of a patient-side slave robot and a master control console. The slave robot has three or four robotic arms that manipulate a stereo endoscope and dexterous surgical instruments such as scissors and needle holders. The surgeon sits at the master control station and grasps handles attached to two dexterous master manipulator arms, which are capable of exerting limited amounts of force feedback to the surgeon. The surgeon's hand motions are sensed by the master manipulators, and the motions are mimicked by the slave manipulators. A variety of control modes may be selected by means of foot pedals on the master console and

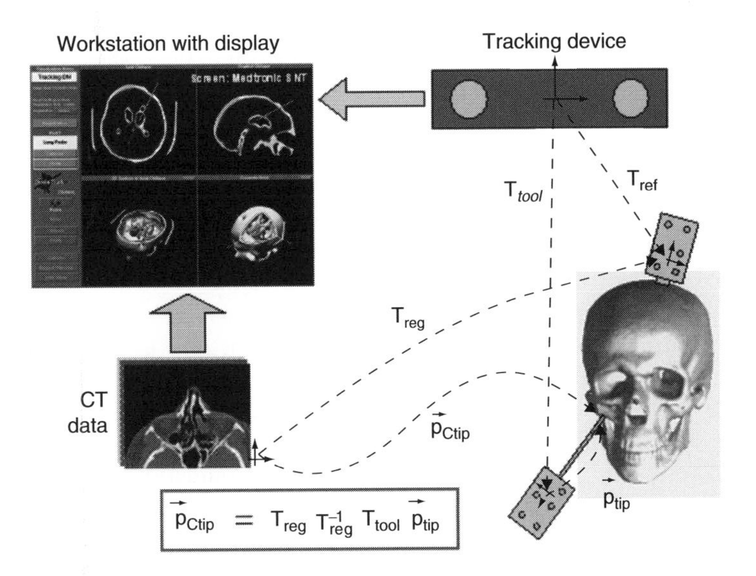

FIGURE 18.9 A typical surgical navigation system showing key coordinate transformations. After registration, the system computes $\vec{p}_{Ctip}$, the position in image coordinates corresponding to the current position of the pointer tip and uses this information to update a display.

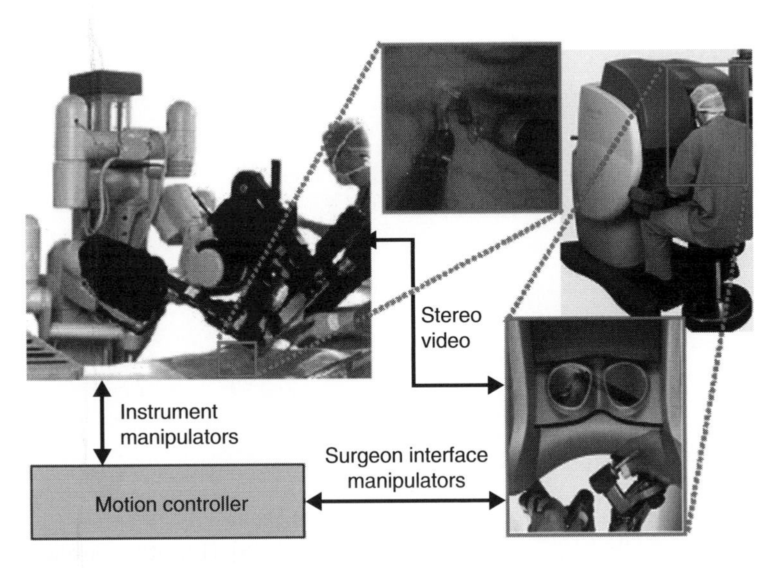

FIGURE 18.10 Architecture of a typical telesurgical system. Photos: Intuitive Surgical Systems.

used for such purposes as determining which slave arms are associated with the hand controllers. Stereo video is transmitted from the endoscope to a pair of high-quality video monitors in the master control station, thus providing high-fidelity stereo visualization of the surgical site. The display and master manipulators are arranged so that it appears to the surgeon that the surgical instruments (inside the patient) are in the same position as his or her hands inside the master control console. Other telesurgical systems employ the same basic architecture, although there are many differences in implementation. For example, many systems [26, 116] use more conventional stereo TV set displays that use polarizing glasses or liquid crystal display shutter glasses to multiplex left and right eye images. Some surgeons find this arrangement more comfortable for long-duration procedures, although much of the immersive feel of the daVinci is lost. Similarly, research systems incorporate many different mechanical designs for the patient-side slave robots.

A primary advantage promised by telesurgical robots for minimally invasive surgery is their ability to permit the surgeon to perform dexterous manipulation of instruments and tissues inside the patient's body. A major theme in current research has been development of highly dexterous, miniaturized robotic end-effectors suitable for this purpose. Some examples are shown in Figure 18.11. Many systems [20, 25, 115, 117, 118] have used cable-actuated tools. One drawback of this approach is that it becomes increasingly difficult to provide high strength and dexterity as the mechanisms get smaller and smaller. This has led various groups to investigate alternatives. For example, several groups have explored microhydraulic systems [119, 120]. At Johns Hopkins, we have explored another approach, illustrated in Figure 18.11(b) and (c), using parallel superelastic spines to produce snakelike end-effectors [119]. Although most current surgical robots employ manipulator arms to position tools within the patient's body, with wrist-like mechanisms to provide distal dexterity, there has been some work on systems with a greater degree of autonomous motion capability [121–123].

Although teleoperation has many advantages, especially for high-dexterity robotic manipulation inside the patient's body, it also has some drawbacks. The amount of equipment required is large, since both master and slave manipulators are needed. The surgeon is frequently somewhat removed from the patient because he or she is sitting at a master control station and may have a reduced overall awareness of the surgical situation.

Consequently, several groups, including our own, have developed an alternative approach based on hands-on admittance control, in which the robot moves in response to forces exerted by the surgeon directly on the robot's end-effector or on a handle attached to the robot. Our early experiences with Robodoc® [83] and other surgical robots [51, 124] showed us that surgeons find this form of control very convenient and natural for surgical tasks. Two notable uses of cooperative control are the Imperial College Acrobot™ orthopedic system [85] [Figure 18.8(c) and (d)] and the Johns Hopkins Steady Hand microsurgery system [125] (Figure 18.12). Although cooperative control is usually limited to precise positioning tasks, it can also provide force scaling via the use of two force sensors: one to sense the surgeon's input and another to measure tool-to-tissue interaction forces and then move the robot in response to a scaled difference between these forces [66, 125].

Other groups have developed completely free-hand instruments that sense and actively cancel physiologic tremor [129, 130]. The main advantage of this approach is that it requires the least change in normal operating room procedure. The

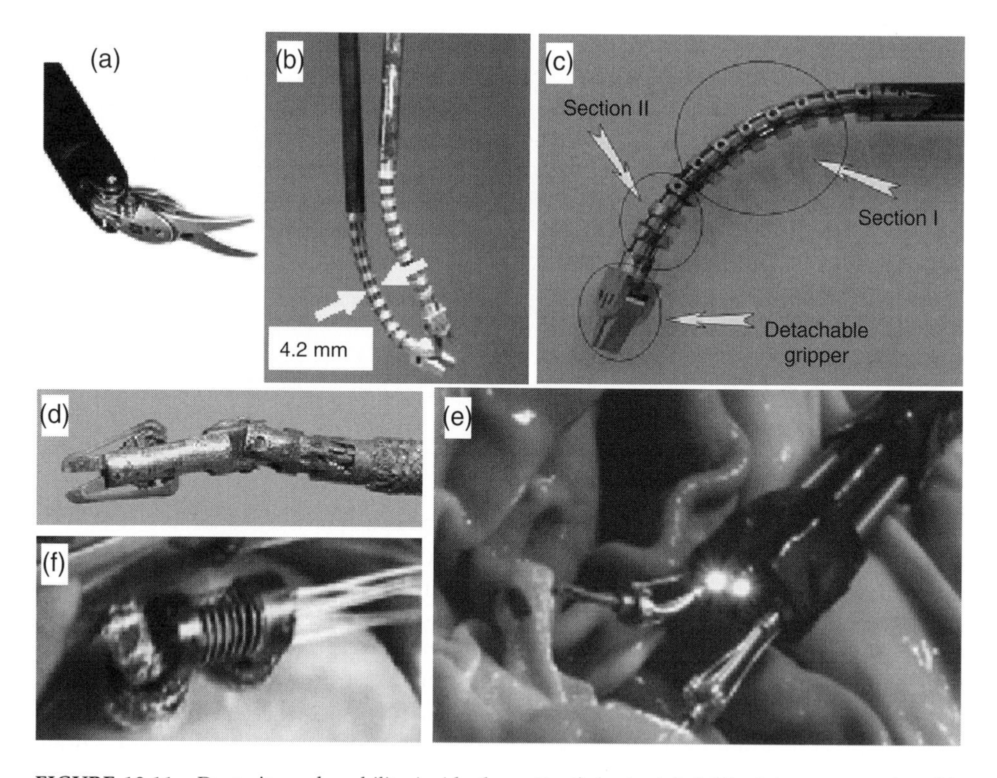

FIGURE 18.11 Dexterity and mobility inside the patient's body. (a) daVinci dexterous wrist with typical surgical instrument. Courtesy Intuitive Surgical. (b, c) 4.2-mm diameter Johns Hopkins University/Columbia University snake manipulator [132, 133]. (d) Five degree-of-freedom, 3 mm-diameter microcatheter robot [112, 117]. (e) Dexterous robot for endogastric surgery [118]. (f) Mobile Heart Lander robot for crawling across the heart [123].

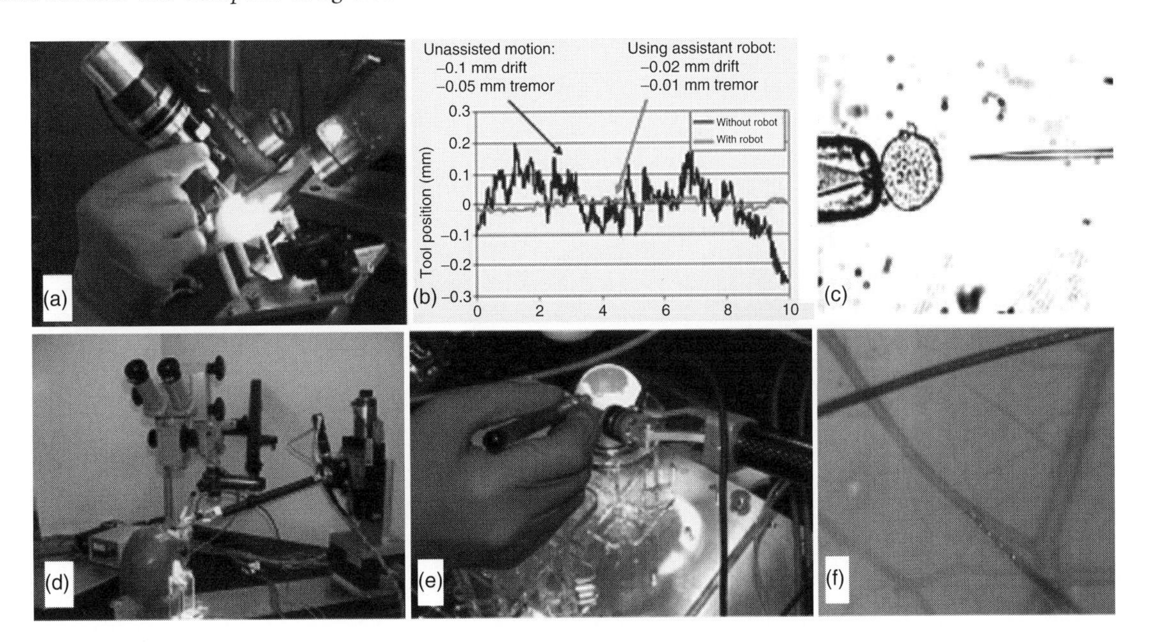

FIGURE 18.12 Johns Hopkins University Steady Hand cooperative manipulation systems for microsurgery. (a) First-generation system [125], here used to demonstrate fenestration of the stapes bone for an otology application [67]. (b) Comparative motion tremor with freehand instrument manipulation and steady-hand robot manipulation [126]. (c) Steady-hand micro-injections into mouse embryos [127]. (d) Newer-generation steady-hand robot for eye surgery [128]. (e, f) Evaluation on chick embryos.

surgeon uses the tremor-reducing tool just as he or she would use any other instrument. The challenges are instrument ergonomics (mostly size and weight) and precise motion performance, which is still not as good as that of fully robotic devices.

One problem commonly encountered in all forms of medical robotics is the difficulty of maintaining a desired relationship between an instrument held by the robot and moving patient anatomy. Broadly speaking, there are two approaches to solving this problem. The first approach [52, 133] is to sense the relative motion, most commonly with computer vision or some other form of imaging device, and then move the robot. The second approach [83, 86, 134, 135] is to attach the robot's base firmly to the patient's anatomy, so that it rides with the patient. This approach is especially common in orthopedics but may be applied in other areas such as otolaryngology, neurosurgery, or ophthalmology, where a good attachment point is available.

18.4.4 Auxiliary Surgeon Supports

Although attention is often focused on robotic systems that directly extend the surgeon's ability to manipulate surgical instruments, many of the most successful robotic applications in surgery have focused on auxiliary tasks such as patient positioning [136], surgical instrument delivery [137, 138] and laparoscopic camera positioning [51, 52, 139]. In fact, the AESOP® laparoscopic camera surgery system [46, 140] (formerly distributed by Computer Motion, Goleta, CA) was one of the first widely deployed surgical robots.

18.4.5 Remote Telesurgery and Telementoring

The possibility for using master–slave telesurgery systems to perform procedures in which the surgeon and patient are separated by very long distances has long been recognized [141, 142]. Commonly considered applications include space exploration, military combat care, and provision of care in sparsely populated areas. A number of research groups have developed experimental systems over the years [20, 143–148]. A major milestone was achieved by Marescaux et al. in 2001 with successful performance of a trans-Atlantic laparoscopic cholecystectomy [149]. Subsequent work has included efforts by Anvari et al. to develop a practical system for deployment in Canada [150, 151].

There has also been significant interest in using telesurgical technology to provide remote (or on-site) mentoring, in which an expert surgeon advises a less-experienced surgeon in carrying out a procedure [152–154]. Although in some ways similar to more conventional telesurgery, this form of telementoring can introduce some additional challenges. In particular, protocols may be needed to enable the expert and trainee surgeon to trade off control of a surgical robot or otherwise to work cooperatively during completion of the case.

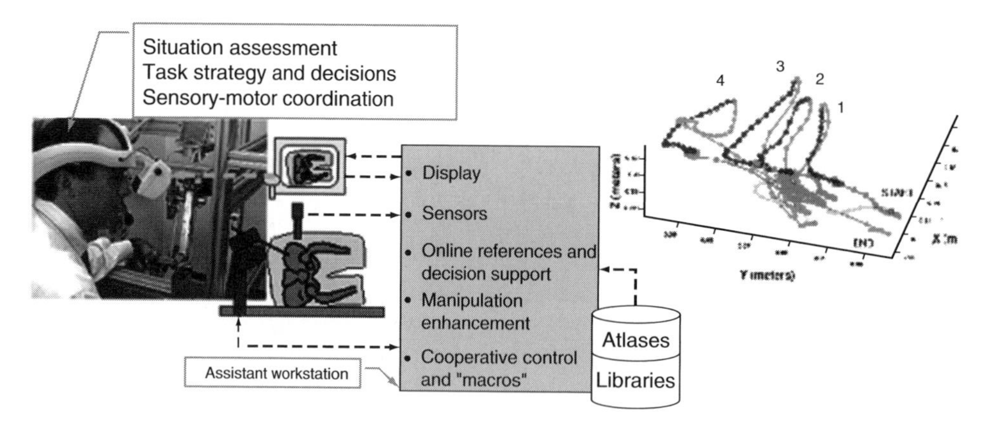

FIGURE 18.13 (a) Functional architecture of a typical intelligent surgical assistant. (b) Segmented trace of daVinci hand motions during a suturing procedure [155].

18.4.6 Toward Intelligent Surgical Assistance

Although one goal of both teleoperation and hands-on control in a surgeon extender system is to enable the surgeon to directly control the motion of the robot, the fact that a computer is actually meditating between the surgeon's command input and the robot's actual motion can create many more possibilities. The simplest is a safety barrier or no fly zone, in which the robot's tool is constrained from entering certain portions of its workspace. More sophisticated versions include virtual springs, dampers, or complex kinematic constraints that help a surgeon align a tool, maintain a desired force, or perform similar tasks. This concept has many names, of which virtual fixtures seems to be the most popular [156–160]. The Acrobot system shown in Figure 18.8(c) and (d) represents a successful clinical application using virtual barriers to limit the motion of cutting tool.

A number of groups [161–164] are exploring extensions of the virtual fixtures concept to active cooperative control, in which the surgeon and robot share or trade off control of the robot during a surgical task or subtask. As the ability of computers to model and follow along surgical tasks [138, 155] improves, these modes will become more and more important in surgical assistant applications. Figure 18.13(a) shows the functional architecture of a typical surgical assistant workstation being developed at Johns Hopkins University. Figure 18.13(b) illustrates initial efforts to develop automatic motion segmentation tools to distinguish the different steps in a suturing procedure.

18.5 Summary

Medical robotics and CIIM are still relatively young fields. Nevertheless, they have grown remarkably, especially in the past 5 to 8 years, as clinical systems have been deployed and as more researchers enter them. This short chapter has only provided a brief introduction to some of the main areas of research and practice, and our treatment has necessarily skipped over important research and groups working in the field. To those who may have been left out, we extend our sincere apologies and hope that readers of this chapter will be motivated to pursue further reading, perhaps starting with books such as those of references [165, 166], recent journal special issues such as those found in references [114, 167], or any of the many conference proceedings in the field.

By coupling information to action in ways that were not possible before, these systems have the potential to fundamentally change the practice of interventional medicine. Enough progress has been made in all of the architectural elements shown in Figure 18.2 so that clinically useful systems can indeed be deployed. However, further advances are still needed across the board in the modeling and analysis required for medical robotic applications, for the interface technologies required to relate the data world to the physical world of patients and clinicians, and to the system science that makes it possible to put everything together safely, robustly, and efficiently. It is our belief that this research is best done in interdisciplinary teams motivated by important applications. Our experience has been that building a strong researcher-surgeon-industry team is one of the most challenging, but also one of the most rewarding, aspects of medical robotics and CIIM research. The only greater satisfaction is the knowledge that the results of such teamwork can have a very direct impact on patients' health. This is a challenging area, but it is worth it.

18.6 Exercises

1. Develop an outline for evaluating alternative approaches to a surgical system or application, including such factors as cost, safety, effectiveness of pain relief, accuracy, time. For each such criterion, include:

a. Short definition or explanation of the criterion
b. Short discussion of how that criterion should be assessed (e.g., units of measure, means of gathering information)
c. Short discussion of how important each criterion is to each relevant group affected (patient, surgeon, hospital administrator, insurance company, employer, etc.)

2. Use your outline to evaluate robotic versus navigationally assisted versus conventional manual joint reconstruction surgery.
3. Use your outline to evaluate robotically assisted versus conventional minimally invasive surgery.
4. Consider the surgical navigation system shown in Figure 18.9. The expression given shows the calculation of the CT coordinates corresponding to the position of the pointer tip:

$$\vec{\mathbf{p}}_{\mathrm{Ctip}} = \mathbf{T}_{\mathrm{reg}}\mathbf{T}_{\mathrm{ref}}^{-1}\mathbf{T}_{\mathrm{tool}}\vec{\mathbf{p}}_{\mathrm{tip}}$$

Suppose, now, that each of the constituent expressions is subject to some small error:

$$\begin{aligned}\mathbf{T}^*_{reg} &= \mathbf{T}_{reg}\Delta\mathbf{T}_{reg}\\ \mathbf{T}^*_{ref} &= \mathbf{T}_{ref}\Delta\mathbf{T}_{ref}\\ \mathbf{T}^*_{tool} &= \mathbf{T}_{tool}\Delta\mathbf{T}_{tool}\\ \vec{\mathbf{p}}^*_{tip} &= \vec{\mathbf{p}}_{tip} + \Delta\vec{\mathbf{p}}_{tip}.\end{aligned}$$

We will assume that the errors are small so that the following approximations are valid:

$$\Delta\mathbf{T}_{reg}\cdot\vec{\mathbf{v}} \approx \vec{\mathbf{v}} + \vec{\alpha}_{reg}\times\vec{\mathbf{v}} + \vec{\varepsilon}_{reg},$$
where $\vec{\alpha}_{reg}$ and $\vec{\varepsilon}_{reg}$ are small vectors

with similar conventions for the other quantities.

a. Define $\vec{\mathbf{p}}_{tt} = \mathbf{T}_{Ctip}\cdot\vec{\mathbf{p}}_{tip}$. Write an expression estimating the error in the computed value of $\vec{\mathbf{p}}_{tt}$. That is, write an expression for

$$\Delta\vec{\mathbf{p}}_{tt} = \mathbf{T}_{tool}\Delta\mathbf{T}_{tool}(\vec{\mathbf{p}}_{tip} + \vec{\varepsilon}_{tip}) - \mathbf{T}_{tool}\vec{\mathbf{p}}_{tip}.$$

b. Define $\vec{\mathbf{p}}_{rt} = \mathbf{T}_{ref}^{-1}\vec{\mathbf{p}}_{tt}$. Write an expression estimating the error in the computed value of $\Delta\vec{\mathbf{p}}_{rt}$ (Hint, use your previous result as a start).
c. Write an expression estimating the error in the computed value of $\vec{\mathbf{p}}_{Ctip}$.
d. Suppose that value errors given above for $\Delta\mathbf{T}_{tool}$ and $\Delta\mathbf{T}_{ref}$ correspond to random measurement errors. Suppose, further, that the tracker has developed an unsuspected systematic error, such that if the reported value of a tracked frame A is $\mathbf{T}_A$ then the actual value is $\mathbf{T}^*_A = \Delta\mathbf{T}_{err}\mathbf{T}_A$. How does this affect your answer to question c? Justify your answer.

18.7 References and Bibliography

1. R. H. Taylor et al. *Computer-Integrated Surgery.* MIT Press, 1996.
2. R. H. Taylor. A perspective on medical robotics. *IEEE Proceedings.* 94:1652–1664, 2006.
3. R. H. Taylor and L. Joskowicz. Computer-integrated surgery and medical robotics. In M. Kutz (Ed.). *Standard Handbook of Biomedical Engineering and Design.* McGraw Hill, 2003:29.3–29.45.
4. R. H. Taylor and D. Stoianovici. Medical robotics in computer-integrated surgery. *IEEE Trans. Rob. Autom.* 19:765–781, 2003.
5. J. B. Maintz and M. A. Viergever. A survey of medical image registration. *Med. Image Anal.* 2:1–37, 1998.
6. S. Lavallee. Registration for computer-integrated surgery: Methodology, state of the art. In R. H. Taylor et al. (Eds.). *Computer-Integrated Surgery.* MIT Press. 77–98, 1996.
7. P. J. Besl and N. D. McKay. A method for registration of 3-D shapes. *IEEE Transactions on Pattern Analysis and Machine Intelligence.* 14:239–256, 1992.
8. R. J. Maciunas. *Interactive Image-Guided Neurosurgery:* American Association of Neurological Surgeons, 1993.
9. A. DiGioia et al. *Computer and Robotic Assisted Knee and Hip Surgery.* Oxford Press, 2004.
10. J. B. Stiehl, W. H. Konerman, and R. G. Haaker. *Navigation and Robotics in Total Joint and Spine Surgery.* Springer, 2003.
11. R. Metson, R. E. Gliklich, and M. Cosenza. A comparison of image guidance systems for sinus surgery. *Laryngoscope* 108:1164–1170, 1998.
12. F. L. Chassat. Experimental protocol for accuracy evaluation of 6-d localizers for computer-integrated surgery: Application to four optical localizers. In *Medical Image Computing and Computer-Aided Interventions (MICCAI) Lecture Notes in Computer Science.* 277–284, 1998.
13. Q. Li et al. Effect of optical digitizer selection on the application accuracy of a surgical localization system. *Comput. Aided Surg.* 4:314–321, 1999.
14. J. Troccaz, M. Peshkin, and B. L. Davies. The use of localizers, robots, and synergistic devices in CAS. *Proc. First Joint Conference of CVRMed and MRCAS, vol. 1205.* Springer, 727–729, 1997. J. Troccaz, E. Grimson, and R. Mosges (Eds.). *Lecture Notes in Computer Science.*
15. M. A. Peshkin et al. Cobot architecture. *IEEE Trans. Rob. Autom.* 17:377–390, 2001.
16. R. H. Taylor et al. Augmentation of human precision in computer-integrated surgery. *Innovation et Technologie en Biologie et Medicine.* 13:450–459, 1992.
17. B. L. Davies et al. A clinically applied robot for prostatectomies. In *Computer Integrated Surgery: Technology and Clinical Applications.* MIT Press. 593–601, 1996.

18. R. H. Taylor et al. A telerobotic assistant for laparoscopic surgery. In R. Taylor et al. (Eds.). *Computer-Integrated Surgery.* MIT Press. 581–592, 1996.
19. D. Stoianovici et al. A modular surgical robotic system for image-guided percutaneous procedures. In W. Wells and R. Kikinis (Eds.). *Medical Image Computing and Computer-Assisted Interventions (MICCAI-98)*, vol. 1496. *Lecture Notes in Computer Science.* Springer. 404–410, 1998.
20. M. Mitsuishi et al. A telemicrosurgery system with colocated view and operation points and rotational-force-feedback-free master manipulator. *Proc. 2nd Int. Symp. on Medical Robotics and Computer Assisted Surgery.* MRCAS '95 Symposium. 111–118, 1995.
21. S. DiMaio et al. Design of an prostate needle placement robot in MRI scanner. *IEEE International Conference on Biomedical Robotics.* 2006.
22. A. Krieger et al. A hybrid method for 6-DOF tracking of MRI-compatible robotic interventional devices. In *IEEE International Conference on Robotics and Automation.* 2006.
23. K. Chinzei, R. Gassert, and E. Burdet. Workshop on MRI/fMRI compatible robot technology—A critical tool for neuroscience and image guided intervention. In *IEEE Int. Conference on Robotics and Automation.* 2006.
24. E. Hempel et al. An MRI-compatible surgical robot for precise radiological interventions. *Comput. Aided Surg.* 8:180–91, 2003.
25. K. Harada et al. Micro manipulators for intrauterine fetal surgery in an open MRI. In *IEEE International Conference on Robotics and Automation (ICRA).* 504–509, 2005.
26. D. F. Louw et al. Surgical robotics: A review and neurosurgical prototype development. *Neurosurgery.* 54:525–537, 2004.
27. D. Stoianovici et al. MRI-guided robot for prostate interventions. In *Society for Minimally Invasive Therapy (SMIT) 18th Annual Conference.* 2006.
28. B. Davies. A discussion of safety issues for medical robots. In R. Taylor et al. (Eds.). *Computer-Integrated Surgery.* MIT Press, 287–296, 1996.
29. F. Picard, J. Moody, and A. DiGioia. Clinical classification of CAOS systems. In *Computer and Robotic Assisted Knee and Hip Surgery.* Oxford University Press. 43–48, 2004.
30. E. M. Frets et al. A frameless stereotaxic operating microscope for neurosurgery. *IEEE Trans. Biomed. Eng.* 36:608–617, 1989.
31. D. W. Roberts et al. The sonic digitizing microscope. In R. J. Maciunas (Ed.). *Interactive Image-Guided Neurosurgery, USA.* American Association of Neurological Surgeons, 1993.
32. A. P. King et al. Stereo augmented reality in the surgical microscope. *Presence: Teleoperators and Virtual Environments.* 9:360–368, 2000.
33. T. Sasama et al. A novel laser guidance system for alignment of linear surgical tools: Its principles and performance evaluation as a man-machine system. *5th International Conference on Medical Image Computing and Computer-Assisted Intervention, vol. 2489. Lecture Notes In Computer Science.* Springer-Verlag. 125–132, 2002.
34. G. S. Fischer et al. MRI guided needle insertion–comparison of four techniques. *Annual Scientific Conference of the Society of Interventional Radiology.* 2006.
35. D. R. Uecker et al. A speech-directed multi-modal man-machine interface for robotically enhanced surgery. *First Int. Symp. on Medical Robotics and Computer Assisted Surgery (MRCAS 94).* 176–183, 1994.
36. P. K. Gupta. *A Method to Enhance Microsurgical Tactile Perception and Performance Through the Use of Auditory Sensory Perception.* [master's thesis]. Johns Hopkins University, 2001.
37. R. A. Abovitz and A. E. Quaid. The future use of networked haptic learning information systems in computer-assisted surgery. In A. Digioia and L. Nolte (Eds.). *Proc. CAOS USA 2001*, Pittsburgh: CAOS International. 337–338, 2001.
38. O. Gerovich, P. Marayong, and A. M. Okamura. The effect of visual and haptic feedback on computer-assisted needle insertion. *Comput. Aided Surg.* 9:243–249, 2004.
39. A. M. Okamura. Methods for haptic feedback in teleoperated robot-assisted surgery. *Industrial Robot.* 31:499–508, 2004.
40. A. E. Quaid and R. A. Abovitz. Haptic information displays for computer-assisted surgery. *IEEE International Conference on Robotics and Automation.* 2092–2097, 2002.
41. T. Akinbiyi et al. Dynamic augmented reality for sensory substitution in robot-assisted surgical systems. *28th Annual International Conference of the IEEE Engineering in Medicine and Biology Society.* 567–570, 2006.
42. M. Blackwell et al. An image overlay system for medical data visualization. *Med. Image Anal.* 4:67–72, 2000.
43. G. Fichtinger et al. Image overlay guidance for needle insertion on CT scanner. *IEEE Trans. Biomed. Eng.* 52:1415–1424, 2005.
44. G. S. Fischer et al. Musculoskeletal needle placement with MRI image overlay guidance. *Annual Meeting of the International Society for Computer Assisted Surgery.* Montreal, Canada. 158–160, 2006.
45. J. Leven et al. DaVinci canvas: A telerobotic surgical system with integrated, robot-assisted, laparoscopic ultrasound capability. In *MICCAI.* 811–818, 2005.
46. L. Mettler, M. Ibrahim, and W. Jonat. One year of experience working with the aid of a robotic assistant (the voice-controlled optic holder AESOP) in gynaecological endoscopic surgery. *Hum. Reprod.* 13:2748–2750, 1998.

47. H. Reichenspurner et al. Use of the voice-controlled and computer-assisted surgical system zeus for endoscopic coronary artery surgery bypass grafting, *J. Thorac. Cardiovasc. Surg.* 118:1999.
48. R. Sturges and S. Laowattana. A voice-actuated, tendon-controlled device for endoscopy. In R. H. Taylor et al. (Eds.). *Computer-Integrated Surgery.* MIT Press, 1996.
49. R. G. Confer and R. C. Bainbridge. Voice control in the microsurgical suite. In *Proc. of the Voice I/O Systems Applications Conference '84.* 1984.
50. A. Nishikawa et al. FAce MOUSE: A novel human-machine interface for controlling the position of a laparoscope. *IEEE Trans. Rob. Autom.* 19:818–824, 2003.
51. R. H. Taylor et al. A telerobotic assistant for laparoscopic surgery. In *IEEE EMBS Magazine Special Issue on Robotics in Surgery.* 279–291, 1995.
52. A. Krupa et al. Autonomous 3D positioning of surgical instruments in robotized laparoscopic surgery using visual servoing. *IEEE Trans. Rob. Autom.* 19:842–853, 2003.
53. G. Fischer et al. Ischemia and force sensing surgical instruments for augmenting available surgeon information. In *IEEE International Conference on Biomedical Robotics and Biomechatronics—BioRob 2006.* 2006.
54. G. S. Fischer et al. Intraoperative ischemia sensing surgical instruments. *International Conference on Complex Medical Engineering.* 2005.
55. G. Fischer et al. An intra-operative system for relating ischemic damage to retraction forces. *BMES.* 2005.
56. A. Morimoto et al. Force sensor for laparoscopic babcock. In K. S. Morgan et al. (Eds.). *Medicine Meets Virtual Reality.* IOS Press. 354–361, 1997.
57. A. Bicchi et al. A sensorised minimally invasive surgery tool for detecting tissutal elastic properties. *Proc. of the IEEE International Conference on Robotics and Automation.* 884–888, 1996.
58. B. Poulose et al. Human versus robotic organ retraction during laparoscopic Nissen fundoplication. In W. Wells and R. Kikinis (Eds.). *Medical Image Computing and Computer-assisted Interventions (MICCAI-98) Lecture Notes in Computer Science.* 1496:197–206, 1998.
59. P. K. Poulose et al. Human vs robotic organ retraction during laparoscopic Nissen fundoplication. *Surg. Endosc.* 13:461–465, 1999.
60. S. Prasad et al. A modular 2-DOF force-sensing instrument for laparoscopic surgery. *Conference on Medical Image Computing and Computer Assisted Intervention, Montreal.* 279–286, 2003.
61. P. Gupta, P. Jensen, and E. de Juan. Quantification of tactile sensation during retinal microsurgery. *MICCAI99: The Second International Conference on Medical Image Computing and Computer-Assisted Intervention.* 1999.
62. P. Gupta, P. Jensen, and E. de Juan. Surgical forces and tactile perception during retinal microsurgery. In C. Taylor and A. Colchester (Eds.). *MICCAI,* vol. 1679. Springer-Verlag, 1999:1218–1225.
63. J. Rosen et al. The BlueDRAGON—A system for measuring the kinematics and the dynamics of minimally invasive surgical tools *in vivo. IEEE International Conference on Robotics and Automation.* 1876–1881, 2002.
64. J. Rosen, B. Hannaford, M. MacFarlane and M. Sinanan. Force controlled and teleoperated endoscopic grasper for minimally invasive surgery—Experimental performance evaluation. *IEEE Trans. Biomed. Eng.* 46:1212–1221, 1999.
65. A. Menciassi et al. Force sensing microinstrument for measuring tissue properties and pulse in microsurgery. *IEEE/ASME Transactions on Mechatronics.* 8:10–17, 2003.
66. P. J. Berkelman et al. A miniature microsurgical instrument tip force sensor for enhanced force feedback during robot-assisted manipulation. *IEEE Trans. Rob. Autom.* 19:917–922, 2003.
67. D. L. Rothbaum et al. Robot-assisted stapedotomy: Micropick fenestration of the stapes footplate. *Otolaryngol. Head Neck Surg.* 127:417–426, 2002.
68. R. D. Howe et al. Remote palpation technology. *IEEE Eng. Med. Biol.* 14(3):318–323, 1995.
69. R. Beasly and R. Howe. Tactile tracking of arteries in robotic surgery. *IEEE International Conference on Robotics and Automation.* 3801–3806, 2002.
70. M. P. Ottensmeyer and J. K. Salisbury. *In vivo* data acquisition instrument for solid organ mechanical property measurement. *Proceedings of the Medical Image Computing and Computer-Assisted Intervention 4th International Conference.* 975–982, 2001.
71. I. Brouwer et al. Measuring *in vivo* animal soft tissue properties for haptic modeling in surgical simulation. In J. D. Westwood (Ed.). *Medicine Meets Virtual Reality.* IOS Press. 69–74, 2001.
72. S. B. Solomon et al. Robotically driven interventions: A method of using CT fluoroscopy without radiation exposure to the physician. *Radiology.* 225:277–82, 2002.
73. A. Patriciu et al. Robotic kidney and spine percutaneous procedures using a new laser-based CT registration method. *Proceedings to Medical Image Computing and Computer-Assisted Intervention,* vol. 2208. *Lecture Notes in Computer Science.* Springer-Verlag. 249–257, 2001.
74. T. Lozano-Pérez. Robot programming. *Proceedings of the IEEE.* 71, 1983.
75. P. Kazanzides et al. Development of open source software for computer-assisted intervention systems. *ISC/NAMIC/MICCAI Workshop on Open-Source Software.* 2005. Also available online at *Insight Journal* http://hdl.handle.net/1926/46.
76. A. Kapoor, A. Deguet, and P. Kazanzides. Software components and frameworks for medical robot control. *Proc.*

IEEE Intl. Conf. on Robotics and Automation, 2006. In press.

77. P. Kazanzides et al. System architecture and toolkits for image-guided intervention systems. In *Medicine Meets Virtual Reality 14.* 2006.
78. C. Maurer et al. Registration of head volume images using implantable fiducial markers. *IEEE Trans. Med. Imaging.* 16:447–462, 1997.
79. N. G. Levensen and C. S. Turner. An investigation of the Therac-25 accidents. *Computer.* 26:18–41, 1993.
80. R. E. McDermott, R. J. Mikulak, and M. R. Beauregard. *The Basics of FMEA.* Quality Resources, 1996.
81. P. Pott, H. Scharf, and M. Schwarz. Today's state of the art in surgical robotics. *Comput. Aided Surg.* 10:101–132, 2005.
82. B. Mittelstadt et al. The evolution of a surgical robot from prototype to human clinical use. In R. H. Taylor et al. (Eds.). *Computer-Integrated Surgery.* MIT Press. 397–407, 1996.
83. R. H. Taylor et al. An image-directed robotic system for precise orthopaedic surgery. *IEEE Trans. Rob. Autom.* 10:261–275, 1994.
84. J. L. Garbini et al. Robotic instrumentation in total knee arthroplasty. *Proc. 33rd Annual Meeting, Orthopaedic Research Society.* 413, 1987.
85. M. Jakopec et al. The first clinical application of a hands-on robotic knee surgery system, *Comput. Aided Surg.* 6:329–339, 2001.
86. D. S. Kwon et al. The mechanism and the registration method of a surgical robot for hip arthroplasty. *IEEE International Conference on Robotics and Automation.* 1889–2949, 2002.
87. N. M. N. Sugita et al. Development of a computer-integrated minimally invasive surgical system for knee arthroplasty. *IEEE/RAS-EMBS International Conference. Biomedical Robotics and Biomechatronics.* 323–328, 2006.
88. S. Marcacci et al. Computer-assisted knee arthroplasty. In R. H. Taylor et al. (Eds.). *Computer-Integrated Surgery.* MIT Press. 417–423, 1996.
89. W. Siebert and S. Mai. One-year clinical experience using the robot system CASPAR for TKR. In A. Digioia and L. Nolte (Eds.). *Proc. CAOS USA 2001.* CAOS International. 141–142, 2001.
90. W. Bargar et al. Robodoc multi-center trial: An interim report. *Proc. 2nd Int. Symp. on Medical Robotics and Computer Assisted Surgery.* 208–214, 1995.
91. H. Skibbe et al. Revision THR using the ROBODOC system. *CAOS/USA '99.* 110–111, 1999.
92. U. Wiesel et al. Total knee replacement using the Robodoc system. *Proc. First Annual Meeting of CAOS International.* Davos, 88, 2001.
93. H. Paul et al. Development of a surgical robot for cementless total hip arthroplasty. *Clinical Orthop. Relat. Res.* 285:57–66, 1992.
94. P. Kazanzides et al. An integrated system for cementless hip replacement. *IEEE Eng. Med. Biol.* 14:307–313, 1995.
95. S. Cohan. ROBODOC achieves pinless registration. *Ind. Rob.* 28:381–386, 2001.
96. J. Zuhars and T. Hsia. Nonhomogeneous material milling using a robot manipulator with force controlled velocity. *IEEE Intl. Conf. on Robotics and Automation, vol. 2.* 1461–1467, 1995.
97. Y. S. Kwoh et al. A robot with improved absolute positioning accuracy for CT guided stereotactic brain surgery. *IEEE Trans. Biomed. Eng.* 35:153–160, 1988.
98. A. L. Benabid et al. Computer-driven robot for stereotactic surgery connected to CT scan and magnetic resonance imaging. Technological design and preliminary results. *Appl. Neurophysiol.* 50:153–154, 1987.
99. K. Masamune et al. Development of an MRI-compatible needle insertion manipulator for stereotactic neurosurgery. *J. Image Guid. Surg.* 1:242–248, 1995.
100. Q. Li et al. The application accuracy of the NeuroMate robot—A quantitative comparison with frameless and frame-based surgical localization systems. *Comput. Assist. Surg.* 7:90–98, 2002.
101. K. Masamune et al. System for robotically assisted percutaneous procedures with computed tomography guidance. *J. Comput. Assist. Surg.* 6:370–383, 2001.
102. A. Jain et al. Fluoroscope tracking fiducial. *Med. Phys.* 32:3185–3198, 2005.
103. R. C. Susil et al. Transrectal prostate biopsy and fiducial marker placement in a standard 1.5T MRI scanner. *J. Urol.* 175:113–120, 2006.
104. K. Chinzei, R. Kikinis, and F. A. Jolesz. MR compatibility of mechatronic devices: Design criteria. *Second International Conference on Medical Image Computing and Computer-Assisted Intervention,* vol. 1679. *Lecture Notes in Computer Science.* 1020–1030, 1999.
105. E. Watanabe et. al. 3D digitizer (neuronavigator): New equipment for computed tomography-guided stereotaxic surgery. *Surg. Neurol.* 27:543–547, 1987.
106. L. Adams et al. Orientation aid for head and neck surgeons. *Innovation et Technologie en Biologie et Medicine.* 14:409–424, 1992.
107. C. VanderKolk et al. An interactive 3D-CT surgical localizer for craniofacial surgery. In A. Montoya (Ed.). *Craniofacial Surgery.* 25, 1992.
108. C. B. Cutting, F. L. Bookstein, and R. H. Taylor. Applications of simulation, morphometrics and robotics in craniofacial surgery. In R. H. Taylor et al. (Eds.). *Computer-Integrated Surgery.* MIT Press. 641–662, 1996.
109. P. S. Schenker, H. O. Das, and R. Timothy. Development of a new high-dexterity manipulator for robot-assisted microsurgery. *Proceedings of SPIE—The International Society for Optical Engineering: Telemanipulator and Telepresence Technologies vol. 2351.* 191–198, 1995.

110. M. C. Cavusoglu et al. Robotics for telesurgery: Second generation Berkeley/UCSF laparoscopic telesurgical workstation and looking towards the future applications. *Ind. Rob.* 30(1):22–29, 2003.
111. S. E. Salcudean, S. Ku, and G. Bell. Performance measurement in scaled teleoperation for microsurgery. *Proc. First Joint Conference of CVRMed and MRCAS.* Springer. 1205:789–798, 1997. J. Troccaz, E. Grimson, and R. Mosges (Eds.). *Lecture Notes in Computer Science.*
112. K. Ikuta, T. Hasegawa, and S. Daifu. Hyper redundant miniature manipulator hyper finger for remote minimally invasive surgery in deep area. *IEEE Conference on Robotics and Automation.* 1098–1102, 2003.
113. K. Hongo et al. NeuRobot: Telecontrolled micromanipulator system for minimally invasive microneurosurgery-Preliminary results. *Neurosurgery.* 51:985–988, 2002.
114. T. Kanade, B. Davies, and C. Riviere. Special issue on medical robotics. *IEEE Proc.* 94, 2006.
115. G. S. Guthart and J.K. Salisbury. The intuitive telesurgery system: Overview and application. *Proc. of the IEEE International Conference on Robotics and Automation (ICRA2000).* 2000.
116. J. Marescaux and F. Rubino. The ZEUS robotic system: Experimental and clinical applications. *Surg. Clin. North Am.* 83:1305–1315, 2003.
117. K. Ikuta, K. Yamamoto, and K. Sasaki. Development of remote microsurgery robot and new surgical procedure for deep and narrow space. *IEEE Conference on Robotics and Automation.* 1103–1108, 2003.
118. N. Suzuki, M. Hayashibe, and A. Hattori. Development of a downsized master-slave surgical robot system for intragastic surgery. *ICRA Surgical Robotics Workshop. IEEE Robotics and Automation Society,* 2005.
119. K. Ikuta et al. Multi-degree of freedom hydraulic pressure driven safety active catheter. *Proceedings of the 2006 IEEE International Conference on Robotics and Automation.* 4161–4166, 2006.
120. L. Ascari et al. A new active microendoscope for exploring the sub-arachnoid space in the spinal cord. *IEEE Conference on Robotics and Automation,* 2:2657–2662, 2003.
121. S. W. Grundfest, J. W. Burdick, and A. B. Slatkin. The development of a robotic endoscope. *IEEE International Conference on Robotics and Automation.* 162–171, 1995.
122. C. Stefanini, A. Menciassi, and P. Dario. Modeling and experiments on a legged microrobot locomoting in a tubular, compliant and slippery environment. *Int. J. Rob. Res.* 25:551–560, 2006.
123. N. Patronik et al. The HeartLander: A novel epicardial crawling robot for myocardial injections. *Proceedings of the 19th International Congress of Computer Assisted Radiology and Surgery,* vol. 1281C. Elsevier. 735–739, 2005.
124. T. M. Goradia, R. H. Taylor, and L. M. Auer. Robot-assisted minimally invasive neurosurgical procedures: First experimental experience. *Proc. First Joint Conference of CVRMed and MRCAS.* 1205:319–322, 1997. J. Troccaz, E. Grimson, and R. Mosges (Eds.). *Lecture Notes in Computer Science.*
125. R. Taylor et al. Steady-hand robotic system for microsurgical augmentation. *Int. J. Rob. Res.* 18:1201–1210, 1999.
126. M. A. Gomez-Blanco, C. N. Riviere, and P. K. Khosla. Intraoperative tremor monitoring for vitreoretinal microsurgery. *Proc. Medicine Meets Virtual Reality.* 8:99–101, 2000.
127. A. Kapoor, R. Kumar, and R. Taylor. Simple biomanipulation tasks with a steady hand cooperative manipulator. *Proceedings of the Sixth International Conference on Medical Image Computing and Computer Assisted Intervention—MICCAI 2003.* Springer. 1:141–148, 2003. *Lecture Notes in Computer Science (Vol. 2878).*
128. I. Iordachita et al. Steady-hand manipulator for retinal surgery. In K. Cleary (Ed.). *MICCAI Workshop on Medical Robotics.* Copenhagen. 66–73, 2006.
129. C. V. Riviere, R. S. Rader, and N. V. Thakor. Adaptive real-time cancelling of physiological tremor for microsurgery. *Proc. 2nd Int. Symp. on Medical Robotics and Computer Assisted Surgery (MRCAS).* 1995.
130. W. T. Ang and C. N. Riviere. Neural network methods for error canceling in human-machine manipulation. *22nd Annu. Conf. IEEE Eng. Med. Biol. Soc., Istanbul, 2001.* 3462–3465.
131. N. Simaan, R. Taylor, and P. Flint. High dexterity snake-like robotic slaves for minimally invasive telesurgery of the throat. *Int. Symp. on Medical Image Computing and Computer-Assisted Interventions.* Springer. 2:17–24, 2004.
132. N. Simaan et al. Robotic surgery of the upper airways: Addressing the challenges of dexterity enhancement in confined spaces. In R. Faust (Ed.). *Robotics in Surgery: History, Current, and Future Applications.* Nova Science Publishing. 261–280, 2007.
133. S. Xu et al. 3D motion tracking of pulmonary lesions using CT fluoroscopy images for robotically assisted lung biopsy. *SPIE International Society of Optical Engineering.* 5367:394–402, 2004.
134. M. Shoham et al. Bone-mounted miniature robot for surgical procedures: Concept and clinical applications. *IEEE Trans. Rob. Autom.* 19:893–901, 2003.
135. C. Plaskos et al. Praxiteles: A miniature bone-mounted robot for minimal access total knee arthroplasty. *Int. J. Med. Rob. Comput. Assist. Surg.* 1:67–79, 2005.
136. J. A. McEwen et al. Development and initial clinical evaluation of pre-robotic and robotic retraction systems for surgery. *Proc. Second Workshop on Medical and Health Care Robotics.* 91–101, 1989.

137. A. Kochan. Scalpel please, robot: Penelope's debut in the operating room. *Ind. Rob.* 32:449–451, 2005.
138. F. Miyawaki et al. Scrub nurse robot system-intra-operative motion analysis of a scrub nurse and timed-automata-based model for surgery. *IEEE Trans. Ind. Electronics.* 52:1227–1235, 2005.
139. E. Begin, M. Gagner, and R. Hurteau. A robotic camera for laparoscopic surgery: Conception and experimental results. *Surg. Laparosc. Endosc.* 5:1995.
140. J. M. Sackier and Y. Wang. Robotically assisted laparoscopic surgery. From concept to development. *Surg. Endosc.* 8:63–66, 1994.
141. R. Satava. Robotics, telepresence, and virtual reality: A critical analysis fo the future of surgery. *Minimally Invasive Therapy.* 1:357–363, 1992.
142. R. Satava. Virtual reality, telesurgery, and the new world order of medicine. *J. Image Guid. Surg.* 1:12–16, 1995.
143. P. Green et al. Telepresence: Advanced teleoperator technology of minimally invasive surgery [abstract]. *Surg. Endosc.* 6:1992.
144. M. Mitsuishi et al. Remote ultrasound diagnostic system. *Proc. IEEE Conf. on Robotics and Automation. Seoul.* 1567–1574, 2001.
145. D. d. Cunha et al. The MIDSTEP system for ultrasound guided remote telesurgery. *IEEE EMBS.* IEEE. 20/3: 1266–1269, 1998.
146. B. R. Lee et al. TELEPAKY: A new robotic system for active remote telesurgery. *Lancet.* 1999.
147. J. Bauer et al. Remote percutaneous renal access using a new automated telesurgical robotic system. *Telemed. J. E. Health.* 7:341–46, 2001.
148. D. Frimberger et al. Telerobotische Chirurgie zwischen Baltimore und München. *Der Urologe A.* 41:489–492, 2002.
149. J. Marescaux et al. Transatlantic robot-assisted telesurgery. *Nature.* 413:379–380, 2001.
150. M. Anvari et al. The impact of latency on surgical precision and task completion during robotic-assisted remote telepresence surgery. *Comput. Aided Surg.* 10:93–99, 2005.
151. L. Dotto. Application-revolutionary telemedicine techniques, 2006. Summary online article about Anvari's remote telesurgery work, http://www.haivision.com/downloads/CSCmas.pdf.
152. L. Kavoussi et al. Telerobotic-assisted laparoscopic surgery: Initial laboratory and clinical experience. *Urology.* 44:15–19, 1994.
153. E. Hanly et al. Mentoring console improves collaboration and teaching in surgical robotics. *J. Laparoendosc Adv. Surg. Tech.* 16(5):445–451, 2006.
154. B. Herman et al. Telerobotic surgery creates opportunity for augmented reality surgery. *Telemed. J. E. Health.* 11:203, 2005.
155. H. C. Lin et al. Automatic detection and segmentation of robot-assisted surgical motions. *MICCAI.* 3749:802–810, 2005. *Lecture Notes in Computer Science.*
156. L. B. Rosenberg. Virtual fixtures: Perceptual tools for telerobotic manipulation. *Proceedings of IEEE Virtual Reality International Symposium.* 76–82,1993.
157. S. Park, R. D. Howe, and D. F. Torchiana. Virtual fixtures for robotic cardiac surgery. *Fourth International Conference on Medical Image Computing and Computer-Assisted Intervention,* 2001.
158. M. Li and A. M. Okamura. Recognition of operator motions for real-time assistance using virtual fixtures. *11th International Symposium on Haptic Interfaces for Virtual Environment and Teleoperator Systems.* 125–131, 2003.
159. P. Marayong A. Bettini, and A. Okamura. Effect of virtual fixture compliance on human-machine cooperative manipulation. *EEE/RSJ International Conference on Intelligent Robots and Systems,* 2002. 1089–1095.
160. M. Li and R. H. Taylor. Spatial motion constraints in medical robots using virtual fixtures generated by anatomy. *IEEE Conf. on Robotics and Automation.* 1270–1275, 2004.
161. H. Mayer, I. Nagy, and A. Knoll. Skill transfer and learning by demonstration in a realistic scenario of laparoscopic surgery. *IEEE International Conference on Humanoids.* 2003.
162. D. Kragic et al. Human-machine collaborative systems for microsurgical applications. B. Siciliano, O. Khatib, and F. C. A. Groen (Eds.). *International Journal of Robotics Research.* 24:731–745, 2005.
163. M. Li, M. Ishii, and R. H. Taylor. Spatial motion constraints in medical robot using virtual fixtures generated by anatomy. *IEEE Trans. Rob.* 23(1):4–19, 2007.
164. A. Kapoor, M. Li, and R. H. Taylor. Constrained control for surgical assistant robots. *IEEE Int. Conference on Robotics and Automation.* 2006:231–236.
165. R. H. Taylor et al. *Computer Integrated Surgery.* MIT Press. 1996.
166. R. Faust. *Robotics in Surgery: History, Current, and Future Applications.* Nova Science Publishing, 2007.
167. R. H. Taylor, J. Troccaz, and P. Dario. Special issue on medical robotics. *IEEE Trans. Rob. Autom.* 19:2003.

19

Functional Techniques for Brain Magnetic Resonance Imaging

Dr. Sirong Chen,[1,2]
Dr. Kai-Ming Au Yeung,[1] and
Dr. Gladys Goh Lo[1]
[1]*Hong Kong Sanatorium & Hospital*
[2]*University of Sydney*

19.1 Introduction

19.1.1 Background

The increased use of computerized processing and analysis techniques in medical imaging modalities, along with the rapid advance in information technology has resulted in significant advancement in medicine and health care [1–4]. These advances have given rise to new 2D, 3D, and multidimensional imaging modalities such as computerized axial tomography (CAT or CT), magnetic resonance imaging (MRI), single photon emission computed tomography (SPECT), positron emission tomography (PET), and fused imaging modalities such as SPECT/CT, PET/CT, and their clinical significance in the diagnosis and treatment of disease is overwhelming.

The introduction of clinical MRI in the 1980s completed the modern revolution of neuroradiology. MRI, unlike CT, depends on the physical phenomenon of the nuclei spinning to provide biomedical information. MRI is free from the radiation and is therefore a relatively safe technique. It can characterize and discriminate various tissues by using their physical and biochemical properties such as water, iron, fat, and blood products. Since calcium emits no signal on MRI, tissues surrounded by bone, including the posterior fossa and spine, can be imaged without beam-artifact such as occur in CT scan. With its high spatial and tissue contrast resolution, MRI can also provide excellent delineation of anatomic structures. The ability to obtain multiplane images with equivalent resolution without moving the patient offers special advantages for diagnosis and radiation and surgical treatment planning. In addition, MRI contrast agents are well tolerated, with fewer allergic reactions and nephrotoxic effects compared to CT contrast agents. The application of MRI in various aspects of medical science enormously benefits patients, medical practitioners, and scientists. Therefore MRI has been widely accepted in the medical community since its first use on humans by Dr. Raymond Damadian in 1977 [5].

The progress in MRI technology has moved very rapidly and the pace shows little sign of slowing. Major advances in MR technology and particularly in magnetic field gradient designs have created a new popularity in the application of high-speed MRI. The recent development of integrating high-field magnet (3.0T) into MR machines gives rise to higher spatial and temporal resolution in both functional and anatomic imaging. Software improvements include novel pulse sequence designs and advanced image processing techniques to speed up reconstruction and to correct for patient motion, etc. Dynamic contrast-enhanced MRI techniques for evaluating soft tissue masses or cervical lymph nodes can help to differentiate nonmalignant tissue from malignant tumors [6–10]. Because structural imaging techniques alone are inadequate for defining the cerebral function that is increasingly important in clinical assessment, functional imaging techniques were investigated to improve the evaluation of various pathologic processes [7–14].

Functional imaging techniques other than MRI, like nuclear medicine techniques, involve ionizing radiation and have disadvantages from imprecise anatomic localization due to poor spatial and temporal resolution [15, 16]. Functional MRI can produce an image with spatial resolution of less than 1mm apart, whereas the latest commercial PET scanners can resolve images of structures within 4mm of each other. With the continuous support of increasingly powerful computer and advanced image processing methods, MRI allows for not only qualitative but also quantitative analysis, which is extremely valuable for understanding disease and planning treatment. Given its higher spatial and temporal resolution, lack of ionizing radiation, and lower cost, functional MRI may become more widely used than other existing imaging modalities for clinical brain function mapping. Therefore, under the goal of providing an integrated view of brain structure, chemistry, and physiology, MRI occupies a central and currently irreplaceable position in the field of diagnostic brain imaging in spite of many exciting and promising emerging functional techniques.

19.1.2 Overview of Brain Magnetic Resonance Imaging

MRI provides multiplanar large field-of-view images of the body with excellent soft tissue contrast but without ionizing radiation, making it a pivotal imaging modality in the evaluation of brain disease [17, 18]. This noninvasive imaging technique requires a strong and homogeneous external magnetic field (1.5 to 3T). During the examination, the transmitting coils emit radiofrequency (RF) magnetic field to excite the biological tissues; then the excited tissues would release the energy (the magnetic signals), which would be detected by the receiving coils. The received magnetic signals are then transmitted to the computer, where the magnetic signals are reconstructed to MR images. The intensity of MR signals depends on T1, T2 relaxation time and proton density. Usually high-resolution anatomic brain MRI will make use of T1-weighted, T2-weighted, proton density-weighted, or fluid attenuated inversion recovery (FLAIR) sequences. Recent development of functional MRI techniques has also expanded the clinical brain MRI application because of its unequivocally established relationship with physiologic function, energy metabolism, and localized blood supply [19–23]. Based on detecting the changes of diverse physiologic parameters, functional brain MRI can potentially differentiate the pathologic tissue from normal brain tissue.

Functional brain MRI techniques comprise of diffusion-weighted imaging (DWI) for cerebral water molecule mobility assessment, perfusion-weighted imaging (PWI) for cerebral tissue perfusion assessment, MR spectroscopy (MRS) for cerebral chemical metabolites assessment, and blood oxygenation level dependent (BOLD) technique for functional brain localization or lateralization (fMRI). The combination of functional and anatomic information affords a new means of understanding the origin and temporal sequences of various brain diseases.

The material in this chapter was presented at an introductory level for those who have little experience in brain MRI. In addition to the overview of brain MRI, different functional MRI techniques will be covered in the following subsections. After reading this chapter, readers will be inspired by the potential use of MRI techniques for unraveling the mysteries of the human psyche and brain.

19.2 Diffusion-Weighted Magnetic Resonance Imaging in Brain

In addition to T1, T2 relaxation time and proton density, molecular motion due to pulsatile flow and convective or diffusion processes also contribute to MR signal intensity. Diffusion is an important process to MRI or nuclear magnetic resonance (NMR) imaging. In the diffusion MRI, regions of the brain are depicted not only based on physical properties, such as T2 relaxation time and spin density, but also on local characteristics of water molecule diffusion [24, 25]. DWI is used to probe random microscopic motion of water protons, that is, the Brownian motion, on a pixel basis [26]. Shifts of the water protons between tissue compartments are related to physical or anatomic constraints including permeability of the cell membrane, osmolarity of the tissue fluid, and active transportation, all of which have an impact on the extent of proton mobility, or diffusivity. This property makes DWI a powerful tool for the diagnosis of diseases involving alteration in water mobility, such as ischemia, multiple sclerosis (MS), dyslexia, schizophrenia, trauma, and various intracranial abnormalities [27–30].

19.2.1 Basic Principles

The physical model of diffusion is one of successive small, random steps since the size and direction of each step is unrelated to the preceding ones (Figure 19.1a). After a large number of individual steps, there will be, on average, no net displacement of a molecule from its starting location. However, there will be a region around the starting location where the molecule could be expected at any one time (Figure 19.1b). If τ is the average time between steps, t is the total diffusion time, then, after a large number of steps, the mean-squared (expected) displacement from the starting location, $\overline{\Delta x^2}$, is directly related to the mean-squared displacement of each step, $\overline{d^2}$:

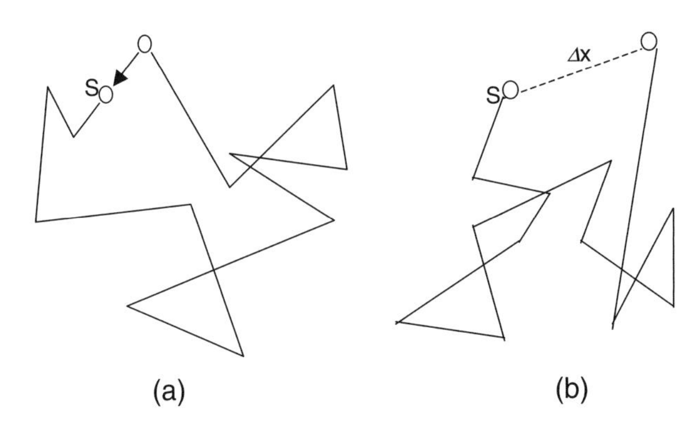

FIGURE 19.1 (a) The physical model of random diffusion movement. (b) There will be an increasing large region in which the molecule can be expected at any one time.

$$\overline{\Delta x^2} = \left(\frac{\overline{d^2}}{\tau}\right) t = 2Dt, \quad (19.1)$$

where $D(= \overline{d^2}/2\tau)$, the constant of proportionality, is the free diffusion coefficient term. This equation applies to the inter-diffusion of molecules of a single type and can be measured experimentally by the use of NMR excitation or radioactive isotope labeling.

To understand how diffusion can lead to net displacement, assume a gradient of concentration, *dC/dx*. According to Fick's first law of diffusion, the net flux, *J*, or net movement across the plane (of area *S*) per time τ can be expressed as

$$J = -SD\frac{dC}{dx}, \quad (19.2)$$

which describes that diffusion acts to even out concentrations. Because diffusion acts to eliminate any concentration differences, there needs to be a mechanism to maintain unequal chemical solutions of the living systems. This is possible through the use of membrane barriers surrounding individual cells or organelles.

Most DWI sequences are based on the spin-echo (SE) Stejska-Tanner sequence. Diffusion sensitization is typically added to SE sequence by applying a pair of pulsed magnetic field gradients as shown in Figure 19.2. After the initial 90° pulse (dephasing pulse), the ensemble of transverse spins rapidly gets out of phase. Then after the 180° pulse (rephrasing pulse), the dephased spins would match the pulsed gradient again. If the position of the spins does not change between the two pulsed gradients, the second pulse would generate a similar spatially dependent precession rate variation. Therefore, stationary water molecules would be fully rephrased. However, moving water molecules would cause signal loss owing to incomplete rephrasing, that is, diffusion between the two gradients would reduce the NMR signal intensity. The diffusion decay factor is normally formed as *exp* (*-bD*), where *D* is the diffusion coefficient of water molecule and *b* is a constant defined as

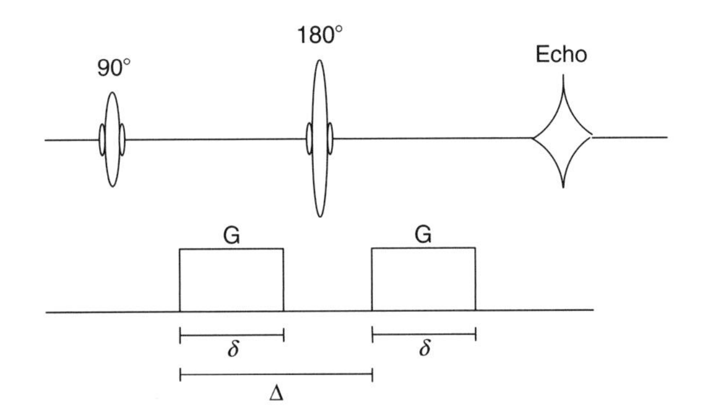

FIGURE 19.2 A typical SE diffusion MRI sequence.

$$b = \gamma^2 G^2 \delta^2 (\Delta - \delta/3) \quad (19.3)$$

where γ is the proton gyromagnetic ratio, and *G*, δ, and Δ are the magnitude, duration, and interval of the pulsed gradient pair, respectively (Figure 19.2). Raw images obtained with a high *b* value are often called diffusion-weighted images.

Since most DWI sequences are based on SE sequence, diffusion-weighted images usually retain some properties of the SE image (i.e., with T2 contribution). When T2 elevation dominates (T2 shine-through effect), the diffusion-weighted image shows hyperintensity in the presence of rapid or facilitated diffusion. Therefore, diffusion image (not DWI), free from T2 contribution, is desired. An alternative solution is to create two (or more) images with different diffusion gradients, which is the only difference between the two DWI sequences, then the ratio of the two images, *S*1 and *S*2, would be:

$$\frac{S_1}{S_2} = \frac{\exp[-b_1 D]}{\exp[-b_2 D]} = \exp[(b_2 - b_1)D], \quad (19.4)$$

where b_1 and b_2 are the *b* factors of the corresponding sequences. Taking the logarithm of the ratio image and dividing the result of each pixel by the known quantity (b_2–b_1) would generate an image of the diffusion coefficient (*D*).

In biological tissues, factors other than diffusion contribute to the signal loss, such as vessel flow, cerebrospinal fluid flow or restriction due to organelles, and cell membrane or fiber packing. Therefore, the apparent diffusion coefficient (ADC), which combines the effects of capillary perfusion and water diffusion in the extracellular space, is preferred [31].

Normally, there are two kinds of diffusion in the living system. Diffusion within tissues that has a random microstructure or unrestricted media will have diffusion equal in all directions, called isotropic diffusion. On the other hand, diffusion within regions that has restricted movement of molecules refers to anisotropic diffusion. The diffusion within the nerve fiber is anisotropic diffusion.

Despite its well-established clinical advantage, DWI is not without pitfalls. High diffusion gradients make the images very sensitive to other types of motion, particularly patients' motion and blood flow. Therefore, development of effective motion suppression methods is very important. Signal averaging, restraining holders, and sedation can help to suppress bulk motions, such as breathing and jerking [32]. Motion artifacts can largely be eliminated with the use of diffusion-weighted echo-planar MRI whose acquisition time is within 150 msec [33].

19.2.2 Clinical Applications

The clinical application of DWI began in the last decade with the demonstration of its capabilities for depicting the anatomy of the white matter fiber tracts in the brain. With its quantitative evaluation based on the diffusion tensor, DWI has a role in the assessment of brain maturation and white matter diseases in the fetus, neonate, and child. Figure 19.3 demonstrates the brain DWI of a normal volunteer, depicting the size and course of major white matter pathways. In the adult, the white matter tracts (association, projection, and commissural white matter pathways) of both peripheral and central nervous system (CNS) could be mapped using the diffusion tensor imaging (DTI). Figure 19.4 shows the displacement (rather than destruction) of the right white matter tracts of a patient with right frontal meningioma, which is very important for planning the operation. MS is an inflammatory disease of the CNS, leading to a progressive decline of motor and sensory functions, and eventually permanent disability. In the setting of MS, diffusion increases in plaques due to demyelination. Therefore, the ADC or D values of the normal-appearing white matter are increased compared to the control values. In addition, the D value of the normal-appearing grey matter is increased as well, which correlates with the cognitive deficit. Diffusion change might be a more sensitive marker in predicting the progression of MS disease than conventional imaging findings [34].

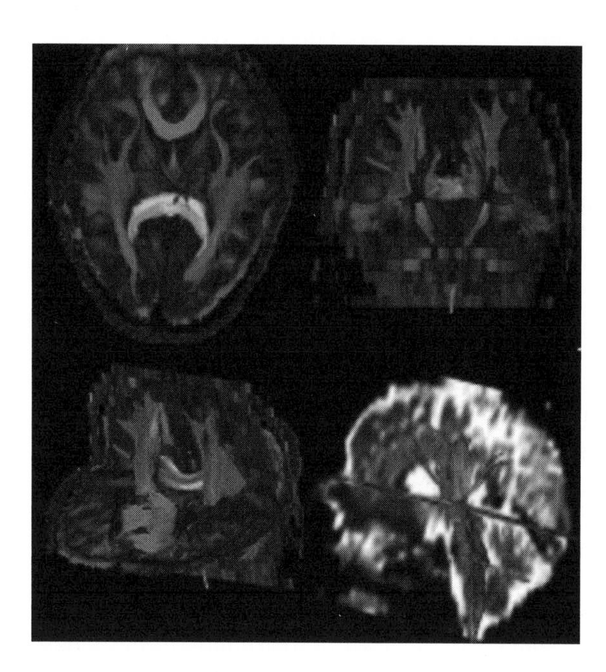

FIGURE 19.3 The diffusion tensor tractography (direction-encoded color map) of the normal volunteer identifying the major association, commissural, and projection pathways in the brain. The different colors of the white matter tracts represent their different directions/orientations. For a more detailed view of this figure, please visit our companion site at: http://books.elsevier.com/companions/9780123735836.

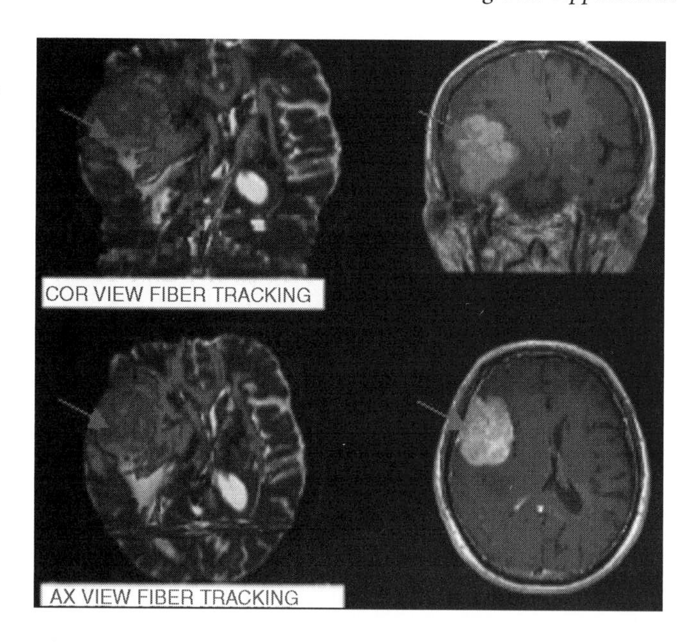

FIGURE 19.4 The DTI (fiber tracking) of a patient with right frontal meningioma, depicting the displacement (rather than destruction) of the right white matter tracts. For a more detailed view of this figure, please visit our companion site at: http://books.elsevier.com/companions/9780123735836.

Cerebrovascular disease is one of the leading causes of death worldwide. Of the 60% of patients who survive a first stroke, only 10% recover completely, while the majority have permanent disability, with many requiring institutional care. The most important and widely used clinical application of DWI is in the field of cerebrovascular disease, which includes cerebral ischemia and infarction. Patients with signs or symptoms of cerebral ischemia are now evaluated by DWI, which is much more sensitive than CT. As shown in Figure 19.5, the DWI

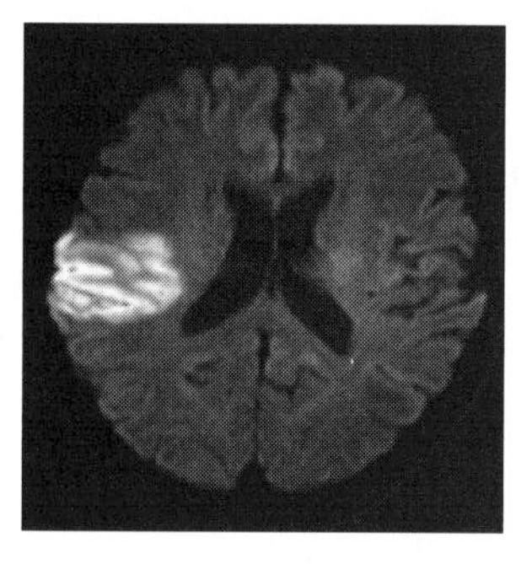

FIGURE 19.5 The diffusion-weighted image shows marked hyperintensity in the acute infarction at the right cerebral hemisphere.

demonstrates marked hyperintensity at the right cerebral hemisphere, indicating acute infarction. DWI could provide one of the earliest demonstrations of ischemic lesions. In one study [27], during the first hour following middle cerebral artery occlusion in a cat, changes could be obtained on DWI when no abnormality was found on T1- or T2-weighted images. T2 signal abnormality did not become evident until 6 to 8 hours or as late as 12 hours after the ischemic event [35]. The ADC decreases in the acute phase but it will ultimately increase in the chronic phase of the stroke as a result of increased local water content [36].

DWI could differentiate arachnoid cyst and epidermoid, which is a difficult task for conventional MRI because both arachnoid cyst and epidermoid show hyperintensity on T2-weighted images (Figure 19.6a,b). As shown in Figure 19.6c,d, arachnoid cysts show increased diffusion; whereas epidermoid has relative restricted diffusion. DWI is also suitable for determining glioma grade and regions of active tumor growth [37]. Furthermore, it is helpful to assess the tumor response to therapy, possible therapy failure, and therapy complications such as radiation necrosis.

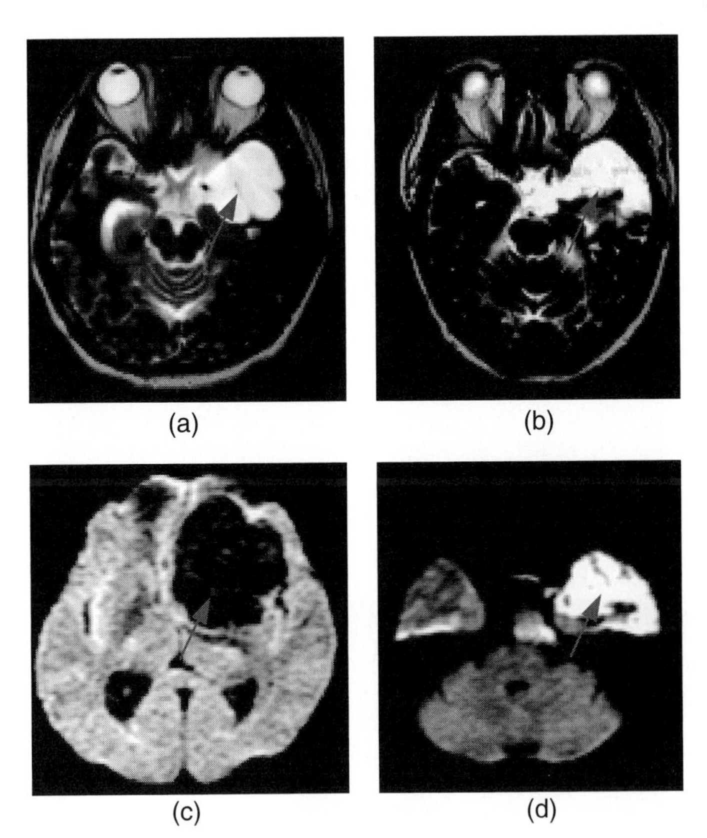

FIGURE 19.6 T2-weighted images of (a) arachnoid cyst and (b) epidermoid, which have similar signal intensity and appearance. (c) DWI of arachnoid cyst shows unrestricted diffusion. (d) DWI of epidermoid depicts restricted diffusion.

19.3 Magnetic Resonance Perfusion Imaging in Brain

The perfusion-weighted MR imaging of brain tissue is widely used. The term perfusion refers to the passage of blood in the brain capillaries, most often related to contrast agent from an arterial supply to venous drainage through the cerebral microcirculation. PWI is sensitive to microscopic tissue-level blood flow. The PWI techniques may or may not require intravenous administration of an MR contrast agent. Most current clinical experience is with contrast-based technique. Quantitative imaging of cerebral perfusion has been commonly performed with nuclear medicine techniques. Advantages of PWI over nuclear medicine techniques include relatively lower cost, higher spatial resolution, and easier comparison with anatomic images.

19.3.1 Basic Principles

MR perfusion imaging is based on monitoring the passage of a nondiffusible tracer, such as the paramagnetic gadolinium-diethyltriamine pentaacetic acid (GdDTPA), within the brain tissue dynamically. The paramagnetic contrast agent (tracer) has both T1 shortened effect through dipole-dipole interaction (routinely associated with contrast-enhancement) and T2 (T2*) shortened effect through magnetic susceptibility effect. If the blood-brain barrier is intact, the contrast agent in the CNS can only produce enhancement of the blood pool. Nevertheless, the magnetic susceptibility effect due to the magnetic field gradient between the lumen of the vessel and surrounding tissue would cause loss of coherence of spins, whose effect is more dominant than T1 shortened effect. The susceptibility effect is most significant during the first passage of contrast bolus through the brain. Gd-DTPA is a suitable contrast agent for PWI in view of its availability and T2 (T2*) susceptibility effect [38]. The mechanism causing shortened T2 (T2*) is very efficient in the capillaries due to their large intravoxel dispersion and relatively large surface area (compared with arteries and veins). The T2 (T2*) MR signal drop of a brain region depends on both the vascular contrast concentration and the concentration in small (3 to 10 μm) vessels [39, 40], and is therefore served as the relative perfusion to that region.

PWI could be performed using the SE or gradient-echo pulse sequence. The sensitivity of the two acquisition methods for evaluating pathologic brain abnormalities may be similar. However, SE-based PWI is more accurate than gradient-echo measurement for representing the capillary perfusion in the human brain [41] because the later technique may incorporate extensive artifacts from the large cerebral vessels [42].

PWI technique is capable of providing quantitative as well as qualitative assessment of brain microcirculation. By using rapid imaging techniques to resolve the tissue transit of intravenously administered contrast agents and applying tracer kinetic modeling techniques to the acquired dynamic perfusion images, the diagnostically valuable parameters, such as the cerebral blood volume (CBV), cerebral blood flow (CBF), and mean transit time (MTT), could be estimated. To apply tracer kinetic analysis for the measurement of these physiologic parameters, a relationship between the PWI signal intensity and local contrast concentration should be determined to generate the contrast concentration-time curve of the region of interest (ROI). Analysis of these curves by accounting for the kinetics and compartmentalization of contrast delivery to tissue as well as how these factors affect MR signal intensity, will allow calculation of the above-mentioned tissue microcirculatory perfusion parameters.

19.3.2 Quantitative Analysis

By repeated imaging in a short interval (typically 1 to 2 seconds), one can observe magnetic susceptibility effect within the brain tissue during the first pass of the contrast bolus as a transient signal drop from the baseline shown in Figure 19.7. The degree of signal intensity drop depends on the contrast concentration, relative blood volume, and a number of hemodynamic parameters. Villringer et al. [43] experimentally verified that the relationship between the signal intensity and T2 relaxation rate change ($\Delta R2 = 1/\Delta T2$) could be approximated by

$$S(t) = S_0 e^{[-TE(\Delta R2(t))]} \tag{19.5}$$

where S_0 and $S(t)$ are the signal intensities at baseline and time t, and TE is the echo time. Hence, MR signal intensity-time curve can be readily converted into $\Delta R2$-time curve:

$$\Delta R2(t) = -\frac{\ln(S(t)/S_0)}{TE} \tag{19.6}$$

Both theoretic and empiric data [44] have shown a linear relationship between tissue contrast concentration ($C_t(t)$) and T2 relaxation rate change:

$$\Delta R2(t) = k_2 C_t(t) \tag{19.7}$$

where k_2 is a constant that depends on tissue type, magnetic field strength, and MR pulse sequence [40]. Therefore, the contrast concentration-time curve in brain tissue ($C_t(t)$) can be measured by mapping the MR signal intensity dynamically through

$$C_t(t) = \frac{-\ln(S(t)/S_0)}{k_2 TE} \tag{19.8}$$

With the measurement of $C_t(t)$, CBV can be determined by the ratio of the areas under the tissue and arterial concentration-time curves within a given ROI:

$$CBV = \frac{\int_0^\infty C_t(t)dt}{\int_0^\infty I(t)dt} \tag{19.9}$$

where $I(t)$ is the arterial input function (AIF). Since directly measuring the AIF by blood sampling is very invasive, relative CBV (rCBV) approximated by the integral of the tissue concentration-time curve, is therefore often used. Recirculation in blood vessels always occurs before complete washout of the contrast is finished [45, 46]. By modeling the measured $C_t(t)$ using the gamma-variate function and proper curve fitting methods, such as the nonlinear regression methods, the tissue concentration-time curve could be corrected. MTT (one of the hemodynamic parameters) is the average time taken for the contrast passing through the tissue following bolus injection. It could be evaluated for each pixel in the concentration-time curve, referring to the time to the centroid. By central volume principle:

$$CBF = \frac{CBV}{MTT} \tag{19.10}$$

CBF could be calculated.

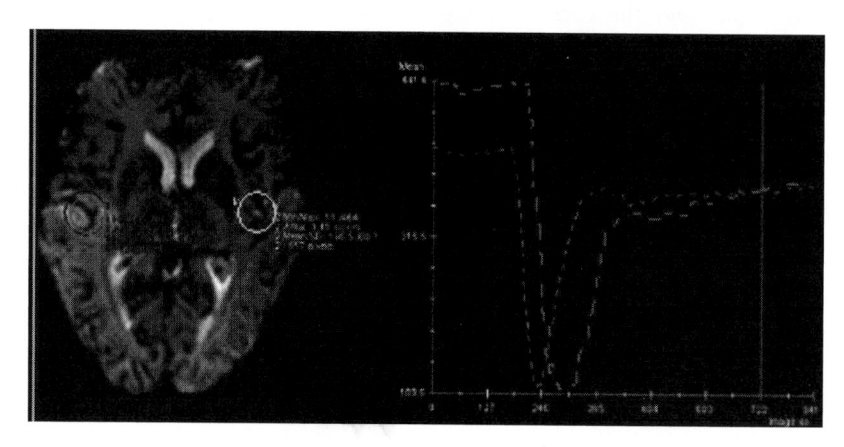

FIGURE 19.7 A frame of bolus-tracking perfusion-weighted imaging and MR signal intensity-time curves of two ROIs.

Although CBV could be obtained approximately by the integral of the tissue concentration-time curve, CBF determination, which is associated with the flow-weighted tissue impulse response, is limited by the fact that an ideal contrast bolus has a duration of 0 seconds, whereas in clinical reality, contrast injection is of relatively long duration. Accordingly, the corrected $C_t(t)$ should be equal to the convolution of flow-weighted impulse response model with the AIF. Usually, the arterial contrast concentration in vessels directly supplying the brain tissue could be approximated by dynamic MRI measurements of the carotid artery. Given the high spatial and temporal resolution, MRI could provide sufficient information to characterize both $I(t)$ and $C_t(t)$ curves. Then, deconvolution of the gamma-variate corrected $C_t(t)$ with the image-derived AIF from the carotid artery could be applied to estimate the model parameters including CBF.

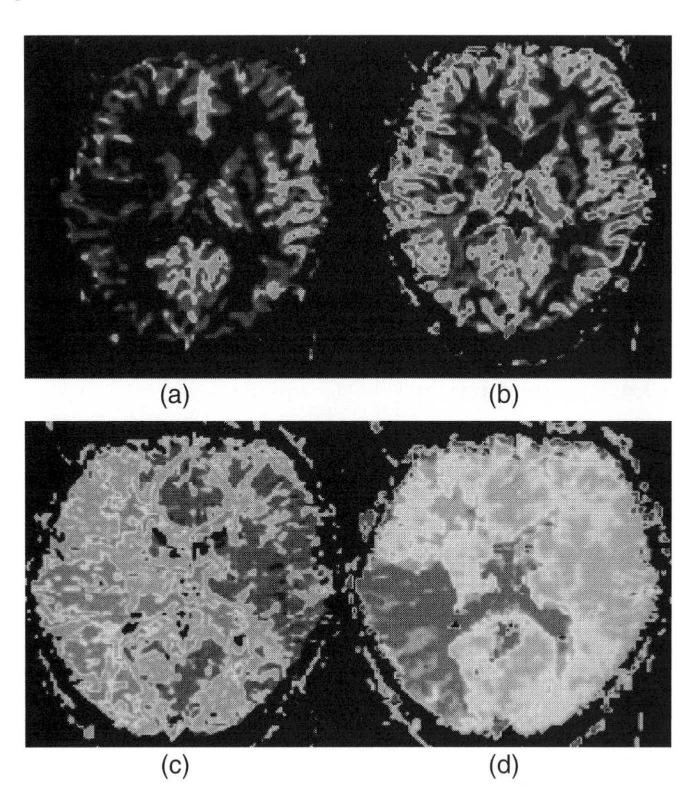

FIGURE 19.8 Maps of (a) CBF, (b) CBV, (c) MTT, and (d) TTP obtained from the perfusion-weighted images of a patient with acute infarct. The MTT and TTP maps demonstrate elevated signal intensity at the posterior right temporal lobe. The CBF map depicts reduced blood flow not only at the corresponding region but also at the right frontal lobe (to a lesser degree). For a more detailed view of this figure, please visit our companion site at: http://books.elsevier.com/companions/9780123735836.

19.3.3 Clinical Applications

Vascular disease of the brain, with the associated tissue ischemia and infarction, is a major health problem. Using the dynamic susceptibility-weighted bolus-tracking method to measure tissue perfusion, brain PWI has a significant impact in the diagnosis of ischemic diseases and treatment of affected patients [41]. PWI could also identify areas where blood flow and oxygen supply are compromised to such an extent that sequent tissue damage is imminent. From PET studies, CBF is known to be very important for evaluating tissue survival. Furthermore, MTT is roughly inversely proportional to the perfusion pressure, therefore, prolonged MTT represents an important indicator of the hemodynamic impairment after acute stroke. Figure 19.8 shows the maps of (a) CBF, (b) CBV, (c) MTT, and (d) time to peak (TTP) obtained from the perfusion-weighted images of a patient with acute infarct. The MTT and TTP maps demonstrate elevated signal intensity, suggestive of prolonged MTT and TTP, at the posterior right temporal lobe. The CBF map depicts reduced blood flow not only at the corresponding infarct region but also at the right frontal lobe (to a lesser degree). Figure 19.9 gives the PWI of a patient with old stroke and the corresponding CBV map, demonstrating the focal reduced blood volume in the right brain due to old infarction. The signal intensity-time curve of the infarct region shows less reduced signal compared to that of the normal brain region.

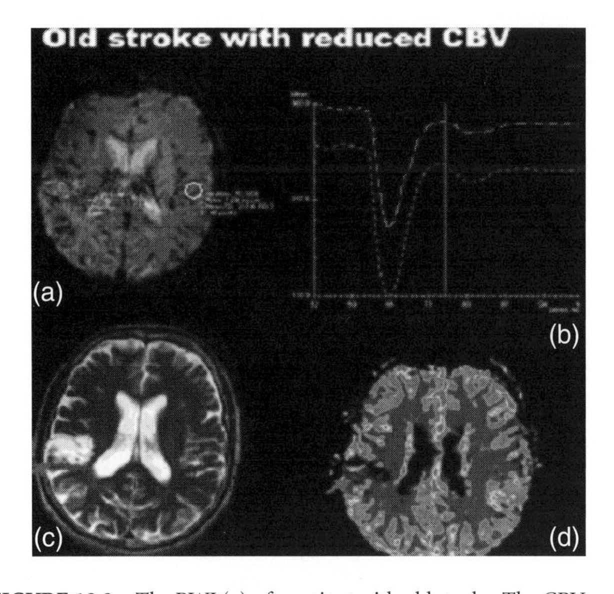

FIGURE 19.9 The PWI (a) of a patient with old stroke. The CBV map (d) demonstrates the focal reduced blood volume in the right brain due to old infarction. The signal intensity-time curve (b) of the infarct region denoted by red dashed line displays less reduced signal. For a more detailed view of this figure, please visit our companion site at: http://books.elsevier.com/companions/9780123735836.

In addition to assessing cerebral ischemia, PWI can help to select patients who will benefit most from treatment such as thrombolysis [47]. The mismatch area between the DWI and PWI will give rise to the penumbra area or salvageable area. Differentiation between brain abscesses and cystic brain tumors such as high-grade gliomas and metastases is often difficult with conventional MRI. Tumor angiogenesis by release of humoral factors (vascular endothelial growth factor) is one of the hallmarks of tumor growth and also constitutes the target of novel approaches to treat human neoplasm. PWI could be used to assess the size, density, and integrity of tumor microvessels, which allows the differentiation of pyogenic brain abscess from cystic brain tumor, making it a strong additional imaging modality in the early diagnosis of these two entities [48].

19.4 Functional Magnetic Resonance Imaging Using BOLD Techniques

19.4.1 Introduction

In the last decade, much emphasis on neuro-MRI has been beyond the confine of anatomy and pathology. fMRI not only refines the management of various brain disorders but also shows a novel insight into normal and abnormal cognition and behavior. Further, it is one of the noninvasive and powerful methods for presurgical mapping of functional cortical areas in relation to underlying brain lesions and surgically important anatomy (areas that include sensorimotor, visual, language, or even memory centers). While fMRI is evolving rapidly, it will be a useful tool in understanding the neurobiology of many neuropsychiatric disorders.

The brain is made up of a large number of specialized regions that have extensive primary functional activities and responses to different specific neural stimulations such as vision, hearing, movement, or sensation. The structure of the microcirculation in a tissue is distinctively characteristic of that tissue and varies among regions with specialized functionality. There can be flow heterogeneity within tissues, reflecting local metabolic demands or patterns of growth. The fraction of blood in tissue provides an approximate estimate of the metabolic rate [49]. It was found that neuronal activity induces focal increase in CBF [50], CBV [51], and blood oxygenation [52]. Conventional fMRI techniques are CBV-based and CBF-based methods. Currently, the most widely used fMRI is the BOLD technique. To make use of the BOLD technique correctly, suitable models describing the underlying physiologic processes contributing to measured BOLD signal changes, including contribution from changes in CBF, CBV, cerebral metabolic rate for oxygen (CMRO2), and cerebral metabolic rate for glucose (CMRGlu), should be developed [53].

19.4.2 Basic Principles

The use of quantitative tools for clinical evaluation and scientific research aids in the understanding of normal and pathologic brain functions. It is well established that CBF, CMRO2, and CMRGlu are tightly coupled in the normal resting state [54]. However, during focal activation, CBF and CMRO2 are discordant: CBF and CMRGlu increase up to 50% [55, 56], whereas CMRO2 increases only 5% [52, 57]. The rate of oxygen delivery to the activated brain increases in proportion to CBF, whereas there is little or no increase in the rate of oxygen consumption, resulting in increased capillary-venous oxyhemoglobin content but relatively decreased deoxyhemoglobin content during activation. Deoxyhemoglobin is paramagnetic and functions as an endogenous intravascular magnetic susceptibility contrast agent. It increases the magnetic field strength in red blood cells and thus creates a microscopic intravascular field gradient. This gradient, although weak, degrades spin-phase coherence on gradient-weighted sequence and attenuates the signal intensity. Therefore, the relatively decreased deoxyhemoglobin during the brain activation will lead to hyperintensity on MRI (the signal change is around 2%–8%). fMRI techniques are based on imaging sequences sensitive to the changes of the magnetic properties of blood related to the focal cerebral microcirculation changes during brain activation [58–60]. Actually, the principles of fMRI are similar to those of PWI except that the magnetic susceptibility of PWI is from the contrast agent (exogenous), whereas for fMRI it is from deoxyhemoglobin in the microcirculation (endogenous). Hence, movies of brain activity could be revealed as patients perform various tasks or are exposed to various stimuli such as visual, sensory, or motor stimulation. Figure 19.10 shows the BOLD images of a normal volunteer during (a) visual stimulation and (b) during motor cortex activation by asking the subject to touch the thumb to individual fingers in a predetermined sequence. With the support of powerful computer hardware and advanced image processing techniques, BOLD images could be generated on site and superimposed on the high-resolution structural MRI in different planes, as shown in Figure 19.11.

The implementation of BOLD fMRI includes the design of task paradigm, MR data acquisition, generation of functional maps by statistical analysis, and postprocessing. A statistical threshold is used to discriminate the inactive brain regions from the active brain regions responding to the condition of the paradigm. As shown in Figure 19.11, the results of activation analysis are then registered to the high-resolution structural images for more accurate evaluation of the brain regions involved in the activation task. In fMRI data analysis, differentiating the noise-induced signal from brain activity-induced signal is achieved by test hypothesis, including the null and alternate hypotheses. The null hypothesis states that the acquired MR signal is generated by noise; whereas the alternate hypothesis states that it is generated by brain activity. Test statistics such as the Student t test, correlation analysis, and Kolmogorov-Smirnov statistics are widely used.

fMRI involves no ionizing radiation or external injection, therefore, it has been increasingly used in clinical practice for the evaluation of brain diseases. The invasive techniques involving electrode-placement or intracarotid Wada test, along with the PET scan, have gained less popularity. On the other hand, PET scan still retains the advantage of being able to identify the brain receptors activated by neurotransmitters, abused drugs, or potential treatment compounds [61, 62].

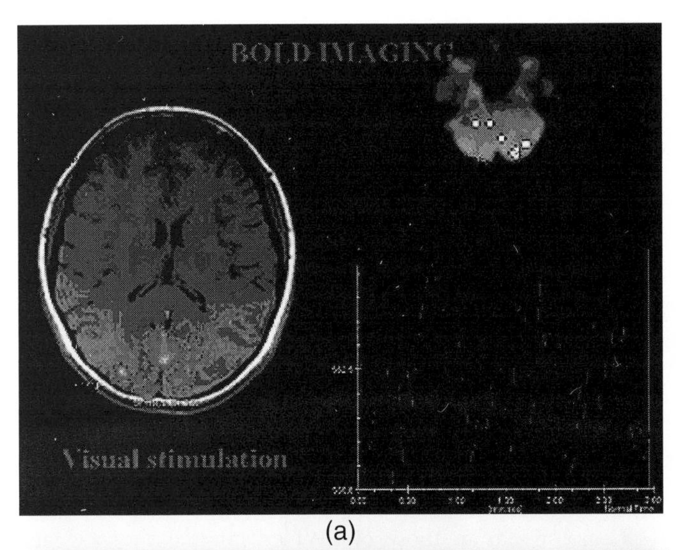

(a)

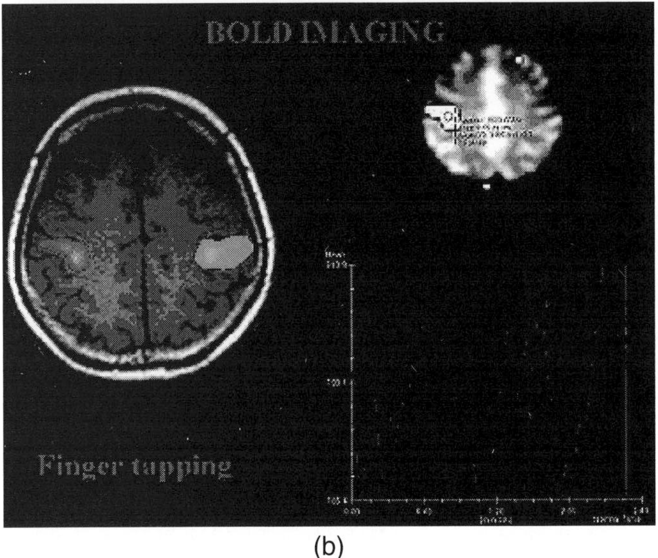

(b)

FIGURE 19.10 BOLD images of the activations of (a) occipital cortex during visual stimulation and (b) motor cortex. The signal intensity-time curves show increased BOLD signals during stimulation. For a more detailed view of this figure, please visit our companion site at: http://books.elsevier.com/companions/9780123735836.

19.5 Clinical Magnetic Resonance Spectroscopy in Brain

NMRS has attracted much attention in recent years and has become an important tool to study the biochemical aspect of brain disorders. It provides a noninvasive *in vivo* measurement of serial biochemical and metabolic changes in various brain diseases and characterizes biochemical components of normal and abnormal brain tissue. Recent development of spatial localization methods makes possible the combined presentation of the biochemical or metabolic information obtained from MRS and the anatomic information provided by conventional MRI. This integration has widened the horizon in understanding the origin and characteristics of brain diseases.

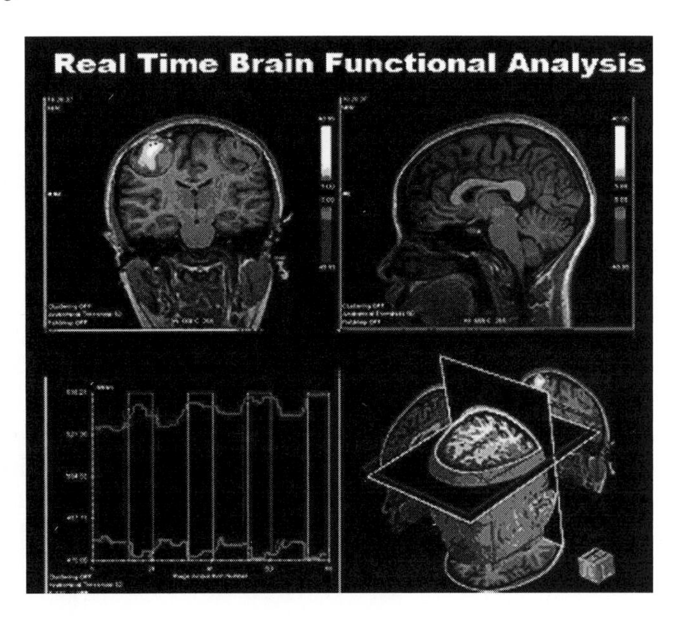

FIGURE 19.11 Real-time BOLD image (top left) of the cortical activation of specific center, which could be superimposed on the multiplaner high-resolution structural MRI (bottom right). The BOLD signal intensity-time curves of the two ROIs demonstrate significant difference. For a more detailed view of this figure, please visit our companion site at: http://books.elsevier.com/companions/9780123735836.

19.5.1 Basic Principles

Before using the MRS technique, the choice of RF coils, the nucleus to be monitored (i.e., H-1 or P-31), and the method of spatial localization should be decided. Localization can be achieved by employing RF gradient, static B_0 gradient, pulsed spatial gradient, or a combination of gradients similar to those currently used in MRI. The proton (H-1) is widely used for MRS because of its high natural abundance (100%) of organic substances and high nuclear magnetic sensitivity. Additionally, diagnostically resolvable hydrogen MR spectra may be obtained with existing units (1.5T or above) and standard head coils. Proton MRS (1H-MRS) could provide noninvasive assessment of changes in brain metabolism underlying several brain diseases since almost all metabolites contain hydrogen atoms [63]. Typically, changes in levels of N-acetylaspartate (NAA), choline (Cho), and creatine (Cr) are evaluated. Figure 19.12 illustrates the normal 1H-MRS spectra of NAA, Cho, and Cr of different ROIs. Other metabolites such as lactate and glutamine/glutamate could also be monitored.

If a proton is placed in an external static magnetic field (B_0), the frequency (f) of the proton processing around the strong external magnetic field is well defined by the Larmor equation:

$$f = \frac{\gamma B_0}{2\pi} \tag{19.11}$$

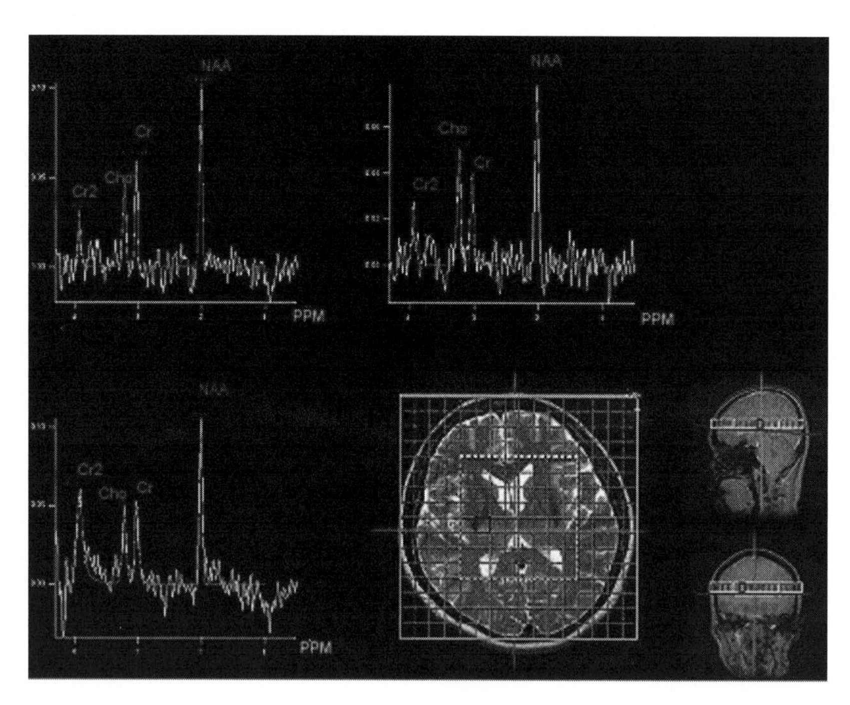

FIGURE 19.12 Normal proton MRS spectra of NAA, Cho, and Cr of different brain regions.

where γ is the gyromagnetic ratio. Proctor and Yu [64] found that the chemical environment would produce a small change of the Larmor resonance frequency (f) (so-called chemical shift), which is the result of tiny magnetic fields created by the orbiting electron clouds around the nucleus opposing the external magnetic field. Therefore, the magnetic field experienced by the hydrogen nucleus is slightly less than B_0, which is called shielding effect. Protein and fat molecules have stronger nuclear shielding effect than water and soft tissue. Consequently, water and fat protons would have distinct resonance peaks on the MRS spectrum. The chemical shift is expressed as parts per million (ppm). For NAA, Cho, and Cr, the values are 2ppm, 3.2ppm, and 3.03ppm, respectively. Results obtained in healthy control subjects would serve as the reference for reliable identification and quantification of metabolites concentration in CNS. The technical problems of using 1H-MRS include water suppression, shimming localization, editing, quantitiation, and spectra interpretation.

19.5.2 Clinical Applications

Currently, the most common diagnostic indication for MRS is to characterize suspected cerebral tumor recurrence from post-irradiation necrosis, infections, degenerative brain disorders, hepatic encephalopathy, ischemia, and demyelination. MRS is a potential tool for capturing specific metabolic profiles and offering a differential diagnosis. In addition, spectroscopic mapping allows visualization of different metabolite concentrations and distribution within lesions.

Brain damage in Alzheimer's disease (AD) and mild cognitive impairment (MCI) is widespread, with involvement of large portions of the neocortex and subcortical white matter. With the measurement of NAA, Cr, myo-inositol, and Cho, 1H-MRS could characterize the white matter biochemical profiles of MCI and patients with early AD [65, 66]. Since neuronal damage is already evident and widespread in individuals with MCI before the onset of clinical dementia [67], biochemical changes can be observed by 1H-MRS in the preclinical period. People with higher Cho/Cr ratios have a higher risk to develop dementia or AD [68]. In patients with AD, a reduction of NAA/tCr is present. Reduced NAA levels suggest neuronal loss or dysfunction in the observed region. The observed regional metabolic alteration reflects the characteristic neurologic symptoms in AD (dementia) and mirrors the disease progress over time [69].

1H-MRS could evaluate brain infarction. Normal brain shows no detectable lactate, while brain with infarction shows the continued presence of lactate and substantial reduction of NAA, Cr, and Cho in the infarct area, as shown in Figure 19.13. This is primarily the result of diminished cell density. The presence of lactate indicates increased anaerobic glycolysis due to ischemia [70].

1H-MRS also demonstrates its clinical usefulness in classifying brain tumor type and grade, monitoring response to therapy and progression to higher grade, and determining tumor extent for treatment planning. The metabolite concentration and pH value in human brain tumors differ significantly from those in normal brains. The NAA/Cr, NAA/Cho,

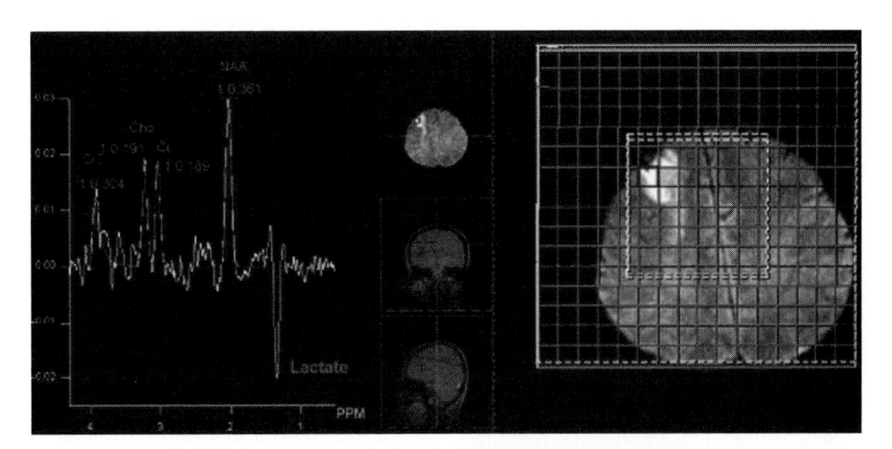

FIGURE 19.13 The MRS of a patient with brain infarction.

and Cr/Cho ratios can distinguish normal brain from gliomas, low-grade astrocytoma from high-grade group, and tumor recurrence from postirradiation necrosis. Figure 19.14 shows the MRS of a patient who underwent radiation for a brain tumor. The spectrum of postirradiation necrosis demonstrates reduced Cho, Cr, and NAA levels as well as an inverted doublet at 1.3 ppm compatible with lactate. Figure 19.15 demonstrates a patient with glioblastoma multiforme. There is increased lactate and a reduced NAA level as well as increased free lipid and Cho levels that are different from those found with necrosis. 1H-MRS could also be used to stage metastatic brain tumors [71].

Brain abscess and brain tumor may have similar clinical presentations. Additionally, the differential diagnosis of brain abscess versus cystic or necrotic tumor may be difficult based on CT or MRI findings. However, the strategies of management for abscess and neoplasm are very different; therefore, it is especially imperative to have a correct diagnosis before any surgical intervention. Spectral patterns of 1H-MRS permit differentiation of brain abscess from necrotic or cystic tumor [72]. For the cerebral abscess, there are various resonances attributed to lactate, valine, alanine, leucine, acetate, and succinate, whereas there is only one resonance attributed to lactate in the cerebral tumor. 1H-MRS provides valuable information on tumor biochemistry, which is an important complement to conventional radiology.

19.5.3 Conclusion

MRS of the brain can provide interpretable spectra for measuring tissue metabolites and can increase confidence in the diagnosis and treatment of brain lesions. Recent recommendation

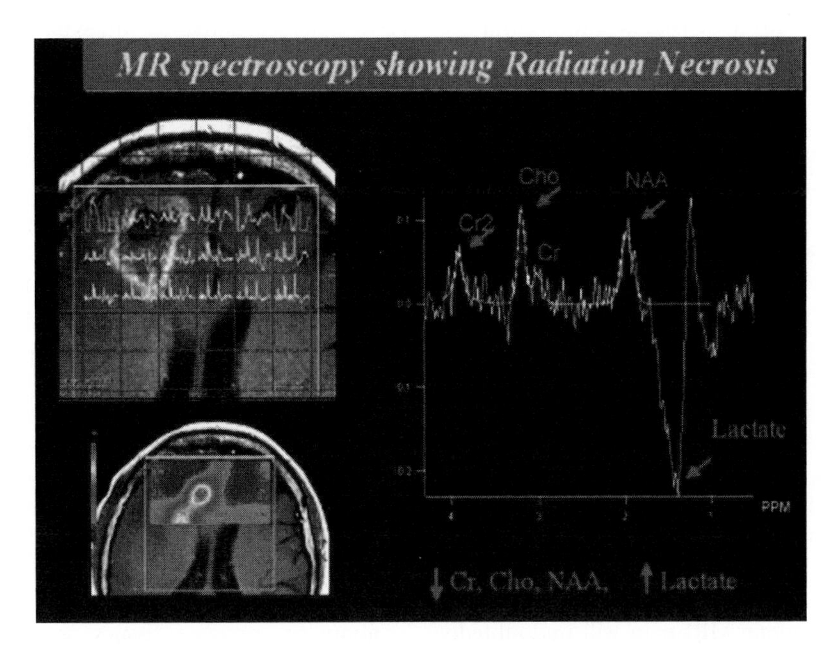

FIGURE 19.14 The MRS of a patient with postirradiation necrosis.

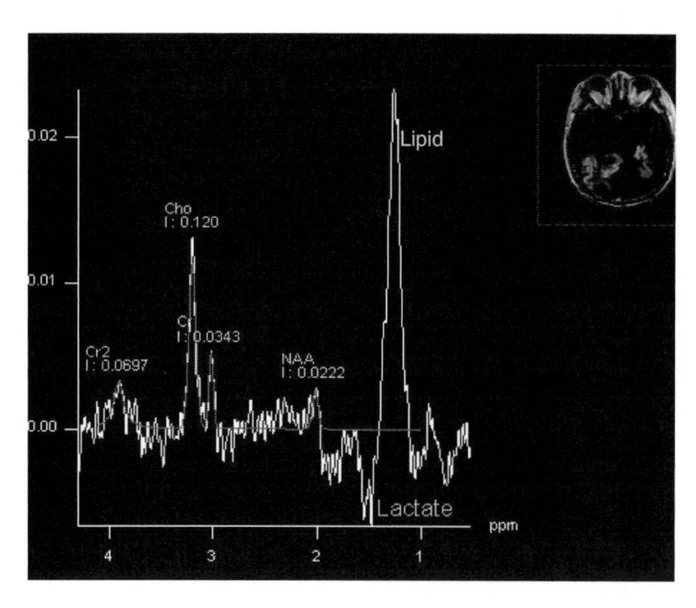

FIGURE 19.15 The MRS of a patient with glioblastoma multiforme.

for monitoring MS treatment suggests an increasing role of MRS [73]. Ongoing improvement in the equipment and pulse sequence design may make cerebral MRS more practical in the near future. There will always be the need for the development and validation of new and better MRS methods including P-31.

19.6 Summary

Conventional MRI has been used successfully in neuroradiology. The advanced techniques in functional MR imaging such as PWI, DWI, fMRI, and MRS provide a more complete picture of structural and functional brain abnormalities [74]. Functional MRI promises to be a valuable tool in the management of patients with a broad spectrum of brain disorders. Clinical applications of functional MRI will continue to expand in the presence of unlimited innovative technical development of both hardware and software. As computers become more powerful, the advanced information technologies of rapid 3D or even multidimensional image processing, graphic display, and multimodality data integration/fusion will result in break-throughs in neuroradiology. These developments would not only improve the diagnostic image quality but also the visualization of real-time brain activities. The role of functional MRI may be limited only by our imagination.

PET, SPECT, MRI, and the electroencephalogram are noninvasive functional imaging techniques that can measure biological activity and reveal the living human brain at work. The full potential of functional imaging has not yet been appreciated, let alone harnessed. The ability to fuse metabolic PET or SPECT image with an anatomic MRI map will undoubtedly prove valuable for routine clinical use in the future.

Acknowledgments

The authors would express their great gratitude to Mr. Raymond Lee for collecting the clinical images in this chapter and are grateful to the support from ARC, PolyU/UGC grant.

19.7 Exercises

1. List the main advantages of functional MRI compared to other existing medical imaging modalities.
2. Specify why and how "diffusion" could be used in MRI. Give the most common application of DWI in the detection of brain diseases.
3. How to quantitatively interpret the MR signal on perfusion images and what are the physiological parameters used for the quantitative estimation on PWI?
4. Describe the usefulness of BOLD fMRI in neuroradiology. How to quantitatively analyze the signal change on fMRI during different types of stimulation?
5. State the underlying principle for the "chemical shift" and list the common metabolites evaluated by 1H-MRS.

19.8 References and Bibliography

1. K. F. Kamm. The future of digital imaging. *Br. J. Radiol.* 70:S145–152, 1997.
2. G. L. Allan and J. Zylinski. The teaching of computer programming and digital image processing in radiography. *Int. J. Med. Inform.* 50(1–3):139–143, 1998.
3. R. Mattheus. Medical imaging part of multi-media. *Int. J. Cardiac Imaging.* 11(Suppl 3):187–190, 1995.
4. S. Wirth et al. [PACS: Storage and retrieval of digital radiological image data]. *Radiologe.* 45(8):690–697, 2005.
5. C. M. Carroll-Callahan and L. A. Andersson. MRI: Are you playing your system like a fiddle or a Stradivarius? Where we are headed and how to keep up. *Radiol. Manage.* 26(2):36–41; Quiz 42–44, 2004.
6. G. V. Shah et al. Newer MR imaging techniques for head and neck. *Magn. Reson. Imaging Clin. North Am.* 11(3):449–469, 2003.
7. M. Libicher, G. W. Kauffmann, and W. Hosch. Dynamic contrast-enhanced MRI for evaluation of cardiac tumors. *Eur. Radiol.* 16(8):1858–1859, 2006.
8. A. Semiz Oysu et al. Dynamic contrast-enhanced MRI in the differentiation of posttreatment fibrosis from recurrent carcinoma of the head and neck. *Clin. Imaging.* 29(5):307–312, 2005.
9. C. Siegel. Dynamic contrast-enhanced MRI in normal and abnormal prostate tissues as defined by biopsy, MRI, and 3D MRSI. *J. Urol.* 175(4):1366, 2006.

10. N. Tuncbilek, H. M. Karakas, and O. O. Okten. Dynamic contrast enhanced MRI in the differential diagnosis of soft tissue tumors. *Eur. J. Radiol.* 53(3):500–505, 2005.
11. K. Tatsuro. [A brief introduction about functional MRI]. *Nippon Ronen Igakkai Zasshi.* 43(1):1–6, 2006.
12. J. J. Pekar. A brief introduction to functional MRI. *IEEE Eng. Med. Biol. Mag.* 25(2):24–26, 2006.
13. A. Wismuller et al. Cluster analysis of dynamic cerebral contrast-enhanced perfusion MRI time-series. *IEEE Trans. Med. Imaging.* 25(1):62–73, 2006.
14. A. S. Doria et al. Dynamic contrast-enhanced MRI quantification of synovium microcirculation in experimental arthritis. *AJR Am. J. Roentgenol.* 186(4):1165–1171, 2006.
15. K. Kikuchi et al. Measurement of cerebral hemodynamics with perfusion-weighted MR imaging: comparison with pre- and post-acetazolamide 133Xe-SPECT in occlusive carotid disease. *AJNR Am. J. Neuroradiol.* 22(2):248–254, 2001.
16. J. O. Karonen et al. Diffusion and perfusion MR imaging in acute ischemic stroke: A comparison to SPECT. *Comput. Methods Programs Biomed.* 66(1):125–128, 2001.
17. P. C. Brugger and D. Prayer. [Fetal MRI of pathological brain development]. *Radiologe.* 46(2):112–119, 2006.
18. S. B. Peterman et al. Nuclear magnetic resonance imaging (NMR), (MRI), of brain stem tumours. *Neuroradiology.* 27(3):202–207, 1985.
19. Y. Liu et al. Cerebral hemodynamics in human acute ischemic stroke: A study with diffusion- and perfusion-weighted magnetic resonance imaging and SPECT. *J. Cereb. Blood Flow Metab.* 20(6):910–920, 2000.
20. Y. Kuwabara et al. [Pre and post operative evaluation of the perfusion reserve by acetazolamide 99mTc-HMPAO SPECT in patients with chronic occlusive cerebral arteries: A comparative study with PET]. *Kaku Igaku.* 31(9): 1039–1050, 1994.
21. A. Vlasenko et al. Comparative quantitation of cerebral blood volume: SPECT versus PET. *J. Nucl. Med.* 38(6): 919–924, 1997.
22. B. B. Chin et al. Hemodynamic indices of myocardial dysfunction correlate with dipyridamole thallium-201 SPECT. *J. Nucl. Med.* 37(5):723–729, 1996.
23. Y. R. Tran Dinh et al. Cerebral postischemic hyperperfusion assessed by Xenon-133 SPECT. *J. Nucl. Med.* 38(4):602–607, 1997.
24. A. Hiwatashi et al. Hypointensity on diffusion-weighted MRI of the brain related to T2 shortening and susceptibility effects. *AJR Am. J. Roentgenol.* 181(6):1705–1709, 2003.
25. D. Le Bihan. Looking into the functional architecture of the brain with diffusion MRI. *Nat. Rev. Neurosci.* 4(6):469–480, 2003.
26. R. Bammer. Basic principles of diffusion-weighted imaging. *Eur. J. Radiol.* 45(3):169–184, 2003.
27. M. E. Moseley et al. Diffusion-weighted MR imaging of acute stroke: Correlation with T2-weighted and magnetic susceptibility-enhanced MR imaging in cats. *AJNR Am. J. Neuroradiol.* 11(3):423–429, 1990.
28. M. Nakahara, K. Ericson, and B. M. Bellander. Diffusion-weighted MR and apparent diffusion coefficient in the evaluation of severe brain injury. *Acta. Radiol.* 42(4): 365–369, 2001.
29. R. Bammer and F. Fazekas. Diffusion imaging in multiple sclerosis. *Neuroimaging Clin. North Am.* 12(1):71–106, 2002.
30. T. Klingberg et al. Microstructure of temporo-parietal white matter as a basis for reading ability: Evidence from diffusion tensor magnetic resonance imaging. *Neuron.* 25(2):493–500, 2000.
31. D. Le Bihan et al. Separation of diffusion and perfusion in intravoxel incoherent motion MR imaging. *Radiology.* 168(2):497–505, 1988.
32. C. Thomsen, O. Henriksen, and P. Ring. *In vivo* measurement of water self diffusion in the human brain by magnetic resonance imaging. *Acta. Radiol.* 28(3):353–361, 1987.
33. R. Turner, D. Le Bihan, and A. S. Chesnick. Echo-planar imaging of diffusion and perfusion. *Magn. Reson. Med.* 19(2):247–253, 1991.
34. M. Mascalchi et al. Diffusion-weighted MR of the brain: Methodology and clinical application. *Radiol. Med. (Torino).* 109(3):155–197, 2005.
35. A. D. Elster and D. M. Moody. Early cerebral infarction: Gadopentetate dimeglumine enhancement. *Radiology.* 177(3):627–632, 1990.
36. L. Gray and J. MacFall. Overview of diffusion imaging. *Magn. Reson. Imaging Clin. North Am.* 6(1):125–138, 1998.
37. M. Hartmann, S. Heiland, and K. Sartor. [Functional MRI procedures in the diagnosis of brain tumors: Perfusion- and diffusion-weighted imaging]. *Rofo.* 174(8):955–964, 2002.
38. M. H. Lev and B. R. Rosen. Clinical applications of intracranial perfusion MR imaging. *Neuroimaging Clin. North Am.* 9(2):309–331, 1999.
39. J. L. Boxerman et al. MR contrast due to intravascular magnetic susceptibility perturbations. *Magn. Reson. Med.* 34(4):555–566, 1995.
40. B. R. Rosen et al. Perfusion imaging with NMR contrast agents. *Magn. Reson. Med.* 14(2):249–265, 1990.
41. O. Speck et al. Perfusion MRI of the human brain with dynamic susceptibility contrast: Gradient-echo versus spin-echo techniques. *J. Magn. Reson. Imaging.* 12(3):381–387, 2000.
42. C. R. Fisel et al. MR contrast due to microscopically heterogeneous magnetic susceptibility: Numerical simulations and applications to cerebral physiology. *Magn. Reson. Med.* 17(2):336–347, 1991.

43. A. Villringer et al. Dynamic imaging with lanthanide chelates in normal brain: Contrast due to magnetic susceptibility effects. *Magn. Reson. Med.* 6(2):164–174, 1988.
44. B. R. Rosen, J. W. Belliveau, and D. Chien. Perfusion imaging by nuclear magnetic resonance. *Magn. Reson. Q.* 5(4):263–281, 1989.
45. C. F. Starmer and D. O. Clark. Computer computations of cardiac output using the gamma function. *J. Appl. Physiol.* 28(2):219–220, 1970.
46. H. K. Thompson Jr. et al. Indicator transit time considered as a gamma variate. *Circ. Res.* 14:502–515, 1964.
47. C. B. Grandin. Assessment of brain perfusion with MRI: Methodology and application to acute stroke. *Neuroradiology.* 45(11):755–766, 2003.
48. C. Erdogan et al. Brain abscess and cystic brain tumor: Discrimination with dynamic susceptibility contrast perfusion-weighted MRI. *J. Comput. Assist. Tomogr.* 29(5):663–667, 2005.
49. J. B. Bassingthwaighte. Microcirculatory considerations in NMR flow imaging. *Magn. Reson. Med.* 14(2):172–178, 1990.
50. P. T. Fox et al. Mapping human visual cortex with positron emission tomography. *Nature.* 323(6091):806–809, 1986.
51. J. W. Belliveau et al. Functional mapping of the human visual cortex by magnetic resonance imaging. *Science.* 254(5032):716–719, 1991.
52. P. T. Fox et al. Nonoxidative glucose consumption during focal physiologic neural activity. *Science.* 241(4864): 462–464, 1988.
53. D. G. Nair. About being BOLD. *Brain Res. Brain Res. Rev.* 50(2):229–243, 2005.
54. J. C. Baron et al. Local interrelationships of cerebral oxygen consumption and glucose utilization in normal subjects and in ischemic stroke patients: A positron tomography study. *J. Cereb. Blood Flow Metab.* 4(2): 140–149, 1984.
55. P. T. Fox et al. A noninvasive approach to quantitative functional brain mapping with H2 (15)O and positron emission tomography. *J. Cereb. Blood Flow Metab.* 4(3): 329–333, 1984.
56. J. H. Greenberg et al. Metabolic mapping of functional activity in human subjects with the [18F]fluorodeoxyglucose technique. *Science.* 212(4495):678–680, 1981.
57. P. T. Fox and M. E. Raichle. Focal physiological uncoupling of cerebral blood flow and oxidative metabolism during somatosensory stimulation in human subjects. *Proc. Natl. Acad. Sci. U S A.* 83(4):1140–1144, 1986.
58. S. Naruse. [Functional MRI of the brain]. *Rinsho Shinkeigaku.* 35(12):1345–1350, 1995.
59. L. Hertz-Pannier et al. Brain functional MRI: physiological, technical, and methodological bases, and clinical applications. *J. Radiol.* 81(6 suppl):717–730, 2000.
60. C. Tanaka et al. [Theory and clinical application of functional MRI: 3D-functional brain mapping]. *Nippon Rinsho.* 55(7):1660–1665, 1997.
61. M. Itoh et al. [*In vivo* visualization of neurotransmitter function in the human brain by PET]. *No To Hattatsu.* 27(2):146–152, 1995.
62. M. Kato, T. Taniwaki, and Y. Kuwabara. [The advantages and limitations of brain function analyses by PET]. *Rinsho Shinkeigaku.* 40(12):1274–1276, 2000.
63. A. P. Burlina et al. MR spectroscopy: A powerful tool for investigating brain function and neurological diseases. *Neurochem. Res.* 25(9–10):1365–1372, 2000.
64. W. G. Proctor and F. C. Yu. The dependence of a nuclear magnetic resonance frequency upon chemical compound. *Phys. Rev.* 70(5):717, 1950.
65. M. Catani et al. (1)H-MR spectroscopy differentiates mild cognitive impairment from normal brain aging. *Neuroreport.* 12(11):2315–2317, 2001.
66. R. E. Jung et al. Biochemical markers of intelligence: A proton MR spectroscopy study of normal human brain. *Proc. Biol. Sci.* 266(1426):1375–1379, 1999.
67. A. Falini et al. A whole brain MR spectroscopy study from patients with Alzheimer's disease and mild cognitive impairment. *Neuroimage.* 26(4):1159–1163, 2005.
68. T. den Heijer et al. MR spectroscopy of brain white matter in the prediction of dementia. *Neurology.* 66(4):540–544, 2006.
69. W. Block et al. *In vivo* proton MR-spectroscopy of the human brain: Assessment of N-acetylaspartate (NAA) reduction as a marker for neurodegeneration. *Amino Acids.* 23(1–3):317–323, 2002.
70. J. H. Duijn et al. Human brain infarction: Proton MR spectroscopy. *Radiology.* 183(3):711–718, 1992.
71. P. E. Sijens et al. 1H MR spectroscopy detection of lipids and lactate in metastatic brain tumors. *N.M.R. Biomed.* 9(2):65–71, 1996.
72. S. H. Kim et al. Brain abscess and brain tumor: Discrimination with *in vivo* H-1 MR spectroscopy. *Radiology.* 204(1):239–245, 1997.
73. S. D. Rand, R. Prost, and S. J. Li. Proton MR spectroscopy of the brain. *Neuroimaging Clin. North Am.* 9(2):379–395, 1999.
74. A. H. Hoon Jr. Neuroimaging in cerebral palsy: Patterns of brain dysgenesis and injury. *J. Child Neurol.* 20(12): 936–939, 2005.

20

Molecular Imaging in Cancer

Prof. Kristine Glunde,
Dr. Catherine A. Foss, and
Prof. Zaver M. Bhujwalla
The Johns Hopkins University School of Medicine

20.1 Introduction

Cancer imaging has traditionally consisted of answering the questions of where and how big are the tumors. These answers are determined by conducting planar X-ray, computed tomography (CT), ultrasound (US), or magnetic resonance imaging (MRI) scans (Figure 20.1). The scan results then guide the surgeon about where to cut or the radiation oncologist about where to direct the radiation beam. Additionally, the progress of chemotherapy and radiation therapy are typically monitored by assessing whether or not existing tumors are shrinking in size by comparison of pre- and post-treatment images. Today, cancer is understood to be a genetic disease where tumor cells are characterized by unique molecular marker expression profiles. In contrast with standard anatomic imaging to answer where and how big, cancer imaging is increasingly focusing on detecting unique molecular markers to help characterize the identity, extent, and progression of specific neoplastic disease.

Molecular imaging can be defined as "the non-invasive visualization of molecular processes" [1]. Molecular imaging uses many of the same modalities mentioned previously (X-ray, CT, MRI, US) in addition to fluorescence microscopy and endoscopy as well as nuclear scanning techniques like positron-emission tomography (PET) and single-photon emission tomography (SPECT). The study of cancer encompasses many disciplines from the basic sciences, which explore the genetic and biochemical origins and progression of neoplastic disease, to clinical practice, where cancer detection and treatment take place. Molecular imaging in cancer is of use in basic research and clinical management of cancer as scientists use the techniques to validate hypotheses and potential treatments in preclinical models and clinicians use these methods to noninvasively detect, characterize, and follow treatment in living patients. Examples provided in the following sections in this chapter illustrate how molecular imaging is currently being used in the study of cancer.

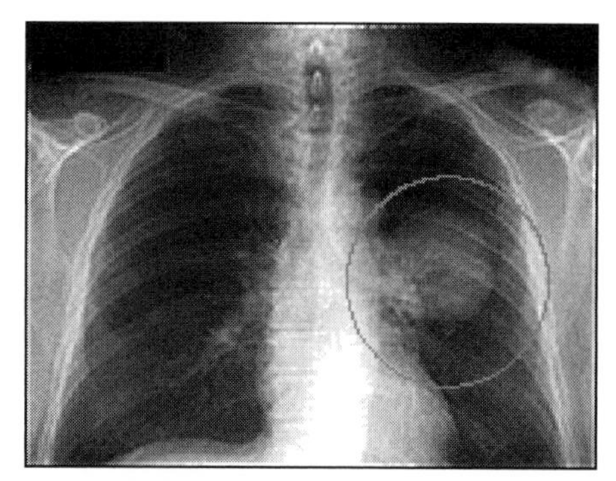

Planar X-ray showing lung tumor.

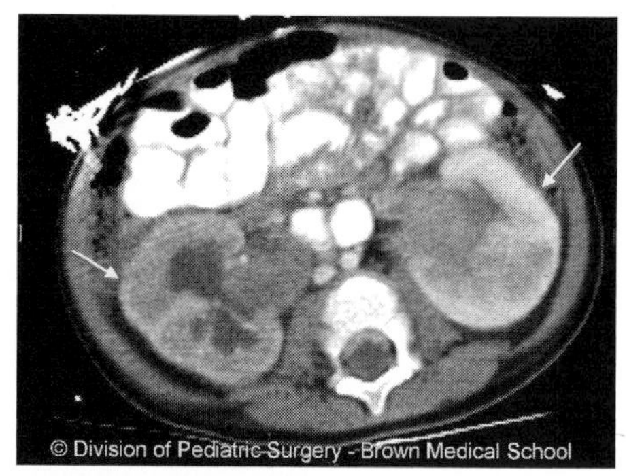

CT scan showing Wilm's tumor.

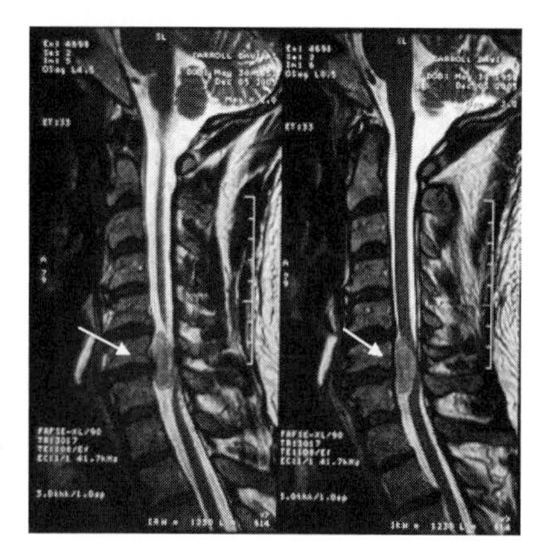

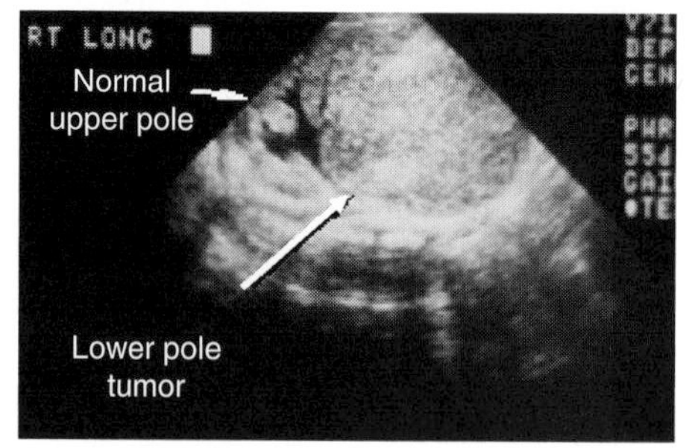

Left, MRI scan showing a spine tumor.
Above, US scan showing metastatic lung cancer.

FIGURE 20.1 Planar X-ray, computed tomography, magnetic resonance imaging, and ultrasound scans showing anatomic features of cancer in patients. Circles and arrows indicate the location of each tumor.

20.2 Imaging of Gene Expression

Cancer is initiated by and progresses through genetic changes. Initial genetic changes allow cancer cells to evade biologic programs that regulate and limit cellular growth under normal conditions. As tumor growth continues, the unique physiologic microenvironment that cancer cells are exposed to in solid tumors influences tumor progression, aggressiveness, and response to treatment. Therefore, imaging of cancer-related genes has been developed to study cancer initiation and progression *in vivo*, and allow for visualization of gene expression in the context of the unique features of solid tumors both in real time and dynamically over a period of time. The ability to image gene expression has helped tremendously to study key pathways in tumor models *in vivo*. Researchers can now study genetic processes in the biologically complex setting of whole tumors in living animals. Molecular imaging of gene expression also has the potential of being translated to the clinic to achieve specific detection of the genetic make-up of a particular tumor and to monitor gene delivery to the tumor in future gene therapies.

Molecular imaging can be applied to visualize gene expression-related processes such as promoter activity or transcriptional activity, among other processes, in living organisms *in vivo*. To detect these processes, it is necessary to couple the expression of reporter genes to the process under investigation. Researchers have designed reporter genes for each of the available imaging modalities such as optical, MR, and nuclear imaging methods. For optical imaging, the reporter genes typically used are luciferase genes [2] for bioluminescence imaging and fluorescent protein genes [3] for fluorescence imaging. For PET and SPECT imaging, the herpes simplex virus thymidine kinase (HSVtk) genes [4] or the sodium iodide symporter (NIS) [5] are frequently used. The ferritin gene [6] or chemical exchange saturation transfer (CEST) reporters [7] are available for imaging of gene expression using MRI. It is also possible to use several of these reporter genes in the same system for multimodality imaging where different imaging modalities can either be used in combined scanners or sequentially. For example, a triple fusion reporter gene has been reported that allows *in vivo* multimodality imaging of bioluminescence, fluorescence, and PET imaging [8]. The principle for imaging gene expression is generally outlined in Figure 20.2. The reporter gene coupled to the promoter or gene under investigation is first delivered within cells, which can be achieved by means of transfection agents in live cell applications, or systemically by viruses or other vehicles such as liposomes in tumors *in vivo* (Figure 20.2).

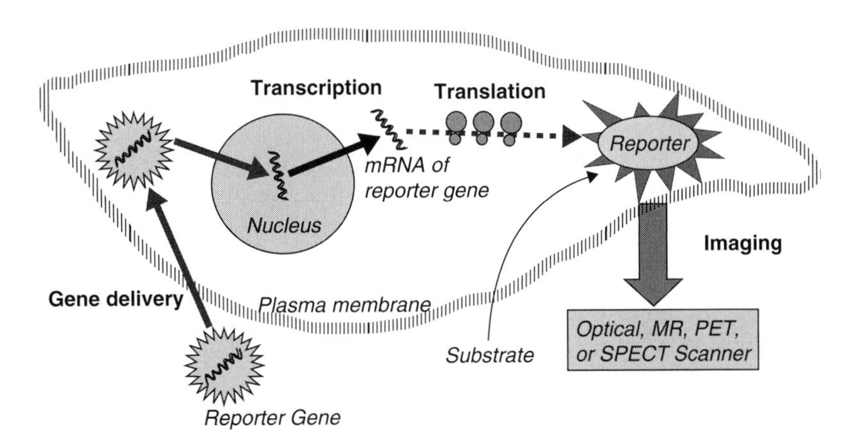

FIGURE 20.2 Principle for imaging of gene expression. The reporter gene is delivered into the cell, transcribed, translated, and imaged directly or following the action of an enzyme on a substrate.

The cellular machinery for transcription and translation produces the corresponding mRNA and protein of the reporter gene, and the latter can be detected by an imaging modality such as optical, MR, PET, or SPECT imaging either directly or in conjunction with a substrate that is activated by the reporter (Figure 20.2).

To study the promoter activity of a cancer-relevant gene *in vivo*, reporter genes are placed under the control of the promoter of a cancer-relevant gene. To visualize the final product of gene expression, which is the encoded protein, a reporter gene is positioned in-frame with the gene of interest in solid tumors of animal models *in vivo* to produce a fusion protein that can be imaged. Such fusion proteins, however, can be functionally altered because they contain additional protein units that are attached to the native protein. To avoid this, an internal ribosome entry site (IRES) coding sequence can be introduced between the reporter gene and the gene under investigation. Transcription of such an IRES-containing construct results in bicistronic mRNA, which ribosomes translate into two separate proteins: the reporter protein and the unaltered protein of interest. Imaging of gene expression has helped researchers delineate mechanistic and functional aspects of oncogenes such as *myc* and tumor suppressor genes such as *p53* [9, 10]. Examples will be given in Section 20.2.1

20.2.1 Optical Imaging of Gene Expression

Optical imaging of gene expression *in vivo* requires that gene expression be coupled to the emission of light photons that can travel through tissue. As demonstrated in Figure 20.3, these light photons should lie within the spectral window of low tissue autofluorescence, photon attenuation, and light scattering, in the spectral region between 650 nm and 950 nm [11]. Molecules such as water and oxygenated and deoxygenated hemoglobin, which are highly abundant in biologic tissues, absorb a lot of light in the lower portion of the visible

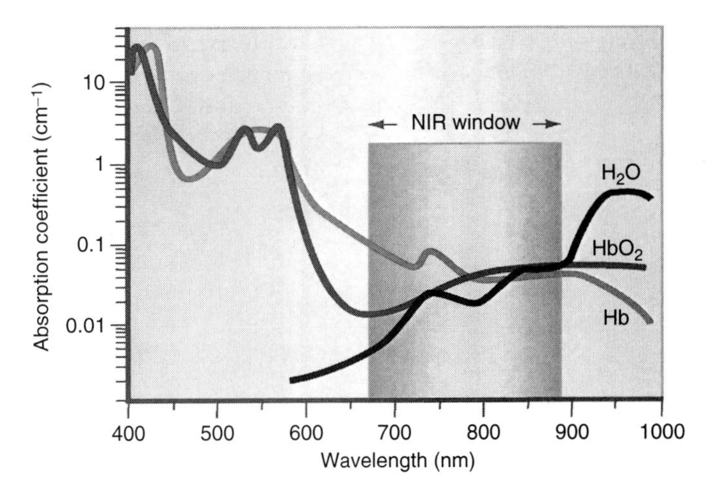

FIGURE 20.3 Absorption of light versus wavelength. Given the decreased absorption of light in the near-infrared (NIR) region compared with visible light (400–650 nm) and infrared light (>900 nm), tissue penetration of NIR photons may be up to 10–15 cm. Adapted from Mahmood and Weissleder [11].

spectrum as well as in the infrared region of the spectrum (Figure 20.3).

Luciferase genes, which are used for bioluminescence imaging [2], occur naturally in some insects, enabling them to glow in the dark. The most useful luciferase genes for molecular imaging have been found in the firefly (*Photinus pyralis*), *Renilla*, green or red click beetle (*Pyrophorus plagiophthalamus*), and *Gaussia*. Bioluminescence imaging of luciferase reporters has the advantages of being robust and cost-effective and provides high signal-to-noise levels. However, gene expression imaging of luciferase bioluminescence requires the administration of a substrate such as luciferin because luciferases are enzymes that catalyze the oxidation of this substrate, and visible light photons are a product of this reaction, producing bioluminescence.

Therefore, bioluminescence imaging depends on substrate pharmacokinetics. Fluorescent protein genes [3], which are used for fluorescence imaging of gene expression, do not require substrates, because the fluorescent proteins themselves emit light photons following excitation with light of a shorter wavelength than the emitted light, as demonstrated in Figure 20.4. Green fluorescent protein genes were first cloned from the jellyfish *Aequorea victoria*, and since then numerous mutants and novel monomeric fluorescent proteins with various spectral properties have been generated as shown in Figure 20.4 [3]. For imaging of gene expression *in vivo*, the generation of red-shifted fluorescent protein variants was important because it improved the signal-to-noise ratio due to decreased tissue autofluorescence, photon attenuation, and light scattering in this region of the visible spectrum, as shown in Figure 20.3. Fluorescent proteins, however, only achieve a relatively low light photon output compared to luciferase bioluminescence, and therefore molecular imaging of fluorescent proteins *in vivo* suffers from relatively poor sensitivity.

In this section, we will give two examples of how researchers have used molecular imaging to study cancer-related gene expression *in vivo*: the *myc* oncogene and the *p53* tumor suppressor gene. The *myc* oncogene is one of the most commonly activated oncogenes in liver cancers, which are often refractory to clinical treatment. Therefore, basic research studies and future clinical imaging of *myc* function and *myc* gene expression are of great interest. Dynamic long-term *in vivo* bioluminescence imaging was applied in a study of transgenic mice that had conditional expression of doxycycline-inducible *myc* proto-oncogene and were also transgenic for firefly luciferase in subcutaneous liver tumors [12]. As shown in Figure 20.5, *myc* oncogene inactivation resulted in tumor regression and dormancy as long as *myc* remained inactive, indicated by decreased bioluminescence. When *myc* expression was reactivated, tumor growth was reactivated as well (Figure 20.5b), and the neoplastic features of previously differentiated hepatocytes and biliary cells were immediately restored [12]. This study demonstrated that *myc* inactivation resulted in tumor regression, and *myc* reactivation caused malignant expansion of previously dormant liver cancer cells [12]. Bioluminescence imaging of gene expression enabled direct visualization of *myc* oncogene inactivation and reactivation in this study, demonstrating the

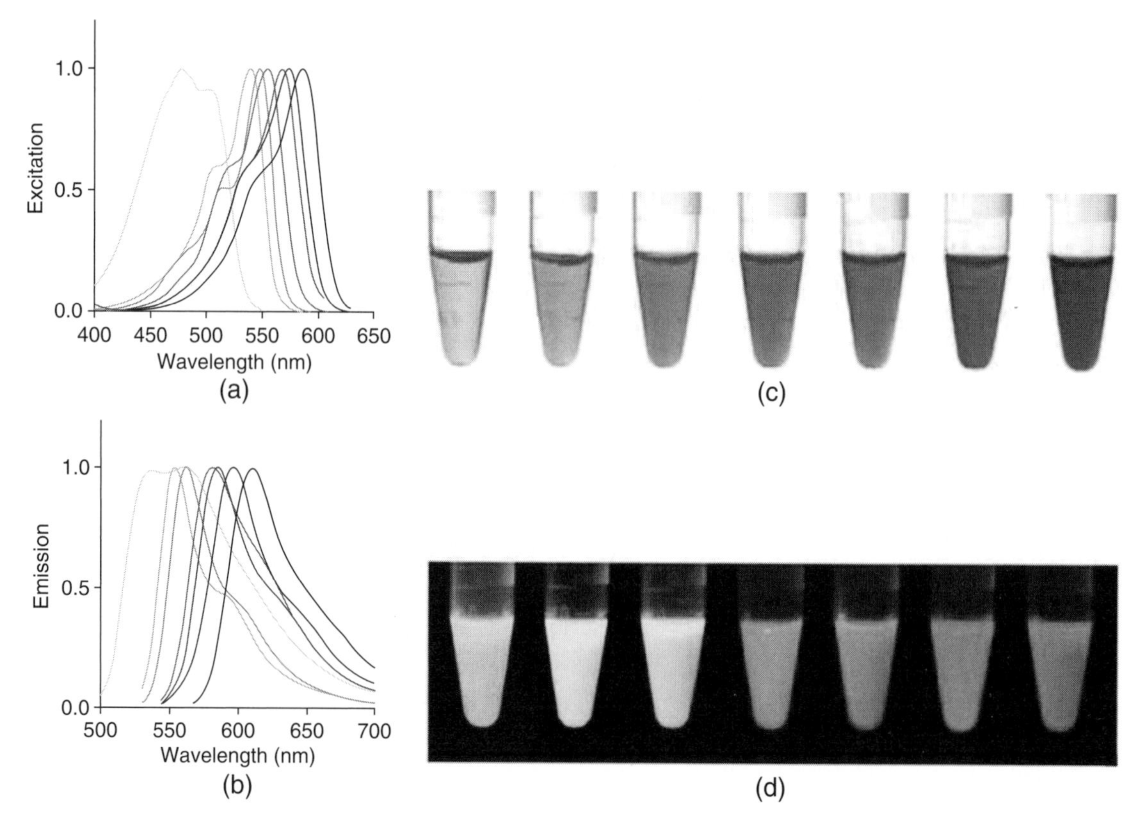

FIGURE 20.4 (a) Excitation and (b) emission spectra are shown as solid or dashed lines for monomeric variants and as a dotted line for dTomato and tdTomato, with colors corresponding to the color of each variant. Purified proteins (from left to right, mHoneydew, mBanana, mOrange, tdTomato, mTangerine, mStrawberry, and mCherry) are shown in (c) visible light and (d) fluorescence. The fluorescence image is a composite of several images with excitation ranging from 480 nm to 560 nm. Adapted from Shaner et al. [3]. For a more detailed view of this figure, please visit our companion site at: http://books.elsevier.com/companions/9780123735836.

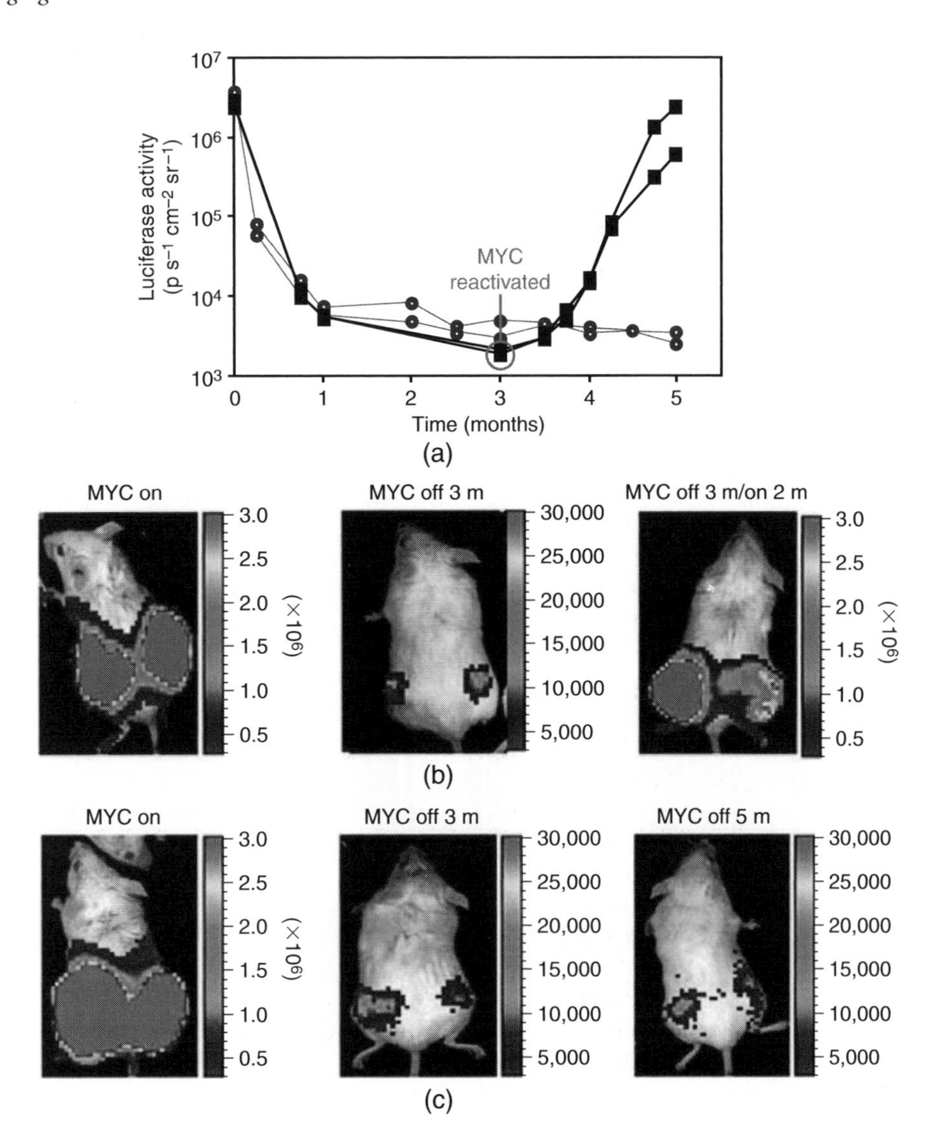

FIGURE 20.5 (a) Kinetics of tumor regression using *in vivo* bioluminescence imaging of luciferase-labeled liver tumors. Transplanted tumors undergo rapid regression but residual, persisting luciferase activity remains at the site of tumor growth. Upon *myc* reactivation, tumor growth reoccurred. For visualization of tumor growth, a pseudocolor image representing luciferase light intensity is superimposed over a greyscale reference image of the representative animals in each treatment group: squares, *myc* on, then *myc* off, and finally *myc* on; circles, *myc* on then *myc* off. Luciferase activity is measured in photons/cm^2/s per steradian (p cm^{-2} s^{-1} sr^{-1}). (b) Representative images for a mouse where *myc* is on (left); *myc* is on and then off for 3 months (3 m) (center); and *myc* is on, off for 3 months, and then reactivated for 2 months (right). (c) A representative control mouse is represented for the same time points: *myc* on (left); *myc* on then off for 3 months (center); and *myc* remains off for 5 months (right). Data are representative of five different experiments with 1 to 10 animals in each group. Adapted from Shachaf et al. [12]. For a more detailed view of this figure, please visit our companion site at: http://books.elsevier.com/companions/9780123735836.

impact of optical imaging of gene expression *in vivo* in cancer research.

The *p53* tumor suppressor gene is mutated in approximately 50% of all human cancers. *p53* plays a key role in cell cycle regulation and apoptosis following DNA damage, and it functions as a sequence-specific transcription factor. *p53* mutations and *p53* deficiency also contribute to an aggressive and chemotherapy- or radiotherapy-resistant cancer phenotype. Therefore, much research has been devoted to studying the function and molecular regulation of *p53* in cancer development and response to radiation therapy. An example of how *p53* gene expression was optically imaged using bioluminescence imaging will be outlined next. The transcriptional activity of *p53* was noninvasively evaluated *in vivo* in a transgenic mouse model conditionally expressing the firefly luciferase gene, which was placed under the control of the *p53*-responsive P2 promoter for the murine double minute 2 *(mdm2)* gene [9]. *p53* acts as a sequence-specific transcription factor modulating the expression of target genes, one of which is *mdm2*. Following DNA damage caused by ionizing radiation, *p53* is stabilized, and its transcriptional function is activated, which leads to up-regulation of genes containing a *p53*-responsive P2 promoter, such as the P2 promoter firefly luciferase construct in this study. Following exposure to ionizing radiation, the *in vivo p53* transcriptional activity was dynamically visualized by bioluminescence imaging for up to 14 hours in transgenic mice containing this construct. *p53* transcriptional activity following radiation displayed a distinct oscillatory pattern as demonstrated in Figure 20.6, confirming *p53* transcriptional oscillations previously observed in cultured cells [9]. Such *in vivo* bioluminescence imaging studies of *mdm2*-P2-promoter-bioluminescence mice will help assess the *p53* response *in vivo* following systemic administration of novel therapeutic *p53* inhibitors or agents modulating the response to ionizing radiation [9].

20.2.2 Nuclear Imaging of Gene Expression

Nuclear imaging using SPECT or PET relies on reporter genes that result in intracellular trapping or selective retention of radiolabeled substrates. Nuclear imaging is highly sensitive and can detect femtomolar concentrations of radiotracer quantitatively. Because SPECT and PET are inherently tomographic, it is feasible to use these modalities for imaging of gene expression. However, expensive instrumentation, availability of in-house radiopharmaceutical drug production, and dependence on tracer pharmacokinetics are some disadvantages for these molecular imaging modalities when imaging gene expression. Nuclear imaging of gene expression can be achieved with the herpes simplex virus

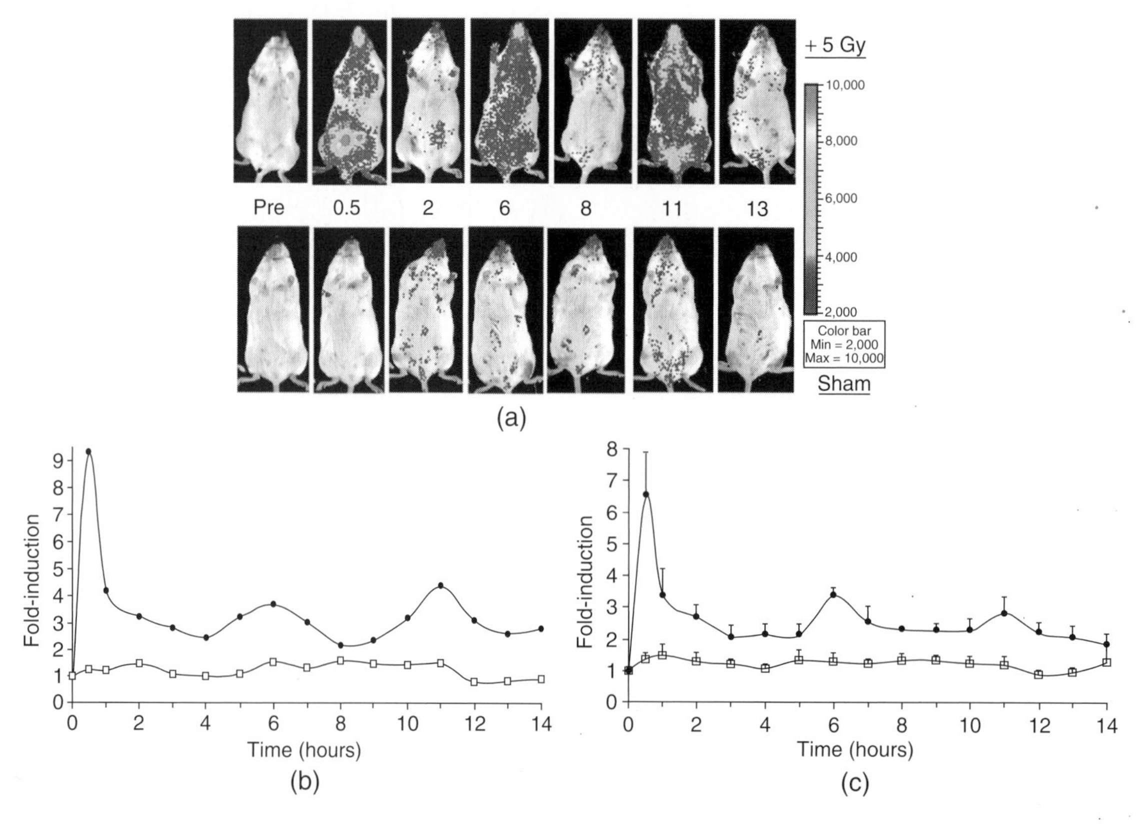

FIGURE 20.6 (a) Oscillatory behavior of *p53* following total body irradiation. MDM2-luciferase mice were radiated with 5 Gy total body irradiation or sham irradiated and then followed with serial bioluminescence imaging scans: before radiation, 0.5 hour, 1 hour, and then every hour until 14 hours after radiation. Top, serial images from one representative irradiated animal; bottom, serial images from one representative control animal. (b) Quantification of bioluminescence induction for the abdominal region of interest (ROI). Fold-induction above baseline for the abdominal ROI for the irradiated (black circles) or control (white squares) animals depicted in (a). (c) Quantification of the bioluminescence induction for the abdominal ROI. Points, average of five irradiated (black circles) and four control animals (white squares); bars, standard error. Adapted from Wiener et al. [9]. For a more detailed view of this figure, please visit our companion site at: http://books.elsevier.com/companions/9780123735836.

HSVtk gene [4] or with NIS [5] as reporter genes. Nuclear imaging of HSVtk relies on the enzymatic modification of a radiolabeled substrate, which leads to selective retention of the converted substrate in reporter cells. Substrates used for nuclear HSVtk reporter imaging are [^{18}F] 9-[4-fluoro-3-(hydrox-ymethyl) butyl] guanine ([^{18}F]FHBG), or [^{124}I]/[^{131}I] 2′-fluoro-2′-deoxy-5-iodouracil-β-D-arabino furanoside ([^{124}I]/[^{131}I]FIAU) and analogous compounds. Intracellularly expressed HSVtk phosphorylates these compounds so that they are trapped in the HSVtk-expressing cells. The HSVtk gene can also be used in gene-directed enzyme prodrug therapy, which is a two-step therapeutic approach for cancer gene therapy. In this therapy, the transgene HSVtk is first delivered into and expressed in the tumor. Then the prodrug ganciclovir is administered and is selectively activated by the expressed HSVtk enzyme. The antiviral drug ganciclovir is nontoxic to cells, but is a potent antitumor agent when phosphorylated in the presence of HSVtk expression to its triphosphate. The use of NIS as a reporter gene requires intravenous injection of free [^{124}I] NaI, [^{131}I]NaI, or [^{99m}Tc]pertechnetate. The NIS is a transmembrane protein capable of transporting iodide into the cells, and thus, iodide or pertechnate will be trapped in cancer cells that express the NIS reporter. We will explain in the following two preclinical research examples how nuclear molecular imaging of gene expression is used to study cancer.

Telomerase is an important tumor marker because it remains active in cancer cells and leads to continued proliferation, while it is inactive in most differentiated cells [5]. More than 85% of all solid tumors contain high telomerase activity, which makes telomerase an attractive target for molecular imaging of cancer. Novel therapeutic approaches to target telomerase in cancer treatment are also being developed. The expression pattern of telomerase promoter fragments in mice was imaged using PET *in vivo* [5] as shown in Figure 20.7. For this purpose researchers generated two different recombinant adenoviruses that each contained promoter fragments from either the RNA component of telomerase, which is human telomerase RNA (hTR), or the catalytic protein component of telomerase, which is human telomerase reverse transcriptase (hTERT), to drive the expression of the NIS PET reporter gene [5]. They found that both of these telomerase gene promoter fragment constructs resulted in cancer-specific expression of the NIS transgene, which was visualized by PET imaging (Figure 20.7). This example shows how PET imaging of the NIS reporter gene can be used to develop novel methods to measure telomerase activity in tumors. In the future, this basic research can lead to clinical applications, including the use of these telomerase gene promoter fragments for therapeutic transgene expression in gene delivery vectors [5].

A unique example of using the HSVtk reporter gene was demonstrated in a novel hybrid vector in which ligand-directed tumor targeting was combined with molecular PET reporter imaging [13]. In this report, researchers combined single-stranded phages (P) that display short peptides that bind to an integrin to achieve tumor targeting with recombinant adeno-associated virus (AAV), which is required to achieve gene delivery into the cancer cells, to form chimeric viral particles (AAVP) [13]. These AAVP also contained the HSVtk gene for molecular PET imaging using 2′-[^{18}F]-fluoro-2′-deoxy-1-β-D-arabino-furanosyl-5-ethyl-uracil ([^{18}F]FEAU) or the luciferase gene for molecular bioluminescence imaging. Comparison of treatment with nontargeted or scrambled control AAVP-HSVtk particles with the integrin-targeted RGD-4C AAVP-HSVtk particles demonstrated improved tumor targeting, specificity, and efficacy of gene delivery [13] as evident in Figure 20.8. Since HSVtk also serves as a suicide gene when combined with ganciclovir (GCV) administration [13], improved therapeutic response following GCV administration was observed in the RGD-4C AAVP-HSVtk particle-treated animals as well [13] (Figure 20.8).

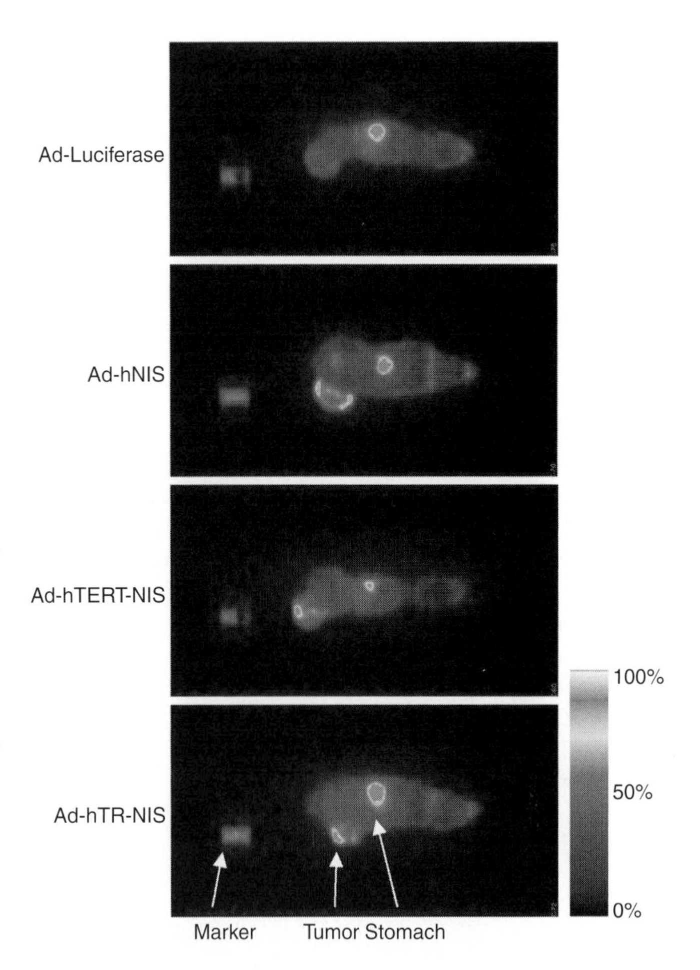

FIGURE 20.7 PET imaging of A2780 tumor-bearing BALB/c nu/nu mice after injection of adenoviruses. The tumors were injected with the various recombinant adenoviruses (5×10^8 plaque-forming units, three mice/group) and scanned 72 hours later after injection of Na^{124}I. The data were acquired for 1 hour. Single 0.5-mm coronal slices of the 30–60 minute time frame are shown. hTERT, human telomerase reverse transcriptase; hTR, human telomerase RNA; NIS, Na/I symporter. Adapted from Groot-Wassink et al. [5]. For a more detailed view of this figure, please visit our companion site at: http://books.elsevier.com/companions/9780123735836.

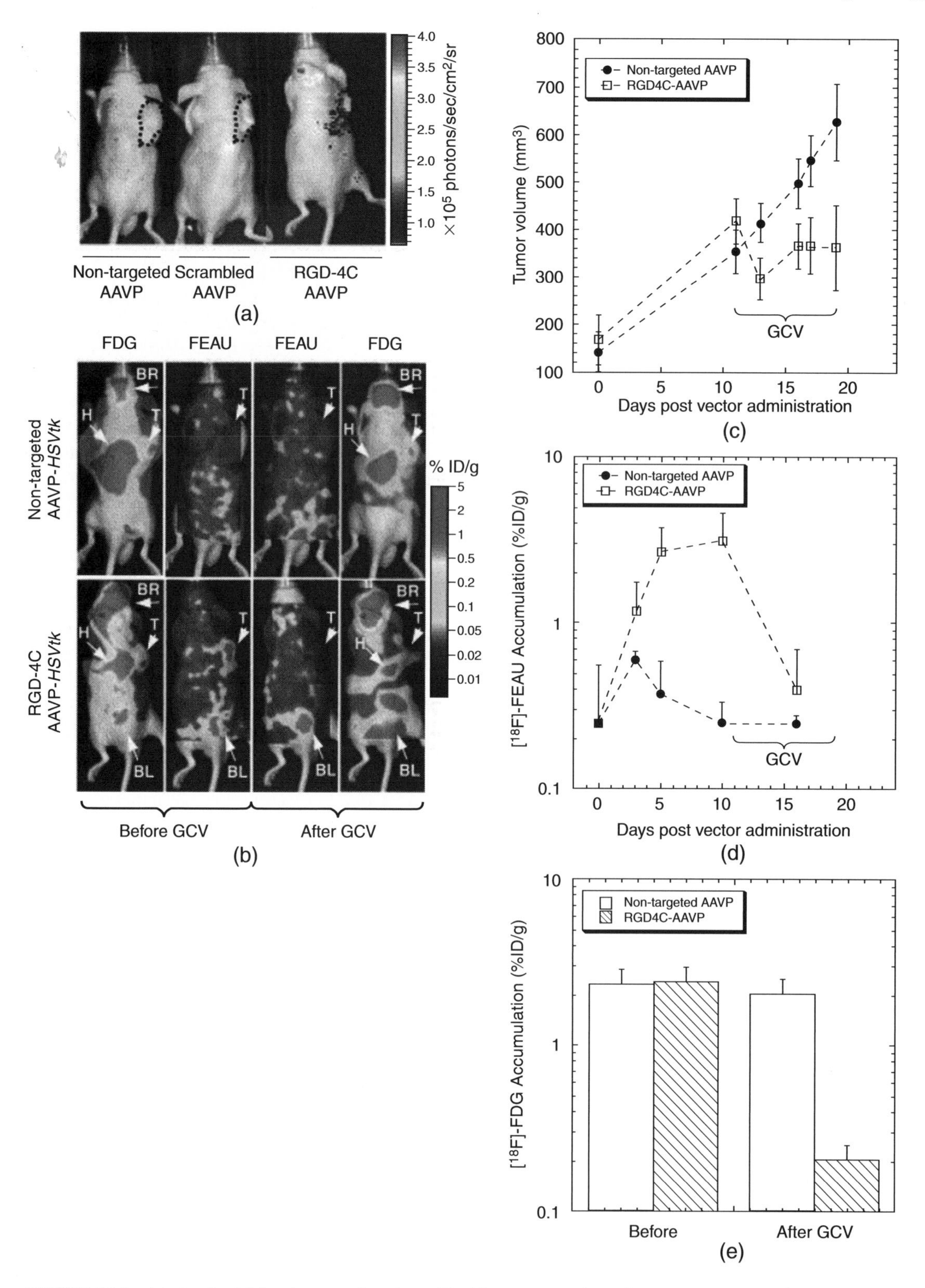

FIGURE 20.8 (a) *In vivo* bioluminescent imaging of luciferase expression after systemic AAVP delivery. Nude mice bearing DU145-derived tumor xenografts received an intravenous single dose of either RGD-4C AAVP-Luciferase (5×10^{11} TU) or control (nontargeted AAVP-Luc or scrambled RGD-4C AAVP-Luciferase). Ten days later, bioluminescence imaging of tumor-bearing mice was performed. (b) Multitracer PET imaging in tumor-bearing mice after systemic delivery of RGD-4C AAVP-HSVtk. Nude mice bearing DU145-derived tumor xenografts (n = 9 tumor-bearing mice per cohort) received an intravenous single dose (5×10^{11} TU) of RGD-4C AAVP-HSVtk or nontargeted AAVP-HSVtk. PET images with [^{18}F]FDG and [^{18}F]FEAU obtained before and

20.2.3 Magnetic Resonance Imaging of Gene Expression

Several different approaches have been pursued to generate MR contrast for molecular imaging of gene expression, most of which require addition of substrate or additional contrast agent. The ferritin MR reporter gene has been developed to generate MR contrast without additional substrate [6]. Cellular ferritin reporter expression provides contrast by transiently lowering the intracellular iron concentration, leading to a physiologic compensation mechanism that triggers cellular iron uptake. The overexpressed ferritin protein, which is a ubiquitously expressed iron-binding protein, will store excess intracellular iron. Ferritin shortens T_1 and T_2 relaxation times in MRI. In a recent study, researchers analyzed the longitudinal (R_1), and transverse (R_2) relaxation maps in ferrritin expressing C6 glioma xenografts in nude mice as demonstrated in Figure 20.9 [6]. Both R_1 and R_2 relaxation were increased in ferritin overexpressing tumors (Figure 20.9) [6]. In this study, a tetracycline (TET)-activated system was used to perform MRI on comparable tumor xenografts with and without ferritin expression [6]. To validate their results using fluorescence imaging studies and biochemical analyses, researchers constructed a cassette of multimodality reporter genes containing the ferritin gene tagged with hemagglutinin (HA) as well as enhanced green fluorescent protein (EGFP) under the control of TET. This process is outlined in Figure 20.9 [6]. The advantage of this novel MR-detectable reporter gene is that no exogenous administration of contrast-generating substrate is required.

20.3 Receptor Imaging

One of the most promising targets of *in vivo* molecular imaging is the cellular receptor. Receptors are defined as proteins that receive and respond to stimuli from other proteins, hormones, or small molecules. Receptors make good molecular targets because their expression is frequently tissue- and/or disease-specific and they often have high selectivity for their ligands. High-affinity ($\leq$ 5 nM binding affinity) and highly specific ligands are chosen for use as probes for their given receptor. Receptors may be located within a membrane (e.g., plasma, nuclear, golgi, mitochondrial) or within the cytoplasm.

The key to selecting an appropriate receptor target for oncologic imaging is similar to that for selecting good real estate: location, location, location! First, the receptor must be expressed either only within the cancerous tissue (the ideal situation) or be greatly overexpressed within the tumor with low expression in normal tissues. This is critical to differentiating neoplastic and normal tissue. How does one identify a good receptor target for the neoplasm of choice? Today, the emerging field of proteomics is the preferred way to identify selectively expressed receptors for a given cell type. Proteomics can be defined as the qualitative and quantitative comparison of proteomes [proteome = PROTEin matched to its genOME] under different conditions to further unravel biologic processes (http://ca.expasy.org). This method works by collecting many tissue samples from both healthy and affected patients and comparing gene and/or protein expression levels within those tissues. Look for receptors that (1) are highly expressed in the tissue you wish to image/study (i.e., prostate carcinoma), (2) are highly expressed in a large majority of patients with the given tumor type (if you want to develop a universal screening process for patients to detect the selected type of tumor), and (3) receptors with known ligands.

Ligand selection is as important as receptor target identification. Ligand selection is also about choosing the right location. Criteria for selecting the right ligand are as follows:

- High affinity ($\leq$ 5 nM binding affinity) for the receptor and high selectivity for the target receptor over other receptors and receptor subtypes
- The ligand must be amenable to labeling with an optical, MR, nuclear, or US opaque moiety that will generate measurable contrast signal within the appropriate scanning device. The addition of the contrast moiety must not significantly alter ligand binding affinity or specificity for its receptor. The addition of a contrast-enhancing group to the receptor ligand makes it a molecular probe.
- The probe should have characteristics such that it can endure a general route of administration (for example intravenously), arrive at and quickly bind to the target receptors in the target tissue, and quickly wash out of nontarget tissues that lack the target receptor. In this way, the probe contrast is detected largely at the site of interest (location).
- The probe should also display favorable biodistribution characteristics where the only nonspecific probe uptake occurs at sites of metabolic processing such as the kidney and bladder or in the gastrointestinal tract for hepatobiliary clearance.
- Finally, ligands should be chosen to maximize the two points listed above by matching the ligand choice to the location of the receptor.

This last criterion is the most complicated. The tissue as well as subcellular location of the receptor will also determine what type

after GCV treatment are presented. T, tumor; H, heart; BR, brain; BL, bladder. Calibration scales are provided in (a) and (b). Superimposition of PET on photographic images of representative tumor-bearing mice was performed to simplify the interpretation of [^{18}F]FDG and [^{18}F]FEAU biodistribution. (c) Growth curves of individual tumor xenografts after AAVP administration. (d) Temporal dynamics of HSVtk gene expression as assessed by repetitive PET imaging with [^{18}F]FEAU at different days post-AAVP administration. (e) Changes in tumor viability before and after GCV therapy as assessed with [^{18}F]FDG PET. Error bars in (c) through (e) represent standard deviations. Adapted from Hajitou et al. [13]. For a more detailed view of this figure, please visit our companion site at: http://books.elsevier.com/companions/9780123735836.

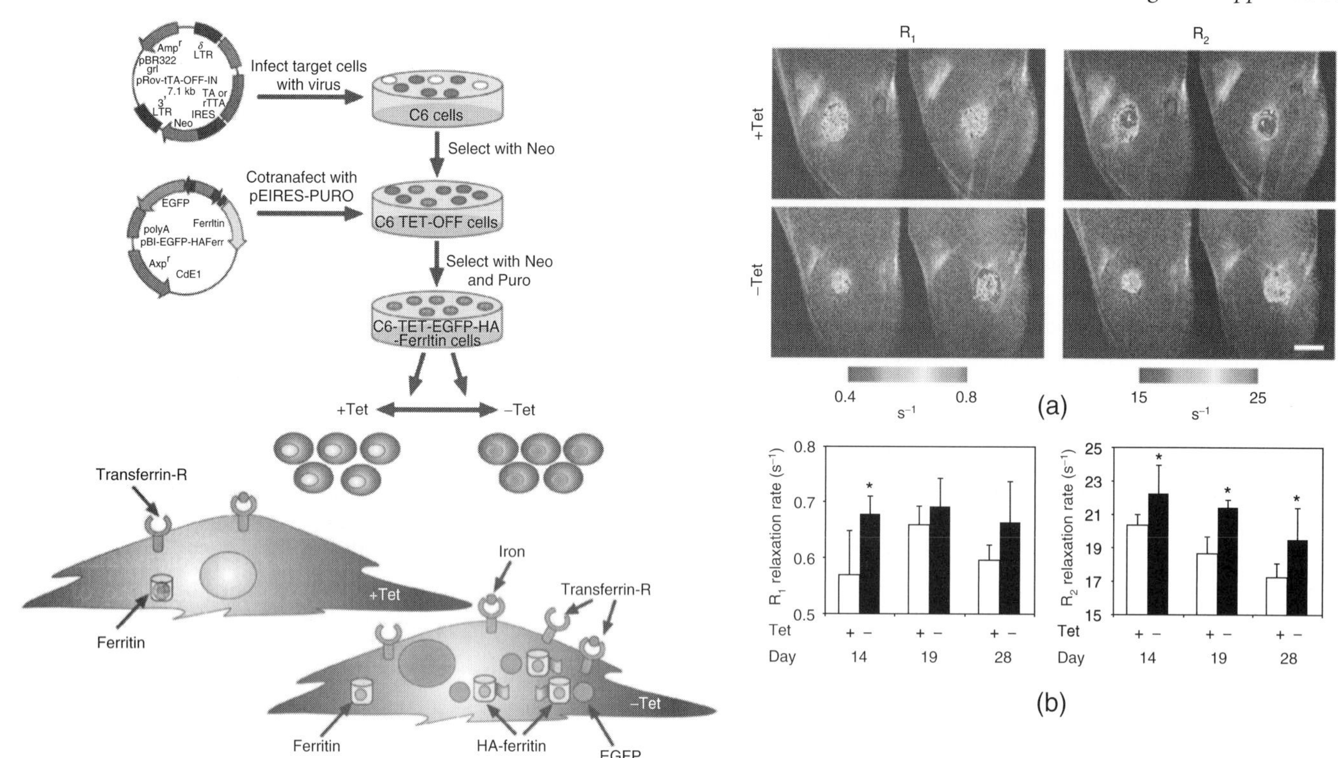

FIGURE 20.9 (Left) Schematic of how C6 cells were infected with viruses carrying the TET transactivator (tTA) under a constitutive promoter (pRev-tTA-OFF-IN). The cells were then transfected to express TET-EGFP-HA-ferritin using a bidirectional vector (pBI-EGFP-HA-Ferr vector). Selected clones showed overexpression of EGFP and HA-tagged ferritin, both of which were tightly suppressed by administration of TET (+Tet). (Right) *In vivo* MRI detection of switchable ferritin expression in C6 tumor xenografts in the hind limb of nude mice generated from C6 cell clones stably expressing TET-EGFP-HA-ferritin. TET and sucrose (or sucrose only for –Tet) were supplied in drinking water, starting 2 days before inoculation. (a) R_1 and R_2 maps of tumor regions overlaid on the MR images are shown for two representative mice from each group. (b) R_1 and R_2 values (mean $\pm$ SD) at the tumor region in the presence (ferritin off; n = 7) or absence (ferritin on; n = 4) of TET in drinking water. $^*P < 0.05$: two-tailed unpaired Student *t* test. Scalebar = 2.5 mm. Adapted from Cohen et al. [6]. For a more detailed view of this figure, please visit our companion site at: http://books.elsevier.com/companions/9780123735836.

of ligand will be suitable for imaging. For example, if monoclonal antibodies have been generated against a particular receptor that is overexpressed in cancerous tissue (i.e., α-methylacylCoA racemase [AMACR] in prostate carcinoma [14, 15]), the antibody will only be suitable as a probe if it can access its receptor *in vivo*. This is only possible if the receptor is located in the plasma membrane and the antibody recognizes an extracellular epitope. In the case of AMACR-directed antibodies, the AMACR protein is expressed in the mitochondria and peroxisomes, which are intracellular. Antibodies do not enter cells passively or nonspecifically, so in the case of an intracellular receptor target, a successful probe must be able to enter the cell and interact with the receptor in its native environment. Currently, the search for a suitable small-molecule ligand able to bind the AMACR receptor tightly and specifically *in vivo* is under way.

20.3.1 Nuclear Receptor Imaging

Nuclear receptor imaging involves the use of either PET or SPECT. Each has its own strengths and weaknesses. Table 20.1 illustrates some of the more important features of both, including availability, cost of the isotope, ligand limitations, and sensitivity.

Most larger hospitals possess their own SPECT scanner(s) and can order either the free isotope for custom labeling or preformulated tracers from a radiopharmacy. Currently, the vast majority of SPECT scans that take place every day around the world are nonspecific scans such as [^{67}Ga]Ga citrate (nonspecific tumor uptake), [^{201}Tl]TlCl (myocardial heart uptake), and [^{99m}Tc]MDP (bone scans). However, the use of receptor-targeted probes in clinical SPECT is gaining acceptance. Three examples of now commonly used receptor-based SPECT tracers for tumor detection include [^{111}In]OctreoScan® for the detection of somatostatin receptor-expressing tumors [16], [^{111}In]Zevalin for the detection of CD20-positive lymphomas [17], and [^{111}In]ProstaScint for detecting prostate-specific membrane antigen optimized-expressing prostate cancer [18]. Figure 20.10 illustrates (PSMA) results seen in clinical scans with these tracers.

The three examples cited above are all labeled with In-111 and comprise peptide (OctreoScan) and antibody probes (Zevalin and ProstaScint). In the case of antibody probes, their target

TABLE 20.1 SPECT vs. PET in clinical imaging

SPECT (clinical)	PET (clinical)
\$200-\$500K (scanner)	\$1 × 10^6 (scanner)
Generator-produced isotopes \$50-\$2,500/ dose	Cyclotron-produced isotopes \$2–5 × 10^6 (cyclotron)
[^{99m}Tc], [^{123}I], [^{111}In], [^{67}Ga]	[^{11}C], [^{13}N], [^{18}F], [^{124}I], [^{64}Cu], [^{62}Cu], [^{86}Y], [^{94m}Tc]
Mostly chelation chemistry	Physiologic tracers (mostly direct covalent attachment)
Largely qualitative	Quantitative
1-1.5 cm resolution	0.4-0.8 cm resolution

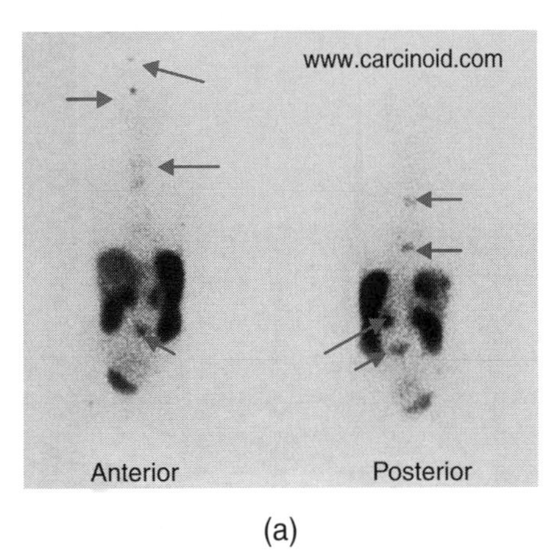

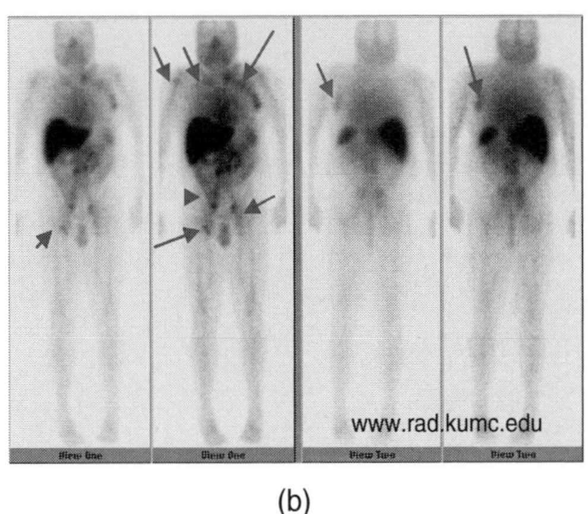

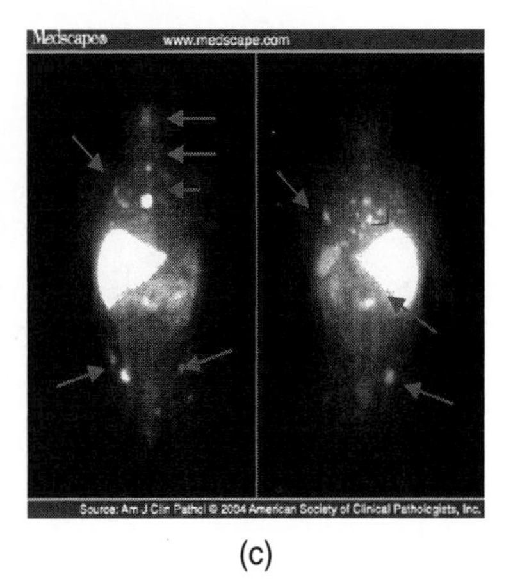

(a) (b) (c)

FIGURE 20.10 (a) [^{111}In]OctreoScan showing somaostatin receptor-positive metastatic cancer (arrows). (b) [^{111}In]Zavalin scan showing CD20+ recurrence on non-Hodgkin's lymphoma (arrows). (c): [^{111}In]ProstaScint scan showing PSMA+ metastatic prostate cancer (arrows).

receptor affinity is very high, but their biodistribution includes substantial nontarget uptake in both the liver and spleen. However, given sufficient uptake and clearance time (2 to 6 days), these probes do effectively show the locations where high densities of their target receptors are located and whether those tumors will respond to that specific targeted therapy.

Current trends favor the development of small-molecule probes for receptor imaging that possess both high affinity and specificity for their targets and display favorable pharmacokinetics. Pharmacokinetics are essentially the speed and tissue distribution that describe the movement of a probe through the body. Small hydrophilic molecules ($\leq$ 1 kDa) generally display excellent pharmacokinetic properties such as quick wash-in rates and quick nontarget wash-out rates. These typically also clear through the urine and so do not display nonspecific liver or GI uptake. An example of a small-molecule probe with these attributes is [^{18}F]DCFBC, a 391 Da PSMA inhibitor probe under development for use in primary and metastatic prostate cancer detection using PET. Currently, this probe has been tested in preclinical models of prostate cancer in mice (R. Mease et al., patent pending). Other small-molecule probes for PSMA expression using SPECT are also being developed.

20.3.2 Magnetic Resonance Receptor Imaging

Recently, novel receptor probes that take advantage of inherently high resolution of MRI and couple it with detection of specific biochemical targets have been developed. Contrast enhancement using MRI is typically induced by altering the relaxation rate constants of the abundant water signal in tissue. This is done typically by using either chelated Gd^{3+} to create T_1 (spin-lattice relaxation time) positive contrast or superparamagnetic Fe_2O_3 particles to create T_2 (spin-spin relaxation time) negative contrast. Generally, probes that are amenable to labeling with radiometals for SPECT or PET can also be labeled with Gd^{3+} for use in MRI. The difference is that the sensitivity of detection for MRI is considerably lower than that of either SPECT or PET. Therefore, MRI probes must either contain many chelated Gd^{3+} ions or the density of receptor targets must be high. Additionally, these probes can only be used for cell surface receptors.

An example of successful receptor imaging using T_1 contrast MRI has been accomplished using a biotinylated herceptin antibody to target Her2/neu expressing breast cancer tumors [19]. Once the antibody has had sufficient time to bind and clear from nonspecific tissues, a Gd^{3+}-chelated avidin probe is injected and binds specifically to the biotin present on the herceptin antibody [19]. This generates positive contrast, enabling imaging of Her2/neu expressing tumors. This has been achieved in preclinical mouse models as demonstrated in Figure 20.11 [19].

Another example of receptor imaging using MRI is imaging the folate receptor with superparamagnetic iron-containing dendrimers decorated with folate ligands [20]. This employs negative T_2 contrast and has been successfully demonstrated in

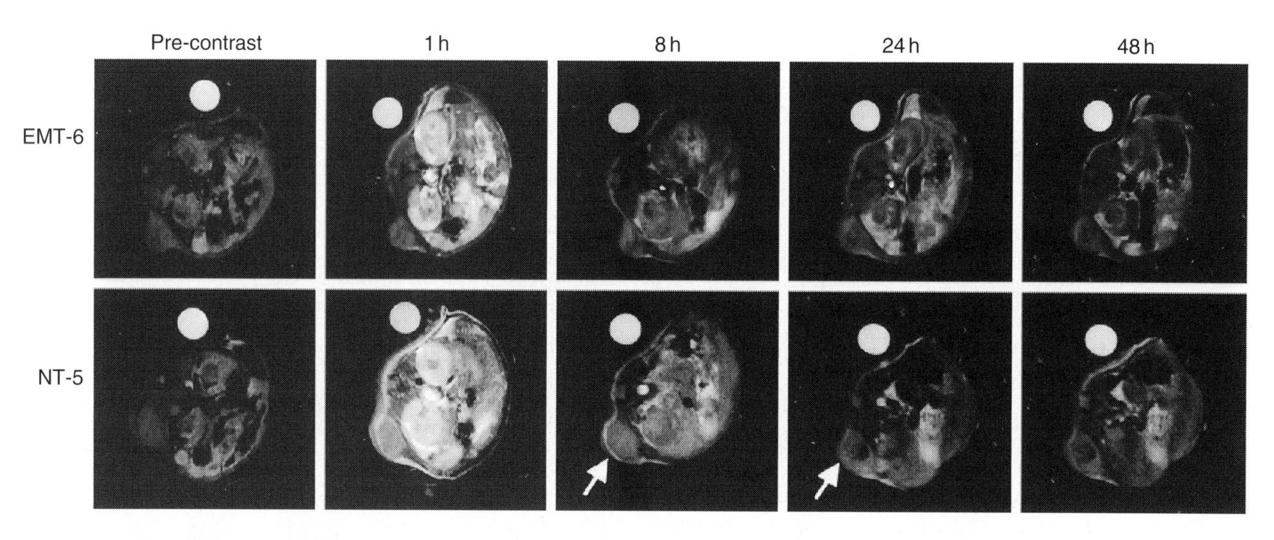

FIGURE 20.11 MR T_1 weighted images of control EMT-6 and NT-5 tumors obtained before administration of the contrast agent (avidin-GdDTPA conjugate) and at 1, 8, 24, and 48 hours after contrast. Arrows show enhanced signal from the tumor at the 8- and 24-hour time points for the HER-2/neu-expressing NT-5 tumor. Adapted from Artemov et al. [19].

imaging folate receptor overexpressing tumors in preclinical tumor models. Many types of cancer overexpress the folate receptor and this method may provide a high-resolution noninvasive method for detecting cancers that express this molecular marker.

20.3.3 Optical Receptor Imaging

Optical receptor imaging uses fluorescently labeled probes and is traditionally used to detect receptors in either cell culture or tissue sections. Immunofluorescence is a widely used technique to detect well-known cancer markers in biopsy specimens to help characterize cancer type and stage. An emerging aspect of this field is in using fluorescently labeled probes to image receptors *in vivo*. Cooled charge-coupled device cameras are now available within small animal imagers. These animal scanners have the ability to excite fluorophores with various frequencies of light and then capture their emission in either planar or tomographic mode (http://www.xenogen.com/wt/page/imaging, http://www.visenmedical.com/technologies/index.html). These *in vivo* scanners are used in rodent preclinical studies only because of the large attenuation observed in animal tissue even at shallow depths. Beyond 1 cm in tissue depth, the signal becomes unusable.

Both antibody and small molecule probes have been conjugated to fluorophores emitting from the green through near-infrared (NIR) wavelengths. NIR light passes through tissue with less attenuation than shorter wavelengths and so these dyes are becoming the standard choice for fluorescence imaging (see Figure 20.3). Attenuation is the critical limitation of this technique and optical receptor imaging is only just emerging as a preclinical modality to study cancer and cancer therapy.

One example of optical receptor imaging is the use of fluorescently labeled cyclic arginine-glycine-aspartic acid (RGD) peptides to study $\alpha_v\beta_3$ integrin expression in tumor neovasculature. The $\alpha_v\beta_3$ integrin subtype has been found to be overexpressed in a variety of tumor types and in tumor neovasculature. Conjugation of a cyclic RGD peptide, a known high-affinity ligand of this receptor subtype, to the Cy-7 NIR dye has been done to visualize metastatic ovarian tumor models in mice as a proof-of-principle demonstration of targeted receptor imaging [21].

Additional clinical applications for optical receptor imaging lie in using a confocal endomicroscope, which can be maneuvered into body cavities to image epithelial surfaces [22, 23]. It can also potentially be used during surgery to aid the surgeon in determining where the margins are located and where to stop cutting.

20.3.4 Ultrasound Imaging of Receptors

To generate contrast for receptor imaging in ultrasound applications, microbubbles have recently been used, typically perfluorohydrocarbon gas in hydrophobic vesicles [24, 25]. Microbubbles enhance the US echo by creating backscatter because they expand and contract when being exposed to US beams of any frequency [25]. Researchers have attached US contrast-generating microbubbles to the peptide arginine-arginine-leucine, which can specifically be found in the tumor vasculature [26]. Avidin-biotin binding was used to attach microbubbles to the targeting peptide [26]. This US contrast agent was tested preclinically and generated vasculature-specific contrast in US images of a prostate tumor xenograft, as demonstrated in Figure 20.12 [26].

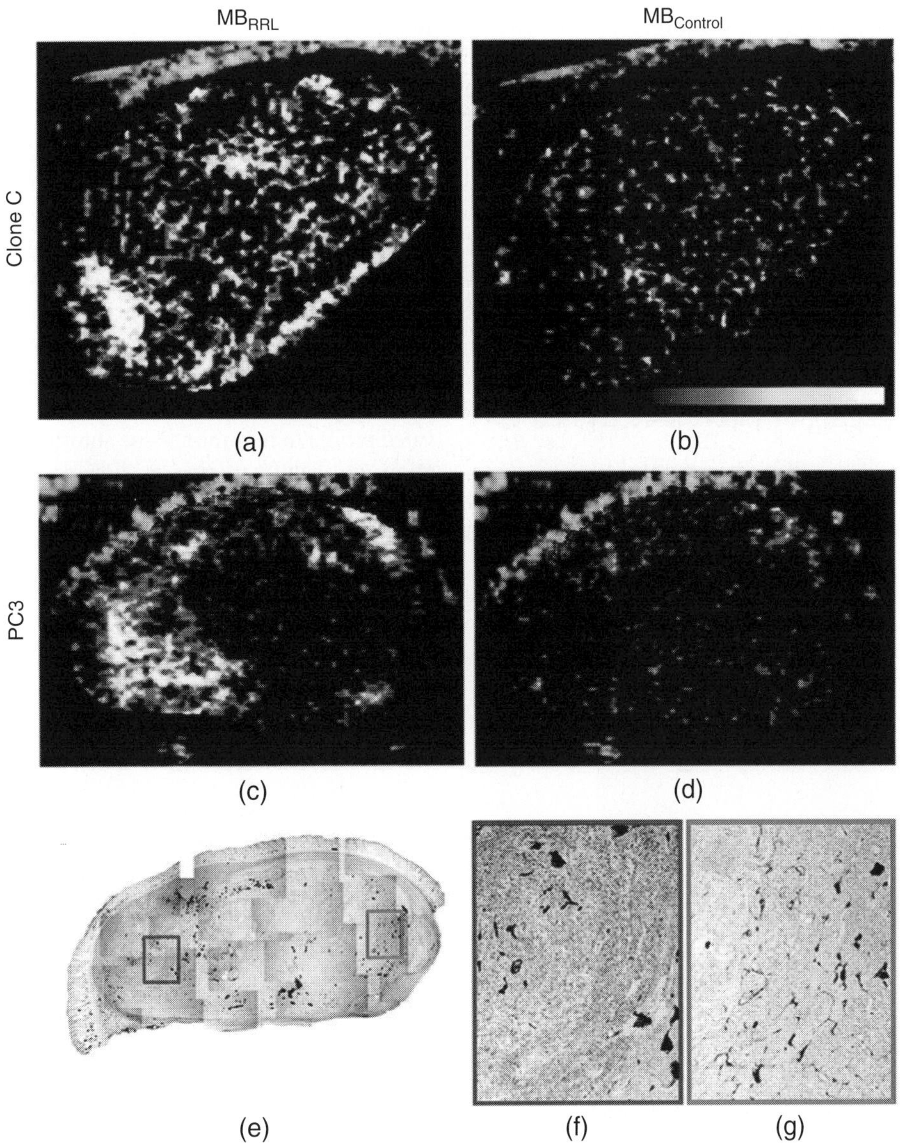

FIGURE 20.12 (a) Background-subtracted, color-coded US image taken 120 seconds after injection of microbubbles (MBs) conjugated to RRL (MB_{RRL}) into a mouse bearing a Clone C tumor. Within the colored areas, gradations from red to orange to yellow to white denote greater signal enhancement by contrast material. Non-color coded portions are not background subtracted and do not influence the videointensity data. MB_{RRL} resulted in greater contrast enhancement. (b) Corresponding image for MBs conjugated to a glycine control peptide ($MB_{Control}$) in the same mouse as A. (c) and (d) Similar ultrasound images as in (a) and (b), but from a mouse with a PC3 tumor. (e) Collage of high-resolution photomicrographs taken of a midline PC3 tumor section immunohistochemically stained for factor VIII, showing localization of the microvasculature predominantly to the periphery of the tumor. Cells are counterstained with hematoxylin. Some expected shrinkage has occurred secondary to formalin fixation. Original magnification ×20. Adapted from Weller et al. [26]. For a more detailed view of this figure, please visit our companion site at: http://books.elsevier.com/companions/9780123735836.

20.4 Enzyme-Activated Probes

Enzyme-activated probes are smart contrast agents that become detectable following enzymatic cleavage. Thus, the contrast agent needs to be a substrate for the enzyme activity to be monitored by molecular imaging. This substrate should only be detectable by molecular imaging once the enzyme under investigation has modified it enzymatically. Several such enzymatic substrates have been developed for optical imaging. They provide the possibility of delivering fluorophores that are quenched in the substrate because they are in close proximity. Once the enzyme cleaves the substrate, the previously quenched fluorophores are released and start fluorescing. Examples are discussed in Section 20.4.1. For MRI and magnetic resonance spectroscopy (MRS), enzyme-activated probes have been developed in the context of molecular imaging of gene expression, as discussed in Section 20.2. β-galactosidase, which can be used as a reporter for molecular imaging of gene expression, requires the addition of a substrate for molecular imaging by MRI or MRS. Some reporter genes for nuclear imaging also require administration of a labeled substrate for molecular imaging, as described in Section 20.2.

20.4.1 Optical Imaging of Enzyme-Activated Probes

Optical imaging of enzyme-activated probes has been most heavily explored in molecular imaging of enzymes that are overexpressed in cancers. Several different proteases, which are enzymes that cleave proteins at specific sites based on a specific amino acid sequence, participate in degrading and remodeling the extracellular matrix (ECM) and basement membranes. The ECM is a tissue-structuring meshwork of several types of structural proteins consisting mainly of collagens, laminin, and fibronectin. ECM degradation and remodeling facilitate cancer invasion, metastasis, and angiogenesis. The overexpression of thiol proteases such as cathepsin B and matrix metalloproteases (MMPs) such as MMP-2 (gelatinase) has been associated with tumor aggressiveness and poor clinical outcome in several cancers [27, 28]. Researchers have developed molecular imaging of protease activity in tumors. Most frequently they use NIR optical imaging of protease-activated probes to minimize tissue autofluorescence, photon attenuation, and light scattering, as discussed in Section 20.2.1 (Figure 20.3). These probes contain quenched fluorophores, which are released from a carrier after cleavage of the probe by an enzyme,

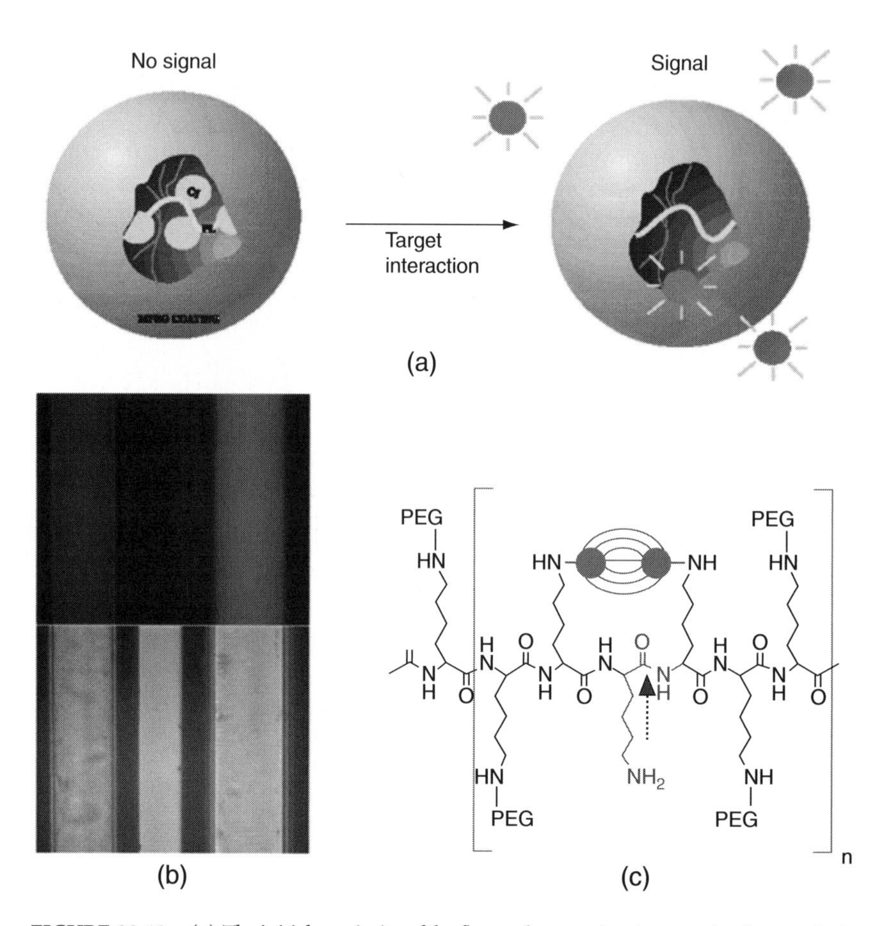

FIGURE 20.13 (a) The initial proximity of the fluorophore molecules to each other results in signal quenching. (b) NIR fluorescence image (top) and bright light image (bottom) of non-activated C-PGC (left) and activated probe (right). Fluorophore concentration: 0.17 M. Image acquisition time: 30 seconds. Excitation: 670 nm, emission: 700 nm. Note the difference in signal intensity between enzyme-activated and unactivated probe. (c) Chemical structure of repeating graft copolymer segment indicating quenching of Cy5.5 and enzymatic degradation site (green arrow). Adapted from Weissleder et al. [29]. For a more detailed view of this figure, please visit our companion site at: http://books.elsevier.com/companions/9780123735836.

and then a fluorescent signal is detected, as depicted in Figure 20.13 [29]. A well-tolerated nontoxic nonimmunogenic synthetic graft copolymer consisting of poly-L-lysine (PL) sterically protected by multiple methoxypolyethylene glycol (MPEG) side chains was used as a delivery vehicle of quenched fluorophores to tumors [29]. Each PL backbone contained an average of 92 MPEG molecules and 11 molecules of Cy5.5 yielding $(Cy5.5)_{11}$-PL-$MPEG_{92}$ (abbreviated as C-PGC) [29].

These smart optical contrast agents have been developed for MMP-2 detection [27] and for NIR optical imaging of cathepsin B-sensitive probes in breast tumor models [28], as shown in Figure 20.14. Activation of the cathepsin B probe was observed in both well-differentiated and highly invasive metastatic tumors. However, the fluorescence signal intensity was higher in the metastatic tumor model, which was consistent with higher cathepsin B protein expression or activity in the more metastatic tumor [28]. This cathepsin B-activated smart contrast agent for optical imaging has great potential for being translated to clinical applications to discern breast cancers with high proteolytic activity that may, as a result, be more metastatic. This may be possible in the future because optical imaging scanners are being tested in clinical trials.

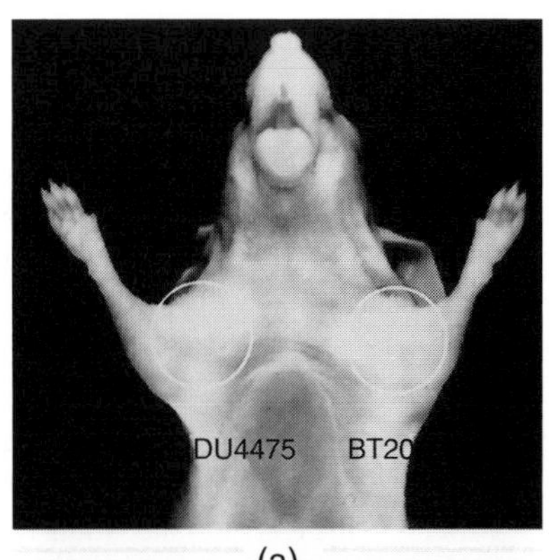

(a)

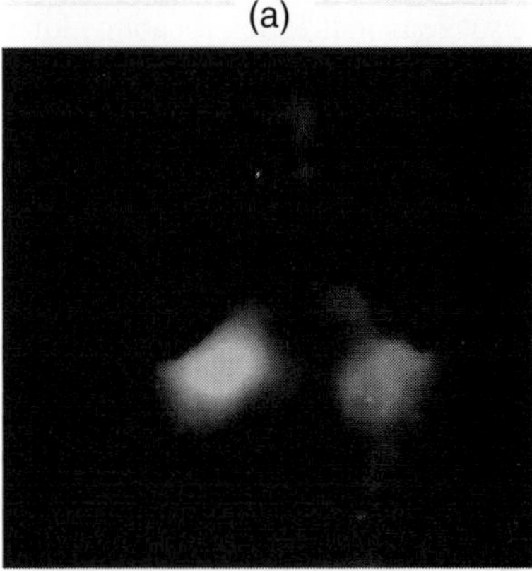

(b)

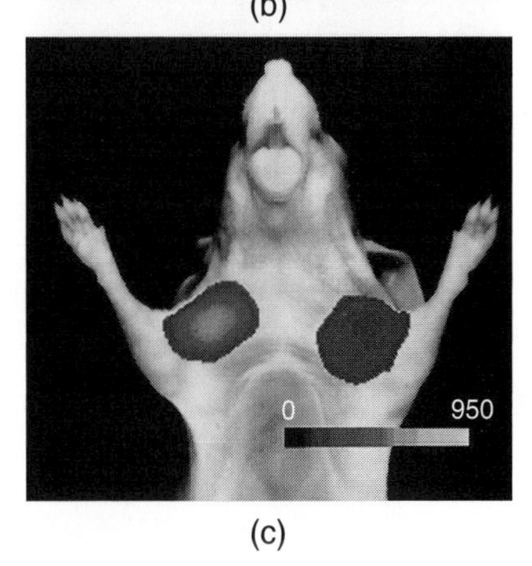

(c)

FIGURE 20.14 NIR fluorescence imaging 24 hours after intravenous injection of the cathepsin-B-sensitive autoquenched probe in a representative animal. (a) Light image; (b) raw NIR fluorescence image; and (c) color encoded NIR fluorescence signal (arbitrary units of NIR fluorescence intensity) superimposed on light image. The highly invasive breast adenocarcinoma (DU4475) was implanted on the right of the chest and the well-differentiated adenocarcinoma (BT20) on the left. Note the higher fluorescent signal depicted on the highly invasive breast lesion (b, c). Adapted from Bremer et al. [28]. For a more detailed view of this figure, please visit our companion site at: http://books.elsevier.com/companions/9780123735836.

20.5 Metabolic Imaging

Metabolic imaging can be performed using probe-free contrast where endogenous molecules are imaged or by using isotope-labeled metabolite precursors that are being imaged. Nuclear metabolic imaging requires administration of radiolabeled tracer compounds that are analogs of endogenous metabolites. Proton and ^{31}P MRS can be used to detect endogenous metabolite levels, while ^{13}C MRS has been applied for metabolic tracer studies following administration of ^{13}C-labeled substrates. Typical metabolic features found in most tumors are the activation of glycolysis as well as active suppression of the tricarboxylic acid cycle [30]. Increased total choline levels are also typical of tumors and occur primarily as a result of increased phosphocholine levels [31], which have been linked to oncogenic *ras* signaling [32] and oncogenic transformation [33].

20.5.1 Nuclear Metabolic Imaging

Nuclear metabolic imaging is the most prolific and widely practiced form of molecular imaging. Every day, hundreds of patients around the world are scanned using [^{18}F]FDG PET or PET-CT to determine the extent of their cancer. Fluorodeoxyglucose (FDG) is a glucose analog that is taken up by all cells but is taken up in much higher quantity by cancer cells that have high glycolytic activity. Most tumors display higher glucose uptake as well as higher glycolytic rates and therefore sequester a larger amount of FDG than surrounding normal tissues. [^{18}F]FDG PET is a highly sensitive method to detect rapidly growing tumors noninvasively because of both overexpression of glucose transporters and presence of the glycolytic enzyme hexokinase.

Tumors typically also display a higher rate of DNA synthesis due to increased proliferation as well as increased protein and lipid metabolism. These traits can be targeted by using radiolabeled probes such as [^{18}F]FLT (fluoro-L-thymidine), which is incorporated into replicating DNA following deoxyribosylation; [^{11}C]choline, which accumulates in tumors due to increased choline phospholipid metabolism in cancer cells [31]; and [^{11}C]methionine, which measures total protein synthesis. Tracers such as [^{11}C]choline are becoming even more useful as they are also taken up in slower-growing tumor types, which contain elevated levels of phosphocholine but exhibit a lower glycolytic rate and thus lower [^{18}F]FDG uptake than other tumor types. Evidence suggests that elevated choline kinase and phosphocholine levels are associated with an aggressive phenotype and may indicate a higher propensity to metastasize [31, 34]. Figures 20.15 and 20.16 show representative scans of [^{18}F]FDG and [^{11}C]choline PET, illustrating the differences between metabolic imaging of increased glycolytic rates, proliferation, and increased choline phospholipid metabolism in tumors. The ability to characterize a tumor's metabolic profile noninvasively is valuable in staging and selecting an appropriate therapy.

20.5.2 Metabolic Magnetic Resonance Spectroscopy

Proton MRS can be used to detect endogenous tumor lactate levels because tumors have an elevated glycolytic activity. Although ^{1}H MRS of lactate has been used as a tumor marker, contamination from the lipid signal as well as variability in its clearance make it a less reliable marker than elevated total choline. Proton MRS is frequently used to detect elevated tumor total choline levels, which are detected as a single overlapping MRS signal consisting of the endogenous choline-containing molecules in tumors: glycerophosphocholine, phosphocholine, and free choline. Figure 20.17 demonstrates how increased total choline levels in an MRS image overlap with the contrast-enhancing region in a postcontrast T_1-weighted image of a tumor in a woman with breast cancer [35]. In the clinic, metabolic molecular imaging of choline in cancers is currently being used as an adjunct for diagnosis of primary malignant tumors in brain, prostate, and breast [36]. Proton MRS of the total choline signal is useful clinically in treatment planning for radiation therapy and in assessing treatment response in brain, breast, and prostate [36].

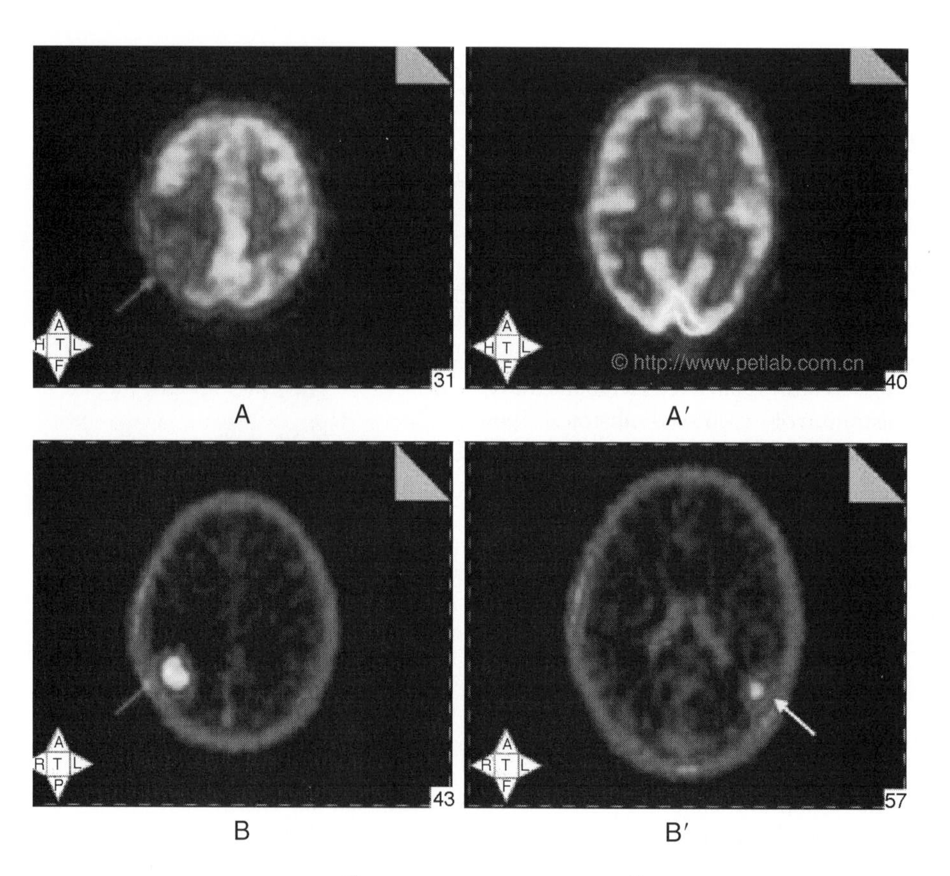

FIGURE 20.15 Comparison of [^{18}F]FDG PET (A and A′) and [^{11}C]choline PET (B and B′) in two patients with brain tumors. The tumor (red arrow) in A and B has a high glycolytic rate and high choline levels. The tumor in A′ and B′ contains high choline levels, but a relatively low glycolytic rate. For a more detailed view of this figure, please visit our companion site at: http://books.elsevier.com/companions/9780123735836.

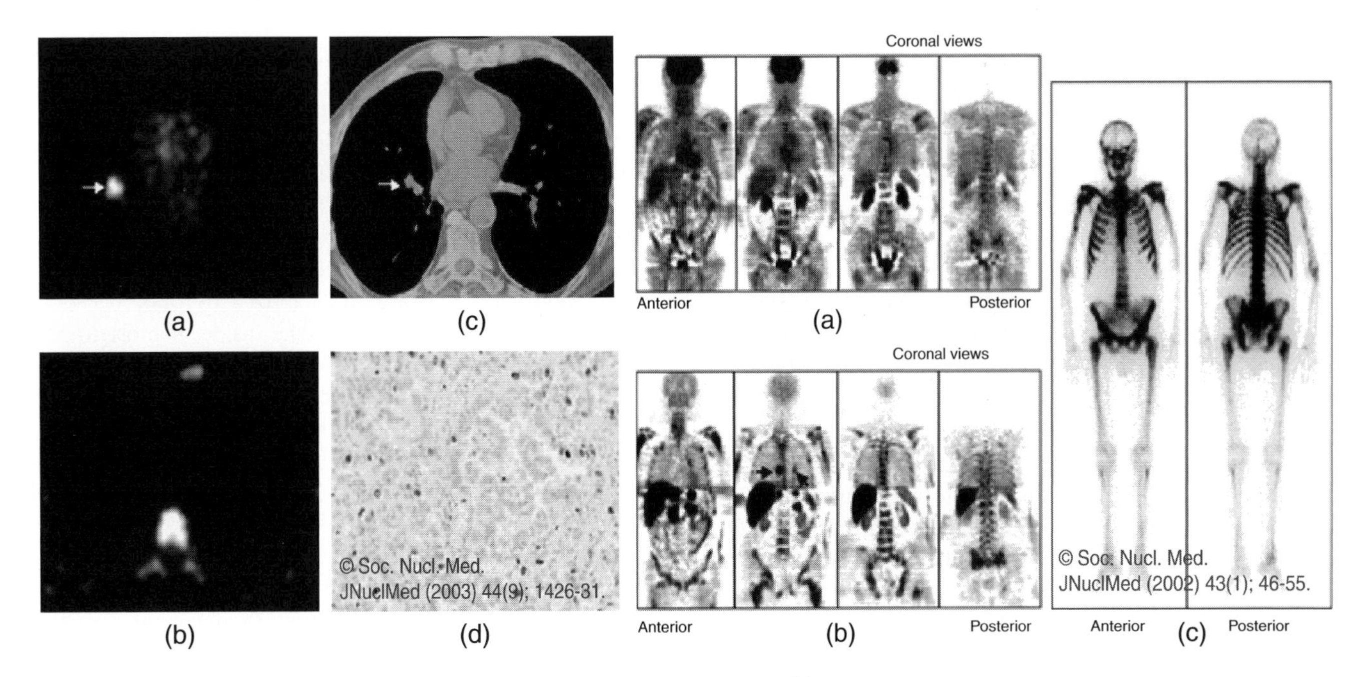

FIGURE 20.16 (Left) [^{18}F]FDG PET of a lung tumor (a, arrow) compared with [^{18}F]FLT PET (b), indicating an increased glycolytic rate with no increase in proliferation. (Right) [^{18}F]FDG PET showing metastatic prostate cancer (a) with clearer uptake using [^{11}C]methionine PET (b, arrow).

20.6 Imaging of Permeability, Perfusion, and Blood Flow

Angiogenesis plays an important role in tumor growth because once a tumor reaches a certain size, it cannot grow any further unless it recruits new blood vessels that supply the tumor with nutrients and oxygen. The formation of new blood vessels, also referred to as neovascularization, is therefore necessary for tumor progression. This process involves multiple angiogenic factors that induce existing vasculature in the surrounding tissue to migrate toward the cancer cells and establish neovasculature [37]. Multiple anti-angiogenic factors also exist and the net outcome is the balance between pro- and anti-angiogenic factors [38]. Vascular endothelial growth factor (VEGF) is an example of a potent angiogenic agent that induces angiogenesis as well as increases the permeability of vasculature [38]. Tumor vasculature can be imaged by PET, contrast-enhanced MRI, CT, and US. In addition, imaging can provide functional parameters such as blood flow, blood volume, and permeability in preclinical models and in patients [24]. These functional imaging techniques measure tumor perfusion and microvascular permeability by detecting the pharmacokinetic behavior of intravenously administered contrast agents, which are carried in the blood and distributed in the tumor tissue [24].

20.6.1 Magnetic Resonance Imaging and Magnetic Resonance Spectroscopy of Permeability, Perfusion, and Blood Flow

In preclinical studies, tumor vascular volume and permeability can be characterized by the use of a macromolecular contrast agent consisting of albumin bound to gadolinium-diethyltriamine pentaacetic acid (GdDTPA). The GdDTPA on the albumin results in a reduction of T_1 of water that can be quantified to derive maps of vascular volume and permeability. MRI studies using this agent, which is approximately 90 kDa in size, of human breast and prostate cancer xenografts [39] revealed that the vasculature of metastatic tumors was significantly more permeable than the vasculature of nonmetastatic tumors [39]. These researchers also observed that increased vascular endothelial growth factor (VEGF) expression levels correlated with the increased permeability of these tumors [39]. Macromolecular contrast agents have also been used to detect the effect of anti-angiogenic treatment [40]. Low molecular weight GdDTPA-based contrast agents such as Magnevist^{TM} are routinely used in the clinic to identify tumors by contrast-enhancement. Low molecular weight agents also provide values of relative perfusion of tumors [40].

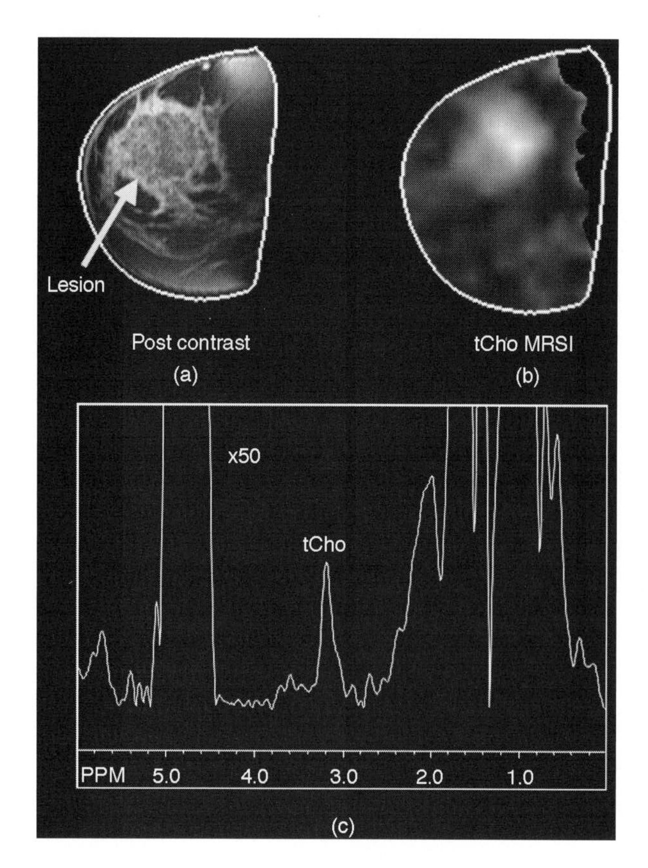

FIGURE 20.17 (a) Postcontrast T_1-weighted images of a breast lesion. (b) Proton MRSI image of total choline-containing compounds (tCho). (c) Magnified (×50) representative ^1H MR spectrum from within the tumor demonstrating elevated tCho (signal-to-noise ratio of 10.6) levels within the tumor. Adapted from Jacobs et al. [35].

20.6.2 Nuclear Imaging of Perfusion and Blood Flow

Measurement of perfusion and blood flow in nuclear imaging is done principally to account for blood pool uptake in images (so as not to be confused with tumor uptake) and also is critical for molecular modeling to determine the input function for a given probe (see chapter sections 5.2 and 6.4). Currently, anti-vasculature treatments are being developed to kill tumors, and methods are needed to measure the progress of these therapies and validate new treatments.

Methods for measuring blood flow include [^{15}O]water PET, [^{11}C]CO [41] and [^{11}C]CO_2 PET as well as [^{133}Xe]Xe SPECT and [^{99m}Tc]-labeled human serum albumin SPECT. Measurements are taken during the first minute or two following tracer injection to capture the first-pass kinetics of the tracer. In this way, hypervascular areas associated with tumor growth can be detected [42] as well as hypovascular areas associated with large, necrotic tumors or obstruction of major vessels. In the case of measuring the effects of anti-angiogenic or anti-vascular agent therapy, blood flow measurements are taken before and following therapy and the region of interest surrounding the tumor is quantitated to determine the effects of the treatment [43, 44].

Perfusion can be measured using [^{64}Cu]PTSM PET and virtually all of the above listed probes when acquiring data following probe equilibrium. The intent is to capture data after the first few passes of the tracer in the body and then quantitate the deposition of the tracer into the tissue. Perfusion in tumors is critical since blood vessels inside tumors tend to be tortuous and leaky and are characterized by poor perfusion into the tumor itself. This is why the inside of most tumors become necrotic and hypoxic after growing beyond a certain size. The effectiveness of many chemotherapeutics and radiation therapy is directly related to tumor perfusion since cytotoxic drugs need to be delivered within the tumor to be effective against the cancer cells, and radiation therapy is more effective in well-oxygenated tumors. Therefore, a number of tracers have been developed to facilitate measurement of tumor perfusion. [^{64}Cu]PTSM is now being used in the clinic on a trial basis to gauge tumor perfusion in an effort to determine whether tumors are likely to respond to radiation therapy prior to initiation of treatment [42]. This has the potential to spare patients damaging ionizing radiation therapy if their tumors turn out to be poor candidates for such treatment.

20.6.3 Ultrasound Imaging of Perfusion and Blood Flow

The tumor vasculature can be visualized by unenhanced ultrasound imaging. Ultrasound imaging can detect vascular features in tumors with a high anatomical resolution of 40 μm to 200 μm vessel diameters [24]. The velocity of blood flow within an evaluated vessel can be measured by color Doppler ultrasound. This technique assigns color to the measured blood vessel based on a scale that is proportional to the velocity of the blood flow [24].

20.7 Imaging of the Tumor Microenvironment

Solid tumors consist of tumor cells that are surrounded, and influenced, by other cells, such as fibroblasts and vascular endothelial cells. The extracellular matrix (ECM) is a meshwork of proteins with tumor-specific physiological environments, such as hypoxia and acidic extracellular pH, which characterize the tumor microenvironment. This tumor stroma plays key roles in processes ultimately leading to tumor progression, such as tumor angiogenesis [45], lymphangiogenesis [46], inflammation [47], and invasion and metastasis [48]. Because the tumor vasculature is chaotic, easily collapses, and sometimes cannot meet the oxygen and substrate demands of the cancer cells, tumors contain hypoxic and acidic physiological environments [49]. Tumor hypoxia is associated with poor prognosis, and resistance to radiation and chemotherapy

[50]. Hypoxia also triggers angiogenesis as the expression of VEGF, which is regulated by a hypoxia response element in its promoter region [37]. For all these reasons, researchers have attempted to image tumor hypoxia with different molecular imaging modalities. Several potential therapeutic targets in the tumor stroma have been discovered with the help of novel imaging techniques, which help investigate the interactions between cancer cells and the tumor microenvironment noninvasively and *in vivo* [51].

20.7.1 Nuclear Imaging of the Tumor Microenvironment

The ability to image hypoxia noninvasively is a boon to both researchers and physicians. Several positron-emitting radiotracers have been developed to facilitate this purpose. By far, the most commonly used are [^{18}F]-fluoromisonidazole ([^{18}F]FMISO) and [^{64}Cu]-copper(II)-diacetyl-bis(N(4)-methylthiosemicarbazone) ([^{64}Cu]ATSM). Both of these tracers are dependent on the concentration of NADH inside the cell. Once the tracer is reduced, it becomes trapped in the cytosol and accumulates, while tracer diffusing into well-oxygenated tissues diffuses back out and clears from the body. Physicians have used both [^{18}F]FMISO and [^{64}Cu]ATSM to probe the hypoxic content of tumors prior to treatment in an effort to determine whether PET scans with these agents have any predictive value of the efficacy of certain therapies or for patient prognosis [52]. For example, studies in patients with head and neck cancer who were scanned using [^{18}F]FMISO PET prior to treatment established that patients who had hypoxic tumors had a more favorable response to tirapazamine, a hypoxia-targeting drug, than to standard cytotoxic therapy alone, whereas patients who did not have hypoxic tumors did not benefit from the tirapazamine [53]. Also, in an attempt to transiently oxygenate a tumor, hyperthermia treatments were tested and progress was monitored using [^{64}Cu]ATSM in mouse models of breast cancer [54]. The researchers found that hyperthermic treatment decreased uptake of the tracer while pO_2 increased.

Similarly, [^{18}F]FDG PET clearly shows whether solid tumors are homogeneous in their glycolytic activity and whether there are pockets of necrosis. This is useful to help in staging and treatment planning to determine whether the entire tumor needs treatment or only the viable rim. A combination of blood flow analysis, hypoxia scanning, and [^{18}F]FDG PET gives a clear picture of the microenvironment of a tumor as blood flow and perfusion relate to hypoxia/normoxia and tissue viability. These data are important in understanding tumor dynamics.

20.7.2 Magnetic Resonance Imaging and Magnetic Resonance Spectroscopy of the Tumor Microenvironment

Parameters characterizing the ECM have been measured by molecular imaging using MRI methods. Peritumoral interstitial convection and lymphatic drain have been measured *in vivo* using macromolecular contrast-enhanced MRI [55]. These parameters provide an index of the ECM integrity and significant differences in vascular and ECM transport were observed in two breast cancer models with different invasiveness. These MRI findings suggest that a combination of increased invasiveness and reduced extracellular matrix integrity may increase lymph node metastasis [56]. Contrast-enhanced MRI has also been used to visualize the spatial distribution of tumor interstitial fluid pressure *in vivo* using contrast-enhanced magnetic resonance imaging [57].

20.7.3 Optical Imaging of the Tumor Microenvironment

Researchers currently apply optical imaging techniques such as differential interference contrast (DIC) microscopy and second harmonic generation microscopy [58] to study the tumor microenvironment, which is characterized by an interaction of tumor cells, stromal cells, and the ECM. Optical imaging using DIC optics generates imaging contrast by means of different optical path length gradients passing through a Nomarski prism [59]. Such DIC microscopy has been used by researchers to dynamically track cell-induced matrix remodeling [59]. Second harmonic generation (SHG) microscopy is a nonlinear optical process that requires an environment without a center of symmetry, such as an interfacial region, to produce a signal. Researchers have used this optical contrast mechanism to image endogenous structural proteins such as collagen-rich layers within the dermis of the mouse ear [60]. It is possible to visualize collagen fibers in melanoma xenografts grown in a window chamber in severe combined immunodeficient mice as shown in Figure 20.18 [61]. These researchers were able to detect collagen fibers *in vivo* without adding contrast agent [61]. Such SGH imaging of tumoral fibrillar collagen provided estimates of the relative diffusive hindrance in tumors and demonstrated that enzymatic modification of tumor collagen by relaxin can improve diffusive transport in tumors, which is important for drug delivery processes in tumors.

20.8 Multimodality Imaging

The most powerful form of *in vivo* oncologic imaging combines molecular, functional, and anatomic imaging. Measuring a specific chemical signal leads ideally to a focused area of increased contrast, where the density of the target molecule is highest. Since most molecular probes are also subject to metabolic processing and breakdown within the body, it is crucial to have a co-registered anatomic image of the subject to verify that the regions of increased contrast correspond to the site(s)

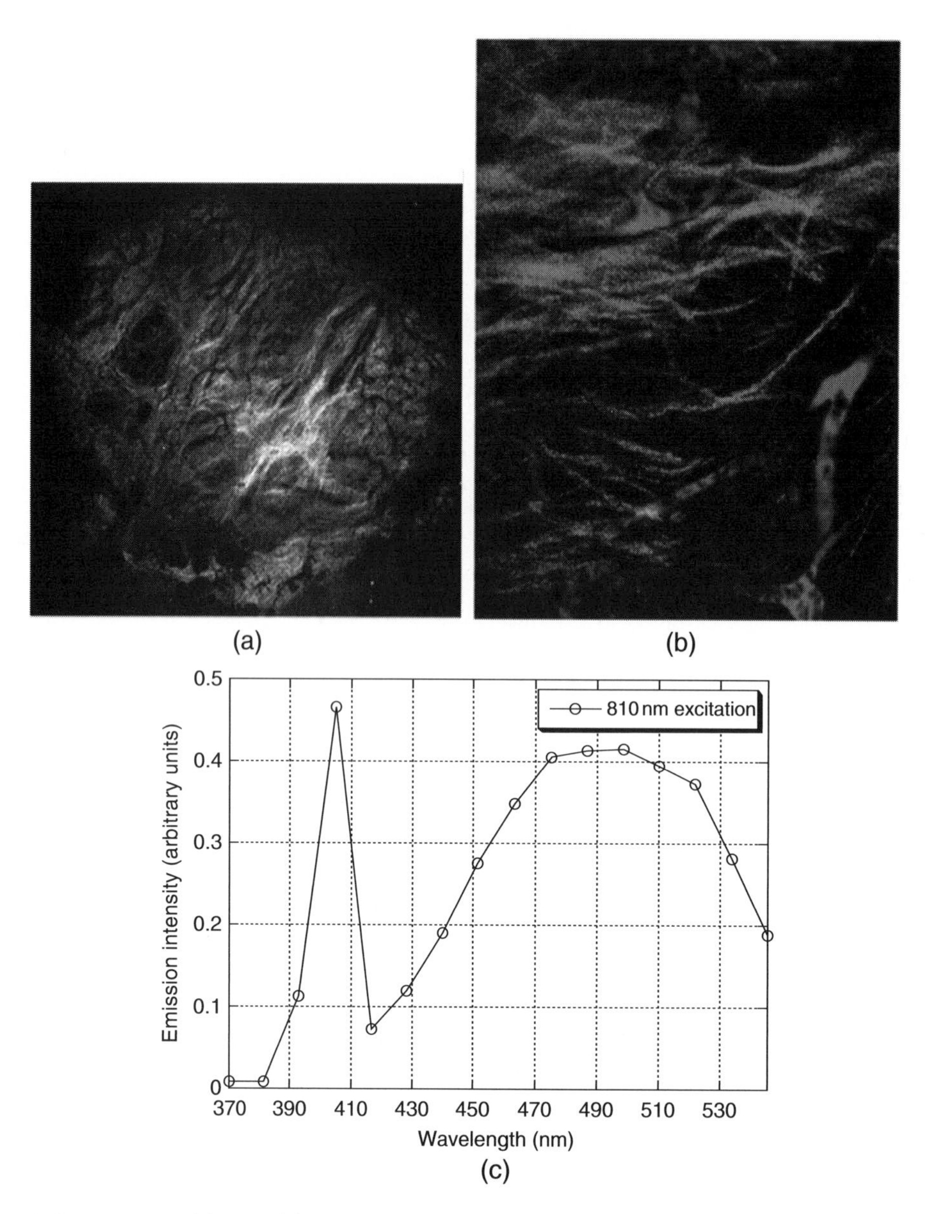

FIGURE 20.18 (a) Second-harmonic signal in a Mu89 melanoma grown in the dorsal skinfold chamber of a severe combined immunodeficient mouse. This image was a montage of 12 separate images, each of which was a maximum intensity projection of 5 images obtained at 20-μm steps. The image shown is 6.6 mm in width. (b) Second-harmonic signal with highlighted vessels. Vessels were highlighted with an intravenous injection of 0.1 ml tetramethylrhodamine-dextran (10 mg/ml; red pseudocolor). SHG signal, green pseudocolor. There was no colocalization of SHG signal with the borders of blood vessels. The image shown is 275 μm in width. (c) Average spectra of light generated with 810 nm excitation of an approximately 0.25-mm^2 region of a Mu89 melanoma in the dorsal skinfold chamber of an immunodeficient mouse. Adapted from Brown et al. [61]. For a more detailed view of this figure, please visit our companion site at: http://books.elsevier.com/companions/9780123735836.

of the tumor and as an anatomic aid to the surgeon or radiation oncologist where to cut. Figure 20.19 illustrates how combined SPECT-CT helps pinpoint the location of a prostate tumor and metastases where SPECT or CT alone would each yield only half of the picture.

Useful combinations of multimodality imaging pair a molecular technique that possesses high sensitivity with an anatomic technique that produces a high-resolution image of the body's soft and hard tissues. Each modality has its own strengths and weaknesses regarding sensitivity and resolution.

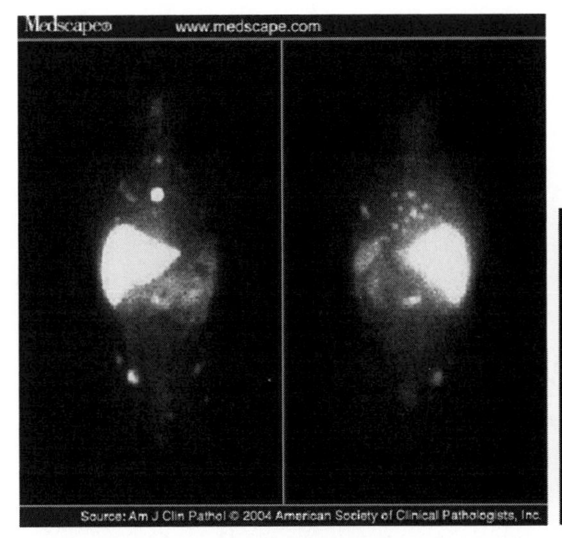

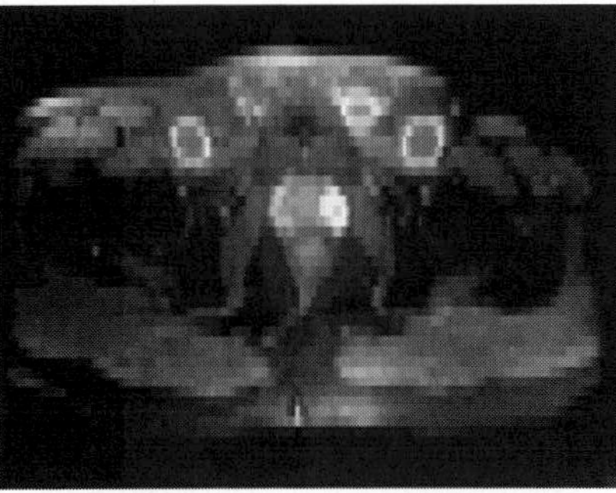

FIGURE 20.19 SPECT versus SPECT-CT in combined PSMA receptor and anatomic prostate cancer imaging. (Left) SPECT scan of a patient receiving ProstaScint™ showing the presence of multiple metastases as well as liver and spleen uptake. (Right) Fused SPECT-CT images in a different patient also receiving ProstaScint™ as well as major blood vessels in the imaged region (yellow arrows). For a more detailed view of this figure, please visit our companion site at: http://books.elsevier.com/companions/9780123735836.

Choosing which modality to use is always a balance among what the target molecule is, whether it has a ligand capable of being labeled, the resolution needed for the application, and the size of the subject. Table 20.2 illustrates the differences in sensitivity and resolution for each modality to help choose which one is right for your application.

Multimodality imaging has been performed in several preclinical studies, one of which will be described in the following section. In this study, researchers combined MRI and optical imaging to characterize the relationship between hypoxia and vascular parameters in prostate cancer xenografts in mice. The spatial distribution of hypoxia in these tumors was detected by imaging fluorescence from cells stably expressing green fluorescent protein under control of a hypoxia response element [62]. Vascular volume and permeability were measured using macromolecular contrast-enhanced MRI [62]. Combined macromolecular contrast-enhanced MRI and optical images demonstrated that vascular volume was low in fluorescing hypoxic tumor regions, which were also frequently permeable [62] as shown in Figure 20.20. The researchers concluded that these observations were consistent with the finding that hypoxia regulates VEGF expression by a hypoxia response element in the VEGF promoter region [37], and thus impacts vascular parameters such as vascular volume and permeability.

TABLE 20.2 Imaging modalities and their respective sensitivities, contrast agents, and applications

Molecular imaging in intact species: methods and agents						
Sensitivity	Modality	Agents	H	R	Primary uses	Examples
	• *Optical*					
pM	FMT	Fluorescent proteins		X	Gene expression, tagging superficial structures	GFP, RFP, NIRF probes
	BLI	Luciferin		X	Gene expression, therapeutic monitoring	fLuc rLuc
	• *Nuclear*					
nM	SPECT	^{99m}Tc, $^{123/5}$I, ^{111}In	X	X	Site-selectivity, protein labeling	^{99m}Tc-annex in V, ^{123}I-A85380
	PET	^{11}C, ^{18}F, ^{124}I, $^{64/62/60}$Cu	X	X	Site-selectivity, gene expression, drug development	^{11}C-RAC, ^{124}I-FIAU, ^{64}Cu-ATSM
	• *MRI*					
μM	MRS	Endogenous metabolites	X	X	CNS, prostate, heart, breast	NAA, Cr, Cho, Glx, mI, ^{31}P
	Contrast agents	Gd, Mn, FeO		X	Cell trafficking, enzymatic activation	Poly-L-lysine, dendrimers, MION
	• *Ultrasound*					
(10 μm)	Contrast agents	Perfluorinated microbubbles		X	Drug delivery, gene transfection	Human albumin (Optison)

H=human, R=rodent

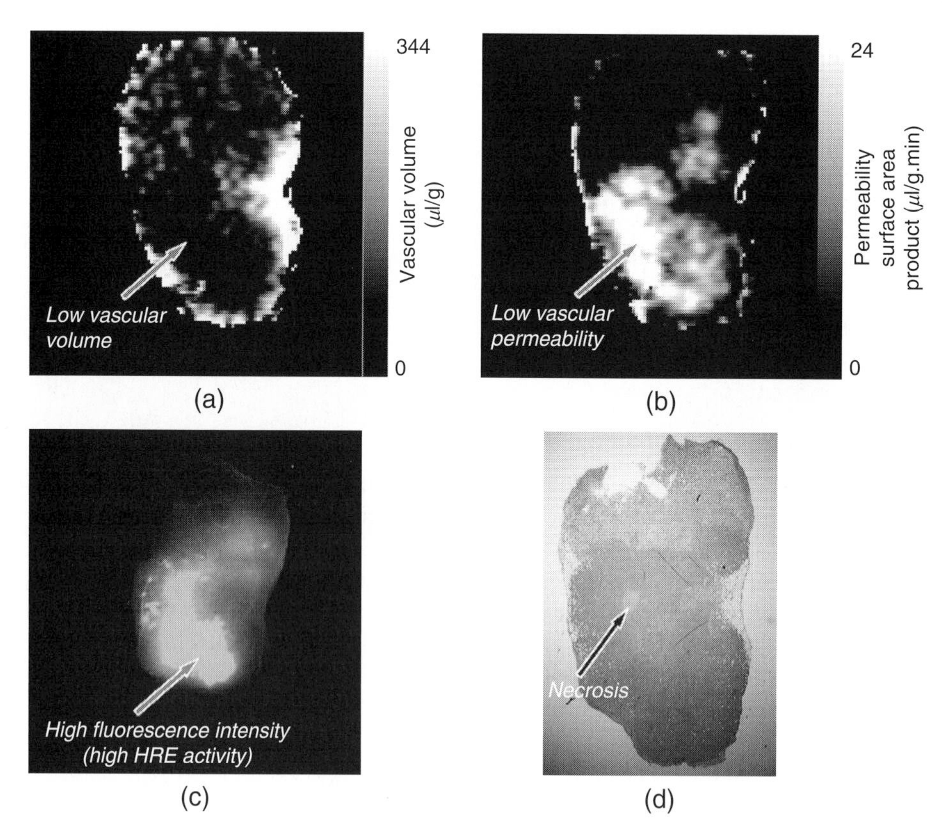

FIGURE 20.20 Maps of (a) vascular volume (VV) and (b) permeability surface area product (PSP) obtained from a central slice of a PC-3 prostate tumor xenograft (180 mm^3) expressing enhanced green fluorescent protein (EGFP) under the control of a hypoxia response element (HRE). VV ranged from 0 to 344 μl/g and PSP from 0 to 24 μl/g min. (c) Fluorescent microscopy of a fresh tissue slice obtained from the imaged slice, using a Nikon TS100-F microscope ($\times$1 objective) with a wavelength of 512 nm. (d) Hematoxylin and eosin stained, 5-μm-thick section from the central MRI slice. The region exhibiting EGFP consisted of viable cells. The less dense staining in the upper part of the section is due to uneven sectioning. The only area of dying cells was in a small necrotic focus (black arrow). Adapted from Raman et al. [62]. For a more detailed view of this figure, please visit our companion site at: http://books.elsevier.com/companions/9780123735836.

20.9 Summary

Cancer is a multifaceted disease that requires individual characterization for each tumor but also shares common characteristics. It is important to be able to identify and visualize molecular markers that occur either collectively or individually in various cancers. Such ability is valuable in the quest for the development and application of more potent molecular targeted cancer therapies that kill malignant tissue while sparing normal tissue. As multimodality imaging instruments become increasingly available, a combined molecular-functional-anatomic imaging approach will become more commonplace for preclinical and clinical investigations and will play an integral role in characterizing tumors to select and validate treatment, screen for sensitivity, and monitor treatment.

Acknowledgments

The authors gratefully acknowledge funding from National Institutes of Health (NIH) grants P50 CA103175 (JHU ICMIC Program) and R24 CA92871 (JHU SAIRP).

20.10 Exercises

1. Please list the advantages of performing molecular imaging in cancer diagnosis over purely anatomic imaging.
2. Which aspects of molecular imaging are most useful in the clinic and why?
3. What biologic processes can be visualized using molecular imaging of gene expression?

4. What reporter genes are available for molecular imaging of gene expression, what imaging modalities are required to visualize each reporter gene, and what are the advantages and disadvantages of these reporter genes?
5. What are the optimal features of a receptor that make it feasible for molecular receptor imaging in tumors, and why should the perfect receptor have these features?
6. What are smart contrast agents?
7. What imaging modality has been used most frequently in molecular imaging of smart contrast agents, and why?
8. Compare the advantages and disadvantages of the various imaging modalities used in molecular imaging.

20.11 References and Bibliography

1. R. G. Blasberg and J. G. Tjuvajev. Molecular-genetic imaging: Current and future perspectives. *J. Clin. Invest.* 111(11):1620–1629, 2003.
2. J. F. Rodriguez et al. Expression of the firefly luciferase gene in vaccinia virus: A highly sensitive gene marker to follow virus dissemination in tissues of infected animals. *Proc. Natl. Acad. Sci. USA.* 85(5):1667–1671, 1988.
3. N. C. Shaner et al. Improved monomeric red, orange and yellow fluorescent proteins derived from Discosoma sp. red fluorescent protein. *Nat. Biotechnol.* 22(12):1567–1572, 2004.
4. J. G. Tjuvajev et al. Noninvasive imaging of herpes virus thymidine kinase gene transfer and expression: A potential method for monitoring clinical gene therapy. *Cancer Res.* 56(18):4087–4095, 1996.
5. T. Groot-Wassink et al. Noninvasive imaging of the transcriptional activities of human telomerase promoter fragments in mice. *Cancer Res.* 64(14):4906–4911, 2004.
6. B. Cohen et al. Ferritin as an endogenous MRI reporter for noninvasive imaging of gene expression in C6 glioma tumors. *Neoplasia.* 7(2):109–117, 2005.
7. A. A. Gilad et al. Artificial reporter gene providing MRI contrast *in vivo* based on chemical exchange. Proceedings of the Fourteenth Scientific Meeting of the International Society for Magnetic Resonance in Medicine 2006: Abstract #100.
8. P. Ray et al. Imaging tri-fusion multimodality reporter gene expression in living subjects. *Cancer Res.* 64(4): 1323–1330, 2004.
9. D. A. Hamstra et al. Real-time evaluation of p53 oscillatory behavior *in vivo* using bioluminescent imaging. *Cancer Res.* 66(15):7482–7489, 2006.
10. W. Wang, S. H. Kim, and W. S. El-Deiry. Small-molecule modulators of p53 family signaling and antitumor effects in p53-deficient human colon tumor xenografts. *Proc. Natl. Acad. Sci. USA.* 103(29):11003–11008, 2006.
11. U. Mahmood and R. Weissleder. Near-infrared optical imaging of proteases in cancer. *Mol. Cancer Ther.* 2(5):489–496, 2003.
12. C. M. Shachaf et al. MYC inactivation uncovers pluripotent differentiation and tumour dormancy in hepatocellular cancer. *Nature.* 431(7012):1112–1117, 2004.
13. A. Hajitou et al. A hybrid vector for ligand-directed tumor targeting and molecular imaging. *Cell.* 125(2):385–398, 2006.
14. B. M. Carswell et al. Detection of prostate cancer by alpha-methylacyl coa racemase (P504S) in needle biopsy specimens previously reported as negative for malignancy. *Histopathology.* 48(6):668–673, 2006.
15. B. P. Adley and X. J. Yang. Application of alpha-methylacyl coenzyme A racemase immunohistochemistry in the diagnosis of prostate cancer: A review. *Anal. Quant. Cytol. Histol.* 28(1):1–13, 2006.
16. R. E. Weiner and M. L. Thakur. Radiolabeled peptides in oncology: Role in diagnosis and treatment. *Biodrugs.* 19(3):145–163, 2005.
17. C. P. Theuer et al. Radioimmunotherapy of non-Hodgkin's lymphoma: Clinical development of the Zevalin regimen. *Biotechnol. Annu. Rev.* 10:265–295, 2004.
18. D. Yao et al. The utility of monoclonal antibodies in the imaging of prostate cancer. *Semin. Urol. Oncol.* 20(3):211–218, 2002.
19. D. Artemov et al. Magnetic resonance molecular imaging of the HER-2/neu receptor. *Cancer Res.* 63(11):2723–2727, 2003.
20. E. C. Wiener et al. Imaging folate binding protein expression with MRI. *Acad. Radiol.* 9(suppl 2):S316–S319, 2002.
21. Z. H. Jin et al. Noninvasive optical imaging of ovarian metastases using Cy5-labeled RAFT-c(-rgdfk-)4. *Mol. Imaging.* 5(3):188–197, 2006.
22. S. Ito et al. Principle and clinical usefulness of the infrared fluorescence endoscopy. *J. Med. Invest.* 53(1–2):1–8, 2006.
23. A. L. Polglase, W. J. Mclaren, and P. M. Delaney. Pentax confocal endomicroscope: A novel imaging device for *in vivo* histology of the upper and lower gastrointestinal tract. *Expert Rev. Med. Devices.* 3(5):549–556, 2006.
24. H. E. Daldrup-Link, G. H. Simon, and R. C. Brasch. Imaging of tumor angiogenesis: Current approaches and future prospects. *Curr. Pharm. Des.* 12(21):2661–2672, 2006.
25. V. R. Stewart and P. S. Sidhu. New directions in ultrasound: Microbubble contrast. *Br. J. Radiol.* 79(939): 188–194, 2006.
26. G. E. Weller et al. Ultrasonic imaging of tumor angiogenesis using contrast microbubbles targeted via the tumor-binding peptide arginine-arginine-leucine. *Cancer Res.* 65(2):533–539, 2005.
27. C. Bremer et al. Optical imaging of matrix metalloproteinase-2 activity in tumors: Feasibility study in a mouse model. *Radiology.* 221(2):523–529, 2001.

28. C. Bremer et al. Imaging of differential protease expression in breast cancers for detection of aggressive tumor phenotypes. *Radiology.* 222(3):814–818, 2002.
29. R. Weissleder et al. *In vivo* imaging of tumors with protease-activated near-infrared fluorescent probes. *Nat. Biotechnol.* 17(4):375–378, 1999.
30. J. W. Kim et al. HIF-1-mediated expression of pyruvate dehydrogenase kinase: A metabolic switch required for cellular adaptation to hypoxia. *Cell Metab.* 3(3):177–185, 2006.
31. E. Ackerstaff, K. Glunde, and Z. M. Bhujwalla. Choline phospholipid metabolism: A target in cancer cells? *J. Cell Biochem.* 90(3):525–533, 2003.
32. A. Ramirez de Molina et al. Regulation of choline kinase activity by Ras proteins involves Ral-GDS and PI3K. *Oncogene.* 21(6):937–946, 2002.
33. E. O. Aboagye and Z. M. Bhujwalla. Malignant transformation alters membrane choline phospholipid metabolism of human mammary epithelial cells. *Cancer Res.* 59(1):80–84, 1999.
34. K. Glunde et al. RNA interference-mediated choline kinase suppression in breast cancer cells induces differentiation and reduces proliferation. *Cancer Res.* 65(23): 11034–11043, 2005.
35. M. A. Jacobs et al. Proton magnetic resonance spectroscopic imaging of human breast cancer: A preliminary study. *J. Magn. Reson. Imaging.* 19(1):68–75, 2004.
36. K. Glunde, M. A. Jacobs, and Z. M. Bhujwalla. Choline metabolism in cancer: Implications for diagnosis and therapy. *Expert Rev. Mol. Diagn.* 6(6):821–829, 2006.
37. R. Haubner and H. J. Wester. Radiolabeled tracers for imaging of tumor angiogenesis and evaluation of anti-angiogenic therapies. *Curr. Pharm. Des.* 10(13):1439–1455, 2004.
38. Z. Huang and S. D. Bao. Roles of main pro- and anti-angiogenic factors in tumor angiogenesis. *World J. Gastroenterol.* 10(4):463–470, 2004.
39. Z. M. Bhujwalla et al. Vascular differences detected by MRI for metastatic versus nonmetastatic breast and prostate cancer xenografts. *Neoplasia.* 3(2):143–153, 2001.
40. A. R. Padhani. MRI for assessing antivascular cancer treatments. *Br. J. Radiol.* 76(spec. no. 1):S60–S80, 2003.
41. K. D. Miller et al. Randomized phase II trial of the anti-angiogenic potential of doxorubicin and docetaxel; primary chemotherapy as biomarker discovery laboratory. *Breast Cancer Res.* Treat. 89(2):187–197, 2005.
42. J. S. Lewis et al. Copper-64-pyruvaldehyde-bis(N(4)-methylthiosemicarbazone) for the prevention of tumor growth at wound sites following laparoscopic surgery: monitoring therapy response with micropet and magnetic resonance imaging. *Cancer Res.* 62(2):445–449, 2002.
43. P. Kunz et al. Angiopoietin-2 overexpression in Morris hepatoma results in increased tumor perfusion and induction of critical angiogenesis-promoting genes. *J. Nucl. Med.* 47(9):1515–1524, 2006.
44. M. A. Flower et al. 62Cu-PTSM and PET used for the assessment of angiotensin II-induced blood flow changes in patients with colorectal liver metastases. *Eur. J. Nucl. Med.* 28(1):99–103, 2001.
45. M. Sund, L. Xie, and R. Kalluri. The contribution of vascular basement membranes and extracellular matrix to the mechanics of tumor angiogenesis. *Apmis.* 112(7–8):450–462, 2004.
46. W. P. Li and C. J. Anderson. Imaging matrix metalloproteinase expression in tumors. *QJ. Nucl. Med.* 47(3): 201–208, 2003.
47. J. W. Pollard. Tumour-educated macrophages promote tumour progression and metastasis. *Nat. Rev. Cancer.* 4(1):71–78, 2004.
48. N. Wernert. The multiple roles of tumour stroma. *Virchows Arch.* 430(6):433–443, 1997.
49. R. K. Jain. Normalization of tumor vasculature: An emerging concept in antiangiogenic therapy. *Science.* 307(5706): 58–62, 2005.
50. J. Czernin, W. A. Weber, and H. R. Herschman. Molecular imaging in the development of cancer therapeutics. *Annu. Rev. Med.* 57:99–118, 2006.
51. M. M. Mueller and N. E. Fusenig. Friends or foes: Bipolar effects of the tumour stroma in cancer. *Nat. Rev. Cancer.* 4(11):839–849, 2004.
52. J. G. Rajendran et al. Hypoxia imaging-directed radiation treatment planning. *Eur. J. Nucl. Med. Mol. Imaging.* 33(suppl 13):44–53, 2006.
53. D. Rischin et al. Prognostic significance of [18F]-misonidazole positron emission tomography-detected tumor hypoxia in patients with advanced head and neck cancer randomly assigned to chemoradiation with or without tirapazamine: A substudy of Trans-Tasman Radiation Oncology Group Study 98.02. *J. Clin. Oncol.* 24(13): 2098–2104, 2006.
54. R. J. Myerson et al. Monitoring the effect of mild hyperthermia on tumour hypoxia by Cu-ATSM PET scanning. *Int. J. Hyperthermia.* 22(2):93–115, 2006.
55. H. Dafni et al. Overexpression of vascular endothelial growth factor 165 drives peritumor interstitial convection and induces lymphatic drain: Magnetic resonance imaging, confocal microscopy, and histological tracking of triple-labeled albumin. *Cancer Res.* 62(22):6731–6739, 2002.
56. A. P. Pathak et al. Lymph node metastasis in breast cancer xenografts is associated with increased regions of extravascular drain, lymphatic vessel area, and invasive phenotype. *Cancer Res.* 66(10):5151–5158, 2006.
57. Y. Hassid et al. Noninvasive magnetic resonance imaging of transport and interstitial fluid pressure in ectopic human lung tumors. *Cancer Res.* 66(8):4159–4166, 2006.

58. P. Friedl. Dynamic imaging of cellular interactions with extracellular matrix. *Histochem. Cell Biol.* 122(3):183–190, 2004.
59. W. M. Petroll and L. Ma. Direct, dynamic assessment of cell-matrix interactions inside fibrillar collagen lattices. *Cell Motil. Cytoskeleton.* 55(4):254–264, 2003.
60. P. J. Campagnola et al. Three-dimensional high-resolution second-harmonic generation imaging of endogenous structural proteins in biological tissues. *Biophys. J.* 82(1 Pt 1):493–508, 2002.
61. E. Brown et al. Dynamic imaging of collagen and its modulation in tumors *in vivo* using second-harmonic generation. *Nat. Med.* 9(6):796–800, 2003.
62. V. Raman et al. Characterizing vascular variables in hypoxic regions: A combined magnetic resonance and optical imaging study of a human prostate cancer model. *Cancer Res.* 66(20):9929–9936, 2006.

21

Molecular Imaging in Biology and Pharmacology

Prof. Sung-Cheng Huang,
Prof. Anna M. Wu, and
Prof. Jorge R. Barrio
*University of California,
Los Angeles (UCLA)*

21.1 Introduction and Background

Recently, there has been tremendous progress in biology and medicine, in particular in expanding our understanding of biology and disease at the cellular and molecular levels. In parallel, there has been dramatic progress in noninvasive biological/medical imaging. It is now possible to examine *in vivo* specific biochemical pathways and gene expression in cells. Specifically, a labeled molecule, called a tracer, can be used as a sensor/marker to indicate the activity level of a specific gene, enzyme, biochemical pathway, or other cellular process at the molecular level. This approach is commonly referred to as *molecular imaging*. Molecular imaging already plays an important role in disease diagnosis, treatment selection, and monitoring of treatment response in patients. Furthermore, molecular imaging provides critical information for drug development by allowing investigators to understand the underlying biology. It allows the efficacy of drugs to be easily assessed *in vivo* and is expected to have a large impact on future drug evaluation and development [1–3].

The field of molecular imaging involves expertise from many basic sciences, including physics, chemistry, engineering, etc., in addition to biology and medicine. Accelerating advances in these scientific and technical areas in recent years have created significant momentum and led to the flourishing state of molecular imaging today [1–3]. Information technology has played a key role in molecular imaging, and its importance is expected to increase in the future. The objectives of this chapter are to introduce the basic elements of molecular imaging, describe recent advances, provide examples that highlight the role of information technology, and introduce examples of applications that illustrate how molecular imaging is used in biology, medicine, and pharmacology.

21.1.1 Basic Elements and New Developments in Molecular Imaging

Molecular imaging generally involves the administration of a labeled molecule (called a tracer or labeled probe) in the body, the measurement of the tracer with an imaging device, and the analysis and interpretation of the measured images to give biologically meaningful information. Thus, from a technical point of view, the basic components of molecular imaging are synthesis of labeled tracers, tracer measurement instrumentation, and image analysis.

In the area of tracer development, numerous new compounds are continually being made available to follow various biological processes or to bind to various receptors or enzymes. There are three major types of tracers in terms of the imaging device to monitor their body distribution/kinetics:

- Positron emitters
- Single photon emitters
- Optical light emitters

They have different advantages and limitations, but positron-emitting tracers are experiencing the most rapid rate of growth, because they can be natural compounds with one atom in the molecule replaced by a positron-emitting radioisotope (e.g., O-15, C-11, N-13, F-18) or with the molecular structure slightly altered to include one of the positron-emitting radionuclides but maintaining similar biological/biochemical properties. In this chapter, we will focus mainly on this type of tracer, whose associated imaging modality is positron emission tomography (PET) [4]. Readers are referred to other chapters in this book or other books for details of other types of tracers. Among the imaging devices for the three types of tracers, PET can give the most quantitative measure of the concentration level of a tracer and thus can provide the most accurate biological information in local tissue regions.

The development/preparation/synthesis of PET compounds also has experienced rapid growth. Commonly used positron-emitting radionuclides have relatively short half-lives ($t\frac{1}{2}$) (from 70 seconds for Rb-82 to 110 minutes for F-18) and are typically produced using a cyclotron. After such radioisotopes are produced, they need to be incorporated quickly into specific molecules to make the compounds as PET tracers. The development of small self-shielded cyclotrons (called baby cyclotrons) has made it possible to have cyclotrons installed on-site in most hospitals or medical centers. The developments of new targets and automated synthesis units have facilitated the process of incorporating the cyclotron-produced positron-emitting radioisotope into desired molecules to produce a usable PET tracer. With these developments, fewer people are needed to support the routine operation of tracer preparation, and higher-quality tracers can be made widely available reliably.

In the area of imaging devices, the development of the PET scanner in the early 1970s represented a major milestone [4]. It uses scintillation detectors and coincidence detection to detect positron annihilation events, and mathematical tomographic reconstruction [5–8] to determine the 3D distribution of the positron-emitting tracers. Using tomographic reconstruction, the generated image gives not only a relative distribution of the tracer, but also a quantitatively accurate measurement of the tracer concentration in local regions in absolute physical units. This ability to measure tracer concentration *in vivo* without invasive procedures is an amazing accomplishment (Figure 21.1).

Numerous developments in scintillation crystals, detection electronics, detector and scanner designs, and reconstruction algorithms have improved detection sensitivity and image signal-to-noise ratio and have achieved spatial resolutions of less than 1.75 mm *full width at half maximum* (FWHM) (see Sections 21.5, 21.6, and 21.7). The temporal resolution for imaging distribution changes over time has also been dramatically improved (Section 21.7). Recent developments in PET/computed tomography (CT) and time-of-flight PET have further improved the image quality and incorporation of anatomical information.

In addition to improving image quality through image processing and image reconstruction, data/image analysis plays an important role in linking the physical measurement of tracer distribution to biological information, which is what the end users of molecular imaging are primarily interested in. The use of tracers for biological studies has been around for a long time, and the basic principles of tracer kinetics have been used in many practical procedures for measuring various physiological functions (e.g., cardiac output, cerebral blood perfusion, glomerular filtration rate). Since PET images provide regional tracer concentration in absolute units, tracer principles can be applied rigorously. Compartmental models, commonly used in physiological studies, have been adopted for molecular imaging. Analysis has also been simplified by the use of the blood time activity of the tracer as the input function to describe the tissue tracer kinetics obtained from dynamic PET images. This approach has improved the robustness of model parameter estimations and has become a common method for interpreting PET images in terms of biological information [9]. The robustness of the parameter estimation for some studies has been further improved by the use of the kinetics in reference tissue to replace the input function [10, 11]. These improvements in robustness plus the incorporation of the imaging process into the model and the development of graphical analysis methods have allowed the generation of parametric images, in which the image value is a biological parameter expressed in absolute biological units (e.g., mg/g/min for reaction rates, nl/g for receptor density).

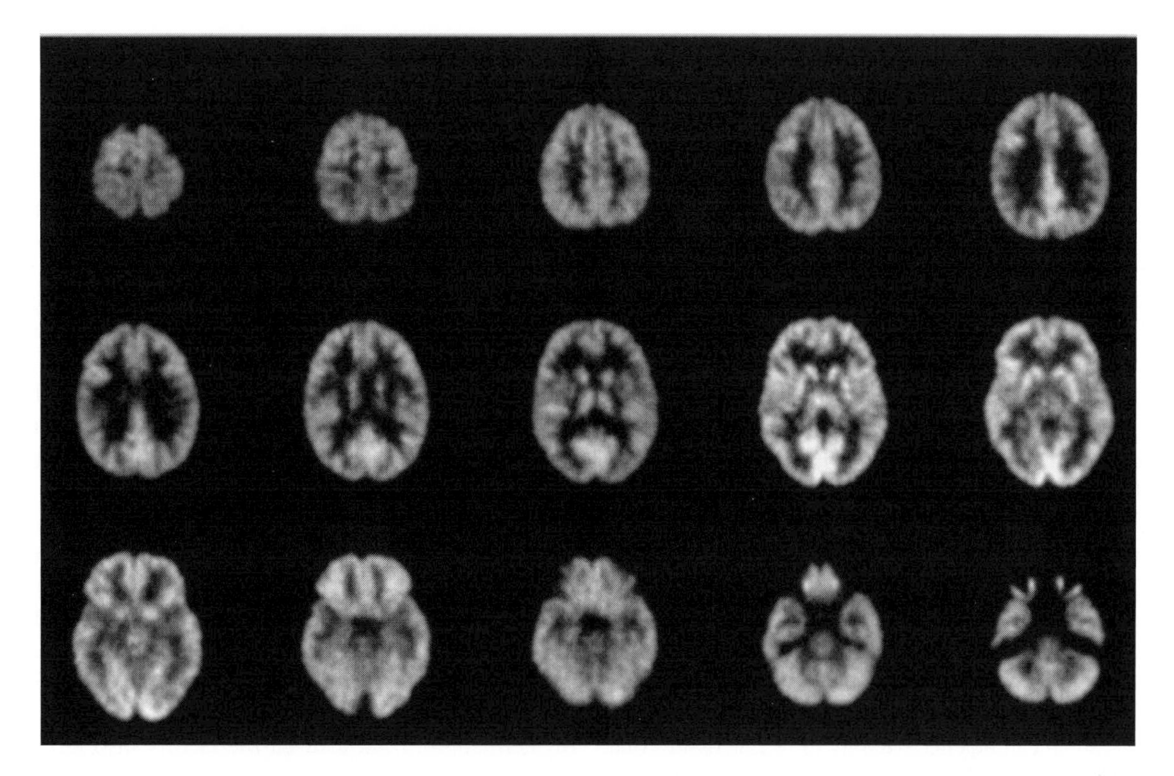

FIGURE 21.1 PET images of the brain of a normal subject. The images are transaxial cross sections from the top to the bottom of the brain. The tracer used in the imaging study was 2-deoxy-2-[F-18]fluoro-D-glucose (fluorodeoxyglucose, FDG) given intravenously as a bolus. Brightness in the images indicates the utilization rate of the glucose, which is related directly to neurological function in brain tissues. Various gray-matter substructures of the brain are clearly delineated on the images.

Recent developments in image coregistration and spatial normalization have allowed the easy merging of functional and anatomical information.

However, technical advances alone would not have created the current large impact of molecular imaging had there not been parallel and equally impressive advances in biology and medicine.

21.1.2 Recent Developments in Biology and Pharmaceuticals

Biology and medicine are being revolutionized by the profusion of information flowing from highly successful research initiatives in genomics, proteomics, systems biology, and informatics. Completion of the sequence of the human genome has provided the foundation for the identification of genes and proteins and, most importantly, for deciphering their roles and interactions in health and disease. Comprehensive methods are now available for analysis of the DNA, RNA, and protein in a variety of samples, which can be cells and tissues from laboratory studies, animal models, or clinical specimens. At the DNA level, high-throughput DNA sequencing, rapid analysis of single nucleotide polymorphisms (SNPs), and the use of competitive genome hybridization (CGH) for genome-wide surveys are examples of some of the techniques that can be used to detect the existence of genetic variations. Of equal importance is whether or not individual genes are active. Expression profiling of gene activity is most commonly conducted by microarray analysis for the detection of the presence of the specific RNA transcripts from each gene. Finally, it is the expression and activity of the protein product of each gene that truly represents gene function.

The comprehensive analysis of the structure, function, expression, and interactions of proteins has been accelerated by the field of proteomics, which combines rapid, high-throughput analysis of proteins using mass spectrometry for peptide fingerprinting and sequencing, modern electrophoretic techniques for identifying protein modifications, structural proteomics (employing X-ray crystallography and nuclear magnetic resonance methodology), and functional assays to document protein-protein interactions.

This wealth of information has required new approaches and even new disciplines in order to analyze and interpret data. The most important shift has been away from the intensive study of a single gene, single protein, or single metabolic or signaling pathway, toward more global, integrated strategies.

Systems biology has provided a framework for measuring and cataloging gene expression, cell signaling, metabolic signatures, and other meta-information that informs us of the biological state of cells or tissues. Modeling of biological systems allows the identification of key nodes or components that can represent targets for detection as well as sites for therapeutic intervention.

As an example, in oncology, the *epidermal growth factor receptor* (EGFR) family of receptor tyrosine kinases has long been known to play a key role in the cancerous growth of cells of epithelial origin [12]. Years of work have led to the identification of gene amplification of family members such as EGFR and HER2 in tumors, culminating in the identification, development, and ultimately clinical approval of pharmaceuticals that target the presence and action of these very specific biomarkers. Successful drugs targeting these specific growth-signaling molecules include classic small molecule pharmaceuticals (gefitinib, erlotinib), as well as biotherapeutics in the form of engineered antibodies (trastuzumab, cetuximab, panitumumab). Furthermore, it has become clear that the molecular characterization of tumors is key to understanding their susceptibility to molecularly targeted therapies. For example, trastuzumab can be effective only in tumor tissues that overexpress the Her2/c-erb-B2 growth factor receptor, and resistance to gefitinib is a function of mutations in the EGFR kinase region. Some key information can be extracted directly from tumor biopsies. However, it is becoming equally imperative to develop methods to monitor the genotype and phenotype of tumor tissues in living organisms, including patients, in order to assess biological potential, to predict drug susceptibility, and to monitor response to targeted therapies.

Fortunately, the same information that facilitates identification of novel targets for therapy also provides potential targets for molecular imaging probe development. Molecular imaging can be based on the presence of elevated levels of receptors, enzymes, kinases, etc., at the target tissue. For example, elevated expression of glucose transporters and hexokinases leads to increased phosphorylation and retention of [^{18}F]-fluorodeoxyglucose (FDG) in metabolically active tumor cells or activated immune cells during infection and inflammation. Increased expression of thymidine kinase leads to trapping of [^{18}F]-fluorothymidine in highly proliferative cells in tumors, bone marrow, and other tissues. Both are readily detectable by PET/microPET. Overexpression of specific proteases (such as cathepsins) in tissues can be assessed by using self-quenched peptide probes that contain appropriate cleavage sequences and by using fluorescence imaging detection. Antibodies specific for a spectrum of cell-surface markers can be conjugated to a variety of signal-generating moieties, including radionuclides, magnetic particles, fluorescent dyes, bioluminescent enzymes, ultrasound bubbles, etc., to generate molecularly specific imaging probes. Molecular imaging provides a window into development, health, and disease processes in the living organism.

21.2 Considerations for Quantitative Molecular Imaging

In most cases of molecular imaging, quantitative biological information is expected. In order to provide quantitative information, issues related to tracer, biology, imaging device characteristics, study procedure, and data/image processing all need to be carefully coordinated or considered to provide the needed information. For example, for measuring the transport/reaction rate of a dynamic process, not only must the tracer follow the biological process of interest, but its kinetics must be significantly influenced by rate changes in the process of interest. In the terminology of tracer methods, this means that the uptake/clearance of tracer should be limited by the key step of the dynamic process of interest. If the change is reflected significantly in the absolute amount of tracer uptake, the imaging device would need to provide accurate measurement of the concentration of the tracer in tissue, and the time of imaging would need to include the period when the tracer uptake reaches high levels compared with the background. Since the tracer, biology, and imaging device properties cannot be described by a finite set of variables like indexes of merit, optimization of all the factors involved is difficult. Instead, the factors are usually considered one or a couple at a time, with the others fixed. So, usually a suboptimal and practical study would be performed. For consideration of imaging devices, one is referred to published articles or books on biomedical imaging instrumentations. (Section 21.3 contains a comprehensive discussion of tracer/probe design and selection.)

We now address issues related to study procedure and tracer kinetics. Measurements from the imaging device are assumed to be able to provide accurate information on the concentration level of tracer in local tissues like those provided by PET. To interpret acquired images in terms of biology, molecular imaging utilizes tracer methods, in which the physical measurement needs to be processed to give the biological information. Many confounding factors could potentially affect the tracer concentration in tissue. These factors include:

- Specific biochemical processes that the tracer probes
- Amount of dose administered
- Route of tracer administration
- Blood perfusion of the tissue of interest
- Body size of the subject
- Systemic condition of the body (uptake of tracer in other tissue/organs, excretion rate of the tracer, body fat content, biochemical reaction/metabolism of the tracer in the body, etc.)
- Endogenous substrate/transmitter competition
- Nonspecific biochemical environment in tissue
- Biochemical properties of labeled metabolites in plasma
- Vascular volume in tissue
- Time of uptake post–tracer administration

A common approach to reduce the effects of confounding factors is the use of the blood time activity curve (TAC) of the tracer, usually called the "input function." Since the amount of tracer in local tissue is delivered through blood perfusion and is cleared primarily through venous blood flow, the tracer kinetics in local tissue regions can be isolated from considerations in the rest of the body if the activity level of the tracer in blood is measured. This simplifies greatly the problem of extracting biological information from measured kinetics in local regions. So, in the following subsection, issues related to the input function are first addressed. Modeling for extracting biological information from local tissue kinetics will be discussed afterward.

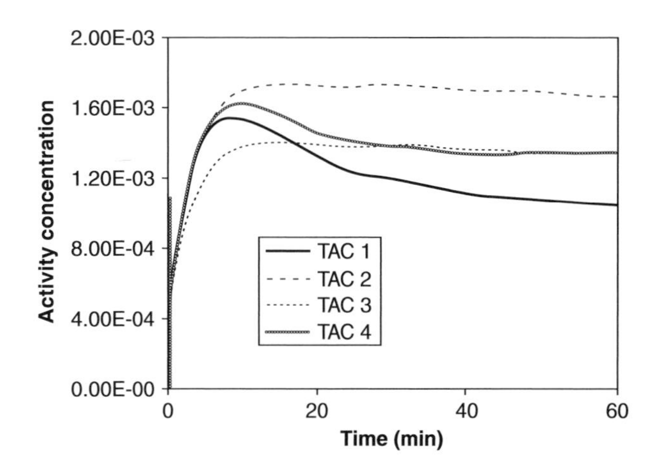

FIGURE 21.2 Tissue time activity curves (TACs) that illustrate the effects of the input function and tissue biological parameters.

21.2.1 Input Function

The input function can be viewed as a summary of the tracer administration and the kinetics of the whole body that influence the kinetics in local tissues. In other words, if the input function is known, one can focus only on the kinetics of the tracer in the local tissue of interest. That is, one does not need to consider the kinetics of the tracer in the rest of the body, so far as the estimation of the biological information in the local tissue is concerned. For example, change in the kidney clearance of a hydrophilic tracer is reflected on the time course of the blood concentration level of the tracer. The tracer kinetics in other tissue regions (e.g., brain, muscle) would be affected also (even though local biological condition remains the same), but only as a consequence of the change in the time course of the tracer concentration in blood, that is, changes in the input function.

In addition to changes in the clearance rate of the tracer, the way the tracer is administered to the body can affect directly the shape of the input function. Usually, the tracer in molecular imaging studies is administered as a bolus through intravenous (IV) injection. If the injection speed is slowed down (i.e., as a slow bolus), the input function would have a wider early peak. If the tracer administration is through constant infusion or orally, there may not be an early peak at all. As a result, the kinetics in local tissue would be different, because local tissue kinetics are dependent on the input function and the tissue's biological condition. This does not imply that the method/route of tracer administration as reflected in the shape of the input function is of no significance. The input function, when available, simplifies the relationship and thus the considerations. For example, for extracting biological information from local tissue kinetics, one needs to be concerned only about how changes in tissue biological parameters affect the tissue kinetics. Figure 21.2 illustrates that the tissue kinetics could change due to either biological changes in tissue or changes in the input functions. It shows that the kinetics are sensitive to both factors; and for comparable tracer levels in tissue at certain times, they could correspond to two significantly different biological states because they have two different input functions. This illustrates the importance of knowing the input function for interpreting the tracer kinetic data.

In Figure 21.2, the tracer in tissue is assumed to follow a compartmental model like that for FDG (see the figure in the next subsection). TAC 1, TAC 2, and TAC 3 have the same input function but correspond to three different tissue biological states (i.e., model parameters), which are reflected in different shapes of the kinetics. TAC 4 has the same tissue parameters as does TAC 1, but the input function is different. Over the time period from 30 to 60 minutes, the activity levels in tissue for TAC 3 and TAC 4 are comparable, although they correspond to two different biological states—illustrating the importance of knowing the input function for interpreting the tracer kinetic data.

21.2.2 Physiological/Biological Model

As stated earlier and as the basis of tracer methods, tissue kinetics are related to the biological condition in the local tissue. Considering the complexity of structure, composition, and function in tissue, the relationship between tracer kinetics in tissue and the biological condition can be quite complicated. However, for the time scale of the kinetic measurements taken during most molecular imaging studies, the relationship can be simplified and described by compartmental models that consist of interconnected compartments. For each compartment in the model, the tracer concentration is constant, and the rate of transport between compartments is linearly related to the concentration of the originating compartment through the use of a rate constant (not time varying). The values of these rate constants in the model are used to define a particular biological condition in tissue. Changes in the biological condition are represented as changes in the values of these rate constants.

In Figure 21.3(a) is depicted a comprehensive model of tracer delivery and tissue trapping/binding. Figure 21.3(b) is

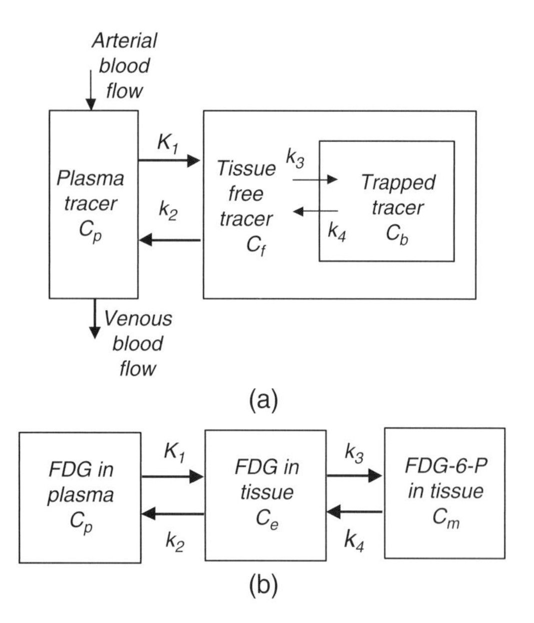

FIGURE 21.3 Models of tracer kinetics in local tissue region. C_p = plasma concentration; C_f = free concentration; C_b = bound concentration; C_e = concentration of unmetabolized (extravascular) tracer; C_m = concentration of metabolized tracer.

a compartmental model derived from panel (a), when the delivery of tracer to tissue is not limited by blood perfusion. The model in panel (b) is commonly used for describing the tissue kinetics of FDG in the brain and myocardium. The relationship among the concentration levels of all the compartments can be easily described mathematically using linear differential equations, which can be easily solved and usually have good mathematical properties. The simplicity of this type of model for describing the relationship between kinetics and biological condition in tissue is very desirable, and thus accounts for its popularity in many tracer studies.

To illustrate the procedure of setting up a compartmental model for describing the kinetics of a tracer in tissue, we will go through the steps evolving from biology to a model, using the kinetics of F-18–labeled FDG, an analogue of glucose, as an example. Tracer is delivered to tissue through blood perfusion ("arterial blood flow" in Figure 21.3[a]). While the majority of the tracer molecules pass through the capillary and leave the tissue via venous blood flow (at the left in Figure 21.3[a]), a fraction is transported across the capillary and cell membrane into cells through a carrier-facilitated transport system (e.g., GLU1 and GLU3 in the brain tissue), as indicated by the arrow K_1. Some tracer molecules will undoubtedly be transported back to the vasculature (arrow k_2) and cleared away; some will be phosphorylated to FDG-6-P (as indicated by arrow k_3) by the cellular enzyme hexokinase. Unlike glucose-6-P, which continues to be converted to glucose-1-P and moved onto the glycolytic pathway, FDG-6-P is not a substrate for the isomerase in the next step and is accumulated in cells. So, a simple compartment model, as shown in Figure 21.3(a) can be used to described the kinetics of FDG in tissue. The transport step k_4 shown in the model indicates that the accumulated FDG-6-P can be dephosphorylated slowly back to the unphosphorylated form. Furthermore, if the perfusion rate is significantly larger than the transport rate into tissue (i.e., $F >> K_1$), the tissue kinetics are not flow-limited and the model can be further reduced to the form shown in Figure 21.3(b). With this model, the kinetics of F-18 activity in tissue ($C_i(t)$) can be described by the following equation [13]:

$$C_i(t) = \left(\frac{K_1}{\alpha_2 - \alpha_1} [(k_3 + k_4 - \alpha_1) \exp(-\alpha_1 t) - (k_3 + k_4 - \alpha_2) \exp(-\alpha_2 t)] \right) \otimes C_p(t) + v_b C_p(t)$$

with

$$\alpha_1, \alpha_2 = 0.5^* \left[(k_2 + k_3 + k_4) \mp \sqrt{(k_2 + k_3 + k_4)^2 - 4k_2 k_4} \right] \tag{21.1}$$

where C_p is the input function, K_1, k_2, k_3, and k_4 are the rate constants of the model shown in Figure 21.3(b); the symbol $\otimes$ denotes the mathematical operation of convolution; and the last term of the equation accounts for the activity in the vasculature in tissue, with v_b denoting the vascular volume in tissue. So, based on the equation, the input function, and the measurement of the tissue kinetics, one can use nonlinear regression [9, 14] to estimate the values of the rate constants, from which the net uptake constant of FDG from blood to tissue can be determined as $K_1 k_3/(k_2 + k_3)$ [13].

However, the transport and phosphorylation of FDG go through the same biochemical pathway as does glucose. Therefore, if the efficiencies of these carrier/enzymatic reactions for FDG relative to those of glucose are known or can be calibrated, the utilization rate of glucose in tissue can be estimated from the accumulation rate of FDG. The calibration constant between FDG and glucose through these steps is commonly referred to as the *lumped constant* (LC), a term coined originally by Sokoloff [15] for deoxyglucose (DG) and later adopted for FDG [13, 16].

The example in Figure 21.3 illustrates the modeling process for a tracer that follows a dynamic process/reaction in tissue. The reaction can be a natural cellular biochemical pathway (like glucose utilization) or it can be used to indicate another cellular process/state, like the use of FHBG (fluorohydroxymethylbutyl-guanine) for gene expression [17–19]. In either case, the important information that molecular imaging provides is the transport/uptake constant (i.e., $K_1 k_3/(k_2 + k_3)$) of the tracer in tissue that reflects the overall underlying biological process involved. Instead of going through nonlinear regression for estimation of the individual rate constants and then calculating the uptake rate, the Patlak analysis can provide the uptake constant directly using a graphical method [20] and is commonly used for this type of study.

Another type of tracer/probe commonly used is for determination of the density of receptors/enzymes in tissue. An example of this type of tracer/probe is FDDNP (fluoroethyl(-methyl)amino-2naphthyl ethylidene malononitrile) for assay of the density of amyloid plague and neurofibrillary tangles (NFTs) in tissue for assessment of pathological processes related to Alzheimer's disease (see Section 21.4). The model for this type of tracer is similar to that for tracers of a dynamic process, except that the k_3 step is proportional to the binding affinity of the tracer to the receptor site and the density of the receptors, and k_4 represents the dissociation rate of the tracer from the binding site. The ratio of k_3 to k_4 is thus related directly to the ratio of receptor density and the dissociation constant of the tracer for receptor binding and is commonly called the *binding potential* [21]. Theoretically, if one knows the value of the dissociation constant, the receptor density can be easily obtained from the binding potential. Due to the uncertainty of the biochemical environment, the receptors are exposed to tissue *in vivo,* accurate estimation of the dissociation constant of the binding tracer is difficult, and the binding potential is usually used instead as an index of receptor density.

The primary parameter of interest for this type of study is thus the ratio k_3/k_4. Although the values of k_3,k_4, or the ratio can be obtained from nonlinear regression of the model to the measured tissue kinetics, there is a simple graphical method, the Logan plot [22], that can give the distribution volume (V_d) of the tracer in tissue relative to that in blood. This distribution volume, according to the model, is equal to $(K_1/k_2)(1 + k_3/k_4)$. If a tissue region does not have any specific receptor that the tracer will bind to, the value of V_d will simply be K_1/k_2. Assuming that this K_1/k_2 ratio is uniform over all tissue regions, the ratio k_3/k_4 can be obtained by the following equation:

$$k_3/k_4 = (V_d)/(V_d)_{\text{ref}} - 1.$$

Moreover, it has been shown that the ratio of distribution volumes in this equation can be determined easily with the use of the kinetics of the reference tissue as the input function in applying the Logan plot [11]. With this approach, one does not even need to determine the input function from blood samples. However, for this type of simplified analysis, one needs to keep in mind the underlying assumptions of a uniform K_1/k_2 ratio and the existence of a reference tissue that is devoid of specific receptors for the tracer/probe used.

A unique feature of measuring receptor density is the binding saturability of the binding sites on receptors. The available number of binding sites for the tracer/probe can be affected by endogenous ligands that also bind to the same receptor. Therefore, different tracers/probes that have different binding affinity to the same receptor could give different density results due to their differing competitiveness with the endogenous ligands. Binding saturability can also be caused by exogenous ligands that are introduced externally via IV or oral administration [23, 24]. In fact, one can repeat the molecular imaging study at multiple levels of externally administered ligands to determine both the receptor density ($B_{\max}$) and the dissociation constant (K_D), similar to the *in vitro* receptor assay procedure that uses the Scatchard plot to determine $B_{\max}$ and K_D simultaneously [25].

In the discussion of tracer methods so far, the tracers/probes are assumed to have ideal properties for the specific biological processes. However, many practical tracers/probes have properties that require some modification of the model or the input function. One common finding among many tracers/probes is that they are metabolized peripherally (e.g., in the liver), and some labeled metabolites would circulate in the blood along with the original tracer/probe. Depending on the chemical property of the labeled metabolites, methods or models to account for the effects of the metabolites could have a wide range of complexities. Some examples are the metabolism of O-15 oxygen to O-15 water [26–28] (for measurement of oxygen metabolic rate) and the metabolism of FDOPA ([^{18}F]fluorodopa) to OMFD (3-*O*-methyl-[^{18}F]FDOPA) (for measurement of presynaptic dopaminergic function) [29]. (Readers are referred to the cited references for details.) Other issues frequently encountered include the transport/uptake of tracer/probe in red blood cells, and nonspecific bindings of tracers/probes in tissue.

In summary, since the interpretation of molecular imaging results is based on tracer methods, it is important to understand the general characteristics of these methods for assessing biological functions in tissues/cells. Generally, the biological information that tracer methods provide can be grouped into two categories:

1. Information about the transport/reaction rates of a dynamic process
2. Information on the density/capacity of macromolecules (e.g., receptors, enzymes, amyloid plaques) in tissue

Even though compartmental models are used for both categories, they have different characteristics, and different analysis techniques are usually used.

21.3 Design/Development of Molecular Imaging Probes

21.3.1 Chemical Probes (Small Molecules)

The introduction of radioactive molecules into living organisms has become one of the preferred methods for the study of biological systems. Introduction of tracer probes, due to the extremely low mass of the probe, produces very minimal disturbances in the steady state of the living system under investigation. Even though tracer techniques have led to the development of useful clinical applications (e.g., conventional nuclear medicine techniques), the development of PET has led the way in the development and use of quantitative assays of local biochemical and pharmacological processes in humans.

An important property of molecular imaging probes used with PET is that compounds labeled with positron-emitting radioisotopes can be prepared with high specific activity (e.g., >>1 Ci (curies)/mmol) so that the process to be measured is not perturbed.

Another important property of molecular imaging probes is that with cyclotron-produced positron-emitting radioisotopes of carbon (C-11; $t\frac{1}{2} = 20.38$ min), nitrogen (N-13; $t\frac{1}{2} = 9.96$ min), oxygen (O-15; $t\frac{1}{2} = 2.03$ min), and fluorine (F-18; $t\frac{1}{2} = 109.72$ min), true molecular imaging probes matching the strict requirements of enzyme and/or receptor targets can be designed. For example, labeling with the most common positron-emitting radioisotopes, carbon-11, nitrogen-13, and oxygen-15, renders compounds biochemically indistinguishable from their natural counterparts; furthermore, fluorine-18 can be used to provide labeled substrate analogues (e.g., FDG, 6-fluoro-L-DOPA, FDOPA) or pharmacological agents (F-18–labeled neurotransmitter receptor ligands) to trace biochemical or pharmacological processes in a predictable manner. Because of its small size and the strength of the C-F bond, the fluorine atom is commonly used to replace H or OH on a molecule. This modification allows favorable interactions of the new molecule (e.g., molecular imaging probe or drug) with the target (e.g., enzyme, receptor) to occur without steric hindrance. Moreover, the presence of fluorine may, with target enzymes, specifically block subsequent reactions of a given substrate analogue (e.g., FDG vs. glucose).

The positron emitters carbon-11, nitrogen-13, oxygen-15, and fluorine-18 constitute the only externally detectable forms of carbon, nitrogen, oxygen, and fluorine, respectively. Thus, the dynamic course of radioactive emission can be readily quantitated with PET, permitting the application of tracer kinetic techniques for the measurement of substrate concentrations, reaction rates, and receptor binding in various tissues.

It is then important to recognize that specific probes are designed and used to target the process to be investigated (e.g., glucose metabolism, neurotransmitter synthesis, neurotransmitter reuptake, postsynaptic receptor binding, protein synthesis, gene expression). Based on the tissue (e.g., enzymes, receptors, pathological deposition), molecular imaging probes can be divided into a few large groups:

1. *Probes based on enzyme-mediated transformations.* These are designed as specific substrates of the enzymes to be targeted in tissue. Examples include FDG for targeting hexokinase as the initial step in glucose metabolism; FDOPA for targeting aromatic amino acid decarboxylase; and fluorothymidine (FLT) for targeting thymidine kinase, the first step for the incorporation of thymidine in DNA.
2. *Probes based on stoichiometric binding interactions.* These include most receptor ligand interactions and detection of brain pathological deposition in neurodegenerative diseases. Examples include receptor ligands for neuroreceptors and FDDNP for amyloid aggregates in brain tissue.
3. *Probes for determination of perfusion* (e.g., [N-13]ammonia). Perfusion probes have no specific structural requirements, except for their high vascular membrane permeability without specific macromolecular targets in tissue.
4. *Probes targeting specific transporters.* Examples include radiolabeled aminoacids for sodium-dependent transporters for detection of cancer (e.g., FDOPA as a brain tumor marker).

The general criteria for selecting and using molecular imaging probes have been recently described [30]. In brief, to select a molecular imaging probe to measure a specific process or assess organ function, the probe should meet the following criteria:

1. The probe has target specificity. Ideally, it should be restricted to the target process.
2. The probe has high membrane permeability to reach target areas.
3. Trapping of the labeled molecule or labeled reaction product occurs in a slow turnover pool.
4. Analogues specific to one biochemical pathway are used to isolate one or a few steps of the process. Thus, the kinetics of only the administered compound is represented in the measured data.
5. The molecular probe has high affinity for its tissue target and is rapidly cleared from nonspecific areas.
6. The blood pool is rapidly cleared of the molecular imaging probe to reduce blood pool background at the tissue target.
7. The probe is not or is slowly metabolized so that it is the only or primary chemical entity in the blood.
8. The probe has high specific activity to trace the process under investigation without exerting mass effects on the target molecule.
9. The probe has low nonspecific binding to increase target specificity and target-to-background ratios >> 1.
10. The molecular imaging probe has a small number of transport and biochemical reaction steps to allow tracer kinetic modeling to establish quantitative parameters for imaging determination.

21.3.2 Biological Probes (Antibodies, Peptides, Aptamers)

Labeling of biological molecules for use as tracers has a long history, which parallels that of chemical (small molecule) probes. Proteins and peptides are capable of highly specific interactions, and living organisms capitalize on these recognition properties by using specific receptors to sense the presence of ligands. For example, peptide hormones and neurotransmitters and their cognate receptors embody the recognition of short *peptide* sequences, whereas for cytokines, growth factors, and their corresponding receptors, recognition of specific *proteins* controls key cellular events. To take this a step further, researchers have developed somatostatin analogues such as octreotide for therapeutic modulation of somatostatin receptors (SSTr's). Furthermore, radiolabeling of octreotide and related molecules provides imaging probes for neuroendocrine tumors that overexpress SSTr [31]. Antibodies that recognize tumor-associated antigens, such as prostate-specific membrane antigen, have been developed into imaging agents (Prostascint^{TM}, capromab pendetide) for detection of cancer based on cell-surface biomarkers [32].

The development of biological probes for molecular imaging is set to rapidly expand due to the following recent advances:

1. Burgeoning knowledge allowing the identification of informative biomarkers (including cell-surface biomarkers) arising from genomics, proteomics, and bioinformatics research (as described in Section 21.1.2)
2. Widespread availability of PET scanners in hospitals and clinics and of small-animal PET (microPET) scanners in academic, biotech, and pharmaceutical company research facilities
3. Expanding availability of "exotic" PET radionuclides (such as ^{64}Cu, ^{68}Ga, ^{76}Br, ^{86}Y, and ^{124}I) combined with robust, general methods for conjugation to biological molecules. Particularly important has been the improved availability of positron-emitting nuclides with longer half-lives (e.g., ^{64}Cu and ^{124}I) compatible with the kinetics of biological processes *in vivo* (protein distribution and clearance, cell trafficking) [33, 34]. Furthermore, there is significant interest in the radiopharmaceutical field in codevelopment of pairs of imaging/therapeutic radionuclides (^{124}I/^{131}I, ^{86}Y/^{90}Y, ^{64}Cu/^{67}Cu) for attachment to molecular targeting agents. This is a growing area where quantitative imaging, biomathematical modeling, and dose estimation will play especially important roles.
4. Powerful platforms for generation of biological molecules (antibodies, peptides, aptamers) with any desired binding specificity

Advances in ligand generation are probably the most significant for the development of molecular imaging probes. Antibodies represent a classic example from nature in which the immune system has devised a mechanism for the generation of molecules with a broad range of binding properties. Specifically, binding diversity is encoded in genomic DNA, in the form of multiple copies of antibody variable-region genes. Combinatorial rearrangements of the variable regions, along with joining and diversity gene segments, result in a diverse repertoire of antibody binding specificities expressed by the B lymphocytes of the immune system. When a mouse, or human, is exposed to a foreign antigen, the lymphocytes producing the corresponding antibody are triggered to expand and produce the appropriate binding agent (antibody). This process has been captured by monoclonal antibody technology, in which immunization of a mouse with an antigen or protein of interest can be followed by routine isolation of a high-affinity antibody with the desired specificity.

Antibody technology has been further accelerated by the replacement of traditional mouse monoclonal antibodies with antibodies isolated using microbial or *in vitro* display methods. A classic example is phage display, in which an antibody binding site is "displayed" on an individual phage (bacterial virus) by fusion of the antibody variable genes to the phage coat protein [35]. A diverse library of phage (ca10^9–10^{10} different binders) is generated, and the individual phage exhibiting the desired specificity is selected by "panning" on the target antigen. This shortens the time required for identifying specific binders from months down to weeks. Variations include yeast display and mammalian cell display [36]. Alternatively, microbes and cells can be eliminated from the process entirely, by employing rapid *in vitro* techniques such as ribosome display or *in vitro* compartmentation [37].

Similar technologies also allow the rapid identification of novel peptides with unique binding properties. Libraries of linear or constrained peptides (in which loops are linked by a disulfide bridge) have been constructed in phage display systems and screened for novel binding properties. A further advance has been the use of alternative small protein scaffolds (intermediate in size between peptides and antibody variable domains), for generation of libraries of diverse binders [38]. The use of peptides, small proteins, and antibodies allows the identification of binding agents that recognize a variety of target shapes, from grooves, pockets, and indentations to flat surfaces. Finally, aptamers represent another broad class of ligands, derived from nucleic acids instead of polypeptides [39]. RNA and DNA molecules can adopt defined three-dimensional structures and exhibit exquisitely specific binding to proteins and other targets. *In vitro* selection using the SELEX method (Systematic Evolution of Ligands by EXponential enrichment) has allowed generation of aptamers that can bind to a variety of biological targets. Nucleic acid chemistry then allows production of derivatives with the stability required for *in vivo* use.

Following is a general outline of the steps required for development of a biological probe for molecular imaging, along with associated considerations at each step:

1. *Target selection.* Most biological molecular probes have been directed toward cell-surface targets. Cell surfaces in

living organisms are readily accessible to properly designed biologicals, whereas delivery to intracellular targets requires additional strategies to facilitate cell uptake. Moreover, biologicals are biodegradable, so their metabolism and fate are difficult to control once inside the cell. Depending on the nature of the target, either aptamers, peptides, or antibodies can be selected to provide a variety of sizes and shapes for the binding site.

2. *Generation of molecular probe.* As previously described in detail, powerful methods are available for routine isolation of antibodies, peptides, or aptamers with any binding specificity. These methods can also be used to improve the affinity and refine the specificity of binders once initial candidates are identified.
3. *Production.* Peptides and aptamers can be synthesized chemically in bulk once the desired sequence is identified by the display and library methods outlined above. Automated peptide or nucleic acid synthesizers are available that can also incorporate nonstandard amino acids or nucleosides and nucleotide backbones to enhance the properties of peptides and aptamers. Antibodies are proteins, and while there are systems for *in vitro* synthesis, mainstream methods of production involve microbial or mammalian cell expression. Since cells are employed to synthesize these proteins, the process takes longer, but the methods are standardized and widespread.
4. *Radiolabeling.* Methods for routine labeling of proteins, peptides, and aptamers are widely available. Different chemistries for radio-iodination can result in linkages that are labile or stable, suiting different applications. A variety of bifunctional chelating agents have been developed for conjugation to biomolecules, to allow labeling with radiometals. It is also possible to directly engineer chelating sites into proteins and peptides using cysteine residues or hexahistidine sequences. An added benefit is that site-specific, stoichiometric radiolabeling can be enabled. An important benefit of using large biomolecules is that the addition of a small radioactive tag is unlikely to impact the biological and binding properties of the probe, in contrast to chemical modification of small molecule probes.
5. *Pharmacokinetics and disposition in vivo.* As is the case for all molecular imaging probes, whether small molecules or biomolecules, the distribution, targeting, clearance, and metabolism of the imaging agent are of utmost importance to understand. Protein engineering can be used to modulate pharmacokinetics; the *in vivo* properties of peptides and aptamers can also be modified chemically—for example, by conjugation to polyethylene glycol (PEG).

In summary, all the elements are in place for the expanded use of biological molecules as probes for molecular imaging.

21.4 Molecular Imaging of Beta-Amyloid and Neurofibrillary Tangles

21.4.1 Brief Review of Molecular Probes for Beta-Amyloid Imaging

Alzheimer's disease (AD) is characterized by a progressive loss of cognitive function with neuronal loss and with β-amyloid senile plaques (SPs) and neurofibrillary tangles (NFTs) as the pathological hallmarks of the disease [40]. In 1990 more than 4 million Americans were diagnosed with AD, making it the most common form of dementia [41]; and assuming that a cure is not found, the number of AD patients has been extrapolated to quintuple by 2040 [42]. It is important to note that the definitive diagnosis of AD can only be made based on postmortem histopathological examination of brain tissue and detection of NFTs and SPs [43]. These neuropathological aggregates have become important imaging targets for early detection of the disease. Imaging probes specific for these aggregates have also become surrogate markers for monitoring the effectiveness of therapeutic interventions aimed at removing these aggregates from the brain.

A priori, monomeric peptides (e.g., radiolabeled β-amyloid peptides [44, 45]) or the monoclonal antibodies directed against them (see e.g., [46]) can be used as molecular imaging probes for amyloid plaques (SPs) present in the brains of AD patients. These radiolabeled peptides can orient themselves in the highly ordered arrangement of peptide monomers and thus accumulate as the result [47]. A very important limitation, however, is the very poor permeability of these peptides to cross the blood-brain barrier, which has limited their possible use for *in vivo* imaging. The development of β-amyloid–specific small molecule imaging agents with improved brain entry has been more successful. Target recognition for these ligands may be provided by the cross-β sheet structure in the core of the fibril anchored together with hydrogen bonds from the peptide backbones and with π stacking of the aromatic amino acid residues, as well as electrostatic interactions [48]. Some of these compounds have structural features similar to Congo Red, a histological dye used for *in vitro* detection of amyloid-like structures. Examples of Congo Red–related structures include X-34 [49], methoxy-X04 [50], and BSB ((trans, trans),-1-bromo-2,5-bis-(3-hydroxycarbonyl-4-hydroxy)styrylbenzene) [51].

Another histological dye, thioflavin T, has provided the framework for the development of thioflavin T–related molecular imaging probes, like 2-[(4′-methylamino)phenyl]-6-hydroxybenzothiazole (PIB) [52], imidazo[1,2-a]pyridine derivatives [53, 54], benzoxazole derivatives such as IBOX (2-(4′-dimethylaminophenyl)-6-iodobenzoxazole) [55] and styrylbenzoxazoles [56], and 4-methylamino-4′-hydroxystylbene [57]. Other types of probes are based on aromatic or heteroaromatic polycyclic moieties such as fluorene [58], acridine in BF-108 [59], and [F-18]FDDNP [60].

Only three of these molecular imaging probes have been applied to *in vivo* imaging of AD brain with PET. [F-18]FDDNP was the first molecular imaging probe reported to be effective in the visualization of neuropathology in the living brain of AD patients [60–62]. Specifically, [F-18]FDDNP labels β-amyloid and NFTs both *in vitro* and *in vivo* [63] and has proven useful to follow reliably the neuropathological progression of the disease in the living brain [64]. [F-18]FDDNP accumulates significantly in several cortical areas (medial and lateral temporal lobe, parietal lobe, frontal lobe), with the highest increases in the medial temporal lobe of AD patients as a result of Aβ and NFT deposition in this area. Since the medial temporal lobe is associated with initial pathology formation [65], [F-18]FDDNP offers an excellent opportunity for early detection (i.e., patients at risk and patients with mild cognitive impairment). An attempt to image Aβ aggregates in living human subjects was reported by Klunk et al. [66] using the hydroxylated benzothiazole aniline derivative ([C-11] 2-(4-methylaminophenyl)-6-hydroxybenzothiazole ([C-11] 6-OH-BTA or [C-11] PIB). In AD patients, [C-11]PIB retention is most prominently increased in frontal cortex (standard uptake value [SUV] = 1.56) and parietal areas (SUV = 1.45). Temporal accumulation (1.26) is low, similar to that of the pons (1.31), known to lack Aβ aggregates [66]. [C-11]PIB retention was equivalent in AD patients and control subjects for areas known to have no or minimum amyloid deposition (e.g., white matter, pons, cerebellum). Verhoeff et al. [67] also reported human PET data with a novel hydroxylated stilbene derivative, 4-[C-11]methylamino-4′-hydroxystylbene ([C-11]SB-13), with promising results.

21.4.2 *In Vitro* Characterization of FDDNP

In vitro binding is a necessary, but not sufficient condition, to validate the potential value of a molecular imaging probe for *in vivo* use. Normally, determinations of *in vitro* binding are a prelude to *in vivo* utilization of molecular imaging probes, and these results provide a line of evidence as to the possible usefulness of a probe when used *in vivo*. For example, the binding constant of [F-18]FDDNP to synthetic Aβ(1–40) fibrils was initially determined [61] in 0.25% ethanol in phosphate buffered saline (PBS) to be in the subnanomolar range, indicating its potential utility *in vivo*. Radioactive binding assays with brain homogenates from AD and normal control patients, performed in 1% ethanol in PBS, confirmed the high-affinity binding of [F-18]FDDNP to *ex vivo* SPs and NFTs. The resulting Scatchard plot of [F-18]FDDNP binding in AD homogenates yielded a K_D value of 0.75 nM and a B_{max} value of 144 nM with the brain sample studied [63]. Confocal fluorescence microscopy and immunohistochemistry were used to correlate the distribution of radiofluorinated [F-18]FDDNP in digital autoradiograms of AD brain specimens. Digital autoradiography of AD brain specimens using [F-18]FDDNP in 1% ethanol in saline [63, 68] revealed its binding in the temporal and parietal cortices matching the immunohistochemistry of adjacent slices and the pattern of SP and NFT distribution. There was no appreciable binding of [F-18]FDDNP to homogenates from age-matched control brains. The high-affinity binding and B_{max} for [F-18]FDDNP sites in AD brain homogenates fulfills the requirement for *in vivo* PET visualization of probe binding to brain receptor sites if it is assumed that the binding sites on SPs are analogous to the receptor model of binding [47, 62].

The apparent K_D value for [F-18]FDDNP in the low nanomolar range is also consistent with the specific labeling of SPs and NFTs, as microscopically evident by the fluorescence images and the gross pattern of binding observed with digital autoradiography, wide-field fluorescence microscopy, and immunostaining. It has also been shown that FDDNP is able to label all neuropathological aggregates—generically named amyloids—having cross-β sheets as a secondary structure element the protofilaments [68], including SPs, NFTs (paired-helical tau filaments), and prion aggregates. As such, FDDNP parallels results with thioflavin T in the same brain tissue. FDDNP labeling also predicts birefringence in Congo Red–stained histological sections with high reliability [68]. These studies with FDDNP have also been extended to living patients.

21.4.3 *In Vivo* Imaging of Beta-Amyloid and Neurofibrillary Tangles in Alzheimer's Disease

The availability of animal models of β-amyloid deposition in brain has opened a new avenue for *in vivo* imaging research with microPET. These models offer an invaluable opportunity for testing new potential molecular imaging probes for amyloid aggregates and also for evaluation of new anti-aggregation therapies. For example, [F-18]FDDNP binds to the β-amyloid–rich areas of the rat brain (frontal cortex and hippocampus) in a triple transgenic rat model of β-amyloid deposition [69].

In humans, the spatial pattern of SP and NFT distribution in neocortex depends on the severity of disease [65]. As an example, the deposition of β-amyloid SPs follows distinctive spatial and temporal patterns, for which Braak and Braak [65] have proposed three stages. In stage A, neocortical areas of temporal lobe and orbitofrontal cortex first develop SP deposits. In stage B, these deposits become denser in the same regions but also spread into the rest of the frontal lobe and into the parietal lobe. Finally, in stage C, SPs have invaded the whole neocortex. This indicates that both pathology load (SPs and NFTs) in a specific region and the pattern of pathology distribution in the brain are important factors to consider for diagnostic work with *in vivo* imaging techniques targeting these pathologies.

AD is today almost exclusively diagnosed based on clinical symptoms. However, the diagnosis of probable AD can be made in advanced stages of the disease only when the death toll of neurons in the central nervous system is already heavy and widespread [65, 70]. At that point, therapeutic interventions are mostly palliative. Therefore, new imaging tools for early diagnosis are essential, and this is a key for effective therapeutic interventions. Studies in living subjects have indeed demonstrated that [F-18]FDDNP-PET can differentiate mild cognitive impairment from normal aging and AD [64, 71] and has proven useful in early detection of neurodegeneration. These findings also suggest the utility of FDDNP-PET to monitor the effects of anti-aggregation drug candidates. Autopsy evaluation in one patient showed that the regional pattern of FDDNP *in vivo* distribution measured with PET was consistent with plaque and tangle accumulation patterns. Because FDDNP binds both plaques and tangles, regional binding patterns may be helpful in differentiating early AD from normal aging, nonamnestic mild cognitive impairment, or other forms of dementia.

Initial FDDNP studies of frontotemporal dementia show binding in frontal and temporal but not parietal regions, suggesting that FDDNP labels regional tau pathology and thus differentiates these two dementia forms according to binding patterns. Future studies will determine whether combining several informative imaging techniques will improve diagnostic accuracy and whether the benefits of using multiple scans outweigh the added costs.

21.5 Molecular Imaging Using Antibody Probes

21.5.1 Imaging Cell-Surface Phenotype

Intact antibodies, or immunoglobulins, are large (150,000 daltons), complex, glycosylated proteins comprising four polypeptide chains with intra- and interchain disulfide bridges. Nonetheless, they are adept at penetrating and permeating tissues *in vivo*, to provide the first line of recognition and defense against infection. As previously discussed, antibody technologies have allowed the development of countless monoclonal antibodies, including a large subset that recognize cell-surface proteins. Indeed, the techniques of flow cytometry and fluorescence-activated cell sorting (FACS) are heavily dependent on monoclonal antibodies that recognize cell-surface biomarkers. Important classes of markers include the human CD antigens that define the differentiation and activation state of immune cells, and the growing list of well-characterized tumor-associated antigens.

Monoclonal antibodies have become laboratory workhorses, widely employed in research and diagnostic methods, including enzyme-linked immunosorbent assays (ELISAs), radioimmunoassays, Western blots, immunohistochemical staining, immunofluorescence, and flow cytometry and FACS. A number of years ago, efforts to extend the utility of antibodies to targeted delivery *in vivo* led to the first examples of radioimmunoscintigraphy. Pioneering work of Goldenberg and colleagues showed that administration of a ^{131}I-radiolabeled murine monoclonal antibody with specificity for *carcinoembryonic antigen* (CEA) allowed detection of CEA-expressing colon cancer in patients by external scanning. Over the subsequent years, radiolabeled antibody conjugates have been developed and approved for immunoscintigraphy in humans, including CEA-Scan™, Oncoscint™, and Prostascint™.

21.5.2 Optimization of Antibodies for *In Vivo* Targeting

Radiolabeled antibodies for clinical imaging have not been widely accepted, and one reason may be the suboptimal properties of the antibody component. The first-generation tracers are based on murine antibodies, which can be highly immunogenic in humans. Initial attempts at implementing antibody-based imaging have also been hampered by pharmacokinetics that are far from ideal. In particular, native antibodies exhibit prolonged blood and whole-body retention, as is fitting for their role as defense agents. However, when coupled to a radionuclide for use as an imaging agent, slow blood clearance results in high background activity, which can persist for days.

Antibody engineering has provided solutions to most of these issues. Reduction of immunogenicity has been achieved by replacing the murine components of monoclonal antibodies with the corresponding human domains, to yield chimeric and humanized antibodies. More recently, techniques such as phage display (described in Section 21.3.2) and the development of transgenic mice carrying germline human antibody genes have streamlined the process of directly producing fully human antibodies of any specificity. The pharmacokinetics of recombinant antibodies and fragments have systematically been explored by evaluation of cognate fragments that differ in molecular weight, valency, and presence or absence of native antibody domains. For example, an antibody that recognizes CEA, a well-characterized biomarker in human colon cancer, was used as the starting point for the generation of a series of engineered antibody fragments [72]. These included monovalent single-chain antibodies (scFv, 25 kDa) consisting solely of the variable regions of the parental antibody joined by a peptide linker. Larger, bivalent fragments were also evaluated, including diabodies (55 kDa), minibodies (80 kDa), scFv-Fc fragments (110 kDa), and intact chimeric antibodies (150 kDa). Evaluation of the tumor targeting, distribution, and clearance of radiolabeled anti-CEA fragments in athymic mice bearing human colorectal carcinoma xenografts showed the following:

1. Bi- or higher valency is required for good tumor retention.
2. Engineered proteins with molecular weights below about 60 kDa (the threshold for first-pass renal clearance) clear

the circulation rapidly, resulting in low background but also limiting the tumor uptake.

3. Larger fragments such as minibody and scFv-Fc clear more slowly and attain higher levels of tumor activity.
4. The targeting and clearance properties of scFv-Fc fragments (and intact antibodies) can be tailored by introducing specific mutations into the Fc region to modulate blood half-life.

Smaller fragments, such as the diabody and minibody, labeled with ^{64}Cu or ^{124}I, produced excellent tumor images by microPET scanning, within a few hours of systemic administration [73, 74]. Ongoing clinical imaging studies will delineate which format(s) are optimal for tumor detection in patients [75].

An important question is whether the results obtained in the CEA antigen system can be extended to other cell-surface biomarkers. Minibodies and scFv-Fc fragments were produced from the anti-Her2 therapeutic antibody, trastuzumab. When radiolabeled with ^{64}Cu, these engineered antibody fragments demonstrated excellent localization to Her2-expressing breast cancer xenografts in mice, as shown by microPET imaging [76]. More recently, the approach has been confirmed by production of anti-CD20 minibodies based on the CD20-specific therapeutic antibody, rituximab [77], and a diabody and minibody specific for a novel cell-surface biomarker in prostate cancer, prostate stem cell antigen (PSCA) (see Figure 21.4) [78]. Thus, antibody engineering provides a platform for producing molecular imaging agents that can recognize any cell-surface target.

21.5.3 Measurement of Target Expression

Antibody recognition is based on stoichiometric binding interactions and does not offer the versatility of some small molecule probes, which can assess perfusion, transport, enzyme activity, and other biological processes. Instead, antibody imaging can provide a readout on the presence or absence of a specific molecular biomarker. The spatial distribution of an antibody probe can provide an indication of cells in an unusual location (e.g., a tumor). The cell-surface phenotype (e.g., CEA-positive, CD20-positive) can indicate the tissue of origin. Presence or absence of cell-surface markers can also provide an indication of the biology of the cells in a tissue—for example, indicating responsiveness to growth factors or cytokines or providing a readout on the activation state of immune cells.

One particularly useful path is the codevelopment of antibody imaging agents and matching antibody therapeutics. Ongoing studies are focusing on the measurement of specific molecular targets. For example, *in vivo* imaging using a ^{64}Cu-radiolabeled trastuzumab antibody fragment can be used to distinguish two breast cancer tumor xenografts in a mouse, based on whether the tumor cell lines express high (MCF-7/HER2) or low (MDA-MB-231) levels of the target antigen, Her2 (Figure 21.4(b)). In the future, quantitative assessment of the level of expression of cell-surface targets by molecular imaging may prove to be an important determinant for selection of targeted therapies.

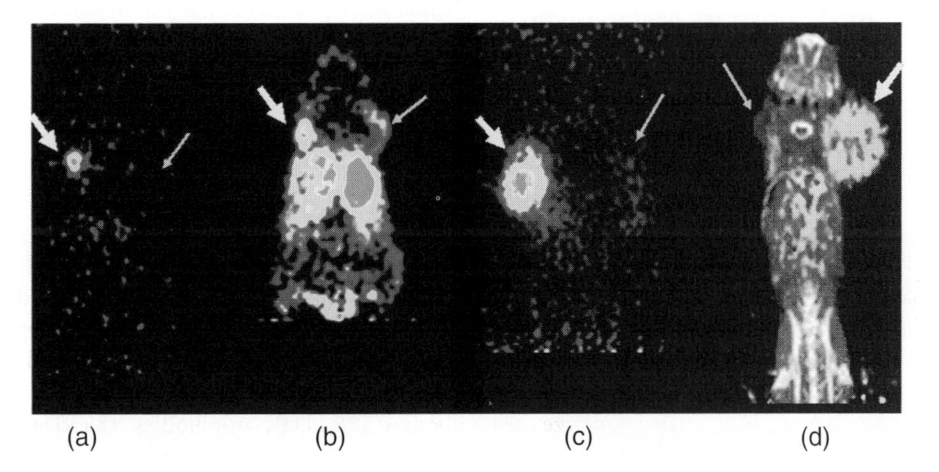

FIGURE 21.4 Antibodies of any specificity can readily be engineered for *in vivo* targeting and imaging by microPET. In each panel, the antigen-positive xenograft is indicated by the thick yellow arrow, and the antigen-negative control tumor is shown by the thin orange arrow. All images were acquired at 18–21 hours postinjection; coronal slices are shown. (a) CEA imaging using I-124 cT84.66 diabody. (b) Her2 imaging using Cu-64-DOTA (tetraazacyclo-dodecane-tetraacidic acid) trastuzumab scFv-Fc DM. (c) CD20 imaging using I-124 rituximab minibody. (d) PSCA imaging using I-124 hu2B3 minibody. For a more detailed view of this figure, please visit our companion site at: http://books.elsevier.com/companions/9780123735836.

21.5.4 Monitoring Response to Therapy

Radiolabeled antibodies and fragments can also be used as molecular imaging probes for following the response of cells or tissues to therapeutic intervention. A simple case would be to use an antibody that recognizes a tumor-specific marker as a means to quantitate tumor burden; upon therapy, a decrease in signal would indicate a loss of viable tumor cells. Such an approach would need to be validated; for example, if the antigen is particularly stable, it might remain at the tumor site even if therapy were effective. More sophisticated implementations of antibody imaging can be based on knowledge of the molecular actions of specific drugs. For example, Smith-Jones [79] used trastuzumab $(Fab')_2$ fragments radiolabeled with Ga-68 to assess the effect of the drug 17-AAG *in vivo* by microPET imaging. This drug interferes with the action of hsp90, a chaperone protein that protects Her2 from degradation. Treatment of a xenograft-bearing mouse with 17-AAG resulted in a reduction of Her2 signal at the tumor. Thus, using an antibody probe to monitor the downstream effect on a cell-surface biomarker provided confirmatory evidence of the mechanism of action of the drug *in vivo.*

21.6 Some Other Molecular Imaging Applications

A few other molecular imaging examples are also interesting, and represent certain types of application in biology and pharmacology. They are described briefly.

21.6.1 *In Vivo* Regional Substrate Metabolism in Human Brain

Human brain tissue has a unique metabolic characteristic. It is known that glucose taken up from plasma is the primary metabolic substrate used in brain tissue to generate the energy for its cellular functions. Normally, the glucose molecules in brain cells go through aerobic metabolism in the Krebs cycle and are converted to carbon dioxide and water, using six molecules of oxygen for each glucose molecule. So, the whole-brain utilization rates of oxygen and glucose are normally around 6:1, in terms of molar quantities. However, it is unclear whether this oxygen-to-glucose utilization ratio (OGR) holds uniformly for both gray and white matter, especially under pathological conditions. For example, for brain trauma patients, a better understanding of regional metabolic changes is expected to help treatment. A molecular imaging study addressing these questions has been performed by Bergsneider et al. [80–84] at the UCLA Brain Injury Research Center and the UCLA Nuclear Medicine Clinic. The study involved the use of a multiple of dynamic PET sessions, using O-15 carbon monoxide (for blood volume), O-15 water (for blood perfusion), O-15 oxygen (for oxygen utilization), and FDG (for glucose utilization). The study was performed within one week of brain injury. (Readers are referred to the cited references for detailed study procedures.) Some parametric images of cerebral blood flow (CBF), oxygen extraction fraction (OEF), and cerebral metabolic rate of oxygen (CMRO) in a normal subject are shown in Figure 21.5, and the average values of these parameters in normal subjects and in subjects with traumatic brain injury (TBI) are summarized in Table 21.1.

In general, the results show that in TBI subjects, the brain tissue not adjacent to the contusion region has normal blood perfusion and normal OEF, but the glucose and oxygen utilization rate in the gray matter are both reduced. The OGR value has some particular implication for the metabolic state in tissue. OGR = 6 indicates that the mechanism in mitochondria for aerobic metabolism is functioning well. If OGR < 6, some glucose probably goes through anaerobic metabolism, and lactate is produced (potentially lowering the tissue pH).

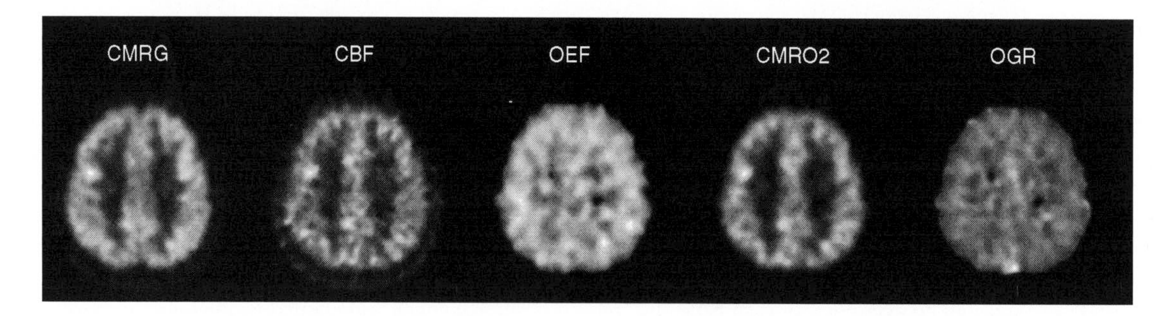

FIGURE 21.5 Parametric images from a dynamic brain PET study in a normal human subject, using O-15 CO, O-15 oxygen, O-15 water, and FDG tracers, sequentially in a single session. The dynamic O-15 water PET images were used to generate the CBF image; O-15 CO and O-15 oxygen images were used along with the CBF image to generate the CMRO2 and OEF images; the FDG image was used to calculate the image for cerebral metabolic rate of glucose (CMRG). The OGR image was generated as the ratio of the CMRO2 and CMRG images. Figure provided by Dr. H. M. Wu, Department of Molecular and Medical Pharmacology, UCLA School of Medicine.

TABLE 12.1 Metabolic parameters in human brain tissues*

	Normal ($n = 16$)	TBI ($n = 11$)
CMRG (mg/min/100g)	4.31 ± 0.69	3.29 ± 0.58
	1.87 ± 0.36	1.97 ± 0.25
CBF (m/min/g)	0.44 ± 0.09	0.43 ± 0.08
	0.22 ± 0.06	0.22 ± 0.04
OEF	0.40 ± 0.06	0.38 ± 0.06
	0.31 ± 0.05	0.30 ± 0.07
CMRO (ml/min/100g)	3.18 ± 0.47	2.44 ± 0.54
	1.21 ± 0.14	0.97 ± 0.18
OGR	6.03 ± 1.04	6.01 ± 1.09
	5.37 ± 1.00	3.99 ± 0.77

*Numbers are mean ± standard deviation, with the top number of each group for gray-matter regions, and the bottom number for white-matter regions. The values for TBI excluded regions with contusion. Values in this table are taken from Wu et al. [83].

If this couples with a reduced oxygen utilization rate, one could question whether the oxygen delivery to tissue is limited. On the other hand, if OGR > 6, substrates (e.g., lactate, ketone bodies) in addition to glucose are being used. It could indicate that oxygen delivery is not limited but that glucose transport or metabolism is altered or that the tissue prefers other substrates in that pathological state. Full interpretation of this, however, requires consideration of additional information, such as of perfusion, OEF, and arterial-venous differences (in concentration) of various substrates. Nevertheless, results from the molecular imaging study provided new biological information that is difficult to obtain by other methods available today.

21.6.2 Cell Proliferation Rate in Mouse Tumor

FLT is an analogue of thymidine involved in DNA synthesis during the S-phase of cell division. It has been used in PET studies to examine the cell proliferation rate in tumors [85–91] to grade tumor aggressiveness or to show the efficacy of a treatment protocol. The animal study we will describe shows an application of FLT-PET and FDG-PET for investigating tumor response post–irradiation treatment. The study involved tumor implantation via *transgenic adenocarcinoma of mouse prostate* (TRAMP), irradiation, and multiple imaging longitudinally. The study was a collaboration between Huang's and McBride's laboratories at UCLA [92–94]. (Readers are referred to the cited references for detailed study procedures.)

Figure 21.6 summarizes the results. Within one week post-irradiation, FDG uptake on irradiated tumor was usually higher than that of the reference. In two weeks, FDG uptake was lower than that of the reference. Thereafter, FDG uptakes on active growing areas of the tumors on both sides were comparable. For FLT, uptake in the first week was lower in the irradiated tumor, but in the second week and thereafter, FLT uptake was comparable to that of the reference. In other words, cell proliferation rate in tumor post-irradiation was suppressed initially but fully recovered thereafter. Glucose utilization rate, on the other hand, was enhanced immediately after irradiation but went through a suppressed period afterward (in the second week) before recovering to its reference level. The results are consistent with an oxidative metabolism impairment immediately following irradiation, possibly due to mitochondrial damage, resulting in increases in anaerobic glycolysis (increase in FDG uptake). While oxidative metabolism is impaired, DNA replication and cell proliferation are decreased. The discordant FDG and FLT uptake changes following irradiation suggest caution in the interpretation of FDG-PET scan images in cancer patients. FLT-PET is probably more useful for treatment efficacy assessment. The study is currently ongoing at UCLA to examine the optimal dosage fractionation in radiation treatment of cancer.

21.6.3 Measurement of Murine Cardiovascular Physiology

Historically, many physiological parameters were first determined in humans using tracer methods and were later confirmed by other, more sophisticated or more direct measurements. Some parameters, like blood perfusion and cardiac output (CO), continue to be measured clinically in patients by tracer methods due to the methods' simple procedures and robustness [95]. For small animals such as mice, the cardiac function is also an important physiological consideration in the study of drug response but is usually very difficult to measure due to their small body size (~28 gm), small total blood volume (~2 ml), and fast heart rate (~400 beats/min). At the UCLA Nuclear Medicine Clinic, a small-animal PET procedure has been developed that can provide information on cardiac function in mice conveniently and noninvasively, along with other biological information in a single imaging session [96, 97]. Figure 21.7 shows typical image data of the study. With such a study procedure, the cardiac output was found to average 20.4 ± 3.4 ml/min in anesthetized mice, with a stroke volume (SV) of 45.0 ± 6.9 μl. The results are comparable to those obtained with a much more elaborate and tedious procedure using magnetic resonance imaging (MRI) [98] and demonstrate the capability of molecular imaging for measuring cardiovascular function in small-animal models. When cardiac stress was simulated with dobutamine injection, CO and SV were increased by ~75% and ~23%, respectively.

21.7 Summary and Future Perspectives

Many new technical and biological developments are being worked on by many investigators in many institutions. A few of them are summarized below.

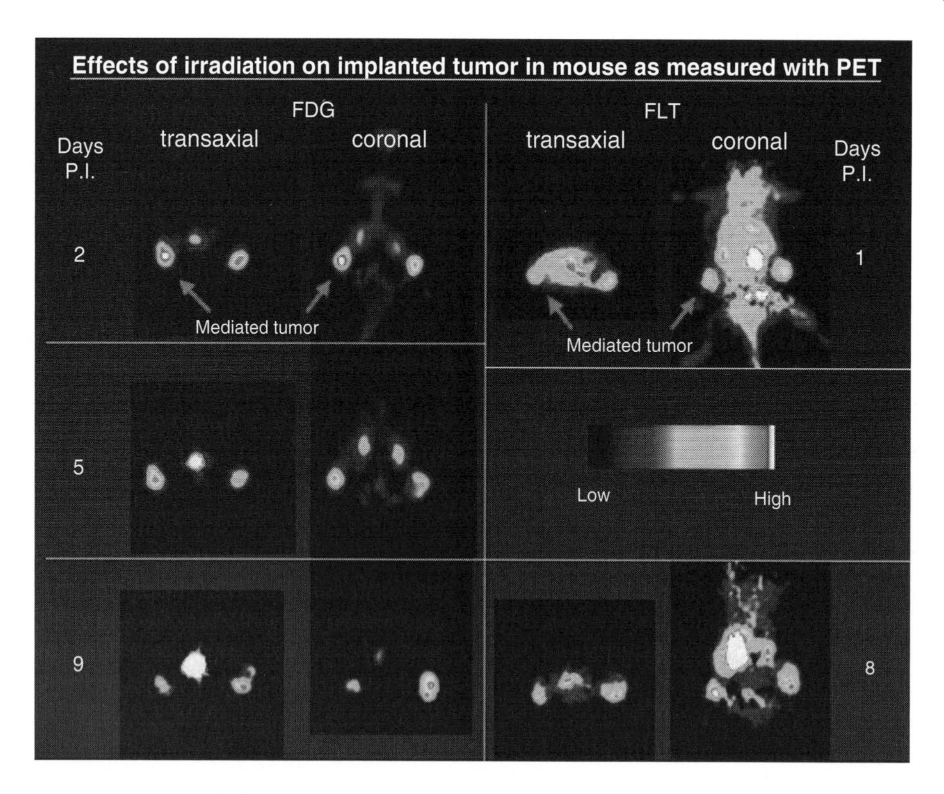

FIGURE 21.6 Multiple microPET images (using FDG and FLT) of a mouse (C57BL/6) at various days post-irradiation (25 Gy) applied to an implanted tumor (TRAMP). The tumor on the other side did not receive radiation and was used as a control. The images are the transaxial (left) and coronal (right) sectional images of the mouse through the center of the tumors. They were scanned at 60 minutes after tracer injection (IV bolus through the tail vein). FDG uptakes at days 2 and 5 post-irradiation (P.I.) are seen to have increased (compared with that of the control tumor) until day 9 P.I., when it decreased. FLT uptake at day 1 P.I. is seen to be lower than that in the control tumor but got back to a comparable level at day 8 P.I. For a more detailed view of this figure, please visit our companion site at: http://books.elsevier.com/companions/9780123735836.

21.7.1 Optical Imaging, Small-Animal Single Photon Emission Computed Tomography, and Microfluidic Blood Samplers

Optical imaging, though still facing a number of hurdles to becoming a quantitative measurement tool, has some favorable characteristics for biological studies. The imaging device required is relatively simple. It does not involve radioactivity, and it does not require a cyclotron to make the tracer. The group led by Dr. Arion Chatzioannou at UCLA is developing a device that combines optical imaging with small-animal PET. The new device will have complementary information from both imaging modalities and is expected to add additional information for molecular imaging.

While the spatial resolution of PET is limited by the positron range of the labeling radioisotope, the resolution of pinhole single photon emission CT (SPECT) can be changed by adjusting the imaging geometry and the distances between the pinhole, the object, and the detector plane. Thus, it can in principle have a much higher spatial resolution than PET. The potential has been demonstrated in some prototype scanners [99–102]. While issues related to radiation dose level, detection efficiency, and photon attenuation still need to be addressed, the spatial resolution that this type of small-animal SPECT scanner can provide is impressive.

As discussed earlier in this chapter, the availability of the input function is very important for image quantification in terms of biological information. The current method of using manually drawn blood samples is tedious, labor intensive, and difficult to use for small animals like mice. Nanotechnology (e.g., microfluidic devices) is promising to relieve some of these difficulties. A prototype microfluidic blood sampler has already been demonstrated [103, 104]. It takes a very small and precise amount of blood for each sample and can be operated by computer control. Coupled with nanotechnologies for chemical assay [105, 106], this type of device can further simplify the procedures for determining red blood cell uptake and labeled metabolites and is expected to make quantitative molecular imaging more convenient and practical to perform.

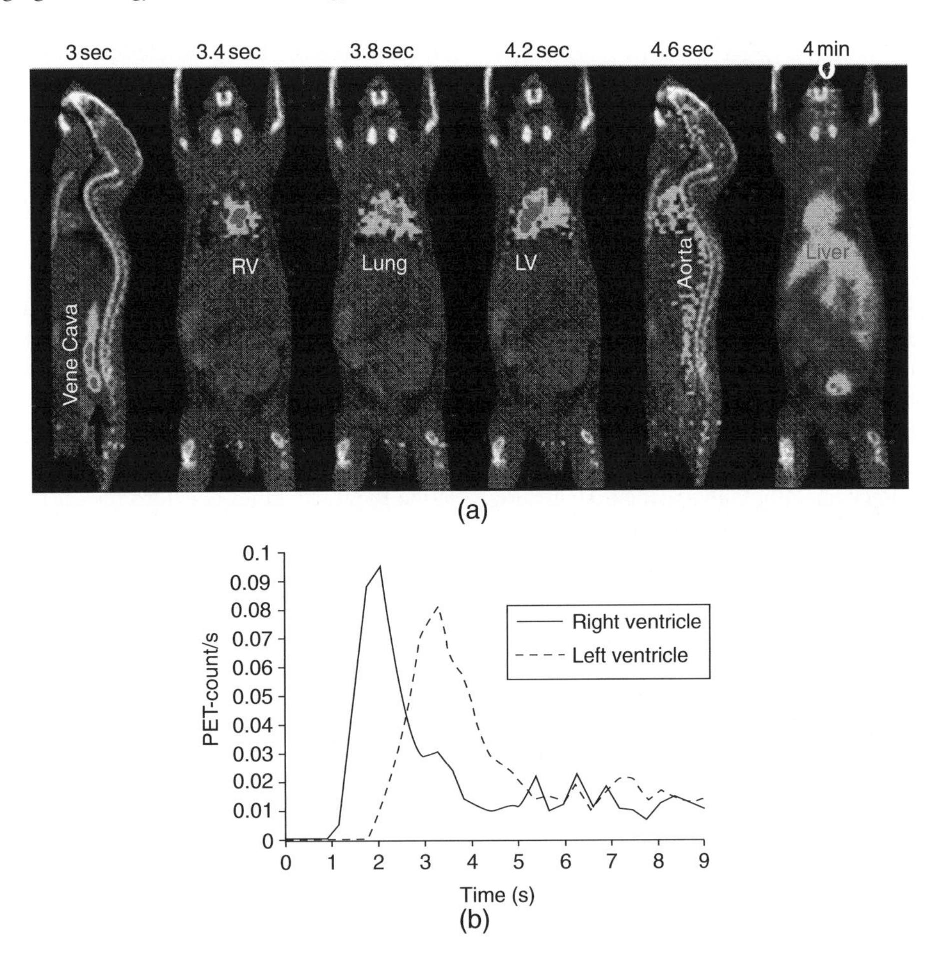

FIGURE 21.7 (a) Short-time-frame PET images (coronal and sagittal sections) of a mouse immediately after a bolus of FDG injected IV in the tail vein. The PET images (in color) were fused over the x-CT images (in black and white) of the animal. The injected tracer first appeared in the vena cava, then in the right ventricle (RV), the lung, the left ventricle (LV), and the aorta in sequence in less than five seconds. Afterward, the tracer was delivered to various organ tissues in the body. (b) Time activity curves obtained from regions of interest defined on the RV and LV. The kinetics can be used to calculate the transit time through the pulmonary system and the cardiac output of the animal [96, 97]. Figure provided by Dr. H. M. Wu, Department of Molecular and Medical Pharmacology, UCLA School of Medicine. For a more detailed view of this figure, please visit our companion site at: http://books.elsevier.com/companions/9780123735836.

21.7.2 Automated Image/Data Analysis

Currently, one of the major time-consuming steps in a molecular imaging procedure is the data/image analysis to convert the measured physical quantities to biological information. It usually involves multiple steps of manual interaction, and the result is susceptible to many error sources. Hence, many investigators simply skip the analysis and settle for visual examination of images that reflect only the distribution of tracer concentration. This greatly compromises the advantages of quantitative molecular imaging. The development of parametric imaging to facilitate the conversion of tracer concentration to biological information has helped ease somewhat the difficulty of image quantitation. In brain imaging, extensive effort has been devoted to image warping and alignment to standardize spatially the images to a common space [107, 108] to allow easy extraction of region-of-interest values and statistical evaluation (i.e., SPM2 [109, 110], NeuroSTAT [111, 112], AIR [113], and others [114]). In cardiac studies, people have also adopted polar maps as a standard [115]. However, this has not been generalized to the whole body.

Much more can be done to automate the analysis procedure, especially for small-animal studies. We expect that the procedure of image/data analysis will be facilitated in the future to become as routine as blood tests in clinical laboratories in hospitals today. Ideally, a biology/pharmacology investigator will no longer need to be involved directly in image/data analysis processing and will simply receive a summary report after a molecular imaging experiment.

21.7.3 Virtual Experimentation

Animal experiments are critical to biomedical and pharmacological studies, but they are usually very time-consuming and costly to perform. If one has a clear idea of the expected outcome before a set of experiments is performed, one can reduce the number of experiments and obtain the results faster. This applies to molecular imaging studies as well. With the tremendous advancement in computational methodology and computers, computation can be used to reduce the number of necessary experiments. In fact, this has been recognized in many scientific fields. Computation has already played an important role in many areas. New terms like "computational fluid dynamics," "computational biology," and "computer aided design" have become popular. For molecular imaging, computational methods are useful in many different ways. For example, Monte Carlo simulation of photon emission and detection can be used before a prototype imaging device is built. Whole-body tracer kinetics for a tracer with certain chemical and biological properties can be simulated to see whether the tracer is suited for certain application before the tracer is actually synthesized.

When an investigator decides that an experiment is needed, computer simulation of the kinetics in all organ tissues in the whole body can be used to guide the protocol design of the experiment. An Internet-based software system, KIS (Kinetic Imaging System) [116], has been developed at the Department of Molecular and Medical Pharmacology at UCLA to address the general need in this direction. The system currently consists of four modules—glossary, virtual experimentation, image analysis, and kinetic model fitting, as shown in Figure 21.8. The virtual experimentation module is focused on simulating the tracer kinetics in various organs in a mouse and can generate a set of dynamic PET images of a mouse as if it were from a real small-animal PET scanner, for any selected tracer of

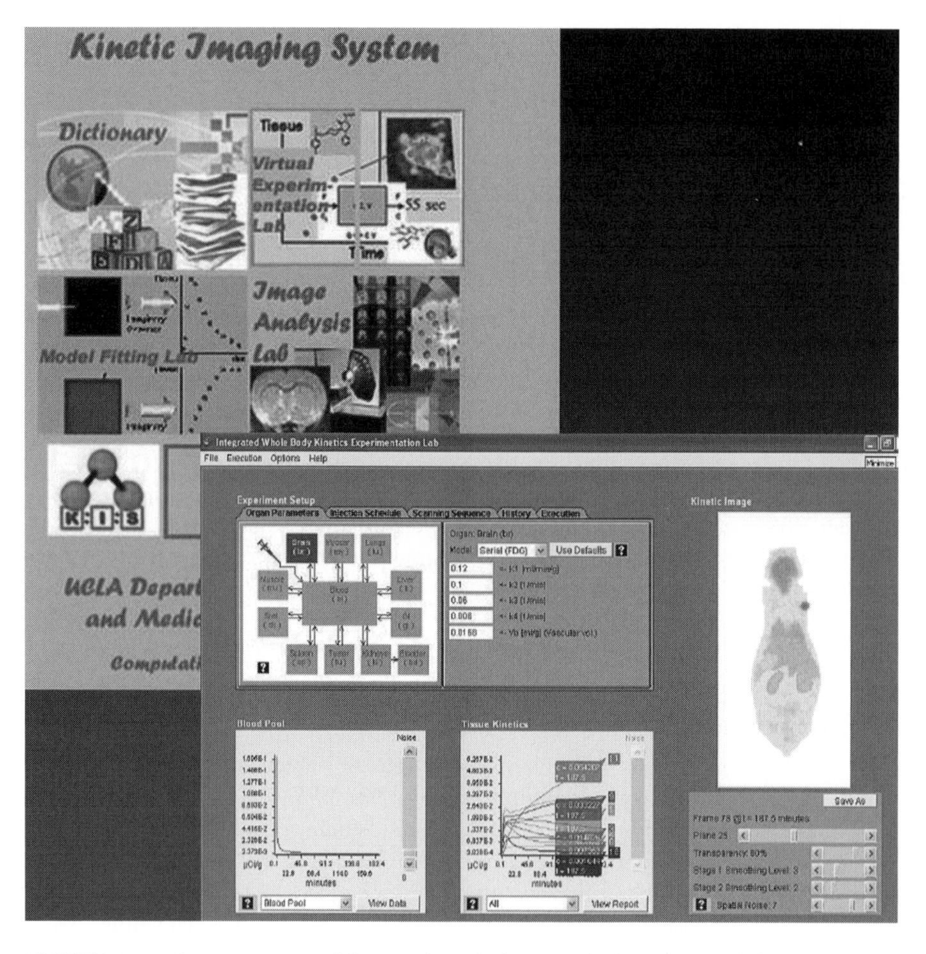

FIGURE 21.8 Screen capture of the opening window panel of KIS (upper left background) and that of the whole-body kinetics simulation panel of the virtual experimentation module in KIS. For a set of kinetic parameters specified by the user for a tracer in various organs of the body, virtual experimentation can generate the time activity curves in the body organs and corresponding dynamic mouse PET images, which simulate the real PET images if the tracer were actually given to an experimental mouse. For a more detailed view of this figure, please visit our companion site at: http://books.elsevier.com/companions/9780123735836.

assumed chemical/biological properties. It is clear that the potential capability of such systems is enormous, and the current software system can be greatly expanded. For example, virtual experimentation can be made to resemble more closely the true physiology in a mouse and to provide the drug kinetics. The expansion can also be directed to simulation of tracer/drug kinetics in other animals and in humans.

21.8 Exercises

1. List and elaborate on the major components of molecular imaging.
2. Elaborate in general terms why molecular imaging can provide biological information *in vivo*.
3. State the rationale for the need of an input function for biological quantification of molecular images.
4. What are the two most commonly used graphical analysis methods in molecular imaging? What are their differences and how would one choose one over the other for any specific study?
5. Discuss the general criteria for selecting and using chemical probes (small molecules) for molecular imaging.
6. What are the major steps/considerations required for development of a biological probe (antibodies, peptides, aptamers) for molecular imaging?

21.9 References and Bibliography

1. M. E. Phelps. The merging of biology and imaging into molecular imaging. *J. Nucl. Med.* 41:661–681, 2000.
2. M. E. Phelps. Molecular imaging with positron emission tomography. *Annual Review of Nuclear and Particle Science.* 52:303–338, 2002.
3. M. E. Phelps. *PET: Molecular Imaging and Its Biological Applications.* Springer, 2004.
4. M. E. Phelps et al. Application of annihilation coincidence detection to transaxial reconstruction tomography. *J. Nucl. Med.*16:210–224, 1975.
5. G. N. Hounsfield. Computerized transverse axial scanning (tomography). Part I. Description of system. *Brit. J. Radiol.* 46:1016–1022, 1973.
6. R. N. Bracewell and A. C. Riddle. Inversion of fan-beam scans in radioastronomy. *Astrophys. J.* 150:427–434, 1967.
7. A. C. Kak and M. Slaney. *Principles of Computerized Tomographic Imaging.* IEEE Press, 1988:327.
8. S. R. Deans. *The Radon Transform and Some of Its Applications.* John Wiley & Sons, 1983:289.
9. S. C. Huang and M. E. Phelps. Principles of tracer kinetic modeling in positron emision tomography and autoradiography. In M. E. Phelps, J. Mazziotta, and H. R. Schelbert (Eds.). *Positron Emission Tomography and Autoradiography.* Raven Press. 287–346, 1985.
10. A. Lammertsma. Simplified reference tissue model for PET receptor studies. *Neruoimage.* 4:153–158, 1996.
11. J. Logan et al. Distribution volume ratios without blood sampling from graphical analysis of PET data. *J. Cereb. Blood Flow Metab.* 16:834–840, 1996.
12. A. Citri and Y. Yarden. EGF-ERBB signalling: Towards the systems level. *Nat. Rev. Mol. Cell. Biol.* 7(7):505–516, 2006.
13. S. C. Huang et al. Noninvasive determination of local cerebral metabolic rate of glucose in man. *Am. J. Physiol.* 238:E69–E82, 1980.
14. J. Bard. *Nonlinear Parameter Estimation.* Academic Press, 1974.
15. L. Sokoloff et al. The [C-14]deoxyglucose method for the measurement of local cerebral glucose utilization: Theory, procedure, and normal values in the conscious and anesthetized albino rat. *J. Neuro. Chem.* 28:897–916, 1977.
16. M. E. Phelps et al. Tomographic measurement of regional cerebral glucose metabolic rate in man with (F-18)fluorodeoxyglucose: Validation of method. *Ann. Neurol.* 6:371–388, 1979.
17. S. Gambhir et al. Imaging gene expression: Principles and assays. *J. Nucl. Cardiol.* 6:219–233, 1999.
18. S. Gambhir et al. Imaging transgene expression with radionuclide imaging technologies. *Neoplasia.* 2:118–138, 2000.
19. L. Green et al. Tracer kinetic modeling of FHBG in mice imaged with microPET for quantitation of reporter gene expression. *J. Nucl. Med.* 41(5 suppl):228, 2000.
20. C. Patlak and R. G. Blasberg. Graphical evaluation of blood-to-brain transfer constants from multiple-time uptake data. *J. Cereb. Blood Flow Metab.* 3:1–7, 1983.
21. M. A. Mintun et al. A quantitative model for the in vivo assessment of drug binding sites with positron emission tomography. *Ann Neurol.* 15(3):217–227, 1984.
22. J. Logan et al. Graphical analysis of reversible radioligand binding from time-activity measurements applied to [N-[C-11]-methyl]-(-)-cocaine PET studies in human subjects. *J. Cereb. Blood Flow Metab.* 10(5):740–747, 1990.
23. S. C. Huang, J. R. Barrio, and M. E. Phelps. Neuroreceptor assay with PET: Equilibrium versus dynamic approaches. *J. Cereb. Blood Flow Metabol.* 6:515–521, 1986.
24. M. P. Kung and H. F. Kung. Mass effect of injected dose in small rodent imaging by SPECT and PET. *Nucl. Med. Biol.* 32:673–678, 2005.
25. W. C. Eckelman (Ed.). *Receptor Binding Radiotracers,* vols. I and II. CRC Press, 1982.
26. M. A. Mintun et al. Brain oxygen utilization measured with O-15 radiotracers and positron emission tomography. *J. Nucl. Med.* 25(2):177–187, 1984.
27. N. Hattori et al. Accuracy of a method using short inhalation of O-15-O2 for measuring cerebral oxygen extrac-

tion fraction with PET in healthy humans. *J. Nucl. Med.* 45:765–770, 2004.
28. S. Huang et al. Modeling approach for separating blood time-activity curves in positron emission tomographic studies. *Phys. Med. Biol.* 36:749–761, 1991.
29. S. Huang et al. Kinetics and modeling of 6-[F-18]fluoro-L-DOPA in human positron emission tomographic studies. *J. Cereb. Blood Flow Metab.* 11:898–913, 1991.
30. J. R. Barrio. PET: Molecular imaging and its biological applications. In M. E. Phelps (Ed.). *The Molecular Basis of Disease.* Springer-Verlag. 270–320, 2004.
31. W. W. de Herder et al. Neuroendocrine tumors and somatostatin: Imaging techniques. *J. Endocrinol. Invest.* 28(11 suppl):132–136, 2005.
32. N. H. Bander. Technology insight: Monoclonal antibody imaging of prostate cancer. *Nat. Clin. Pract. Urol.* 3(4):216–225, 2006.
33. P. McQuade et al. Positron-emitting isotopes produced on biomedical cyclotrons. *Curr. Med. Chem.* 12(7):807–818, 2005.
34. I. Verel, G. W. Visser, and G. A. van Dongen. The promise of immuno-PET in radioimmunotherapy. *J. Nucl. Med.* 46(Suppl 1):164S–171S, 2005.
35. G. P. Smith and V. A. Petrenko. Phage display. *Chem. Rev.* 97:391–410, 1997.
36. H. R. Hoogenboom. Selecting and screening recombinant antibody libraries. *Nat. Biotechnol.* 23(9):1105–1116, 2005.
37. A. Rothe, R. J. Hosse, and B. E. Power. Ribosome display for improved biotherapeutic molecules. *Expert Opin. Biol. Ther.* 6(2):177–187, 2006.
38. A. Rothe, R. J. Hosse, and B. E. Power. *In vitro* display technologies reveal novel biopharmaceutics. *Faseb J.* 20(10):1599–1610, 2006.
39. J. F. Lee, G. M. Stovall, and A. D. Ellington. Aptamer therapeutics advance. *Curr. Opin. Chem. Biol.* 10(3): 282–289, 2006.
40. J. C. Vickers et al. The cause of neuronal degeneration in Alzheimer's disease. *Prog. Neurobiol.* 60(2):139–165, 2000.
41. D. A. Evans. Estimated prevalence of Alzheimer's disease in the United States. *Milbank Q.* 68(2):267–289, 1990.
42. K. Iqbal. *Alzheimer's Disease: Basic Mechanisms, Diagnosis and Therapeutic Strategies.* Wiley, 1991.
43. M. Ball et al. Consensus recommendations for the postmortem diagnosis of Alzheimer's disease. *Neurobiol. Aging.* 18(Suppl 4):s1–s2, 1997.
44. A. Kurihara and W. M. Pardridge. Abeta(1–40) peptide radiopharmaceuticals for brain amyloid imaging: In chelation, conjugation to poly(ethylene glycol)-biotin linkers, and autoradiography with Alzheimer's disease brain sections. *Bioconjug. Chem.* 11(3):380–386, 2000.
45. H. J. Lee et al. Imaging brain amyloid of Alzheimer disease in vivo in transgenic mice with an Abeta peptide radiopharmaceutical. *J. Cereb. Blood Flow Metab.* 22(2): 223–231, 2002.
46. R. P. Friedland et al. Development of an anti-A beta monoclonal antibody for *in vivo* imaging of amyloid angiopathy in Alzheimer's disease. *Mol. Neurobiol.* 9(1–3):107–113, 1004.
47. K. Shoghi-Jadid et al. Exploring a mathematical model for the kinetics of beta-amyloid molecular imaging probes through a critical analysis of plaque pathology. *Mol. Imaging Biol.* 8(3):151–162, 2006.
48. O. S. Makin et al. Molecular basis for amyloid fibril formation and stability. *Proc. Natl. Acad. Sci. U S A.* 102(2):315–320, 2005.
49. S. D. Styren et al. X-34, a fluorescent derivative of Congo red: A novel histochemical stain for Alzheimer's disease pathology. *J. Histochem. Cytochem.* 48(9):1223–1232, 2000.
50. W. E. Klunk et al. Imaging Abeta plaques in living transgenic mice with multiphoton microscopy and methoxy-X04, a systemically administered Congo red derivative. *J. Neuropathol. Exp. Neurol.* 61(9):797–805, 2002.
51. C. W. Lee et al. Isomerization of (Z,Z) to (E,E)1-bromo-2,5-bis-(3-hydroxycarbonyl-4-hydroxy)styrylbenzene in strong base: Probes for amyloid plaques in the brain. *J. Med. Chem.* 44(14):2270–2275, 2001.
52. C. A. Mathis et al. Synthesis and evaluation of 11C-labeled 6-substituted 2-arylbenzothiazoles as amyloid imaging agents. *J. Med. Chem.* 46(13):2740–2754, 2003.
53. M. P. Kung et al. IMPY: An improved thioflavin-T derivative for *in vivo* labeling of beta-amyloid plaques. *Brain Res.* 956(2):202–210, 2002.
54. L. Cai et al. Synthesis and evaluation of two 18F-labeled 6-iodo-2-(4′-N,N-dimethylamino)phenylimidazo[1,2-a] pyridine derivatives as prospective radioligands for beta-amyloid in Alzheimer's disease. *J. Med. Chem.* 47(9): 2208–2218, 2004.
55. Z. P. Zhuang et al. IBOX(2-(4′-dimethylaminophenyl)-6-iodobenzoxazole): A ligand for imaging amyloid plaques in the brain. *Nucl. Med. Biol.* 28(8):887–894, 2001.
56. N. Okamura et al. Styrylbenzoxazole derivatives for *in vivo* imaging of amyloid plaques in the brain. *J. Neurosci.* 24(10):2535–2541, 2004.
57. M. Ono et al. 11C-labeled stilbene derivatives as Abeta-aggregate-specific PET imaging agents for Alzheimer's disease. *Nucl. Med. Biol.* 30(6):565–571, 2003.
58. C. W. Lee et al. Dimethylamino-fluorenes: Ligands for detecting beta-amyloid plaques in the brain. *Nucl. Med. Biol.* 30(6):573–580, 2003.
59. T. Suemoto et al. *In vivo* labeling of amyloid with BF-108. *Neurosci. Res.* 48(1):65–74, 2004.
60. J. R. Barrio et al. PET imaging of tangles and plaques in Alzheimer's disease. *J. Nucl. Med.* 40(Suppl S):284, 1999.
61. E. D. Agdeppa et al. Binding characteristics of radiofluorinated 6-dialkylamino-2-naphthylethylidene derivatives as positron emission tomography imaging probes for beta-

amyloid plaques in Alzheimer's disease. *J. Neurosci.* 21(24):RC189, 2001.
62. K. Shoghi-Jadid et al. Localization of neurofibrillary tangles and beta-amyloid plaques in the brains of living patients with Alzheimer disease. *Am. J. Geriatr. Psychiatry.* 10(1):24–35, 2002.
63. E. D. Agdeppa et al. 2-Dialkylamino-6-acylmalononitrile substituted naphthalenes (DDNP analogs): Novel diagnostic and therapeutic tools in Alzheimer's disease. *Mol. Imaging Biol.* 5(6):404–417, 2003.
64. G. W. Small et al. Positron emission tomography scanning of cerebral amyloid and tau deposits in mild cognitive impairment. *N. Engl. J. Med.* 355:2652–2663, 2006.
65. H. Braak and E. Braak. Neuropathological stageing of Alzheimer-related changes. *Acta Neuropathol. (Berl.).* 82(4):239–259, 1991.
66. W. E. Klunk et al. Imaging brain amyloid in Alzheimer's disease with Pittsburgh compound-B. *Ann. Neurol.* 55(3):306–319, 2004.
67. N. P. Verhoeff et al. *In vivo* imaging of Alzheimer disease beta-amyloid with [11C]SB-13 PET. *Am. J. Geriatr. Psychiatry.* 12(6):584–595, 2004.
68. L. M. Smid et al. The 2,6-disubstituted naphthalene derivative FDDNP labeling reliably predicts Congo red birefringence of protein deposits in brain sections of selected human neurodegenerative diseases. *Brain Pathol.* 16(2):124–130, 2006.
69. V. Kepe et al. [F-18]MicoPET imaging of β-amyloid deposits in the living brain of triple transgenic rat model of β-amyloid deposition. *Mol. Imaging Biol.* 7(2):105, 2005.
70. J. L. Price and J. C. Morris. Tangles and plaques in nondemented aging and "preclinical" Alzheimer's disease. *Ann. Neurol.* 45(3):358–368, 1999.
71. V. Kepe et al. Serotonin 1A receptors in the living brain of Alzheimer's disease patients. *Proc. Natl. Acad. Sci. USA.* 103(3):702–707, 2006.
72. A. M. Wu and P. D. Senter. Arming antibodies: Prospects and challenges for immunoconjugates. *Nat. Biotechnol.* 23(9):1137–1146, 2005.
73. A. M. Wu et al. High-resolution microPET imaging of carcinoembryonic antigen-positive xenografts by using a copper-64-labeled engineered antibody fragment. *Proc. Natl. Acad. Sci. USA.* 97(15):8495–8500, 2000.
74. G. Sundaresan et al. 124I-labeled engineered anti-CEA minibodies and diabodies allow high-contrast, antigen-specific small-animal PET imaging of xenografts in athymic mice. *J. Nucl. Med.* 44(12):1962–1969, 2003.
75. J. Y. Wong et al. Pilot trial evaluating an 123I-labeled 80-kilodalton engineered anticarcinoembryonic antigen antibody fragment (cT84.66 minibody) in patients with colorectal cancer. *Clin. Cancer Res.* 10(15):5014–5021, 2004.
76. T. Olafsen et al. Optimizing radiolabeled engineered anti-p185HER2 antibody fragments for *in vivo* imaging. *Cancer Res.* 65(13):5907–5916, 2005.
77. T. Olafsen et al. MicroPET imaging of CD20 lymphoma xenografts using engineered antibody fragments. *J. Nuc. Med.* 47:33P, 2006.
78. J. V. Leyton et al. An anti–prostate stem cell antigen (PSCA) diabody for prostate cancer targeting. *Mol. Imaging Biol.* 8:91, 2006.
79. P. M. Smith-Jones et al. Imaging the pharmacodynamics of HER2 degradation in response to Hsp90 inhibitors. *Nat. Biotechnol.* 22(6):701–706, 2004.
80. M. Bergsneider et al. Metabolic recovery following human traumatic brain injury based on FDG-PET: Time course and relationship to neurological disability. *J. Head Trauma Rehabil.* 16(2):135–148, 2001.
81. M. Bergsneider et al. Dissociation of cerebral glucose metabolism and level of consciousness during the period of metabolic depression following human traumatic brain injury. *J. Neurotrauma.* 17:389–401, 2000.
82. P. Vespa et al. Metabolic crisis without brain ischemia is common after traumatic brain injury: A combined microdialysis and positron emission tomography study. *J. Cereb. Blood Flow and Metab.* 25(6):763–774, 2005.
83. H. M. Wu et al. Selective metabolic reduction in gray matter acutely following human traumatic brain injury. *J. Neurotrauma.* 21(2):149–161, 2004.
84. H. M. Wu et al. Subcorticle white matter metabolic changes remote from focal hemorrhagic lesions suggest diffuse injury following human traumatic brain injury (TBI). *Neurosurgery.* 55(6):1306–1317, 2004.
85. A. Shields et al. Imaging proliferation *in vivo* with [F-18]FLT and positron emission tomography. *Nat. Med.* 4(11):1334–1336, 1998.
86. A. Shields, P. T. Ho, and J. R. Grierson. The role of imaging in the development of oncologic agents. *J. Clin. Pharmacol.* 39(Suppl):40–44, 1999.
87. J. Toyohara et al. Basis of FLT as a cell proliferation marker: Comparative uptake studies with [3H]thymidine and [3H]arabinothymidine, and cell-analysis in 22 asynchronously growing tumor cell lines. *Nucl. Med. Biol.* 29(3): 281–287, 2002.
88. W. Chen et al. Imaging proliferation in brain tumors with 18F-FLT PET: Comparison with 18F-FDG. *J. Nucl. Med.* 46(6):945–952, 2005.
89. D. C. Cobben et al. Is 18F-3′-fluoro-3′-deoxy-L-thymidine useful for the staging and restaging of non–small cell lung cancer? *J. Nucl. Med.* 45(10):1677–1682, 2004.
90. D. A. Mankoff, A. F. Shields, and K. A. Krohn. PET imaging of cellular proliferation. *Radiol. Clin. North Am.* 43(1):153–167, 2005.
91. D. Mankoff, F. Dehdashti, and A. F. Shields. Characterizing tumors using metabolic imaging: PET imaging of cellular proliferation and steroid receptors. *Neoplasia.* 2(1):71–88, 2000.

92. S. Huang et al. Post-irradiation temporal changes in glucose metabolism and cell proliferation in implanted murine tumors as measured by FDG and FLT PET. *J. Nucl. Med.* 43 suppl:25P, 2002.
93. M. Pan et al. Evaluation of dose response cell proliferation changes in mouse tumors after irradiation: Comparison between image-derived analysis modalities. *J. Nucl. Med.* 47:199P, 2006.
94. M. H. Pan et al. Kinetics of F-18-FLT and F-18 FDG in mouse tuors after irradiation. *J. Nucl. Med.* 46:383P, 2005.
95. H. W. Strauss. *Cardiovascular Nuclear Medicine.* Mosby. 442, 1980.
96. M. C. Kreissl et al. Non-invasive measurement of cardiovascular function in mice with ultra-high temporal resolution small animal PET. *J. Nucl. Med.* 47:974–980, 2006.
97. H. M. Wu et al. First-pass angiography in mice using FDG-PET—A new method of deriving the cardiovascular transit time without the need of region of interest drawing. *IEEE Trans. Nucl. Sci.* 52:1311–1315, 2005.
98. F. Wiesmann et al. Dobutamine-stress magnetic resonance microimaging in mice: Acute changes of cardiac geometry and function in normal and failing murine hearts. *Circ. Res.* 88:563–569, 2001.
99. F. J. Beekman and B. Vastenhouw. Design and simulation of a high-resolution stationary SPECT system for small animals. *Phys. Med. Biol.* 49:4579–4592, 2004.
100. S. R. Meikle et al. A prototype coded aperture detector for small animal SPECT. *IEEE Trans. Nucl. Sci.* 49(5):2167–2171, 2002.
101. S. R. Meikle et al. *Performance Evaluation of a Multipinhole Small Animal SPECT System.* IEEE MIC conference record, 2003.
102. F. J. Beekman et al. U-SPECT-I: A novel system for submillimeter-resolution tomography with radiolabeled molecules in mice. *J. Nucl. Med.* 46(7):1194–1200, 2005.
103. H. M. Wu et al. An integrated microfluidic blood sampler for determination of blood input function in quantitative mouse microPET studies. *IEEE MIC, 2005.*
104. H. M. Wu et al. Application of integrated microfluidics to small animal positron emission tomography to obtain fully quantitative *in vivo* imaging. *J. Nucl. Med.* 47:55P, 2006.
105. M. Adler et al. Detection of femtogram amounts of biogenic amines using self-assembled DNA-protein nanostructures. *Nature Methods.* 2:147–149, 2005.
106. D. N. Breslauer, P. J. Lee, and L. P. Lee. Microfluidics-based systems biology. *Mol. BioSyst.* 2:97–112, 2006.
107. J. Mazziotta et al. A probabilistic atlas of the human brain: Theory and rationale for its development. *NeuroImage.* 2:89–101, 1995.
108. A. Evans et al. Three-dimensional correlative imaging: Applications in human brain mapping. In R. Thatcher et al. (Eds.). *Functional Neuroimaging: Technical Foundations.* 145–162, 1994.
109. K. Friston et al. Spatial registration and normalization of images. *Hum. Brain Mapp.* 2:165–189, 1995.
110. J. Ashburner and K. J. Friston. Nonlinear spatial normalization using basis function. *Hum. Brain Mapp.* 7:254–266, 1999.
111. S. Minoshima et al. Stereotactic PET atlas of the human brain: Aid for visual interpretation of functional brain images. *J. Nucl. Med.* 35:949–954, 1994.
112. S. Minoshima et al. Anatomic standardization: Linear scaling and nonlinear warping of functional brain images. *J. Nucl. Med.* 35:1528–1537, 1994.
113. R. P. Woods et al. Automated image registration I. General methods and intrasubject, intramodality validation. *J. Compt. Assist. Tomogr.* 22:141–154, 1998.
114. K. P. Lin et al. A general technique for inter-study registration of multifunction and multimodality images. *IEEE Trans. Nucl. Sci.* 41:2850–2855, 1994.
115. S. G. Nekolla et al. Reproducibility of polar map generation and assessment of defect severity and extent assessment in myocardial perfusion imaging using positron emission tomography. *Eur. J. Nucl. Med.* 25:1313–1321, 1998.
116. S. C. Huang et al. An Internet-based Kinetic Imaging System (KIS) for MicroPET. *Mol. Imaging Biol.* 7:330–341, 2005.

22

From Telemedicine to Ubiquitous M-Health: The Evolution of E-Health Systems

Dr. Dejan Rašković,[1]
Dr. Aleksandar Milenković,[2]
Prof. Piet C. De Groen,[3] and
Dr. Emil Jovanov[2]
[1]*University of Alaska*
[2]*University of Alabama, Huntsville*
[3]*Mayo Clinic*

22.1 Introduction

Existing health care systems are designed to react on illness and are optimized to manage illness. The widespread use of communication and information technologies has facilitated the delivery of medical services at a distance, which is known as *telemedicine* [1]. Ranging from tele- and videoconferencing to robotic surgery, telemedicine has extended the reach of medical services from elite medical institutions to remote villages in Finland and isolated Greek islands. New approaches have also forced changes in clinical practices, the most notable being the introduction of electronic medical records and information and communications technology. This new paradigm is known as *eHealth* [2, 3].

Recent developments in sensors, wearable computing, and ubiquitous communications have the potential of providing clinicians and users with tools and environments to gather physiological data over extended periods of time. This emerging concept is known as *m* (mobile)-*health* and represents the evolution of eHealth systems from traditional desktop telemedicine platforms to wireless and mobile configurations [4–6].

The main enabling technological trends for m-health systems include:

- Increased communication and computation capabilities of cell phones
- The new generation of power-efficient processors and communication controllers
- Revolutionary changes in microelectromechanical systems (MEMSs) and nano-sensor technologies enabling embedded and implanted biomedical sensors in frequently used objects in homes and offices

This chapter outlines enabling technologies and the taxonomy of m-health applications and introduces the ultimate concept in unobtrusive system organization for "anytime, anywhere" monitoring—wireless body area networks (WBANs) of intelligent wireless sensors. We discuss system integration and implementation issues, future trends, and possible applications.

22.2 Overview of M-Health Systems

22.2.1 Introduction

We divide m-health systems into two broad categories: WBAN-based systems and systems based on "smart clothes" (Section 22.2.3). However, as the section about smart clothes will discuss, these do not represent two completely disjointed sets of systems. We also present examples of wearable biomedical sensors to provide enough insight into different types of sensors and to serve as an illustration of our taxonomy.

Examples of m-health systems and sensors are grouped according to the medical condition they apply or cannot apply to. We will discuss current and past commercial and research projects related to mobile monitoring of health conditions. Wireless telemetry has been available for a few decades, but wireless intelligent sensors capable of real-time signal processing have been developed only recently. Therefore, most of the work related to wireless intelligent sensors that will be discussed in this section represents research projects.

22.2.1.1 Cardiopulmonary Monitoring

Medtronic offers the Reveal® Plus Insertable Loop Recorder [78], developed in collaboration with the Division of Cardiology, University of Western Ontario. It provides up to 14 months of monitoring and data acquisition of critical cardiac events. Up to 40 minutes of history can be stored after an episode. This device weighs 17 g, with an approximate volume of 8 mL. A previously recorded episode can be uploaded on demand to the computer for analysis.

The implantable EndoSure MEMS blood pressure sensor from CardioMEMS [9, 10] was originally developed at Georgia Tech. It was the first implantable pressure sensor that combined wireless and MEMS technology to receive approval from the U.S. Food and Drug Administration (FDA). The device is implanted during aneurysm repair; at the same time that a graft is placed in the aneurysm sac, the sensor is inserted into the sac, from within which it will take pressure readings; the readings can be transmitted from the sensor to an external device using radiofrequency (RF) scavenging techniques.

Scientists at the d'Arbeloff Laboratory for Information Systems and Technology at the Massachusetts Institute of Technology (MIT) have developed a ring sensor that continuously monitors heart rate using a photoplethysmograph (PPG) signal and sends data wirelessly to a host computer [11]. Shaped like a ring, the device can be worn on a finger.

Researchers at MIT and Massachusetts General Hospital have developed a behind-the-ear PPG-based sensor that uses a modified hydrostatic-based oscillometric method. It employs a MEMS accelerometer to reliably measure height [12]. Philips Research Europe has developed a system based on the 802.15.4 standard of the Institute of Electrical and Electronics Engineers (IEEE) that enables continuous cuffless blood pressure estimation using an electrocardiography (ECG) dry sensor worn on the waist and a behind-ear PPG sensor [13].

Researchers at the University of Alabama in Huntsville developed a system for stress level assessment based on heart rate variability measurements [14]. The system performs synchronous measures of individual heart rate during prolonged stressful training. Data are stored locally (for up to 60 hours) and collected wirelessly from the entire group of users using mobile gateways.

22.2.1.2 Diabetes Control

A typical example of a commercial system is Symphony™ Diabetes Management [15] from Sontra Medical Corporation. Sontra offers a patch sensor that will continuously extract interstitial fluid, draw the analytes into the sensor, and measure and calculate the blood glucose concentration. The results are calculated and wirelessly sent to the receiver every 3.8 seconds. Currently, the system is used for only glucose measurements, but the company plans to add sensors to measure other analytes as well. DexCom [16] and Medtronic [17, 18], among others, currently offer systems based on subcutaneous sensors that can be worn for up to 72 hours before replacement is needed. DexCom also has reported results for patients with surgically implanted long-term glucose level sensors.

A "skin breakdown detection" device is intended for use by people suffering from diabetes [19]. The device is worn in the shoe and records temperature, pressure, and humidity under the heel and metatarsal heads. The data are periodically evaluated offline to detect abnormal conditions that may lead to skin breakdown; the goal is to prevent formation of foot ulcers, which in a patient with advanced diabetes may lead to amputation.

Medtronic MiniMed is an example of efforts to develop an artificial pancreas for diabetes patients. The first step in the development is an insulin pump (e.g., MiniMed Paradigm®) that disperses insulin based on the results of blood glucose measurements [18].

LG Electronics and Healthpia [20] have developed the LG KP8400 cell phone that features a built-in glucose monitor. The cell phone is currently commercially available in some countries in Asia and is awaiting approval from the FDA. The user places a drop of blood on a test strip and inserts the strip into a slot in the cell phone. The glucose measurement is automatically sent to a caregiver.

22.2.1.3 Brain and Muscle Activity Recording/Stimulation

Researchers from the University of Washington, Caltech, and Case Western Reserve University have developed an implantable microcomputer [21] capable of recording nerve and muscle signals from small animals during their normal activity. They use flexible metallic needles to collect signals from nerve bundles and micromachined silicon probes to record activity of neural assemblies. However, to collect signals from individual neurons, scientists from the University of Washington are working on silicon MEMS probes that will mimic the performance of glass capillaries used on constrained animals. The implantable device consists of variable-gain amplifiers, a system-on-a-chip microcontroller, and a high-density memory. In this project, researchers decided to avoid antennas and charge pumps needed for RF-powered devices, due to size constraints of the implantable device. Instead, they plan to employ thin-film batteries.

At the University of Michigan's Center for Wireless Integrated Microsystems, researchers developed a BiCMOS (bipolar complementary metal-oxide semiconductor) wireless stimulator chip [22], to be used in conjunction with micromachined passive stimulating microprobes. This design allows wireless and stand-alone operation for unlimited time, since the chip uses a 4 MHz carrier signal to receive both data and power through inductive coupling. Total power dissipation of the chip is less than 10 mW, and its surface area is about 13 mm^2.

Another example of an RF-powered intelligent sensor is a miniature implantable wireless neural recording device [23, 24] developed at the University of California, Los Angeles. This device records and transmits neural signals. The device size is less than 1 cm^2, and power dissipation has been measured at 13.8 mW. Tests have shown that the transmitting range is up to 0.5 m and that the demodulated signal is highly correlated with the original signals in the range between 5 and 1.5 mV.

Medtronic has developed the Activa Therapy Deep Brain Stimulator, a surgically implanted device similar to a cardiac pacemaker, to block the brain signals associated with dystonia, Parkinson's disease, and essential tremor. It delivers carefully controlled electrical stimulation to targeted areas within the brain [25].

22.2.1.4 Gastrointestinal Monitoring

Researchers from the Universities of Glasgow, Edinburgh, and Strathclyde are developing a capsule [26] traversing the gastrointestinal tract (part of the Integrated Diagnostics for Environmental and Analytical Systems [IDEAS] project). The capsule-based sensor gathers data that cannot be collected using traditional endoscopy. The device is battery powered and integrates sensors, processing, and RF bidirectional communication onto a single piece of silicon (current device size is 32×11.5 mm) [27, 28].

Given Imaging offers the commercially available Given® Diagnostic System [29, 30]. A disposable imaging capsule is swallowed by a patient and passes through the gastrointestinal tract while wirelessly transmitting images to a receiver worn on a belt. The images are received through an array of antennas; the antennas are used also to determine the exact location of the capsule.

22.2.1.5 Heterogeneous Sensor Systems

The U.S. Army Research Institute of Environmental Medicine (USARIEM) and the U.S. Army Medical Research and Materiel Command (USAMMRC) led the Warfighter Physiological Status Monitoring (WPSM) project [31, 32]. The experimental system in development for that project includes sensors for heart rate, metabolic energy cost of walking, core and skin temperatures, geolocation, and activity/inactivity. Data collected by various sensors are transmitted wirelessly to a hub (worn on a soldier's belt) through a low-power personal area network (PAN). Sensors are expected to be low-cost and disposable, capable of collecting data for up to a few weeks. Aggregated data can be stored or forwarded to a warfighter's digital fighting system, command center, or, in the future, the Internet. The final system is expected to be able to predict the critical aspects of a soldier's performance under extreme conditions.

The careTrends™ system, offered by Sensitron, uses a combination of Bluetooth and IEEE 802.11b transmission to send patient vital signs data from point of care to a server [34]. Currently, the company offers monitoring of blood pressure, pulse, temperature, weight, oxygen saturation, and respiration rate; measurements are uploaded wirelessly to a Patient Communication Unit. The caregiver can use the handheld unit to input pain scores, view and manage test results, and communicate with a careTrends access point.

Cleveland Medical Devices Inc. markets Crystal Monitor [35] as a lightweight programmable wireless physiological monitor, capable of viewing and recording electroencephalography (EEG), ECG, electromyography (EMG), electro-oculography (EOG), pulse oximetry oxygen saturation (SpO_2), and other signals. Collected data are wirelessly transferred to a personal computer (PC) up to 50 feet away, using the 2.4 GHz Industry, Science, and Medicine (ISM) band. The device can operate continuously for up to 12 hours on two AA batteries. In addition, it uses a removable secure digital (SD) card to store over 60 hours of patient data for unattended monitoring.

Equivital Limited has developed the Equivital™ system for continuous monitoring and storage of physiological life signs, to be used by the military, emergency services, first responders, performance sports, and general health care. The system allows for real-time or offline analysis of the data and incorporates the sensors for monitoring heart rate, respiratory rate, user's motion and position, temperature, and G shocks caused by falls and heavy impacts. It also provides a rudimentary cognitive response from the user to assess the user's consciousness and awareness [36].

The 3G wireless cellular data system can be used for direct transmission of all patient data (video, medical images, ECG signals, etc.) [37].

22.2.2 Taxonomy of M-Health Systems

We introduce two taxonomy groups. One classifies personal medical devices based upon their usage, while the other deals with their implementation. The taxonomies allow us to abstract from a particular application and devise design principles that hold for all applications that fall within a given category. Our taxonomy groups will include some devices not covered in the examples listed in the previous section. The functionality of these devices is either well known or goes beyond the scope of this paper—defibrillators/pacemakers, hearing aid devices, artificial hearts, artificial limbs, artificial eyes, drug pumps, etc. We include them in the taxonomy for completeness and to show possible areas for future medical applications of wearable computing.

22.2.2.1 Description of Taxonomy of Usage

Our taxonomy of usage for personal medical devices is shown in Figure 22.1. The taxonomy places devices in a 2D space defined by two axes: capability and mode of wearability. A third axis could also be introduced, to address duration of the required service. We see four broad categories along the capability axis: recording/transmitting, processing, correcting, and replacing. We define these functions as follows:

- *Recording/transmitting:* Devices that store or send relevant signals and data from the patient but do not evaluate the signals (except for signal conditioning) in any manner or provide feedback to the patient. The signals are evaluated offline.
- *Processing:* Devices that process relevant signals and provide immediate feedback to the patient about his or her current condition. This feedback may or may not be continuous, as in the case of an ECG monitor that provides alerts of impending cardiac events. These processing devices may also store the signals so that they may be further processed offline, just like the recording devices.
- *Correcting:* Devices that provide appropriate stimuli directly to a malfunctioning organ in order to correct its behavior.
- *Replacing:* Devices that replace an organ entirely (prostheses).

There may be another category in the future, somewhat related to the correcting category, but going a step further: Devices that train the body in some way but then can be removed, analogous to braces for the teeth. For the purpose of this taxonomy, we treat as implanted all the devices that are inside the user's body, even if they were not inserted surgically (i.e., even if they were swallowed, inserted subcutaneously by the user, etc.). In the not-too-distant future, devices may be small enough to be introduced into the body by other means, such as inhalation. Some devices straddle the border between categories. (For more information, see Raskovic et al. [33].)

22.2.2.2 Description of Taxonomy of Implementations

The taxonomy of implementations classifies systems according to the system mobility (represented in Figure 22.2). It characterizes systems along two axes: patient mobility and gateway availability. We use this taxonomy to emphasize the fact that the development of medical monitoring equipment constantly goes in the direction of increased patient mobility and the flexibility of positioning of external monitoring equipment. In the past, patient monitoring has been performed in hospitals or labs with patients strapped to the fixed monitoring system and not able to move about freely. With technological advances, wearable monitors allow patients to walk around the hospital. The ultimate goal would be normal patient mobility while patients are monitored as they go about their everyday routine with the miniaturized monitoring equipment concealed on their persons.

In most modern systems, data are not only provided to the user, but at the same time forwarded to a hospital information or telemedical system. Therefore, it is necessary to have access points (or gateways) to medical networks. One of the main

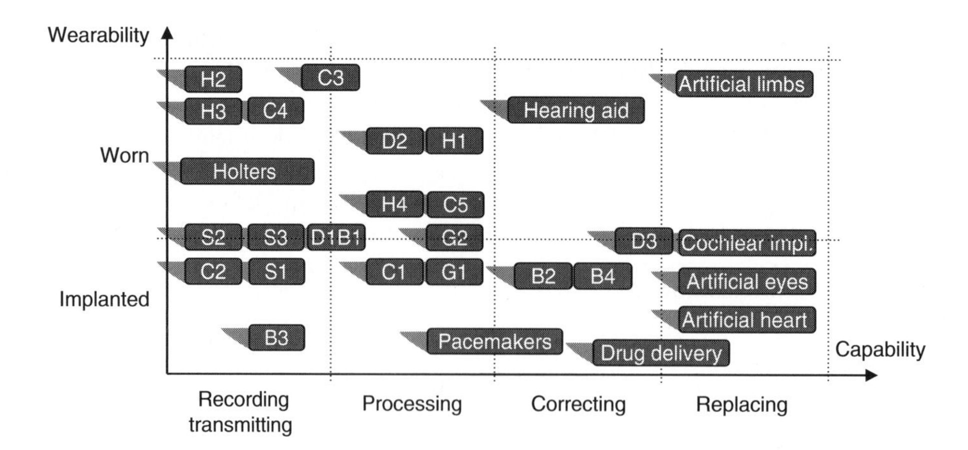

FIGURE 22.1 Taxonomy of usage. Adapted from Raskovic et al. [33].

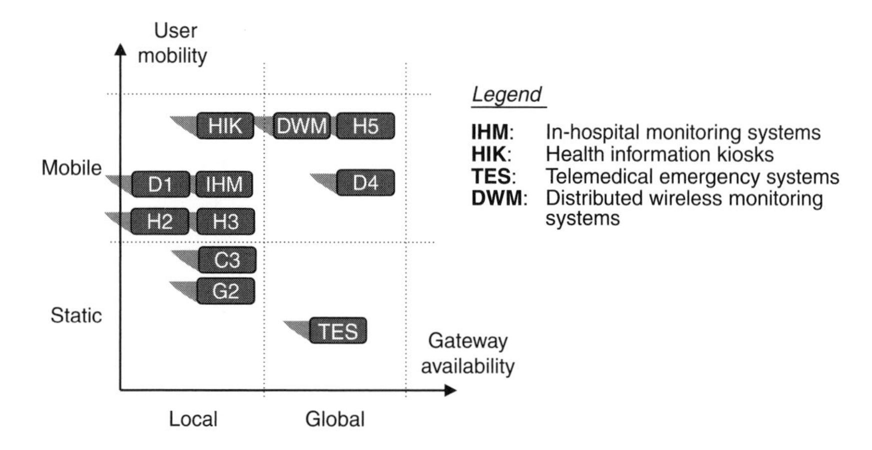

FIGURE 22.2 Taxonomy of implementations. Adapted from Raskovic et al. [33].

system design issues is availability of these gateways, which, depending on the application, can be implemented as networks of fixed gateways within a medical institution, health kiosks throughout the city, or global access points, in the case of satellite-based systems. The smaller the size of the monitoring device, the more limited will be the available power for signal transmission. One design solution is to keep communication distances short but to allow access points to move around and connect with individual monitors. Alternatively a limited set of access points are available to monitor devices that move in and out of reach of the access points. Such solutions keep power consumption by the monitor low, at the cost of delayed data delivery.

We define two broad categories along each of the axes of patient mobility and gateway availability: The patient can be either *mobile* or *static*, while the gateway availability can be either *local* or *global*. A moving gateway or a group of local gateways will be treated as a global gateway. The ultimate goal of ubiquitous personal health monitoring systems is maximal user mobility. Using these axes, we divide systems into four categories and provide some typical examples:

- *Local gateway/static patient* (LGSP) is typical of older monitoring systems, such as fixed bedside hospital monitors.
- *Local gateway/mobile patient* (LGMP) is usually used in wireless in-hospital or home monitoring systems and allows patient mobility within the range of the network of access points. Another possible future application of LGMP is in terms of health information kiosks that could be used to collect data from personal monitors at different locations in town.
- *Global gateway/static patient* (GGSP) is exemplified by an emergency response vehicle globally connected with a medical network, while patients are incapacitated.
- *Global gateway/mobile patient* (GGMP) allows patients to move freely over large areas, made possible by distributed wireless monitoring systems. These systems employ either cell phone infrastructures or mobile gateways as access points.

22.2.3 Smart Clothes

Smart clothes are extensions to networks of independent wireless medical sensors. Smart clothes are not yet widely employed, but once current problems with fabrication and usability are solved and their costs come down, wider acceptance is expected. Smart clothes were first applied for monitoring of patients, athletes, and high-risk workers, but applications for everyday life followed very soon afterward [38].

Smart clothes can be used to detect biomechanical (e.g., respiration, body movement, posture monitoring), bioelectrical (ECG, EMG, EEG, etc.), temperature, and other parameters. The IEEE Engineering in Medicine and Biology Society (EMBS) Technical Committee for Wearable Biomedical Sensors and Systems [39] considers smart clothes the core of a wearable biomedical system, because they are convenient, personal, and in close proximity to the source of most biomedical signals. In addition, they can be worn with little chance of disclosing the possible medical conditions of their users.

In their simplest form, smart clothes provide only interconnectivity among sensors, electrodes, and external electronics. However, if smart clothes are to become truly wearable m-health systems, electronics need to be embedded into clothing as well. Ideally, an entire smart clothes system should wirelessly communicate with electronic devices that the wearers would normally use. For example, a smart phone can be used to display the current state of the smart clothes system, issue warnings, and relay information to higher tiers of a medical system.

Several major challenges need to be addressed successfully before smart clothes can be widely employed:

- *Wearability*. Many of the characteristics that define wearability (unobtrusiveness, stretchability, washability, etc.)

are in direct conflict with the requirements for the increased functionality of smart clothes (more sensors embedded in clothing, integrated processing and display, etc.).

- *Interconnectivity.* Sensors, electrodes, and conductive yarns are sometimes built in different technologies. This can create difficulties in interfacing them electrically and can also compromise the flexibility of the yarn because of differences in mechanical properties.
- *Motion and other artifact suppression.* Signal integrity is crucial in any medical system and becomes even more important in the case of wearable systems. Smart clothes, by their very nature, are prone to different types of artifacts (interference of other personal electronic devices, difficulties with propagation of signals near and through the human body, etc.), among which motion artifacts are especially prominent. Motion and other artifacts can be reduced or completely removed through careful design of sensors and interconnections, sensor redundancy, signal conditioning, and signal processing. However, requirements for increased signal integrity often conflict with requirements related to acceptable levels of comfort while wearing smart clothes.

A number of commercial products are available, such as the Sensatex SmartShirt System [40, 41], the VivoMetrics LifeShirt System [42], and Information Society Technologies' WEALTHY (WEarable heALTH care sYstem) [43]. Universities are also performing a range of research activities in this area [44–48].

22.3 M-Health Based on Wireless Body Area Networks

22.3.1 General Concepts

Recent technological advances in sensors, integrated circuits, and wireless networking facilitate wireless sensor networks that are deeply embedded in their native environments. Wireless sensor networks are highly suitable for many applications, such as habitat monitoring [49], machine health monitoring and guidance, traffic pattern monitoring and navigation, plant monitoring in agriculture [50], and infrastructure monitoring. The current technological and economic trends will enable new generations of wireless sensor networks with more compact and lighter sensor nodes, more processing power, and more storage capacity. In addition, the ongoing proliferation of wireless sensor networks across many application domains will result in a significant cost reduction.

One of the most promising application domains is health monitoring [51], and within health care, WBANs in particular are emerging as promising enabling technologies to implement m-health. A WBAN for health monitoring consists of multiple sensor nodes that can measure and report the user's physiological state. A WBAN for health monitoring may also feature active devices for control of the user's physiological state—for example, some WBAN nodes may be responsible for drug delivery. These sensor nodes are strategically placed on the human body. The exact location and attachment of the sensor nodes on the human body depend on the sensor type, size, and weight. Sensors can be worn as stand-alone devices or can be built into jewelry, applied as tiny patches on the skin, hidden in the user's clothes or shoes, or even implanted in the user's body. Each node in the WBAN is typically capable of sensing, sampling, processing, and wirelessly communicating one or more physiological signals. The exact number and type of physiological signals to be measured, processed, and reported depends on end-user application and may include a subset of the following physiological sensors:

- An ECG sensor for monitoring heart activity
- An EMG sensor for monitoring muscle activity
- An EEG sensor for monitoring brain electrical activity
- A PPG sensor for monitoring pulse and blood oxygen saturation
- A cuff-based pressure sensor for monitoring blood pressure
- A resistive or piezoelectric chest belt sensor for monitoring respiration
- A galvanic skin response (GSR) sensor for monitoring autonomous nervous system arousal
- A blood glucose level sensor
- A thermistor for monitoring body temperature

In addition to these sensors, a WBAN for health monitoring may include sensors that can help determine the user's location, discriminate among the user's states (e.g., lying, sitting, walking, running), or estimate the type and level of the user's physical activity. These sensors typically include the following:

- A localization sensor (e.g., a global positioning system [GPS])
- A tilt sensor for monitoring trunk position
- A gyroscope-based sensor for gait-phase detection
- Accelerometer-based motion sensors on extremities to estimate type and level of users' activities
- A "smart sock" or an insole sensor to count steps and/or delineate phases and distribution of forces during individual steps

Environmental conditions may often influence the user's physiological state (e.g., it has been shown that blood pressure may depend on the subject's ambient temperature) or accuracy of the sensors (e.g., background light may influence the readings from PPG sensors). Consequently, WBANs may benefit from integrating the third group of sensors, which provide information about environmental conditions such as humidity, light, ambient temperature, atmospheric pressure, and noise.

All technological trends and the ability to measure a wide variety of physiologically important signals indicate that WBANs are well positioned to become a key component in providing continual, unobtrusive, and affordable monitoring in health care.

22.3.2 System Architecture and Organization

Typically, a WBAN will form the lowest tier (Tier 1) of a multitiered medical information system for health monitoring. Figure 22.3 illustrates a general system architecture of a medical monitoring information system that includes a personal server at Tier 2 and a series of medical servers at Tier 3. The exact system architecture and the number of system tiers depend predominantly on target applications, available infrastructure, and type and number of users. Though we focus on the health monitoring system described in Figure 22.3, we will identify possible alternatives to this type of system organization.

The WBAN in Figure 22.3 includes one heart sensor and two motion sensors, one attached at a wrist and the other to an ankle. One possible target application for such a WBAN is fitness monitoring—helping to track duration, type, and intensity of regular daily exercise. A similar system can be used for monitoring of cardiac patients during a rehabilitation period at home. The heart sensor can operate in multiple modes reporting either (1) a raw ECG signal (from one or multiple channels), (2) time-stamped heart beats, or (3) averaged heart rate over a certain period of time. The motion sensors, each equipped with a 3D accelerometer, can also operate in several modes, reporting either (1) raw acceleration signals for *x*-, *y*-, and *z*-axes, (2) extracted features (e.g., time-stamped steps or phases of a step), or (3) an estimated level of activity (e.g., activity-induced energy expenditure [AEE] over a certain period of time). The sensor nodes (together with a network coordinator) attached to a personal server compose the WBAN. Upon configuration, the WBAN continually performs sensing, sampling, and signal processing. Sensors wait for command and control messages from the WBAN coordinator and report continual sensor readings or events of interest as they occur.

Tier 2 encompasses the personal server, which is responsible for a number of tasks, providing a transparent interface to the wireless sensor nodes, an interface to the user, and an interface to the medical server. The interface to the WBAN includes network configuration and management. Network configuration encompasses the following tasks: sensor node registration (type and number of sensors), initialization (e.g., specifying sampling frequency and mode of operation), customization (e.g., running user-specific calibration or user-specific signal-processing-procedure upload), and setup of a secure communication (security key exchange). Once the WBAN network is configured, the personal server manages the network and takes care of channel sharing, time synchronization, data retrieval and processing, and fusion of the data. Based on synergy of information from multiple physiological, location, activity, and environmental sensors, the personal server can determine users' states and their health status; in addition, the personal server can provide feedback through a user-friendly and intuitive graphical or audio user interface. Finally, if a communication channel to the medical server is available, the personal

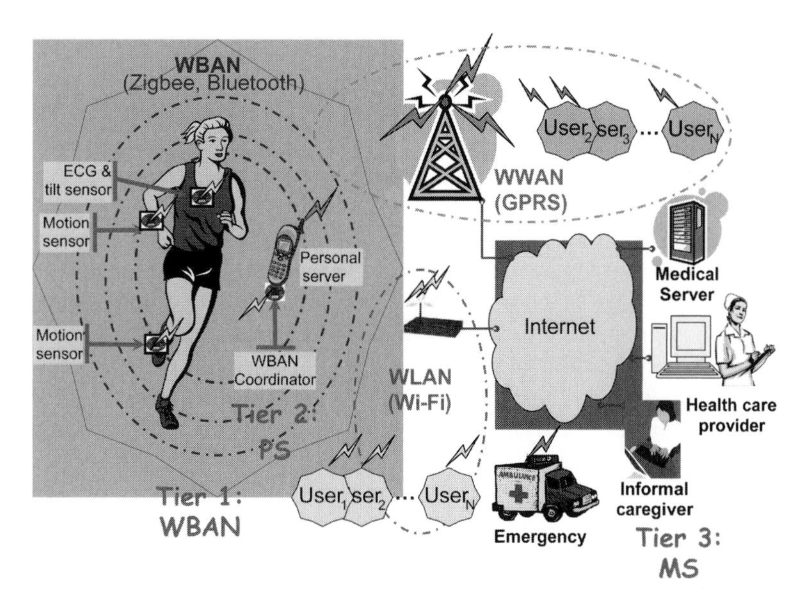

FIGURE 22.3 A multitiered health monitoring system based on WBAN. PS = personal server; MS = medical server; GPRS = general packet radio service.

server can establish a secure link to the medical server and send condensed or detailed reports about users' health status. These reports can be processed, displayed, and integrated into users' medical records. However, if a link between the personal server and the medical server is not available, the personal server should be able to store the data locally and initiate uploads when a link becomes available. Depending on the use scenario, the personal server can run on a smart phone (as illustrated in Figure 22.3), on a personal digital assistant (PDA) enabled by a wireless wide area network (WWAN), or on a home PC.

Tier 3 includes a medical server accessed via the Internet. In addition to the medical server, the last tier may encompass other servers, such as informal caregivers, commercial health care providers, and even emergency services. The medical server keeps electronic medical records of registered users and provides various services to the users, medical personnel, and informal caregivers. It is the responsibility of the medical server to authenticate users, accept health monitoring session uploads, format and insert the session data into corresponding medical records, analyze the data patterns, recognize serious health anomalies in order to contact emergency caregivers, and forward new instructions to the users, such as physician-prescribed exercises. Patients' physicians can access the data from their offices via the Internet and examine the data to ensure that the patients are within expected health metrics (in terms of heart rate, blood pressure, activity) and that they are responding to a given treatment or performing prescribed exercises. A server agent may inspect the uploaded data and create an alert in the case of a potential medical condition.

The large amount of data collected through these services can also be utilized for knowledge discovery through data mining. Integration of the collected data into research databases and quantitative analysis of conditions and patterns likely will prove invaluable to researchers trying to link symptoms and diagnoses with historical changes in health status, physiological data, or other parameters (e.g., gender, age, weight). In a similar way, a WBAN–personal server–medical server infrastructure could significantly contribute to monitoring and study of drug therapy effects.

22.3.3 Applications of Wireless Body Area Networks

WBANs can be used in a number of applications, from fitness/exercise monitoring of healthy users, to monitoring of patients with chronic or impeding medical conditions in hospitals and ambulatory settings, to early detection of disease, to emergency care. Table 22.1 lists medical conditions and their corresponding relevant physiological signals.

For each medical condition, a series of WBAN solutions can be devised; it is not our intention to cover a broad series of medical conditions. Instead, we opt to present a hypothetical case study of a representative condition in a patient recovering from a heart attack, to illustrate the usefulness of WBAN-based health monitoring systems. We discuss many common problems that patients face after a heart attack and describe how our system can be used to address these problems; in addition, we will show how WBANs would provide advantages over typical present-day solutions.

TABLE 22.1 Medical conditions and suggested minimal configurations of wireless body area network (WBAN)

Medical condition	WBAN Sensors
Cardiac arrhythmias/heart failure	Heart rate/ECG, blood pressure, activity
Asthma	Respiration rate, peak flow, oxygen saturation
Cardiac rehabilitation	Heart rate/ECG, activity, environmental sensors
Postoperative rehabilitation	Heart rate/ECG, temperature, activity
Diabetes	Blood glucose level, activity, temperature
Obesity/weight loss programs	Heart rate, smart scale, activity (accelerometers)
Epilepsy	EEG, gait (gyroscope, accelerometers)
Parkinson's disease	Gait, tremor, activity (gyroscope, accelerometers)

22.3.3.1 Case Example

Peter Petrovich is recovering from a heart attack. After release from the hospital, he attended supervised cardiac rehabilitation for several weeks. His recovery process is going well, and Peter is to continue a prescribed exercise regimen at home. However, the unsupervised rehabilitation at home does not go well for Peter. He does not follow the exercise regimen as prescribed. He exercises but does not truthfully disclose to the treating health care providers the minimal intensity and duration of his exercise. As a result, Peter's recovery is slower than expected, which raises concerns among his health care providers about his health status: Is the damage to Peter's heart greater than initially suspected, or does he not follow medical advice? His physician has no quantitative way to verify his adherence to the exercise program.

A WBAN-based health monitoring system offers a solution for Peter and all persons undergoing cardiac rehabilitation at home, as well as for the health care providers. Peter is equipped with a WBAN-based ambulatory health monitoring system. Tiny electronic inertial sensors measure movement on extremities and the number of steps Peter takes, while electrodes on the chest measure Peter's heart activity. The WBAN provides continual reporting of heart rate and AEE. The time and duration of his normal and exercise activities are recorded, and the level of intensity of the exercise can be determined by calculating an estimate of energy expenditure from the motion sensors. The information is available on Peter's smart phone, which acts as his personal server. The personal server may also assist Peter in his exercise efforts: It may alert him that he has not initiated or is not reaching his intended goals,

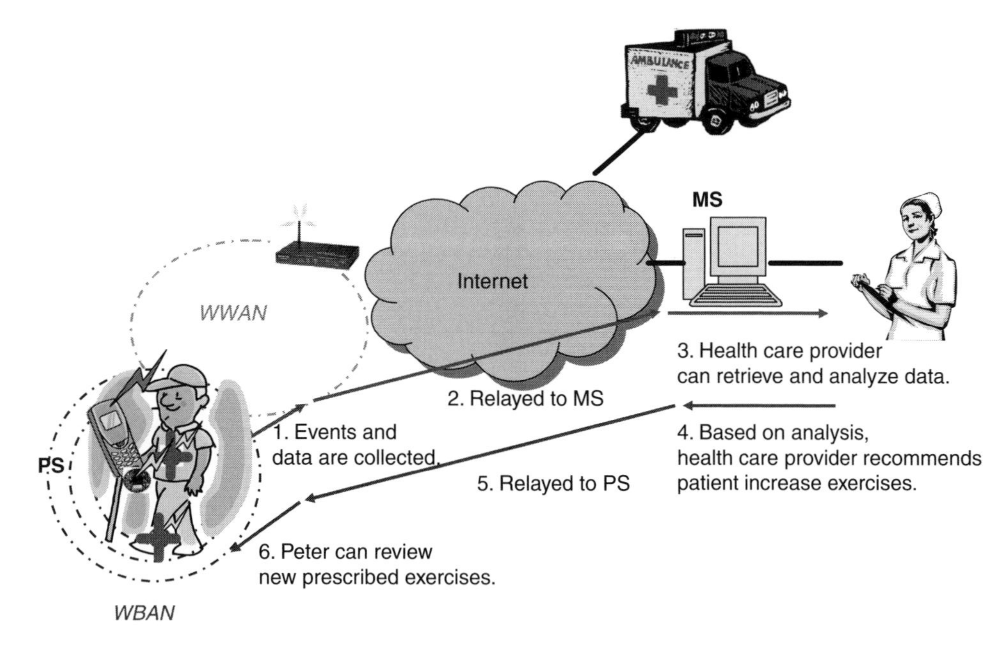

FIGURE 22.4 Example of data flow in the proposed WBAN health care monitoring system.

or it may generate warnings in case of excessive exercise (e.g., heart rate above the maximum threshold for a person of his age, weight, and condition).

Through the Internet or cell phone–connected server, his health care providers can collect and review all data, verify that Peter is exercising regularly, issue new prescribed exercises, adjust data threshold values, and schedule office visits. Peter's description of his progress continues to be important, but his health care providers no longer need to rely on only subjective descriptions. Instead, they have an objective and quantitative dataset of his level and duration of exercise. In addition, Peter's parameters of heart rate variability provide a direct measure of his physiological response to the exercise, serving as an in-home stress test. Substituting these remote stress tests and data collection for in-office tests, Peter's health care providers reduce the number of office visits. This decreases health care costs and makes better use of the health care providers' time. In urgent cases, however, the personal server can directly contact Emergency Medical Services (EMS) if the user subscribes to this service. Figure 22.4 illustrates one possible data flow.

22.4 Wireless Intelligent Sensors for M-Health

Each WBAN sensor node typically performs four basic tasks:

1. Sensing and sampling of relevant physiological or environmental signals
2. Digital signal processing of input signals (e.g., filtering, feature extraction, data compression)
3. On-sensor data buffering
4. Wireless communication with the personal server

Consequently, a WBAN node encompasses the following physical resources, shown in Figure 22.5:

- Sensor devices
- Signal conditioning circuitry
- Analog-to-digital converter circuitry
- Processing units
- Memory
- Communication input/output (I/O) devices (e.g., radio interfaces) and power supply

In addition to the monitoring function, a sensor node may include actuators, capable of changing or reacting to the user's state. For instance, a WBAN sensor node may include a drug delivery pump that is automatically activated once certain conditions are met; a blood glucose sensor may be augmented with actuators that control dosage of insulin. Another example of an acting sensor node is an EEG sensor augmented with actuators for electrical neural stimulation to prevent the development of epileptic seizures.

The actual hardware organization of each sensor node is greatly influenced by the main design requirements for the WBAN, such as functionality, wearability, ease of deployment/durability, reliability of communications, security, and interoperability.

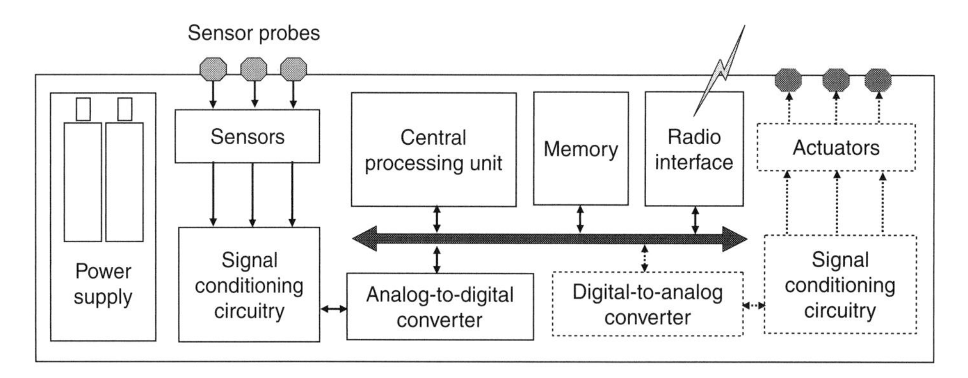

FIGURE 22.5 WBAN node architecture.

- *Functionality.* The end-user application determines (i) the number and type of vital statistics the WBAN needs to provide, (ii) the required precision and accuracy of sensor readings, and (iii) the sample frequency and frequency of reported data. For example, a fitness monitoring application targeting healthy users may not require ECG and may also tolerate possible loss of heart beat messages. However, arrhythmia monitoring applications require a very precise heart beat stream or even raw ECG. Possible loss of heart beat events is unacceptable, as it may result in false alarms or missed events.
- *Wearability.* To achieve noninvasive and unobtrusive continuous health monitoring, WBAN sensors should be lightweight and have a small form factor, so they can be built into clothes or applied as tiny patches on the skin. The size and weight of current sensor platforms are determined predominantly by the size and weight of batteries. However, a battery's capacity is directly proportional to its size. It is desirable to provide an extended period of operation without the need for battery replacements because multiple sensors requiring frequent battery changes will hamper users' acceptance of wearable systems. In addition, longer battery life will decrease WBAN operational costs. Consequently, energy efficiency is one of the key design requirements for WBAN sensor nodes, as it improves both wearability and user compliance. Electrophysiological signals rely on contact electrodes with gels used to reduce contact resistance. It is known that prolonged wear of resistive electrodes can result in skin irritation. A number of other problems can arise. For example, if the contact gel dries out, the signal quality likely deteriorates. Worse is if the electrode pulls away from the skin completely. These problems can be overcome by using recently introduced insulated noncontact bio-electrodes [52].
- *Deployment and durability.* The ideal location of specific WBAN sensors is still an open issue—for instance, for activity research, the research community is investigating a minimal set of motion sensors and their placement that will enable almost perfect discrimination of the user's states. Sensor attachment is also a critical factor, since the movement of loosely attached sensors creates spurious oscillations after an abrupt movement; such signal artifacts can generate false events or mask real events. The sensor nodes also need to be robust and durable, so that environmental conditions and time will not influence sensor readings.
- *Reliable communication* is of utmost importance for medical applications that rely on WBANs. The communication requirements of different medical sensors vary with required sampling rates, from less than 1 Hz to 1,000 Hz. One approach to improve reliability is to move beyond telemetry by performing on-sensor signal processing. For example, instead of transferring raw data from an ECG sensor, one could perform feature extraction on the sensor and transfer only the information about an event (e.g., QRS features and the corresponding time stamp of the R peak). In addition to reducing heavy demands for the communication channel, the reduced communication requirements decrease total energy consumption and consequently increase battery life. A careful trade-off between communication and computation is crucial for optimal system design.
- *Security.* Another important issue is overall system security. The problem of security arises at all three tiers of a WBAN-based telemedical system. At the lowest level, wireless medical sensors must meet privacy requirements mandated by law for all medical devices and must guarantee data integrity. Though security key establishment, authentication, and data integrity are challenging tasks in resource-constrained medical sensors, the relatively small number of nodes in a typical WBAN and short communication ranges make these requirements achievable [53].
- *Interoperability.* Wireless medical sensors should allow users to easily assemble a robust WBAN depending on the user's state of health. Standards that specify interoperability of wireless medical sensors will promote vendor competition and eventually result in more affordable systems.

22.4.1 Sensor Architecture

Physical sensors are devices that detect and convert natural physical quantities into analog signals (voltages and currents). Electrophysiological signals, such as ECG, EEG, EMG, and GSR, are sensed directly through contact or contactless electrodes attached to certain parts of the human body. Parameters of physical quantities such as blood pressure, blood glucose level, and body motion are converted into electrical signals using corresponding transducers. For example, MEMS-based accelerometers attached to the human body convert acceleration measured on that location into an electrical signal. Often these electrical signals need to be conditioned before sampling. Signal conditioning circuits amplify signals that are too weak (e.g., ECG signals are in millivolts, and these signals are typically amplified into the range of volts before sampling). Other signal conditioning circuits may reduce the signal level or reduce the frequency range of the signal through filtering. Finally, analog signals are converted into corresponding digital code that can be further processed. These three functions—sensing (with transducing), signal conditioning, and analog-to-digital conversion (ADC)—are typically implemented by multiple integrated circuits, but current trends are toward integration of all these functions into a single-chip solution.

The WBAN physical sensors must satisfy a key requirement: to be unobtrusive and easily deployable. They also need to have a stable function over long time periods and be easy to calibrate. Sensor characteristics such as accuracy, resolution, sampling rate, and the number of channels depend on health monitoring applications. Table 22.2 shows for a series of typical WBAN sensors min-max sampling rate, min-max resolution, number of channels, type of sensor probes, and preferred sensor location.

22.4.1.1 Computing

Processing resources on a WBAN node include one or more processors/microcontrollers. They are responsible for coordinating sampling activities; preprocessing sampled data (e.g., filtering); performing feature extraction; managing local memory resources; and initializing, controlling, and managing WBAN communication. In order to meet strict requirements for small size and weight, the WBAN sensor nodes have limited processing and storage resources. Processing and storage requirements of a WBAN node vary greatly depending on physiological signals (type, resolution, sampling rate) and WBAN application requirements. For example, a WBAN node equipped with one or more foot switches poses minimal requirements for processing power; similarly, transmission of a raw ECG signal does not require significant processing power. But a heart sensor featuring morphological ECG analysis requires higher processing power.

WBAN nodes must have enough storage resources for temporary data buffers to accommodate for lost messages and intermittent communication. The size of these buffers is determined by allowed event latency and available memory capacity. Event latency requirements define the maximum propagation delay from the moment an event has been detected on a WBAN node until the moment the personal server application has received that event. For example, a

TABLE 22.2 Physiological signals: sampling rates, precision typical for wearable health monitoring applications, and likely locations of deployment

Physiological parameter	Sampling rate (Hz) (min–max)	Precision (bits) (min–max)	Channels (min–max)	Type of sensing device	Placement location
ECG (per channel)	(100–1000)	(12–24)	(1–3)	Electrodes	Chest
EMG	(125–1000)	(12–24)	(1–8)	Electrodes	Muscles
EEG	(125–1000)	(12–24)	(1–8)	Electrodes	Head
PPG	(100–1000)	(12–16)	1	Photodiode	Ear or finger
Blood pressure	(100–1000)	(12–24)	1	Pressure cuff	Arm or finger
Respiration	(25–100)	(8–16)	1	Elastic chest belt or electrodes	Chest
Blood glucose	<0.01	(8–16)	1	Chemical	Skin
GSR	(50–250)	(8–16)	1	Electrodes	Fingers
Skin temperature	<1 in 60 sec	(16–24)	1	Thermistor probe	Wrist/arm
Localization	(0.01–10)	(80–120)	1	GPS receiver	Personal server (PS)
Gait	(25–100)	(16–32)	(1–3)	Inertial gyroscope	Chest
Activity	(25–100)	(12–24)	3	Accelerometers	Chest, extremities
Steps	(2–100)	(1–16)	(1–8)	Mechanical foot switch	Shoe insole
Humidity	<1 in 60 sec	(12–16)	1	—	Attached to PS
Light	<1 in 60 sec	(12–16)	1	—	Attached to PS
Ambient temperature	<1 in 60 sec	(12–16)	1	—	Attached to PS
Atmospheric pressure	<1 in 60 sec	(12–16)	1	—	Attached to PS
Ambient noise	<1 in 60 sec	(12–24)	1		Attached to PS

WBAN node or multiple nodes monitoring posture of an elderly person must notify the personal server that a fall has been detected within a couple of seconds, so that the personal server may create an alert event for emergency services or home health care providers. Contrary to this, a WBAN application targeting monitoring of physical activity and exercise of a healthy user does not pose strict requirements for event delay propagation. The data upload does not need to be in real time, and it can be done once a day. However, even in this case, available memory capacity imposes limitation on total operating time if we do not want to lose any data (see Example #3 in Section 22.4.1.4).

TABLE 22.3 Zigbee vs. bluetooth: comparison of main characteristics

Parameter	Zigbee	Bluetooth
Frequency band	2.4 GHz	2.4 GHz
Modulation technique	Direct sequence spread spectrum (DSSS)	Frequency hopping spread spectrum (FHSS)
Protocol stack size	4–32 KB	250 KB
Battery changes	Rare	Intended for frequent recharges
Max bandwidth	250 Kb/s	750 Kb/s
Max range	Up to 70 m	1–100 m
Typical network join time	30 ms	3 sec
Network size	65536	8

22.4.1.2 Wireless Communication

The radio interface of a WBAN node must be able to receive command and calibration messages from the network coordinator and to transmit sensor readings, extracted events, and status messages to the network coordinator. Emerging wireless standards and the expected proliferation of large-scale wireless sensor networks enable continual advances in radio interfaces—each new generation of radio devices provides higher bit rates at lower cost and energy consumption, with higher levels of integration and miniaturization. System designers need to estimate the required application bandwidth. In general, bandwidth depends on the number and type of sensor signals, their sampling frequency, and sample sizes. The required communication bandwidth may be estimated as follows:

$$SBW = \sum_{i=1}^{N} \sum_{j=1}^{Nch_i} Fs_i \cdot SS_i \cdot Rov_i,$$

where

- SBW is the total required system bandwidth (without communication protocol overhead),
- N is the total number of monitored signals in the system (i.e., the body area network),
- Nch_i is the number of channels of the signal i,
- FS_i is the sampling frequency of the signal i, and
- SS_i is a sample size of the signal i.
- Rovi is the recorded message overheard with the signal i.

The WBAN communication may feature a custom wireless protocol or a wireless PAN based on IEEE standard 802.15.4 (Zigbee) or 802.15.1 (Bluetooth). ZigBee has been developed for control and home automation applications; has a low data rate, low power consumption, and short latency; and supports short packet devices and a large number of devices in the network. Bluetooth, on the other hand, uses a higher data rate and higher power consumption and works with large packet devices. Table 22.3 shows the main characteristics of Zigbee and Bluetooth.

22.4.1.3 Putting Everything Together: The Actis System

In the spirit of the system architecture previously described, a prototype WBAN for health monitoring has been developed [54, 55]. Figure 22.6 shows the prototype components. The prototype includes two activity sensors (ActiS), an integrated ECG and tilt sensor (eActiS), and a personal server. Each sensor node includes a custom application-specific board and uses the Tmote sky platform for processing and 802.15.4-compliant wireless communication [56, 57]. The personal server runs on either a laptop computer or a wide local area network/ WWAN–enabled handheld pocket PC. The network coordinator with wireless ZigBee interface is implemented on another Tmote sky module that connects to the personal server through

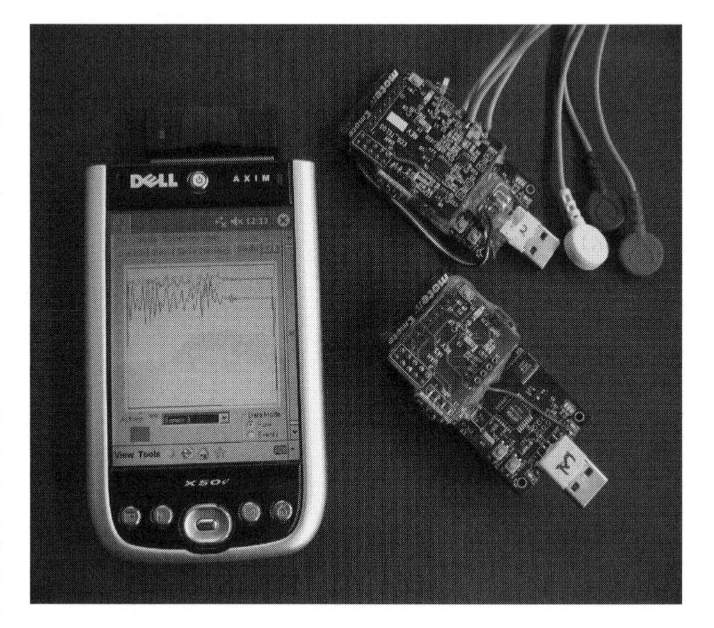

FIGURE 22.6 Prototype WBAN. From left to right: the personal server with network coordinator, ECG sensor with electrodes, and a motion sensor.

a universal serial bus (USB) interface. For an alternative setting, a custom network coordinator has been developed featuring a ZigBee wireless interface, an advanced RISC (reduced instruction set computer) machine (ARM) processor, and a compact flash interface to the personal server. More details about prototype architecture, implementation issues, communication protocol, and software architecture can be found in [54, 55].

22.4.1.4 Examples

Example #1: Calculate the min-max bandwidth requirements [B/sec] of a heart sensor that streams a subject's three-channel ECG signal.

Solution: Bandwidth = [Sampling rate] * [Number of bytes per sample] * [Number of channels] = [100, 1000] * 2 * 3 = [600 B/sec, 6000 B/sec]

Example #2: Calculate min-max bandwidth requirements of a heart sensor that streams RR intervals.

Solution: Normal heart rate is in the range of 30 to 240 bpm. The heart sensor detects each heart beat and time-stamps it. Therefore, an R-peak event is represented by an associated time-stamp. If we assume a common time tick of 1/32 kHz = 31.25 ms, the first step is to determine the number of bits needed for each time-stamp. For the maximum heart rate of 240 bpm, the number of clock ticks is (60/240) * 32,000 = 8,000. For the minimum heart rate of 30 bpm, the number of clock ticks is (60/30) * 32,000 = 64,000. To represent this range (8,000–64,000), a two-byte unsigned integer will suffice. Consequently, the required min-max bandwidth is in the range of (30 bpm * 2 B)/60 = 1 byte/sec, and (240 bpm * 2 B)/60 = 8 B/sec.

Example #3: Consider a WBAN node with a three-level memory hierarchy, including on-chip local random-access memory (RAM), on-chip or on-board flash memory, and an external flash disk. Assume that available memory capacity for data buffering is as follows:

M_1 = RAM capacity, 5 KB (40 Kb)
M_2 = flash memory capacity, 4Mb
M_3 = flash disk capacity, 1 GB (8 Gb)

What are expected operating times for an ECG sensor node transmitting (a) a single channel ECG signal (sampling rate = [100, 1000], resolution = 16 bits) and (b) RR intervals for each level of the memory hierarchy?

Solution: The system operating time (OT) can be determined as:

$$OT = M_i / BW_i,$$

where M_i is the given memory capacity and BW_i is the amount of data to be stored in memory (sampling_rate * sample_length).

(a)

$$OT_1(\text{RAM}) = M_1/BW_{\text{ECG}} = 40\,\text{Kb}/[1{,}600..16{,}000]\text{bps} = [2.56..25.6]\text{s}$$

$$OT_2\ (\text{flash\_memory}) = M_2/BW_{\text{ECG}} = 4\,\text{Mb}/[1{,}600..16{,}000]\text{bps} \approx [0.72..7.2]\text{hours}$$

$$OT_3(\text{flash\_disk}) = M_3/BW_{\text{ECG}} = 8\,\text{Gb}/[1{,}600..16{,}000]\text{bps} \approx [2.5..25]\text{days}$$

(b)

$$OT_1(\text{RAM}) = M_1/BW_{\text{RR}} = 40\,\text{Kb}/[8..64]\text{bps} \approx [10..85]\ \text{min}$$

$$OT_2(\text{flash\_memory}) = M_2/BW_{\text{RR}} = 4\,\text{Mb}/[8..64]\text{bps} \approx [0.75..6]\text{days}$$

$$OT_3(\text{flash\_disk}) = M_3/BW_{\text{ECG}} = 8\,\text{Gb}/[8..64]\text{bps} \approx [4.25..42]\text{years}$$

22.5 Wireless Mobile Devices for M-Health

The WBAN personal server application can run on wireless handheld devices, such as smart phones (which tend to be voice-centric devices with PDA-like data capabilities) or WWAN-enabled PDAs or personal communicators (which tend to be data-centric devices with voice capabilities). In home monitoring settings, the personal server application may also run on a PC. Each new generation of wireless handheld devices includes more processing power, more storage, and longer battery life, so that their capabilities meet the requirements of the personal server application. Consequently, the focus of this section is on personal server application requirements.

The personal server provides user interface, controls the WBAN, fuses data and events, and creates unique session archive files. It begins a health monitoring session by wirelessly configuring sensor parameters, such as sampling rate, selection of the type of physiological signal of interest, and specification of events of interest. Sensors, in turn, transmit pertinent event messages to the personal server. The personal server must aggregate the multiple data streams, create session files, and archive the information in the patient database. Real-time feedback is provided through the user interface. The user can self-monitor vital signs and be notified of any detected warnings or alerts.

The user interface must provide seamless control of the WBAN, implementing all the necessary control over it, such as node identification, sensor configuration [53], sensor calibration, visual real-time data capture, and graphical presentation of events, alerts, and health status.

Sensor node identification requires a method for uniquely identifying a single sensor node to associate the node with a specific function during a health monitoring session. For example, a motion sensor placed on the arm performs an entirely different function than a motion sensor placed on the leg. Because two motion sensors are otherwise indistinguishable, it is necessary to identify which sensor should function as an arm-motion sensor and which should function as a leg-motion sensor. The personal server application will typically guide a user through the process of sensor mounting and setup.

Another important function is sensor calibration, which can be permanent (once in a lifetime) or session specific (e.g., activity sensors on the leg may require an initial calibration of the default orientation on the body).

The personal server is solely responsible for collecting data and events from the WBAN. Each sensor node in the network samples, collects, and processes data. Depending on the type of sensor and the degree of processing specified at the time of configuration, a variety of events will be reported to the personal server. An event log is created by aggregating event messages from all the sensors in the WBAN; the log must then be inserted into a session archive file. The personal server must recognize events as they are received and make decisions based on the nature and severity of the event. Normally, neither R-peak nor heart beat events create alerts and only are recorded in the event log. However, the personal server will recognize when the corresponding heart rate exceeds predetermined threshold values; in that case, it alerts the user that the heart rate has exceeded the target range.

Even in a deployed system where intelligent sensors analyze and process raw data and transmit application event messages, there may be cases where it is necessary to transmit raw data samples. Such cases become apparent when considering a deployed ECG monitor. When embedded signal processing routines detect an arrhythmic event, the node should send an event message to the personal server, which will then be relayed to the appropriate medical server. The medical server, in turn, will provide an alert to the patient's physician. However, a missed heart beat can also be caused by electrode movement. Therefore, it would be useful to augment this event with actual recording of the fragment of unprocessed ECG sensor data. The recording can be used by a physician to evaluate the type and exact nature of the event or to dismiss it as a recording artifact. In such a case, the embedded sensor will begin streaming the real-time data to the personal server for a predefined time period.

22.6 Next-Generation M-Health Systems

To gain wider acceptance, the next generation of m-health systems will have to address several challenges that today limit the usefulness of such systems. Those challenges include the availability of infrastructure and bandwidth in current and new wireless networks, miniaturization of medical sensors, convenience, and standardization of communication protocols and interfaces between medical and nonmedical devices.

With the advances in wireless mobile devices, their usage in the m-health context becomes more practical. A smart phone (see section 22.5) seems to be the most frequently considered candidate for a future personal multimedia hub. In addition to handling all personal multimedia needs, it could serve as a central part of m-health systems by taking over the tasks of medical sensor coordination, monitoring, archiving, and reporting. The answer to the question of why to use a phone as a personal server seems to be an easy one. Recent advances in computational and storage capacity, the dramatic increase in the available wireless bandwidth, and advances in screen technologies have made it possible to turn a once simple device into a convenient do-it-all gadget. In addition, it appears that everyone has a cell phone now, or will have one in the near future. While in 1991 there were only about 16 million cell phone subscribers worldwide, by 2005 the number of subscribers had grown to 2.14 billion [58]. Worldwide, cellular subscribers are expected to top 3.2 billion in 2010 and to continue to grow in numbers. In the United States, the number of land lines reached almost 193 million in 2000 and has been declining since. At the same time, the number of cell phone subscribers went from 5.3 million in 1990 to 202 million in 2005 [59].

22.6.1 Wireless Cellular Technologies for M-Health Systems

Despite the usual issues of concern whenever medical systems are considered (security, reliability, latency, physical size), one of the biggest obstacles to widespread use of cellular networks in telemedical systems is the lack of bandwidth and performance.

There is some discrepancy in the way different groups classify current and future wireless cellular technologies into generations (1G, 2G, 3G, and 4G) [60, 61]. We will give a short description of some of the most important technologies available today or currently being developed.

22.6.1.1 First and Second Generations

1G was simply the first generation of analog mobile phones, oriented exclusively to voice communication.

2G, the second generation, replaced 1G analog mobile phones. 2G phones were intended primarily for digital transmission of voice. Three systems were developed: Pacific Digital Cellular (PDC, widespread in Japan), Interim Standard 95 (IS-95) and IS-136 in the United States, and the Global System for Mobile Communications (GSM, widespread in Europe).

22.6.1.2 Generations 2.5 and 2.75

2.5G was generally reserved for General Packet Radio Service (GPRS) (see Figure 22.3), delivered as a network overlay for

GSM, Code Division Multiple Access (CDMA), and Time Division Multiple Access (TDMA) networks. While the basic GSM service allowed data rates of up to 9.6 kbps only, GPRS is capable of having 14 kbps per channel (after protocol and error-correction overhead). GPRS can combine up to eight channels, bringing the total to over 100 kbps in theory. For the first time, the focus of a wireless phone service was primarily on data transmission. The protocol architecture of the backbone network is based on the Internet protocol (IP), which can be supplemented by the transmission control protocol (TCP) for reliability or the user datagram protocol (UDP) for applications that do not require that level of robustness. The most important feature for end users was the always-on connectivity of GPRS. The user could remain continuously connected but the network resources and bandwidth were used (and the user charged) only when data were transmitted. GPRS offered multiple services—web browsing, transfer of still images and video clips, document sharing and remote collaborative working, etc. All of these services allowed the emergence of the first usable medical systems based on global availability.

2.75G referred to Enhanced Data for GSM Evolution (EDGE). If the GPRS protocol was the first step toward 3G, the EDGE protocol was the second one. In theory, EDGE offered data rates of up to 384 kbps, with the actual data rates being much lower. The real attractiveness of EDGE was in its ability to work on the existing GSM spectrum. The EDGE protocol was adopted mostly by mobile operators in countries where the allocation of spectrum for the 3G systems was delayed (e.g., the United States).

22.6.1.3 Third Generation

In 2000, European mobile phone operators spent well over $100 billion on 3G spectrum licenses. In September 2006, the U.S. Federal Communications Commission (FCC) completed the auction of licenses for one of the bands (AWS-1) of Advanced Wireless Services. The total amount raised from the auction was close to $14 billion [62]. High fees and the necessity of building entirely new infrastructure delayed the introduction of 3G systems, except in some Asian countries (Japan, South Korea) where fees were almost nonexistent. The main air interface for the third generation is wideband CDMA (W-CDMA). Two services using W-CDMA are the Universal Mobile Telecommunications System (UMTS, a GSM successor) and Freedom of Mobile Multimedia Access (FOMA, implemented in Japan). The International Telecommunication Union (ITU) approved UMTS as a part of the ITU-R M.1457 recommendation. UMTS provides up to 2 Mbps indoors (a low-mobility condition), up to 384 kbps outside (at the speed of slow-moving pedestrians), and up to 144 kbps for fast-moving mobile phones. The general idea behind IMT (International Mobile Telecommunications)-2000/UMTS is to have a unified, seamless operation through the combined use of pico- and microcells indoors and in urban areas, macrocells in outdoor and rural areas, and satellite networks when necessary.

In the early days of defining the 3G systems, video telephony was envisioned as the killer application. However, music downloading was the most frequently used service among the early adopters of these systems. It has been shown that 3G systems can easily support the amount of data required for medical applications [37].

22.6.1.4 Beyond 3G

Post-3G requirements and classifications become blurred. Instead of making a clear distinction among systems, many use the term "3G and beyond" or B3G (beyond 3G) to include both 3G and 4G systems. As expected, some of the key requirements of new systems are increased bandwidth, stable performance, and quality of service (QoS). However, the emphasis seems to be even more on providing a generalized access network that will allow internetworking between different access systems in terms of horizontal and vertical handover [63].

The following standards and technologies are by some considered as 4G systems, while others treat them only as the first steps toward the "real" 4G systems:

- *High Speed Downlink Packet Access* (HSDPA) is designed for data rates of up to 14.4 Mbps and features lower delays of approximately 100 ms. More importantly, in its initial implementation, it is capable of delivering average throughput rates of about 1 Mbps. Cingular Wireless (since 2005) and a number of other companies (since 2006) have offered HSDPA on a commercial basis. The peak network rates are expected to reach 7.2 Mbps by 2008.
- *WiMax*, originally based on the IEEE 802.16 specification, is designed to deliver up to 70 Mbps over a 50 km radius. The IEEE 802.16-2004 standard was developed for an unlicensed band (5.8 GHz) and intended primarily for local connectivity. The IEEE 802.16e-2005 added support for mobile radio operation. One aspect of WiMax that currently limits its usability is the fact that scheduling becomes inefficient if a large number of users are present in the same sector.

22.6.2 Future Trends and Obstacles

Most people agree that to be considered "truly" 4G, the system has to be capable of achieving up to 100 Mbps when the user is stationary (indoor/urban) and up to 1 Gbps when the user is moving (outdoors). 4G systems are data- and visual-centric. In addition, a 4G system is expected to have the following features:

- Use of IP version 6, which increases the number of addresses, eliminates the need for network address

translation (NAT) devices, and makes it possible to use concepts and applications developed for other devices for easier integration into global systems
- Multiple antennas at the transmitter and the receiver, to sustain the increased data rate
- Support for *pervasive networking* (handover), as is being defined by the IEEE 802.21 standard

Some of the envisioned features beyond 4G include:

- Being able to smell the environment of the other person on the phone
- Communicating without emitting any voice (lip movement recognition)

Having in mind the current trends for increasing the data rates in new wireless mobile systems, it seems that the most important issues to be resolved prior to widespread adoption of m-health medical systems are:

- The need for further miniaturization of medical sensors in order to increase user comfort and reduce power consumption and the possibility of unwanted disclosure of the patient's condition
- The lack of standardization and the ability to interface existing medical equipment with new communication systems
- The security of data during transmission and while stored in the user's personal hub (smart phone)
- Users' and caregivers' confidence in m-health systems

22.7 Summary

M-health is becoming a major technological trend for ambulatory and prolonged physiological monitoring. It has the potential to shift the paradigm of health care from reactive to proactive, from disease management to disease prevention. System developers will still have to resolve a number of issues. The most important are (a) wearability and compliance, (b) system integration, (c) standardization of protocols and procedures, (d) seamless system integration, and (e) data mining of huge datasets.

In this chapter we have discussed the current state of technology, existing systems, and the main issues to provide system designers a feel for the "landscape" of the design space.

22.8 Exercises

1. Prepare a survey of relevant sensing techniques for noninvasive *blood glucose monitoring* used in commercial systems and research prototypes. Discuss their accuracy, system design, and suitability for wearable applications.

   Do the same for *blood pressure monitoring*.
2. Wireless interfaces consume the most energy in WBAN sensor platforms. To reduce energy requirements (and consequently improve wearability), on-platform compression of biomechanical and bioelectrical signals can be employed. Prepare a survey of the existing approaches for compression of these signals and discuss their suitability for on-platform implementation.
3. A microcontroller system is performing the following task:

   16-bit samples have a frequency f_{ADC} and are processed and stored in internal memory. After 16 samples are collected, they are sent using an external wireless interface operating at 200 kbps. Data are encapsulated into a simple frame format:

Header (preamble + sync)	Sample 0	Sample 1	...	Sample 15	Check sum
6 bytes	2 bytes	2 bytes		2 bytes	2 bytes

The microcontroller also keeps a software real-time clock with a 500 μs precision. The microcontroller is running at 8 MHz (main clock), and the internal ADC is using the same clock. It takes:

- 8 clock cycles to sample 16-bit data
- 13 clock cycles to convert it
- 14 clock cycles to process each sample and store it to memory
- 12 clock cycles to update the real-time clock
- 12 clock cycles to prepare a byte and send it to the external wireless interface

It takes 6 ms to wake up the wireless interface and to begin transmission. Assume that the microcontroller can wake up instantaneously and that the previously given processing times include interrupt overhead (if appropriate). The current consumption of the microcontroller and the wireless interface is as follows:

Mode	Current consumption
Active mode, ADC off	2 mA
Active mode, ADC on	4 mA
Sleep mode	2 μA
Wireless interface, active mode	15 mA

If the system is running on batteries that have a capacity of 2,000 mAh, calculate the maximum sampling frequency (f_{ADC}) such that the calculated battery life is at least 6 months (180 days).

4. Calculate the required bandwidth of a heart sensor that reports both raw ECG (single channel), with a sampling frequency of 100 Hz, and heart rate, assuming average heart rate of 72 bpm. Heart beat stream includes 4-byte timestamps.
5. Calculate the required bandwidth for ECG and EEG monitoring. The system features three channels of ECG, four channels of EEG, and a tilt sensor. ECG and EEG signals are sampled at 125 and 250 Hz, respectively, and the tilt sensor is sampled once every 10 seconds and saved as a one-byte status/position.
6. How many users with ECG monitors can we simultaneously monitor in a network with effective data bandwidth of 100 kbps? Each ECG monitor records three channels of ECG with a sampling frequency of 500 Hz.
7. What is the expected operation time of a monitoring system with one ECG channel and two sensors with 3D accelerometers? Sampling frequency of the ECG is 250 Hz, sampling frequency of accelerometers is 40 Hz, and the personal server uses a 512 KB flash memory card.
8. What is the effective duty cycle of the system in Question 7 assuming constant message size of 50 bytes with effective payload of 25 B and 250 kbps wireless communication bandwidth?
9. What is the expected battery life assuming that the system is powered with 2 AAA batteries with capacity of 750 mAh. Average power supply current in active mode is 1 mA and during wireless communication 20 mA. The system features TDMA protocol with 50 ms time slots for each sensor. Each sensor listens throughout the master time slot (50 ms) and transmits in its own time slot. Average transmission time is 10 ms.

22.9 References and Bibliography

1. C. S. Pattichis et al. Wireless telemedicine systems: An overview. *IEEE Antennas and Propagation Magazine.* 44(2):143–153, 2002.
2. H. Oh et al. What is eHealth? (3): A systematic review of published definitions. *World Hosp. Health Serv.* 41(1):32–40, 2005.
3. G. Eysenbach. What is e-health? *J. Med. Internet Res.* 3(2):e20, 2001. http://www.jmir.org/2001/2/e20
4. R. S. H. Istepanian, E. Jovanov, and Y. T. Zhang. Guest editorial introduction to the special section on m-health: Beyond seamless mobility and global wireless health-care connectivity. *IEEE Trans. Inf. Technol. Biomed.* 8(4): 405–414, 2004.
5. E. Jovanov. Wireless technology and system integration in body area networks for mHealth applications. *Proceedings of the 27th Annual International Conference of the IEEE Engineering in Medicine and Biology Society,* 2005.
6. A. Milenkovic, C. Otto, and E. Jovanov. Wireless sensor networks for personal health monitoring: Issues and an implementation. *Computer Communications* (Special issue: Wireless Sensor Networks: Performance, Reliability, Security, and Beyond). 29(13/14):2521–2533, 2006.
7. A. Krahn et al. Recording that elusive rhythm. *Can. Med. Assoc. J.* 161:1424–1425, 1999.
8. Medtronic. *Reveal® Plus Insertable Loop Recorder (ILR).* http://www.medtronic.com/physician/reveal/index.html
9. Georgia Tech News Release. *New CardioMEMS Device Helps Aneurysm Patients.* http://www.gatech.edu/newsroom/release.php?id=846
10. CardioMEMS. *Cardio Micro Sensor.* http://www.cardiomems.com
11. S. Rhee, B.-H. Yang, and H. H. Asada. Artifact-resistant power-efficient design of finger-ring plethysmographic sensor. *IEEE Trans. Biomed. Eng.* 48:795–805, 2001.
12. P. A. Shaltis, A. Reisner, and H. H. Asada. Wearable, cuff-less PPG-based blood pressure monitor with novel height sensor. *Proc. 28th IEEE EMBS Annu. Int. Conf.,* 908–911, 2006.
13. J. Espina et al. Wireless body sensor network for continuous cuff-less blood pressure monitoring. *Proc. 28th IEEE EMBS Annu. Int. Conf.,* 11–15, 2006.
14. E. Jovanov et al. Stress monitoring using a distributed wireless intelligent sensor system. *IEEE Eng Med. Biol. Mag.,* 2003.
15. Sontra Medical Corporation. *Symphony Diabetes Management System.* http://www.sontra.com
16. DexCom. *DexCom STS System.* http://www.dexcom.com/html/dexcom_products_sts_starter_kit.html
17. J. A. Tamada, M. Lesho, and M. J. Tierney. Keeping watch on glucose. *IEEE Spectrum* 39(4):52–57, 2002.
18. Medtronic MiniMed. *MiniMed Paradigm® 522 or 722 Insulin Pump.* http://www.minimed.com/products/insulinpumps/components/insulinpump.html
19. R. E. Morley, Jr. et al. In-shoe multisensory data acquisition system. *IEEE Trans. Biomed. Eng.* 48:815–820, 2001.
20. HealthPia. *Diabetes Phone.* http://www.healthpia.us/products.asp
21. C. Diorio and J. Mavoori. Computer electronics meet animal brains. *IEEE Computer.* 38(1):69–75, 2003.
22. M. Ghovanloo and K. Najafi. A BiCMOS wireless stimulator chip for micromachined stimulating microprobes. *Second Joint EMBS/BMES Conference,* 2113–2114, 2002.
23. P. S. Motta and J. W. Judy. Multielectrode microprobes for deep brain stimulation fabricated with a customizable 3-D electroplating process. *IEEE Trans. Biomed. Eng.* 52(5):923–933, 2005.
24. P. Irazoqui-Pastor, I. Mody, and J. W. Judy. Transcutaneous RF-powered neural recording device. *Second Joint EMBS/BMES Conference,* 2105–2106, 2002.
25. Medtronic. *Activa Therapy.* http://www.medtronic.com/neuro/brainpacemaker/index.html

26. A. Astaras et al. A miniature integrated electronics sensor capsule for real-time monitoring of the gastrointestinal tract (ideas). *ICBME 2002: The Bio-Era: New Challenges, New Frontiers*, 2002.
27. E. A. Johannessen et al. Implementation of radiotelemetry in a lab-in-a-pill format. *Lab on a Chip.* 6(1):39–45, 2006.
28. E. A. Johannessen et al. Implementation of multichannel sensors for remote biomedical measurements in a microsystems format. *IEEE Trans. Biomed. Eng.* 51(3):525–535, 2004.
29. Given Imaging. *PillCam™ SB Capsule Endoscopy.* http://www.givenimaging.com/Cultures/en-US/Given/English/Products/CapsuleEndoscopy
30. B. Lewis and P. Swain. Capsule endoscopy in the evaluation of patients with suspected small intestinal bleeding: The results of a pilot study. *Gastrointest. Endosc.* 56(3): 349–353, 2002.
31. R. W. Hoyt et al. Combat medical informatics: Present and future. *AMIA 2002 Annual Symposium*, 2002.
32. J. Obusek. Warfighter physiological status monitoring. *The Warrior.* 6–8, 2001.
33. D. Raskovic, T. Martin, and E. Jovanov. Medical monitoring applications for wearable computing. *Computer Journal.* 47(4), 2004.
34. Sensitron. *The CareTrends Systems.* http://www.sensitron.-net/US/caretrends/wirelessEnab.html
35. Cleveland Medical Devices Inc. *Crystal Monitor® 20–B.* http://www.clevemed.com/home/home_medical_p_crystalmonitor20_b.php
36. Equivital Limited. *Physiological Life Signs Monitoring.* http://www.equivital.co.uk
37. Y. Chu and A. Ganz. A mobile teletrauma system using 3G networks. *IEEE Trans. Inf. Technol. Biomed.* 8(4):456–462, 2004.
38. Adidas. *Addidas_1 Intelligence Level 1.1.* http://www.adidas.com
39. P. Bonato et al. IEEE EMBS Technical Committee on Wearable Biomedical Sensors and Systems: Position paper. *Proc. Int. Workshop on Wearable and Implantable Body Sensor Networks*, 212–214, 2006.
40. Sensatex. *Smart Shirt System.* http://www.sensatex.com
41. Georgia Institute of Technology. *School of Textile and Fiber Engineering*, 2003. http://www.tfe.gatech.edu
42. VivoMetrics. *LifeShirt System.* http://www.vivometrics.com/site/system.html
43. Wealthy-IST. *Wearable Health Care System.* http://www.wealthy-ist.com
44. Virginia Tech E-Textiles Laboratory [website]. http://www.ccm.ece.vt.edu/etextiles
45. J. Edmison et al. An e-textile system for motion analysis. *Proc. Int. Workshop on New Generation of Wearable Computers for eHealth*, 215–223, 2003 (invited paper).
46. Y. Zhang et al. A health-shirt using e-textile materials for the continuous and cuffless monitoring of arterial blood pressure. *Proc. 3rd IEEE-EMBS International Summer School and Symposium on Medical Devices and Biosensors*, 69–86, 2006.
47. R. Paradiso and D. de Rossi. Advances in textile technologies for unobtrusive monitoring of vital parameters and movements. *Proc. 28th IEEE EMBS Annu. Int. Conf.*, 392–395, 2006.
48. R. Paradiso, G. Loriga, and N. A. Taccini. Wearable health care system based on knitted integrated sensors. *IEEE Trans. Inf. Technol. Biomed.* 9(3):337–344, 2005.
49. R. Szewczyk et al. Habitat monitoring with sensor networks. *Communications of the ACM.* 47:34–40, 2004.
50. J. Burrell, T. Brooke, and R. Beckwith. Vineyard computing: Sensor networks in agricultural production. *IEEE Pervasive Computing.* 3:38–45, 2004.
51. E. Jovanov et al. A wireless body area network of intelligent motion sensors for computer assisted physical rehabilitation. *J. Neuroengineering Rehabil.* 2, 2005.
52. Quasar [website]. http://www.quasar-usa.com
53. S. Warren and E. Jovanov. The need for rules of engagement applied to wireless body area networks. *IEEE Consumer Communications and Networking Conference CCNC2006*, 2006.
54. A. Milenkovic, C. Otto, and E. Jovanov. Wireless sensor networks for personal health monitoring: Issues and an implementation. *Computer Communications.* 29:2521–2533, 2006.
55. C. Otto et al. System architecture of a wireless body area sensor network for ubiquitous health monitoring. *Journal of Mobile Multimedia.* 1:307–326, 2006.
56. Moteiv [website]. http://www.moteiv.com
57. J. Polastre, R. Szewczyk, and D. Culler. *Telos: Enabling Ultra-Low Power Wireless Research.* http://www.moteiv.com
58. *USA Today.* Recycled cell phones help drive Third World wireless boom. August 20, 2006. http://www.usatoday.com/tech/wireless/phones/2006-08-20-cellphone-recycling_x.htm
59. eTForecasts. *Cellular Subscriber Forecast By Country.* http://www.etforecasts.com/products/ES_cellular.htm
60. G. Aggelou. *Mobile Ad Hoc Networks: From Wireless LANs to 4G Networks.* McGraw-Hill, 2006.
61. P. Rysavy. *Mobile Broadband EDGE, HSPA, and LTE.* White paper prepared for 3GAmericas, 2006.
62. U.S. Federal Communications Commission. *Auction of Advanced Wireless Services Licenses Closes.* Public Notice DA 06–1882, 2006.
63. IEEE. *Handover and Interoperability Between Heterogeneous Network Types.* http://www.ieee802.org/21

23

Multimedia for Future Health—Smart Medical Home

Dr. Jinman Kim,[1]
Dr. Zhiyong Wang,[1]
Dr. Tom Weidong Cai,[1] and
Prof. David Dagan Feng[1,2]
[1] *University of Sydney*
[2] *Hong Kong Polytechnic University*

23.1 Introduction

With recent advances in multimedia technology, its impact toward information technology in biomedicine is ever increasing [1–3]. Multimedia technologies are enabling more comprehensive and intuitive uptake of information in a wide range of fields that have a direct impact on our life, particularly in entertainment, education, work, and health. Systems and services have been developed to harness the advantages of multimedia technology, which ranges from video-conferencing, online shopping in virtual environments, video-on-demand services and E-learning to remote healthcare [1, 4, 5]. The core components behind these multimedia technologies are human-centered multimedia services, which combine many fields of information technology including computing, telecommunication, databases, mobile devices, sensors, and virtual/augmented reality systems. Human-centered multimedia services are built upon three key research pillars as shown in Figure 23.1. These are (1) human-computer interaction (HCI); (2) multimedia delivery; and (3) multimedia data management. HCI (e.g., via the use of keyboard/mouse input devices) is the initial component of the multimedia information flow with the responsibility of generating outputs by interpreting inputs from the users. Multimedia delivery systems (e.g., the Internet) are responsible for transparent information delivery (e.g., streaming video) from sources to destinations. Finally,

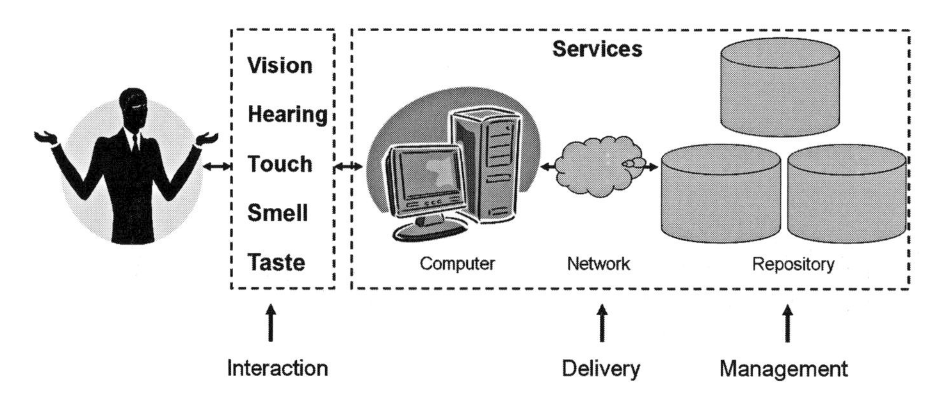

FIGURE 23.1 Illustration of the three pillars of human-centered multimedia systems: (1) interaction; (2) delivery; and (3) management.

the multimedia data management components facilitate information access (e.g., browsing, retrieval, and indexing).

One area of biomedicine that has seen rapid transition and great benefit from state-of-the-art developments in multimedia technology is the smart medical home, known also as smart houses [3, 6–10], and often considered to be the hub of future health care [11]. The smart medical home is a subcomponent of the concept of a smart home [3, 12, 13]. This notion of smart home was first introduced in the early 1980s with the proposal to integrate intelligent implementations of consumer electronic devices, electrical equipment, and security devices for the purpose of automation of domestic tasks, easy communication, and human-friendly control, as well as safety [3, 13]. Smart homes include devices that have automatic functions and systems that can be remotely controlled by the user with the primary objective of enhancing comfort, saving energy, and increasing security for the residents of the house. These developments have found applications in the field of enhancing the medical capabilities of homes for people with medical conditions and special needs.

The aim of a smart medical home is analogous to that of smart homes, namely to create an integrated system of affordable, easy to use, intelligent health care tools for consumers in their home [6, 14]. The smart medical home has recently seen significant research and development [3, 9, 14–16]. This is attributed to the trend of consumers increasingly taking control of their own health care. This trend is evident from the shift observable in medical treatment plans, which are increasingly moving from a hospital-based to a patient-centered system [17]. The same phenomenon is observable in the increased use of the Internet to search for health-related information and in the billions of dollars being spent annually on alternative and nontraditional health products [9, 16, 18, 19]. More significantly, the health care system may not be able to cope with the impending influx of new patients as the population continues to age. It is expected that by the year 2020, the 65-year-and-older population in the United States will reach 53 million, an increase of 18 million from 2000 [3]. Thus, the need for technologies that are able to complement the health care system while enabling people to live healthier, longer lives in their own home is becoming critical. With the rapid expansion of networking and information technologies into our daily lives through such technologies as the Internet, mobile phones, and interactive digital television (DTV), the acceptance level for potential smart medical home technologies is at an all-time high. The needs of our aging population will accelerate the movement and awareness of self-care and wellness and will irreversibly alter traditional doctor-patient relationships [3, 7, 20].

This chapter presents latest research and development in multimedia technologies and the transition of these technologies into health care products for the smart medical home. It is subdivided into two parts: (1) enabling multimedia technologies and (2) applications involving multimedia technologies in biomedicine. In the first part, a general introduction to multimedia technologies is presented, continuing into a discussion of the visual, audio, and other emerging media components for HCI in section two. This is followed by multimedia content management in section three. The technologies for delivering this multimedia content are presented in section four. The second part starts with a general description of biomedical technologies that either have already found or are finding their way into smart medical homes. Section five introduces and gives examples of developments in smart medical homes, with emphasis on enabling multimedia technologies. This is followed by sections six to eight which present the major applications used in medical homes of telemedicine (monitoring, consultation, etc.), sensors (wearable and stand-alone devices), and computer-assistance technologies (medication advisor, decision support, etc.). Seamless integration of these different multimedia technologies is necessary for medical devices used in a smart medical home. A more in-depth discussion of the biomedical information technology topics of telemedicine and wearable medical devices in biomedicine can be found in Chapter 22. However, to fully appreciate the contents of this chapter, these topics will be briefly covered, with emphasis on their application to the smart medical home and their use of multimedia technologies. Section nine discusses potential applications of virtual/augmented reality, followed by the developments in patient awareness toward biomedical multimedia technologies in section ten. Finally, a summary of the chapter is given in section eleven.

23.2 Multimedia for Human-Computer Interaction

The aim of HCI is to mimic human-human interactions. Of course a complete picture of how human beings interact with the real world is not yet available to us, and this remains one of the greatest scientific challenges. Interacting with the computer is essentially the first step toward manipulating and using digital information. Given the ever-increasing role of computers in society, HCI has become increasingly important in our daily lives [21]. However, interaction between the human and computer via the use of traditional input devices that are often the combination of mouse, keyboard, joystick, or remote control, is far less flexible than spontaneous human-human interaction. The constraints derived from these devices have become even more restrictive with the emergence of techniques such as virtual/augmented reality [22, 23] and wearable computers [10, 15, 24].

In general, human-human interaction consists of all five basic senses of human cognition: vision, hearing, smell, taste, and touch [25]. The ultimate aim of HCI is to make use of all natural human actions, such as facial expressions, body movement, speech, and eye gaze, in communicating with the computer, which interprets these and generates outputs that are understandable by the human operators. Vision and speech are two of the most dominant senses, and hence they will be focused on in greater depth in the sections below, followed by a review of emerging technologies for other sensing modalities.

23.2.1 Visual Information Processing

Visual information refers to what a human perceives through his or her eyes or information captured by optical cameras. A key contributor in the field of visual information processing is face recognition technology [26], as used in such applications as security and surveillance systems [27], gesture recognition [28], lip reading for the deaf [29], and optical character recognition [30]. One of the main limitations of current visual information processing systems is the need to apply constraints to the users, such as the need to wear gloves to ease hand tracking for gesture recognition systems to provide enough information for the computation to make use of the data.

The other key research area in visual information processing is the way that computers present users with visual information. Computer graphics and visualization have greatly contributed to this issue, providing approaches that include stereo or multiple-view image analysis, 3D reconstruction, view synthesis and rendering, 3D displays, graph drawing, etc. Computer graphics and visualization, as a field, aims to produce realistic representation and visual information of data in 2D, 3D, or in greater dimensions, through the use of mathematical models and algorithms [4, 31, 32]. These include but are not limited to ray tracing, texture-based rendering, and illustrative rendering. Computer graphics and visualization techniques have already been widely applied to a large variety of domains including public transport, biology, social science, and archaeology. These research areas focus on facilitating information comprehension for the users by means of virtual environments, generally to permit users to capture computational information through their visual senses. In general, visual information processing requires intensive computation and therefore continuous research on how to achieve efficient computation is of great importance. Chapter 9 covers the applications of data visualization and display of digital medical images.

23.2.2 Speech Processing

Apart from visual information, another dominant modality in human communication is sound, with speech in particular. The two main areas for speech processing in HCI are speech recognition and speech synthesis. The concept of speech recognition is for a computer program to acquire analog signal (speech) from a microphone, convert it to a digital waveform, and process it to search for a matching wave (recognition) [33, 34]. The conversion requires sophisticated algorithms that compare the input with a database of known words. Once the words are recognized, these words are often represented in digital text format. Automatic speech recognition (ASR) has been a research topic for decades [34] and tremendous advancements have been made. Many prototypes (e.g., Sphinx project II [35]) and commercial systems (e.g., IBM ViaVoice [36]) are now readily available. Most speech recognition systems are based on statistical models of the acoustic features of spoken words and of natural language. Therefore, it is error prone because of the diversity of different speakers, the physical or emotional change of a speaker, and different physical environments (e.g., noise). This issue has been partially addressed by the introduction of training and adaptive tuning algorithms, which require an initial training session but allow continuous updates to the software model of the user's speech through user-conducted error checking and corrections. Most current ASR systems constrain users to special training, a special speaking style (e.g., prepared vs. spontaneous and discrete vs. continuous), and known physical environments.

Speech synthesis is a process of converting unrestricted text into speech and communicating information to users. As reviewed in Breen [37], speech synthesis systems in general operate via the following steps. First, the text is converted into a symbolic linguistic/phonologic description. Second, the phonologic component converts the set of orthographic symbols into a set of distinctive features or sounds (i.e., phonemes) depending on the phonologic model. The phoneme is the most popular form of phonologic representation and the set of phonemes of a language can be understood as the smallest segments of sounds that can be distinguished by their contrast within words. Finally, this abstract symbolic description is transformed into an acoustic signal. The success of speech synthesis has been beneficial to many applications such as automatic telephone banking and taxi booking. A trend in speech synthesis is to reinforce the message with paralinguistic cues so that communication moods and other content beyond the text itself can be delivered. This area of study is called expressive text-to-speech synthesis, with interest in non-emotional expressive speaking styles growing in recent years. It has been recognized that depending on the domain and the target group of speech applications, different expressive styles are required. For example, expressing suspense and global storytelling style is essential to storytelling applications. Theune et al. [38] proposed to generate expressive speech for storytelling applications through a set of prosodic rules extracted from human storytellers' speech. Recently, IBM [33] introduced an expressive text-to-speech engine that can be directed, via text markup, to use a variety of expressive styles including questioning, contrastive emphasis, and conveying good and bad news for American English.

23.2.3 Emerging Sensing Modalities

Motivated by the tremendous need to explore better HCI paradigms, there has been a growing interest in developing innovative sensing modalities [39–43]. Besides the vision and speech senses for HCI, computers can also simulate tactile sensing, which enables the feeling of realism through the use of haptic devices, for example, in virtual reality (VR) [41]. Haptic interfaces allow users to input commands into the computer by means of hand movements and provide users with tactile and force feedback that is consistent with what the user is viewing, thus providing users with senses to manipulate 3D virtual objects with respect to features such as shape, weight, surface texture, and temperature [42]. Haptic interfaces provide the opportunity for complex yet potentially more intuitive means of interacting with a computer, and this ability has been widely explored in medical applications [42, 43].

Another sensing modality that has seen an exponential increase in research interest is the monitoring of brain electrical activity (via electroencephalogram). Brain activity can be monitored noninvasively from the surface of the scalp and can be harnessed to directly control a computer [40]. The hands-free nature of such HCI is potentially useful in situations where hands are needed for other tasks, such as in aircraft piloting. Such sensing modality is also of paramount importance for physically disabled patients, as it allows them to interact with the latest information technologies.

23.2.4 Virtual/Augmented Reality

VR, formerly known as a visually coupled system, is a concept that aims to integrate all the sensing technologies seamlessly and allow users to gain more realistic experience in a physically and perceptually appropriate manner [22]. This approach is generally believed to be the next generation of HCI [44], as it leverages the multimodal nature of human-human interaction to facilitate multimedia computing without the need for specialized training. One of the first multimodal HCI systems can be credited to Bolt [45]. His put-that-there system fused spoken input and magnetically tracked 3D hand gestures using a frame-based integration architecture. The system was used for the simple management of a limited set of virtual objects such as the selection of objects, modification of object properties, and object relocation. Even though the natural feel of the interaction was hindered by the limitations of the technology at the time, put-that-there has remained the inspiration of all modern multimodal interfaces. A comprehensive review of multimodal HCI can be found in Sharma et al. [21] and Pantic et al. [46].

More advanced VR systems demand support from advanced multimedia technologies that include computer graphics, visualization, speech recognition, and haptic interaction. This demand is clearly illustrated by Schreer et al. [4], where the state of 3D multimedia technologies including 3D video reconstruction and rendering and 3D audio processing have been reviewed for their applications to VR as well as to telepresence. Recent trends in VR systems have resulted in less dependence on special wearable HCI devices (i.e., head-mounted displays and sensory gloves) and a move toward larger scale 3D displays and systems that minimize HCI requirements [4]. This has further facilitated the development of more immersive VR systems, leading to the development of augmented reality (AR) systems, which differ in that the visualized information and real-world visual objects co-exist in the same user interaction space [23].

23.3 Multimedia Content Management

Creating and publishing digital multimedia content today is easier than ever before, at both individual and organizational levels. Every individual in the world is a potential content producer who is capable of creating digital content that can be easily distributed and published. The ease of content production is accelerating its growth, and thus leading to problems in content management and content identification. Health care is a domain that stands to significantly benefit from enhancements in content management [47]. In a typical health care system, patient records consisting of multimedia content (images, audio, etc.) are stored electronically, and this content needs to be searchable by physicians. Thus, with growth in patient records, the ability to manage these contents so that searching can be performed efficiently is becoming increasingly important. To efficiently use the advantages of multimedia data, it is essential to develop intelligent approaches to processing and managing these data and indexing their content.

23.3.1 Multimedia Content Analysis

Multimedia data are typically annotated manually with textual descriptions and then stored in a database management system (DBMS) which controls access to these data [48–50]. However, such a solution was found to possess serious constraints when applied to multimedia content management [51]. The problems associated with manual annotation are that it is labor intensive and subjective to the operators and traditional textual annotation is limited in its ability to contextualize multimedia contents. For instance, it is not feasible to contextualize the texture visual feature of each image by keywords. Therefore, in the early 1990s, content-based retrieval was proposed to resolve these issues and allow users to access multimedia data based on their perceptual content.

As multimedia data have different formats and characteristics (e.g., image, video, and audio), different approaches are used to contextualize their contents. The ever-improved field of data processing and analysis, as described in Chapter 7, has greatly contributed to the growth of visual feature extraction. Visual features such as color, shape, and texture are extracted to characterize image and video content [52]. For example, the color histogram is used to represent the color distribution of a given image, shape for object contours, and texture for visual patterns (e.g., stripes) [53–55]. Additionally, motion information can be exploited to contextualize the movements of objects and cameras in the categorization of videos [56].

In the medical domain, visual features such as shape and texture have been used for medical image retrieval [47]. Due to the advances in image processing and possible inclusion of prior knowledge, content-based image retrieval (CBIR) has a great potential in medical image database applications. Current developments of CBIR of medical images can be found in Chapter 4.

Much as with visual features, audio feature extraction has benefited significantly from advances in audio processing that are enabling the use of, for example, loudness and harmonicity as features to characterize audio content [57], with a good example being the automated classification of music genres [58].

23.3.2 Multimedia Content Description Interface

The great potential of multimedia retrieval has attracted much interest from a large number of researchers. Therefore, many feature extraction approaches have been proposed to characterize both the perceptual and conceptual contents of multimedia data. Meanwhile, no systematic way has been found to exchange the features and to model multimedia content through these features, which may result in proprietary solutions in multimedia content access. Motivated by such a demand, the motion picture experts group (MPEG) in 1996 initiated MPEG-7 to look into the issues of providing interoperable descriptions to bridge multimedia content and its consumption and facilitate multimedia content access [59]. Unlike previous MPEG standards that target the compression and reproduction of the data itself, MPEG-7 is geared towards enhancing that data that describes the context and contents of the multimedia data, the so-called metadata.

MPEG-7 descriptions are intended to provide extensible metadata solutions for a wide range of applications where content description can be at different levels of abstraction from the low-level (automatic and statistical features) to the representation of high-level features that convey semantic meaning. In addition, highly structured MPEG-7 descriptions support the combination of low-level and high-level features in a single description. Examples of content-based access with MPEG-7 include finding information using spoken queries, hand-drawn images, and query by humming [59], as well as the personalized service of TV news [60, 61]. MPEG-7 has also been adopted to facilitate the management, delivery, and access of medical data as demonstrated by Rege et al. [62] in whose study human brain images were annotated and used to capture the semantic information such that both the retrieval tasks and answering domain-specific complex queries can be supported for image-guided neurosurgery. Similarly, electroencephalogram images were organized efficiently by combining textual information and low-level image information with MPEG-7 [63]. In a recent study, Cuggia et al. [64] introduced an integration of MPEG-7 with existing medical standards to manage digital audiovisual medical resources.

23.4 Multimedia Delivery

Multimedia data places considerable demand on computing resources and subsequently presents formidable technical difficulties for storage, networking, and computing infrastructures. Compression techniques are critical to deliver multimedia content to a wide range of communities. Telecommunication technologies offer valuable opportunities for distributing multimedia data. Undoubtedly, the Internet is one of the most popular manners, though it was not initially designed for multimedia services. Wireless sensor networks have also recently enabled a paradigm shift in the science of monitoring applications such as weather and soil moisture [65]. In this section, we will introduce a new potentially useful delivery system, the digital TV, which enables the distribution of multimedia content as well as interactive data. It is expected that almost every nation is in the process of transforming analog TV to digital TV [66]. We will also introduce the MPEG technology, which is the most routinely used multimedia delivery standard.

23.4.1 Digital Television

The TV has long been an integral part of our life, whether as an entertainment appliance or as a window to the world. Its legacy role of high penetration media-delivery in households will be further emphasized in the DTV era because DTV via satellite, cable, and terrestrial broadcasting provides much better picture and sound quality as well as more opportunities for applications such as interactive services and data-casting. A TV that receives its signal digitally is no longer just a passive box that displays pictures and sound. A DTV that is properly equipped can be a powerful and interactive computer with similarities to networked desktop PCs. Access to such TV has even been extended to handheld devices such as mobile phones and personal digital assistants (PDAs). These advantages will significantly enrich the viewing experience and take it beyond its current dimensions in addition to fostering a number of novel applications (such as the ability to access health information and medical consultation) in the field of home health care and mobile health systems (m-health) [67]. Cable TV networks have also been used to transmit alarms, emergency calls and biomedical data and to provide home telecare interactively [68].

Traditionally, users follow TV program schedules to watch their favorite shows. This situation has changed since the videocassette recorder was introduced, allowing users to record programs and watch them at the user's convenience. Increases in modern computing and storage capacity are bringing users more improved services such as video-on-demand and digital recording.

23.4.2 Multimedia Compression

Today, we can enjoy digital music almost anywhere with music players becoming smaller and able to store increasing numbers of songs. In addition, digital video can be watched through the Internet, DTV, and even hand-held devices. Behind all these successful multimedia products are standards that provide interoperability and thus generate a marketplace in which consumer equipment manufacturers can produce competitive yet conformant products.

In terms of static images, the joint photographic experts group (JPEG) standard has achieved enormous success [69], as is evident from the use of JPEG in all digital cameras to store pictures. To accommodate advances in multimedia technology, JPEG has evolved into JPEG2000. JPEG2000 uses wavelet compression as the core technique [70] to develop a new image coding standard for different types of images (e.g., bi-level, gray-level, and color) with different characteristics (e.g., natural, scientific, medical, and text), and thus allowing different imaging models (e.g., real-time transmission and image library archival) preferably within a unified and integrated system. The advantages of JPEG2000 have spurred its application in biomedicine. For instance, Khademi and Krishman [71] successfully used JPEG 2000 for robust and real-time digital mammogram compression with efficient database access and remote access to digital libraries that was shown to reduce the time required during diagnosis.

Different standards of MPEG which include MPEG-1, MPEG-2, and MPEG-4 have been proposed for different applications, a decision which forms a large part of the MPEG format's popularity. MPEG-1 (issued in 1991), entitled "Coding of Moving Pictures and Associated Audio at Up to About 1.5 Mbps" is the first standard by the MPEG and is intended for medium-quality (e.g., VHS quality) and medium bit-rate video and audio compression. MPEG-1 organizes audio coding schemes in three layers, simply called layer-1, layer-2, and layer-3. Encoder complexity and performance (sound quality per bit rate) progressively increase from layer-1 to layer-3. Each audio layer extends the features of the layer with the lower number. The popular MP3 file format is an abbreviation for MPEG-1 layer-3, which set the stage for the ongoing revolution in distributing digital music.

MPEG-2 (issued in 1994) [72], entitled "Generic Coding of Moving Pictures and Associated Audio" was designed to support more coding schemes, a wider range of bitrates, and more choice in video resolution (e.g., high definition TV). Although MPEG-2 systems have video and audio specifications that are largely based on the MPEG-1 specifications, MPEG-2 provides higher picture quality by using higher data rates. Advanced audio coding is added to provide a significant performance increase over backward-compatible audio. MPEG-2 tries to be a generic coding standard for a wide range of applications by comprising a large set of tools to meet the requirements of various applications. The tool sets are characterized in terms of profiles and levels that are defined to provide coding solutions with appropriate complexity as well as to limit the memory and computational requirements for various applications. For example, set-top boxes of standard definition TV and high definition TV correspond to the implementation of different profiles/levels of MPEG-2 standard.

MPEG-4 (issued in 1999), entitled "Coding of Audio-Visual Objects" is the latest video coding standard and is designed to move from the pixel-based to an object-based approach [56]. This is achieved by embodying a significant conceptual jump in audiovisual content representation and object-based modeling and thus enabling an audiovisual scene to be built as a composition of independent objects with their own coding, features, and behaviors. Temporal and spatial dependencies between objects can also be described with a binary format for scene description. In summary, the object-based coding approach in MPEG-4 allows for hybrid natural and synthetic coding, content-based interaction and reuse, content-based coding (e.g., text coding tools for text objects and 3D model coding for 3D objects), and universal access, which has the potential to revolutionize the way users create, reuse, access, and consume multimedia content. The power and advantages of the object-based representation also make MPEG-4 a standard that can be applied to applications ranging from low bit-rate personal mobile communications to high-quality studio production.

23.4.3 Multimedia Framework

As mentioned in the sections above, MPEG has played a key role in developing the standards for multimedia-enabled products and applications. However, in general, each multimedia entity needs to communicate with its environment, clients, and applications. A solution is required to offer users transparent and interpretative consumption and delivery of rich multimedia content so that the coding and metadata standards can be linked with access technologies, rights and protection mechanisms, adoption technology, and a standardized event reporting mechanism under the umbrella of a complete multimedia framework. Furthermore, the need for a solution also stems from the fact that the universal availability of digital networks, particularly the Internet, is changing traditional business models, adding the new possibility of the electronic trade of digital content in addition to the trade of physical goods. The aim of the framework is to fill the gaps in the multimedia delivery chain and to create seamless and universal delivery of multimedia. As such, MPEG-21 (issued in 2000) was introduced to provide a complete framework for delivering and managing multimedia content throughout the chain, encompassing content creation, production, delivery, personalization, consumption, presentation, and trade, to meet its vision "to enable transparent and augmented use of multimedia resources across a wide range of networks and services" [73].

MPEG-21 proposed a new distribution entity termed the digital item (DI) for use in interaction with all users in a distributed multimedia system, where a user is any entity that interacts with the MPEG-21 environment or that makes use of a DI. Such users include individuals, consumers, communities, organizations, corporations, consortia, governments, and other standards bodies and initiatives around the world, and their roles are identified specifically with regard to their relationship to another user for a certain interaction. In particular, content management, intellectual property management, and content adaptation are regulated to handle different service classes. MPEG-21 is thus a major step forward in multimedia standards. It collects the technologies to create an interoperable infrastructure for transparent and protected digital media consumption and delivery. Many application domains have benefited from adopting MPEG-21. MPEG-21 has been used to establish personalized video systems [61] and in backpack journalism scenarios [74]. Together with MPEG-7, MPEG-21 can also be leveraged to create distributed multimedia databases [75]. Successful applications of MPEG-21 have also been reported in the medical and health care domains. The intellectual property management and protection function of MPEG-21 was employed to provide accurate audit trails to authenticate appropriate access to medical information (e.g., patient records) that is shared nationally in England [76]. It has been shown that MPEG-21 can be used to implement information architecture for electronic health records and features of MPEG-7 such as universal accessibility and interoperability will make the architecture highly interoperable in both existing health care systems and different multimedia systems [77].

23.5 Smart Medical Home

The primary aim of medical homes is to develop an integrated health system that is personalized to an individual's home. This technology will allow consumers, in the privacy and comfort of their own homes, to maintain health, detect the onset of disease, and manage symptoms. The data collected 24 hours a day, 7 days a week inside the home will augment the data collected by physicians and hospitals. The data collection modules in the home start with the measurement of traditional vital signs (blood pressure, pulse, respiration) and work to include measurement of new vital signs such as gait, behavior patterns, sleep patterns, general exercise, and rehabilitation exercises [3, 7]. The smart medical home has the potential to delay or partially remove the dependence on retirement nursing homes and thereby extend the person's quality of life. Incorporating smart medical devices into homes can potentially make a strong and positive impact on the lives of persons with physical disabilities and those with chronic diseases [7, 78, 79]. Clinical studies have demonstrated that the use of medical devices in the patient's home can identify adverse trends in clinical signs early and reduce time spent in the hospital [79]. This is made possible by combining multimedia technologies such as networked care systems (telemedicine) with integrated sensors that monitor clinical signs, medication reminders, health education, and daily logs. Figure 23.2 illustrates an example of the multimedia-enabled components that make up a smart medical home.

23.5.1 Recent Projects in the Smart Medical Home

Numerous medical devices and systems have been designed and developed for the home environment with the purpose of

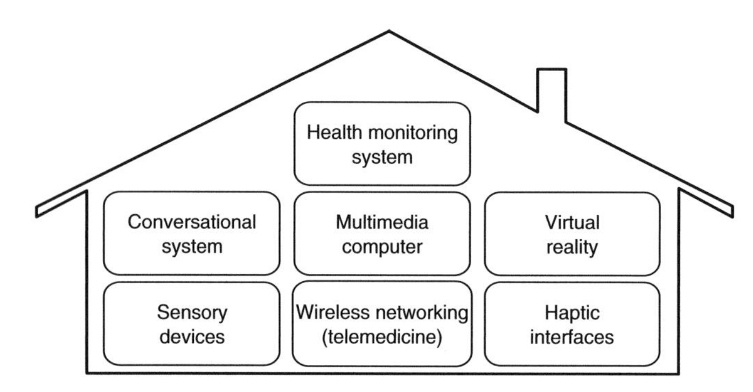

FIGURE 23.2 Primary components for the smart medical home using multimedia technologies. These components all share the purpose of improving health care and quality of life of the consumers. In a typical scenario, the health information collected from the sensors and conversational system in the medical home is transmitted (wireless networking) to physicians in the hospital. The received information is then augmented with already existing patient information and can be used by the physician for diagnosis and consultation with the patient via teleconsultation (telemedicine) using a multimedia-enabled computer with haptic controls. Treatment prescribed for the patient maybe through the use of virtual reality system that is designed for surgical rehabilitation.

providing health benefits to the residents. There are many large projects in the field of smart medical homes that bring together knowledge and expertise from many different disciplines of engineering and biomedicine. Table 23.1 presents selected projects involved with smart medical homes that have made significant research contributions and continue to expand its aim of establishing a smart medical home.

The AgeLab at MIT, [e1] in Table 23.1, aims at meeting the needs of the aging population to invent the future of healthy and active living. Smart medical advisor is a wireless system that uses the consumer's health information to help with food purchasing decisions. Another project titled adaptive devices for independent living is an assistive device to deliver personal information, basic health care, support, and critical assistance for older adults and people living with a degenerative condition. Employing a combination of state-of-the-art technologies, these devices are being designed to be user-friendly so even those with limited technology skills can benefit from their design in their own home.

The smart medical home project at Rochester University's center for future health, [e2] in Table 23.1, is a controlled environment in which research, preconcept testing, concept testing, pilots, and prototype testing are done. The center's overall goal is to develop and integrate a personal health system for the home so that all technologies work seamlessly and allow consumers, in the privacy of their own homes, to maintain health, detect the onset of disease, and manage disease. Selected projects for smart medical homes include smart bandages, wearable medical devices, motion understanding—investigating the ability to learn human motion for medical diagnosis, conversational medical advisor—a system that has conversational interface for consumer use (i.e., medication management), and other medical home components such as networking and decision support.

The medical automation research center at the University of Virginia, [e3] in Table 23.1, has a research and development project working on smart house technologies. These are passive and unobtrusive technologies for monitoring elders' activities, designed with privacy and security in mind. Some of the monitoring technologies developed are a sleep monitoring system that uses noninvasive sensors to gather sleep and physiologic signals and a smart in-home monitoring system that is composed of a suite of low-cost, noninvasive sensors (strictly no cameras or microphones), and a data logging and communications module, in addition to an integrated data management system, linked to the Internet for the purpose of lessening the burden of the caregivers and increasing quality of life for the elders.

These projects are varied in their technical aspects, audience, and management aspects (from single institutional research to international collaboration), and hence provide a broad

TABLE 23.1 Selected homepages of projects involved with the smart medical home

Project Name	Institution	Features	Homepage
[e1] AgeLab	Massachusetts Institute of Technology	Developing systems for aging populations, such as medical advisors and adaptive devices for independent living	http://web.mit.edu/agelab
[e2] Smart Medical Home	Center for Future Health, University of Rochester	Integration of health technologies for home, with projects based on motion understanding, a conversational medical advisor, smart bandages, and others	http://www.futurehealth.rochester.edu/smart_home
[e3] SmartHouse Technologies	Medical Automation Research Center, Virginia University	Health technology for elders that emphasizes health monitoring, including smart in-home monitoring and sleep monitoring	https://smarthouse.med.virginia.edu
[e4] Personal Ambulatory Monitoring	E-Health Research Center, CSIRO	Investigating chronic disease management solutions from home by monitoring patient's vital signs	http://www.e-hrc.net

overview and great insights into the requirements and prospects necessary for the smart medical home.

23.6 Telemedicine in the Smart Medical Home

Telemedicine refers to the use of multimedia technologies that consist of audio, visual, and network for use in medical diagnosis, treatment, and patient care. This is made possible through the exchange of health information, which allows the provision of health care services across geographic, time, social, and cultural barriers between patients and physicians [80–83]. There has been significant growth in the field of clinical applications for telemedicine. This includes, for example, teleconsultation [84, 85], teleradiology [86, 87], and remote patient monitoring [88].

Teleconsultation and remote patient monitoring are fundamental technologies for a successful smart medical home. As an example, for patients undergoing chronic disease management, it is necessary to consult physicians on a regular basis [78]. Health monitoring technologies can potentially reduce the need for the patient to physically meet the physician. Rather, the practitioner could perform teleconsultation via video conferencing, with the necessary medical information shared through remote monitoring systems [88]. Another technology that is often found together with telemedicine for use in the smart medical home is a special sensory device that is used to aid communications such as gesture recognition [89] and speech recognition [6]. Greater details regarding sensory devices and health monitoring systems are discussed in the following section.

23.7 Sensory Devices and Health Monitoring

Due to a continuous decline in size, costs, and power consumption of sensory devices, it is now common to find sensors embedded in different places and objects, such as home appliances/furniture [7, 9] or wearable items like wristbands [81], jewelry [89], and clothing [24, 90, 91]. This section will discuss sensor technology and its application to the smart medical home.

23.7.1 Wearable Devices in Health Care

Wearable devices can be broadly defined as mobile electronic devices that can be unobtrusively embedded in the user's outfit as part of clothing or as an accessory [91, 92]. These devices are made up of three main components: sensors that measure vital health signs; computing hardware that processes, displays, and transmits information from the sensors; and clothing that acts as the supporting element and cosmetic exterior of the device [91]. These mobile devices are fundamentally designed to be operated and accessed without interfering with the user's daily living activities [91, 93]. There has been much discussion about the development of wearable devices for medical applications. Examples include discussion of sensors to measure vital signs in patients with dementia [93], evaluation of response to stressful training situations [94], and the rehabilitation of patients with stroke and heart disease [95]. These studies and others have demonstrated that wearable devices have the potential to become integral components of a modern health care system, as they can provide alternative options and solutions to numerous medical and social requirements [10, 15, 96]. These devices not only improve the provision of health care to enhance the quality of life of the chronically ill and the disabled, but also have proven to be financially rewarding by saving the health service money via hospitalization reductions, either through prevention or by helping provide appropriate means for independent living.

Instead of measuring health signs, a study by Starner et al. [89] introduced a wearable device for use in measuring the tremor of a patient's hand as the user makes a gesture. A pendant that consists of a small camera was designed to be worn by a patient, such that the pendant records and interprets hand gestures that are performed by the wearer. The measurement of tremors, particularly in older populations, was shown to be beneficial in detecting the signs of various medical conditions.

23.7.2 Health Monitoring Systems

There is a clear need for health monitoring systems to form a part of the smart medical home, thereby providing both the monitoring of the occupant's vital health signs and the ability to react to changing health signs [7, 9, 80, 81, 94, 96–99]. These systems are necessary and include a monitoring system that automatically alerts the hospital staff when patients' vital conditions are abnormal and a patient tracking system that is used to track patients who may require immediate assistance. In a typical health monitoring setup, sensors are used to measure patient vital signs and feed information to the monitoring system.

A monitoring system that measures patient blood pressure and records an electrocardiogram was presented by Hung et al. [80]. Here, the novelty was the use of a mobile phone as the monitoring interface using the wireless application protocol such that Web-enabled mobile phones could be used as a health monitoring device. In another study, Anliker et al. [81] presented an advanced care and alert portable telemedical monitor, which is a wearable medical monitoring and alerting system designed for cardiac/repository patients. This system combines the measurement of multiple vital signs, online analysis, and cellular communication to a telemedicine center in a wearable wrist device that is unobtrusive for everyday use. In another paper, Korhenen et al. [9] reported on a wireless wellness project (WWM) that aims to develop a prototype system for homes supporting ubiquitous computing applications for wellness management and home automation. The WWM focused on building a home network where multiple simple household and health monitoring devices were connected. An Internet-based information and support system for patient home recovery after coronary artery bypass graft surgery was presented in the work by Brennan et al. [99]. This system was designed to extend the scope of nursing services to patients from hospital through home, in addition to providing information and support that are tailored to individual patients' needs during recovery in a timely manner.

23.8 Speech Recognition and Conversational Systems

Speech and language processing systems that enable users to communicate with computers using conversational speech are expected to greatly improve our health care system through the development of an easier and more efficient manner to communicate and interact with computers [6, 100–102].

23.8.1 Speech Recognition in Medical Applications

Speech recognition systems provide computers with the ability to identify spoken words and phrases, thereby allowing their use as an interface to command and control computer programs [100] as well as to provide a means of control for physically impaired people [103]. Recent studies have shown that the quality of speech recognition and its usability are continuously being improved, with increases in speech recognition systems being adopted in medical applications [100, 101, 104]. In the medical field, these innovations have primarily been incorporated into dictation systems for the development of reporting systems [104]. For example, speech recognition systems are steadily replacing conventional transcription services in hospitals, primarily driven by the potential increase in operational productivity (shorter time for examination) and reduction in time required to index the reports into the hospital's information system [100]. The ability for the user to create his or her vocabulary dictionary in speech recognition systems has increased their usability and acceptance in the medical field, which enables the addition of medical vocabulary and technical medical terms. A key factor in the rapid growth of speech recognition systems has been the introduction of picture archiving and communication systems (PACS) [105]. PACS enables immediate availability of medical images, which has greatly increased the time between the availability of these images and their corresponding reports [100].

23.8.2 Conversational Human-Computer Interface Systems

Conversational systems differ from speech recognition systems in that conversational systems not only recognize the words input by the user but they also attempt to interpret the meaning of these words (i.e., understand the spoken dialogue of the user). Dialogue systems have found various applications, for example, intelligent dialogue systems [106] and problem solving assistant and speech translational systems [107], which demonstrated the use of a mobile device that recognized user input in Japanese, and output an English translation of the input. In another study, Polifroni et al. [108] reported the use of dialogue input to allow users to search information from the Web so that the interaction with the computer was natural and flexible.

The adaptation of speech recognition and conversational HCI systems in the smart medical home domain has created several novel applications. In particular, the medication advisor project proposed by Ferguson et al. [6] is an intelligent assistant that interacts with its users via conversational natural language, with the purpose of providing the users with information and advice regarding their prescription medication. This study has shown that the dialogue system between human and computer has the potential to aid people in managing their medication, and that such systems can find many other smart medical home applications that can improve and enhance the lives of people.

23.9 Multimedia Technologies for Patient Education and Care

Informing patients about diagnosis, surgery, and treatment is necessary and often has a significant impact on patient awareness of medical procedures [16, 99, 109–113]. The use of multimedia to facilitate patient awareness is becoming widespread, with the advent of visual, audio, interactive, and Internet content that compliments traditional paper-based information. The acceptance of multimedia usage and the relay of information from these media are an important aspect of patient awareness. Two topics are discussed in this section: use of multimedia content for patient education and awareness and multimedia technologies for reducing patient anxiety in regards to surgery.

23.9.1 Multimedia for Patient Education and Awareness

The use of multimedia has been an inseparable tool in education as it enables the presentation of visual, audio, and interactive information. The use of multimedia computers for patient education is becoming widespread in the health care industry from the increase in use and acceptance of computers by patients and improvements in technology modules that have eased comfort and usage of these educational systems [109, 114, 115]. This is becoming more important as patients require increased access to medical information to help facilitate decisions about their health care. Krishna et al. [116] discussed a trial on the effect of an Internet-enabled interactive multimedia asthma education program, with results indicating a significant increase in asthma knowledge for both patients and caregivers. Furthermore, this study also found that this knowledge resulted in a reduction of symptom days and emergency department visits. Fagerlin et al. [117] developed and evaluated a multimedia education program for patients with prostate cancer. The program was carried out on a stand-alone personal computer and participations used a selection of on-screen buttons that controlled interaction and navigation with an assistant providing instructions and support. The study reported that the majority of the information needs were fulfilled; however, like many other studies, the information provided was not completely satisfactory [115].

Another area that has benefited from the use of multimedia is the delivery of information to patients and their families in the hospital waiting room. These waiting rooms typically have a large amount of information in print media forms (pamphlets and posters) regarding preoperative procedures. Often there is also a nurse, social worker, physician, or volunteer present to assist patients and families with this information and with understanding the progress of patients through talking in person or over the telephone. Such practice of information delivery is low-tech and can potentially benefit from the use of multimedia technologies [118]. Wheeler et al. [111] presented the use of videotapes containing information regarding the use of antibiotics that was shown to improve both patient awareness and understanding of the drugs. Wheeler et al. further stated that the conventional paper media (pamphlets) were not read by patients, reinforcing the advantages of using multimedia contents. In another study, Oermann et al. [119] demonstrated the advantage from the use of multimedia content in informing patients. Here, educational videos were also used to provide instructions in the clinical waiting area. They concluded that such video instructions can be an effective and efficient teaching intervention for health information.

23.9.2 Multimedia for Reducing Patient Anxiety

Multimedia technologies have been used as a tool for reducing patient anxiety [112, 120–122]. The use of multimedia has the potential to reduce the need for sedatives to help patients relax during surgery [123] and thereby benefit patients. Different methods have been used to alleviate anxiety and fear during the operative periods, ranging from friendly hospitals, multimedia-embedded operating theatre environment, and the availability of relevant information and explanations. It is common to find music being played in the operating theatre during surgery [120, 121]. Music was shown to reduce patient anxiety by lessening the unfamiliar noise and auditory stimuli that occur during surgery and to reduce anxiety before surgery. Even though the use of multimedia has advantages for patients, there have been many reported studies that suggest these aids to patients can cause distraction to physicians and nurses during surgery, primarily with regard to hearing vital sign measures and potential conflicts in communications with staff when music is used in the operating room. Therefore, it is important to take the above issues into consideration when multimedia systems are designed.

Apart from music, Man et al. [112] presented the effect of intraoperative video on patient anxiety with results indicating ease in patient anxiety during operation and in overall improvement of comfort and satisfaction. In this study, patients undergoing surgery were equipped with specially designed glasses that included liquid crystal display screens and audio used to playback video.

23.10 Multimedia Operating Theater and Virtual Reality

Computer-driven simulations of operating theater in VR are articulating huge interests from both education and clinical environments [124–127]. The ability to utilize multimedia components to create real-time simulation of surgical procedures in a controlled and realistic setting (i.e., a virtual theater) provides the operating surgeon with the ability to perform preoperative planning and practice, surgical education, and training. Such systems have demonstrated usefulness where the outcomes of surgery are often determined by the technical skills of the surgeon. In particular, the use of virtual/augmented reality for surgical planning and education has great potential to evolve for medical home uses (i.e., virtual simulations that are targeted to increase emotional comfort of older and disabled patients at home) [3, 128, 129].

23.10.1 Multimedia Operating Theatre

The utilization of computers in surgical procedures is often referred to as computer assisted/aided surgery and/or computer-integrated surgery (CIS) [130]. Recent studies have shown that the use of CIS has the potential to contribute to cost cutting in health care by allowing fewer staff to perform the same surgery in less time than with traditional methods [131]. Refer to Chapter 18 for further details about CIS and other related topics.

Multimedia technologies play an essential role in the development of CIS systems. In a typical CIS, these systems consist of image viewing (two-dimension/three-dimension) software, a telecommunication system with video/audio conferencing, and an interactive user interface that uses input devices that provide haptic feedback, such as a joystick device [132, 133]. Other multimedia devices have also been used such as the use of remote visualization in the operating theatre [126], where a projection of the images is displayed and controlled by the surgeon using a joystick. Gering et al. [134] reported a camera and position sensory device that was embedded into a surgical cutting knife, where the captured video was used to navigate through a volume rendering of the patient's data in relation to the knife. There has also been introduction of robotics that are controlled using computers for use in surgery [135].

23.10.2 Virtual Reality for Medical Teaching and Training

Recent studies in biomedical information technologies demonstrated the capability of using VR in surgical procedures as a tool for simulation and training [124, 136–139]. As described in Section 23.2.4, VR can be used to create an immersive environment that is made up of multimedia contents that simulate realistic conditions of surgery [125]. One of the key potentials of this multimedia technology is its application to aid in education by allowing interactive training of surgical procedures [139, 140]. VR can compliment traditional approaches to teaching of surgical skills which usually involve a *see and do* approach [141]. It has been demonstrated on a number of applications such as for simulating a vascular reconstruction in a virtual operating theatre [125] and for neurosurgery planning in virtual workbench [138]. An alternative use of VR in medical education is described by Johnsen et al. [142]. Here, the experience in interaction between patients and physicians was simulated through the use of virtual characters. These life-size characters were projected to screens and were used to interact using gestures and speech.

VR in biomedicine is not only restricted to teaching and training, but can also be used for numerous health care applications. Flynn et al. [115] reported a virtual reality system that has been used to enhance recovery of skilled arm and hand movements after stroke. Particularly with redundant input, VR systems enable physically or cognitively handicapped people to access computers.

23.11 Summary

This chapter has discussed the advances in multimedia technologies and their applications to the smart medical home (and hospitals). The core technologies covered include human-computer interaction, multimedia content management, and multimedia delivery, followed by their impact on the development of medical applications for smart medical homes (and hospitals), including telemedicine, sensory devices, speech and conversational systems, patient education and care, operating theatres, and virtual reality. Multimedia technologies have already demonstrated their usefulness, and they will continue to expand their usefulness in health care and in improving the quality of people's lives.

Acknowledgments

The authors are grateful to the support from ARC, PolyU/UGC grants.

23.12 Exercises

1. Analyze the influence of MPEG standards in health care systems for medical homes.
2. State and discuss the most significant differences between MPEG-4 and MPEG-1/2.
3. Describe a multimedia-enabled health care system or product and identify what enabling multimedia technologies have been adopted.
4. Telemedicine is a term used to describe the telecommunications technology for medical diagnosis and patient care when the provider and client are separated by distance. List three or more telemedicine applications and their similarities and differences.
5. What are the advantages and disadvantages of a virtual operating theatre in the training of physicians?
6. Multimedia is often used to reduce patient anxiety in both the waiting room and operating theatre. What other benefits does multimedia have in these environments for patients?
7. Wearable devices are different from other medical devices in that they must not only be designed to aid in the user's health, but also be comfortable and unobtrusive. What are the guidelines to follow in the design of wearable medical devices?

23.13 References and Bibliography

1. D. D. Feng. Guest editorial: Multimedia information technology in biomedicine. *IEEE Trans. Inf. Technol. Biomed.* 4(2):85–87, 2000.
2. E. H. Shortliffe et al. *Medical Informatics—Computer Applications in Health Care and Biomedicine.* 2nd ed. Springer, 2000.
3. D. H. Stefanov, Z. Bien, and W.-C. Bang. The smart house for older persons and persons with physical disabilities: Structure, technology arrangements, and perspectives. *IEEE Trans. Neural Syst. Rehabil. Eng.* 12(2):228–250, 2004.

4. O. Schreer, P. Kauff, and T. Sikora. *3D Videocommunication: Algorithms, Concepts and Real-Time Systems in Human Centred Communication.* John Wiley & Sons, 2005.
5. X. Lu et al. Construction of multimedia courseware and web-based E-learning courses of "biomedical materials." In *Proc. IEEE EMBS.* 2886–2889, 2005.
6. G. Ferguson et al. The Medication Advisor project: Preliminary report. CS Dept., U. Rochester, 2002.
7. M. Chan et al. Smart house automation system for the elderly and the disabled. *IEEE Proc. Conf on Systems, Man and Cybernetics.* 1586–1589, 1995.
8. C. Kidd et al. The aware home: A living laboratory for ubiquitous computing research. *Proc. Workshop on Cooperative Buildings Co. Build.* 1999.
9. I. Korhonen, J. Parkka, and M. Van Gils. Health monitoring in the home of the future. *IEEE Eng. Med. Biol. Mag.* 22(3):66–73, 2003.
10. P. Ooi, G. Culjak, and E. Lawrence. Wireless and wearable overview: Stages of growth theory in medical technology applications. *Proc. Int. Conf. Mobile Business.* 528–536, 2005.
11. K. Kowalenko. Home as the hub of health care. In *IEEE Spectrum.* 3–5, 2006.
12. F. Moraes et al. Using the CAN protocol and reconfigurable computing technology for web-based smart house automation. *Symp. Integrated Circuits and Systems Design.* 38–43, 2001.
13. H. B. Stauffer. Smart enabling system for home automation. *IEEE Trans. Consumer Electronics.* 37(2):29–25, 1991.
14. S. Brownsell, G. Williams, and D. A. Bradley. Information strategies in achieving an integrated home care environment. In *Proc. IEEE Conf. EMBS.* 1224, 1999.
15. K. Hung, Y. T. Zhang, and B. Tai. Wearable medical devices for tele-home healthcare. *IEEE Proc. Engineering in Medicine and Biology Society, EMBC 2004.* 5384–5387, 2004.
16. A. D. Fisk, A. L. Mykityshyn, and W. A. Rogers. Learning to use a home medical device: Mediating age-related differences with training. *Human Factors.* 44(3):354–364, 2002.
17. H. K. Huang. Special issue on multimedia applications in health care. *IEEE Multimedia.* 4(2), 1997.
18. J. Allen, G. Ferguson, and A. Stent. An architecture for more realistic conversational systems. *Proc. Int. Conf. Intelligent User Interfaces.* ACM Press. 1–8, 2001
19. S. Park and S. Jayaraman. Enhancing the quality of life through wearable technology. *IEEE Eng. Med. Biol. Mag.* 22(3):41–48, 2003.
20. A. Melenhorst, W. A. Rogers, and E. C. Caylor. The use of communication technologies by older adults: Exploring the benefits from the users perspective. *Proc. HFES Annual Meeting.* 221–225, 2001.
21. R. Sharma, V. I. Pavlovic, and T. S. Huang. Toward multimodal human-computer interface. *Proc. IEEE.* 86(5):853–869, 1998.
22. J. A. Adam. Virtual reality is for real. *IEEE Spectrum.* 30 (10):22–29, 1993.
23. R. Azuma et al. Recent advances in augmented reality. *IEEE Comput. Graph. Appl.* 21(6):34–47, 2001.
24. S. Mann. Wearable computing: A first step toward personal imaging. *IEEE Computer.* 30(2):25–32, 1997.
25. L. Barfield. *Design for New Media: Interaction Design for Multimedia and the Web.* Pearson Addison Wesley, 2004.
26. S. Li et al. *Handbook of Face Recognition.* Springer, 2005.
27. M. Valera and S. A. Velastin. Intelligent distributed surveillance systems: A review. *IEEE Proc. of Vision, Image and Signal Processing.* 152(2):192–204, 2005.
28. D. Moore. *Vision-Based Recognition of Actions Using Context.* [dissertation]. Atlanta: Georgia Institute of Technology, 2000.
29. S. L. Wang, W. H. Lau, and S. H. Leung. Automatic lipreading with limited training data. *Int. Conf. Pattern Recognition.* Hong Kong. 881–884, 2006.
30. R. Plarnondon and S. N. Srihari. Online and off-line handwriting recognition: A comprehensive survey. *IEEE Trans. Pattern. Anal. Mach. Intell.* 22(1):63–84, 2000.
31. S. R. Buss. *3D Computer Graphics: A Mathematical Introduction with OpenGL.* Cambridge University Press, 2003.
32. I. G. Tollis et al. *Graph Drawing: Algorithms for the Visualization of Graphs.* Prentice Hall, 1998.
33. J. F. Pitrelli et al. The IBM expressive text-to-speech synthesis system for American English. *IEEE Trans. on Audio, Speech and Language Processing.* 14 (4):1099–1108, 2006.
34. L. Rabiner and B. Juang. *Fundamentals of Speech Recognition.* Prentice Hall, 1993.
35. Sphinx project II Available from: http://cmusphinx.sourceforge.net/.
36. IBM ViaVoice. Available from: http://www–3.ibm.com/software/speech/.
37. A. Breen. Speech synthesis models: A review. *J. Electronics and Communication Engineering.* 4(1):19–31, 1992.
38. M. Theune et al. Generating expressive speech for storytelling applications. *IEEE Trans. on Audio, Speech and Language Processing.* 14(4):1137–1144, 2006.
39. M. Bergamsco. Haptic interfaces: The study of force and tactile feedback systems. *IEEE Int. Workshop on Robot and Human Communication,* 1995.
40. H. Lusted et al. Controlling computers with neural signals. *Scientific American.* 96:82–87, 1996.
41. M. A. Srinivasan and C. Basdogan. Haptics in virtual environments: Taxonomy, research status, and challenges. *Computer and Graphics.* 21 (4):393–404, 1997.
42. M. L. McLaughlin. Simulating the sense of touch in virtual environments: Applications in the health sciences. *Digital Media: Transformations in Human Communication.* Peter Lang Publishers, 265–274, 2006.
43. Haptics user interfaces for multimedia systems. *IEEE Multimedia.* 13(3), 2006.

44. A. Jaimes and N. Sebe. Multimodal human computer interaction: A survey. *ICCV Workshop on HCI.* Beijing China. 1–15, 2005.
45. R. A. Bolt. Put that there: Voice and gesture at the graphics interface. *ACM Computer Graphics.* 14(3):262–270, 1980.
46. M. Pantic et al. Toward an affect-sensitive multimodal human-computer interaction. *Proc. IEEE.* 91(9): 1370–1390, 2003.
47. H. Muller et al. A review of content-based image retrieval systems in medical applications: Clinical benefits and future directions. *Int. J. Medical Informatics.* 73(1):1–23, 2004.
48. N. S. Chang and K. S. Fu. Query by pictorial example. *IEEE Trans. Software Engineering.* 6(6):519–524, 1980.
49. S. K. Chang and A. Hsu. An intelligent image database system. *IEEE Trans. Software Engineering.* 14 (5):681–688, 1988.
50. S. F. Chang et al. Next-generation content representation, creation, and searching for new media applications in education. *Proc. IEEE.* 86(5):884–904, 1998.
51. D. Ritendra, L. Jia, and Z.W. James. Content-based image retrieval: Approaches and trends of the new age. In *ACM SIGMM Int. Workshop on Multimedia Information Retrieval.* 253–262, 2005.
52. D. Feng, W. C. Siu, and H. J. Zhang. *Multimedia Information Retrieval and Management—Technological Fundamentals and Applications.* Springer-Verlag, 2003.
53. T. R. Reed and H. D. Buf. A review of recent texture segmentation and feature extraction techniques. *Computer Vision, Graphics, and Image Processing: Image Understanding.* 57(3): 359–379, 1993.
54. J. Zhang and T.-N. Tan. Brief review of invariant texture analysis methods. *Pattern Recognition.* 35(3):735–747, 2002.
55. M. Chantler and L. V. Gool. Special issue on texture analysis and synthesis. *Int. J. Comput. Vision.* 62(1–2):2005.
56. T. Ebrahimi and F. Pereira. *The MPEG–4 Book.* Prentice Hall PTR, 2002.
57. I. Mierswa and K. Morik. Automatic feature extraction for classifying audio data. *Machine Learning.* 58(2–3):2004.
58. N. Scaringella, G. Zoia, and D. Mlynek. Automatic genre classification of music content: A survey. *IEEE Signal Processing.* 133–141, 2006.
59. B. S. Manjunath, P. Salembier, and T. Sikora. *Introduction to MPEG-7: Multimedia Content Description Interface.* Wiley, 2002.
60. M. Rovira et al. A MPEG-7 based personalized recommendation system for digital TV. *IEEE Int. Conf. Multimedia and Expo (ICME2004).* 823–826, 2004.
61. B. Tseng et al. Using MPEG-7 and MPEG-21 for personalizing video. *IEEE Multimedia.* 11 (1):42–52, 2004.
62. M. Rege et al. Using MPEG-7 to build a human brain image database for image guided neurosurgery. *SPIE Int. Symp. Medical Imaging.* 512–519, 2005.
63. L.-S.P. et al. A medical EEG/video multimedia content description system. *Int. Symp. Intelligent Signal Processing and Communication Systems.* 592–595, 2004.
64. M. Cuggia, F. Mougin, and P. L. Beux. Indexing method of digital audiovisual medical resources with semantic web integration. *Int. J. Med. Inform.* 74:169–177, 2005.
65. J. Gehrke and L. Liu. Introduction to sensor-network applications. *IEEE Internet Computing.* 10(2):2006.
66. R. M. Rast. The dawn of digital TV. *IEEE Spectrum.* 24 (10):26–31, 2005.
67. B. Gunter. *Digital Health: Meeting Patient and Professional Needs Online.* Lawrence Erlbaum Associates, 2005.
68. R.-G. Lee et al. Home telecare system using cable television plants: An experimental field trial. *IEEE Trans. Inf. Technol. Biomed.* 4 (1):37–44, 2000.
69. M. Ghanbari. *Standard Codecs: Image Compression to Advanced Video Coding.* IEE Press, 2003.
70. T. Acharya and P.-S. Tsai. *JPEG2000 Standard for Image Compression: Concepts, Algorithms, and VLSI Architectures.* Wiley-Interscience, 2005.
71. A. Khademi and S. Krishnan. Comparison of JPEG2000 and other lossless compression schemes for digital mammograms. *IEEE Conf. Engineering in Medicine and Biology,* 2005.
72. B. Haskell et al. *Digital Video: An Introduction to MPEG-2.* Chapman & Hall, 1997.
73. I. S. Bernett. *The MPEG-21 Book.* Wiley, 2006.
74. G. Drury and I. Burnett. MPEG-21 in backpack journalism scenario. *IEEE Multimedia.* 12(4):24–32, 2003.
75. H. Kosch. *Distributed Multimedia Database Technologies Supported by {MPEG-7} and MPEG-21.* CRC Press, 2004.
76. G. A. Brox. MPEG-21 as an access control tool for the national health service care records service. *J. Telemed. Telecare.* 11(suppl 1):23–25, 2005.
77. M. Alexander, R. Clarke, and C. W. Johnson. An information architecture for electronic health records using MPEG-21 multimedia framework. *National Health Informatics Conferences.* 128–135, 2005.
78. T. Bodenheimer et al. Patient self-management of chronic disease in primary care. *JAMA.* 288(19):2469–2475, 2002.
79. B. G. Celler, N. H. Lovell, and J. Basilakis. Using information technology to improve the management of chronic disease. *Med. J. Australia.* 179(5):242–246, 2003.
80. K. Hung and Y.-T. Zhang. Implementation of a WAP-based telemedicine system for patient monitoring. *IEEE Trans. Inf. Technol. Biomed.* 7(2):101–107, 2003.
81. U. Anliker et al. AMON: A wearable multiparameter medical monitoring and alert system. *IEEE Trans. Inf. Technol. Biomed.* 8(4):415–427, 2004.
82. J. E. Cabral Jr. and K. Yongmin. Multimedia systems for telemedicine and their communications requirements. *IEEE Communications.* 34(7):20–27, 1996.

83. M. Hameed and A. Al-Taei. Telemedicine needs for multimedia and integrated services digital network (ISDN). *ICSC Congress on Computational Intelligence Methods and Applications*. 1–4, 2005.
84. E. Kyriacou et al. Multi-purpose healthcare telemedicine systems with mobile communication link support. *Biomed. Eng.Online*. 2(1):7, 2003.
85. R. W. Jones et al. The AIDMAN project—A telemedicine approach to cardiology investigation, referral and outpatient care. *J. Telemed. Telecare*. 6:32–34, 2000.
86. N. W. John et al. Bringing 3D to teleradiology. *IEEE Int. Conf. Information Visualization*. 4–9, 2000.
87. C. Xinhua and H. K. Huang. Current status and future advances of digital radiography and PACS. *IEEE Eng. Med. Biol. Mag.* 19(5):80–88, 2000.
88. P. Seung-Hun et al. Real-time monitoring of patients on remote sites. *Proc. IEEE Conf. EMBC*. 1321–1325, 1998.
89. T. Starner et al. The gesture pendant: A self-illuminating, wearable, infrared computer vision system for home automation control and medical monitoring. *Symp. Wearable Computers*. 87–94, 2000.
90. P. Bonato. Wearable sensors/systems and their impact on biomedical engineering. *IEEE Eng. Med. Biol. Mag.* 22(3):18–20, 2003.
91. C. Glaros and D.I. Fotiadis. Wearable devices in healthcare. In *Studies in Fuzziness and Soft Computing: Intelligent Paradigms for Healthcare Enterprises*. Springer. 237–264, 2005.
92. G. Troster. The agenda of wearable healthcare. In S. Schattauer (Ed.). *IMIA Yearbook of Medical Informatics: Ubiquitous Health Care Systems*. Schattauer. 125–138, 2004.
93. C.-C. Lin et al. A wireless healthcare service system for elderly with dementia. *IEEE Trans. Inf. Technol. Biomed.* 10(4):696–704, 2006.
94. E. Jovanov et al. Stress monitoring using a distributed wireless intelligent sensor system. *IEEE Eng. Med. Biol. Mag.* 22(3):49–55, 2003.
95. J. M. Winters and Y. Wang. Wearable sensors and telerehabilitation. *IEEE Eng. Med. Biol. Mag.* 22(3):56–65, 2003.
96. T. Martin, E. Jovanov, and D. Raskovic. Issues in wearable computing for medical monitoring applications: A case study of a wearable ECG monitoring device. In *Int. Symp. Wearable Computers*. 43–49, 2000.
97. M. Murero, G. D'Ancona, and H. Karamanoukian. Use of the internet by patients before and after cardiac surgery: An interdisciplinary telephone survey. *J. Med. Internet Res.* 3(3):e27, 2001.
98. V. C. Protopappas et al. An ultrasound wearable system for the monitoring and acceleration of fracture healing in long bones. *IEEE Trans. Biomed. Eng.* 52(9):1597–1608, 2005.
99. P. F. Brennan et al. HeartCare: An Internet-based information and support system for patient home recovery after coronary artery bypass graft (CABG) surgery. *J. Adv. Nurs.* 35(5):699–708, 2001.
100. K. S. White. Speech recognition implementation in radiology. *Pediatr. Radiol.* 35(9):841–846, 2005.
101. M. A. Grasso. The long-term adoption of speech recognition in medical applications. *IEEE Symp. Computer-Based Medical Systems*. 257–262, 2003.
102. J. F. Allen et al. Toward conversational human-computer interaction. *AI Mag.* 22(4):27–38, 2001.
103. M. Fried-Oken. Voice recognition device as a computer interface for motor and speech impaired people. *Arch. Phys. Med. Rehabil.* 10:678–681, 1985.
104. A. J. Gutierrez, M. E. Mullins, and R. A. Novelline. Impact of PACS and voice-recognition reporting on the education of radiology residents. *J. Digit. Imaging.* 18(2):100–108, 2005.
105. T. Warfel and P. Chang. Integrating dictation with PACS to eliminate paper. *J. Digit. Imaging.* 17(1):37–44, 2004.
106. D. Litman and S. Silliman. ITSPOKE: An intelligent tutoring spoken dialogue system. *Proc. Conf. Human Language Technology.* 2004.
107. T. Takezawa et al. A Japanese-to-English speech translation system: ATR–MATRIX. *Proc. ICSLP.* 957–960, 1998.
108. J. Polifroni, G. Chung, and S. Seneff. Towards the automatic generation of mixed-initiative dialogue systems from web content. *Proc. EUROSPEECH.* 193–196, 2003.
109. A. D. Beischer et al. The role of multimedia in patient education for total hip replacement surgery. *J. Bone Joint Surg.* 84–B(SUPP III):290, 2002.
110. J. Littlefield et al. A multimedia patient simulation for teaching and assessing endodontic diagnosis. *J. Dent. Educ.* 67(6):669–677, 2003.
111. J. G. Wheeler et al. Impact of a waiting room videotape message on parent attitudes toward pediatric antibiotic use. *Pediatrics.* 108(3):591–596, 2001.
112. A. K. Y. Man et al. The effect of intra-operative video on patient anxiety. *Anaesthesia.* 58(1):64–68, 2003.
113. R. S. Perocchia et al. Raising awareness of on-line cancer information: Helping providers empower patients. *J. Health Commun.* 10(0):157–172, 2005.
114. J. L. Wofford, E. D. Smith, and D. P. Miller. The multimedia computer for office-based patient education: A systematic review. *Patient Educ. Couns.* 59(2):148–157, 2005.
115. D. Flynn et al. The utility of a multimedia education program for prostate cancer patients: A formative evaluation. *Br. J. Cancer.* 91:855–860, 2004.
116. S. Krishna et al. Internet-enabled interactive multimedia asthma education program: A randomized trial. *Pediatrics.* 111(3):503–510, 2003.
117. A. Fagerlin et al. Patient education materials about the treatment of early-stage prostate cancer: A critical review. *Ann. Intern. Med.* 140(9):721–728, 2004.

118. R. Hyde, F. Bryden, and A. J. Asbury. How would patients prefer to spend the waiting time before their operations? *Anaesthesia.* 53(2):192–195, 1998.
119. M. H. Oermann, S. A. Webb, and J. A. Ashare. Outcomes of videotape instruction in clinic waiting area. *Orthop. Nurs.* 22(2):102–105, 2003.
120. S.-M. Wang et al. Music and preoperative anxiety: A randomized, controlled study. *Anesth. Analg.* 94(6): 1489–1494, 2002.
121. Y. Ullmann et al. The sounds of music in the operating room. *Injury.* In press.
122. H. W. Chiu et al. Using heart rate variability analysis to assess the effect of music therapy on anxiety reduction of patients. *IEEE Proc. Conf. Computers in Cardiology.* p. 469–472, 2003.
123. P. White. Pharmacologic and clinical aspects of preoperative medication. *Anesth. Analg.* 65(9):963–974, 1986.
124. W. M. Wysocki et al. Surgery, surgical education and surgical diagnostic procedures in the digital era. *Med. Sci. Monit.* 9:RA69–RA75, 2003.
125. R. G. Belleman and P. M. A. Sloot. Simulated vascular reconstruction in a virtual operating theatre. *Comp. Assist. Radiol. Surg.* 1230:986–992, 2001.
126. N. W. John. High performance visualization in a hospital operating theatre. *Proc. Theory and Practice of Computer Graphics.* 170–175, 2003.
127. G. Graschew et al. Interactive telemedicine in the operating theatre of the future. *J. Telemed. Telecare.* 6:20–24, 2000.
128. A. S. Merians et al. Virtual reality-augmented rehabilitation for patients following stroke. *Phys. Ther.* 82(9): 898–915, 2002.
129. L. Xun et al. Integration of augmented reality and assistive devices for post-stroke hand opening rehabilitation. *IEEE Proc. Conf. Engineering in Medicine and Biology Society.* 6855–6858, 2005.
130. D. Stoianovici et al. A modular surgical robotic system for image guided percutaneous procedures. In *Lecture Notes in Computer Science: Medical Image Computing and Computer-Assisted Interventation.* 404–410, 1998.
131. S. Grange et al. M/ORIS: A medical/operating room interaction system. *Proc. ACM Multimodal Interfaces (ICMI).* 159–166, 2004.
132. A. Rosset et al. Navigating the fifth dimension: Innovative interface for multidimensional multimodality image navigation. *Radiographics.* 26(1):299–308, 2006.
133. L. Kim and S. H. Park. Haptic interaction and volume modeling techniques for realistic dental simulation. *The Visual Computer.* 22(2):90–98, 2006.
134. D. Gering et al. An integrated visualization system for surgical planning and guidance using image fusion and an open MR. *J. Magn. Reson. Imaging.* 13:967–975, 2001.
135. L. Joskowicz and R. H. Taylor. Computers in imaging and guided surgery. *IEEE Computing in Science & Engineering.* 3(5):65–72, 2001.
136. L.-P. Nolte et al. A new approach to computer-aided spine surgery: Fluoroscopy-based surgical navigation. *Eur. Spine J.* V9(7): S078–S088, 2000.
137. Z. Qingsong, K. C. Keong, and N. W. Sing. Interactive surgical planning using context based volume visualization techniques. *Proc. Int. Workshop Medical Imaging and Augmented Reality.* 21–25, 2001.
138. C. G. Guan et al. Volume-based tumor neurosurgery planning in the virtual workbench. *Proc. IEEE Symp. Virtual Reality.* 167–173, 1998.
139. A. Neubauer et al. Advanced virtual endoscopic pituitary surgery. *IEEE Trans. Visualization and Computer Graphics.* 11(5):497–507, 2005.
140. T. M. Krummel. Surgical simulation and virtual reality: The coming revolution. *Ann. Surg.* 228(5):635–637, 1998.
141. J. Shah et al. Simulation in urology: A role for virtual reality? *BJU Int.* 88(7):661–665, 2001.
142. K. Johnsen et al. Experiences in using immersive virtual characters to educate medical communication skills. *IEEE Proc. Virtual Reality.* 179–186, 2005.

Index

A

B

C

D

E

F

G

H

I

J

K

L

M

N

Q

R

S

X

Y

Z